Sets of Numbers

Natural numbers	$N = \{1, 2, 3, 4, 5, \ldots\}$ (counting numbers)
Whole numbers	$W = \{0, 1, 2, 3, 4, 5, \ldots\}$
Integers	$Z = \{\ldots, -3, -2, -1, 0, 1, 2, 3, \ldots\}$
Rational numbers	$Q = \left\{ \dfrac{p}{q} \,\middle|\, p, q \in Z, q \neq 0 \right\}$
Irrational numbers	$I = \{x \mid x \in R, x \text{ is not a rational number}\}$
Real numbers	$R = \{x \mid x \text{ corresponds to a point on the number line}\}$
Complex numbers	$C = \{a + bi \mid a, b \in R\}$

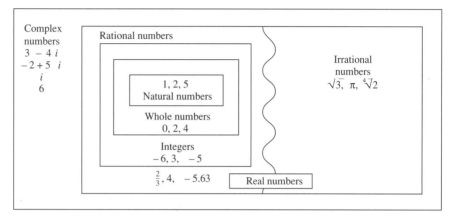

$$N \subset W \subset Z \subset Q \subset R \subset C$$

Properties of Real Numbers

Commutative	**Associative**	**Distributive**
$a + b = b + a$	$a + (b + c) = (a + b) + c$	$a \cdot (b + c) = a \cdot b + a \cdot c$
$a \cdot b = b \cdot a$	$a \cdot (b \cdot c) = (a \cdot b) \cdot c$	$(a + b) \cdot c = a \cdot c + b \cdot c$

Identities

There is a unique real number called the **_additive identity,_** represented by 0, which has the property that $a + 0 = 0 + a = a$ for all $a \in R$.

There is a unique real number called the **_multiplicative identity,_** represented by 1, which has the property that $a \cdot 1 = 1 \cdot a = a$ for all $a \in R$.

Inverses

Each real number a has a unique **_additive inverse,_** represented by $-a$, which has the property that $a + (-a) = (-a) + (a) = 0$.

Each real number a, except 0, has a unique **_multiplicative inverse,_** represented by $\dfrac{1}{a}$, which has the property that $a\left(\dfrac{1}{a}\right) = \left(\dfrac{1}{a}\right)a = 1$.

Closure Properties

The sum of two real numbers is a real number.

The product of two real numbers is a real number.

www.brookscole.com

www.brookscole.com is the World Wide Web site for Thomson Brooks/Cole and is your direct source to dozens of online resources.

At www.brookscole.com you can find out about supplements, demonstration software, and student resources. You can also send e-mail to many of our authors and preview new publications and exciting new technologies.

www.brookscole.com
Changing the way the world learns®

Understanding
Intermediate Algebra

A Course for College Students   **Sixth Edition**

Lewis Hirsch
Rutgers University

Arthur Goodman
Queens College,
City University of New York

Australia • Brazil • Canada • Mexico • Singapore • Spain
United Kingdom • United States

THOMSON

BROOKS/COLE

Understanding Intermediate Algebra: A Course for College Students, 6e
Lewis Hirsch, Arthur Goodman

Executive Editor: *Jennifer Laugier*
Assistant Editor: *Rebecca Subity*
Editorial Assistant: *Christina Ho*
Technology Project Manager: *Sarah Woicicki*
Marketing Manager: *Greta Kleinert*
Marketing Assistant: *Brian Smith*
Marketing Communications Manager: *Bryan Vann*
Project Manager, Editorial Production: *Cheryll Linthicum*
Creative Director: *Rob Hugel*
Art Director: *Vernon T. Boes*
Print Buyer: *Rebecca Cross*

Permissions Editor: *Joohee Lee*
Production Service: *Hearthside Publishing Services*
Text Designer: *Kim Rokusek*
Photo Researcher: *Gretchen Miller*
Copy Editor: *Barbara Willette*
Illustrator: *Jade Myers*
Cover Designer: *Katherine Minerva*
Cover Art: *Judith Harkness*
Cover Printer: *Phoenix Color Corp*
Compositor: *Better Graphics*
Printer: *Courier Kendallville*

Thomson Higher Education
10 Davis Drive
Belmont, CA 94002-3098
USA

Library of Congress Control Number: 2005932071

ISBN 0-534-41795-7

For more information about our products, contact us at:
Thomson Learning Academic Resource Center
1-800-423-0563
For permission to use material from this text or product, submit a request online at http://www.thomsonrights.com.
Any additional questions about permissions can be submitted by e-mail to thomsonrights@thomson.com.

Preface to the Instructor

Purpose

Understanding Intermediate Algebra, 6th edition, is our attempt to offer an intermediate algebra textbook which reflects our philosophy—that students can *understand* what they are doing in algebra and why.

We offer a view of algebra that takes every opportunity to explain why things are done in a certain way, to show how concepts or topics are related, and to show how supposedly "new" topics are actually just new applications of concepts already learned.

This text assumes a knowledge of elementary algebra. However, we realize that students arrive in intermediate algebra with a diversity of previous mathematical experiences. Some may have recently learned elementary algebra but to varying degrees; some may have learned elementary algebra well but have been away from it for a while. In this spirit, we have tried to develop a text flexible enough to meet the needs of this heterogeneous group by providing explanations for what we regard as key elementary concepts that are crucial to the understanding of the more difficult intermediate algebra concepts. The less well prepared students certainly benefit from these explanations; the better prepared students appreciate this second look as a way to help them fit the seemingly unrelated parts of algebra together. The instructor has the choice of discussing this material or leaving it for students to review on their own.

Pedagogy

We believe that a student can successfully learn and understand algebra by mastering the basic concepts and being able to apply them to a wide variety of situations. Thus, each section begins by relating the topic to be discussed to material previously learned. In this way the students can see algebra as a whole rather than as a series of isolated ideas. Many sections begin with a real-life application to motivate the topic to be discussed.

Basic concepts, rules, and definitions are motivated and explained via numerical, analytic (algebraic), and graphical examples.

Concepts are developed in a series of carefully constructed illustrative examples. Through the course of these examples, we compare and contrast related ideas, helping the student to understand the sometimes subtle distinctions among various situations. In addition, these examples strengthen a student's understanding of how this "new" idea fits into the overall picture.

Every opportunity has been taken to point out common errors often made by students and to explain the misconception that often leads to a particular error.

Basic rules and/or procedures are highlighted so that students can find important ideas quickly and easily.

Features

Examples   The various steps in the solutions to examples are explained in detail. Many steps appear with annotations (highlighted in color) that involve the student in the solution. These comments explain how and why a solution is proceeding in a certain way.

Exercises   There are over 6,500 exercises, including calculator exercises. Not only have the exercise sets been matched odd/even, but they have also been designed so that, in many situations, successive odd-numbered exercises compare and contrast subtle differences in applying the concepts covered in the section. Additionally, variety has been added to the exercise sets so that the student must be alert as to what the problem is asking. For example, the exercise set for Section 6.6, which deals primarily with solving rational equations, also contains some exercises on adding rational expressions.

Study Skills   One of the main sources of students' difficulties is that they do not know how to study algebra. In this regard we offer a totally unique feature. The preface to the student and each section in the first four chapters conclude with a Study Skill. This is a brief paragraph discussing some aspect of studying algebra, doing homework, or preparing for or taking exams. Our students have indicated that they found the Study Skills very helpful. The study skills appearing in this text are part of a collection of general mathematics study skills developed by Lewis R. Hirsch and Mary C. Hudspeth at the Pennsylvania State University.

Questions for Thought   Almost every exercise set contains Questions for Thought, which offer the student an opportunity to *think* critically about various algebraic ideas. They may be asked to compare and contrast related ideas, or to examine an incorrect solution and explain why the solution is wrong. The Questions for Thought are intended to be answered in complete sentences and in grammatically correct English. The Questions for Thought were originally designed for having students write across the curriculum, and can be used by instructors for this purpose.

Margin Comments   Margin comments have been added where appropriate in order to involve the students more actively in the learning process as they read the text. A margin comment will usually seek to emphasize a point made in the text presentation or ask a question requiring the student to focus attention on a particularly crucial aspect of the discussion.

Different Perspectives   Different Perspectives boxes appear wherever there is an opportunity to highlight the connection between algebraic and geometric aspects of the same concept. In this way the student is encouraged to think about mathematical ideas from more than one point of view.

Thinking Out Loud   Thinking Out Loud is a feature in which the solution to certain examples is presented in a question-and-answer format so that students can see examples of the thought processes involved in approaching and solving new or unfamiliar problems. This helps students develop more appropriate problem-solving strategies.

Mini-Review   Most sections contain a Mini-Review, which consists of exercises that allow students to periodically review important topics as well as help them prepare for the material to come. These Mini-Reviews afford the student additional opportunity to see new topics within the framework of what they have already learned.

Functions   The concept of functions is introduced early, starting at an elementary level in Chapter 3 and spiraling throughout the text; more sophisticated function concepts are covered in Chapter 9.

Chapter Summary   Each chapter contains a chapter summary describing the basic concepts in the chapter. Each point listed in the summary is accompanied by an example illustrating the concept or procedure.

Review Exercises   Each chapter contains a set of chapter review exercises and a chapter practice test. Additionally, there are four cumulative review exercise sets and four cumulative practice tests following Chapters 3, 6, 9, and 11. These offer the student more opportunities to practice choosing the appropriate procedure in a variety of situations.

Answer Section   The answer section contains answers to all the odd-numbered exercises, as well as to *all* review exercises and practice test problems. The answer to each verbal problem contains a description of what the variable(s) represent and the equation (or system of equations) used to solve it. In addition, the answers to the cumulative review exercises and cumulative practice tests contain a reference to the section in which the relevant material is covered.

New to the Sixth Edition

- The THINKING OUT LOUD feature has been expanded to more examples in order to more fully support stronger active learning for the students.
- A new STUDY SKILL has been added to the STUDY SKILLS feature. The latest study skill encourages and instructs students on how to form study groups. Author and reviewer experience indicates that students who participate in study groups enjoy more success and have a more positive experience in their math courses.
- In response to reviewer feedback, Interval Notation has been added to this edition. It is introduced in Section 2.3, First Degree Inequalities.
- Many of the exercises have been rewritten and numerous applications have been updated.
- 40 HOURS PER WEEK OF ONLINE TUTORING are available to your students with the purchase of a new textbook. Students can access VMentor ™ tutoring service through the iLrn ™ Tutorial packaged free with the text.
- Icons point students to material contained on the Brooks/Cole Website and on the Interactive Video Skillbuilder CD-ROM that accompanies the text.

Teaching Tools for the Instructor

Complete Solutions Manual (0534-421717)   The Complete Solutions Manual provides worked-out solutions to all of the problems in the text.

Text-Specific Videotapes (0534-421725)   These text-specific videotape sets, available at no charge to qualified adopters of the text, feature 10- to 20-minute problem-solving lessons that cover each section of every chapter.

Test Bank (0534-421733)   The Test Bank includes multiple tests per chapter as well as final exams. The tests are made up of a combination of multiple-choice, free-response, true/false, and fill-in-the-blank questions.

iLrn Instructor Version™ (0534-421792)   Providing instructors and students with unsurpassed control, variety, and all-in-one utility, iLrn™ is a powerful and fully integrated teaching and learning system. iLrn ties together five fundamental learning activities: diagnostics, tutorials, homework, quizzing, and testing. Easy to use, iLrn offers instructors complete control when creating assessments in which they can draw from the wealth of exercises provided or create their own questions. iLrn features the greatest variety of problem types – allowing instructors to assess the way they teach! A real timesaver for instructors, iLrn offers automatic grading of homework, quizzes, and tests with results flowing directly into the gradebook. The auto-enrollment feature also saves time with course set up as students self-enroll into the course gradebook. iLrn provides seamless integration with Blackboard™ and WebCT™.

TLE Labs—The ideal upgrade for any iLrn™ Tutorial!   Each of the 15 **TLE Labs**, scheduled one per week in a traditional semester course, introduce and explore key concepts. **iLrn Tutorial** reinforces those concepts with unlimited practice. With the opportunity to explore these core concepts interactively at their own pace, students are solidly prepared for the work of the traditional course. **TLE Labs** correlate to the key concepts of the text and the course in general. All labs are developed with seven key components: Introduction • Tutorial • Examples • Summary • Practice & Problems • Extra Practice • Self Check. Each of these lesson components can be accessed easily from any other component and in any order after the lesson is activated.

In addition to the printed textbook, students receive everything they need to succeed in the course: an online version of the text, access to the **TLE Labs**, text-specific interactive tutorials, and access to **vMentor**™—all in the same, single, unified environment. Ask your Thomson representative about **TLE Labs!** They're a great value for your students.

College Success Factors Index   Developed by Edmond Hallberg, Kaylene Hallberg, and Loren Sauer

Help your students assess their college success needs! This unique online assessment system allows students to identify the behaviors and attitudes that will help them succeed in college. Students can take the 80-statement survey, based on eight important areas that have been proven to correlate with college success, in a password-protected area right from their own computer. A special section for instructors offers information and advice on how to administer, interpret, and report the data. Contact your Thomson Brooks/Cole representative for packaging information.

JoinIn™ on TurningPoint®   Thomson Brooks/Cole is pleased to offer you book-specific **JoinIn**™ content for electronic response systems tailored to **Understanding Intermediate Algebra**. You can transform your classroom and assess your students' progress with instant in-class quizzes and polls. Our exclusive agreement to offer **TurningPoint**® software lets you pose book-specific questions and display students' answers seamlessly within the Microsoft® PowerPoint® slides of your own lecture, in conjunction with the "clicker" hardware of your choice. Enhance how your students interact with you, your lecture, and each other. Contact your local Thomson representative to learn more.

For the Student

Student Solutions Manual (0534-421709)   The Student Solutions Manual provides worked-out solutions to the odd-numbered problems in the text.

Thomson Brooks/Cole Mathematics Web site   http://mathematics.brooks cole.com   Visit us on the web for access to a wealth of free learning resources, including graphing calculator tutorials, tools to address math anxiety, Historical Notes, an extensive glossary of math terms, career information, and more.

iLrn™ Tutorial Student Version (0534-421628)   Featuring a variety of approaches that connect with all types of learners, iLrn™ Tutorial offers text-specific tutorials that require no set up by instructors. Students can begin exploring active examples from the text by using the access code packaged free with a new book. iLrn Tutorial supports students with explanations from the text, examples, step-by-step problem solving help, unlimited practice, and chapter-by-chapter video lessons. With this self-paced system, students can even check their comprehension along the way by taking quizzes and receiving feedback. If they still are having trouble, students can easily access vMentor™ for online help from a live math instructor. Students can ask any question and get personalized help through the interactive whiteboard and using their computer microphones to speak with the instructor. While designed for self-study, instructors can also assign the individual tutorial exercises.

vMentor™...live, online tutoring   Packaged FREE with every text! Accessed seamlessly through **iLrn Tutorial**, **vMentor** provides tutorial help that can substantially improve student performance, increase test scores, and enhance technical aptitude. Your students will have access, via the Web, to highly qualified tutors with thorough knowledge of our textbooks. When students get stuck on a particular problem or concept, they need only log on to **vMentor**, where they can talk (using their own computer microphones) to **vMentor** tutors who will skillfully guide them through the problem using the interactive whiteboard for illustration. Brooks/Cole also offers **Elluminate *Live!***, an online virtual classroom environment that is customizable and easy to use. **Elluminate *Live!*** keeps students engaged with full two-way audio, instant messaging, and an interactive whiteboard—all in one intuitive, graphical interface. For information about obtaining a **Elluminate *Live!*** site license, please contact your Thomson representative. *For proprietary, college, and university adopters only. For additional information please consult your local Thomson representative.*

Interactive Video Skillbuilder CD-ROM (0534-421636)   Think of it as portable office hours! The Interactive Video Skillbuilder CD-ROM contains video instruction covering each chapter of the text. The problems worked during each video lesson are shown first so that students can try working them before watching the solution. To help students evaluate their progress, each section contains a 10-question Web quiz (the results of which can be emailed to the instructor) and each chapter contains a chapter test, with the answer to each problem on each test. A new learning tool on this CD-ROM is a graphing calculator tutorial for precalculus and college algebra, featuring examples, exercises, and video tutorials. Also new, English/Spanish closed caption translations can be selected to display along with the video instruction. This CD-ROM also features MathCue tutorial and testing software. Keyed to the text, MathCue offers these components:

- MathCue Skill Builder—Presents problems to solve, evaluates answers and tutors students by displaying complete solutions with step-by-step explanations.
- MathCue Quiz—Allows students to generate large numbers of quiz problems keyed to problem types from each section of the book.
- MathCue Chapter Test—Also provides large numbers of problems keyed to problem types from each chapter.
- MathCue Solution Finder–This unique tool allows students to enter their own basic problems and receive step-by-step help as if they were working with a tutor.
- Score reports for any MathCue session can be printed and handed in for credit or extra-credit.
- Print or e-mail score reports—Score reports for any MathCue session can be printed or sent to instructors via MathCue's secure e-mail score system.

Acknowledgments

The authors would like to thank the following reviewers of the fifth edition for their helpful comments and suggestions:

Laurette Blakey Foster, Prairie View A&M University
Holly Broesamle, Oakland Community College
John Burghduff, Kingwood College
Mary Jane Ferguson, Houston Community College
Gary R. Grohs, Elgin Community College

Bob Harbison, Blinn College
Robert M. Kaufmann, University of Alabama, Birmingham
Barbara Kurt, St. Louis Community College
Richard Penn, Montgomery College
Charles Wheeler, Montgomery College

We would also like to thank the following people who reviewed the book for the sixth edition:

Joe Howe, St. Charles Community College
David George, The Ohio State University
John Taylor, Shoreline Community College
Elena Kravchuk, The University of Alabama at Birmingham
Jadwiga Weyant, Edmonds Community College
Gary Grohs, Elgin Community College
Amy Main, Blinn College

The production of a textbook is a collaborative effort. We must thank our editor, Jennifer Laugier for her support and encouragement; project manager Cheryll Linthicum for her coordination of the entire production; Anne Seitz of Hearthside Publishing Services for her handling of the manuscript; assistant editor Rebecca Subity for her help in preparing the ancillaries package that accompanies the text; editorial assistant Christina Ho for her administrative support, and Ian Crewe of St. Francis School for his assistance in checking solutions.

Finally, we would like to thank our wives, Cindy and Sora, and our families for their constant support and encouragement.

Preface to the Student

This text is designed to help you understand algebra. We are convinced that if you understand what you are doing and why, you will be a much better algebra student. (Our students who have used previous editions of this book seem to agree with us.) This does not mean that after reading each section you will understand all the concepts clearly. Much of what you learn comes through the course of doing lots and lots of exercises and seeing for yourself exactly what is involved in completing an exercise. However, if you read the textbook carefully and take good notes in class, you will find algebra not quite so menacing.

Here are a few suggestions for using this textbook:

- Always read the textbook with a pencil and paper in hand. Reading mathematics is not like reading other subjects. *You* must be involved in the learning process. Work out the examples along with the textbook and *think* about what you are reading. Make sure you understand what is being done and why.
- You must work homework exercises on a daily basis. While attending class and listening to your instructor are important, do not mistake understanding someone else's work for the ability to do the work yourself. (Think about watching someone else driving a car, as opposed to driving yourself.) Make sure *you* know how to do the exercises.
- Read the Study Skills which follow this preface and appear at the end of each section in the first four chapters. They discuss the best ways to use the textbook and your notes. They also offer a variety of suggestions on how to study, do homework, and prepare for and take tests. There is a complete list of the Study Skills (with page references) on the inside front cover of the book.
- Do not get discouraged if you have difficulty with some topics. Certain topics may not be absolutely clear the first time you see them. Be persistent. We all need time to absorb new ideas and become familiar with them. What was initially difficult will become less so as you spend more time with a subject. Keep at it and you will see that you are making steady progress.

Using a Calculator

A calculator is a very powerful tool for doing calculations quickly and accurately. When you sit down to work on a math problem you should have the tools you will need—pencil, paper, and calculator. Nevertheless, a calculator cannot substitute for understanding what you are doing and why.

In general, it is a good idea to work out as much of an exercise as you can on paper, and try to save any calculator computations for the last step. If you need to use a calculator for the intermediate steps in the solution to a problem, then be sure to write down the results of those steps. Writing the steps down helps you understand the problem and makes it easier to spot errors and check that the calculator answer makes sense (and that you used the calculator correctly); see Study Skill topic "Estimation" on page xvii.

Knowing how and *when* to use a calculator for a problem can make a difference in the accuracy of your answer as well. In this text, unless otherwise specified, all equations require exact answers (answers that are not rounded). Suppose, for example, you need to enter $\frac{33}{7}$ in your calculator before using it in your calculations to solve an equation. If you round $\frac{33}{7}$ to 4.71, you will automatically introduce a rounding error in your calculations. (How much this error is magnified and reflected in the solution will depend on the operations performed on, or with, this number.) Your solution will not be exact. Even if you enter $\frac{33}{7}$ in your calculator as $\boxed{3}\ \boxed{3}\ \boxed{\div}\ \boxed{7}\ \boxed{=}$, your calculator will round the number before performing computations; the result will be a more accurate answer, but it still may not be exact.

Consider another example concerning exact solutions. Suppose the solution to an equation is $\frac{5}{13}$. If the problem is solved correctly and all numbers are entered accurately, you should arrive at the answer **0.384615385** using your calculator. While this answer is accurate to 9 decimal places, it is still a rounded answer, not the exact answer $\frac{5}{13}$. If you try checking this answer in the original equation, the answer may not check exactly.

To get the most out of your calculator, you need to become familiar with its various capabilities. One way to do this is to read the "Getting Started" sections, which deal with how the calculator operates.

We list the following general advice for using your calculator while working on algebra problems:

1. When an exact answer such as $\sqrt{3}$ or $\frac{3}{7}$ is requested or required, the calculator should *not* be used except possibly to *check* the accuracy of your computations.
2. When solving most problems, work out as much of the problem as possible on paper and save the calculator computations for the last step. This way you will be able to check the logic of the steps in your solution, spot errors, and check that the calculator answer makes sense. This also reduces errors due to rounding.
3. For some problems, you may not be able to save the calculator for the last step—the problem may require that you use the calculator for many stages of the problem. Again, you should try to write down as many steps as you can so you can check over your work.
4. You *should* use your calculator to check all computations including checking your solution whenever possible. (If your answer is exact, keep in mind that the calculator may give a rounded answer.)

Strengthening Your Study Skills

Studying Algebra—How Often?

In most college courses, you are typically expected to spend 2–4 hours studying outside of class for every hour spent in class.

It is especially important that you spend this amount of time studying algebra, since you must both acquire and *perfect* skills; and, as most of you who play a musical instrument or participate seriously in athletics already should know, it takes time and lots of practice to develop and perfect a skill.

It is also important that you distribute your studying over time. That is, do not try to do all your studying in 1, 2, or even 3 days, and then skip studying the other days. You will find that understanding algebra and acquiring the necessary skills are much easier if you spread your studying out over the week, doing a little each day. If you study in this way, you will need less time to study just before exams.

In addition, if your study sessions are more than 1 hour long, it is a good idea to take a 10-minute break within every hour you spend reading math or working exercises. The break helps to clear your mind, and allows you to think more clearly.

Previewing Material

Before you attend your next class, preview the material to be covered beforehand. First, skim the section to be covered, look at the headings, and try to guess what the sections will be about. Then read the material over carefully.

You will find that when you read over the material before you go to class, you will be able to follow the instructor more easily, things will make more sense, and you will learn the material more quickly. Now, if there was something you did not understand when you previewed the material, the teacher will be able to answer your questions *before* you work your assignment at home.

What to Do First

Before you attempt algebra exercises, either for homework or for practicing your skills, it is important to review the relevant portions of your notes and text.

Memorizing a bunch of seemingly unrelated algebraic steps to follow in an example may serve you initially, but in the long run (most likely before Chapter 3), your memory will be overburdened—you will tend to confuse examples and/or forget steps.

Reviewing the material before doing exercises makes each solution you go through more meaningful. The better you understand the concepts underlying the exercise, the easier the material becomes, and the less likely you are to confuse examples or forget steps.

When reviewing the material, take the time to *think about what you are reading*. Try not to get frustrated if it takes you an hour or so to read and understand a few pages of a math text—that time will be well spent. As you read your text and your notes, think about the concepts being discussed: **(a)** how they relate to previous concepts covered, and **(b)** how the examples illustrate the concepts being discussed. More than likely, worked-out examples will follow verbal material, so look carefully at these examples and try to understand why each step in the solution is taken. When you finish reading, take a few minutes and think about what you have just read.

Doing Exercises

After you have finished reviewing the appropriate material, you should be ready to do the relevant exercises. Although your ultimate goal is to be able to work out the exercises accurately *and* quickly, when you are working out exercises on a topic that is new to you it is a good idea to take your time and think about what you are doing while you are doing it.

Think about how the exercises you are doing illustrate the concepts you have reviewed. Think about the steps you are taking and ask yourself why you are proceeding in this particular way and not some other. Why this technique or step and not a different one?

Do not worry about speed now. If you take the time at home to think about what you are doing, the material becomes more understandable and easier to remember. You will then be less likely to "do the wrong thing" in an exercise. The more complex-looking exercises are less likely to throw you. In addition, if you think about these things in advance, you will need much less time to think about them during an exam, and so you will have more time to work out the problems.

Once you believe you thoroughly understand what you are doing and why, you may work on increasing your speed.

Reading Directions

One important but frequently overlooked aspect of an algebraic problem is the verbal instructions. Sometimes these instructions are given in a single word, such as "simplify" or "solve" (occasionally it takes more time to understand the instructions than it takes to do the exercise). The verbal instructions tell us what we are expected to do, so make sure you read the instructions carefully and understand what is being asked.

Two examples may look the same, but the instructions may be asking you to do two different things. For example:

Identify the following property:

$$a + (b + c) = (a + b) + c$$

versus

Verify the following property by replacing the variables with numbers:

$$a + (b + c) = (a + b) + c$$

On the other hand, two different examples may have the same instructions but require you to do different things. For example:

$$\text{Evaluate:} \quad 2(3 - 8) \qquad \text{versus} \qquad \text{Evaluate:} \quad 2(3)(-8)$$

You are asked to evaluate both expressions, but the solutions require different steps.

It is a good idea to familiarize yourself with the various ways the same basic instructions can be worded. In any case, always look at an example carefully and ask yourself what is being asked and what needs to be done, *before you do it.*

Comparing and Contrasting Examples

When learning most things for the first time, we may easily get confused and treat things that are different as though they were the same because they "look" similar. Algebraic notation can be especially confusing because of the detail involved. Move or change one symbol in an expression and the entire expression is different; change one word in a verbal problem and the whole problem has a new meaning.

It is important that you become capable of making these distinctions. The best way to do this is by comparing and contrasting examples and concepts that look almost identical, but are not. It is also important that you ask yourself in what ways these things are similar and in what ways they differ. For example, as you may remember from elementary algebra, the commutative property of addition is similar in some respects to the commutative property of multiplication, but different from it in others. Also, the two expressions $3 + 2 \cdot 4$ and $3 \cdot 2 + 4$ look similar, but are actually very different.

When you are working out exercises (or reading a concept), ask yourself, "What examples or concepts are similar to those I am now doing? In what ways are they similar? How do I recognize the differences?" Doing this while you are working the exercises will help to prevent you from making careless errors later on.

Forming a Study Group

There has been much research indicating that students who study in groups generally do better in Math courses than do students who study alone. Consequently, you might form a study group for both doing homework and studying for tests.

A study group should ideally be three or four students of varying abilities. Students who know the material better increase their understanding when they explain it to others. One student's strength may be another student's weakness, so you can help each other. Studying together can also make doing homework and studying for tests less stressful and more fun.

Contents

4 Equations of a Line and Linear Systems in Two Variables   199

5 Polynomial Expressions and Functions   263

6 Rational Expressions and Functions   319

7 Exponents and Radicals   387

8 Quadratic Functions and Equations   455

Appendixes

A Sets   721

B Conic Sections   727

C Sequences and Series: The Binomial Theorem   747

1

The Fundamental Concepts

© Masterfile Royalty-Free

In elementary algebra we developed certain skills, and learned a number of basic concepts and principles and how to apply them in various situations.

In intermediate algebra we will develop our skills to a higher level and, although we will learn new concepts, to a great extent we will draw on the same basic notions covered in elementary algebra. This requires that we understand the fundamental principles of elementary algebra and how to recognize and apply these principles. Therefore, we will begin with a review of the elementary concepts and will eventually generalize them to the more complex cases in later chapters.

Basic Definitions: The Real Numbers and the Real Number Line

We begin this section with a quick summary of set notation. (A more thorough review of set notation appears in Appendix A.)

Set Notation

A *set* is a well-defined collection of objects. One way to designate a set is to *list* the members or elements of the set within braces; for example, we may write $A = \{4, 6, 8, 10\}$. When the set we want to list is infinite, we list a few elements followed by a comma and three dots. For example, the set of numbers we use for counting is called the set of **natural numbers,** which is usually denoted by the letter N:

$$N = \{1, 2, 3, 4, 5, \ldots\}$$

When we include the number 0 with this set it is called the set of **whole numbers** and is denoted by W:

$$W = \{0, 1, 2, 3, 4, 5, \ldots\}$$

Alternatively, we write a set by using **set-builder notation** (see Appendix A); for example, the set A listed above can be written using set-builder notation as $A = \{x \mid x \text{ is an even number between 2 and 12}\}$. (Note that, unless we indicate otherwise, when we say *between* and *in between*, we do *not* include the first and last numbers.)

The symbol used to designate that an object is a member of a particular set is "$\in$"; hence $6 \in A$ is a symbolic way of writing that 6 is an element of the set A. The set B is a **subset** of A, written $B \subset A$, if all elements of B are also in A. For example, $N \subset W$ since the set of whole numbers "contains" the set of natural numbers. A set with no members is called the **empty set** or the **null set** and is symbolized by "$\varnothing$".

For example, the set $H = \{x \mid x \text{ is both even and odd}\}$ has no elements, and we would write $H = \varnothing$.

We can create new sets from existing sets by the set "operations" of *union* and *intersection*. The **union** of two sets A and B, written $A \cup B$, is made up of the elements in A or in B, or in both A and B. For example, if $S = \{a, b, c\}$ and $T = \{a, d, f\}$, then $S \cup T = \{a, b, c, d, f\}$. The **intersection** of two sets A and B, written $A \cap B$, is made up of all elements common to both A and B. For the sets S and T defined above, $S \cap T = \{a\}$.

Sums, Terms, Products, and Factors

Sum is the word we use with addition. In an expression involving a sum, the quantities to be added in the sum are called the **terms.** The symbol used to indicate a sum is the familiar "$+$" sign.

Product is the word we use with multiplication. In an expression involving a product, the numbers being multiplied are called **factors.**

The most frequent error made by students in algebra is that of confusing terms with factors and factors with terms. There are things that can be done with factors that cannot be done with terms, and vice versa. For example,

$$\frac{x\cancel{y}}{\cancel{x}}$$ *The x is a* factor *of the numerator and* therefore *can be reduced with the factor of x in the denominator.*

but

$$\frac{x + y}{x}$$ *The x is a* term *of the numerator and therefore* cannot *be reduced with the factor of x in the denominator.*

If *a* and *b* are natural numbers, then saying that *a* is a ***multiple*** of *b* is equivalent to saying that *b* is a ***factor*** of *a*.

> 30 is a multiple of 6   because 6 is a factor of 30.
> 63 is a multiple of 7   because 7 is a factor of 63.

Thus, a factor of *n* is a number that divides exactly into *n*, whereas a multiple of *n* is a number that is exactly divisible by *n*.

The set of *prime numbers* is an important subset of the natural numbers.

DEFINITION

A ***prime number*** is a natural number, greater than 1, that is divisible only by itself and 1.

A natural number greater than 1 that is not prime is called ***composite.***

In other words, a prime number is a number whose only factors are itself and 1.

For example, the numbers 7 and 11 are prime numbers because they are not divisible by any numbers other than themselves and 1. The number 12 is composite (not prime) because it is divisible by other numbers, such as 3 and 4.

Every composite number can be uniquely written as a product of prime factors. For example,

$$70 = 2 \cdot 5 \cdot 7$$

EXAMPLE 1

Write the number 54 as a product of prime factors.

Solution

We start by writing the number as a product of two factors (any two factors we recognize) and continue factoring until *all* the factors are prime numbers.

We have many choices for our first two factors. We will illustrate just two of the possible paths to the answer.

The order of the factors is unimportant.

$$54 = 2 \cdot 27 \quad \textit{Factor 27.} \qquad \text{or} \qquad 54 = 9 \cdot 6 \quad \textit{Factor 9 and factor 6.}$$
$$= 2 \cdot 3 \cdot 9 \quad \textit{Factor 9.} \qquad\qquad\qquad = 3 \cdot 3 \cdot 2 \cdot 3$$
$$= \boxed{2 \cdot 3 \cdot 3 \cdot 3} \qquad\qquad\qquad\qquad = \boxed{2 \cdot 3 \cdot 3 \cdot 3}$$

No matter which factors you decide to start with, the final answer (since it involves *prime* factors only) will be the same.

The Real Numbers

We can represent whole numbers on the number line as follows:

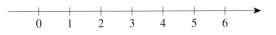

As we remember from elementary algebra, we cannot find the solution to the equation

$$x + 2 = 1$$

if we are restricted to whole number solutions, because there is no whole number that, when increased by 2, will yield 1. Thus, we introduce the ***integers,*** the set $\{\ldots, -3, -2, -1, 0, 1, 2, 3, \ldots\}$, designated by Z, to find the solution, $x = -1$.

The integers can be represented on the number line by extending the whole number line to the left of 0 and labeling the unit marks as follows:

The integers seem to be a satisfactory system until we try to solve the equation

$$3x = 1$$

which has no solution in the set of integers (which integer multiplied by 3 will yield 1?). This requires us to create new numbers called ***rational numbers*** or *fractions* in order to find the solution, $x = \frac{1}{3}$.

The set of rational numbers is usually designated by the letter Q. This set is difficult to list, primarily because no matter where you start, there is no *next* rational number. (What is the "first" fraction after 0?) Instead, we use set-builder notation to describe the set of rational numbers:

$$Q = \left\{ \frac{p}{q} \;\middle|\; p, q \in Z, \text{ and } q \neq 0 \right\}$$

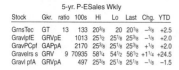
In words, this says that the set of rational numbers, Q, is the set of all numbers that can be represented as fractions, or quotients of integers, provided the denominator is not equal to 0.

$$8 = \frac{8}{1} \qquad \frac{-3}{7} \qquad 0 = \frac{0}{3} \qquad 0.73 = \frac{73}{100}$$

Rational numbers: Both fractions and decimals are used in this stock market quote.

Notice that integers are also rational numbers since they *can* be represented as quotients of integers.

Now we associate the rational numbers with units on the number line and with points *between* the units as follows:

But we will find that there are still points on the number line that cannot be labeled with a rational number. The numbers associated with these points are called *irrational numbers.*

To get a better idea of what irrational numbers look like, let's first examine rational numbers in decimal form (often called *decimal fractions*). To convert a fraction to its decimal form, we divide the numerator by the denominator. However, converting from the decimal form to a fraction is not quite so straightforward. We recognize that the decimal 0.73 is equal to $\frac{73}{100}$. Similarly, we recognize from experience that the decimal 0.33333 (where the bar above the last 3 indicates that the 3 repeats forever) is equal to the fraction $\frac{1}{3}$.

On the other hand, it is highly unlikely that we would recognize the decimal 0.407407$\overline{407}$ as being equal to the fraction $\frac{11}{27}$. (Divide 27 into 11 and verify that you get 0.407407$\overline{407}$.)

It turns out that if a decimal terminates (the decimal ends) or if a decimal is repeating (the same group of digits in the decimal is repeated infinitely), then the decimal represents a rational number.

This leaves us with nonterminating and nonrepeating decimals, which are *not* rational numbers. In other words, such a decimal cannot be represented as the quotient of two integers. This set of nonrepeating, nonterminating decimals is called the set of ***irrational numbers,*** and we shall designate the set by the letter *I*.

It is necessary for us to consider irrational numbers because just as the whole numbers were insufficient to fill all our needs, the rational numbers do not quite do the job either. When we try to solve the equation

$$x^2 = 9$$

(we are looking for a number that when multiplied by itself gives a product of 9) we will fairly quickly come up with two answers:

$$3 \quad [since \; 3 \cdot 3 = 9] \qquad and \qquad -3 \quad [since \; (-3)(-3) = 9]$$

These are called ***square roots*** of 9. Thus, we see that both 3 and -3 are square roots of 9.

When we try to solve

$$x^2 = 2$$

A square with sides equal to 1 foot has diagonal equal to $\sqrt{2}$ feet

however, the answer does not come as easily. We could say our answer is $\sqrt{2}$ or $-\sqrt{2}$, but what are those numbers exactly? Which decimal number, when squared, will give us 2? It turns out that, if we try to find the answer by trial and error (using a calculator to do the multiplication would help), we can get closer and closer to 2 but we will never get 2 exactly. For example, if we try (1.4)(1.4), we get 1.96, so we see that 1.4 is too small. If we try (1.5)(1.5), we get 2.25, and we see that 1.5 is too big. If we continue in this way we can get better and better approximations to a square root of 2. We might reach the approximate answer 1.414235, but (1.414235)(1.414235) = 2.0000606 (rounded off to seven decimal places). In fact, no matter how many places we get for x, the decimal never stops (because the square of x is never exactly 2) and never repeats. This implies that the square root of 2 (written $\sqrt{2}$) is an irrational number. Other examples of irrational numbers are $\sqrt{7}$, $\sqrt{15}$, $-\sqrt{5}$, and π.

The important thing for us to recognize is that the irrational numbers also represent points on the number line. If we take all the rational numbers together with all the irrational numbers (both positive and negative), we get all the points on the number line. This set is called the set of ***real numbers*** and is usually designated by the letter *R*:

$$R = \{x \mid x \text{ corresponds to a point on the number line}\}$$

The real number line is shown in Figure 1.1, with some specified points indicated.

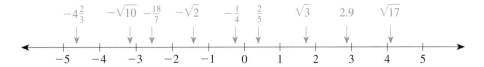

Figure 1.1
Real numbers

We use set notation to describe the relationship among the sets discussed above, as shown in Figure 1.2. From now on, unless we are told otherwise, we will assume that the set of real numbers serves as our basic frame of reference.

$$N \subset W \subset Z \subset Q \subset R$$

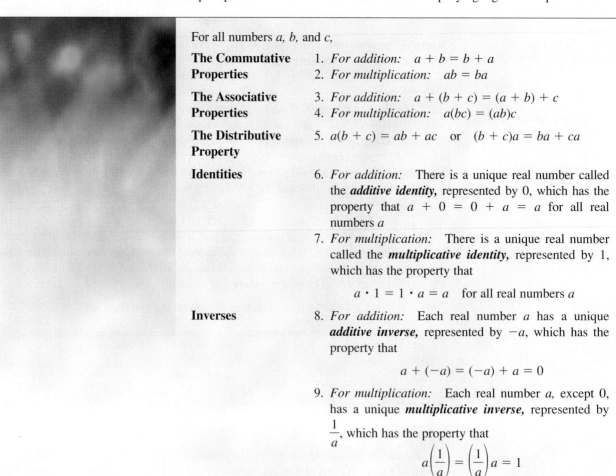

Figure 1.2
Relationships among subsets of real numbers

Properties of Real Numbers

The real numbers, along with the operations of addition ($+$) and multiplication ($\cdot$), obey the 11 properties listed in the following box. Most of these properties are straightforward and may seem trivial. Nevertheless, we shall see that these 11 basic properties are quite powerful and allow us to do much in simplifying algebraic expressions.

For all numbers a, b, and c,

The Commutative Properties
1. *For addition:* $a + b = b + a$
2. *For multiplication:* $ab = ba$

The Associative Properties
3. *For addition:* $a + (b + c) = (a + b) + c$
4. *For multiplication:* $a(bc) = (ab)c$

The Distributive Property
5. $a(b + c) = ab + ac$ or $(b + c)a = ba + ca$

Identities
6. *For addition:* There is a unique real number called the **additive identity,** represented by 0, which has the property that $a + 0 = 0 + a = a$ for all real numbers a
7. *For multiplication:* There is a unique real number called the **multiplicative identity,** represented by 1, which has the property that
$$a \cdot 1 = 1 \cdot a = a \quad \text{for all real numbers } a$$

Inverses
8. *For addition:* Each real number a has a unique **additive inverse,** represented by $-a$, which has the property that
$$a + (-a) = (-a) + a = 0$$
9. *For multiplication:* Each real number a, except 0, has a unique **multiplicative inverse,** represented by $\dfrac{1}{a}$, which has the property that
$$a\left(\frac{1}{a}\right) = \left(\frac{1}{a}\right)a = 1$$

Closure Properties
10. *For addition:* The sum of two real numbers is a real number.
11. *For multiplication:* The product of two real numbers is a real number

Some observations about these properties are in order.

The Commutative Properties:

$$a + b = b + a \quad \text{and} \quad ab = ba$$

Observe first that there are two commutative properties—one for each operation. Note the similarities and differences between them. Essentially, the commutative properties indicate that in adding (or multiplying) a pair of numbers, order is unimportant. For example:

$$x + 2 = 2 + x$$
$$(a + 8) + 4 = 4 + (a + 8) \qquad \textit{Note the order:} \quad \textit{(a + 8) is interchanged with 4.}$$

The Associative Properties:

$$(a + b) + c = a + (b + c) \qquad \text{and} \qquad (ab)c = a(bc)$$

For the associative properties, the order of the variables remains the same, but the grouping changes. The associative properties indicate that how you *group* terms for addition or how you group factors for multiplication is unimportant. Consider the following examples:

*Associative property of **addition***

(a) $(x + 5) + 4 = x + (5 + 4)$ *Note that the order of the terms remains the same; only the grouping is changed.*

(b) $(x + y) + (z + q) = x + (y + z) + q$ *The associative property can be generalized to include more than one pair of grouping symbols.*

The Distributive Property:

$$a(b + c) = ab + ac \qquad \text{or} \qquad (b + c)a = ba + ca$$

The associative and commutative properties involve a single operation; that is, addition and multiplication each have their own associative and commutative properties. On the other hand, the distributive property involves both operations together. For example:

(a) $3(x + 2) = 3 \cdot x + 3 \cdot 2 = 3x + 6$

(b) $xy(a + b + c) = xy \cdot a + xy \cdot b + xy \cdot c$ *Note that we can generalize the distributive property to more than two terms.*
$$= xya + xyb + xyc$$

(c) $5x + 10 = 5(x + 2)$ *The distributive property also yields a method for factoring as well as multiplying.*

(d) $(3x + 7)(x + 4) = (3x + 7)x + (3x + 7)4$

(e) $a(b - c) = ab - ac$ *This is another variation of the distributive property using subtraction.*

Example **(d)** illustrates that the variables in the properties may stand for more complex expressions such as $3x + 7$, as well as for a single letter or number.

$$a \quad \cdot (b + c) = \quad a \quad \cdot b + \quad a \quad \cdot c$$
$$(3x + 7) \cdot (x + 4) = (3x + 7) \cdot x + (3x + 7) \cdot 4$$

Identities

Zero is the additive identity; in other words, *adding* 0 to any number will not change the number. The multiplicative identity is 1; *multiplying* any number by 1 will not change the number. Note the similarities and differences between definitions. Each identity serves the same purpose with respect to its operation.

Inverses

Additive inverses are also called **opposites** (or **negatives**), and multiplicative inverses are also called **reciprocals.** Again, note the similarities and differences between definitions:

The *additive* inverse of *x* is that number which, when *added* to *x,* yields the *additive* identity, 0. The *multiplicative* inverse of *x* is that number which, when *multiplied* by *x,* yields the *multiplicative* identity, 1.

Closure Properties

When we state that a set is closed under an operation, we mean that when we perform the operation on two elements in the set, the result will be an element in the set. For example, we could start with the set of whole numbers and develop a system of whole numbers that has the associative, commutative, distributive, and identity properties (there are no inverses). This system would be closed under addition and multiplication; that is, sums and products of whole numbers are again whole numbers.

EXAMPLE 2

Name the property that the statement illustrates. If the statement is not true for all real numbers, write "false."

(a) $(5x)(yz) = 5(xyz)$
(b) $(x + 7) + 9 = x + (7 + 9)$
(c) $(x - y + 3)(a + b) = (x - y + 3)a + (x - y + 3)b$
(d) $8a + 0 = 8a$
(e) $4[10(7)] = (4 \cdot 10)(4 \cdot 7)$

Solution

(a) $(5x)(yz) = 5(xyz)$ *Associative property of multiplication*
(b) $(x + 7) + 9 = x + (7 + 9)$ *Associative property of addition*
(c) $(x - y + 3)(a + b)$
   $= (x - y + 3)a + (x - y + 3)b$ *Distributive property*
(d) $8a + 0 = 8a$ *Additive identity*
(e) $4[10(7)] = (4 \cdot 10)(4 \cdot 7)$ *False*

As we stated previously, the 11 basic properties are powerful because many of our techniques for simplifying expressions can be derived from them. For example, the following theorem can be derived from the 11 properties just discussed. (A theorem is a statement of an important fact that can be proven.)

**THEOREM:
THE MULTIPLICATION
PROPERTY OF ZERO**

For all real numbers *a,*

$$a \cdot 0 = 0 \cdot a = 0$$

The proof of this theorem is discussed in the Questions for Thought at the end of Exercises 1.1. We will discuss more theorems that can be derived from the basic properties when we discuss integer operations in the next section.

Order and the Real Number Line

The number line provides us with a simple way of defining the idea of numerical "order." For example, 5 is less than 7 because 5 is to the left of 7 on the number line.

In general, we define a to be "less than" b if a is to the left of b on the number line. The symbol we use for "less than" is "<".

> $a < b$ means that a is to the left of b on the number line.

$2 < 6$ is the symbolic statement for "2 is less than 6," which *means* that 2 is to the left of 6 on the number line.

Similarly, the symbol ">" is used for the expression "greater than." Thus, $a > b$ means a is to the right of b on the number line. The symbols "<" and ">" are called *inequality symbols.*

The next box contains a list of all our equality and inequality symbols, what each means, and an example of a *true* statement using each symbol.

Equality and Inequality Symbols			
	$a = b$	a "is equal to" b	$8 + 6 = 9 + 5$
	$a \neq b$	a "is not equal to" b	$4 + 5 \neq 10$
	$a < b$	a "is less than" b	$3 < 8$
	$a \leq b$	a "is less than *or* equal to" b	$3 \leq 8$
	$a > b$	a "is greater than" b	$7 - 2 > 4$
	$a \geq b$	a "is greater than *or* equal to" b	$6 \geq 6$

Note that $a \leq b$ means that either $a < b$ or $a = b$, and similarly, $a \geq b$ means $a > b$ or $a = b$.

Inequalities using the "<" and ">" symbols are called ***strict inequalities,*** whereas inequalities using the "≤" and "≥" symbols are called ***weak inequalities.***

If we are asked to "graph" a set on the number line, it means that we want to indicate those points on the real number line that are in the set. If we are graphing real-number inequalities, we use a heavy *line* (actually half-line or ray) as follows:

Strict inequalities:

$x > a$

a

For strict inequalities we place an **empty circle** *at the end point to indicate that the* **endpoint** *is* **excluded.**

$x < a$

a

Weak inequalities:

$x \geq a$

a

For weak inequalities we place a **filled-in circle** *at the endpoint to indicate that the* **endpoint is included.**

$x \leq a$

a

In all four cases above, the set of concern is the real numbers. The heavy line indicates that starting at or near *a,* all points on the line are included.

Algebraically, the symbol $<$ has the following meaning:

> $a < b$ means that $b - a$ is a positive number.

The symbol $\approx$ means "approximately equal to."

Hence $3 < 8$, *since* $8 - 3 = 5$ is positive; we can also see that $\sqrt{2} < 2$ *since* $2 - \sqrt{2} \approx 2 - 1.414 = 0.586$ is positive.

By the algebraic definitions of "$<$" and "$>$",

> $a > 0$ means that a is positive   and $a < 0$ means that a is negative.

DIFFERENT PERSPECTIVES: Inequalities

Geometric Description

$a < b$ means that a is to the left of b on the number line.

Algebraic Description

$a < b$ means that $b - a$ is a positive number.

The **double inequality,** $a < x < b$, is used to indicate "betweenness." For example,

$-3 < x < 6$ means that x is between -3 and 6 and is read "x is greater than -3 and less than 6."

We read the expression or variable in the middle first, then the left-hand number, and then the right-hand number.

The double inequality is actually a combination of two inequalities that must be satisfied simultaneously. That is, $a < x < b$ is actually a combination of the two inequalities

$$a < x \quad \textbf{and} \quad x < b$$

where *x satisfies* **both** *inequalities at the same time.* For example,

$3 < x < 7$ means that both $3 < x$ and $x < 7$.

For the double inequality $a < x < b$ to make sense, a must be less than b.

Unlike single inequalities, which have one endpoint (they begin at a point and go forever in one direction), double inequalities have two endpoints.

With double inequalities we use the same convention for representing endpoints on the number line as with single inequalities: an empty circle at the endpoint for strict inequalities ("$<$" and "$>$"), and a filled-in circle for weak inequalities ("$\le$" and "$\ge$"). Thus, the graph of $-2 \le x < 5$ would look like this:

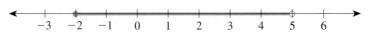

| EXAMPLE 3 | Graph the following inequalities on the number line. |

(a) $\{x \mid x > -1\}$ (b) $\{s \mid -3 \le s \le 6\}$ (c) $\{z \mid -5 \le z < 3, z \in Z\}$

Solution   **(a)** $\{x \mid x > -1\}$ *This is the set of all real numbers that are strictly greater than -1.*

Note the empty circle to indicate that the point -1 is excluded.

(b) $\{s \mid -3 \le s \le 6\}$ *This is the set of all (real) numbers lying between -3 and 6,*
including -3 and 6.

Note that both -3 and 6 are included (filled-in circles at -3 and 6).

(c) $\{z \mid -5 \le z < 3, z \in Z\}$

Note that we are limited to the integers rather than the real numbers, and that 3 is not included.

STUDY SKILLS 1.1

Coping with Getting Stuck

All of us have had the frustrating experience of getting stuck on a problem; sometimes even the simple problems can give us difficulty.

Perhaps you do not know how to begin; or you are stuck halfway through an exercise and are at a loss as to how to continue; or your answer and the book's answer do not seem to match. (Do not assume the book's solutions are 100% correct—we are only human, even if we are math teachers. But do be sure to check that you have copied the exercise accurately.)

Assuming you have reviewed all the relevant material beforehand, be sure you have spent enough time on the problem. Some people take one look at a problem and simply give up without giving the problem much thought. This is not what we regard as "getting stuck," since it is giving up before even having started.

If you find after a reasonable amount of time,

effort, and *thought,* that you are still not getting anywhere, and if you have looked back through your notes and textbook and still have no clue as to what to do, try to find exercises (with answers in the back) that you can do that are similar to the one on which you are stuck. Analyze what you did to arrive at the solution and try to apply those principles to the problem you are finding difficult. If you have difficulty with those similar problems as well, you may have missed something in your notes or in the textbook. Reread the material and try again. If you are still not successful, go on to different problems or take a break and come back to it later.

If you are still stuck, wait until the next day. Sometimes a good night's rest is helpful. Finally, if you are still stuck after reading the material, see your teacher (or tutor) as soon as possible.

EXERCISES 1.1

In Exercises 1–10, list the elements in the specified set.

1. $\{a \mid a \in W, a$ is greater than or equal to 3 and less than 12$\}$

2. $\{t \mid t \in W, t$ is greater than 30 and less than or equal to 42$\}$

3. $\{n \mid n \in W, n$ is greater than 8 and less than 6$\}$

4. $\{n \mid n \in N$ and n is a multiple of 8$\}$

5. $\{n \mid n$ is a prime number between 40 and 50, $n \in N\}$

6. $\{n \mid n$ is a composite number between 40 and 50, $n \in N\}$

7. $\{t \mid t \in W$ and t is a multiple of 7$\}$

8. $\{t \mid t \in W$ and t is a multiple of 3 but not of 6$\}$

9. $\{n \mid n$ is a factor of 54$\}$

10. $\{t \mid t$ is a factor of 36$\}$

In Exercises 11–16, let

$A = \{0, 1, 2, 3, 4, 5, 6\}$
$B = \{x \mid x \in W, x$ is a multiple of 3, and x is less than 36$\}$
$C = \{p \mid p$ is a prime number$\}$
$D = \{x \mid x$ is between 6 and 14, $x \in N\}$

List the elements in the indicated set.

11. $A \cap B$

12. $A \cap C$

13. $A \cup B$

14. $B \cap C$

15. $A \cup D$

16. $A \cap D$

In Exercises 17–24, write the number as a product of prime factors. If the number is prime, say so.

17. 66

18. 78

19. 128

20. 144

21. 61

22. 51

23. 91

24. 1,050

In Exercises 25–36, indicate whether the statement is true or false.

25. $-8 \in Z$

26. $0 \in Q$

27. Every whole number is an integer.

28. $-\dfrac{4}{7} \in Q$

29. $1.8 \in Q$

30. Every real number is a rational number.

31. $\sqrt{19} \in Q$

32. $\sqrt{19} \in R$

33. $W \subset N$

34. $Z \subset Q$

35. $Q \subset R$

36. $R \subset Q$

In Exercises 37–40, list all appropriate ordering symbols. Choose from $<$, $>$, $\leq$, $\geq$, $=$, and $\neq$.

37. $6 \cdot 2$ _____ $6 + 2$

38. $4 + 1$ _____ $4 \cdot 1$

39. $3 \cdot 0$ _____ $3 + 0$

40. $18 - 6$ _____ $6 \cdot 2$

In Exercises 41–56, graph the set on the number line. Unless otherwise indicated, all numbers are real numbers.

41. $\{x \mid x < 4\}$

42. $\{x \mid x > 4\}$

43. $\{a \mid a \leq -3\}$

44. $\{a \mid a \geq -3\}$

45. $\{y \mid y \geq -5\}$

46. $\{y \mid y \leq -4\}$

47. $\{y \mid -2 < y < 5, y \in Z\}$

48. $\{y \mid -5 < y < 2, y \in W\}$

49. $\{r \mid 5 \leq r \leq 10\}$

50. $\{r \mid -10 \leq r \leq 5\}$

51. $\{z \mid -3 < z \leq 0, z \in Z\}$

52. $\{n \mid 0 < n \leq 4, n \in N\}$

53. $\{x \mid -4 < x \leq -7\}$

54. $\{y \mid -5 \leq y \leq 8\}$

55. $\{a \mid -3 < a < 3\}$

56. $\{a \mid -4 \leq a \leq 4\}$

In Exercises 57–60, let

$$A = \{x \mid x \geq -3, x \in Z\}$$
$$B = \{x \mid x \leq -3, x \in Z\}$$
$$C = \{x \mid -4 \leq x \leq 6, x \in Z\}$$
$$D = \{x \mid 1 \leq x < 9, x \in Z\}$$

Graph the set on the number line and then describe it using set notation.

57. $C \cup D$

58. $A \cap B$

59. $A \cap D$

60. $B \cap C$

In Exercises 61–66, let

$$A = \{x \mid x \geq -3\}$$
$$B = \{x \mid x < -3\}$$
$$C = \{x \mid -4 \leq x \leq 6\}$$
$$D = \{x \mid 1 \leq x < 9\}$$

Graph the set on the number line and then describe it using set notation.

61. $A \cup B$

62. $A \cap B$

63. $C \cap D$

64. $C \cup D$

65. $A \cup C$

66. $A \cap C$

In Exercises 67–92, if the statement is true for all real numbers, name the property that the statement illustrates; otherwise, write "false."

67. $(5 + 3) + 7 = 5 + (3 + 7)$

68. $(5 + 3) + 7 = 7 + (5 + 3)$

69. $(x + y)z = (y + x)z$

70. $(x + y)z = z(x + y)$

71. $(x + y)z = xz + yz$

72. $(xy)z = x(yz)$

73. $10 - (7 - 3) = (10 - 7) - 3$

74. $a - (b - c) = (a - b) - c$

75. $(x + 3)(x + 2) = (x + 3)x + (x + 3)2$

76. $(x + 3)(x + 2) = x(x + 2) + 3(x + 2)$

77. $(a + b) \cdot 0 = a + b$

78. $(a + b) + 0 = a + b$

79. $x[(x + 3)(x + 2)] = [x(x + 3)](x + 2)$

80. $x\left(\dfrac{1}{x}\right) = 1 \qquad (x \neq 0)$

81. $(m + 3)\left(\dfrac{1}{m + 3}\right) = 1 \qquad (m \neq -3)$

82. $36 \cdot (12 \cdot 3) = (36 \cdot 12) \cdot 3$

83. $(r + s) + [-(r + s)] = 0$

84. $(a + b)[-(a + b)] = 0$

85. $(a + b)(e + f) = (a + b)(f + e)$

86. $abef = abfe$

87. $2x + 3x = (2 + 3)x$

88. $(a + b) + (-a + b) = 0$

89. The product of two integers is an integer.

90. The difference of two whole numbers is a whole number.

91. The quotient of two integers is an integer.

92. The difference of two integers is an integer.

❓ Questions for Thought

93. The following is a method of representing an infinitely repeating decimal as a fraction. We will show that $0.\underline{578}578578$ is a rational number.

Let $x = 0.578\overline{578578}$ *Multiply both sides of the equation by 1,000.*

Then $1,000x = 578.\overline{578578}$

Now subtract $x = 0.\overline{578578}$ from $1,000x = 578.\overline{578578}$:

$$1,000x = 578.\overline{578578}$$
$$-\qquad x = \quad 0.\overline{578578}$$

Notice that the decimal portions of the numbers match up exactly.

Therefore, $999x = 578$

$$x = \frac{578}{999}$$

Our decimal $x = 0.578\overline{578578}$ is $\frac{578}{999}$ and is therefore a rational number. Note that the equation $1,000x = 578.\overline{578578}$ is found by multiplying both sides of the equation $x = 0.578\overline{578578}$ by 1,000 (since 1,000 is the power of 10 needed to move the decimal point over and match up the infinitely repeating decimal portions of the numbers).

   Try this method of representing an infinitely repeating decimal as a fraction for $0.674\overline{674674}$. Try it for $0.929\overline{292}$.

94. Are both forms of the distributive property necessary? Could we derive one form from the other by using the other properties?

95. Is there a commutative or associative property for subtraction? Explain your answer.

96. Is there a commutative or associative property for division? Explain your answer.

97. Supply the reason for each step in the following proof for the multiplication property of zero. (We have already supplied the property for step 3, which will be discussed in Section 1.6.)

1. $a \cdot 0 = a \cdot (0 + 0)$

2. $a \cdot 0 = a \cdot 0 + a \cdot 0$

3. $-(a \cdot 0) + a \cdot 0 = -(a \cdot 0) + (a \cdot 0 + a \cdot 0)$ Addition property of equality

4. $0 = -(a \cdot 0) + (a \cdot 0 + a \cdot 0)$

5. $0 = [-(a \cdot 0) + (a \cdot 0)] + a \cdot 0$

6. $0 = 0 + a \cdot 0$

7. $0 = a \cdot 0$

1.2   Operations with Real Numbers

In Section 1.1, we discussed the property of additive inverses. We mentioned that the additive inverse of a number is also called the opposite or negative of the number. We represent an additive inverse by putting a negative sign before a number (hence the term *negative*). Thus, $-x$ is the additive inverse of x; by definition, it is the number that when added to x, will yield 0.

-3 is the additive inverse of 3   because $(-3) + 3 = 0$

6 is the additive inverse of -6 because $6 + (-6) = 0$

In the last example, note that 6 is the additive inverse of -6. This illustrates that the additive inverse of a number is not necessarily negative. The *additive inverse of a number is a number opposite in sign* (hence the term *opposite*). Symbolically we have

THEOREM

$$-(-x) = x$$

In words, the opposite of $-x$ is x.

The following theorem can be derived from the real number properties and will be useful to us later on.

THEOREM

$$(-1)x = -x$$

Absolute Value

Geometrically, the **absolute value** of a number is its distance from 0 on the number line. The absolute value of x is symbolized as $|x|$. Hence,

$|-4| = 4$ since -4 is 4 units away from 0 on the number line.

$|4| = 4$ since 4 is 4 units away from 0 on the number line.

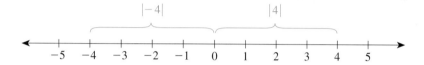

Algebraically, we define absolute value as follows:

$$|x| = \begin{cases} x & \text{if } x \geq 0 \\ -x & \text{if } x < 0 \end{cases}$$

This notation will be discussed in more detail in a later chapter. For now we read it as follows: If x is positive or 0, then the absolute value of x is x. If x is less than 0 (that is, negative), then the absolute value of x is $-x$, which is positive. Consequently, $|x|$ *can never be negative.* For example, by the algebraic definition of absolute value, we have

$$|-8| = -(-8) \qquad \textit{Since } -8 < 0$$
$$= 8 \qquad \textit{By the theorem above: } -(-x) = x$$
$$|0| = 0 \qquad \textit{Since } 0 \geq 0$$
$$|6| = 6 \qquad \textit{Since } 6 \geq 0$$

We introduce the concept of absolute value here so that we may refer back to whole numbers in defining operations with signed numbers.

Addition of Real Numbers

Recall from elementary algebra that we defined addition of signed numbers as moving a specified distance either right (if positive) or left (if negative) on the number line. Hence, $(-8) + (+3)$ can be represented on the number line as follows:

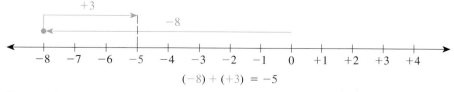

$$(-8) + (+3) = -5$$

As a result, we developed the following rules for adding signed numbers:

| **Addition of Signed Numbers** | 1. When adding two numbers with the same sign, add their absolute values and keep their common sign.
2. When adding two numbers with opposite signs, subtract the smaller absolute value from the larger, and keep the sign of the number with the larger absolute value. |

As usual, we use the symbol "+" to indicate addition, and would represent an addition problem as follows:

$$(-4) + (+3) \qquad \textit{The sum of } -4 \textit{ and } +3$$

EXAMPLE 1

Find the following sums.

(a) $(-6) + (-11)$
(b) $(+7) + (-18)$
(c) $(+9) + (-3)$
(d) $(-4.2) + (+6.3) + (-2.8) + (-5.9)$

Solution

(a) $(-6) + (-11)$
$$= -(6 + 11) = \boxed{-17}$$
↑
and keep the negative sign.

Both numbers have the same signs. Since $|-6| = 6$ and $|-11| = 11$, we add 6 and 11 to get 17

(b) $(+7) + (-18)$
$$= -(18 - 7) = \boxed{-11}$$
↑
and keep the negative sign since $|-18| > |+7|$.

The two numbers have opposite signs. Since $|+7| = 7$ and $|-18| = 18$, we find the difference in absolute values

(c) $(+9) + (-3) = +(9 - 3) = \boxed{+6}$

(d) $(-4.2) + (+6.3) + (-2.8) + (-5.9) = (+2.1) + (-2.8) + (-5.9)$
$$= (-0.7) + (-5.9)$$
$$= \boxed{-6.6}$$

Actually, we could derive the rules for adding signed numbers from the 11 properties given in Section 1.1. For example, to add -4 and -6:

$$(-4) + (-6) = (-1)4 + (-1)6 \qquad \textit{By theorem: } -x = (-1)x$$
$$= (-1)(4 + 6) \qquad \textit{Distributive property}$$
$$= (-1)(10) \qquad \textit{Addition of whole numbers}$$
$$= -10 \qquad \textit{Again, since } -1(x) = -x$$

The rules for addition allow us to shorten this process and should be used instead. The above steps demonstrate the power of the 11 real number properties to define addition of signed numbers.

Subtraction

To find $7 - 4$, we are looking for the number that when added to 4 will yield 7.

Just as with whole numbers, subtraction of real numbers is defined in terms of addition. To find $a - b$, you look for a number that, when added to b, will yield a.

We define subtraction as follows:

$$a - b = a + (-b)$$

Hence, subtracting b is adding the additive inverse of b. Since the additive inverse changes the sign of a number, **to subtract b we change the sign of b and add.**

EXAMPLE 2

Perform the indicated operations.

(a) $(-8) - (+4)$

(b) $(+7) - (-2)$

(c) $(-16) - (-7)$

(d) $(-8.41) + (-6.23) - (-4.9) - (-12)$

(e) $-6 - 3 + 5 - 4 + 7$

Solution

(a) $(-8) - (+4) = (-8) + (-4)$ *Change sign of +4 and add.*

$\qquad = \boxed{-12}$

(b) $(+7) - (-2) = (+7) + (+2)$ *Change −2 to +2 and add.*

$\qquad = \boxed{+9}$

(c) $(-16) - (-7) = (-16) + (+7)$ *Change −7 to +7 and add.*

$\qquad = \boxed{-9}$

(d) For this example it would be easier to use a calculator. If you are using a scientific calculator, we can enter the sign of the number using the ⌑+/−⌑ key *after we enter the number.* We enter the following:

⌑ (⌑ ⌑ 8.41 ⌑ ⌑ +/− ⌑ ⌑) ⌑ ⌑ + ⌑ ⌑ (⌑ ⌑ 6.23 ⌑ ⌑ +/− ⌑ ⌑) ⌑ ⌑ − ⌑ ⌑ (⌑ ⌑ 4.9 ⌑ ⌑ +/− ⌑ ⌑) ⌑ ⌑ − ⌑ ⌑ (⌑ ⌑ 12 ⌑ ⌑ +/− ⌑ ⌑) ⌑ ⌑ = ⌑

to get the answer ⌑ 2.26 ⌑.

Since the calculator reads the ⌑+/−⌑ as part of the number, we do not need to use parentheses. Instead, we can press the following keys:

⌑ 8.41 ⌑ ⌑ +/− ⌑ ⌑ + ⌑ ⌑ 6.23 ⌑ ⌑ +/− ⌑ ⌑ − ⌑ ⌑ 4.9 ⌑ ⌑ +/− ⌑ ⌑ − ⌑ ⌑ 12 ⌑ ⌑ +/− ⌑ ⌑ = ⌑

If you use a graphing calculator to compute this expression, you may enter the expression exactly as you read it. However, be careful that you distinguish between the symbol used for subtraction, ⌑−⌑ (found above the ⌑+⌑ key), and the symbol used to indicate the opposite, or additive inverse, of a number ⌑(−)⌑. Note that the negative symbol is shorter and a bit higher up than the subtraction symbol. For graphing calculators we enter the following:

⌑ (−) ⌑ ⌑ 8.41 ⌑ ⌑ + ⌑ ⌑ (−) ⌑ ⌑ 6.23 ⌑ ⌑ − ⌑ ⌑ (−) ⌑ ⌑ 4.9 ⌑ ⌑ − ⌑ ⌑ (−) ⌑ ⌑ 12 ⌑ ⌑ ENTER ⌑

Again we get the answer ⌑ 2.26 ⌑.

(e) $-6 - 3 + 5 - 4 + 7$

We often drop the addition symbol between signed numbers we are adding. In this case addition is understood: We add (-6), (-3), $(+5)$, (-4), and $(+7)$.

$$-6 - 3 + 5 - 4 + 7$$
$$= -9 + 5 - 4 + 7 \qquad \text{Since } (-6) + (-3) = -9$$
$$= -4 - 4 + 7 \qquad \text{Since } (-9) + (+5) = -4$$
$$= -8 + 7 \qquad \text{Since } (-4) + (-4) = -8$$
$$= \boxed{-1}$$

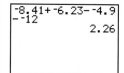

A typical graphing calculator screen for 1(d)

Note that since the graphing calculator reads ⌑(−)⌑ *as part of the number, we do not need to use parentheses.*

Multiplication

In elementary algebra, we derived the rules for multiplying signed numbers by viewing multiplication as repeated addition. Thus, for example,

$$3(-2) = (-2) + (-2) + (-2) = -6$$

hence, the product of a positive number and a negative number is negative.

Rules for Multiplying Two Signed Numbers	1. If two numbers have the same sign, then their product is positive. 2. If two numbers have opposite signs, then their product is negative.

Again, we can use the properties to derive the rules (see the Question for Thought at the end of Exercises 1.2). In general, we can derive the following theorems:

THEOREM	1. $(-a)b = a(-b)$ 2. $a(-b) = -(ab)$ 3. $(-a)(-b) = ab$

EXAMPLE 3   Perform the operations.

(a) $(-3)(+8)$ (b) $(-6)(-9)$ (c) $(-7)(+2)(-3)(-1)$

Solution   (a) $(-3)(+8) = \boxed{-24}$

(b) $(-6)(-9) = \boxed{+54}$

(c) $(-7)(+2)(-3)(-1) = (-14)(-3)(-1)$

$$= (+42)(-1) = \boxed{-42}$$

When multiplying more than two signed numbers together, we can figure out the sign of the product quickly as follows:

If there is an *even* number of negative factors in a product, the product will be positive; otherwise, the product will be negative.

For example, $(-2)(-3)(+1)(-2)(-4)$ will be *positive* because there are four (an even number of) negative factors in the product.

Division

Division is defined in terms of multiplication. To find $\frac{a}{b}$, we look for the number that, when multiplied by b, will yield a. For example, $\frac{6}{3} = 2$ because $2 \cdot 3 = 6$.

Just as we defined subtraction as adding the additive inverse, we define division as multiplication by the multiplicative inverse:

$$\frac{a}{b} = a\left(\frac{1}{b}\right)$$

As a result, the sign rules for division are the same as those for multiplication.

Rules for Division of One Signed Number by Another	1. If the signs are the same, the quotient is positive. 2. If the signs are opposite, the quotient is negative.

What happens if we try to divide a nonzero number by 0? Let's assume $\frac{6}{0} = x$; then this means $x \cdot 0 = 6$. But since multiplication by 0 will only yield 0, we end up with $0 = 6$, which is of course false. Thus, for any *nonzero* number a, $\frac{a}{0}$ will yield the contradiction $a = 0$. Therefore, division by 0 is *undefined.* On the other hand, $\frac{0}{a} = 0$ since $0 \cdot a = 0$.

If we try to divide 0 by 0, we run into problems for another reason. Let's suppose $\frac{0}{0}$ represents some number, call it r. Then this means $r \cdot 0 = 0$. But this is true for all numbers r. This means that any number will work; that is, $\frac{0}{0}$ is not unique. Therefore, we say that $\frac{0}{0}$ is indeterminate (there is no unique answer).

> If a is any nonzero number, then
>
> $$\frac{a}{0} \text{ is undefined}$$
>
> $$\frac{0}{a} = 0$$
>
> $$\frac{0}{0} \text{ is indeterminate}$$

EXAMPLE 4

Perform the operations.

(a) $(-6) \div (-3)$ **(b)** $\dfrac{-21}{+7}$ **(c)** $\dfrac{-5}{0}$

Solution

(a) $(-6) \div (-3) = \boxed{+2}$

(b) $\dfrac{-21}{+7} = \boxed{-3}$

(c) $\dfrac{-5}{0}$ is $\boxed{\text{undefined}}$.

Exponents

Multiplication by a positive integer is shorthand for repeated addition (for example, $5 + 5 + 5 + 5 = 4 \cdot 5$). Similarly, exponential notation is shorthand for repeated multiplication. For example, we would write $5 \cdot 5 \cdot 5 \cdot 5$ as 5^4 using exponential notation. We formally define exponential notation in the box below.

Exponential Notation	$x^n = x \cdot x \cdot x \cdot x \cdot \ \cdots \ \cdot x$ where the factor x occurs n times. In x^n, x is called the ***base*** and n is called the ***exponent.*** A natural number exponent tells how many times x occurs as a factor in the product. The expression x^n is called a ***power,*** the nth power of x.

Note that $x = x^1$; that is, a missing exponent is assumed to be 1.

The expression x^2 is read "x squared" and x^3 is read "x cubed."

To compute or evaluate an expression with a numerical base means to multiply out the expression.

EXAMPLE 5

Write the following without exponents and then evaluate.

(a) 3^5 **(b)** $(-2)^4$ **(c)** -2^4

Solution

(a) $3^5 = 3 \cdot 3 \cdot 3 \cdot 3 \cdot 3$ *Written without exponents*

$= \boxed{243}$ *Evaluated*

(b) $(-2)^4 = (-2)(-2)(-2)(-2)$ *Written without exponents*

$= \boxed{+16}$ *Evaluated*

(c) $-2^4 = -(2 \cdot 2 \cdot 2 \cdot 2) = \boxed{-16}$ *Note: Only "2" is raised to the fourth power.*

Note the differences between parts (b) *and* (c).

Multiple Operations

If there is more than one operation to perform in an expression, there must be agreement as to which operation should be performed first—an order of operations. For example, given the expression $3 + 2 \cdot 4$, if addition were performed first, the value would be 20; if multiplication were performed first, the value would be 11. To rectify this ambiguity, the order of operations given in the next box is agreed upon.

Order of Operations

1. Start by performing operations in the grouping symbols, innermost first.
2. Calculate powers (and roots) in any order.
3. Perform multiplication and division, working left to right.
4. Perform addition and subtraction, working left to right.

EXAMPLE 6

Perform the operations.

(a) $38 - 5^2$ **(b)** $7^2 + 3^2 - (2 + 3)^2$ **(c)** $3 + 7 - 2[5 - 3(9 - 8)]$

Solution

(a) $38 - 5^2$ *Powers first; you are subtracting 5^2.*

$= 38 - 25$

$= \boxed{13}$

(b) $7^2 + 3^2 - (2 + 3)^2$ *Addition in parentheses*

$= 7^2 + 3^2 - (5)^2$ *Then powers*

$= 49 + 9 - 25$ *Next we add.*

$= 58 - 25$

$= \boxed{33}$

(c) $3 + 7 - 2[5 - 3(9 - 8)]$ *Innermost grouping symbol first*

$= 3 + 7 - 2[5 - 3(1)]$ *Then multiplication in brackets*

$= 3 + 7 - 2[5 - 3]$ *Then subtraction in brackets*

$= 3 + 7 - 2[2]$ *Next we multiply.*

$= 3 + 7 - 4$

$= \boxed{6}$

Expressions involving operations with integers must be read very carefully. For example,

$(-3)(-2)$ is a multiplication problem.
$(-3) - (2)$ is a subtraction problem.

Keep in mind that grouping symbols do not represent operations in themselves. It is the symbol between *pairs* of grouping symbols that indicates the operation to be performed (the lack of a symbol between pairs of grouping symbols indicates multiplication). This may seem obvious to you as presented above in its simplified form, but is not as obvious in an expression such as

$$7 - (-7)(-7) = 7 - (49) = -42$$

Two other expressions that are distinct and yet frequently confused concern integers raised to powers. For example,

$$-2^4 = -(2 \cdot 2 \cdot 2 \cdot 2) = -16 \quad \text{and} \quad (-2)^4 = (-2)(-2)(-2)(-2) = +16$$

Keep in mind that the exponent applies only to that which is to its immediate left. In the expression $3 \cdot 2^4$, we raise 2 to the fourth power and then multiply by 3 to get 48. Similarly, in -2^4, we raise 2 to the fourth power and then use the negative sign to get -16.

A bar or line appearing above or below an expression, such as a fraction bar, is treated as a grouping symbol. For example, in simplifying the expression

$$\frac{9 - 5 \cdot 3}{2 \cdot 4 - 10}$$

we perform all operations above the fraction bar and all operations below the fraction bar separately to get $\dfrac{9 - 15}{8 - 10} = \dfrac{-6}{-2}$, *before* performing the division indicated by the fraction bar to get 3.

Absolute value symbols have the priority of grouping symbols; that is, we must perform all operations within absolute value symbols before evaluating the absolute value of an expression.

EXAMPLE 7

Perform the operations.

(a) $\left(-\dfrac{3}{4}\right)(-8) - (-5)(-1)$ **(b)** $-6(-3 - 2) - 3^2$

(c) $-6(-3 - 2)(-3)^2$ **(d)** $\dfrac{4[-5 - 3(-2)^2]}{-6 - 7(-3 - 2) + 5}$

(e) $|-5 - 7| - |7 - 5|$

Solution   (a) $\left(-\dfrac{3}{4}\right)(-8) - (-5)(-1)$ *Multiply first.*

$= (+6) - (+5)$ *Then subtract.*

$= 6 - 5$

$= \boxed{1}$

(b) $-6(-3 - 2) - 3^2$ *Operations in parentheses first.*

$= -6(-5) - 3^2$ *Then powers.*

$= -6(-5) - 9$ *Next multiply.*

$= 30 - 9$ *Subtract.*

$= \boxed{21}$

(c) $-6(-3 - 2)(-3)^2$ *Parentheses first*

$= -6(-5)(-3)^2$ *Then powers*

$= -6(-5)(+9)$ *Multiply.*

$= \boxed{270}$

Note the differences between parts (b) and (c).

(d) $\dfrac{4[-5 - 3(-2)^2]}{-6 - 7(-3 - 2) + 5} = \dfrac{4[-5 - 3(+4)]}{-6 - 7(-5) + 5}$

$= \dfrac{4[-5 - 12]}{-6 + 35 + 5}$

$= \dfrac{4(-17)}{34} = \dfrac{-68}{34} = \boxed{-2}$

(e) $|-5 - 7| - |7 - 5|$ *Perform operations in absolute value symbols first.*

$= |-12| - |2|$ *Then find absolute values.*

$= 12 - 2$ *Then subtract.*

$= \boxed{10}$

If you use the calculator to evaluate a fractional expression in which the numerator and/or the denominator contains more than one term or factor, *you must enclose the numerator and/or the denominator in parentheses.* For example, if you were to try to compute $\dfrac{3 + 4}{6 + 2}$ by entering $3 + 4 \div 6 + 2$, the calculator, following the order of operations, interprets this expression as $3 + \dfrac{4}{6} + 2$. Instead, you must enter $\dfrac{3 + 4}{6 + 2}$ as $(3 + 4) \div (6 + 2)$.

For the same reason, if you try to compute $\dfrac{4}{3 \cdot 5}$ by entering this expression as $4 \div 3 \cdot 5$, the calculator will interpret this as $\dfrac{4}{3} \cdot 5$ (leftmost operation first), so you must enter it as $4 \div (3 \times 5)$.

```
(3+4)/(6+2)
            .875
4/(3*5)
      .2666666667
```

Entering $\dfrac{3 + 4}{6 + 2}$ and $\dfrac{4}{3 \cdot 5}$ on the graphing calculator

EXAMPLE 8

Evaluate the numerical expression $\dfrac{-5.2 + 3.85}{6.7 - 8.4}$ correct to four decimal places.

Solution

The calculator should be used to evaluate this expression. We enter the expression

$\dfrac{-5.2 + 3.85}{6.7 - 8.4}$ as $(-5.2 + 3.85) \div (6.7 - 8.4)$ in a scientific calculator, using the following keys:

to get the answer ⟨ .7941 ⟩ correct to four places.

EXAMPLE 9

Evaluate the numerical expression

$$3\left(\frac{2}{9}\right)^2 + 4\left(\frac{2}{9}\right) + 5$$

(a) Find the exact answer.

(b) Find the answer rounded to two decimal places using a calculator.

Solution

(a) $3\left(\dfrac{2}{9}\right)^2 + 4\left(\dfrac{2}{9}\right) + 5$ *Square $\dfrac{2}{9}$ first to get $\left(\dfrac{2}{9}\right)\left(\dfrac{2}{9}\right) = \dfrac{4}{81}$.*

$= 3\left(\dfrac{4}{81}\right) + 4\left(\dfrac{2}{9}\right) + 5$ *Multiply.*

$= \dfrac{4}{27} + \dfrac{8}{9} + 5$ *Change to equivalent fractions.*

$= \dfrac{4}{27} + \dfrac{24}{27} + \dfrac{135}{27}$

$= \boxed{\dfrac{163}{27} \quad \text{or} \quad 6\dfrac{1}{27}}$

If your calculator does not have an x^2 key but has a y^x key, you can follow the same sequence only substitute y^x 2 for x^2.

(b) Using a scientific calculator, we would press the following keys:

⟨ 3 ⟩ ⟨ × ⟩ ⟨ (⟩ ⟨ 2 ⟩ ⟨ ÷ ⟩ ⟨ 9 ⟩ ⟨) ⟩ ⟨ x^2 ⟩ ⟨ + ⟩ ⟨ 4 ⟩ ⟨ × ⟩ ⟨ (⟩ ⟨ 2 ⟩ ⟨ ÷ ⟩ ⟨ 9 ⟩ ⟨) ⟩ ⟨ + ⟩ ⟨ 5 ⟩ ⟨ = ⟩

to get ⟨ 6.037037 ⟩ to six places, which rounded to two places, is ⟨ 6.04 ⟩.

Compare the answers to parts **(a)** and **(b).**

Substitution

Applications of algebra very often require us to substitute numerical values of variables and then to perform the operations with the substituted values. In such a situation, we will be asked to evaluate an expression given values for the variables.

EXAMPLE 10

Given $a = -3$, $b = -2$, $c = 4$, and $d = 0$, evaluate the following.

(a) $a^2 - 2ab + b^2$ **(b)** $\dfrac{(4a^2 + 3c)d}{6}$

Solution

It is usually a good idea to enclose the values being substituted in parentheses. This helps us to avoid confusing the original operations.

(a) $a^2 - 2ab + b^2 = \quad a^2 - 2 \cdot \quad a \quad \cdot \quad b \quad + \quad b^2$

$$= (-3)^2 - 2 \cdot (-3) \cdot (-2) + (-2)^2$$

$$= 9 - 2(6) + 4$$

$$= 9 - 12 + 4 = \boxed{1}$$

(b) $\dfrac{(4a^2 + 3c)d}{6} = \dfrac{[4 \cdot a^2 + 3 \cdot c] \cdot d}{6}$

$$= \frac{[4(-3)^2 + 3(4)](0)}{6} \qquad \textit{Multiplication by 0 yields 0.}$$

$$= \frac{0}{6} = \boxed{0}$$

EXAMPLE 11

Using population data from 1983 to 2002 presented in Figure 1.3, we can estimate the total population of the United States by the linear equation $P = 2.9375t - 5593$, or by the quadratic equation $P = 0.0341215t^2 - 133.036t + 129868$, where P is the population in millions in year t.

U.S. Population: 1983-2002

[In thousands, except percent (180,671 represents 180,671,000). Estimates as of July 1. Total population includes Armed Forces abroad; civilian population excludes Armed Forces. For bases of estimates, see text of this section.]

Year	Total Population	Percent change[1]	Resident population	Civilian population
1983	234,307	0.91	233,792	232,097
1984	236,348	0.87	235,825	234,110
1985	238,466	0.90	237,924	236,219
1986	240,651	0.92	240,133	238,412
1987	242,804	0.89	242,289	240,550
1988	245,021	0.91	244,499	242,817
1989	247,342	0.95	246,819	245,131
1990	250,132	1.13	249,623	247,938
1991	253,493	1.34	252,981	251,370
1992	256,894	1.34	256,514	254,929
1993	260,255	1.31	259,919	258,446
1994	263,436	1.22	263,126	261,714
1995	266,557	1.18	266,278	264,927
1996	269,667	1.17	269,394	268,108
1997	272,912	1.20	272,647	271,394
1998	276,115	1.17	275,854	274,633
1999	279,295	1.15	279,040	277,841
2000	282,388	1.11	282,178	280,954
2001	285,321	1.04	285,094	283,879
2002	288,205	1.01	287,974	286,728

[1]Percent change from immediate preceding year.

Source: U.S. Census Bureau, *Current Population Reports*, P25-802 and P25-1095; Table CO-EST2001-12-00-Time Series of Intercensal State Population Estimates: April 1, 1990 to April 1, 2000 ; published 11 April 2002;<http://eire.census .gov/ popest/data/counties/tables/CO-EST2001-12/CO-EST2001-12-00.php>; Table NA-2003EST-01-Monthly Population Estimates for the United States: April 1, 2000 to March 1, 2004 ; published: 28 May 2004; <http://eire.census.gov/popest/data/ national/tables/NA-EST2003-01.php>. and unpublished data.

Figure 1.3
U.S. Population: 1983–2002

(a) Compute the total population in 2003 predicted by each equation to the nearest thousand.

(b) The actual population in 2003 was 291,049,000. Which equation gives the better estimate for the actual population in 2003?

Solution

(a) We compute the resident population of the United States by substituting 2003 into each equation using a calculator. For the linear equation we press the keys

$\boxed{2.9375}$ $\boxed{\times}$ $\boxed{2003}$ $\boxed{-}$ $\boxed{5593}$ to get $\boxed{290.8125}$, which is 290,812,500.

For the quadratic equation we press the keys $\boxed{.0341215}$ $\boxed{\times}$ $\boxed{2003}$ $\boxed{x^2}$ $\boxed{-}$ $\boxed{133.036}$ $\boxed{\times}$ $\boxed{2003}$ $\boxed{+}$ $\boxed{129868}$ to get $\boxed{292.6571}$ or 292,657,000 rounded to the nearest thousand.

(b) We are interested in which predicted value comes closer to the actual value, so we look at the differences between the predicted values and the actual value. Since it is of no interest whether the predicted estimate is above or below the actual value, we look at the absolute value of the differences. For the linear equation, the difference is $|290{,}812{,}500 - 291{,}049{,}000| = 236{,}500$. For the quadratic equation we have $|292{,}657{,}000 - 291{,}049{,}000| = 1{,}608{,}000$. We can see that the linear equation yields a smaller difference in absolute value and therefore gives a better predicted value.

EXAMPLE 12

Given $s_e = s_y \sqrt{1 - r^2}$, $s_y = 1.2$, and $r = 0.42$, find s_e rounded to two places.

Solution

$$s_e = s_y \sqrt{1 - r^2} \qquad \textit{Substitue } s_y = 1.2 \textit{ and } r = 0.42.$$
$$= 1.2 \sqrt{1 - (0.42)^2}$$

Given the nature of the computation we are required to perform for this problem, and that we are allowed to round the answer, we can see that this problem requires the use of the calculator. Using a scientific calculator, we press the following keys:

$\boxed{1.2}$ $\boxed{\times}$ $\boxed{(}$ $\boxed{1}$ $\boxed{-}$ $\boxed{.42}$ $\boxed{x^2}$ $\boxed{)}$ $\boxed{}$ $\boxed{=}$

to get $\boxed{1.089029}$ (to six places), which, rounded to two places, is

$$\boxed{1.09}$$

STUDY SKILLS 1.2

Seeking Help from Your Instructor

In the last Study Skill we discussed some ideas dealing with how to cope if you get stuck on a particular problem. However, sometimes you may feel that the difficulty you are having is more substantial than just the inability to do a problem.

If you find yourself in such a situation it is important to recognize that one of the keys to success in a mathematics class is knowing when and where to get the help you need.

Your first step should always be to make an honest effort to review your class notes and the relevant text material. If this does not answer your question, then you should try to speak to your instructor after class. Sometimes just a word or two from your instructor can clarify a point that was interfering with your understanding of the material.

If this does not work then you need to get additional help outside the class. Your instructor has most likely informed you of his or her office hours. These are the times that your instructor is available to give you individualized attention. Going to office hours affords you the opportunity to explain your difficulty to your instructor in much greater detail than is normally possible by asking questions during class. This enables your instructor to understand your difficulty much more clearly, and therefore to give you a more complete and meaningful response.

In order to make your visit as productive as possible, it is best that you formulate your questions clearly beforehand. (It is a good idea to write them down.) Just telling your instructor "I don't understand" does not help your instructor in assisting you.

One additional point: Some students feel that they are a burden on their instructor when they come to office hours. This is not true. Your instructor is there to help you and is pleased that you are making the extra effort to learn the material.

EXERCISES 1.2

In Exercises 1–82, evaluate each of the expressions, if possible.

1. $-3 + 8$

2. $-8 + 3$

3. $-3 - 8$

4. $8 - (-3)$

5. $-3(-8)$

6. $-8(3)$

7. $1.692 - 3.965 + 8.754$

8. $-8.236 - 12.257 + 4.3$

9. $-3 - 4 - 5$

10. $7 - 5 - 6$

11. $-3 - 4(5)$

12. $7 - 5(-6)$

13. $-3(-4)(-5)$

14. $7(-5)(-6)$

15. $3(-4 - 5)$

16. $7(-5 - 6)$

17. $8 - 4 \cdot 3 - 7$

18. $(9 - 6)(2 - 5)$

19. $8 - (4 \cdot 3 - 7)$

20. $9 - (6 \cdot 2 - 5)$

21. $(8 - 4)(3 - 7)$

22. $9 - 6(2 - 5)$

23. $(2.6)^2 - |7.8 - 13.69|$

24. $|-2.94 - 3.8| + (3.9)^2$

25. $\dfrac{-20}{-5}$

26. $\dfrac{-28}{7}$

27. $\dfrac{-5 - 11}{-9 + 4}$

28. $\dfrac{-8 - 7}{-2 - 7}$

29. $\dfrac{-10 - 2 - 4}{-2}$

30. $\dfrac{-9 - 3 - 6}{-3}$

31. $\dfrac{-10 - (2 - 4)}{-2}$

32. $\dfrac{-9 - (3 - 6)}{-3}$

33. $\dfrac{-10 - 2(-4)}{-2}$

34. $\dfrac{-9 - 3(-6)}{-3}$

35. $\dfrac{-4(-3)(-6)}{-4(-3) - 6}$

36. $\dfrac{-2(-4)(-6)}{-2(-4) - 6}$

37. $8 - 3(5 - 1)$

38. $9 - 5(4 - 1)$

39. $-7 - 2(4 - 6)$

40. $-6 - 3(2 - 5)$

41. $7 + 2[4 + 3(4 + 1)]$

42. $6 + 3[1 + 2(3 + 1)]$

43. $7 - 2[4 - 3(4 - 1)]$

44. $6 - 3[1 - 2(3 - 1)]$

45. $8 - \left(\dfrac{10}{-5}\right)$

46. $-6 - \left(\dfrac{-8}{-4}\right)$

47. $8\left(\dfrac{10}{-5}\right)$

48. $-6\left(\dfrac{-8}{-4}\right)$

49. $\dfrac{12}{-4} - \left(\dfrac{10}{-2}\right)$

50. $\dfrac{18}{-3} - \left(\dfrac{12}{-6}\right)$

51. $\dfrac{-8 + 2}{4 - 6} - \dfrac{6 - 11}{-3 - 2}$

52. $\dfrac{13 - (-5)}{-3 - 3} - \dfrac{13 - (-5)}{-3(-3)}$

53. $-12 - \dfrac{6 - 2(-3)}{-3}$

54. $18 + \dfrac{7 - 4(-2)}{-5}$

55. $(-6)^2$

56. -6^2

57. $2^2 + 3^2 + 4^2$

58. $(2 + 3 + 4)^2$

59. $2(5)^2$

60. $(2 \cdot 5)^2$

61. -3^4

62. $(-3)^4$

63. $-8 - 2(-4)^2$

64. $10 - 3(2)^3$

65. $2(-5)(-6)^2$

66. $3(-4)(-5)^2$

67. $2(-5) - 6^2$

68. $3(-4) - 5^2$

69. $\dfrac{5[-8 - 3(-2)^2]}{-6 - 6 - 2}$

70. $\dfrac{5[-12 - 2(-3)^2]}{-4 - 6 - 2}$

71. $-3 - 2[-4 - 3(-2 - 1)]$

72. $-5 - 4[3 - 4(-3 - 2)]$

73. $-3\{5 - 3[2 - 6(3 - 5)]\}$

74. $-5\{6 - 2[3 - 2(-2 - 2)]\}$

75. $|3 - 8| - |3| - |-8|$

76. $|2 - 9| - |-2| - |9|$

77. $|-3 - 2 - 4| - (2 - 3)^3$

78. $|-4 - 5 - 1| - (5 - 6)^3$

79. $(-3 - 2)^2 - (-3 + 2)^2$

80. $(-5 - 4)^2 - (-5 + 1)^2$

81. $\dfrac{42.2}{1.63 - (2.1)(5.8)}$

82. $\dfrac{-9.6}{8.49 - (2.3)^2}$

83. Richard has a balance of $1,542.75 in his checkbook. If his balance falls below $1,000.00 any time during the month, the bank deducts $10.00 from his account. He wrote the following checks and made the following deposits, which were posted by the bank in the following order: checks for $234.25, $134.82, and $68.05 on 8/12, a deposit of $352.00 on 8/14, checks for $274.15 on 8/15 and $392.24 on 8/16, a deposit of $1,241.00 on 8/21, and checks for $382.15, $112.12, $142.55, and $155.00 on 8/24. How much did he have in his account?

84. As of August 28, 2000, the lowest temperature ever recorded was in Vostok, Antarctica, at $-89.2°C$, on July 21, 1983. The highest temperature ever recorded was in Death Valley, CA, on July 10, 1913, at $56.7°C$. Find the difference between these extreme temperatures.

85. The highest point on earth is Mt. Everest in the Himalayan Mountains in Nepal, at 29,035 feet above sea level. The lowest point on the earth is on the shore of the Dead Sea at the Israel–Jordan border, recently recorded at $-1,300$ feet. Find the difference between the highest and lowest points on earth.

86. On July 12, Colleen bought 220 shares of stock selling for $112.50 per share. On September 4, when the stock went down to $82.25 per share, she considered selling all shares. She decided not to sell at that time, but did sell all her shares when the stock moved back up to $94.50 on October 7. Not counting the commissions and fees, how much did she lose when she sold her stock in October? How much would she have lost if she had sold her stock on September 4? How much less did she lose as a result of waiting to sell her stocks in October rather than selling in September?

In Exercises 87–88, evaluate the numerical expressions by (a) finding the exact answer and (b) finding the answer rounded to two decimal places using a calculator.

87. $6\left(\dfrac{3}{7}\right)^2 + 4\left(\dfrac{3}{7}\right) - 3$

88. $5\left(\dfrac{-5}{6}\right)^2 + 11\left(\dfrac{-5}{6}\right) + 8$

In Exercises 89–98, evaluate the expressions for $x = -2$, $y = -3$, and $z = 5$.

89. $x + y + z$

90. $x(y + z)$

91. xyz

92. $xy + z$

93. $-x^2 - 4x + 2$

94. $-z^2 + 3z + 1$

95. $|xy - z|$

96. $|x - y - z|$

97. $\dfrac{3x^2y - x^3y^2}{3x - 2y}$

98. $\dfrac{3x^2y + x^2y^2}{3x + 2y}$

99. Given $s_e = s_y\sqrt{1 - r^2}$, find s_e rounded to two places if $s_y = 2.3$ and $r = 0.74$.

100. Given $t = \dfrac{\overline{X} - a}{\dfrac{s_x}{\sqrt{n}}}$, compute t to two places, for

    (a) $\overline{X} = 78$, $a = 70$, $s_x = 6.5$, $n = 15$

    (b) $\overline{X} = 46$, $a = 70$, $s_x = 6.5$, $n = 15$

101. Given $\sigma_r = \sqrt{\dfrac{1 - \rho^2}{n - 1}}$, compute σ_r to three places, for

    (a) $\rho = 0.72$, $n = 100$

    (b) $\rho = 0.64$, $n = 50$

102. The *harmonic mean* of n positive numbers, $X_1, X_2, X_3, \ldots, X_n$, is defined as follows:

$$h = \dfrac{n}{\dfrac{1}{X_1} + \dfrac{1}{X_2} + \dfrac{1}{X_3} + \cdots + \dfrac{1}{X_n}}$$

Find the harmonic mean of 2, 4, 5, 4, 9, 2, and 6, both exact and rounded to four places.

103. Based on population data from 1960 to 2002, the resident population of the United States can be estimated by the linear equation $P = 2.533t - 4{,}786$, where P is the population in millions at year t.
    (a) Compute the resident population for 2002 predicted by the equation.

    (b) The actual population in 2002 was 288 million. Find the difference between the actual population in 2002 and the population predicted by the equation for 2002.

104. The national health expenditures (the amount of money spent on health care) for the United States from 1980 to 2002 can be estimated by the equation

$$E = 58(t - 1980) + 146.8$$

where E is the amount spent in billions of dollars during year t.

    (a) Use this equation to determine how much was spent in the United States on health care in 1998, 1999, 2000, and 2001.

    (b) Compute the differences between expenditures for each pair of consecutive years found in part **(a)**, and then compute the percent increase (from the preceding year) for each year from 1998 to 2001.

105. The national health expenditures for hospital care only (the amount of money spent on hospital care) for the United States from 1980 to 2002 can be estimated by the equation

$$E = 16.64(t - 1980) + 91.36$$

where E is the amount spent in billions of dollars during year t.

    (a) Use this equation to determine how much was spent in the United States on hospital care in 1998, 1999, 2000, and 2001.

    (b) Compute the differences between hospital care expenditures for each pair of consecutive years found in part **(a)**, and then compute the percent increase (from the preceding year) for each year from 1998 to 2001.

106. The following table from the Bureau of Labor Statistics gives the percentage of females over the age of 16 in the civilian population that is employed in January of each of the years listed.

Year	1993	1994	1995	1996	1997	1998
Employment rate	53.7%	54.9%	55.5%	55.5%	56.5%	57.1%

Based on these data, a statistician suggests that the following formula can be used to approximate the female employment rate R given year t (in January of the years 1993 through 1998):

$$R = -0.025(t - 1,990)^2 + 0.9(t - 1,990) + 51.4$$

(a) Use this formula to compute R for each of the years 1993 through 1998.

(b) Compare the values collected by the Department of Labor with the values obtained from the formula, and comment about the accuracy of the formula.

 Question for Thought

107. Supply the reason for each of the following steps:

$$(-3)(4) = [(-1)(3)](4)$$
$$= (-1)[(3)(4)]$$
$$= (-1)[12]$$
$$= -12$$

1.3   Algebraic Expressions

Suppose you deposit $5,000 in a bank that pays interest at the effective rate of 7.5% simple yearly interest, and you want to determine how much money you will have in the bank at the end of the year. You may approach this problem by first computing the interest you would earn after 1 year, and then adding this interest to the original amount you deposited (called the *principal*) as follows:

Amount in the bank at the end of the year = Principal + Interest
= Principal + Principal × Rate
= $5,000 + $5,000 × 0.075
= $5,000 + $375
= $5,375

So if you deposited $5,000 in this bank, you would have $5,375 at the end of the year.

Notice the number of computations you had to perform to come up with your answer. Is there a simpler way to compute the amount you would have after 1 year?

Let's look at this process using symbolic notation. If we let P stand for the principal, then we can rewrite the amount at the end of the year as

Amount in the bank at the end of the year = Principal + Interest
= Principal + Principal × Rate
= $P + P \times 0.075$
= $P + 0.075P$

The distributive property allows us to factor out P to get

$$P + .075P = (1 + 0.075)P$$
$$= 1.075P$$

This says that to answer the question, all we need to do is multiply the principal by 1.075. Note that our use of symbolic notation has not only given us a more convenient (simpler) way to compute the answer, but also has given a general way to compute that answer for *any* principal. The expression $1.075P$ is an elementary example of what we intend to discuss for the remainder of this section: *simplifying* very basic algebraic expressions.

A *variable* is a symbol, usually a letter, that stands for a number (or set of numbers). In algebra, there are primarily two ways variables are used. One use of a variable is to represent a particular number (or set of numbers) whose value(s) needs to be found. Equations are examples of this type of variable use.

A second use of variables is to describe a general relationship involving numbers and/or arithmetic operations, such as in the commutative property:

$$a + b = b + a$$

Variables represent unknown quantities. A *constant,* on the other hand, is a symbol whose value is fixed, such as 2, -6, 8.33, or π.

DEFINITION	An *algebraic expression* is an expression consisting of constants, variables, grouping symbols, and symbols of operations arranged according to the rules of algebra.

Our goal in this section is to review how to simplify algebraic expressions. That is, given a basic set of guidelines and the real number properties, we will take algebraic expressions and change them into simpler equivalent expressions. By *equivalent expressions,* we mean expressions that represent the same number for all valid replacements of the variables in the expression. For example, the expression $3x + 7x$ is equivalent to the expression $10x$ because when we substitute any number for x, the computed value of $3x + 7x$ will be the same as the computed value of $10x$.

Products

Recall the definition of exponential notation:

$$x^n = x \cdot x \cdot \cdots \cdot x \qquad \text{The factor x occurs n times.}$$

The natural number n is called the *exponent,* x is called the *base,* and x^n is called the *power.* Thus,

$$5a^4b^2 = 5aaaabb \qquad (-3x)^3 = (-3x)(-3x)(-3x)$$

DEFINITION	The numerical factor of a term is called the *numerical coefficient* or simply the *coefficient.*

For example:

The coefficient of $3x^2y$ is 3.
The coefficient of $-5a^4b^2$ is -5.
The coefficient of z^3y is understood to be 1.

We will call the nonnumerical factors in a term the *literal part.* For example:

The literal part of $3x^2y$ is x^2y.

To multiply two powers with the same base, such as $x^5 \cdot x^4$, we can write out the following:

$$x^5 \cdot x^4 = (x \cdot x \cdot x \cdot x \cdot x)(x \cdot x \cdot x \cdot x) \qquad \textit{Write } x^5 \textit{ and } x^4 \textit{ without exponents.}$$

$$= x \cdot x \cdot x \cdot x \cdot x \cdot x \cdot x \cdot x \cdot x \qquad \textit{Then count the x's.}$$

$$= x^9$$

We can see that multiplying two powers with the same base is a matter of counting the number of times x appears as a factor. This gives us the first rule of exponents.

The First Rule of Exponents	$x^n x^m = x^{n+m}$

Thus, to *multiply* two powers of the *same base,* we simply keep the base and *add* the exponents.

The first rule can be generalized to include more than two powers. For example,

$$x^p x^q x^r x^m = x^{p+q+r+m}$$

When asked to *simplify* an expression involving exponents, we should write the expression with bases and exponents occurring as few times as possible.

EXAMPLE 1

Simplify each of the following.

(a) $x^2 x^7 x$ **(b)** $(-2)^3(-2)^2$

Solution

(a) $x^2 x^7 x = x^{2+7+1}$ *Remember that* $x = x^1$.

$$= \boxed{x^{10}}$$

(b) $(-2)^3(-2)^2 = (-2)^{3+2}$

$$= \boxed{(-2)^5} \qquad \textit{Evaluated, this answer is } \boxed{-32}.$$

To find a product such as $(3x^3 y)(-4xy^2)$, we *could* proceed as follows:

$(3x^3 y)(-4xy^2)$

$= (3)[(x^3 y)(-4x)](y^2)$ *Associative property of multiplication (Notice that only the grouping symbols were changed.)*

$= (3)[(-4x)(x^3 y)](y^2)$ *Commutative property of multiplication*

$= (3)(-4)(x \cdot x^3)(y \cdot y^2)$ *Associative property of multiplication*

$= -12x^4 y^3$ *Multiplication and the first rule of exponents*

We stated above that we *could* proceed in this way; however, we usually do *not* proceed in this manner unless we need to *prove* that $(3x^3 y)(-4xy^2)$ is equal to $-12x^4 y^3$.

Actually, to find a *product* such as $(3x^3 y)(-4xy^2)$ quickly, we can ignore the original order and grouping of the variables and constants. Therefore, we just multiply the coefficients and then multiply the powers of each variable using the first rule of exponents.

$$(3x^3 y)(-4xy^2) \qquad \textit{Reorder and regroup.}$$

$$= 3(-4)x^3 \cdot x \cdot y \cdot y^2 \qquad \textit{Multiply.}$$

$$= -12x^4 y^3$$

The statement that "we can ignore the original order and grouping of variables and constants" is, in fact, a restatement of the commutative and associative properties of multiplication.

EXAMPLE 2

Multiply the following.

(a) $(-7a^2b^3c^2)(-2a^3b^4)(-3ac^5)$ **(b)** $-(3x)^4$ **(c)** $(-3x)^4$

Solution

(a) $(-7a^2b^3c^2)(-2a^3b^4)(-3ac^5)$ *First we reorder and regroup.*

$= (-7)(-2)(-3)(a^2a^3a)(b^3b^4)(c^2c^5)$ *Then multiply.*

$= \boxed{-42a^6b^7c^7}$

(b) $-(3x)^4 = -(3x)(3x)(3x)(3x)$ *Rewrite without exponents, reorder, and regroup.*

$= -(3 \cdot 3 \cdot 3 \cdot 3 \cdot x \cdot x \cdot x \cdot x)$

$= \boxed{-81x^4}$

(c) $(-3x)^4 = (-3x)(-3x)(-3x)(-3x)$ *Rewrite without exponents, reorder, and regroup.*

$= (-3)(-3)(-3)(-3)x \cdot x \cdot x \cdot x$

$= \boxed{+81x^4}$

Note the differences between parts (b) and (c).

Combining Terms

We discussed the product of expressions; now we will discuss how we *combine* or add and subtract terms. You probably remember from elementary algebra that you can add *"like terms,"* terms with identical literal parts, but you cannot add *"unlike terms."* What allows us to add like terms is the distributive property:

$$ba + ca = (b + c)a$$

For example,

$$5x + 7x = (5 + 7)x   \text{*Distributive property*}$$
$$= 12x$$

$$8ab - 17ab + 4ab = (8 - 17 + 4)ab   \text{*Distributive property*}$$
$$= -5ab$$

EXAMPLE 3

Combine the following.

(a) $2x^2 - 7x^2$ **(b)** $5x - 2y - 4x - 5y$ **(c)** $4xy^2 - 4y^2x + 3x^2y$

Solution

We will use the distributive property to demonstrate how it is used in combining terms. This step, however, should be done mentally.

(a) $2x^2 - 7x^2 = (2 - 7)x^2$ *This step is usually done mentally.*

$= \boxed{-5x^2}$

(b) $5x - 2y - 4x - 5y = (5 - 4)x + (-2 - 5)y$ *Note that we combine only "like" terms.*

$= 1x - 7y$

$= \boxed{x - 7y}$ *A coefficient of 1 is understood.*

(c) $4xy^2 - 4y^2x + 3x^2y = (4 - 4)xy^2 + 3x^2y$ *Note that $xy^2 = y^2x$ by the commutative*
$$= 0xy^2 + 3x^2y$$ *property and they are therefore "like" terms.*

$$= \boxed{3x^2y}$$ *Since $0xy^2 = 0$*

In summary, to combine "like" terms, add their numerical coefficients.

Removing Grouping Symbols

If we wanted to compute a numerical expression such as $7(9 + 4)$, we would evaluate it as follows:

$$7(9 + 4) = 7(13) = 91$$

We perform the operations within parentheses first.

However, in multiplying algebraic expressions with variables, as in the product $3x(x + 4)$, we cannot simplify within the parentheses. Instead, we multiply using the distributive property:

$$a(b + c) = ab + ac \quad \text{or} \quad (a + b)c = ac + bc$$

Verbally stated, multiplication distributes over addition.

Multiplication distributes over subtraction as well:

$$a(b - c) = ab - ac$$

For example,

$$7(9 + 4) = 7 \cdot 9 + 7 \cdot 4 = 63 + 28 = 91$$ *Note that we still arrive at the same answer as $7(13)$.*

$$3x(x + 4) = (3x)(x) + (3x)(4) = 3x^2 + 12x$$
$$-2(x + y - 3) = (-2)x + (-2)y + (-2)(-3) = -2x - 2y + 6$$

When a negative sign immediately precedes a grouping symbol, as in

$$-(x + 4 - 3y)$$

we may interpret this as the negative of the quantity $x + 4 - 3y$. In the previous section, we found that we can rewrite $-a$ as $(-1)a$. We do the same to $-(x + 4 - 3y)$ to remove the grouping symbol:

$$-(x + 4 - 3y) = (-1)(x + 4 - 3y)$$ *Since $-a = (-1)a$*
Then we use the distributive property.

$$= (-1)(x) + (-1)(4) + (-1)(-3y)$$
$$= -x - 4 + 3y$$

Thus, we can interpret $-(x + 4 - 3y)$ as multiplying the quantity $x + 4 - 3y$ by -1. When subtracting a quantity within grouping symbols, change the sign of *each term* within the grouping symbols.

If a positive sign precedes a grouping symbol, as in

$$+(x + 4 - 7y)$$

we interpret it as $+1(x + 4 - 7y)$. Hence,

$$+(x + 4 - 7y) = +1(x + 4 - 7y)$$
$$= (+1)x + (+1)(+4) + (+1)(-7y)$$
$$= x + 4 - 7y$$

Note that the signs of the terms within the grouping symbols remain unchanged.

EXAMPLE 4

Perform the following operations.

(a) $-5(2x - 3y + 5)$ **(b)** $-3x(-a + b)$

(c) $-3x - (a + b)$ **(d)** $3a - [5a - 3(2a - 1)]$

Solution

(a) $-5(2x - 3y + 5) = -5(2x) - 5(-3y) - 5(5)$ *Distributive property*

$$= \boxed{-10x + 15y - 25}$$

(b) $-3x(-a + b) = (-3x)(-a) - 3x(b)$ *Distributive property*

$$= \boxed{3ax - 3bx}$$ *(We prefer to write factors of terms in alphabetical order; hence, we write 3ax rather than 3xa.)*

Compare parts (b) and (c) of this example.

(c) $-3x - (a + b) = \boxed{-3x - a - b}$ *Since you are subtracting a + b, change the sign of each term.*

(d) $3a - [5a - 3(2a - 1)] = 3a - [5a - 6a + 3]$ *Simplify in brackets first (multiply 2a − 1 by −3). Then combine terms in brackets.*

$$= 3a - [-a + 3]$$ *Next, remove brackets.*

$$= 3a + a - 3$$ *Then combine terms.*

$$= \boxed{4a - 3}$$

EXAMPLE 5

Based on data from the U.S. Census Bureau, from 1985 to 1996, total enrollment in all public and private schools in the United States can be estimated by the equation

$$E = 0.8104(t - 1{,}985) + 56.65$$

where E is the total enrollment in all schools in millions during year t.

(a) Simplify the expression on the right side of the equation.

(b) Compute the estimated enrollment for the year 1997 using the original equation and then with the simplified version completed in part **(a).**

Solution

(a) We simplify the right side of the equation using a calculator as follows:

$$E = 0.8104(t - 1{,}985) + 56.65$$ *Distribute 0.8104.*

$$= 0.8104t - 1{,}608.644 + 56.65$$ *Add −1,608.644 and 56.65 to get*

$$E = 0.8104t - 1{,}551.994$$

Hence the original equation can be written as $\boxed{E = 0.8104t - 1{,}551.994}$.

(b) We compute the enrollments by substituting 1,997 for t for each equation:

Note the differences in computing the answer using each form of the same equation. Which was easier?

$$E = 0.8104(t - 1{,}985) + 56.65 \qquad E = 0.8104t - 1{,}551.994$$

$$= 0.8104(1{,}997 - 1{,}985) + 56.65 \qquad\qquad = 0.8104(1{,}997) - 1{,}551.994$$

$$= 0.8104(12) + 56.65 = 66.3748 \qquad\qquad = 66.3748$$

We can see that we arrive at the same value for each equation as expected: 66.3748, which is $\boxed{66{,}374{,}800 \text{ students}}$.

EXAMPLE 6

In terms of x, find the area of the shaded region bounded by rectangles in Figure 1.4.

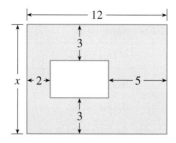

Figure 1.4

Solution   We first observe that the shaded area can be found by subtracting the area of the inner rectangle from the area of the outer rectangle:

$$\text{Area of shaded region} = \text{Area}_{\text{outer rectangle}} - \text{Area}_{\text{inner rectangle}}$$

The area of the outer rectangle is simple enough:

$$A_{\text{outer rectangle}} = \text{Length} \times \text{Width}$$
$$= 12x$$

The area of the inner rectangle can be found by observing that the length of the inner rectangle is $12 - 5 - 2 = 5$, and the width is $x - 3 - 3 = x - 6$. Hence

$$A_{\text{inner rectangle}} = \text{Length} \times \text{Width}$$
$$= 5(x - 6)$$

The area of the shaded region is therefore

$$A_{\text{shaded region}} = A_{\text{outer rectangle}} - A_{\text{inner rectangle}}$$
$$= 12x \qquad\quad -5(x - 6) \qquad \textit{First we multiply } x - 6 \textit{ by } -5.$$
$$= 12x - 5x + 30$$
$$= \boxed{7x + 30}$$

STUDY SKILLS 1.3

Reviewing Old Material

One of the most difficult aspects of learning algebra is that each skill and concept depends on those previously learned. If you have not acquired a certain skill or learned a particular concept well enough, this will more than likely affect your ability to learn the next skill or concept.

Thus, even though you have finished a topic that was particularly difficult for you, you should not breathe too big a sigh of relief. Eventually you will have to learn that topic well to be able to understand subsequent topics. It is important that you try to master all skills and understand all concepts.

Whether or not you have had difficulty with a topic, you should be constantly reviewing previous material as you continue to learn new subject matter. Reviewing helps to give you a perspective on the material you have covered. It helps you to tie the different topics together and makes them *all* more meaningful. Some statement you read 3 weeks ago, and which may have seemed very abstract then, suddenly becomes simple and obvious in the light of all you now know.

Since many problems require you to draw on the skills you have developed previously, it is important for you to review so that you will not forget or confuse them. You will be surprised to find how much constant reviewing aids in learning new material.

When working the exercises, always try to work out some exercises from earlier chapters or sections. Try to include some review exercises at every study session, or at least at every other session. Take the time to reread the text material in previous chapters. When you review, think about how the material you are reviewing relates to the topic you are presently learning.

EXERCISES 1.3

In Exercises 1–68, simplify the expression as completely as possible.

1. $6x + 2x$

2. $5a + 4a$

3. $6x(2x)$

4. $5a(4a)$

5. $2x - 6x$

6. $5a - 4a$

7. $2x(-6x)$

8. $5a(-4a)$

9. $3m - 4m - 5m$

10. $-8y(-3y)(-2y)$

11. $3m(-4m)(-5m)$

12. $-8y - 3y - 2y$

13. $-2t^2 - 3t^2 - 4t^2$

14. $-5z^3 - 2z^3 - 4z^3$

15. $-2t^2(-3t^2)(-4t^2)$

16. $-5z^3(-2z^3)(-4z^3)$

17. $2x + 3y + 5z$

18. $3s + 5t + 6u$

19. $2x(3y)(5z)$

20. $3s(5t)(6u)$

21. $x^3 + x^2 + 2x$

22. $a^5 + 4a^3 + a^2$

23. $x^3(x^2)(2x)$

24. $a^5(4a^3)(a^2)$

25. $-5x(3xy) - 2x^2y$

26. $6r(-2r^2)(-4r^3t)$

27. $-5x(3xy)(-2x^2y)$

28. $6r(-2r^2t) - 4r^3t$

29. $2x^2 + 3x - 5 - x^2 - x - 1$

30. $-7t^3 - 4t^2 - 8 - t^3 - 5t^2 + 2$

31. $10x^2y - 6xy^2 + x^2y - xy^2$

32. $8a^2b^2 - 5a^2b^3 - a^2b^3 + a^2b^2$

33. $3(m + 3n) + 3(2m + n)$

34. $5(2u + 3w) + 4(u + 3w)$

35. $6(a - 2b) - 4(a + b)$

36. $7(3p - q) - 3(p + 2q)$

37. $8(2c - d) - (10c + 8d)$

38. $10(y - z) - (4y + 10z)$

39. $x(x - y) + y(y - x)$

40. $w(w^2 - 4) + w^2(w + 3)$

41. $a^2(a + 3b) - a(a^2 + 3ab)$

42. $t^3(t^2 - 3t) - t^2(t^3 - 3t)$

43. $5a^2bc(-2ab^2)(-4bc^2)$

44. $-2xyz(-4x^3y)(6yz^2)$

45. $(2x)^3(3x)^2$

46. $(5a)^2(2a)^4$

47. $2x^3(3x)^2$

48. $5a^2(2a)^4$

49. $(-2x)^5(x^6)$

50. $(-3a)^4a^8$

51. $(-2x)^4 - (2x)^4$

52. $(-5a)^2 - (5a)^2$

53. $(-2x)^3 - (2x)^3$

54. $(-5a)^3 - (5a)^3$

55. $4b - 5(b - 2)$

56. $8z - 9(z - 3)$

57. $8t - 3[t - 4(t + 1)]$

58. $7y - 5[y - 6(y - 1)]$

59. $a - 4[a - 4(a - 4)]$

60. $c - 6[c - 6(c - 6)]$

61. $x + x[x + 3(x - 3)]$

62. $y^2 + y[y + y(y - 4)]$

63. $x - \{y - 3[x - 2(y - x)]\}$

64. $y - 2\{x - [z - (x - y)]\}$

65. $3x + 2y[x + y(x - 3y) - y^2]$

66. $6r + 2t[r - t(r - 5t) - r^2]$

67. $6s^2 - [st - s(t + 5s) - s^2]$

68. $9z^3 - [wz - z(w - 4z^2) - z^3]$

In Exercises 69–72, **(a)** *without simplifying, evaluate the expression for x = 2.82 and y = 7.25;* **(b)** *simplify and then evaluate the expression for x = 2.82 and y = 7.25.*

69. $3x - 15y - 11x$

70. $-2x + 5y - 3x - 5y$

71. $2(3x - 4) - 5(3x + 8)$

72. $5(3x - 4y) - 2(5x - 10y)$

73. The number of physicians in the United States from 1980 to 2000 can be estimated by the equation

$$P = 16.13(t - 1980) + 449.2$$

where P is the number of physicians in thousands during year t.

(a) Simplify the expression on the right of the equation.

(b) Compute the estimated number of physicians for 2006 using the original equation and then with the simplified version completed in part **(a)**.

74. The number of dentists in the United States from 1980 to 2000 can be estimated by the equation

$$D = 2.24(t - 1980) + 122.4$$

where D is the number of dentists in thousands during year t.

(a) Simplify the expression on the right side of the equation.

(b) Compute the estimated number of dentists for the year 2006 using the original equation and then with the simplified version completed in part **(a)**.

In Exercises 75–78, find the area of the shaded region.

75.

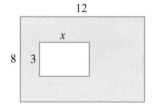

76.

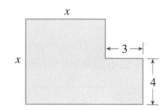

77.

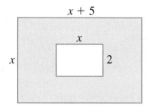

78.

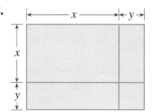

1.4   Translating Phrases and Sentences into Algebraic Form

In this section we focus our attention on translating phrases and sentences into their algebraic form, and leave the formulation and solution of entire problems to later chapters.

Many students look for *key words* to help them quickly determine which algebraic symbols to use in translating an English phrase or sentence. However, the key words do not usually indicate how the symbols should be put together to yield an accurate translation of the words. To be able to put the symbols together in a meaningful way, we must understand how the key words are being used in the context of the given verbal expression.

Let's look at two verbal expressions that use the same key words and yet do not have the same meaning:

1. Five times the sum of six and four
2. The sum of five times six and four

Both expressions contain the following key words:

"five"	the number 5	(5)
"times"	meaning multiply	($\cdot$)
"sum"	meaning addition	($+$)
"six"	the number 6	(6)
"four"	the number 4	(4)

How should the symbols be put together?

Expression 1 says: five times . . . .

Five times what? . . . *the sum*

Thus, we must *first* find the sum of 6 and 4, before multiplying by 5.

Expression 2 says: the sum of . . . .

The sum of what *and* what? . . . *the sum of* 5 *times* 6 **and** 4

Thus, we must *first* compute 5 times 6 before we can determine what the sum is.

Translating the two verbal expressions into algebraic form, we get:

1. $5(6 + 4) = 5 \cdot 10 = 50$ *Five times the sum of six and four*
2. $5 \cdot 6 + 4 = 30 + 4 = 34$ *The sum of five times six and four*

Thus, in comparing expressions 1 and 2, we can see that the simple change of word positions in a phrase can change the meaning quite a bit.

EXAMPLE 1

Translate each of the following into algebraic form.

(a) Nine more than 5 times a number

(b) Seven less than 5 times a number is 26.

(c) The sum of two numbers is 6 less than their product.

Solution

(a) "Nine more than" means we are going to add 9 to something. To what? To "5 *times a number.*"

If we represent the number by n (you are free to choose the letter you like), we get

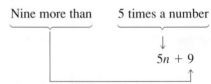

Thus, our answer is

$$\boxed{5n + 9}$$

Of course, $9 + 5n$ is also correct.

(b) "Seven less than" means we are going to subtract 7 from something. From what? From "5 *times a number.*"

The word *is* translates to "=".
If we represent the number by *s*, we get

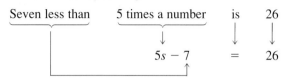

Thus, our final answer is

$$5s - 7 = 26$$ *Note that* $7 - 5s = 26$ *is* **not** *a correct translation.*

(c) The sentence mentions two numbers, so let's call them *x* and *y*.

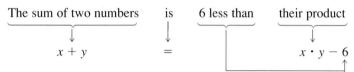

Thus, our translation is

$$x + y = xy - 65$$

The next example again illustrates the need to understand the meaning of the words in a problem clearly to be able to put them together properly.

EXAMPLE 2

Translate each of the following into algebraic form.

(a) The sum of the squares of two consecutive integers

(b) The square of the sum of two consecutive integers

Solution

Some examples of two consecutive integers are 5 and 6, 12 and 13, and −11, and −10.

(a) Let *x* represent the first of the two consecutive integers. Then $x + 1$ represents the second of the two consecutive integers.

The "sum of the squares" means that we must *first* square each one of the two consecutive integers and *then* add the results. Thus, our translation is

$$x^2 + (x + 1)^2$$

(b) We let *x* represent the first of the two consecutive integers. Then $x + 1$ represents the second of the two consecutive integers.

The "square of the sum" means that we must *first* add the two numbers and *then* square the result. Thus, our translation becomes

$$(x + x + 1)^2 \quad \text{or} \quad \boxed{(2x + 1)^2}$$

Even though both phrases in Example 2 contained the words *square* and *sum,* those words alone do not give us the meaning of the entire phrase. We still need to analyze how the words fit together.

While it is necessary to recognize that particular words and phrases imply specific arithmetic operations, and to understand what the problem means as a whole, there is still one more important skill necessary to successfully translate verbal expressions into algebraic ones. This additional skill is the ability to apply your basic general knowledge and common sense to a particular problem.

Let's look at how these various components blend together in a problem.

EXAMPLE 3

Suppose Harry averages $25 an hour working as a programmer for his company and $45 an hour as a consultant in his spare time.

(a) How much does Harry make working 40 hours for his company?

(b) In terms of x, how much does Harry make working x hours for his company?

(c) How much does Harry make if he does 40 hours of consulting?

(d) If Harry works x hours for the company and y hours consulting, how much does he make altogether, in terms of x and y?

(e) If Harry still works x hours for the company and he works a total of 20 hours in both jobs, how much does he make altogether, in terms of x?

Solution

(a) How much does Harry make working 40 hours for the company?

The problem assumes that you understand that if a person is paid a certain amount per hour (pay rate), and works for a number of hours, then the total pay would be found by multiplying the pay rate by the number of hours:

$$\text{Total pay} = (\text{Pay rate per hour}) \cdot (\text{Number of hours})$$

Hence, Harry makes $(25)(40) = \boxed{\$1{,}000}$ by working 40 hours for the company at $25 per hour.

(b) In terms of x, how much does Harry make working x hours for his company?

Again, we find the total pay by multiplying the rate per hour by the number of hours. Hence,

$$\text{Harry makes } (25) \cdot x = \boxed{25x \text{ dollars}}$$

(c) How much does Harry make if he does 40 hours of consulting?

$$\text{Harry makes } (45)(40) = \boxed{\$1{,}800}$$

(d) If Harry works x hours for the company and y hours consulting, how much does he make altogether, in terms of x and y?

Now we are required to find Harry's total pay for the two jobs. His total is the sum of the incomes from each job, or

$$\begin{pmatrix} \text{Pay from} \\ \text{company} \end{pmatrix} + \begin{pmatrix} \text{Pay from} \\ \text{consulting} \end{pmatrix} = \text{Total pay}$$

$$\begin{pmatrix} \text{Rate from} \\ \text{company} \end{pmatrix} \cdot \begin{pmatrix} \text{Number of} \\ \text{hours working} \\ \text{for company} \end{pmatrix} + \begin{pmatrix} \text{Rate from} \\ \text{consulting} \end{pmatrix} \cdot \begin{pmatrix} \text{Number} \\ \text{of hours} \\ \text{consulting} \end{pmatrix} = \text{Total pay}$$

Hence, we have

$$25 \cdot x \qquad + \qquad 45 \cdot y \qquad = \text{Total pay}$$

Thus, his total pay, in terms of x and y, is

$$\boxed{25x + 45y \text{ dollars}}$$

(e) This part is similar to part **(d),** except that we are given only one of the number of hours worked. Let's use the same analysis as in part **(d)** and see what we can find at this point.

$$\begin{pmatrix} \text{Pay from} \\ \text{company} \end{pmatrix} \qquad + \qquad \begin{pmatrix} \text{Pay from} \\ \text{consulting} \end{pmatrix} \qquad = \text{Total pay}$$

$$\begin{pmatrix} \text{Rate from} \\ \text{company} \end{pmatrix} \cdot \begin{pmatrix} \text{Number of} \\ \text{hours working} \\ \text{for company} \end{pmatrix} + \begin{pmatrix} \text{Rate from} \\ \text{consulting} \end{pmatrix} \cdot \begin{pmatrix} \text{Number} \\ \text{of hours} \\ \text{consulting} \end{pmatrix} = \text{Total pay}$$

Hence, we have

$$25 \cdot x \qquad\qquad + \qquad\qquad 45 \cdot ? \qquad\qquad = \text{Total pay}$$

What remains is to fill in the question mark.

Although we are not given the number of hours consulting, we are given that Harry worked a total of 20 hours. Therefore, if he worked a total of 20 hours and x of those hours are spent working for the company, then the remaining hours must be spent consulting. Hence,

$$(\text{Total hours}) - \begin{pmatrix} \text{Number of} \\ \text{hours working} \\ \text{for company} \end{pmatrix} = \begin{pmatrix} \text{Number} \\ \text{of hours} \\ \text{consulting} \end{pmatrix}$$

$$20 \qquad - \qquad x \qquad = \begin{pmatrix} \text{Number} \\ \text{of hours} \\ \text{consulting} \end{pmatrix}$$

Since the number of hours consulting is $20 - x$, we can fill in the rest of the previous expression:

$$\begin{pmatrix} \text{Rate from} \\ \text{company} \end{pmatrix} \cdot \begin{pmatrix} \text{Number of} \\ \text{hours working} \\ \text{for company} \end{pmatrix} + \begin{pmatrix} \text{Rate from} \\ \text{consulting} \end{pmatrix} \cdot \begin{pmatrix} \text{Number} \\ \text{of hours} \\ \text{consulting} \end{pmatrix} = \text{Total pay}$$

$$25 \cdot x \qquad\qquad + \qquad\qquad 45 \cdot ? \qquad\qquad = \text{Total pay}$$

$$25 \cdot x \qquad\qquad + \qquad\qquad 45 \cdot (20 - x) \qquad\qquad = \text{Total pay}$$

Harry's total pay is $25x + 45(20 - x)$, which we simplify:

$$25x + 45(20 - x) = 25x + 900 - 45x$$

$$= \boxed{900 - 20x \text{ dollars}}$$

EXAMPLE 4

The length of a rectangle is 5 more than twice its width.

(a) Express the *area* of the rectangle in terms of one variable.

(b) Express the *perimeter* of the rectangle in terms of one variable.

Solution

We draw a picture of a rectangle. Since the length is expressed in terms of the width, we let

Width $= x$, then Length $= 2x + 5$ *The length is 5 more than twice the width.*

Then we label the rectangle as shown in Figure 1.5.

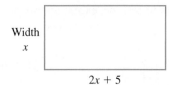

Width
x

$2x + 5$
Length

Figure 1.5

(a) The area, A, of a rectangle, is given by

$$A = (\text{Length})(\text{Width})$$
$$= (2x + 5)(x)$$
$$= \boxed{2x^2 + 5x}$$

(b) The perimeter, P, of a rectangle is the distance around the rectangle. Hence,

$$P = 2W + 2L \quad \textit{Where W = width and L = length}$$

Therefore, in terms of x,

$$P = 2 \cdot x + 2(2x + 5) \quad \textit{Then simplify.}$$
$$= 2x + 4x + 10$$
$$= \boxed{6x + 10}$$

EXAMPLE 5

The Ace Car Rental Company charges $30 per day plus a mileage charge of $0.15 per mile; the Better Car Rental Company charges $28 per day and $0.18 per mile; the Cheap Car Rental Company charges a flat fee of $40 per day with no mileage charge.

(a) How much would it cost you to rent a car from each of the companies if you plan to rent it for 3 days and drive a total of 150 miles?

(b) Express the cost of renting a car from each company if you plan to rent the car for d days and drive m miles.

(c) Assume you plan to rent the car for 3 days and plan to drive *approximately* 200 miles. Discuss which company gives you the best price.

Solution

(a) The Ace Car Rental Company charges $30 per day and $0.15 per mile. A car rented for 3 days and driven 150 miles would cost

$$C_{\text{Ace}} = 30 \times 3 + 0.15 \times 150$$
$$= 90 + 22.50 = \$112.50$$

The Better Car Rental Company charges $28 per day and $0.18 per mile. Renting the car for 3 days and driving a total of 150 miles would cost

$$C_{\text{Better}} = 28 \times 3 + 0.18 \times 150$$
$$= 84 + 27 = \$111$$

The Cheap Car Rental Company charges $40 per day and no mileage charge. Renting the car for 3 days and driving a total of 150 miles would cost

$$C_{\text{Cheap}} = 40 \times 3$$
$$= \$120$$

(b) As we can see by the numerical examples above, the cost of renting a car from each company is found by multiplying the mileage rate by the number of miles, and adding the resulting quantity to the product of the number of days and the daily charge.

The cost of renting a car from each company for d days and driving m miles would be

$$C_{\text{Ace}} = 30d + 0.15m \qquad C_{\text{Better}} = 28d + 0.18m \qquad C_{\text{Cheap}} = 40d$$

Note that m is missing from the equation for C_{Cheap} because there is no mileage charge.

(c) The cost to rent a car from each company for 3 days and driving m miles would be

$$C_{\text{Ace}} = 90 + 0.15m \qquad C_{\text{Better}} = 84 + 0.18m \qquad C_{\text{Cheap}} = 120$$

We plan to drive *approximately* 200 miles; the actual mileage may be more or less than 200 miles. Let's make a chart and see what happens as we drive more or less than the 200 miles.

Number of miles (m)	$90 + 0.15m$	$= C_{\text{Ace}}$	$84 + 0.18m$	$= C_{\text{Better}}$	$120 = C_{\text{Cheap}}$
150 miles	$90 + 0.15 \times 150 = \$112.50$		$84 + 0.18 \times 150 = \$111$		\$120
199 miles	$90 + 0.15 \times 199 = \$119.85$		$84 + 0.18 \times 199 = \$119.82$		\$120
200 miles	$90 + 0.15 \times 200 = \$120$		$84 + 0.18 \times 200 = \$120$		\$120
201 miles	$90 + 0.15 \times 201 = \$120.15$		$84 + 0.18 \times 201 = \$120.18$		\$120
250 miles	$90 + 0.15 \times 250 = \$127.50$		$84 + 0.18 \times 250 = \$129$		\$120

The chart indicates that if we think we may drive less than 200 miles, then the Better Car Rental Company gives us the best price. On the other hand, if we think we may drive more than 200 miles, the Cheap Car Rental Company gives the best price.

EXAMPLE 6

A stock loses 30% of its previous day's value on September 21. The next day, September 22, it gains 30% of its previous day's value.

(a) If you have \$2,400 worth of that stock on September 20, how much is it worth at the end of the day on September 22?

(b) If you have \$$x$ worth of that stock on September 20, how much is it worth at the end of the day on September 22?

Solution

A 30% gain will not offset a 30% loss. Keep in mind the differences between percent and actual values: The actual value of a 30% change is dependent on the previous value, and the previous values are different on September 20 and September 21.

(a) A 30% loss means that on September 21, the stock is worth 70% of its value from September 20 (subtract the percent loss from 100%). To find the percent of an amount we multiply the value by the decimal equivalent of the percent, 70%, which is 0.70. Hence the stock valued at \$2,400 on September 20 is worth $(0.70)(2,400) = \$1,680$ on September 21.

On September 22, the stock gains 30% *of its September 21 value*. This means the stock is worth 130% of its September 21 value (add the percent gain to 100%). We convert 130% to a decimal and multiply it by the new value, \$1,680, to get

$$(1.30)(1,680) = \boxed{\$2,184}.$$

Hence if the stock was worth \$2,400 on September 21, then the value of the stock on September 22 is \$2,184.

(b) A 30% loss means that on September 21, the stock is worth 70% of its value from September 20. To find the percent of an amount we multiply the value by the decimal equivalent of the percent, 70%, which is 0.70. Hence the stock valued at \$$x$ on September 20 is worth $0.70x$ on September 21.

On September 22, the stock gains 30% *of its September 21 value*. This means the stock is worth 130% of its September 21 value. We convert 130% to a decimal and multiply it by the new value, 0.70x, to get

$$\text{Value on September 22} = 130\% \text{ of the value on September 21}$$
$$= (1.3)(0.70x)$$
$$= 0.91x$$

Hence if the stock was worth $x on September 21, then the value of the stock on September 22 is $\boxed{\$0.91x}$.

STUDY SKILLS 1.4

Reflecting

When you have finished reading or doing examples, it is always a good idea to take a few minutes to think about what you have just covered. Think about how the examples relate to the verbal material, and how the material just covered relates to what you have learned previously. How are the examples and concepts you have just covered similar to or different from those you have already learned?

EXERCISES 1.4

In Exercises 1–8, translate the given phrase or sentence. Indicate clearly what each variable represents.

1. Eight more than a number

2. Eight less than a number

3. Three less than twice a number

4. Three more than twice a number

5. Four more than three times a number is seven less than the number.

6. Nine less than twice a number is six more than twice the number.

7. The sum of two numbers is one more than their product.

8. The quotient of two numbers is five less than their sum.

In Exercises 9–18, represent all the numbers in the exercise in terms of one variable.

9. There are two numbers, the larger of which is 5 more than twice the smaller.

10. There are two numbers, the smaller of which is 12 less than 3 times the larger.

11. There are three numbers. The middle number is 3 times the smallest, and the largest number is 12 more than the middle number.

12. There are three numbers. The largest number is 5 more than 8 times the smallest, and the middle number is 10 less than the largest number.

13. Represent two consecutive integers.

14. Represent two consecutive odd integers.

15. Represent three consecutive even integers.

16. Represent three consecutive odd integers.

17. Represent the sum of the cubes of two consecutive integers.

18. Represent the cube of the sum of two consecutive integers.

In Exercises 19–22, represent the numbers in terms of one variable.

19. The sum of two numbers is 40.

20. The sum of two numbers is 29.

21. The sum of three numbers is 100. One of the numbers is twice one of the other numbers.

22. The sum of three numbers is 82. One of the numbers is 9 more than one of the others.

23. The length of a rectangle is 3 times its width. Represent its area and perimeter in terms of one variable.

24. The length of a rectangle is 6 less than 4 times its width. Represent its area and perimeter in terms of one variable.

25. The length of the first side of a triangle is twice the length of its second side, and the length of its third side is 4 more than the length of its second side. Express its perimeter in terms of one variable.

26. The length of the first side of a triangle is 2 more than 3 times the length of its third side, and the length of its second side is 5 more than the length of its third side. Express its perimeter in terms of one variable.

27. A collection of coins consists of 12 nickels, 9 dimes, and 10 quarters.

    (a) How many coins are there?

    (b) What is the value of the coins?

28. The length of a rectangular plot of land is 20 meters and its width is 12 meters. Heavy-duty fence for the length costs $5 per meter, and regular fence for the width costs $2 per meter.

    (a) How much regular fence is needed?

    (b) How much will the regular fence cost?

    (c) How much heavy-duty fence is needed?

    (d) How much will the heavy-duty fence cost?

    (e) What will the total cost for fencing the plot of land be?

29. Repeat Exercise 27 for n nickels, d dimes, and q quarters.

30. Flight attendant A serves six meals per minute for 15 minutes, while flight attendant B serves eight meals per minute for 20 minutes.

    (a) How many meals does flight attendant A serve?

    (b) How many meals does flight attendant B serve?

    (c) How many meals do they serve altogether?

31. Repeat Exercise 28 if the width is w meters and the length is 3 times the width.

32. Repeat Exercise 30 if flight attendant A serves n meals per minute for 18 minutes while flight attendant B serves nine meals per minute for t minutes.

33. A collection of 20 coins consists of nickels and dimes. Represent the value of all the coins using one variable.

34. A collection of 37 coins consists of nickels, dimes, and quarters. There are 10 more dimes than nickels. Represent the value of all the coins using one variable.

35. A bank pays 6% simple yearly interest. Suppose you deposit $1,500 in this bank.

    (a) How much money would you have accumulated in the bank at the end of the first year?

(b) How much would you have accumulated at the end of the second year?

(c) Write a formula to determine how much money is in the bank at the end of the nth year.

(d) Use this formula to find how much will be in the bank at the end of 15 years.

36. A certificate of deposit for a bank pays 5.5% simple yearly interest. Suppose you deposit $2,500 in this bank.

(a) How much money would you have accumulated in the bank at the end of the first year?

(b) How much would you have accumulated at the end of the second year?

(c) Write a formula to determine how much money is in the bank at the end of the nth year.

(d) Use this formula to find how much will be in the bank at the end of 12 years.

37. A long-distance phone company has two plans available: The Standard Plan has a charge of 10¢ per minute for long-distance calls anywhere in the continental United States with no monthly fee; the Saver Plan charges 6¢ per minute and has a monthly fee of $9.00.

(a) How much would it cost under each plan for 2 hours of long-distance calls in the continental United States per month? One hour per month?

(b) Express the monthly cost of each plan if you use t minutes in long-distance calls in the continental United States per month.

(c) Discuss when one plan would be "better" than the other. Argue for each plan and use mathematics to back up your arguments as to *when* one plan would be better than the other.

38. A cellular phone company offers two basic plans for the poor executive: plan A and plan B. Plan A is a monthly service charge of $25, and a charge of 25¢ per minute for telephone airtime; plan B is a monthly service charge of $40, and a charge of 20¢ per minute for telephone airtime.

(a) How much would it cost you per month for each plan if you use airtime on the average of 2 hours per month? One hour per month?

(b) Express the monthly cost for each plan if you use airtime of t minutes per month.

(c) Discuss when one plan would be "better" than the other. Use your best English in arguing for each plan and use mathematics to back up your arguments as to *when* one plan would be better for you than the other.

39. A department store has a 25% discount sale on all of its furniture. There is a sales tax of 5% on the sale price.

(a) Including tax, how much would a piece of furniture selling originally for $1,250 actually cost a customer during the sale?

(b) Including tax, how much would a piece of furniture selling originally for $x actually cost a customer during the sale?

40. In 2006, the local real estate tax for an area is 1.5% of the assessed value of a home. The assessed value of a home is computed at 90% of its market value, when the local government assessed the homes in 2005.

(a) If the market value of a home is $220,000 in 2005, how much are the annual real estate taxes?

(b) If the market value of a home is $x in 2005, how much are the annual real estate taxes?

41. A stock loses 50% of its previous day's value on October 29. The next day, October 30, it gains 50% of its previous day's value.

   (a) If you have $2,800 worth of that stock on October 28, how much is it worth at the end of the day on October 30?

   (b) If you have $x worth of that stock on October 28, how much is it worth at the end of the day on October 30?

42. Suppose you bought stock A at 28 ($28 per share) and stock B at 77 ($77 per share). The next day stock A falls 5 points to 23 and stock B rises 5 points to 82.

   (a) If you originally bought $2,000 worth of stock A and $2,000 worth of stock B, how much are your stocks worth the day after you bought them?

   (b) If you originally bought $x worth of stock A and $y worth of stock B, how much are your stocks worth the day after you bought them?

   (c) If you originally bought $x worth of stock A and $x worth of stock B, how much are your stocks worth the day after you bought them?

1.5 First-Degree Equations and Inequalities

First-Degree Equations

An equation is a symbolic statement of equality. That is, rather than writing "Twice a number is four less than the number," we write $2x = x - 4$. Our goal is to find the solutions to a given equation. By **solution** we mean the value or values of the variable that make the algebraic statement true.

In some cases we may have an equation that is always true. For example, $x + 6 = x + 8 - 2$ is *true for all values of the variable for which it is defined.* We call such equations **identities.** On the other hand, there are equations for which there is no solution, such as $x + 2 = x + 1$. Such an equation is called a **contradiction.** Equations that are true for some values of the variable and false for other values are called **conditional.** The conditional equation $3x - 2 = 5x - 1$ is true when x is $-\frac{1}{2}$ and false when x is any other value. We will have more to say about identities and contradictions in Chapter 2. For now we will focus on conditional equations.

EXAMPLE 1

Determine whether the values $x = 1.6$, -1.6, and 1.4 satisfy the equation $3x - 4(x - 5) = 18.6$.

Solution

If we substitute a value for the variable into an equation and it makes the equation true, we say that the value *satisfies* the equation. To determine whether $x = 1.6$ satisfies the equation $3x - 4(x - 5) = 18.6$, we substitute 1.6 for x in the equation and check to see whether the left and the right sides of the equation agree.

$$3x - 4(x - 5) = 18.6 \qquad \textit{Substitute } x = 1.6.$$
$$3(1.6) - 4(1.6 - 5) \stackrel{?}{=} 18.6$$
$$3(1.6) - 4(-3.4) \stackrel{?}{=} 18.6$$
$$4.8 + 13.6 \stackrel{?}{=} 18.6$$
$$18.4 \neq 18.6$$

Since the values on both sides of the equation do not agree

$$\boxed{x = 1.6 \text{ does } not \text{ satisfy the equation}}$$

To evaluate the same expression for $x = -1.6$, we have

$$3x - 4(x - 5) = 18.6 \qquad \textit{Substitute } x = -1.6.$$
$$3(-1.6) - 4(-1.6 - 5) \overset{?}{=} 18.6$$
$$3(-1.6) - 4(-6.6) \overset{?}{=} 18.6$$
$$21.6 \neq 18.6$$

$\boxed{x = -1.6 \text{ does } not \text{ satisfy the equation}}$, since both sides of the equation do not agree.

We will use a calculator to evaluate the left side of the equation for $x = 1.4$. We type in the following keys:

After pressing $\boxed{=}$, we get $\boxed{\mathsf{18.6}}$, which means both sides of the equation agree and therefore,

$$\boxed{x = 1.4 \textit{ does} \text{ satisfy the equation}}$$

Properties of Equality

If we consider the following four equations:

$$5a - 2(a - 3) = 2a + 10$$
$$5a - 2a + 6 = 2a + 10$$
$$3a + 6 = 2a + 10$$
$$a = 4$$

we can easily check that $a = 4$ is a solution to each of them. (Check this!) In fact, as we shall soon see, $a = 4$ is the only solution of these equations.

DEFINITION   Equations that have exactly the same set of solutions are called *equivalent equations.*

Of the four equations above, the solution to the last one was the most obvious. An equation of this form (that is, $a = 4$) is often called an *obvious equation.*

To develop a systematic method for solving equations, we would like to be able to transform an equation into successively simpler *equivalent* ones, finally resulting in an obvious equation. Developing such a method depends on recognizing the basic properties of equations.

If $a = b$, then a and b are interchangeable. Any expression containing a can be rewritten with b replacing a. Thus, if $a = b$, the expression $a + 7$ can be rewritten as $b + 7$ and we have $a + 7 = b + 7$. Another way of saying this is that if $a = b$, we can "add 7 to both sides of the equation," obtaining $a + 7 = b + 7$, without disturbing the equality. Consequently, recognizing what the equal sign means, we can see that equalities have the properties listed in the next box.

Properties of Equality

For all real numbers a, b, and c,

Addition Property of Equality	1. If $a = b$, then $a + c = b + c$.
Subtraction Property of Equality	2. If $a = b$, then $a - c = b - c$.
Multiplication Property of Equality	3. If $a = b$, then $a \cdot c = b \cdot c$.
Division Property of Equality	4. If $a = b$, then $\dfrac{a}{c} = \dfrac{b}{c}$, $c \neq 0$.

We can verbally summarize these properties with regard to equations as shown in the following box.

We will see in Section 2.2 why multiplying by zero will not yield an equivalent equation.

> An equation can be transformed into an equivalent equation by adding or subtracting the same quantity on both sides of the equation, or by multiplying or dividing both sides of the equation by the same *nonzero* quantity.

Solving First-Degree Equations

As we progress through this textbook we will develop methods to solve various kinds of equations. As we shall see, ***the method of solution we choose will very much depend on the kind of equation we are trying to solve.*** Therefore, when we begin to solve an equation, we must first try to determine what kind of equation it is.

In this section we restrict our attention to a specific kind of equation.

DEFINITION   A *first-degree equation* is an equation that can be put in the form

$$ax + b = 0$$

where a and b are constants ($a \neq 0$).

In other words, a first-degree equation is an equation in which the variable appears to the first power only. For example,

$$5x + 3 = 0, \quad 3x + 7 = x - 2, \quad \text{and} \quad 8 - 2x = 1$$

are all first-degree equations, whereas

$$2x^2 = 4x + 3$$

is a second-degree equation. (The degree of an equation is not always obvious and we will have more to say about this a bit later on in the text.)

Our method of solution for a first-degree equation in one variable consists of applying the properties of equality to transform the original equation into progressively simpler *equivalent* equations until we eventually obtain an obvious equation. Recall that an obvious equation is an equation of the form $x = \#$ or $\# = x$. In other words, in solving a *first-degree* equation,

> Our goal is to isolate the variable on one side of the equation.

EXAMPLE 2   Solve for a: $3a - 5 = -11$

Solution   To isolate a, we must deal with the 3 multiplying a and the 5 being subtracted. Since the 3 is multiplying the a, to eliminate 3 we use the inverse operation and divide by 3. However, dividing by 3 means we must divide both sides of the equation by 3, which forces us to work with fractions. Consequently, it is a better strategy to leave the division for last.

We proceed as follows:

$$3a - 5 = -11 \qquad \text{\textit{First we add +5 to both sides of the equation.}}$$

$$3a - 5 \;\boxed{+ 5} = -11 \;\boxed{+ 5}$$

$$3a = -6 \qquad \text{\textit{Then divide both sides of the equation by 3.}}$$

$$\frac{3a}{3} = \frac{-6}{3}$$

$$\boxed{a = -2}$$

Check: $a = -2$: $\quad 3a - 5 = -11$

$$3(-2) - 5 \overset{?}{=} -11$$

$$-6 - 5 \overset{?}{=} -11$$

$$-11 \overset{\checkmark}{=} -11$$

EXAMPLE 3

Solve for x: $\quad 12 - 4x = 8x + 3$

Solution

We want to isolate x on one side of the equation. To do this, we collect all the x terms on one side of the equation and all the non-x terms on the other side. The side on which we choose to isolate x is irrelevant.

$$12 - 4x = 8x + 3 \qquad \textit{We decide to isolate x on the right.}$$
$$\textit{Add 4x to both sides of the equation.}$$

$$12 - 4x \boxed{+ 4x} = 8x + 3 \boxed{+ 4x} \qquad \textit{Combine like terms.}$$

$$12 = 12x + 3 \qquad \textit{Next add } -3 \textit{ to both sides of the equation.}$$

$$12 \boxed{- 3} = 12x + 3 \boxed{- 3}$$

$$9 = 12x \qquad \textit{Finally, divide both sides of the equation by 12.}$$

$$\frac{9}{12} = \frac{12x}{12} \qquad \textit{Reduce the fraction to get}$$

$$\boxed{\frac{3}{4} = x}$$

Check: $x = \frac{3}{4}$: $\quad 12 - 4x = 8x + 3$

$$12 - 4\left(\frac{3}{4}\right) \overset{?}{=} 8\left(\frac{3}{4}\right) + 3$$

$$12 - 3 \overset{?}{=} 6 + 3$$

$$9 \overset{\checkmark}{=} 9$$

Often it is necessary to simplify an equation before proceeding to the actual solution.

EXAMPLE 4

Solve for t: $\quad 2.5(7.3 - t) = 2.24 - (3.5 + 1.5t)$

Solution

We begin by simplifying each side of the equation as completely as possible (with the help of a calculator):

$$2.5(7.3 - t) = 2.24 - (3.5 + 1.5t) \qquad \textit{Remove parentheses.}$$

$$18.25 - 2.5t = 2.24 - 3.5 - 1.5t \qquad \textit{Where possible, combine terms on each side.}$$

$$18.25 - 2.5t = -1.26 - 1.5t \qquad \textit{Add 1.5t to each side.}$$

$$18.25 - 2.5t \boxed{+ 1.5t} = -1.26 - 1.5t \boxed{+ 1.5t}$$

$$18.25 - t = -1.26$$

Subtract 18.25 from each side.

$$18.25 \;\boxed{-\; 18.25}\; - t = -1.26 \;\boxed{-\; 18.25}$$

$$-t = -19.51$$

We are not finished yet. We want t (not −t) so we divide both sides of the equation by the coefficient of t, which is −1.

$$\frac{-t}{-1} = \frac{-19.51}{-1}$$

$$\boxed{t = 19.51}$$

The check is left to the student.

EXAMPLE 5

Solve for x: $\dfrac{x}{2} + 10 = \dfrac{4x}{3}$

Solution

Recall from elementary algebra that to solve this equation we would first "clear the denominators" by multiplying each side of the equation by the *least common denominator*, or the LCD (the smallest number that is divisible by all the denominators in the equation). In this case, the LCD is $2 \cdot 3 = 6$.

$$\frac{x}{2} + 10 = \frac{4x}{3}$$

Multiply each side by 6 (the LCD).

$$6\left(\frac{x}{2} + 10\right) = 6\left(\frac{4x}{3}\right)$$

Distribute 6 on the left side.

$$\frac{6}{1} \cdot \frac{x}{2} + 6 \cdot 10 = \frac{6}{1} \cdot \frac{4x}{3}$$

$$\frac{\overset{3}{\cancel{6}}}{1} \cdot \frac{x}{\cancel{2}} + 6 \cdot 10 = \frac{\overset{2}{\cancel{6}}}{1} \cdot \frac{4x}{\cancel{3}}$$

$$3x + 60 = 8x$$

Now we have a simpler equation to solve. Subtract 3x from each side.

$$3x + 60 \;\boxed{-\; 3x}\; = 8x \;\boxed{-\; 3x}$$

$$60 = 5x$$

Divide both sides by 5.

$$\boxed{12 = x}$$

The check is left to the student.

EXAMPLE 6

The federal budget surplus (or deficit) from 1992 to 1998 can be estimated by the equation

$$F = 58.599(t - 1{,}992) - 314.73$$

where F is the federal surplus (or deficit if F is negative) in billions of dollars in year t.

(a) What was the federal surplus in 1996?

(b) During what year (between 1992 and 1998) was the federal *deficit* \$50 billion?

Solution

(a) Computing the surplus (or deficit) for a given year is just a matter of substituting for t and computing F:

$$F = 58.599(t - 1{,}992) - 314.73$$

Substitute t = 1,996.

$$= 58.599(1{,}996 - 1{,}992) - 314.73$$

$$= 58.599(4) - 314.73$$

Use a calculator to get

$$= -80.334$$

Hence, according to this equation, there was a deficit of about $80.3 billion in 1996.

(b) On the other hand, finding the year given the federal surplus (deficit) requires that we substitute -50 (a 50-billion-dollar deficit) for F and then solve the equation for t:

$$F = 58.599(t - 1,992) - 314.73$$ *Substitute $F = -50$.*

$$-50 = 58.599(t - 1,992) - 314.73$$ *Now solve the equation for t. First simplify the right-hand side: distribute 58.599.*

$$-50 = 58.599t - 116,729.208 - 314.73$$

$$-50 = 58.599t - 117,043.938$$ *Add 117,043.938 to each side.*

$$-50 \;\boxed{+\; 117{,}043.938}\; = 58.599t - 117{,}043.938 \;\boxed{+\; 117{,}043.938}$$

$$116,993.938 = 58.599t$$ *Divide both sides by 58.599.*

$$\frac{116,993.938}{58.599} = t$$

$$1,996.517654 = t$$

Hence the federal deficit was $50 billion during 1996.

Literal Equations

When we solve an equation, we will not always get a numerical answer. Solving a conditional equation with one variable will usually give us a numerical solution. However, the situation is different in the case of an equation with more than one variable.

An equation that contains more than one variable (letter) is called a ***literal equation.*** In certain special cases it is called a *formula.* A literal equation in which one of the variables is totally isolated on one side of the equation is said to be solved ***explicitly*** for that variable.

For example, the following three equations are alternative forms of exactly the same equation:

$$y = -3x + 7$$ Solved explicitly for y

$$x = \frac{-y + 7}{3}$$ Solved explicitly for x

$$3x + y = 7$$ Not solved explicitly for either variable

When we are asked to solve a literal equation we must be told which variable we are to solve for explicitly.

EXAMPLE 7   Solve the equation: $4x - 7y = 12$

(a) Explicitly for x **(b)** Explicitly for y

Solution   **(a)** Solving explicitly for x means we want x isolated on one side of the equation.

$$4x - 7y = 12$$ *Add 7y to both sides.*

$$4x - 7y \;\boxed{+\; 7y}\; = 12 \;\boxed{+\; 7y}$$

$$4x = 7y + 12$$ *Then divide both sides by 4.*

$$\frac{4x}{4} = \frac{7y + 12}{4}$$

$$\boxed{x = \frac{7y + 12}{4} \quad \text{or} \quad x = \frac{7}{4}y + 3}$$

(b) Solving explicitly for y means we want y isolated on one side of the equation.

$$4x - 7y = 12 \qquad \textit{Add } -4x \textit{ to both sides.}$$

$$4x - 7y \;\boxed{-\,4x} = 12 \;\boxed{-\,4x}$$

$$-7y = -4x + 12 \qquad \textit{Then divide both sides by } -7.$$

$$\frac{-7y}{-7} = \frac{-4x + 12}{-7}$$

$$\boxed{y = \frac{-4x + 12}{-7} \quad \text{or} \quad y = \frac{4}{7}x - \frac{12}{7}}$$

If we isolated y on the right side we would get $y = \dfrac{4x - 12}{7}$. Do you see that this answer is equivalent to $y = \dfrac{-4x + 12}{-7}$? We will discuss this in Section 6.1.

First-Degree Inequalities

DEFINITION

A *first-degree inequality* is an inequality that can be put in the form $ax + b < 0$, where a and b are constants with $a \neq 0$. (The $<$ symbol can be replaced with $>$, $\leq$, or $\geq$.)

As with equations, *solving inequalities* means to find the values of the variable that make the algebraic statement true. As with equations, we may have an inequality that is *true for all values of the variable,* such as $x + 6 < x + 8$. We call these inequalities *identities.* An inequality for which there is no solution, for example $x + 2 > x + 3$, is called a *contradiction.* Inequalities that are true for some values of the variable and false for other values are called *conditional.*

Two inequalities are *equivalent* if they have identical solutions. Our goal will be to take more complicated inequalities such as $2 - x - 3 < 5 - 2x$ and transform them to equivalent "obvious inequalities" such as $x < 6$. For this we will need the properties of inequalities listed below.

Keep in mind that while conditional first-degree equations have single solutions, the solutions to conditional first-degree inequalities are infinite sets.

Properties of Inequalities

For a, b, and $c \in R$, if $a < b$, then

1. $a + c < b + c$

2. $a - c < b - c$

3. $ac < bc$ when c is positive; $ac > bc$ when c is negative

4. $\dfrac{a}{c} < \dfrac{b}{c}$ when c is positive; $\dfrac{a}{c} > \dfrac{b}{c}$ when c is negative

(The "$<$" symbol can be replaced with "$>$", "$\geq$", or "$\leq$".) We can verbally summarize these properties as follows:

To produce an equivalent inequality, we may add (subtract) the same quantity to (from) both sides of an inequality, or multiply (divide) both sides of an inequality by the same *positive* quantity. On the other hand, we must reverse the inequality symbol to produce an equivalent inequality if we multiply (divide) both sides of an inequality by the same *negative* quantity.

EXAMPLE 8

Solve the following inequality for x. $3x - 5 < 7x + 2$

Solution

As with equations, our goal is to isolate the variable on one side of the inequality.

$$3x - 5 < 7x + 2$$ *Subtract 7x from each side of the inequality.*

$$3x \;\boxed{-\,7x}\; - 5 < 7x \;\boxed{-\,7x}\; + 2$$ *The inequality symbol is not reversed when we subtract from each side.*

$$-4x - 5 < 2$$ *Add 5 to each side.*

$$-4x - 5 \;\boxed{+\,5}\; < 2 \;\boxed{+\,5}$$

$$-4x < 7$$ *Divide each side by −4.*

$$\downarrow$$

$$\frac{-4x}{-4} > \frac{7}{-4}$$

$$\boxed{\,x > \frac{-7}{4}\,}$$

The inequality symbol is reversed.

Visualizing the solutions of an inequality on the real number line can be helpful. To sketch the graph of the solution set for this example (see Figure 1.6), we use the notation introduced in Section 1.1.

Note that $-\dfrac{7}{4}$ is between −2 and −1.

Figure 1.6

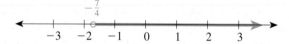

The empty circle indicates that the point $-\dfrac{7}{4}$ is to be excluded.

EXAMPLE 9

Solve for a and sketch the solution set on the real number line.

$$5a - 3 \le -9 - (4 - 3a)$$

Solution

$$5a - 3 \le -9 - (4 - 3a)$$ *Simplify the right side of the inequality.*

$$5a - 3 \le -9 - 4 + 3a$$

$$5a - 3 \le -13 + 3a$$ *Then apply the inequality properties.*

$$5a - 3 \;\boxed{-\,3a}\; \le -13 + 3a \;\boxed{-\,3a}$$

$$2a - 3 \le -13$$

$$2a - 3 \;\boxed{+\,3}\; \le -13 \;\boxed{+\,3}$$

$$2a \le -10$$

$$\frac{2a}{2} \le \frac{-10}{2}$$

$$\boxed{\,a \le -5\,}$$

We graph the solution set on the number line as shown in Figure 1.7.

Figure 1.7

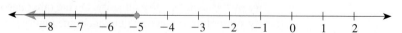

Note that a filled circle indicates that the point −5 is to be included.

STUDY SKILLS 1.5

Checking Your Work and Estimation

We develop confidence in what we do by knowing that we are right. One way to check to see if we are right is to look at the answers usually provided in the back of the book. However, few algebra textbooks provide *all* the answers. And of course, answers are not provided during exams, when we need the confidence most.

Isn't it frustrating to find out that you incorrectly worked a problem on an exam, and then to discover that you would easily have seen your error had you just taken the time to check over your work? Therefore, you should know how to check your answers.

The method of checking your work should be different from the method used in the solution. In this way you are more likely to discover any errors you might have made. If you simply rework the problem the same way, you cannot be sure you did not make the same mistake twice.

Ideally, the checking method should be quicker than the method for solving the problem (athough this is not always possible).

Learn how to check your answers, and practice checking your homework exercises as you do them.

Also, as you work out exercises and solve problems, it is very important to be constantly aware of the reasonableness of your answer so that you do not propose impossible solutions to problems. This is particularly true when solving verbal problems. It is obvious that if in a certain problem, x represents the number of 25-lb boxes, then an answer of $x = -7$ is ridiculous. However, it is also unreasonable to get an answer of $x = 8.3$. Why?

Sometimes recognizing an impossible answer is more subtle. For example, if a total of $5,000 is split into two investments and x represents one of the investments, then $x = \$6,500$ is impossible. Why?

Sometimes students are lulled into a false sense of security, when they use a calculator to do computations. While it is true that the calculator does not make computational mistakes, you need to be sure that you have chosen to do the correct computations. In addition, you may inadvertently hit the wrong operations key or put a decimal point in the wrong place. Having an estimate of the correct answer will make it much easier to recognize these types of errors.

EXERCISES 1.5

In Exercises 1–8, determine whether the listed values of the variable satisfy the equation.

1. $5(x - 3) - (x + 2) = -5$; $x = 0, 3, 5$

2. $7(x + 1) - (x - 2) = -3$; $x = -2, 0, 2$

3. $3(a - 5) + 2(1 - a) = -11$; $a = -3, 0, 2$

4. $4(t + 3) - (7 - t) = 1$; $t = -1, 1, 4$

5. $x^2 - 4x = 5$; $x = -1, 2, 5$

6. $6x - x^2 = 0$; $x = -4, 3, 6$

7. $a(a + 8) = a + 8$; $a = -1, 1, 3$

8. $(a + 4)^2 = a$; $a = -8, 2, 8$

In Exercises 9–50, solve the equation.

9. $2x - 7 = 5$

10. $3a - 5 = 19$

11. $4y + 2 = -1$

12. $7z + 6 = -2$

13. $m + 3 = 3 - m$

14. $m - 3 = 3 - m$

15. $3t - 5 = 5t - 13$

16. $8w - 9 = 13w + 21$

17. $11 - 3y = 38$

18. $9 - 7y = 16$

19. $5s + 2 = 3s - 7$

20. $40t + 15 = 22t - 3$

21. $3.24x - 5.2 = 7.74x + 0.3$

22. $7.24x - 12.3 = 0.5 + 4.84x$

23. $5 - 21x = 3x - 16$

24. $4 - 3b = 5b + 7$

25. $3x - 12 = 12 - 3x$

26. $5y - 7 = 7 - 5y$

27. $3x - 12 = -12 - 3x$

28. $5y - 7 = -7 - 5y$

29. $2t + 1 = 7 - t$

30. $6 - 5t = t - 12$

31. $3(x - 1) = 2(x + 1)$

32. $3(a + 5) = 4(a + 5) + (a + 5)$

33. $3(x + 1) + x = 2(x + 3)$

34. $4(x - 3) - 2x = 4(x - 1)$

35. $0.06x + 0.0725(22{,}500 - x) = 1{,}500$

36. $6.5(2.5x - 4) = -2.5(5x + 5.34)$

37. $\dfrac{x}{2} - 1 = 5$

38. $\dfrac{2x}{3} + 1 = 6$

39. $\dfrac{2x}{3} + 2 = \dfrac{x}{2}$

40. $\dfrac{3x}{2} - 4 = \dfrac{x}{3}$

41. $5x - \dfrac{2}{3} = \dfrac{x}{4}$

42. $4x - \dfrac{1}{2} = \dfrac{2}{3}$

43. $6x - \dfrac{2}{5} = \dfrac{3x}{4}$

44. $7x - \dfrac{7}{8} = \dfrac{x}{4}$

45. $\dfrac{3x}{5} - \dfrac{2}{3} = 5x$

46. $\dfrac{2x}{7} + \dfrac{3}{4} = \dfrac{5x}{3}$

47. $\dfrac{3x - 2}{4} = 5$

48. $\dfrac{5x - 3}{2} = 8$

49. $\dfrac{7x - 1}{2} = \dfrac{2}{3}$

50. $\dfrac{3x + 1}{5} = \dfrac{3}{8}$

51. The median value of a home in the United States from 1940 to 2000 can be estimated by the equation

$$V = 1{,}481(t - 1940) + 28{,}871$$

where V is the median value in thousands of dollars at year t.

   (a) What was the median value of a home in the United States in 2000?

   (b) In what year (between 1940 and 2000) was the median value of a home $100,000?

52. The national health expenditures for hospital care only (how much was spent on hospital care) for the United States from 1980 to 2002 can be estimated by the equation

$$E = 16.64(t - 1980) + 91.36$$

where E is the amount spent in billions of dollars during year t.

   (a) Use this equation to determine how much was spent in the United States on hospital care in 1985.

   (b) In what year (between 1980 and 2002) was 280 billion dollars spent in the United States on hospital care?

In Exercises 53–58, solve each of the following equations explicitly for the indicated variable.

53. $5x + 7y = 4$ for x

54. $5x + 7y = 4$ for y

55. $2x - 9y = 11$ for y

56. $2x - 9y = 11$ for x

57. $2(x - y) = 3x + 4$ for x

58. $2(x - y) = 3x + 4$ for y

In Exercises 59–70, solve the inequality and graph the solution on the number line.

59. $2x - 5 < 7x - 2$

60. $3x + 4 \geq 6x - 3$

61. $4x - 25 < x - 8$

62. $2 - (x + 5) \geq x + 2$

63. $3 + (2x - 1) < 4x - 6$

64. $3x - (2x - 1) < x + 5$

65. $3x - (2x - 7) \leq 4x + 6$

66. $5 - (4x - 3) \geq 2 - (x + 1)$

67. $6x - (3x + 1) \geq 2x + (2x - 3)$

68. $4.23x - 8.5 > 7.3x + 2$

69. $3.4 - (2x - 5.8) \leq 2.6x + 4$

70. $8.2 - (3.1x - 1.6) > 4.25x - 7.1$

CHAPTER 1 SUMMARY

After having completed this chapter, you should be able to:

1. Distinguish between natural numbers (N), whole numbers (W), the integers (Z), the rational (Q) and irrational (I) numbers, and understand the makeup of the set of real numbers (R) (Section 1.1).

2. Graph inequalities on the number line (Section 1.1).

For example:

(a) On the number line, the graph of $\{x \mid -5 < x \leq 3, x \in Z\}$ is

(b) On the number line, the graph of $\{x \mid -2 < x \leq 3\}$ is

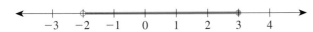

Note the empty circle at -2 and the filled circle at 3.

3. Identify the properties of the real numbers (Section 1.1).

For example: The statement

$$(x + 4) + (x - 6) = (4 + x) + (x - 6)$$

illustrates the ***commutative property of addition.***

4. Perform operations on real numbers (Section 1.2).

For example: Perform the given operatons.

(a) $-6 - 2[5 - (7 - 8)]$ *Perform operations inside parentheses first.*

$= -6 - 2[5 - (-1)]$ *Then simplify in brackets.*

$= -6 - 2[5 + 1]$

$= -6 - 2[6]$

$= -6 - 12$

$= -18$

(b) $\dfrac{-5 + 4(2 - 3)}{-5(3 - 3)}$ *Perform operations inside parentheses.*

$= \dfrac{-5 + 4(-1)}{-5(0)}$

$= \dfrac{-5 + 4(-1)}{0}$

Undefined   *Since division by 0 is undefined*

5. Substitute real numbers for variables in expressions or formulas and evaluate the numerical expression (Section 1.2).

For example: Given $S = s_0 + v_0 t - 16t^2$, find S given $s_0 = 50.4$, $v_0 = 25.2$, and $t = 1.5$.

Solution: Using a calculator, we enter the numbers and operations as shown below.

$$\boxed{50.4}\; \boxed{+}\; \boxed{25.2}\; \boxed{\times}\; \boxed{1.5}\; \boxed{-}\; \boxed{16}\; \boxed{\times}\; \boxed{1.5}\; \boxed{x^2}\; \boxed{=}$$

The solution is $s = 52.2$.

6.   Perform operations on and simplify algebraic expressions (Section 1.3).

For example: Perform the operations and simplify the following.

(a) $3x(2x - y) - 2x(x + y)$ *First use the distributive property.*

$\quad = 6x^2 - 3xy - 2x^2 - 2xy$ *Combine like terms.*

$\quad = 4x^2 - 5xy$

(b) $2a - [3a - 5(2a - 1)]$ *Perform operations inside brackets.*

$\quad = 2a - [3a - 10a + 5]$

$\quad = 2a - [-7a + 5]$ *Then subtract (change the sign of each term in brackets).*

$\quad = 2a + 7a - 5$

$\quad = 9a - 5$

7.   Translate verbal sentences into algebraic statements (Section 1.4).

For example: If Jack has n nickels, q quarters, and d dimes, express the number of coins and the value of his coins in terms of n, q, and d.

Solution: If Jack has n nickels, q quarters, and d dimes, then he has $n + q + d$ coins.

The value of his nickels is $5n$ cents.

The value of his quarters is $25q$ cents.

The value of his dimes is $10d$ cents.

So the total value of all his coins is

$$5n + 25q + 10d \text{ cents}$$

8.   Solve and check a first-degree equation in one variable (Section 1.5).

For example: Solve for x: $2x - 10 = \dfrac{x}{3}$

Solution:

$$2x - 10 = \frac{x}{3}$$ *We "eliminate" the denominators by multiplying each side of the equation by 3.*

$$3(2x - 10) = 3\left(\frac{x}{3}\right)$$ *Multiply out the left side; simplify the right side.*

$$6x - 30 = x$$ *Subtract x from each side.*

$$5x - 30 = 0$$ *Add 30 to each side.*

$$5x = 30$$ *Divide both sides by 5.*

$$x = 6$$

Check: $x = 6$: $2(6) - 10 \overset{?}{=} \dfrac{6}{3}$

$\qquad\qquad\qquad 12 - 10 \overset{\checkmark}{=} 2$

9. Solve a first-degree inequality and graph the solution set (Section 1.5).

For example: Solve for *x*: $x - (3x + 5) \geq 2x$

Solution:

$$x - (3x + 5) \geq 2x \qquad \textit{Simplify the left side of the inequality.}$$
$$x - 3x - 5 \geq 2x$$
$$-2x - 5 \geq 2x \qquad \textit{Add 2x to each side of the inequality.}$$
$$-5 \geq 4x \qquad \textit{Divide each side by 4.}$$
$$-\frac{5}{4} \geq x$$

The sketch of the graph appears in Figure 1.8.

Figure 1.8

10. Solve simple literal equations explicitly for a given variable (Section 1.5).

For example: Given $8x - 3y = a + 2y$, solve explicitly for *y*.

Solution: Since we are solving explicitly for *y*, we collect all the *y* terms on one side of the equation and all the non-*y* terms on the other side.

$$8x - 3y = a + 2y \qquad \textit{Add 3y to each side of the equation.}$$
$$8x - 3y \boxed{+ 3y} = a + 2y \boxed{+ 3y} \qquad \textit{Combine terms on each side where possible.}$$
$$8x = a + 5y \qquad \textit{Subtract a from each side.}$$
$$8x \boxed{- a} = a + 5y \boxed{- a} \qquad \textit{Simplify each side.}$$
$$8x - a = 5y \qquad \textit{Divide each side by 5.}$$
$$\frac{8x - a}{5} = y$$

CHAPTER 1 REVIEW EXERCISES

In Exercises 1–6, you are given

$$A = \{x \mid x \leq 4, x \in N\} \qquad B = \{a \mid a > 5, a \in W\}$$
$$C = \{a, r, b, s, e\} \qquad D = \{b, c, f, g\}$$

List the elements in the following sets.

1. *A*

2. *B*

3. $C \cap D$

4. $C \cup D$

5. $A \cup B$

6. $A \cap B$

In Exercises 7–12, you are given

$$A = \{a \mid a \in W, \text{ and } a \text{ is a factor of } 12\}$$
$$B = \{b \mid b \in W, \text{ and } b \text{ is a multiple of } 12\}$$
$$C = \{c \mid c \in W, \text{ and } c \text{ is a multiple of } 6\}$$

List the elements in the following sets.

7. A

8. B

9. $A \cap B$

10. $A \cup B$

11. $B \cap C$

12. $A \cap C$

In Exercises 13–14, you are given

$$A = \{x \mid -2 < x \le 4, x \in Z\}$$
$$B = \{y \mid 3 \le y \le 12, y \in Z\}$$

Describe the sets using set notation.

13. $A \cap B$

14. $A \cup B$

In Exercises 15–20, graph the sets on the real number line.

15. $\{x \mid x \le 4\}$

16. $\{x \mid x > 4\}$

17. $\{b \mid -8 \le b \le 5\}$

18. $\{b \mid -8 < b < 5\}$

19. $\{a \mid -2 < a \le 4\}$

20. $\{b \mid -2 \le b < 4\}$

In Exercises 21–28, determine whether the statement is true or false.

21. $\dfrac{1}{2} \in Z$

22. $\sqrt{9} \in Z$

23. $0.47\overline{6476} \in Q$

24. $\pi \in R$

25. $\pi \in Q$

26. $6\dfrac{2}{7} \in Q$

27. $I \subset R$

28. A rational number is an integer.

In Exercises 29–36, if the statement is true, state the property illustrated by the statement. If the statement is not true, write "false."

29. $(x - 4) + 3 = 3 + (x - 4)$

30. $(5 - x)7 = 7(5 - x)$

31. $(x - 4) \cdot 3 = x \cdot 3 - 4 \cdot 3$

32. $(x + 5)(y - 2) = x(y - 2) + 5(y - 2)$

33. $\left(\dfrac{1}{2x}\right)2x = 1 \quad (x \ne 0)$

34. $(5x + 2) + 0 = 5x + 2$

35. $(3x - 2) + (4x - 1) = 3x - (2 + 4x) - 1$

36. $3 \cdot 2(x - 1) = 6(2x - 2)$

In Exercises 37–54, perform the operations.

37. $(-2) + (-3) - (-4) + (-5)$

38. $(-6) - (-2) + (-3) - (+4)$

39. $6 - 2 + 5 - 8 - 9$

40. $-7 - 2 - 3 + 8 - 5$

41. $(-2)(-3)(-5)$

42. $(-7)(-3)(-5)(+2)$

43. $(-2)^6$

44. -2^6

45. $(-2) - (-3)^2$

46. $-5 - (-2)^3$

47. $(-6 - 3)(-2 - 5)$

48. $(-4 + 7)(-3 - 2)$

49. $5 - 3[2 - (4 - 8) + 7]$

50. $-6 + 2[5 - (3 - 9)]$

51. $5 - \{2 + 3[6 - 4(5 - 9)] - 2\}$

52. $-7 + \{3 - 6[12 - 5(9 - 11)]\}$

53. $\dfrac{4[5 - 3(8 - 12)]}{-6 - 2(5 - 6)}$

54. $\dfrac{6 - [2 - (3 - 5)]}{7 + [5 + 6(6 - 8)]}$

In Exercises 55–60, evaluate, given $x = -2$, $y = -1$, and $z = 0$.

55. $x^2 - 2xy + y^2$

56. $2x^2 + 3xy - 3y^2$

57. $|x - y| - (|x| - |y|)$

58. $|x + y| + |x| - |y|$

59. $\dfrac{2x^2y^3 + 3y^2}{zx^2y}$

60. $\dfrac{(5xy^3 + x^4)z}{x^2y}$

61. Given

$$t = \frac{\overline{X} - a}{\frac{s_x}{\sqrt{n}}}$$

compute t to two places, for $\overline{X} = 100$, $a = 95$, $s_x = 7.1$, and $n = 30$.

62. Given

$$\sigma_r = \sqrt{\frac{1 - \rho^2}{n - 1}}$$

compute σ_r to two places, for $\rho = 0.65$ and $n = 80$.

In Exercises 63–79, perform the operations and express your answer in simplest form.

63. $(2x + y)(-3x^2y)$

64. $(-6ab^2)(-3a^2b)(-2b)$

65. $(2xy^2)^2(-3x)^2$

66. $(-3r^2s^2)^3(-2r^2)$

67. $3x - 2y - 4x + 5y - 3x$

68. $-6a + 2b + 3a - 4b - 5a$

69. $-2r^2s + 5rs^2 - 3sr^2 - 4s^2r$

70. $-5x^2y^3 + 6xy^2 + 2x^2y^3 - 8y^2x$

71. $(y - 5) - (y - 4)$

72. $3a(a - b + c)$

73. $5rs(2r + 3s)$

74. $2x - 3(x - 4)$

75. $7y - 9(2x + 1)$

76. $3a - [5 - (a - 4)]$

77. $-2r + 3[s - 2(s - 6)]$

78. $5x - \{3x + 2[x - 3(5 - x)]\}$

79. $7 - 3\{y - 2[y - 4(y - 1)]\}$

80. The median value of a home in the United States from 1940 to 2000 can be estimated by the equation

$$V = 1,481(t - 1,940) + 28,871$$

where C is the median cost in thousands of dollars at year t.

   (a) Simplify the expression on the right-hand side of the equation.

   (b) Compute the median value of a home for 1993 using the original equation and then with the simplified version completed in part **(a)**.

In Exercises 81–90, translate the statements algebraically.

81. Five less than the product of two numbers is 3 more than their sum.

82. Eight more than the sum of a number and itself is 3 less than the product of the number and itself.

83. The sum of the first two of three consecutive odd integers is 5 less than the third.

84. The product of the last two of three consecutive even integers is 8 more than 10 times the first.

85. The length of a rectangle is 5 less than 4 times its width. Express its area and perimeter in terms of one variable.

86. The length of a rectangle is 5 more than 3 times its width. Express its area and perimeter in terms of one variable.

87. The sum of the squares of two numbers is 8 more than the product of the two numbers.

88. The square of the sum of two numbers is 8 more than the product of the two numbers.

89. Xavier has 40 coins in nickels and dimes. Express the total value of his coins in terms of the number of dimes.

90. Cassandra makes $18 an hour as a garage mechanic and $10 an hour as a typist. If she works a total of 30 hours, and she works x of these hours as a mechanic, express the total amount she makes in terms of x.

In Exercises 91–107, solve the equation.

91. $5x - 2 = -2$

92. $3x - 4 = 8$

93. $3x - 5 = 2x + 6$

94. $3y - 5 = 8y - 9$

95. $11x + 2 = 6x - 3$

96. $7x + 8 = 3x$

97. $5(a - 3) = 2(a - 4)$

98. $7 - (2 - b) = 3b - 4(b - 1)$

99. $6(q - 4) + 2(q + 5) = 8q - 19$

100. $3(z - 7) + 2z - 4 = 5(z - 5)$

101. $3x - 5x = 7x - 4x$

102. $2a + 5a = 7a - 3a$

103. $x - \dfrac{2}{3} = 2x + 4$

104. $\dfrac{2x - 1}{3} = \dfrac{1}{2}$

105. $\dfrac{x - 3}{5} = x + 1$

106. $2y - \dfrac{1}{3} = y + \dfrac{1}{2}$

107. $3.2x + 0.14 = 1.4x - 21.46$

108. The number of physicians in the United States from 1980 to 2000 can be estimated by the equation

$$P = 16.13(t - 1980) + 449.2$$

where P is the number of physicians in thousands during year t.

(a) Use this equation to estimate the number of physicians in the United States in 1995.

(b) In what year (between 1980 and 1997) were there 600,000 physicians in the United States?

In Exercises 109–112, solve each equation explicitly for the indicated variable.

109. $7x + 8y = 22$ for y

110. $3a - 5b = a + 2$ for a

111. $3x - 6y = 2 - 4x + 2y$ for y

112. $3a - 5b = \dfrac{a + 2}{2}$ for a

In Exercises 113–118, solve each inequality and graph the solution on the number line.

113. $3x + 12 < 2x - 9$

114. $5x - 8 \geq 3x - 6$

115. $5 - (2x - 7) > 3x - 15$

116. $3 - (8x + 12) \leq 9x + 24$

117. $3.2 - (5.3x - 1.2) \leq 3.4x - 1.8$

118. $8.2x - 8.3 + 1.4x > 5 - (3x - 2.4)$

CHAPTER 1 PRACTICE TEST ⊙

1. Given

$A = \{x \mid x$ is a prime factor of 210$\}$

$B = \{y \mid y$ is a prime number less than 25$\}$

(a) List the elements in $A \cap B$.

(b) List the elements in $A \cup B$.

2. Indicate whether each of the following is true or false:

(a) $\sqrt{16} \in I$

(b) $2 \in Q$

(c) $-\dfrac{3}{4} \in Z$

3. Graph the following sets on the real number line.

(a) $\{a \mid a > 4\}$

(b) $\{x \mid -3 \le x < 10\}$

4. If the statement is true for all real numbers, state the property that the given statement illustrates. If the statement is not true for all real numbers, write "false."

(a) $7(xy + z) = 7xy + z$

(b) $(x + y) + 3 = 3 + (x + y)$

5. Evaluate the following:

(a) $-3 - (-6) + (-4) - (-9)$

(b) $(-7)^2 - (-6)(-2)(-3)$

(c) $\lvert 3 - 8 \rvert - \lvert 5 - 9 \rvert$

(d) $6 - 5[-2 - (7 - 9)]$

6. Evaluate the following, given $x = -2$ and $y = -3$:

(a) $(x - y)^2$

(b) $\dfrac{x^2 - y^2}{x^2 - 2xy - y^2}$

7. Simplify the following:

(a) $(5x^3y^2)(-2x^2y)(-xy^2)$

(b) $3rs^2 - 5r^2s - 4rs^2 - 7rs$

(c) $2a - 3(a - 2) - (6 - a)$

(d) $7r - \{3 + 2[s - (r - 2s)]\}$

8. The length of a rectangle is 8 less than 3 times its width. Express its perimeter in terms of one variable.

9. Wallace has 34 coins in nickels and dimes. If x represents the number of dimes, express the number of nickels in terms of x. In terms of x, what is the total value of his coins?

10. Solve for x.

(a) $3x - 2 = 5x + 8$

(b) $2(x - 4) = 5x - 7$

(c) $\dfrac{2x + 1}{3} = \dfrac{1}{4}$

(d) $3x - (5x - 4) \le 2x + 8$

(e) $2.1x - 5 > 5.3x - 4.9$

11. Solve explicitly for t. $3s - 5t = 2t - 4$

STUDY SKILLS 1.6

Making Study Cards

Study cards are $3'' \times 5''$ or $5'' \times 8''$ index cards that contain summary information needed for convenient review. The process of making study cards is a learning experience in itself. We will discuss how to use study cards in the next chapter. For now, we will cover three types of cards: the definition/principle card, the warning card, and the quiz card.

The *definition/principle (D/P) cards* contain a single definition, concept, or rule for a particular topic.

The front of each D/P card should contain the following:

1. A heading of a few words.
2. The definition, concept, or rule accurately recorded.
3. If possible, a restatement of the definition, concept, or rule in your own words.

The back of the card should contain examples illustrating the idea on the front of the card.

Here is an example of a D/P card.

FRONT

The Distributive Property
$a(b + c) = ab + ac$
or
$(b + c)a = ba + ca$
Multiply <u>each</u> term by a.

BACK

(1) $3x(x + 2y) = 3x(x) + 3x(2y)$
$= 3x^2 + 6xy$
(2) $-2x(3x - y) = -2x(3x) - 2x(-y)$
$= -6x^2 + 2xy$
(3) $(2x + y)(x + y)$
$= 2x(x + y) + y(x + y)$

Warning (W) cards contain errors that you may be making consistently on homework, quizzes, or exams, or those common errors pointed out by your teacher or your textbook. The front of the warning card should contain the word WARNING; the back of the card should contain an example of both the correct way an example should be done and the common error. Be sure to label clearly which solution is correct and which is not.

For example:

FRONT

WARNING
EXPONENTS
An exponent refers only to the factor immediately to the left of the exponent.

BACK

EXAMPLES
$2 \cdot 3^2 = 2 \cdot 3 \cdot 3$ NOT $(2 \cdot 3)(2 \cdot 3)$
$(-3)^2 = (-3)(-3) = 9$
↑ Parentheses mean -3 is the factor to be squared.
BUT
$-3^2 = -3 \cdot 3 = -9$
↑ The factor being squared here is 3, not -3.

Quiz cards are another type of study card. They will be used to help us construct practice tests. For now, go through your textbook and pick out a few of the odd-numbered exercises (just the problem) from each section, putting one or two problems on one side of each card. Make sure that you copy the *instructions* as well as the problem accurately. On the back of the card, write down the exercise number and section of the book where the problem was found. For example:

FRONT

Translate the given sentence. Indicate clearly what each variable represents.
Four more than three times a number is seven less than the number.

BACK

Exercise 5
Section 1.4

2

Equations and Inequalities

© John Foxx/Alamy

Equations as Mathematical Models

Let's suppose you are nearing the end of a course in which the final exam counts as 40% of your course grade. Your average grade for all but the final exam is 73, and you want to determine the minimum score needed on your final exam to get a grade of 80 for the course.

There are several ways we can approach this problem, but let's make sure we understand the relationship between the final exam grade and the course grade; that is, how does the final exam affect your already established grade of 73? In this example, the final exam is 40% of the course grade, and your average going into the final is 73; hence 60% of your grade is 73. If we let the final exam score be x, we would compute the course grade as follows:

Course grade $= (0.60)(73) + 0.40x$

$\qquad = 43.8 + 0.40x$

The course grade in this case is called a *weighted average* because it is computed by adding the products of grades with their percentage weights. Now that we have an equation or a *mathematical model* describing the relationship between the final exam score and the course grade, we can try to guess the value of x and check to see how close we come to our goal of a course grade of 80.

If you get a 95 on the final, then you would have

$$43.8 + 0.40(95) = 81.8$$

as a course grade. This is higher than 80.

Suppose you get 85 as a final exam score; then you would have

$$43.8 + 0.40(85) = 77.8$$

as a course grade. This is lower than we want.

Suppose you get 90 as a final exam score; then you would have

$$43.8 + 0.40(90) = 79.8$$

as a course grade. This is again too low (remember, we want the course grade to be 80). This guess-and-check method is quite uncertain and can take some time. How long would it take you to check other possible final course grades (and what you would need on the final to attain them) as well?

In Section 1.4, we began to construct equations that describe the relationships between quantities discussed in word problems. A **mathematical model** is a mathematical description of a real-life situation. For example, the equation we discussed above, Course grade $= 43.8 + 0.40x$, is a mathematical model describing the relationship between course grade and final exam score x for a student with a 73 average before the final exam. In this section we will focus on mathematical models. In the next section we will turn our attention to the algebraic procedures used in solving equations.

Let's look at some more models.

If you haven't done so already, we suggest that you review Section 1.4 and 1.5 in preparation for this section.

EXAMPLE 1

Cynthia is paid a base salary of $180 a week plus a 5% commission on her gross sales.

(a) Create a model which relates her weekly pay W to her gross sales g.
(b) In order to pay her living expenses, she has to earn at least $400 a week. How much does she have to sell to earn enough for living expenses?

Solution

(a) To develop a model describing the relationship between Cynthia's weekly pay and her gross sales, we need to understand the relationship between the quantities and/or the variables under discussion. It is often helpful to substitute values for the variable quantities: as we examine the results, we should be able to spot a pattern to help us describe this relationship in general terms using symbolic notation. (Though this may seem too elementary a step for this problem, we encourage you to get in the habit of developing this general approach as a way to attack more difficult problems.)

To begin, let's note that if Cynthia sells $500 worth of merchandise one week, her commission is 5% of $500, found by multiplying $0.05 \times 500 = 25$ dollars. So her weekly pay would be $180 + $25 = $205. Let's try one or two more gross sales figures and compute her weekly pay:

$$\text{Weekly pay} = \text{Base salary} + \text{Commission}$$

Weekly pay with sales of $1,000 = \quad $180 \quad + 0.05 \times $1,000 = $230

Weekly pay with sales of $2,000 = \quad $180 \quad + 0.05 \times $2,000 = $280

Hence, if we let W = her weekly pay, then W is related to her gross sales, g, in the following way:

$$W = 180 + 0.05g$$

(b) To determine what her gross sales must be to earn $400 a week, we substitute $W = 400$ in the equation $W = 180 + .05g$, and find g as follows:

$$W = 180 + 0.05g \qquad \textit{Substitute W = 400.}$$
$$400 = 180 + 0.05g \qquad \textit{Now solve for g; first add −180 to each side.}$$
$$220 = 0.05g \qquad \textit{Divide both sides by 0.05.}$$
$$\frac{220}{0.05} = \frac{0.05g}{0.05}$$
$$4,400 = g$$

Hence she should sell

$$\boxed{\$4,400 \text{ worth of merchandise per week}}$$

to have a weekly income of $400. We can check this value by observing that $W = 180 + 0.05(4,400) = 180 + 220 = $400.

EXAMPLE 2

The length of a rectangle is 3 more than 4 times its width.

(a) Create a model expressing the perimeter in terms of its width.
(b) Determine the dimensions of the rectangle if the perimeter is to be 50 m.

Solution

(a) A diagram is especially useful in a problem like this. Note that the length is described in terms of the width. Let

$$w = \text{Width}$$

Then

$$4w + 3 = \text{Length} \qquad \textit{The example tells us that the length of the rectangle is "3 more than 4 times its width."}$$

The rectangle is drawn in Figure 2.1.

w

$4w + 3$

Figure 2.1

Recall that the formula for the perimeter of a rectangle is

$$P = 2w + 2L \qquad \textit{Where w = width and L = length}$$

With P as the perimeter and w as the width, we substitute $L = 4w + 3$ to get

$$P = 2w + 2(4w + 3) \qquad \textit{Which can be simplified to}$$
$$= 2w + 8w + 6$$

$$P = 10w + 6 \qquad \textit{This is our model.}$$

If we are given the width, then from this model we can determine the length and the perimeter.

To get a sense of how the variables w, L, and P are related, we constructed the table on the left for a few values of w.

w	$L = 4w + 3$	$P = 10w + 6$
1	7	16
2	11	26
3	15	36
4	19	46
5	23	56

(b) To find the dimensions of the rectangle when the perimeter is 50 m, we first substitute $P = 50$ in the equation $P = 10w + 6$, and solve for w:

$$P = 10w + 6 \qquad \textit{Substitute P = 50.}$$
$$50 = 10w + 6 \qquad \textit{Solve for w; add } -6 \textit{ to each side.}$$
$$44 = 10w \qquad \textit{Divide both sides by 10}$$
$$\frac{44}{10} = w$$

Hence, for the perimeter to be 50 m, the width must be $\frac{44}{10} = 4.4$ m, and therefore the length must be $L = 4(4.4) + 3 = 20.6$ m. Hence the dimensions are

$$\boxed{4.4 \text{ m} \times 20.6 \text{ m}}.$$

EXAMPLE 3

Jonas has $15,000 to invest. He decides to invest some of it in a long-term certificate that yields an annual interest rate of 7% and the rest in a short-term certificate that yields an annual interest of 5%.

(a) Create a model that expresses the amount of interest earned in terms of the amount invested in the long-term certificate.

(b) Use this model to explain how the interest is affected by changes in the distribution of the $15,000 investment. Determine the maximum and minimum interest he could earn at the end of the year.

(c) Determine how much he should invest in each certificate if he wants to earn $1,100 in interest this year.

Solution

(a) As we discussed in Chapter 1, the amount of interest, I, earned in 1 year if P dollars is invested at a rate of $r\%$ (simple interest) per year is computed as

$$\text{Interest} = (\text{Principal})(\text{Rate})$$
$$I = P \cdot r \qquad \textit{Where r is written as a decimal}$$

For example, if $2,000 is invested at 7% per year, then the interest earned in 1 year is

$$I = (2,000)(0.07) = \$140$$

Based on the information given in this example, let

$$x = \text{The amount invested at } 7\%$$

Then

$$15,000 - x = \text{The amount invested at 5\%}$$

Since the total amount invested is $15,000 (total minus part equals remainder)

Let T be the total interest earned. We construct an equation that involves the yearly interest collected on each certificate:

Total interest = Interest on the 7% certificate + Interest on the 5% certificate

$$T = (0.07)(\text{Amount invested at 7\%}) + (0.05)(\text{Amount invested at 5\%})$$

$$T = \qquad 0.07x \qquad + \qquad 0.05(15,000 - x)$$

Hence our model is

$$T = 0.07x + 0.05(15,000 - x) \qquad \textit{Which is simplified to}$$

$$= 0.07x + 750 - 0.05x$$

$$\boxed{T = 0.02x + 750}$$

(b) What is the significance of the 0.02 and the 750 in the equation we found in part **(a)**? Let's examine this model by trying to understand how the relationship between the variables and constants explains what is occurring in this real-life situation. The model we have obtained is

$$T = 0.02x + 750$$

where T is the total amount of interest collected from both investments and x is *that portion of the $15,000 invested in the 7% certificate.*

First note that if Jonas invests nothing in the 7% certificate, then he is setting $x = 0$ in the equation for T, obtaining $T = 0.02(0) + 750 = \$750$. This occurs because $x = 0$ means all the money goes into the 5% certificate and 5% of $15,000 is $750. This is the minimum interest Jonas would get from his investment.

Where does the 0.02 come from? Consider this: If Jonas puts all the money into the 5% certificate, then he would earn $750 (i.e., $0.05 \times 15,000$) as his total annual interest. *For every dollar transferred from the 5% certificate to the 7% certificate, he gains 2% more interest on that dollar.* If x dollars are transferred, he gains 2% of x dollars, hence the interest equals $750 + 0.02x$. If all the money is put into the 7% certificate, Jonas gets $750 + 0.02(15,000) = 750 + 300 = \$1,050$ in interest. This is the maximum amount of interest he would get for his investment.

(c) How much should he put into the long-term certificate to get interest totaling $1,100? We can reason that since the most he can get from the investment is $1,050, he can never get $1,100.

If we tried to find the solution using our model, we would substitute $T = 1,100$ in the equation $T = 0.02x + 750$:

$$T = 0.02x + 750 \qquad \textit{Substitute } T = 1,100.$$

$$1,100 = 0.02x + 750 \qquad \textit{Solve for x: add } -750 \textit{ to each side.}$$

$$350 = 0.02x \qquad \textit{Divide both sides by 0.02}$$

$$\frac{350}{0.02} = 17,500 = x$$

We find that we do have an answer, but does it make sense? x is the amount invested at 7%, but $17,500 is more than the total amount invested in both certificates.

$$\boxed{\text{Therefore he can never earn \$1,100 in interest with those investments}}$$

| EXAMPLE 4 | Jean can stuff 30 envelopes per hour, and Bob can stuff 45 envelopes per hour. |

(a) Create a model that expresses the number of envelopes, *E*, they can stuff working together in terms of the number of hours, *t,* they work together.

(b) Create a model that expresses the number of envelopes, *E,* they can stuff working together if Bob starts 1 hour after Jean. Discuss how the two models differ.

Solution   **(a)** Let's carefully analyze how we construct this model.

THINKING OUT LOUD

What do we need to find?

An equation expressing the number of envelopes stuffed by Jean and Bob working together in terms of the number of hours *t* they work together.

In general, how do we find the number of envelopes stuffed during a given time period?

Let's look at Jean alone. She can stuff 30 envelopes per hour. Therefore, in 1 hour, she can stuff 30 envelopes; in 2 hr, she can stuff $30 \times 2 = 60$ envelopes; in 3 hr, she can stuff $30 \times 3 = 90$ envelopes, etc. We can see by the pattern that in *t* hours, she can stuff $30t$ envelopes; hence $E_J = 30t$ is the model for the number *Jean* can stuff for *t* hours. In the same way, since Bob's rate is 45 envelopes per hour, the model for Bob is $E_B = 45t$.

How do we find the number of envelopes they can stuff working together?

The equation for Jean, $E_J = 30t$, gives the number of envelopes stuffed by Jean in *t* hours. Likewise, the equation for Bob, $E_B = 45t,$ gives the number of envelopes stuffed by Bob in *t* hours. Together they stuff $E_J + E_B = 30t + 45t$ envelopes. Hence we have

$$E = E_J + E_B = 30t + 45t = 75t$$

Thus, the model is: $E = 75t$.

In general we see that the relationship between the amount of work done and the rate at which the work is being done is as follows:

$$\text{Amount of work done} = (\text{Work rate}) \cdot (\text{Time})$$

or

$$W = r \cdot t$$

The model we developed for Jean and Bob working together is $E = 75t$. We could have approached this more intuitively by saying that if Jean can stuff 30 envelopes an hour and Bob can stuff 45 envelopes per hour, then together they can stuff $30 + 45 = 75$ envelopes an hour. Hence our model will be

$$E = 75t$$

which is equivalent to the model we found before. Why bother with the first method at all? It will become apparent in part **(b)** of the example.

(b) How long would it take if Bob started an hour after Jean? Because we are starting with different rates at different hours, we cannot simply add the rates as we did above in the second method. Instead we will apply the logic of method 1: First we let

$$t = \text{Number of hours Jean spends stuffing envelopes}$$

Then

$$t - 1 = \text{Number of hours Bob spends stuffing envelopes}$$
Since Bob starts an hour later, he is working one hour less than Jean.

Thus we have

$$\left(\begin{array}{c}\text{Amount of work} \\ \text{done by Jean}\end{array}\right) + \left(\begin{array}{c}\text{Amount of work} \\ \text{done by Bob}\end{array}\right) = \left(\begin{array}{c}\text{Total amount of} \\ \text{work done}\end{array}\right)$$

$$(\text{Jean's rate}) \cdot (\text{Jean's time}) + (\text{Bob's rate}) \cdot (\text{Bob's time}) = E$$

$$30 \cdot t \qquad + \qquad 45 \cdot (t - 1) \qquad = E$$

Hence the model is

$$E = 30t + 45(t - 1)$$

which simplifies to

$$E = 30t + 45t - 45$$

or

$$E = 75t - 45$$

Compare the two models. Note that the second model, $E = 75t - 45$, differs from the first model, $E = 75t$, by 45; that is, the second model will always produce 45 less envelopes than the first model for any value of t, the amount of time they work together. This is because in the second model, Bob worked one hour less; and since Bob stuffs 45 envelopes per hour, 45 less are being stuffed.

STUDY SKILLS 2.1

Preparing for Exams: Is Doing Homework Enough?

At some point you will probably take an algebra exam. More than likely, your time will be limited and you will not be allowed to refer to any books or notes during the test.

Working problems at home may help you to develop your skills and to better understand the material, but there is still no guarantee that you can demonstrate the same high level of performance during a test as you may be showing on your homework. There are several reasons for this:

1. Unlike homework, exams must be completed during a limited time.
2. The fact that you are being assigned a grade may make you anxious and therefore more prone to careless errors.
3. Homework usually covers a limited amount of material, while exams usually cover much

more material. This increases the chance of confusing or forgetting skills, concepts, and rules.
4. Your books and notes *are* available as a guide while working homework exercises.

Even if you do not deliberately go through your textbook or notes while working on homework exercises, the fact that you know what sections the exercises are from, and what your last lecture was about, cues you in on how the exercises are to be solved. You may not realize how much you depend on these cues and may be at a loss when an exam does not provide them for you.

If you believe that you understand the material and/or you do well on your homework, but your exam grades just do not seem to be as high as you think they should be, then the study skills discussed in this chapter should be helpful.

EXERCISES 2.1

1. Tamara's math teacher counts the final exam as 25% of her course grade, and Tamara's average grade for all but the final exam is 82. Devise a mathematical model (an equation) that shows how Tamara's course grade is related to her final exam score.

   (a) Use this model to find the minimum score she would need on her final exam to get a course grade of 80.

   (b) Determine the minimum score she would need on her final exam to get a course grade of 90.

2. Lisa's math teacher counts the final exam as 45% of her course grade, and Lisa's average grade for all but the final exam is 82. Devise a mathematical model that shows how Lisa's course grade is related to her final exam score.

   (a) Use this model to find the minimum score she would need on her final exam to get a course grade of 80.

   (b) Determine the minimum score she would need on her final exam to get a grade of 90 for the course.

3. Carla's math teacher counts the final exam as 25% of her course grade, and Carla's average grade for all but the final exam is 78. Devise a mathematical model that shows how Carla's course grade is related to her final exam score. Use this model to determine the minimum score she would need on her final exam to get a grade of 80 for the course.

4. Chuck's math teacher counts the final exam as 55% of his course grade, and Chuck's average grade for all but the final exam is 65. Devise a mathematical model that shows how Chuck's course grade is related to his final exam score. Use this model to find the minimum score he would need on his final exam to get a course grade of 80.

5. Suppose that a salesperson is paid a commission of 9% of his gross sales. Develop an equation relating his commission to his gross sales and use this equation to find the weekly gross sales he needs to earn a commission of $600 a week.

6. Suppose that a salesperson is paid a base salary of $100 plus a commission of 5% of her gross sales. Develop an equation relating her pay to her gross sales and use this equation to determine the weekly gross sales she needs to earn $400 a week.

7. On October 1, 2004, the euro (European dollar) was worth $1.24 (U.S. dollars). Express the value of x euros in U.S. dollars. Use the model to find the value of 300 euros in U.S. dollars.

8. On October 1, 2004, the euro (European dollar) was worth $1.24 (U.S. dollars). Express the value of x U.S. dollars in terms of euros. Use the model to find the value of $300 in euros.

9. On October 1, 2004, the British pound was worth $1.7984 (U.S. dollars). Express the value of *x* British pounds in terms of U.S. dollars. Use the model to find the value of 400 British pounds in U.S. dollars.

10. On October 1, 2004, the U.S. dollar was worth 110.42 Japanese yen. Express the value of *x* U.S. dollars in terms of the Japanese yen. Use the model to estimate the value of $200 in yen.

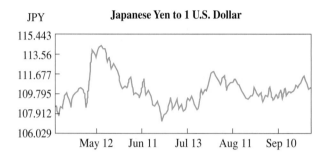

11. Suppose you are given the choice of two pay plans: strictly commission or a base salary plus commission. The strictly commission plan pays 9% of your gross sales, whereas the base salary plus commission plan pays you a base salary of $120 and a commission of 6% of your gross sales. Create a mathematical model that describes each plan and use these models to determine *when* one plan is better than the other.

12. A cell phone company offers two basic plans for the poor executive: plan A and plan B. Plan A is a monthly service charge of $25, and a charge of 25¢ a minute for telephone airtime; plan B is a monthly service charge of $40, and a charge of 20¢ a minute for telephone airtime. Create a mathematical model that describes each plan and use these models to determine *when* one plan is better than the other.

13. The length of a rectangle is 1 more than 3 times its width. Express the perimeter in terms of its width, and use this model to find the dimensions of the rectangle if the perimeter is to be 80 in.

14. The length of the first side of a triangle is twice the length of the second side, and the length of the third side is 5 more than the length of the second side. Express the perimeter of the triangle in terms of the length of its second side, and use this model to find the lengths of the three sides if the perimeter is to be 100 in.

15. Anna has decided to invest in a high-risk certificate that yields an annual interest rate of 8.3%. Express the amount of interest earned in terms of the amount she invests in this certificate and use this model to find how much she should invest if she wants to earn $1,000 in interest this year.

16. Maria has some money to invest and decides to invest in a low-risk certificate that yields an annual interest rate of 4.6%. Express the amount of interest earned in terms of the amount she invests in this certificate and use this model to determine how much interest she would receive if she invested $12,500.

17. John has $12,500 to invest. He decides to invest some of it in a long-term certificate that yields an annual interest rate of 8%, and the rest in a short-term certificate that yields an annual interest rate of 5%. Express the amount of interest earned in terms of the amount invested in the long-term certificate and use this model to explain how the interest is affected by changes in the distribution of the $12,500 investment. Determine the maximum and minimum interest he could earn at the end of the year.

18. An LGP39 printer can print 9 copies per minute, whereas a GHJ27 printer can print only 4 copies per minute. Express the total number of copies made in terms of time for each machine and for both machines working together. Determine how long it takes for 142 pages to be printed by each machine alone, and for both machines working together.

19. An LGP39 printer can print 9 copies per minute, whereas a GHJ27 printer can print only 4 copies per minute. Express the total number of copies made in terms of time (for the whole job) for both machines working together, if the slower model starts 5 minutes after the faster model starts. Use this model to find how long it takes for 142 pages to be printed by both machines.

20. A computer discount store held an end-of-summer sale on two types of computers; 58 computers were sold. If one type sold for $600 and the other type sold for $850, express the amount of money collected on the sale of the 58 computers in terms of the number of $600 computers sold. Use the model to determine how many $600 computers were sold if $40,300 was collected on the sale of all the computers.

21. For the computer sale of Exercise 20, express the amount of money collected on the sale of the 58 computers in terms of the number of $850 computers sold. Use the model to determine how many $850 computers were sold if $40,300 was collected on the sale of all computers.

22. On October 29, a stock loses 30% of its previous day's value. The next day, October 30, it gains 30% of its previous day's value. Express the value of the stock on October 30 in terms of its value on October 28. Use this model to determine the value of the stock on October 28, if the stock is worth $2,500 on October 30.

23. Suppose you bought shares of stock at 61 ($61 per share) yesterday, and today the stock falls to 47. If the total value of all your shares yesterday was x, express the value of the stock today in terms of x. Use this model to determine the value of the stock yesterday if the stock is worth $2,500 today.

24. Arthur can process 300 forms per hour, whereas Lewis can process 200 forms per hour. Express the total number of forms processed in terms of time (for the whole job) for both Arthur and Lewis working together. Use this model to find how long it takes to process 3,000 forms by both working together.

25. Arthur and Lewis of Exercise 24 worked together, but Arthur started a half-hour after Lewis. Express the total number of forms processed in terms of time (for the whole job). Use this model to find how long it takes for 3,000 forms to be processed if Arthur starts a half-hour after Lewis.

26. A truck carries a load of 50 boxes; some are 20-lb boxes, and the rest are 25-lb boxes. Devise a mathematical model that shows how the total weight of boxes in the truck is related to the number of 20-lb boxes. Use this model to find how many of each type of box there is if the total weight of boxes is 1,075 lb.

27. A merchant wishes to purchase a shipment of 24 clock radios. Simple AM models cost $35 each, whereas AM/FM models cost $50 each. In addition, there is a delivery charge of $70 for the shipment. Express the total cost of all radios together in terms of the number of AM radios. Use this model to determine how many of each type of radio he purchased if the shipment cost $1,000.

28. A plumber charges $32 per hour for her time and $15 per hour for her assistant's time. On a certain job the assistant works alone for 2 hours doing preparatory work, then the plumber and her assistant complete the job together. Express the total cost of the job in terms of the amount of time the plumber worked. Use this model to determine how many hours the plumber worked if the bill for the job was $312.

2.2 First-Degree Equations and Applications

In the previous section we focused on constructing a few mathematical models and applied some basic equation-solving strategies to answer questions about these models. In this section we begin with a formal discussion of the fundamentals of solving equations in order to solidify the foundation on which we can build later. In addition, we will examine more equations as well as continue to explore other models.

An *equation* is a mathematical statement that two algebraic expressions are equal. Let's begin by examining the three equations

1. $5(x + 2) - 5x - 15 = 0$
2. $5(x + 2) - 5x - 10 = 0$
3. $5(x + 2) - 15 = 0$

These equations may look very similar at first glance. However, if we examine them more carefully, we see that there are very fundamental differences. If we simplify the left side of each equation, we obtain the following:

1. $5(x + 2) - 5x - 15 = 0$
   $5x + 10 - 5x - 15 = 0$
   $-5 = 0$
2. $5(x + 2) - 5x - 10 = 0$
   $5x + 10 - 5x - 10 = 0$
   $0 = 0$
3. $5(x + 2) - 15 = 0$
   $5x + 10 - 15 = 0$
   $5x - 5 = 0$

After simplifying each equation, we can see that:

Since $-5 \neq 0$, equation 1 is false.

DEFINITION An equation that is always false, regardless of the value of the variable, is called a *contradiction.*

Since $0 = 0$, equation 2 is true for all values of x.

DEFINITION An equation that is always true, regardless of the value of the variable, is called an *identity.*

Equation 3 is neither true nor false. Its truth depends on the value of the variable. If x is equal to 1, the equation is true; otherwise, it is false.

DEFINITION An equation whose truth depends on the value of the variable is called a *conditional equation.*

EXAMPLE 1

Determine whether each of the following is an identity or a contradiction.

(a) $x(x - 3) - (5x + 7) = 2x(x + 3) - 7(2x - 1) - x^2 - 14$

(b) $3x + 6 - (3x - 2) = 5 + 4x + 4(3 - x)$

Solution

Clearly, in their present form we cannot tell what types of equations we have. Our first step in dealing with an equation is always to simplify each side as completely as possible using the methods outlined in Chapter 1.

(a) $x(x - 3) - (5x + 7) = 2x(x + 3) - 7(2x - 1) - x^2 - 14$

Multiply out using the distributive property; combine like terms.

$$x^2 - 3x - 5x - 7 = 2x^2 + 6x - 14x + 7 - x^2 - 14$$
$$x^2 - 8x - 7 = x^2 - 8x - 7$$

Thus, we see that the equation is an $\boxed{\text{identity}}$.

The fact that an equation contains a variable does not necessarily mean it is conditional.

(b) $3x + 6 - (3x - 2) = 5 + 4x + 4(3 - x)$
$$3x + 6 - 3x + 2 = 5 + 4x + 12 - 4x$$
$$8 = 17$$

Thus, we see that the equation is a $\boxed{\text{contradiction}}$.

In the case of a conditional equation, we are usually interested in finding those values that make it true.

DEFINITION

A value that makes an equation true is called a ***solution*** of the equation. We also say that the value ***satisfies*** the equation.

EXAMPLE 2

Determine whether each of the values listed satisfies the given equation.

$$4 + 3(w - 1) = 5(w + 1) - (1 - w) \quad \text{for } w = 1, -1$$

Solution

To check whether a particular value satisfies an equation, we replace each occurrence of the variable, on both sides, by the proposed value.

Check: $w = 1$: $\quad 4 + 3(w - 1) = 5(w + 1) - (1 - w)$
$$4 + 3(1 - 1) \overset{?}{=} 5(1 + 1) - (1 - 1)$$
$$4 + 3(0) \overset{?}{=} 5(2) - 0$$
$$4 + 0 \overset{?}{=} 10 - 0$$
$$4 \neq 10$$

$\boxed{\text{Therefore, } w = 1 \text{ is } not \text{ a solution}}$.

Be careful when you substitute $w = -1$: Watch your computations.

Check: $w = -1$: $\quad 4 + 3(w - 1) = 5(w + 1) - (1 - w)$
$$4 + 3(-1 - 1) \overset{?}{=} 5(-1 + 1) - [1 - (-1)]$$
$$4 + 3(-2) \overset{?}{=} 5(0) - (1 + 1)$$
$$4 - 6 \overset{?}{=} 0 - 2$$
$$-2 \overset{\checkmark}{=} -2$$

$\boxed{\text{Therefore, } w = -1 \text{ } is \text{ a solution}}$.

Note that the equation is conditional since it is neither always true nor always false.

DEFINITION A *first-degree equation in one variable* is an equation that can be put in the form $ax + b = 0$, where a and b are constants and $a \neq 0$.

This definition was given in Chapter 1. In Chapter 3 we will see why a first-degree equation is also called a linear equation.

Equations that have identical solutions are called **equivalent equations.** The equations

$$3x - [5 + 2(x - 1)] = 2x + 1$$
$$x - 3 = 2x + 1$$
$$x = -4$$

are equivalent since they each have exactly the same solution, -4. Our goal is to take an equation and, with the help of a few properties, gradually change the given equation into an equivalent *obvious equation,* an equation of the form $x = a$ (or $a = x$), where x is the variable for which we are solving. We will need the following properties of equality.

The Substitution Property of Equality

If two quantities are equal, then one quantity can be substituted for the other.

The substitution property states that if $a = b$, then a can be used to replace b and vice versa. Hence, the statement "if $a = b$, then $a + c = b + c$" is simply a particular application of the substitution property, as are the other properties repeated here from Section 1.5.

Properties of Equality

Addition Property of Equality	1. If $a = b$, then $a + c = b + c$.
Subtraction Property of Equality	2. If $a = b$, then $a - c = b - c$.
Multiplication Property of Equality	3. If $a = b$, then $a \cdot c = b \cdot c$.
Division Property of Equality	4. If $a = b$, then $\dfrac{a}{c} = \dfrac{b}{c}$, $c \neq 0$.

We summarize these properties with regard to equations as shown in the following box.

An equation can be transformed into an equivalent equation by adding or subtracting the same quantity on both sides of the equation, or by multiplying or dividing both sides of the equation by the same *nonzero* quantity.

A comment about the multiplication property is in order here. While property 3 as stated is correct, if we apply the multiplication property with $c = 0$, we may not obtain an equivalent equation. For example, the equation $x + 3 = 5$ is a conditional equation whose only solution is $x = 2$. If we apply the multiplication property with $c = 0$ (that is, we multiply both sides of the equation by 0), we get

$$x + 3 = 5$$
$$0(x + 3) = 0(5)$$
$$0 = 0 \qquad \textit{Which is an identity}$$

Multiplying both sides of the equation by 0 has not given us an equivalent equation. To ensure that whenever we apply property 3 we will obtain an equivalent equation, we must restrict the use of property 3 to the case when $c \neq 0$.

Actually we could have consolidated this list of four properties into two properties. Since subtraction is defined in terms of addition and division is defined in terms of multiplication, all we really need are the addition and multiplication properties.

We list three additional properties to complete our discussion.

More Properties of Equalty		
Reflexive Property of Equality	1. $a = a$	
Symmetric Property of Equality	2. If $a = b$, then $b = a$.	
Transitive Property of Equality	3. If $a = b$ and $b = c$, then $a = c$.	

The first four properties of equality given on page 77 give us four ways to transform an equation into an equivalent equation. Our goal is to apply the preceding transformations to a given equation until we have reduced it to an obvious equation.

In general, the strategy we use is described in the box.

> 1. First simplify each side of an equation as much as possible. If there are fractions in the equation, use the multiplication property to "clear the denominators."
> 2. Using the addition (and/or subtraction) property, collect all terms containing the variable for which we are solving on one side of the equation.
> 3. Use the division (or multiplication) property to isolate the variable and get the obvious equation.
> 4. Finally, check all solutions in the original equation.

EXAMPLE 3

Solve for t: $3(t - 4) - 4(t - 3) = t + 3 - (t - 2)$

Solution We begin by simplifying each side of the equation as completely as possible.

$$3(t - 4) - 4(t - 3) = t + 3 - (t - 2)$$ *Remove parentheses.*

$$3t - 12 - 4t + 12 = t + 3 - t + 2$$ *Then combine like terms.*

$$-t = 5$$ *We are not finished yet. We want t alone so we divide both sides of the equation by the coefficient of t, which is −1.*

$$\frac{-t}{-1} = \frac{5}{-1}$$

$$\boxed{t = -5}$$

The check is left to the student.

EXAMPLE 4

Solve for q: $\dfrac{q}{5} + \dfrac{5 - q}{3} = 2$

Solution We clear the denominators by multiplying both sides of the equation by the least common denominator, which is $5 \cdot 3 = 15$. Multiplying both sides by 15 means multiplying each term on both sides by 15.

$$\frac{15}{1} \cdot \frac{q}{5} + \frac{15}{1} \cdot \frac{5-q}{3} = 15 \cdot 2 \qquad \textit{Next, simplify the left side.}$$

$$\frac{\overset{3}{\cancel{15}}}{1} \cdot \frac{q}{5} + \frac{\overset{5}{\cancel{15}}}{1} \cdot \frac{5-q}{\cancel{3}} = 15 \cdot 2$$

$$3q + 5(5 - q) = 30$$

$$3q + 25 - 5q = 30$$

$$25 - 2q = 30 \qquad \textit{Add} -25 \textit{ to each side.}$$

$$-2q = 5$$

$$\boxed{q = -\frac{5}{2}}$$

The check is left to the student.

EXAMPLE 5

Solve for y: $3(y - 1) - 4(y + 2) = 11 - y$

Solution

$$3(y - 1) - 4(y + 2) = 11 - y$$

$$3y - 3 - 4y - 8 = 11 - y$$

$$-y - 11 = 11 - y \qquad \textit{Add} +y \textit{ to both sides.}$$

$$-11 = 11$$

Since $-11 = 11$ is always false, and is equivalent to our original equation, our original equation is a *contradiction* (this means there is no value of y that makes the equation true) and therefore there is $\boxed{\text{no solution}}$.

When we solve an equation that is in fact a contradiction, the variable drops out entirely and we get an equation that says that two unequal numbers are equal—which is always false.

EXAMPLE 6

Solve for s: $8 - (s - 1) = -s + 9$

Solution

$$8 - (s - 1) = -s + 9 \qquad \textit{Simplify the left side.}$$

$$8 - s + 1 = -s + 9$$

$$9 - s = -s + 9 \qquad \textit{Add} +s \textit{ to both sides.}$$

$$9 = 9$$

Since $9 = 9$ is always true and is equivalent to our original equation, our original equation is an *identity* (this means every value of s will make the equation true) and we say the equation is

$$\boxed{\text{true for all real numbers}}$$

Actually, you might already have recognized the third line of the solution as an identity.

When we solve an equation that is in fact an identity, the variable drops out entirely and we get an equation that says that a number is equal to itself—which is always true.

We have seen that if a first-degree equation is conditional, then it has a unique solution, which we can find by using the outline given previously. Otherwise, it must be either an identity or a contradiction.

Let's now apply some of the skills we have acquired in solving equations.

EXAMPLE 7

A discount shoe outlet collected $1,348 on the sale of 62 pairs of sneakers. Some of the sneakers sold were women's sneakers selling for $24 per pair and the remainder were girls' sneakers selling for $20 per pair. How many of each type were sold?

Solution

At this point in our development we will restrict ourselves to solutions involving one variable. In Chapter 4 we will discuss a two-variable approach to this type of problem.

Let

$$x = \text{Number of pairs of women's sneakers sold}$$

Then

$$62 - x = \text{Number of pairs of girls' sneakers sold}$$ *Because there are 62 pairs sold altogether (total minus part equals remainder)*

Our equation involves the amount of money collected on the sale of the sneakers (that is, the *value* of the sneakers sold). We compute the amount as follows:

$$\begin{pmatrix}\text{Amount collected from} \\ \text{sale of all women's sneakers}\end{pmatrix} + \begin{pmatrix}\text{Amount collected from} \\ \text{sale of all girls' sneakers}\end{pmatrix} = \begin{pmatrix}\text{Total} \\ \text{collected}\end{pmatrix}$$

$$\begin{pmatrix}\text{\# of pairs} \\ \text{of women's} \\ \text{sneakers} \\ \text{sold}\end{pmatrix} \cdot \begin{pmatrix}\text{Cost of} \\ \text{1 pair} \\ \text{of women's} \\ \text{sneakers}\end{pmatrix} + \begin{pmatrix}\text{\# of pairs} \\ \text{of girls'} \\ \text{sneakers} \\ \text{sold}\end{pmatrix} \cdot \begin{pmatrix}\text{Cost of} \\ \text{1 pair} \\ \text{of girls'} \\ \text{sneakers}\end{pmatrix} = 1,348$$

$$x \cdot 24 \qquad\qquad + \qquad (62 - x) \cdot 20 \qquad\qquad = 1,348$$

Then our equation is

$$24x + 20(62 - x) = 1,348$$

We now solve the equation:

$$24x + 1,240 - 20x = 1,348$$
$$4x + 1,240 = 1,348$$
$$4x = 108$$
$$x = 27$$

Thus, there were $x = \boxed{27 \text{ pairs of women's sneakers sold}}$

and $62 - 27 = \boxed{35 \text{ pairs of girls' sneakers sold}}$

Check: $27 + 35 = 62$

$24(27) + 20(35) = 648 + 700 = 1,348$

Look back at Example 3 in Section 2.1 and compare it with Example 7 above. Note that, although on the surface they seem to have nothing to do with each other, the equations we obtained are structurally very similar. Such problems are often called *value problems.*

The basic elements of the solutions to these examples are outlined in the box.

Outline of Strategy for Solving Verbal Problems	1. *Read the problem carefully,* as many times as is necessary to understand what the problem is saying and what it is asking.

1. *Read the problem carefully,* as many times as is necessary to understand what the problem is saying and what it is asking.

2. *Use diagrams* whenever you think it will make the given information clearer.

3. *Find the underlying relationship or formula* relevant to the given problem. Ask whether there is some underlying relationship or formula you need to know. If not, then the words of the problem themselves give the required relationship.

4. Clearly *identify the unknown quantity* (or quantities) in the problem, and label it (them) using one variable.

5. By using the underlying formula or relationship in the problem, *write an equation* involving the unknown quantity (or quantities). (This is the *crucial* and probably most difficult step.)

6. *Solve* the equation.

7. *Answer the question.* Make sure you have answered the question that was asked.

8. *Check* the answer(s) in the original words of the problem.

EXAMPLE 8

Jane begins a 20-mile race at 9:00 A.M., running at an average speed of 10 mph. One hour later her brother leaves the starting line on a motorcycle and follows her route at the rate of 40 mph. At what time does he catch up to her?

Solution

A simple diagram may help us visualize the problem. Figure 2.2 emphasizes the fact that when Jane's brother overtakes her, they have both traveled the *same distance.*

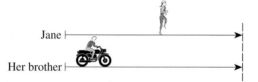

Figure 2.2
Diagram for Example 8

The basic idea in the problem is that if you travel at a constant rate of speed, then

$$\text{Distance} = (\text{Rate})(\text{Time})$$

or

$$d = rt$$

As indicated above, we have

$$d_{\text{Jane}} = d_{\text{brother}}$$

Using the formula $d = rt$, we get

$$r_{\text{Jane}} \cdot t_{\text{Jane}} = r_{\text{brother}} \cdot t_{\text{brother}}$$

The information given in the problem tells us $r_{\text{Jane}} = 10$ and $r_{\text{brother}} = 40$, so our problem becomes

$$10 \cdot t_{\text{Jane}} = 40 \cdot t_{\text{brother}}$$

Since we are told that Jane's brother left 1 hour later, let

$$t = \text{Number of hours Jane runs until her brother catches her}$$

Then

$t - 1 =$ Number of hours Jane's brother rides until he catches up   *He leaves 1 hour later so he travels 1 hour less.*

Thus, our equation becomes

$$10t = 40(t - 1)$$

$$10t = 40t - 40$$

$$-30t = -40$$

$$t = \frac{-40}{-30} = \frac{4}{3} = 1\frac{1}{3} \text{ hours} = 1 \text{ hr } 20 \text{ min}$$

Jane's brother reached her 1 hour and 20 minutes after *she* began the race, at

$$\boxed{10.20 \text{ A.M.}}$$

Check: Jane started at 9:00 A.M. and met her brother at 10:20 A.M. Thus, she ran for 1 hour and 20 minutes ($1\frac{1}{3}$ hours). Traveling for $1\frac{1}{3}$ hours at 10 mph, Jane travels

$$\left(1\frac{1}{3}\right)10 = \left(\frac{4}{3}\right)10 = \frac{40}{3} = 13\frac{1}{3} \text{ miles}$$

Jane's brother started an hour after Jane, at 10:00 A.M., and met his sister at 10:20 A.M. Therefore, her brother traveled for 20 minutes $\left(\frac{1}{3} \text{ hour}\right)$. Traveling for $\frac{1}{3}$ hour at 40 mph, her brother travels

$$\left(\frac{1}{3}\right)40 = \frac{40}{3} = 13\frac{1}{3} \text{ miles}$$

Since the distances are the same, the answer checks.

Note that we could have let $t =$ time for Jane's brother to catch up, and then let $t + 1 =$ the time Jane runs until her brother catches up. While the value for t will not be the same, the answer to the problem remains unchanged.

EXAMPLE 9

Jean can process 25 forms per hour and Bob can process 15 forms per hour.

(a) How long would it take them to process 710 forms working together?

(b) How long would it take if Bob started 2 hours after Jean?

Solution

As with Example 8, we are dealing with a rate (a quantity per unit time). We created a model for a similar example in Section 2.1 (Example 4). The basic relationship between the amount of work done and the rate at which the work is being done is

$$\text{Amount of work done} = (\text{Work rate}) \cdot (\text{time})$$

or

$$W = r \cdot t$$

(a) How long will it take Jean and Bob to process 710 forms? The model was constructed by noting that the total amount of work done is the sum of the amount of work done by each individual:

$$\begin{pmatrix} \text{Amount of work} \\ \text{done by Jean} \end{pmatrix} + \begin{pmatrix} \text{Amount of work} \\ \text{done by Bob} \end{pmatrix} = \begin{pmatrix} \text{Total amount} \\ \text{of work done} \end{pmatrix}$$

$$(\text{Jean's rate}) \cdot (\text{Jean's time}) + (\text{Bob's rate}) \cdot (\text{Bob's time}) = \qquad F$$

We are given Jean's and Bob's rate: Since Jean and Bob are working the same amount of time, we can let t = number of hours Jean works = number of hours Bob works. Thus, the model becomes

$$25 \cdot t \qquad + \qquad 15 \cdot t \qquad = \qquad F$$

Hence the model is

$$F = 25t + 15t$$

which simplifies to

$$F = 40t$$

where F is the number of forms processed and t is the number of hours it takes for them to process the forms working together.

Since we want to know how long it takes for them to process 710 forms working together, we are looking for t when F is 710.

$$F = 40t \qquad \textit{Substitute F = 710 and solve for t.}$$
$$710 = 40t$$
$$\frac{710}{40} = t$$
$$17\tfrac{3}{4} = t$$

It takes $\boxed{17\tfrac{3}{4} \text{ hours}}$ for Bob and Jean to process 710 forms.

(b) How long would it take if Bob started 2 hours after Jean? We developed the model in the same way:

$$\begin{pmatrix} \text{Amount of work} \\ \text{done by Jean} \end{pmatrix} \quad + \quad \begin{pmatrix} \text{Amount of work} \\ \text{done by Bob} \end{pmatrix} \quad = \quad \begin{pmatrix} \text{Total amount} \\ \text{of work done} \end{pmatrix}$$

$$\text{(Jean's rate)} \cdot \text{(Jean's time)} + \text{(Bob's rate)} \cdot \text{(Bob's time)} = \qquad F$$

We are given Jean's and Bob's rates. Since Bob starts 2 hours later than Jean, he works 2 hours less than Jean. Thus, if we let t = number of hours Jean works, then $t - 2$ = number of hours Bob works. The model becomes

$$25 \cdot t \qquad + \qquad 15 \cdot (t - 2) \qquad = \qquad F$$

Hence the model is

$$F = 25t + 15(t - 2) \qquad \textit{Which simplifies to}$$
$$F = 25t + 15t - 30$$
$$F = 40t - 30$$

where F is the number of forms processed and t is the number of hours it takes for them to process the forms working together when Bob starts 2 hours later. Since we want to know how long it takes for them to process 710 forms, we are looking for t when F is 710 with this model.

$$F = 40t - 30 \qquad \textit{Substitute F = 710 and solve for t.}$$
$$710 = 40t - 30$$
$$740 = 40t$$
$$\frac{740}{40} = t$$

Hence $t = 18\tfrac{1}{2}$

So it takes them $\boxed{18\tfrac{1}{2} \text{ hours}}$ to process 710 forms if Bob starts 2 hours later.

The check is left to the student.

One final comment: Do not get discouraged if you do not always get a complete solution to every problem. As you do more problems, you will get better at solving problems. Make an honest attempt to solve the problems and keep a written record of your work so that you can go back over it to see exactly where you encounter any difficulties.

STUDY SKILLS 2.2

Preparing for Exams: When to Study

When to start studying and how to distribute your time studying for exams is as important as how to study. To begin with, "pulling all-nighters" (staying up all night to study just prior to an exam) seldom works. As with athletic or musical skills, algebraic skills cannot be developed overnight. In addition, without an adequate amount of rest, you will not have the clear head you need for working on an algebra exam. It is usually best to start studying early—from $1\frac{1}{2}$ to 2 weeks before the exam. In this way you have the time to perfect your skills and, if you run into a problem, you can consult your teacher to get an answer in time to include it as part of your studying.

It is also a good idea to distribute your study sessions over a period of time. That is, instead of putting in 6 hours in one day and none the next two days, put in 2 hours each day over the three days. You will find that not only will your studying be less boring, but also you will retain more with less effort.

As we mentioned before, your study activity should be varied during a study session. It is also a good idea to take short breaks and relax. A study "hour" could consist of 50 minutes of studying and a 10-minute break.

EXERCISES 2.2

In Exercises 1–8, determine whether the equation is an identity or a contradiction.

1. $x - (x - 3) = 4$

2. $3a - 3(a - 2) = -6$

3. $4(w - 2) = 4w - 8$

4. $4(w - 2) = 4w - 2$

5. $2(y + 1) - (y - 5) = 4(y - 2) - 3(y - 5)$

6. $3(y - 1) - (3y - 1) = 2y - 1 - (2y + 1)$

7. $6(2y + 1) - 4(3y - 1) = y - (y + 10)$

8. $2(y - 4) + 2(4 - y) = 5(7 - y) + 5(y - 7)$

In Exercises 9–12, determine whether the listed values of the variable satisfy the equation.

9. $5(x - 3) - (x + 2) = 8 - x; \quad x = -0.08, 0, 5$

10. $7(x + 1) - (x - 2) = x - 1: \quad x = -2, 0.3, 2$

11. $x^2 - 5x = 5 - x; \quad x = -5, -\frac{1}{2}, 5$

12. $6x - x^2 = x - 6; \quad x = -6, -0.6, 6$

In Exercises 13–42, solve the equation by algebraic methods.

13. $3x - 12 = 12 - 3x$

14. $5y + 7 = 7 - 5y$

15. $3x - 12 = -12 + 3x$

16. $5y - 7 = -7 - 5y$

17. $2t + 1 = 7 + t$

18. $6 - 5t = t + 12$

19. $2(x - 1) = 2(x + 1)$

20. $3(a + 5) = 4(a + 5) - (a + 5)$

21. $3(x + 1) - x = 2(x + 3)$

22. $4(x - 3) - 2x = 2(x - 1)$

23. $5(2t - 1) - 3t = 5 - 7t$

24. $6(3q - 4) + 5q - 2 = 5(4q + 1) + 3q - 7$

25. $6(3 - z) + 2(4z - 5) = z - (3 - z)$

26. $9 - 5(x + 4) = -4(x + 1) - (x + 7)$

27. $4(3x - 1) - 5(3x - 2) = 2(x + 3) - 5x$

28. $7z - 3(4 - z) = 5(2z) + 12$

29. $x(x - 2) - 15 = x(x + 5) - 3(x + 5)$

30. $a(a + 4) - 6(a + 4) = a(a - 4) + 6(a - 4)$

31. $6x - [2 - (x - 1)] = x - 5(x + 1)$

32. $3x - [5 - 2(x + 1)] = x - 7(x + 2)$

33. $3 - [4t + 5(2t - 1) - 3t] = 3(4 - t)$

34. $5d - [6d + 4(5 - d)] = 3 - [2d - (2d - 1)]$

35. $\dfrac{x}{2} - \dfrac{3x}{5} = 4$

36. $\dfrac{2a}{3} - \dfrac{1}{2} = a$

37. $\dfrac{y}{2} - \dfrac{4 - y}{3} = 4$

38. $\dfrac{5a}{2} - a = 2a + \dfrac{1}{2}$

39. $\dfrac{2y - 1}{4} = y - 3$

40. $\dfrac{t}{3} - \dfrac{t - 5}{2} = 6$

41. $\dfrac{3a}{2} - \dfrac{1}{3} = \dfrac{a + 5}{4}$

42. $\dfrac{t + 5}{2} - \dfrac{t}{3} = \dfrac{1}{2} + t$

Solve each of the following problems algebraically. Be sure to clearly label what the variable represents.

43. One number is 5 more than twice another, and their sum is 23. What are the numbers?

44. One number is 6 less than 7 times another. If the sum of the two numbers is -6, find the numbers.

45. The sum of four consecutive integers is 1 more than the third. Find them.

46. Four consecutive odd integers are such that the product of the first and third is 48 less than the product of the second and fourth. Find them.

47. On October 1, 2004, the euro (European dollar) was worth $1.24 (U.S. dollars). What is the value of 500 Euros in U.S. dollars?

48. On October 1, 2004, the U.S. dollar was worth 110.42 Japanese yen. What is the value of $600 in yen?

49. On January 5, a stock loses 25% of its previous day's value. The next day, January 6, it gains 25% of its previous day's value. If the stock is worth $2,500 on January 6, what was it worth on January 4?

50. Suppose you bought shares of stock at 75 ($75 per share) yesterday, and today the stock rises to 82.

   (a) What is the value of the stock today if it was worth $3,000 yesterday?

   (b) What *was* the value of the stock yesterday if the value of the stock *today* is $3,000?

51. The length of a rectangular garden is twice its width. If the garden is fenced in with 42 m of fencing, what are the dimensions of the garden?

52. The width of a rectangular garden is 8 feet less than its length. If the garden is fenced in with 72 feet of fencing, what are the dimensions of the garden?

53. The first side of a triangle is 5 cm less than the second side, and the third side is twice as long as the first side. If the perimeter of the triangle is 33 cm, find the lengths of the sides of the triangle.

54. The shortest side of a triangle is 8 in. less than the medium side, and the longest side is 4 in. longer than 3 times the shortest side. If the perimeter is 27 in., find the length of the shortest side.

55. The length of a rectangle is 1 more than 3 times the width. If the width is increased by 2 and the length is doubled, the new perimeter is 3 less than 5 times the original length. Find the original dimensions.

56. The width of a rectangle is 12 less than twice the length. If the length is increased by 3 and the width is doubled, the new perimeter is 6 less than 6 times the original length. What are the original dimensions?

57. A salesperson works strictly on commission of 8% of his gross sales. Determine his weekly gross sales required to earn $600 a week.

58. Suppose that a salesperson is paid a base salary of $200 a week plus 5% commission on his gross sales. Determine his weekly gross sales required to earn $600 a week.

59. Suppose you are nearing the end of a course in which the teacher counts the final exam as 45% of your course grade. Your average grade for all but the final exam is 72. Determine the minimum score needed on your final exam to get a grade of 80 for the course.

60. Suppose you are nearing the end of a course in which the teacher counts the final exam as 45% of your course grade. Your average grade for all but the final exam is 72. Determine the minimum score needed on your final exam to get a course grade of 90.

61. Bernard has $164 in $1, $5, and $10 bills. If he has 25 bills altogether and he has one more $10 bill than $5 bills, how many of each type of bill does he have?

62. Cheryl buys a total of 100 stamps at the post office for $20.56. If she bought 5¢, 15¢, and 37¢ stamps only, and there were ten more 37¢ stamps than 5¢ stamps, how many of each did she buy?

63. A truck carries a load of 50 packages; some are 20-lb packages and the rest are 25-lb packages. If the total weight of all packages is 1,075 lb, how many of each type are on the truck?

64. A truck carries a load of 85 boxes; some are 30-lb boxes, some are 25-lb boxes, and some are 20-lb boxes. If there are 3 times as many 30-lb boxes as 20-lb boxes, and the total weight of all the boxes is 2,245 lb, how many of each type of box are in the truck?

65. Yan wants to buy 200 shares of stock in an Internet company. She buys some shares at $18.825 per share, and the rest when the price has gone up to $20.375 per share. If she spent a total of $3,889 for all the shares, how many shares did she buy at each price?

66. A manufacturer orders a batch of 8,000 screws for $128. The batch contained two types of screws: regular screws costing $0.015 each and magnetic screws costing $0.0175 each. How many magnetic screws were ordered?

67. Maggie works in a shoe store. She earns a salary of $300 per week plus a commission of $3 on each pair of boots and $2 on each pair of shoes she sells. During a certain week she made 70 sales and earned a total income of $468. How many pairs of shoes did she sell?

68. A merchant wishes to purchase a shipment of clock radios. Simple FM models cost $20 each, whereas AM/FM models cost $30 each. In addition, there is a delivery charge of $70. If he spends $610 on 24 clock radios, how many of each type did he buy?

69. A gourmet food shop wants to mix some $4-per-pound coffee beans with 30 pounds of $6-per-pound coffee beans to produce a mixture that will sell for $5.20 per pound. How much $4-per-pound coffee should be used?

70. A toy store sells puzzles for $8 each and games for $19 each. A man spends $595 buying a certain number of puzzles and ten more games than puzzles. How many puzzles were bought?

71. Orchestra seats to a certain Broadway show are $96 each and balcony seats are $76 each. If a theater club spends $5,016 on the purchase of 56 seats, how many orchestra seats were purchased?

72. Orchestra seats to the off-Broadway show *Iwog* are $60 each and balcony seats are $28 each. If 156 tickets were sold for the matinee performance grossing $4,816, how many of each type of ticket were sold?

73. A plumber charges $42 per hour for her time and $24 per hour for her assistant's time. On a certain job the assistant works alone for 2 hours doing preparatory work, then the plumber and her assistant complete the job together. If the total bill for the job was $444, how many hours did the plumber work?

74. The plumber and assistant of Exercise 73 complete another job in which the assistant does the preparatory work alone and then the plumber completes the job alone. If the job took 9 hours and the bill came to $240, how many hours did each work?

75. At 3:00 P.M., two cars are 345 miles apart and are traveling toward each other. If one car is traveling at 55 mph and the other at 60 mph, at what time will they meet?

76. Two cars leave from the same spot at the same time and travel in opposite directions. If one car travels at 50 mph and the other travels at 60 mph, how long will it be until they are 495 miles apart?

77. Two cars leave the same spot at the same time and travel in opposite directions. If one car is traveling at 35 kph and the other at 50 kph, how long will it take for them to be 595 km apart?

78. Alice left her house at 7:00 A.M. for work. Her husband found her pocketbook at home at 8:00 A.M., so he jumped into his car to try to catch up with her. If Alice travels to work at 30 mph and her husband is traveling at 75 mph, can he catch up to her before she gets to work if her office is 60 miles from her house? If he does, what time does he catch up to her?

79. How long would it take someone jogging at 17 kph to overtake someone walking at 7 kph with a 3-hour head start?

80. Repeat Exercise 79 if the walker walks at the rate of 6 kph for 3 hours and then walks at 7 kph.

81. A person drives to a convention at the rate of 48 kph and returns home at the rate of 54 kph. If the total driving time is 17 hours, how far away is the convention?

82. A person can drive from town A to town B at a certain rate of speed in 6 hours. If he increases his speed by 20 kph he can make the trip in 4 hours. How far is it from town A to town B?

83. A relay race requires each team of two contestants to complete a 124-km course. The first person runs part of the course, then the second person completes the remainder of the course by bicycle. One team covers the running section at 18 kph and the bicycle section at 50 kph in a total of 6 hours. How long did it take to complete the running section? How long is the running section of the course?

84. A bike race consists of two segments whose total length is 110 km. The first segment is covered at 15 kph and takes 2 hours longer to complete than the second segment, which is covered at 25 kph. How long is each segment?

85. A manager needs 5,125 copies of a document. A new duplicating machine can make 50 copies per minute, whereas an older model can make 35 copies per minute. The older machine begins making copies but breaks down before completing the job and is replaced by the new machine, which completes the job. If the total time for the job is 1 hour and 50 minutes, how many copies did the older machine make?

86. A new computer printer can print 30 pages per minute and an older model can print 20 pages per minute. The older printer begins printing a 1,190-page report but breaks down before completing the job. The newer printer then completes the job. If the total printing time for the report was 47 minutes, how many pages did the older printer print?

87. An experienced worker can process 60 items per hour; a new worker can process 30 per hour. How many hours will it take to complete a job of processing 6,750 items if they work together?

88. Repeat Exercise 87 if the experienced worker works alone for 12 hours and then the new worker completes the job alone.

89. Repeat Exercise 87 if the experienced worker begins the job of processing the 6,750 items 3 hours before being joined by the new worker.

90. Repeat Exercise 87 if the new worker begins the job of processing the 6,750 items 3 hours before being joined by the experienced worker.

91. An experienced assembly-line worker can package 80 items per hour, whereas a trainee can package 48 items per hour. A trainee begins packaging a stack of 656 items at 10:00 A.M. and is joined by the experienced worker 3 hours later. At what time will they finish packaging the stack?

92. A college professor is grading 80 exam papers. He grades the papers at a certain rate for 3 hours. Then fatigue sets in and he completes the grading in 2 more hours at a rate of 5 exam papers per hour slower. How fast was he grading the papers for the first 3 hours?

2.3   First-Degree Inequalities and Applications

In Section 2.2 we started by discussing the various types of first-degree equations and discovered that not all first-degree equations have unique solutions.

A similar analysis applies to inequalities:

An inequality may be a *contradiction,* such as $x - 3 > x$. (Why is this always false?)

Or it may be an *identity,* such as $x + 1 > x$. (Why is this always true?)

Or it may be *conditional,* such as $x + 1 > 3$.

EXAMPLE 1   Determine whether the following is an identity or a contradiction:

$$7x - (3x - 4) < 5x + (4 - x)$$

Solution   As with some equations, we may not be able to tell in its present form what type of inequality we have. Thus, our first step is again to simplify each side as completely as possible.

$$7x - (3x - 4) < 5x + (4 - x)$$ *Remove parentheses by the distributive property and combine terms.*
$$7x - 3x + 4 < 5x + 4 - x$$
$$4x + 4 < 4x + 4$$

Thus, we see that $\boxed{\text{the inequality is a contradiction.}}$ (Why?)

As with equations, the values that make an inequality true are called *solutions* of the inequality, and we say that those values *satisfy* the inequality. But whereas conditional first-degree equations have single-value solutions, the solutions to conditional first-degree inequalities are infinite sets. Even the solution to the double inequality $0 < x < 1$ is the infinite set of numbers that lie between 0 and 1.

EXAMPLE 2

Determine whether each of the values listed satisfies the given inequality.

$$8x + 2(x - 4) \geq 12 \qquad \text{for } x = -1, 2, 4$$

Solution

We replace each occurrence of the variable, on both sides of the inequality, by the proposed value.

Check: $x = -1$:

$$8x + 2(x - 4) \geq 12$$
$$8(-1) + 2(-1 - 4) \overset{?}{\geq} 12$$
$$-8 + 2(-5) \overset{?}{\geq} 12$$
$$-18 \not\geq 12$$

Therefore, $\boxed{x = -1 \text{ is } not \text{ a solution}}$.

Check: $x = 2$:

$$8x + 2(x - 4) \geq 12$$
$$8(2) + 2(2 - 4) \overset{?}{\geq} 12$$
$$16 + 2(-2) \overset{?}{\geq} 12$$
$$12 \overset{\checkmark}{\geq} 12$$

Therefore, $\boxed{x = 2 \text{ is a solution}}$.

Remember: A number is equal to itself and is therefore greater than or equal to itself.

Check: $x = 4$:

$$8x + 2(x - 4) \geq 12$$
$$8(4) + 2(4 - 4) \overset{?}{\geq} 12$$
$$32 + 2(0) \overset{?}{\geq} 12$$
$$32 \overset{\checkmark}{\geq} 12$$

Therefore, $\boxed{x = 4 \text{ does satisfy the inequality}}$.

Properties of Inequalities

Inequalities that have exactly the same set of solutions are called *equivalent inequalities*. As with equations, it will be our goal to take an inequality and transform it into successively simpler *equivalent* inequalities, finally resulting in an *obvious inequality*—one that has the variable isolated on one side of the inequality or in the middle of a double inequality.

The properties of inequalities will help us to systematically achieve our goal. These properties are a bit different from the properties of equalities. Keeping in mind that $a < b$ means that a is to the left of b on the real number line, we can see that adding or subtracting the same quantity on both sides of an inequality *preserves* the direction (or

sense) of the inequality. For example,

$$-2 < 1$$
$$-2 + 4 \;?\; 1 + 4 \quad \text{\textit{Add 4 to both sides.}}$$
$$2 < 5$$

$$-2 < 0$$
$$-2 - 4 \;?\; 0 - 4 \quad \text{\textit{Subtract 4 from both sides.}}$$
$$-6 < -4$$

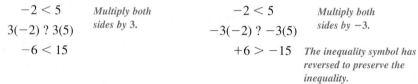

However, multiplication and division are somewhat different.

Multiplying by a positive number   **Multiplying by a negative number**

$$-2 < 5$$
$$3(-2) \;?\; 3(5) \quad \text{\textit{Multiply both sides by 3.}}$$
$$-6 < 15$$

$$-2 < 5$$
$$-3(-2) \;?\; -3(5) \quad \text{\textit{Multiply both sides by }} -3.$$
$$+6 > -15 \quad \text{\textit{The inequality symbol has reversed to preserve the inequality.}}$$

These numerical examples illustrate what the situation is in general. We also need to point out that the division and multiplication properties must be the same. After all, $\dfrac{a}{c} = \left(\dfrac{1}{c}\right)a$, so that dividing by c is the same as multiplying by $\dfrac{1}{c}$. If c is positive, so is $\dfrac{1}{c}$; if c is negative, so is $\dfrac{1}{c}$.

Properties of Inequalities

To obtain an equivalent inequality:
1. We may add or subtract the same quantity on each side of an inequality, and the inequality symbol remains the same.
2. We may multiply or divide each member of an inequality by a *positive* quantity, and the inequality symbol remains the same.
3. We may multiply or divide each member of an inequality by a *negative* quantity, and the inequality symbol is *reversed*.

Algebraically, we can write these properties as follows (for properties and theorems throughout this book, the "$<$" symbol can be replaced by "$>$," "$\geq$," or "$\leq$"):

Properties of Inequalities

If $a < b$, then

$$a + c < b + c$$
$$a - c < b - c$$

$ac < bc$ when c is positive; $ac > bc$ when c is negative

$\dfrac{a}{c} < \dfrac{b}{c}$ when c is positive; $\dfrac{a}{c} > \dfrac{b}{c}$ when c is negative

---

Solving First-Degree Inequalities

DEFINITION   A *first-degree inequality* is an inequality that can be put in the form

$$ax + b < 0$$

where a and b are constants and $a \neq 0$.

The procedure we have outlined for solving first-degree equations works equally well for first-degree inequalities. However, we must exercise a bit more care, as Example 3 illustrates.

EXAMPLE 3	Solve the following inequality for x.

$$9 - 5(x - 3) \geq 14$$

Solution   As with an equation, our goal is to isolate the variable on one side of the inequality.

$9 - 5(x - 3) \geq 14$ *First simplify the left side of the inequality.*

$9 - 5x + 15 \geq 14$

$24 - 5x \geq 14$ *Then apply the inequality properties.*

$24 - 5x \boxed{-24} \geq 14 \boxed{-24}$

$-5x \geq -10$ *Divide both sides by -5.*

$\downarrow$

$\dfrac{-5x}{-5} \leq \dfrac{-10}{-5}$ *Remember that when we divide both sides of an inequality by a negative number, we must reverse the inequality symbol.*

$x \leq 2$

Visualizing the solutions of an inequality on the real number line can be helpful. To sketch the graph of the solution set for this example (see Figure 2.3), we use the notation introduced in Chapter 1.

Figure 2.3
Solution set for Example 3

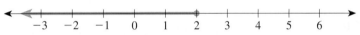

Remember, a **filled-in** *circle indicates that the point 2 is to be* **included** *as part of the solution.*

This solution means that any number less than or equal to 2 should satisfy the original inequality. How do we check this? It is impossible to check every number less than or equal to 2.

We would like to check that the endpoint, 2, is correct and that the inequality symbol is facing in the right direction. In general, the endpoint of our answer is the solution to the equation corresponding to the original inequality. That is, the endpoint, 2, is the solution to the *equation,* $9 - 5(x - 3) = 14$.

Thus, we check by replacing x with 2 in the original inequality to determine whether both sides of the inequality yield the same value. This tells us that the endpoint, 2, is correct.

To determine whether our inequality symbol is facing the right direction, we check by using a number less than 2 to see that it does satisfy the inequality. (This you can often do mentally.)

Checks:

Treating the inequality as an equation, we check to see what happens when $x = 2$:

$9 - 5(x - 3) = 14$

$9 - 5(2 - 3) \stackrel{?}{=} 14$

$9 - 5(-1) \stackrel{?}{=} 14$

$9 + 5 \stackrel{\checkmark}{=} 14$

Therefore, our *endpoint* 2, is correct.

We can choose any number less then 2—say, $x = 0$. This should satisfy the inequality.

$9 - 5(x - 3) \geq 14$

$9 - 5(0 - 3) \stackrel{?}{\geq} 14$

$9 - 5(-3) \stackrel{?}{\geq} 14$

$9 + 15 \stackrel{\checkmark}{\geq} 14$

Therefore, the direction of our inequality is correct.

The *double inequality, a < x < b,* is used to indicate "betweenness." For example,

$$-7 < x < 1$$

means that x is between -7 and 1 and is read "x is greater than -7 and less than 1." *Note that we read the middle variable first, then the left-hand number, and then the right-hand number.*

The inequality symbols separate the double inequality into three parts called *members:*

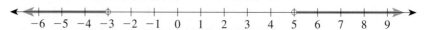

$$a \quad < \quad x \quad < \quad b$$

Left member   Middle member   Right member

Recall that the double inequality is actually a combination of two inequalities that must be satisfied simultaneously. That is,

$$a < x < b$$

is actually a combination of the two inequalities

$$a < x \quad \text{and} \quad x < b$$

where *x satisfies both inequalities at the same time.* In other words, x must be *simultaneously* greater than a *and* less than b.

We have to be careful to avoid putting two single inequalities together into one double inequality where either it is not required or it does not make sense. For example, if we have a region such as that shown in Figure 2.4, where $x < -3$ *or* $x > 5$, we cannot use the double inequality notation because x *cannot* satisfy both inequalities simultaneously (x is not "between" -3 and 5).

Figure 2.4
$x < -3$ *or* $x > 5$

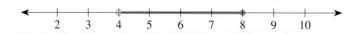

When a double inequality is written in its obvious form ($a < x < b$, where a and b are constants and x is the variable), the left-hand constant, a, must be the smaller of the two constants; this way the double inequality reflects the relative positions occupied by the members on the number line (see Figure 2.5).

Figure 2.5
$4 < x \leq 8$

EXAMPLE 4

Which of the following do not make sense?

(a) $4 < x < 9$ **(b)** $-5 < x < -7$

(c) $-5 < z > 4$ **(d)** $-2 \geq y > -5$

Solution

(a) $4 < x < 9$ is satisfied by $x = 5$ and therefore

$$\boxed{\text{makes sense}}$$

(b) $-5 < x < -7$

$$\boxed{\text{does not make sense}}$$

because there is no number that is both greater than -5 and, at the same time, less than -7.

(c) $-5 < z > 4$ is $\boxed{\text{not acceptable notation}}$

Remember that double inequalities are used to indicate "betweenness."

(d) $-2 \geq y > -5$ is satisfied by $y = -3$ and

$$\boxed{\text{does make sense}}$$

However, the standard practice is to write the inequality with the smaller number as the left member. Hence, it is usually written as

$$-5 < y \leq -2$$

As stated previously, the double inequality

$$-7 \leq 2x + 3 < 9$$

is actually two inequalities:

$$-7 \leq 2x + 3 \quad \text{and} \quad 2x + 3 < 9$$

We *could* solve them both, proceeding as follows:

$$-7 \boxed{-3} \leq 2x + 3 \boxed{-3} \qquad \text{and} \qquad 2x + 3 \boxed{-3} < 9 \boxed{-3}$$

$$-10 \leq 2x \qquad \text{and} \qquad 2x < 6$$

$$\frac{-10}{2} \leq \frac{2x}{2} \qquad \text{and} \qquad \frac{2x}{2} < \frac{6}{2}$$

$$-5 \leq x \qquad \text{and} \qquad x < 3$$

Since x has to satisfy both inequalities simultaneously, we would rewrite the two inequalities as one double inequality:

$$-5 \leq x < 3$$

On the other hand, notice that we applied exactly the same transformation to each inequality at the same time. So it would be more convenient to leave the double inequality as it is and to *apply the same transformation to each **member** of the inequality*. Again the goal is to isolate x, but for double inequalities, x will be the middle member.

$$-7 \leq 2x + 3 < 9$$ *To isolate x in the middle, first subtract 3 from all three members.*

$$-7 \boxed{-3} \leq 2x + 3 \boxed{-3} < 9 \boxed{-3}$$

$$-10 \leq 2x < 6$$ *Then divide all three members by 2.*

$$\frac{-10}{2} \leq \frac{2x}{2} < \frac{6}{2}$$

$$-5 \leq x < 3$$

We should check that the endpoints, -5 and 3, are correct, and then check a value between the endpoints. We leave this to the student.

EXAMPLE 5

Solve for x and sketch the solution set on the real number line:

$$-1 < 7 - 4x \leq 9$$

Solution   In solving a double inequality, we want to isolate the variable as the *middle member* of the inequality.

$$-1 \boxed{-7} < 7 - 4x \boxed{-7} \le 9 \boxed{-7}$$

Subtract 7 from each member of the inequality.

$$-8 < -4x \le 2$$

$$\frac{-8}{-4} > \frac{-4x}{-4} \ge \frac{2}{-4}$$

Divide each member of the inequality by −4. Note that both inequality symbols reverse when we divide by a negative number. Now simplify fractions.

$$2 > x \ge -\frac{1}{2}$$

Rewirte the double inequality in standard form.

$$\boxed{-\frac{1}{2} \le x < 2}$$

The sketch of this solution set is shown in Figure 2.6.

Figure 2.6
Solution set for Example 5

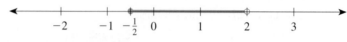

Note the filled circle at $-\frac{1}{2}$ and the empty circle at 2.

We leave the check to the student.

As in the case with equations, if we find that in the process of solving an inequality the variable drops out entirely, then we have either an identity or a contradiction.

Interval Notation

By an interval, we mean a section of the number line.

Another way to express sets of numbers described by inequalities is to use interval notation. Interval notation is a convenient and compact way of representing intervals on the number line. We will begin with bounded intervals—that is, intervals that have two endpoints.

We use parentheses to indicate that an endpoint is *not* included and brackets to indicate that an endpoint *is* included. Hence, for $a < b$, we have the following.

Interval Notation: Bounded Intervals

Set Notation		Interval Notation	Line Graph
$\{x \mid a \le x \le b\}$	is written as	$[a, b]$, called the *closed interval from a to b.*	
$\{x \mid a < x < b\}$	is written as	(a, b), called the *open interval from a to b.*	
$\{x \mid a \le x < b\}$	is written as	$[a, b)$, called a *half-open interval: closed at a and open at b.*	
$\{x \mid a < x \le b\}$	is written as	$(a, b]$, called a *half-open interval: open at a and closed at b.*	

The smaller number is always written to the left of the larger. Unfortunately, the open interval (a, b) uses the same notation as the ordered pair (a, b). It should always be clear, however, from the context of a problem whether we are talking about an interval or an ordered pair.

When expressing unbounded intervals, lines, or half-lines, we use the infinity symbol, ∞ or $-\infty$, with interval notation, as follows.

Interval Notation: Unbounded Intervals

Set Notation	Interval Notation	Line Graph
$\{x \mid x \geq a\}$	is written as $[a, \infty)$	
$\{x \mid x > a\}$	is written as (a, ∞)	
$\{x \mid x \leq a\}$	is written as $(-\infty, a]$	
$\{x \mid x < a\}$	is written as $(-\infty, a)$	

The symbols ∞ and $-\infty$ do not represent numbers; they are simply symbols to indicate that the interval goes on forever, or increases (or decreases) without bound. Therefore, we always write a parenthesis next to the ∞ symbol.

Remember that whenever we use interval notation, we are working within the framework of the real number system. The heavy line on the graph indicates that all points on the line are included.

EXAMPLE 6

Graph the following inequalities on the number line, and express the set using interval notation.

(a) $\{x \mid x > -3\}$ **(b)** $\{s \mid s \leq 4\}$ **(c)** $\{x \mid -2 \leq x \leq 6\}$

(d) $\{x \mid -2 < x < 6\}$ **(e)** $\{t \mid -2 < t \leq 6\}$ **(f)** All Reals

Solution

(a) $\{x \mid x > -3\}$ which is $(-3, \infty)$

(b) $\{s \mid s \leq 4\}$ which is $(-\infty, 4]$

(c) $\{x \mid -2 \leq x \leq 6\}$ which is $[-2, 6]$

(d) $\{x \mid -2 < x < 6\}$ which is $(-2, 6)$

(e) $\{t \mid -2 < t \leq 6\}$ which is $(-2, 6]$

(f) All Reals which is $(-\infty, \infty)$

EXAMPLE 7

Solve for the variable. Express your answer using interval notation.

(a) $5x + 8(20 - x) \geq 2(x - 10)$ **(b)** $0 \leq \dfrac{5}{9}(F - 32) < 100$

Solution

(a) $5x + 8(20 - x) \geq 2(x - 10)$ *Simplify each side.*

$5x + 160 - 8x \geq 2x - 20$

$160 - 3x \geq 2x - 20$ *Now apply the inequality properties.*

$-5x \geq -180$ *Divide both sides by −5.*

$x \leq 36$ *Note that the inequality symbol is reversed.*

Using interval notation, the answer is $(-\infty, 36]$.

(b)

$$0 \le \frac{5}{9}(F - 32) < 100$$

$$0 \le \frac{5}{9}(F - 32) < 100 \qquad \textit{Multiply each member by } \tfrac{9}{5}.$$

$$\frac{9}{5}(0) \le \frac{9}{5} \cdot \frac{5}{9}(F - 32) < \frac{9}{5}(100)$$

$$0 \le F - 32 < 180 \qquad \textit{Add 32 to each member.}$$

$$32 \le F < 212 \quad \text{or, using interval notation,} \quad [32, 212)$$

The outline suggested in Section 2.2 for solving verbal problems applies equally well to problems that give rise to inequalities. A few examples will serve to illustrate this idea.

EXAMPLE 8

What numbers satisfy the condition that "3 less than 4 times the number is less than 29"?

Solution

In translating this problem algebraically, we must be careful to distinguish the phrase "3 less than" from the phrase "3 is less than." The phrase "3 less than" indicates that we are subtracting 3 from some quantity—it is not a statement of inequality. An example of this would be the phrase "3 less than 8," which gets translated as $8 - 3$.

On the other hand, the phrase "3 is less than 8" is a statement of inequality, which is translated as $3 < 8$.

Let $x = $ a number satisfying the condition given in the example.

The given condition can be translated as follows:

$$\underbrace{\text{3 less than 4 times the number}}_{4x - 3} \quad \underbrace{\text{is less than}}_{<} \quad \underbrace{29}_{29}$$

Thus, our inequality is

$$4x - 3 < 29$$

Now we solve this inequality:

$$4x - 3 \;\boxed{+ 3} \;< 29 \;\boxed{+ 3}$$

$$4x < 32$$

$$\frac{4x}{4} < \frac{32}{4}$$

$$x < 8$$

Check:

Check the endpoint $x = 8$:

$$4x - 3 = 29$$
$$4(8) - 3 \stackrel{?}{=} 29$$
$$32 - 3 \stackrel{?}{=} 29$$
$$29 \stackrel{\checkmark}{=} 29$$

The endpoint, 8, is correct.

Check any value less than 8, say 7:

$$4x - 3 < 29$$
$$4(7) - 3 \stackrel{?}{<} 29$$
$$28 - 3 \stackrel{?}{<} 29$$
$$25 \stackrel{\checkmark}{<} 29$$

The direction of the inequality symbol is correct.

Thus,

$$\boxed{\text{any number less than 8}}$$

satisfies the condition of this problem.

EXAMPLE 9

The perimeter of a rectangle is to be at least 48 cm and no greater than 72 cm. If the width is 9 cm, what is the range of possible values for the length?

Solution

The rectangle, with length L, is shown in Figure 2.7. The perimeter, P, is $2L + 2(9)$ or

$$P = 2L + 18$$

9

Figure 2.7
Rectangle for Example 7

L

The problem tells us that we require

$$48 \leq \text{Perimeter} \leq 72$$

Since the perimeter is $2L + 18$, the inequality becomes

$$48 \leq 2L + 18 \leq 72$$

Now we solve the inequality:

$$48 \;\boxed{-18}\; \leq 2L + 18 \;\boxed{-18}\; \leq 72 \;\boxed{-18}$$

$$30 \leq 2L \leq 54$$

$$\frac{30}{2} \leq \frac{2L}{2} \leq \frac{54}{2}$$

$$15 \leq L \leq 27$$

Thus, in order for the perimeter to be between 48 and 72 cm (inclusive), the length must range between

$$\boxed{15 \text{ cm and } 27 \text{ cm, inclusive}}$$

The check is left to the student.

EXAMPLE 10

A printer is determining the cost of producing a pamphlet. There will be a certain number of color pages that cost 13¢ each, and 28 more than that number of black and white pages that cost 5¢ each. In addition, there is a 25¢ per pamphlet cover charge. Create a model which expresses the cost of the job in terms of the number of color pages. If the cost of the pamphlet is to be no more than $3.50, determine how many pages can be put in the pamphlet.

Solution   Let

$$c = \text{Number of color pages}$$

Then

$$c + 28 = \text{Number of black and white pages}$$ *There are 28 more black and white pages than color pages.*

Our model expresses the cost of printing each pamphlet. We let T be the total cost. (We will write our equation in cents to simplify the arithmetic.)

$$\left(\begin{array}{c}\text{Cost of printing} \\ \text{color pages}\end{array}\right) + \left(\begin{array}{c}\text{Cost of printing} \\ \text{black and white pages}\end{array}\right) + \left(\begin{array}{c}\text{Cover} \\ \text{charge}\end{array}\right) = \left(\begin{array}{c}\text{Total} \\ \text{Cost}\end{array}\right)$$

$$\left(\begin{array}{c}\text{Number} \\ \text{of color} \\ \text{pages}\end{array}\right) \cdot \left(\begin{array}{c}\text{Cost of} \\ \text{1 color} \\ \text{page}\end{array}\right) + \left(\begin{array}{c}\text{Number of} \\ \text{black and} \\ \text{white pages}\end{array}\right) \cdot \left(\begin{array}{c}\text{Cost of} \\ \text{1 b/w} \\ \text{page}\end{array}\right) + \left(\begin{array}{c}\text{Cover} \\ \text{charge}\end{array}\right) = \left(\begin{array}{c}\text{Total} \\ \text{cost}\end{array}\right)$$

$$c \cdot 13 \quad + \quad (c + 28) \cdot 5 \quad + \quad 25 \quad = \quad T$$

Hence

$$T = 13c + 5(c + 28) + 25 \qquad \textit{Simplify.}$$
$$= 13c + 5c + 140 + 25$$

$$\boxed{T = 18c + 165}$$ *This equation represents how the total cost, T, is related to c, the number of color pages.*

We are being asked to find how many pages can be in the pamphlet if the cost T is to be no more than \$3.50. Since T is in cents, we can write this as $T \le 350$. We have the model $T = 18c + 165$, so we can find c, the number of color pages, when $T \le 350$ as follows:

$$T \le 350 \qquad \textit{Substitute } T = 18c + 165.$$
$$18c + 165 \le 350 \qquad \textit{Solve for c.}$$
$$18c \le 185$$
$$c \le \frac{185}{18} \approx 10.28$$

Since c must be a whole number, $c \le 10.28$ means that c (the number of color pages) can be at most 10. Keep in mind we are looking for the *total* number of pages. First we note that if c is the number of color pages, then the number of black and white pages is $c + 28$. Hence the total number of pages, in terms of c is $c + (c + 28) = 2c + 28$. Since there cannot be more than 10 color pages, there cannot be more than $2(10) + 28 = 48$ pages altogether if the cost of the pamphlet is to be no more than \$3.50.

EXAMPLE 11

An aide in the mathematics department gets paid \$8 per hour for clerical work and \$15 per hour for tutoring. If she wants to work a total of 20 hours and earn at least \$265, what is the maximum number of hours she can spend on clerical work?

Solution   This problem is similar to the value problems covered in Section 2.2, except that it requires an inequality. Let's take a closer look at the problem.

If the aide works the whole 20 hours on clerical work, then she would be paid $20(8) = \$160$, certainly less than her goal. So we know that she has to spend less than 20 hours on clerical work.

On the other hand, if she spends 0 hours on clerical work, then she would spend all 20 hours tutoring and would make $20(15) = \$300$, more than enough for her goal. We can see that her goal of earning at least $265 *can* be realized by working between 0 and 20 hours on clerical work.

Now let's set up the problem: Let

$$x = \text{Number of hours spent on clerical work}$$

Then

$$20 - x = \text{Number of hours spent tutoring}$$

$$\begin{pmatrix} \text{Amount earned} \\ \text{from clerical work} \end{pmatrix} + \begin{pmatrix} \text{Amount} \\ \text{earned from} \\ \text{tutoring} \end{pmatrix} = \begin{pmatrix} \text{Total} \\ \text{amount} \\ \text{earned} \end{pmatrix}$$

$$\begin{pmatrix} \text{Hourly rate} \\ \text{for} \\ \text{clerical} \\ \text{work} \end{pmatrix} \cdot \begin{pmatrix} \text{Number of} \\ \text{hours spent} \\ \text{on clerical} \\ \text{work} \end{pmatrix} + \begin{pmatrix} \text{Hourly rate} \\ \text{for} \\ \text{tutoring} \end{pmatrix} \cdot \begin{pmatrix} \text{Number of} \\ \text{hours spent} \\ \text{on tutoring} \end{pmatrix} \geq 265$$

$$8 \quad \cdot \quad x \quad + \quad 15 \quad \cdot \quad (20 - x) \quad \geq 265$$

Hence, the inequality is

$$8x + 15(20 - x) \geq 265$$

Now we solve the inequality:

$$8x + 15(20 - x) \geq 265$$
$$8x + 300 - 15x \geq 265$$
$$300 - 7x \geq 265$$
$$-7x \geq -35$$
$$x \leq 5 \qquad \text{\textit{Note that we are dividing by} -5 \textit{and must therefore}}$$
$$\text{\textit{reverse the inequality symbol.}}$$

Thus, the number of hours spent on clerical work must be less than or equal to 5, or *the maximum number of hours the aide can spend on clerical work is*

$$\boxed{5 \text{ hours}}$$

if she is to make at least $265.

Preparing for Exams: Study Activities

If you are going to learn algebra well enough to be able to demonstrate high levels of performance on exams, then you must concern yourself with developing your skills in algebraic manipulation and understanding what you are doing and why you are doing it.

Many students concentrate only on skills and resort to memorizing the procedures for algebraic manipulations. This may work for quizzes or a test covering just a few topics. For exams covering a chapter or more, this can be quite a burden on your memory. Eventually, interference occurs and problems and procedures get confused. If you find yourself doing well on quizzes but not on longer exams, this may be your problem.

Concentrating on understanding what a method is and why it works is important. Neither the teacher nor the textbook can cover every possible way in which a particular concept may present itself in a problem. If you understand the concept, you should be able to recognize it in any problem. But again, if you concentrate only on understanding concepts and not on developing skills, you may find yourself prone to making careless and costly errors under the pressure of an exam.

In order to achieve both skill development and conceptual understanding, your studying should include four activities: (**1**) practicing problems, (**2**) reviewing your notes and textbook, (**3**) drilling with study cards, and (**4**) reflecting on the material and exercises.

Rather than doing any one of these activities over a long period of time, it is best to do a little of the first three activities during a study session and save some time for reflection at the end of the session.

EXERCISES 2.3

In Exercises 1–8, determine whether the inequality is an identity or a contradiction.

1. $x + 1 > x - 1$

2. $a - 7 \leq a + 3$

3. $2x - (x - 3) \geq x + 5$

4. $2x - (x - 3) < x + 5$

5. $-6 \leq r - (r + 6) < 3$

6. $-6 < r - (r + 6) < 3$

7. $0 < x - (x - 3) \leq 5$

8. $-7 \leq a - (a + 3) < -3$

In Exercises 9–14, determine whether the listed values of the variable satisfy the inequality.

9. $4 + 3u > 10; \quad u = -3.4, 2$

10. $7 + 2u \leq 15; \quad u = 4, 5.8$

11. $6 - 4y \leq 8; \quad y = -2.5, -1$

12. $8 - 5y \geq -3; \quad y = -1, 1.7$

13. $-3 \leq 3 - (4 - z) < 3; \quad z = -2, 4.6$

14. $-4 < 8 - 2(1 - r) \leq 4; \quad r = -1.4, 5$

In Exercises 15–22, determine whether the inequality makes sense. Put those that do make sense in standard form if they are not already in standard form.

15. $5 < x < 8$

16. $-5 < z < -8$

17. $-6 > w \geq -8$

18. $-6 \leq a < -5$

19. $-3 < a < -2$

20. $-4 < b < 3$

21. $2 < x < -3$

22. $2 > x > -3$

In Exercises 23–28, express the inequality using interval notation.

23. $-2 < x < 6$

24. $0 \leq x \leq 5$

25. $x < 7$

26. $x > 3$

27. $-12 < x \leq -8$

28. $x \leq -9$

In Exercises 29–36, solve each inequality.

29. $5y \geq 2 - 3y$

30. $5y < 7y - 2$

31. $\dfrac{2}{3}x - 4 \leq 3x + \dfrac{1}{2}$

32. $\dfrac{2x - 3}{4} > x + 1$

33. $3x + 4 \geq 2x - 5$

34. $5x - 7 > 2x + 6$

35. $11 - 3y < 38$

36. $9 - 7y \geq 16$

In Exercises 37–56, solve each inequality. Express your answer using interval notation and sketch the graph of the solution set on the real number line.

37. $3a - 8 < 8 - 3a$

38. $5y - 4 \geq 4 - 5y$

39. $3t - 5 > 5t - 13$

40. $8w - 9 < 13w + 21$

41. $2y - 3 \leq 5y + 7$

42. $4 - 3t \geq 13 - t$

43. $5(a - 2) - 7a > 2a + 10$

44. $4(m + 7) - 5 < 20$

45. $\dfrac{3x + 2}{5} < x + 4$

46. $\dfrac{2}{5}x + \dfrac{1}{3} > x - 1$

47. $7 - 5(t - 2) \leq 2$

48. $4 + 3(1 - a) \geq -8$

49. $2(y + 3) + 3(y - 4) < 3(y + 2) + 2(y + 1)$

50. $5(x - 2) - 3(x + 4) \geq 2x - 22$

51. $1 \leq c + 3 < 5$

52. $-3 < r - 3 \leq 1$

53. $6 < 2k - 1 \leq 11$

54. $11 \leq 3z + 2 < 15$

55. $-1 < 4 - t < 3$

56. $-3 \leq 6 - t \leq -1$

In Exercises 57–68, solve each inequality.

57. $1 \leq 8 - 3t \leq 12$

58. $0 < 9 - 5t < 29$

59. $2 < 5 - 3(x + 1) < 17$

60. $13 < 7 - 2(x - 3) \leq 31$

61. $-2 < x - 5 \leq 1$

62. $-4 < k - 3 < 2$

63. $-2 \leq 5 - x < 1$

64. $-4 < 3 - k < 2$

65. $1 \leq 1 - (5z - 2) \leq 11$

66. $0 < 2 - (3w - 4) < 6$

67. $0.39 \leq 0.72x - 1.5 < 8.1$

68. $7.55 < 0.75x - 2.5 \leq 8.5$

Solve each of the following problems algebraically. Be sure to clearly label what each variable represents.

69. What numbers satisfy the condition "4 less than 3 times the number is less than 17"?

70. What numbers satisfy the condition "5 less than 4 times the number is greater than 19"?

71. If 12 more than 6 times a number is greater than 3 times the number, how large must the number be?

72. If 9 less than 5 times a number is less than twice the number, how small must the number be?

73. What is the maximum length of the side of a square that has a maximum perimeter of 72 ft?

74. What is the minimum length of the side of a square that has a minimum perimeter of 94 ft?

75. The width of a rectangle is 8 cm. If the perimeter is to be at least 80 cm, what must the length be?

76. If the width of a rectangle is 10 m and the perimeter is not to exceed 120 m, how large can the length be?

77. The length of a rectangle is 18 in. If the perimeter is to be at least 50 in. but not greater than 70 in., what is the range of values for the width?

78. The length of a rectangle is twice the width. If the perimeter is to be at least 80 ft but not more than 100 ft, what is the range of values for the width of the rectangle?

79. The length of a rectangle is 5 in. more than 3 times the width. If the width varies from 12 in. to 16 in., what is the range of values for the perimeter?

80. The length of a rectangle is 3 ft less than twice its width. If the width varies from 6 to 13 ft, what is the range of values for the length? What is the range of values for the perimeter?

81. The medium side of a triangle is 2 cm longer than the shortest side, and the longest side is twice as long as the shortest side. If the perimeter of the triangle is to be at least 30 cm and no more than 50 cm, what is the range of values for the shortest side?

82. The first side of a triangle is 3 times the second side and the third side is 5 m more than the second side. If the perimeter is to be between 35 and 50 m, what is the range of values for each side?

83. The body-mass index, BMI, is a measure of height for weight that indicates weight status.

$$BMI = \frac{703W}{H^2}$$

where W is weight in pounds and H is height in inches.

   For adults over 20, a BMI between 18.5 and 24.9 is considered healthy. To the nearest pound, what weight range gives a healthy BMI for

   (a) 68 inches

   (b) Your height.

84. To get the maximum benefit from exercising, you should try to keep your heart rate within the range known as the target heart rate, or THR. The THR lies between 60% and 80% of your maximum heart rate, where your maximum heart rate is estimated by subtracting your age from 220 if you are a male, or subtracting your age from 226 if you are a female.

   (a) Construct a mathematical model for the THR for males and a model for females. To the nearest whole number, find the THR range.

   (b) for a male of age 28

   (c) a female of age 55

   (d) your age.

85. An organization wants to sell tickets to a concert. It plans on selling 300 reserved-seat tickets and 150 tickets at the door. The price of a reserved-seat ticket is to be $2 more than a ticket at the door. If the organization wants to collect at least $3,750, what is the minimum price it can charge for a reserved-seat ticket?

86. Repeat Exercise 85 if the price of a reserved-seat ticket is to be $3 more than a ticket at the door.

87. Romi has 40 coins in nickels and dimes. What is the maximum number of dimes he should have to have no more than $2.85?

88. Bridgett has 50 coins in nickels and dimes. What is the minimum number of dimes she should have to have at least $3.45?

89. Carolyn makes two types of stuffed animals: stuffed panda bears and stuffed elephants. She charges $20 for the bears and $25 for the elephants. If she decides to make 24 stuffed animals this week, what is the minimum number of elephants she should make if she is to gross at least $575 for the 24 animals?

90. Jules enjoys spending time as a teacher, although he is paid only $120 a day. On the other hand, he receives $180 a day for what he considers to be a less enjoyable job as a car mechanic. If he plans to work 25 days this month, what is the minimum number of days he should work as a mechanic if he is to make at least $3,900?

91. Ian makes $20 an hour when he tutors and averages $10 an hour when he works at a local restaurant. Out of 30 hours working time, what is the minimum number of hours he should tutor if he is to make at least $650?

92. Barry makes $5 for each case of chocolate bars he sells and $7 for each case of hard candy. If he is supposed to sell 75 cases of either type, what is the minimum number of cases of hard candy he should sell to make at least $530?

93. Steve makes 25¢ on every regular hot dog he sells from his stand and 45¢ on every super-dog he sells. Steve just bought a new sign for his stand for $38.60. What is the minimum number of super-dogs he must sell to make enough to pay for his new sign by the 100th sale of either dog?

94. Repeat Exercise 93 and determine the minimum number of super-dogs Steve must sell by his 100th sale (of either dog), if his new sign cost $52.50.

95. Zach wants to invest his money in the stock market. He decides on two stock issues that both sell at the same price. Stock A is a "safe," conservative stock that pays an annual dividend of $2 per share. Stock B, on the other hand, is a more risky investment but pays an annual dividend of $3 per share. Zach wants to invest enough money to buy 1,000 shares of either stock, but he does not want to take the risk of buying all stock B. He decides that he would like to get at least $2,400 annual dividends from his stock investment. What is the minimum number of shares of stock B he should buy to ensure a $2,400 dividend annually?

Questions for Thought

96. Describe what is *wrong* (if anything) with the following:

(a) $3x < -6$

$$\frac{3x}{3} \overset{?}{<} \frac{-6}{3}$$

$$x \overset{?}{>} -2$$

(b) $-2x > 4$

$$\frac{-2x}{-2} \overset{?}{>} \frac{4}{-2}$$

$$x \overset{?}{<} 2$$

(c) $-2x > 4$

$$\frac{-2x}{-2} \overset{?}{>} \frac{4}{-2}$$

$$x \overset{?}{>} -2$$

(d) $x + 2 \leq 5$

$$x + 2 - 2 \leq 5 - 2$$

$$x \overset{?}{\geq} 3$$

97. What is the difference between a conditional inequality and an identity?

98. What is the difference between a contradiction and an identity?

99. Discuss the differences between the properties of equations and the properties of inequalities.

100. What does it mean to say that two inequalities are equivalent?

101. Describe the difference between the method of checking an equation and that of checking an inequality.

2.4 Absolute-Value Equations and Inequalities

In Chapter 1 we defined the absolute value of x, $|x|$, *geometrically* as follows:

$|x|$ is the distance on the real number line between x and 0.

In this section we will draw on this idea of the absolute value of an expression being its distance to 0 to solve first-degree equations and inequalities involving absolute values.

EXAMPLE 1   Solve for x: $|x| = 4$

Solution   We will begin solving absolute-value equations by using the geometric definition of absolute value to interpret the equation on the real number line. We will then replace the absolute-value equation with two equations *without* absolute values.

The equation $|x| = 4$ means that the distance from x to 0 is 4. In other words, x is located 4 units away from 0. Therefore, x must be 4 or -4, as illustrated in Figure 2.8.

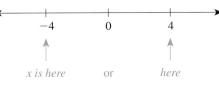

Figure 2.8
$|x| = 4$

Therefore, $\boxed{x = -4 \quad \text{or} \quad x = 4}$

EXAMPLE 2

Solution

Solve for a: $|3a - 1| = 5$

$3a - 1$ must be 5 units away from 0, as shown in Figure 2.9.

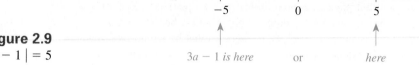

Figure 2.9
$|3a - 1| = 5$

Therefore,

$$3a - 1 = -5 \quad \text{or} \quad 3a - 1 = 5$$
$$3a = -4 \qquad\qquad 3a = 6$$
$$\boxed{a = -\frac{4}{3}} \qquad\qquad \boxed{a = 2}$$

Check: $a = -\dfrac{4}{3}$:

$$|3a - 1| = 5$$
$$\left|3\left(-\tfrac{4}{3}\right) - 1\right| \overset{?}{=} 5$$
$$|-4 - 1| \overset{?}{=} 5$$
$$|-5| \overset{\checkmark}{=} 5$$

Check: $a = 2$:

$$|3a - 1| = 5$$
$$|3(2) - 1| \overset{?}{=} 5$$
$$|6 - 1| \overset{?}{=} 5$$
$$|5| \overset{\checkmark}{=} 5$$

Now that we have seen the geometric (number-line) interpretation of absolute-value equations, we can see that:

> For $a \geq 0$,
>
> $$|x| = a \qquad \text{is equivalent to} \qquad x = a \quad \text{or} \quad x = -a.$$

Hence,

$$|x| = 7 \quad \text{yields two equations (and therefore two answers):}$$
$$x = 7 \quad \text{or} \quad x = -7$$

Similarly, $|2t - 5| = 7$ yields two equations:

$$2t - 5 = 7 \quad \text{or} \quad 2t - 5 = -7 \qquad \textit{We solve for } t \textit{ to find the}$$
$$t = 6 \quad \text{or} \qquad\qquad t = -1 \qquad \textit{two answers.}$$

EXAMPLE 3

Solution

Solve for t: $|2t + 3| - 4 = 3$

To apply the analysis we have used thus far, we need to have the absolute value isolated on one side of the equation.

$$|2t + 3| - 4 \;\boxed{+ 4}\; = 3 \;\boxed{+ 4}\; \qquad \textit{Add 4 to both sides.}$$

$$|2t + 3| = 7$$

This means that the expression $2t + 3$ must be 7 units away from 0, or, equivalently,

$$2t + 3 = 7 \quad \text{or} \quad 2t + 3 = -7$$
$$2t = 4 \qquad\qquad 2t = -10$$
$$\boxed{t = 2} \quad \text{or} \quad \boxed{t = -5}$$

Check: $t = 2$: $|2t + 3| - 4 = 3$ **Check:** $t = -5$: $|2t + 3| - 4 = 3$

$|2(2) + 3| - 4 \stackrel{?}{=} 3$ $\qquad$ $|2(-5) + 3| - 4 \stackrel{?}{=} 3$

$|4 + 3| - 4 \stackrel{?}{=} 3$ $\qquad$ $|-10 + 3| - 4 \stackrel{?}{=} 3$

$|7| - 4 \stackrel{?}{=} 3$ $\qquad$ $|-7| - 4 \stackrel{?}{=} 3$

$7 - 4 \stackrel{\checkmark}{=} 3$ $\qquad$ $7 - 4 \stackrel{\checkmark}{=} 3$

EXAMPLE 4

Solve for y: $|y - 3| = -2$

Solution

Be careful! Do not blindly apply a procedure where it is not appropriate. This example requires that the absolute value, which is a distance, be negative. But distance is always measured in *nonnegative* units. The absolute value of an expression can never be negative; therefore, there are

$$\boxed{\text{no solutions}}.$$

EXAMPLE 5

Solve for x: $|2x + 5| = |x - 2|$

Solution

For the absolute value of two expressions to be equal, they must both be the same distance from 0. This can happen if the two expressions are equal *or* if they are negatives of each other. For instance, if $|a| = |b|$ then a and b can both be 6, $|6| = |6|$, or a can be 6 and b can be -6 or vice versa, $|6| = |-6|$.

Thus, our absolute-value equation will be true if

$$2x + 5 = x - 2 \quad \text{or} \quad 2x + 5 = -(x - 2) \qquad \textit{Note that we must take the negative}$$
$$\boxed{x = -7} \qquad\qquad 2x + 5 = -x + 2 \qquad \textit{of the entire expression.}$$
$$3x = -3$$
$$\boxed{x = -1}$$

The check is left to the student.

The same kind of analysis allows us to solve absolute-value inequalities as well. Again, our goal is to replace the absolute-value inequality with an inequality or inequalities *without* absolute values.

EXAMPLE 6

Solve for x: $|x| < 3$

Solution

This inequality means that x must be less than 3 units away from 0. Therefore, x must fall between -3 and 3, as illustrated in Figure 2.10.

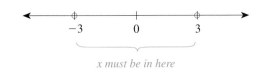

Figure 2.10

$|x| < 3$

We can write the solution set as the double inequality

$$-3 < x < 3 \qquad \text{or using interval notation,} \quad (-3, 3)$$

EXAMPLE 7

Solve for n and sketch the solution on the real number line:

$$|2n - 1| < 5$$

Solution

Proceeding as we did in the last example, we require that $2n - 1$ be less than 5 units away from 0. Therefore, $2n - 1$ must lie between -5 and 5, as shown in Figure 2.11.

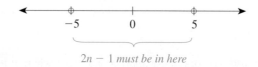

Figure 2.11
$|2n - 1| < 5$

$2n - 1$ *must be in here*

We write this as the double inequality:

$$-5 < 2n - 1 < 5 \qquad \textit{Now solve for n.}$$

$$-5 \;\boxed{+1}\; < 2n - 1 \;\boxed{+1}\; < 5 \;\boxed{+1}$$

$$-4 < \qquad 2n \qquad < 6$$

$$\frac{-4}{2} < \qquad \frac{2n}{2} \qquad < \frac{6}{2}$$

$$-2 < n < 3 \qquad \text{or using interval notation; } (-2, 3)$$

Keep in mind that this solution means that for the expression $2n - 1$ to be less than 5 units away from 0, n must be between -2 and 3, as shown in Figure 2.12.

Figure 2.12
Solution set for
Example 7

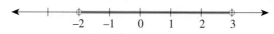

Note that the empty circles at the endpoints indicate a strict inequality.

The check is left to the student.

EXAMPLE 8

Solve for y: $\quad |y| > 2$

Solution

This inequality means that y must be more than 2 units away from 0. Therefore, y must be to the left of -2, or to the right of 2, as indicated in Figure 2.13.

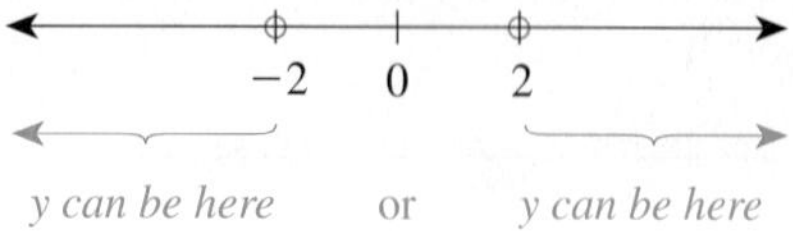

Figure 2.13
$|y| > 2$

y can be here or *y can be here*

We must write the solution set as two separate inequalities:

$$y < -2 \quad \text{or} \quad y > 2$$

This answer is composed of two intervals. To express the answer using interval notation, we usually write the two intervals joined with the *union* symbol ∪ as follows:

$(-\infty, -2) \cup (2, \infty)$. See Appendix *A* for more information on the union symbol, ∪.

EXAMPLE 9

Solve for w and sketch the solution on the real number line:

$$|\,3w - 2\,| > 4$$

Solution

This inequality says that the expression $3w - 2$ must be more than 4 units away from 0. Therefore, $3w - 2$ must be to the left of -4 or to the right of 4, as shown in Figure 2.14.

Figure 2.14
$|\,3w - 2\,| > 4$

$3w - 2$ *can be here*   or $3w - 2$ *can be here*

We can write the solution set as *two* separate inequalities:

$$3w - 2 < -4 \qquad\qquad \text{or} \qquad\qquad 3w - 2 > 4$$

$$3w - 2 \;\boxed{+\ 2}\; < -4 \;\boxed{+\ 2}\; \qquad \text{or} \qquad 3w - 2 \;\boxed{+\ 2}\; > 4 \;\boxed{+\ 2}\;$$

$$3w < -2 \qquad\qquad\qquad\qquad 3w > 6$$

$$\boxed{\; w < -\frac{2}{3} \qquad \text{or} \qquad w > 2 \;}$$

Using interval notation, we express the answer as $\left(-\infty,\ -\frac{2}{3}\right) \cup (2,\ \infty)$

On the number line the solution set appears as shown in Figure 2.15.

Figure 2.15
Solution set for
Example 9

With regard to solving absolute-value *inequalities*, we can summarize the previous discussion as follows:

For $a > 0$.

$$|\,x\,| < a \quad \text{is equivalent to} \quad -a < x < a$$

whereas

$$|\,x\,| > a \quad \text{is equivalent to} \quad x < -a \quad \text{or} \quad x > a$$

Note the differences between the two types of inequalities and how they should be handled.

EXAMPLE 10

Solve each of the following for x. Express your answer using interval notation and sketch the solution on the real number line:

(a) $|\,5x - 3\,| \le 7$ **(b)** $|\,5x - 3\,| > 7$

Solution

(a) $|\,5x - 3\,| \le 7$ is equivalent to $-7 \le 5x - 3 \le 7$.

$$-7 \;\boxed{+\ 3}\; \le 5x - 3 \;\boxed{+\ 3}\; \le 7 \;\boxed{+\ 3}\;$$

$$-4 \le \quad 5x \quad \le 10$$

$$-\frac{4}{5} \le \quad \frac{5x}{5} \quad \le \frac{10}{5}$$

$$-\frac{4}{5} \le x \le 2$$

The solution is $\boxed{\left[-\dfrac{4}{5}, 2\right]}$. The solution set is shown in Figure 2.16.

Figure 2.16
Solution set for
Example 10(**a**)

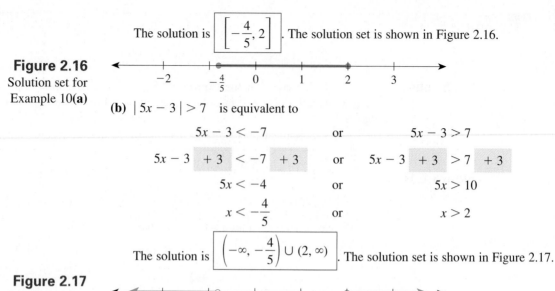

(b) $\left|5x - 3\right| > 7$ is equivalent to

$$5x - 3 < -7 \qquad \text{or} \qquad 5x - 3 > 7$$

$$5x - 3 \;\boxed{+\,3}\; < -7 \;\boxed{+\,3}\; \qquad \text{or} \qquad 5x - 3 \;\boxed{+\,3}\; > 7 \;\boxed{+\,3}\;$$

$$5x < -4 \qquad \text{or} \qquad 5x > 10$$

$$x < -\dfrac{4}{5} \qquad \text{or} \qquad x > 2$$

The solution is $\boxed{\left(-\infty, -\dfrac{4}{5}\right) \cup (2, \infty)}$. The solution set is shown in Figure 2.17.

Figure 2.17
Solution set for
Example 10(**b**)

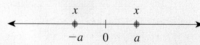

Notice the similarities and the differences between parts **(a)** and **(b)**. We cannot use a double inequality to describe the solution in part **(b)**.

DIFFERENT PERSPECTIVES: Absolute-Value Equations and Inequalities

GEOMETRIC DESCRIPTION

The absolute value of x, $\left|x\right|$, is its distance from 0 on the number line.
For $a > 0$,
$\left|x\right| = a$ means x is a units from 0 on the number line:

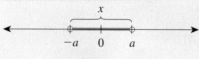

$\left|x\right| < a$ means x is less than a units from 0 on the number line:

$\left|x\right| > a$ means x is more than a units from 0 on the number line:

ALGEBRAIC DESCRIPTION

The absolute value of x, $\left|x\right|$, is defined algebraically as follows.
For $a > 0$,
$\left|x\right| = a$ means $x = a$ or $x = -a$

$\left|x\right| < a$ means $-a < x < a$

$\left|x\right| > a$ means $x < -a$ or $x > a$

EXAMPLE 11

Solve for t:

(a) $\left|t + 3\right| < -2$ **(b)** $\left|t - 5\right| > -1$

Solution

As we have mentioned before, look at a problem carefully; don't just blindly plunge ahead.

(a) Since absolute value means distance from 0, this part of this example is asking that a distance be less than -2, which is impossible. (Distance is always measured in nonnegative units.) Thus, the inequality in part **(a)** is a contradiction and has $\boxed{\text{no solutions}}$.

Using interval notation, we can write all real numbers as $(-\infty, \infty)$.

(b) Part **(b)** is asking that a distance be greater than -1, which is always true (a distance is *always* greater than or equal to 0). Therefore, part **(b)** is an identity and all real numbers satisfy the inequality.

EXAMPLE 12

A lumber company cuts boards for building construction and claims that the lengths of the boards are accurate to within $\}116\}$ of an inch of their specified length. What is the range of lengths for an 8-ft piece of lumber from this company?

Solution

If boards are cut to within $\frac{1}{16}$ of an inch of their specified length, this means that the difference between the actual length of the board, 96 in. $(= 12 \times 8)$, must be less than or equal to $\frac{1}{16}$ in. Letting $x =$ the actual length, we can write this as an absolute-value inequality:

$$\left| x - 96 \right| \leq \frac{1}{16} \qquad \textit{Which can be solved as follows}$$

Rewrite without the absolute value symbols:

$$-\frac{1}{16} \leq \quad x - 96 \quad \leq \frac{1}{16} \qquad \textit{Add 96 to each member.}$$

$$-\frac{1}{16} + 96 \leq x - 96 + 96 \leq 96 + \frac{1}{16} + 96$$

$$95\frac{15}{16} \leq \quad x \quad \leq 96\frac{1}{16}$$

Hence the actual length of an 8-ft board can be anywhere from $95\frac{15}{16}$ in. to $96\frac{1}{16}$ in.

EXAMPLE 13

An economic forecasting corporation predicts that in the next year, the price of oil will stay within 5% of its current price. According to this prediction, what will the price range be for a barrel of oil if the current price for a barrel is $30?

Solution

The company is predicting that the difference between current price and the future price should remain within 5% of its current value. Letting $F =$ the future price, we can write this as an absolute-value inequality:

We could just as well have used $\left| F - 30 \right|$.

$$\left| \text{Current price} - F \right| \leq 5\% \text{ of the current price} \qquad \textit{Substitute the current}$$
$$\textit{price of \$30.}$$
$$\left| 30 - F \right| \leq 0.05(30)$$

$$\left| 30 - F \right| \leq 1.50$$

We start by rewriting the inequality without the absolute value symbols:

$$-1.50 \leq 30 - F \leq 1.50 \qquad \textit{Subtract 30 from each member.}$$
$$-31.50 \leq \quad -F \quad \leq -28.50 \qquad \textit{Divide each member by } -1.$$
$$31.50 \geq \quad F \quad \geq 28.50 \qquad \textit{Rewrite the double inequality in standard form.}$$
$$28.50 \leq \quad F \quad \leq 31.50$$

Hence the actual price range predicted for the following year is between $28.50 and $31.50.

STUDY SKILLS 2.4

Preparing for Exams: Using Study Cards

The process of making up study cards is a learning experience in itself. Study cards are convenient to use—you can carry them along with you and use them for review in between classes or as you wait for a bus.

Use the (D/P and W) cards as follows:

1. Look at the heading of a card and, covering the rest of the card, see if you can remember what the rest of the card says.

2. Continue this process with the remaining cards. Pull out cards you know well and put them aside, but review them from time to time. Study cards that you do not know.

3. Shuffle the cards so that they are in random order and repeat the process again from the beginning.

4. As you go through the cards, ask yourself the following questions (where appropriate):

   a. When do I use this rule, method, or principle?

   b. What are the differences and similarities between problems?

   c. What are some examples of the definitions or concepts?

   d. What concept is illustrated by the problem?

   e. Why does this process work?

   f. Is there a way to check this problem?

EXERCISES 2.4

In Exercises 1–42, solve the absolute-value equation or inequality.

1. $|x| = 4$
2. $|a| = 6$
3. $|x| < 4$
4. $|a| > 6$

5. $|x| > 4$
6. $|a| < 6$
7. $|x| \leq 4$
8. $|a| \geq 6$

9. $|x| \geq 4$
10. $|a| \leq 6$
11. $|x| = -4$
12. $|a| = -6$

13. $|x| > -4$
14. $|a| \leq -6$

15. $|x| \leq -4$
16. $|a| > -6$

17. $|t - 3| = 2$
18. $|y| + 2 = 7$

19. $|t| - 3 = 2$
20. $|y + 2| = 7$

21. $|5 - n| = 1$
22. $|3 - y| = 2$

23. $|a - 5| < 3$
24. $|a + 1| > 4$

25. $|a - 1| \geq 2$
26. $|a + 7| \leq 3$

27. $|2a - 5| < -1$
28. $|2a - 9| \geq -5$

29. $|2a| - 5 < -1$
30. $|2a| - 9 \geq -5$

31. $|3x - 2| - 3 = 1$
32. $|5t - 4| + 4 = 3$

33. $|3x - 2| + 3 = 1$
34. $|5t - 4| - 4 = 3$

35. $|2(x - 1) + 7| = 5$
36. $|3(y + 2) - 10| = 4$

37. $|3 - a| \leq 2$
38. $|6 - m| > 5$

39. $|5 - 2a| > 1$
40. $|7 - 2a| \leq 3$

41. $|4(x - 1) - 5x| < 4$
42. $|3(y + 2) - 7y| \geq 6$

In Exercises 43–52, express your answer to inequalities using interval notation, and sketch the solution set of the equation or inequality on the real number line.

43. $|x - 1| = 5$

44. $|a + 3| = 4$

45. $|3 - x| \leq 2$

46. $|5 - a| \geq 3$

47. $|2x + 7| > 1$

48. $|2a - 9| < 5$

49. $|4x - 5| < 3$

50. $|5a - 4| < 3$

51. $|3 - 4x| < 1$

52. $|4 - 3x| < 1$

In Exercises 53–62, solve the absolute-value equation.

53. $|5t - 1| = |4t + 3|$

54. $|3t + 2| = |2t - 5|$

55. $|4r - 3| = |2r + 9|$

56. $|5r + 6| = |3r - 12|$

57. $|a - 5| = |2 - a|$

58. $|4 - n| = |n - 6|$

59. $|3x - 4| = |4x - 3|$

60. $|2x - 5| = |5 - 2x|$

61. $|x + 1| = |x - 1|$

62. $|t - 3| = |t + 3|$

63. The Yadav Heating and Plumbing Company subcontracts out the manufacture of thermostats to another company. The Yadav Company requires the subcontractor to produce thermostats accurate to 0.75°F. It tests the thermostats by putting them in a room and taking a reading when the temperature is a true value of 40°F, 50°F, and 80°F. What is the allowable range of thermostat readings for each temperature?

64. A weather forecaster predicts the day's high temperature and claims that she is always accurate to within 5°F. What is the acceptable range of temperature values for a forecast high of 79°F?

65. A political pollster predicts that in 2 days, the mayor of a certain city will win with 64% of the votes. If the pollster's projected margin of error for this prediction is 6 percentage points, give the range of values of the mayor's proportion of votes predicted by this pollster.

66. In a certain state, it was predicted that next year's sales tax should not change by more than 8% of the current sales tax. Currently, the sales tax in this state is 6%. Based on this prediction, what is the range of possible values of the sales tax for next year?

? Questions for Thought

67. Verbally describe the differences between the following two inequalities:

$$|x| < a \quad \text{and} \quad |x| > a$$

Describe the approach you would use in solving them.

68. Discuss what is ***wrong*** (if anything) with the solution to the following problem.

Solve for x: $\quad |3x + 4| - 9 = 2$

$$3x + 4 - 9 = 2 \quad \text{or} \quad 3x + 4 - 9 = -2$$
$$3x - 5 = 2 \quad \text{or} \quad 3x - 5 = -2$$
$$3x = 7 \quad \text{or} \quad 3x = 3$$
$$x = \frac{7}{3} \quad \text{or} \quad x = 1$$

STUDY SKILLS 2.5

Preparing for Exams: Reviewing Your Notes and Textbook; Reflecting

Another activity we suggest as an important facet of studying for exams is to review your notes and text. Your notes are a summary of the information you believed was important at the time you wrote it down. In the process of reviewing your notes and textbook you may turn up something you missed. A gap in your understanding may get filled and consequently give more meaning to some of the definitions, rules, and concepts on your study cards (and make them easier to remember). Perhaps you will understand a shortcut that you missed the first time around.

Reviewing the explanations or problems in the textbook *and* your notes gives you a better perspective and helps to tie the material together. Concepts will begin to make more sense when you review and think about how they are interrelated. It is also important to practice review problems so that you will not forget those skills you have already learned. Do not forget to review old homework exercises, quizzes, and exams—especially those problems that were incorrectly done. Review problems also offer an excellent opportunity to work on your speed as well as your accuracy.

We discussed the importance of reflecting on the material you are reading and the exercises you are doing. Your thinking time is usually limited during an exam, and you want to be able to anticipate variations in problems and to make sure that careless errors will be minimized at that time. For this reason it is a good idea to try to think about possible problems ahead of time. In areas where you tend to get confused, make the distinctions that exist as clear as possible.

As you review material, ask yourself the study questions given in Study Skills 1.1 and 2.4. Also look at the Questions for Thought at the end of most of the exercise sets and ask yourself those questions as well.

CHAPTER 2 SUMMARY

After having completed this chapter, you should be able to:

1. Create and use a mathematical model that describes a relationship between variables discussed in an application. (Section 2.1)

    For example: A cell phone company charges a flat fee of $10 a month and a fee of 18¢ per minute for airtime. Express the monthly cost of using a cell phone in terms of the number of minutes of airtime. Determine how much airtime you can allow yourself if you have budgeted $35 a month for the cell phone.

    Solution: If your monthly cell phone airtime is 100 minutes, your cost would be

    $$C = \$10 + (100 \times 0.18) = \$28 \text{ per month}$$

    Hence you find the cost by adding 10 to the product of 0.18 and the number of minutes of airtime, or, if we let C be the monthly cost, and t be the number of minutes airtime, we have

    $$C = 10 + 0.18t$$

    To find how much airtime is allowed for $35, we substitute $C = 35$ in the equation $C = 10 + 0.18t$, and solve for t:

$C = 10 + 0.18t$	*Substitute $C = 35$.*
$35 = 10 + 0.18t$	*Solve for t; subtract **10** from each side.*
$25 = 0.18t$	*Divide both sides by **0.18***
$\dfrac{25}{0.18} \approx 138.9 = t$	

    Hence, with this plan, to stay under $35 per month, one cannot spend more than 138 minutes on the phone per month.

2. Understand and recognize the basic types of equations and inequalities: conditional, identity, and contradiction; and the properties of equations and inequalities (Sections 2.2 and 2.3).

3. Solve (algebraically) and check a first-degree equation in one variable (Section 2.2).

For example: Solve for x: $5(x - 3) - (8x - 1) = 3(x - 2) - 8(x + 2)$

Solution:

$$5(x - 3) - (8x - 1) = 3(x - 2) - 8(x + 2)$$
$$5x - 15 - 8x + 1 = 3x - 6 - 8x - 16$$

Simplify both sides of the equation, then apply equality properties.

$$-3x - 14 = -5x - 22$$ *Add 5x to both sides of the equation.*
$$2x - 14 = -22$$ *Add 14 to both sides of the equation.*
$$2x = -8$$ *Divide both sides of the equation by 2.*
$$x = -4$$

Check: $x = -4$: $5(-4 - 3) - [8(-4) - 1] \stackrel{2}{=} 3(-4 - 2) - 8(-4 + 2)$
$$5(-7) - (-32 - 1) \stackrel{2}{=} 3(-6) - 8(-2)$$
$$-35 - (-33) \stackrel{2}{=} -18 + 16$$
$$-2 \stackrel{\checkmark}{=} -2$$

4. Solve (algebraically), check, and sketch the solution set of a first-degree inequality in one variable (Section 2.3).

For example: Solve for x. $7 - 3(x + 2) < 22$

Solution: $7 - 3(x + 2) < 22$ *Simplify where necessary.*
$$7 - 3x - 6 < 22$$
$$1 - 3x < 22$$ *Then apply inequality properties. Add -1 to both sides of the inequality.*
$$-3x < 21$$
$$\frac{-3x}{-3} > \frac{21}{-3}$$ *Dividing by a negative number reverses the inequality symbol.*
$$x > -7$$

Check: Check the endpoint $x = -7$:   Pick a value of x greater than -7. Let $x = -2$:

$$7 - 3(x + 2) = 22 \qquad\qquad 7 - 3(x + 2) < 22$$
$$7 - 3(-7 + 2) \stackrel{2}{=} 22 \qquad\qquad 7 - 3(-2 + 2) \stackrel{?}{<} 22$$
$$7 - 3(-5) \stackrel{2}{=} 22 \qquad\qquad 7 - 3(0) \stackrel{?}{<} 22$$
$$7 + 15 \stackrel{2}{=} 22 \qquad\qquad 7 - 0 \stackrel{\checkmark}{<} 22$$
$$22 \stackrel{\checkmark}{=} 22$$

The solution set on the real number line is shown here:

5. Use interval notation to express solutions to inequalities. (Sections 2.3 and 2.4)

For example: Using interval notation.

(a) We can write $-3 \le x$ as $[-3, \infty)$.

(b) We can write $x < 2$ as $(-\infty, 2)$

(c) We can write $-5 \le x \le 8$ as $[-5, 8]$.

(d) We can write $x < -3$ or $x \ge 5$ as $(-\infty, -3) \cup [5, \infty)$

6. Solve verbal problems algebraically by using first-degree equations or inequalities in one variable (Sections 2.2 and 2.3).

For example: John spends a total of $19 on 12 batteries. If alkaline batteries cost $2 each and regular batteries cost $1 each, how many of each type did he buy?

Solution: Let

$$x = \text{Number of alkaline batteries purchased}$$

Then

$$12 - x = \text{Number of regular batteries purchased}$$

We compute the total cost as follows:

$$\underbrace{2x}_{\substack{\textit{Cost of} \\ \textit{alkaline batteries}}} + \underbrace{1(12 - x)}_{\substack{\textit{Cost of} \\ \textit{regular batteries}}} = \underbrace{19}_{\textit{Total cost}}$$

Our equation is
$$2x + (12 - x) = 19$$
$$2x + 12 - x = 19$$
$$x + 12 = 19$$
$$x = 7$$

John purchased 7 alkaline batteries and $12 - 7 = 5$ regular batteries.

7. Solve absolute-value equations and inequalities of the first degree (Section 2.4).

For example:

(a) Solve for *t:* $\left| 3t - 5 \right| = 8$

Solution: $\left| 3t - 5 \right| = 8$ means that $3t - 5$ is 8 units away from 0. Therefore,

$$3t - 5 = 8 \quad \text{or} \quad 3t - 5 = -8$$
$$3t = 13 \quad \text{or} \quad 3t = -3$$
$$t = \frac{13}{3} \quad \text{or} \quad t = -1$$

(b) Solve for *a:* $\left| 2a - 3 \right| \leq 7$

Solution: $\left| 2a - 3 \right| \leq 7$ means that $2a - 3$ is less than or equal to 7 units away from 0 on the real number line. Therefore,

$$2a - 3 \text{ is between } -7 \text{ and } 7, \text{ inclusively.}$$

Hence our inequality becomes:

$$-7 \leq 2a - 3 \leq 7 \qquad \textit{Add +3 to each member.}$$

$$-7 \boxed{+3} \leq 2a - 3 \boxed{+3} \leq 7 \boxed{+3}$$

$$-4 \leq 2a \leq 10 \qquad \textit{Divide each member by 2.}$$
$$\frac{-4}{2} \leq \frac{2a}{2} \leq \frac{10}{2}$$
$$-2 \leq a \leq 5$$

which, using interval notation, is $[-2, 5]$

(c) Solve for *x:* $\left| 8 - 5x \right| > 2$

Solution: $\left| 8 - 5x \right| > 2$ means that $8 - 5x$ is more than 2 units away from 0 on the real number line. Therefore,

$$8 - 5x < -2 \quad \text{or} \quad 8 - 5x > 2$$

Solve each inequality.

$$8 - 5x < -2 \quad \text{or} \quad 8 - 5x > 2 \qquad \textit{First add } -8 \textit{ to both sides of each inequality.}$$

$$-5x < -10 \quad \text{or} \quad -5x > -6 \qquad \textit{Divide each side of each inequality by } -5.$$

$$x > 2 \quad \text{or} \quad x < \frac{6}{5} \qquad \textit{Note that we reverse both inequality symbols.}$$

which, using interval notation, is expressed as $\left(-\infty, \frac{6}{5}\right) \cup (2, \infty)$

CHAPTER 2 REVIEW EXERCISES

1. The Bedding Store contracts out to a delivery service, which charges the store $15 for delivering single beds and $20 for delivering larger beds. The service delivered 24 beds yesterday. Construct an equation relating the amount of money collected by the delivery service in terms of the number of single beds delivered for The Bedding Store yesterday. Use the equation to determine the number of each type of bed delivered if the delivery service collected $415.

2. A salesperson is paid a base salary of $120 plus a commission of 6% of her gross sales. Construct an equation relating her pay to her gross sales and determine the weekly gross sales she needs to earn $500 a week.

3. The length of a rectangle is 2 less than 4 times its width. Express the perimeter in terms of its width. Use the equation to determine the dimensions of the rectangle if the perimeter is to be 100 in.

4. Carol averages $10 an hour grading homework papers for the mathematics department, and $25 an hour tutoring for the mathematics department. She can work no more than 25 hours a week. Express the amount of money she can earn in terms of the number of hours she grades homework papers and use this model to explain how the amount she makes is affected by changes in the number of hours she tutors. Determine the maximum number of hours she can grade homework and still make at least $500.

In Exercises 5–32, solve the equation.

5. $3x - 4(x - 2) = 2x - 5$

6. $\dfrac{2x - 3}{4} + \dfrac{x}{2} = 4$

7. $\dfrac{x}{2} - \dfrac{3}{2} = \dfrac{x}{5} + 1$

8. $4x - 2 = 2(x - 5) + 2x$

9. $3x - 5x = 7x - 4x$

10. $2a + 5a = 7a - 3a$

11. $6(8 - a) = 3(a - 4) + 2(a - 1)$

12. $3(a - 4) - 2(a - 4) = a - 4$

13. $7[x - 3(x - 3)] = 4x - 2$

14. $5[x - 2(x + 5) + 3] = 3x - 5$

15. $5\{x - [2 - (x - 3)]\} = x - 2$

16. $2 - \{x - 6[x - (4 - x)]\} = 3(x + 2)$

17. $|x| = 3$

18. $|a| = -5$

19. $|x| = -3$

20. $|y| = 7$

21. $|2x| = 8$

22. $|-3x| = 5$

23. $|a + 1| = 4$

24. $|a - 3| = 5$

25. $|4z + 5| = 0$

26. $|7x + 2| = 6$

27. $|2y - 5| - 8 = -3$

28. $|4x - 9| + 5 = 7$

29. $|x - 5| = |x - 1|$

30. $|a + 3| = |a + 1|$

31. $|2t - 4| = |t - 2|$

32. $|2a - 5| = |a - 3|$

In Exercises 33–48, solve the inequality.

33. $3x - 4 \leq 5$

34. $2x \leq x - 3$

35. $5x - 4 \leq 2x$

36. $3x - 2 > 5x - 6$

37. $5z + 4 > 2z - 1$

38. $3y - 2 > -y + 3$

39. $5(s - 4) < 3(s - 4)$

40. $2(t - 3) + 4 \leq 2t - 7$

41. $3(r - 2) - 5(r - 1) \geq -2r - 1$

42. $5(x - 7) - 2(x - 1) < 3(x - 11)$

43. $-3 \leq 2x - 1 \leq 5$

44. $5 \leq 3x + 2 \leq 20$

45. $-5 \leq 6 - 2x \leq 7$

46. $4 \leq 3 - x \leq 12$

47. $-3 \leq 3(x - 4) \leq 6$

48. $-5 \leq 5(x - 3) \leq 6$

In Exercises 49–70, solve the inequality. Express your answer using interval notation and sketch a graph of its solution set if it is conditional.

49. $3[a - 3(a + 2)] > 0$

50. $2 - (b - 4) \leq 5 - b$

51. $5 - [2 - 3(2 - x)] < -3x + 1$

52. $5 - [2 - 3(x - 2)] > -3x + 1$

53. $3 - \{2 - 5[q - (2 - q)]\} \leq 3[q - (q - 5)]$

54. $2\{x - [3 - (5 - x)]\} > 2 - x$

55. $|x| < 4$

56. $|x| > 4$

57. $|s| \geq 5$

58. $|x| \leq 5$

59. $|t| < 0$

60. $|t| \leq 0$

61. $|t - 1| < 2$

62. $|t - 1| > 2$

63. $|a - 6| \geq 3$

64. $|a - 6| \leq 3$

65. $|r + 9| \leq 4$

66. $|r - 9| \geq 4$

67. $|2x - 1| \geq 2$

68. $|2x - 1| > 2$

69. $|3x - 2| < 4$

70. $|3x - 2| > 4$

In Exercises 71–76, solve the inequality.

71. $|2x - 3| \leq 5$

72. $|2x - 3| \geq 5$

73. $|3x + 5| + 2 > 7$

74. $|3x + 5| + 2 < 7$

75. $|3 - 2x| + 8 \leq 4$

76. $|3 - 2x| - 8 \leq 4$

In Exercises 77–90, solve algebraically.

77. If 3 times a number is 4 less than 4 times the number, what is the number?

78. If 5 times the sum of a number and 6 is 2 less than the number, what is the number?

79. On October 12, the U.S. dollar was worth 114.05 Japanese yen. What is the value of $800 in yen?

80. A stock loses 15% of its previous day's value on April 5. The next day, April 6, it gains 20% of its previous day's value. If the stock is worth $4,500 on April 6, what was it worth on April 4?

81. A salesperson works strictly on commission of 9% on his gross sales. Determine his weekly gross sales required to earn $650 a week.

82. Suppose that a salesperson is paid a base salary of $100 a week plus a 5% commission on his gross sales. Determine his weekly gross sales necessary to earn $600 a week.

83. Suppose you are nearing the end of a course in which the teacher counts the final exam as 40% of your course grade. Your average grade for all but the final exam is 74. Determine the minimum score needed on your final exam to get a grade of 80 for the course.

84. Suppose you are nearing the end of a course in which the teacher counts the final exam as 45% of your course grade. Your average grade for all but the final exam is 78. Determine the minimum score needed on your final exam to get a course grade of 80.

85. Thirty packages were delivered. Some were 8-lb packages and the rest were 5-lb packages. If the total weight of all the packages was 186 lb, how many of each type of package were delivered?

86. Forty-five packages were in a storeroom. Some weighed 8 lb and the rest weighed 5 lb. If the total weight of all the packages was 276 lb, how many of each type of package was in the storeroom?

87. A bedding store contracts out to a delivery service, which charges the store $10 for delivering single beds and $15 for delivering larger beds. Yesterday, the service delivered a total of 23 beds and collected $295. How many of each type of bed were delivered?

88. A plumber charges $65 an hour for his services and $40 an hour for his assistant's services. On a certain job the assistant did the preparatory work alone and then the plumber completed the job alone. The total bill for the job came to $369.50: $27 for parts and the rest for 7 hours' total labor. How many hours did each work on the job?

89. Bob averages $20 an hour editing textbooks and $25 an hour tutoring. If he is working 30 hours this week, what is the minimum number of hours he must tutor to make at least $660?

90. Refer to Exercise 89. If Bob works 40 hours, what is the maximum number of hours he can tutor if he is to make no more than $925?

91. An optical mark reader is being used to grade 8,500 grade sheets. A machine that reads 200 sheets per minute begins the job but breaks down before completing all the grade sheets. A new machine, which reads 300 sheets per minute, then completes the job. If it took a total of 30 minutes to read all the grade sheets, how long did the first machine work before it broke down?

92. A commercial Web-site company charges $875 for designing and setting up a basic Web site and $2,050 for a deluxe Web site, which incudes 2 months of technical support. During a certain month the company bills $23,950 for 22 Web sites. How many Web sites of each type were designed and set up?

93. A Web-site manager notices that on a certain day her Web site experiences 18,576 "hits." She separates the hits into three categories: Category 1: Hits that spend less than 1 minute on the Web site. Category 2: Hits that spend between 1 minute and 5 minutes on the Web site. Category 3: Hits that spend more than 5 minutes on the Web site. There were twice as many hits in category 2 as in category 3, and the total number in cataegory 1 is equal to the numbers in categories 2 and 3 together. How many hits were in each category?

Preparing for Exams: Using Quiz Cards

A few days before the exam, select an appropriate number of problems from the quiz cards, old exams, or quizzes, and make up a practice test for yourself. You may need the advice of your teacher as to the number of problems and the amount of time to allow yourself for the test. If they are available, old quizzes and exams may help guide you.

Now find a quiet, well-lit place with no distractions, set your clock for the appropriate time limit (the same as your class exam will be), and take the test. Pretend it is a real test; that is, do not leave your seat or look at your notes, books, or answers until your time is up. (Before giving yourself a test you may want to refer to the next chapter's Study Skills discussions on taking exams.)

When your time is up, stop; you may now look up the answers and grade yourself. If you are making errors, check over what you are doing wrong. Find the section where those problem types are covered, review the material, and try more problems of that type.

If you do not finish your practice test on time, you should definitely work on your speed. Remember that speed, as well as accuracy, is important on most exams.

Think about what you were doing as you took your test. You may want to change your test-taking strategy or reread the next chapter's discussion on taking exams. If you were not satisfied with your performance, and you have time after the review, give yourself another practice test.

CHAPTER 2 PRACTICE TEST

1. Linda has $5,000 to invest. She invests part in a savings account paying 3.5% simple interest, and the rest in a certificate of deposit paying 6%. Create a mathematical model that expresses the total amount of interest earned yearly in terms of the amount of money invested in the savings account, and use this model to estimate how much is in the savings account if the total annual interest from both accounts is $220.

Solve the following equations.

2. $7x - 9 = 5x + 2$

3. $3x - 4(x - 5) = 7(x - 2) - 8(x - 1)$

4. $6a - [2 - 3(a - 4)] = 5(a - 10)$

Solve the following inequalities.

5. $3x + 5 \leq 5x - 3$

6. $3x - 2[x - 3(1 - x)] > 5$

7. $-3 < 5 - 2x < 4$

Solve the following. Express answers to inequalities using interval notation and graph the solution set on the real number line.

8. $|2x - 7| + 2 = 9$

9. $|3x - 4| < 5$

10. $|5x - 7| \geq 2$

11. If the euro (European dollar) is worth $1.20 (U.S. dollars), what is the value of 500 euros in U.S. dollars?

12. A truck carries a load of 93 boxes; some weigh 35 kg and the rest weigh 45 kg. If the total weight of all the boxes is 3,465 kg, how many of each type are in the truck?

13. How long would it take a jogger traveling at 8 mph to catch up with a person walking 3 mph with a 2-hour head start?

14. Ken has 32 coins in his pocket in nickels and dimes. What is the minimum number of dimes if he has at least $2.65?

3

Graphing Straight Lines and Functions

© image 100/Alamy

In this chapter we introduce the concepts of relations and functions, which play an important role in mathematics and its applications. We will pay particular attention to drawing, understanding, and interpreting the graphs of certain functions and the relationships they represent.

Suppose that a researcher collects data in the following table, which indicates the number of calories burned, *c*, during a brisk walk that lasts *m* minutes.

m	2	4	6	8	10
c	12	23	32	40	53

Based on these data, a researcher conjectures that the number of calories burned during a brisk walk and the number of minutes the walk lasts are related by the equation $c = 5m + 2$. In Section 3.2 we will examine this equation in more detail to analyze this relationship and determine how well this equation correlates with the data obtained. We will find that having a "picture" of this equation allows us to understand this relationship better. The ideas we develop in this chapter play an important role in performing this type of analysis and obtaining such a picture, called the graph of the equation.

The Rectangular Coordinate System and Graphing Straight Lines

Let's begin by considering a first-degree equation in two variables, such as

$$2x + y = 6$$

What does it mean to have a solution to this equation? A moment's thought reveals that a single solution to such an equation consists of a *pair* of numbers. We need an *x*-value and a *y*-value to produce *one* solution to this equation. For example, the pair of numbers $x = 1$ and $y = 4$ makes the equation true:

$$2x + y = 6 \qquad \textit{Substitute } x = 1, y = 4.$$
$$2(1) + 4 \overset{?}{=} 6$$
$$6 \overset{\checkmark}{=} 6$$

The pair $x = 2$ and $y = 2$ also satisfies the equation:

$$2x + y = 6 \qquad \textit{Substitute } x = 2, y = 2.$$
$$2(2) + 2 \overset{?}{=} 6$$
$$6 \overset{\checkmark}{=} 6$$

However, the pair $x = 3$ and $y = 1$ does *not* satisfy the equation:

$$2x + y = 6 \qquad \textit{Substitute } x = 3, y = 1.$$
$$2(3) + 1 \overset{?}{=} 6$$
$$7 \neq 6$$

Thus, we see that the equation $2x + y = 6$ is neither always true nor always false. Nevertheless, this equation has infinitely many solutions. In fact, we can produce as many solutions to this equation as we like by simply picking *any* value for one of the variables and solving for the other variable. (Remember that a single solution to this equation consists of two numbers.)

For instance, to obtain a solution to $2x + y = 6$ we can choose $x = 3$ and solve for y:

$$2x + y = 6 \qquad \textit{Substitute } x = 3.$$
$$2(3) + y = 6$$
$$6 + y = 6$$
$$y = 0 \qquad \textit{Thus, } x = 3 \textit{ and } y = 0 \textit{ is a solution.}$$

Or, we can choose $y = 6$ and solve for x:

$$2x + y = 6 \qquad \textit{Substitute } y = 6.$$
$$2x + 6 = 6$$
$$2x = 0$$
$$x = 0 \qquad \textit{Thus, } x = 0 \textit{ and } y = 6 \textit{ is a solution.}$$

x	y
0	6
1	4
2	2
3	0

One way to keep track of the solutions to $2x + y = 6$ that we have found thus far is to make a table, as shown in the margin. Another way of listing the solutions is by using ***ordered pair*** notation, which records the pair of numbers $x = 1$ and $y = 4$, for example, as $(1, 4)$. That is, an ordered pair of numbers is of the form (x, y), where the first number (sometimes called the first component or ***abscissa***) is the x-value and the second number or component (sometimes called the ***ordinate***) is the y-value. These are called *ordered pairs* for obvious reasons—the order of the numbers matters. As we saw earlier, the pair $(1, 4)$ satisfies the equation $2x + y = 6$, whereas the pair $(4, 1)$ does not.

As we have already pointed out, we can produce as many solutions as we want to a first-degree equation in two variables by simply choosing a value for one of the variables and solving for the other. How then can we exhibit all the solutions if there are infinitely many of them? Since we cannot list all of them, we develop an alternative method for displaying the solution set.

You are familiar with the two-dimensional coordinate system called a ***rectangular*** or ***Cartesian coordinate system*** (named after the French mathematician and philosopher René Descartes, 1596–1650). It is obtained by taking two number lines, one horizontal and one vertical, perpendicular to each other at their respective zero points. This is illustrated in Figure 3.1.

The horizontal axis is the x-axis; the vertical axis is the y-axis. The axes intersect at the origin.

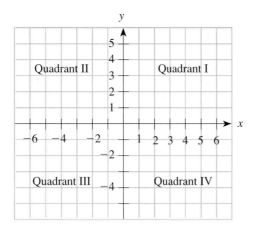

Figure 3.1
Rectangular coordinate
system

The x- and y-axes divide the plane into four parts, called ***quadrants.*** These are numbered in a conventional way, as indicated in the figure.

The x- and y-axes are referred to as the ***coordinate axes.*** The first and second members of the ordered pair are often called the x-coordinate and y-coordinate, respectively. *Note that points **on** the coordinate axes are not considered as being **in** any of the quadrants.*

This coordinate system allows us to associate a point in the plane with each ordered pair (x, y).

To plot (graph) the point associated with an ordered pair (x, y), we start at the origin and move $|x|$ units to the right if x is positive, to the left if x is negative, and then $|y|$ units up if y is positive, down if y is negative.

The point at which we arrive is the graph of the ordered pair (x, y):

$$(x, \quad y)$$

$$\uparrow \quad \uparrow$$

Tells you   *Tells you*
right/left   *up/down*

EXAMPLE 1

Plot (graph) the points with coordinates $(3, 4)$, $(-2, 1)$, $(-3, -5)$, $(2, -3)$, $(5, 0)$, and $(0, 4)$ on the rectangular coordinate system.

Solution

To graph the point $(3, 4)$ we start at the origin and move 3 units to the right and then 4 units up. (Or, we could first move 4 units up and then 3 units to the right.) In a similar manner, we plot all the points as shown in Figure 3.2.

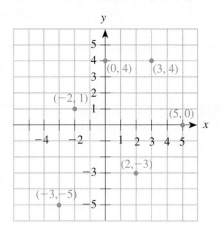

Figure 3.2
Solution for Example 1

We call particular attention to the points $(5, 0)$ and $(0, 4)$. Notice that the point $(5, 0)$ is on the x-axis because its *y-coordinate* is 0; the point $(0, 4)$ is on the y-axis because its *x-coordinate* is 0.

Just as every ordered pair is associated with a point, so too is every point associated with an ordered pair. If we want to see which ordered pair is associated with a specific point, P, we construct a perpendicular line from P to the x-axis. The point at which the perpendicular intersects the x-axis is the x-coordinate of the ordered pair associated with P. Similarly, we construct a perpendicular from P to the y-axis to find the y-coordinate of the ordered pair. Figure 3.3 illustrates this process.

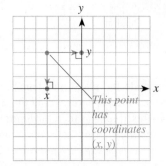

Figure 3.3
Determining the coordinates
of a point

Thus, we have a one-to-one correspondence between the points in the plane and ordered pairs of real numbers; that is, every point on the plane is identified by a unique pair of real numbers, and every pair of real numbers corresponds to a unique point on the plane. For this reason we frequently say "the point (x, y)" rather than "the ordered pair (x, y)."

With this coordinate system in hand, we can now return to the question, "How can we exhibit the solution set of the equation $2x + y = 6$?" We have seen that the pairs of numbers in the accompanying table satisfy $2x + y = 6$. If we write them in ordered pair notation, we get the ordered pairs given in the table on the left (remember that x is the first coordinate and y the second):

x	y	(x, y)
0	6	$(0, 6)$
1	4	$(1, 4)$
2	2	$(2, 2)$
3	0	$(3, 0)$

Now we can plot these points, as shown in Figure 3.4. The figure strongly suggests that we "connect the dots" and draw a straight line. In doing so, we are making two statements: first, that every ordered pair that satisfies the equation is a point on the line, and second, that every point on the line has coordinates that satisfy the equation.

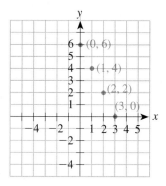

Figure 3.4

We will find the following definition useful.

DEFINITION

The *graph* of an equation is the set of all points whose coordinates satisfy the equation.

Thus, in drawing the straight line in Figure 3.5, we are saying that the graph of the equation $2x + y = 6$ is the straight line. This graph is now the "picture" of the solution set to the equation $2x + y = 6$.

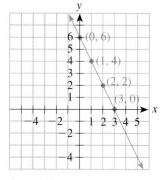

Figure 3.5
The graph of the equation
$2x + y = 6$

The following theorem, whose proof we will discuss further, generalizes this entire discussion.

THEOREM

The graph of an equation of the form $Ax + By = C$, where A and B are *not* both equal to 0, is a straight line.

It is for this reason that an equation of the form $Ax + By = C$ is often called a *linear equation*. This particular form, $Ax + By = C$, is called the *general form* of a linear equation. We postpone the proof of this theorem until Section 4.2; however, we can put this theorem to immediate use.

Basically, this theorem tells us that if we have a first-degree equation in two variables, we know that its graph is a straight line. For example, if we want to graph the equation $2x - y = 8$, the theorem tells us that the graph is going to be a straight line. Since two points determine a straight line, all we need to find are two points that satisfy the equation.

As we mentioned earlier in this section, we can find as many points—that is, generate as many solutions to this equation—as we please, by simply picking a value for one of the variables and solving for the other variable. For example, for the equation $2x - y = 8$, let's pick two values for x and then find the y-value associated with each x-value.

	Let $x = 2$:	$2x - y = 8$	**Let** $x = 5$:	$2x - y = 8$
		$2(2) - y = 8$		$2(5) - y = 8$
		$4 - y = 8$		$10 - y = 8$
		$-y = 4$		$-y = -2$
		$y = -4$		$y = 2$

Hence, two solutions are $(2, -4)$ and $(5, 2)$.

We would plot the two points and draw a line through them, as indicated in Figure 3.6. It is a good idea to check your graph by finding a third solution to the equation and checking to see that it lies on the line.

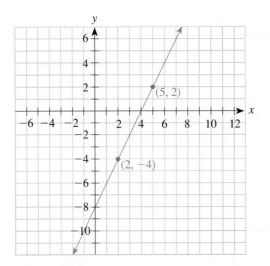

Figure 3.6
The graph of $2x - y = 8$

We see that we can graph a linear equation by arbitrarily finding two points. However, throughout our work in graphing there are certain points to which we want to pay particular attention.

DEFINITION   The *x-intercepts* of a graph are the *x*-value of the points where the graph crosses the *x*-axis. The *y-intercepts* of a graph are the *y*-values of the points where the graph crosses the *y*-axis.

If we look back at the graph of $2x - y = 8$ (Figure 3.6), we can see that the *x*-intercept is 4 and the *y*-intercept is -8.

Since the graph crosses the *x-axis* when $y = 0$ (why?),

the x-intercept of a graph occurs when $y = 0$.

Similarly, since the graph crosses the *y-axis* when $x = 0$ (why?),

the y-intercept of a graph occurs when $x = 0$.

Whenever possible (and practical), we label the *x*- and *y*-intercepts of a graph.

EXAMPLE 2

Sketch the graph of the following equation and label the intercepts.

$$-2x + 3y = 15$$

Solution

The graph of $-2x + 3y = 15$ will be a straight line. Thus, we need find only two points—the intercepts:

$-2x + 3y = 15$ *To find the x-intercept, set y = 0 and solve for x.*

$-2x + 3(0) = 15$

$-2x = 15$

$x = -\dfrac{15}{2}$ *The x-intercept is $-\dfrac{15}{2}$. Hence, the graph crosses the x-axis at $\left(-\dfrac{15}{2}, 0\right)$.*

$-2x + 3y = 15$ *To find the y-intercept, set x = 0 and solve for y.*

$-2(0) + 3y = 15$

$3y = 15$

$y = 5$ *The y-intercept is 5. Hence, the graph crosses the y-axis at (0, 5).*

Now we can sketch the graph (see Figure 3.7).

It is easier to locate $-\dfrac{15}{2}$ if we rewrite it as -7.5.

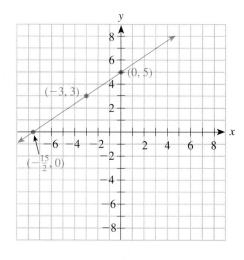

	x	**y**
x-intercept	$-\frac{15}{2}$	0
y-intercept	0	5
check	-3	3

Figure 3.7
The graph of
$-2x + 3y = 15$

We check with a third point. Letting $y = 3$, we find that $x = -3$. From the graph we can see that $(-3, 3)$ is on the line.

The method we have outlined in Example 2 is called the ***intercept method*** for graphing a straight line.

While the intercept method is the preferred method to use, there are occasions when it does not quite do the job.

EXAMPLE 3

Sketch the graph of $y = -2x$.

Solution

Again, since this is a first-degree equation in two variables, we know that the graph is going to be a straight line. We will find the intercepts.

To find the *x*-intercept, we set $y = 0$ and solve for *x*:

$$y = -2x$$

$$0 = -2x$$

$$0 = x \qquad \textit{The x-intercept is 0.}$$

But $(0, 0)$, as the origin, is on the y-axis as well, and so 0 is also the y-intercept. Since we get only one point from our search for the intercepts, we must find another point on the line.

Again, we simply choose a convenient value for x or y. We choose $x = 1$:

$$y = -2x$$
$$y = -2(1)$$
$$y = -2 \qquad \textit{Another point on the line is } (1, -2).$$

Now we can sketch the graph (see Figure 3.8). We use the point $(-1, 2)$ as a check for this graph.

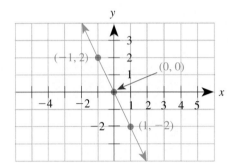

x	y
0	0
1	-2
-1	2

Figure 3.8
The graph of $y = -2x$

DIFFERENT PERSPECTIVES: Intercepts

Consider the geometric and algebraic interpretations of the x- and y-intercepts of the graphs of $3x - 2y = 6$.

Graphical Interpretation

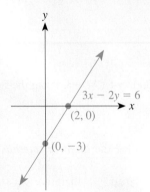

Algebraic Interpretation

To find the x-intercept, set $y = 0$ and solve for x:

$$3x - 2y = 6$$
$$3x - 2(0) = 6$$
$$3x = 6$$
$$x = 2$$

To find the y-intercept, set $x = 0$ and solve for y:

$$3x - 2y = 6$$
$$3(0) - 2y = 6$$
$$-2y = 6$$
$$y = -3$$

The line crosses the x-axis at the point $(2, 0)$. $\longleftrightarrow$ The x-intercept is 2.

The line crosses the y-axis at the point $(0, -3)$. $\longleftrightarrow$ The y-intercept is -3.

EXAMPLE 4

Sketch the graphs of:

(a) $x = -3$

(b) $y = 2$

Solution

First we must keep in mind that we are working in a two-dimensional coordinate system. Recall that we defined the standard form of a first-degree equation in two variables to be $Ax + By = C$, with A and B not both equal to 0.

(a) The equation $x = -3$ is of this form with $A = 1$, $B = 0$, and $C = -3$:

$$Ax + \quad By = C$$
$$1 \cdot x + 0 \cdot y = -3$$

x	y
−3	−2
−3	1
−3	5

Try substituting various values for x and y in the equation $1 \cdot x + 0 \cdot y = -3$ and you will find that it does not matter what value we substitute for y, the x-value must still be -3, as indicated in the accompanying table.

For an ordered pair to satisfy the equation $x = -3$, its x-coordinate must be -3. The equation $x = -3$ places no condition whatsoever on the y-coordinate—the y-coordinate can be any real number. Hence, the graph is a vertical line 3 units to the left of the y-axis, as illustrated in Figure 3.9(a).

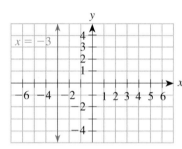

Figure 3.9a
The graph of $x = -3$

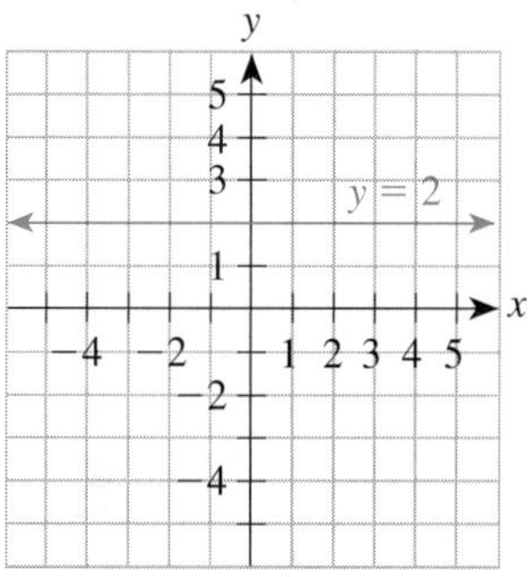

Figure 3.9b
The graph of $y = 2$

(b) Similarly, we can rewrite the equation $y = 2$ in general form with $A = 0$, $B = 1$, and $C = 2$:

$$Ax + \quad By = C$$
$$0 \cdot x + 1 \cdot y = 2$$

We can substitute numbers for x and find that no matter what value is substituted for x, the y-value will always be 2.

For an ordered pair to satisfy the equation $y = 2$, its y-coordinate must be 2, but the equation places no condition on the x-coordinate. Hence, the graph is a horizontal line 2 units above the x-axis (see Figure 3.9b).

EXAMPLE 5

A salesperson earns $320 per week plus a commission of $0.50 per item sold. During a typical week, she sells a maximum of 900 items.

(a) Write an equation expressing the salesperson's weekly income, I, in terms of n, the number of items sold.

(b) Sketch a graph of this equation.

(c) Pick a point on the line, estimate its coordinates, and explain what it tells you about the salesperson's weekly income.

Solution   (a) If the salesperson were to sell 240 items during a particular week, the income for that week would be

$$I = 320 + 240(0.50)$$

We multiply the number of items by 0.50 and add the $320.

$$= \$440$$

In general, then, the weekly income I will be

$I = 320 + 0.50n$ dollars   where n is the number of items sold

(b) This equation is a first-degree equation in two variables, so its graph is a straight line. The equation is written in terms of n and I rather than x and y, so we will assign the variable n to the horizontal axis and the variable I to the vertical axis. Hence ordered pairs will be of the form (n, I) rather than (x, y). To sketch the graph properly, we should create appropriate scales for the axes. Given the fact that the salesperson can sell from 0 to 900 items, we label the horizontal n-axis from 0 to 900. For the vertical I-axis, we note that since the maximum she sells is 900 items, the salesperson's maximum weekly income will be $I = 320 + 0.5(900) = \$770$. So the vertical axis should reach at least $770. For convenience of scaling, we will make it $800 (see Figure 3.10).

We must find two ordered pairs satisfying the equation. If $n = 0$, then $I = 320 + 0.5(0) = 320$. Hence we plot the ordered pair $(0, 320)$. (This means that if she sold no items, she still makes $320.) As we saw above, if $n = 900$, then $I = \$770$. So we may plot the ordered pair $(900, 770)$, and graph the line (see Figure 3.10).

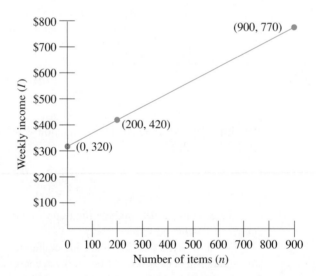

Figure 3.10

(c) This graph represents the relationship between the salesperson's weekly income and the number of items she sells. Every point on the graph gives a pair of numbers that are specific values of the weekly income for each number of items sold that week. For example, in Figure 3.10 we have chosen a point on the line and estimate its coordinates to be $(200, 420)$: This tells us that for a week during which the salesperson sells 200 items, she makes an income of $420.

DIFFERENT PERSPECTIVES: The Graph of an Equation

An equation and its graph offer two ways of looking at a relationship.

Algebraic Description

An equation such as $y = 3x - 5$ describes a relationship in which the y-value is 5 less than 3 times the x-value. Ordered pairs such as $(1, -2)$, $(0, -5)$, and $(3, 4)$ satisfy this relationship and hence are solutions to the equation.

Graphical Description

The graph of an equation gives a pictorial representation of the relationship. The following is the graph of the equation $y = 3x - 5$.

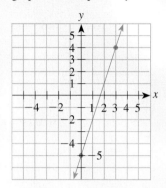

Every point (x, y) on the graph also gives a pair of numbers that satisfies the equation, and every pair of numbers that satisfies the equation corresponds to a point on the graph. We can see that the points $(1, -2)$, $(0, -5)$, and $(3, 4)$ are on the graph and by substituting we can check that they satisfy the equation $y = 3x - 5$.

Current technology can facilitate some of these graphing procedures. This becomes particularly useful for more complicated graphs. We illustrate these capabilities for the straight line $y = 2x + 6$ in the standard $[-10, 10]$ by $[-10, 10]$ window shown in Figure 3.11a. Some calculators, such as the TI-83 Plus, have a TABLE option, which allows us to examine a table of x and y-values for this equation. (See Figure 3.11b.)

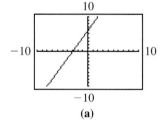

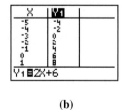

Figure 3.11
The graph (a) and a table of values (b) for the equation $y = 2x + 6$

(a)

(b)

STUDY SKILLS 3.1

Taking an Algebra Exam: Just Before the Exam

You will need to concentrate and think clearly during the exam. For this reason it is important that you get plenty of rest the night before the exam and that you have adequate nourishment.

It is *not* a good idea to study up until the last possible moment. You may find something that you missed and become anxious because there is not enough time to learn it. Then rather than simply missing a problem or

two on the exam, the anxiety may affect your performance on the entire exam. It is better to stop studying sometime before the exam and do something else. You could, however, review formulas you need to remember and warnings (common errors you want to avoid) just before the exam.

Also, be sure to give yourself plenty of time to get to the exam.

EXERCISES 3.1

In Exercises 1–8, determine whether the given ordered pair satisfies the equation.

1. $3x - 5y = 17$;   (4, 1)

2. $3x - 5y = 17$; $(-6, -7)$

3. $4y - 3x = 7$; $(1, -1)$

4. $4y - 3x = 20$; $(2, -4)$

5. $2x + 3y = 2$; $\left(\dfrac{3}{2}, -\dfrac{1}{3}\right)$

6. $5x - 4y = 0$; $\left(\dfrac{1}{10}, \dfrac{1}{8}\right)$

7. $\dfrac{2}{3}x - \dfrac{1}{4}y = 1$;   (6, 12)

8. $\dfrac{3}{4}y - \dfrac{4}{5}x = 4$;   (20, 16)

9. $y = 5$;   (2, 5)

10. $x = 4$;   (3, 7)

In Exercises 11–16, fill in the missing component of the given ordered pairs for the equations:

11. $x + y = 8$: $(-1, \ ), (0, \ ), (1, \ ), (\ , -2), (\ , 0), (\ , 4)$

12. $x - 2y = 5$: $(-3, \ ), (-1, \ ), (0, \ ), (\ , -3), (\ , -1), (\ , 0)$

13. $5x + 4y = 20$: $(-2, \ ), (0, \ ), (4, \ ), (\ , -5), (\ , 0), (\ , 4)$

14. $3x + 7y = 15$: $(-7, \ ), (0, \ ), (5, \ ), (\ , -3), (\ , 0), (\ , 2)$

15. $\dfrac{x}{3} + \dfrac{y}{4} = 1$: $(-3, \ ), (0, \ ), (3, \ ), (\ , -4), (\ , 0), (\ , 4)$

16. $\dfrac{2x}{7} - \dfrac{y}{2} = 2$: $(-7, \ ), (0, \ ), (7, \ ), (\ , -2), (\ , 0), (\ , 2)$

In Exercises 17–30, find the x- and y-intercepts of the graphs of the given equations.

17. $x + y = 6$

18. $y + x = 5$

19. $x - y = 6$

20. $y - x = 5$

21. $y - x = 6$

22. $x - y = 5$

23. $2x + 4y = 12$

24. $3x + 6y = 12$

25. $y = -\dfrac{4}{3}x + 4$

26. $y = -\dfrac{3}{2}x + 3$

27. $y = \dfrac{3}{5}x - 3$

28. $y = \dfrac{5}{3}x + 5$

29. $2y - 3x = 7$

30. $2x - 3y = 8$

In Exercises 31–58, sketch the graphs of the given equations. Label the x- and y-intercepts.

31. $4x + 3y = 0$

32. $5x - 2y = 0$

33. $y = x$

34. $y = -x$

35. $\dfrac{x}{2} - \dfrac{y}{3} = 1$

36. $\dfrac{y}{3} + \dfrac{x}{5} = 1$

37. $y = 3x - 1$

38. $y = -2x + 3$

39. $y = -\dfrac{2}{3}x + 4$

40. $y = \dfrac{3}{5}x - 6$

41. $5x - 4y = 20$

42. $3x + 6y = 18$

43. $5x - 7y = 30$

44. $3x + 8y = 15$

45. $5x + 7y = 30$

46. $3x - 8y = 16$

47. $x = 5$

48. $y = -3$

49. $-\dfrac{3}{4}x + y = 2$

50. $\dfrac{5}{3}x - y = 6$

51. $\dfrac{3}{4}x - y = 2$

52. $-\dfrac{5}{3}x + y = 6$

53. $y + 5 = 0$ **54.** $x - 7 = 0$

55. $5x - 4y = 0$ **56.** $3x + 6y = 0$

57. $5x - 4 = 0$ **58.** $3 + 6y = 0$

59. The number of physicians in the United States from 1980 to 2000 can be estimated by the equation

$$P = 16.13t - 31{,}488.2$$

where P is the number of physicians in thousands during year t. Sketch a graph of this equation using the vertical axis for P and the horizontal axis for t restricted to values between 1980 and 2000.

60. The number of dentists in the United States from 1980 to 2000 can be estimated by the equation

$$D = 2.24(t - 1980) + 122.4$$

where D is the number of dentists in thousands during year t. Sketch a graph of this equation using the vertical axis for D and the horizontal axis for t restricted to values between 1980 and 2000.

61. A salesperson earns $220 per week plus a commission of 5% of the value of all the items she sells. She can sell a maximum of $20,000 worth of items during the week.
  (a) Write an equation expressing the salesperson's weekly salary, S, in terms of v, the value of the items she sells weekly.
  (b) Sketch a graph of this equation.
  (c) Pick a point on the line, estimate its coordinates, and explain what it tells you about the salesperson's weekly income.

62. A cell phone company charges $30 a month and 21¢ a minute for each minute of airtime. Assume that the maximum amount of airtime you will use per month is 500 minutes.
  (a) Write an equation expressing the cost, C, in terms of t, the time you spend on air per month.
  (b) Sketch a graph of this equation.
  (c) Pick a point on the line, estimate its coordinates, and explain what it tells you about the monthly costs under this plan.

63. At 11:00 A.M. John is 260 miles away from home. He continues to drive away from home at an average speed of 52 miles per hour for the next 6 hours.
  (a) Write an equation that expresses his distance, d, from home in terms of the number of hours, t, he has driven since 11:00 A.M.
  (b) Sketch a graph of this equation.
  (c) Pick a point on the line, estimate its coordinates, and explain what it tells you about John's distance from home.

64. A bank charges its checking account customers a monthly fee of $8.50 plus a charge of $0.30 per check.
  (a) Write an equation expressing the total monthly charge, C, in terms of the number of checks processed, n.
  (b) Sketch a graph of this equation.
  (c) Pick a point on the line, estimate its coordinates, and explain what it tells you about the total monthly checking charge.

65. A car rental company charges $29 per day plus a mileage charge of $0.14 per mile.
  (a) Write an equation expressing the total rental charge, C, for a 1-day rental in terms of the number of miles driven, n.
  (b) Sketch a graph of this equation.

    (c) Pick a point on the line, estimate its coordinates, and explain what it tells you about the total rental charge.

66. A telephone company charges a monthly fee of $17.50 plus a charge of $0.08 per local call.
    (a) Write an equation expressing the total monthly charge, *C,* in terms of the number of local calls made, *n.*
    (b) Sketch a graph of this equation.
    (c) Pick a point on the line, estimate its coordinates, and explain what it tells you about the total monthly telephone charge.

? Questions for Thought

67. Describe the *x*- and *y*-intercepts of a graph both geometrically and algebraically.

68. Sketch the graph of $2s + 3t = 6$ on a set of coordinate axes with *t* being the horizontal axis and *s* the vertical axis.

69. Repeat Exercise 68 with the axes reversed. Does reversing the labeling of the axes affect the graph? How?

70. Although we usually use the same scale on both the horizontal and vertical axes, sometimes it is necessary (or more convenient) to use different scales for each axis. Use appropriate scales on the *x*- and *y*-axes to sketch a graph of the following:
    (a) $y = 200x + 400$ **(b)** $y = 0.04x$

3.2 Graphs and Equations

In the previous section we discussed how to graph the line $Ax + By = C$, by using the equation to generate ordered pairs, plot the corresponding points, and draw the line. In this section we will do the opposite; that is, given a graph we will have to identify ordered pairs that lie on the graph. (In the next chapter we will do more—given a linear graph, we will construct its equation.) We do not need to confine our discussion to linear graphs.

    A graph describes a relationship between variables. The graph in Figure 3.12 gives us values of *y* for various values of *x*, and values of *x* for various values of *y*. For example, we can see that the point $(-2, 3)$ is *on the graph*, hence we state that $y = 3$ when $x = -2$ (or $x = -2$ when $y = 3$).

When we want to focus on a particular point on the graph, we often exaggerate this point with a heavy dot, such as the dot on $(-2, 3)$. The arrowhead at either end of a graph indicates that the graph continues indefinitely in the direction the arrow is pointing.

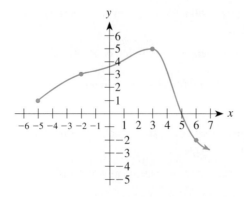

Figure 3.12

    If we are asked to fill in the missing component of the "point" $(3, \ )$ lying on the graph, we are being asked to find the *y*-coordinate on the graph whose *x*-coordinate is 3. To do this we locate $x = 3$ *on the graph,* and identify the y-coordinate of that point. If we can put a grid on the graph (Figure 3.13a), then we can easily see that on the graph, the y-coordinate is 5 when $x = 3$; or, the point $(3, 5)$ lies on the graph.

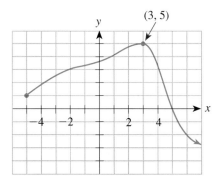

Figure 3.13a

If we cannot put a grid on the graph then we have to locate the y-value by locating the x-value 3 on the x-axis (see Figure 3.13b) and move vertically toward the graph until we meet the graph. Then move horizontally toward the y-axis until we meet it. Read off the y-value on the y-axis. This will give us the value $y = 5$.

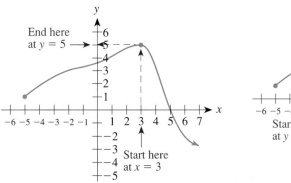

Figure 3.13b

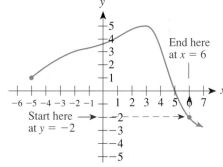

Figure 3.13c

On the other hand, if we are asked to find the x-value when $y = -2$, we would start on the y-axis (see Figure 3.13c) and move horizontally toward the graph until we meet the graph. Then move vertically toward the x-axis until we meet it. Read off the x-value on the x-axis. This will give us the value $x = 6$. Hence when $y = -2$, $x = 6$, or the ordered pair $(6, -2)$ satisfies the equation of the graph.

EXAMPLE 1

Using the graph in Figure 3.14, determine the following:

(a) Find y when $x = -3$, and 4.
(b) Find x when $y = 4$.
(c) Fill in the missing component of the "point" $(6, \ )$ lying on the graph.

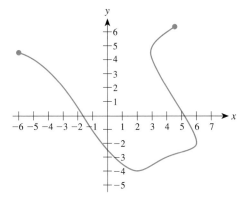

Figure 3.14

Solution   **(a)** To find y when $x = -3$, we start on the x-axis at $x = -3$, project vertically until we meet the graph, then move horizontally until we meet the y-axis, and read off the value on the y-axis: $y = 2$. See Figure 3.15a. Hence $y = 2$ when $x = -3$.

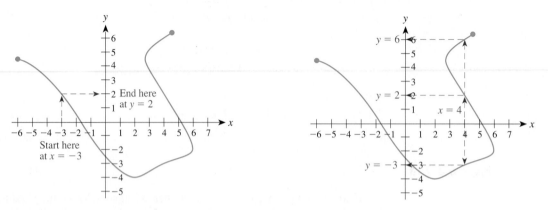

Figure 3.15a   **Figure 3.15b**

To find y when $x = 4$, we start on the x-axis at $x = 4$ and project vertically until we meet the graph. In this case, however, note that we meet the graph in three places. This means that there are three points on the graph where the x-coordinate has the value 4. For each point on the graph, we move horizontally until we meet the y-axis, and read off the values on the y-axis. We get three answers: $y = -3$, $y = 2$, and $y = 6$. See Figure 3.15b. Hence when $x = 4$, $y = -3$, 2, and 6.

(b) To find x when $y = 4$, we start on the y-axis at $y = 4$ and project horizontally until we meet the graph. In this case, however, note that we meet the graph in two places. This means that there are two points on the graph where the y-coordinate has the value 4. For each point on the graph, we move vertically until we meet the x-axis, and read off the values on the x-axis. We get two answers: $x = -5$ and $x = 3$. See Figure 3.16 in red. Hence when $y = 4$, $x = -5$ and 3.

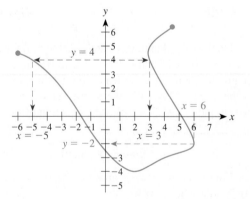

Figure 3.16

(c) Filling in the missing component of $(6, \quad)$ means we are looking for the value of y when $x = 6$. To find y when $x = 6$, we start on the x-axis at $x = 6$, project vertically until we meet the graph, then move horizontally until we meet the y-axis, and read off the value on the y-axis: $y = -2$. See Figure 3.16 in green. Hence $(6, -2)$ lies on the graph.

If the graph can also be expressed as an equation, then, as we have seen in Section 3.1, finding ordered pairs of points on the graph is equivalent to finding ordered pairs that satisfy the equation of the graph. Given a graph and its equation, the following are equivalent ways of expressing the same idea:

"$y = 5$ when $x = 3$," or

"The point $(3, 5)$ lies on the graph," or

"The ordered pair $(3, 5)$ satisfies the equation of the graph."

EXAMPLE 2

Figure 3.17 shows the graph of the equation $y = x^2 - 9$.

(a) Find y for $x = 2$ by using the graph, and verify the value using the equation of the graph.

(b) Identify the x- and y-intercepts of the graph, and verify the intercepts using the equation of the graph.

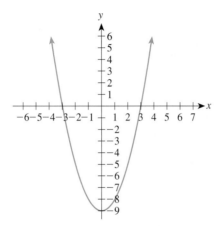

Figure 3.17
The graph of $y = x^2 - 9$

Solution

(a) To find y when $x = 2$ by using the graph, we start on the x-axis at $x = 2$, project vertically until we meet the graph. Next, move horizontally until we meet the y-axis, and read off the value on the y-axis. We get $y = -5$ (see Figure 3.18).

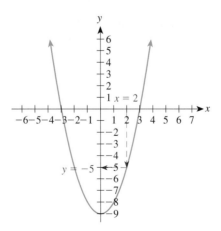

Figure 3.18

Hence, the point $(2, -5)$ lies on the graph. Since we are told that the equation $y = x^2 - 9$ is the equation of the graph, we can verify that the point $(2, -5)$ lies on the graph by checking that the equation is satisfied by the ordered pair $(2, -5)$.

$$y = x^2 - 9 \qquad \text{\textit{Substitute }} x = 2 \text{ \textit{and }} y = -5 \text{ \textit{in the equation.}}$$

$$-5 \overset{?}{=} 2^2 - 9$$

$$-5 \overset{\checkmark}{=} 4 - 9 \qquad \text{\textit{Hence the point }} (2, -5) \text{ \textit{does lie on the graph.}}$$

(b) The y-intercept is the value of y where the graph crosses the y-axis. We can see that the graph crosses the y-axis when $y = -9$. Hence the y-intercept is -9.

The graph crosses the y-axis when $x = 0$; the point $(0, -9)$ lies on the graph. We verify that $(0, -9)$ is the y-intercept:

$$y = x^2 - 9 \qquad \textit{Substitute } x = 0 \textit{ and } y = -9 \textit{ in the equation.}$$
$$-9 \overset{?}{=} -0^2 - 9$$
$$-9 \overset{\checkmark}{=} -9 \qquad \textit{Hence the point } (0, -9) \textit{ lies on the graph.}$$

An x-intercept is a value of x where the graph crosses the x-axis. We can see that the graph crosses the x-axis twice: when $x = -3$ and when $x = 3$. Hence the x-intercepts are -3 and 3.

The graph crosses the x-axis when $y = 0$; the points $(-3, 0)$ and $(3, 0)$ lie on the graph. We verify that $(-3, 0)$ is an x-intercept:

$$y = x^2 - 9 \qquad \textit{Substitute } x = -3 \textit{ and } y = 0 \textit{ in the equation.}$$
$$0 \overset{?}{=} (-3)^2 - 9$$
$$0 \overset{\checkmark}{=} 9 - 9 \qquad \textit{Hence the point } (-3, 0) \textit{ lies on the graph.}$$

We leave it to the student to verify algebraically that $(3, 0)$ is also an x-intercept.

EXAMPLE 3

Consider the graph in Figure 3.19.

(a) Determine the smallest and largest x-coordinate of any point on the graph.
(b) Determine the smallest and largest y-coordinate of any point on the graph.

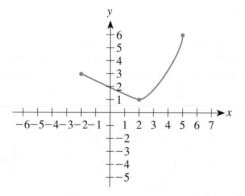

Figure 3.19

Solution

To find the *smallest* x-coordinate on this graph, we are looking for the x-coordinate of the *leftmost* point on the graph. We can see that the value of x at the leftmost point occurs at the point $(-2, 3)$, when $x = -2$. Hence the smallest x-coordinate is -2. The *largest* x-coordinate on this graph is the x-coordinate of the *rightmost* point on the graph. This occurs at the point $(5, 6)$. Therefore, the largest x-coordinate is 5.

To find the *smallest* y-coordinate on this graph, we are looking for the y-coordinate of the *lowest* point on the graph. We can see that the value of y at the lowest point occurs at the point $(2, 1)$, when $y = 1$. Hence the smallest y-coordinate on the graph is 1. The *largest* y-coordinate on this graph is the y-coordinate of the *highest* point on the graph. This occurs at the point $(5, 6)$. Therefore, the largest y-coordinate is 6.

EXAMPLE 4

A bank charges its checking account customers a monthly fee of $7.50 plus a charge of $0.30 per check.

(a) Write an equation expressing the total monthly charge, C, in terms of the number of checks processed, n.
(b) Sketch a graph of this equation.
(c) Use the graph to estimate the total monthly charge if 40 checks are processed.
(d) Use the graph to determine how many checks were processed if the total monthly charge was $15.00.

Solution

(a) If you were to write 20 checks, the total monthly charge would be

$$C = 7.50 + 20(0.30) \qquad \textit{We multiply the number of checks by **0.30** and add **\$7.50**.}$$
$$= \$13.50$$

We can see that for n checks, the total monthly charge C would be

$$\boxed{C = 7.50 + 0.30n}$$

(b) The equation is written in terms of n and C rather than x and y, so we will assign the variable n to the horizontal axis and the variable C to the vertical axis. Hence ordered pairs will be of the form (n, C) rather than (x, y). We scale the horizontal axis n from 0 to 50, and the vertical C-axis from 0 to 25.

We find two ordered pairs satisfying the equation: If $n = 0$, then we have $C = 7.5 + 0.3(0) = 7.5$. Hence we plot the point $(0, 7.5)$. (This means that if no checks were processed, the charge is $7.50.) If the number of checks is $n = 50$, then $C = 7.50 + 0.30(50) = 22.50$. So we may plot the point $(50, 22.5)$ and draw the line (see Figure 3.20a).

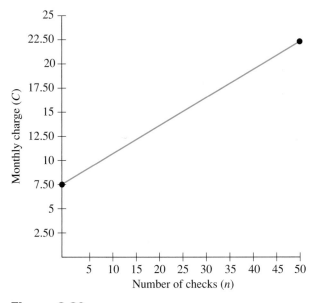

Figure 3.20a

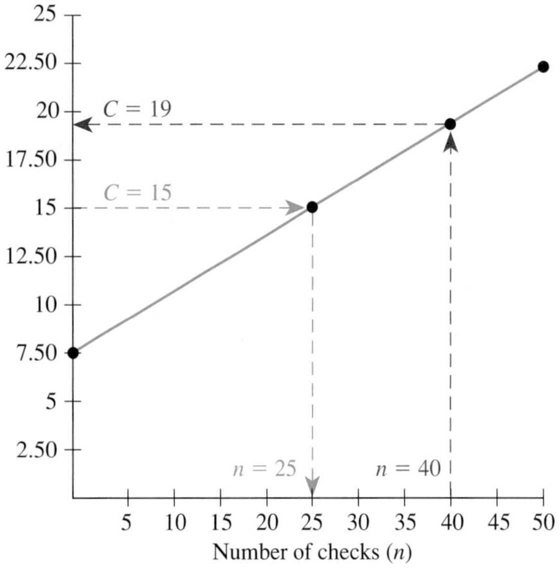

Figure 3.20b

Find C when $n = 40$ using the equation $C = 7.50 + 0.30n$.

(c) Finding the monthly charge for 40 checks means we are looking for C when $n = 40$. We start on the horizontal axis at $n = 40$ and project vertically until we meet the graph. Then we move horizontally until we meet the vertical axis, and read off the value on the C-axis to get approximately \$19.00. See Figure 3.20b. Hence the total monthly cost for processing 40 checks is about \$19.00.

Find n when C = 15 using the same equation.

(d) Finding the number of checks processed given a monthly charge of \$15.00 means we are looking for n when $C = 15.00$. We start on the vertical axis at $C = 15$ and project horizontally until we meet the graph. Then we move vertically until we meet the horizontal axis, and read off the value on the n-axis to get (approximately) 25. See Figure 3.20b. Hence the number of checks processed for a total monthly cost of \$15.00 is about 25.

Let's now return to the question posed at the beginning of this chapter.

EXAMPLE 5

Suppose that a researcher collects data in the following table, which indicates the number of calories burned, c, during a brisk walk that lasts m minutes.

m	2	4	6	8	10
c	12	23	32	40	53

Based on these data, a researcher conjectures that the number of calories c burned during a brisk walk and the number of minutes are related by the equation $c = 5m + 2$.

(a) Graph the data given in the chart using m as the horizontal axis and c as the vertical axis, and graph the equation on the same set of coordinate axes.

(b) Determine how well the proposed equation agrees with the observed data.

Solution

We recognize that $c = 5m + 2$ is a linear equation and therefore its graph is a straight line. To graph the equation $c = 5m + 2$, we see that if $m = 0$, then $c = 5(0) + 2 = 2$. Hence we plot the ordered pair $(0, 2)$. If $m = 10$, then $c = 5(10) + 2 = 52$. So we may plot the ordered pair $(10, 52)$ and graph the line (see Figure 3.21). The graph of the data appears in Figure 3.21 as well.

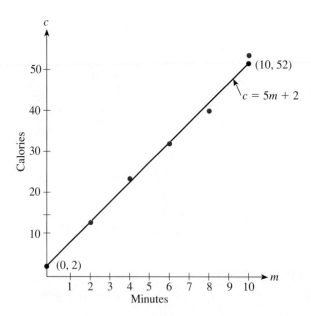

Figure 3.21

The data points do seem to "come close" to the line. We reproduced the table of data on the next page and added a row of c values predicted by the equation

$c = 5m + 2$. The following table illustrates numerically how close the equation pedicts the recorded actual number of calories, c.

m	2	4	6	8	10
Actual c	12	23	32	40	53
$c = 5m + 2$	12	22	32	42	52

STUDY SKILLS 3.2

Taking an Algebra Exam: Beginning the Exam

At the exam, make sure that you listen carefully to the instructions given by your instructor or the proctor.

As soon as you are allowed to begin, jot down the formulas you think you might need, and write some key words (warnings) to remind you to avoid common errors or errors you have made previously. Writing down the formulas first will relieve you of the burden of worrying about whether you will remember them when you need to, thus allowing you to concentrate more.

You should refer back to the relevant warnings as you go through the exam to make sure you avoid these errors.

Remember to read the directions carefully.

EXERCISES 3.2

1. Using the graph below, determine the following:

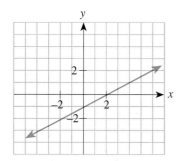

    (a) Find y when $x = -4$ and 4.
    (b) Find x when $y = 2$.

2. Using the graph below, determine the following:

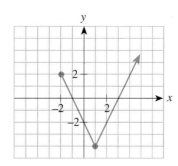

    (a) Find y when $x = 3$ and -2.
    (b) Find x when $y = -4$.

3. Using the graph below, determine the following:

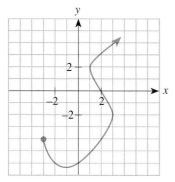

    (a) Find y when $x = -3$ and 2.
    (b) Find x when $y = 2$ and -6.

4. Using the graph below, determine the following:

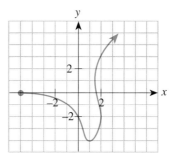

    (a) Find y when $x = -5$ and 2.
    (b) Find x when $y = -2$ and 3.

Exercises 5–8 refer to the figure given below.

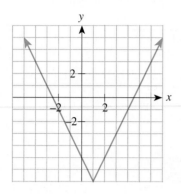

5. Find the value of y that satisfies the equation of the graph when $x = -2$.

6. Find the value of x that satisfies the equation of the graph when $y = 3$.

7. Find the value of x that satisfies the equation of the graph when $y = -3$.

8. Find the value of y that satisfies the equation of the graph when $x = 2$.

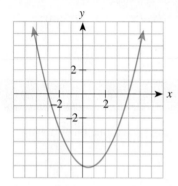

9. For the figure above, fill in the missing coordinates for the points that lie on the graph.

    (a) $(3, \ )$ **(b)** $(-4, \ )$

    (c) $(\ , -3)$ **(d)** $(\ , 4)$

10. For the figure above, fill in the missing coordinates for the ordered pairs that satisfy the equation of the graph.

    (a) $(\ , -5)$ **(b)** $(\ , 0)$

    (c) $(0, \ )$ **(d)** $(1, \ )$

11. The following is the graph of the equation $x^2 + y^2 = 13$.

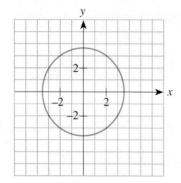

    (a) Find y for $x = 2$ by using the graph, and verify the value(s) using the equation of the graph.

    (b) Find x for $y = -3$ by using the graph, and verify the value(s) using the equation of the graph.

12. The following is the graph of the equation $16x^2 + 4y^2 = 16$.

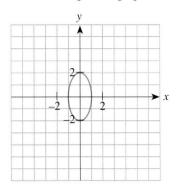

(a) Find y for $x = 0$ by using the graph, and verify the value(s) using the equation of the graph.

(b) Find x for $y = 0$ by using the graph, and verify the value(s) using the equation of the graph.

13. The following is the graph of the equation $x + 2y = 8$.

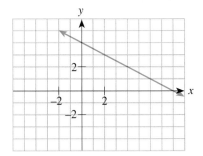

(a) Find the y-intercept using the graph, and verify your answers using the equation of the graph.

(b) Find the x-intercept using the graph, and verify your answer using the equation of the graph.

14. The following is the graph of the equation $4y^2 - 4x = 16$.

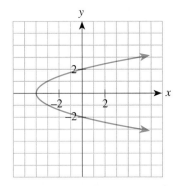

(a) Find the y-intercepts using the graph, and verify your answer using the equation of the graph.

(b) Find the x-intercept using the graph, and verify your answer using the equation of the graph.

15. The following is the graph of the equation $y = \sqrt{x} + 4$.

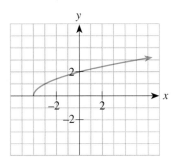

(a) Find the y-intercept using the graph, and verify your answer using the equation of the graph.

(b) Find the x-intercept using the graph, and verify your answer using the equation of the graph.

(c) Determine the smallest and largest x-coordinate of any point on the graph.

(d) Determine the smallest and largest y-coordinate of any point on the graph.

16. The following is the graph of the equation $y = |x - 1|$.

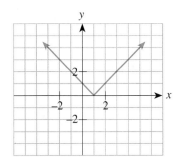

(a) Find the y-intercept using the graph, and verify your answer using the equation of the graph.

(b) Find the x-intercept using the graph, and verify your answer using the equation of the graph.

(c) Determine the smallest and largest x-coordinate of any point on the graph.

(d) Determine the smallest and largest y-coordinate of any point on the graph.

17. Given the accompanying graph.

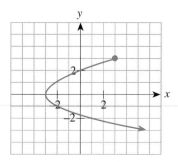

    (a) Determine the smallest and largest x-coordinate of any point on the graph.

    (b) Determine the smallest and largest y-coordinate of any point on the graph.

18. Given the accompanying graph.

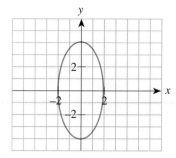

    (a) Determine the smallest and largest x-coordinate of any point on the graph.

    (b) Determine the smallest and largest y-coordinate of any point on the graph.

19. The following table gives the cost of tuition and fees for public four-year colleges in the United States during the years 1977 to 2004.

Year	1977	1981	1985	1989	1993	1997	2001	2004
Cost	617	804	1,228	1,578	2,334	2,975	3,487	4,694

Based on these data, a statistician suggests that the following formula can be used to approximate the cost C of tuition and fees for public four-year colleges for year t given in the table:

$$C = 145t - 286,400$$

    (a) Graph the data given in the chart using the horizontal axis for t and the vertical axis for C, and graph the given equation on the same set of coordinate axes.

    (b) Determine how well the proposed equation agrees with the observed data.

20. The following table gives the cost of tuition and fees (in dollars) for public two-year colleges in the United States during the years 1977 to 2004.

Year	1977	1981	1985	1989	1993	1997	2001	2004
Cost	283	391	584	799	1,116	1,465	1,642	1,905

Based on these data, a statistician suggests that the following formula can be used to approximate the cost of tuition and fees for public two-year colleges, C, for year t given in the table:

$$C = 62.4t - 123,250$$

    (a) Graph the data given in the chart using the horizontal axis for t and the vertical axis for C, and graph the given equation on the same set of coordinate axes.

    (b) Determine how well the proposed equation agrees with the observed data.

21. The table below contains the life expectancy at birth for men and women in the United States from 1950 to 2001:

Year	1950	1960	1970	1980	1985	1990	1995	1998	2001
Life Expectancy	68.2	69.7	70.8	73.7	74.7	75.4	75.8	76.7	77.2

The life expectancy at birth from 1950 to 2001 can be estimated by the equation

$$L = 0.182t + 67.9$$

L is the life expectancy during year t, where $t = 0$ corresponds to 1950, $t = 10$ corresponds to 1960, etc. Sketch a graph of this equation using the vertical axis for L and the horizontal axis for t, restricted for values between 1950 and 2001 (National Center for Health Statistics, 2003).

22. The table below contains the death rates (per 100,000 resident population) for motor-vehicle-related injuries for men and women in the United States from 1970 to 2000:

Year	1970	1980	1990	1995	2000
Life Expectancy	27.6	22.3	18.5	16.3	15.4

The death rates per 100,000 resident population for motor-vehicle-related injuries from 1970 to 2000 can be estimated by the equation

$$D = -0.4t + 27$$

D is the death rate during year t, where $t = 0$ corresponds to 1970, $t = 10$ corresponds to 1980, etc. Sketch a graph of this equation using the vertical axis for D and the horizontal axis for t, restricted for values between 1970 and 2000 (National Center for Health Statistics, 2003).

23. A salesperson is paid a commission salary of 10% of her gross sales. Her weekly gross sales rarely exceed $10,000.

   (a) Write an equation expressing her weekly commission, C, in terms of her gross sales, g.

   (b) Sketch a graph of this equation.

   (c) Use the graph to estimate her commission for the week in which her gross sales were $4,000.

   (d) Use the graph to determine her gross sales for the week in which she makes $850 in commissions.

24. A salesperson is paid a base salary of $100 plus a commission of 8% of her gross sales.

   (a) Write an equation expressing her weekly income, I, in terms of her gross sales, g.

   (b) Sketch a graph of this equation.

   (c) Use the graph to estimate her income for the week in which her gross sales were $4,000.

   (d) Use the graph to determine her gross sales for the week in which her income is $850.

25. A bank charges its checking account customers a monthly fee of $10.00 plus a charge of $0.25 per check.

   (a) Write an equation expressing the total monthly charge, C, in terms of the number of checks processed, n.

   (b) Sketch a graph of this equation.

   (c) Use the graph to estimate the total monthly charge if 50 checks are processed.

   (d) Use the graph to determine how many checks were processed if the total monthly charge was $15.00.

26. Suppose you are given the choice of two salary plans: a strictly commission salary or a base salary plus commission. The strictly commission salary pays 5% of your gross sales, while the base salary plus commission pays you a weekly base salary of $120 and a commission of 3% of your gross sales.

(a) Write an equation *for each plan* expressing the weekly income, *I*, in terms of weekly gross sales, *g*.

(b) Sketch a graph of each equation on the same set of coordinate axes.

(c) Use the graph to determine under what conditions one plan is better than the other.

Relations and Functions: Basic Concepts

The concept of a function is one of the most important in mathematics. We often hear or read statements such as "Insurance rates are a function of a person's age" or "Crop yields are a function of the weather." We understand these statements to mean that one of the quantities is dependent on the other.

We often find that two things are related to each other by some type of rule or correspondence. For example:

To each person there corresponds a Social Security number.

The distance *d* you travel at an average speed of 50 miles per hour is related to *h*, the number of hours you travel, according to the equation $d = 50h$.

To each positive number there correspond two square roots.

Not all correspondences can be expressed by some simple mathematical formula. Sometimes a graph can be a very efficient way to describe a relationship between two quantities. In the remainder of this chapter we develop two ideas: We make precise the idea of one quantity being dependent on another, and demonstrate the various ways we can express specific types of relationships.

DEFINITION

A *relation* is a correspondence between two sets in which to each element of the first set there is associated or assigned one or more elements of the second set. The first set is called the *domain* and the second set is called the *range.*

A relation is simply a rule by which we decide how to match up or associate elements from one set with elements from another. For example, students are usually assigned an identification number, as illustrated below:

Names of students (*Domain*)		*Student numbers* (*Range*)	
John Jones	$\longrightarrow$	#17345	
Sam Klass	$\longrightarrow$	#65734	*The arrow leads from the domain to the range.*
Carol Kane	$\longrightarrow$	#75664	

The relation shown above describes the assignment of numbers to names. The domain is the set of names, {John Jones, Sam Klass, Carol Kane}, and the range is the set of numbers assigned to the three names, {17345, 65734, 75664}. *The relation is the assignment.*

A relation can have many elements of the domain assigned to many elements of the range. For example, we can describe the following relation, by which we associate with each person his or her telephone number(s):

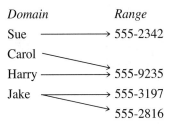

Note that Jake has two telephone numbers, while the number 555-9235 is associated with two people, Carol and Harry.

Let's consider another example. Let the domain be the set $A = \{a, b, c\}$ and the range be the set $B = \{1, 2, 3\}$. Four of the many possible relations between the two sets are shown below.

Relation R	*Relation S*

Domain   *Range*   *Domain*   *Range*

$a \longrightarrow 1$ $a \longrightarrow 2$

$b \longrightarrow 2$ $b \longrightarrow 3$

$c \longrightarrow 3$ $c \longrightarrow 1$

Relation T	*Relation U*

Domain   *Range*   *Domain*   *Range*

$a \longrightarrow 1$ $a \longrightarrow 1$

$b \quad\quad 2$ $b \quad\quad 2$

$c \longrightarrow 3$ $c \longrightarrow 3$

In relation *R,* a is assigned 1, b is assigned 2, and c is assigned 3.

In relation *S,* a is assigned 2, b is assigned 3, and c is assigned 1.

In relation *T,* a is assigned 1, b is assigned 1, c is assigned 2, and c is also assigned 3.

In relation *U,* a is assigned 1, a is also assigned 2, b is assigned 3, and c is assigned 3.

Thus, we can describe any type of correspondence we want as long as we assign to each element of the domain an element or elements of the range.

An alternative way to describe a relation is to use ordered pairs. For example, instead of writing the correspondence of relation *R:*

$$a \longrightarrow 1$$
$$b \longrightarrow 2$$
$$c \longrightarrow 3$$

we can write the correspondence without arrows as $(a, 1)$, $(b, 2)$, $(c, 3)$. Thus, the set $\{(a, 1), (b, 2), (c, 3)\}$ describes the relation *R* between the set *A* and the set *B*.

When using ordered pair notation, *we will always assume that the first member of the ordered pair is an element of the domain and the second member is the associated element of the range.* Thus, the first component, *x,* of the ordered pair (x, y) is an element of the domain. The second component, *y,* is an element of the range such that *y* is assigned to *x.*

We usually call the variable representing possible values of the domain the **independent variable** and the variable representing possible values of the range the **dependent variable.**

We can rewrite the other previous examples as follows:

Relation *S:* $\{(a, 2), (b, 3), (c, 1)\}$ *Note that this is all you need to describe the relation.*

Relation *T:* $\{(a, 1), (b, 1), (c, 2), (c, 3)\}$

Relation *U:* $\{(a, 1), (a, 2), (b, 3), (c, 3)\}$

We define a relation using ordered pair notation as follows:

ALTERNATE DEFINITION

A *relation* is a set of ordered pairs (x, y). The set of x-values is called the ***domain*** and the set of y-values is called the ***range.***

If you haven't done so already, you may want to review set notation in Appendix A.

For example,

The set of ordered pairs $R = \{(8, 2), (6, -3), (5, 7), (5, -3)\}$ is the relation between the sets $\{5, 6, 8\}$ and $\{2, -3, 7\}$ where $\{5, 6, 8\}$ is the domain and $\{2, -3, 7\}$ is the range; 8 is assigned 2, 6 is assigned -3, 5 is assigned 7, and 5 is also assigned -3.

Another example:

$S = \{(a, b), (a, c), (b, c), (c, a)\}$ is a set of ordered pairs describing the relation between set $\{a, b, c\}$ and itself. Thus, $\{a, b, c\}$ is both the domain and the range: a is assigned b and c, b is assigned c, and c is assigned a.

If the domain and range are infinite, we cannot write out each element assignment, so instead we can use set-builder notation to describe the relation(ship) between the variables. For example,

$$S = \{(x, y) \mid y = x + 5, x \text{ is a real number}\}$$

is a relation between the set of real numbers and itself (both the domain and range are the set of real numbers). Each value of x is assigned a value, y, that is 5 more than x. Thus,

3 is assigned value 8 because $8 = 3 + 5$ *or* $(3, 8) \in S$.

9 is assigned value 14 because $14 = 9 + 5$ *or* $(9, 14) \in S$.

-10 is assigned value -5 because $-5 = -10 + 5$ *or* $(-10, -5) \in S$.

We can drop the set notation and simply write $y = x + 5$ to represent the relation

$$\{(x, y) \mid y = x + 5, \quad x \text{ is a real number}\}$$

When we write a relation as an equation in x and y, we assume that x represents the independent variable and y represents the dependent variable. *We also assume that the domain is the set of real numbers that will yield real-number values for y.* (The range is often more difficult to identify.)

For example, suppose a relation is defined by

$$y = \frac{1}{x}$$

Then x can be any real number except 0 because 0 will produce an undefined y-value. Therefore, the domain (allowable values of x) is the set of all real numbers except 0. We can also write the domain as $\{x \mid x \neq 0\}$.

EXAMPLE 1

Find the domains of the relations defined by each of the following:

(a) $y = \dfrac{3x}{2x + 1}$ **(b)** $y = -2x + 7$

Solution

(a) We are looking for the domain of the relation defined by the given equation, that is, the set of allowable values of x (values of x that will yield real-number values of y).

By our previous experiences with rational expressions, we know that a sum, difference, product, or quotient of real numbers is a real number *except* when the divisor of a quotient is 0 (in which case the quotient is undefined). Therefore, the denominator, $2x + 1$, cannot be 0, or x cannot be $-\frac{1}{2}$.

Thus, the domain is $\boxed{\left\{ x \mid x \neq -\frac{1}{2} \right\}}$.

(b) $y = -2x + 7$. Any real value substituted for x will produce a real value for y.

Therefore, the domain is $\boxed{\text{all real numbers}}$.

Functions

Let's consider a real-life relation described by the telephone keypad illustrated in Figure 3.22.

Figure 3.22
A typical telephone keypad

It is very common today to see a business advertise its phone number by using something like

"To get an over-the-phone price quotation on a new car, dial BUY CARS."

Anyone wanting to call this number can easily decode BUY CARS as the number 289-2277. This is because the telephone keypad sets up a correspondence between the set of letters of the alphabet and the set of numbers $\{2, 3, 4, 5, 6, 7, 8, 9\}$. The correspondence between the two sets is that indicated by the telephone keypad—that is, the letters A, B, and C correspond to the number 2; the letters D, E, and F correspond to the number 3; and so forth.

What is important to note is that although the telephone keypad sets up a correspondence between letters and numbers, as indicated in the figure, it can be viewed two ways. There is a qualitative difference between the letter → number correspondence and the reverse number → letter correspondence: Each letter corresponds to exactly one number, but each number corresponds to 3 letters. Consequently, a phone number such as 438-2253 cannot be *uniquely* encoded into words. The number 438-2253 can be interpreted as GET CAKE or IF U BAKE.

$$
\begin{array}{ccccccc}
4 & 3 & 8 & 2 & 2 & 5 & 3 \\
\downarrow & \downarrow & \downarrow & \downarrow & \downarrow & \downarrow & \downarrow \\
G & E & T & C & A & K & E
\end{array}
\quad \text{or} \quad
\begin{array}{ccccccc}
4 & 3 & 8 & 2 & 2 & 5 & 3 \\
\downarrow & \downarrow & \downarrow & \downarrow & \downarrow & \downarrow & \downarrow \\
I & F & U & B & A & K & E
\end{array}
$$

Mathematically it is important for us to distinguish among relations that assign a *unique* range element to each domain element (such as the letter → number correspondence on the telephone) and those that do not.

DEFINITION   A *function* is a correspondence between two sets such that to each element of the first set (the domain) there is assigned *exactly* one element of the second set (the range).

Thus, for a relation to be a function, we cannot assign more than one element of the range to an element of the domain. However, in a function we can still assign an element of the range to more than one element of the domain.

For example, the relation

$$3 \longrightarrow 6$$
$$2 \longrightarrow 5$$
$$7 \longrightarrow 4$$

which is the set $\{(3, 6), (2, 5), (7, 4)\}$, *is* a function because each element in the domain $\{3, 2, 7\}$ is assigned only one element in the range $\{6, 5, 4\}$.

Another example is the relation

$$3 \longrightarrow 6$$
$$2 \nearrow$$
$$7 \longrightarrow 4$$

which is the set $\{(3, 6), (2, 6), (7, 4)\}$. Even though the range element 6 corresponds to two elements of the domain, 3 and 2, this relation *is still* a function because *each element in the domain, $\{3, 2, 7\}$, is associated with one element of the range, $\{4, 6\}$*: 3 is assigned only one value, 6; 2 is assigned only one value, 6; and 7 is assigned only one value, 4.

On the other hand, the relation

$$3 \longrightarrow 6$$
$$7 \longrightarrow 4$$

which is the set $\{(3, 6), (3, 4), (7, 4)\}$, is *not* a function, because the domain element 3 is assigned two elements of the range, 6 and 4.

The following is a useful metaphor: We can describe a function as a machine into which you throw elements of the domain, and out of which come elements of the range (see Figure 3.23).

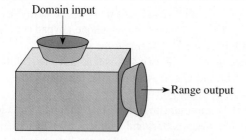

Domain input

Range output

Figure 3.23
Function machine

Of the relations R, S, T, and U given on p. 145, which are functions?

Consider the function $\{(3, 5), (2, 5), (7, 4)\}$. Note that if we put 3 into the machine, out comes 5; if we put 2 into the machine, out comes 5; and if we put 7 into the machine, out comes 4.

If we tried to use the function machine for the relation $\{(3, 5), (3, 4), (2, 6)\}$, what would happen when we throw in 3? If we throw in 3, we cannot be sure whether we will get 5 or 4. This ambiguity is precisely why this relation is *not* a function.

Remember

All functions are relations, but not all relations are functions. Functions are a special kind of relation such that to every element in the *domain* there is assigned *exactly one* element in the *range*.

EXAMPLE 2

Indicate the domain of the following relations and determine whether each relation is a function.

(a) $\{(3, 5), (4, 2), (3, 6), (5, 7)\}$ **(b)** $\{(7, -2), (6, -4), (5, 8), (-4, 8)\}$

Solution

(a) The domain is $\{3, 4, 5\}$.

The relation is $\boxed{\text{not a function}}$ since 3 is assigned more than one element of the range (3 is assigned both 5 and 6).

(b) The domain is $\{-4, 5, 6, 7\}$.

The relation $\boxed{\text{is a function}}$ since each element in the domain is assigned no more than one element of the range.

We can now give an alternative definition of a function.

DEFINITION

A *function* is a relation in which no two distinct ordered pairs have the same first coordinate.

A function can also be defined by an equation. Unless the contrary is stated, x will always represent values from the domain, and y will always represent values in the range.

EXAMPLE 3

Explain why the equation $y = 3x - 5$ defines y *as a function of x*.

Solution

As indicated previously, a function requires that exactly one range value corresponds to each domain value. In other words, there corresponds exactly one y-value to each x-value. The given equation $y = 3x - 5$ tells us that to obtain the range value y, we take the domain value x, multiply it by 3, and then subtract 5. For each x-value the equation produces a unique y-value, and so this equation does define y as a function of x.

EXAMPLE 4

Does the equation $y^2 = 3x - 5$ define y as a function of x?

Solution

If we choose $x = 7$, we get the equation

$$y^2 = 3x - 5 \qquad \textit{Substitute } x = 7.$$
$$y^2 = 3(7) - 5$$
$$y^2 = 16 \qquad \textit{We recognize that this equation has \textbf{two} solutions.}$$
$$y = \pm 4$$

The fact that there are two y-values that correspond to $x = 7$ means that the relation $y^2 = 3x - 5$ is not a function. In fact, if we choose *any* number in its domain other than $x = \frac{5}{3}$, there will be two y-values associated with it. However, to demonstrate that this equation does not define y as a function of x, all we need find is a single x-value that has two y-values associated with it.

EXAMPLE 5

Does the equation $y = 3x^2 - 5$ define y as a function of x?

Solution

Let's examine this function by making a table for various values of x:

x	-3	-2	-1	0	1	2	3
y	22	7	-2	-5	-2	7	22

From this table, we can see y-values that are associated with two different x-values. For example, we can see that both $x = -3$ and $x = 3$ give a y-value of 22. The fact that two different x-values (in the domain) give the same y-value (in the range) is *not* a problem as far as the equation defining y as a function of x is concerned. A function merely requires that to each x-value there corresponds one and only one y-value. In this equation we take the x-value, square it, multiply that result by 3, and then subtract 5. For each x-value (input) there is only one y-value (output), and so this equation *does* define y as a function of x.

EXAMPLE 6

Find the domain of the function $y = \sqrt{2x - 6}$.

Solution

Let's analyze this question carefully, step by step, to develop a strategy for its solution.

THINKING OUT LOUD

What do we need to find?

The domain of the function $y = \sqrt{2x - 6}$

What do we mean by the domain of a function?

According to our ground rules, the domain of a function is the set of real numbers x for which the corresponding y-value is also a real number.

Are there any values of x that must be excluded?

Since the square root of a negative number is undefined in the real number system, we must *exclude* any value of x for which $2x - 6 < 0$. Equivalently, the domain *includes* all values of x for which $2x - 6 \geq 0$.

How can I restate the problem in simpler terms?

Solve the inequality $2x - 6 \geq 0$.

If we substitute $x = 1$ in the expression $\sqrt{2x - 6}$, we get $\sqrt{2(1) - 6} = \sqrt{-4}$, which is undefined in the real number system. There is no real number whose square is equal to -4.

$$2x - 6 \geq 0$$
$$2x \geq 6$$
$$x \geq 3$$

Thus the domain is $\boxed{\{x \mid x \geq 3\}}$. Notice that any x-value less than 3 will not give a real number for y.

We began this section by describing the process of defining a relation or function as a set of ordered pairs. Since a graph is nothing more than the picture of a particular set of ordered pairs, we can actually define a relation or a function by means of a graph. Let's consider the graph in Figure 3.24.

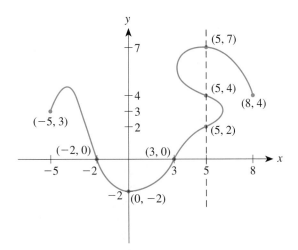

Figure 3.24
The graph of a relation

Each point on the graph corresponds to a pair of numbers (x, y). For example, we observe that the point $(-5, 3)$ is on the graph. This means that if we choose $x = -5$, the corresponding y-value is 3. We can see that the graph crosses the x-axis at $x = -2$ and $x = 3$. This means that there are two domain values, -2 and 3, that give the range value $y = 0$.

However, we also observe that the points $(5, 2)$, $(5, 4)$, and $(5, 7)$ are on the graph. This means that if we choose $x = 5$, there are three possible y-values: y can be 2, 4, or 7. This is exactly what a function is not allowed to do, and so this graph is not the graph of a function.

How can we look at a graph and determine whether it is the graph of a function? According to the definition of a function, if any x-value in the domain has more than one y-value associated with it, the relation is not a function. Using the graph in Figure 3.24 as an example, we see that the vertical line passing through $x = 5$ passes through three points on the graph. This means that there are three y-values corresponding to $x = 5$, which in turn means that the graph cannot be the graph of a function. This condition is usually very easy to check and is called the *vertical line test*.

The Vertical Line Test

If any vertical line intersects a graph in more than one point, then the graph does not represent y as a function of x.

EXAMPLE 7

Determine which of the following relations are functions, by the vertical line test.

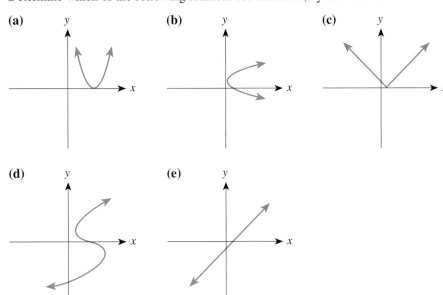

Solution The relations in parts **(a)**, **(c)**, and **(e)** are functions, since any vertical line will intersect the graphs at only one point.

The relations in parts **(b)** and **(d)** are not functions, since some vertical lines will intersect the graphs at more than one point, as shown in the accompanying figures.

(b)

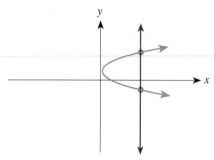

This vertical line intersects the graph at two points; therefore, this relation is not a function.

(d)

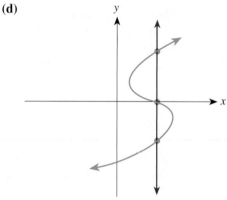

This vertical line intersects the graph at three points; therefore, this relation is not a function.

DIFFERENT PERSPECTIVES: The Vertical Line Test

Consider the graphical and algebraic descriptions of a function.

Graphical Description

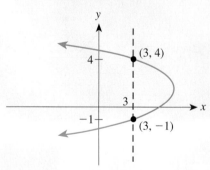

If any vertical line intersects a graph in more than one point, then the graph cannot be the graph of a function.

Algebraic Description

For an equation to define y as a function of x, each x-value must correspond to exactly one y-value. As the graph at the left illustrates, if a graph fails the vertical line test, then it follows that there is at least one x-value to which there correspond at least two y-values. In particular, since the points $(3, -1)$ and $(3, 4)$ are on the graph, there are two y-values that correspond to $x = 3$. This means that if we had the equation of this graph, substituting $x = 3$ would give us *two* y-values satisfying the equation, which violates the definition of a function.

One last comment about the graph in Figure 3.24: We can also use the graph to "read off" the domain and range of the relation from the graph. If we project vertically from the graph to the *x*-axis, we can see that every *x*-value between $x = -5$ and $x = 8$ has at least one *y*-value associated with it, and therefore the domain is $\{x \mid -5 \le x \le 8\}$; see Figure 3.25a.

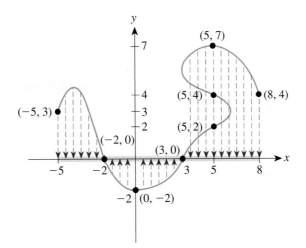

Figure 3.25a
Finding the domain
from a graph

Similarly, if we project horizontally from the graph to the *y*-axis, we can see that every *y*-value between $y = -2$ and $y = 7$ is associated with at least one *x*-value, and so the range is $\{y \mid -2 \le y \le 7\}$; see Figure 3.25b.

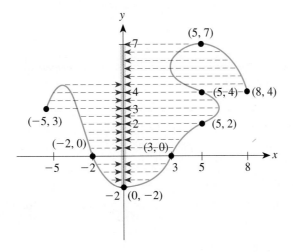

Figure 3.25b
Finding the range
from a graph

We have also used an arrow diagram to represent a function. Although a table and an arrow diagram are virtually the same, a table is more commonly used.

This is just a first illustration that much useful information about a relationship can be extracted from its graph.

In summary, we have seen that we can express a relation (or function) in four possible ways: We can use a table, or use a set of ordered pairs, or write an equation, or draw a graph. As we proceed through the book we will see how each of these modes of representing a function gives its own useful information about the function.

DIFFERENT PERSPECTIVES : Representing Functions

The following are four different ways of representing the same function.

Algebraic Description

We can represent a function by an equation:

$$y = 8 - x^2 \quad \text{for } x = -3, -1, 0, 2, 3$$

Ordered Pair Description

We can represent a function by a set of ordered pairs:

$$\{(-3, -1), (-1, 7), (0, 8), (2, 4), (3, -1)\}$$

Table Description

We can represent a function by a table of values:

x	-3	-1	0	2	3
y	-1	7	8	4	-1

Graphical Description

We can represent a function by a graph:

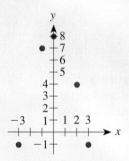

Functions are frequently used to describe real-life relationships, as illustrated in the next example.

EXAMPLE 8

Michelle wants to build a rectangular pen against her house with 50 ft of fencing. Only three sides of the pen are enclosed with fencing, and one of the sides adjacent to the house is x ft (see Figure 3.26).

(a) Express the *area* of the pen in terms of x.

(b) Determine the area of the pen when $x = 5, 10, 20,$ and 24 ft.

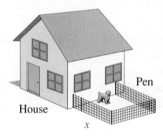

Figure 3.26

Solution   Let's analyze this question carefully, step by step, in order to develop a strategy for its solution.

THINKING OUT LOUD

What are we being asked to do?	To express the area of a rectangular pen in terms of x
How do we find the area of a rectangle?	Multiply its length by its width.
Are we given any dimensions of the rectangle?	We are given that one side adjacent to the house is x (and therefore the other adjacent side is x). We are not given the length of the other side.
To find the area of the rectangle, we need the length of the other side. Is there any given information that can help us determine the length of the missing side?	We are given that the total length of the fence is 50 ft. We can see that the entire length of the fence consists of the three sides: two are labeled x, and the third side, the side for which we are looking, is missing.

At this point we need to recognize that since the entire length of the fence is 50 ft, if we subtract the two x's from 50, we will get the missing side:

$$\text{The missing side} = 50 - x - x = 50 - 2x$$

We now label the figure as shown in Figure 3.27.

Figure 3.27

House

Pen x

$50 - 2x$

Hence the area is

$$\boxed{A = x(50 - 2x)} \quad \text{or} \quad \boxed{A = 50x - 2x^2}$$

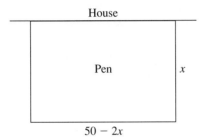

Finding the areas using the table function of a graphing calculator

(b) To find the area for $x = 5, 10, 20$ and 24, we simply substitute the values for x into either formula to compute the corresponding values for A: We will use the formula $A = x(50 - 2x)$ and make a table for the four area values corresponding to the given values of x:

x	$A = x(50 - 2x)$
5	$5(50 - 2 \cdot 5)\ \ = 200$ sq ft
10	$10(50 - 2 \cdot 10) = 300$ sq ft
20	$20(50 - 2 \cdot 20) = 200$ sq ft
24	$24(50 - 2 \cdot 24) = \ \ 48$ sq ft

STUDY SKILLS 3.3

Taking an Algebra Exam: What to Do First

Not all exams are arranged in ascending order of difficulty (from easiest to most difficult). Since time is usually an important factor, you do not want to spend too much time working on a few problems that you find difficult and then find that you do not have enough time to solve the problems that are easier for you. Therefore, it is strongly recommended that you first look over the exam and then follow the order given below:

1. Start with the problems that you know how to solve quickly.
2. Then go back and work on problems that you know how to solve but that take longer.
3. Then work on those problems that you find more difficult, but for which you have a general idea of how to proceed.
4. Finally, divide the remaining time between doing the problems you find most difficult and checking

your solutions. Do not forget to check the warnings you wrote down at the beginning of the exam.

You probably should not be spending a lot of time on any single problem. To determine the average amount of time you should be spending on a problem, divide the amount of time given for the exam by the number of problems on the exam. For example, if the exam lasts for 50 minutes and there are 20 problems, you should spend an average of $\frac{50}{20} = 2\frac{1}{2}$ minutes per problem. Remember, this is just an estimate. You should spend less time on "quick" problems (or those worth fewer points), and more time on the more difficult problems (or those worth more points). As you work the problems be aware of the time; if half the time is gone, you should have completed about half of the exam.

EXERCISES 3.3

In Exercises 1–4, write the relation depicted in the diagram using ordered pair notation.

1. 3 ⟶ 9
8 ⟶ 7
7 ⟶ 2

2. 4 ⟶ 8
2 ⟶ 7
6

3. 3 ⟶ *a*
8 ⟶ *b*
−1 ⟶ *c*

4. 2 ⟶ 1
3 ⟶ 2
0 ⟶ 5

In Exercises 5–8, determine the domain and range.

5. $\{(3, 2), (4, 2), (5, 3)\}$

6. $\{(-1, 5), (-2, 6), (-3, 7)\}$

7. $\{(3, -2), (-2, 3), (3, -1), (4, 3)\}$

8. $\{(5, -1), (-1, 5)\}$

In Exercises 9–24, determine the domain.

9. $y = \dfrac{3}{x}$

10. $y = \dfrac{5}{x + 1}$

11. $y = 3x - 7$

12. $y = 2x - 3$

13. $y = \dfrac{4x}{2x + 3}$

14. $y = \dfrac{5x}{3x - 2}$

15. $y = x^2 - 3x + 4$

16. $y = 3x^2 - 5x + 2$

17. $y = \sqrt{x - 4}$

18. $y = \sqrt{x + 3}$

19. $y = \sqrt{5 - 4x}$

20. $y = \sqrt{6 - 5x}$

21. $y = 5 - \sqrt{4x}$

22. $y = 6 - \sqrt{5x}$

23. $y = \dfrac{x}{\sqrt{x - 3}}$

24. $y = \dfrac{2x}{\sqrt{3 - x}}$

In Exercises 25–44, determine which of the relations are functions.

25. 2 ────→ 6
3 ──╱
4 ────→ 1

26. 5 ────→ 7
4 ──╱
2 ────→ 1

27. 6 ────→ 3
╲──→ 1
5 ────→ 4

28. 3 ────→ 9
5 ⇄ 4
╲──→ 6

29. $\{(6, 3), (5, 3)\}$

30. $\{(8, -2), (7, -2)\}$

31. $\{(3, 1), (3, 2)\}$

32. $\{(5, 1), (5, 7)\}$

33. $\{(6, 5), (5, 6)\}$

34. $\{(7, -2), (-2, 7)\}$

35. $\{(3, 1), (5, 2), (6, 2)\}$

36. $\{(6, 2), (5, 8), (6, 3)\}$

37. $\{(9, -1), (6, -2), (9, 3)\}$

38. $\{(7, 4), (5, -2), (3, -2)\}$

39. $y = x + 3$

40. $2y = x + 4$

41. $x^2 + y^2 = 81$

42. $x^2 + 3y^2 = 9$

43. $x = y^2 - 4$

44. $y = x^2 + 2$

In Exercises 45–52, determine which of the relations are functions.

45.

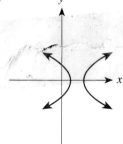

46.

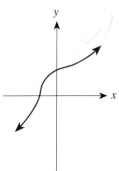

47.

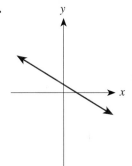

48.

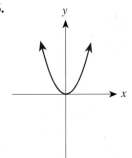

49.

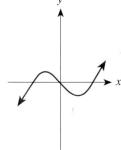

50.

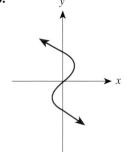

51.

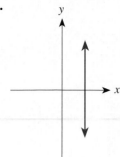

52.

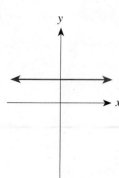

In Exercises 53–62, use the following graph to answer the question or complete the ordered pair.

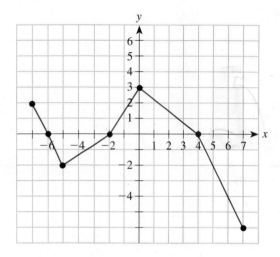

53. If $x = 4$, then $y = $ _____.

54. If $x = -5$, then $y = $ _____.

55. (, -6)

56. (-7,)

57. For how many values of x is $y = 0$? What are they?

58. For how many values of y is $x = 0$? What are they?

59. For how many values of y is $x = 3$?

60. For how many values of x is $y = 2$?

61. What is the domain of this function?

62. What is the range of this function?

In Exercises 63–72, use the following graph to answer the question or complete the ordered pair.

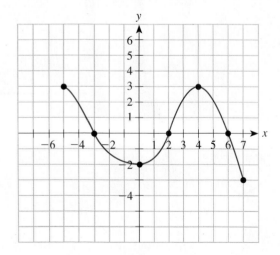

63. If $x = 4$, then $y =$ _____.

64. If $x = -5$, then $y =$ _____.

65. $(\ , -3)$

66. $(-4,\ )$

67. For how many values of x is $y = 0$? What are they?

68. For how many values of y is $x = 0$? What are they?

69. For how many values of y is $x = 3$?

70. For how many values of x is $y = 2$?

71. What is the domain of this function?

72. What is the range of this function?

In Exercises 73–78, use the given graph to determine the domain and range of the relation or function.

73.

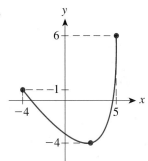

74.

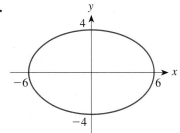

75.

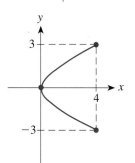

76.

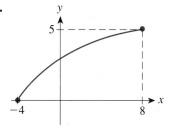

77.

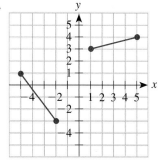

78.

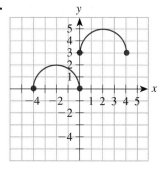

79. Express the area of a square as a function of the length, s, of its side.

80. The length of a rectangle is 5 cm less than twice its width, w. Express the area of the rectangle as a function of w.

81. Express the area of the shaded region of the accompanying figure as a function of a. Based on the dimensions given in the figure, what are the possible values for a?

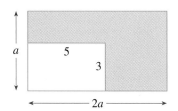

82. Express the area of the shaded region of the accompanying figure as a function of t.

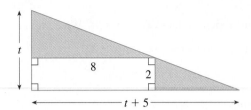

83. A car rental agency charges a flat fee of $29 per day plus a mileage charge of $0.22 per mile. Express the cost, C, of a 4-day rental as a function of m, the number of miles driven.

84. A person drives 3 hours at r mph, and then drives 15 mph faster for 5 additional hours. Express the total distance covered, d, as a function of r.

85. A retailer buys 10 shirts that cost d dollars each and 12 shirts that cost $5 more per shirt. Express the total cost of all the shirts as a function of d.

86. A farmer sets up a fence to enclose and subdivide a rectangular garden as indicated in the accompanying figure.

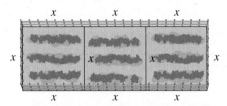

To provide greater security, heavy-duty fencing that costs $10 per foot is used for the outside boundary, whereas regular fencing that costs $8 per foot is used for the two dividers. Express the total cost to set up this enclosure, C, as a function of x.

87. A company's production cost, C, in dollars is a function of n, the number of items produced, according to the formula $C = 0.02n^2 + 100n + 10,000$. Find the cost to produce 100, 200, 500, and 1,000 items.

88. Muriel's monthly income, I, is computed as $I = 0.03(x - 8,000) + 2,000$, where x is the number of items that she sells that month. Compute her monthly income if the monthly sales are 2,500; 10,000; 20,000; and 30,000 items.

89. A cellular phone company charges a flat monthly fee of $29.95 plus $0.33 per minute of airtime.

  (a) Express the monthly cellular phone charge, C, as a function of m, the number of minutes of airtime.

  (b) Find the monthly cellular phone charge with airtimes of 75 minutes, 180 minutes, and 300 minutes.

90. A building contractor charges a flat fee of $2,450 for planning and permits, plus a construction fee of $475 per square meter of construction.

  (a) Express the contractor's charge, C, as a function of x, the number of square meters of construction.

  (b) Compute the cost to construct an area of 90 square meters, 115 square meters, 140 square meters, and 175 square meters.

91. The relationship between the monthly profit (P) generated from all two-bedroom units and the rent (r) charged for each unit in an apartment complex is described in the graph on page 161.

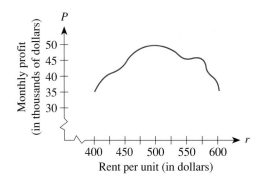

(a) Is this the graph of a function?

(b) What are its domain and range?

(c) What rent per unit generates the maximum profit? What is the maximum profit?

 Questions for Thought

92. State in words what makes a relation a function.

93. In Example 8 we found that the area of a rectangular pen built alongside a house is given by $A = x(50 - 2x)$. This describes the area A as a function of the length of one of the sides, x. Given that this function represents a real-life situation, what is the domain of this function?

3.4 # Function Notation

In the previous section we discussed real-valued functions such as $y = x + 2$ and $y = 3x^2 - 2$. Because these equations are explicitly solved for y, we can say that y is a function of x, or that y is dependent on the values of x (hence, y is called the dependent variable). We now introduce a special notation for functions. The statement "y is a function of x" can be written symbolically as

$$y = f(x) \quad \textit{f(x) is read "f of x" and denotes the value of f at x.}$$

We can bypass the y variable completely and describe an expression such as $3x^2 - 2$ as being a function of x by writing

$$f(x) = 3x^2 - 2 \quad \textit{The expression } 3x^2 - 2 \textit{ is a function of x.}$$

The letters f, g, and h are used most frequently for functions. Other examples of functions are as follows:

$$f(x) = x^2 - 2x + 5$$
$$g(x) = 2 - 5x$$
$$h(a) = \frac{a}{a + 1}$$

Note that the parentheses in this notation are *not* used as grouping symbols—that is, $f(x)$ is *not* the product of f and x. The parentheses are being used to specify the independent variable.

The notation $f(x)$ is a useful shorthand for evaluating expressions or substituting variables. For example, if we want to know the value of the function $f(x)$ when $x = 3$, we write $f(3)$.

Using the vocabulary introduced in the previous section, we can say that 3 is the **input** *and f(3) is the* **output**. *In general, x is the input and f(x) is the output.*

If $f(x) = 7x - 4$, and we want to find the function value when $x = 3$, then

$$f(3) = 7(3) - 4 = 21 - 4 = 17 \qquad \textit{Thus, f(3) = 17.}$$

In the same way, $f(-5)$ is the value of $f(x)$ when $x = -5$:

$$f(-5) = 7(-5) - 4 = -35 - 4 = -39 \qquad \textit{Thus, f(-5) = -39.}$$

Thus,

> $f(a)$ is the value of $f(x)$ when a is substituted for x in $f(x)$.

For example, if we have $g(x) = 2x^2 - 4x + 5$, then $g(3)$ is the value of $g(x) = 2x^2 - 4x + 5$ when x is replaced by 3. Hence,

$$g(3) = 2(3)^2 - 4(3) + 5 \qquad \textit{g(3) is g(x) evaluated when x = 3.}$$
$$= 2(9) - 4(3) + 5$$
$$= 11$$

We are not necessarily restricted to numbers as possible replacements for x. For example, if $f(x) = 7x - 4$, then

$$f(a) = 7a - 4 \qquad \textit{Replace x with a in f(x).}$$
$$f(z) = 7z - 4 \qquad \textit{Replace x with z in f(x).}$$

EXAMPLE 1

If $f(x) = 3x - 5$ and $g(x) = 2x^2 - 3x + 1$, find each of the following:

(a) $f(-2)$ (b) $g(4)$ (c) $g(-3)$ (d) $g(r)$

Solution

(a) Evaluate $f(x)$ for $x = -2$:

$$f(-2) = 3(-2) - 5 = -11$$

> Thus, $f(-2) = -11$

Remember, g(4) = 21 tells us that the point (4, 21) is on the graph of y = g(x).

(b) Substitute 4 for x in $g(x)$:

$$g(4) = 2(4)^2 - 3(4) + 1 = 2 \cdot 16 - 3 \cdot 4 + 1 = 21$$

> Thus, $g(4) = 21$

(c) Evaluate $g(x)$ for $x = -3$:

$$g(-3) = 2(-3)^2 - 3(-3) + 1 = 2 \cdot 9 + 9 + 1 = 28$$

> Thus, $g(-3) = 28$

(d) Substitute r for x in $g(x)$:

$$g(r) = 2r^2 - 3r + 1$$

> Thus, $g(r) = 2r^2 - 3r + 1$

EXAMPLE 2

Given $f(x) = 3x - 1$, find each of the following:

(a) $f(a + 1)$ (b) $f(2x - 1)$

Solution   **(a)** $f(a + 1)$ is the expression we get when we substitute $a + 1$ for x in $f(x)$. Since

$$f(x) = 3x - 1 \qquad \text{\textit{Substituting } a + 1 \textit{ for } x, \textit{ we get}}$$

$$f(a + 1) = 3(a + 1) - 1 \qquad \text{\textit{Then simplify.}}$$

$$= 3a + 3 - 1$$

$$= 3a + 2$$

$$\boxed{\text{Thus, } f(a + 1) = 3a + 2}$$

(b) $f(2x - 1)$ is the expression we get when we substitute $2x - 1$ for x in $f(x)$. Since

$$f(x) = 3x - 1 \qquad \text{\textit{Substituting } 2x - 1 \textit{ for } x, \textit{ we get}}$$

$$f(2x - 1) = 3(2x - 1) - 1$$

$$= 6x - 3 - 1 \qquad \text{\textit{Then simplify.}}$$

$$= 6x - 4$$

$$\boxed{\text{Thus, } f(2x - 1) = 6x - 4}$$

EXAMPLE 3

Given $f(x) = 4x - 1$, find each of the following:

(a) $f(x) + 2$ **(b)** $f(x + 2)$ **(c)** $f(x) + f(2)$ **(d)** $f(x + 2) - f(x)$

Solution   **(a)** $f(x) + 2$ *Means add 2 to f(x); since f(x) = 4x − 1,*

$$= \underbrace{4x - 1}_{f(x)} + 2$$

$$= 4x + 1$$

$$\boxed{\text{Hence, } f(x) + 2 = 4x + 1}$$

(b) $f(x + 2)$ means to substitute $x + 2$ for x in $f(x)$. Since $f(x) = 4x - 1$, we have

$$f(x) = 4x - 1$$

$$f(x + 2) = 4(x + 2) - 1$$

$$= 4x + 8 - 1$$

$$= 4x + 7$$

$$\boxed{\text{Hence, } f(x + 2) = 4x + 7}$$

Note $f(x) + 2$ and $f(x + 2)$ are not the same. In this example, $f(x) + 2 = 4x + 1$ while $f(x + 2) = 4x + 7$.

(c) $f(x) + f(2)$ is the sum of two expressions: $f(x)$ and $f(2)$.

$$f(x) + f(2) = \underbrace{4x - 1}_{f(x)} + \underbrace{4(2) - 1}_{f(2)}$$

$$= 4x - 1 + 8 - 1$$

$$= 4x + 6$$

$$\boxed{\text{Hence, } f(x) + f(2) = 4x + 6}$$

Note the differences between part **(c)** and parts **(a)** and **(b)**. You cannot simply add $f(2)$ to $f(x)$ to get $f(x + 2)$. In general, $f(a + b) \neq f(a) + f(b)$.

(d) First find $f(x + 2)$; we found in part **(b)** that $f(x + 2) = 4x + 7$. Then

$$f(x + 2) - f(x) = \underbrace{4x + 7}_{f(x+2)} - \underbrace{(4x - 1)}_{f(x)} \quad \textit{Note the parentheses around } 4x - 1.$$

$$= 4x + 7 - 4x + 1$$

$$= 8$$

> Hence, $f(x + 2) - f(x) = 8$

A function does not have to be defined algebraically; it can be defined by a graph and the functional values can be determined from this graph.

Interpreting $f(x)$ from a Graph

Let's carefully examine the graph in Figure 3.28a and review what it means for a point (x, y) to lie on a graph and how, given the graph of a relationship, we can determine y given x.

We can see in Figure 3.28a that the point $(5, -6)$ is on the graph. This means that the graph passes through a point that is 5 horizontal units to the right of the y-axis and 6 vertical units below the x-axis. If we need to find the y-value associated with a particular x-value, we start at the x-value on the x-axis, project vertically upward or downward to the graph, and then project horizontally to the y-axis and read this y-value. Hence if $x = 3$, then we see that for this graph, $y = 1$. Notice that at $x = 3$, the "height" of the graph (the vertical distance from the x-axis) is 1; likewise, if $x = -4, y = 3$.

The x-coordinate gives the horizontal distance from the y-axis, and the y-coordinate gives the vertical distance from the x-axis.

The x-coordinate is often called the "directed" distance from the y-axis, meaning that $|x|$ is the distance, and the sign of x gives the direction. Similarly, the y-coordinate is the directed distance from the x-axis.

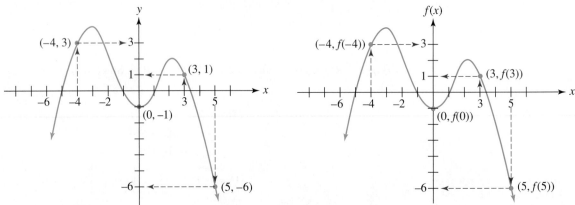

Figure 3.28a   **Figure 3.28b**

Returning to the graph in Figure 3.28a, using function notation, we can label the graph $y = f(x)$. Then, instead of writing $y = 3$ when $x = -4$, we can write $f(-4) = 3$. Similarly, $y = -1$ when $x = 0$ can be written as $f(0) = -1$. Figure 3.28b repeats Figure 3.28a with the vertical axis labeled $f(x)$ instead of y.

Referring to Figure 3.28b, let's summarize the equivalence of these alternatives. See Table 3.1.

Table 3.1

(x, y) **Notation**	$y = f(x)$ **Notation**	$(x, f(x))$ **Notation**
$(-4, 3)$	$3 = f(-4)$	$(-4, f(-4))$
$(3, 1)$	$1 = f(3)$	$(3, f(3))$
$(0, -1)$	$-1 = f(0)$	$(0, f(0))$

Again, $f(x)$ is the y-value at x, and gives the "height" of the graph (the vertical distance above or below the x-axis) at x. To be more precise, $f(x)$ is the directed distance of the graph from the x-axis at x. Hence, $f(a)$ is the directed distance of the graph from the x-axis at $x = a$; the directed distance of the graph from the x-axis at $x = b$ is $f(b)$. See Figure 3.29.

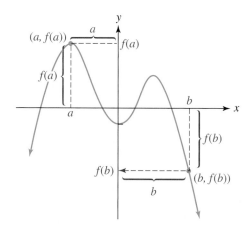

Figure 3.29

EXAMPLE 4

Consider the function defined by the graph in Figure 3.30 to find each of the following.

(a) $f(6)$ **(b)** $f(-3)$ **(c)** $f(0)$ **(d)** For what values of x is $f(x) = 0$?

(e) Which is larger, $f(-3)$ or $f(3)$?

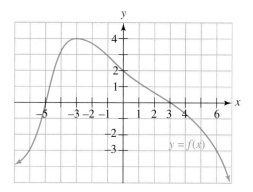

Figure 3.30

Solution

(a) To find $f(6)$, we must find the y-value that corresponds to $x = 6$. Looking at the graph, we can see that when $x = 6$, the corresponding y-value is $y = -3$. Therefore, $f(6) = -3$.

(b) Similarly, from the graph we see that when $x = -3$, the corresponding y-value is $y = 4$. Therefore, $f(-3) = 4$.

(c) To find $f(0)$, we must determine where the graph crosses the y-axis (because when $x = 0$ the point must be on the y-axis). From the graph we can see that $y = 2$ when $x = 0$, hence $f(0) = 2$.

(d) We are looking for those x-values for which $y = 0$ [remember that $f(x)$ is just another name for y]. But $y = 0$ means that the point must be on the x-axis. Therefore, from the graph we see that $f(-5) = 0$ and $f(3) = 0$, so the x-values that make $f(x) = 0$ are -5 and 3.

(e) While it is true that 3 is larger than -3, this question asks you to compare the y or f values, not the x-values. We cannot answer this question until we determine what the function f actually does to x. We see on the graph that when $x = -3$, the corresponding y-value is $y = 4$; hence $f(-3) = 4$. Similarly, the y-value corresponding to $x = 3$ is $y = 0$; hence $f(3) = 0$. So $f(-3) = 4$ is larger than $f(3) = 0$.

At the beginning of this section we learned that algebraically, $f(a + 3)$ and $f(a) + 3$ are not the same. Keeping in mind that $f(a)$ is the "height" of the graph at a, we can geometrically demonstate the differences between the two quantities. See Figure 3.31.

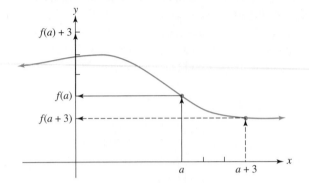

Figure 3.31
Comparing
$f(a) + 3$ and $f(a + 3)$

Notice that finding $f(a) + 3$ requires us first to find $f(a)$. To find $f(a)$, we first locate a on the x-axis, move vertically up until we intersect the graph, and then project horizontally onto the y-axis. To find $f(a) + 3$, we move vertically up three more units. Using the vocabulary introduced earlier this is equivalent to using a as the input and then adding 3 to the output, $f(a)$.

On the other hand, finding $f(a + 3)$ requires us first to find $a + 3$, move vertically up until we intersect the graph, and then project horizontally onto the y-axis. This is the same as using $a + 3$ as the input and then finding the output, $f(a + 3)$. Notice the difference between the two expressions in their values as well as the procedures used to find each expression.

DIFFERENT PERSPECTIVES: Function Notation

Consider the geometric and algebraic interpretation of function notation. Let's consider the function $y = f(x) = x^2 - x - 6$, whose graph appears below.

Graphical Description

$y = f(4) = 6$

The graph of $y = f(x) = x^2 - x - 6$

To find $f(4)$ *using the graph,* we are looking for the y-value that corresponds to $x = 4$. We can see by the graph that when $x = 4$, the corresponding y value is $y = 6$. Hence $f(4) = 6$.

Algebraic Description
To find $f(4)$ we substitute 4 for x in $f(x) = x^2 - x - 6$:

$$f(4) = 4^2 - 4 - 6 = 16 - 4 - 6 = 6$$

Hence $f(4) = 6$.

STUDY SKILLS 3.4

Taking an Algebra Exam: Dealing with Panic

In the first two chapters of this textbook we have given you advice on how to learn algebra and discussed how to prepare for an algebra exam. If you followed this advice and put the proper amount of time to good use, you should feel fairly confident and less anxious about the exam. But you may still find during the course of the exam that you are suddenly stuck or you draw a blank. This may lead you to panic and say irrational things like "I'm stuck. . . . I can't do this problem. . . . I can't do any of these problems. . . . I'm going to fail this test." Your heart may start to beat faster and your breath may quicken. You are entering a panic cycle.

These statements are irrational. Getting stuck on a few problems does not mean that you cannot do any algebra. These statements only serve to interfere with your concentration on the exam itself. How can you think about solving a problem while you are telling yourself that you cannot? The increased heart and breath rate are part of this cycle.

What we would like to do is break this cycle. What we recommend that you do is first put aside the exam and silently say to yourself **STOP!** Then try to relax, clear your mind, and encourage yourself by saying to yourself things such as "This is only one (or a few) problems, not the whole test" or "I've done problems like this before, so I'll get the solution soon." (Haven't you ever talked to yourself this way before?)

Now take some slow deep breaths and search for some problems that you know how to solve and start with those. Build your concentration and confidence slowly with more problems. When you are through with the problems you can complete, go back to the ones on which you were stuck. If you have the time, take a few minutes and rest your head on your desk, and then try again. But make sure you have checked the problems you have completed.

EXERCISES 3.4

In Exercises 1–12, given $f(x) = 2x - 3$, $g(x) = 3x^2 - x + 1$, and $h(x) = \sqrt{x + 5}$, find:

1. $f(0)$ **2.** $g(0)$ **3.** $g(2)$ **4.** $f(2)$

5. $g(-2)$ **6.** $f(-2)$ **7.** $h(3)$ **8.** $h(4)$

9. $h(-3)$ **10.** $h(-4)$ **11.** $h(a)$ **12.** $g(a)$

In Exercises 13–24, given $f(x) = x^2 + 2$ and $g(x) = 2x - 3$, find (and simplify):

13. $f(5) + f(2)$ **14.** $f(5 + 2)$ **15.** $f(6 - 4)$ **16.** $f(6) - f(4)$

17. $g(x + 2)$ **18.** $g(x) + g(2)$ **19.** $f(2x)$ **20.** $g(3x)$

21. $g(3x + 2)$ **22.** $3g(x) + 2$ **23.** $g(x + 2) - g(x)$ **24.** $g(x + 5) - g(x)$

In Exercises 25–34, given $f(x) = x^2 + 2x - 3$. find (and simplify):

25. $f(-2) + f(3)$ **26.** $f(-2 + 3)$ **27.** $f(x) - 4$ **28.** $f(x) + 3$

29. $f(x) - f(4)$ **30.** $f(x) + f(3)$ **31.** $f(3x)$ **32.** $2f(x)$

33. $3f(x)$ **34.** $f(2x)$

In Exercises 35–40, given $g(x) = \dfrac{x}{x + 5}$, find (and simplify):

35. $g(-3)$ **36.** $g(10)$ **37.** $g(0)$

38. $g(-5)$ **39.** $g(x + 5)$ **40.** $g(x - 5)$

41. Use the following graph of $y = g(x)$ to find

    (a) $g(-5)$ **(b)** $g(-2)$ **(c)** $g(0)$

    (d) $g(1)$ **(e)** $g(3)$

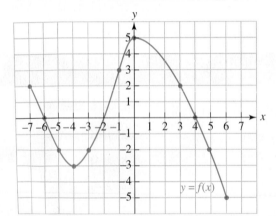

42. Use the following graph of $y = h(x)$ to find

    (a) $h(-2)$ **(b)** $h(6)$ **(c)** $h(0)$

    (d) $h(4)$ **(e)** $h(-4)$

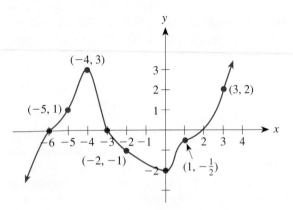

43. Use the following graph of $y = f(x)$ to find

    (a) $f(-1)$ **(b)** $f(0)$ **(c)** $f(4)$

    (d) For what value(s) of x is $f(x) = -2$?

    (e) For how many value(s) of x is $f(x) = 4$?

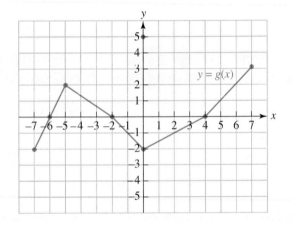

44. Use the following graph of $y = g(x)$ to find

    (a) $g(-1)$ **(b)** $g(0)$ **(c)** $g(4)$

    (d) For what value(s) of x is $g(x) = -2$?

    (e) For what value(s) of x is $g(x) = 4$?

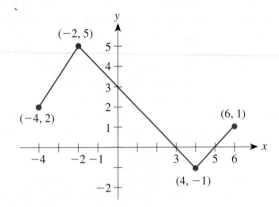

45. A car rental agency charges a flat fee of $29 per day plus a mileage charge of $0.14 per mile. Express the charge, C, for a 3-day rental as a function of m, the number of miles driven.

46. The length of a rectangle is 8 more than twice its width w. Express the perimeter, P, of the rectangle as a function of w.

47. The width of a rectangle is 12 less than three times its length, L. Express the area, A, of the rectangle as a function of L.

48. A cellular phone company charges a flat monthly fee of $33 per month plus an airtime fee of $0.54 per minute of phone usage. Express the monthly cellular phone bill, C, as a function of m, the number of minutes of phone usage.

49. A ticket club has 600 concert tickets to sell. It sells x tickets at a price of $30 per ticket, and all the remaining tickets at $24 per ticket. If all the tickets are sold, express the total amount collected, C, as a function of x.

50. A woman drives for a total of 10 hours. She drives for t hours at a rate of 45 mph, and the remaining time at a rate of 55 mph. Express the total distance covered, d, as a function of t.

51. Two copying machines are working on a project. Machine A can produce 20 copies per minute; machine B can produce 22 copies per minute. If machine A works for m minutes, and machine B works for 35 minutes longer than machine A, express the total number, N, of copies made as a function of m.

52. Two workers are packing items on an assembly line. One worker packages 30 items per hour for h hours and the second worker packages 35 items per hour for 3 hours longer than the first worker. Express the total number, N, of items packaged by the two workers as a function of h.

53. An electrician charges $65 per hour for her time and $45 per hour for her assistant's time. She works for h hours on a job and her assistant works 2 hours less than she does. Express the total amount, A, they earn as a function of h.

54. A plumber charges $62 per hour for his time and $47 per hour for his assistant's time. On a particular job that takes a total of 8 hours, the plumber works alone for h hours and the assistant then works alone for the remainder of the time. Express the total amount, A, earned by the plumber and his assistant as a function of h.

55. Lamont has part-time jobs as a tutor and as a clerk. He earns $30 per hour as a tutor and $9.35 per hour as a clerk. During a particular week, he works a total of 15 hours tutoring and clerking. Express his weekly income, I, as a function of t, the number of hours he tutors.

56. Cheryl owns a gourmet food shop. During a certain week she sells p pounds of a $5.35-per-pound coffee blend and 12 fewer pounds of a $6.85 coffee blend. Express the total coffee revenue, R, for the week as a function of p.

3.5   Interpreting Graphs

In the first four sections of this chapter, we discussed the rectangular coordinate system and how we can get a picture of the solutions of an equation from its graph. In this section we will discuss how graphs themselves are legitimate mathematical tools that can help us understand or evaluate quantifiable relationships.

Most of us have had some experience with graphs outside a mathematics class: Graphs may accompany a newspaper or magazine article, or appear in a textbook. A graph is usually used to give us a convenient picture illustrating some relationship between two quantities that allows us to make, summarize, or clarify a point. The graph also gives more detail as to the nature of the relationship under discussion. In mathematics, the graph is a tool that helps us to visualize a trend or relationship between two quantities. While an equation is useful for finding exact values, a graph not only can be used to get reasonable estimates of these values but also has the added advantage of allowing us to "see" the nature of the relationship.

EXAMPLE 1

The graph in Figure 3.32 on page 170 illustrates how the profit a company expects on the sale of a certain item depends on its selling price. The horizontal axis is labeled s and represents the selling price of the item (in dollars). The vertical axis is labeled P and represents the profit (in thousands of dollars) that the company earns. Use the information given in this graph to describe how the profit relates to the selling price.

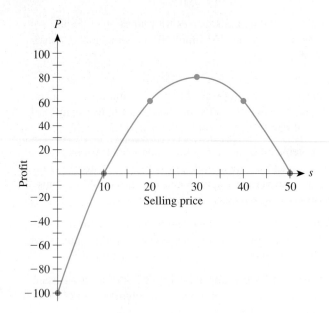

Figure 3.32
The graph for Example 1

Solution   Examining the graph, we can draw the following conclusions:

1. The point $(0, -100)$ is on the graph. This means that when the selling price is $0 ($s = 0$), the profit is $-\$100,000$ ($P = -100$). We can interpret this negative profit to mean that if the item were given away free, the company would have a *loss* of \$100,000. This loss may be due to fixed costs such as rent and taxes, which must be paid regardless of what the selling price is or how many items are sold.

2. As the selling price starts to increase, the profit increases as well. The point $(10, 0)$ on the graph tells us that when the selling price is \$10 ($s = 10$), the profit is \$0 ($P = 0$). (This is often called the *break-even point.*)

3. As the selling price continues to increase, the profit continues to increase as well. The highest point on the graph is $(30, 80)$. This highest point corresponds to the *maximum profit* the company can earn, which is \$80,000.

4. As the selling price increases beyond \$30 the profit decreases until, when the selling price is \$50, the profit is again \$0. This can be explained by the fact that once the selling price gets too large, fewer people will buy the item, thus decreasing the profit.

At a glance we can see that the trend is for the profit to increase as the selling price increases until it reaches the particular selling price at which the profit is the highest. Then, as the selling price continues to increase, the profit decreases until there is \$0 profit. Thus, this graph gives a "snapshot" of the relationship between the selling price and the expected profit.

EXAMPLE 2

A psychologist developed a treatment for reducing certain compulsive behaviors. In order to collect preliminary evidence of the effectiveness of her treatment, she had 35 of her clients keep diaries, recording the number of times they exhibited their compulsive behavior daily. They did this for a total of 12 weeks (before, during, and after treatment), and then they reported the same information again a year later. Figure 3.33 on page 171 shows the average daily number of compulsive behaviors exhibited by the group during the 12 weeks, and 1 year later (for 2 weeks). Assuming that the changes were a direct result of the therapy and nothing else, what can you conclude about the effectiveness of the treatment?

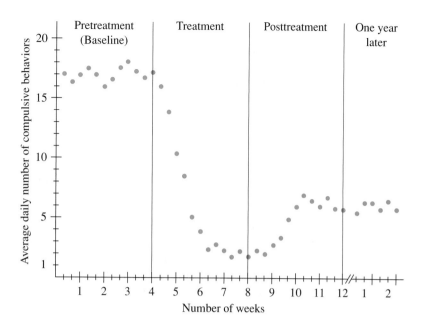

Figure 3.33

Solution   The graph is split into four sections: before the treatment (called the baseline), during the treatment, after the treatment, and a year later. During the first 4 weeks, we can see that the average daily number of compulsive behaviors exhibited by clients was between 16 and 18. During therapy, the next 4 weeks, the average number of compulsive behaviors decreased steadily until it fell in the range of about 2 to 3 at the end of the treatment. For the first week following the treatment, the number of compulsive behaviors remained about the same as during the last week of the treatment phase. Then the number of compulsive behaviors slowly increased during the next week until it reached a level between 6 and 7. A year later, the number remained at the same level as it had during the end of the posttreatment period.

One may conclude that although the treatment was not successful in eliminating the response entirely, it certainly reduced the number of occurrences of the compulsive behavior by about 60–66%.

EXAMPLE 3

Jani is traveling on business in her car. Suppose that the graph in Figure 3.34 describes her day's travel. The horizontal axis is labeled t and represents the number of hours since Jani began her trip. The vertical axis is labeled d and represents the distance (in miles) that Jani is from her home. Use the information given in this graph to describe her day as best you can.

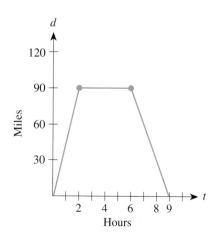

Figure 3.34
The graph for Example 3

Solution   Examining the graph, we can draw the following conclusions:

1. The point $(0, 0)$ is on the graph. This means that when $t = 0$, $d = 0$. Keeping in mind that t is the number of hours since Jani began her trip, $t = 0$ means that Jani has not yet begun her trip; since d is the number of miles that Jani is from her home, $d = 0$ means that she is 0 miles from her home. In other words, Jani is starting her trip from home.
2. The line is rising from $(0, 0)$ to $(2, 90)$, which means that during this 2-hour period Jani's distance from home is *increasing*.
3. The point $(2, 90)$ is on the graph, which means that when $t = 2$, $d = 90$. In other words, after 2 hours, Jani is 90 miles from home.
4. The next portion of the graph is horizontal, which means that the distance is not changing. The distance from home remains 90 miles for the next 4 hours. Possibly Jani is at a business meeting that lasts for 4 hours.
5. The line is falling from $(6, 90)$ to $(9, 0)$, which means that during this 3-hour period Jani's distance from home is *decreasing*.
6. The point $(9, 0)$ means that after 9 hours Jani's distance from home is again 0. After 9 hours Jani has returned to her home.

Actually, we can glean a bit more information from this graph. When Jani starts her trip, we can see that she covers a distance of 90 miles in 2 hours; hence, her average speed during her first 2 hours of travel is

$$\text{Averaged speed} = \frac{90 \text{ miles}}{2 \text{ hours}} = 45 \text{ mph}$$

On her return trip home (the 3-hour period from $t = 6$ to $t = 9$), Jani covers a distance of 90 miles in 3 hours, so her average speed during her return trip home is 30 mph.

The main point here is that the graph allows us to see at a glance some important aspects of Jani's trip.

EXAMPLE 4   Let's suppose that we want to determine how much of a medication is in the bloodstream at various times after the medication is taken. We might take a series of blood samples from a patient as follows: The first sample is taken before a particular medication is administered; subsequent blood samples are taken at 20-minute intervals after the patient receives the medication, and the amount of medication in the blood at each time is determined.

By designating the horizontal axis, labeled t, as the time in minutes after the medication is taken, and the vertical axis, labeled A, as the amount of medication (in milligrams) present in the blood, we can represent the data collected as points on a graph (see Figure 3.35a). If we draw line segments connecting the points (see Figure 3.35b), then the resulting graph allows us to analyze the relationship between the elapsed time and the amount of medication in the bloodstream.

Use the graph to describe the connections between the time elapsed since the medication was taken and the amount of medication present in the bloodstream.

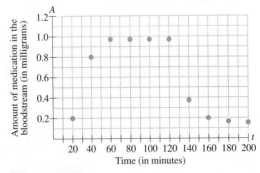

Figure 3.35a

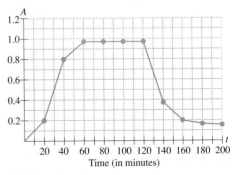

Figure 3.35b

Solution   If we examine the graph in Figure 3.35b, we can see that over the first 20-minute interval, the amount of medication in the blood rises from 0 mg to 0.2 mg. In other words, the medication is entering the blood at an average rate of

$$\frac{0.2 \text{ mg}}{20 \text{ min}} = 0.01 \frac{\text{mg}}{\text{min}}$$

During the next 20-minute interval, the amount of medication in the blood rises from 0.2 mg to 0.8 mg; thus, the amount of medication in the blood increases by 0.6 mg in 20 minutes. In other words, the medication is entering the blood at an average rate of

$$\frac{0.6 \text{ mg}}{20 \text{ min}} = 0.03 \frac{\text{mg}}{\text{min}}$$

Based on this analysis, we can say that during the second 20-minute interval, the medication is entering the bloodstream at 3 times the rate it did during the first 20 minutes.

Now that we have an idea of how to interpret the steepness of the line, we can identify the following trends:

1. The medication enters the blood relatively slowly during the first 20 minutes, then more quickly during the 20–40-minute time interval.
2. During the 40–60-minute time interval, the medication is still entering the blood, but the *rate* at which the medication is entering the bloodstream is slower than in the second interval. (The third line segment is less steep than the second line segment.)
3. During the next three 20-minute intervals (60–120 minutes), the line segment is horizontal, which indicates that no more medication is entering (or leaving) the bloodstream.
4. During the 120–140-minute time interval, the line is falling relatively steeply. This tells us that the amount of medication in the bloodstream is *decreasing* at a comparatively rapid rate.
5. During the next three 20-minute intervals, the line segments are also falling but are progressively less steep. This tells us that the amount of medication in the bloodstream is continuing to decrease but at a slower rate during each time period.

Figure 3.36 serves to summarize this analysis and interpretation of the graph.

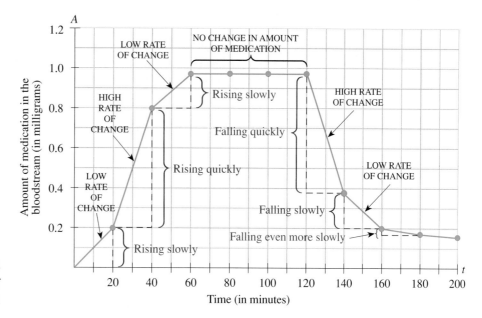

Figure 3.36

Interpreting the rate of change from a graph

Keep in mind that our analysis and interpretation are dependent on the fact that we connected the data points with straight line segments. To actually justify such an assumption in the real world, we would need to collect more data.

EXAMPLE 5

The forestry service located a species of deer and monitored the size of the population for 15 years. The graph of the relationship between the size of the population and time is illustrated by Figure 3.37. Discuss the population growth illustrated by the graph.

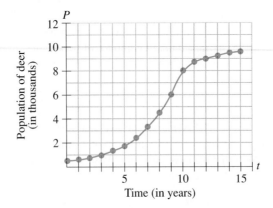

Figure 3.37

Solution

Look at the graph to get an overall picture of how the population increases. Divide the horizontal axis into 5-year intervals, and look at the population increases for each interval. Note that for the first 5-year interval, the graph rises a bit more than 1 unit (representing a population increase of over 1,000); and for the next 5-year interval, the graph rises a bit more than 6 units (representing a population increase of 6,000). For the last 5-year interval, the graph rises $1\frac{1}{2}$ units. We can restate these trends in general terms as follows: The population grows slowly during the first 5 years, then grows rapidly between the 5th and the 10th years. Between the 10th and the 15th years, the population growth tapers off or slows down (or the growth rate slows down during the last 5 years).

STUDY SKILLS 3.5

Taking an Algebra Exam: A Few Other Comments about Exams

Do not forget to check over all your work, as we have suggested on numerous occasions. Reread all directions and make sure that you have answered all the questions as directed.

If you are required to show your work (such as for partial credit), make sure that your work is neat. Do not forget to put your answers where directed or at least indicate your answers clearly by putting a box or circle around them. For multiple-choice tests be sure you have filled in the correct space.

One other bit of advice: Some students are unnerved when they see others finishing the exam early. They begin to believe that there may be something wrong with themselves because they are still working on the exam. They should not be concerned that some students can do the work quickly and that others leave the exam early, not because the exam was easy for them, but because they give up.

In any case, do not be in a hurry to leave the exam. If you are given 1 hour for the exam then take the entire hour. If you have followed the suggestions in this chapter such as checking your work, etc., and you still have time left over, relax for a few minutes and then go back and check your work again.

EXERCISES 3.5

1. The telephone company monitored telephone usage in a small town by recording the number of calls made every 2 hours during a 24-hour period. The following graph illustrates the level of telephone usage during this period. The horizontal

axis, labeled *t,* represents the hour of the day: $0 = $ midnight, $1 = 1$ A.M., $14 = 2$ P.M., etc. The vertical axis, labeled *n,* represents the number of phone calls (in thousands).

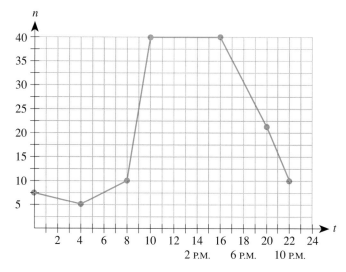

(a) How many calls are made at midnight?

(b) At what time is the number of phone calls a minimum?

(c) How many phone calls are made at 6:00 P.M.?

(d) During what period(s) of time is the number of phone calls decreasing?

(e) At what time is the number of phone calls a maximum?

(f) During what period of time does the number of phone calls remain constant?

2. The graph on the right illustrates the level of electrical power usage in a small town during a 1-year period. The horizontal axis, labeled *m,* represents the month of the year: 1–2 represents January; 2–3, February; . . . ; 12–13, December. The vertical axis, labeled *K,* represents the number of megawatts of electrical power being used.

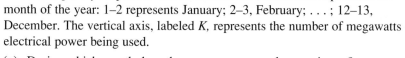

(a) During which month does the power usage reach a maximum?

(b) During which month does the power usage reach a minimum?

(c) During which months does the power usage increase?

(d) During which months does the power usage decrease?

3. The following graph shows the percentage of unemployment rate in the United States from 1994 to 2004 (seasonally adjusted).

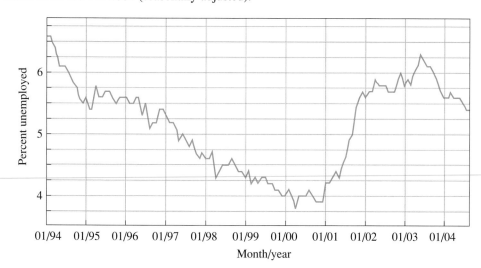

(a) What was the unemployment rate in January 1996?

(b) During what years was the unemployment rate less than 5%.

(c) Verbally describe the general trend shown by the graph from 1994 to 2004.

4. The following graph shows the number employed (in the thousands) by the U.S. Postal Service from 1994 to 2004 (seasonally adjusted).

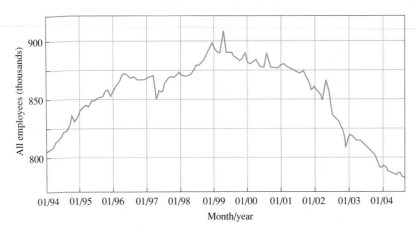

(a) How many were employed by the U.S. Postal Service in January 1996?

(b) During what years was the number of employees above 850,000?

(c) Verbally describe the general trend shown by the graph from 1994 to 2004.

5. The following graph, published by the Department of Labor Statistics, shows the amount of energy consumed, produced, and traded in the United States from 1980 to 1998, in quadrillion British thermal units (Btu). (A Btu is the amount of energy required to raise the temperature of 1 pound of water 1°F at or near 39.2°F.)

Energy Production, Trade, and Consumption: 1980 to 1998

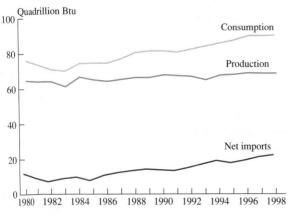

Source: Chart prepared by U.S. Census Bureau.

(a) How many Btu were consumed, produced, and imported in the United States in 1990?

(b) Verbally describe the general trend shown by the graphs for each of production, consumption, and imports from 1980 to 1998.

6. A psychologist developed a treatment for reducing certain inappropriate social behaviors in children. In order to collect preliminary evidence of the effectiveness of her treatment, she observed 20 of her clients interacting socially, for 45 minutes a day, recording the number of times they exhibited inappropriate social behaviors. She did this for a total of 9 weeks (before, during, and after treatment). The accompanying graph shows the average daily number of inappropriate social behaviors exhibited by the group during the 9 weeks. Assuming that

the changes were a direct result of the therapy and nothing else, what can you conclude about the effectiveness of the treatment?

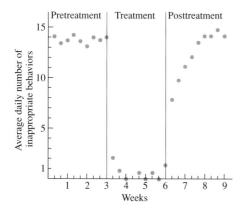

7. The following graph illustrates the relationship between air temperature and altitude. The horizontal axis, labeled *a*, represents the altitude (in kilometers). The vertical axis, labeled *T*, represents the temperature (in degree Celsius).

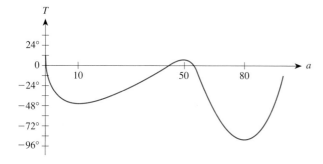

(a) It is commonly believed that the temperature drops steadily as the altitude increases. Does the graph confirm or deny this belief?

(b) If a temperature-measuring device is sent aloft, between what altitudes will the temperature be decreasing?

(c) As the device rises from an altitude of 45 km to an altitude of 55 km, how is the temperature changing?

8. The following graph illustrates the level of a certain substance in the blood after a medication containing this substance is taken. The horizontal axis, labeled *m*, represents the number of minutes after the medication is taken. The vertical axis, labeled *A*, represents the amount of the substance, measured in milligrams, present in the blood.

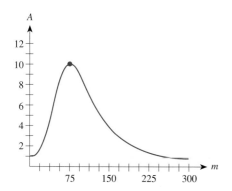

(a) How many milligrams of the substance are present in the blood before any medication is taken? This is the body's normal level of this substance. [*Hint:* Find *A* when *m* = 0.]

(b) How long does it take until the amount of this substance reaches its maximum level in the blood?

(c) After the medication is taken, how long does it take for the level of the substance to return to normal?

9. In a psychology experiment, students were asked to memorize a list of nonsense syllables (meaningless three-letter words such as "ogu," "bir," or "gar"). After successfully demonstrating that they had memorized the entire list, all were retested at various time intervals afterward. The graph of the relationship between their (average) learning score and the retesting time is shown in the figure. The horizontal axis, labeled t, represents the time at which the students were retested; the vertical axis, labeled p, represents the group's average percent score. This is often called a *forgetting curve*.

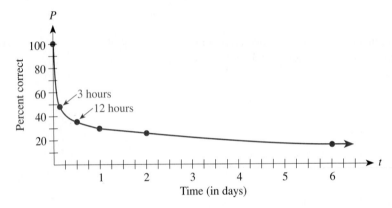

(a) Find the average score 1 day after the students learned the material.

(b) How long will it take for them to remember only 25% of what they had learned (or how long does it take them to forget 75% of what they learned)?

(c) Discuss the relationship between time and forgetting illustrated by the graph. Include in your discussion how quickly forgetting occurs as time passes.

10. Students were given a list of words to memorize. Each day they were given a new list to remember and were tested the following day only on the previous day's list. This continued for 15 days. The following is a graph of the approximate relationship between the percentage recalled on a list and number of different previous lists memorized.

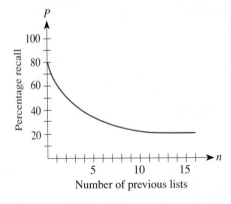

(a) Approximately what percentage of the words were recalled on the seventh list? (This means we want the percentage recall when the number of previous lists is 6.)

(b) *Interference* is the term used to describe the negative effects new learning can have on previous learning. Discuss the relationship between percentage recall and the number of previous lists memorized. What can you conclude about interference in this case?

11. The figure shows learning curves for massed and spaced practice. Students practiced keeping a pointer on a moving target. Each trial lasted 30 seconds. Some students had a 15-second rest period between trials (massed practice) and others a 45-second rest (spaced practice). (Keep in mind that these are actually two graphs: the graph of the massed practice group and the graph of the spaced practice group.)

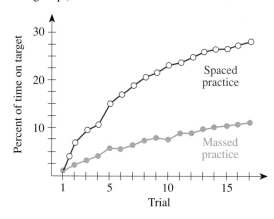

   (a) For the massed practice group, approximately how many trials does it take before they can score 10%?

   (b) For the spaced practice group, approximately how many trials does it take before they can score 10%?

   (c) Based on the graphs or learning curves given, what conclusions can you draw about spaced versus massed practice for this type of skill?

12. Two groups of students memorized a list of nonsense syllables; we will designate them the awake group and the asleep group. Each group was divided into smaller subgroups in which each subgroup was tested on the same words at differing time intervals; that is, one subgroup was tested 1 hour later, another was tested 2 hours later, etc. Students in the asleep group were asleep between the initial learning phase and the final testing phase; students in the awake group were awake between the learning and testing phases. The graph shows the results of the experiment.

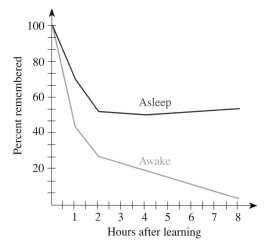

   (a) Compare the two groups by the graph. Which group exhibited less forgetting? If the results of this graph were true for all learning, what conclusions could you draw about when to study?

   (b) Look at the line segment covering only students tested at hour 1. How do these two (tested at 1 hour) groups compare?

   (c) Look at the graph for the awake group only. Describe, in general, how quickly forgetting is occurring as time passes.

   (d) Look at the graph for the awake group only. Use the graph to describe specifically the rates of forgetting (how quickly forgetting is occurring) for each segment.

13. Kyle is taking a business trip. Suppose that the following graph describes his day's travel. The horizontal axis is labeled t and represents the number of hours since Kyle began his trip. The vertical axis is labeled d and represents the distance (in miles) that Kyle is from his home. Use the information given in this graph to describe his day as best you can, and to determine his rate of speed and his direction.

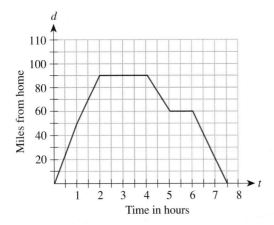

14. The graph below is called the *logistic growth curve*. It shows the growth of a population over time. The horizontal dashed line appearing on the graph is called the *carrying capacity* of the environment, which is defined to be the number of individuals in a population that the environment can support over an indefinite amount of time. The horizontal axis represents time and the vertical axis represents the number present in the population. Use the graph to discuss how population size changes over time. Discuss possible reasons why the graph looks this way.

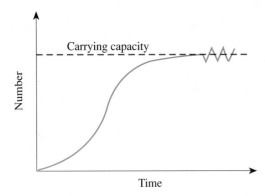

15. A psychologist wanted to test three different procedures or treatments to reduce aggressive behavior of 5-year-old children: treatment A, treatment B, and treatment C. An "aggressive child" was assigned to each treatment, which was administered twice a week for 4 weeks. Before starting treatment, the child was taken to a room where he or she was observed with a group of other children in a playroom for a half-hour. The psychologist's assistant recorded the number of times the child undergoing treatment exhibited aggressive behaviors. The child attended these play sessions after each treatment session, and continued attending

the play sessions after the 4-week treatment was completed. The assistant continued recording the aggressive behaviors.

The accompanying graph shows the results both during and after each of the three different treatments. The horizontal axis is the time, labeled *t,* in days during and following treatment starting at $t = 0$, the first play session before treatment. The vertical axis is labeled *b* and represents the number of aggressive behaviors exhibited during each play session.

Discuss what the graph tells you about the effectiveness of the therapies relative to each other. Which therapy has the "best" results?

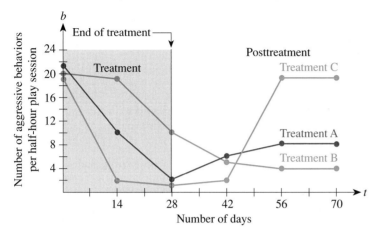

16. A biologist performed the following experiment with two paramecium species: *P. cyntia* and *P. soraria.*

    (a) He first grew the two species under identical conditions in separate containers, as shown in the figure. How do the two compare? What conclusion can you make about one species versus the other?

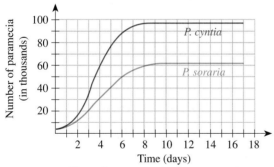

Two cultures grown in separate containers

    (b) The biologist then grew the two together in the same container, with the results shown in the following figure. Grown together, how do the two now compare? What conclusions can you make?

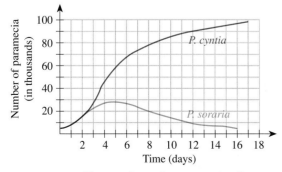

The same two cultures grown together
in the same container

CHAPTER 3 SUMMARY

After completing this chapter, you should be able to:

1. Sketch the graph of a first-degree equation in two variables (Section 3.1).

For example: Sketch the graph of $6y - 3x = 12$. Label the intercepts.

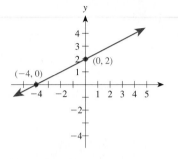

Figure 3.38

To find the x-intercept:	*To find the y-intercept:*
Set y = 0 and solve for x.	*Set x = 0 and solve for y.*
$6y - 3x = 12$	$6y - 3x = 12$
$6(0) - 3x = 12$	$6y - 3(0) = 12$
$-3x = 12$	$6y = 12$
$x = -4$	$y = 2$

The graph crosses the x-axis at $(-4, 0)$ and the y-axis at $(0, 2)$.

The graph of $6y - 3x = 12$ is shown in Figure 3.38. Finding a check point is left to the student.

2. Be able to extract information about an equation from its graph (Section 3.2).

For example: Figure 3.39 shows the graph of an equation:

(a) Using the graph to find y when $x = 4$, we start on the x-axis at $x = 4$ and project vertically until we meet the graph. Move horizontally until we meet the y-axis, and read off the value on the y-axis to get $y = -7$ (see Figure 3.40). Hence $y = -7$ when $x = 4$, or, equivalently, we can say the "point" $(4, -7)$ lies on the graph, or the ordered pair $(4, -7)$ satisfies the equation of the graph.

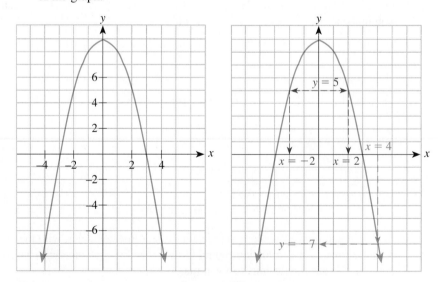

Figure 3.39   **Figure 3.40**

(b) To find x when $y = 5$ by using the graph, we start on the y-axis at $y = 5$ and project horizontally until we meet the graph. Note that we meet the graph in two places. This means that there are two points on the graph where the y-coordinate has the value 5. For each point on the graph, we move vertically until we meet the x-axis, and read off the values on the x-axis. We get two answers: $x = -2$, and $x = 2$ (see Figure 3.40). Hence the points $(-2, 5)$ and $(2, 5)$ lie on the graph.

(c) We can also see on the graph that the y-intercept is 9 (and therefore $(0, 9)$ lies on the graph), and the x-intercepts are -3 and 3 (and therefore $(-3, 0)$ and $(3, 0)$ lie on the graph).

(d) If we know that the equation of the graph given above is $y = 9 - x^2$, then we can verify that the points found in parts **(a), (b),** and **(c)** above lie on the graph by substituting the ordered pairs into the equation of the graph. For example, we verify that the point $(-2, 5)$ lies on the graph using the equation:

$$y = 9 - x^2 \qquad \text{\textit{Substitute } } x = -2 \text{ \textit{and} } y = 5 \text{ \textit{in the equation.}}$$
$$5 \overset{?}{=} 9 - (-2)^2$$
$$5 \overset{\checkmark}{=} 9 - 4 \qquad \text{\textit{Hence the point} } (-2, 5) \text{ \textit{lies on the graph.}}$$

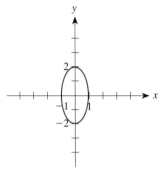

Figure 3.41
The graph of $4x^2 + y^2 = 4$

3. Understand the meaning of a relation and identify its domain and range (Section 3.3).

For example:

(a) The relation $\{(2, -3), (3, 4), (-2, -3)\}$ has domain $\{-2, 2, 3\}$ and range $\{-3, 4\}$. The relation assigns -3 to 2, 4 to 3, and -3 to -2.

(b) The relation $4x^2 + y^2 = 4$ has the graph shown in Figure 3.41. By looking at the graph, we see that the domain is $\{x \mid -1 \le x \le 1\}$ and the range is $\{y \mid -2 \le y \le 2\}$.

(c) The relation described by $y = \dfrac{2}{x - 1}$ has domain $\{x \mid x \ne 1\}$ because $x = 1$ is the only value of x that produces either an undefined or a nonreal value for y.

4. Understand the meaning of a function (Section 3.3).

For example:

(a) The relation $\{(2, -3), (2, 4)\}$ is not a function because the x-value, 2, is assigned two y-values, $y = -3$ and $y = 4$.

(b) $\{(3, 5), (2, 5)\}$ is a function because no x-value is assigned more than one y-value.

(c) Consider the relation described by the following graph:

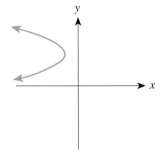

In the next figure we apply the vertical line test:

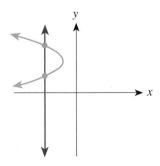

Since the graph of the relation intersects a vertical line at more than one point, this relation is not a function.

5. Use function notation (Section 3.4).

For example:
(a) If $f(x) = 3x^2 - 4$, then

$$f(2) = 3(2)^2 - 4 = 3 \cdot 4 - 4 = 8$$
$$f(-1) = 3(-1)^2 - 4 = 3 \cdot 1 - 4 = -1$$
$$f(s) = 3s^2 - 4$$

(b) If $f(x) = 2x - 1$ and $g(x) = x^2 + 3$, then

$$f(x + 2) = 2(x + 2) - 1 = 2x + 4 - 1 = 2x + 3$$
$$f(x) + 2 = (2x - 1) + 2 = 2x + 1$$
$$f(x) + f(2) = (2x - 1) + (2 \cdot 2 - 1) = 2x - 1 + 4 - 1 = 2x + 2$$
$$g(2x) = (2x)^2 + 3 = 4x^2 + 3$$

(c) Using the graph $y = f(x)$ given in Figure 3.42, we observe the following:

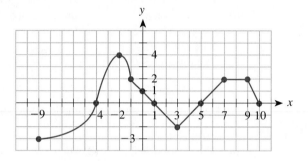

Figure 3.42

(i) $f(7) = 2$ (ii) $f(-9) = -3$ (iii) $f(-4) = 0$ (iv) $f(8) = 2$
(v) $f(-2)$ is greater than $f(2)$ [because $f(-2)$ is positive, whereas $f(2)$ is negative].
(vi) The graph of $y = f(x)$ has x-intercepts at $x = -4, 1, 5$, and 10.

6. Interpret data presented as a graph (Section 3.5).

For example: After a group of students memorized a list of random five-digit numbers, they were tested on their ability to recall the same list of numbers at regular hourly intervals for 5 hours. The graph in Figure 3.43 shows the average percent correct scores for the group at each testing point.

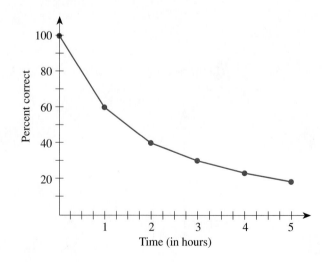

Figure 3.43

Some of the observations we can make are:

1. During the first hour, 40% of the numbers were forgotten. More numbers were forgotten during the first hour than during any other hour in the 5-hour period pictured in the graph.

2. During the third hour (that is, between time = 2 and time = 3), 10% of the numbers were forgotten.

3. At the end of this 5-hour period, 20% of the numbers were still remembered.

CHAPTER 3 REVIEW EXERCISES

In Exercises 1–26, sketch the graph of the equation. Label the intercepts.

1. $2x + y = 6$

2. $2x + y = -6$

3. $2x - 6y = -6$

4. $6y - 2x = 6$

5. $5x - 3y = 10$

6. $3x - 5y = 10$

7. $2x + 5y = 7$

8. $3x + 4y = 5$

9. $3x - 8y = 11$

10. $3x + 8y = 11$

11. $5x + 7y = 21$

12. $4x - 5y = 20$

13. $y = x$

14. $y = -x$

15. $y = -2x$

16. $x = 3y$

17. $y = \dfrac{2}{3}x + 2$

18. $y = -\dfrac{1}{2}x - 3$

19. $\dfrac{x}{3} + \dfrac{y}{2} = 12$

20. $\dfrac{x}{4} - \dfrac{y}{2} = 16$

21. $x - 2y = 8$

22. $x - 2y = 0$

23. $x - 2 = 0$

24. $y + 2 = 0$

25. $2y = 5$

26. $3x = 4$

27. Using the graph in Figure 3.44, determine the following:

   (a) Find y when $x = -2$ and 2.

   (b) Find x when $y = 3$.

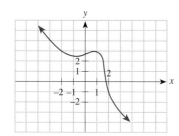

Figure 3.44

28. Using the graph in Figure 3.45, determine the following:

   (a) Find y when $x = -3$.

   (b) Find x when $y = 2$ and -5.

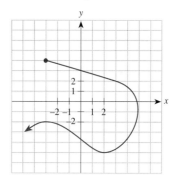

Figure 3.45

Exercises 29–32 refer to the figure given below.

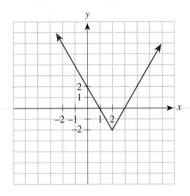

29. Find the value of y that satisfies the equation of the graph given above when $x = -2$.

30. Fill in the missing coordinate for the points that lie on the graph.
   (a) $(5, \ )$ **(b)** $(\ , 4)$

31. Find the x-intercepts.

32. Find the y-intercepts.

Exercises 33–34 refer to the graph of the equation $x^2 + y^2 = 20$ in the figure below.

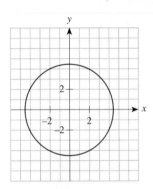

33. Find y for $x = 2$ by using the graph, and verify the value(s) using the equation of the graph.

34. Find x for $y = 4$ by using the graph, and verify the value(s) using the equation of the graph.

35. A salesperson is paid a base salary of $150 plus a commission of 5% of her gross sales.
   (a) Write an equation expressing her weekly income, I, in terms of her gross sales, g.
   (b) Sketch a graph of this equation.
   (c) Use the graph to estimate her income for the week in which her gross sales were $5,000.
   (d) Use the graph to determine her gross sales for the week in which her income was $750.

36. A car rental company charges $30.00 a day plus a mileage charge of 20¢ a mile.
   (a) Write an equation expressing the daily rental costs, C, of a car in terms of the number of miles driven, m.
   (b) Sketch a graph of this equation.
   (c) Use the graph to estimate the cost of driving 200 miles in a day.
   (d) Use the graph to determine how many miles can be driven if the costs are not to exceed $65.

In Exercises 37–42, identify the domain of the function.

37. $y = 2x + 1$ **38.** $2y = 3x$ **39.** $y = \sqrt{4 - x}$ **40.** $y = \sqrt{x + 3}$

41. $y = \dfrac{3x}{x + 2}$ **42.** $y = \dfrac{x}{2x + 1}$

In Exercises 43–50, determine whether the given relation is a function.

43. $\{(2, -5), (3, 8), (4, -5)\}$ **44.** $\{(6, 2), (5, 1), (6, 8)\}$

45. $\{(3, -1), (4, 2), (4, 7)\}$ **46.** $\{(-3, 1), (1, -3)\}$

47. $y = 2x + 3$ **48.** $2y = 3x - 1$

49. $y = x^2 - 4$ **50.** $x = y^2 - 4$

In Exercises 51–54, determine which of the graphs of relations represent functions.

51.

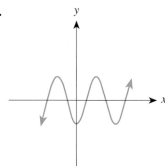

52.

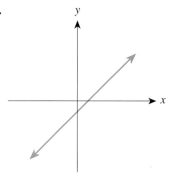

53.

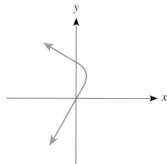

54.

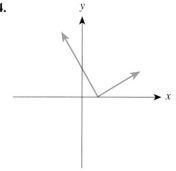

In Exercises 55–62, find the domain and range of the given relation or function;
also indicate whether the relation is a function.

55. $\{(-2, 3), (0, 5), (3, 7), (5, 4)\}$ **56.** $\{(-3, -8), (0, 0), (-3, 7), (2, 10)\}$

57. $\{(4, 9), (6, 9), (1, 9), (8, 9)\}$ **58.** $\{(6, -5), (6, 2), (6, 0), (6, -3)\}$

59.

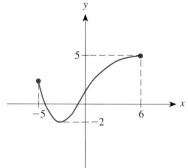

60.

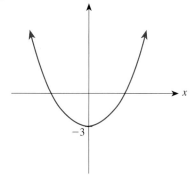

61.

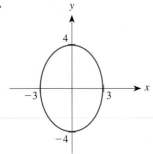

62.

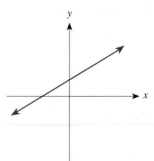

In Exercises 63–72, evaluate the functions at the given values. If a value is not in the domain of the function, state so.

63. $f(x) = 3x + 5$; $f(-1), f(0), f(1), f(2)$

64. $g(x) = 5 - 4x$; $g(-1), g(0), g(1), g(2)$

65. $f(x) = 2x^2 - 3x + 2$; $f(-1), f(0), f(1), f(2)$

66. $g(x) = 3x^3 + 2x - 3$; $g(-1), g(0), g(1), g(2)$

67. $h(x) = \sqrt{x - 5}$; $h(6), h(5), h(4)$

68. $g(x) = \sqrt{5 - 3x}$; $g(1), g(2), g(-1)$

69. $h(x) = \dfrac{x - 1}{x + 3}$; $h(1), h(3), h(-3)$

70. $h(x) = \dfrac{x + 1}{x}$; $h(-1), h(0), h(4)$

71. $f(x) = 2x^2 + 4x - 1$; $f(a), f(z)$

72. $f(a) = 2a^2 - 4a + 2$; $f(x), f(z)$

In Exercises 73–84, $f(x) = 5x + 2$ and $g(x) = 6 - x$. Find:

73. $f(x + 2)$

74. $f(x + 3)$

75. $f(x) + 2$

76. $f(x) + 3$

77. $f(x) + f(2)$

78. $f(x) + f(3)$

79. $g(x + 2)$

80. $g(x + 3)$

81. $g(2x)$

82. $g(3x)$

83. $2g(x)$

84. $3g(x)$

85. Use the graph of $y = f(x)$ given in Figure 3.46 to answer the following questions.
   (a) Find $f(-6)$.   **(b)** Find $f(0)$.   **(c)** Find $f(-4)$.
   (d) Find $f(6)$.   **(e)** Which is smaller, $f(4)$ or $f(5)$?
   (f) Identify the x-intercepts of the graph.

86. Use the graph of $y = h(x)$ in Figure 3.47 to answer the following questions:
   (a) $h(-2)$ **(b)** $h(6)$ **(c)** $h(0)$ **(d)** $h(4)$
   (e) $h(-4)$ **(f)** For what value(s) of x is $h(x) = 0$?

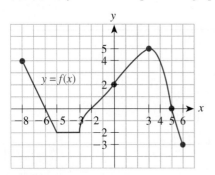

Figure 3.46
The graph of $y = f(x)$

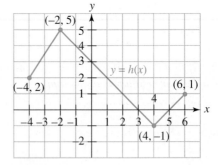

Figure 3.47
The graph of $y = h(x)$

87. Rasheed is taking a business trip. Suppose that the graph on the next page describes his day's travel.

   The horizontal axis is labeled t and represents the number of hours since Rasheed began his trip. The vertical axis is labeled d and represents the distance (in miles) that Rasheed is from his home.

   Use the information given in this graph to describe his day as best you can.

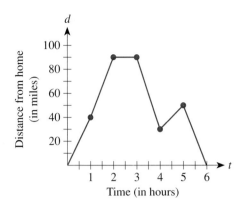

88. At time $t = 0$, an object is dropped from the top of a building, and we record the distance, d, the object is from the top of the building after each half-second elapses. We plot the graph of the relationship as shown on the right. The horizontal axis is the time, t, in seconds, and the vertical axis is the distance, d, in feet.

    Discuss the relationship between the distance the object falls and the amount of time it is in the air. What happens to the speed of the object as it drops?

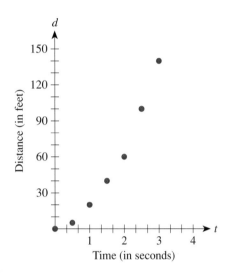

89. An office equipment leasing company rents out computer equipment. It charges a delivery fee of $30 plus a fee of $42 each day the computer is rented. Express the total charges, C, for a rental as a function of n, the number of days the computer is rented.

90. An airline is offering a special promotion to join its frequent flyer program. If you join the program, you get a joining bonus of 5,000 miles plus 2,500 miles for each shuttle flight you take. Express the mileage, M, you earn for joining the program and taking s shuttle flights as a function of s.

91. The following graph shows the number of people employed in the retail trade (in the thousands) in the United States from 1994 to 2004 (seasonally adjusted).
  (a) How many were employed in the retail trade in January of 1996?
  (b) During what years was the number of employees above 15 million?
  (c) Verbally describe the general trend shown by the graph from 1994 to 2004.

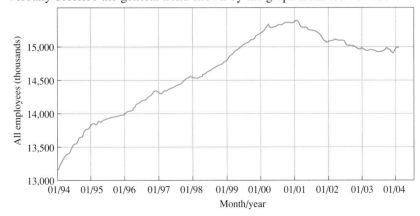

CHAPTER 3 PRACTICE TEST ⊙

1. Graph the following using the intercept method:
   (a) $3x - 5y = 30$ (b) $x - 7 = 0$

2. The graph of the equation $y = \sqrt{x} + 4$ appears in the figure below. Use the graph to find y for $x = 5$, and verify the value(s) using the equation of the graph.

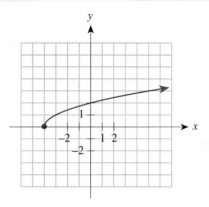

3. Identify the domain and range of each of the following relations:
   (a) $\{(2, -3), (2, 5), (3, 5), (4, 6)\}$
   (b)

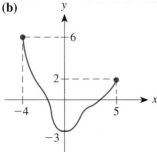

4. Identify the domain of each of the following functions:

   (a) $y = \sqrt{x - 4}$ (b) $y = \dfrac{x}{3x - 4}$

5. Identify which of the following relations are functions:
   (a) $\{(2, 5)\}, (2, 4)\}$ (b) $\{(3, 2), (4, 3), (5, 2)\}$ (c) $x = 3y^2$

   (d)

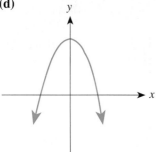

6. Given $f(x) = 3x^2 - 4$, $g(x) = \sqrt{2x + 1}$, and $h(x) = 5x - 3$, find:
   (a) $g(2)$ (b) $f(-3)$ (c) $h(x - 2)$
   (d) $f(x^2)$ (e) $f(x) - f(5)$

7. Use the graph of $y = f(x)$ given in Figure 3.48 to answer the following questions.

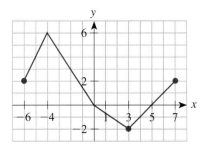

Figure 3.48

(a) Find $f(-6)$ (b) Find $f(3)$ (c) Find $f(0)$ (d) Find $f(7)$
(e) Which is the largest, $f(-4)$, $f(3)$, or $f(5)$?
(f) Find all x such that $f(x) = 0$.

8. The graph in Figure 3.49 illustrates the blood level, L, in milligrams of a certain substance m minutes after a medication is taken.

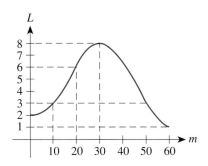

Figure 3.49

(a) How long does it take for the substance to reach its maximum level in the bloodstream?
(b) What is the maximum level of the substance in the bloodstream?
(c) During which of the 10-minute intervals indicated along the m-axis does the substance level increase the most? By how much does it increase?

CHAPTERS 1–3 CUMULATIVE REVIEW

In Exercises 1–4,

$$A = \{x \mid x \text{ is a prime number; } 7 < x < 41\}$$
$$B = \{y \mid y \text{ is a multiple of 3; } 6 \le y \le 42\}$$
$$C = \{z \mid z \text{ is a multiple of 5; } 6 < z < 40\}$$

List the elements in the following sets:

1. A

2. $A \cap B$

3. $A \cap C$

4. $A \cup C$

In Exercises 5–6, answer true or false.

5. $\dfrac{6}{2} \in Q$

6. $\sqrt{7} \in Q$

In Exercises 7–8, graph the following sets:

7. $\{x \mid -2 \le x < 8\}$

8. $\{x \mid 2 \le x < 8, \ x \in Z\}$

In Exercises 9–10, list the real number property illustrated by the statement.

9. The sum of two real numbers is a real number.

10. $\left(\dfrac{1}{x-3}\right)(x-3) = 1 \quad (x \ne 3)$

In Exercises 11–14, perform the operations.

11. $-2 + 3 - 4 - 8 + 5$

12. $(-2)(4) - (-3)(-3)$

13. $(-9 + 3)(-4 - 2)$

14. $6 - \{3 - 5[2 - (3 - 9)]\}$

15. Evaluate the expressions given $x = -4$, $y = 2$, and $z = 0$.

   (a) $|y - x| - (y - x)$

   (b) $\dfrac{3x^2y - 2x}{3xz}$

In Exercises 16–17, perform the operations and express your answer in simplest form.

16. $-3x(x^2 - 2x + 3)$

17. $(-2xy)^2(-3x^2y)$

18. The number of motor vehicle registrations in the United States from 1980 to 1997 can be estimated by the equation

$$M = 3.058(t - 1,980) + 156.35$$

where M is the number of motor vehicle registrations (in millions) during year t.
   (a) Simplify the right-hand side of the equation.
   (b) Use both the simplified and the original version of this equation to compute the number of motor vehicle registrations in the United States in 1995.

In Exercises 19–22, translate the statements algebraically.

19. Three more than the product of a number and 8

20. The product of the first two of three consecutive integers less the third consecutive integer

21. The length of a rectangle is 3 less than 4 times its width. Express its area and perimeter in terms of its width.

22. Harold has 35 coins in dimes and nickels; express the value of his coins in terms of the number of nickels.

In Exercises 23–30, solve the equation.

23. $4x - 3 = 5x + 8$

24. $5(a - 1) = 5(a + 2)$

25. $2x + 3 - (x - 2) = 5 - (x - 4)$

26. $3x - 2 + 2(x - 1) = 1 - 2(3 - x)$

27. $3(x - 4) + 2(3 - x) = 5[x - (2 - 3x)]$

28. $5(x - 2) - 2(3x + 1) = 2(x - 4)$

29. $|2x + 1| = 9$

30. $|3x - 2| = |x - 1|$

In Exercises 31–38, solve the inequalities and graph the solutions on the number line.

31. $3x - 5 \leq x - 8$

32. $3 - 5x > 5 + 3x$

33. $6(x - 2) + (2x - 3) < 2(4x + 1)$

34. $-5 < 3 - 2x \leq 9$

35. $|x - 5| < 4$

36. $|2x + 1| > 7$

37. $|5 - 4x| > 7$

38. $|8 - 3x| \leq 8$

39. John's math teacher counts the final exam as 40% of his course grade, and John's average for all but the final exam is 63. Construct a mathematical model that shows how John's course grade is related to his final exam score. Use this model to *determine* the minimum score he would need on his final exam to get a course grade of 70.

40. A salesperson is paid a base salary of $250 a week plus a 2% commission on her gross sales. Construct a mathematical model that shows how her weekly salary is related to her gross sales. Use this model to determine her weekly gross sales required to earn $500 a week.

In Exercises 41–48, solve the problem algebraically.

41. If 4 times a number is 3 less than twice the number, what is the number?

42. Cindy has 25 coins in quarters and dimes, totaling $3.25. How many of each coin does she have?

43. A man went to a store and bought 28 packages of paper plates for $33. Some of the packages contained 300 plates and cost $2 per package, and the rest of the packages contained 100 plates and cost $1 per package. How many plates did he buy altogether?

44. A handyman charges $14 an hour to paint a house and his assistant gets $7 an hour. If the bill for labor for painting a house comes to $567, how many hours did the assistant paint if altogether they worked a total of 58 hours?

45. A GL-70 printer can print 80 characters per second, whereas a VF-44 printer can print 120 characters per second. How long would it take a VF-44 printer to print a document that takes 30 minutes for the GL-70 to print?

46. A company installed a photocopy machine that can make 90 copies per minute. For a certain job it had to make 9,750 copies. The job was begun on the new machine, but after a while the machine broke down. The remaining copies were made by the old machine, which makes 60 copies per minute. If the total job took 2 hours, how many copies did the new machine make?

47. The length of a rectangle is 2 more than 3 times its width. If the width varies from 5 to 12 ft, what is the range of values of the perimeter?

48. Charles gets $12 an hour as a tutor and $15 an hour as a mechanic. If he is to work a total of 30 hours, what is the minimum number of hours he should work as a mechanic if he wants to make at least $399?

Exercises 49–50 refer to the graph of the equation $x + y^2 = 4$ in the figure below.

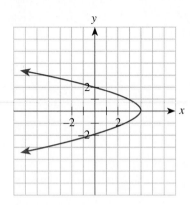

49. Find x for $y = 3$ by using the graph, and verify the value(s) using the equation of the graph.

50. Find y for $x = -5$ by using the graph, and verify the value(s) using the equation of the graph.

In Exercises 51–54, find the domain and range of each relation and determine whether it is a function.

51. $\{(2, 3), (4, -1), (7, 5)\}$

52. $\{(1, 3), (-2, 3), (5, 6)\}$

53. $\{(4, 2), (3, 9), (4, 7)\}$

54. $\{(-1, 1), (-2, 4), (1, 1), (2, 4)\}$

In Exercises 55–58, determine the domain of the given function.

55. $y = x^2 - x + 3$

56. $y = \sqrt{x - 3}$

57. $y = \dfrac{x}{5x - 4}$

58. $y = \dfrac{\sqrt{x + 4}}{x - 3}$

In Exercises 59–62, determine which are graphs of functions.

59.

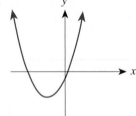

60.

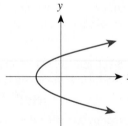

61.

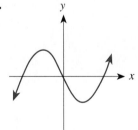

62.

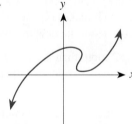

In Exercises 63–74, if $f(x) = x + \dfrac{1}{x}$, $g(x) = \dfrac{1}{x^2}$, and $h(x) = 3x + 1$, find:

63. $f(-3)$

64. $g\left(\dfrac{3}{4}\right)$

65. $h(x^2)$

66. $g(x^2)$

67. $f(a + 3)$

68. $f(a) + 3$

69. $f(3a)$

70. $3f(3a)$

71. $f(1) + g(-1)$

72. $h(8) + 6$

73. $f(7) + h\left(-\dfrac{1}{3}\right)$

74. $h\left(\dfrac{5}{2}\right)$

In Exercises 75–80, sketch a graph of the equation using the intercept method.

75. $5x - 4y = 10$

76. $3x - 2y = 12$

77. $5y = 3x - 10$

78. $5x = 3y - 10$

79. $3x - 2 = 8$

80. $2y + 4 = 6$

81. A biologist takes a culture of bacteria and observes it growing in a certain environment. The accompanying graph illustrates the population growth of a colony of bacteria over time.

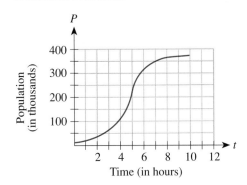

(a) Estimate the size of the population 2 hours after she takes the culture.

(b) Estimate when the population will reach 200,000.

(c) When is the culture growing most rapidly?

(d) When does the population growth begin to level off?

82. A salesperson earns a weekly salary of $275 plus a commission of 2.5% of sales. Express the weekly income, I, as a function of s, the weekly sales.

83. An interior designer charges a flat fee of $350 plus an hourly fee of $75. Express the total charges, C, on a project that takes the designer h hours to complete.

In Exercises 84–85, determine the domain and range of each function from its graph.

84.

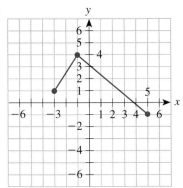

85.

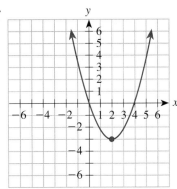

86. Use the accompanying figure to answer the following questions.

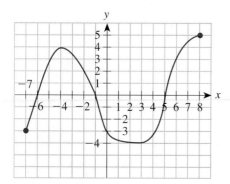

(a) Find $f(-7)$.   (b) Find $f(0)$.   (c) Find $f(8)$.   (d) Find $f(3)$.

(e) Which is smaller, $f(-4)$ or $f(1)$?

(f) Identify the x-intercepts of the graph of $f(x)$.

87. Jonas averages $15 grading homework papers, and $25 an hour tutoring. He can work no more than 20 hours a week. Express the amount of money he can make as a function of the number of hours he grades homework papers. Graph this function. Use this graph to estimate how much he makes if he works 5 hours grading homework papers this week. Use this graph to estimate how many hours he should be grading homework papers if he wants to make $380 this week.

CHAPTERS 1–3 CUMULATIVE PRACTICE TEST

1. Given the sets

$$A = \{a \mid a \text{ is a multiple of } 4, \quad 3 < a < 21\}$$
$$B = \{x \mid x \text{ is a multiple of } 2, \quad 0 < x < 24\}$$

find:
(a) $A \cap B$
(b) $A \cup B$

2. What real number property is illustrated by the following statement?

$$(x + 2y)(x + 3y) = x(x + 3y) + 2y(x + 3y)$$

3. Evaluate the following:
(a) $(-2)(-2) - (-2)^2(2)$
(b) $5 - \{6 + [2 - 3(4 - 9)]\}$

4. Given $x = -4$ and $y = -5$, evaluate $\dfrac{x^2 - 3xy + 2y^2}{x - y}$.

5. Solve the following equations:
(a) $3x - 2 = 5x + 4$
(b) $3(x - 5) - 2(x - 5) = 3 - (5 - x)$
(c) $|x - 3| = 8$

6. Solve each of the following inequalities and graph the solution set on the number line.

(a) $3 - (2x - 4) < 5 - (7 - 2x)$

(b) $|3x - 2| < 5$

(c) $|5 - 4x| \geq 1$

7. A truck is carrying 170 packages weighing a total of 3,140 lb. If each package weighs either 10 lb or 30 lb, how many of each weight package are on the truck?

8. How long would it take a car traveling at 55 mph to overtake a car traveling at 40 mph with a 1-hour head start?

9. Evan is taking a business trip. Suppose that the following graph describes his day's travel. The horizontal axis is labeled t and represents the number of hours since Evan began his trip. The vertical axis is labeled d and represents the distance (in miles) that Evan is from his home.

  Use the information given in this graph to describe his day as best you can.

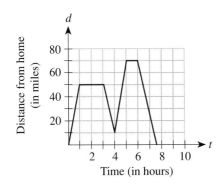

10. Sketch the graph of $3x - 4y = 18$ using the intercept method.

11. The graph of the equation $9x^2 + 16y^2 = 144$ is shown here. Identify the x-intercepts by using the graph, and verify the value(s) using the equation of the graph.

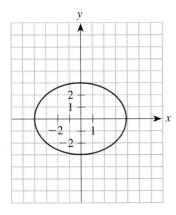

12. Find the domain of each of the following functions:

(a) $f(x) = \dfrac{x}{x - 9}$

(b) $g(x) = \sqrt{2x - 3}$

13. Given $f(x) = \dfrac{x - 2}{x + 1}$ and $g(x) = 4x - 1$, find:

(a) $f(-5)$ **(b)** $g(x^2)$

(c) $f(x + 2)$ **(d)** $f(x) + 2$

(e) $g(5x)$ **(f)** $5g(x)$

(g) $f(-1) + g(0)$ **(h)** $g(-1) - g(1) - 1$

14. Use the accompanying figure to answer the following questions.

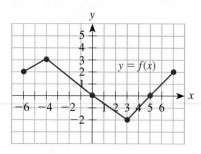

(a) Find $f(-6)$.
(b) Find $f(3)$.
(c) Find $f(0)$.
(d) Find $f(-4)$.
(e) Which is larger, $f(-3)$ or $f(2)$?
(f) Identify the x-intercepts of the graph of $f(x)$.
(g) Identify the domain and range.

15. A salesperson is paid a base salary of $200 a week plus a 10% commission on his gross sales. Express his weekly salary as a function of his gross sales. Sketch a graph of this function and use the graph to determine his weekly gross sales required to earn $600 a week.

4

Equations of a Line and Linear Systems in Two Variables

STUDY SKILLS

© Thinkstock/Alamy

Straight Lines and Slope

Recall that the graph of a first-degree equation in two variables is a straight line. We can state this algebraically: The graph of an equation of the form $Ax + By = C$ (where A and B are not both 0) is a straight line. For this reason, such an equation is called a ***linear equation.*** We also saw that (provided $B \neq 0$) we can solve such an equation explicitly for y, and such an equation defines y as a function of x. This type of function is called a ***linear function.***

EXAMPLE 1

Solution

We have also found a third check point, $(-10, 14)$.

Figure 4.1
The graph of
$F = 1.8C + 32$

The Celsius (C) and Fahrenheit (F) temperature scales are related according to the equation $F = 1.8C + 32$. Sketch the graph of this relationship.

The equation $F = 1.8C + 32$ is a first-degree equation in two variables. We must first decide which variables should be represented along the horizontal axis and the vertical axis. We choose to let the horizontal axis represent the Celsius temperature, C, and the vertical axis to represent the Fahrenheit temperature, F. Recognizing this as a first-degree equation in two variables, we know that the graph will be a straight line. Using the intercept method discussed in Section 3.1, we find that the graph crosses the C-axis at approximately $(-17.8, 0)$ and crosses the F-axis at $(0, 32)$. Using these two points, we obtain the graph in Figure 4.1.

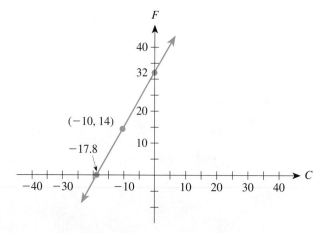

This graph certainly satisfies the vertical line test, which reinforces the fact that such a linear equation defines F as a function of C.

In graphing the equation $F = 1.8C + 32$ of Example 1, we are looking for all ordered pairs that satisfy the particular condition that "the F-coordinate is 32 more than 1.8 times the C-coordinate." The straight line we obtained in Example 1 is the graph of all the ordered pairs that satisfy this particular condition.

In general, a first-degree equation in two variables can be viewed as a condition on the two variables. If we let the variables x and y represent the first and second coordinates of points in a rectangular coordinate system, then certain points satisfy the condition and others do not. The set of all points that do satisfy this condition—that is, the graph of a first-degree equation in two variables—is a straight line.

What if we now reverse this situation? Suppose we have a straight-line graph. How do we find its equation? The importance of this question is illustrated by the following example.

EXAMPLE 2

The following table contains employee data for a certain manufacturing company.

Years of Operation	x	1	2	3	4	5	6	7	8	9	10
Number of Employees	y	26	29	34	38	44	48	53	59	62	67

In this table, x represents the number of years the company has been operating, and y represents the number of employees working at the firm. Sketch a graph of these data and use the graph to guess how many employees will be working at this firm in year 11.

Solution

In Figure 4.2(a), we have plotted the points corresponding to the ordered pairs given in the table. We observe that the data points appear to fall "approximately" along a straight line, as indicated in Figure 4.2(b). If the trend indicated by these data continues, then from the graph we can "guess" that in year 11 (that is, when $x = 11$), the company would have 72 employees ($y = 72$).

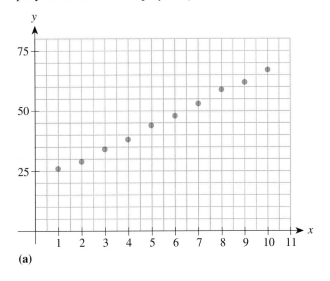

(a)

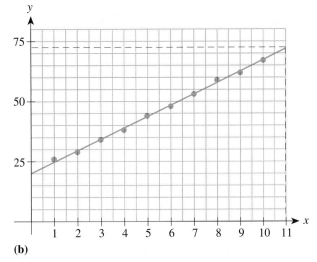

Figure 4.2
The points corresponding
to the data given in
Example 2

(b)

If we knew that the equation of this approximation line were $y = 4.7x + 20$, then we could simply substitute $x = 11$ to obtain $y = 4.7(11) + 20 = 71.7$. This tells us that after 11 years the company will have approximately 71.7 employees, which we would round to 72 employees.

As Example 2 indicates, having an equation of a straight line gives us easy access to x and y values that correspond to points that lie on the line. This section and the next are devoted to the question, How do we determine an equation of a line whose graph is given? Suppose we are given the graph of a straight line and we want to produce an equation for this straight line. Keeping in mind that an equation of a line is the condition that the points on the line must satisfy, we must naturally ask what condition must the points on the given line satisfy.

Look at Figure 4.3. From basic geometry we find that triangles ABC and DEF are similar triangles. ($\angle BAC \cong \angle EDF$ because $\overline{AC}$ and $\overline{DF}$ are parallel, and $\angle ABC \cong \angle DEF$ since $\overline{BC}$ and $\overline{EF}$ are parallel. So corresponding angles are equal.) Therefore, their corresponding sides are in proportion. That is,

$$\frac{|BC|}{|AC|} = \frac{|EF|}{|DF|}$$ *Note:* $|BC|$ *means the length of line segment BC.*

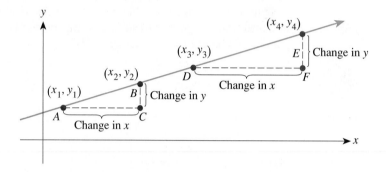

Figure 4.3

In other words, as we move from any point on a nonvertical line to any other point on the line, the ratio of the change in the y-coordinates of the points to the change in the x-coordinates of the points is constant. This fact is what we are looking for—a condition that all points on a line must satisfy.

The remainder of this section is devoted to further development of this idea. In the next section we will return to answer the question raised earlier about how to obtain an equation for a line when we have its graph.

We define the following:

DEFINITION   Let $P_1(x_1, y_1)$ and $P_2(x_2, y_2)$ be any two points on a nonvertical line L. The *slope* of the line L, denoted by m, is given by

$$m = \frac{y_2 - y_1}{x_2 - x_1} \qquad (x_1 \neq x_2)$$

Note that this definition uses what we saw in Figure 4.3, that the ratio of the change in y to the change in x is *independent* of the points chosen; that is, for any two points on a particular line, the ratio will remain the same.

For example, let's see how we would find the slope of the line passing through the points $(2, -3)$ and $(8, 1)$. Although it is not always necessary, it is usually helpful to draw a diagram illustrating the given information (see Figure 4.4).

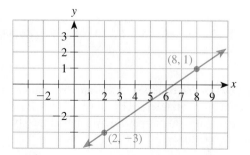

Figure 4.4

The formula for the slope of a line is

$$m = \frac{y_2 - y_1}{x_2 - x_1}$$

but which of our points is (x_1, y_1) and which is (x_2, y_2)? The fact of the matter is that it does not make any difference which we call the "first" point and which we call the "second" point, as long as we are consistent for both the x- and y-coordinates.

We can let $P_1(x_1, y_1) = (2, -3)$ and $P_2(x_2, y_2) = (8, 1)$ and we get

$$m = \frac{y_2 - y_1}{x_2 - x_1} = \frac{1 - (-3)}{8 - 2} = \frac{4}{6} = \frac{2}{3}$$

or we can let $P_1(x_1, y_1) = (8, 1)$ and $P_2(x_2, y_2) = (2, -3)$ and we get

$$m = \frac{y_2 - y_1}{x_2 - x_1} = \frac{-3 - 1}{2 - 8} = \frac{-4}{-6} = \frac{2}{3}$$

Thus, the slope is $m = \frac{2}{3}$.

EXAMPLE 3

Compute the slope of the line, and sketch the line passing through each pair of points.

(a) $P(8, 3)$ and $Q(4, 6)$ **(b)** $R(2, -5)$ and $S(5, -2)$

Solution

(a) Let $(x_1, y_1) = (8, 3)$ and $(x_2, y_2) = (4, 6)$. Then the slope formula gives

$$m = \frac{y_2 - y_1}{x_2 - x_1} = \frac{6 - 3}{4 - 8} = \frac{3}{-4}$$

Therefore, $\boxed{\text{the slope is } -\dfrac{3}{4} = -0.75}$.

(b) Let $(x_1, y_1) = (2, -5)$ and $(x_2, y_2) = (5, -2)$. The slope formula gives

$$m = \frac{y_2 - y_1}{x_2 - x_1} = \frac{-2 - (-5)}{5 - 2} = \frac{3}{3} = 1$$

Therefore, $\boxed{\text{the slope is } 1}$.

Figure 4.5 illustrates the lines passing through each pair of points.

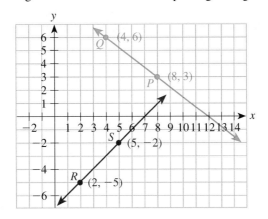

Figure 4.5

Note that if we start at any point on the line through P and Q and we move 4 units to the right, then we must also move 3 units down to get back on the line. Alternatively, we can think of the slope as $\dfrac{-0.75}{1}$, and thus if we move 1 unit to the right from a point on the line, then we must also move down 0.75 unit to get back on the line.

Similarly, we note that if we start at any point on the line through R and S and we move 1 unit to the right, then we must also move 1 unit up to get back on the the line. We have more to say about this idea a bit later in this section.

EXAMPLE 4

Find the slope of the line whose equation is $y = \dfrac{3}{2}x - 4$.

Solution

In the next section we will find a quick way to arrive at the answer to this question. But for now, to find the slope of a line, we need two points on the line. Since the slope is independent of the points chosen, we can arbitrarily choose any two points that satisfy the equation and therefore lie on the line.

If $x = 0$ then $y = \dfrac{3}{2}(0) - 4 = -4$; thus $(0, -4)$ is one point on the line.

If $x = 4$ then $y = \dfrac{3}{2}(4) - 4 = 6 - 4 = 2$; thus $(4, 2)$ is another point on the line.

Using the slope formula, we have

$$m = \frac{2 - (-4)}{4 - 0} = \frac{6}{4} = \frac{3}{2} = 1.5$$

Hence, the slope of the line whose equation is $y = \dfrac{3}{2}x - 4$ is $\boxed{\dfrac{3}{2}}$.

If we choose two other points that satisfy the equation, for example $(2, -1)$ and $(10, 11)$ (check for yourself that these points satisfy the equation), we will get the same result because the slope of a line is independent of the points chosen on the line:

$$m = \frac{-1 - 11}{2 - 10} = \frac{-12}{-8} = \frac{3}{2} = 1.5$$

The graph of the equation $y = \dfrac{3}{2}x - 4$ appears in Figure 4.6.

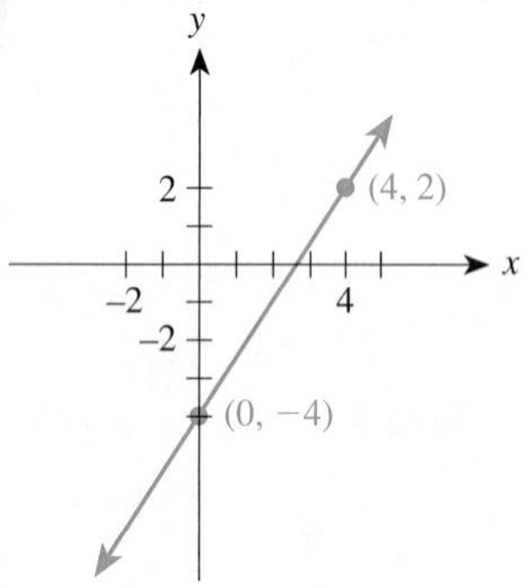

Figure 4.6
The graph of $y = \frac{3}{2}x - 4$

It is important to note that in the definition of the slope of a line, we have specified that the line be nonvertical. The reason for this is that for a vertical line, all x-coordinates are the same; we would be forced to divide by 0 in the computation of the slope.

For example, if we *try* to compute the slope of the vertical line passing through the points $(1, -4)$ and $(1, 2)$ (see Figure 4.7), we get

$$m = \frac{2 - (-4)}{1 - 1} = \frac{6}{0}$$

which is undefined. (This is why, in the definition of the slope, we specified $x_1 \neq x_2$.)

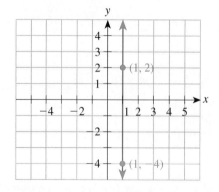

Figure 4.7
The vertical line through $(1, -4)$ and $(1, 2)$

Thus, we find that the following is true:

> The slope of a vertical line is undefined.

Let us now examine what this number, the slope, tells us about a line. Recall that *whenever we describe a graph, we describe it for increasing values of x (that is, moving from left to right)*.

The line in Figure 4.8(a) is rising (as we move from left to right), while the line in Figure 4.8(b) is falling (as we move from left to right).

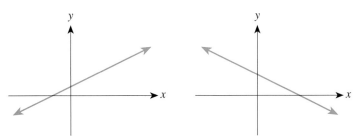

The y values are increasing as we move from left to right.

(a) This line is rising.

The y values are decreasing as we move from left to right.

(b) This line is falling.

Figure 4.8

The slope of a line is a number that tells us the *rate* at which its *y* values are increasing or decreasing. In other words, it is a measure of the steepness of a line. For example, if the slope of a line is $\frac{2}{5}$, this tells us that

$$m = \frac{2}{5} = \frac{\text{Change in } y}{\text{Change in } x}$$

This means that a 5-unit change in *x* gives a 2-unit change in *y*. Therefore, the line has the *steepness* shown in Figure 4.9.

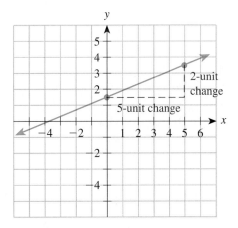

Figure 4.9
A line whose slope is $\frac{2}{5}$

A line whose slope is $\frac{5}{2}$ has the steepness indicated in Figure 4.10.

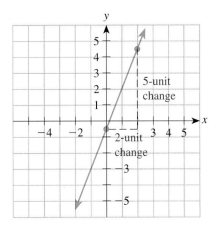

Figure 4.10
A line with slope $\frac{5}{2}$

Figure 4.11 shows what happens when we vary the slope of a rising line passing through the point (2, 3). Notice that the greater the slope, the steeper the line. We can see that, as we move from *left to right,*

> A line with positive slope rises.

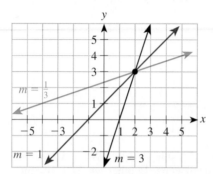

Figure 4.11

What about lines with negative slope? Let's look at a line with a slope of $-\dfrac{3}{4}$ passing through (3,1). We can view $m = -\dfrac{3}{4}$ as $\dfrac{-3}{4}$ or $\dfrac{3}{-4}$. A slope of $\dfrac{-3}{4}$ means that a 4-unit change in x gives a -3-unit change in y. A slope of $\dfrac{3}{-4}$ means that a -4-unit change in x gives a 3-unit change in y. Thus, we can draw the line as shown in Figures 4.12a and 4.12b. In both cases we get the same figure.

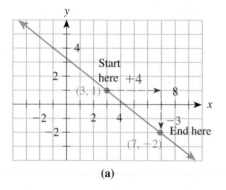 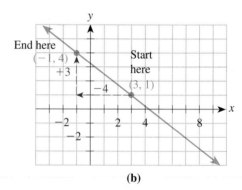

Figure 4.12
Line passing through the point (3, 1) with slope $-\dfrac{3}{4}$

      (a)   (b)

We can see that, as we move from *left to right,*

> A line with negtive slope falls.

Figure 4.13 illustrates that the greater the *absolute value* of the slope, the steeper the line.

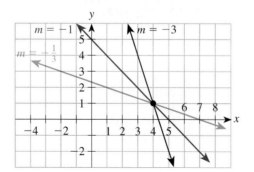

Figure 4.13

EXAMPLE 5

What is the slope of the line through the points $(1, 4)$ and $(3, 4)$?

Solution $m = \dfrac{4 - 4}{3 - 1} = \dfrac{0}{2} = 0$

This line is horizontal, as shown in Figure 4.14.

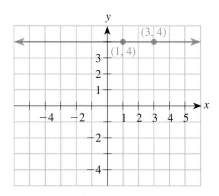

Figure 4.14

Any horizontal line, for which the y-coordinates will all be equal, will have slope equal to 0. It is perfectly reasonable that a horizontal line should have a slope equal to 0, since it has *no steepness*.

> A horizontal line has zero slope.

Do not confuse a slope of 0 with undefined slope.
A line with slope 0 has no steepness (it is horizontal).
A line with undefined slope has "infinite" steepness (it is vertical).

It is fairly intuitive that lines that have the same slope and therefore the same steepness are parallel. Conversely, lines that are parallel have the same steepness and hence the same slope. This fact is the subject of the following theorem.

THEOREM   Two distinct lines L_1 and L_2 with slopes m_1 and m_2, respectively, are parallel if and only if $m_1 = m_2$.

EXAMPLE 6

Show that the points $P(2, 4)$, $Q(8, 0)$, $R(3, -3)$, and $S(-3, 1)$ form the vertices of a parallelogram.

Solution   We first plot the given points as shown in Figure 4.15.

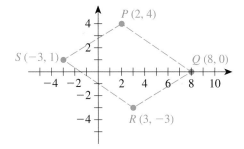

Figure 4.15

A parallelogram is a quadrilateral (a four-sided polygon) whose opposite sides are parallel. Thus, we must show that $\overline{PQ}$ is parallel to $\overline{SR}$ and that $\overline{SP}$ is parallel to $\overline{RQ}$. By the previous theorem, we can find out whether those sides are parallel by examining their slopes. Thus, we compute the slopes of each side:

$$m_{\overline{PQ}} = \frac{4-0}{2-8} = \frac{4}{-6} = -\frac{2}{3} \qquad m_{\overline{SR}} = \frac{1-(-3)}{-3-3} = \frac{4}{-6} = -\frac{2}{3}$$

$$m_{\overline{SP}} = \frac{1-4}{-3-2} = \frac{-3}{-5} = \frac{3}{5} \qquad m_{\overline{RQ}} = \frac{-3-0}{3-8} = \frac{-3}{-5} = \frac{3}{5}$$

Since the slopes of the opposite sides are the same, the opposite sides are parallel and therefore ⟨ the figure is a parallelogram ⟩.

If two lines pass through a single point and have the same slope, then they must be the same line. We can restate this as follows:

> A point and a slope determine a line.

Thus, just as we can draw a line by knowing two points, we should also be able to draw a line given one point and the slope of the line.

EXAMPLE 7

(a) Graph the line passing through the point $(3, 1)$ with slope $= \frac{2}{5}$.

(b) Graph the line passing through the point $(-2, 1)$ with slope $= -4$.

Solution

(a) First we plot the point $(3, 1)$. Since the line must have slope $m = \frac{2}{5}$, this means that

$$m = \frac{\text{Change in } y}{\text{Change in } x} = \frac{2}{5}$$

or that for every 5 units we move off the line to the right (change in x equal to $+5$), we must move 2 units up to return to the line (change in y equal to $+2$). Hence, if we start at $(3, 1)$, count 5 units right and then 2 units up, we arrive at another point on the line. We draw a line through the two points, as illustrated in Figure 4.16.

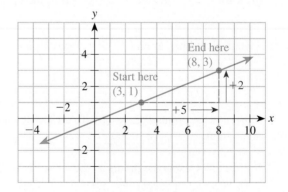

Figure 4.16

Notice that if we decided to move 10 units right (change in x equal to $+10$), then we would have to move 4 units up to return to the line (change in y equal to $+4$) and make the ratio $m = \frac{4}{10} = \frac{2}{5}$. On the other hand, if we moved 5 units *left* (change in x equal to -5), then we would have to move 2 units *down* (change in x equal to -2) to keep $m = \frac{-2}{-5} = \frac{2}{5}$.

(b) To graph the line passing through $(-2, 1)$ with slope $= -4$, we start by plotting the point $(-2, 1)$. To find the next point, we note that since the slope, m, is -4, we can rewrite -4 as

$$\frac{-4}{1} = \frac{\text{Change in } y}{\text{Change in } x}$$

Therefore, for every 1 unit we move off the line to the right (change in *x* equal to $+1$), we must travel *down* 4 units (change in *y* equal to -4) before we find another point on the line. Thus, we start at $(-2, 1)$, move 1 unit right and 4 units down to find another point, and then draw a line through the two points (see Figure 4.17).

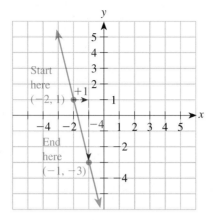

Figure 4.17

Example 7 demonstrates that given a point and a slope, we can geometrically determine the graph of the line. In the next section we will find that given the same information, we can (algebraically) determine the equation of the line.

We have already discussed parallel lines. What about the slopes of perpendicular lines?

EXAMPLE 8

Sketch the lines with slopes $\frac{6}{7}$ and $-\frac{7}{6}$ that pass through the point $(10, 4)$.

Solution

$m = \frac{6}{7}$ means a 7-unit change (increase) in *x* produces a 6-unit change (increase) in *y*.

$m = -\frac{7}{6}$ means a -6-unit change (decrease) in *x* produces a 7-unit change (increase) in *y*.

Let's sketch two lines with slopes $\frac{6}{7}$ and $-\frac{7}{6}$ (see Figure 4.18).

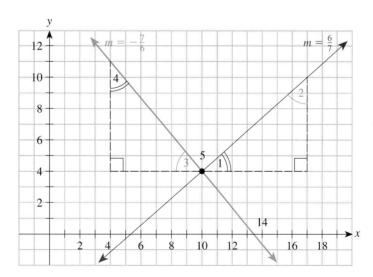

Figure 4.18

On the figure we have drawn in two triangles to help us visualize the following:

We know that $\angle 1 + \angle 2 = 90°$.

Since these two triangles are congruent (compare the lengths of their sides), we also know that $\angle 2 = \angle 3$ (they are both opposite the side of the triangle whose length is 7).

Therefore, $\angle 1 + \angle 3 = 90°$.

But then $\angle 5$ must also be $90°$ (because $\angle 1 + \angle 3 + \angle 5 = 180°$) and so the two lines are perpendicular.

$$\text{Note that } \left(\frac{6}{7}\right)\left(-\frac{7}{6}\right) = -1.$$

Example 8 is a particular case of the following theorem.

THEOREM   Two nonvertical lines L_1 and L_2 with slopes m_1 and m_2, respectively, are perpendicular if and only if $m_1 \cdot m_2 = -1$.

$$\text{If } m_1 \cdot m_2 = -1, \text{ we can write } m_2 = -\frac{1}{m_1} \quad \text{or} \quad m_1 = -\frac{1}{m_2}.$$

For this reason we often say that the slopes of nonvertical perpendicular lines are "negative reciprocals" of each other.

We should note that throughout the previous discussion we have insisted on nonvertical lines. While it is true that a horizontal line and a vertical line are perpendicular, the slope of a vertical line is undefined and so is not covered by the last theorem.

EXAMPLE 9

Given two points on each of the lines L_1, L_2, L_3, and L_4, compute the slope of each line and determine whether any two of the lines are parallel or perpendicular.

L_1:   (1, 3) and (4, 5) L_2:   (−2, 5) and (0, 2)

L_3:   (−3, 0) and (0, 2) L_4:   (2, 1) and (−1, 3)

Solution

$$m_1 = \frac{5 - 3}{4 - 1} = \frac{2}{3}$$

$$m_2 = \frac{2 - 5}{0 - (-2)} = \frac{-3}{2} = -\frac{3}{2}$$

$$m_3 = \frac{2 - 0}{0 - (-3)} = \frac{2}{3}$$

$$m_4 = \frac{3 - 1}{-1 - 2} = \frac{2}{-3} = -\frac{2}{3}$$

Note that L_4 is neither parallel nor perpendicular to L_1, L_2, or L_3.

Since $m_1 = m_3$, $\boxed{L_1 \text{ and } L_3 \text{ are parallel}}$.

Since $m_2 = -\dfrac{1}{m_1}$ and $m_2 = -\dfrac{1}{m_3}$, $\boxed{L_2 \text{ is perpendicular to } L_1 \text{ and } L_3}$.

EXAMPLE 10   Find a value for c so that the line passing through the points (3, 4) and (−1, c) has slope $\frac{1}{2}$.

Solution   Using the formula for the slope of a line, we determine that the slope is

$$\frac{c - 4}{-1 - 3} = \frac{c - 4}{-4}$$

and we want this to be equal to $\frac{1}{2}$. The equation is

$$\frac{c - 4}{-4} = \frac{1}{2} \qquad \textit{Multiply both sides by } -4.$$

$$\frac{-4}{1} \cdot \frac{c - 4}{-4} = \frac{1}{2} \cdot \frac{-4}{1}$$

$$c - 4 = -2$$

$$\boxed{c = 2}$$

EXAMPLE 11

An engineer has specified that sewage pipe for a certain building must have a 3% drop in grade.

(a) How much vertical clearance must a builder allow for a sewage pipe that is to carry waste from a point in a building to a point in the street that is 300 feet away in a horizontal direction?

(b) How long does this pipe have to be (to the nearest tenth of a foot)?

Solution

(a) Figure 4.19 illustrates the given situation.

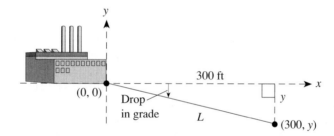

Figure 4.19

To simplify the computation, we have set up the situation as if the pipe were situated at the origin. As the diagram illustrates, we need the horizontal distance (the x-coordinate) to be 300 ft and we want to find the vertical clearance (the y-coordinate) that will give a 3% drop in grade. A 3% drop in grade (or downgrade) means that the ratio of the vertical distance to the horizontal distance is -0.03. In effect, we require that the slope of the line segment L be -0.03. We therefore have

$$\frac{y - 0}{300 - 0} = -0.03 \qquad \textit{Now we solve for y.}$$

$$y = -0.03(300) = -9$$

Recall that the Pythagorean theorem states that in a right triangle, $c^2 = a^2 + b^2$, where c is the length of the hypotenuse, and a and b are the lengths of the legs.

Therefore, the builder must set the pipe so that the street end of the pipe is 9 feet lower than the building end.

(b) In Figure 4.19 we have labeled the length of the pipe L. Since we have a right triangle, we can use the Pythagorean theorem to find the length L.

$$L^2 = 9^2 + 300^2$$

$$L^2 = 90{,}081 \qquad \textit{Find the square root of 90,081 on a calculator.}$$

$$L = \boxed{300.1 \text{ ft}} \qquad \textit{Rounded to the nearest tenth}$$

Interpreting the Slope of the Graph of a Line

EXAMPLE 12

In Example 1 we indicated that the Celsius and Fahrenheit temperature scales are related by the equation $F = 1.8C + 32$. Use the graph found in Example 1 to describe this relationship.

Solution

The graph of this equation is repeated in Figure 4.20.

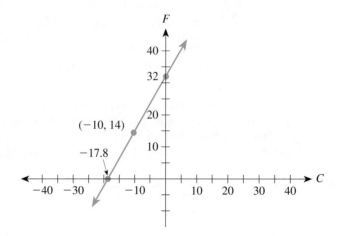

Figure 4.20
The graph of
$F = 1.8C + 32$

We see some points on the graph with which we may already be familiar—for example, the temperature at which water freezes is 0°C or 32°F. In other words, 0°C is equivalent to 32°F and so the point (0, 32) is on the graph. We can see that the line is rising, which means that as the Celsius temperature increases, so does the Fahrenheit temperature, but at what rate?

Recall that the slope of a line tells us the ratio of the vertical change to the horizontal change as we move along the line. Hence, the slope is a *rate* that tells us how much one quantity changes with respect to another. For example, by its definition, a slope of 3 means that

This means that

$$\frac{\text{Change in } y}{\text{Change in } x} = 3$$

Change in $y = 3 \cdot$ (Change in x)

Hence, as we move along the line, vertical change in position will be 3 times any horizontal change in position. We have pointed out that a slope is a characteristic of a line that does not change regardless of where we are on the line. Hence, we can state that a straight line always increases (or decreases) *at the same rate*.

Using the two points (0, 32) and (−10, 14), we compute the slope of the line to be $\frac{9}{5}$ or, equivalently, 1.8. This slope means that as C increases by 5 degrees, F increases by 9 degrees. Alternatively, we can rewrite $\frac{9}{5}$ as $\frac{9/5}{1}$ and say that every 1-unit increase in the horizontal (C) direction is accompanied by a 1.8-unit increase in the vertical (F) direction. We interpret this to mean that a 1-degree change in the Celsius temperature is equivalent to a 1.8-degree change in the Fahrenheit temperature. Since the slope of a line is constant, this statement is true for *all* values of C and F.

The slope is the change in F over the change in C.

EXAMPLE 13

The average normal body temperature of a human being is approximately 98.6°F or 37°C. If Samantha has a temperature of 39°C, is this cause for concern? (Something like this actually did happen to one of the authors. His daughter looked feverish, and he wanted to take her temperature, but the only thermometer available was in °C. While he was familiar with how to interpret °F, he was not accustomed to interpreting a thermometer reading in °C.)

Solution   Samantha's temperature is 2° above normal on the Celsius scale. On the Fahrenheit scale, 2 degrees above normal is 100.6°F, which in most cases is usually interpreted as a slight fever, and no cause for alarm. However, a 2-degree change on the Celsius scale means something different.

We could substitute 39°C into the formula above, converting Celsius to Fahrenheit, and determine whether there is cause for concern based on the °F temperature. However, we pointed out in Example 12 that the slope of the line $F = 1.8C + 32$ tells us that a 1-degree increase in °C is approximately equivalent to a 2-degree increase in °F. Therefore, a 2-degree increase in °C is equivalent to an increase of approximately 4 degrees Fahrenheit. This means that Samantha has a temperature of approximately 102.6°F [the exact value is $98.6 + 2(1.8) = 102.2°F$]. This is a cause for some concern.

In Example 12 the slope tells us how much the Fahrenheit temperature changes in relation to Celsius temperature. (Fahrenheit temperature increases 1.8° for each 1° increase in Celsius.) In general, the slope gives us the *rate* at which the vertical units change in comparison to the change in horizontal units. As with interpreting trends, how the slope is interpreted depends on what quantities the axes represent.

EXAMPLE 14

Use the graph of Jani's trip described in Example 3 of Section 3.5, which we repeat here as Figure 4.21, to interpret the slope of each line segment.

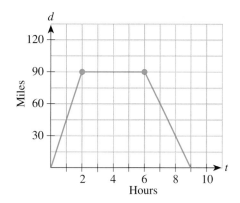

Figure 4.21

Solution   Looking at the graph in Figure 4.21, we note that the horizontal axis is time (in hours) and the vertical axis is distance (in miles). Hence, the slope—being the change in vertical units compared to horizontal units—is interpreted in this example as the change in distance compared to the change in time:

$$m = \frac{\text{Change in distance}}{\text{Change in time}}$$

But we define speed as $\frac{\text{Distance}}{\text{Time}}$. Hence, we can use the slope of each line segment to find Jani's speed for each line segment, or time interval. The slope for the first line segment covering the first two hours is $m_1 = \frac{90}{2} = 45$. Hence, her speed in the first leg of her journey is 45 miles per hour. For the second line segment, between hour 2 and hour 6, $m_2 = 0$. Hence, her speed during that time period is 0 miles per hour. For the last line segment, between hour 6 and hour 9, $m_3 = -\frac{90}{3} = -30$.

For the first line segment, the distance from Jani's home was increasing, so the slope was positive. For the last line segment, however, her average speed was still 30 mph, as she did travel a distance of 90 miles over 3 hours, but the negative sign indicates that her distance from her home *decreased* during that time span. Hence, the slope for this graph gives us both the speed and the direction Jani is traveling to and from her home. We can say that for this graph the absolute value of the slope is the *speed,* and the sign of the slope tells us whether the distance from her home is increasing (positive slope) or decreasing (negative slope).

Let's return to Example 4 of Section 3.5. Figure 4.22 is the graph of the amount of medication (in milligrams), *m,* entering the bloodstream of a patient at time *t* (at 20-minute intervals) after the medication is administered.

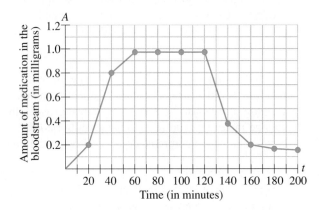

Figure 4.22

The graph is made up of several line segments, each of which has a slope. What do the slopes of the line segments tell us? Let's look at the time interval between 40 and 60 minutes. We note that this segment joins the points (40, 0.8) and (60, 0.97) (we estimated 0.97 on the graph). The slope of that line segment is

$$\frac{0.97 - 0.8 \text{ mg}}{60 - 40 \text{ min}} = \frac{0.17 \text{ mg}}{20 \text{ min}} = 0.0085 \text{ mg/min}$$

Hence, we can see that the slope tells us the *rate* at which the medication enters (or leaves) the bloodstream. Line segments with different slopes indicate different rates. Figure 4.23 shows the slopes, or rates, for each time interval or line segment. Compare this graph with Figure 3.36 on page 173.

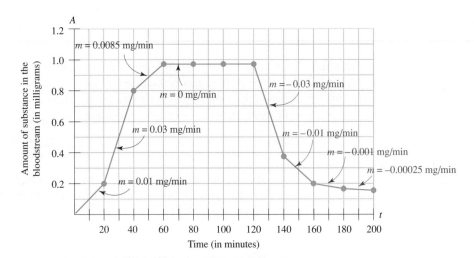

Figure 4.23

For the first 20 minutes, the rate at which the medication enters the bloodstream is 0.01 mg/min. Notice that the slope is positive (this line segment is rising), indicating that more medication is in the bloodstream at the end of the 20-minute interval than at the beginning. For the next 20 minutes, the rate is 0.03 mg/min. Notice that *m* is larger for the second 20 minutes—that is, medication is entering the bloodstream at a faster rate, and therefore the line is steeper. For the 40–60-minute time interval, the line is less steep as the *rate* at which medication is entering the bloodstream is slowing down

($m = 0.0085$ mg/min). From 60 to 120 minutes the slope of the line segments is 0, which means that no additional medication is entering (or leaving) the bloodstream.

From 120 to 140 minutes, the slope is -0.03 mg/min. The slope is negative (the line segment is falling); this means that there is less medication in the bloodstream at the end of the 20-minute interval than at the beginning. Hence medication is *leaving* the bloodstream—leaving at a rate of 0.03 mg/min. From 140 to 160 minutes, the slope of the line segment is -0.01 mg/min. Since the slope is negative, the medication is still leaving the bloodstream, but since $0.01 < 0.03$, it is leaving at a slower rate than during the previous 20 minutes. Note that the 140–160-minute interval segment is less steep than the 120–140-minute interval segment.

STUDY SKILLS 4.1

Reviewing Your Exam: Diagnosing Your Strengths and Weaknesses

Your exam will be a useful tool in helping you to determine what topics, skills, or concepts you need to work on in preparation for the next topic, or in preparation for future exams. After you get your exam back, you should review it carefully: Examine what you did correctly and the problems you missed.

Don't quickly gloss over your errors and assume that any errors were minor or careless. Students often mistakenly label many of their errors as "careless," when in fact they are a result of not clearly understanding a certain concept or procedure. Be honest with yourself. Don't delude yourself into thinking that all errors are careless. Ask yourself the following questions about your errors:

Did I understand the directions or perhaps misinterpret them?

Did I understand the topic to which the question relates?

Did I misuse a rule or property?

Did I make an arithmetic error?

Look over the entire exam. Did you consistently make the same type of error throughout the exam? Did you consistently miss problems covering a particular topic or concept? You should try to follow your work and see what you were doing on the exam. If you think you have a problem understanding a concept, topic, or approach to a problem, you should immediately seek help from your teacher or tutor, and reread relevant portions of your textbook.

EXERCISES 4.1

In Exercises 1–6, sketch the line through the given pair of points and compute its slope.

1. $(1, -2)$ and $(-3, 1)$

2. $(2, -3)$ and $(-1, 4)$

3. $(0, 2)$ and $(2, 0)$

4. $(5, 0)$ and $(0, 5)$

5. $(-3, -4)$ and $(-2, -5)$

6. $(-4, -3)$ and $(-2, -1)$

In Exercises 7–18, compute the slope of the line passing through the given pair of points.

7. $(2, 4)$ and $(-3, 4)$

8. $(1, -5)$ and $(1, 3)$

9. $(4, 2)$ and $(4, -3)$

10. $(-5, 1)$ and $(3, 1)$

11. (a, a) and (b, b) $(a \neq b)$

12. (a, b) and (b, a) $(a \neq b)$

13. (a, a^2) and (b, b^2) $(a \neq b)$

14. (a, b^2) and (b, a^2) $(a \neq b)$

15. $(0.7, -0.2)$ and $(0.06, 0.14)$

16. $(1250, 22)$ and $(700, 12)$

17. $(-0.2, 7)$ and $(0.14, 0.06)$

18. $(22, 1250)$ and $(12, 700)$

In Exercises 19–22, estimate the slope of the line.

19.

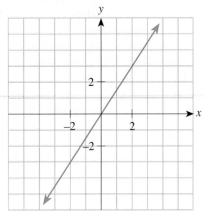

20.

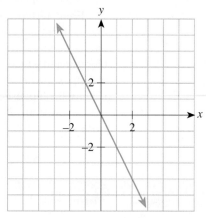

21.

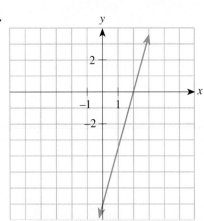

22.

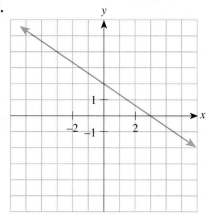

23. Find the slope of the line that crosses the *x*-axis at $x = 5$ and the *y*-axis at $y = -3$.

24. Find the slope of the line that crosses the *y*-axis at $y = 7$ and the *x*-axis at $x = -4$.

In Exercises 25–36, sketch the graph of the line L that contains the given point and has the given slope m.

25. $(1, 3)$, $m = 2$

26. $(2, 5)$, $m = 1$

27. $(1, 3)$, $m = -2$

28. $(2, 5)$, $m = -1$

29. $(0, 3)$, $m = -\dfrac{1}{4}$

30. $(0, 3)$, $m = -4$

31. $(4, 0)$, $m = \dfrac{2}{5}$

32. $(-4, 0)$, $m = -\dfrac{2}{5}$

33. $(2, 5)$, $m = 0$

34. $(-1, 3)$,   undefined slope

35. $(2, 5)$,   undefined slope

36. $(-1, 3)$, $m = 0$

In Exercises 37–42, find the slope of the line by picking two points on the line.

37. $y = 2x - 7$

38. $y = -\dfrac{1}{3}x + 4$

39. $y = -0.4x + 5$

40. $y = 5x - 3$

41. $2y - 3x = 8$

42. $3x + 4y = 7$

In Exercises 43–48, determine whether the line passing through the points P_1 and P_2 is parallel to or perpendicular to (or neither parallel to nor perpendicular to) the line passing through the points P_3 and P_4.

43. $P_1(1, 2)$, $P_2(3, 4)$, $P_3(-1, -2)$, $P_4(-3, -4)$ **44.** $P_1(5, 6)$, $P_2(7, 8)$, $P_3(-5, -6)$, $P_4(-7, -8)$

45. $P_1(0, 4)$, $P_2(-1, 2)$, $P_3(-3, 5)$, $P_4(1, 7)$ **46.** $P_1(2, 3)$, $P_2(3, 0)$, $P_3(-2, -5)$, $P_4(1, -6)$

47. $P_1(3, 5)$, $P_2(-2, 5)$, $P_3(1, 4)$, $P_4(1, -2)$

48. $P_1(a, b)$, $P_2(b, a)$, $P_3(c, d)$, $P_4(-d, -c)$ $(a \neq b, c \neq d)$

49. Find a number h so that the line passing through the points $(4, -2)$ and $(1, h)$ has slope -5.

50. Find a number k so that the line passing through the points $(k, 2)$ and $(-3, -1)$ has slope 4.

51. Using slopes, show that the points $(0, 0)$, $(2, 1)$, $(-2, 5)$, and $(0, 6)$ are the vertices of a parallelogram.

52. Using slopes, show that the points $(-3, 1)$, $(-7, 4)$, $(0, 5)$, and $(-4, 8)$ are the vertices of a rectangle.

53. Using slopes, show that the points $(-3, 2)$, $(-1, 6)$, and $(3, 4)$ are the vertices of a right triangle.

54. Using slopes, show that the points $(1, 1)$, $(3, 5)$, $(-1, 6)$, and $(-5, -2)$ are the vertices of a trapezoid.

55. A highway engineer specifies that a certain section of roadway covering a horizontal distance of 2 km should have an upgrade of 8%. Compute the change in elevation of this section of the road.

56. An escalator with an upgrade of 5% carries people through a vertical distance of 80 ft. Through what horizontal distance does this escalator carry passengers?

57. A cable car is carrying passengers down a hill at a downgrade of 12%. If the passengers descend 1,250 ft, through what horizontal distance have they traveled?

58. A wire is attached from a window to a pole standing 60 ft away from a building. If the slope of the wire is -0.6, how far below window level is the wire attached to the pole?

59. Suppose that in a rectangular coordinate system the horizontal axis represents the amount of time that has passed (in seconds), and the vertical axis represents the distance that a sprinter covers (in meters). How would you interpret the slope of the line segment joining the points $(2, 15)$ and $(8, 80)$?

60. Suppose that in a rectangular coordinate system the horizontal axis represents the number of items a company produces, and the vertical axis represents the profit (in dollars) the company earns on the production of that many items. How would you interpret the slope of the line segment joining the points $(0, 0)$ and $(100, 875)$?

61. Suppose that in a rectangular coordinate system the horizontal axis, labeled t, represents the hour of a particular afternoon, and the vertical axis, labeled F, represents the temperature in °F. How would you interpret the slope of the line segment joining the points $(1, 56)$ and $(5, 42)$?

62. Suppose that in a rectangular coordinate system the horizontal axis, labeled n, represents the year, and the vertical axis, labeled P, represents the population of a small town. How would you interpret the slope of the line segment joining the points $(1980, 8972)$ and $(1990, 7456)$?

63. A company manufactures monitors. The accompanying figure shows a graph of the relationship between the wholesale price of a monitor, p, in dollars, and the company's daily profit, P, in thousands of dollars. The horizontal axis represents the wholesale price of a monitor and the vertical axis represents the daily profit. For the price range given in the figure, the actual data points are plotted. These data points can be approximated by the given line. Estimate the slope of the line. What does the slope of the line tell you about this relationship?

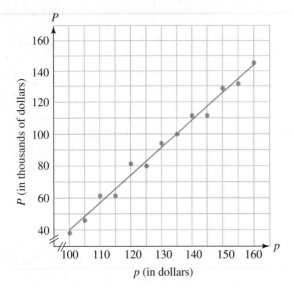

64. The figure below shows a graph of the relationship between the daily cost, C, of electricity to keep a particular home at a constant $72°$, and the average daily outdoor temperature, T, in July. The horizontal axis represents the average daily temperature and the vertical axis represents the daily cost of electricity. For the temperature range given in the figure, the actual data points are plotted. These data points can be approximated by the given line. Estimate the slope of the line. What does the slope of the line tell you about this relationship?

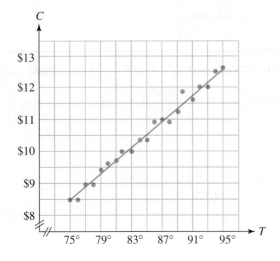

❓ Questions for Thought

65. How could you use the idea of slope to determine whether the three points $(-2, -1)$, $(0, 4)$, and $(2, 9)$ all lie on the same line (are collinear)?

66. How would you describe a line with positive slope? Negative slope? Zero slope? Undefined slope?

67. Suppose we do not insist on the units along the x- and y-axes being the same. Sketch the graph of the line with slope 3 passing through the point $(1, 2)$ if the units on the x-axis are twice as large as the units on the y-axis.

68. Repeat Exercise 67 if the units on the y-axis are twice as large as the units on the x-axis.

69. Sketch the graphs of $y = 2x$, $y = 2x + 3$, $y = 2x + 5$, $y = 2x - 3$, and $y = 2x - 5$ on the same set of coordinate axes. Determine the slope of each line from its graph. Can you draw any conclusions about determining the slope of a line from its equation?

 ## Mini-Review

70. *Simplify.* $(3x^4)^2(6x^3)$

71. *Simplify.* $2xy(3x - 5y) - 6y(2x^2 + xy)$

72. *Evaluate for $x = 3$ and $y = -2$.* $|5y + x| - y^2$

73. *Solve for t.* $\dfrac{t}{3} - 2 = 5$

 4.2

Equations of a Line and Linear Functions as Mathematical Models

In our previous discussions we have stressed the idea that a first-degree equation in two variables is a *condition* that an ordered pair must satisfy. We have seen that the set of all points that satisfy such an equation is a straight line.

We are now prepared to reverse the question. Suppose we "specify a line." How can we find its equation? In other words, what condition must all the points on the line satisfy?

To answer this question we must first understand what is meant by the phrase "specify a line." We know that a line is determined by two points, but we also saw in the previous section that, for a nonvertical line, if we know one point on the line and its slope, we can find as many additional points on the line as we choose. Therefore, equivalently we can say that to specify a nonvertical line we specify a point on the line and its slope.

Let us suppose we have a line passing through a given point (x_1, y_1) with slope m (see Figure 4.24). To determine whether another point (x, y) is on this line, we must check whether the points (x_1, y_1) and (x, y) give us the required slope of m.

$$\text{If } \frac{y - y_1}{x - x_1} = m, \text{ then } (x, y) \text{ is on the line.}$$

$$\text{If } \frac{y - y_1}{x - x_1} \neq m, \text{ then } (x, y) \text{ is not on the line.}$$

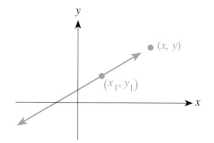

Figure 4.24

Thus, the equation

$$\frac{y - y_1}{x - x_1} = m$$

is the condition that a point (x, y) must satisfy for it to be on this line.

If we multiply both sides of the equation by $x - x_1$, we get the relationship given in the box.

Point-Slope Form of the Equation of a Straight Line	An equation of the line with slope m passing through (x_1, y_1) is $$y - y_1 = m(x - x_1)$$

EXAMPLE 1   Write an equation of the line with slope $\frac{2}{3}$ that passes through $(-3, 4)$.

Solution   The given point $(-3, 4)$ corresponds to (x_1, y_1) and $m = \frac{2}{3}$.

$$y - y_1 = m(x - x_1)$$

$$y - 4 = \frac{2}{3}[x - (-3)]$$

$$\boxed{y - 4 = \frac{2}{3}(x + 3)}$$

This is the equation of the line shown in Figure 4.25.

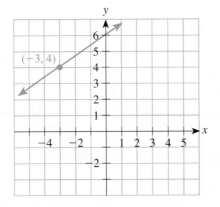

Figure 4.25
Line with slope $\frac{2}{3}$ passing through $(-3, 4)$

The point–slope form allows us to write an equation of a nonvertical line if we know *any* point on the line and its slope. What if the given point happens to be where the graph of the line crosses the y-axis? That is, suppose the line has slope m and y-intercept equal to b. Then the graph passes through the point $(0, b)$. Using the point–slope form, we can write its equation as

$$y - y_1 = m(x - x_1)$$
$$y - b = m(x - 0)$$
$$y - b = mx$$
$$y = mx + b$$

This last form is called the **slope–intercept form** of the equation of a straight line.

Slope–Intercept Form of the Equation of a Straight Line	An equation of the line with slope m and y-intercept b is $$y = mx + b$$

EXAMPLE 2

Write an equation of the line with slope 4 and y-intercept -3, and graph the equation.

Solution   Since we are given the slope and y-intercept, it is appropriate to use the slope–intercept form.

$$y = mx + b \qquad \textit{We are given m = 4 and b = −3.}$$
$$y = 4x + (-3)$$
$$\boxed{y = 4x - 3}$$

We could also use the point-slope form with $m = 4$ and $(x_1, y_1) = (0, -3)$.

$$y - y_1 = m(x - x_1)$$
$$y - (-3) = 4(x - 0)$$
$$y + 3 = 4x$$
$$\boxed{y = 4x - 3}$$

We graph the equation as we graphed equations in the previous section, given the slope and a point. Since the point is the y-intercept, we start at the y-axis at -3. Since the slope is $4 = \frac{4}{1}$, we move right 1 unit and up 4 units to find another point on the line and then draw the line (see Figure 4.26).

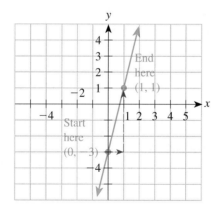

Figure 4.26

EXAMPLE 3

Find the slopes of the lines with the following equations:

(a) $y = -3x + 5$ **(b)** $3y + 5x = 8$

Solution   **(a)** Perhaps the most useful feature of the slope–intercept form is the fact that when the equation of a straight line is written in this form, it is easy to "read off" the slope (as well as the y-intercept):

$$y = -3x + 5$$
$$\qquad \uparrow \qquad \uparrow$$
$$y = \ mx + b$$

Therefore, $\boxed{\text{the slope is } -3}$.

(b) To "read off" the slope of a line from its equation, the equation must be *exactly* in slope–intercept form. That is, the equation must be in the form $y = mx + b$.

$$3y + 5x = 8 \qquad \text{\textit{We solve for y.}}$$
$$3y = -5x + 8$$
$$y = \frac{-5}{3}x + \frac{8}{3}$$

Therefore, the slope is $-\dfrac{5}{3}$.

In Section 3.1, we stated that the graph of a first-degree equation in two variables is a straight line. In fact, the following statement is also true.

Any nonvertical line can be represented by an equation of the form $y = mx + b$.

EXAMPLE 4

Write an equation of the line passing through the points $(1, -2)$ and $(-3, -5)$.

Solution   The graph of the line appears in Figure 4.27.

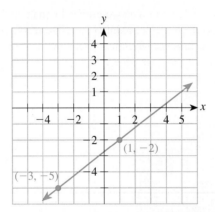

Figure 4.27

Whether we decide to write an equation for this line using the point–slope form or the slope–intercept form, we must know the slope of the line:

$$m = \frac{-2 - (-5)}{1 - (-3)} = \frac{3}{4}$$

We can now write an equation using the point–slope form *or* the slope–intercept form.

Using point–slope form: If we choose to use the point–slope form, we have another choice to make. We can choose to use either $(1, -2)$ or $(-3, -5)$ as our given point (x_1, y_1).

Using $(1, -2)$: $y - (-2) = \dfrac{3}{4}(x - 1)$ or $\boxed{y + 2 = \dfrac{3}{4}(x - 1)}$

Using $(-3, -5)$: $y - (-5) = \dfrac{3}{4}[x - (-3)]$ or $\boxed{y + 5 = \dfrac{3}{4}(x + 3)}$

As we shall see in a moment, although these two equations may look different, they are in fact equivalent.

Using slope–intercept form: The slope–intercept form is $y = mx + b$. Since we know $m = \frac{3}{4}$, we can write

$$y = \frac{3}{4}x + b$$

We do not yet know the value of b since we do not know the y-intercept. However, we do know that the points $(1, -2)$ and $(-3, -5)$ are on the line and so must satisfy the equation. Therefore, if we substitute one of these points, say $(1, -2)$, into the equation $y = \frac{3}{4}x + b$, we can solve for b.

$$y = \frac{3}{4}x + b \qquad \textit{Substitute } (1, -2).$$

$$-2 = \frac{3}{4}(1) + b$$

$$-2 = \frac{3}{4} + b$$

$$-2 - \frac{3}{4} = b$$

$$-\frac{11}{4} = b$$

Thus, the equation is

$$\boxed{y = \frac{3}{4}x - \frac{11}{4}}$$

While the three answers we have obtained look different, they are in fact equivalent. If we take our first two answers and solve for y we get:

First answer

$$y + 2 = \frac{3}{4}(x - 1)$$

$$y + 2 = \frac{3}{4}x - \frac{3}{4}$$

$$y = \frac{3}{4}x - \frac{3}{4} - 2$$

$$y = \frac{3}{4}x - \frac{11}{4}$$

Second answer

$$y + 5 = \frac{3}{4}(x + 3)$$

$$y + 5 = \frac{3}{4}x + \frac{9}{4}$$

$$y = \frac{3}{4}x + \frac{9}{4} - 5$$

$$y = \frac{3}{4}x - \frac{11}{4}$$

and so all three answers are equivalent.

Using the slope–intercept form in the example required the most work, but it had the advantage of giving us a uniform answer. *For that reason, we will write all our final answers in slope–intercept form.*

Nevertheless, unless there are instructions to the contrary, you may use whichever form *you* find most convenient to obtain an equation of a line. As the last example illustrates, any time you are asked to write an equation of a nonvertical line, you may use either the point–slope or slope–intercept form. The given information in each problem will determine which form is more efficient to use.

EXAMPLE 5

Solution

Write an equation of the line passing through the point $(2, 3)$ and parallel to the line whose equation is $3x - 6y = 12$.

The example becomes much clearer if we draw a diagram illustrating exactly what is being asked. Figure 4.28 illustrates that we are being asked to find the equation of the red line.

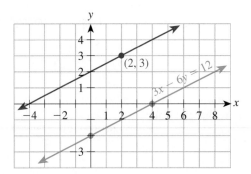

Figure 4.28

Let's analyze this question carefully, step by step, to develop a strategy for its solution.

THINKING OUT LOUD

What do we need to find?	The equation of a line satisfying certain conditions
What information is needed to write an equation of a line?	A point on the line and the slope of the line
What information is given in the problem?	The point is given, along with the equation of a line parallel to the line whose equation we want to find.
How do we find the slope of a line given the equation of a line parallel to it?	The slopes of parallel lines are equal. If we find the slope of the parallel line, then we have the slope of the line whose equation we want to find.
How do we find the slope of a line whose equation is given?	As we saw in Example 3, the easiest way is to put the given equation in slope–intercept form.

Based on this analysis, we take the equation of the parallel line, put it in slope–intercept form, and then "read off" the slope:

$$3x - 6y = 12 \qquad \textit{We solve explicitly for y.}$$
$$-6y = -3x + 12$$
$$y = \frac{-3}{-6}x + \frac{12}{-6}$$
$$y = \frac{1}{2}x - 2$$

Therefore the slope of the parallel line is $\frac{1}{2}$.

Now we can use the point–slope form to write the equation:

$$y - y_1 = m(x - x_1) \qquad \textit{The given point is (2, 3); } m = \frac{1}{2}.$$
$$y - 3 = \frac{1}{2}(x - 2) \text{ or } \boxed{y = \frac{1}{2}x + 2}$$

Notice that in this solution we used both the slope–intercept and point–slope forms. We used the slope–intercept form because it was the most efficient way to find the slope of the parallel line, and then we used the point–slope form because it was the most efficient method to write the equation.

EXAMPLE 6

Write an equation of the line passing through each pair of points:

(a) $(2, 5)$ and $(-3, 5)$ **(b)** $(1, 3)$ and $(1, -4)$

Solution

(a) The slope of the line is

$$m = \frac{5 - 5}{2 - (-3)} = \frac{0}{5} = 0$$

Using the point–slope form with the point $(2, 5)$, we get

$$y - 5 = 0(x - 2)$$
$$y - 5 = 0$$
$$\boxed{y = 5}$$

Alternatively, we may recognize at the outset that the line passing through the points $(2, 5)$ and $(-3, 5)$ is a horizontal line 5 units above the x-axis (see Figure 4.29). As we saw in Section 3.1, the equation of this line is $y = 5$.

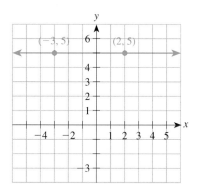

Figure 4.29

(b) If we attempt to compute the slope of this line, we get

$$m = \frac{3 - (-4)}{1 - 1} = \frac{7}{0} \quad \text{which is undefined!}$$

Therefore, we cannot use the point–slope or slope–intercept form to write an equation. (Both these forms require the line to have a slope. That is why in the discussion leading up to obtaining those forms we always specified a "nonvertical line.")

However, once we recognize that this is a vertical line 1 unit to the right of the y-axis (see Figure 4.30), we can write its equation as

$$\boxed{x = 1}$$

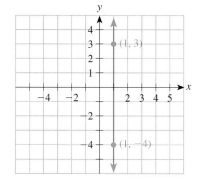

Figure 4.30

EXAMPLE 7

A manufacturer determines that the relationship between the profit earned, P, and the number of items produced, x, is linear. Suppose the profit is \$1,500 on 45 items and \$2,500 on 65 items.

(a) Write an equation relating P and x.

(b) What would the expected profit be if 100 items were produced?

Solution

Keep in mind that labeling the points (x, P) means that the horizontal axis is the x-axis and the vertical axis is the P-axis.

(a) The fact that the relationship is linear means we can write a first-degree equation in two variables, x and P. It is natural to assume that the profit, P, depends on the number of items produced, x. Thus, x is the independent variable and P is the dependent variable. When using ordered pairs, the accepted practice is to assign the independent variable as the first component, and the dependent variable as the second. Hence we will label the points (x, P). We can write the information we are given as two points: $(45, 1500)$ and $(65, 2500)$. Thus, the slope of the line is

$$m = \frac{2,500 - 1,500}{65 - 45} = \frac{1,000}{20} = 50$$

We can now write the equation using the point–slope form with the point $(45, 1500)$ and $m = 50$ (keep in mind that we are using P instead of y):

$$P - 1,500 = 50(x - 45)$$

Or, if we put this in slope–intercept form, we have

$$\boxed{P = 50x - 750}$$

It is worthwhile to note that in real-life applications, the slope of a line often has a practical significance. In this example, there are actual units attached to the numbers when we compute the slope:

$$m = \frac{\$2,500 - \$1,500}{65 \text{ items} - 45 \text{ items}} = \frac{\$1,000}{20 \text{ items}} = \frac{\$50}{\text{item}} = \$50 \text{ per item}$$

Thus, the significance of the slope is that it tells us the amount of profit earned *per item*. Looking at the equation $P = 50x - 750$, we now recognize that it says the *total profit* is \$50 per item minus \$750. In fact, we recognize that profit depends on the number of items: Using the language of functions we can state "P is a function of x," or write this symbolically as $P = P(x)$.

(b) We want the expected profit if 100 items are produced. In other words, we want to find P when $x = 100$, or, in the language of functions, we want to find $P(100)$:

$$P = P(x) = 50x - 750 \qquad \textit{To find P(100), substitute x = 100 in P(x) to get}$$
$$= 50(100) - 750$$
$$= 5,000 - 750$$
$$= \$4,250$$

Thus, the expected profit if 100 items are produced is $\boxed{\$4,250}$.

It is often helpful to graph the equation to visualize the relationship between the variables (see Figure 4.31).

Having a graph allows us to approximate an answer or check its reasonableness and accuracy.

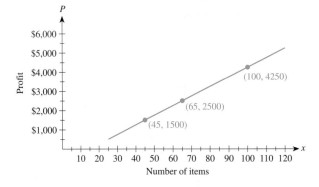

Figure 4.31

EXAMPLE 8

An automobile dealership recognizes that there is an approximately linear relationship between the number of cars and trucks sold each week. During a certain week, 42 cars and 18 trucks were sold, whereas during another week 35 cars and 16 trucks were sold. Find and graph a function relating the number c of cars and the number t of trucks sold each week.

Solution

It is unlikely that the number of cars sold *depends* on the number of trucks sold, or vice versa. Therefore, we may choose to make c the independent variable and t the dependent variable [in which case we will record the ordered pairs as (c, t), and the horizontal axis will be labeled c and the vertical axis will be labeled t], or we may do the reverse. We illustrate both approaches.

Approach 1 We record the ordered pairs as (c, t). The horizontal axis is c and the vertical axis is t.

Since we are told that the relationship is approximately linear, we will use the point–slope form for the equation of a line with the points $(42, 18)$ and $(35, 16)$. The slope of the line is

$$m = \frac{t_2 - t_1}{c_2 - c_1} = \frac{18 - 16}{42 - 35} = \frac{2}{7}$$

We can now write the equation of the line as

$$t - 18 = \frac{2}{7}(c - 42) \quad \text{or equivalently} \quad t = \frac{2}{7}c + 6$$

Approach 2 We record the ordered pairs as (t, c).

We use the point–slope form for the equation of a line with the points $(18, 42)$ and $(16, 35)$. The slope of the line is

$$m = \frac{c_2 - c_1}{t_2 - t_1} = \frac{42 - 35}{18 - 16} = \frac{7}{2}$$

We can now write the equation of the line as

$$c - 42 = \frac{7}{2}(t - 18) \quad \text{or equivalently} \quad c = \frac{7}{2}t - 21$$

Note that both approaches produce functions that describe the relationship between c and t. Figure 4.32 shows the graphs of these functions.

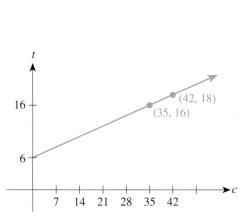

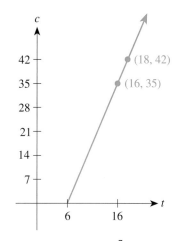

Figure 4.32

(a) The graph of $t = \frac{2}{7}c + 6$

(b) The graph of $c = \frac{7}{2}t - 21$

Note that we have drawn the graphs in the first quadrant only, since it makes no sense for either c or t to be negative. Both graphs give us information about the relationship between c and t; they differ simply in terms of the perspective from which we view the relationship.

EXAMPLE 9

A company manufactures optical scanners. Figure 4.33 shows a graph of the relationship between the selling price, d, of a certain model scanner, in dollars, and the company's daily profit, P, on that model, in thousands of dollars. The horizontal axis represents the selling price of the scanner, and the vertical axis represents the daily profit on that model. For the price range given in the figure, the actual data points are plotted. These data points can be approximated by the given line. Find an equation of the line.

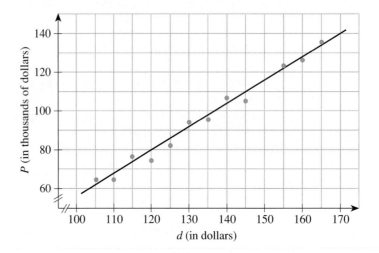

Figure 4.33

Solution

Here is a case in which we are given the graph of a line and we need to find its equation. We could proceed by estimating the slope of the line (as we have done in the previous section) and identifying a point through which the line passes. Another approach would be to identify two points on the line and find the equation as we have done in the previous examples. We will take the latter approach.

Which points do we use to construct an equation of the line? First, we note that the points we choose must be *on* the line: We cannot pick data points if they do not lie on the line. Second, the best choices of points are those that require the least estimation—those that lie on the grid if possible. See Figure 4.34.

For this example, we choose the points (120, 80) and (145, 110).

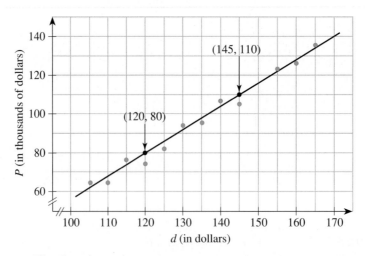

Figure 4.34

The slope is

$$\frac{110 - 80}{145 - 120} = \frac{30}{25} = 1.2$$

Hence, using the point (120, 80), the equation is

$$P - 80 = 1.2(d - 120)$$

which simplifies to

$$P = 1.2d - 64 \quad \textit{Remember that P is in thousands of dollars.}$$

We should check with a point (keeping in mind that our equation is only as good as the accuracy of the points we chose). If we let the price p be 150, then the equation yields a daily profit of

$$P = 1.2(150) - 64 = 116 \quad \textit{which means a profit of \$116,000}$$

Hence, the point (150, 116) should lie on (or very close to) the line if the equation is accurate. We can see by the figure that the point does lie on or close the line.

One final comment: In this section we have shown that the equation of a straight line is a first-degree equation in x and y. However, we have still not proven the theorem stated in Section 2.1 that the graph of a first-degree equation in two variables is a straight line. A proof of this fact is outlined in Exercise 73.

STUDY SKILLS 4.2

Reviewing Your Exam: Checking Your Understanding

If you carefully looked over your exam and you believe that you understand the material and what you did wrong, then do the following:

Copy the problems over on a clean sheet of paper and rework the problems without your text, notes, or exam. When you are finished, check to see whether your answers are correct. If your answers are correct, try to find problems in the text similar to those problems and work the new problems on a clean sheet of paper (again without notes, text, or exam). If some of your new answers are incorrect then you may have

learned how to solve your test problems, but you probably have not thoroughly learned the topic being tested. You may need to repeat these steps several times until you are confident you understand your errors.

If any case, you should keep your exams (with the correct answers) because they are a good source of information for future studying. You may want to record errors that you consistently made on the exam on your warning cards: Types of exam problems can be used when you make up your quiz cards (see Study Skills 1.6).

EXERCISES 4.2

In Exercises 1–28, write an equation of the line L satisfying the given conditions.
Where possible, express your answer in slope–intercept form.

1. L has slope 5 and passes through the point $(1, -3)$.

2. L has slope -4 and passes through the point $(-2, 4)$.

3. L has slope -3 and passes through the point $(-5, 2)$.

4. L has slope 2 and passes through the point $(-1, 6)$.

5. L passes through $(6, 1)$ and has slope $\frac{2}{3}$.

6. L passes through $(10, 3)$ and has slope $\frac{3}{5}$.

7. L passes through $(4, 0)$ and has slope $-\frac{1}{2}$.

8. L passes through $(0, 4)$ and has slope $-\frac{1}{2}$.

9. L has slope $\frac{3}{4}$ and passes through $(0, 5)$.

10. L has slope $\frac{4}{3}$ and passes through $(5, 0)$.

11. L has slope 0 and passes through $(-3, -4)$.

12. L has undefined slope and passes through $(-3, -4)$.

13. L has undefined slope and passes through $(-4, 7)$.

14. L has slope 0 and passes through $(-4, 7)$.

15. L passes through the points $(2, 3)$ and $(5, 7)$.

16. L passes through the points $(1, 4)$ and $(3, 8)$.

17. L passes through the points $(-2, -1)$ and $(-3, -5)$.

18. L passes through the points $(-3, -2)$ and $(-1, -4)$.

19. L passes through the points $(2, 3)$ and $(0, 5)$.

20. L passes through the points $(0, -1)$ and $(3, 1)$.

21. L has slope 4 and crosses the y-axis at $y = 6$.

22. L has slope 4 and crosses the x-axis at $x = 6$.

23. L has slope -2 and crosses the x-axis at $x = -3$.

24. L has slope $-\frac{1}{5}$ and crosses the y-axis at $y = -2$.

25. L passes through the points $(2, 3)$ and $(6, 3)$.

26. L passes through the points $(3, 2)$ and $(3, 6)$.

27. L is vertical and passes through $(-2, -4)$.

28. L is horizontal and passes through $(-5, -3)$.

In Exercises 29–44, write an equation of the line L satisfying the given conditions.

29. L crosses the x-axis at $x = -3$ and the y-axis at $y = 2$.

30. L crosses the y-axis at $y = -5$ and the x-axis at $x = -2$.

31. L passes through $(2, 2)$ and is parallel to $y = 3x + 7$.

32. L passes through $(2, 2)$ and is perpendicular to $y = 3x + 7$.

33. L passes through $(-3, 2)$ and is perpendicular to $y = -\frac{2}{3}x - 1$.

34. L passes through $(-3, 2)$ and is parallel to $y = -\frac{2}{3}x - 1$.

35. L passes through $(0, 0)$ and is perpendicular to $y = x$.

36. L passes through $(0, 0)$ and is parallel to $x + y = 6$.

37. L passes through $(-1, -2)$ and is parallel to $2y - 3x = 12$.

38. L passes through $(-1, -2)$ and is perpendicular to $2y - 3x = 12$.

39. L passes through $(1, -3)$ and is perpendicular to $4x - 3y = 9$.

40. L passes through $(1, -3)$ and is parallel to $4x - 3y = 9$.

41. L is perpendicular to $8x - 5y = 20$ and has the same y-intercept.

42. L is perpendicular to $8x - 5y = 20$ and has the same x-intercept.

43. L passes through the point $(0, 4)$ and is parallel to the line passing through the points $(3, -6)$ and $(-1, 2)$.

44. L passes through the point $(4, 0)$ and is perpendicular to the line passing through the points $(-3, 1)$ and $(2, 6)$.

In Exercises 45–48, identify the line L as horizontal or vertical and write its equation.

45. L passes through $(4, 3)$ and is parallel to the x-axis.

46. L passes through $(4, 3)$ and is parallel to the y-axis.

47. L passes through $(-1, -2)$ and is perpendicular to the x-axis.

48. L passes through $(-1, -2)$ and is perpendicular to the y-axis.

In Exercises 49–52, find the equation of the line.

49.

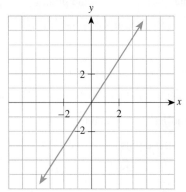

50.

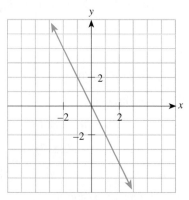

51.

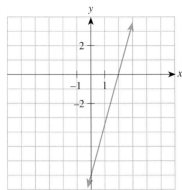

52.

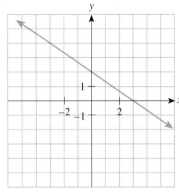

53. The figure shows the graph of the number of retail prescriptions (in the billions)
   P dispensed in the United States from 1992 to 2001. The vertical axis represents
   P, and the horizontal axis represents t. For the time range given, the actual data
   points are plotted. These data points can be approximated by the given line.

   (a) Find the equation of the line.

   (b) Use the equation of the line found in part **(a)** to predict P for 2005.

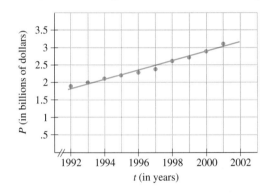

54. The figure shows the graph of the total enrollments in elementary and secondary
   schools, E, in the United States, in millions, for each year t, from 1992 to 2000.
   The vertical axis represents E and the horizontal axis represents t. For the time
   range given in the figure, the actual data points are plotted. These data points can
   be approximated by the given line.

   (a) Find an equation of the line.

   (b) Use the equation found in part **(a)** to predict E for 2004.

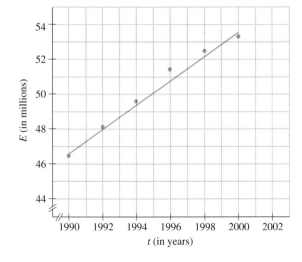

In Exercises 55–62, determine whether the given pairs of lines are perpendicular, parallel, or neither.

55. $3x - 2y = 5$ and $3x - 2y = 6$

56. $5x - 7y = 4$ and $5x + 7y = 4$

57. $2x = 3y - 4$ and $2x + 3y = 4$

58. $6x - 2y = 7$ and $3x - y = 8$

59. $5x + y = 2$ and $5y = x + 3$

60. $2x - y = 8$ and $4x = 2y + 9$

61. $3x - 7y = 1$ and $6x = 14y + 5$

62. $3x + 5y = 2$ and $4 + 10x = 6y$

63. A car rental company charges a flat fee of $29 per day plus a mileage fee of $0.12 per mile. Express C, the cost of a 1-day rental, as a linear function of n, the number of miles driven. What is the slope of the line whose equation you just found?

64. Each month a local utility company charges a flat fee of $12 plus a fee of $0.13 per kilowatt-hour. Express the monthly utility cost, U, as a linear function of k, the number of kilowatt-hours used. What is the slope of the line whose equation you just found?

65. Each month a local phone company charges a flat fee of $23 plus a fee of $0.825 per local call. Express the amount of the monthly phone bill, B, as a linear function of n, the number of local phone calls made. What is the slope of the line whose equation you just found?

66. A typesetter charges a flat fee of $85 as a setup charge, plus a fee of $0.825 per page. Express the cost, C, of a typesetting job as a linear function of p, the number of pages. What is the slope of the line whose equation you just found?

67. A manufacturer determines that the relationship between the profit earned, P, and the number of items produced, x, is linear. If the profit is $200 on 18 items and $2,660 on 100 items, write an equation relating P to x and determine what the expected profit would be if 200 items were produced.

68. Joe found that the relationship between the profit, P, he made on his wood carvings and the number of wood carvings he produced, x, is linear. If he made a profit of $10 on 5 carvings and $90 on 15 carvings, write an equation relating P to x and determine his expected profit if he produced 35 carvings.

69. A psychologist found that the relationship between the scores on two types of personality tests, test A and test B, was perfectly linear. An individual who scored a 35 on test A would score a 75 on test B, and an individual who scored a 15 on test A would score a 35 on test B. What test B score would an individual get if she scored a 40 on test A?

70. A factory foreman found a perfectly linear relationship between the number of defective widgets produced weekly, D, and the total number of overtime hours per week, h, put in by the widget inspectors. When the inspectors put in 100 hours total overtime, 85 defective widgets were found that week; when the inspectors put in 40 hours total overtime, 30 defective widgets were found. How many defective widgets should be found during the week the inspectors put in 120 hours total overtime?

71. A math teacher found that the performance, E, of her students on their first math exam was related linearly to their performance, V, on a video game located in the recreation room. A student who scored a 70 on his math exam scored 35,000 points on the video game. On the other hand, a student who scored an 85 on her exam scored a 20,000 on the video game. Write an equation relating E and V, and predict what exam score a student would have received if he scored 15,000 points on the video game.

72. A physiologist found that the relationship between the length of the right-hand thumb, T, and the length of the left little toe, t, was perfectly linear for a group of hospital residents. For one of the residents, the right-hand thumb was 5 cm and the left little toe was 2.5 cm; for another resident, the right-hand thumb was 7 cm and the left little toe was 2 cm. Write an equation relating T to t and predict the size of a resident's toe if his thumb is 8 cm.

? Questions for Thought

73. We have already proven that if a graph is a straight line, then its equation can be put in the form $y = mx + b$ (or $x = k$ if the line is vertical). In either case, its equation is first-degree. To prove the converse we need to show that any first-degree equation in two variables has as its graph a straight line.

Suppose we have the equation $Ax + By = C$, where $B \neq 0$. Show that *any* two ordered pairs (x_1, y_1) and (x_2, y_2) that satisfy the equation will yield the same slope. [*Hint:* If (x_1, y_1) satisfies the equation $Ax + By = C$, then

$$Ax_1 + By_1 = C$$

Now solve for y_1 and we have

$$y_1 = \frac{C - Ax_1}{B}$$

Therefore, we can write our ordered pair as $\left(x_1, \dfrac{C - Ax_1}{B}\right)$.

Similarly, for (x_2, y_2), we get $\left(x_2, \dfrac{C - Ax_2}{B}\right)$.

[Now show that the slope you get from these two points is independent of x_1 and x_2.]

74. Put the equation of the line $Ax + By = C$ in slope–intercept form. What is the slope of the line (in terms of A, B, and C)?

◆ Mini-Review

75. *Evaluate:* $\dfrac{-4(-2)(-6)}{-4 - 2(-6)}$

76. *Solve:* $8 - 5(x - 3) \le 28$

77. Find the domain of $f(x) = \dfrac{x}{4x - 5}$

78. *Solve for y:* $8x - 3y = 12$

4.3   Linear Systems in Two Variables

It is frequently the case that we are considering a problem that requires us to satisfy two or more different conditions simultaneously. For example, given its particular circumstances, a business may want to choose a method of advertising that minimizes cost but also maximizes exposure.

Each condition can sometimes be represented by an equation in two or more variables. We then seek the numbers (if any) that satisfy all the equations (conditions) simultaneously.

In this section we will examine systems of two linear equations in two variables, and discuss methods that involve manipulation of these equations. In Chapter 10 we will generalize these methods to other systems, as well as discuss other methods for solving linear systems.

Let's begin by considering the following situation.

EXAMPLE 1

As an employee in a sales department, your employer offers you the option of receiving your weekly income in two possible ways: You can either be paid a straight commission of 8% of your gross sales, or you can receive a base pay of $180 per week plus a commission of 6% of your gross sales. Which plan should you choose?

Solution

The answer to this question depends on how much you expect to sell each week. For example, if you expect to sell $3,000 worth of merchandise in a week, you would receive a weekly income of $0.08 \cdot \$3,000 = \$240$ taking the straight commission plan, but $0.06 \cdot \$3,000 + \$180 = \$360$ taking the base pay plus commission plan. On the other hand, if you expect to sell $10,000 worth of merchandise in a week, you would receive a weekly income of $0.08 \cdot \$10,000 = \$800$ taking the straight commission plan, but $0.06 \cdot \$10,000 + \$180 = \$780$ taking the base pay plus commission plan.

Notice that if you expect to sell "a lot" of merchandise, the straight commission plan is more desirable than the base pay plus commission plan. On the other hand, if you do not expect to sell "a lot," the base pay plus commission plan is better. We can write an equation for each plan expressing the weekly income, I, in terms of the gross sales, g, as follows:

$$I = 0.08g \qquad \text{for straight 8\% commission}$$
$$I = 180 + 0.06g \quad \text{for base pay plus 6\% commission}$$

Each of these equations is a first-degree or linear equation in two variables, and hence has a straight line as its graph. We represent g along the horizontal axis, and I along the vertical axis, and graph each line on the same set of axes in Figure 4.35.

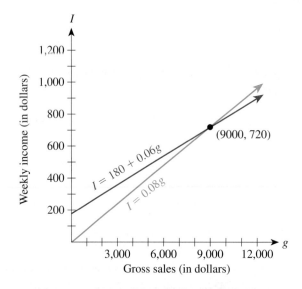

Figure 4.35
Graphs of $I = 180 + 0.06g$
and $I = 0.08g$

The point of intersection of the two lines, (9000, 720), can be determined algebraically. We will do this later in this section.

In the previous section we emphasized the importance of drawing the graph of an equation in order to better understand the relationship between the variables. Notice that the graph of each line tells us the weekly income, I, of each plan based on gross sales, g.

By determining the point of intersection of the lines, we find what the gross sales should be for the incomes yielded by each plan to be the same. For this example, gross sales of $9,000 will yield the same weekly income, $720, for both plans. The point of intersection of the two lines, (9000, 720) is important for our purposes because it tells us when one plan is more advantageous than the other. You can see from the graph in Figure 4.35 that if you sell less than $9,000 worth of merchandise per week, the base pay plus commission plan yields a higher income; if you sell more than $9,000 worth of merchandise per week, the straight commission plan yields a higher income.

Your decision as to what plan to take reduces to whether you think you can average above or below $9,000 in gross sales per week.

The process of solving a system of equations by graphing both equations and observing the point(s) of intersection is called the **graphical method** of solution.

As we saw in the preceding example, finding the point of intersection of two lines was a crucial step in making the decision about which plan to choose. Determining a point of intersection by the graphical method often gives an approximate solution. We would like to develop algebraic methods for finding the *exact* solution to such a problem. We will also see that algebraic solutions can sometimes be very efficient and time-saving. Let's begin by introducing some basic terminology.

DEFINITION Two or more equations considered together are called a **system of equations.** In particular, if the equations are of the first degree, it is called a **linear system.**

Thus,

$$\begin{cases} 2x - 3y = 6 \\ x - y = 1 \end{cases}$$

is an example of a linear system in two variables. This is called a **2 × 2** (read "2 by 2") **system,** since there are two equations and two variables. The "{" indicates that the two equations are to be considered together.

Solving such a *system of equations* means finding all the ordered pairs that satisfy *all* the equations in the system. Keep in mind that *one* solution to the system consists of *two* numbers—an *x*-value and a *y*-value.

We know that a single linear equation has an infinite number of solutions. How many solutions can a 2 × 2 linear system have? Or how many ordered pairs can satisfy two linear equations in two unknowns? Each of the equations in a 2 × 2 linear system is a straight line. Thus, we have the following three possibilities:

Case 1 The lines intersect in exactly one point. The coordinates of the point are the solution to the system.

Such a system is called **consistent** and **independent** (see Figure 4.36a).

Case 2 The lines are parallel and therefore never intersect. There are no solutions to the system.

Such a system is called **inconsistent** (see Figure 4.36b).

Case 3 The lines coincide. All the points that satisfy one of the equations also satisfy the other. Thus, there are infinitely many solutions.

Such a system is called **dependent** (see Figure 4.36c).

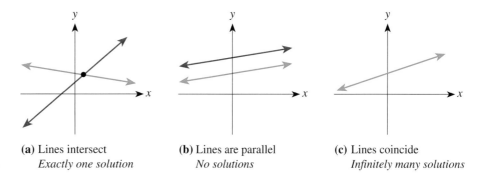

(a) Lines intersect
Exactly one solution

(b) Lines are parallel
No solutions

(c) Lines coincide
Infinitely many solutions

Figure 4.36

The algebraic methods we are about to discuss will allow us to determine which situation we have and, in case 1, to find the unique solution.

The Elimination Method

The *elimination* (or *addition*) *method* is based on the following idea. We already know how to solve a variety of equations involving *one* variable. If we can manipulate our system of equations so that one of the variables is eliminated, we can then solve the resulting equation in one variable.

We illustrate the elimination method with several examples.

EXAMPLE 2

Solve the following system of equations:

$$\begin{cases} x + y = -5 \\ x - y = 9 \end{cases}$$

Solution

In solving an equation, we are used to adding the same number or quantity to both sides. However, the addition property of equality allows us to add equal *quantities* to both sides of an equation. According to the second equation, $x - y$ and 9 are equal quantities; thus, we can just "add" that quantity to both sides of the equation $x + y = -5$, but we add $x - y$ to the left side of $x + y = -5$ and 9 to the right side. Thus, we "add" the two equations together.

$$\begin{array}{r} x + y = -5 \\ x - y = 9 \qquad \textit{Add.} \\ \hline 2x = 4 \end{array}$$

Notice that we have *eliminated* one of the variables and now we have an equation in one variable. We solve for that variable:

$$x = 2$$

To find the other variable, y, we substitute $x = 2$ in either one of the equations and solve for y:

$$\begin{aligned} x - y &= 9 \qquad \textit{Substitute } x = 2 \textit{ in the second equation and solve for y.} \\ 2 - y &= 9 \\ y &= -7 \end{aligned}$$

The solution is $\boxed{(2, -7)}$.

Check: We substitute $x = 2$ and $y = -7$ into both equations:

$$\begin{array}{cc} x + y = -5 & x - y = 9 \\ 2 + (-7) \overset{\checkmark}{=} -5 & 2 - (-7) \overset{\checkmark}{=} 9 \end{array}$$

EXAMPLE 3

Solve the following system of equations:

$$\begin{cases} 2x - 3y = 6 \\ x - y = 1 \end{cases}$$

Solution

We would like to "add" the two equations in such a way that one of the variables is eliminated. To do this we must change the coefficients of either the x- or y-variable so that they are exact opposites. Thus, when we add the two equations, the variables with opposite coefficients will be eliminated.

For example, to eliminate the y-variable, we multiply the second equation by -3. This produces a y-coefficient of $+3$ in the second equation, which is the opposite of -3, the y-coefficient in the first equation.

We proceed as follows:

$$2x - 3y = 6 \quad \textit{As is} \rightarrow \qquad 2x - 3y = 6 \quad \textit{Add the resulting equations.}$$
$$x - y = 1 \quad \textit{Multiply by } -3 \rightarrow \qquad \underline{-3x + 3y = -3}$$
$$-x = 3 \quad \textit{Solve for x.}$$
$$x = -3$$

Now we can substitute $x = -3$ into one of the original equations (we will use the first one) and solve for y:

$$2x - 3y = 6$$
$$2(-3) - 3y = 6$$
$$-6 - 3y = 6$$
$$-3y = 12$$
$$y = -4$$

Thus, the solution is $\boxed{(-3, -4)}$.

Check: We substitute $x = -3$ and $y = -4$ into both equations:

$$2x - 3y = 6 \qquad\qquad x - y = 1$$
$$2(-3) - 3(-4) \overset{?}{=} 6 \qquad -3 - (-4) \overset{?}{=} 1$$
$$-6 + 12 \overset{\checkmark}{=} 6 \qquad\qquad -3 + 4 \overset{\checkmark}{=} 1$$

Note that we could have chosen to eliminate x in Example 3 by multiplying the second equation by -2 and then adding the two equations. The elimination method is also called the *addition method*.

EXAMPLE 4

Solve the following system of equations:

$$\begin{cases} 3x = 4y + 6 \\ 5y = 2x - 4 \end{cases}$$

Solution

To make the elimination process easier to perform, we should first line up "like" variables to make it easier to see how to eliminate one of them.

$$3x = 4y + 6 \qquad \rightarrow \qquad 3x - 4y = 6$$
$$5y = 2x - 4 \qquad \rightarrow \qquad -2x + 5y = -4$$

A system in this form with the variables lined up is said to be in **standard form**.

We choose to eliminate x. To keep the arithmetic as simple as possible, we convert the coefficients of x to $+6$ and -6. Note that 6 is the least common multiple (LCM) of 3 and 2.

$$3x - 4y = 6 \quad \textit{Multiply by 2} \rightarrow \qquad 6x - 8y = 12 \quad \textit{Add the resulting}$$
$$-2x + 5y = -4 \quad \textit{Multiply by 3} \rightarrow \qquad \underline{-6x + 15y = -12} \quad \textit{equations.}$$
$$7y = 0 \quad \textit{Solve for y.}$$
$$y = 0$$

Substitute $y = 0$ into one of the original equations and solve for x:

$$3x = 4y + 6 \qquad \textit{We substitute } y = 0 \textit{ into the first equation.}$$
$$3x = 4(0) + 6$$
$$3x = 6$$
$$x = 2$$

Thus, our solution is $\boxed{(2, 0)}$.

You should check to see that $(2, 0)$ does indeed satisfy both equations.

We can summarize the elimination method as follows.

The Elimination (Addition) Method	1. Put the system of equations in standard form—that is, make sure the variables and constants line vertically. 2. Decide which variable you want to eliminate. 3. Multiply one or both equations by appropriate constants so that the variable you have chosen to eliminate appears with opposite coefficients. 4. Add the resulting equations. 5. Solve the resulting equation in *one* variable. 6. Substitute the value of the variable obtained into one of the original equations, and solve for the other variable. 7. Check your solution in both of the original equations.

EXAMPLE 5

Solve the following system of equations:

$$\begin{cases} 3a - \dfrac{b}{2} = 7 \\[2mm] \dfrac{a}{5} - \dfrac{2b}{3} = 3 \end{cases}$$

Solution   We begin by clearing the system of fractions.

$$3a - \frac{b}{2} = 7 \qquad \textit{Multiply by 2} \rightarrow \qquad 6a - b = 14$$

$$\frac{a}{5} - \frac{2b}{3} = 3 \qquad \textit{Multiply by 15} \rightarrow \qquad 3a - 10b = 45$$

We choose to eliminate *a*:

$$\begin{array}{lll} 6a - b = 14 & \textit{As is} \rightarrow & 6a - b = 14 \qquad \textit{Add.} \\ 3a - 10b = 45 & \textit{Multiply by } -2 \rightarrow & \underline{-6a + 20b = -90} \\ & & \qquad\qquad 19b = -76 \qquad \textit{Solve for b.} \\ & & \qquad\qquad\quad b = -4 \end{array}$$

To obtain the value for *a*, we may substitute $b = -4$ into any of the equations containing *a* and *b*. We substitute $b = -4$ into the equation $6a - b = 14$ (the first of our equations without fractional coefficients):

$$6a - b = 14$$
$$6a - (-4) = 14$$
$$6a + 4 = 14$$
$$6a = 10$$
$$a = \frac{10}{6} = \frac{5}{3}$$

Thus, the solution is $\boxed{a = \dfrac{5}{3},\ b = -4}$.

Check: (in the original equations)

$$3a - \frac{b}{2} = 7 \qquad\qquad \frac{a}{5} - \frac{2b}{3} = 3$$

$$3\left(\frac{5}{3}\right) - \frac{(-4)}{2} \overset{?}{=} 7 \qquad\qquad \frac{\frac{5}{3}}{5} - \frac{2(-4)}{3} \overset{?}{=} 3$$

$$5 + 2 \overset{\checkmark}{=} 7 \qquad\qquad \frac{5}{3} \cdot \frac{1}{5} - \left(\frac{-8}{3}\right) \overset{?}{=} 3$$

$$\frac{1}{3} + \frac{8}{3} \overset{\checkmark}{=} 3$$

The Substitution Method

There is another method we can use to solve a system of equations. While our goal remains the same—to obtain an equation in one variable—our approach will be slightly different. We illustrate the substitution method with several examples.

EXAMPLE 6

Solve the following system of equations:

$$\begin{cases} 5x - 3y = 7 \\ \qquad x = 6y - 4 \end{cases}$$

Solution

We notice that the second equation is solved explicitly for x. Since $x = 6y - 4$, we can substitute $6y - 4$ into the first equation in place of x:

$$5x - 3y = 7 \qquad \textit{Replace each occurrence of x by 6y − 4.}$$
$$5(6y - 4) - 3y = 7 \qquad \textit{Now we have an equation in one variable. Solve for y.}$$
$$30y - 20 - 3y = 7$$
$$27y - 20 = 7$$
$$27y = 27$$
$$y = 1$$

Now we substitute $y = 1$ into the second of our original equations (since it is already solved explicitly for x):

$$x = 6y - 4$$
$$x = 6(1) - 4$$
$$x = 2$$

Thus, the solution is $\boxed{(2,\ 1)}$.

Check:

$$5x - 3y = 7 \qquad\quad x = 6y - 4$$
$$5(2) - 3(1) \overset{?}{=} 7 \qquad\quad 2 \overset{?}{=} 6(1) - 4$$
$$10 - 3 \overset{\checkmark}{=} 7 \qquad\quad 2 \overset{\checkmark}{=} 6 - 4$$

We can summarize the substitution method as follows.

The Substitution Method	1. Solve one of the equations explicitly for one of the variables. 2. Substitute the expression obtained in step 1 into the other equation. 3. Solve the resulting equation in one variable. 4. Substitute the obtained value into one of the original equations (usually the one solved explcitly in step 1) and solve for the other variable. 5. Check the solution.

EXAMPLE 7

Solution

Solve the following system of equations:

$$\begin{cases} \dfrac{5}{2}x + y = 4 \\ 2y + 5x = 10 \end{cases}$$

We solve the first equation explicitly for y:

$$\frac{5}{2}x + y = 4 \quad \rightarrow \quad y = -\frac{5}{2}x + 4$$

Substitute for y in the second equation:

$$2y + 5x = 10 \qquad \textit{Substitute } -\frac{5}{2}x + 4 \textit{ for y.}$$

$$2\left(-\frac{5}{2}x + 4\right) + 5x = 10$$

$$-5x + 8 + 5x = 10$$

$$8 = 10 \qquad \textit{This is a contradiction.}$$

Thus, there are no solutions common to both equations.

If we solve the *second* equation in our system for y, we get

$$y = -\frac{5}{2}x + 5 \qquad \textit{This equation is in the form } y = mx + b.$$

We can see that the two lines never meet (they both have the same slope of $-\frac{5}{2}$ but have different y-intercepts).

This system of equations has no solutions and is therefore $\boxed{\text{inconsistent}}$.

EXAMPLE 8

Solution

Solve the following system of equations:

$$\begin{cases} 6x - 4y = 10 \\ 2y + 5 = 3x \end{cases}$$

If we choose to use the substitution method, then we must solve one of the equations explicitly for one of the variables. Whichever equation and whichever variable we choose, we are forced to work with fractional expressions. (Try it!) In this case the elimination method seems to be easier.

We begin by getting the system in standard form, and then eliminate y:

$$\begin{array}{llll} 6x - 4y = 10 & \rightarrow & 6x - 4y = 10 & \textit{As is} \rightarrow & 6x - 4y = 10 \\ 2y + 5 = 3x & \rightarrow & -3x + 2y = -5 & \textit{Multiply by 2} \rightarrow & \underline{-6x + 4y = -10} \\ & & & & 0 = 0 \end{array}$$

$$\textit{This is an identity.}$$

If we solve both equations for *y*, we obtain

$$y = \frac{3}{2}x - \frac{5}{2}$$

and we see that the lines are identical (they have the same slope *and* the same *y*-intercept). Thus, every ordered pair that satisfies one of the equations also satisfies the other. There are infinitely many solutions. The equations are | dependent | .

The solution set is the set of all the points on the line, that is,

$$\{(x, y) \mid 6x - 4y = 10\}$$

EXAMPLE 9

In Example 1 of this section we constructed two equations to describe two possible income plans:

$$\begin{cases} I = 0.08g \\ I = 180 + 0.06g \end{cases}$$

We graphed the two equations, assigning *g* as the independent variable (represented by the horizontal axis) and *I* as the dependent variable (represented by the vertical axis). Find the point of intersection of the graphs of these two equations.

Solution   The point of intersection of the two graphs lies on both lines and so is an ordered pair that satisfies both equations. Hence, we are being asked to solve the system of equations.

Since both equations are solved explicitly for *I,* we can substitute for *I* in either equation:

$I = 180 + 0.06g$ *Use the first equation to substitute 0.08g for I in the second equation.*

$0.08g = 180 + 0.06g$ *Subtract **0.06g** from both sides to get*

$0.02g = 180$ *Divide each side by **0.02**.*

$$g = \frac{180}{0.02} = 9,000$$

Substitute *g* = 9,000 for *I* into the first equation to get

$$I = 0.08g$$
$$I = 0.08(9,000) = 720$$

Since *g* is the independent variable, the ordered pairs are of the form (*g, I*): The point of intersection is (9000, 720). The graphs of these equations appear in Figure 4.35 on page 234.

Having the ability to solve a system of equations gives us a great deal of flexibility in how we set up our solutions to verbal problems. In many cases we may be able to solve a verbal problem either by using a one-variable approach as we did in Chapters 2 and 3, or by writing a system of equations.

EXAMPLE 10

A stationery store ordered 50 cases of envelopes costing a total of $551.50. Among the 50 cases were some that contained legal-size envelopes and cost $11.95 each, whereas the remaining cases contained letter-size envelopes and cost $9.95 each. How many of each type were ordered?

Solution   *If we want to use more than one variable, then we must have as many independent equations as we have variables.*

$$\text{Let } x = \text{Number of cases of legal-size envelopes}$$
$$\text{Let } y = \text{Number of cases of letter-size envelopes}$$

We must create two equations—the first will relate the *number* of cases, the second will relate the *cost* of the cases.

Our equations are

$$\begin{cases} x + y = 50 \\ 11.95x + 9.95y = 551.50 \end{cases}$$

There are 50 cases altogether.

The total cost is $551.50.

$$x + y = 50 \qquad \textit{Multiply by } -995 \rightarrow \qquad -995x - 995y = -49{,}750$$
$$11.95x + 9.95y = 551.50 \qquad \textit{Multiply by } 100 \rightarrow \qquad \underline{1{,}195x + 995y = 55{,}150}$$
$$200x = 5{,}400$$
$$x = 27$$

Substitute $x = 27$ into the first equation:

$$x + y = 50$$
$$27 + y = 50$$
$$y = 23$$

Hence, there are 27 cases of legal-size envelopes and 23 cases of letter-size envelopes.

The check is left to the student.

EXAMPLE 11

The length of a rectangle is 5 in. more than 4 times the width, and the perimeter is 182 in. Find the dimensions of the rectangle.

Solution   The applications discussed earlier in the text were solved using a one-variable approach. In fact many applications can be solved using either a one-variable or a two-variable approach. We illustrate both approaches in this example.

One-Variable Approach: We draw a rectangle and label the width as *w*.

$$\text{Let } W = \text{Width}$$

Then

$$4W + 5 = \text{Length} \quad \textit{Since the length is 5 more than 4 times the width}$$

We label the length of the rectangle $4W + 5$ (see Figure 4.37).

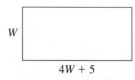

Figure 4.37

The formula for the perimeter of a rectangle is $P = 2L + 2W$, where P is the perimeter, L is the length, and W is the width. Hence

$$P = 2L + 2W$$

Since P = 182 and the width is W and the length is 4W + 5, we have

$$182 = 2(4W + 5) + 2W$$

Which can be simplified to

$$182 = 8W + 10 + 2W$$

$$182 = 10W + 10$$

This is a first-degree equation. We proceed to solve it: Subtract 10 from each side.

$$172 = 10W$$

Divide each side by 10 to get

$$\frac{172}{10} = \frac{10W}{10}$$

$$17.2 = W$$

The width is 17.2 in.

Hence the width $W = 17.2$ in. and the length $L = 4W + 5 = 4(17.2) + 5 = 73.8$ in.

The dimensions of the rectangle are $\boxed{17.2 \text{ in. by } 73.8 \text{ in.}}$

Check:

Is 73.8 (the length) 5 more than 4 times 17.2 (the width)? $73.8 \overset{\checkmark}{=} 4(17.2) + 5$.

Is the perimeter 182 in.? $2(73.8) + 2(17.2) \overset{\checkmark}{=} 182$

Two-Variable Approach: We draw a picture of a rectangle and label the sides using *two* variables, L and W, as shown in Figure 4.38.

Figure 4.38

We create two equations—one relating the sides, and the other relating the perimeter to the sides:

$$L = 4W + 5$$
$$182 = 2L + 2W$$

The length is 5 inches more than 4 times the width.

The perimeter is 182 in.

Using the first equation, we substitute into the second equation and get

$$182 = 2(4W + 5) + 2W$$

This is exactly the same equation we obtained using the one-variable approach. The rest of the solution is as it appears above.

In Examples 10 and 11, neither the one-variable nor the two-variable approach offers any particular advantages. Sometimes, however, as the following example illustrates, the way a problem is stated makes the two-variable approach significantly easier.

EXAMPLE 12

George and Ruth both go into a store to buy blank videotape cassettes. George buys eight 120-minute cassettes and five 160-minute cassettes for a total of $28.87, while Ruth buys six 120-minute cassettes and ten 160-minute cassettes for a total of $37.84. What are the prices of a single 120-minute and a single 160-minute cassette?

Solution   If we try the one-variable approach here we will find that it is difficult to represent both prices in terms of one variable. (Try it!) However, the statement of the problem allows us to use the two-variable approach quite naturally.

Let t = price of a single 120-minute cassette

Let s = price of a single 160-minute cassette

From the statement of the problem, we obtain the following system:

$$8t + 5s = 28.87 \quad \text{\textit{This equation represents George's purchase.}}$$
$$6t + 10s = 37.84 \quad \text{\textit{This equation represents Ruth's purchase.}}$$

We choose to eliminate s.

$$8t + 5s = 28.87 \quad \text{\textit{Multiply by} } -2 \rightarrow \quad -16t - 10s = -57.74$$
$$6t + 10s = 37.84 \quad \text{\textit{As is} } \rightarrow \quad \underline{6t + 10s = 37.84}$$
$$-10t = -19.90$$
$$t = 1.99$$

Substitute $t = 1.99$ into the first equation:

$$8t + 5s = 28.87$$
$$8(1.99) + 5s = 28.87$$
$$15.92 + 5s = 28.87$$
$$5s = 12.95$$
$$s = 2.59$$

A 120-minute cassette costs \$1.99 and a 160-minute cassette costs \$2.59.

The check is left to the student.

STUDY SKILLS 4.3

Preparing for a Comprehensive Final Exam

Many algebra courses require students to take a comprehensive final exam—an exam covering the entire course. Because of the amount of material covered in this type of exam, your preparation should necessarily differ from your preparation for other exams.

To succeed in a comprehensive final, even those students who have been doing well on exams all along will need to take the time to review the ideas and procedures covered earlier in the course. As more material is learned, it becomes easier to forget and/or confuse concepts learned previously.

For example, we often find that, on finals, students manage to correctly solve complex problems from the later chapters, but have difficulty with some of the simpler problems covered in the earlier chapters. This is because the more complex material was most recently learned, whereas some of the "simple" topics, covered much earlier in the course, were forgotten or became obscured by the newer material. Even if you knew the material well earlier in the course, there is

no substitute for timely review of the *entire* course syllabus.

Your studying should begin at least two weeks before the exam. Starting two weeks early means that you will probably be learning new material as you are studying for the final exam. You should consider studying for the final as separate from learning the new material. This means that your total math study time should be increased and divided between learning the new material and reviewing previous material (even the "simple" topics).

Your studying should include the following:

1. Review your class notes. Pay particular attention to those topics you may have found a bit difficult the first time you learned them.
2. If you have been writing out Study Cards, be sure to review them. Pay particular attention to the Warning Cards.
3. Review a selection of homework exercises. Be

sure to choose a wide variety of exercises to cover all the topics in your syllabus, and the various types of exercises within each topic.

4. Review all your class exams and quizzes. Be sure you understand how every single problem is solved, and where you made your errors.

5. Find out whether copies of previous final exams are available for you to examine. Be sure you know the format of the final exam: Will the exam be multiple choice, fill-in, or a combination? Will there be partial credit?

6. A few days before the final exam, make up a practice final exam and follow the directions given in Study Skills 2.6: Using Quiz Cards.

Remember, going over all this material takes time, so be sure to start your studying well in advance of the final exam date.

EXERCISES 4.3

In Exercises 1–38, solve the system of equations. State whether the system is independent, inconsistent, or dependent. Use whichever method you prefer.

1. $\begin{cases} 2x + y = 12 \\ 3x - y = 13 \end{cases}$

2. $\begin{cases} x + 4y = 6 \\ -x + 3y = 8 \end{cases}$

3. $\begin{cases} -x + 5y = 11 \\ x - 2y = -2 \end{cases}$

4. $\begin{cases} 4x + y = 16 \\ 3x - y = 5 \end{cases}$

5. $\begin{cases} 3x - y = 0 \\ 2x + 3y = 11 \end{cases}$

6. $\begin{cases} 5x - y = 13 \\ 3x - 2y = 5 \end{cases}$

7. $\begin{cases} x + 7y = 20 \\ 5x + 2y = 34 \end{cases}$

8. $\begin{cases} -x + 5y = 12 \\ -3x + 4y = 3 \end{cases}$

9. $\begin{cases} 4x + 5y = 0 \\ 2x + 3y = -2 \end{cases}$

10. $\begin{cases} 5x - 3y = 18 \\ 4x - 6y = 0 \end{cases}$

11. $\begin{cases} 2x + 3y = 7 \\ 4x + 6y = 14 \end{cases}$

12. $\begin{cases} 3x - 5y = 4 \\ 6x - 10y = 9 \end{cases}$

13. $\begin{cases} 5x - 6y = 3 \\ 10x - 12y = 5 \end{cases}$

14. $\begin{cases} -2x + 14y = 8 \\ x - 7y = -4 \end{cases}$

15. $\begin{cases} 2x - 3y = 10 \\ 3x - 2y = 15 \end{cases}$

16. $\begin{cases} 2x + 3y = 18 \\ 3x + 2y = 12 \end{cases}$

17. $\begin{cases} y = 2x + 3 \\ 2x + y = -1 \end{cases}$

18. $\begin{cases} x = 3y - 4 \\ 3x + 2y = 10 \end{cases}$

19. $\begin{cases} 6a - 3b = 1 \\ 8a + 5b = 7 \end{cases}$

20. $\begin{cases} 2a - 6b = -4 \\ 5a - 7b = -4 \end{cases}$

21. $\begin{cases} s = 3t - 5 \\ t = 3s - 5 \end{cases}$

22. $\begin{cases} s = 5t - 8 \\ t = 5s - 8 \end{cases}$

23. $\begin{cases} 3m - 2n = 8 \\ 3n = m - 8 \end{cases}$

24. $\begin{cases} 5m - 3n = 2 \\ m - 4 = 2n \end{cases}$

25. $\begin{cases} 3p - 4q = 5 \\ 3q - 4p = -9 \end{cases}$

26. $\begin{cases} 5p + 6q = 1 \\ 5q + 6p = -1 \end{cases}$

27. $\begin{cases} \dfrac{u}{3} - v = 1 \\ u - \dfrac{v}{2} = 5 \end{cases}$

28. $\begin{cases} u - \dfrac{v}{4} = 4 \\ \dfrac{u}{5} - v = -3 \end{cases}$

29. $\begin{cases} \dfrac{w}{4} + \dfrac{z}{6} = 4 \\ \dfrac{w}{2} - \dfrac{z}{3} = 4 \end{cases}$

30. $\begin{cases} \dfrac{w}{6} - \dfrac{z}{5} = 2 \\ \dfrac{w}{2} - \dfrac{z}{10} = 1 \end{cases}$

31. $\begin{cases} \dfrac{x}{6} + \dfrac{y}{8} = \dfrac{3}{4} \\ \dfrac{x}{4} + \dfrac{y}{3} = \dfrac{17}{12} \end{cases}$

32. $\begin{cases} \dfrac{x}{5} + \dfrac{y}{3} = \dfrac{2}{3} \\ \dfrac{x}{10} - \dfrac{y}{4} = \dfrac{3}{4} \end{cases}$

33. $\begin{cases} \dfrac{x + 3}{2} + \dfrac{y - 4}{3} = \dfrac{19}{6} \\ \dfrac{x - 2}{3} + \dfrac{y - 2}{2} = 2 \end{cases}$

34. $\begin{cases} \dfrac{a - 2}{4} - \dfrac{b + 1}{2} = \dfrac{3}{2} \\ \dfrac{a - 3}{3} + \dfrac{b + 1}{4} = \dfrac{25}{4} \end{cases}$

35. $\begin{cases} 0.1x + 0.01y = 0.37 \\ 0.02x + 0.05y = 0.41 \end{cases}$

36. $\begin{cases} 0.3x - 0.7y = 2.93 \\ 0.06x - 0.2y = 0.58 \end{cases}$

37. $\begin{cases} \dfrac{x}{2} + 0.05y = 0.35 \\ 0.3x + \dfrac{y}{4} = 0.65 \end{cases}$

38. $\begin{cases} 0.02x + \dfrac{y}{2} = 0.3 \\ \dfrac{x}{2} - 0.4y = 2.34 \end{cases}$

Solve the following problems algebraically by writing an equation or a system of equations. Clearly label what each variable represents.

39. Susan wants to invest a total of $14,000 so that her yearly interest is $835. If she chooses to invest part in a certificate of deposit paying 5% and the remainder in a corporate bond paying 8%, how much should she invest at each rate?

40. Harry makes two investments. He invests $6,000 more at 8.5% than he invests at 7.5%. If the total yearly interest from both investments is $1,022, how much is invested at each rate?

41. To the nearest cent, how can $10,000 be split into two investments, one paying 10% interest and the other paying 8% interest, so that the yearly interest from the two investments is equal?

42. $20,000 is split into two investments: one paying 5% and the other paying 6.5%. To the nearest cent, how much should be invested in each so that the yearly interest from the 5% investment is double the interest from the 6.5% investment?

43. A person invests money at 6% and at 4%, earning a total yearly interest of $540. Had the amounts invested been reversed, the yearly interest would have been $510. How much was invested altogether?

44. Jake invests in a bond paying 4% and a bond paying 2.5%, earning a total yearly interest of $360. Had the amounts invested been reversed, the yearly interest would have been $400. How much was invested in each bond?

45. The perimeter of a rectangle is 36 cm. If the length is 2 cm more than the width, find the dimensions of the rectangle.

46. The side of a square is 2 less than 3 times the side of an equilateral triangle. If the perimeter of the square is 12 more than twice the perimeter of the triangle, find the lengths of the sides of both figures.

47. Albert and Audrey both go into a camera store to buy film. Albert spends $35.60 on 5 rolls of 35-mm film and 3 rolls of movie film. Audrey spends $43.60 on 3 rolls of 35-mm film and 5 rolls of movie film. What are the costs of a single roll of each type of film?

48. Jim has a part-time job selling newspaper and magazine subscriptions. One week he earns $62.20 by selling 10 newspaper and 6 magazine subscriptions. The following week he earns $79 by selling 12 newspaper and 8 magazine subscriptions. How much does he earn for each newspaper and each magazine subscription he sells?

49. A donut shop sells a box containing 7 cream-filled and 5 jelly donuts for $6.32, and a box containing 4 cream-filled and 8 jelly donuts for $6.08. Find the costs of a single cream-filled and a single jelly donut.

50. A candy shop sells a mixture containing 1 lb of hard candy and 2 lb of chocolates for $15.39, and a mixture containing 2 lb of hard candy and 1 lb of chocolates for $12.93. What are the costs for 1 lb of hard candy and 1 lb of chocolates?

51. An electronics manufacturer produces two types of transmitters. The more expensive model requires 6 hours to manufacture and 3 hours to assemble. The less expensive model requires 5 hours to manufacture and 2 hours to assemble. If the company can allocate 730 hours for manufacturing and 340 hours for assembly, how many of each type can be produced?

52. The mathematics department hires tutors and graders. For the month of October, the department budgets $1,650 for 80 hours of tutoring and 45 hours of grading. In November, the department budgets $1,200 for 60 hours of tutoring and 30 hours of grading. How much does the department pay for each hour of tutoring and for each hour of grading?

53. A couple has a choice of two health insurance plans: one through the husband's employer and one through the wife's employer. The husband's plan will cost $140 per month and will pay for 90% of all medical expenses. The wife's plan will cost $100 per month and will pay for 75% of all medical expenses. At what level of annual medical expenses are the two plans equivalent?

54. A family has a choice of two health insurance plans: one through the father's company and one through the mother's employer. The father's plan will cost $150 per month and will pay for 100% of all medical expenses. The mother's plan will cost $125 per month and will pay for 80% of all medical expenses. At what level of annual medical expenses should the family choose the mother's plan?

55. The Cheapo Car Rental Company charges $29 per day plus $0.12 per mile, whereas the Cut-Rate Car Rental Company charges $22 per day plus $0.15 per mile. On a 1-day rental, how many miles must be driven to make the cost of the two companies the same?

56. An office equipment rental company offers two different leases on a copying machine. One lease agreement costs $75 per month plus $0.024 per copy. A second lease agreement costs $85 per month plus $0.019 per copy. How many copies per month must a customer make for the second lease agreement to be cheaper than the first?

57. The costs, C, that a manufacturer incurs in producing a product are usually divided into two parts: *fixed costs* and *unit costs*. The unit costs are computed by multiplying the number of units manufactured, x, by the cost of manufacturing each unit. The *break-even* point for a company is the number of units the company must sell so that its costs and revenue are equal. Suppose that a company's costs are given by the equation $C = 1.6x + 7{,}200$ and its revenue is given by the equation $R = 2.1x$. Find the break-even point.

58. Suppose that a company's costs are given by the equation $C = 3.5x + 18{,}750$ and its revenue is given by the equation $R = 4.2x$. Find the break-even point. (See Exercise 57.)

59. The telephone company offers two billing plans for local service. Under the first plan, the customer pays a fee of $18 per month plus a per-call fee of $0.11. The second plan charges the customer a fee of $24 per month, which includes 50 free calls plus a per-call fee of $0.09 for each call above 50 per month. How many local calls per month make the two plans equivalent?

60. A cellular phone company offers two billing plans for its monthly service. Under the first plan, the customer pays a fee of $33 per month, which includes 30 minutes of free airtime, plus a fee of $0.57 for each minute of airtime above 30. The second plan charges the customer a monthly fee of $40 per month plus a fee of $0.34 per minute of airtime. How many minutes of phone usage per month make the two plans equivalent?

61. A bank teller receives a deposit of 43 bills totaling $340. If the bills are all $5 and $10 bills, how many of each are there?

62. Dorothy purchases a total of 80 stamps for $25.76. If she bought 34-cent and 25-cent stamps, how many of each did she buy?

63. A car rental agency charges a flat fee plus a mileage rate for a 1-day rental. If the charge for a 1-day rental with 85 miles is $44.30 and the charge for a 1-day rental with 125 miles is $51.50, find the flat fee and the charge per mile.

64. A discount airline has a fixed charge for processing tickets plus a mileage fee. A ticket for a 200-mile trip costs $48, while a ticket for a 300-mile trip costs $61. Find the fixed charge and the charge per mile.

65. A plane can cover a distance of 2,310 km in 6 hours with a tailwind (with the wind) and a distance of 1,530 km in the same time with a headwind (against the wind). Find the speed of the plane and the speed of the wind.

66. On a river, a boat travels a distance of 60 km in 4 hours downstream, but can travel 40 km in 5 hours upstream. Find the speed of the boat and the speed of the river.

67. Find the point (x, y) so that the line passing through the points (x, y) and $(2, 3)$ with a slope of -1 intersects the line passing through (x, y) and $(1, -2)$ with a slope of 2.

68. Find the values of A and B so that the line whose equation is $Ax + By = 8$ passes through the points $(1, -8)$ and $\left(5, \frac{8}{3}\right)$.

69. Between 1990 and 2004, the number of physicians who specialize in dermatology in the United States can be estimated by the equation $P = 205(t - 1,990) + 6,005$, where P is the number of dermatologists during the given year t. Between 1990 and 2004, the number of physicians who specialize in gastroenterology in the United States can be estimated by the equation $P = 390(t - 1,990) + 5,205$, where P is the number of gastroenterologists during the given year t. In what year (between 1990 and 2004) was the number of dermatologists equal to the number of gastroenterologists?

❓ Questions for Thought

70. When asked to solve a system of equations, describe what you look for in deciding whether to use the elimination or substitution method.

71. Solve the following system by the elimination method:

$$\begin{cases} 4x - 9y = 3 \\ 10x - 6y = 7 \end{cases}$$

In using the elimination method, we eliminated a variable and then used *substitution* to find the value of the variable eliminated. A variation of the elimination method is to use the elimination process twice—once for each variable—to solve for each variable. Would this be easier for the system you just solved? Why?

72. Solve the following systems for u and v by first substituting x for $\frac{1}{u}$ and y for $\frac{1}{v}$:

(a) $\begin{cases} \dfrac{1}{u} + \dfrac{1}{v} = 5 \\ \dfrac{1}{u} - \dfrac{1}{v} = 1 \end{cases}$ **(b)** $\begin{cases} \dfrac{2}{u} + \dfrac{1}{v} = 3 \\ \dfrac{6}{u} + \dfrac{1}{v} = 5 \end{cases}$

◆ Mini-Review

73. *Solve:* $4(x - 3) - 6(x - 8) = 10 - (x - 18)$

74. $\dfrac{x}{3} - \dfrac{x}{5} = \dfrac{16}{5}$

75. If $f(x) = 3x^2 - 2x + 1$, find $f(-4)$ and $f(a)$.

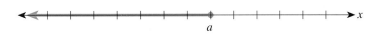

4.4 Graphing Linear Inequalities in Two Variables

Let's begin by reexamining the *number line* graph of a first-degree inequality in *one variable*. As illustrated in Figure 4.39, on the number line, the graph of a first-degree conditional inequality in one variable, $x \le a$, forms a half-line. This half-line has as its endpoint the point that is the solution of the *equation* $x = a$. In other words, on the number line, the graph of the inequality $x \le a$ is bounded by the graph of the equation $x = a$.

Figure 4.39
The graph of $x \le a$ on the number line

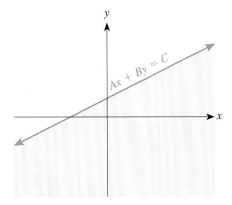

In graphing linear inequalities in *two variables* on the *rectangular coordinate system*, we find an analogous situation. The graph of $Ax + By \le C$ will be a *half-plane* bounded by the graph of the *equation* $Ax + By = C$ (see Figure 4.40).

Figure 4.40
A possible graph of
$Ax + By \le C$

Using what we have learned thus far, we can graph linear inequalities in two variables.

EXAMPLE 1

Sketch the solution set to the inequality $x + y \le 4$.

Solution

We begin by graphing the equation $x + y = 4$. We sketch the graph by the intercept method as outlined in Section 4.1 (see Figure 4.41).

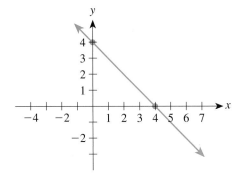

Figure 4.41
Graph of $x + y = 4$

One way to look at this inequality is to solve $x + y \le 4$ for y to get

$$y \le 4 - x$$

We want $y \le 4 - x$. That is, we want every point whose y-coordinate is *less than or equal to* 4 minus its x-coordinate. Since the line in Figure 4.41 is the set of points with each y-coordinate *equal* to 4 minus its x-coordinate, we want those points *below* the line. (Remember that the *y*-coordinate of a point is its "height." Asking for *y* to be less than something means we want the points to be *lower*.) Thus, our graph is as illustrated in Figure 4.42.

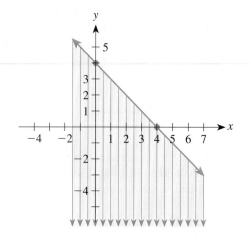

Figure 4.42
The graph of $x + y \leq 4$
(points on or below
the line)

Note that we could also have proceeded as follows:

$$x + y \leq 4 \qquad \textit{Solve for x.}$$
$$x \leq 4 - y$$

This means that we want the *x*-coordinate to be *less than* or equal to 4 minus the *y*-coordinate. Thus, we want the points to the *left* of the line, as indicated in Figure 4.43. Note that this gives us the same graph as in Figure 4.42.

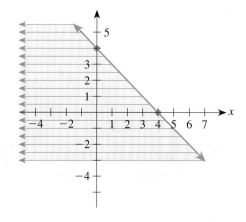

Figure 4.43
The graph of $x + y \leq 4$
(points on or to the left of
the line)

Let's look at another example.

EXAMPLE 2

Solution

Sketch the solution set of $3x - 2y > 10$.

Even though we have a strict inequality, we begin by sketching the graph of the equation $3x - 2y = 10$. The intercepts are $(0, -5)$ and $\left(\frac{10}{3}, 0\right)$, (see Figure 4.44). Note that we drew a *dashed* line instead of a solid line. This is because the inequality is strict and the points on the line are *not* included in the solution set.

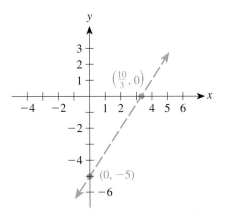

Figure 4.44
The graph of $3x - 2y = 10$

Rather than proceed as we did in the previous example to solve the inequality for x or y, let's think a moment. We know that the graph of the inequality is going to be a half-plane lying either above or below (or to the left or right of, if you like) the graph of the line. We already know how to graph the line. The question that remains is, "To which side of the line will the solutions to the inequality lie?" Either *all* solutions will lie in one part or *all* solutions will lie in the other part.

Therefore, all we need do is pick a convenient point in one of the parts (above or below the line) and substitute it into the inequality. If that point satisfies the inequality, then *all* the points in that region do. If that point does not satisfy the inequality, then *all* the points in the *other* part do.

The point $(0, 0)$ is often a convenient point to choose (note that it lies *above* the line $3x - 2y = 10$):

$$3x - 2y > 10 \qquad \textit{Substitute } (0, 0) \textit{ to see whether it satisfies the inequality.}$$
$$3(0) - 2(0) \overset{?}{>} 10$$
$$0 - 0 \not> 10$$

Since $(0, 0)$ does *not* satisfy the inequality and $(0, 0)$ is located in the region *above* the line, we know that the solution set consists of all points *below* the line (see Figure 4.45).

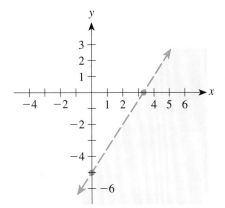

Figure 4.45
The graph of
$3x - 2y > 10$

The last two examples lead us to the following outline, which can be used to graph linear inequalities in two variables.

**Outline for Graphing
Linear Inequalities**

1. Sketch the graph of the equation that is the boundary of the solution set. If the inequality is *weak* (involves ≤ and ≥), draw a solid line. If the inequality is *strict* (involves < or >), draw a dashed line.
2. Choose a convenient test point (a point not on the line) and substitute it into the inequality.
3. If the test point satisfies the inequality, then shade the region that contains the test point. Otherwise, shade the region on the other side of the line.

EXAMPLE 3

A nutritional products company wants to create a dietary supplement that must contain at most 5 milligrams (mg) of potassium. Food A contains 0.35 mg of potassium per gram of the food substance. Food B contains 0.65 mg of potassium per gram of the food substance. Write an inequality describing the amounts of food A and food B that can be used in the supplement and sketch the graph of this inequality.

Solution

If we let x be the number of grams of food A used, then because each gram of food A contains 0.35 mg of potassium, the number of milligrams of potassium that would be contributed to the supplement is $0.35x$. Similarly, if we let y be the number of grams of food B used, then the number of milligrams of potassium that would be contributed is $0.65y$. Therefore, since the supplement must contain at most 5 mg of potassium, we have the inequality

$$0.35x + 0.65y \le 5$$

Using the outline given above, we obtain the graph in Figure 4.46.

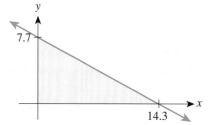

Figure 4.46
The graph of
$0.35x + 0.65y \le 5$

It makes no sense for either x or y to be negative in this situation, so we have shaded only the region in the first quadrant.

The points (1, 3.7) and (3.5, 2.8) are in the shaded region, which means that, for example, the company could use 1 gram of food A together with 3.7 grams of food B, or 3.5 grams of food A together with 2.8 grams of food B for the supplement and still stay at or below the maximum allowable potassium content. For this reason, the shaded region that contains all the solutions to the inequality is often called the *feasible set.*

EXAMPLE 4

In a rectangular coordinate system, sketch the graphs of:

(a) $x \ge -2$ **(b)** $y < 5$

Solution

(a) The graph of $x = -2$ is a vertical line 2 units to the left of the y-axis. Since we want x to be *greater than or equal to* -2, we shade the region to the *right* of the line, as indicated in Figure 4.47.

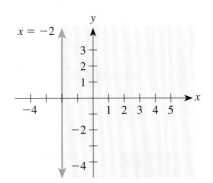

Figure 4.47
The graph of $x \geq -2$

(b) The graph of $y = 5$ is a horizontal line 5 units above the x-axis. Since we want y to be *less than* 5, we shade the region below the line (see Figure 4.48).

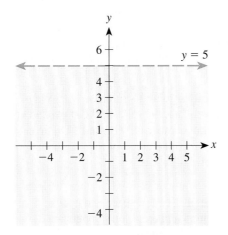

Figure 4.48
The graph of $y < 5$

Notice that the example instructions said "In a rectangular coordinate system. . . ." Without such instructions we might have just sketched the solution sets as inequalities in *one variable* on the number line.

EXERCISES 4.4

Sketch the solution set to each of the following inequalities on a rectangular coordinate system.

1. $y \leq x + 3$

2. $x > y + 3$

3. $x > y - 2$

4. $y \leq x - 2$

5. $x + y < 3$

6. $x + y \geq 4$

7. $x + y \geq 3$

8. $x + y < 4$

9. $2x + y \leq 6$

10. $x + 4y > 8$

11. $x + 2y > 6$

12. $4x + y \leq 8$

13. $3x + 2y \geq 12$

14. $5x + 3y < 30$

15. $2x + 5y < 10$

16. $4y + 3x \geq 24$

17. $2x + 5y > 10$

18. $4y + 3x \leq 24$

19. $2x + 5y \geq 10$

20. $4y + 3x < 24$

21. $2x - 5y \geq 10$

22. $4y - 3x < 24$

23. $y \leq x$

24. $x > -y$

25. $2x - y < 4$

26. $x - 2y \geq 4$

27. $4x - y \geq 8$

28. $x - 4y < 8$

29. $3x - 4y > 12$

30. $4x - 3y \leq 12$

31. $7x - 3y < 15$

32. $8x + 3y > 18$

33. $7x + 3y > 15$

34. $8x + 3y \leq 18$

35. $\dfrac{x}{2} + \dfrac{y}{3} < 4$

36. $\dfrac{x}{3} - \dfrac{y}{2} > 5$

37. $\dfrac{x}{3} - \dfrac{y}{4} \geq 3$ **38.** $\dfrac{x}{5} + \dfrac{y}{2} \leq 2$ **39.** $y < 3$ **40.** $y > -1$

41. $x \geq -2$ **42.** $x < 5$ **43.** $x < -2$ **44.** $x > 5$

45. $y < 0$ **46.** $y \geq 0$ **47.** $x \leq 0$ **48.** $x < 0$

49. $x < \dfrac{1}{2}$ **50.** $y \geq \dfrac{2}{3}$ **51.** $\dfrac{y}{2} > 0$ **52.** $\dfrac{x}{3} < 0$

53. The perimeter of a rectangle of length x and width y cannot exceed 80 ft. Write an inequality that represents this situation and sketch the graph of the solution set. List three possible feasible points in the solution set.

54. The perimeter of a rectangle of length x and width y must be at least 15 m. Write an inequality that represents this situation and sketch the graph of the solution set. List three feasible points in the solution set.

55. You are shopping at a discount clothes outlet where all shirts are priced at $20 and all pants are priced at $28. Suppose you buy s shirts and p pairs of pants. Write an inequality that represents the number of shirts and pants you can purchase on a maximum budget of $200. Sketch the graph of this inequality and list three feasible points in the solution set.

56. A wholesale buyer for an audio equipment store is going to purchase MP3 players and CD players. She will buy m MP3 players at $90 each and d CD players at $80 each. Write an inequality that represents the number of MP3 and CD players she can purchase for a maximum of $5,000. Sketch the graph of this inequality and list three feasible points in the solution set.

57. A warehouse with 2,500 sq ft of floor space is going to be used to store refrigerators and air conditioners. Each refrigerator uses 12 sq ft of space and each air conditioner uses 8 sq ft of space. If r refrigerators and a air conditioners are stored, write an inequality representing this situation. Sketch the graph of this inequality and list three feasible points in the solution set.

58. A director of personnel at a department store is going to hire clerks and salespeople. Clerks get a salary of $400 per week, whereas salespeople get a salary of $475 per week. The director will hire c clerks and s salespeople but must not exceed a total weekly payroll of $10,000. Write an inequality representing this situation and list three feasible points in the solution set.

59. A bank must be able to serve at least 90 customers per hour on average. Regular clerks can serve an average of 15 customers per hour, whereas special service clerks can serve an average of 10 customers per hour. Write an inequality representing the number of regular and special service clerks required to meet the bank's needs. Sketch the graph of the solution set and find three feasible points.

60. A meals-on-wheels program can provide two types of truck. One type can deliver 800 meals per day and the other type can deliver 1,200 meals per day. If a public agency can afford to contract for at most 21,000 meals, write an inequality representing how many of each type of truck can be contracted for. Sketch the graph of the solution set and find three feasible points.

 Mini-Review

61. *Evaluate:* $\left| -6 - 3 \right| - \left| 6 \right| - \left| 3 \right|$ **62.** *Solve for x:* $2x - 3 \leq 5x + 4$

63. The following statement illustrates which property of real numbers?

$$3\left(\dfrac{1}{2}\right) + 4\left(\dfrac{1}{2}\right) = (3 + 4)\left(\dfrac{1}{2}\right)$$

64. *Solve for x:* $|5 - 2x| = 5$

65. How long will it take for Bobby running 6 mph to overtake Linda running 5 mph if Linda had an hour head start?

66. The new DVX model machine can process 40 items per minute, whereas the old DVX model can process 32 items per minute. How long will it take to complete a job of processing 3,960 items if both machines are operating together?

CHAPTER 4 SUMMARY

After completing this chapter you should be able to:

1. Use the definition of slope to compute the slope of a line when two points on the line are known (Section 4.1).

For example: Compute the slope of the line passing through the points $(-1, 3)$ and $(2, -5)$.

$$m = \frac{y_2 - y_1}{x_2 - x_1} = \frac{3 - (-5)}{-1 - 2} = \frac{8}{-3} = -\frac{8}{3}$$

2. Recognize that lines with equal slopes are parallel and that lines whose slopes are negative reciprocals of each other are perpendicular (Section 4.1).

For example:

If line L_1 has slope $\frac{2}{5}$, then any line parallel to L_1 must have slope $\frac{2}{5}$.

If line L_2 has slope $\frac{3}{7}$, then any line perpendicular to L_2 must have slope $-\frac{7}{3}$.

3. Write an equation of a line using either

$$\text{the point–slope form} \quad y - y_1 = m(x - x_1)$$

or

$$\text{the slope–intercept form} \quad y = mx + b$$

(Section 4.2).

For example: Write an equation of the line passing through the points $(-2, 1)$ and $(1, 2)$.

We first find the slope:

$$m = \frac{y_2 - y_1}{x_2 - x_1} = \frac{1 - 2}{-2 - 1} = \frac{-1}{-3} = \frac{1}{3} \quad \text{or} \quad y = \frac{1}{3}x + \frac{5}{3}$$

The point–slope form is $y - y_1 = m(x - x_1)$. We can use either $(-2, 1)$ or $(1, 2)$ as our given point (x_1, y_1). Using $(1, 2)$ as the given point, we get

$$y - 2 = \frac{1}{3}(x - 1)$$

4. Find the slope of a line by putting its equation in slope–intercept form (Section 4.2).

For example: Find the slope of the line whose equation is $5x + 4y = 7$.

We can find the slope by putting the equation in slope–intercept form—that is, in the form $y = mx + b$:

$$5x + 4y = 7$$
$$4y = -5x + 7$$
$$y = -\frac{5}{4}x + \frac{7}{4}$$

Therefore, we can "read off" the slope, which is $-\frac{5}{4}$.

5. Write an equation of a line satisfying certain conditions (Section 4.2).

For example: Write an equation of the line that passes through the point $(3, -2)$ and is perpendicular to $2x - 7y = 11$.

We first find the slope of $2x - 7y = 11$ by putting it in slope–intercept form:

$$2x - 7y = 11$$
$$-7y = -2x + 11$$
$$y = \frac{2}{7}x - \frac{11}{7} \qquad \textit{The slope of this line is } \frac{2}{7}.$$

Since the required line is perpendicular to $2x - 7y = 11$, it will have slope $-\frac{7}{2}$. We can now write the equation using point–slope form:

$$y - y_1 = m(x - x_1)$$
$$y - (-2) = -\frac{7}{2}(x - 3)$$
$$y + 2 = -\frac{7}{2}(x - 3) \quad \text{or} \quad y = -\frac{7}{2}x + \frac{17}{2}$$

6. Solve a 2 × 2 system of linear equations by using either the elimination or the substitution method (Section 4.3).

For example: Solve for x and y:

$$\begin{cases} 3x - 4y = 10 \\ 4x - 5y = 13 \end{cases}$$

Solution: We use the elimination method. We choose to eliminate y.

$$\begin{cases} 3x - 4y = 10 & \textit{Multiply by } -5 \rightarrow & -15x + 20y = -50 \\ 4x - 5y = 13 & \textit{Multiply by } 4 \rightarrow & \underline{16x - 20y = 52} \\ & & x = 2 \end{cases}$$

Substitute $x = 2$ into the first equation:

$$3x - 4y = 10$$
$$3(2) - 4y = 10$$
$$6 - 4y = 10$$
$$-4y = 4$$
$$y = -1$$

The solution is $(2, -1)$. Check in both equations.

7. Solve a variety of verbal problems that give rise to a linear system of equations (Section 4.3).

For example: A candy shop sells a mixture containing 1 lb of regular jelly beans and $\frac{1}{2}$ lb of gourmet jelly beans for $2.75, and a mixture containing $\frac{1}{2}$ lb of regular jelly beans and 1 lb of gourmet jelly beans for $3.25. What is the cost per pound of each type of jelly bean?

Solution:

Let r = the cost per pound of regular jelly beans
Let g = the cost per pound of gourmet jelly beans

Then the given information translates into the following system of equations:

$$\begin{cases} 1r + \dfrac{1}{2}g = 2.75 \quad \textit{Represents the cost of the first mixture} \\ \dfrac{1}{2}r + 1g = 3.25 \quad \textit{Represents the cost of the second mixture} \end{cases}$$

Multiplying both sides of each equation by 2 yields the following system:

$$\begin{cases} 2r + g = 5.5 \quad \textit{As is} \rightarrow \\ r + 2g = 6.5 \quad \textit{Multiply by } -2 \rightarrow \end{cases}$$

$$\begin{array}{r} 2r + g = 5.5 \\ -2r - 4g = -13 \qquad \textit{Add.} \\ \hline -3g = -7.5 \\ g = 2.5 \end{array}$$

To find r, substitute 2.5 for g in $2r + g = 5.5$:

$$2r + g = 5.5$$
$$2r + 2.5 = 5.5$$
$$2r = 3$$
$$r = 1.5$$

The cost of regular jelly beans (r) is \$1.50 per pound and the cost of gourmet jelly beans (g) is \$2.50 per pound.

8. Sketch the graph of an inequality in two variables (Section 4.4).

For example: Sketch the graph of $x - 3y \le 6$.

We first sketch the graph of $x - 3y = 6$, whose intercepts are $(6, 0)$ and $(0, -2)$, using a solid line. Now we use $(0, 0)$ as a test point:

$$x - 3y \le 6$$
$$0 - 3(0) \overset{?}{\le} 6$$
$$0 - 0 \overset{?}{\le} 6$$
$$0 \overset{\checkmark}{\le} 6$$

Since $(0, 0)$ satisfies the inequality, we shade the region containing $(0, 0)$, as indicated in the figure.

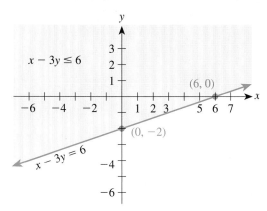

9. Use two-variable linear inequalities and their graphs to represent and visualize real-life situations (Section 4.4).

For example: A furniture manufacturer has a warehouse with 8,400 sq ft of storage space, which will be used to store recliners and couches. Each recliner requires 15 sq ft of space and each couch requires 32 sq ft. Write an inequality describing the number of recliners and couches that can be stored in this area and sketch the graph of this inequality.

Solution: If we let r be the number of recliners, then since each recliner requires 15 sq ft of space, all the recliners will require $15r$ sq ft of space. Similarly, if we let c be the number of couches, then the number of square feet required for all the couches is $32c$. Since the total storage space available is 8,400 sq ft, we have the inequality

$$15r + 32c \le 8{,}400$$

Using the outline for graphing a linear inequality, we obtain the graph in Figure 4.49. Since it makes no sense for either r or c to be negative, we have shaded only the region in the first quadrant.

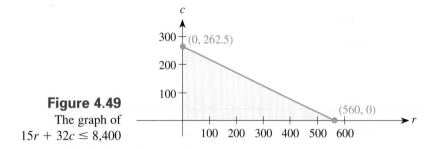

Figure 4.49
The graph of
$15r + 32c \le 8{,}400$

The points (300, 100) and (100, 200) are in the shaded region, which means that, for example, 300 recliners and 100 couches, or 100 recliners and 200 couches could be stored. These are examples of feasible points. It is also important to note that although a point such as (50.3, 71.5) is in this region, it is not a feasible point for this problem, because it is impossible to have a fraction of a couch or recliner.

CHAPTER 4 REVIEW EXERCISES

In Exercises 1–18, find the slope of the line satisfying the given condition(s).

1. Passing through the points $(-1, 0)$ and $(3, -2)$

2. Passing through the points $(6, -3)$ and $(-4, 3)$

3. Its equation is $y = 3x - 5$.

4. Its equation is $3y = 2x + 1$.

5. Its equation is $4y - 3x = 1$.

6. Its equation is $x = 3y - 5$.

7. Parallel to the line passing through the points $(3, 5)$ and $(1, 4)$

8. Parallel to the line passing through the points $(2, 4)$ and $(5, 0)$

9. Perpendicular to the line passing through the points $(4, 7)$ and $(4, 9)$

10. Perpendicular to the line passing through the points $(3, 5)$ and $(1, 4)$

11. Passing through the points $(-7, 6)$ and $(2, 6)$

12. Passing through the points $(3, 5)$ and $(3, -2)$

13. Parallel to the line whose equation is $y = 3x - 7$

14. Perpendicular to the line whose equation is $y = 5x + 1$

15. Perpendicular to the line whose equation is $3y - 5x + 6 = 0$

16. Parallel to the line whose equation is $6x - 4y - 9 = 0$

17. Parallel to the line whose equation is $x = 3$

18. Perpendicular to the line whose equation is $x = 4$

*In Exercises 19–22, find the value(s) of **a** that satisfy the given conditions.*

19. The line through the points $(4, a)$ and $(1, 2)$ has slope 4.

20. The line through the points $(a, 3)$ and $(2, 5)$ has slope 1.

21. The line through the points $(-3, a)$ and $(0, 3)$ is parallel to the line through the points $(7, a)$ and $(0, 0)$.

22. The line through the points $(2, a)$ and $(0, 5)$ is perpendicular to the line through the points $(0, -1)$ and $(a, 4)$.

In Exercises 23–44, write an equation of the line satisfying the given conditions.

23. The line passes through the points $(-2, 3)$ and $(1, -4)$.

24. The line passes through the points $(-1, -4)$ and $(0, -2)$.

25. The line passes through the points $(3, 5)$ and $(-3, -5)$.

26. The line passes through the points $(2, -6)$ and $(-2, 6)$.

27. The line passes through the point $(2, 5)$ and has slope $\frac{2}{5}$.

28. The line passes through the point $(0, 4)$ and has slope -4.

29. The line passes through the point $(4, 7)$ and has slope 5.

30. The line passes through the point $(3, 8)$ and has slope $-\frac{3}{4}$.

31. The line has slope 5 and y-intercept 3.

32. The line has slope -3 and y-intercept -4.

33. The horizontal line passes through the point $(2, 3)$.

34. The horizontal line passes through the point $(-3, -4)$.

35. The line passes through the point $(0, 0)$ and is parallel to the line $y = \frac{3}{2}x - 1$.

36. The line passes through the point $(-3, 0)$ and is perpendicular to the line $y = -5x + \frac{3}{7}$.

37. The line is perpendicular to $2y - 5x = 1$ and passes through the point $(0, 6)$.

38. The line is parallel to $6x - 7y + 3 = 0$ and passes through the point $(5, 5)$.

39. The line is perpendicular to $3x = -5y$ and passes through the point $(0, 0)$.

40. The line is parallel to $3x = -5y$ and passes through the point $(0, 0)$.

41. The line crosses the x-axis at $x = 3$ and the y-axis at $y = -5$.

42. The line has x-intercept 5 and y-intercept 8.

43. The line is parallel to $3x - 2y = 5$ and has the same y-intercept as $5y = x + 3$.

44. The line is parallel to $2x + 5y = 4$ and has the same y-intercept as $2y = x - 4$.

45. A manufacturer found that the relationship between his profit, P, and the number of gadgets produced, x, is linear. If he makes \$12,000 by producing 250 gadgets and \$20,000 by producing 300 gadgets, write an equation relating P to x and predict how much he would make if he produced 400 gadgets.

46. A psychologist found that the relationship between scores on two tests, test A and test B, was perfectly linear. Joe scored 32 on test A and 70 on test B; Sue scored 45 on test A and 96 on test B. If Jake scored 40 on test A, what would he score on test B? If Charles scored 80 on test B, what would he score on test A?

47. The figure below shows the graph of the number of physicians in the United States from 1980 to 2000, P, in thousands, for each year t. The horizontal axis represents t and the vertical axis represents P. For the time range given in the figure, the actual data points are plotted. These data points can be approximated by the given line.

(a) Find an equation of the line.

(b) Use the equation found in part **(a)** to predict P in 2004.

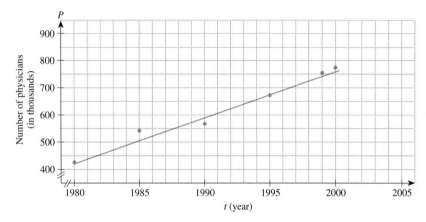

48. The figure below shows the graph of the number of active dentists in the United States from 1980 to 2000, D, in thousands, for each year t. The horizontal axis represents t and the vertical axis represents D. For the time range given in the figure, the actual data points are plotted. These data points can be approximated by the given line.

(a) Find an equation of the line.

(b) Use the equation found in part **(a)** above to predict D in 2004.

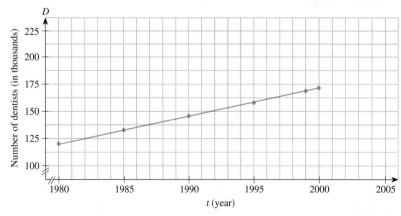

In Exercises 49–56, solve the systems of equations.

49. $\begin{cases} x - y = 4 \\ 2x - 3y = 7 \end{cases}$

50. $\begin{cases} 4x + 5y = 2 \\ 7x + 6y = 9 \end{cases}$

51. $\begin{cases} \dfrac{x}{6} - \dfrac{y}{4} = \dfrac{4}{3} \\ \dfrac{x}{5} - \dfrac{y}{2} = \dfrac{8}{5} \end{cases}$

52. $\begin{cases} \dfrac{2x}{3} + \dfrac{3y}{4} = \dfrac{7}{24} \\ \dfrac{6x}{5} - \dfrac{3y}{2} = \dfrac{1}{10} \end{cases}$

53. $\begin{cases} 3x - \dfrac{y}{4} = 2 \\ 6x - \dfrac{y}{2} = 4 \end{cases}$ **54.** $\begin{cases} \dfrac{x}{6} + y = 1 \\ \dfrac{x}{3} + 2y = 4 \end{cases}$ **55.** $\begin{cases} x = 2y - 3 \\ y = 3x + 2 \end{cases}$ **56.** $\begin{cases} 2s - 5t = 6 \\ 4t + 3s = 8 \end{cases}$

57. A total of $8,500 is split into two investments. Part is invested in a certificate of deposit paying 4.75% per year and the rest is invested in a bond paying 6.65% per year. If the annual interest from the two investments is $512.05, how much is invested at each rate?

58. Tom goes into a bakery and orders 3 lb of bread and 5 lb of cookies for a total of $22.02. Sarah buys 2 lb of bread and 3 lb of cookies in the same bakery for a total of $13.43. What are the prices per pound for bread and for cookies?

In Exercises 59–68, graph the inequality on a rectangular coordinate system.

59. $y - 2x < 4$

60. $y + 2x < 4$

61. $2y - 3x > 6$

62. $2y + 3x < 12$

63. $5y - 8x \le 20$

64. $2y - 7x \ge 14$

65. $\dfrac{x}{2} + \dfrac{y}{3} \ge 6$

66. $\dfrac{x}{5} - \dfrac{y}{3} < 15$

67. $y < 5$

68. $x \le -1$

69. The perimeter of a rectangle of length x and width y must exceed 100 ft. Write an inequality that represents this situation and sketch the graph of the solution set. List three feasible points in the solution set.

70. You are shopping at a discount clothes outlet where all ties are priced at $9 and all belts are priced at $6. Suppose you buy t ties and b belts. Write an inequality that represents the number of ties and belts you can purchase on a maximum budget of $72. Sketch the graph of this inequality and list three feasible points in the solution set.

CHAPTER 4 PRACTICE TEST 💿

1. Sketch the graph of each of the following equations in a rectangular coordinate system. Be sure to label the intercepts.

(a) $4x + 3y - 24 = 0$ (b) $2x = 8$

(c) $y = \frac{1}{2}x - 6$

2. Find the slope of the line satisfying the given conditions:

(a) Passing through $(3, 9)$ and $(-2, 4)$ (b) Equation is $3x - 2y = 8$

3. Find the value of a if a line with slope 2 passes through the points $(a, 2)$ and $(2, 5)$.

4. Write the equation of the line satisfying the given conditions.

(a) The line passes through points $(2, -3)$ and $(3, 5)$.

(b) The line passes through $(1, 0)$ with slope -4.

(c) The line passes through $(2, 5)$ with y-intercept 3.

(d) The line passes through $(2, -3)$ and is parallel to $x + 3y = 8$.

(e) The line passes through $(2, -3)$ and is perpendicular to $x + 3y = 8$.

(f) The horizontal line passes through $(3, -1)$.

5. A psychologist finds that the relationship between scores on two tests, test A and test B, is perfectly linear. A person scoring 60 on test A scores 90 on test B; someone scoring 80 on test A scores 150 on test B. What should a person receive on test B if he or she scores 85 on test A?

6. Find an equation of the line given in the figure below.

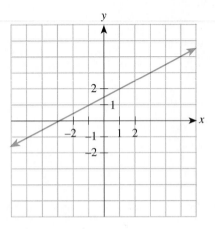

7. Solve the following systems of simultaneous equations:

(a) $\begin{cases} 2x + 3y = 1 \\ 3x + 4y = 4 \end{cases}$

(b) $\begin{cases} \dfrac{a}{3} + \dfrac{b}{2} = 2 \\ a = \dfrac{b}{3} - 5 \end{cases}$

8. Jerry invested \$3,500 in two savings certificates. One yields $8\frac{1}{2}\%$ annual interest and the other yields 9% annual interest. How much was invested in each certificate if Jerry receives \$309 in annual interest?

9. Graph the inequality $3x - 8y > 12$.

10. A personnel department is going to hire clerks and salespeople. Clerks receive a salary of \$450 per week, whereas salespeople receive a salary of \$525 per week. The department will hire c clerks and s salespeople but must not exceed a total weekly salary of \$9,200. Write an inequality representing this situation and list three feasible points in the solution set.

5

Polynomial Expressions and Functions

© Dennis MacDonald/Alamy

In business, the demand function gives the price per item, p, in terms of the number of items sold, x. Suppose a company finds that the price p (in terms of the number of items sold for its model GC-5 calculator is related to the number of calculators sold, x (in millions), and is given by the demand function.

$$p = 80 - 4x^2$$

We will return to a similar example in Example 2 of this section.

The manufacturer's revenue is determined by multiplying the number of items sold x by the price per item p. Thus the revenue function R for this calculator manufacturer is

$$R = xp = x(80 - 4x^2) = 80x - 4x^3$$

where R is calculated in millions of dollars.

These demand and revenue functions are examples of polynomial functions. We will see additional applications that give rise to some of the other functions discussed in this chapter.

Polynomial Functions as Mathematical Models

Let's begin by introducing some basic terminology.

DEFINITION   A *monomial* is an algebraic expression that is either a constant or a product of constants and one or more variables with whole-number exponents.

The following are examples of monomials:

$$3x^2y \quad (\text{or } 3xxy) \qquad 5xy^4 \quad (\text{or } 5xyyyy) \qquad \frac{x}{4} \quad (\text{or } \frac{1}{4}x) \qquad 7$$

Note that each is a product of constants and/or variables. On the other hand,

$$3x + 2y \qquad 5x^{1/2} \qquad \frac{3}{x} \qquad x - 4$$

cannot be represented as products of constants and/or variables with whole-number exponents, and are therefore *not* monomials.

DEFINITION   A *polynomial* is a finite sum of monomials.

We may name polynomials by the number of monomials or terms making up the polynomial: A *binomial* is a polynomial consisting of two terms and a *trinomial* is a polynomial consisting of three terms.

Besides the number of terms, we can also classify a polynomial by its *degree*. Before defining the degree of a polynomial, however, we first define the degree of a monomial.

DEFINITION   The *degree of a monomial* is the sum of the exponents of its variables. The degree of a nonzero constant is zero. The degree of the *number* 0 is undefined.

For example:

$3x^4$ has degree 4, since the exponent of its variable is 4.

$5x^3y^6$ has degree 9, since the sum of the exponents of the variables is 9.

$-3x^2y^3z$ has degree 6, since $2 + 3 + 1 = 6$ (remember $z = z^1$).

4 has degree 0 (as we will see in Chapter 7, we can write 4 as $4x^0$).

Now we can define the degree of a polynomial:

DEFINITION   The *degree of a polynomial* is the highest degree of any monomial in it.

EXAMPLE 1   Identify the degree of each of the following:

(a) $3x^5 + 2x^3 + 6$ **(b)** $7x^2y^4 - 3x^7y^5z^2$ **(c)** $2^5x^3y^4$ **(d)** -7

Solution   **(a)** $\underbrace{3x^5}_{\text{Degree 5}} + \underbrace{2x^3}_{\text{Degree 3}} + \underbrace{6}_{\text{Degree 0}}$

$3x^5 + 2x^3 + 6$ has $\boxed{\text{degree 5}}$, since the highest monomial degree is 5.

(b) $\underbrace{7x^2y^4}_{\text{Degree 6}} - \underbrace{3x^7y^5z^2}_{\text{Degree 14}}$

$7x^2y^4 - 3x^7y^5z^2$ has $\boxed{\text{degree 14}}$, since the highest degree of any monomial in it is 14.

(c)   According to the definition of degree, we add the exponents of the *variables;* thus, we ignore the exponent of the constant 2. Hence, $2^5x^3y^4$ has $\boxed{\text{degree 7}}$.

(d) -7 is a constant. Therefore, it has $\boxed{\text{degree 0}}$.

When the number 0 is considered as a polynomial, we call it the *zero polynomial.* Since we can write 0 as $0x^4$ or as $0x^{10}$, you can see that we would have difficulty assigning a degree to the zero polynomial. Hence, *the zero polynomial has no degree.* Thus, whereas a nonzero constant has degree 0, the degree of the number 0 is undefined.

A third way to classify polynomials is by the number of variables. For example, $3x^2y^2 - 2xz$ is a fourth-degree binomial in three variables: *x, y,* and *z;* similarly, $5x^3 - 2x^2 + 3x$ is a third-degree trinomial in one variable.

We usually write polynomials in descending powers; that is, the highest-degree term is written first, followed by the next-highest-degree term, and so on.

Symbolically, we often see polynomials in one variable defined as follows:

DEFINITION   A *polynomial in one variable* is an expression of the form

$$a_nx^n + a_{n-1}x^{n-1} + a_{n-2}x^{n-2} + \cdots + a_2x^2 + a_1x + a_0, \quad a_n \neq 0$$

where the a_i's are real numbers, x is a variable, and n is a nonnegative integer called the *degree* of the polynomial.

Let's examine this definition, especially the notation, more closely. To begin with, $a_0, a_1, a_2, a_3, \ldots$ are simply real-number coefficients: a_2 is the coefficient of x^2, a_4 is the coefficient of x^4, a_n is the coefficient of x^n. The subscript numbers are used to conveniently distinguish the constants from one another. We could have used letters of the English alphabet instead, but then we would have to stop at 26 since there are only 26 letters in the English alphabet.

When a polynomial is written with *descending powers,* it is said to be in *standard form.* The polynomial $5 + 3x - 4x^2$ in standard form is $-4x^2 + 3x + 5$. In this form, by the definition above, $n = 2$, $a_2 = -4$, $a_1 = 3$, and $a_0 = 5$. Note how the subscripts of a conveniently match the exponents of x.

When a polynomial is written with descending powers, we can assume that missing powers of the variable have coefficients of 0. For example:

$$3x^5 - 2x + 3 \quad \text{can be rewritten as} \quad 3x^5 + 0x^4 + 0x^3 + 0x^2 - 2x + 3$$

Then, $n = 5$, $a_5 = 3$, $a_4 = 0$, $a_3 = 0$, $a_2 = 0$, $a_1 = -2$, and $a_0 = 3$.

Notice how the meaning of *degree* is worked into the preceding definition of a polynomial in one variable: The highest power of x is n and therefore the degree of the polynomial is n.

A ***polynomial function*** is a function of the form "$f(x) =$ a polynomial." Numerous real-life situations can be described using polynomial functions, as the next few examples illustrate.

EXAMPLE 2

Suppose that a particular company knows that if the price of an item is x dollars, then it can sell $80,000 - 1,000x^2$ items.

(a) Express the revenue, R, as a function of x.
(b) Use the function R to compute the revenue when the price of an item is $4, $5, and $6.

Solution

(a) *Revenue* refers to the amount of money a business takes in. In a manufacturing business, the revenue earned is computed by multiplying the price charged per item by the number of items sold. For example, if an item sells for $6.35, and 1,200 items were sold, the revenue R will be $R = (6.35)(1,200) = \$7,620$.
Using the given information we have

Revenue = (number of items) × (price per item)

$$\text{Revenue} = (\text{Price per item}) \cdot (\text{Number of items sold})$$
$$R = \quad x \quad \cdot \quad (80,000 - 1,000x^2)$$
$$= 80,000x - 1,000x^3$$

Hence, $\boxed{R = R(x) = 80,000x - 1,000x^3}$. Notice that this function is a third-degree polynomial.

(b) We compute the revenue, $R(x)$ (or R), for various prices of the item x by substituting the values $x = 4$, 5, and 6 in the function $R(x) = 80,000x - 1,000x^3$. Recall that when we use the functional notation, $R(x)$, rather than stating "find $R(x)$ when $x = 4$," we say "find $R(4)$." Hence, $R(4)$ is the revenue when $x = 4$. Similarly, $R(5)$ and $R(6)$ are the revenues when $x = 5$ and $x = 6$, respectively.

$$R(x) = 80,000x - 1,000x^3 \qquad \textit{We find R(4) by substituting 4 for x in R(x).}$$
$$R(4) = 80,000(4) - (1,000)4^3 = 256,000 \quad \textit{Next we find R(5).}$$
$$R(5) = 80,000(5) - (1,000)5^3 = 275,000 \quad \textit{Finally, we find R(6).}$$
$$R(6) = 80,000(6) - (1,000)6^3 = 264,000$$

Thus, when the price of an item is $4, the revenue is $256,000; when the price is $5, the revenue is $275,000; when the price is $6, the revenue is $264,000.

EXAMPLE 3

The base of a rectangular box has a length that is 5 cm more than its width, which is w cm; its height is twice its width. Express the volume of the box as a function of w, and use this function to compute the volume when the width of the box is 15 cm, 25 cm, and 48 cm.

Solution

Figure 5.1 depicts the box with width, length, and height as described in the statement of the example.

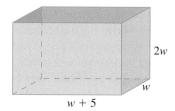

Figure 5.1

If we let V = the volume of the box, then V is found by multiplying its width, length, and height. Thus,

$$V = V(w) = w(2w)(w + 5) = 2w^2(w + 5) = 2w^3 + 10w^2$$

which expresses the volume as a third-degree polynomial function of the width, w.

To compute the volume $V = V(w)$ for $w = 15, 25,$ and 48 cm, we substitute these values of w into the function and evaluate V. In the language of functions, we are finding $V(15)$, $V(25)$, and $V(48)$:

$V(w) = 2w^3 + 10w^2$ *We find V(15) by substituting 15 for w in V(w).*
                                     We can use a calculator to get

$V(15) = 2(15)^3 + 10(15)^2 = 9,000$ *Next we find V(25).*

$V(25) = 2(25)^3 + 10(25)^2 = 37,500$ *Finally we find V(48).*

$V(48) = 2(48)^3 + 10(48)^2 = 244,224$

We can see that when the width is 15 cm, the volume is 9,000 cm³; when the width is 25 cm, the volume is 37,500 cm³; when the width is 48 cm, the volume is 244,224 cm³.

EXAMPLE 4

If an object is thrown straight up from a height of 10 m with an initial velocity of 50 m per second, then its height, h (in meters), above the ground t seconds after the object is thrown is given by the function

$$h = h(t) = 10 + 50t - 4.9t^2$$

(a) Use this function to find the height of the object at 2-second intervals, and describe the motion of the object.

(b) Use the information in part **(a)** to estimate *when* the object reaches its maximum height, and estimate the maximum height of the object.

(c) Estimate when the object will hit the ground.

Solution

(a) To determine the height of the object at 2-second intervals, we create a table of values:

These repetitive computations should be done with a calculator.

t	$h(t) = 10 + 50t - 4.9t^2$	*Height at t Seconds*
0	$h(0) = 10 + 50(0) - 4.9(0)^2$	$= 10$
2	$h(2) = 10 + 50(2) - 4.9(2)^2$	$= 90.4$
4	$h(4) = 10 + 50(4) - 4.9(4)^2$	$= 131.6$
6	$h(6) = 10 + 50(6) - 4.9(6)^2$	$= 133.6$
8	$h(8) = 10 + 50(8) - 4.9(8)^2$	$= 96.4$
10	$h(10) = 10 + 50(10) - 4.9(10)^2$	$= 20$
12	$h(12) = 10 + 50(12) - 4.9(12)^2$	$= -95.6$

Examining the table, we can see that when $t = 2$ (meaning that 2 seconds have elapsed since the object was thrown), $h(t) = 90.4$ (meaning that the object is 90.4 m above the ground). We can see by the table of values that initially (at time

$t = 0$), the height of the object is 10 m. The height continues to grow until "around" the 6th second ($t = 6$), where the highest value of $h(t)$ *in the table* is 133.6 (the object is 133.6 m above the ground), and then $h(t)$ gets smaller.

If we were to add additional data points within the 2-second intervals, we would find that the height of the object seems to increase until it reaches a maximum, and then decreases. If we graph the data using t as the independent variable, we can clearly see this trend (see Figure 5.2).

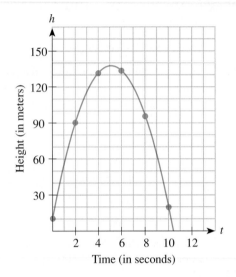

Figure 5.2
The graph of
$h(t) = 10 + 50t - 4.9t^2$

Try to estimate the answers to parts (b) and (c) using the graph. Do your answers agree?

The shape of this graph is called a *parabola*. The graph of a second-degree polynomial function such as $h(t)$ will have the shape of a parabola. We will discuss this in more detail in Chapter 8.

(b) Based on the analysis in part **(a),** we know that the maximum height, or the highest value of $h(t)$, occurs somewhere near $t = 6$ seconds. But does this occur between 4 and 6 seconds or between 6 and 8 seconds? We check the value of $h(t)$ for $t = 5$ to get

$$h(5) = 10 + 50(5) - 4.9(5)^2 = 137.5$$

which is the largest value yet. Hence, the maximum value must lie near $t = 5$. We could improve the estimate by checking the values $h(4.5)$ and $h(5.5)$, but $t = 5$ is a good enough estimate for our purposes. Hence, the maximum height occurs at approximately $t = 5$ seconds. The maximum height at that time is approximately 138 m.

Recall that $h(10) = 20$ means when $t = 10, h(10) = 20$.

(c) We note from the table that $h(10) = 20$, and $h(12) = -95.6$. This tells us that after 10 seconds ($t = 10$), the object was 20 m *above* the ground [$h(t) = 20$], whereas after 12 seconds ($t = 12$), the object would be 95.6 m *below* the ground [$h(t) = -95.6$]. Now since the object cannot travel below ground, this must be interpreted to mean that sometime between 10 and 12 seconds, the object hits the ground.

We are trying to find the value of t when $h(t) = 0$.

We check the value of $h(t)$ for $t = 11$ to get

$$h(11) = 10 + 50(11) - 4.9(11)^2 = -32.9$$

We can now narrow down the time of impact to be between 10 and 11 seconds. If we want to improve our estimate, we can compute $h(10.5)$, the height from the ground when $t = 10.5$:

$$h(10.5) = 10 + 50(10.5) - 4.9(10.5)^2 = -5.225$$

We can continue to examine values between 10 and 10.5 to improve the estimate even more, but for now we see that it takes between 10 and 10.5 seconds for the object to hit the ground.

The remainder of this chapter focuses on the various algebraic operations that can be performed with polynomial expressions and functions.

EXERCISES 5.1

In Exercises 1–12, determine whether the expression is polynomial or not. If it is a polynomial, determine

(a) The number of terms (monomial, binomial, trinomial, etc.)
(b) The degree of the polynomial
(c) The number of variables
(d) The coefficient of each term

1. $5x^2 + 4$

2. $7x^3 - 6xy^4 + 10$

3. 59

4. $-5x$

5. $-25x^{1/3} + 9x^4$

6. $-x^4 + 2x^3 - 6x + 1$

7. $-4xy^3z^7$

8. $3x^5y^7 - 6xyz^3 + x^2 - 4t$

9. $\dfrac{3}{x^2}$

10. $\dfrac{-8b^2}{2b^3 - 5b + 1}$

11. $4 - x - 2x^3 + x^2$

12. 0

13. The following table gives the per-capita poultry consumption (in pounds per year) in the United States from 1970 to 2002.

Year	1970	1975	1980	1985	1990	1995	2000	2002
Number	33.8	32.8	40.8	45.6	56.2	62.1	67.9	70.7

Based on these data, a statistician suggests that the following formula can be used to approximate the value of the per-capita poultry consumption (in pounds per year) P, for year t given in the table, where $t = 0$ represents 1970, $t = 1$ represents 1971, $t = 2$ represents 1972, etc.:

$$P = P(t) = 0.13t^2 + 0.845t + 31.42$$

Use this function to compute the per capita poultry values for each year given in the table. How well does the proposed equation agree with the observed data?

14. The median income of all families (in 2002 dollars) in the United States from 1990 to 2003 can be estimated by the equation

$$M = -0.025t^3 + 0.465t^2 - 1.78t + 40.97$$

where M is the median income (in thousands of dollars) during year t, where $t = 0$ represents 1990, $t = 1$ represents 1991, $t = 2$ represents 1992, etc.

(a) Use this equation to estimate the median income for 1992, 1994, 1996, 2000, and 2003.

(b) Use the information obtained in part **(a)** to estimate during what year the median income reached $40,000.

15. A candy manufacturer finds that when he charges d dollars per pound, he can sell $4,300 - 10d^2$ lb of candy per week.

(a) Express the revenue, R, as a function of d.

(b) Use the function R to compute the revenue when he charges $8, $10, $12, and $14 per pound.

16. An electronics distributor finds that when she charges x dollars per unit, she can sell $1{,}080 - 10x^2$ units.

    (a) Express the revenue, R, as a function of x.

    (b) Use the function R to compute the revenue when she charges $4, $5, and $7 per unit.

17. A hardware manufacturer can sell n items at a price of $2 + 0.45n - 0.001n^2$ dollars per item.

    (a) Express the revenue, R, as a function of n.

    (b) Use the function R to compute the revenue *and price* when 100, 200, and 300 items are sold.

18. A cellular phone company can lease n phones per month at a price (in dollars) of $24 + 0.58n - 0.003n^2$ per phone per month.

    (a) Express the revenue, R, as a function of n.

    (b) Use the function R to compute the revenue *and price* when 50, 100, and 150 phones are leased per month.

19. The base of a closed rectangular box has length equal to 3 times its width, which is w inches; its height is 6 in. less than its width. Express the surface area, S, of the box as a function of w, and use this function to compute the surface area when the width of the box is 20 in., 26 in., and 42 in.

20. The base of a rectangular box has length equal to twice its width, which is w cm; its height is 3 cm less than its width. Express the volume, V, of the box as a function of w, and use this function to compute the volume when the width of the box is 12 cm, 18 cm, and 36 cm.

21. A pharmaceutical company finds that the amount A (in milligrams) of a new medication present in the bloodstream m minutes after the medication is taken can be approximated by the function

$$A = A(m) = 0.28m + 0.39m^2 - 0.02m^3$$

    (a) Use this function to approximate the number of milligrams of the medication present in the bloodstream at 5-minute intervals for the first 30 minutes after it is taken.

    (b) Use the information obtained in part (a) to estimate when the amount of medication in the bloodstream is at a maximum.

22. If an object is thrown straight up from ground level with an initial velocity of 45 feet per second, then its height h (in feet) above the ground t seconds after the object is thrown is given by the function

$$h = h(t) = 45t - 16t^2$$

    (a) Use this function to find the height of the object at half-second intervals.

    (b) Use the information obtained in part (a) to estimate (to the nearest half-second) when the object will hit the ground.

◆ Mini-Review

23. *Evaluate:* $-3 - (-4) + (-5) - (-8)$

24. *Evaluate:* $-3 - (-2)^2 - (4 - 5)$

25. The length of the first side of a triangle is twice the length of the second side; the third side is 24 in. If the perimeter of the triangle is 75 in., find the length of the first and second sides.

26. A carpenter charges $30 per hour for his time and $18 per hour for his assistant's time. On a certain job the assistant works alone for 3 hours, and is then joined by the carpenter so that they complete the job together. If the total bill was $270, how many hours did the carpenter work?

5.2 Polynomials: Sums, Differences, and Products

Adding and Subtracting Polynomials

Addition and subtraction of polynomials is simply a matter of grouping with parentheses (if not supplied in the problem), removing grouping symbols, and combining like terms.

EXAMPLE 1

(a) Add the polynomials $3x^2 - 2x + 5$, $5x^3 - 4$, and $3x + 2$.

(b) Subtract $3x^3 + 5x^2 - 2x - 3$ from $7x - 4$.

Solution

(a) We supply the grouping symbols to distinguish the given polynomials:

$$(3x^2 - 2x + 5) + (5x^3 - 4) + (3x + 2) \qquad \textit{Remove the grouping symbols.}$$
$$= 3x^2 - 2x + 5 + 5x^3 - 4 + 3x + 2 \qquad \textit{Then combine like terms.}$$
$$= \boxed{5x^3 + 3x^2 + x + 3}$$

(b) Subtraction can be tricky. Make sure you understand what is being subtracted:

$$(7x - 4) - (3x^3 + 5x^2 - 2x - 3) \qquad \textit{Note how the polynomials are placed.}$$

$\textit{Remove grouping symbols by distributing}$
$\textit{$-1$ (observe the sign of each term).}$

$$= 7x - 4 - 3x^3 - 5x^2 + 2x + 3 \qquad \textit{Combine like terms.}$$
$$= \boxed{-3x^3 - 5x^2 + 9x - 1}$$

EXAMPLE 2

Perform the given operations and simplify.

(a) $(3x^2 + 2x - 4) + (5x^2 - 3x + 2) - (3x^2 - 2x + 1)$

(b) Subtract the sum of $2x^3 - 3x$ and $5x^2 - x + 4$ from $3x^2 - 2x + 5$.

Solution

(a) $(3x^2 + 2x - 4) + (5x^2 - 3x + 2) - (3x^2 - 2x + 1) \qquad \textit{Removing grouping}$
$\textit{symbols.}$

$$= 3x^2 + 2x - 4 + 5x^2 - 3x + 2 - 3x^2 + 2x - 1 \qquad \textit{Then combine}$$
$\textit{like terms.}$
$$= \boxed{5x^2 + x - 3}$$

(b) We translate the problem as follows:

$$(3x^2 - 2x + 5) - [(2x^3 - 3x) + (5x^2 - x + 4)] \qquad \textit{Remove parentheses in brackets.}$$
$$= (3x^2 - 2x + 5) - [2x^3 - 3x + 5x^2 - x + 4] \qquad \textit{Then combine terms in brackets.}$$
$$= (3x^2 - 2x + 5) - [2x^3 + 5x^2 - 4x + 4] \qquad \textit{Remove remaining grouping}$$
$\textit{symbols.}$

$$= 3x^2 - 2x + 5 - 2x^3 - 5x^2 + 4x - 4 \qquad \textit{We write the answer in standard}$$
$\textit{form.}$
$$= \boxed{-2x^3 - 2x^2 + 2x + 1}$$

Products of Polynomials

The distributive property gives us a procedure for multiplying polynomials. Recall that the property states that multiplication distributes over addition, that is:

$$a(b + c) = a \cdot b + a \cdot c$$

or

$$(b + c)a = b \cdot a + c \cdot a$$

As we discussed in Chapter 1, the real number properties, along with the first rule of exponents, give us a procedure for multiplying monomials by polynomials.

EXAMPLE 3

Perform the operations for each of the following:

(a) $5a^2b(4ab + 3ab^2)$ **(b)** $3x^3y^2(2x^2 - 7y^3)$

Solution

(a) $5a^2b(4ab + 3ab^2)$ *Apply the distributive property.*

$\quad = (5a^2b)(4ab) + (5a^2b)(3ab^2)$ *Then use the first rule of exponents.*

$\quad = \boxed{20a^3b^2 + 15a^3b^3}$

(b) $3x^3y^2(2x^2 - 7y^3) = 3x^3y^2(2x^2) - 3x^3y^2(7y^3)$

$\quad = \boxed{6x^5y^2 - 21x^3y^5}$

Our next step is to demonstrate how to multiply two polynomials of more than one term. We can distribute any polynomial in the same way that we distribute a monomial. For example, in multiplying $(2x + 1)(x + 3)$, we distribute $(x + 3)$ as

$$(2x + 1)(x + 3) = 2x(x + 3) + 1(x + 3)$$
$$(B + C) \cdot A \quad = B \cdot A \quad + C \cdot A$$

It is important for you to understand that a variable may stand not only for numbers, but also for other variables, expressions, or polynomials. Hence, we can let A stand for the binomial $x + 3$ and apply the distributive property above. The problem is still unfinished, for we must now apply the distributive property again and then combine terms:

$(2x + 1)(x + 3)$ *Distribute $x + 3$.*

$\quad = (2x)(x + 3) + 1(x + 3)$ *Apply the distributive property again.*

$\quad = (2x)x + (2x)3 + (1)x + (1)3$

$\quad = 2x^2 + 6x + x + 3$ *Combine like terms.*

$\quad = 2x^2 + 7x + 3$

You may have learned to multiply binomials by some memorized procedure such as **FOIL,** or to multiply polynomials by the vertical method. However, factoring problems will require that you thoroughly understand the logic underlying polynomial multiplication. For this reason, it is important that you realize that polynomial multiplication is derived from the *distributive property.*

EXAMPLE 4

Multiply the following:

(a) $(3a - 2)(2a + 5)$

(b) $(2y^2 + 3y + 1)(y - 5)$

(c) $(3x - 5)^2$

Solution   **(a)** $(3a - 2)(2a + 5) = (3a)(2a + 5) - 2(2a + 5)$

$$= 6a^2 + 15a - 4a - 10$$

$$= \boxed{6a^2 + 11a - 10}$$

(b) $(2y^2 + 3y + 1)(y - 5) = (2y^2)(y - 5) + 3y(y - 5) + 1(y - 5)$

$$= 2y^3 - 10y^2 + 3y^2 - 15y + y - 5$$

$$= \boxed{2y^3 - 7y^2 - 14y - 5}$$

Remember:

$(a - b)^2 \neq a^2 - b^2$

(c) $(3x - 5)^2 = (3x - 5)(3x - 5)$ *First distribute $3x - 5$.*

$$= 3x(3x - 5) - 5(3x - 5)$$

$$= 9x^2 - 15x - 15x + 25$$

$$= \boxed{9x^2 - 30x + 25}$$ *Note that the answer is not $9x^2 - 25$.*

We can find the product of two polynomials by the ***vertical method*** as well; for example, $(x^2 - 3x + 4)(2x^2 - 3)$ can be multiplied as follows:

$$
\begin{array}{r}
x^2 - 3x + 4 \\
2x^2 \qquad - 3 \\
\hline
-3x^2 + 9x - 12 \\
2x^4 - 6x^3 + 8x^2 \qquad\quad \\
\hline
2x^4 - 6x^3 + 5x^2 + 9x - 12
\end{array}
$$

$\leftarrow (-3)(x^2 - 3x + 4)$

$\leftarrow (2x^2)(x^2 - 3x + 4)$

$\uparrow \quad \uparrow \quad \uparrow \quad \uparrow \quad \uparrow$

Like terms lined up in columns

The previous examples illustrate the following general rule for multiplying polynomials.

Rule for Multiplying Polynomials

To multiply two polynomials, multiply each term in the first polynomial by each term in the second polynomial.

As the next example illustrates, polynomial operations can be applied to polynomial functions as well.

EXAMPLE 5

Suppose $f(x) = x - 3$ and $g(x) = x^2 + 3x + 9$. Find:

(a) $g(x) - f(x)$

(b) $f(x) \cdot g(x)$

Solution   **(a)** $g(x) - f(x) = \underbrace{(x^2 + 3x + 9)}_{g(x)} - \underbrace{(x - 3)}_{f(x)}$ *Note that the parentheses around $g(x)$ are optional, whereas the parentheses around $f(x)$ are essential to ensure that we subtract each term of $f(x)$. We remove the parentheses.*

$$= x^2 + 3x + 9 - x + 3$$ *Now combine like terms.*

$$= \boxed{x^2 + 2x + 12}$$

(b) To multiply the two functions, we follow the rule for multiplication of polynomials given in the box.

$$f(x) \cdot g(x) = (x - 3)(x^2 + 3x + 9)$$

Each term in the first set of parentheses multiplies each term in the second.

$$= x^3 + 3x^2 + 9x - 3x^2 - 9x - 27$$

Combine like terms.

$$= x^3 - 27$$

We will have more to say about operations on functions in Chapter 9.

EXAMPLE 6

A circular garden has a radius of 10 ft.

(a) If a brick walkway with a uniform width of x ft is to be built surrounding the garden, express the area of the walkway in terms of x.

(b) Find the area of the walkway if the width of the walkway is 2.5 ft. Round your answer to the nearest tenth.

Solution

(a) We draw a diagram of the circular garden and surrounding brick walkway, labeling the radius of the garden, 10 ft, and the width of the walkway, x ft, as shown in Figure 5.3.

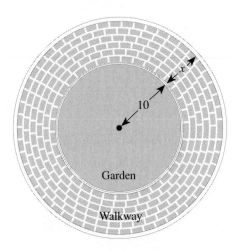

Figure 5.3

To find the area of the brick walkway, we have to find the area of the inner and outer circles, and then subtract the area of the inner circle from the area of the outer circle. Since the area of a circle is πr^2, where r is the radius of the circle, we need to find the radius of each circle.

The radius of the inner circle is given as 10 ft and, as we can see by the figure, the radius of the outer circle must be $(10 + x)$ ft. Thus,

$$\text{Area}_{\text{walkway}} = \text{Area}_{\text{outer circle}} - \text{Area}_{\text{inner circle}}$$

Where R is the radius of the outer circle and r is the radius of the inner circle

$$= \pi R^2 \qquad - \pi r^2$$

$$= \pi(10 + x)^2 - \pi(10)^2$$

Now we perform the operations.

$$= \pi(10 + x)(10 + x) - \pi(10)^2$$

$$= \pi(100 + 20x + x^2) - 100\pi$$

$$= 100\pi + 20\pi x + \pi x^2 - 100\pi$$

Simplify.

$$= (\pi x^2 + 20\pi x) \text{ sq ft}$$

(b) In part **(a)** of this example, we found that the area, A, of the walkway is given by $A = \pi x^2 + 20\pi x$ sq ft, where x is the width of the walkway. We are given that

$x = 2.5$ ft. Thus,

$$A = \pi x^2 + 20\pi x \qquad \textit{Substitute x = 2.5.}$$
$$= \pi(2.5)^2 + 20\pi(2.5) \qquad \textit{We use the } \boxed{\pi} \textit{ key on a calculator to get}$$
$$\approx 176.7 \text{ sq ft}$$

Note that if you approximate π using 3.14, you will get 176.6 sq ft as the answer.

EXERCISES 5.2

In Exercises 1–54, perform the indicated operations.

1. $(3x^2 - 2x + 5) + (2x^2 - 7x + 4)$

2. $(7xy^2 - 3x^2y) - (4xy^2 - 3x)$

3. $(15x^2 - 3xy - 4y^2) - (16x^2 - 3x + 2)$

4. $(2x^2 - 3x) - (-4x + 5) - (3x^2 - 2x + 1)$

5. $(3a^2 - 2ab + 4b^2) + [(3a^2 - b^2) - (3ab)]$

6. $[(5s^2 - 3) - (2st - 4t^2)] - 7s^2 - 3st + 2t^2$

7. $(3a^2 - 2ab + 4b^2) - [(3a^2 - b^2) - (3ab)]$

8. $7s^2 - 3st + 2t^2 - [(5s^2 - 3) - (2st - 4t^2)]$

9. $[(3a^2 - b^2) - (3ab)] - (3a^2 - 2ab + 4b^2)$

10. Subtract $5y^2 - 3y + 2$ from $y - 1$.

11. Add $3xy + 2y^2$ to $-7y^2 - 3y + 4$.

12. Subtract $y - 1$ from $5y^2 - 3y + 2$.

13. Subtract $3x^2 - 2xy + 7y^2$ from $-8x^2 - 5xy + 9y^2$.

14. Find the sum of $3a^2 - 6ab + 16b^2$, $3ab - 6b^2$, and $-5a^2 - 3ab$.

15. Subtract the sum of $2a^2 - 3ab + 4b^2$ and $6a^2 + 2ab - 2b^2$ from $3a^2 - 2b^2$.

16. Subtract the sum $3x + 2y$ and $3y - 2x$ from the sum of $x - y$ and $x + y$.

17. $x^2y(3x^2 - 2xy + 2y^2)$

18. $-3a^2b^2(4a^2b - 3ab + 2a)$

19. $(3x^2 - 2xy + 2y^2)(x^2y)$

20. $-13a^3b^2(5a^2b - 3a^2 + 2)$

21. $(x + 3)(x + 4)$

22. $(y - 4)(y - 8)$

23. $(3x - 1)(2x + 1)$

24. $(2x - 3)(3x + 5)$

25. $(3x + 1)(2x - 1)$

26. $(2x + 3)(3x - 5)$

27. $(3a - b)(a + b)$

28. $(2a - 5b)(a + 2b)$

29. $(2r - s)^2$

30. $(3b - 2)^2$

31. $(2r + s)(2r - s)$

32. $(3b - 2)(3b + 2)$

33. $(3x - 2y)(3x + 2y)$

34. $(2x^2 + y)(2x^2 - y)$

35. $(3x - 2y)^2$

36. $(2x^2 - y)^2$

37. $(x - y)(a - b)$

38. $(a + b)(x + y)$

39. $(2r + 3s)(2a + 3b)$

40. $(4x^2 - 5y)(5x - 3y^2)$

41. $(2y^2 - 3)(y^2 + 1)$

42. $(3a^2 - x^3)(3a^2 + x^3)$

43. $(5x^2 - 3x + 4)(x + 3)$

44. $(5x^2 - 3x + 4)(x - 3)$

45. $(9a^2 + 3a - 5)(3a + 1)$

46. $(7x^2 + 4x - 7)(5x - 2)$

47. $(a^2 - ab + b^2)(a + b)$

48. $(a^2 + ab + b^2)(a - b)$

49. $(x - y - z)^2$

50. $(x + y + z)^2$

51. $(x^2 - 2x + 4)(x + 2)$

52. $(x^2 + 2x + 4)(x - 2)$

53. $(a + b + c + d)(a + b)$

54. $(x^3 + x^2 + x + 1)(x - 1)$

55. The length of a rectangle is 5 more than 3 times its width. If the width of the rectangle is w, express the area and perimeter of the rectangle in terms of w.

56. The length of a rectangle is 1 less than twice its width. If the width of the rectangle is w, express the area and perimeter of the rectangle in terms of w.

57. A circular garden has a radius of 8 ft. If a walkway with a uniform width of x ft is to be built surrounding the garden, express the area of the walkway in terms of x.

58. A circular swimming pool has a radius of 50 ft. If a concrete walkway with a uniform width of x ft surrounds the pool, express the area of the walkway in terms of x.

59. A 20 ft by 60 ft rectangular swimming pool is surrounded by a path of uniform width. If the width of the walkway is x ft, express the area of the walkway in terms of x.

60. A 30 ft by 30 ft square swimming pool is surrounded by a concrete walkway of uniform width. If the width of the walkway is x ft, express the total area of the pool and walkway in terms of x.

In Exercises 61–74, given $f(x) = 2x - 3$, $g(x) = 3x^2 - 5x + 2$, and $h(x) = x^3 - x$, find and simplify.

61. $f(x) - g(x)$

62. $g(x) - f(x)$

63. $h(x) - [f(x) - g(x)]$

64. $f(x) - [g(x) - h(x)]$

65. $f(x) \cdot g(x)$

66. $g(x) \cdot h(x)$

67. $h(x)[f(x) + g(x)]$

68. $f(x)[h(x) - g(x)]$

69. $g(x) + g(2)$

70. $g(x) + 2$

71. $g(x + 2)$

72. $h(x - 1)$

73. $h(x) - 1$

74. $h(x) - h(1)$

? Questions for Thought

75. The rule for multiplying polynomials that was stated in this section is based on the fundamental properties of the real numbers. Supply the real-number property that justifies each of the following steps.

$$
\begin{aligned}
(x + y)(3x - 2y) &= (x + y)(3x) + (x + y)(-2y) \\
&= x(3x) + y(3x) + x(-2y) + y(-2y) \\
&= (3x)x + (3x)y + (-2y)x + (-2y)y \\
&= 3(xx) + 3(xy) - 2(yx) - 2(yy) \\
&= 3x^2 + 3xy - 2yx - 2y^2 \\
&= 3x^2 + 3xy - 2xy - 2y^2 \\
&= 3x^2 + (3 - 2)xy - 2y^2 \\
&= 3x^2 + xy - 2y^2
\end{aligned}
$$

76. Discuss what is ***wrong*** (if anything) with the following:

(a) $(3x^2)(2x^3) \overset{?}{=} 6x^6$ (b) $2x - \{3x - 2[4 - x]\} \overset{?}{=} 2x - \{3x - 8 + 2x\}$

$$\overset{?}{=} 2x - \{5x - 8\}$$

$$\overset{?}{=} -10x^2 + 16x$$

Mini-Review

77. *Evaluate:* $\dfrac{-4 - [4 - 5(2 - 8)]}{4 - 5 \cdot 2 - 8}$

78. A truck carries a load of 60 packages. Some are 25-lb packages and the rest are 30-lb packages. If the total weight of all the packages is 1,680 lb, how many packages of each kind are in the truck?

79. *Solve for x:* $|3 - 2x| \le 5$

80. *Solve for a:* $3a - 8 = 5a - (4 + 2a)$

5.3 General Forms and Special Products

We can now develop some procedures for multiplying polynomials quickly as long as we keep in mind *why* the process works. Our emphasis will be on quick procedures for *binomial* multiplication, since such products occur frequently in algebra. Later in this section, we will generalize the procedures to more complex cases.

General Forms

If we multiplied $(x + a)(x + b)$, we would get the following:

$(x + a)(x + b)$ *Distribute x + b.*

$= x(x + b) + a(x + b)$ *Distribute x and a.*

$= x^2 + bx + ax + ab$ *Factor x from the middle terms.*

$= x^2 + (a + b)x + ab$

Thus, we have:

General Form 1

$$(x + a)(x + b) = x^2 + (a + b)x + ab$$

Verbally stated, when two binomials of the form (Variable + Constant) are multiplied, the product will be a trinomial, where the coefficient of the first-degree term will be the *sum* of the binomial constants and the numerical term will be the *product* of the binomial constants. Knowing this form allows us to quickly multiply binomials of this type mentally.

EXAMPLE 1   (a) $(x + 3)(x + 5)$ (b) $(y - 7)(y + 4)$ (c) $(a - 6)(a - 5)$

Solution $(x + a)(x + b) = x^2 + (a + b)x + ab$

(a) $(x + 3)(x + 5) = x^2 + (3 + 5)x + 3 \cdot 5$

$= \boxed{x^2 + 8x + 15}$

(b) $(y - 7)(y + 4) = y^2 + (-7 + 4)y + (-7)(4)$

$$= \boxed{y^2 - 3y - 28}$$

(c) $(a - 6)(a - 5) = a^2 + (-6 - 5)a + (-6)(-5)$

$$= \boxed{a^2 - 11a + 30}$$

If we multiplied out $(ax + b)(cx + d)$, we would get:

$(ax + b)(cx + d)$ *Distribute cx + d.*

$\quad = ax(cx + d) + b(cx + d)$ *Distribute ax and b.*

$\quad = acx^2 + adx + bcx + bd$ *Factor x from the middle terms.*

$\quad = acx^2 + (ad + bc)x + bd$

Thus, we have:

General Form 2	$(ax + b)(cx + d) = acx^2 + (ad + bc)x + bd$

This form is certainly more complicated than the previous general form. Therefore, rather than trying to memorize this form to perform quick multiplication of binomials, we will present another view of the same procedure for multiplying binomials, called the **FOIL method.**

FOIL stands for

$$\textbf{\textit{F}irst,} \quad \textbf{\textit{O}uter,} \quad \textbf{\textit{I}nner,} \quad \textbf{\textit{L}ast}$$

It is a systematic way to keep track of the terms to be multiplied, demonstrated as follows:

$$(3x + 4)(2x + 5) = (3x)(2x) + (3x)(5) + (4)(2x) + (4)(5)$$
$$= 6x^2 + 15x + 8x + 20$$
$$= 6x^2 + 23x + 20$$

Keep in mind that FOIL is only a name to help us to be systematic in carrying out *binomial* multiplication; *each term of one binomial is still being multiplied by each term of the other.*

EXAMPLE 2

Perform the indicated operations.

(a) $(5x - 4)(2x + 1)$
(b) $(3a + 4)(3a - 4)$
(c) $(3x + 1)^2$

Solution

(a) $(5x - 4)(2x + 1) = (5x)(2x) + (5x)(1) - 4(2x) + (-4)(1)$

$$= 10x^2 + 5x - 8x - 4$$

$$= \boxed{10x^2 - 3x - 4}$$

(b) $(3a + 4)(3a - 4) = (3a)(3a) - (3a)4 + 4(3a) - 4(4)$

$$= 9a^2 - 12a + 12a - 16$$

$$= \boxed{9a^2 - 16}$$

(c) $(3x + 1)^2 = (3x + 1)(3x + 1)$

$$= (3x)(3x) + (3x)(1) + 1(3x) + 1(1)$$

$$= 9x^2 + 3x + 3x + 1$$

$$= \boxed{9x^2 + 6x + 1}$$

You should practice your multiplication skills in the Exercise sets until you can do the multiplications flawlessly in your head.

Special Products

Special products are specific products of binomials that can be derived from the general forms given previously (which, in turn, are derived from the distributive properties).

Special Products	
1. $(a + b)(a - b) = a^2 - b^2$	*Difference of two squares*
2. $(a + b)^2 = a^2 + 2ab + b^2$	*Perfect square of sum*
3. $(a - b)^2 = a^2 - 2ab + b^2$	*Perfect square of difference*

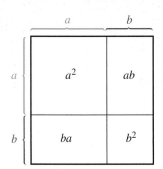

The area of the square with side $a + b$ illustrates the perfect square of a sum:
$(a + b)^2 = a^2 + 2ab + b^2$

Proof:

1. $(a + b)(a - b) = a(a - b) + b(a - b)$

$$= a^2 - ab + ba - b^2$$

$$= a^2 - b^2$$

2. $(a + b)^2 = (a + b)(a + b)$

$$= a(a + b) + b(a + b)$$

$$= a^2 + ab + ba + b^2$$

$$= a^2 + 2ab + b^2$$

Proof of special product 3 is left as an exercise.

Special product 1 is known as the ***difference of two squares,*** whereas special products 2 and 3 are known as ***perfect squares.*** Note the differences between the special products, especially between special products 1 and 3, which are often confused with each other.

The expressions $a + b$ and $a - b$ are called ***conjugates of each other.*** A *conjugate* of a binomial is formed by changing the sign of one of its terms. Special product 1 tells us that the product of conjugates yields the differences of two squares.

Conjugates will be important to us in subsequent chapters, but for now you should understand that $-a - b$ is *not* the conjugate of $a + b$, but rather it is the *negative* of $a + b$, since

$$-(a + b) = -a - b$$

A *conjugate* of $x - 4$ is $x + 4$, but the *negative* of the expression $x - 4$ is $-(x - 4) = -x + 4$, usually written as $4 - x$.

Notice that in the negative of a binomial, the signs of *both* terms are changed, whereas in a conjugate only one sign is changed.

Special products are important in factoring; in many cases, the quickest way to factor an expression is to recognize it as a special product. In addition, recognizing and using special products can reduce the time needed for multiplication.

EXAMPLE 3

Perform the indicated operations.

(a) $(3x + 5)^2$

(b) $(2a - 7b)(2a + 7b)$

(c) $(2a - 7b)^2$

Solution

Remember that
$(A + B)^2 \neq A^2 + B^2.$

All these problems can be worked by using the general forms, or more slowly by using the distributive property. The quickest way, however, is to use the special products.

(a) $(3x + 5)^2$ *This is a perfect square of a sum.*

$= (3x)^2 + 2(3x)(5) + (5)^2$

$= \boxed{9x^2 + 30x + 25}$

(b) $(2a - 7b)(2a + 7b)$ *This is the difference of two squares.*

$= (2a)^2 - (7b)^2$

$= \boxed{4a^2 - 49b^2}$

(c) $(2a - 7b)^2$ *This is a perfect square of a difference.*

$= (2a)^2 - 2(2a)(7b) + (7b)^2$

$= \boxed{4a^2 - 28ab + 49b^2}$

Study the differences between parts **(b)** and **(c)** of Example 3; note their similarities. Keep in mind that when you square a binomial, you will get a middle term in the product. Again, practice these so that you can quickly do them mentally.

Multiple Operations

Let's put together what we have learned so far and examine how to simplify expressions requiring multiple operations with polynomials. We follow the same order of operations discussed in Chapter 1 (that is, parentheses, exponents, multiplication and division, addition and subtraction).

EXAMPLE 4

Perform the indicated operations.

(a) $2x - 3x(x - 5)$ **(b)** $(x - y)^3$

(c) $2x(x - 3)(3x + 1)$ **(d)** $5x^3 - 2x(x - 4)(2x + 3)$

(e) $(a - b)^2 - (a - b)(a + b)$

Solution

(a) $2x - 3x(x - 5)$ *Multiplication before subtraction: First distribute* $-3x$.

$= 2x - 3x^2 + 15x$ *Then combine terms.*

$= \boxed{-3x^2 + 17x}$

(b) $(x - y)^3$ *Rewrite $(x - y)^3$ as a product.*

$= (x - y)(x - y)(x - y)$ *Then multiply two binomials together.*

$= (x - y)(x^2 - 2xy + y^2)$ *Multiply the result by the third binomial.*

$= x^3 - 2x^2y + xy^2 - x^2y + 2xy^2 - y^3$ *Combine like terms.*

$= \boxed{x^3 - 3x^2y + 3xy^2 - y^3}$

(c) $2x(x - 3)(3x + 1)$ *There is less chance of your making an error if you multiply the binomials first.*

$= 2x(3x^2 - 8x - 3)$

$= \boxed{6x^3 - 16x^2 - 6x}$

(d) $5x^3 - 2x(x - 4)(2x + 3)$ *Multiply binomials first.*

$= 5x^3 - 2x(2x^2 - 5x - 12)$ *Distribute $-2x$.*

$= 5x^3 - 4x^3 + 10x^2 + 24x$ *Combine like terms.*

$= \boxed{x^3 + 10x^2 + 24x}$

e. $(a - b)^2 - (a - b)(a + b)$ *$(a - b)^2$ is a perfect square.*

$= a^2 - 2ab + b^2 - (a - b)(a + b)$ *$(a + b)(a - b)$ is the difference of two squares.*

$= a^2 - 2ab + b^2 - (a^2 - b^2)$ *The parentheses around $a^2 - b^2$ are necessary. Next remove parentheses.*

$= a^2 - 2ab + b^2 - a^2 + b^2$

$= \boxed{-2ab + 2b^2}$

Note that in part **(e)**, we multiplied $(a + b)(a - b)$ before we subtracted. It is necessary to retain parentheses around $a^2 - b^2$ since we are subtracting the entire expression $(a^2 - b^2)$.

EXAMPLE 5

Given $g(x) = x^2 - 3x + 1$, find each of the following:

(a) $g(a + 2)$ **(b)** $g(2x)$

(a) $g(a + 2)$ is the expression we get when we substitute $a + 2$ for x in $g(x)$:

$$g(x) = x^2 - 3x + 1$$
$$g(a + 2) = (a + 2)^2 - 3(a + 2) + 1 \qquad \textit{Then simplify.}$$
$$= a^2 + 4a + 4 - 3a - 6 + 1$$
$$= a^2 + a - 1$$

Thus,

$$\boxed{g(a + 2) = a^2 + a - 1}$$

(b) $g(2x)$ is the expression we get when we substitute $2x$ for x in $g(x)$:

$$g(x) = x^2 - 3x + 1$$
$$g(2x) = (2x)^2 - 3(2x) + 1 \qquad \textit{Simplify.}$$
$$= 4x^2 - 6x + 1$$

Thus,

$$\boxed{g(2x) = 4x^2 - 6x + 1}$$

Applying Special Products

Of course, we can apply the special products to binomials containing more complex monomials, as shown in the next example.

EXAMPLE 6

Perform the operations using special products or general forms.

(a) $(2x^2 - 3y)(2x^2 + 3y)$ **(b)** $(3r^3 - 5x^2)^2$

Solution

(a) $(2x^2 - 3y)(2x^2 + 3y) = (2x^2)^2 - (3y)^2$ *Difference of two squares*

$$= \boxed{4x^4 - 9y^2}$$

(b) $(3r^3 - 5x^2)^2 = (3r^3)^2 - 2(3r^3)(5x^2) + (5x^2)^2$ *Perfect square*

$$= \boxed{9r^6 - 30r^3x^2 + 25x^4}$$

General forms and special products allow us to multiply quickly without going through intermediate steps. They can also be applied to more complex expressions. For example, to multiply

$$(x + r + s)(x + r - s)$$

you could simplify the expressions within the parentheses and multiply out using the horizontal or vertical method. On the other hand, if we rewrite this product as $[(x + r) + s][(x + r) - s]$, recognize this problem as the difference of two squares, and apply what we already know about special products, we could reduce our labor for this problem. Again, keep in mind that in equivalent expressions such as special products, the variables can represent polynomials as well as other letters and numbers.

$$(a\ \ + b) \cdot (\ a\ \ - b) =\ \ a^2\ \ - b^2$$
$$[(x + r) + s] \cdot [(x + r) - s] = (x + r)^2 - s^2$$

We finish the problem by applying special product 2 to $(x + r)^2$:

$$(x + r)^2 - s^2 = \boxed{x^2 + 2xr + r^2 - s^2}$$

EXAMPLE 7

Perform the operations using special products: $[x + y + z]^2$

Solution

We are viewing $[x + y + z]^2$ as $[(x + y) + z]^2$. Try regrouping the given expression as $[x + (y + z)]^2$ and verify that you get the same result.

We can view this expression as the square of the sum of two terms, $(x + y)$ and z, and use the special product for the square of a sum:

$$(a\ \ + b)^2 =\ \ a^2\ \ + 2 \cdot a \cdot b + b^2$$
$$[(x + y) + z]^2 = (x + y)^2 + 2(x + y)z + z^2$$ *Apply special product 2 to $(x + y)^2$.*
$$= x^2 + 2xy + y^2 + 2z(x + y) + z^2$$
$$= \boxed{x^2 + 2xy + y^2 + 2xz + 2yz + z^2}$$

The key is to recognize complex expressions as being in special product form.

Multiplication in the manner illustrated in Example 7 is much faster than using the horizontal or vertical method. Again, you should practice these problems because we will return to these forms again in the next sections.

EXAMPLE 8

An open cardboard box is to be made from a 20-in. by 30-in. rectangular sheet of cardboard by cutting out identical squares of side x from each of the corners of the sheet and folding up the sides to form a box, as shown in Figure 5.4. Express the volume of the box in terms of x.

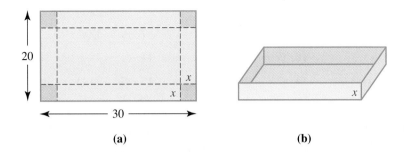

Figure 5.4

(a)   (b)

Solution

THINKING OUT LOUD

What do we need to find?	An expression for the volume of the box in terms of x
How do we find the volume of the box in the figure?	The formula for the volume of a rectangular box is $V = Length \times Width \times Height$. Hence, we need to identify the length, width, and height.
How do we identify the length, width, and height of the box?	First note that the box is constructed by cutting out squares with sides of length x, from the *sheet,* and folding up that sides at the dashed lines. We can therefore see that both the length and width of the *box* are found by subtracting x twice (once from each end) from each dimension. By looking at Figure 5.4(b), we can see that the height of the box is x. See Figure 5.5.

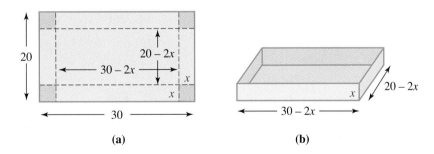

(a)   (b)

Figure 5.5

From Figure 5.5, we see that the length is $30 - 2x$, the width is $20 - 2x$, and the height is x. The volume of the box is therefore

$$V = Length \times Width \times Height$$
$$= (30 - 2x)(20 - 2x)x \qquad \text{\textit{We multiply out the right-hand side:}}$$
$$\text{\textit{First multiply out the binomial.}}$$
$$= (600 - 100x + 4x^2)x$$
$$V = 600x - 100x^2 + 4x^3$$

Hence, the volume (written in standard form) is $V = 4x^3 - 100x^2 + 600x$.

EXERCISES 5.3

Perform the operations and simplify. Use special products wherever possible.

1. $(x + 4)(x + 5)$ **2.** $(a - 2)(a + 4)$ **3.** $(x + 3)(x - 7)$ **4.** $(y - 3)(y - 9)$

5. $(x - 8)(x - 11)$ **6.** $(x - 21)(x + 20)$ **7.** $(x + 5)(x - 6)$ **8.** $(x + 5) + (x - 6)$

9. $x + 5(x - 6)$ **10.** $(x - 7)(x - 4)$ **11.** $(x - 7) - (x - 4)$ **12.** $x - 7(x - 4)$

13. $(2x + 1)(x - 4)$ **14.** $(3a + 2b)(a - b)$ **15.** $(5a - 4)(5a + 4)$ **16.** $(6s - 4)(6s + 4)$

17. $(3z + 5)^2$ **18.** $(7r - 3t)(7r + 3t)$ **19.** $(3z - 5)^2$ **20.** $(7r - 3t)^2$

21. $(3z - 5)(3z + 5)$ **22.** $(7r + 3t)^2$ **23.** $(5r^2 + 3s)(3r + 5s)$ **24.** $(4x^2 - 5)(5x - 4)$

25. $(3s - 2y)^2$ **26.** $(7x - y)(7x - y)$ **27.** $(3y + 10z)^2$ **28.** $(y - 2x)(y + 2x)$

29. $(3y + 10z)(3y - 10z)$ **30.** $(y - 2x)^2$ **31.** $(3a + 2b)(3a + 4b)$ **32.** $(4x - y)(4x + y)$

33. $(8a - 1)^2$ **34.** $(8a - 1)(8a + 1)$ **35.** $(5r - 2s)(5r - 2s)$ **36.** $(2r - 3x)(2r + 7x)$

37. $(7x - 8)(8x - 7)$ **38.** $(5x + 9y)(9x - 5y)$ **39.** $(5t - 3s^2)(5t + 3s)$ **40.** $(2x^2 - 3)(2x + 3)$

41. $(3y^3 - 4x)^2$ **42.** $(2x^2 - 3)^2$ **43.** $(3y^3 - 4x)(3y^3 + 4x)$ **44.** $(7r^2s^2 - 5g)(7r^2s^2 + 5g)$

45. $(2rst - 7xyz)^2$ **46.** $(7r^2 - t^3)^2$ **47.** $(x - 4) - (3x + 1)^2$ **48.** $5x^2 - (2x + 1)(3x + 2)$

49. $(a - b)^2 - (b + a)^2$ **50.** $(3x - 1)(2x + 3) - (x - 4)(x + 1)$

51. $(3a + 2)(-5a)(2a - 1)$ **52.** $(3x - 1)(2x + 3) - x - 4x - 1$

53. $(3a + 2) - 5a(2a - 1)$ **54.** $b^2 - 2b(3b + 2)(2b - 3)$

55. $3r^3 - 3r(r - s)(r + s)$ **56.** $(5r - 2)3r - (5r - 2)(r - 1)$

57. $(3y + 1)(y + 2) - (2y - 3)^2$ **58.** $(x - 2y)^2 - (x - 2y)^2$

59. $(5a - 3b)^3$ **60.** $(x - 3y)^3$

61. $3x(x + 1) - 2x(x + 1)^2$ **62.** $[x - (a + b)][x + (a + b)]$

63. $[(a + b) + 1][(a + b) - 1]$ **64.** $[(x + 2y) - 3z]^2$

65. $[(a - 2b) + 5z]^2$ **66.** $[(x + 2y) - 3z][(x + 2y) + 3z]$

67. $[a - 2b + 5z][a - 2b - 5z]$ **68.** $[2(a + b) + 1][3(a + b) - 5]$

69. $[(a + b) + (2x + 1)][(a + b) - (2x + 1)]$ **70.** $(a - b - 3)(a + b + 3)$

71. $(a^n - 3)(a^n + 3)$ **72.** $(x^n - y^n)(x^n + y^n)$

In Exercises 73–82, given $f(x) = 3x^2 - 2x + 1$ and $g(x) = x^2 - 4$, find

73. $f(x + 2)$ **74.** $g(x - 3)$ **75.** $f(x) + f(2)$ **76.** $g(x) - g(3)$

77. $f(2x)$ **78.** $g(3x)$ **79.** $2f(x)$ **80.** $3g(x)$

81. $3f(x) - 2g(x)$ **82.** $2g(x) + 3f(x)$

83. An open cardboard box is to be made from a 1-ft-square sheet of cardboard by cutting out identical squares of side x from each of the corners of the sheet and folding up the sides to form a box, as shown in Figure 5.4 on page 283. Express the volume of the box in terms of x and simplify.

84. An open cardboard box is to be made from a 12-in. by 8-in. rectangular sheet of cardboard by cutting out identical squares of side x from each of the corners of the sheet and folding up the sides to form a box, as shown in Figure 5.4 on page 283. Express the volume of the box in terms of x and simplify.

In Exercises 85–88, express the shaded area in terms of x.

85.

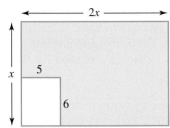

86.

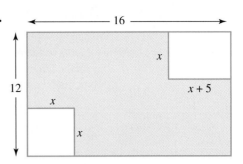

87.

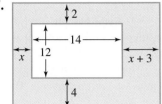

88.

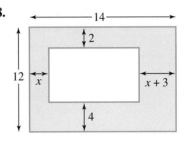

? Questions for Thought

89. *Verbally* describe how to find the product $(x + a)(x + b)$.

90. Discuss what is **wrong** (if anything) with each of the following.
   (a) $(x + y)^2 - (x + y)(x - y) \overset{?}{=} x^2 + 2xy + y^2 - x^2 - y^2$
   $$\overset{?}{=} 2xy$$

   (b) $3x(x - 5)(x + 5) \overset{?}{=} (3x^2 - 15x)(3x^2 + 15x)$
   $$\overset{?}{=} 9x^4 - 225x^2$$

91. Discuss the differences between the expression $(x + y)^2$ and the expression $x^2 + y^2$.

92. Describe the relationships among the following expressions:
$$a + b, \quad a - b, \quad b - a, \quad -a + b, \quad -a - b$$

◆ Mini-Review

93. *Solve for x:* $|3x - 2| > 5$.

94. Two cars leave at the same time from the same location and travel in opposite directions. If one car is traveling at 45 mph and the other is traveling at 55 mph, how long will it be until they are 325 miles apart?

95. The length of a rectangle is 4 more than 5 times its width. If the perimeter of the rectangle is 41.6 in., find the dimensions of the rectangle.

96. Sketch the graphs of the following equations in a rectangular coordinate system.
   (a) $4x - 5y + 20 = 0$ **(b)** $y - 3 = 0$

5.4 Factoring Out the Greatest Common Factor

We can view polynomial multiplication as changing products to sums. In the same way, we can view factoring as changing sums to products. The distributive property gives us the method for multiplying polynomials and it also gives us a method for factoring polynomials:

$$\textit{Multiplying}\longrightarrow$$
$$(a + b)c = a \cdot c + b \cdot c$$
$$\longleftarrow\textit{Factoring}$$

Before we begin to factor polynomials, we should mention that in factoring whole numbers, we were interested in a representation as a product of *whole numbers*. Thus, 12 could be factored as $4 \cdot 3$, $6 \cdot 2$, $3 \cdot 2 \cdot 2$, or $12 \cdot 1$. Even though 12 could be represented as a product of $\frac{1}{3}$ and 36, we did not consider factors that are not whole numbers.

In the same way, we will not consider factors of polynomials with fractional or irrational coefficients; *all polynomial factors should have integer coefficients.* Hence, although $x - 1$ can be represented as the product $\frac{1}{4}(4x - 4)$, we will not consider $\frac{1}{4}$ as a factor of $x - 1$ because it is not an integer.

The first type of factoring we will consider is called **common monomial factoring.** Essentially, we are interested in factoring out the greatest monomial common to each term in a polynomial.

For example, the greatest common monomial factor of $6x^2 + 8x$ is $2x$ because $2x$ is the greatest common factor of *both* terms $6x^2$ and $8x$. Therefore, we can rewrite $6x^2 + 8x$ as $2x \cdot 3x + 2x \cdot 4$ and then factor out $2x$ by the distributive property to get $2x(3x + 4)$. Thus,

$$6x^2 + 8x = 2x \cdot 3x + 2x \cdot 4 = 2x(3x + 4)$$

Keep in mind that this is a two-step process: First you find the greatest common factor; then you determine what remains within the parentheses.

| EXAMPLE 1 | Factor the following as completely as possible. |

(a) $3x^2 + 2x^3 - 4x^5$ **(b)** $24x^2y^3 - 16xy^3 - 8y^4$
(c) $3x^3y^3 - 9xy^4 + 3$ **(d)** $2a^2b - 8ab^2c + 5c^2$

Solution In general, it is probably easiest to begin by first determining the greatest common numerical factor, then the greatest common x factor, then the greatest common y factor, and so forth.

(a) $3x^2 + 2x^3 - 4x^5$

$= \boxed{x^2(3 + 2x - 4x^3)}$ *We factor out x^2 from each term since it is the greatest common factor of $3x^2$, $2x^3$, and $4x^5$.*

Always check your answer by multiplying (you can check each term mentally).

Check: $x^2(3 + 2x - 4x^3) = 3x^2 + 2x^3 - 4x^5$

(b) $24x^2y^3 - 16xy^3 - 8y^4$

$= \boxed{8y^3(3x^2 - 2x - y)}$ *The greatest common factor of $24x^2y^3$, $16xy^3$, and $8y^4$ is $8y^3$.*

Check this answer.

If we had factored part **(b)** as $4y(6x^2y^2 - 4xy^2 - 2y^3)$, it would check when we multiplied it out. But it is not factored *completely;* we have not factored out the *greatest* common factor. We could still factor $2y^2$ from the trinomial. Thus,

$$4y(6x^2y^2 - 4xy^2 - 2y^3) \qquad \textit{Factor } 2y^2 \textit{ from } 6x^2y^2 - 4xy^2 - 2y^3.$$
$$= 4y[2y^2(3x^2 - 2x - y)]$$
$$= 8y^3(3x^2 - 2x - y)$$

Always check that there are no more common factors in the parentheses.

(c) $3x^3y^3 - 9xy^4 + 3$ \qquad *The greatest common factor is 3.*

$$= \boxed{3(x^3y^3 - 3xy^4 + 1)} \qquad \textit{Do not forget to include +1 to hold a place for the 3.}$$

Check: $3(x^3y^3 - 3xy^4 + 1) = 3x^3y^3 - 9xy^4 + 3$

(d) $2a^2b - 8ab^2c + 5c^2$ \qquad *Since there is no common factor of $2a^2b$, $8ab^2c$, and $5c^2$ other than 1, we write:*

$$\boxed{\text{not factorable}}$$

We can generalize common factoring to factoring *polynomials* of more than one term from expressions as follows:

Factor the following: $3x(y - 4) + 2(y - 4)$

Note that $y - 4$ is common to both expressions, $3x(y - 4)$ and $2(y - 4)$, and therefore can be factored out, just as we would factor A from $3xA + 2A$.

$$3x \cdot A \quad + 2 \cdot A \quad = \quad A \cdot (3x + 2)$$
$$3x(y - 4) + 2(y - 4) = (y - 4)(3x + 2)$$

If we read the equation from right to left, we see multiplication by the distributive property.

EXAMPLE 2

Factor the following completely:

(a) $5a(x - 2y) - 3b(x - 2y)$ \qquad **(b)** $3x(x - 3) + 5(x - 3)$

(c) $(x + 2)^2 + (x + 2)$

Solution

(a) $5a(x - 2y) - 3b(x - 2y)$ \qquad *Since $x - 2y$ is common to both terms, $5a(x - 2y)$ and $-3b(x - 2y)$, we can factor $x - 2y$ out, and are left with $5a$ and $-3b$.*

$$= \boxed{(x - 2y)(5a - 3b)}$$

(b) $3x(x - 3) + 5(x - 3)$ \qquad *Factor out $x - 3$ and we are left with $3x$ and $+ 5$.*

$$= \boxed{(x - 3)(3x + 5)}$$

(c) $(x + 2)^2 + (x + 2)$ \qquad *$x + 2$ is the factor common to both $(x + 2)^2$ and $x + 2$. This is like factoring $A^2 + A$ to get $A(A + 1)$.*

$$A^2 \quad + \quad A \quad = \quad A \cdot (\quad A \quad + 1)$$
$$(x + 2)^2 + (x + 2) = (x + 2)[(x + 2) + 1] \qquad \textit{Simplify inside brackets.}$$
$$= (x + 2)[x + 2 + 1]$$
$$= \boxed{(x + 2)(x + 3)}$$

Not all polynomials are conveniently grouped as in Example 2. We often have to take a step or two to put the polynomial in factorable form. For example, in factoring

$$ax + ay + bx + by$$

you can see that there is no common *monomial* we can factor from *all four terms.* However, if we group the terms by pairs and factor the pairs, we can then factor the *binomial* from each group as follows:

$ax + ay + bx + by$ *Group the pairs together.*

$= ax + ay \ + \ bx + by$ *Then factor a from the first pair and b from the second pair.*

$= a(x + y) + b(x + y)$ *Factor x + y from each group.*

$= (x + y)(a + b)$

Grouping and factoring *parts* of a polynomial to factor the polynomial itself is called **factoring by grouping.**

EXAMPLE 3

Factor the following completely.

(a) $7xy + 2y + 7xa + 2a$

(b) $3xb - 2b + 15x - 10$

Solution   **(a)** $7xy + 2y + 7xa + 2a$ *Since there is no common factor of all terms, we group in pairs.*

$= 7xy + 2y \ + \ 7xa + 2a$ *Factor each pair.*

$= y(7x + 2) + a(7x + 2)$ *Factor 7x + 2 from each group.*

$= \boxed{(7x + 2)(y + a)}$

(b) $3xb - 2b + 15x - 10$ *There is no common factor, so group each pair.*

$= 3xb - 2b \ + \ 15x - 10$ *Factor each pair.*

$= b(3x - 2) + 5(3x - 2)$ *Factor 3x - 2 from each group.*

$= \boxed{(3x - 2)(b + 5)}$

A *word of caution:*   Factoring parts of a polynomial, as in

$$3xb - 2b + 15x - 10 = b(3x - 2) + 5(3x - 2)$$

is an intermediate step to guide us in factoring by grouping. It is *not* the factored form of the polynomial. (You would not say 13 is factored as $3 \cdot 3 + 2 \cdot 2$, would you?) Thus, keep in mind that when you are asked to factor a polynomial, the whole polynomial must be represented as a *product* of polynomials.

When a negative sign appears between the pairs of binomials we intend to group, we occasionally have to factor out a negative factor first, to see whether each group contains the same binomial factor. In factoring, $5ac + 20c - 3a - 12$, for example, we would have to factor out -3 in $-3a - 12$:

$5ac + 20c - 3a - 12$ *First separate the pairs.*

$= 5ac + 20c \ - \ 3a - 12$ *Factor out 5c and −3.*

       Be careful. Check signs with multiplication.

           ↓   ↓

$= 5c(a + 4) - 3(a + 4)$ *Factor out a + 4.*

$= \boxed{(a + 4)(5c - 3)}$

EXAMPLE 4

Factor the following completely.

(a) $10xy - 2y + 5x - 1$ **(b)** $6x^2 + 2x - 9x - 3$

(c) $x^3 + x^2 + x + 1$ **(d)** $x^3 + x^2 - x + 1$

Solution   **(a)** $10xy - 2y + 5x - 1$ *Group each pair.*

$\qquad\qquad = 10xy - 2y \;+\; 5x - 1$ *Then factor each pair.*

$\qquad\qquad = 2y(5x - 1) + 1(5x - 1)$

$\qquad\qquad\qquad\qquad \uparrow$

Note: It is helpful to hold this place with the understood factor of 1.

$\qquad\qquad = \boxed{(5x - 1)(2y + 1)}$

Normally we would first com-
bine like terms for this problem,
but here we want to illustrate
a point.

(b) $6x^2 + 2x - 9x - 3 = 6x^2 + 2x \;-\; 9x - 3$ *Factor out 2x from the first pair,*
$\qquad\qquad\qquad\qquad\qquad\qquad\qquad\qquad\qquad$ *−3 from the second pair.*

$\qquad\qquad = 2x(3x + 1) - 3(3x + 1)$

$\qquad\qquad = \boxed{(3x + 1)(2x - 3)}$

(c) $x^3 + x^2 + x + 1 = x^3 + x^2 \;+\; x + 1$

$\qquad\qquad\qquad\qquad\quad = x^2(x + 1) + 1(x + 1)$

$\qquad\qquad\qquad\qquad\quad = \boxed{(x + 1)(x^2 + 1)}$

(d) $x^3 + x^2 - x + 1 = x^2(x + 1) - 1(x - 1)$ *Note that the binomials x − 1 and x + 1 are*
$\qquad\qquad\qquad\qquad\qquad\qquad\qquad\qquad\qquad$ *not identical and therefore this expression*
$\qquad\qquad\qquad\qquad\qquad\qquad\qquad\qquad\qquad$ *cannot be factored by grouping.*

Answer: $\boxed{\text{Not factorable}}$

Note the differences between parts **(c)** and **(d).** Even if we tried grouping in part **(d)** without factoring out -1, we would get $x^2(x + 1) + 1(-x + 1)$, which is still not factorable by grouping, since $x + 1$ and $-x + 1$ are not identical.

EXERCISES 5.4

Factor the following completely.

1. $4x^2 + 2x$

2. $5ab + ab^2$

3. $x^2 + x$

4. $3x^3 + 9$

5. $3xy^2 - 6x^2y^3$

6. $7x^2 - 3xy + 2x$

7. $6x^4 + 9x^3 - 21x^2$

8. $12x^2y^3 - 18x^2$

9. $35x^4y^4z - 15x^3y^5z + 10x^2y^3z^2$

10. $15a^2b - 10ab^2 - 10$

11. $24r^3s^4 - 18r^3s^5 - 6r^2s^3$

12. $15a^2b - 10ab^2$

13. $35a^2b^3 - 21a^3b^2 + 7ab$

14. $15rs^4t - 15r^2s^3 + 10rs^2$

15. $3x(x + 2) + 5(x + 2)$

16. $7r(2s + 1) + 3(2s + 1)$

17. $3x(x + 2) - 5(x + 2)$

18. $7r(2s + 1) - 3(2s + 1)$

19. $2x(x + 3y) + 5y(x + 3y)$

20. $2s(a - b) + 3(a - b)$

21. $3x(x + 4y) - 5y(x + 4y)$

22. $2x(x - 5) + (x - 5)$

23. $(2r + 1)(a - 2) + 5(a - 2)$

24. $(3a + 4)(x - 1) + 2x(x - 1)$

25. $2x(a - 3)^2 + 2(a - 3)^2$

26. $5y(a - b)^2 - 5(a - b)^2$

27. $16a^2(b - 4)^2 - 4a(b - 4)$

28. $3y(y - 2)^2 + (y - 2)$

29. $2x^2 - 8x + 3x - 12$

30. $5a^2 + 10a + 7a + 14$

31. $3x^2 - 12xy + 5xy - 20y^2$

32. $2a^2 + 2ab + 3ab + 3b^2$

33. $7ax - 7bx + 3ay - 3by$

34. $5ra - 5rb + 3sa - 3sb$

35. $7ax + 7bx - 3ay - 3by$

36. $5ra + 5rb - 3sa - 3sb$

37. $7ax - 7bx - 3ay + 3by$

38. $5ra - 5rb - 3sa - 3sb$

39. $7ax - 7bx - 3ay - 3by$

40. $5ra - 5rb - 3sa + 3sb$

41. $2r^2 + 2rs - sr - s^2$

42. $3x^2 + 6x - 4x - 8$

43. $2r^2 + 2rs - sr + s^2$

44. $3x^2 - 4x + 6x - 8$

45. $5a^2 - 5ab - 2ab + 2b^2$

46. $2x^2 + 10x - x - 5$

47. $3a^2 - 6a - a + 2$

48. $5x^2 + 20x + x + 4$

49. $3a^2 - 6a + a - 2$

50. $5x^2 - 20x + x - 4$

51. $3a^2 + 6a - a - 2$

52. $5x^2 + 20x - x + 4$

53. $a^3 + 2a^2 + 4a + 8$

54. $a^3 - 3a^2 + a - 3$

 ## Questions for Thought

55. Verbally describe what is *wrong* (if anything) with the following:

   (a) $9x^2 + 15x + 3 \overset{?}{=} 3(3x^2 + 5x)$ **(b)** $(x - 2)^2 + (x - 2) \overset{?}{=} (x - 2)^3$

56. Factoring by grouping illustrates what property?

57. Complete the following expressions so that they can be factored by grouping:

   (a) $5x + 10y + 3x + ?$ **(b)** $4a - 12y - 5a^2 + ?$

Mini-Review

58. *Solve for a:* $5 - [3 - (a - 2)] = 5a - 2$

59. *Solve for a:* $|9 - 3a| \le 6$

60. *Solve for a:* $|9 - 3a| \ge 6$

61. Sketch the graph of the line with slope $-\frac{2}{3}$ that passes through the point $(4, -1)$.

 5.5

Factoring Trinomials

Factoring $x^2 + Bx + C$

When we multiplied $x + a$ by $x + b$ in Section 5.3, we found that

$$(x + a)(x + b) = x^2 + (a + b)x + ab$$

Note the relationship between the constants a and b in the binomials to be multiplied, and the coefficients of the trinomial product: The x term coefficient is the *sum* of a and b; the numerical term is the product of a and b. Therefore, if the trinomial $x^2 + Bx + C$ could be factored into two binomials, $(x + a)$ and $(x + b)$, then B is the *sum* of a and b, whereas C is the *product* of a and b. Thus, all we need to find are two factors of B that sum to C. For example,

To factor $x^2 + 8x + 12$, we need to find two factors of $+12$ that sum to $+8$.

We first determine the signs of the two factors, arriving at the answer by logical deduction as follows:

The two factors must have the same signs since their product, $+12$, is positive. Both signs must be positive since the sum $+8$ is positive.

If we systematically check the factors of 12 ($1 \cdot 12$; $2 \cdot 6$; $3 \cdot 4$), we arrive at $+6$ and $+2$ as the factors of $+12$ that sum to $+8$. Thus,

$$x^2 + 8x + 12 = (x + 6)(x + 2)$$

EXAMPLE 1

Factor each of the following.

(a) $x^2 - 4x - 12$ (b) $x^2 - 21x + 54$
(c) $a^2 - 10a + 25$ (d) $a^2 - 25$
(e) $3y^3 - 6y^2 - 105y$ (f) $x^2 + 2x + 3$

Solution

(a) $x^2 - 4x - 12$ Find factors of -12 that sum to -4. First consider the signs of the two factors: The signs must be opposite since the product is negative.

Since we know the signs are *opposite,* we ignore the signs and look for two factors of 12 whose *difference* is 4.

Possible candidates are 12 and 1, 6 and 2, and 3 and 4. The pair 6 and 2 yields a difference of 4 and therefore the two factors are 6 and 2. Now, looking at the signs, we must have -6 and $+2$ to sum to -4 as required. Hence,

$$\boxed{x^2 - 4x - 12 = (x - 6)(x + 2)}$$

Always take the time to check your answer by multiplying the factors. Be careful with the signs.

(b) $x^2 - 21x + 54$ Find the two factors of $+54$ that sum to -21. The signs of the two factors must be the same (since $+54$ is positive): Both factors must be negative since their sum, -21, is negative.

Since we know the signs are the same, we ignore the signs and look for two factors of 54 whose *sum* is 21.

Possible candidates are 1 and 54, 2 and 27, 3 and 18, and 6 and 9. The answer is 3 and 18; considering the signs, we must have -3 and -18. Hence,

$$\boxed{x^2 - 21x + 54 = (x - 18)(x - 3)}$$

Check the answer.

(c) $a^2 - 10a + 25$ Find two factors of $+25$ that sum to -10. Since $+25$ is positive, the signs must be the same. Both factors must be negative since their sum is -10.

The factors must be -5 and -5. Hence,

$$\boxed{a^2 - 10a + 25 = (a - 5)(a - 5)}$$

(d) $a^2 - 25$ We can rewrite this as $a^2 + 0a - 25$ to determine that we are seeking two factors of -25 that sum to 0. The signs of the two factors must be opposite (since -25 is negative).

Ignoring the signs for the moment, we are searching for two factors of 25 whose *difference* is 0.

The factors must be -5 and 5. It obviously does not matter which factor has the negative sign:

$$\boxed{a^2 - 25 = (a - 5)(a + 5)}$$ *Note this is a difference of two squares.*

Compare this problem with the one in part (c).

(e) Remember always to look for the greatest common factor first:

$$3y^3 - 6y^2 - 105y = 3y(y^2 - 2y - 35)$$

Now try to factor $y^2 - 2y - 35$: The signs of two factors of -35 are opposite. Therefore, ignoring the signs, what two factors of 35 yield a *difference* of 2? Answer: 7 and 5. Including signs: -7 and $+5$, since they must sum to -2.
Thus,

$$y^2 - 2y - 35 = (y - 7)(y + 5)$$

and

$$3y^3 - 6y^2 - 105y = 3y(y^2 - 2y - 35)$$

$$= \boxed{3y(y - 7)(y + 5)}$$

(f) $x^2 + 2x + 3$ Find two factors of $+3$ that sum to $+2$.

Since $+3$ is positive, the signs of the two factors must be the same, and both factors must be positive since their sum is $+2$. But the only factors of $+3$ are 1 and 3, and $+1$ and $+3$ do not add up to $+2$. Since no factors satisfy these conditions, $x^2 + 2x + 3$ is

$$\boxed{\text{not factorable}}$$

Factoring $Ax^2 + Bx + C$

We saw in Section 5.3 that

$$(ax + b)(cx + d) = acx^2 + (ad + bc)x + bd$$

which is obviously more complex than the previous case; it is complicated by the coefficients of the x terms in the binomials.

Note the relationships between the constants a, b, c, and d, and the coefficients of the trinomial product. Therefore, if $Ax^2 + Bx + C$ were to factor into two binomials, A would be the product of the x coefficients in the binomials ($a \cdot c$); C would be the product of the numerical term coefficients in the binomials ($b \cdot d$); and B would be the interaction (inner + outer) of the four coefficients a, b, c, and d. We will demonstrate the trial-and-error process by example.

EXAMPLE 2

Factor $2x^2 + 5x + 3$.

Solution

Note first that there are no common factors.

The only possible factorization of 2 is $2 \cdot 1$. These must be the coefficients of x in the binomials.

The only possible factorization of 3 is $3 \cdot 1$. These must be the numerical terms in the binomials.

There are two possible answers:

$$(2x + 1)(x + 3) \quad \text{and} \quad (2x + 3)(x + 1)$$

Multiplying out, we get

$$2x^2 + 7x + 3 \quad \text{and} \quad 2x^2 + 5x + 3$$

Note that both first and last terms are identical. The middle term indicates that $(2x + 3)(x + 1)$ is the answer. Hence,

$$2x^2 + 5x + 3 = \boxed{(2x + 3)(x + 1)}$$

EXAMPLE 3

Factor $6x^2 + 19x + 10$ completely.

Solution

Note that there is no common monomial to factor from $6x^2 + 19x + 10$. Since there seem to be many possible combinations, we check out the possibilities as follows:

Possible factorizations of 6 are $6 \cdot 1$ and $2 \cdot 3$.

Possible factorizations of 10 are $10 \cdot 1$ and $5 \cdot 2$.

The possible factorizations of $6x^2 + 19x + 10$ are

$$(6x + 5)(x + 2) = 6x^2 + 17x + 10$$
$$(6x + 2)(x + 5) = 6x^2 + 32x + 10*$$
$$(6x + 10)(x + 1) = 6x^2 + 16x + 10*$$
$$(6x + 1)(x + 10) = 6x^2 + 61x + 10$$
$$(2x + 5)(3x + 2) = 6x^2 + 19x + 10$$
$$(2x + 2)(3x + 5) = 6x^2 + 16x + 10*$$
$$(2x + 10)(3x + 1) = 6x^2 + 32x + 10*$$
$$(2x + 1)(3x + 10) = 6x^2 + 23x + 10$$

Each pair of binomials will yield the same first and last term, but only one combination will yield the correct middle term, $19x$:

$$6x^2 + 19x + 10 = \boxed{(2x + 5)(3x + 2)}$$

Note the following:

1. You should stop when you hit the right combination and then check your answer by multiplying.
2. Take a close look at the second possibility, $(6x + 2)(x + 5)$. This possibility can be factored further into $2(3x + 1)(x + 5)$ by factoring 2 from $6x + 2$. If this were the answer, it would imply that 2 is a common factor of $6x^2 + 19x + 10$ (why?). But clearly, there is no common factor of $6x^2 + 19x + 10$, so we can eliminate $(6x + 2)(x + 5)$ as a possibility. For the same reason, we can eliminate the other possibilities that appear with an asterisk (*).

EXAMPLE 4

Factor $12a^3 + 2a^2 - 4a$.

Solution

$12a^3 + 2a^2 - 4a$ *Factor out the common monomial, 2a, first.*

$= 2a(6a^2 + a - 2)$ *Factor $6a^2 + a - 2$ into $(2a - 1)(3a + 2)$.*

$= \boxed{2a(2a - 1)(3a + 2)}$

The trial-and-error procedure can be quite laborious—especially when the numbers have numerous factors. If you do not go about checking the factors in a systematic manner, it is very easy to miss the correct factors (and their relative positions) and assume the expression cannot be factored. For this reason, we recommend the *factoring by grouping* method, which we now describe.

Factoring Trinomials by Grouping

Another way to factor trinomials is to use grouping. To factor $Ax^2 + Bx + C$ we would do the following:

1. Find the product, AC.
2. Find two factors of AC that sum to B.
3. Rewrite the middle term as a sum of terms whose coefficients are the factors found in step 2.
4. Factor by grouping.

EXAMPLE 5

Factor $12x^2 - 17x - 5$ by grouping.

Solution

1. Find the product, AC:

$$12(-5) = -60$$

2. Find two factors of -60 that add to -17:

$$-20 \text{ and } +3, \quad \text{since} \quad (-20)(+3) = -60 \text{ and } -20 + 3 = -17$$

Note that we could have also written this as $12x^2 + 3x - 20x - 5$.

3. Rewrite the middle term, $-17x$, as $-20x + 3x$. Hence,

$$12x^2 - 17x - 5 = 12x^2 - 20x + 3x - 5$$

4. Factor by grouping:

$$12x^2 - 20x + 3x - 5$$
$$= 4x(3x - 5) + 1(3x - 5)$$
$$= \boxed{(3x - 5)(4x + 1)}$$

Although we still have to look for factors (of a number larger than A or C), the process of factoring by grouping is usually a bit more efficient than the trial-and-error process.

EXAMPLE 6

Factor $20a^2 - 7ab - 6b^2$ completely.

Solution

Find the product, $20(-6) = -120$.

Which two factors of -120 will yield -7 when added? Answer: -15 and $+8$.

Rewrite $-7ab$ as $-15ab + 8ab$.

Thus,

$$20a^2 - 7ab - 6b^2 = 20a^2 - 15ab + 8ab - 6b^2$$
$$= 5a(4a - 3b) + 2b(4a - 3b)$$
$$= \boxed{(4a - 3b)(5a + 2b)}$$

EXAMPLE 7

Factor $5x^2 - 13x + 6$ completely.

Solution

Since $(5)(6) = 30$, we want factors of 30 that sum to -13. Answer: -10 and -3. Hence,

$$5x^2 - 13x + 6 \qquad \text{\textit{Rewrite} } -13x \text{ \textit{as} } -10x - 3x.$$
$$= 5x^2 - 10x - 3x + 6 \qquad \text{\textit{Factor by grouping.}}$$

$$= 5x(x - 2) - 3(x - 2)$$

$$= \boxed{(x - 2)(5x - 3)}$$

> If there are no pairs of factors of AC that sum to the middle term coefficient, then the trinomial does not factor into two binomials.

EXAMPLE 8

Factor the following completely.

(a) $5r^2 - 2rs + 10s^2$ **(b)** $-10z^2 + 11z - 3$

Solution

(a) The factors of $5(10) = 50$ are $1 \cdot 50$, $2 \cdot 25$, and $5 \cdot 10$. Since there are no pairs of factors of $10(5) = 50$ that sum to -2, the polynomial is

$$\boxed{\text{not factorable}}$$

(b) The factors of $(-10)(-3) = 30$ that sum to 11 are $+6$ and $+5$. Hence,

$$-10z^2 + 11z - 3 = -10z^2 + 6z + 5z - 3$$
$$= -2z(5z - 3) + 1(5z - 3)$$
$$= \boxed{(5z - 3)(-2z + 1)}$$

Another way to solve this problem is to factor -1 from the trinomial first:

$$-10z^2 + 11z - 3 = -[10z^2 - 11z + 3] \qquad \textit{Factor by grouping.}$$
$$= -[10z^2 - 6z - 5z + 3]$$
$$= -[2z(5z - 3) - 1(5z - 3)]$$
$$= -[(5z - 3)(2z - 1)]$$

or simply

$$= \boxed{-(5z - 3)(2z - 1)}$$

Look at the two answers given for this same problem. They actually are the same. If you multiply out $-(2z - 1)$, you will get $-2z + 1$, which is the second factor of the first answer. [Remember that $-(ab) = (-a)(b)$.]

Factoring Using Special Products

We have discussed factoring $x^2 - 10x + 25$ and $x^2 - 25$ by using trial and error, and it took a few steps to arrive at each solution. However, if we had recognized these polynomials as forms of special products, we could have cut down our labor a bit. In this section, we will discuss how to recognize and factor special products without resorting to trial-and-error factoring or factoring by grouping.

First, we relist in the box the special products that we have had thus far.

Factoring Special Products

1. $a^2 - b^2 = (a + b)(a - b)$ *Difference between two squares*

2. $a^2 + 2ab + b^2 = (a + b)^2$ *Perfect square of sum*

3. $a^2 - 2ab + b^2 = (a - b)^2$ *Perfect square of difference*

The first polynomial, $a^2 - b^2$, is called the *difference of two squares;* the difference of squares of two terms can be factored into two binomial conjugates. For example, we could rewrite the binomial

$$4x^2 - 9y^2$$

as the difference of two squares

$$(2x)^2 - (3y)^2$$

which factors into

$$(2x - 3y)(2x + 3y)$$

The last two special products are called *perfect squares* (just as $36 = 6^2$ is a perfect square). Notice that the first and last terms of both perfect-square trinomials are perfect squares. For example,

$$9x^2 - 12xy + 4y^2$$
$$= (3x)^2 - 2(3x)(2y) + (2y)^2$$
$$= (3x - 2y)^2$$

If a trinomial is factorable, it can be factored by the methods covered in the previous section. However, if a trinomial can be recognized as a special product, it can be factored quickly and with less effort. For example, when we see a trinomial such as

$$16x^2 - 40xy + 25y^2$$

there are quite a few possible combinations of factors to check if we use trial and error or factoring by grouping. Our experience with special products, however, makes us suspicious when we see two perfect square terms, $16x^2$ and $25y^2$, in the trinomial. We would immediately check to see whether it is a perfect square by choosing

$$(4x - 5y)(4x - 5y)$$

as the first *possible* factorization. Multiplication will confirm our suspicion that this is indeed the proper factorization of $16x^2 - 40xy + 25y^2$.

With regard to the difference of two squares, the only binomials up to this point that factor into two *binomials* are those that are the difference of two squares. Obviously, it is easier to use the difference-of-squares factorization than to introduce a middle term coefficient of 0 as we did in the previous section using the other factoring methods.

To factor quickly, you must recognize the relationships given in the previous box.

EXAMPLE 9

Factor the following completely.

(a) $x^2 - 6x + 9$ **(b)** $9a^2 + 30a + 25$ **(c)** $4x^2 - 25y^2$
(d) $x^4 - y^4$ **(e)** $x^3 + x^2 - x - 1$

Solution

(a) $x^2 - 6x + 9 = x^2 - 2(3)x + 3^2$ *Note the relationships between terms in a perfect-square trinomial.*

$$= \boxed{(x - 3)^2}$$

(b) $9a^2 + 30a + 25 = (3a)^2 + 2(3a)(5) + 5^2$

$$= \boxed{(3a + 5)^2}$$

(c) $4x^2 - 25y^2 = (2x)^2 - (5y)^2$

$$= \boxed{(2x - 5y)(2x + 5y)}$$

(d) $x^4 - y^4$ *There is no common monomial to factor.*
Rewrite as a difference of two squares.

$$= (x^2)^2 - (y^2)^2 \qquad \textit{Factor.}$$

$$= (x^2 - y^2)(x^2 + y^2) \qquad \textit{Do not forget to factor } x^2 - y^2.$$

$$= \boxed{(x - y)(x + y)(x^2 + y^2)}$$

Always check to see whether there is any more factoring to be done.

(e) $x^3 + x^2 - x - 1$ *Factor by grouping.*

$$= x^2(x + 1) - 1(x + 1) \qquad \textit{Then factor out } x + 1.$$

$$= (x + 1)(x^2 - 1) \qquad \textit{Do not forget to factor } x^2 - 1.$$

$$= \boxed{(x + 1)(x - 1)(x + 1) \quad \text{or} \quad (x - 1)(x + 1)^2}$$

Here are two more special products that will be useful.

Factoring the Sum and Difference of Cubes	4. $a^3 - b^3 = (a - b)(a^2 + ab + b^2)$ *Difference of two cubes* 5. $a^3 + b^3 = (a + b)(a^2 - ab + b^2)$ *Sum of two cubes*

Note the similarities and differences between factoring the sum and difference of two cubes. Keep in mind that $a^3 - b^3$ is *not* the same as $(a - b)^3$, which when multiplied out will have middle terms. Also note that either right-hand factor, $a^2 - ab + b^2$ or $a^2 + ab + b^2$, will not have binomial factors. (Do not confuse it with the perfect square $a^2 + 2ab + b^2$, which has binomial factors.)

EXAMPLE 10

Factor the following completely.

(a) $27x^3 - y^3$ **(b)** $8c^3 + 125b^3$

Solution

(a) $27x^3 - y^3$ *Write as a difference of two cubes.*

$$\overset{A^3 - B^3}{}$$

$$= (3x)^3 - y^3 \qquad \textit{Factor.}$$

$$\overset{(A - B)(A^2 + A \cdot B + B^2)}{}$$

$$= (3x - y)[(3x)^2 + (3x)(y) + y^2] \qquad \textit{Simplify inside the brackets.}$$

$$= \boxed{(3x - y)(9x^2 + 3xy + y^2)}$$

(b) $8c^3 + 125b^3$ *Rewrite as a sum of two cubes.*

$$= (2c)^3 + (5b)^3 \qquad \textit{Factor.}$$

$$= (2c + 5b)[(2c)^2 - (2c)(5b) + (5b)^2] \qquad \textit{Simplify inside the brackets.}$$

$$= \boxed{(2c + 5b)(4c^2 - 10cb + 25b^2)}$$

Can you factor $A^2 + B^2$?

Note that although the sum of two *cubes **does*** factor, the sum of two *squares* does ***not.***

In general, we offer the following advice for factoring polynomials.

General Advice for Factoring Polynomials	1. Always factor out the greatest common factor first.
	2. If the polynomial to be factored is a binomial, then it may be a difference of two squares, or a sum or difference of two cubes (remember that a sum of two squares does not factor).
	3. If the polynomial to be factored is a trinomial, then:
	   a. If two of the three terms are perfect squares, the polynomial may be a perfect square.
	   b. Otherwise, the polynomial may be one of the general forms.
	4. If the polynomial to be factored consists of four or more terms, then try factoring by grouping.

EXERCISES 5.5

Factor the following completely.

1. $x^2 + 4x - 45$

2. $x^2 - 10x + 24$

3. $y^2 - 3y - 10$

4. $a^2 - 8ab - 20b^2$

5. $x^2 - 2xy - 15y^2$

6. $r^2 - rs - 2s^2$

7. $r^2 - 81$

8. $a^2 - 49$

9. $x^2 - xy - 6y^2$

10. $a^2 + 4ab - 21b^2$

11. $x^2 - xy + 6y^2$

12. $a^2 + 4ab + 21b^2$

13. $r^2 - 7rs + 12s^2$

14. $a^2 - 12ab + 20b^2$

15. $r^2 - rs - 12s^2$

16. $a^2 + 12ab - 20b^2$

17. $9x^2 - 49y^2$

18. $4a^2 - 81b^2$

19. $15x^2 + 17x + 4$

20. $9a^2 - 6a - 8$

21. $15x^2 - xy - 6y^2$

22. $10a^2 + 23ab - 12b^2$

23. $15x^2 + xy - 6y^2$

24. $10a^2 - 23ab + 12b^2$

25. $10xy + 3x^2 - 25y^2$

26. $13xy + 6y^2 + 6x^2$

27. $10a^2 + 21ab - 10b^2$

28. $7x^2 + 20xy + 12y^2$

29. $25 - 5y - 2y^2$

30. $21 + 8a - 4a^2$

31. $25 - 5y - y^2$

32. $21 - 8a - 4a^2$

33. $2x^3 + 4x^2 - 16x$

34. $3y^3 + 30y^2 + 63y$

35. $x^3 + 3x^2 - 28x$

36. $y^3 - 5y^2 - 36y$

37. $2x(x - 3) + (x - 1)(x + 2)$

38. $2u(u + 1) + (u + 2)(u - 5)$

39. $18ab - 15abx - 18abx^2$

40. $30y^3 + 25y^2 + 5y$

41. $90y^4 - 114y^3 + 36y^2$

42. $54xy^4 + 45x^2y^3 + 6x^3y^2$

43. $6r^4 - r^2 - 2$

44. $6x^4 - x^2 - 12$

45. $3x^2 - 27$

46. $2x^2 - 50$

47. $20xy^5 + 2x^2y^3 - 8x^3y$

48. $6x^2 + 5xy^2 - 6y^4$

49. $108x^3y + 72x^2y^3 - 15xy^5$

50. $25x^4 + 10x^2y^2 + 15y^4$

51. $12a^7b + a^4b - 6ab$

52. $24a^7b - 20a^4b^3 + 4ab^5$

53. $15a^8 + 19a^4 - 10$

54. $4a^4 + 8a^2 + 3$

55. $16x^2 - 9y^2$

56. $x^2 - 6x + 9$

57. $x^2 + 4xy + 4y^2$

58. $9a^2 + 30ab + 25b^2$

59. $x^2 - 4xy + 4y^2$

60. $9a^2 - 30ab + 25b^2$

61. $x^2 - 4xy - 4y^2$

62. $9a^2 - 30ab - 25b^2$

63. $6x^2y^2 - 5xy + 1$

64. $9r^2 + 42rs + 49s^2$

65. $81r^2s^2 - 16$

66. $25x^2y^2 - 4$

67. $9x^2y^2 - 4z^2$

68. $9a^2b^2 - 4$

69. $25a^2 - 4a^2b^2$

70. $1 - 9a^2$

71. $1 - 14a + 49a^2$

72. $125a^3 + b^3$

73. $8a^3 - b^3$

74. $8a^3 - 343b^3$

75. $x^3 + 125y^3$

76. $9x^4 + 24x^2 + 16$

77. $4x^2 + 25y^2$

78. $9x^4 - 24x^2 - 16$

79. $4y^2 - 20y + 10$

80. $45a^3b - 20ab^3$

81. $12a^3c + 36a^2c^2 + 27ac^3$

82. $18x^5 + 24x^3 + 8x$

83. $4x^4 - 81y^4$

84. $x^8 - y^8$

85. $25a^6 + 10a^3b + b^2$

86. $12y^6 - 27y^2$

87. $12y^6 + 27y^2$

88. $12x^4 - 12y^4$

89. $(a + b)^2 - 4$

90. $(3a + b)^2 - (a + b)^2$

91. $a^2 - 2ab + b^2 - 16$

92. $r^2 - 10r + 25 - x^2$

93. $x^2 + 6x + 9 - r^2$

94. $x^2 + 2xy + y^2 - 81$

95. $a^3 + a^2 - 4a - 4$

96. $r^3 - r^2 - 9r + 9$

97. $x^4 + x^3 - x - 1$

98. $8x^4 - 16x^3 - x + 2$

99. If a race has n entrants, the number of possible different first-, second-, and third-place finishes (excluding ties) is given by the formula $P = n^3 - 3n^2 + 2n$. Factor this polynomial.

100. The volume of a cylindrical shell (see the accompanying figure) is given by the formula $V = \pi R^2 h - \pi r^2 h$. Factor this polynomial.

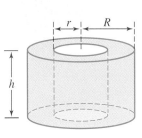

? Questions for Thought

101. In this section we mentioned that if a polynomial does *not* have a common factor, we can eliminate any binomial factors that *do* have common factors. For example, the trinomial $6x^2 - 5x - 6$ has no common numerical factors. Why does this imply that $3x - 3$ could not be a binomial factor of $6x^2 - 5x - 6$?

102. Find all k such that:

(a) $x^2 + kx + 10$ is factorable.   **(b)** $x^2 + kx - 10$ is factorable.

(c) $2x^2 + kx - 3$ is factorable.

103. Find all p such that $x^2 + 3x + p$ is factorable, where $-20 < p < 20$.

104. How would you factor the following?

(a) $3(x + y)^2 - 4(x + y) - 4$ (b) $2a^2 + 4ab + 2b^2 - 7a - 7b + 3$

105. Describe what relationship must exist among terms for a trinomial to factor as a perfect square.

106. Describe the factors of the sum and the differences of two cubes. How do they differ? How are they the same?

107. Discuss the differences between $a^3 - b^3$ and $(a - b)^3$.

108. Factor $x^6 - y^6$ completely.

(a) Begin by using the difference of two squares.

(b) Begin by using the difference of two cubes.

(c) Should the answers to part (a) and (b) be the same? Why or why not?

109. Why is $(a^2b - b)(a + 1)$ *not* a complete factorization of $a^3b + a^2b - ab - b$? What should the answer be?

 Mini-Review

110. Find the slope of the line passing through the points $(2, -6)$ and $(3, 5)$.

111. Find the equation of the line passing through the point $(5, 1)$ and perpendicular to the line $x - 6y = 6$.

112. Sketch the graph of $3x - 6y \leq 12$.

113. A shoe manufacturer finds that during the summer there is a linear relationship between the weekly number of pairs of sandals he sells and the average weekly temperature (in °F). If he sells 20 pairs during the week that the temperature averages 74°F, and 40 pairs during the week that the temperature averages 84°F, how many pairs should he expect to sell the week that the temperature averages 90°F?

5.6 Solving Polynomial Equations by Factoring

We have previously learned how to solve first-degree equations in one variable. In fact, these were polynomial equations of degree 1. Polynomial equations of degree 2 are called *quadratic equations*. We define the *standard form* of a quadratic equation as

$$Ax^2 + Bx + C = 0 \quad \text{(where } A \neq 0)$$

In this standard form, A is the coefficient of the second-degree term, x^2; B is the coefficient of the first-degree term, x; and C is the numerical constant.

As with all other equations, the solutions to quadratic equations are values that when substituted for the variable satisfy the equation. The solutions to a polynomial equation are also called the *roots* of the equation. The factoring techniques we have discussed in this chapter allow us to solve certain quadratic and higher-degree polynomial equations.

We begin by recognizing that if the product of two quantities is 0, then either one or both of the quantities must be 0. Symbolically written, if $a \cdot b = 0$, then $a = 0$, or $b = 0$, or both $a = 0$ and $b = 0$. In mathematics, the word *or* includes the possibility of both. Thus, we can write

The Zero-Product Rule

If $a \cdot b = 0$, then $a = 0$ or $b = 0$.

For example, if we want to find the solution to the equation

$$(x - 2)(x + 3) = 0$$

then either the factor $x - 2$ must equal 0 or the factor $x + 3$ must equal 0. Now we have two first-degree equations:

$$x - 2 = 0 \qquad \text{or} \qquad x + 3 = 0$$
$$\text{Therefore,} \quad x = 2; \qquad \text{Therefore,} \quad x = -3.$$

Hence, our solutions to the equation $(x - 2)(x + 3) = 0$ are $x = 2$ and $x = -3$.

EXAMPLE 1

Solve each of the following equations.

(a) $(3x - 2)(x + 3) = 0$ **(b)** $y(y + 4) = 0$ **(c)** $5x(x - 1)(x + 8) = 0$

Solution

(a) If $(3x - 2)(x + 3) = 0$, then

$$3x - 2 = 0 \quad \text{or} \quad x + 3 = 0 \qquad \textit{Solve each first-degree equation.}$$
$$3x = 2 \qquad\qquad x = -3$$

Hence, $\boxed{x = \dfrac{2}{3} \quad \text{or} \quad x = -3}$.

Let's check the solution $x = -3$:

$$(3x - 2)(x + 3) = 0 \qquad \textit{Substitute } -3 \textit{ for } x.$$
$$[3(-3) - 2](-3 + 3) \stackrel{?}{=} 0$$
$$\uparrow$$

Note that this arithmetic is not necessary once we establish that one of the factors is 0.

$$(-11)(0) \stackrel{\checkmark}{=} 0$$

We leave the other check to the student.

(b) If $y(y + 4) = 0$, then

$$y = 0 \quad \text{or} \quad y + 4 = 0 \qquad \textit{If the product of factors is 0, then at least one of the factors must be 0.}$$

Hence, $\boxed{y = 0 \quad \text{or} \quad y = -4}$.

(c) We can generalize the zero-product rule to more than two factors. We solve the equation $5x(x - 1)(x + 8) = 0$ by setting each factor equal to 0:

$$x = 0 \quad \text{or} \quad x - 1 = 0 \quad \text{or} \quad x + 8 = 0$$

If the product of factors is 0, then at least one of the factors must be 0. We solve each equation. Note that we can ignore the constant factor 5 since $5 \neq 0$.

Hence, $\boxed{x = 0, \quad x = 1, \quad \text{or} \quad x = -8}$.

We mentioned that we can ignore the *factor* 5 because $5 \neq 0$. We would arrive at the same conclusion if we were to first divide both sides of the equation by 5.

If we multiply out the left side in part **(a)** of the previous example, we get $3x^2 + 7x - 6 = 0$, which is a quadratic equation in standard form. If the second-degree expression of a quadratic equation (in standard form) factors into two first-degree factors, we can apply the above principle to solve the quadratic equation. This is called the *factoring method* of solving quadratic equations.

For example, to solve the second-degree equation $5x^2 - 2 = 3x^2 - x + 4$, we first put the equation in standard form:

$$5x^2 - 2 = 3x^2 - x + 4 \qquad \text{\textit{Put in standard form.}}$$

$$2x^2 + x - 6 = 0 \qquad \text{\textit{Factor the left side.}}$$

$$(2x - 3)(x + 2) = 0 \qquad \text{\textit{Since the product is 0, set each factor equal to 0.}}$$

$$2x - 3 = 0 \quad \text{or} \quad x + 2 = 0 \qquad \text{\textit{Then solve each first-degree equation.}}$$

$$2x = 3$$

$$x = \frac{3}{2} \quad \text{or} \qquad x = -2$$

We leave it to the student to check these solutions.

EXAMPLE 2

Solve each of the following equations.

(a) $2x^2 - 3x - 1 = 1$ **(b)** $3a^2 = 5a$

Solution

(a) Do not try to factor first; $2x^2 - 3x - 1 = 1$ is not in standard form. Remember, the principle on which the factoring method is based requires that the product be equal to 0.

$$2x^2 - 3x - 1 = 1 \qquad \text{\textit{First we put the equation in standard form (subtract 1 from both sides of the equation).}}$$

$$2x^2 - 3x - 2 = 0 \qquad \text{\textit{Then factor.}}$$

$$(2x + 1)(x - 2) = 0 \qquad \text{\textit{Then solve each first-degree equation.}}$$

$$2x + 1 = 0 \quad \text{or} \quad x - 2 = 0$$

$$2x = -1 \qquad \qquad x = 2$$

$$\boxed{x = -\frac{1}{2} \quad \text{or} \quad x = 2}$$

(b) Do not make the mistake of dividing both sides of the equation by a; a is a variable that may be 0, and division by 0 is not allowed.

$$3a^2 = 5a \qquad \text{\textit{First put in standard form.}}$$

$$3a^2 - 5a = 0 \qquad \text{\textit{Then factor.}}$$

$$a(3a - 5) = 0 \qquad \text{\textit{Set each factor equal to 0.}}$$

$$a = 0 \quad \text{or} \quad 3a - 5 = 0$$

$$3a = 5$$

$$\boxed{a = 0 \quad \text{or} \quad a = \frac{5}{3}}$$

The same approach can be applied to polynomials of higher degree if they can be factored.

EXAMPLE 3

Solve for x: $x^3 - 4x^2 = 12x$

Solution

This is a third-degree polynomial equation. To use the factoring method, we get one side of the equation to equal to 0, and hope that we can factor.

$$x^3 - 4x^2 = 12x \qquad \text{\textit{Subtract 12x from both sides.}}$$

$$x^3 - 4x^2 - 12x = 0 \qquad \text{\textit{Factor the left side.}}$$

$$x(x^2 - 4x - 12) = 0$$

$$x(x - 6)(x + 2) = 0 \qquad \text{\textit{We use the zero-product rule and set each factor equal to 0.}}$$

$$x = 0 \quad \text{or} \quad x - 6 = 0 \quad \text{or} \quad x + 2 = 0$$

$$\boxed{x = 0 \quad \text{or} \quad x = 6 \quad \text{or} \quad x = -2}$$

The student should check these solutions.

EXAMPLE 4

If $f(x) = 2x^2 + x - 10$, solve each of the following polynomial equations.

(a) $f(x) = 0$ **(b)** $f(x) = -9$

Solution

(a) Solving the polynomial equation $f(x) = 0$ means we want to find the value x that will make $f(x) = 0$. Since $f(x) = 2x^2 + x - 10$, we solve the equation $2x^2 + x - 10 = 0$:

$$f(x) = 2x^2 + x - 10 \qquad \text{\textit{Substitute f(x) = 0 to get}}$$

$$0 = 2x^2 + x - 10 \qquad \text{\textit{Solve for x by factoring.}}$$

$$0 = (x - 2)(2x + 5) \qquad \text{\textit{Set each factor equal to 0.}}$$

$$0 = x - 2 \quad \text{or} \quad 0 = 2x + 5$$

$$\boxed{x = 2 \quad \text{or} \quad x = -\frac{5}{2}}$$

Hence, $f(2) = 0$ and $f\left(-\dfrac{5}{2}\right) = 0$.

(b) Solving the polynomial equation $f(x) = -9$ means we want to find the value x that will make $f(x) = -9$. Hence we solve the equation $2x^2 + x - 10 = -9$:

$$f(x) = 2x^2 + x - 10 \qquad \text{\textit{Substitute f(x) = -9 to get}}$$

$$-9 = 2x^2 + x - 10 \qquad \text{\textit{Put the equation in standard form (Add 9 to each side.)}}$$

$$0 = 2x^2 + x - 1 \qquad \text{\textit{Solve for x by factoring.}}$$

$$0 = (x + 1)(2x - 1) \qquad \text{\textit{Set each factor equal to 0.}}$$

$$0 = x + 1 \quad \text{or} \quad 0 = 2x - 1$$

$$\boxed{x = -1 \quad \text{or} \quad x = \frac{1}{2}}$$

Hence, $f(-1) = -9$ and $f\left(\dfrac{1}{2}\right) = -9$.

We will discuss more general methods for solving quadratic equations in Chapter 8.

EXAMPLE 5

A rectangle has length 4 inches greater than its width. If its area is 77 sq in., what are the dimensions of the rectangle?

Solution

Draw a diagram and label the sides, as indicated in Figure 5.6.

Let x = Width

Figure 5.6

Then

$$x + 4 = \text{Length} \qquad \textit{Since the length is 4 more than the width}$$

The formula for the area of a rectangle is

$$A = (\text{Length})(\text{Width})$$

$$77 = (x + 4)x$$

Solve:

$$77 = x^2 + 4x$$

$$0 = x^2 + 4x - 77$$

$$0 = (x - 7)(x + 11)$$

$$x = 7 \quad \text{or} \quad x = -11 \qquad \textit{We eliminate the negative answer since length is never negative.}$$

Thus, $x = 7$ in. (the width) and $x + 4 = 7 + 4 = 11$ in. (the length).

The dimensions are $\boxed{7 \text{ in. by } 11 \text{ in.}}$.

EXAMPLE 6

A 20 ft by 60 ft rectangular pool is surrounded by a concrete walkway of uniform width. If the total area of the walkway is 516 sq ft, how wide is the walkway?

Solution

Since the walkway is the same width at all points around the pool, we label this width x and our picture is as shown in Figure 5.7.

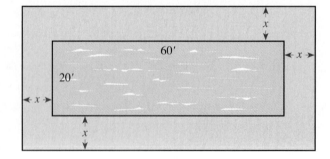

Figure 5.7
Diagram for Example 6

Let's analyze this question carefully, step by step, to develop a strategy for its solution.

THINKING OUT LOUD

What do we need to find?

The width of the walkway

What information are we given?

The dimensions of the pool and the area of the walkway

Is there a connection between what we are given and what we want to find?

From the diagram we see that the area of the outer rectangle equals the area of the pool plus the area of the walkway.

How do we find the area of the outer rectangle?

Since the area of a rectangle is length times width, we need to express the length and width of the outer rectangle in terms of the width x of the walkway.

How do we find x, the width of the walkway?

We can use the area relationship to write an equation and solve for x.

Let's carry out this strategy. We notice that the length of the outer rectangle as you go across from left to right is $x + 60 + x$, which simplifies to $60 + 2x$. Similarly, the width of the outer rectangle from top to bottom is $x + 20 + x$, which simplifies to $20 + 2x$. Thus, in terms of x, the area of the outer rectangle (the pool *and* the walkway) is

$$(20 + 2x)(60 + 2x)$$

The area of the inner rectangle (the pool) is $(60)(20) = 1{,}200$ sq ft, and we are given that the area of the walkway itself is 516 sq ft.

Figure 5.7 gives us the following relationship:

Area of outer rectangle = Area of inner rectangle + Area of walkway
$$(20 + 2x)(60 + 2x) = \qquad 1{,}200 \qquad + \qquad 516$$

We solve the quadratic equation:

$$(20 + 2x)(60 + 2x) = 1{,}200 + 516$$

$$1{,}200 + 160x + 4x^2 = 1{,}716 \qquad \textit{Put in standard form.}$$

$$4x^2 + 160x - 516 = 0 \qquad \textit{Divide both sides of the equation by 4.}$$

$$\frac{4(x^2 + 40x - 129)}{4} = \frac{0}{4}$$

$$x^2 + 40x - 129 = 0 \qquad \textit{Factor.}$$

$$(x - 3)(x + 43) = 0$$

$$x = 3 \quad \text{or} \quad x = -43 \qquad \textit{Eliminate the negative answer. (Why?)}$$

Hence, $x = 3$.

The width of the walkway is $\boxed{3 \text{ ft}}$.

EXAMPLE 7

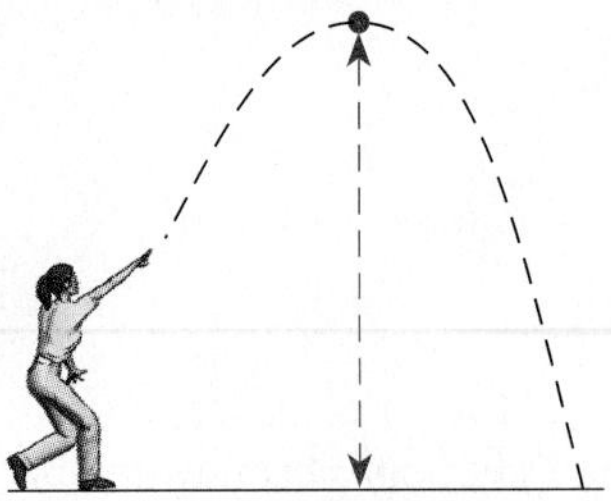

Figure 5.8
Diagram for Example 7

Jenna throws a ball up into the air. The equation

$$h = h(t) = -16t^2 + 30t + 4$$

gives the height, h (in feet), above the ground that the ball reaches t seconds after she throws it (see Figure 5.8).

(a) How far above the ground is the ball exactly 1 second after she throws it? After $\frac{3}{4}$ of a second?

(b) How many seconds will it take for the ball to hit the ground from the time she throws it?

Solution

(a) Since the ball is $h = h(t)$ ft above the ground in t seconds, to find the height of the ball at 1 second, we find $h(1)$:

$$h(t) = -16t^2 + 30t + 4$$
$$h(1) = -16(1)^2 + 30(1) + 4$$
$$= -16 + 30 + 4 = 18$$

The ball is $\boxed{18 \text{ ft above the ground}}$ at 1 second.

At $\frac{3}{4}$ of a second, we find $h\left(\frac{3}{4}\right)$:

$$h(t) = -16t^2 + 30t + 4$$
$$h\left(\tfrac{3}{4}\right) = -16\left(\frac{3}{4}\right)^2 + 30\left(\frac{3}{4}\right) + 4$$
$$= -16\left(\frac{9}{16}\right) + 30\left(\frac{3}{4}\right) + 4$$
$$= -9 + \frac{90}{4} + 4$$
$$= 17\frac{1}{2}$$

Hence, the ball is $\boxed{17\frac{1}{2} \text{ ft above the ground}}$ at $\frac{3}{4}$ of a second.

(b) In this part of the problem, we need to find t, but to find t, we need to know $h(t)$. Since $h(t)$ is the height of the ball above the ground, when the ball hits the ground $h(t)$ must be 0. Hence, we are being asked to find t when $h(t) = 0$.

$$h(t) = -16t^2 + 30t + 4 \qquad \textit{Let } h(t) = 0.$$
$$0 = -16t^2 + 30t + 4 \qquad \textit{Find t: Factor } -2 \textit{ from the right.}$$
$$0 = -2(8t^2 - 15t - 2) \qquad \textit{Divide both sides by } -2.$$
$$\frac{0}{-2} = \frac{-2(8t^2 - 15t - 2)}{-2}$$
$$0 = 8t^2 - 15t - 2$$
$$0 = (8t + 1)(t - 2)$$
$$8t + 1 = 0 \quad \text{or} \quad t - 2 = 0$$
$$8t = -1 \qquad\qquad t = 2$$
$$t = -\frac{1}{8}$$

So $h = 0$ at $t = -\frac{1}{8}$ and at $t = 2$. But $-\frac{1}{8}$ second does not make any sense and we therefore eliminate this answer.

Thus, the ball reaches the ground in ☐ 2 seconds .

Note also that at $t = 0$ seconds, the height, $h(t)$ is 4 ft, indicating that the ball actually starts 4 ft above the ground.

EXERCISES 5.6

In Exercises 1–44, solve the equation.

1. $(x + 2)(x - 3) = 0$

2. $(a - 5)(a + 7) = 0$

3. $0 = (2y - 1)(y - 4)$

4. $0 = (2a + 3)(a - 4)$

5. $x^2 = 25$

6. $a^2 = 81$

7. $x(x - 4) = 0$

8. $y(y + 3) = 0$

9. $12 = x(x - 4)$

10. $15 = y(y + 2)$

11. $5y(y - 7) = 0$

12. $3x(x + 2) = 0$

13. $x^2 - 16 = 0$

14. $a^2 - 225 = 0$

15. $0 = 9c^2 - 16$

16. $0 = 25x^2 - 4$

17. $8a^2 - 18 = 0$

18. $2x^2 = 32$

19. $0 = x^2 - x - 6$

20. $0 = a^2 + 2a - 8$

21. $2y^2 - 3y + 1 = 0$

22. $5a^2 - 14a - 3 = 0$

23. $0 = 8x^2 + 4x - 112$

24. $0 = 6x^2 + 8x - 8$

25. $x^3 + x^2 - 6x = 0$

26. $y^4 - y^2 = 0$

27. $t^4 + 2t^3 + t^2 = 0$

28. $s^3 - s = 0$

29. $20m^3 = 5m$

30. $6b^3 = 9b^4$

31. $x(x - 3) = x^2 - 10$

32. $x(2x - 3) = x^2 + 10$

33. $(t - 4)(t + 1) = (t - 3)(t - 2)$

34. $(a + 2)(a + 5) = (a - 1)(a + 6)$

35. $(2t - 4)(t + 1) = (t - 3)(t - 2)$

36. $(a + 2)(a + 4) = (2a - 1)(a + 6)$

37. $(x + 6)^2 = 16$

38. $(x - 5)^2 = 49$

39. $(3x - 4)^2 = 20 - 24x$

40. $(2x + 3)^2 = 12x + 18$

41. $2(x - 3)(x + 2) = 0$

42. $2(x - 3) + (x + 2) = 0$

43. $2x - 3(x + 2) = 4$

44. $(2x - 3)(x + 2) = 4$

45. If $f(x) = x^2 - 2x - 15$, solve each of the following polynomial equations:

    (a) $f(x) = 0$ **(b)** $f(x) = 9$

46. If $h(x) = 3x^2 + 5x - 2$, solve each of the following polynomial equations:

    (a) $h(x) = 0$ **(b)** $h(x) = 20$

Solve each of the following problems algebraically.

47. Five less than the square of a positive number is 1 more than 5 times the number. Find the number.

48. Seven more than the square of a positive number is 1 less than 6 times the number. Find the number(s).

49. The sum of the squares of two positive numbers is 68. If one of the numbers is 8, find the other number.

50. The sum of the squares of two negative numbers is 30. If one of the numbers is -5, find the other number.

51. Find the numbers such that the square of the sum of the number and 6 is 169.

52. Find the numbers such that the square of the sum of the number and 3 is 100.

53. The sum of a number and its reciprocal is $\frac{13}{6}$. Find the number(s).

54. The sum of a number and its reciprocal is $\frac{25}{12}$. Find the number(s).

55. The sum of a number and twice its reciprocal is 3. Find the number(s).

56. The sum of a number and three times its reciprocal is 4. Find the number(s).

57. The profit in dollars (P) made on a concert is related to the price in dollars (d) of a ticket in the following way:

$$P = 10,000(-d^2 + 12d - 35)$$

(a) How much profit is made by selling tickets for $5?

(b) What must the price of a ticket be to make a profit of $10,000?

58. The hourly cost in dollars (C) of producing sofas in a furniture factory is related to the number of sofas (x) produced in the following way:

$$C(x) = -100x^2 + 1,000x - 1,600$$

(a) How much does it cost per hour to produce 4 sofas?

(b) How many sofas must be produced so that the hourly cost is $800?

59. A man jumps off a diving board into a pool. The equation

$$s = s(t) = -16t^2 + 64$$

gives the distance s (in feet) the man is above the pool t seconds after he jumps.

(a) How high above the pool is the diver after the first second?

(b) How long does it take for him to hit the water?

(c) How high is the diving board? [*Hint:* Let $t = 0$.]

60. Alex throws a ball straight up into the air. The equation

$$s = s(t) = -16t^2 + 80t + 44$$

gives the distance s (in feet) the ball is above the ground t seconds after he throws it.

(a) How high is the ball at $t = 2$ seconds?

(b) How long does it take for the ball to hit the ground?

61. Find the dimensions of a rectangle whose area is 80 sq ft and whose length is 2 ft more than its width.

62. Find the dimensions of a rectangle whose area is 108 sq ft and whose length is 3 ft more than its width.

63. A 20 ft by 55 ft rectangular swimming pool is surrounded by a concrete walkway of uniform width. If the area of the concrete walkway is 400 sq ft, find the width of the walkway.

64. A 21 ft by 21 ft square swimming pool is surrounded by a path of uniform width. If the area of the path is 184 sq ft, find the width of the path.

 Questions for Thought

65. On what property of real numbers is the factoring method based?

66. Discuss what is *wrong* (and why) with the solutions to the following equations:

    (a) Problem: Solve $(x - 3)(x - 4) = 7$.

        Solution: $x - 3 = 7$ or $x - 4 = 7$

                     $x = 10$ or $x = 11$

    (b) Problem: Solve $3x(x - 2) = 0$.

        Solution: $x = 3,$ $x = 0$ or $x = 2$

◆ Mini-Review

67. How long will it take a car traveling at a rate of 52 mph to catch up with a car traveling at a rate of 40 mph with an hour head start?

68. Solve for x. Express your answer using interval notation and graph the solution set on a number line. $-4 \le 3x - 2 < 5$

69. Find the slope of the line whose equation is $3x - 2y = 8$.

70. Write an equation of the line passing through the points $(-4, 0)$ and $(0, 5)$.

5.7   Polynomial Division

In this section we will discuss division of polynomials—in particular, division by binomials and trinomials. We will discuss division by monomials in Chapter 6 within the context of fractional expressions.

    Recall that when we divide numbers using long division, we use the following procedure:

$$
\begin{array}{r}
12 \\
23\overline{)279} \\
23 \\
\hline
49 \\
46 \\
\hline
3
\end{array}
$$

Which is shorthand for

$$
\begin{array}{r}
10 + 2 \\
20 + 3\,\overline{)200 + 70 + 9} \\
200 + 30 \\
\hline
40 + 9 \\
40 + 6 \\
\hline
3
\end{array}
$$

We divide 20 into 200.

We divide 20 into 40.

    First we focus our attention on dividing by 20, but then we multiply the number in the quotient by $20 + 3$, subtract, bring down the next term, and repeat this process. Polynomial division is handled similarly. We will demonstrate by example.

EXAMPLE 1 $(2x^2 + 3x + 3) \div (x + 4)$

        Solution   **1.** Put both polynomials into standard form (arrange terms in descending order of degree).

                                     $x + 4\overline{)2x^2 + 3x + 3}$

2. Divide the highest-degree term of the divisor into the highest-degree term of the dividend:

$$\frac{2x^2}{x} = 2x$$

$$\begin{array}{r} 2x \phantom{{}+3x+3} \\ x+4\overline{)2x^2 + 3x + 3} \end{array}$$

3. Multiply the resulting quotient, $2x$, by the whole divisor:

Multiply.

$$\begin{array}{r} 2x \phantom{{}+3x+3} \\ x+4\overline{)2x^2 + 3x + 3} \\ 2x^2 + 8x \phantom{{}+3} \end{array}$$

$$2x(x+4) = 2x^2 + 8x$$

4. Subtract the result of step 3 from the dividend:

$$\begin{array}{l} 2x^2 + 3x \\ -(2x^2 + 8x) \end{array} \quad \rightarrow \quad \begin{array}{l} 2x^2 + 3x \\ \underline{-2x^2 - 8x} \\ {-5x} \end{array}$$

Change signs of all terms in $2x^2 + 8x$ and add.

$$\begin{array}{r} 2x \phantom{{}+3x+3} \\ x+4\overline{)2x^2 + 3x + 3} \quad \textit{Subtract.} \\ -(2x^2 + 8x) \phantom{{}+3} \\ {-5x} \phantom{{}+3} \end{array}$$

5. Bring down the next term.

$$\begin{array}{r} 2x \phantom{{}+3x+3} \\ x+4\overline{)2x^2 + 3x + 3} \\ -(2x^2 + 8x) \phantom{{}+3} \\ -5x + 3 \end{array}$$

6. Repeat steps 2–5 until no more terms can be brought down.
   (a) Divide x into $-5x$ to get -5
   (b) Multiply -5 by $x + 4$ to get $-5x - 20$.
   (c) Subtract $-5x - 20$ from $-5x + 3$ to get 23.

$$\begin{array}{r} 2x - 5 \\ x+4\overline{)2x^2 + 3x + 3} \\ -(2x^2 + 8x) \phantom{{}+3} \\ -5x + 3 \\ -(-5x - 20) \quad \textit{Subtract.} \\ +\,23 \end{array}$$

7. The remaining term, $+23$, is the remainder.

$$\begin{array}{r} 2x - 5 \\ x+4\overline{)2x^2 + 3x + 3} \\ -(2x^2 + 8x) \phantom{{}+3} \\ -5x + 3 \\ -(-5x - 20) \\ +\,23 \end{array}$$

The process ends when the degree of the remainder is less than the degree of the divisor. In this case, the degree of 23 is 0 and the degree of $x + 4$ is 1.

Answer: $\boxed{2x - 5, \quad \text{Rem. } +23}$

Or, we can rewrite this expression as

$$\boxed{2x - 5 + \frac{23}{x + 4}}$$

We should check the answer in the same way we check long division problems with numbers:

$$\text{Dividend} = (\text{Quotient}) \cdot (\text{Divisor}) + \text{Remainder}$$
$$2x^2 + 3x + 3 \overset{?}{=} (2x - 5)(x + 4) + 23 \qquad \textit{Multiply first.}$$
$$\overset{?}{=} 2x^2 + 3x - 20 + 23$$
$$\overset{\checkmark}{=} 2x^2 + 3x + 3$$

When writing the dividend in standard form, it is a good idea to leave a space between terms whenever a consecutive power is missing, or fill in the power with a coefficient of 0. This stops you from making the error of combining two different powers of x in a subtraction step.

EXAMPLE 2

$(2x^3 + x - 18) \div (x - 2)$

Solution

$$
\begin{array}{r}
2x^2 + 4x + 9 \\
x - 2 \overline{)2x^3 + 0x^2 + x - 18} \\
\underline{-(2x^3 - 4x^2)} \\
4x^2 + x \\
\underline{-(4x^2 - 8x)} \\
9x - 18 \\
\underline{-(9x - 18)} \\
0
\end{array}
$$

Missing power of x in dividend. Write in power with 0 coefficient.

Answer: $\boxed{2x^2 + 4x + 9}$

Check:
$$2x^3 + x - 18 \overset{?}{=} (2x^2 + 4x + 9)(x - 2)$$
$$\overset{?}{=} 2x^3 + 4x^2 + 9x - 4x^2 - 8x - 18$$
$$\overset{\checkmark}{=} 2x^3 + x - 18$$

EXAMPLE 3

$(a^5 + 3a^3 + a^2 + 3a + 7) \div (a^3 + a + 1)$

Solution

$$
\begin{array}{r}
a^2 \quad\quad + 2 \\
a^3 + a + 1 \overline{)a^5 \quad\quad + 3a^3 + a^2 + 3a + 7} \\
\underline{-(a^5 \quad\quad + a^3 + a^2)} \\
2a^3 \quad\quad + 3a + 7 \\
\underline{-(2a^3 \quad\quad + 2a + 2)} \\
a + 5
\end{array}
$$

In this example we left a space for the x^4 term rather than writing $0x^4$.

These last two terms must be brought down.

Answer: $\boxed{a^2 + 2, \quad \text{Rem. } a + 5}$ or $\boxed{a^2 + 2 + \dfrac{a + 5}{a^3 + a + 1}}$

Note that the degree of the remainder must be less than the degree of the divisor.

EXAMPLE 4

Verify that $x + 4$ is a factor of $3x^3 + 10x^2 - 9x - 4$, and apply this result to solve the equation $3x^3 + 10x^2 - 9x - 4 = 0$.

Solution

To verify that $x + 4$ is a factor of $3x^3 + 10x^2 - 9x - 4$, we divide $3x^3 + 10x^2 - 9x - 4$ by $x - 4$ using long division: if the remainder is 0, then $x - 4$ is a factor of $3x^3 + 10x^2 - 9x - 4$.

$$
\begin{array}{r}
3x^2 - 2x - 1 \\
x + 4 \overline{)3x^3 + 10x^2 - 9x - 4} \\
-(3x^3 + 12x^2) \\
\hline
-2x^2 - 9x \\
-(-2x^2 - 8x) \\
\hline
-x - 4 \\
-(-x - 4) \\
\hline
0
\end{array}
$$

Hence $x + 4$ is a factor of $3x^3 + 10x^2 - 9x - 4$ and we can write, $3x^3 + 10x^2 - 9x - 4 = (x + 4)(3x^2 - 2x - 1)$. To solve the equation $3x^3 + 10x^2 - 9x - 4 = 0$, we use the results above and proceed as follows:

$3x^3 + 10x^2 - 9x - 4 = 0$ *Factor the left-hand side.*

$(x + 4)(3x^2 - 2x - 1) = 0$ *We can factor $3x^2 - 2x - 1$ into $(3x + 1)(x - 1)$ and get*

$(x + 4)(3x + 1)(x - 1) = 0$ *Use the zero product rule.*

$x + 4 = 0$ or $3x + 1 = 0$ or $x - 1 = 0$

$$
\boxed{x = -4 \quad \text{or} \quad x = -\frac{1}{3} \quad \text{or} \quad x = 1}
$$

EXERCISES 5.7

Divide the following using polynomial long division.

1. $(x^2 - x - 20) \div (x - 5)$
2. $(3y^2 + 5y + 7) \div (y + 1)$

3. $(3a^2 + 10a + 8) \div (3a + 4)$
4. $(2x^2 + x - 14) \div (2x - 3)$

5. $(21z^2 + z - 16) \div (3z + 1)$
6. $(5a^2 + 16a + 11) \div (5a + 1)$

7. $(a^3 + a^2 + a - 8) \div (a - 1)$
8. $(x^3 + 2x^2 + 2x + 1) \div (x + 1)$

9. $(3x^3 - 11x^2 - 5x + 12) \div (x - 4)$
10. $(y^3 - 2y^2 + 3y - 2) \div (y - 2)$

11. $(4z^2 - 15) \div (2z + 3)$
12. $(4a^2 - 35) \div (2a - 7)$

13. $(x^4 + x^3 - 4x^2 - 3x - 2) \div (x - 2)$
14. $(y^4 + 4y^3 - 2y^2 - 2y + 3) \div (y + 1)$

15. $(-10 + 2y^2 + 5y - 5y^3 + 3y^4) \div (y - 1)$
16. $(-10x^2 + 12 + 9x - 3x^3 + 6x^4) \div (2x - 1)$

17. $(6 - 5a + 2a^4 + 4a^3) \div (a + 2)$
18. $(10x^4 - 6x - 15 + 5x^3) \div (2x + 1)$

19. $(y^3 - 1) \div (y + 1)$
20. $(8a^3 - 1) \div (2a - 1)$

21. $(8a^3 + 1) \div (2a - 1)$
22. $(y^3 + 1) \div (y + 1)$

23. $(2y^5 + 4 - 3y^4 - y^2 + y) \div (y^2 + 1)$
24. $(5x^4 + 1) \div (x - 1)$

25. $(-10 + 12z + 3z^5 + 3z^2 - 4z^3) \div (z^2 + 2z - 1)$

26. $(5y^6 + 17y^2 - 14 + 7y - 9y^3 - 28y^4 + 15y^5) \div (y^2 + 3y - 5)$

27. $(7x^6 + 1 + x - 3x^2 - 14x^4) \div (x^2 - 2)$

28. $(x^4 - y^4) \div (x - y)$

29. Verify that $x - 2$ is a factor of $2x^3 - 3x^2 - 8x + 12$, and apply this result to solve the equation $2x^3 - 3x^2 - 8x + 12 = 0$.

30. Verify that $x - 1$ is a factor of $3x^3 + 5x^2 - 3x - 5$, and apply this result to solve the equation $3x^3 + 5x^2 - 3x - 5 = 0$.

31. Verify that $x - 8$ is a factor of $2x^3 - 17x^2 + 7x + 8$, and apply this result to solve the equation $2x^3 - 17x^2 + 7x + 8 = 0$.

32. Verify that $x + 2$ is a factor of $2x^3 - x^2 - 7x + 6$, and apply this result to solve the equation $2x^3 - x^2 - 7x + 6 = 0$.

 Mini-Review

33. If $f(x) = 5x^2 - 3x + 2$, find $f(4)$ and $f(x^2)$.

CHAPTER 5 SUMMARY

After having completed this chapter you should be able to:

1. Identify the degree and coefficients of the terms in a polynomial, as well as the degree of the polynomial itself (Section 5.1)

For example: The polynomial $5x^6 - 3x + 8$ has *three* terms:

The first term, $5x^6$, has coefficient 5 and degree 6.

The second term, $-3x$, has coefficient -3 and degree 1.

The constant term, 8, has degree 0.

The polynomial has degree 6.

2. Combine and simplify polynomial expressions (Section 5.2).

For example:

(a) $2(3x^2 - x + 3y) - 3(y - x) - (2x^2 - 3y)$
$$= 6x^2 - 2x + 6y - 3y + 3x - 2x^2 + 3y$$
$$= 4x^2 + x + 6y$$

(b) $2x^2y(3xy^2 - 6x^3y^2) - 3x(-2xy^2)(-4x^3y) = 6x^3y^3 - 12x^5y^3 - 24x^5y^3$
$$= 6x^3y^3 - 36x^5y^3$$

3. Multiply polynomials (Section 5.2).

For example:

$(2x - y)(x^2 - 3xy - y^2) = 2x^3 - 6x^2y - 2xy^2 - x^2y + 3xy^2 + y^3$
$$= 2x^3 - 7x^2y + xy^2 + y^3$$

4. Multiply polynomials using general forms and special products (Section 5.3).

For example:

(a) $(3x - 4)^2 = (3x)^2 - 2(3x)(4) + 4^2$
$$= 9x^2 - 24x + 16$$

(b) $[(x - y) - 7][(x - y) + 7] = (x - y)^2 - 7^2$
$$= x^2 - 2xy + y^2 - 49$$

5. Factor various types of polynomials (Sections 5.4, 5.5).

For example:

Common monomial factoring: **(a)** $2x^2 + 7x = x(2x + 7)$

Factoring by grouping: **(b)** $x^2 - 3x - xy + 3y$

$$= x(x - 3) - y(x - 3)$$

$$= (x - 3)(x - y)$$

General forms: **(c)** $y^2 - 10y + 21 = (y - 7)(y - 3)$

   (d) $2x^2 + 7x + 6 = (2x + 3)(x + 2)$

   (e) $6x^5 - 15x^4y + 9x^3y^2$

$$= 3x^3(2x^2 - 5xy + 3y^2)$$

$$= 3x^3(2x - 3y)(x - y)$$

Using special products: **(f)** $16m^2 - 9n^2 = (4m - 3n)(4m + 3n)$ *Difference of two squares*

   (g) $25x^2 - 60x + 36 = (5x - 6)^2$ *Perfect square*

   (h) $x^3 - 8 = x^3 - 2^3$ *Difference of two cubes*

$$= (x - 2)(x^2 + 2x + 4)$$

   (i) $t^2 + 6t + 9 - u^2 = (t + 3)^2 - u^2$ *Group the first three terms.*

$$= (t + 3 - u)(t + 3 + u)$$ *Difference of two squares*

6. Use the factoring method to solve polynomial equations (Section 5.6).

For example: To solve the equation $2x^2 - 30 = 7x$, we proceed as follows:

$$2x^2 - 30 = 7x$$ *Write the quadratic equation in standard form.*

$$2x^2 - 7x - 30 = 0$$ *Factor the left side.*

$$(2x + 5)(x - 6) = 0$$ *Using the zero-product rule, we set each factor equal to 0.*

$$2x + 5 = 0 \quad \text{or} \quad x - 6 = 0$$

$$x = -\frac{5}{2} \quad \text{or} \quad x = 6$$

7. Use long division to divide polynomials (Section 5.7).

For example: $(8x^3 - 28x + 19) \div (2x - 3)$

Solution:

$$
\begin{array}{r}
4x^2 + 6x - 5 \\
2x - 3 \overline{)8x^3 + 0x^2 - 28x + 19} \\
\underline{-(8x^3 - 12x^2)} \\
12x^2 - 28x \\
\underline{-(12x^2 - 18x)} \\
-10x + 19 \\
\underline{-(-10x + 15)} \\
4
\end{array}
$$

Answer: $4x^2 + 6x - 5 + \dfrac{4}{2x - 3}$

8. Solve applications using polynomial equations (Sections 5.1 and 5.6).

CHAPTER 5 REVIEW EXERCISES

In Exercises 1–26, perform the operations and simplify:

1. Subtract the sum of $2x^2 - 3x + 4$ and $5x^2 - 3$ from $2x^3 - 4$.

2. Subtract $2x^3 - 3x^2 - 2$ from the sum of $2x^3 - 3x + 4$ and $3x^2 - 5x + 9$.

3. $3x^2(2xy - 3y + 1)$

4. $5rs^2(3r^2s - 2rs^2 - 3)$

5. $3ab(2a - 3b) - 2a(3ab - 4b^2)$

6. $3xy(2x - 3y + 4) - 2x(5xy - 3y) - 2xy^2$

7. $3x - 2(x - 3) - [x - 2(5 - x)]$

8. $5x - 4\{3 - 2[x - (2 - x)]\}$

9. $(x - 3)(x + 2)$

10. $(x - 5)(x - 8)$

11. $(3y - 2)(2y - 1)$

12. $(5y - 3a)(5y + 3a)$

13. $(3x - 4y)^2$

14. $(5a + 3b)(a - 2b)$

15. $(4x^2 - 5y)^2$

16. $(4a - 3b^2)(4a + 3b^2)$

17. $(7x^2 - 5y^3)(7x^2 + 5y^3)$

18. $(2x - 5)(3x^2 - 2x + 1)$

19. $(5y^2 - 3y + 7)(2y - 3)$

20. $(7x^2 + 3xy - 2y^2)(3x - 2y)$

21. $(2x + 3y + 4)(2x + 3y - 5)$

22. $[(a + b) + (x - y)][(a + b) - (x - y)]$

23. $[(x - y) + 5]^2$

24. $(x - 3 + y)(x - 3 - y)$

25. $(a + b - 4)^2$

26. $(x + y + 5)^2$

In Exercises 27–72, factor the expression completely.

27. $6x^2y - 12xy^2 + 9xy$

28. $15a^2b^2 - 10ab^4 + 5ab^2$

29. $3x(a + b) - 2(a + b)$

30. $5y(y - 1) + 3(y - 1)$

31. $2(a - b)^2 + 3(a - b)$

32. $4(a - b)^2 + 7(a - b)$

33. $5ax - 5a + 3bx - 3b$

34. $2xa + 2xb + 3a + 3b$

35. $5y^2 - 5y + 3y - 3$

36. $14a^2 - 7ab + 6ab - 3b^2$

37. $a^2 + a(a - 10)$

38. $t^3 + t(t - 6)$

39. $2t(t + 2) - (t - 2)(t + 4)$

40. $3m(m - 1) - (2m + 1)(m - 2)$

41. $x^2 - 2x - 35$

42. $a^2 + 5a - 36$

43. $a^2 + 5ab - 14b^2$

44. $y^4 + x^2y^2 - 6x^4$

45. $35a^2 + 17ab + 2b^2$

46. $x^2 - 6xy + 9y^2$

47. $a^2 - 6ab^2 + 9b^4$

48. $2x^3 + 6x^2 - 54x$

49. $3a^3 - 21a^2 + 30a$

50. $5a^3b + 5a^2b^2 - 30ab^3$

51. $2x^3 - 50xy^2$

52. $3y^3 + 24y^2 + 48y$

53. $6x^2 + 5x - 6$

54. $25y^2 - 5y - 12$

55. $8x^3 + 125y^3$

56. $x^3 - 27$

57. $6a^2 - 17ab - 3b^2$

58. $12x^2 + 16xy^2 + 5y^4$

59. $21a^4 + 41a^2b^2 + 10b^4$

60. $54a^4 - 16ab^3$

61. $25x^4 - 40x^2y^2 + 16y^4$

62. $18x^3 + 15x^2y - 18xy^2$

63. $20x^3y - 60x^2y^2 + 45xy^3$

64. $4a^4 - 9b^4$

65. $6a^4b - 8a^3b^2 - 8a^2b^3$

66. $28x^4y - 63x^2y^3$

67. $6x^5 - 10x^3 - 4x$

68. $30x^5y - 85x^3y^2 + 25xy^3$

69. $(a - b)^2 - 4$

70. $(3x - 2)^2 - (5y + 3)^2$

71. $9y^2 + 30y + 25 - 9x^2$

72. $25x^4 - y^2 - 8y - 16$

In Exercises 73–76, find the quotient using long division.

73. $(3x^3 - 4x^2 + 7x - 5) \div (x - 2)$

74. $(x^4 - x - 1) \div (x + 1)$

75. $(8a^3 - 27) \div (2a - 3)$

76. $(y^6 + y^5 + y^4 + y^3 + y^2 + y + 1) \div (y^2 + y + 1)$

In Exercises 77–82, solve the given equation by factoring.

77. $x(x + 5)(3x - 4) = 0$

78. $x^2 - 20 = 8x$

79. $x^4 = 25x^2$

80. $x^3 + 10x^2 - 24x = 0$

81. $(x + 6)(x - 3) = (4x - 2)(x + 4)$

82. $(x + 7)^2 = x + 19$

In Exercises 83–86, given $f(x) = 2x^2 - 5x + 3$ and $g(x) = 3x - 1$, find

83. $f(x)g(x)$

84. $f(x) + g(x)$

85. $f(x + 1)$

86. $g(x) - g(2)$

87. Verify that $x - 3$ is a factor of $2x^3 - x^2 - 18x + 9$, and apply this result to solve the equation $2x^3 - x^2 - 18x + 9 = 0$.

88. Verify that $x + 5$ is a factor of $x^3 + 6x^2 + 3x - 10$, and apply this result to solve the equation $x^3 + 6x^2 + 3x - 10 = 0$.

89. The sum of the squares of the first n natural numbers is given by the formula

$$S(n) = 1^2 + 2^2 + 3^2 + \cdots + n^2 = \tfrac{1}{6}(2n^3 + 3n^2 + n).$$ Factor $S(n)$.

90. The sum of the cubes of the first n natural numbers is given by the formula

$$S(n) = 1^3 + 2^3 + 3^3 + \cdots + n^3 = \tfrac{1}{4}(n^4 + 2n^3 + n^2).$$ Factor $S(n)$.

91. A particular company knows that if the price of an item is c dollars, then it can sell $20{,}000 - 1{,}000c^2$ items.

  (a) Express the revenue, R, as a function of c.

  (b) Use the function R to compute the revenue when the price of an item is \$1, \$2, and \$4.

92. The daily profit, P, made on selling a small toy is related to the price, x, of the toy in the following way:

$$P = P(x) = -1{,}000x^2 + 11{,}000x - 24{,}000$$

  (a) How much daily profit is made selling the toy for \$5?

  (b) What must the price of the toy be to make a daily profit of \$4,000?

CHAPTER 5 PRACTICE TEST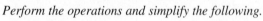

Perform the operations and simplify the following.

1. $(9a^2 + 6ab + 4b^2)(3a - 2b)$

2. $(x - 3)(x + 3) - (x - 3)^2$

3. $(x - y - 2)(x - y + 2)$

4. $(3x + y^2)^2$

Factor the following completely:

5. $10x^3y^2 - 6x^2y^3 + 2xy$

6. $6ax + 15a - 2bx - 5b$

7. $x^2 + 11xy - 60y^2$

8. $10x^2 - 11x - 6$

9. $2x^2 - 3x + 7$

10. $3a^3b - 3ab^3$

11. $r^3s - 10r^2s^2 + 25rs^3$

12. $8a^3 - 1$

13. $(2x + y)^2 - 64$

14. $a^3 + 4a^2 - a - 4$

15. Find the quotient using long division: $(x^3 + 7x - 8) \div (x - 3)$

16. Solve for x.

   (a) $2x^3 + 8x^2 - 24x = 0$

   (b) $(2x - 7)(x - 4) = (4x - 19)(x - 2)$

17. An object is thrown straight up into the air. The equation

$$s = s(t) = -16t^2 + 32t + 48$$

gives the distance, s (in feet), the object is above the ground t seconds after it is thrown.

   (a) How high above the ground is the object at $t = 2$ seconds?

   (b) How long does it take for the ball to hit the ground?

18. Given $f(x) = x^2 - 3x + 2$, find

   (a) $f(5)$ **(b)** $f(x + 5)$

6 Rational Expressions and Functions

We will return to this function in Example 4 of this section.

An environmental consultant for a large factory finds that the cost C (in thousands of dollars) to remove p percent of the pollutants released by the factory can be approximated by the function

$$C(p) = \frac{60p}{100 - p}$$

Understanding how this function behaves would be very useful in determining the monetary investment necessary to make specific improvements in the amount of pollution removed. This cost function $C(p)$ is called a *rational function,* and will be investigated further later in this section.

6.1 Rational Functions

In Chapter 1 we defined the set of rational numbers, Q, in the following way:

$$Q = \left\{ \frac{a}{b} \;\middle|\; a, b \in Z, b \neq 0 \right\}$$

which means that a rational number is any number that can be represented as the quotient of two integers, provided that the denominator is not 0.

Similarly, we define a **rational expression** as a quotient of two polynomials, provided the denominator is not the **zero polynomial.** (Remember that the zero polynomial is simply the number 0.) A **rational function** is defined to be a quotient of two polynomial functions.

DEFINITION

A **rational function** is a function of the form

$$f(x) = \frac{p(x)}{q(x)} \qquad \text{where } p(x) \text{ and } q(x) \text{ are polynomials and } q(x) \neq 0$$

But even if the denominator of a rational expression is not the zero polynomial, we must still be careful about division by 0; a nonzero polynomial could have a value of 0 when certain values are substituted for the variable. For example, the rational expression $\dfrac{3x - 4}{x - 5}$ has the nonzero polynomial $x - 5$ as its denominator. However, since $x - 5$ is equal to 0 when $x = 5$,

$$\frac{3x - 4}{x - 5} \text{ is not defined for } x = 5$$

Alternatively, we can say that the **domain** of the rational function $f(x) = \dfrac{3x - 4}{x - 5}$ is the set of all real numbers *excluding* $x = 5$.

A word about terminology is in order here. While a rational expression is a quotient of polynomials, a **fraction** is *any* quotient of expressions. For example, the following are all examples of fractions:

$$\frac{2x^3 - 3x + 7}{2x^2 + 5}, \qquad \frac{3x - 5}{2x^2 + 8}, \qquad \frac{3x^2 + 2x}{\sqrt{2x - 5}}, \qquad \frac{(2x^2 - 3x + 5)^{1/3}}{2x - 1}$$

The first two fractions are rational expressions. The last two fractions are not rational expressions because they contain expressions that are not polynomials.

EXAMPLE 1

Which value(s) of x must be excluded for each of the following rational expressions?

(a) $\dfrac{6}{2x - 1}$ (b) $\dfrac{x - 4}{5}$

Solution   Essentially, we need to eliminate the values of x that make the denominator equal to 0, and hence the fraction undefined.

(a) $\dfrac{6}{2x - 1}$ is undefined when $2x - 1 = 0$.

Hence $\boxed{\text{we exclude } x = \dfrac{1}{2}}$.

(b) $\dfrac{x - 4}{5}$ $\boxed{\text{is defined for all values of } x}$ because the *denominator* is never equal to 0. Remember that it is permissible for the *numerator* of a fraction to equal 0. If $x = 4$, the fraction becomes $\frac{4-4}{5} = \frac{0}{5}$, which is equal to 0.

EXAMPLE 2   Find the domain of $f(x) = \dfrac{2x + 1}{x^2 - 2x - 8}$.

Solution   Remember that the domain of a function is the set of all values of x that make $f(x)$ defined and real. Hence, we want to find and eliminate the values of x that make the denominator equal to 0 and therefore $f(x)$ undefined. Since $f(x) = \dfrac{2x + 1}{x^2 - 2x - 8}$ is undefined when the denominator is 0, we solve the equation $x^2 - 2x - 8 = 0$:

$$x^2 - 2x - 8 = 0 \qquad \textit{Solve by factoring.}$$
$$(x - 4)(x + 2) = 0 \qquad \textit{Set each factor equal to 0.}$$
$$x - 4 = 0 \quad \text{or} \quad x + 2 = 0$$
$$x = 4 \quad \text{or} \quad x = -2$$

Hence the domain of $f(x)$ is $\boxed{\text{all real numbers except } x = 4 \text{ and } x = -2}$. We can write this domain alternatively as $\{x \mid x \neq 4, x \neq -2\}$.

Just as the integer p may be regarded as a rational number by rewriting it as $\dfrac{p}{1}$, polynomials may also be considered as rational expressions since they can be represented as a quotient of the polynomial and 1.

Let's examine a few situations in which we can use a rational function as a mathematical model.

EXAMPLE 3   The area of a rectangle is 4 sq meters.

(a) Express the length of the rectangle as a function of its width.
(b) Use this function to explain how the length and width for this rectangle are related.

Solution   **(a)** The area of a rectangle is *length* $\times$ *width*. Since the area of the rectangle is 4, we have

$$4 = lw \qquad \textit{Where l = length and w = width}$$

Since we want to express the length as a function of the width, we solve for l to get

$$l = \frac{4}{w} \qquad \textit{We can rewrite this using function notation, replacing l with L(w) (since the length is a function of the width) as}$$

$$L(w) = \frac{4}{w}$$

(b) Let's begin by examining this function algebraically: What does the equation tell us in general about the relationship between the length and width of this rectangle? Consider the equation $L(w) = \dfrac{4}{w}$. What happens to $L(w)$ if we keep changing the numerical value of w, allowing w to get larger and larger? If the denominator of the fraction $\frac{4}{w}$ gets larger and larger, the fraction itself gets smaller and smaller. As w becomes smaller and smaller, the fraction gets larger and larger. Let's construct a table of values for $L(w)$.

Recall from arithmetic that $4 \div \frac{1}{4} = 4 \cdot \frac{4}{1} = 16$. Or you can use your calculator and compute $4 \div \frac{1}{4}$ as $4 \div 0.25$.

Width	Length
w	$\dfrac{4}{w} = L(w)$
$\frac{1}{4}$	$\dfrac{4}{\frac{1}{4}} = 16$
$\frac{1}{2}$	$\dfrac{4}{\frac{1}{2}} = 8$
1	$\frac{4}{1} = 4$
2	$\frac{4}{2} = 2$
4	$\frac{4}{4} = 1$
8	$\frac{4}{8} = \frac{1}{2}$
16	$\frac{4}{16} = \frac{1}{4}$

Note that when $w = 0$, $L(w)$ is undefined (that is, $L(0)$ does not exist), since we are dividing by 0. Observe that as w becomes larger and larger, $L(w)$ becomes smaller and smaller. This table confirms the algebraic relationship we suggested earlier.

Using the table above, we can also plot points and graph this equation to get a "picture" of this relationship (see Figure 6.1).

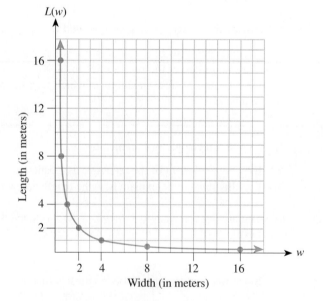

Figure 6.1

The graph of $L(w) = \dfrac{4}{w}$

The graph shows that as w gets bigger and bigger (moving to the right), the $L(w)$ values get smaller and smaller ($L(w)$ gets closer and closer to 0). On the other hand, note that as w gets closer and closer to 0 (moving to the left), the $L(w)$ values get larger and larger ($L(w)$ keeps growing).

EXAMPLE 4

An environmental consultant for a large factory finds that the cost C (in thousands of dollars) to remove p percent of the pollutants released by the factory can be approximated by the function

$$C(p) = \frac{60p}{100 - p}$$

(a) What is the domain of this function?
(b) Create a table of values for this function, and explain what these values mean.
(c) Sketch a graph of this function. Describe the behavior of this cost function. (What is happening to the cost as we get closer and closer to removing 100% of the pollutants?)

Solution

(a) First note that for this function to make sense, p, the percent of pollutants being removed, cannot be less than 0 or greater than 100. We can also see from the equation that p cannot equal 100. (Why?) Hence the domain is $0 \le p < 100$.

(b) The table below shows $C(p)$ for various values of p. $C(p)$ is the cost of removing p percent of the pollutants. For example, $C(50) = 60$ means it costs \$60,000 to remove 50% of the pollutants.

(c) The graph is shown in Figure 6.2.

p	$C(p) = \dfrac{60p}{100 - p} = cost\ (in\ thousands)$
10	$C(10) = \frac{60(10)}{100 - 10} = 6.67$
30	$C(30) = \frac{60(30)}{100 - 30} = 25.71$
50	$C(50) = \frac{60(50)}{100 - 50} = 60$
70	$C(70) = \frac{60(70)}{100 - 70} = 140$
75	$C(75) = \frac{60(75)}{100 - 75} = 180$
80	$C(80) = \frac{60(80)}{100 - 80} = 240$
90	$C(90) = \frac{60(90)}{100 - 90} = 540$
95	$C(95) = \frac{60(95)}{100 - 95} = 1,140$

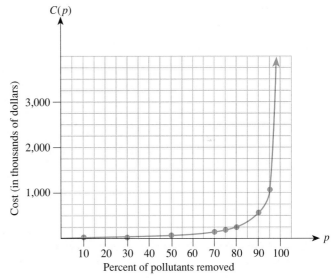

Figure 6.2

We can see that as p, the percent of pollutants removed, increases, the cost to remove this level of pollutants also increases. However, what is interesting to note is that as p increases, the cost increases more and more rapidly. In particular, as p approaches 100, the cost increases enormously. For example, $C(75) = 180$, which means that it costs \$180,000 to remove 75% of the pollutants, and $C(80) = 240$, which means that it costs \$240,000 to remove 80% of the pollutants. Thus, if the factory is already removing 75% of the pollutants, in order to remove an additional 5% of pollutants the factory would require an additional \$60,000.

Note that as the percentage of pollutants removed approaches 100%, the cost becomes prohibitive.

However, $C(90) = 540$ and $C(95) = 1,140$, which means that if the factory is already removing 90% of the pollutants, in order to remove an additional 5% of pollutants, the factory would require an additional \$600,000! The graph shows us that as p approaches 100%, the cost of each additional percent removed is increasing enormously.

EXERCISES 6.1

In Exercises 1–8, what values of the variable(s) should be excluded for the fraction?

1. $\dfrac{x}{x-2}$

2. $\dfrac{x+2}{3y+2}$

3. $\dfrac{3xy^2}{2}$

4. $\dfrac{3x^2z}{5}$

5. $\dfrac{a+b}{5x}$

6. $\dfrac{x^2}{y}$

7. $\dfrac{3xy}{3x-y}$

8. $\dfrac{5ab}{2a-b}$

In Exercises 9–16, find the domain of the given function.

9. $f(x) = \dfrac{3x}{x-8}$

10. $g(x) = \dfrac{x-7}{x+2}$

11. $h(x) = \dfrac{2x+1}{5x-8}$

12. $g(x) = \dfrac{3x-1}{7x-9}$

13. $f(x) = \dfrac{3x-2}{x^2-25}$

14. $h(x) = \dfrac{x-9}{x^2-16}$

15. $f(x) = \dfrac{2x+1}{x^2-x-6}$

16. $g(x) = \dfrac{3x-1}{2x^2-9x-5}$

17. The area of a rectangle is 20 sq ft.

   (a) Express the length of the rectangle as a function of its width.

   (b) Use this function to find the length of this rectangle when the width is 0.5, 1, 2, 4, 5, 10, 15, and 20 ft. Explain how the length and width for this rectangle are related.

18. The area of a rectangle is 100 sq ft.

   (a) Express the length of the rectangle as a function of its width.

   (b) Use this function to find the length of this rectangle when the width is 0.5, 1, 20, 40, 60, 80, 100, and 200 ft. Explain how the length and width for this rectangle are related.

19. A manufacturer finds that the cost C (in dollars) to produce x items is given by $C = 0.4x + 8,000$. It then follows that the average cost per item to produce x items is given by

$$A(x) = \frac{0.4x + 8,000}{x}$$

   (a) Compute $A(5)$, the average per item cost to produce 5 items. Compute $A(10)$, $A(50)$, $A(100)$, $A(200)$, and $A(300)$.

   (b) Use the points found in part **(a)** to sketch a rough graph of this function. Use the graph or a table to explain how the number of items produced is related to the average cost per item.

20. An entomologist uses the function

$$n(d) = \frac{500(d+4)}{0.4d+2}$$

to model the number n of insects present in a colony d days after the initial formation of the colony.

   (a) How many insects are present after 1, 5, 10, 20, 30, 60, and 90 days?

   (b) Use the points found in part **(a)** to sketch a rough graph of this function. Use the graph or a table to explain how the number of insects in the colony changes as time goes by.

21. An ecological study group suggests that under certain conditions the following function relates the number n of deer that can be realistically supported on a acres of foraging land:

$$n(a) = \frac{40a}{0.4a + 8} \qquad \text{for } 0 \leq a \leq 200$$

(a) Compute $n(1)$, $n(10)$, $n(50)$, and $n(100)$, and explain what each of these values means.

(b) Use the points found in part (a) to sketch a rough graph of this function. Use the graph or a table to explain how the number of deer sustained is related to the number of acres of foraging land.

22. A pharmaceutical company finds that when a certain drug is administered to a patient the concentration c of the drug in the bloodstream (in milligrams per liter) h hours after the drug was administered is given by

$$c(h) = \frac{50h}{h^2 + 1}$$

(a) Compute $c(0.25)$, $c(0.5)$, $c(1)$, $c(2)$, $c(6)$, and $c(24)$, and explain what each of these values means.

(b) Use the points found in part (a) to sketch a rough graph of this function. Use the graph or a table to explain how the concentration of the drug in the bloodstream changes as time passes.

23. Suppose that a manufacturer finds that the cost C of managing and storing g gallons of a certain chemical is given by the function

$$C = C(g) = 2g + \frac{5,000}{g}$$

(a) What can the domain of this function be?

(b) Compute $C(1)$, $C(5)$, $C(10)$, $C(20)$, $C(40)$, $C(100)$, $C(200)$, and $C(300)$, and explain what each of these values means.

(c) Use the points found in part (a) to sketch a rough graph of this function. Describe the behavior of this cost function.

24. Suppose that a company finds that the cost per unit C of producing x units is given by the function

$$C = C(x) = 325 + \frac{750}{x^2}$$

(a) What can the domain of this function be?

(b) Compute $C(1)$, $C(3)$, $C(7)$, $C(12)$, $C(15)$, $C(20)$, and $C(30)$, and explain what each of these values means.

(c) Use the points found in part (a) to sketch a rough graph of this function. Describe the behavior of this cost function. What is happening to the cost per unit as the company manufactures more and more units?

6.2   Equivalent Fractions

We know from previous experience with rational expressions that two expressions may look different but may actually be equivalent or represent the same amount. With numerical fractions we can draw pictures to demonstrate that $\frac{2}{3}$ and $\frac{4}{6}$ are equivalent. On the other hand, how do we determine whether $\frac{35x^2}{14x}$, $\frac{5x}{2}$, and $\frac{5x^4}{2x^3}$ are equivalent? With variables involved, we primarily use the Fundamental Principle of Fractions.

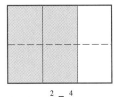

$$\frac{2}{3} = \frac{4}{6}$$

The Fundamental Principle of Fractions	$\dfrac{a \cdot k}{b \cdot k} = \dfrac{a}{b}$ $b, k \neq 0$

This principle says that if we *divide* or *multiply* the numerator and denominator of a fraction by the same nonzero expression, we obtain an equivalent fraction. Reading The Fundamental Principle from left to right is called *reducing fractions to lower terms;* reading it from right to left is called *building fractions to higher terms.*

Reducing Fractions

A fraction *reduced to lowest terms* or *written in simplest form* is a fraction that has no factors (other than ± 1) common to both its numerator and denominator. This requires us to factor both numerator and denominator and then divide out factors common to the numerator and denominator. It is usually quicker to factor out the greatest common factor.

EXAMPLE 1

Express the following in simplest form:

(a) $\dfrac{56}{98}$ (b) $\dfrac{33x^4y^2}{15xy^5}$ (c) $\dfrac{x + y}{x - y}$

Solution

(a) $\dfrac{56}{98} = \dfrac{\not{7} \cdot \not{2} \cdot 2 \cdot 2}{\not{7} \cdot 7 \cdot \not{2}}$ *For convenience, we cross out the factors common to the numerator and denominator and then rewrite the fraction in reduced form.*

$= \boxed{\dfrac{4}{7}}$

Instead of factoring the numerator and denominator completely, we could factor out the greatest common factor and then reduce:

$$\dfrac{56}{98} = \dfrac{4 \cdot \not{14}}{7 \cdot \not{14}}$$ *The greatest common factor is **14**.*

$$= \boxed{\dfrac{4}{7}}$$

(b) $\dfrac{33x^4y^2}{15xy^5} = \dfrac{11 \cdot \not{3} \cdot \not{x} \cdot x \cdot x \cdot x \cdot \not{y} \cdot \not{y}}{5 \cdot \not{3} \cdot \not{x} \cdot \not{y} \cdot \not{y} \cdot y \cdot y \cdot y} = \dfrac{11x^3}{5y^3}$

or

$\dfrac{33x^4y^2}{15xy^5} = \dfrac{11x^3 \cdot \not{3xy^2}}{5y^3 \cdot \not{3xy^2}} = \boxed{\dfrac{11x^3}{5y^3}}$

(c) $\dfrac{x + y}{x - y}$ $\boxed{\text{cannot be reduced}}$.

Remember that the Fundamental Principle of Fractions refers only to common factors. We can divide out common factors but not common terms.

We can now apply the factoring techniques we studied in Chapter 5.

EXAMPLE 2

Express the following in simplest form:

(a) $\dfrac{x^2 - y^2}{(x - y)^2}$ (b) $\dfrac{x - 8}{x^2 - 5x - 24}$

(c) $\dfrac{4x^2 - 4x - 3}{2x^2 - x - 3}$ (d) $\dfrac{16 - x^2}{x^2 - x - 12}$

Solution   **(a)** $\dfrac{x^2 - y^2}{(x - y)^2} = \dfrac{(x - y)(x + y)}{(x - y)(x - y)}$ *First factor; then reduce.*

$$= \dfrac{(x - y)(x + y)}{(x - y)(x - y)}$$

$$= \boxed{\dfrac{x + y}{x - y}}$$

(b) $\dfrac{x - 8}{x^2 - 5x - 24} = \dfrac{x - 8}{(x - 8)(x + 3)}$ *First factor; then reduce.*

$$= \dfrac{x - 8}{(x - 8)(x + 3)}$$

$$= \boxed{\dfrac{1}{x + 3}}$$ *Remember that $\dfrac{x - 8}{x^2 - 5x - 24} = \dfrac{1}{x + 3}$*
for all values of x except 8 and -3. Why?

(c) $\dfrac{4x^2 - 4x - 3}{2x^2 - x - 3} = \dfrac{(2x - 3)(2x + 1)}{(2x - 3)(x + 1)}$ *Factor and reduce.*

$$= \dfrac{(2x - 3)(2x + 1)}{(2x - 3)(x + 1)}$$

$$= \boxed{\dfrac{2x + 1}{x + 1}}$$ *For all values of x except $\tfrac{3}{2}$ and -1*

(d) $\dfrac{16 - x^2}{x^2 - x - 12} = \dfrac{(4 - x)(4 + x)}{(x - 4)(x + 3)}$ *Factor.*

At first glance it appears that there are no common factors, but as we discussed in Chapter 5, $4 - x$ is the negative of $x - 4$; that is,

$$-(x - 4) = -x + 4 = 4 - x$$

Thus, we could factor -1 from *either* $4 - x$ or $x - 4$:

$$\dfrac{(4 - x)(4 + x)}{(x - 4)(x + 3)} = \dfrac{(-1)(x - 4)(4 + x)}{(x - 4)(x + 3)}$$

$$= \boxed{\dfrac{-1(4 + x)}{x + 3} = \dfrac{-4 - x}{x + 3}}$$

A reminder: When we state that two rational expressions or functions are equal, it is always understood that these expressions are equal only for values of x for which both expressions are defined. Thus, in Example 4**(b)**, $\dfrac{x - 8}{x^2 - 5x - 24}$ and $\dfrac{1}{x + 3}$ are equivalent for all values of x *except* $x = 8$ and $x = -3$.

Building Fractions to Higher Terms

In the process of adding fractions with different denominators, we will have to use the Fundamental Principle of Fractions to do the opposite of reducing: ***building to higher terms.*** We demonstrate by example.

EXAMPLE 3

Fill in the question mark to make the two fractions equivalent.

(a) $\dfrac{3x}{5y^2z} = \dfrac{?}{20xy^4z^2}$ **(b)** $\dfrac{2x-3}{x^2-25} = \dfrac{?}{(x-5)(x+5)^2}$

Solution

The Fundamental Principle of Fractions says that two fractions are equivalent if one is obtained by multiplying the denominator and the numerator of the other by the same (nonzero) expression. Thus, all we need to do is to look at what additional factors appear in the denominator of the second fraction and multiply the numerator of the first fraction by these same factors.

(a)

$$\frac{3x}{5y^2z} = \frac{?}{20xy^4z^2}$$

To make $5y^2z$ into $20xy^4z^2$, we need to multiply by $4xy^2z$. Therefore, we must multiply the numerator by $4xy^2z$ as well.

$$\frac{3x(4xy^2z)}{5y^2z(4xy^2z)} = \frac{12x^2y^2z}{20xy^4z^2}$$

The answer is $\boxed{12x^2y^2z}$.

(b)

$$\frac{2x-3}{x^2-25} = \frac{?}{(x-5)(x+5)^2}$$

Factor the denominator of the left fraction.

$$\frac{(2x-3)}{(x-5)(x+5)} = \frac{?}{(x-5)(x+5)^2}$$

Looking at the denominators and moving from left to right, we see that we are missing a factor of $x+5$.

$$\frac{(2x-3)(x+5)}{(x-5)(x+5)(x+5)} = \frac{(2x-3)(x+5)}{(x-5)(x+5)^2}$$

Thus, we must also multiply the numerator by $(x+5)$.

The answer is $\boxed{(2x-3)(x+5) \quad \text{or} \quad 2x^2+7x-15}$.

Signs of Fractions

In a rational expression, a sign may precede any of the following: the numerator, the denominator, or the entire fraction. In general,

$$\frac{a}{b} = -\frac{a}{-b} = -\frac{-a}{b} = \frac{-a}{-b}$$

and

$$-\frac{a}{b} = \frac{-a}{b} = \frac{a}{-b} = -\frac{-a}{-b}$$

You may check that these are equal by letting $a=6$ and $b=3$. Note that if you change *exactly* two of the three signs of a fraction, the result will be an equivalent fraction.

Of the three equivalent forms,

$$-\frac{x}{3} = \frac{-x}{3} = \frac{x}{-3}$$

the first two are usually the preferred forms.

EXERCISES 6.2

In Exercises 1–36, reduce to lowest terms.

1. $\dfrac{3x^2y}{15xy}$

2. $\dfrac{24a^3b^2}{36ab^4}$

3. $\dfrac{16x^2y^3a}{18xy^4a^3}$

4. $\dfrac{87a^2b^3}{57a^5b^9}$

5. $\dfrac{(x-5)(x+4)}{(x-3)(x+4)}$

6. $\dfrac{(x-2)(x+1)}{(x-2)(x+3)}$

7. $\dfrac{(2x-3)(x-5)}{(2x+3)(x+3)}$

8. $\dfrac{(5y-1)(2y-7)}{(5y+1)(3y+2)}$

9. $\dfrac{(2x-3)(x-5)}{(2x-3)(5-x)}$

10. $\dfrac{(5y-1)(3y+2)}{(1-5y)(3y+2)}$

11. $\dfrac{3x^2-3x}{6x^2+18x}$

12. $\dfrac{6a^2-3a}{15a^2-15a}$

13. $\dfrac{x^2-9}{x^2-6x+9}$

14. $\dfrac{x^2-6x+9}{x^2-9}$

15. $\dfrac{w^2-8wz+7z^2}{w^2+8wz+7z^2}$

16. $\dfrac{x^2-4xy+4y^2}{x^2+4xy+4y^2}$

17. $\dfrac{x^2+x-12}{x-3}$

18. $\dfrac{a^2+2a-63}{a-7}$

19. $\dfrac{x^2+x-12}{3-x}$

20. $\dfrac{a^2+2a-63}{7-a}$

21. $\dfrac{3a^2-13a-30}{15a^2+28a+5}$

22. $\dfrac{3a^2-a-2}{2a^2+a-3}$

23. $\dfrac{2x^3-2x^2-12x}{3x^2-6x}$

24. $\dfrac{10a^3-45a^2-25a}{2a^2-10a}$

25. $\dfrac{2x^2}{3x^2-6}$

26. $\dfrac{5a^2}{2a^2-10}$

27. $\dfrac{3a^2-7a-6}{3+5a-2a^2}$

28. $\dfrac{2x^2-3x-2}{2+5x-3x^2}$

29. $\dfrac{6r^2-r-2}{2+r-6r^2}$

30. $\dfrac{15y^2-y-2}{2+y-15y^2}$

31. $\dfrac{a^3+b^3}{(a+b)^3}$

32. $\dfrac{a^3+b^3}{a^2+b^2}$

33. $\dfrac{ax+bx-2ay-2by}{4xy-2x^2}$

34. $\dfrac{4x^2-8x+3}{4x^2-6x+2xy-3y}$

35. $\dfrac{(a+b)^2-(x+y)^2}{a+b+x+y}$

36. $\dfrac{x-1}{x^3+x^2-x-1}$

In Exercises 37–48, fill in the missing expression.

37. $\dfrac{2x}{3y}=\dfrac{?}{9x^2y}$

38. $\dfrac{5a}{3b}=\dfrac{?}{21a^2b}$

39. $\dfrac{5}{3a^2b}=\dfrac{?}{15a^4b^2}$

40. $\dfrac{8}{5x^2y}=\dfrac{?}{20x^3y^3}$

41. $\dfrac{3x}{x-5}=\dfrac{?}{(x-5)(x+5)}$

42. $\dfrac{2z}{z-6}=\dfrac{?}{(z-6)^2}$

43. $\dfrac{a-b}{x+y}=\dfrac{?}{x^2+2xy+y^2}$

44. $\dfrac{x-y}{a-b}=\dfrac{?}{a^2-b^2}$

45. $\dfrac{y-2}{2y-3}=\dfrac{?}{12-5y-2y^2}$

46. $\dfrac{x+1}{2x-5}=\dfrac{?}{15-x-2x^2}$

47. $\dfrac{x-y}{a^2-b^2}=\dfrac{?}{a^2x+a^2y-b^2x-b^2y}$

48. $\dfrac{5x+y}{r-s}=\dfrac{?}{r^2a+r^2b-s^2a-s^2b}$

❓ Questions for Thought

49. Discuss what is **wrong** (if anything) with the following:

(a) $\dfrac{\cancel{a^2}+\cancel{b^2}}{\cancel{a}+\cancel{b}}\overset{?}{=}a+b$

(b) $\dfrac{a^3+b^3}{(a+b)^3}\overset{?}{=}1$

(c) $\dfrac{ax + bx + 2ay + 2by}{a + b} \stackrel{?}{=} \dfrac{(a + b)x + (a + b)2y}{a + b}$

$\stackrel{?}{=} \dfrac{(a + b)x + (a + b)2y}{a + b}$

$\stackrel{?}{=} x + (a + b)2y$

50. Evaluate the following expressions for $y = 0, 1, 2, 3,$ and 4.

$$\dfrac{2y^2 - y - 6}{y^2 - y - 2} \quad \text{and} \quad \dfrac{2y + 3}{y + 1}$$

How do the values compare for each expression?

Mini-Review

51. *Solve for x:* $10(x + 24) - 6x = 400$

52. Write an equation of the line that passes through the point $(-5, 3)$ and is perpendicular to the line whose equation is $4x - 5y = 7$.

53. Given $f(x) = 2x^3 - x^2 + 3$, find $f(x^2)$.

6.3 # Multiplication and Division of Rational Expressions

Multiplication

Now that we have some experience with rational expressions, we can examine the arithmetic operations with these expressions. We begin with multiplication and division, since performing these operations is more straightforward than addition and subtraction.

We define multiplication as follows:

Multiplication of Fractions	
	$\dfrac{a}{b} \cdot \dfrac{c}{d} = \dfrac{a \cdot c}{b \cdot d} \qquad b, d \neq 0$

The product of fractions is defined as the product of the numerators divided by the product of the denominators, provided neither denominator is 0. For example:

$$\dfrac{3x}{y} \cdot \dfrac{4x^2}{7y^3} = \dfrac{(3x)(4x^2)}{(y)(7y^3)} = \dfrac{12x^3}{7y^4}$$

$$\dfrac{x - y}{2x + y} \cdot \dfrac{3x - 2y}{x + y} = \dfrac{(x - y)(3x - 2y)}{(2x + y)(x + y)} = \dfrac{3x^2 - 5xy + 2y^2}{2x^2 + 3xy + y^2}$$

Now that we have defined multiplication, we can take another look at the Fundamental Principle of Fractions. By the multiplicative identity property, we have

$$\dfrac{a}{b} = \dfrac{a}{b} \cdot 1 \qquad \textit{Multiplication by 1 does not change the value of an expression.}$$

$$= \dfrac{a}{b} \cdot \dfrac{k}{k} \qquad \textit{Since } \dfrac{k}{k} = 1, \ (k \neq 0)$$

$$= \dfrac{a \cdot k}{b \cdot k}$$

which is the Fundamental Principle of Fractions. Thus, the Fundamental Principle of Fractions can be viewed as simply stating that multiplying by 1 does not change the value of an expression.

In the previous section we required that our final answer be reduced to lowest terms. In the process of multiplication, factors of each numerator remain in the numerator of the product and factors of each denominator remain in the denominator of the product. Therefore, it is much more efficient to reduce by any common factors before we actually carry out the multiplication.

EXAMPLE 1

Perform the operations. Express your answer in simplest form.

(a) $\dfrac{24x^2y}{13a^3b^2} \cdot \dfrac{26ab^3}{3xy^2}$ **(b)** $\dfrac{x + 2y}{x - y} \cdot \dfrac{y - x}{x - 2y}$

(c) $\left(\dfrac{x^2 + x}{x^2 - 1}\right)\left(\dfrac{x}{x + 2}\right)(x^2 + x - 2)$

Solution

(a) $\dfrac{24x^2y}{13a^3b^2} \cdot \dfrac{26ab^3}{3xy^2} = \dfrac{\overset{8\ x}{\cancel{24x^2y}}}{\underset{a^2}{\cancel{13a^3b^2}}} \cdot \dfrac{\overset{2\ b}{\cancel{26ab^3}}}{\underset{y}{\cancel{3xy^2}}}$

$= \boxed{\dfrac{16bx}{a^2y}}$

(b) $\dfrac{x + 2y}{x - y} \cdot \dfrac{y - x}{x - 2y} = \dfrac{x + 2y}{x - y} \cdot \dfrac{(-1)(x - y)}{x - 2y}$ *Since $y - x = (-1)(x - y)$*

$= \dfrac{x + 2y}{\cancel{x - y}} \cdot \dfrac{-1(\cancel{x - y})}{x - 2y}$ *Then reduce.*

$= \boxed{\dfrac{-(x + 2y)}{x - 2y} = \dfrac{-x - 2y}{x - 2y} = -\dfrac{x + 2y}{x - 2y}}$

Writing in the understood denominator of 1 helps you to avoid errors in reducing fractions.

(c) $\left(\dfrac{x^2 + x}{x^2 - 1}\right)\left(\dfrac{x}{x + 2}\right)(x^2 + x - 2)$ *Factor and reduce.*

$= \dfrac{x(\cancel{x + 1})}{(\cancel{x + 1})(\cancel{x - 1})} \cdot \dfrac{x}{\cancel{x + 2}} \cdot \dfrac{(\cancel{x + 2})(\cancel{x - 1})}{1}$ *Note the 1 in the denominator.*

$= \dfrac{x^2}{1} = \boxed{x^2}$

Division

We can now proceed to formulate the rule for division of fractions. Just as subtraction is the inverse operation of addition, division is the inverse operation of multiplication. That is, $38 \div 2$ is 19 because $19 \cdot 2 = 38$. Thus,

$$38 \div 2 = 19 \quad \text{because} \quad 19 \cdot 2 = 38$$

and

$$\frac{a}{b} \div \frac{c}{d} = \frac{ad}{bc} \quad \text{because} \quad \frac{ad}{bc} \cdot \frac{c}{d} = \frac{a}{b}$$

We rewrite

$$\frac{ad}{bc} \quad \text{as} \quad \frac{a}{b} \cdot \frac{d}{c}$$

Hence, we have the rule for division given in the next box.

| Division of Fractions | $\dfrac{a}{b} \div \dfrac{c}{d} = \dfrac{a}{b} \cdot \dfrac{d}{c}$ $b, c, d \neq 0$ |

Recall from Chapter 1 that the multiplicative inverse of a number is called its *reciprocal*. We usually define reciprocal as follows:

DEFINITION

If $x \neq 0$, the **reciprocal** of x is $\dfrac{1}{x}$.

Thus, the reciprocal of 5 is $\dfrac{1}{5}$.

The reciprocal of $x^2 + 2$ is $\dfrac{1}{x^2 + 2}$.

The reciprocal of $\dfrac{a}{b}$ is $\dfrac{1}{\dfrac{a}{b}}$, which is $1 \div \dfrac{a}{b}$, and, by definition of division,

$$1 \div \dfrac{a}{b} = 1 \cdot \dfrac{b}{a} = \dfrac{b}{a}$$

Hence, if $a \neq 0$ and $b \neq 0$, the reciprocal of $\dfrac{a}{b}$ is $\dfrac{b}{a}$.

We can state the rule for division as follows:

To divide **by** a fraction, multiply by its reciprocal.

EXAMPLE 2

Perform the operations. Express your answer in simplest form.

(a) $\dfrac{16x^3}{9y^4} \div \dfrac{32x^6}{27y^3}$ (b) $\dfrac{3x^2 - 5xy + 2y^2}{2x^2 + 3xy + y^2} \div \dfrac{x - y}{x^2 + xy}$

Solution

(a) $\dfrac{16x^3}{9y^4} \div \dfrac{32x^6}{27y^3}$ *Dividing by $\dfrac{32x^6}{27y^3}$ means multiplying by $\dfrac{27y^3}{32x^6}$.*

$= \dfrac{16x^3}{9y^4} \cdot \dfrac{27y^3}{32x^6}$ *Reduce.*

$= \dfrac{\overset{}{16x^3}}{\underset{y}{9y^4}} \cdot \dfrac{\overset{3}{27y^3}}{\underset{2\,x^3}{32x^6}} = \boxed{\dfrac{3}{2x^3y}}$

(b) $\dfrac{3x^2 - 5xy + 2y^2}{2x^2 + 3xy + y^2} \div \dfrac{x - y}{x^2 + xy}$ *Multiply by the reciprocal of the divisor (the expression you are dividing by).*

$= \dfrac{3x^2 - 5xy + 2y^2}{2x^2 + 3xy + y^2} \cdot \dfrac{x^2 + xy}{x - y}$ *Then factor and reduce.*

$= \dfrac{(3x - 2y)(x - y)}{(2x + y)(x + y)} \cdot \dfrac{x(x + y)}{x - y}$

$= \boxed{\dfrac{x(3x - 2y)}{2x + y}} = \dfrac{3x^2 - 2xy}{2x + y}$

Just as operations on polynomials can be applied to polynomial functions, so too operations on rational expressions can be applied to rational functions.

EXAMPLE 3

Given $f(x) = \dfrac{x + 1}{x - 1}$ and $g(x) = x^2 - 1$. Find

(a) $f(x) \cdot g(x)$ **(b)** $f(x) \div g(x)$

Solution   **(a)** $f(x) \cdot g(x) = \dfrac{x + 1}{x - 1} \cdot (x^2 - 1)$ *Factor and reduce.*

$$= \frac{x + 1}{\cancel{x - 1}} \cdot \frac{(x + 1)(\cancel{x - 1})}{1}$$

$$= \boxed{(x + 1)^2}$$

(b) $f(x) \div g(x) = \dfrac{x + 1}{x - 1} \div (x^2 - 1)$ *Follow the rule for division, and factor.*

$$= \frac{\cancel{x + 1}}{x - 1} \cdot \frac{1}{(\cancel{x + 1})(x - 1)}$$

$$= \boxed{\dfrac{1}{(x - 1)^2}}$$

EXERCISES 6.3

In Exercises 1–40, perform the indicated operations. Reduce all answers to lowest terms.

1. $\dfrac{10a^2b}{3xy^3} \cdot \dfrac{9xy^4}{5a^4b^7}$

2. $\dfrac{32a^2}{7xb^2} \cdot \dfrac{21x^2b^4}{8a^3}$

3. $\dfrac{17a^2b^3}{18yx} \div \dfrac{34a^2}{9xy}$

4. $\dfrac{38xy^2z}{81a^2b^3} \div \dfrac{19ac^2}{27y^2}$

5. $\dfrac{16x^3b}{9a} \cdot 24a^2b^3$

6. $24a^2b^3 \cdot \dfrac{16x^3b}{9a}$

7. $\dfrac{32r^2s^3}{12a^2b} \cdot \left(\dfrac{15ab^2}{16r} \cdot \dfrac{24rs^2}{5ab}\right)$

8. $\dfrac{32x^2c}{12a^2b} \cdot \left(\dfrac{17a^2x^2}{34b^2c} \cdot \dfrac{16x}{9a}\right)$

9. $\dfrac{32r^2s^3}{12a^2b} \div \left(\dfrac{15ab^2}{16r} \cdot \dfrac{24rs^2}{5ab}\right)$

10. $\dfrac{32x^2c}{12a^2b} \div \left(\dfrac{17a^2x^2}{34b^2c} \cdot \dfrac{16x}{9a}\right)$

11. $\dfrac{x^2 - x - 2}{x + 3} \div \dfrac{3x + 9}{2x + 2}$

12. $\dfrac{4x + 8}{3x - 6} \cdot \dfrac{6x - 12}{8x + 16}$

13. $\dfrac{x^2 - x - 2}{x + 3} \cdot \dfrac{3x + 9}{2x + 2}$

14. $\dfrac{4x + 8}{3x - 6} \div \dfrac{6x - 12}{8x + 16}$

15. $\dfrac{5a^3 - 5a^2b}{3a^2 + 3ab} \cdot (a + b)$

16. $\dfrac{x - 3}{2x^2 - 5x - 3} \cdot (8x + 4)$

17. $\dfrac{5a^3 - 5a^2b}{3a^2 + 3ab} \div (a + b)$

18. $(8x + 4) \div \dfrac{x - 3}{2x^2 - 5x - 3}$

19. $\dfrac{2x^2 + 3x - 5}{2x + 5} \cdot \dfrac{1}{1 - x}$

20. $\dfrac{2y^2 - 9y - 5}{2y + 1} \cdot \dfrac{1}{5 - y}$

21. $\dfrac{2a^2 - 7a + 6}{4a^2 - 9} \cdot \dfrac{4a^2 + 12a + 9}{a^2 - a - 2}$

22. $\dfrac{2x^2 - 5x - 12}{3x^2 - 11x - 4} \cdot \dfrac{3x^2 - 14x - 5}{2x^2 - 7x - 15}$

23. $\dfrac{x^2 - y^2}{(x + y)^3} \cdot \dfrac{(x + y)^2}{(x - y)^2}$

24. $\dfrac{2r^2 + rs - 3s^2}{r^2 - s^2} \cdot \dfrac{2r - 2s}{2r^2 + 5rs + 3s^2}$

25. $\dfrac{9x^2 + 3x - 2}{6x^2 - 2x} \div \dfrac{3x + 2}{6x^2}$

26. $\dfrac{2a^2 - 5a - 3}{4a^2 + 2a} \div \dfrac{2a + 1}{4a}$

27. $\dfrac{2x^2 + x - 3}{x^2 - 1} \div \dfrac{2x^2 + 5x + 3}{2 - 2x}$

28. $\dfrac{9x^2 - 9x + 2}{2x - 6x^2} \cdot \dfrac{6x^3}{3x - 2}$

29. $\dfrac{6q^2 - q - 2}{8q^2 + 4q} \cdot \dfrac{8q^2}{6q^2 - 4q}$

30. $\dfrac{4a^2 - 4a - 3}{8a + 4a^2} \cdot \dfrac{16a^2}{4a^2 - 6a}$

31. $\dfrac{m^2 - 10m + 25}{m^2 - 25} \div (5m^2 - 25m)$

32. $\dfrac{2c^3 - 18c^2}{9c^2} \cdot (3c^2 - 9c)$

33. $\left(\dfrac{2t^2 + 3t + 1}{3t^2 - t - 2}\right) \cdot \left(\dfrac{t - 1}{2t - 1}\right) \cdot \left(\dfrac{3t + 2}{t + 1}\right)$

34. $\left(\dfrac{6x^2 + x - 2}{4x^2 - 8x + 3}\right) \cdot \left(\dfrac{x - 1}{3x + 2} \cdot \dfrac{8x - 12}{2x - 2}\right)$

35. $\dfrac{9a^2 + 9a + 2}{3a^2 - 2a - 1} \cdot \left(\dfrac{a - 1}{3a^2 + 4a + 1} \div \dfrac{3a + 2}{a + 1}\right)$

36. $\dfrac{9a^2 + 9a + 2}{3a^2 - 2a - 1} \div \left(\dfrac{a - 1}{3a^2 + 4a + 1} \cdot \dfrac{3a + 2}{a + 1}\right)$

37. $\dfrac{2r^2 - 5r - 3}{6r - 2} \div \left(\dfrac{r - 3}{2} \div \dfrac{3r - 1}{2r + 1}\right)$

38. $\left(\dfrac{2r^2 - 5r - 3}{6r - 2} \div \dfrac{r - 3}{2}\right) \div \dfrac{3r - 1}{2r + 1}$

39. $\dfrac{x^2 - y^2}{y^3 x - x^3 y} \div (x^2 + 2xy + y^2)$

40. $\dfrac{x^2 - y^2}{y^3 x - x^3 y} \cdot (x^2 + 2xy + y^2)$

In Exercises 41–46, use $f(x) = \dfrac{x + 2}{x - 3}$, $g(x) = 2x^2 - 2x - 12$, *and* $h(x) = \dfrac{1}{3x^2 + 6x}$.

41. $f(x) \cdot g(x)$

42. $f(x) \cdot h(x)$

43. $f(x) \div g(x)$

44. $g(x) \div f(x)$

45. $h(x) \div f(x) \cdot g(x)$

46. $xh(x) \cdot g(x)$

 Mini-Review

47. *Solve the following system of equations.*

$$\begin{cases} 6x - 2y = 7 \\ 16x + 3y = 2 \end{cases}$$

48. Express the area of the shaded region of the figure in terms of x.

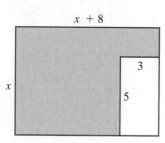

49. What is the degree of the polynomial $-4x^3 + 6x^2 - 5x + 2$?

50. *Solve for x:* $|2x - 3| = |3x + 4|$

6.4   # Sums and Differences of Rational Expressions

The process of adding and subtracting fractions with the same or *common denominators* can be demonstrated by using the distributive property along with the definition of rational multiplication as follows:

$$\frac{a}{c} + \frac{b}{c} = a\left(\frac{1}{c}\right) + b\left(\frac{1}{c}\right) \qquad \textit{Rational multiplication}$$

$$= (a + b)\left(\frac{1}{c}\right) \qquad \textit{The distributive property}$$

$$= \frac{a + b}{c} \qquad \textit{Rational multiplication}$$

Hence, we have the rules stated in the box.

Addition and Subtraction of Rational Expressions	$\dfrac{a}{c} + \dfrac{b}{c} = \dfrac{a + b}{c}$ and $\dfrac{a}{c} - \dfrac{b}{c} = \dfrac{a - b}{c}$

In combining fractions with common denominators, we simply combine the numerators and place this result over the common denominator.

EXAMPLE 1

Perform the operations.

(a) $\dfrac{5y}{x} + \dfrac{3}{x}$ (b) $\dfrac{2x - 3}{x - 2} - \dfrac{x - 1}{x - 2}$

(c) $\dfrac{4x^2 - 33x}{x^2 + x - 2} + \dfrac{2x - 6}{x^2 + x - 2} - \dfrac{(3x - 2)(x - 7)}{x^2 + x - 2}$

Solution

(a) $\dfrac{5y}{x} + \dfrac{3}{x} = \boxed{\dfrac{5y + 3}{x}}$

(b) $\dfrac{2x - 3}{x - 2} - \dfrac{x - 1}{x - 2}$ *Remember, we are subtracting the quantity $x - 1$ and therefore must enclose it in parentheses.*

$= \dfrac{2x - 3 - (x - 1)}{x - 2}$ *Remove parentheses.*

$= \dfrac{2x - 3 - x + 1}{x - 2}$ *Observe the signs of the terms in the numerator. Combine terms in the numerator.*

$= \dfrac{x - 2}{x - 2}$ *Then reduce.*

$= \boxed{1}$

(c) $\dfrac{4x^2 - 33x}{x^2 + x - 2} + \dfrac{2x - 6}{x^2 + x - 2} - \dfrac{(3x - 2)(x - 7)}{x^2 + x - 2}$ *Combine numerators.*

$= \dfrac{4x^2 - 33x + (2x - 6) - (3x - 2)(x - 7)}{x^2 + x - 2}$ *Remember that you are subtracting a product; find the product first.*

$= \dfrac{4x^2 - 33x + 2x - 6 - (3x^2 - 23x + 14)}{x^2 + x - 2}$ *Then subtract.*

$= \dfrac{4x^2 - 33x + 2x - 6 - 3x^2 + 23x - 14}{x^2 + x - 2}$ *Next, combine terms.*

$= \dfrac{x^2 - 8x - 20}{x^2 + x - 2}$ *Factor and Reduce*

$= \dfrac{(x + 2)(x - 10)}{(x - 1)(x + 2)}$

$= \boxed{\dfrac{x - 10}{x - 1}}$

Our last step should be to reduce the answer to lowest terms. Also, keep in mind that we can reduce only common factors, *not* common terms.

The next type of problem to consider is one in which the denominators are different. If the denominators are different, we use the Fundamental Principle of Fractions to build new fractions. The new fractions must be equivalent to the original fractions and all have the same denominator. Once the denominators are the same, we can add the fractions as we did before. The idea is to find an expression that can serve as a common denominator.

This new denominator must be divisible by the original denominators and, for convenience, it should be the "smallest" expression divisible by the original denominators. By "smallest" we mean having the least number of factors. Hence, we call the smallest expression divisible by all denominators in question the **least common denominator,** or **LCD.** To find the LCD, we follow the procedure outlined in the accompanying box.

Procedure for Finding the LCD	1. Factor each denominator *completely.* 2. The LCD consists of the product of each *distinct* factor the *maximum* number of times it appears in any *one* denominator.

EXAMPLE 2

Find the LCD of the following:

(a) $\dfrac{3}{14x^3}, \quad \dfrac{5}{21x^2y^4}, \quad \dfrac{1}{12x^4y^3}$ (b) $\dfrac{3}{x^2 - 9}, \quad \dfrac{x + 1}{x^2 + 6x + 9}$

Solution

(a) The LCD of $\dfrac{3}{14x^3}, \quad \dfrac{5}{21x^2y^4}, \quad$ and $\dfrac{1}{12x^4y^3} \quad$ is found in the following way:

1. Factor each denominator as completely as possible.

$$14x^3 = 7 \cdot 2xxx$$
$$21x^2y^4 = 7 \cdot 3xxyyyy$$
$$12x^4y^3 = 2 \cdot 2 \cdot 3xxxxyyy$$

2. Write each factor the *maximum* (not the total) number of times it appears in any one denominator.

The LCD is $7 \cdot 2 \cdot 2 \cdot 3xxxxyyyy = \boxed{84x^4y^4}$.

7   appears *once* in the first two denominators.
2   appears *twice* in the third denominator and once in the first denominator.
3   appears *once* in the second and third denominators.
x   appears *four* times in the third denominator, three times in the first denominator, and twice in the second denominator.
y   appears *four* times in the second denominator and three times in the third denominator.

Looking at the LCD and the denominators in factored form, note that all three denominators appear within the LCD. This means that the LCD is divisible by all three denominators. Also note that each factor in the LCD is necessary. For example, we cannot drop a factor of *y* or else $21x^2y^4$ will not divide into the LCD; we cannot drop a factor of *x* or $12x^4y^3$ will not divide into the LCD. The fact that there are no "extra" factors makes it the *least* common denominator.

(b) The LCD of $\dfrac{3}{x^2 - 9} \quad$ and $\dfrac{x + 1}{x^2 + 6x + 9} \quad$ is found as follows:

1. Factor each denominator completely.

$$x^2 - 9 = (x + 3)(x - 3)$$
$$x^2 + 6x + 9 = (x + 3)(x + 3)$$

That is, $x - 3$ appears once in the first denominator and $x + 3$ appears twice in the second denominator.

2. Write each factor the maximum number of times it appears in any one denominator.

The LCD is $(x - 3)(x + 3)(x + 3) = \boxed{(x - 3)(x + 3)^2}$.

Now let's see how we can use the LCD and the Fundamental Principle of Fractions to add or subtract fractions with unlike denominators. We can add fractions with unlike denominators by following four general steps.

To Combine Fractions with Different Denominators

1. Find the LCD.
2. Build each fraction to higher terms with the LCD in each denominator.
3. Combine, as with fractions with common denominators.
4. Simplify, if possible

EXAMPLE 3

Perform the indicated operations.

(a) $\dfrac{3}{5x^2 y} - \dfrac{2}{3xy^2} + x$ **(b)** $\dfrac{3x}{x^2 - 9} + \dfrac{5}{2x^3 - 6x^2}$ **(c)** $\dfrac{5x - 10}{x - 4} + \dfrac{3x - 2}{4 - x}$

Solution

(a) First find the LCD:

$$\text{LCD:} \quad 5 \cdot 3 \cdot x^2 y^2$$

Build fractions to higher terms with the LCD, $5 \cdot 3 \cdot x^2 y^2$, in each denominator. Look at $5 \cdot 3 \cdot x^2 y^2$ and determine what factors are missing.

$$\frac{3}{5x^2 y} = \frac{3\ (3y)}{5x^2 y\ (3y)}$$

Multiply numerator and denominator by the missing factors to make $5x^2 y$ into $5 \cdot 3 \cdot x^2 y^2$.

$$\frac{2}{3xy^2} = \frac{2\ (5x)}{3xy^2\ (5x)}$$

Multiply numerator and denominator by the missing factors to make $3xy^2$ into $5 \cdot 3 \cdot x^2 y^2$.

$$\frac{x}{1} = \frac{x\ (5 \cdot 3 \cdot x^2 y^2)}{1\ (5 \cdot 3 \cdot x^2 y^2)}$$

Rewrite x as $\dfrac{x}{1}$ and fill in the missing factors to make 1 into $5 \cdot 3 \cdot x^2 y^2$.

Therefore we have

$$\frac{3}{5x^2 y} - \frac{2}{3xy^2} + x = \frac{3(3y) - 2(5x) + x(5 \cdot 3 \cdot x^2 y^2)}{5 \cdot 3 \cdot x^2 y^2}$$

Place numerators over the LCD and combine.

$$= \boxed{\frac{9y - 10x + 15x^3 y^2}{15x^2 y^2}}$$

Your last step should be to simplify (reduce) your answer. In this case, the answer is already in simplest form.

(b) The LCD is found by factoring the denominators and writing each factor the maximum number of times it appears in any denominator:

$$x^2 - 9 = (x - 3)(x + 3)$$
$$2x^3 - 6x^2 = 2x^2(x - 3)$$

Thus, the LCD is $2x^2(x - 3)(x + 3)$.

$$\frac{3x}{x^2 - 9} + \frac{5}{2x^3 - 6x^2}$$ *Build fractions by filling in missing factors.*

$$= \frac{3x(2x^2)}{(x - 3)(x + 3)(2x^2)} + \frac{5(x + 3)}{2x^2(x - 3)(x + 3)}$$ *Combine, as with like fractions.*

$$= \frac{6x^3 + 5(x + 3)}{2x^2(x - 3)(x + 3)}$$

At this point you may be tempted to reduce the x^2 with x^3 and/or the $(x + 3)$ in the numerator with the $(x + 3)$ in the denominator. However, neither x^3 nor $(x + 3)$ is a factor of the numerator and thus we cannot reduce the fraction at this point. Therefore the final answer is

$$= \boxed{\frac{6x^3 + 5x + 15}{2x^2(x - 3)(x + 3)}}$$

(c) $$\frac{5x - 10}{x - 4} + \frac{3x - 2}{4 - x}$$ *Note: Since $x - 4$ is the negative of $4 - x$, we multiply the numerator and denominator of the second fraction by -1.*

$$= \frac{5x - 10}{x - 4} + \frac{(3x - 2)(-1)}{(4 - x)(-1)}$$

$$= \frac{5x - 10}{x - 4} + \frac{(3x - 2)(-1)}{x - 4}$$ *Now the denominators are the same.*

$$= \frac{5x - 10 + (3x - 2)(-1)}{x - 4}$$ *Combine numerators; place results above the common denominator.*

$$= \frac{5x - 10 - 3x + 2}{x - 4} = \frac{2x - 8}{x - 4} = \frac{2(x - 4)}{x - 4}$$ *Factor and reduce.*

$$= \boxed{2}$$

EXAMPLE 4

Perform the indicated operations. $\dfrac{3}{x^2 - 9} - \dfrac{x + 1}{x^2 + 6x + 9}$

Solution   We found the LCD of these two rational expressions previously in Example 2(**b**).

$$\frac{3}{x^2 - 9} - \frac{x + 1}{x^2 + 6x + 9} = \frac{3}{(x + 3)(x - 3)} - \frac{x + 1}{(x + 3)(x + 3)}$$ *The LCD is $(x - 3)(x + 3)^2$. Build fractions by filling in missing factors.*

$$= \frac{3(x + 3)}{(x - 3)(x + 3)^2} - \frac{(x + 1)(x - 3)}{(x - 3)(x + 3)^2}$$ *Combine.*

$$= \frac{3(x + 3) - (x + 1)(x - 3)}{(x - 3)(x + 3)^2}$$ *Simplify the numerator.*

$$= \frac{(3x + 9) - (x^2 - 2x - 3)}{(x - 3)(x + 3)^2}$$

$$= \frac{3x + 9 - x^2 + 2x + 3}{(x - 3)(x + 3)^2}$$

$$= \boxed{\frac{-x^2 + 5x + 12}{(x - 3)(x + 3)^2}}$$

EXAMPLE 5

If $f(x) = \dfrac{2}{x + 1}$, find and simplify $f(x + h) - f(x)$.

Solution   We begin by finding $f(x + h)$:

$$f(x) = \frac{2}{x + 1} \qquad \textit{f(x + h) is found by substituting x + h for x.}$$

$$f(x + h) = \frac{2}{x + h + 1} \qquad \textit{Next, we write out the expression f(x + h) − f(x).}$$

$$f(x + h) - f(x) = \frac{2}{x + h + 1} - \frac{2}{x + 1} \qquad \begin{array}{l}\textit{Next, find the LCD and build the fractions.}\\ \textit{LCD} = (x + h + 1)(x + 1)\end{array}$$

$$= \frac{2(x + 1)}{(x + h + 1)(x + 1)} - \frac{2(x + h + 1)}{(x + h + 1)(x + 1)} \qquad \begin{array}{l}\textit{Combine the}\\ \textit{numerators.}\end{array}$$

$$= \frac{2(x + 1) - 2(x + h + 1)}{(x + h + 1)(x + 1)}$$

$$= \frac{2x + 2 - 2x - 2h - 2}{(x + h + 1)(x + 1)}$$

$$= \boxed{\frac{-2h}{(x + h + 1)(x + 1)}}$$

Dividing a Polynomial by a Monomial

We discussed polynomial division in Chapter 5 and postponed discussing division by monomials until now because the process typically requires us to rewrite the quotient as a fraction.

EXAMPLE 6   Find the quotient: $(6x^5y^2 - 9x^3y + 12x^2) \div 6x^2y$

Solution   Division by monomials can be accomplished in two ways, depending on what type of answer is being sought. Both methods require us to rewrite the quotient as a fraction.

Method 1:   Just as we can combine several fractions that have the same denominator into a single fraction, we can also write a single fraction with several terms in the numerator as a sum or difference of several fractions.

$$\frac{6x^5y^2 - 9x^3y + 12x^2}{6x^2y} = \frac{6x^5y^2}{6x^2y} - \frac{9x^3y}{6x^2y} + \frac{12x^2}{6x^2y} \qquad \begin{array}{l}\textit{Notice that we are doing the reverse}\\ \textit{of combining fractions.}\end{array}$$

$$= \frac{\overset{x^3\ y}{\cancel{6x^5y^2}}}{\cancel{6x^2y}} - \frac{\overset{3x}{\cancel{9x^3y}}}{\underset{2}{\cancel{6x^2y}}} + \frac{\overset{2}{\cancel{12x^2}}}{\cancel{6x^2y}} \qquad \textit{Reduce each fraction.}$$

$$= \boxed{x^3y - \frac{3x}{2} + \frac{2}{y}}$$

Method 2: An alternative approach is to reduce the fraction as discussed in Section 6.2:

$$\frac{6x^5y^2 - 9x^3y + 12x^2}{6x^2y}$$

Factor and reduce.

$$= \frac{\cancel{3x^2}(2x^3y^2 - 3xy + 4)}{\cancel{6x^2}y}$$

$$= \boxed{\frac{2x^3y^2 - 3xy + 4}{2y}}$$

The answers obtained by the two methods may look different but they are in fact equal. Try combining the fractions in the answer obtained by method 1 into a single fraction and see whether you get the answer produced by method 2.

EXERCISES 6.4

In Exercises 1–60, perform the indicated operations. Express your answer in simplest form.

1. $\dfrac{x + 2}{x} - \dfrac{2 - x}{x}$

2. $\dfrac{x - 4}{2x} + \dfrac{x + 4}{2x}$

3. $\dfrac{10a}{a - b} - \dfrac{10b}{a - b}$

4. $\dfrac{5x}{x + 2} + \dfrac{10}{x + 2}$

5. $\dfrac{3a}{3a^2 + a - 2} - \dfrac{2}{3a^2 + a - 2}$

6. $\dfrac{x}{3x^2 + x - 2} + \dfrac{1}{3x^2 + x - 2}$

7. $\dfrac{x^2}{x^2 - y^2} + \dfrac{y^2}{x^2 - y^2} - \dfrac{2xy}{x^2 - y^2}$

8. $\dfrac{a^2}{a^2 - b^2} + \dfrac{b^2}{a^2 - b^2} + \dfrac{2ab}{b^2 - a^2}$

9. $\dfrac{5x}{x + 3} + \dfrac{3x}{x + 3} + 4$

10. $\dfrac{3t}{t - 2} + \dfrac{2}{t - 2} + 3t$

11. $\dfrac{3x}{2y^2} - \dfrac{4x^2}{9y}$

12. $\dfrac{24}{5x^2y} + \dfrac{14}{45xy^3}$

13. $\dfrac{7y}{6x^2} - \dfrac{8x^2}{9y^2}$

14. $\dfrac{5a}{4b^2} - \dfrac{16b}{25a^2}$

15. $\dfrac{7y}{6x^2} \cdot \dfrac{8x^2}{9y^2}$

16. $\dfrac{5a}{4b^2} \div \dfrac{16b}{24a^2}$

17. $\dfrac{36a^2}{b^2c} + \dfrac{24}{bc^3} - \dfrac{3}{7bc}$

18. $\dfrac{3x}{2a^2b} + \dfrac{2y}{3ab^2} - \dfrac{3}{a^3b^2}$

19. $\dfrac{3}{x} + \dfrac{x}{x + 2}$

20. $\dfrac{5}{y} + \dfrac{y}{y - 3}$

21. $\dfrac{a}{a - b} - \dfrac{b}{a}$

22. $\dfrac{2}{x - 7} - \dfrac{7}{x}$

23. $\dfrac{5}{x + 7} - \dfrac{2}{x - 3}$

24. $\dfrac{3}{x - 2} + \dfrac{4}{x + 5}$

25. $\dfrac{2}{x - 7} + \dfrac{3x + 1}{x + 2}$

26. $\dfrac{3}{a + 4} + \dfrac{5a + 1}{a - 2}$

27. $\dfrac{2r + s}{r - s} - \dfrac{r - 2s}{r + s}$

28. $\dfrac{3x + y}{x + y} - \dfrac{2x + y}{x - y}$

29. $\dfrac{4}{x - 4} + \dfrac{x}{4 - x}$

30. $\dfrac{a}{a - b} + \dfrac{b}{b - a}$

31. $\dfrac{7a + 3}{2a - 1} + \dfrac{5a + 4}{1 - 2a}$

32. $\dfrac{3x + 1}{x - 2} + \dfrac{2x + 3}{2 - x}$

33. $\dfrac{a}{a - b} - \dfrac{b}{a^2 - b^2}$

34. $\dfrac{x}{x - y} + \dfrac{y}{(x - y)^2}$

35. $\dfrac{a}{a - b} \div \dfrac{b}{a^2 - b^2}$

36. $\dfrac{x}{x - y} \cdot \dfrac{y}{(x - y)^2}$

37. $\dfrac{2y + 1}{y + 2} + \dfrac{3y}{y + 3} + y$

38. $\dfrac{2x + 1}{x + 3} + \dfrac{3x}{x - 2} + 2x$

39. $\dfrac{5x + 1}{x + 2} + \dfrac{2x + 6}{x^2 + 5x + 6}$

40. $\dfrac{2a + 3}{a^2 - 2a - 8} - \dfrac{a - 5}{a - 4}$

41. $\dfrac{5x + 1}{x + 2} \div \dfrac{2x + 6}{x^2 + 5x + 6}$

42. $\dfrac{2a + 3}{a^2 - 2a - 8} \div \dfrac{a - 5}{a - 4}$

43. $\dfrac{x + 1}{x^2 - 3x} + \dfrac{x - 2}{x^2 - 6x + 9}$

44. $\dfrac{x - 2}{3x^2 - 12x} - \dfrac{3}{x^2 - 8x + 16}$

45. $\dfrac{a + 3}{a - 3} + \dfrac{a}{a + 4} - \dfrac{3}{a^2 + a - 12}$

46. $\dfrac{x + 6}{x - 3} + \dfrac{x}{x + 2} - \dfrac{10}{x^2 - x - 6}$

47. $\dfrac{y + 7}{y + 5} - \dfrac{y}{y - 3} + \dfrac{16}{y^2 + 2y - 15}$

48. $\dfrac{3x + 4}{x + 4} + \dfrac{3x}{x - 2} + \dfrac{4}{x^2 + 2x - 8}$

49. $\dfrac{3x - 4}{x - 5} + \dfrac{4x}{10 + 3x - x^2}$

50. $\dfrac{2r - 5}{r - 2} + \dfrac{3r}{6 - r - r^2}$

51. $\dfrac{2a - 1}{a^2 + a - 6} + \dfrac{a + 2}{a^2 - 2a - 15} - \dfrac{a + 1}{a^2 - 7a + 10}$

52. $\dfrac{3y + 2}{y^2 - 2y + 1} - \dfrac{7y - 3}{y^2 - 1} + \dfrac{5}{y + 1}$

53. $\dfrac{5a}{3a - 1} + \dfrac{2a + 1}{5a + 2} + 5a + 1$

54. $\dfrac{3x}{x - 1} + \dfrac{x + 3}{x - 2} + 2x - 3$

55. $\dfrac{2r + xs}{r + s} + \dfrac{2s + xr}{r + s}$

56. $\dfrac{ax + ay}{a + b} + \dfrac{bx + by}{a + b}$

57. $\dfrac{r^2}{r^3 - s^3} + \dfrac{rs}{r^3 - s^3} + \dfrac{s^2}{r^3 - s^3}$

58. $\dfrac{a^2}{a^3 + 8b^3} - \dfrac{2ab}{a^3 + 8b^3} + \dfrac{4b^2}{a^3 + 8b^3}$

59. $\dfrac{2s + t}{s^3 - t^3} + \dfrac{3s}{s^2 + st + t^2}$

60. $\dfrac{3x + 1}{x^3 + 8} + \dfrac{2x + 1}{x^2 - 2x + 4}$

In Exercises 61–66, find the quotient by each of the following methods:

(a) *Express as a single fraction reduced to lowest terms.*

(b) *Rewrite the fraction as sums or differences of fractions with the same denominator (and then simplify each fraction).*

61. $\dfrac{x^2 + 4x}{4x}$

62. $\dfrac{a^3 - 6a^2}{3a}$

63. $\dfrac{15x^3 y^2 - 10x^2 y^3}{5x^2 y^2}$

64. $\dfrac{18r^2 s^6 + 24r^3 s^4}{6r^2 s^3}$

65. $\dfrac{6m^2 n - 4m^3 n^2 - 9mn}{15mn^2}$

66. $\dfrac{10u^2 v^3 - 15u^4 v + 20uv^3}{25u^2 v^3}$

In Exercises 67–74, use $f(x) = \dfrac{x - 2}{x + 2}$, $g(x) = \dfrac{2x}{x^2 - 4}$, and $h(x) = \dfrac{x}{x - 2}$.

67. $f(x) + g(x)$ **68.** $f(x) \cdot g(x)$ **69.** $f(x) \cdot h(x)$ **70.** $f(x) - h(x)$

71. $g(x) + h(x)$ **72.** $g(x) \div h(x)$ **73.** $h(x) \div g(x)$ **74.** $f(x) - g(x) + h(x)$

In Exercises 75–78, find and simplify $f(x + h) - f(x)$.

75. $f(x) = \dfrac{3}{x}$

76. $f(x) = \dfrac{1}{x - 6}$

77. $f(x) = \dfrac{x}{x-3}$

78. $f(x) = \dfrac{2x}{x+4}$

 Question for Thought

79. Use the calculator to evaluate the following for $x = -1, -\frac{1}{2}, 0, \frac{1}{2}, 1,$ and 2; discuss your results.

$$\frac{3x+5}{x-1} + \frac{x+7}{1-x}$$

Mini-Review

80. *Solve for x:* $\dfrac{x}{3} - \dfrac{x}{4} = 4$

81. Find the slope of the line passing through the points $(-4, 2)$ and $(5, -2)$.

82. *Simplify:* $(x-3)^2 - (x+3)^2$

83. Sketch the graph of the solution set of $2x - 3y \geq 12$.

6.5   Mixed Operations and Complex Fractions

Now we are in the position to perform mixed operations with fractions.

EXAMPLE 1

Perform the indicated operations and simplify.

$$\left(\frac{x+6}{x+2} - \frac{4}{x}\right) \div \frac{x^2-16}{x^2-4}$$

Solution $\left(\dfrac{x+6}{x+2} - \dfrac{4}{x}\right) \div \dfrac{x^2-16}{x^2-4}$ *Combine fractions in parentheses. The LCD is $x(x+2)$.*

$$= \left(\frac{x(x+6)}{x(x+2)} - \frac{4(x+2)}{x(x+2)}\right) \div \frac{x^2-16}{x^2-4}$$

$$= \frac{x(x+6) - 4(x+2)}{x(x+2)} \div \frac{x^2-16}{x^2-4}$$

$$= \frac{x^2 + 6x - 4x - 8}{x(x+2)} \div \frac{x^2-16}{x^2-4}$$

$$= \frac{x^2 + 2x - 8}{x(x+2)} \div \frac{x^2-16}{x^2-4}$$ *Next, we follow the rule for division.*

$$= \frac{x^2 + 2x - 8}{x(x+2)} \cdot \frac{x^2-4}{x^2-16}$$ *Factor and reduce.*

$$= \frac{\cancel{(x+4)}(x-2)}{x\cancel{(x+2)}} \cdot \frac{\cancel{(x+2)}(x-2)}{\cancel{(x+4)}(x-4)}$$

$$= \boxed{\frac{(x-2)^2}{x(x-4)}} \quad \text{or} \quad \boxed{\frac{x^2 - 4x + 4}{x^2 - 4x}}$$

Another way of expressing a quotient of fractions is to use a large fraction bar instead of the quotient sign, $\div$. Thus, we can rewrite

$$\frac{a}{b} \div \frac{c}{d} \quad \text{as} \quad \frac{\dfrac{a}{b}}{\dfrac{c}{d}}$$

Quotients written this way (fractions within fractions) are called **complex fractions.** This notation is often more convenient than using the quotient sign. It allows us to demonstrate that the "multiply by the reciprocal of the divisor" rule is simply an application of the Fundamental Principle of Fractions:

$$\frac{\dfrac{a}{b}}{\dfrac{c}{d}} = \frac{\dfrac{a}{b} \cdot \dfrac{d}{c}}{\dfrac{c}{d} \cdot \dfrac{d}{c}}$$

Multiply numerator and denominator of the complex fraction by $\dfrac{d}{c}$ (Fundamental Principle of Fractions).

$$= \frac{\dfrac{a}{b} \cdot \dfrac{d}{c}}{\dfrac{\cancel{c}}{\cancel{d}} \cdot \dfrac{\cancel{d}}{\cancel{c}}} = \frac{\dfrac{a}{b} \cdot \dfrac{d}{c}}{1}$$

Reduce.

$$= \frac{a}{b} \cdot \frac{d}{c}$$

Thus, an expression such as

$$\frac{\dfrac{3a^2b}{2xy^3}}{\dfrac{9ab^2}{8x^2y}}$$

can be rewritten as

$$\frac{3a^2b}{2xy^3} \div \frac{9ab^2}{8x^2y} = \frac{3a^2b}{2xy^3} \cdot \frac{8x^2y}{9ab^2}$$

$$= \frac{\overset{a}{\cancel{3a^2b}}}{\underset{y^2}{\cancel{2xy^3}}} \cdot \frac{\overset{4x}{\cancel{8x^2y}}}{\underset{3 \quad b}{\cancel{9ab^2}}} = \frac{4ax}{3by^2}$$

EXAMPLE 2

Solution

If $f(x) = \dfrac{2-x}{x+1}$,   find $f\left(\dfrac{2}{3}\right)$.

$$f\left(\frac{2}{3}\right) = \frac{2 - \dfrac{2}{3}}{\dfrac{2}{3} + 1}$$

This is a complex fraction that needs to be simplified. We offer two methods of solution. The first method treats the complex fraction as a multiple-operation problem similar to Example 1. That is, we treat the complex fraction as if it were written as

$$\left(2 - \frac{2}{3}\right) \div \left(\frac{2}{3} + 1\right)$$

Hence, we combine the fractions in the numerator of the complex fraction, combine the fractions in the denominator of the complex fraction, and then we will have a quotient of two fractions.

Method 1: $\dfrac{2 - \dfrac{2}{3}}{\dfrac{2}{3} + 1}$ *Combine the fractions in the numerator and denominator of the complex fraction. The LCD in both cases is 3.*

$$= \dfrac{\dfrac{6}{3} - \dfrac{2}{3}}{\dfrac{2}{3} + \dfrac{3}{3}}$$

$$= \dfrac{\dfrac{4}{3}}{\dfrac{5}{3}}$$ *We now have the quotient of two fractions. We follow the rule for division; multiply by the reciprocal of the divisor.*

$$= \dfrac{4}{\cancel{3}} \cdot \dfrac{\cancel{3}}{5} = \boxed{\dfrac{4}{5}}$$

The second method attempts to immediately clear the denominators *within* the complex fraction. To do this, we apply the Fundamental Principle by multiplying the numerator and denominator of the complex fraction by the LCD of *all* the simple fractions in the complex fraction.

Method 2: $\dfrac{2 - \dfrac{2}{3}}{\dfrac{2}{3} + 1}$ *Find the LCD of all the simple fractions, which is 3. Next multiply the numerator and denominator of the complex fraction by 3.*

$$= \dfrac{\left(2 - \dfrac{2}{3}\right) \cdot \dfrac{3}{1}}{\left(\dfrac{2}{3} + 1\right) \cdot \dfrac{3}{1}}$$ *Now use the distributive property in the numerator and denominator.*

$$= \dfrac{2 \cdot 3 - \dfrac{2}{\cancel{3}} \cdot \dfrac{\cancel{3}}{1}}{\dfrac{2}{\cancel{3}} \cdot \dfrac{\cancel{3}}{1} + 1 \cdot 3}$$ *Reduce where appropriate.*

$$= \dfrac{6 - 2}{2 + 3} = \boxed{\dfrac{4}{5}}$$

Either method can be used to simplify any complex fraction. There are situations for which method 1 is simpler and others for which method 2 is simpler.

It is a worthwhile exercise to compute the value of this complex fraction using a calculator. You may find that entering this computation correctly on a calculator requires extra care.

EXAMPLE 3

Express the following as a simple fraction reduced to lowest terms:

$$\dfrac{1 - \dfrac{2}{x}}{1 + \dfrac{2}{x} - \dfrac{8}{x^2}}$$

Solution   We again illustrate both methods.

Method 1: $\dfrac{1 - \dfrac{2}{x}}{1 + \dfrac{2}{x} - \dfrac{8}{x^2}}$

Combine the fractions in the numerator and denominator of the complex fraction.

The LCD for the numerator is x; the LCD for the denominator is x^2.

$= \dfrac{\dfrac{x}{x} - \dfrac{2}{x}}{\dfrac{x^2}{x^2} + \dfrac{2x}{x^2} - \dfrac{8}{x^2}}$

$= \dfrac{\dfrac{x - 2}{x}}{\dfrac{x^2 + 2x - 8}{x^2}}$

We now have the quotient of two fractions. Now follow the rule for division: Multiply by the reciprocal of the divisor.

$= \dfrac{x - 2}{x} \cdot \dfrac{x^2}{x^2 - 2x - 8}$ *Factor and reduce.*

$= \dfrac{x - 2}{x} \cdot \dfrac{\overset{x}{x^2}}{(x + 4)(x - 2)}$

$= \boxed{\dfrac{x}{x + 4}}$

Method 2: $\dfrac{1 - \dfrac{2}{x}}{1 + \dfrac{2}{x} - \dfrac{8}{x^2}}$

We multiply the numerator and denominator of the complex fraction by the LCD of all denominators of the simple fractions, which is x^2.

$= \dfrac{\left(1 - \dfrac{2}{x}\right) \cdot \dfrac{x^2}{1}}{\left(1 + \dfrac{2}{x} - \dfrac{8}{x^2}\right) \cdot \dfrac{x^2}{1}}$

Next, use the distributive property in the numerator and denominator.

$= \dfrac{x^2 - \dfrac{2}{x} \cdot \dfrac{x^2}{1}}{x^2 + \dfrac{2}{x} \cdot \dfrac{x^2}{1} - \dfrac{8}{x^2} \cdot \dfrac{x^2}{1}}$

$= \dfrac{x^2 - 2x}{x^2 + 2x - 8}$ *Factor and reduce.*

$= \dfrac{x(x - 2)}{(x + 4)(x - 2)} = \boxed{\dfrac{x}{x + 4}}$

EXAMPLE 4

If $f(x) = \dfrac{3}{x}$, simplify $\dfrac{f(x + 5) - f(x)}{5}$.

Solution   Looking carefully at the given expression, we recognize that we are simply computing the difference between two expressions, $f(x + 5)$ and $f(x)$, and then dividing this difference by 5. Since we are given $f(x)$, our first step is to find $f(x + 5)$:

$$f(x) = \frac{3}{x} \qquad \textit{To find } f(x + 5), \textit{ substitute } x + 5 \textit{ for } x \textit{ in } \frac{3}{x}.$$

$$f(x + 5) = \frac{3}{x + 5}$$

Now that we have $f(x + 5)$, we can find $\dfrac{f(x + 5) - f(x)}{5}$:

$$\frac{f(x + 5) - f(x)}{5} = \frac{\overbrace{\dfrac{3}{x + 5}}^{f(x+5)} - \overbrace{\dfrac{3}{x}}^{f(x)}}{5}$$

We choose to simplify the complex fraction using method 2. We multiply the numerator and denominator of the complex fraction by the LCD of all the simple fractions, which is $x(x + 5)$.

$$= \frac{\left(\dfrac{3}{x + 5} - \dfrac{3}{x}\right)\dfrac{x(x + 5)}{1}}{5 \cdot x(x + 5)}$$

$$= \frac{\dfrac{3}{\cancel{x+5}} \cdot \dfrac{x(\cancel{x+5})}{1} - \dfrac{3}{\cancel{x}} \cdot \dfrac{\cancel{x}(x + 5)}{1}}{5 \cdot x(x + 5)} \qquad \textit{Simplify.}$$

$$= \frac{3x - 3(x + 5)}{5x(x + 5)}$$

Note the necessary parentheses around $x + 5$. Also note that you cannot reduce by a factor of $(x + 5)$ because it is not a factor of the numerator. Now simplify the numerator and reduce the fraction.

$$= \frac{3x - 3x - 15}{5x(x + 5)} = \frac{-15}{5x(x + 5)} = \boxed{\frac{-3}{x(x + 5)}}$$

EXERCISES 6.5

In Exercises 1–30, express each of the following as a simple fraction reduced to lowest terms.

1. $\dfrac{\dfrac{3}{xy^2}}{\dfrac{15}{x^2 y}}$

2. $\dfrac{\dfrac{5}{3a^2 b}}{\dfrac{25}{9ab^2}}$

3. $\dfrac{\dfrac{3}{x - y}}{\dfrac{x - y}{3}}$

4. $\dfrac{\dfrac{5}{x + y}}{\dfrac{5}{x - y}}$

5. $\dfrac{a - \dfrac{1}{3}}{\dfrac{9a^2 - 1}{3a}}$

6. $\dfrac{1 - \dfrac{1}{x^2}}{\dfrac{x - 1}{x}}$

7. $\dfrac{x + \dfrac{2}{x^2}}{\dfrac{1}{x} + 2}$

8. $\left(\dfrac{1}{x} + \dfrac{1}{y}\right) \div \dfrac{3}{xy^2}$

9. $\left(1 - \dfrac{4}{x^2}\right) \div \left(\dfrac{1}{x} - \dfrac{2}{x^2}\right)$

10. $\dfrac{x - \dfrac{3}{y^2}}{\dfrac{2}{x} - \dfrac{3}{y^2}}$

11. $\dfrac{1 - \dfrac{5}{y}}{y + 3 - \dfrac{40}{y}}$

12. $\dfrac{3 + \dfrac{9}{x}}{x - 7 - \dfrac{30}{x}}$

13. $\dfrac{1 - \dfrac{4}{z} + \dfrac{4}{z^2}}{\dfrac{1}{z^2} - \dfrac{2}{z^3}}$

14. $\dfrac{2 + \dfrac{3}{x} - \dfrac{2}{x^2}}{\dfrac{2}{x^2} - \dfrac{1}{x^3}}$

15. $\dfrac{\dfrac{4}{y^2} - \dfrac{12}{xy} + \dfrac{9}{x^2}}{\dfrac{4}{y^2} - \dfrac{9}{x^2}}$

16. $\dfrac{\dfrac{9}{t} + \dfrac{12}{s} + \dfrac{4t}{s^2}}{\dfrac{9}{t} + \dfrac{9}{s} + \dfrac{2t}{s^2}}$

17. $\dfrac{\dfrac{2x}{y} + 7 + \dfrac{5y}{x}}{3x + 2y - \dfrac{y^2}{x}}$

18. $\dfrac{\dfrac{6a}{b} - 5 - \dfrac{6b}{a}}{2 - \dfrac{3b}{a}}$

19. $\dfrac{1 - \dfrac{3}{x} - \dfrac{10}{x^2}}{4 + \dfrac{8}{x}}$

20. $\dfrac{5 + \dfrac{35}{x}}{1 + \dfrac{3}{x} - \dfrac{28}{x^2}}$

21. $\dfrac{9 - \dfrac{25}{t^2}}{6 + \dfrac{10}{t}}$

22. $\dfrac{16 - \dfrac{49}{v^2}}{12 + \dfrac{21}{v}}$

23. $\dfrac{\dfrac{1}{x - 3} - \dfrac{1}{x}}{3}$

24. $\dfrac{\dfrac{1}{a + 5} - \dfrac{1}{a}}{5}$

25. $\dfrac{\dfrac{h}{h + 1} - \dfrac{h}{h - 1}}{2}$

26. $\dfrac{\dfrac{y + 4}{y + 2} - \dfrac{y}{y - 2}}{4}$

27. $\left(3 + \dfrac{1}{2x - 1}\right) \div \left(5 + \dfrac{x}{2x - 1}\right)$

28. $\left(a + \dfrac{2a}{a - 1}\right) \div \left(a - \dfrac{2a}{a - 1}\right)$

29. $\dfrac{\dfrac{4}{x + 3} - \dfrac{4}{x}}{3}$

30. $\dfrac{\dfrac{7}{x + h} - \dfrac{7}{x}}{h}$

In Exercises 31–44, use $f(x) = \dfrac{2x}{x - 3}$, $g(x) = -\dfrac{1}{x}$, *and* $h(x) = \dfrac{x + 2}{x}$ *to find and simplify the given expression.*

31. $f\left(\dfrac{2}{3}\right)$

32. $g\left(\dfrac{3}{4}\right)$

33. $g\left(\dfrac{2}{x}\right)$

34. $f\left(\dfrac{3}{x}\right)$

35. $\dfrac{f(x + 3) - f(x)}{3}$

36. $\dfrac{g(x - 4) - g(x)}{4}$

37. $\dfrac{h(x - 2) - h(x)}{2}$

38. $\dfrac{g(x) - g(5)}{x - 5}$

39. $\dfrac{f(x) + g(x)}{h(x)}$

40. $\dfrac{g(x) + h(x)}{f(x)}$

41. $\dfrac{g(x) + 1}{h(x)}$

42. $\dfrac{h(x) - 1}{g(x)}$

43. $\dfrac{f(x) - h(x)}{3}$

44. $\dfrac{g(x) - f(x)}{3}$

◆ Mini-Review

45. *Solve for x:* $(x - 4)(x + 1) = (x - 3)(x - 2)$

46. *Solve for x:* $(2x - 4)(x + 1) = (x - 3)(x - 2)$

47. Write an equation of the line passing through the point $(-4, 5)$ that is parallel to the line whose equation is $3x - 2y = 5$.

48. *Simplify:* $x - [5 - 3(x - 5)]$

6.6 Fractional Equations and Inequalities

Up to now we have been concerned with performing operations with fractional expressions. In this section we will discuss how to approach fractional equations and inequalities.

In Chapter 2 we developed a strategy for solving first-degree equations. In Section 6.4 we discussed the idea of the least common denominator (LCD). Our method for solving fractional equations and inequalities will use both of these ideas, as illustrated in the following examples.

EXAMPLE 1

Solve for x: $\dfrac{x}{4} - \dfrac{x - 3}{3} = \dfrac{x + 3}{6}$

Solution

Rather than solve this equation directly, we would much prefer to solve an equivalent equation without fractions. The idea of an LCD that we used in Section 6.4 can be used to accomplish this.

Since we are dealing with an equation, we can multiply both sides of the equation by any nonzero quantity we choose. If we multiply the entire equation by a number that is divisible by 3, 4, and 6, we will "eliminate" the denominators. The LCD is exactly the smallest such quantity. This process is called ***clearing the denominators.***

We multiply both sides of the equation by LCD of all the denominators in the equation, which is 12.

$$\frac{x}{4} - \frac{x - 3}{3} = \frac{x + 3}{6}$$

$$12\left(\frac{x}{4} - \frac{x - 3}{3}\right) = 12\left(\frac{x + 3}{6}\right) \qquad \textit{Each fraction is multiplied by 12.}$$

$$\frac{12}{1} \cdot \frac{x}{4} - \frac{12}{1} \cdot \frac{x - 3}{3} = \frac{12}{1} \cdot \frac{x + 3}{6} \qquad \textit{Reduce.}$$

$$\frac{\cancel{12}^{3}}{1} \cdot \frac{x}{\cancel{4}} - \frac{\cancel{12}^{4}}{1} \cdot \frac{x - 3}{\cancel{3}} = \frac{\cancel{12}^{2}}{1} \cdot \frac{x + 3}{\cancel{6}}$$

$$3x - 4(x - 3) = 2(x + 3)$$

Note that the parentheses around $x - 3$ and $x + 3$ are necessary.

$$3x - 4x + 12 = 2x + 6$$
$$-x + 12 = 2x + 6$$
$$6 = 3x$$
$$\boxed{2 = x}$$

The check is left to the student.

EXAMPLE 2

Solve for *x*: $\dfrac{4}{x} + \dfrac{5}{2} = \dfrac{8}{x}$

Solution   We follow the procedure described in Example 1.

$$\frac{4}{x} + \frac{5}{2} = \frac{8}{x}$$

Multiply both sides of the equation by the LCD, which is $2x$.

$$2x\left(\frac{4}{x} + \frac{5}{2}\right) = 2x\left(\frac{8}{x}\right)$$

$$\frac{2x}{1} \cdot \frac{4}{x} + \frac{2x}{1} \cdot \frac{5}{2} = \frac{2x}{1} \cdot \frac{8}{x}$$

We clear the denominators.

$$\frac{2\cancel{x}}{1} \cdot \frac{4}{\cancel{x}} + \frac{\cancel{2}x}{1} \cdot \frac{5}{\cancel{2}} = \frac{2\cancel{x}}{1} \cdot \frac{8}{\cancel{x}}$$

$$8 + 5x = 16$$
$$5x = 8$$

You should continue to check all solutions.

$$\boxed{x = \frac{8}{5} = 1.6}$$

When a variable does not appear in any denominator, clearing the denominators is the first step to take in solving fractional inequalities as well.

EXAMPLE 3

Solve for *q*: $\dfrac{q}{5} + \dfrac{5 - q}{3} \leq 2$

Solution   We clear the denominators by multiplying both sides of the inequality by the LCD, which is $5 \cdot 3 = 15$.

$$\frac{15}{1} \cdot \frac{q}{5} + \frac{15}{1} \cdot \frac{5 - q}{3} \leq 15 \cdot 2$$

$$\frac{\cancel{15}^{3}}{1} \cdot \frac{q}{\cancel{5}} + \frac{\cancel{15}^{5}}{1} \cdot \frac{5 - q}{\cancel{3}} \leq 15 \cdot 2$$

$$3q + 5(5 - q) \leq 30$$
$$3q + 25 - 5q \leq 30$$
$$25 - 2q \leq 30$$
$$-2q \leq 5$$

$$\boxed{q \geq -\frac{5}{2}}$$

Do not forget to reverse the inequality symbol when dividing by a negative number.

EXAMPLE 4

Solve for a: $\dfrac{5}{2a} + \dfrac{3}{a-2} = \dfrac{7}{10a}$

Solution

We clear the denominator by multiplying both sides of the equation by the LCD, which is $10a(a-2)$.

$$\frac{5}{2a} + \frac{3}{a-2} = \frac{7}{10a}$$

Each fraction is multiplied by the LCD.

$$\left(\frac{10a(a-2)}{1}\right)\frac{5}{2a} + \left(\frac{10a(a-2)}{1}\right)\frac{3}{a-2} = \left(\frac{10a(a-2)}{1}\right)\frac{7}{10a}$$

Reduce.

$$\frac{\overset{5}{\cancel{10}a(a-2)}}{1} \cdot \frac{5}{\cancel{2a}} + \frac{10a(\cancel{a-2})}{1} \cdot \frac{3}{\cancel{a-2}} = \frac{\cancel{10}a(a-2)}{1} \cdot \frac{7}{\cancel{10a}}$$

$$25(a-2) + 30a = 7(a-2)$$
$$25a - 50 + 30a = 7a - 14$$
$$55a - 50 = 7a - 14$$
$$48a = 36$$

Use your calculator to check your answer.

$$a = \frac{36}{48} = \boxed{\frac{3}{4}}$$

As the next example shows, solving fractional equations requires some care.

EXAMPLE 5

Solve for x: $\dfrac{7}{x+5} + 2 = \dfrac{2-x}{x+5}$

Solution

Again we clear the denominator by multiplying both sides of the equation by the LCD, which is $x+5$:

$$\frac{7}{x+5} + 2 = \frac{2-x}{x+5}$$

First multiply both sides of the equation by x + 5 and reduce.

$$(\cancel{x+5})\left(\frac{7}{\cancel{x+5}}\right) + 2(x+5) = (\cancel{x+5})\left(\frac{2-x}{\cancel{x+5}}\right)$$
$$7 + 2(x+5) = 2 - x$$
$$7 + 2x + 10 = 2 - x$$
$$2x + 17 = 2 - x$$
$$3x = -15$$
$$x = -5$$

Check: $x = -5$:

$$\frac{7}{x+5} + 2 = \frac{2-x}{x+5}$$

$$\frac{7}{-5+5} + 2 \overset{?}{=} \frac{2-(-5)}{-5+5}$$

$$\frac{7}{0} + 2 \neq \frac{7}{0}$$

Since we are never allowed to divide by 0, $\frac{7}{0}$ is undefined. Therefore, $x = -5$ is *not* a solution.

Have we made an error? No. As we pointed out in Chapter 2, if we multiply an equation by 0 we may no longer have an equivalent equation (that is, an equation with the same solution set). This is exactly what has happened here.

We multiplied the original equation by $x + 5$ to clear the denominators, but if $x = -5$, then $x + 5$ is equal to 0, and so we have multiplied the original equation by 0. The resulting equation, for which $x = -5$ is a solution, is not equivalent to the original equation, for which $x = -5$ is not even an allowable value.

Our logic tells us that $x = -5$ is the only possible solution. Since $x = -5$ does not satisfy the original equation, the original equation has

$$\boxed{\text{no solutions}}$$

Example 5 shows that when we multiply an equation by a variable quantity that might be equal to 0, we *must* check our answer(s) in the original equation. This check is not optional, but rather a necessary step in the solution. We are not checking for errors—we are checking to see whether we have obtained a valid solution.

Another way of saying this is that we look at our original equation in Example 5 and ask, "What are the possible replacement values for x?" We must disqualify $x = -5$ from the outset since it requires division by 0, which is undefined.

EXAMPLE 6

Solve for t: $\dfrac{4}{t} - \dfrac{5}{t + 3} = 1$

Solution   We will use the LCD, which is $t(t + 3)$ to clear the denominators.

$$\frac{4}{t} - \frac{5}{t + 3} = 1 \qquad \textit{Note that } t = 0 \textit{ and}$$
$$\textit{t} = -3 \textit{ must be excluded.}$$

$$t(t + 3) \cdot \frac{4}{t} - t(t + 3) \cdot \frac{5}{t + 3} = t(t + 3) \cdot 1$$

$$4(t + 3) - 5t = t(t + 3)$$

$$4t + 12 - 5t = t^2 + 3t$$

$$-t + 12 = t^2 + 3t \qquad \textit{We recognize this as a quadratic equation.}$$
$$\textit{We get the equation into standard form.}$$

$$0 = t^2 + 4t - 12$$

$$0 = (t + 6)(t - 2)$$

$$t + 6 = 0 \quad \text{or} \quad t - 2 = 0$$

$$\boxed{t = -6 \quad \text{or} \quad t = 2}$$

The check is left to the student.

EXAMPLE 7

Solve for t: $\dfrac{5}{t + 3} - \dfrac{4}{3t} = \dfrac{7}{t^2 + 3t}$

Solution   We will use the LCD to clear the denominators. To find the LCD we want each denominator in factored form.

$$\frac{5}{t + 3} - \frac{4}{3t} = \frac{7}{t(t + 3)} \qquad LCD = 3t(t + 3)$$

$$3t(t + 3) \cdot \frac{5}{t + 3} - 3t(t + 3) \cdot \frac{4}{3t} = 3t(t + 3) \cdot \frac{7}{t(t + 3)}$$

$$15t - 4(t + 3) = 21$$

$$15t - 4t - 12 = 21$$

$$11t - 12 = 21$$
$$11t = 33$$
$$\boxed{t = 3}$$

The check is left to the student.

EXAMPLE 8

Combine and simplify: $\dfrac{2}{t} + \dfrac{3}{t+2} - \dfrac{4}{5t}$

Solution

It is very important to recognize the difference between this example and the previous one. Example 7 was an equation. The multiplication property of *equality* allowed us to "clear the denominators." This example, on the other hand, is an expression, *not* an equation. Therefore, the multiplication property of equality does not apply; we cannot "clear the denominators."

Instead, we are being asked to combine three fractions. We will use the LCD again in this example, but this time to *build* the fractions. Following the outline used in Section 6.4, we proceed as follows:

$$\frac{2}{t} + \frac{3}{t+2} - \frac{4}{5t} \qquad LCD = 5t(t+2)$$

$$= \frac{2 \cdot 5(t+2)}{5t(t+2)} + \frac{3 \cdot 5t}{5t(t+2)} - \frac{4(t+2)}{5t(t+2)}$$

$$= \frac{10(t+2) + 15t - 4(t+2)}{5t(t+2)}$$

$$= \frac{10t + 20 + 15t - 4t - 8}{5t(t+2)} = \boxed{\frac{21t + 12}{5t(t+2)}}$$

Let's examine an application that involves a rational or fractional equation.

EXAMPLE 9

An audio company finds that when n MPEG players are produced, the average cost per player is given by the equation

$$C = \frac{46n + 12{,}006}{n}$$

If the company wants the average cost per player to be $55, how many MPEG players should be manufactured?

Solution

To answer this question, we substitute $C = 55$ in the given equation and then solve for n:

$$C = \frac{46n + 12{,}006}{n} \qquad \textit{Substitute } C = 55.$$

$$55 = \frac{46n + 12{,}006}{n} \qquad \textit{Multiply both sides by n.}$$

$$55n = 46n + 12{,}006$$

$$9n = 12{,}006$$

$$n = \frac{12{,}006}{9} = \boxed{1334} \quad \textit{MPEG players}$$

EXERCISES 6.6

In Exercises 1–70, if the exercise is an equation or inequality, solve it. If it is an expression, perform the indicated operations and simplify.

1. $\dfrac{x}{3} - \dfrac{x}{2} + \dfrac{x}{4} = 1$

2. $\dfrac{a}{4} - \dfrac{a}{5} - \dfrac{a}{10} = 1$

3. $\dfrac{t}{3} + \dfrac{t}{5} < \dfrac{t}{6} - 11$

4. $\dfrac{n}{6} - \dfrac{n}{9} > \dfrac{n}{3} - 5$

5. $\dfrac{a-1}{6} + \dfrac{a+1}{10} = a - 3$

6. $\dfrac{w-3}{8} + \dfrac{w+4}{12} = w - 4$

7. $\dfrac{y-5}{2} = \dfrac{y-2}{5}$

8. $\dfrac{z-2}{7} = \dfrac{z-7}{2}$

9. $\dfrac{y-5}{2} \le \dfrac{y-2}{5} + 3$

10. $\dfrac{z-2}{7} \ge \dfrac{z-7}{2} - 5$

11. $\dfrac{x-3}{4} - \dfrac{x-4}{3} = 2$

12. $\dfrac{x-3}{5} - \dfrac{x+2}{6} = 1$

13. $\dfrac{x-3}{4} - \dfrac{x-4}{3} \ge 2$

14. $\dfrac{x-3}{5} - \dfrac{x+2}{6} \le 1$

15. $\dfrac{3x+11}{6} - \dfrac{2x+1}{3} = x + 5$

16. $\dfrac{7x+1}{4} - \dfrac{3x+1}{8} = \dfrac{x+2}{2}$

17. $\dfrac{x-3}{5} - \dfrac{3x+1}{4} < 8$

18. $\dfrac{5x+1}{2} - \dfrac{x+1}{4} > 2$

19. $\dfrac{5}{x} - \dfrac{1}{2} = \dfrac{3}{x}$

20. $\dfrac{4}{3x} + \dfrac{9}{x} - \dfrac{6}{5}$

21. $\dfrac{4}{x} - \dfrac{1}{5} + \dfrac{7}{2x}$

22. $\dfrac{2}{x} + \dfrac{1}{3} = \dfrac{5}{x}$

23. $\dfrac{1}{t-3} + \dfrac{2}{t} = \dfrac{5}{3t}$

24. $\dfrac{2}{y+2} + \dfrac{3}{y} = \dfrac{5}{2y}$

25. $\dfrac{6}{a-3} - \dfrac{3}{8} = \dfrac{21}{4a-12}$

26. $\dfrac{5}{z-1} - \dfrac{7}{z} = \dfrac{3}{2z-2}$

27. $\dfrac{7}{x-5} + 2 = \dfrac{x+2}{x-5}$

28. $\dfrac{9}{x+2} - 3 = \dfrac{x+11}{x+2}$

29. $\dfrac{4}{y^2 - 2y} - \dfrac{3}{2y} = \dfrac{17}{6y}$

30. $\dfrac{3}{z^2 - 4z} + \dfrac{7}{4z} - \dfrac{3}{8}$

31. $\dfrac{2x}{x-5} - \dfrac{2x+1}{x+2} = \dfrac{3}{x+2}$

32. $\dfrac{3a}{a-4} - \dfrac{3a+1}{a+3} = \dfrac{12}{a-4}$

33. $\dfrac{5}{y^2 + 3y} - \dfrac{4}{3y} + \dfrac{1}{2}$

34. $\dfrac{9}{z^2 + 5z} + \dfrac{6}{5z} = \dfrac{3}{10z}$

35. $\dfrac{9}{x^2 + 4x} = \dfrac{6}{x^2 + 2x}$

36. $\dfrac{6}{t^2 - 3t} = \dfrac{12}{t^2 - 9t}$

37. $x + \dfrac{1}{x} = 2$

38. $a - \dfrac{2}{a} = \dfrac{7}{3}$

39. $\dfrac{3x+2}{x^2 - 4x - 5} + \dfrac{x-4}{x+1} = \dfrac{x}{x-5}$

40. $\dfrac{3x+2}{x^2 - 4x - 5} + \dfrac{x-4}{x+1} - \dfrac{x}{x-5}$

41. $\dfrac{5x+1}{x^2-4} - \dfrac{2x+3}{x+2} + \dfrac{2x+3}{x-2}$

42. $\dfrac{2x-3}{x+2} - \dfrac{5x+1}{x^2-4} = \dfrac{2x+3}{x-2}$

43. $\dfrac{1}{x^2-x-2} + \dfrac{2}{x^2-1} = \dfrac{1}{x^2-3x+2}$

44. $\dfrac{3}{a^2+3a-10} + \dfrac{12}{a^2-2a} = \dfrac{4}{a^2+5a}$

45. $\dfrac{1}{x-4} - \dfrac{5}{x+2} = \dfrac{6}{x^2-2x-8}$

46. $\dfrac{3}{3x-2} - \dfrac{7}{x+1} = \dfrac{5}{3x^2+x-2}$

47. $\dfrac{n}{3n+2} + \dfrac{6}{9n^2-4} - \dfrac{2}{3n-2}$

48. $\dfrac{n}{2n-3} + \dfrac{2}{2n+3} - \dfrac{5n}{4n^2-9}$

49. $\dfrac{1}{3n+4} + \dfrac{8}{9n^2-16} = \dfrac{1}{3n-4}$

50. $\dfrac{5}{2n+1} + \dfrac{3}{2n-1} = \dfrac{22}{4n^2-1}$

51. $\dfrac{4}{2x-1} + \dfrac{2}{x+3} = \dfrac{5}{2x^2+5x-3}$

52. $\dfrac{7}{3a+2} + \dfrac{4}{a+5} = \dfrac{8}{3a^2+17a+10}$

53. $\dfrac{6}{x} - \dfrac{2}{x-1} = 1$

54. $\dfrac{8}{x+3} + \dfrac{14}{x} = 1$

55. $\dfrac{x}{x-1} = \dfrac{2x}{x+1}$

56. $\dfrac{3x}{x+5} = \dfrac{x}{x+3}$

57. $\dfrac{2x}{x+2} = x-1$

58. $\dfrac{2}{x-1} = 4-x$

59. $\dfrac{5}{x} + \dfrac{9}{x+2} = 4$

60. $\dfrac{3}{x-2} - \dfrac{7}{x} = 6$

61. $\dfrac{6}{3a+5} - \dfrac{2}{a-4} = \dfrac{10}{3a^2-7a-20}$

62. $\dfrac{5}{y-2} - \dfrac{3}{2y-1} = \dfrac{4}{2y^2-5y+2}$

63. $\dfrac{2x+3}{x^2-x-2} - \dfrac{x+4}{x^2+3x+2} = \dfrac{x}{x^2-4}$

64. $\dfrac{3x-1}{x^2-3x-10} - \dfrac{1}{x+5} = \dfrac{2x+1}{x^2-3x-10}$

65. $\dfrac{3}{x^2-x-6} + \dfrac{2}{2x^2-5x-3} = \dfrac{5}{2x^2+5x+2}$

66. $\dfrac{10}{2a^2-a-15} - \dfrac{5}{3a^2-4a-15} = \dfrac{3}{6a^2+25a+25}$

67. $\dfrac{2x+1}{x^2+x-2} + \dfrac{4x}{x^2-1} = \dfrac{15x-1}{x^2+3x+2}$

68. $\dfrac{x+1}{x^2-2x-15} - \dfrac{x+2}{x^2-25} + \dfrac{1}{x-5}$

69. $\dfrac{4}{4x^2-9} - \dfrac{5}{4x^2-8x+3} = \dfrac{8}{4x^2+4x-3}$

70. $\dfrac{6}{9y^2-1} - \dfrac{4}{9y^2+9y+2} = \dfrac{9}{9y^2+3y-2}$

71. A DVD manufacturer finds that when n DVD players are produced, the average cost per player is given by the equation

$$C = \frac{48n + 22{,}000}{n}$$

If the manufacturer wants the average cost per player to be \$55, how many DVD players should be manufactured?

72. A satellite radio manufacturer finds that when x radios are produced, the average cost per radio is given by the equation

$$C = \frac{28x + 15{,}000}{x}$$

If the manufacturer wants the average cost per radio to be \$32, how many radios should be manufactured?

In Exercises 73–78, given $f(x) = \dfrac{3x+1}{(x-2)^2}$ and $g(x) = \dfrac{x+5}{x-2}$, express the following as a single simplified fraction.

73. $f(x) + g(x)$

74. $f(x) - g(x)$

75. $2f(x) - 3g(x)$

76. $3f(x) + 2g(x)$

77. $\dfrac{f(x)}{g(x)}$

78. $\dfrac{g(x)}{f(x)}$

Mini-Review

79. *Simplify:* $3x^2y(2x^3y^2)$ **80.** *Simplify:* $3x^2y(2x^3 + y^2)$

81. Give an example of a function $f(x)$ for which $f(x + 2) \neq f(x) + f(2)$.

82. Find the domain of $f(x) = \dfrac{1}{\sqrt{3x - 5}}$.

6.7   Literal Equations

In this section we elaborate on literal equations, which we first discussed in Section 1.5.

A literal equation that has a "real-life" interpretation is often called a ***formula***. For instance, the area, A, of a trapezoid (see Figure 6.3) is given by the formula

$$A = \frac{1}{2}h(b_1 + b_2)$$

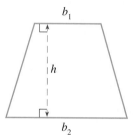

Figure 6.3
Trapezoid

This formula is quite useful if we are given h, b_1, and b_2 and want to compute A. However, if we are given A, h, and b_1 and we want to compute b_2, we would much prefer to have a formula solved explicitly for b_2.

EXAMPLE 1   Solve explicitly for b_2: $A = \frac{1}{2}h(b_1 + b_2)$

Solution   We offer two approaches to illustrate a point.

Method 1: $A = \dfrac{1}{2}h(b_1 + b_2)$ *Multiply the equation by 2 to clear the denominator.*

$2A = h(b_1 + b_2)$ *Apply the distributive property.*

$2A = hb_1 + hb_2$ *Then isolate b_2 (subtract hb_1 from both sides).*

$2A - hb_1 = hb_2$ *Divide both sides by h.*

$\dfrac{2A - hb_1}{h} = \dfrac{hb_2}{h}$

$\boxed{\dfrac{2A - hb_1}{h} = b_2}$

Method 2: $A = \dfrac{1}{2}h(b_1 + b_2)$ *Again clear the denominator.*

$2A = h(b_1 + b_2)$ *Divide both sides by h.*

$\dfrac{2A}{h} = \dfrac{h(b_1 + b_2)}{h}$

$\dfrac{2A}{h} = b_1 + b_2$ *Isolate b_2 (subtract b_1 from both sides).*

$\boxed{\dfrac{2A}{h} - b_1 = b_2}$

Both methods are equally correct. What is important is that you recognize that the two answers are equivalent. If we combine the second answer into a single fraction we get

$$\frac{2A}{h} - b_1 = \frac{2A}{h} - \frac{hb_1}{h} = \frac{2A - hb_1}{h}$$

which is the first answer. Keep this in mind when you check your answers with those in the answer section in the back of the book.

The same procedure can be applied to solving literal *inequalities* as well.

EXAMPLE 2

Solve explicitly for *a*: $a - 2c \leq 6a + 7d$

Solution
$$a - 2c \leq 6a + 7d$$
Add 2c to both sides of the inequality.

$$a - 2c \boxed{+ 2c} \leq 6a + 7d \boxed{+ 2c}$$

$$a \leq 6a + 7d + 2c$$
Do not stop here! The variable a must appear on only one side of the inequality.

$$a \boxed{- 6a} \leq 6a + 7d + 2c \boxed{- 6a}$$

$$-5a \leq 7d + 2c$$
$$\downarrow$$

$$\frac{-5a}{-5} \geq \frac{7d + 2c}{-5}$$
Remember that when we divide both sides of an inequality by a negative number, we reverse the inequality symbol.

$$\boxed{a \geq -\frac{7d + 2c}{5}}$$

EXAMPLE 3

Solve for *t*: $at + 9 = 3t + b$

Solution
We want to get the *t* terms on one side of the equation and the non-*t* terms on the other side.

$$at + 9 = 3t + b$$
First add −3t to both sides of the equation.

$$at + 9 \boxed{- 3t} = 3t + b \boxed{- 3t}$$

$$at + 9 - 3t = b$$
Then subtract 9 from both sides.

$$at - 3t + 9 \boxed{- 9} = b \boxed{- 9}$$

$$at - 3t = b - 9$$
To isolate t, we factor out the t on the left side.

$$t(a - 3) = b - 9$$
Then divide both sides by a − 3.

$$\frac{t(a - 3)}{a - 3} = \frac{b - 9}{a - 3}$$

$$\boxed{t = \frac{b - 9}{a - 3}}$$

Since we cannot divide by 0, this solution assumes that $a \neq 3$.

Notice that in Example 3 we had to factor out *t* from $at - 3t$. This may seem like a new step that we have not seen before, but actually this step was implicit in the equations we solved prior to this section. For example, in solving

$$7t - 9 = 3t + 8$$

we subtract $3t$ from both sides and add 9 to both sides so our equation would look like this:

$$7t - 9 \quad \boxed{- 3t + 9} \quad = 3t + 8 \quad \boxed{- 3t + 9}$$

$$7t - 3t = 8 + 9$$

We normally combine like terms, $7t - 3t = 4t$ and $8 + 9 = 17$, and simplify our equation:

$$4t = 17$$

If you think back to how we are allowed to combine terms (by the distributive property), you will see that the factoring step is implicit in combining terms:

$$7t - 3t = (7 - 3)t \qquad \textit{Distributive property}$$
$$= 4t$$

So this procedure was not new; you have actually been doing it all along.

EXAMPLE 4

Solution

Solve for u: $y = \dfrac{u + 1}{u + 2}$

$$y = \frac{u + 1}{u + 2} \qquad \textit{Begin by multiplying both sides by } u + 2 \textit{ to clear the denominator.}$$

$$(u + 2)y = \frac{u + 1}{u + 2} \cdot \frac{u + 2}{1}$$

$$(u + 2)y = u + 1 \qquad \textit{Apply the distributive property.}$$

$$uy + 2y = u + 1$$

$$uy + 2y \quad \boxed{- u - 2y} \quad = u + 1 \quad \boxed{- u - 2y} \qquad \textit{We collect all the u terms on one side, non-u terms on the other side.}$$

$$uy - u = 1 - 2y \qquad \textit{Factor out u on the left side.}$$

$$u(y - 1) = 1 - 2y \qquad \textit{Then divide both sides of the equation by } y - 1.$$

$$\frac{u(y - 1)}{y - 1} = \frac{1 - 2y}{y - 1}$$

$$\boxed{u = \frac{1 - 2y}{y - 1}}$$

This solution assumes that $y \neq 1$ and $u \neq -2$. Why?

EXERCISES 6.7

Solve each of the following equations or inequalities explicitly for the indicated variable.

1. $5x + 7y = 4$ for x

2. $5x + 7y = 4$ for y

3. $2x - 9y = 11$ for y

4. $2x - 9y = 11$ for x

5. $w + 4z - 1 = 2w - z + 3$ for w

6. $w + 4z - 1 = 2w - z + 3$ for z

7. $2(6r - 5t) > 5(2r + t)$ for r

8. $2(6r - 5t) > 5(2r + t)$ for t

9. $3m - 4n + 6p = 5n + 2p - 8$ for n

10. $3m - 4n + 6p = 5n + 2p - 8$ for m

11. $\dfrac{a}{5} - \dfrac{b}{3} = \dfrac{a}{2} - \dfrac{b}{6}$ for a

12. $\dfrac{c}{12} - \dfrac{d}{8} = \dfrac{c}{4} - \dfrac{d}{6}$ for d

13. $\dfrac{x + y}{3} - \dfrac{x}{2} + \dfrac{y}{6} = 3(x - y)$ for x

14. $\dfrac{w - z}{10} - \dfrac{w}{5} + \dfrac{z}{4} = 2(z + w)$ for z

15. $ax + b = cx + d$ for x

16. $p - rz = q - tz$ for z

17. $3x + 2y - 5 = ax + by + 1$ for x

18. $3x + 2y - 5 = ax + by + 1$ for y

19. $(x + 3)(y + 7) = a$ for x

20. $(3x - 2)(2y - 1) = b$ for y

21. $y = \dfrac{u - 1}{u + 1}$ for u

22. $z = \dfrac{w - 2}{w + 3}$ for w

23. $x = \dfrac{2t - 3}{3t - 5}$ for t

24. $u = \dfrac{3y - 4}{2y - 5}$ for y

Each of the following is a formula from mathematics or the physical or social sciences. Solve each formula for the indicated variable.

25. $A = \dfrac{1}{2}bh$ for b

26. $A = \dfrac{1}{2}bh$ for h

27. $A = \dfrac{1}{2}h(b_1 + b_2)$ for b_1

28. $A = \dfrac{1}{2}h(b_1 + b_2)$ for h

29. $A = P(1 + rt)$ for r

30. $A = P(1 + rt)$ for P

31. $C = \dfrac{5}{9}(F - 32)$ for F

32. $F = \dfrac{9}{5}C + 32$ for C

33. $\dfrac{P_1}{V_1} = \dfrac{P_2}{V_2}$ for P_2

34. $\dfrac{P_1}{V_1} = \dfrac{P_2}{V_2}$ for V_2

35. $S = s_0 + v_0 t + \dfrac{1}{2}gt^2$ for g

36. $S = s_0 + v_0 t + \dfrac{1}{2}gt^2$ for v_0

37. $\dfrac{x - \mu}{s} < 1.96$ for x (assume $s > 0$)

38. $\dfrac{x - \mu}{s} < 1.96$ for μ (assume $s > 0$)

39. $\dfrac{1}{f} = \dfrac{1}{f_1} + \dfrac{1}{f_2}$ for f_1

40. $\dfrac{1}{f} = \dfrac{1}{f_1} + \dfrac{1}{f_2}$ for f

41. $S = 2\pi r^2 + 2\pi rh$ for h

42. $S = 2LH + 2LW + 2WH$ for W

Mini-Review

43. *Solve for x:* $(x - 3)^2 = 4$

44. *Multiply and simplify:* $(x - 2)^3$

45. An auto repair shop spends a total of $940 on 18 car batteries. Some were heavy-duty batteries costing $60 each and the rest were regular batteries costing $50 each. How many of each were brought?

46. Use the following graph of $f(x)$ to determine its domain and range, and find $f(-3)$.

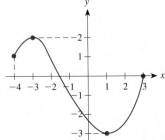

Applications: Rational Functions and Equations as Mathematical Models

As we have seen on numerous occasions, functions and equations can be used to model real-life situations. In this section, we pay particular attention to situations that give rise to functions and equations that involve rational expressions.

The outline suggested in Chapter 2 for solving verbal problems applies equally well to problems that give rise to fractional equations and inequalities. We will repeat this adivce in the next box for your reference.

Outline of Strategy for Solving Verbal Problems	1. *Read the problem carefully,* as many times as is necessary to understand what the problem is saying and what it is asking. 2. *Use diagrams* whenever you think it will make the given information clearer. 3. *Find the underlying relationship or formula* relevant to the given problem. Ask whether there is some underlying relationship or formula you need to know. If not, then the words of the problem themselves give the required relationship. 4. Clearly *identify the unknown quantity* (or quantities) in the problem, and label it (them) using one variable. 5. By using the underlying formula or relationship in the problem, *write an equation* involving the unknown quantity (or quantities). 6. *Solve* the equation. 7. *Answer the question.* Make sure you have answered the question that was asked. 8. *Check* the answer(s) in the original words of the problem.

EXAMPLE 1

Marc drove from his house to Jodi's house. He had driven over the speed limit for one-third of the distance, when he was stopped and given a ticket for speeding. He then drove under the speed limit for one-half the total distance until his car ran out of gas. He walked the remaining 5 miles to Jodi's house. What is the distance from Marc's house to Jodi's house?

Solution

At first glance, this may look like a distance–rate problem, but actually the problem involves only distance. First, draw a picture to represent the distances traveled and label the distances, as shown in Figure 6.4.

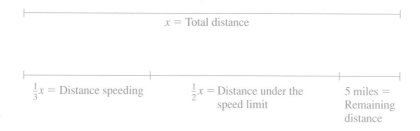

Figure 6.4

The figure illustrates the relationships among the distances traveled under each condition. It gives us the following equation:

$$\frac{1}{3}x + \frac{1}{2}x + 5 = x$$

We solve this equation by first multiplying the equation by the LCD of the fractions, which is 6:

$$\overset{2}{\cancel{6}}\left(\frac{1}{\cancel{3}}x\right) + \overset{3}{\cancel{6}}\left(\frac{1}{\cancel{2}}x\right) + 6 \cdot 5 = 6 \cdot x$$

$$2x + 3x + 30 = 6x$$

$$5x + 30 = 6x \qquad \textit{Subtract 5x from both sides.}$$

$$30 = x$$

$$x = \boxed{30 \text{ miles}}$$

Check: One-third the distance speeding $= \dfrac{1}{3}(30) = 10$ miles

One-half the distance under the limit $= \dfrac{1}{2}(30) = 15$ miles

Remaining distance $= 5$ miles

Total $= 30$ miles

An alternative, shorter way to solve the same problem is to consider the fact that Marc has already traveled $\frac{1}{2} + \frac{1}{3}$, or $\frac{5}{6}$ of the distance before he runs out of gas. So he still has $\frac{1}{6}$ the total distance to walk. This remaining distance is given as 5 miles. Hence, we have

$$\frac{1}{6} \text{ the total distance} = 5 \text{ miles} \quad \text{or} \quad \frac{1}{6}x = 5$$

Multiplying both sides of the equation by 6, we have

$$x = 30 \text{ miles}$$

Ratio and proportion problems are fairly straightforward to set up, as long as you remember to match up the corresponding units.

EXAMPLE 2

In a local district, the ratio of Democrats to Republicans is 5 to 7. If there are 2,100 Republicans in the district, how many Democrats are there?

Solution

First, you should realize that the ratio of a to b is simply the fraction $\dfrac{a}{b}$ reduced to lowest terms. Hence, to *find* the ratio of the number of Democrats to the number of Republicans, we would create the fraction

$$\frac{\text{Number of Democrats}}{\text{Number of Republicans}}$$

and reduce it to get $\frac{5}{7}$.

We let $x =$ number of Democrats and set up our equation with two fractions, being careful to match up the units:

$$\frac{5 \text{ Democrats}}{7 \text{ Republicans}} = \frac{x \text{ Democrats}}{2,100 \text{ Republicans}} \quad \text{or} \quad \frac{5}{7} = \frac{x}{2,100}$$

Multiply both sides of the equation by 2,100 to get

$$1,500 = x$$

Hence, there are $\boxed{1,500 \text{ Democrats}}$ in that district.

EXAMPLE 3

If 1 kg is 2.2 lb, how many kilograms are in 106 lb?

Solution   Again, we match up the units to check that the fractions are set up properly. Let $x =$ the number of kilograms in 106 lb.

$$\frac{x \text{ kg}}{106 \text{ lb}} = \frac{1 \text{ kg}}{2.2 \text{ lb}}$$

$$\frac{x}{106} = \frac{1}{2.2}$$

Since we are solving the x, we multiply both sides of the equation by 106:

$$x = \frac{106}{2.2} = \boxed{48.18 \text{ kg}} \quad \text{(rounded to two decimal places)}$$

EXAMPLE 4

If the ratio of Republicans to Democrats in a district is 4 to 9, and there is a total of 2,210 Republicans and Democrats in that district, how many Republicans are there?

Solution   Let $x =$ number of Republicans. Then

$$2{,}210 - x = \text{Number of Democrats} \qquad \textit{Total minus part = remainder}$$

Hence, our equation becomes

$$\frac{4}{9} = \frac{x}{2{,}210 - x}$$

Multiplying both sides of the equation by $9(2{,}210 - x)$, we get

$$\cancel{9}(2{,}210 - x)\left(\frac{4}{\cancel{9}}\right) = \left(\frac{x}{\cancel{2{,}210 - x}}\right)(9)(\cancel{2{,}210 - x})$$

$$4(2{,}210 - x) = 9x$$

$$8{,}840 - 4x = 9x$$

$$8{,}840 = 13x$$

$$680 = x$$

Thus, there are $\boxed{680 \text{ Republicans}}$ in the district.

We leave the check to the student.

EXAMPLE 5

How many liters each of 35%* and 60% alcohol solutions must be mixed together to make 40 liters of a 53% alcohol solution?

Solution   We offer both a one- and a two-variable approach.

One-Variable Approach:

We can let $x =$ amount of 35% solution.

Then $40 - x =$ amount of 60% solution. Why?

We can visualize this problem as shown in Figure 6.5.

*In this textbook, when we refer to percent we mean percent by volume.

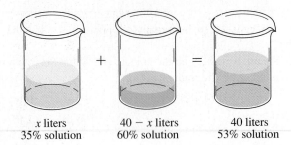

Figure 6.5

x liters
35% solution

$40 - x$ liters
60% solution

40 liters
53% solution

To write an equation to solve this problem, we need to differentiate between the amount of *solution* we have and the amount of *actual alcohol* in the solution. For instance, if we have 100 liters of a 30% alcohol solution, the amount of actual alcohol is

$$0.30(100) = 30 \text{ liters}$$

The basic relationship we use in this example is that the amounts of alcohol in each solution add up to the total amount of alcohol in the final solution.

Thus, our equation is

$$0.35x \quad + \quad 0.60(40 - x) \quad = \quad 0.53(40)$$

Amount of alcohol in the 35% solution + *Amount of alcohol in the 60% solution* = *Total amount of alcohol in final solution*

Since $0.35 = \frac{35}{100}$, $0.60 = \frac{60}{100}$, and $0.53 = \frac{53}{100}$, we multiply both sides of the equation by 100 to clear the decimals:

$$35x + 60(40 - x) = 53(40)$$
$$35x + 2{,}400 - 60x = 2{,}120$$
$$-25x + 2{,}400 = 2{,}120$$
$$-25x = -280$$
$$x = \frac{-280}{-25} = 11.2$$

> We must mix 11.2 liters of 35% alcohol solution with $40 - 11.2 = 28.8$ liters of 60% alcohol solution

Check: $11.2 + 28.8 \overset{\checkmark}{=} 40$ $\qquad 0.35(11.2) + 0.60(28.8) \overset{?}{=} 0.53(40)$

$$3.92 + 17.28 \overset{\checkmark}{=} 21.2$$

Two-Variable Approach: We can also visualize the problem as indicated in Figure 6.6.

Let x = amount of 35% alcohol solution

Let y = amount of 60% alcohol solution

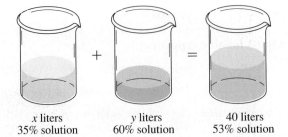

Figure 6.6

Diagram for two-variable approach

x liters
35% solution

y liters
60% solution

40 liters
53% solution

We create two equations—one relating the amounts of *solution,* the other relating the amounts of *pure alcohol.* We obtain the following system of equations:

$$\begin{cases} x + y = 40 & \text{\textit{Amounts of solution}} \\ 0.35x + 0.60y = 0.53(40) & \text{\textit{Amounts of pure alcohol}} \end{cases}$$

We rewrite the first equation, solving for y to get $y = 40 - x$. We then substitute $y = 40 - x$ into the second equation and get

$$0.35x + 0.60(40 - x) = 0.53(40)$$

This is exactly the same equation we obtained using the one-variable approach. The rest of the solution is as it appears for the one-variable approach.

EXAMPLE 6

Mrs. Stone invests a certain amount of money in a bank account paying 5.1% interest per year, and $3,500 more than this amount in a mutual fund paying 8.2% interest per year. If the annual income from the two investments is $905.45, how much is invested at each rate?

Solution   Let's analyze this problem.

THINKING OUT LOUD

What do we need to find?

How much was invested at 5.1% and 8.2%.

What information are we given?

We are given that she invested $3,500 more in the 8.2% investment than in the 5.1% investment, and that the annual income (interest) from the investments is $905.45.

How is the given information related to what we need to find?

We first focus on the annual interest. We know that the amount of interest, I, earned in 1 year if P dollars is invested at a rate of r% per year is computed as

$$I = P \cdot r \quad \textbf{\textit{Interest} = (Principal)(Rate)}$$

For example, if $1,000 is invested at 8% per year, then the interest earned in 1 year is

$$I = (1,000)(0.08) = \$80.$$

How do we apply $I = Pr$ to this problem?

5.1% written as a decimal is 0.051; 8.2% is 0.082.

Let's say she invested x amount at 5.1%: her interest for that investment is then $0.051x$. But if she invested x dollars at 5.1%, then she must have invested $x + 3,500$ at 8.2% (Remember, she invested $3,500 more in the 8.2% investment). Her interest for the 8.2% investment is then $0.082(x + 3,500)$. We can now write an equation involving the interest.

Hence, we let

$$x = \text{Amount invested at 5.1\%}$$

Then

$$x + 3,500 = \text{Amount invested at 8.2\%}$$

Our equation is

$$0.051x \quad + \quad 0.082(x + 3{,}500) = \quad\quad 905.45$$

| Interest received from the 5.1% investment | + | Interest received from the 8.2% investment | = | Total interest received from both investments |

Multiplying both sides of the equation by 1,000 clears the decimals:

$$51x + 82(x + 3{,}500) = 905{,}450$$
$$51x + 82x + 287{,}000 = 905{,}450$$
$$133x = 618{,}450$$
$$x = 4{,}650$$

Thus, there is

$$\boxed{\$4{,}650 \text{ invested at } 5.1\%}$$

and $3,500 more, or

$$\boxed{\$8{,}150 \text{ invested at } 8.2\%}$$

Using a calculator gives us the advantage of skipping the step of "clearing decimals." But you still need to write down the equation steps as you proceed to solve the equation. That is, we can start with

$$0.051x + 0.082(x + 3{,}500) = 905.45 \qquad \text{\textit{Simplify the left side using your}}$$
$$0.051x + 0.082x + 287 = 905.45 \qquad \text{\textit{calculator.}}$$
$$0.133x + 287 = 905.45 \qquad \text{\textit{Isolate x using the calculator}}$$
$$0.133x = 618.45 \qquad \text{\textit{for your computations.}}$$
$$x = 4{,}650$$

Check: $8{,}150 \overset{\checkmark}{=} 4{,}650 + 3{,}500$ $0.051(4{,}650) + 0.082(8{,}150) \overset{?}{=} 905.45$
$$237.15 + 668.30 \overset{\checkmark}{=} 905.45$$

The computations in the check can easily be done by calculator.

EXAMPLE 7

Solution

Greg can paint a house in 20 hours, whereas Mark can paint the same house in 40 hours. How long will it take them to paint the house if they work together?

To solve this problem algebraically we need to assume that Greg and Mark work in a totally cooperative manner and that they work at a constant rate.

This example is similar to the work problem in Section 2.2 (Example 9), in which we were given a rate and needed to know the time it took to get a certain amount done. In general we know that the relationship between the amount of work done and the rate at which the work is being done is

$$\text{Amount of work done} = (\text{Work rate}) \cdot (\text{time}) \qquad \text{or} \qquad W = rt$$

What makes this example slightly different from the version in Section 2.2 is that the rates are stated differently in this problem. We can rewrite the rates given in this problem as follows: To say that it takes Greg 20 hours to paint a house means that Greg can complete $\frac{1}{20}$ of a house in 1 hour, or that his rate is $\frac{1}{20}$ of a house per hour.

Notice that $\frac{1}{20}$ house per hour means that in 3 hours Greg paints $\frac{3}{20}$ of a house, and in 20 hours, he completes $\frac{20}{20} = 1$ house. In general, the portion of the job completed = (rate) $\times$ (time).

Since Mark can paint the same house in 40 hours, his rate is $\frac{1}{40}$ of a house per hour. Now that we have the rates in more familiar form, we can approach this example as we did Example 9 of Section 2.2.

$$\left(\begin{array}{c}\text{Portion of the job}\\\text{completed by Greg}\end{array}\right) \quad + \quad \left(\begin{array}{c}\text{Portion of the job}\\\text{completed by Mark}\end{array}\right) \quad = \quad \left(\begin{array}{c}1 \text{ complete}\\\text{job}\end{array}\right)$$

(Greg's rate) · (Greg's time) + (Mark's rate) · (Mark's time) = 1 complete house

We are given Greg's and Mark's rates. Since Greg and Mark are working the same amount of time, we can let x = number of hours Greg and Mark work together. Thus the equation becomes

$$\frac{1}{20} \cdot x \quad\quad + \quad\quad \frac{1}{40} \cdot x \quad\quad = 1$$

or

$$\frac{x}{20} + \frac{x}{40} = 1$$

Multiply by the LCD, which is 40:

$$\frac{40}{1} \cdot \frac{x}{20} + \frac{40}{1} \cdot \frac{x}{40} = 40 \cdot 1$$

$$2x + x = 40$$

$$3x = 40$$

$$x = \frac{40}{3} = 13\tfrac{1}{3} \text{ hours} = \boxed{13 \text{ hours and 20 minutes}}$$

The check is left to the student.

EXAMPLE 8

It takes Nadia 8 hours to complete a particular job. Suppose Victor begins working on the job with Nadia, and he then quits after 2 hours. Nadia takes 3 more hours to complete the job working alone. How long would it take Victor to do the job by himself?

Solution

This problem is similar to Example 7, but requires more analysis.

Since Nadia can complete the job in 8 hours, her rate is $\frac{1}{8}$ of a job per hour. Similarly, if we let x = the amount of time it takes Victor to complete the job, then his rate is $\frac{1}{x}$ of a job per hour.

They start by working together for 2 hours. This means that for the first 2 hours, Nadia completes

$$W = r \cdot t$$

$$W = \left(\frac{1}{8}\right)(2) = \frac{1}{4} \text{ of the job}$$

For the same first 2 hours, Victor completes

$$W = r \cdot t$$

$$W = \left(\frac{1}{x}\right)(2) = \frac{2}{x} \text{ of the job}$$

Hence after the first 2 hours $\frac{1}{4} + \frac{2}{x}$ of the job is complete.

After Victor quits, it takes Nadia 3 hours to complete the job working alone. Since her rate is $\frac{1}{8}$ of a job per hour, during the last 3 hours she completes

$$W = rt$$

$$W = \left(\frac{1}{8}\right)(3) = \frac{3}{8} \text{ of the job}$$

We can now add the portions of the job completed by each person over each period of time:

$$\underbrace{\frac{1}{4}}_{\substack{\text{Portion completed} \\ \text{by Nadia during} \\ \text{the first 2 hours}}} + \underbrace{\frac{2}{x}}_{\substack{\text{Portion completed} \\ \text{by Victor during} \\ \text{the first 2 hours}}} + \underbrace{\frac{3}{8}}_{\substack{\text{Portion completed} \\ \text{by Nadia during} \\ \text{the last 3 hours}}} = \underbrace{1}_{\text{1 complete job}}$$

We solve for *x:*

$$\frac{1}{4} + \frac{2}{x} + \frac{3}{8} = 1 \qquad \textit{Multiply both sides by the LCD, 8x.}$$

$$8x\left(\frac{1}{4} + \frac{2}{x} + \frac{3}{8}\right) = 8x \cdot 1 \qquad \textit{Perform the operations on each side and simplify.}$$

$$8x \cdot \frac{1}{4} + 8x \cdot \frac{2}{x} + 8x \cdot \frac{3}{8} = 8x$$

$$2x + 16 + 3x = 8x$$

$$5x + 16 = 8x \qquad \textit{Solve for x to get}$$

$$16 = 3x$$

$$\frac{16}{3} = x$$

Hence working alone, it would take Victor $\frac{16}{3} = 5\frac{1}{3}$ hours.

EXAMPLE 9

Solution

A train can make a 480-km trip in the same time that a car can make a 320-km trip. If the train travels 40 kph faster than the car, how long does it take the car to make its trip?

We will use the relationship $d = r \cdot t$ in the form $t = \frac{d}{r}$. Since we are told that the times for the train and car are the same, we have

$$t_{\text{train}} = t_{\text{car}}$$

$$\frac{d_{\text{train}}}{r_{\text{train}}} = \frac{d_{\text{car}}}{r_{\text{car}}}$$

Let r = rate for car. Then $r + 40$ = rate for train. Thus, our equation is

$$\frac{480}{r + 40} = \frac{320}{r} \qquad LCD = r(r + 40)$$

$$r(r + 40) \cdot \frac{480}{r + 40} = r(r + 40) \cdot \frac{320}{r}$$

$$480r = 320(r + 40)$$

$$480r = 320r + 12{,}800$$

$$160r = 12{,}800$$

$$r = 80$$

Keep in mind that the problem asks for the *time* it takes the car to make the 320-km trip, which is

$$t = \frac{320}{80} = \boxed{4 \text{ hours}}$$

Check: The rate of the train is $80 + 40 = 120$ kph.
The time for the train to make its 480-kmp trip is

$$t = \frac{480}{120} = 4 \text{ hours}$$

Let's turn to the situation described at the beginning of this chapter.

EXAMPLE 10

An environmental consultant for a large factory finds that the cost C (in thousands of dollars) to remove p percent of the pollutants released by the factory can be approximated by the function

$$C(p) = \frac{60p}{100 - p}$$

(a) What is the cost of removing 70% of the pollutants?

(b) If the company determines that it can afford to allocate $200,000 to remove pollutants for the factory, what percentage of the pollutants can be removed?

(c) Suppose the company puts A (in thousands) toward removing pollutants. Express the percentage of pollutants that can be removed as a function (in terms) of A, and use this function to answer part **(b).**

Solution

(a) We are given the percentage p of pollutants to be removed, so we simply substitute for p in the given cost function. In other words, we are being asked to computer $C(70)$.

$$C(p) = \frac{60p}{100 - p} \qquad \textit{Substitute } p = 70.$$

$$C(70) = \frac{60(70)}{100 - 70} = \frac{60(70)}{30} = 140$$

Thus, the cost of removing 70% of the pollutants is $140,000.

(b) We are given the cost, $C(p)$, and want to find p. Since $C(p)$ is the cost in thousands, we have $C(p) = 200$. Hence, we need to solve the *rational equation* $C(p) = 200$:

$$C(p) = \frac{60p}{100 - p} \qquad \textit{Set } C(p) = 200 \textit{ and find p.}$$

$$200 = \frac{60p}{100 - p} \qquad \textit{Multiply both sides of the equation by } 100 - p.$$

$$(100 - p)200 = \frac{60p}{100 - p}(100 - p)$$

$$200(100 - p) = 60p$$

$$20,000 - 200p = 60p$$

$$20,000 = 260p$$

$$\frac{20,000}{260} \approx 76.92 = p$$

Hence, rounding to the nearest tenth, 76.9% of the pollutants can be removed for $200,000.

(c) Since $C(p)$ and A are both expressed in thousands, we have $C(p) = A$.

$$C(p) = \frac{60p}{100 - p} \qquad \textit{Set } C(p) = A \textit{ and solve for p:}$$

$$A = \frac{60p}{100 - p} \qquad \textit{Multiply both sides of the equation by } 100 - p.$$

$$(100 - p)A = \frac{60p}{100 - p}(100 - p)$$

$$A(100 - p) = 60p \qquad \textit{Distribute A.}$$

$$100A - pA = 60p \qquad \textit{Collect terms containing p.}$$

$$100A = pA + 60p \qquad \textit{Factor out p.}$$

$$100A = p(A + 60) \qquad \textit{Isolate p: Divide both sides by A + 60.}$$

$$\frac{100A}{A + 60} = p$$

Hence, $p = p(A) = \dfrac{100A}{A + 60}$ gives us the percent of pollutants that can be removed for A thousand dollars.

We can use this function to computer the answer to part **(b)** by substituting $A = 200$ in the function $p(A)$.

$$p = P(A) = \frac{100A}{A + 60} \qquad \textit{Substitute A = 200.}$$

$$= \frac{100(200)}{200 + 60} \approx 76.92 \qquad \textit{Which agrees with our answer to part (b)}$$

EXERCISES 6.8

Solve each of the following problems algebraically. That is, set up an equation or inequality and solve it. Be sure to label clearly what the variable represents.

1. If three-fourths of a number is 7 less than two-fifths of the number, what is the number? Let x = number; $\frac{3}{4}x = \frac{2}{5}x - 7$; $x = -20$

2. If 5 more than five-sixths of a number is 3 less than two-thirds of the number, what is the number? $5 + \frac{5}{6}x = \frac{2}{3}x - 3$; number is -48

3. The ratio of men to women in a certain graduating class is 7 to 9. If there are 810 women in the class, how many men are in the class? Let x = # of men; $\frac{7}{9} = \frac{x}{810}$; 630 men

4. In a certain town, the ratio of dogs to cats is 8 to 11. If there is a total of 2,812 cats and dogs, how many cats are there? $\frac{8}{11} = \frac{2,812 - x}{x}$; 1,628 cats

5. The ratio of two positive numbers is 5 to 12. Find the two numbers, if one number is 21 less than the other. Let x = the larger number; $\frac{5}{12} = \frac{x - 21}{x}$; two numbers are 36 and 15

6. The ratio of two positive numbers is 7 to 9. Find the two numbers, if one number is 4 more than the other. $\frac{7}{9} = \frac{x}{x + 4}$; numbers are 14 and 18

7. If 1 in. is 2.54 cm, how many inches are there in 52 cm? (Round your answer to two decimal places.) Let x = # inches in 52 cm; $\frac{1}{2.54} = \frac{x}{52}$; 20.47 in.

8. If 1 kg is 2.2 lb, how many pounds are there in 17 kg? (Round your answer to two decimal places.) $\frac{1}{2.2} = \frac{17}{x}$; 37.4 lb in 17 kg

9. If one euro (European dollar) is worth $1.24 (U.S. dollars), how much is $312 worth in euros? Let x = # euros; $\frac{1}{1.24} = \frac{x}{312}$; 251.61 euros

10. If one U.S. dollar is worth 110.42 Japanese yen, what is the U.S. dollar value of 5,000 yen? Let x = # U.S. dollars $\frac{1}{110.42} = \frac{x}{5,000}$; $45.28

11. On planet G, the units of currency are the droogs, the dreeps, and the dribbles. If 5 droogs are equivalent to 4 dreeps, and 7 dreeps are equivalent to 25 dribbles, how many dribbles are in 28 droogs? Let x = # of dribbles; $\frac{x}{28} = \frac{25}{7} \cdot \frac{4}{5}$; 80 dribbles

12. On planet P, length is measured in wings, wongs, and wytes. If 14 wings are equivalent to 9 wytes, and 9 wongs are equivalent to 7 wings, how many wongs make a wyte? $\frac{9}{14} = \frac{1}{x}$; $\frac{14}{9}$ wings in 1 wyte; $\frac{9}{7} = \frac{x}{14/9}$; 2 wongs make 1 wyte.

13. The first side of a triangle is one-half the second side; the third side is 2 cm more than the second side. If the perimeter is 22 cm, find the lengths of the sides. Let x = second side; $P = 22 = \frac{1}{2}x + x + x + 2$; lengths are 4, 8, and 10 cm

14. The width of a rectangle is three-fifths the length. If the perimeter is 80 in., find the dimensions of the rectangle. $80 = \frac{6}{5}x + 2x$; 15 in. by 25 in.

15. The perimeter of a rectangle is 50 cm. If its length is $2\frac{1}{2}$ times its width, find its dimensions. Let x = width; $50 = 2(2.5x + x)$; $x = \frac{50}{7}$ cm (width), length = $\frac{125}{7}$ cm

16. The shortest side of a triangle is two-thirds its medium side, and the longest side is 5 ft longer than the shortest side. If the perimeter is $23\frac{2}{3}$ ft, find the length of each side. $\frac{71}{3} = x + \frac{2}{3}x + \frac{2}{3}x + 5$; 8 ft by $\frac{16}{3}$ ft by $\frac{31}{3}$ ft

17. Mike was walking from the ballfield when one-quarter of the way home he decided to take a cab. If he was 6 miles from his home when he decided to take a cab, how far is his home from the ballfield? Let x = distance from home; $\frac{3x}{4} = 6$; 8 miles

18. Half the height of a radio tower is painted blue and $\frac{1}{5}$ the height is painted red. If 22 ft remain unpainted, how tall is the tower? $\frac{1}{2}x + \frac{1}{5}x + 22 = x$; $73\frac{1}{3}$ ft.

19. Valerie walks from her home to her school. One-fifth of the way there she finds a nickel. One-quarter of the rest of the way she finds a dime. If she is still 2 blocks from school when she finds the dime, how many blocks did she walk from her home to school? (Assume that all blocks are equal in length.) Let x = distance from home to school; $x - \frac{1}{5}x - \frac{1}{4}\left(\frac{4}{5}x\right) = 2$; $3\frac{1}{3}$ blocks

20. Raju walks from his home to his school. One-fifth of the way there he finds a nickel. One quarter of the rest of the way there he finds a dime. If he is still 3 blocks from school when he finds the dime, how far did he walk from his home to his school? (Assume that all blocks are equal in length.) $\frac{1}{5}x + \frac{1}{5}x + 3 = x$; 5 blocks

21. A law of physics states that if an electrical circuit has three resistors in parallel, then the reciprocal of the total resistance of the circuit is the sum of the reciprocals of the individual resistances. As a formula, we have

$$\frac{1}{R} = \frac{1}{R_1} + \frac{1}{R_2} + \frac{1}{R_3}$$

where R_1, R_2, and R_3 are the individual resistances measured in ohms, and R is the total resistance measured in ohms (see the accompanying figure).

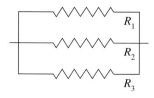

In such a circuit, the total resistance is $1\frac{1}{4}$ ohms, and two of the resistors are 2 ohms and 5 ohms. What is the third resistance? $\frac{4}{5} = \frac{1}{2} + \frac{1}{5} + \frac{1}{R_3}$; $R_3 = 10$ ohms

22. Einstein's theory of relativity states that velocities must be added according to the formula

$$v = \frac{v_1 + v_2}{1 + \frac{v_1 v_2}{c^2}}$$

where v is the resulting velocity, v_1 and v_2 are the velocities to be added, and c is the speed of light. If v_1 and v_2 are both one-half the speed of light, what is the resultant velocity? $v = \frac{4}{5}c$

23. Cindy has invested in two interest-bearing investments: a certificate of deposit that yields 6% interest per year, and a bond that yields 10% interest. She invested $3,000 more in the certificate than in the bond. If she receives a total of $580 in annual interest, how much is invested in each venture?

Let x be invested in bond;
$0.10x + 0.06(x + 3,000) = 580$;
$2,500 in bond; $5,500 in CD

24. Nick invested his money in two money-market certificates: One yields 4% interest per year and the other, higher-risk, certificate yields 8% per year. He invested $3,520 more in the certificate yielding 4% than in the other certificate. How much did he invest in each certificate if he receives a total of $1,100.80 in annual interest from the two certificates?

$0.08x + 0.04(3,520 + x) = 1,100.80$;
8%: $8,000, 4%: $11,520

25. Arthur has a total of $25,000 saved in two banks: One bank gives $5\frac{1}{2}\%$ yearly interest, and the other gives 7% interest. If his total yearly interest from both banks is $1,465, how much was saved in each bank?

Let x be invested at $5\frac{1}{2}\%$;
$0.055x + 0.07(25,000 - x) = 1,465$;
$19,000 at $5\frac{1}{2}\%$, $6,000 at 7%

26. Patrick has invested $15,000 in two bonds: One bond yields 9% annual interest and the other yields $8\frac{3}{4}\%$ annual interest. How much is invested in each certificate if the combined yearly interest from both bonds is $1,340?

$0.09x + 0.0875(15,000 - x) = 1,340$;
9%: $11,000, $8\frac{3}{4}\%$: $4,000

27. Lew wants to invest $18,000 in two bonds. One yields $8\frac{1}{2}\%$ annual interest and the other yields 11% annual interest. How much should he invest in each bond if he wants to receive a combined annual interest rate of 10%?

Let x be invested at $8\frac{1}{2}\%$;
$0.085x + 0.11(18,000 - x) =$
$0.10(18,000)$;
$7,200 at $8\frac{1}{2}\%$, $10,800 at 11%

28. Joan saved $21,000 in two banks. One bank gives $4\frac{3}{4}\%$ annual interest, and the other gives $5\frac{1}{2}\%$ annual interest. How much did she save in each bank if she received a combined annual interest rate of 5% on her savings?

$0.0475x + 0.055(21,000 - x) =$
$0.05(21,000)$; $4\frac{3}{4}\%$: $14,000, $5\frac{1}{2}\%$: $7,000

29. It takes Carol 3 hours to paint a small room and it takes John 5 hours to paint the same room. Working together, how long would it take for them to finish the room?

Let x = number of hours working together; $\frac{x}{3} + \frac{x}{5} = 1$; $1\frac{7}{8}$ hours

30. Pipe A can fill a pool with water in 3 days, and pipe B can fill the same pool in 2 days. If both pipes were used, how long would it take to fill the pool?

$\frac{x}{3} + \frac{x}{2} = 1$; $1\frac{1}{5}$ days

31. The Quickie Cleaning Service can clean a certain office building in 30 hours, while the Super-Quickie Cleaning Service can do the same job in 20 hours. If the two services work together, how long will it take to get the building clean?

Let x = number of hours working together; $\frac{x}{30} + \frac{x}{20} = 1$; 12 hours

32. The Echo brand candy corn machine can make enough candy corn to fill a supermarket order in 16 hours. On the other hand, the newer Echo 2 machine can make enough to fill the same order in 13 hours. How many hours would it take to make enough candy corn to fill the supermarket order if both machines were working together?

$\frac{x}{16} + \frac{x}{13} = 1$; $7\frac{5}{29}$ hours

33. A bricklayer can complete a wall in $2\frac{2}{3}$ days, whereas his assistant can do the same size wall in 5 days. Working together, how long will it take for them to complete the wall?

Let x = time working together;
$\frac{x}{2\frac{2}{3}} + \frac{x}{5} = 1$; $1\frac{17}{23}$ days

34. A clerk can process 500 forms in $1\frac{2}{5}$ days. It takes $2\frac{1}{2}$ days for another clerk to process the same 500 forms. If both clerks work together, how long will it take them to process the 500 forms?

$\frac{x}{7/5} + \frac{x}{5/2} = 1$; $\frac{35}{39}$ day

35. Repeat Exercise 31 if the Quickie Cleaning Service works alone for 10 hours and is then replaced by the Super-Quickie Cleaning Service.

Let x = # hours for the Super-Quickie Service to finish cleaning; $\frac{10}{30} + \frac{x}{20} = 1$; $x = \frac{40}{3}$; Total time = $23\frac{1}{3}$ hours

36. Repeat Exercise 31 if the Super-Quickie Cleaning Service works alone for 10 hours and is then replaced by the Quickie Cleaning Service. $\frac{10}{20} + \frac{x}{30} = 1$; 25 hours together

37. When a bathtub faucet is turned on (and the drain is shut), it can fill a tub in 10 minutes; when the drain is open, a full tub can *empty* in 15 minutes. How long would it take for the bathtub to fill if the water were turned on with the drain left open? [*Hint:* Let the rate at which the bathtub drains be negative.]

Let x = number of minutes to overflow; $\frac{x}{10} - \frac{x}{15} = 1$; 30 minutes

38. How long would it take for the bathtub of Exercise 37 to fill if a full tub can empty in 6 minutes when the drain is opened?

$\frac{x}{10} - \frac{x}{6} = 1;$
No solution; the tub will not fill.

39. Aaron takes 10 hours to complete a particular job, and Kimberly takes 8 hours to complete the same job. Aaron begins work on the job by himself at noon. At 3 P.M. Kimberly joins him. What time do they complete the job?

Let x = # hours they work together;
$\frac{3}{10} + \frac{x}{10} + \frac{x}{8} = 1;$
$x = 3.11$, at 6:07 P.M.

40. It takes 4 hours for David and Malika to complete a particular job working together. If David could have completed the job himself in 8 hours, how long would it have taken Malika to complete the job working alone?

$\frac{4}{8} + \frac{4}{x} = 1;$ 8 hours

41. Lori takes 6 hours to mow her lawn alone. Suppose she starts the job at 10 A.M. and an hour later Megan shows up with her mower and helps her finish at 1 P.M. How long would it have taken Megan to do the job by herself?

Let x = the number of hours it takes Megan to do the job herself;
$\frac{1}{6} + \frac{2}{6} + \frac{2}{x} = 1;$ 4 hours

42. What time would Megan and Lori of Exercise 41 finish mowing the lawn if they both started at 10 A.M.?

$\frac{x}{6} + \frac{x}{4} = 1; x = 2.4$ hr; 12:24

43. Bill can ride 10 km in the same time that Jill can ride 15 km. If Jill rides 10 kph faster than Bill, how fast does Bill ride?

Let x = Bill's rate; $\frac{10}{x} = \frac{15}{x+10};$ 20 kph

44. A car can make a 200-mile trip in the same time that a bike can make a 60-mile trip. If the car travels 35 mph faster than the bike, how long does it take the bike to make its trip?

$\frac{200}{x+35} = \frac{60}{x};$ 4 hours

45. Marla drove 600 miles on an interstate highway. Her speed was 50 mph except on a part of the highway under construction, where her speed was 20 mph. If her total driving time was 14 hours, how many miles did she drive at the slower speed?

Let x = number of miles at slow speed;
$\frac{600-x}{50} + \frac{x}{20} = 14; 66\frac{2}{3}$ miles

46. Susan decided to run to the store, which is 8 km from her house. She ran at a rate of 7 kph for part of the way and then walked at a rate of 3 kph the rest of the way. If the total trip took 2 hours, how many km did she run?

$7x + 3(2 - x) = 8;$ 3.5 km

47. How many ounces of a 20% solution of alcohol must be mixed with 5 oz of a 50% solution to get a mixture of 30% alcohol?

Let x = number of ounces of 20% alcohol; $0.20x + 0.50(5) = 0.30(x + 5);$ 10 oz of 20% alcohol

48. How many ounces of a 35% solution of sulfuric acid (and distilled water) must be mixed with 12 oz of a 20% solution to get a 30% solution of sulfuric acid?

$0.35x + 0.20(12) = 0.30(x + 12);$ 24 oz

49. A chemist has a 30% solution of alcohol and a 75% solution of alcohol. How much of each should be mixed together to get 80 ml of a 50% mixture?

Let x = number of ml of 30% solution; $0.30x + 0.75(80 - x) = 0.50(80);$ $44\frac{4}{9} \approx 44.44$ ml of 30% solution, $35\frac{5}{9} \approx 35.56$ ml of 75% solution

50. A chemist has a 45% mixture of hydrochloric acid and an 80% mixture of the same acid. How much of each should be mixed together to get 60 ml of a 60% solution of hydrochloric acid? $0.45x + 0.80(60 - x) = 0.60(60);$ 45%: 34.3 ml, 80%: 25.7 ml

51. Two alloys, one 40% iron and the other 60% iron, are to be melted down to form another alloy. How much of each alloy should be melted down and mixed to form 80 tons of an alloy that is 55% iron?

Let x = number of tons of 40% iron; $0.40x + 0.60(80 - x) = 0.55(80);$ $x = 20$ tons of 40% iron; 60 tons of 60% iron

52. Two alloys, one 28% iron and the other 82% iron, are to be melted down to form another alloy. How much of each alloy should be melted down and mixed to form 108 tons of an alloy that is 40% iron? $0.28x + 0.82(108 - x) = 0.40(108);$ 28%: 84 tons, 82%: 24 tons

53. Abner has 2 liters of a 60% solution of alcohol. How much pure alcohol must he add to have an 80% solution?

Let x = number of liters of pure alcohol; $0.60(2) + 1(x) = 0.80(x + 2);$ 2 liters of pure alcohol

54. Susan has 2 liters of a 60% solution of alcohol. How much pure water should she add to have a 40% solution? $0.60(2) + 0(x) = 0.40(2 + x);$ 1 liter

55. Jack's radiator has a 3-gallon capacity. His radiator is filled to capacity with a 30% mixture of antifreeze and water. He drains off some of the old solution and refills the radiator to capacity with pure water to get a 20% mixture. How much did he drain off?

Let x = number of gallons drained off = water added; $0.30(3 - x) + 0(x) = 0.20(3);$ 1 gallon

56. Jack's car radiator has a 3-gallon capacity. His radiator is filled to capacity with a 30% mixture of antifreeze and water. He drains off some of the old solution and refills the radiator to capacity with pure antifreeze to get a 45% mixture. How much did he drain off? $0.30(3 - x) + 1(x) = 0.45(3)$; 0.64 gallon

57. Advance tickets were sold for a concert for $25.00 each. At the door, tickets were $30.50 each. How many advance tickets were sold if 3,600 tickets were sold, netting $97,700?

Let x = number of advance tickets sold; $25x + 30.50(3,600 - x) = 97,700$; 2,200 advance tickets

58. Children's tickets at a movie sold for $4.50, whereas adult tickets sold for $7.25. On a single weekday the movie *Spiders* took in $15,650. If 2,500 tickets were sold that day, how many of each type were sold?

$4.50x + 7.25(2,500 - x) = 15,650$; children's: 900, adult: 1,600

59. Sam had a bunch of nickels, dimes, and quarters totaling $20. If he had five more dimes than nickels and twice as many quarters as nickels, how many of each coin did he have?

Let x = number of nickels; $0.05x + 0.10(x + 5) + 0.25(2x) = 20.00$; 30 nickels, 35 dimes, 60 quarters

60. Jake had 205 coins, consisting of nickels, dimes, and quarters. If the total value of the coins was $22 and there were three times as many dimes as nickels, how many of each coin did he have?

$5x + 10(3x) + 25(205 - 4x) = 2,200$; nickels: 45, dimes: 135, quarters: 25

61. Orchestra seating at a play sold for $25.00, balcony tickets sold for $20.50, and general admission tickets sold for $16.00. For a single showing, there were twice as many general admission tickets sold as orchestra tickets. If there were 900 tickets sold and the total gross for the showing was $17,325, how many of each ticket were sold?

Let x = number of orchestra tickets; $25x + 16(2x) + 20.5(900 - 3x) = 17,325$; 250 orchestra seats, 500 general admission, 150 balcony

62. April saved a total of $19,000 in three banks: the First National Bank, the First Federal Bank, and the Fidelity Bank. The First National Bank gave $4\frac{3}{4}\%$ yearly interest, First Federal gave 5% yearly interest, and Fidelity gave $5\frac{1}{2}\%$ yearly interest. April had twice as much money saved in First Federal as she had saved in Fidelity. How much did she have saved in each bank if she received $932.50 in yearly interest for all her savings combined?

$0.055x + 0.05(2x) + 0.0475(19,000 - 3x) = 932.50$; Fidelity: $2,400, 1st Federal: $4,800, 1st National: $11,800

63. Ari's math teacher assigns course grades in the following way: Two exams are given, each worth 20% of the course grade; quizzes and homework together are worth 20% of the course grade; and the final exam is worth 40% of the course grade. Ari received a grade of 85 on the first exam, 65 on the second exam, and had a quiz and homework combined average of 72. What score must he get on the final exam to receive a grade of at least 80 for the course?

Let x = score on final exam; $0.20(85) + 0.20(65) + 0.20(72) + 0.40(x) \geq 80$; he needs at least an 89

64. Repeat Exercise 63 if Ari received the same exam and quiz and homework scores but wanted to find the score he would need on the final exam to receive a grade of at least 70 for the course.

$0.2(85) + 0.2(65) + 0.2(72) + 0.4x \geq 70$; at least 64

65. Tamara had $20,000 to invest. She decided to invest part of it in a high-risk bond that yields 8.2% annual interest, and the rest in a savings bank yielding 3.9% per year. What is the least amount she should invest in the high-risk bond if she wants to receive at least $1,000 interest per year?

Let x = amount invested in high-risk bond; $0.082x + 0.039(20,000 - x) \geq 1,000$; $x \approx 5,116.28$; at least $5,116.28 invested in the high-risk bond

66. Repeat Exercise 65 if Tamara wants to receive at least $2,000 a year interest.

$0.082x + 0.039(20,000 - x) \geq 2,000$; no solution, since $x > \$20,000$

67. A teacher assigns course grades in the following way: Each of three exams is worth 20%, and the final exam is worth 40%. If Chen's exam grades are 85, 92, and 86, what is the lowest he could score on the final exam for him to have an average of at least 90? Let x = score on the final exam; $0.20(85) + 0.20(92) + 0.20(86) + 0.40x \geq 90$; $x \geq 93.5$; at least 93.5

68. Repeat Exercise 67 if Chen receives the same exam grades, but wants to find the lowest he could score on the final exam and still have a course average of at least 80. $0.2(85) + 0.2(92) + 0.2(86) + 0.4x \geq 80$; at least 68.5

69. When two resistors, one of 6 ohms and one of x ohms, are connected in parallel, the resulting circuit has a combined resistance R given by

$$R(x) = \frac{6x}{6 + x}$$

(a) What should the resistance of x be for the resultant circuit to have a combined resistance of 3 ohms?   6 ohms

(b) What should the resistance of x be for the resultant circuit to have a combined resistance of A ohms? $\frac{6A}{6 - A}$

70. It takes Tamika 2 hours to complete a job. If it takes x hours for Juan to do the same job alone, then the number of hours T it takes for the job to be completed when Tamika and Juan work together is given by

$$T(x) = \frac{2 + x}{2x}$$

(a) How long would it take Juan to complete the job working alone if together they could complete the job in $1\frac{1}{2}$ hours?   1 hr

(b) How long would it take Juan to complete the job working alone if together they could complete the job in 1 hour?   2 hr

71. A manufacturer finds that the cost C (in dollars) to produce x items is given by $C = 0.2x + 6,000$. It then follows that the average cost per unit to produce x items is given by

$$A(x) = \frac{0.2x + 6,000}{x}$$

(a) The manufacturer determines that the most profit can be made if the average cost per unit is $100. How many items should be produced if the average per unit cost is $100?   61 ($x = 60.12$)

(b) How many items should be produced if the average per unit cost is T? $\frac{6,000}{T - 0.2}$

72. A state environmental control department finds that the function

$$C(p) = \frac{180p}{100 - p}$$

serves as a good model for the cost C (in millions of dollars) to remove pollutants in the state's lakes and rivers.

(a) If the state allocates $20,000,000 to remove pollutants from its rivers and lakes, what percentage of the pollutants can be removed?   10%

(b) Suppose the state allocates A (in millions) toward removing pollutants. Express the percentage of pollutants that can be removed as a function (in terms) of A. $\frac{100A}{180 + A}$

❓ Question for Thought

73. The distance from Philadelphia to State College is 200 miles. Joe wants to travel the distance in 4 hours, and therefore he figures that he must average 50 mph for the trip. His car develops problems, however, so he can travel only 25 mph. Halfway there, the problem suddenly disappears. Joe now figures that for him to make the total trip within the 4-hour limit, he must average 75 mph for the rest of the trip. What is wrong with this logic?

Joe was traveling at 25 mph for the first half of the trip (100 miles). Hence, the first half of his trip took $\frac{100}{25} = 4$ hours. So it does not matter how fast he travels the second half; he used up the 4 hours he had for the trip.

Mini-Review

74. *Simplify:* $5 - \{2x - 3[x - 2(x - 3)]\}$ $-5x + 23$ **75.** *Factor completely:* $3x^3y - 6x^2y^2 + 3xy^3$ $3xy(x - y)^2$

76. *Factor completely:* $(x - a)^2 - 9$ $(x - a - 3)(x - a + 3)$ **77.** *Solve for x and graph the solution set:* $|2x + 8| \le 10$
$-9 \le x \le 1$. See Answer Section for graph.

CHAPTER 6 SUMMARY

After having completed this chapter you should be able to:

1. Identify a rational function and its domain (Section 6.1).

For example: The function $f(x) = \dfrac{3x - 7}{2 - 5x}$ is a rational function with domain
$x \ne \frac{2}{5}$.

2. Apply the Fundamental Principle of Fractions to reduce fractions (Section 6.2).

For example: $\dfrac{2x^2 - 8x}{x^2 - 16} = \dfrac{2x(x - 4)}{(x + 4)(x - 4)} = \dfrac{2x}{x + 4}$

3. Multiply and divide rational expressions (Section 6.3).

For example:

(a) $\dfrac{x^2 - 3x + 2}{4x - 8} \cdot \dfrac{8x}{x^3 - x^2} = \dfrac{(x - 2)(x - 1)}{4(x - 2)} \cdot \dfrac{8x}{x^2(x - 1)}$

$$= \dfrac{(x - 2)(x - 1)}{4(x - 2)} \cdot \dfrac{\overset{2}{8x}}{x^2(x - 1)} = \dfrac{2}{x}$$

(b) $\dfrac{x^3y - xy^3}{4x + 4y} \div (x^2 - xy) = \dfrac{x^3y - xy^3}{4x + 4y} \cdot \dfrac{1}{x^2 - xy}$ *Use the rule for division.*
Factor and reduce.

$$= \dfrac{xy(x + y)(x - y)}{4(x + y)} \cdot \dfrac{1}{x(x - y)}$$

$$= \dfrac{xy(x + y)(x - y)}{4(x + y)} \cdot \dfrac{1}{x(x - y)}$$

$$= \dfrac{y}{4}$$

4. Find the least common denominator for several fractions (Section 6.4).

For example: Find the LCD of

$$\dfrac{5}{2x^2}, \quad \dfrac{3}{x^2 + 2x}, \quad \dfrac{2}{x^2 - 4}$$

Factor each denominator: $2x^2, \quad x(x + 2), \quad (x + 2)(x - 2)$

The distinct factors are $2, \quad x, \quad x + 2, \quad x - 2$

The LCD is $2x^2(x + 2)(x - 2)$.

5. Use the Fundamental Principle to combine fractions with unlike denominators
(Section 6.4).

For example:

$$\dfrac{5}{2x^2} - \dfrac{3}{x^2 + 2x} + \dfrac{2}{x^2 - 4}$$

$$= \frac{5}{2x^2} - \frac{3}{x(x+2)} + \frac{2}{(x+2)(x-2)} \qquad LCD:\ = 2x^2(x+2)(x-2)$$

$$= \frac{5(x+2)(x-2)}{2x^2(x+2)(x-2)} - \frac{3(2x)(x-2)}{2x^2(x+2)(x-2)} + \frac{2(2x^2)}{2x^2(x+2)(x-2)}$$

$$= \frac{5(x^2-4) - 6x(x-2) + 4x^2}{2x^2(x+2)(x-2)}$$

$$= \frac{5x^2 - 20 - 6x^2 + 12x + 4x^2}{2x^2(x+2)(x-2)}$$

$$= \frac{3x^2 + 12x - 20}{2x^2(x+2)(x-2)}$$

6. Simplify complex fractions (Section 6.5).

For example:

$$\frac{\dfrac{1}{s} + \dfrac{6}{t}}{\dfrac{1}{s^2} - \dfrac{36}{t^2}} = \frac{\left(\dfrac{1}{s} + \dfrac{6}{t}\right)(s^2 t^2)}{\left(\dfrac{1}{s^2} - \dfrac{36}{t^2}\right)(s^2 t^2)}$$ *Multiply numerator and denominator by the LCD of all simple fractions: $s^2 t^2$.*

$$= \frac{st^2 + 6s^2 t}{t^2 - 36s^2}$$

$$= \frac{st(t + 6s)}{(t - 6s)(t + 6s)}$$

$$= \frac{st}{t - 6s}$$

7. Use the LCD to solve fractional equations (Section 6.6).

For example: Solve for x.

$$\frac{6}{x+3} - \frac{3}{x+6} = \frac{4}{x+3}$$ *Multiply both sides of the equation by $(x+3)(x+6)$.*

$$(x+3)(x+6) \cdot \frac{6}{x+3} - (x+3)(x+6) \cdot \frac{3}{x+6} = (x+3)(x+6) \cdot \frac{4}{x+3}$$

$$6(x+6) - 3(x+3) = 4(x+6)$$

$$6x + 36 - 3x - 9 = 4x + 24$$

$$3x + 27 = 4x + 24$$

$$3 = x \qquad \text{*Be sure to check that the solution is valid.*}$$

Check: $\dfrac{6}{3+3} - \dfrac{3}{3+6} \overset{?}{=} \dfrac{4}{3+3}$

$$\frac{6}{6} - \frac{3}{9} \overset{?}{=} \frac{4}{6}$$

$$1 - \frac{1}{3} \overset{\checkmark}{=} \frac{2}{3}$$

8. Solve a literal equation explicitly for a specified variable (Section 6.7).

For example: Solve for t.

$$3t - 7s = at - xs + 1 \qquad \text{*Collect t terms on one side and non-t terms on the other side of the equation.*}$$

$$3t - 7s \ \boxed{-\ at} \ = at - xs + 1 \ \boxed{-\ at} \qquad \text{*Subtract at from both sides.*}$$

$$3t - at - 7s = -xs + 1 \qquad \text{*Add 7s to both sides.*}$$

$$3t - at - 7s \boxed{+ 7s} = -xs + 1 \boxed{+ 7s}$$

$$3t - at = 7s - xs + 1 \qquad \textit{Factor t from the left side.}$$

$$t(3 - a) = 7s - xs + 1 \qquad \textit{Divide both sides of the equation by } 3 - a.$$

$$t = \frac{7s - xs + 1}{3 - a} \qquad a \neq 3$$

9. Solve verbal problems that give rise to fractional equations (Section 6.8).

For example: A clerk can complete a job in 4 hours. If another clerk can complete the same job in 2 hours, how long would it take them to complete the job if they both worked together?

Let x be the number of hours it takes them to complete the job together.

Then $\dfrac{x}{4}$ represents the portion of the job completed by the first clerk in x hours.

$\dfrac{x}{2}$ represents the portion of the job completed by the second clerk in x hours.

Therefore, our equation is

$$\frac{x}{4} + \frac{x}{2} = 1 \qquad \textit{Multiply both sides of the equation by } 4.$$

$$x + 2x = 4$$

$$3x = 4$$

$$x = \frac{4}{3} = 1\frac{1}{3}$$

Hence, working together, they will complete the job in $1\frac{1}{3}$ hours or 1 hour and 20 minutes.

CHAPTER 6 REVIEW EXERCISES

In Exercises 1–4, find the domain of the given function.

1. $f(x) = \dfrac{2x - 1}{x + 5}$ $\{x \mid x \neq -5\}$

2. $g(x) = \dfrac{x - 6}{3x + 4}$ $\{x \mid x \neq -\frac{4}{3}\}$

3. $h(x) = \dfrac{1}{3x^2 - 11x - 4}$ $\{x \mid x \neq -\frac{1}{3}, 4\}$

4. $g(x) = \dfrac{x - 4}{2x^2 + 13x - 7}$ $\{x \mid x \neq -7, \frac{1}{2}\}$

In Exercises 5–12, reduce to lowest terms.

5. $\dfrac{4x^2y^3}{16xy^5}$ $\frac{x}{4y^2}$

6. $\dfrac{12a^2b^4}{16a^5bc^2}$ $\frac{3b^3}{4a^3c^2}$

7. $\dfrac{x^2 + 2x - 8}{x^2 + 3x - 10}$ $\frac{x + 4}{x + 5}$

8. $\dfrac{6x^2 - 7x - 3}{(2x - 3)^2}$ $\frac{3x + 1}{2x - 3}$

9. $\dfrac{x^4 - 2x^3 + 3x^2}{x^2}$ $x^2 - 2x + 3$

10. $\dfrac{12a^3b^4 - 16ab^3 + 18a}{4a^2b}$ $\frac{6a^2b^4 - 8b^3 + 9}{2ab}$

11. $\dfrac{5xa - 7a + 5xb - 7c}{3xa - 2a + 3xb - 2b}$ $\frac{5x - 7}{3x - 2}$

12. $\dfrac{8a^3 - b^3}{4a^2 + 2ab + b^2}$ $2a - b$

In Exercises 13–54, perform the indicated operations. Express your answer in simplest form.

13. $\dfrac{4x^2y^3z^2}{12xy^4} \cdot \dfrac{24xy^5}{16xy}$ $\dfrac{xy^3z^2}{2}$

14. $\dfrac{4x^2y^3z^2}{12xy^4} \div \dfrac{24xy^5}{16xy}$ $\dfrac{2xz^2}{9y^5}$

15. $\dfrac{5}{3x^2y} + \dfrac{1}{3x^2y}$ $\dfrac{2}{x^2y}$

16. $\dfrac{3}{2a^2b} - \dfrac{3a}{2a^3b}$ 0

17. $\dfrac{3x}{x-1} + \dfrac{3}{x-1}$ $\dfrac{3(x+1)}{x-1}$

18. $\dfrac{2a}{a-3} - \dfrac{6}{a-3}$ 2

19. $\dfrac{2x^2}{x^2+x-6} + \dfrac{2x}{x^2+x-6} - \dfrac{12}{x^2+x-6}$ 2

20. $\dfrac{3a^2}{a^2-a-6} - \dfrac{3a}{a^2-a-6} - \dfrac{18}{a^2-a-6}$ 3

21. $\dfrac{5}{3a^2b} - \dfrac{7}{4ab^4}$ $\dfrac{20b^3-21a}{12a^2b^4}$

22. $\dfrac{7}{3xy^2} + \dfrac{2}{5xy} - \dfrac{4}{30xy}$ $\dfrac{4y+35}{15xy^2}$

23. $\dfrac{3x+1}{2x^2} - \dfrac{3x-2}{5x}$ $\dfrac{-6x^2+19x+5}{10x^2}$

24. $\dfrac{7x-1}{3x} + \dfrac{5x+2}{6x^2}$ $\dfrac{14x^2+3x+2}{6x^2}$

25. $\dfrac{x-7}{5-x} + \dfrac{3x+3}{x-5}$ $\dfrac{2(x+5)}{x-5}$

26. $\dfrac{3a+1}{a-4} + \dfrac{2a+3}{4-a}$ $\dfrac{a-2}{a-4}$

27. $\dfrac{x^2+x-6}{x+4} \cdot \dfrac{2x^2+8x}{x^2+x-6}$ $2x$

28. $\dfrac{a^2+2a-15}{4a^2+8a} \cdot \dfrac{2a^2+4a}{a+5}$ $\dfrac{a-3}{2}$

29. $\dfrac{a^2-2ab+b^2}{a+b} \div \dfrac{(a-b)^3}{a+b}$ $\dfrac{1}{a-b}$

30. $\dfrac{r^2-rs-2s^2}{2r-s} \div \dfrac{12r^2-24rs}{8r^2-4rs}$ $\dfrac{r+s}{3}$

31. $\dfrac{3x}{2x+3} - \dfrac{5}{x-4}$ $\dfrac{3x^2-22x-15}{(2x+3)(x-4)}$

32. $\dfrac{5a}{a-1} + \dfrac{3a}{2a+1}$ $\dfrac{13a^2+2a}{(a-1)(2a+1)}$

33. $\dfrac{3x-2}{2x-7} + \dfrac{5x+2}{2x-3}$ $\dfrac{16x^2-44x-8}{(2x-7)(2x-3)}$

34. $\dfrac{7x+1}{x+3} - \dfrac{2x+3}{x-2}$ $\dfrac{5x^2-22x-11}{(x-2)(x+3)}$

35. $\dfrac{5a}{a^2-3a} + \dfrac{2}{4a^3+4a^2}$ $\dfrac{10a^3+10a^2+a-3}{2a^2(a+1)(a-3)}$

36. $\dfrac{3}{x^2+2x-8} - \dfrac{5}{x^2-x-2}$ $\dfrac{-2x-17}{(x+4)(x-2)(x+1)}$

37. $\dfrac{5}{x^2-4x+4} + \dfrac{3}{x^2-4}$ $\dfrac{8x+4}{(x-2)^2(x+2)}$

38. $\dfrac{3}{a^2-6a+9} - \dfrac{5}{a^2-9}$ $\dfrac{-2a+24}{(a+3)(a-3)^2}$

39. $\dfrac{2x}{7x^2-14x-21} + \dfrac{2x}{14x-42}$ $\dfrac{x(x+3)}{7(x-3)(x+1)}$

40. $\dfrac{5}{x-y} + \dfrac{3}{y^2-x^2}$ $\dfrac{5x+5y-3}{x^2-y^2}$

41. $\dfrac{5x}{x-2} + \dfrac{3x}{x+2} - \dfrac{2x+3}{x^2-4}$ $\dfrac{8x^2+2x-3}{x^2-4}$

42. $\dfrac{2x}{x-5} - \dfrac{3x}{x+5} + \dfrac{x+1}{x^2-25}$ $\dfrac{-x^2+26x+1}{x^2-25}$

43. $\left(\dfrac{2x+y}{5x^2y-xy^2}\right)\left(\dfrac{25x^2-y^2}{10x^2+3xy-y^2}\right)\left(\dfrac{5x^2-xy}{5x+y}\right)$ $\dfrac{1}{y}$

44. $\left(\dfrac{3a+2b}{2a^2+3ab}\right)\left(\dfrac{2a^2+ab-3b^2}{9a^2-4b^2}\right)\left(\dfrac{3a^2b^2-2ab^3}{a-b}\right)$ b^2

45. $\dfrac{4x+11}{x^2+x-6} - \dfrac{x+2}{x^2+4x+3}$ $\dfrac{3(x^2+5x+5)}{(x+3)(x-2)(x+1)}$

46. $\dfrac{2x+1}{2x^2+13x+15} - \dfrac{x-1}{2x^2-3x-9}$ See Answer Section

47. $\dfrac{5x}{x^2-x-2} + \dfrac{4x+3}{x^3+x^2} - \dfrac{x-6}{x^3-2x^2}$ $\dfrac{5x+3}{(x-2)(x+1)}$

48. $\dfrac{2r}{r^2-rs-2s^2} + \dfrac{2s}{r^2-3rs+2s^2} + \dfrac{3}{r^2-s^2}$ See Answer Section

49. $\dfrac{4x^2+12x+9}{8x^3+27} \cdot \dfrac{12x^3-18x^2+27x}{4x^2-9}$ $\dfrac{3x}{2x-3}$

50. $\dfrac{4x^2-12x+9}{8x^3-27} \div \dfrac{4x^2-9}{12x^3+18x^2+27x}$ $\dfrac{3x}{2x+3}$

51. $4x \div \left(\dfrac{8x^2-8xy}{2ax+bx-2ay-by} \div \dfrac{2ax+2bx+3ay+3by}{2a^2+3ab+b^2}\right)$ $\dfrac{2x+3y}{2}$

52. $\left(\dfrac{8x^2-8xy}{2ax+bx-2ay-by} \cdot \dfrac{2a^2+3ab+b^2}{2ax+2bx+3ay+3by}\right) \div 4x$ $\dfrac{2}{2x+3y}$

53. $\left(\dfrac{x}{2} + \dfrac{3}{x}\right) \cdot \dfrac{x+1}{x}$ $\dfrac{(x^2+6)(x+1)}{2x^2}$

54. $\left(\dfrac{x}{y} - \dfrac{y}{x}\right) \div \dfrac{x^2-y^2}{x^2y^2}$ xy

In Exercises 55–60, express the complex fraction as a simple fraction reduced to lowest terms.

55. $\dfrac{\dfrac{3x^2y}{2ab}}{\dfrac{9x}{16a^2}}$ $\dfrac{8axy}{3b}$

56. $\dfrac{\dfrac{x}{x-y}}{\dfrac{y}{x-y}}$ $\dfrac{x}{y}$

57. $\dfrac{\dfrac{3}{a}-\dfrac{2}{a}}{\dfrac{5}{a}}$ $\dfrac{1}{5}$

58. $\dfrac{\dfrac{2}{x^2}+\dfrac{1}{x}}{\dfrac{4}{x^2}-1}$ $\dfrac{1}{2-x}$

59. $\dfrac{\dfrac{3}{b+1}+2}{\dfrac{2}{b-1}+b}$ $\dfrac{(b-1)(2b+5)}{b^3+b+2}$

60. $\dfrac{\dfrac{z}{z-2}+\dfrac{1}{z+1}}{\dfrac{3z}{z^2-z-2}}$ $\dfrac{z^2+2z-2}{3z}$

In Exercises 61–64, let $f(x)=\dfrac{x+3}{x-2}$ and $g(x)=\dfrac{1}{x-4}$.

61. Find (and simplify) $f(x)-g(x)$. $\dfrac{x^2-2x-10}{(x-2)(x-4)}$

62. Find (and simplify) $g(x+h)-g(x)$. $\dfrac{-h}{(x+h-4)(x-4)}$

63. Evaluate $g\left(\tfrac{2}{3}\right)$ $-\tfrac{3}{10}$

64. Evaluate $f\left(\tfrac{3}{5}\right)$ $-\tfrac{18}{7}$

In Exercises 65–82, solve the equations or inequalities.

65. $\dfrac{x}{3}+\dfrac{x-1}{2}=\dfrac{7}{6}$ $x=2$

66. $\dfrac{x}{2}-\dfrac{1}{3}=\dfrac{2x+3}{6}$ $x=5$

67. $\dfrac{x}{5}-\dfrac{x+1}{3}<\dfrac{1}{3}$ $x>-5$

68. $\dfrac{x}{2}-\dfrac{1}{3}\ge\dfrac{2x+3}{6}$ $x\ge5$

69. $\dfrac{5}{x}-\dfrac{1}{3}=\dfrac{11}{3x}$ $x=4$

70. $\dfrac{7}{x}-\dfrac{3}{2x}=5$ $x=\tfrac{11}{10}$

71. $\dfrac{x+1}{3}-\dfrac{x}{2}>4$ $x<-22$

72. $\dfrac{2-x}{14}\le\dfrac{x+1}{7}+2$ $x\ge-\tfrac{28}{3}$

73. $-\dfrac{7}{x}+1=-13$ $x=\tfrac{1}{2}$

74. $\dfrac{3}{x-1}+4=\dfrac{5}{x-1}$ $x=\tfrac{3}{2}$

75. $\dfrac{5}{x-2}-1=0$ $x=7$

76. $\dfrac{2}{x-3}+1=3$ $x=4$

77. $\dfrac{x-2}{5}-\dfrac{3-x}{15}>\dfrac{1}{9}$ $x>\tfrac{8}{3}$

78. $\dfrac{2x+1}{3}+\dfrac{x}{4}\le\dfrac{5}{6}$ $x\le\tfrac{6}{11}$

79. $\dfrac{7}{x-1}+4=\dfrac{x+6}{x-1}$ No solution

80. $\dfrac{3}{x-2}+5=\dfrac{1+x}{x-2}$ No solution

81. $\dfrac{4x+1}{x^2-x-6}=\dfrac{2}{x-3}+\dfrac{5}{x+2}$ $x=4$

82. $\dfrac{5}{x+3}+\dfrac{2}{x}=\dfrac{x-12}{x^2+3x}$ No solution

In Exercises 83–92, solve the equations explicitly for the given variable.

83. $5x-3y=2x+7y$ for x $x=\dfrac{10y}{3}$

84. $5x-3y=2x+7y$ for y $y=\dfrac{3x}{10}$

85. $3xy=2xy+4$ for y $y=\dfrac{4}{x}$

86. $5ab-2a=3ab+2b$ for b $b=\dfrac{a}{a-1}$

87. $\dfrac{2x+1}{y}=x$ for y $y=\dfrac{2x+1}{x}$

88. $\dfrac{ax+b}{c}=a$ for a $a=\dfrac{b}{c-x}$

89. $\dfrac{ax+b}{cx+d}=y$ for x $x=\dfrac{dy-b}{a-cy}$

90. $\dfrac{2s+3}{3s-2}=r$ for s $s=\dfrac{3+2r}{3r-2}$

91. $\dfrac{1}{a}+\dfrac{1}{b}+\dfrac{1}{c}=\dfrac{1}{d}$ for b $b=\dfrac{adc}{ac-cd-ad}$

92. $\dfrac{a}{c}-\dfrac{b}{d}=e$ for c $c=\dfrac{ad}{b+de}$

Solve each of the following problems algebraically.

93. If 1 in. is 2.54 cm, how many inches are there in 1 cm? Let $x=$ # of inches in 1 cm; $\dfrac{2.54\text{ cm}}{1\text{ in.}}=\dfrac{1}{x}$, $x=\dfrac{1}{2.54}\approx0.39$ in.

94. The ratio of Democrats to Republicans in a district is 4 to 3. If there is a total of 1,890 Democrats and Republicans, how many Democrats are there? Let $x=$ # of Democrats; $\dfrac{4}{3}=\dfrac{x}{1,890-x}$; 1,080 Democrats

95. Carol walked from her aunt's to her father's house. Halfway there she picked up her brother Arnold and one-third of the rest of the way there she picked up her sister Julie. The three of them walked a distance of $1\frac{1}{2}$ miles before arriving at their father's house. How far did Carol walk?

Let total distance $= x$; $\frac{1}{2}x + \frac{1}{3}\left(\frac{1}{2}x\right) + \frac{3}{2} = x$; 4.5 miles

96. José takes the same amount of time to walk 3 miles as it takes Carlos to ride his bike 5 miles. If Carlos travels 5 miles per hour faster than José, what is Carlos' rate of speed?

Let $r =$ Carlos' rate; $\frac{3}{r-5} = \frac{5}{r}$; $12\frac{1}{2}$ mph

97. It takes Charles $2\frac{1}{2}$ days to refinish a room and it takes Ellen $2\frac{1}{3}$ days to refinish the same room. How long would it take them to refinish the room if they worked together?

Let $x =$ number of days required by both; $\frac{x}{2\frac{1}{2}} + \frac{x}{2\frac{1}{3}} = 1$; $1\frac{6}{29}$ days

98. It takes Jerry twice as long as Sue to clean up the kitchen. If Sue can clean up the kitchen in $\frac{1}{3}$ hour, how long would it take to clean up the kitchen if she works together with Jerry?

Let $x =$ amount of time to clean kitchen together; $\frac{x}{\frac{1}{3}} + \frac{x}{\frac{2}{3}} = 1$; $\frac{2}{9}$ hour

99. Tali takes 6 hours to complete a particular job. Suppose she and John begin working on the job together, and after 2 hours Tali leaves. If John completes the job himself an hour later, how long would it have taken John to complete the same job if he worked alone?

Let $x = $ # hours it takes John to complete the job alone; $\frac{2}{6} + \frac{3}{x} = 1$; $4\frac{1}{2}$ hours

100. It takes 3 hours for two people to complete a particular job working together. If one of them could have completed the job alone in 8 hours, how long would it take the other to complete the job working alone?

Let $x = $ # hours it takes for the other to complete the job; $\frac{3}{8} + \frac{3}{x} = 1$; $4\frac{4}{5}$ hours

101. How much of a 35% solution of alcohol should a chemist mix with 5 liters of a 70% alcohol solution to get a solution that is 60% alcohol?

Let $x =$ amount of 35% alcohol; $0.70(5) + 0.35x = 0.60(x + 5)$; 2 liters

102. How much pure water should be added to 3 liters of a 65% solution of sulfuric acid to dilute it to 30%?

Let $x =$ quantity of pure water; $0(x) + 0.65(3) = 0.30(x + 3)$; $x = 3.5$ liters

103. Children's tickets at a movie theater sold for $3.50, and adult tickets sold for $6.25. On a single weekday the movie *Fred* took in $4,970. If 980 tickets were sold that day, how many of each type were sold?

Let $x =$ number of children's tickets; $3.50x + 6.25(980 - x) = 4,970.00$; 420 children's tickets, 560 adult tickets

104. Peg has 82 coins in pennies, nickels, and dimes, totaling $3.32. If she has 4 times as many nickels as dimes, how many of each coin does she have?

Let $x =$ number of dimes; $0.05(4x) + 0.10x + 0.01(82 - 5x) = 3.32$; 40 nickels, 10 dimes, and 32 pennies

105. A pharmaceutical company finds that when a certain drug is administered to a patient, the concentration c of the drug in the bloodstream (in milligrams per liter) h hours after the drug was administered is given by

$$c(h) = \frac{30.4h}{h^2 + 2}$$

(a) Compute $c(0.25)$, $c(0.5)$, $c(1)$, $c(2)$, $c(4)$, and $c(10)$, and explain what each of these values means.

(b) Use the points found in part **(a)** to sketch a rough graph of the function. Use the graph or a table to explain how the concentration of the drug in the bloodstream changes as time passes. See Answer Section

$c(0.25) = 3.68$; $c(0.5) = 6.76$; $c(1) = 10.13$; $c(2) = 10.13$; $c(4) = 6.76$; $c(10) = 2.98$

106. A manufacturer finds that the cost C (in dollars) to produce x items is given by $C = 0.3x + 2,500$. It then follows that the average cost per unit $A(x)$ to produce x items is given by

$$A(x) = \frac{0.3x + 2,500}{x}$$

(a) The manufacturer determines that the most profit can be made if the average cost per unit is $12.50. How many items should be produced if the average per unit cost is $12.50? 205 items

(b) How many items should be produced if the average per unit cost is $A? $\frac{2,500}{A - 0.3}$ items

CHAPTER 6 PRACTICE TEST ⊙

1. Find the domain of $f(x) = \dfrac{3x + 5}{2x + 5}$. $\left\{x \mid x \neq \frac{5}{2}\right\}$

2. Express the following in simplest form.

(a) $\dfrac{24x^2y^4}{64x^3y}$ $\dfrac{3y^3}{8x}$

(b) $\dfrac{x^2 - 9}{x^2 - 6x + 9}$ $\dfrac{x + 3}{x - 3}$

(c) $\dfrac{6x^3 - 9x^2 - 6x}{5x^3 - 10x^2}$ $\dfrac{3(2x + 1)}{5x}$

3. Perform the indicated operations and express your answer in simplest form.

(a) $\dfrac{4xy^3}{5ab^4} \cdot \dfrac{15}{16x^4y^5}$ $\dfrac{3}{4x^3y^2ab^4}$

(b) $\dfrac{3x}{18y^2} + \dfrac{5}{8x^2y}$ $\dfrac{4x^3 + 15y}{24x^2y^2}$

(c) $\dfrac{r^2 - rs - 2s^2}{2s^2 + 4rs} \div \dfrac{r - 2s}{4s^2 + 8rs}$ $2(r + s)$

(d) $\dfrac{9x - 2}{4x - 3} + \dfrac{x + 4}{3 - 4x}$ 2

(e) $\dfrac{3x}{x^2 - 4} + \dfrac{4}{x^2 - 5x + 6} - \dfrac{2x}{x^2 - x - 6}$ See Answer Section

(f) $\left(\dfrac{3}{x} - \dfrac{2}{x + 1}\right) \div \dfrac{1}{x + 1}$ $\dfrac{x + 3}{x}$

4. Express as a simple fraction reduced to lowest terms.
$$\dfrac{\dfrac{3}{x + 1} - 2}{\dfrac{5}{x} + 1}$$ $\dfrac{x(1 - 2x)}{(x + 1)(x + 5)}$

5. If $f(x) = \dfrac{3x + 1}{x + 4}$, evaluate $f\left(-\frac{3}{5}\right)$. $-\frac{4}{17}$

6. Solve the following:

(a) $\dfrac{2x - 4}{6} - \dfrac{5x - 2}{3} < 5$ $x > \frac{-15}{4}$

(b) $\dfrac{3}{x - 8} = 2 - \dfrac{5 - x}{x - 8}$ No solution

(c) $\dfrac{3}{2x + 1} + \dfrac{4}{2x - 1} = \dfrac{29}{4x^2 - 1}$ $x = 2$

7. Solve explicitly for x:
$$y = \dfrac{x - 2}{2x + 1}$$ $x = \dfrac{-y - 2}{2y - 1}$

8. How much of a 30% alcohol solution must be mixed with 8 liters of a 45% solution to get a mixture that is 42% alcohol?   Let x = amount of 30% alcohol solution; $0.30(x) + 0.45(8) = 0.42(x + 8)$; 2 liters

9. It takes Jackie $3\frac{1}{2}$ hours to complete a job and it takes 2 hours for Eleanor to do the same job. How long would it take for them to complete the job working together?   Let x = time required working together; $\dfrac{x}{3\frac{1}{2}} + \dfrac{x}{2} = 1$; $x = 1\frac{3}{11}$ hours

10. A manufacturer finds that the cost C (in dollars) to produce x items is given by $C = 0.25x + 3{,}200$. It then follows that the average cost per unit $A(x)$ to produce x items is given by
$$A(x) = \dfrac{0.25x + 3{,}200}{x}$$

The manufacturer determines that the most profit can be made if the average cost per unit is $18. How many items should be produced if the average per unit cost is $18?   180 items

CHAPTERS 4–6 CUMULATIVE REVIEW

In Exercises 1–4, sketch the graph of each equation in a rectangular coordinate system. See Answer Section for graphs.

1. $3y - 5x + 9 = 0$ (§4.1)

2. $y = -\dfrac{2}{3}x - 1$ (§4.1)

3. $y = -5$ (§4.1)

4. $x = -5$ (§4.1)

5. On the same coordinate system, sketch the graphs of the lines with slopes 2 and -2 passing through the point $(2, 3)$. (§4.1)

6. On the same coordinate system, sketch the graphs of the lines with slopes $\frac{3}{4}$ and $-\frac{3}{4}$ passing through the point $(3, 2)$. (§4.1)

In Exercises 7–12, find the slope of the line satisfying the given condition(s).

7. Passing through points $(2, -3)$ and $(3, 5)$ $m = 8$ (§4.1)

8. Passing through points $(5, -4)$ and $(-2, 3)$ $m = -1$ (§4.1)

9. Equation is $y = 5x - 8$ $m = 5$ (§4.1)

10. Equation is $3x - 2y = 11$ $m = \frac{3}{2}$ (§4.1)

11. Parallel to the line passing through the points $(2, -1)$ and $(6, 4)$ $m = \frac{5}{4}$ (§4.1)

12. Perpendicular to the line passing through the points $(2, -3)$ and $(4, 2)$ $m = -\frac{2}{5}$ (§4.1)

*In Exercises 13–14, find the values of **a** satisfying the given condition(s).*

13. The line through the points $(2, a)$ and $(a, -2)$ has slope 3. $a = 1$ (§4.1)

14. The line through the points $(a, 2)$ and $(1, 3)$ is parallel to the line through the points $(4, -1)$ and $(5, 1)$. $a = \frac{1}{2}$ (§4.1)

In Exercises 15–22, find the equation of the line satisfying the given conditions.

15. The line passes through the point $(-2, 7)$ and has slope 3. $y = 3x + 13$ (§4.2)

16. The line passes through the point $(2, 5)$ and has slope -2. $y = -2x + 9$ (§4.2)

17. The line passes through the points $(2, 7)$ and $(3, 1)$. $y = -6x + 19$ (§4.2)

18. The line passes through the points $(3, 5)$ and $(4, 5)$. $y = 5$ (§4.2)

19. The line has y-intercept 2 and slope 4. $y = 4x + 2$ (§4.2)

20. The line has x-intercept 3 and slope -1. $y = -x + 3$ (§4.2)

21. The line passes through the point $(2, -3)$ and is parallel to the line $3x + 5y = 4$. $y = -\frac{3}{5}x - \frac{9}{5}$ (§4.2)

22. The line passes through the point $(2, -3)$ and is perpendicular to the line $2y = 4x + 1$. $y = -\frac{1}{2}x - 2$ (§4.2)

23. Estimate the equation of the line given in the figure. $y = \frac{3}{2}x + 1$ (§4.2)

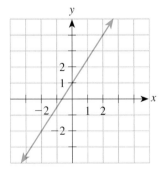

24. The owner of a clothing store found that the relationship between his daily profit on bathing suits (P) and the daily temperature (T) during the summer months is linear. If he makes \$450 when the temperature is 86°F and \$325 when the temperature is 80°F, write an equation relating P to T, and predict his daily profit on a 90°F day. $P = \frac{125}{6}T - \frac{4{,}025}{3}$; \$533.33 (§4.2)

In Exercises 25–32, solve the given system of equations.

25. $\begin{cases} 3x - 2y = 8 \\ 5x + y = 9 \end{cases}$ $x = 2, y = -1$ (§4.3)

26. $\begin{cases} 4x - 3y = 2 \\ 6x - 5y = 3 \end{cases}$ $x = \frac{1}{2}, y = 0$ (§4.3)

27. $\begin{cases} 7u + 5v = 23 \\ 8u + 9v = 23 \end{cases}$ $u = 4, v = -1$ (§4.3)

28. $\begin{cases} -10s + 7t = -6 \\ -4s + 6t = -4 \end{cases}$ $s = \frac{1}{4}, t = -\frac{1}{2}$ (§4.3)

29. $\begin{cases} 2m = 3n - 5 \\ 3n = 2m - 5 \end{cases}$ Inconsistent (§4.3)

30. $\begin{cases} 5w = 4v + 7 \\ 4v = 5w - 7 \end{cases}$ Dependent (§4.3)

31. $\begin{cases} \dfrac{2}{3}y - \dfrac{1}{2}x = 6 \\ \dfrac{4}{5}y - \dfrac{3}{4}x = 6 \end{cases}$ $x = 8, y = 15$ (§4.3)

32. $\begin{cases} \dfrac{x}{4} + \dfrac{5y}{6} = -\dfrac{11}{12} \\ \dfrac{5x}{3} + \dfrac{y}{2} = 4 \end{cases}$ $x = 3, y = -2$ (§4.3)

In Exercises 33–36, graph the inequality on the rectangular coordinate system. See Answer Section for graphs

33. $2x + 3y \geq 18$ (§4.4)

34. $7x - y < 14$ (§4.4)

35. $2y < 5x - 20$ (§4.4)

36. $2y \leq 4$ (§4.4)

In Exercises 37–38, identify the degree of the polynomial.

37. $5xy^2 - 2x^2y^3 + 2$ 5 (§5.1)

38. 3 0 (§5.1)

In Exercises 39–50, perform the operations and express your answer in simplest form.

39. $(3x^2 + 2x - 4) - (2x^2 - 3x + 5)$ $x^2 + 5x - 9$ (§5.2)

40. $-3x(2x - 5y + 4)$ $-6x^2 + 15xy - 12x$ (§5.2)

41. $(x - y)(x - 3y)$ $x^2 - 4xy + 3y^2$ (§5.2)

42. $(3x - 2y)(x - y)$ $3x^2 - 5xy + 2y^2$ (§5.2)

43. $(x + y - 2)(x + y)$ $x^2 + 2xy + y^2 - 2x - 2y$ (§5.2)

44. $(2a + b)(4a^2 - 2ab + b^2)$ $8a^3 + b^3$ (§5.2)

45. $(3x - 5y)^2$ $9x^2 - 30xy + 25y^2$ (§5.3)

46. $(2m + 3n)(2m - 3n)$ $4m^2 - 9n^2$ (§5.3)

47. $(3x - 5y)(3x + 5y)$ $9x^2 - 25y^2$ (§5.3)

48. $(2m + 3n)^2$ $4m^2 + 12mn + 9n^2$ (§5.3)

49. $(2x + y - 3)(2x + y + 3)$ $4x^2 + 4xy + y^2 - 9$ (§5.3)

50. $(x - 2y + 3)^2$ $x^2 - 4xy + 4y^2 + 6x - 12y + 9$ (§5.3)

In Exercises 51–72, factor as completely as possible.

51. $x^2 - 5x - 24$ $(x - 8)(x + 3)$ (§5.5)

52. $y^2 - 4x^2$ $(y - 2x)(y + 2x)$ (§5.5)

53. $y^2 - 12xy + 35x^2$ $(y - 7x)(y - 5x)$ (§5.5)

54. $10a^2 - 3ab - b^2$ $(5a + b)(2a - b)$ (§5.5)

55. $4y^2 + 16yz + 15z^2$ $(2y + 3z)(2y + 5z)$ (§5.5)

56. $9x^2 - 25z^2$ $(3x - 5z)(3x + 5z)$ (§5.5)

57. $25a^2 + 20ab + 4b^2$ $(5a + 2b)^2$ (§5.5)

58. $4x^2 - 12x - 9$ Not factorable (§5.5)

59. $36x^2 - 9$ $9(2x - 1)(2x + 1)$ (§5.5)

60. $25x^2 - 30xz + 9z^2$ $(5x - 3z)^2$ (§5.5)

61. $25a^2 + 20ab - 4b^2$ Not factorable (§5.5)

62. $12y^3 - 16y^2 - 3y$ $y(6y + 1)(2y - 3)$ (§5.5)

63. $3y^3 + 5y^2 - 2y$ $y(3y - 1)(y + 2)$ (§5.5)

64. $25a^4 + 10a^2b^2 - 8b^4$ $(5a^2 - 2b^2)(5a^2 + 4b^2)$ (§5.5)

65. $49a^4 - 14a^2z - 3z^2$ $(7a^2 - 3z)(7a^2 + z)$ (§5.5)

66. $18a^5b - 9a^3b^2 - 2ab^3$ $ab(6a^2 + b)(3a^2 - 2b)$ (§5.5)

67. $(x - y)^2 - 16$ $(x - y - 4)(x - y + 4)$ (§5.5)

68. $(a - 2b)^2 - 25$ $(a - 2b - 5)(a - 2b + 5)$ (§5.5)

69. $16a^4 - b^4$ $(4a^2 + b^2)(2a - b)(2a + b)$ (§5.5)

70. $(x + y)^2 + 3(x + y) + 2$ $(x + y + 2)(x + y + 1)$ (§5.5)

71. $8a^3 + 125b^3$ $\ (2a + 5b)(4a^2 - 10ab + 25b^2)$ (§5.5)

72. $x^3 - 25x - x^2 + 25$ $\ (x - 1)(x + 5)(x - 5)$ (§5.5)

In Exercises 73–78, solve the given equation.

73. $x^2 - x = 12$ $\ x = -3, 4$ (§5.6)

74. $2t^2 + 5 = 7t$ $\ t = 1, \frac{5}{2}$ (§5.6)

75. $(a + 3)(a - 5) = 9$ $\ a = -4, 6$ (§5.6)

76. $(3c + 1)(2c - 5) = (6c - 1)(c + 2)$ $\ c = -\frac{1}{8}$ (§5.6)

77. $(r + 4)^2 = 36$ $\ r = 2, -10$ (§5.6)

78. $3x^2 = 5x$ $\ x = 0, \frac{5}{3}$ (§5.6)

For Exercises 79–80, if $f(x) = x^2 - 5x + 4$, solve the given polynomial equation.

79. $f(x) = 0$ $\ x = 1, 4$

80. $f(x) = 4$ $\ x = 0, 5$

In Exercises 81–84, find the quotients.

81. $(x^2 + 2x + 3) \div (x + 4)$ $\ x - 2, R\ 11$ (§5.7)

82. $(2x^2 + 7x - 1) \div (2x + 1)$ $\ x + 3, R\ -4$ (§5.7)

83. $(4x^3 + 3x + 1) \div (2x + 3)$ $\ 2x^2 - 3x + 6, R\ -17$ (§5.7)

84. $(x^4 + 2) \div (x^2 + 2x + 1)$ $\ x^2 - 2x + 3, R\ -4x - 1$ (§5.7)

In Exercises 85–86, find the domain of the function.

85. $f(x) = \dfrac{3x - 1}{2x + 3}$ $\ \left\{x \mid x \neq -\frac{3}{2}\right\}$ (§6.1)

86. $g(x) = \dfrac{x - 7}{3x^2 - 5x - 2}$ $\ \left\{x \mid x \neq -\frac{1}{3}, 2\right\}$ (§6.1)

In Exercises 87–90, reduce to lowest terms.

87. $\dfrac{18a^3b^2}{16a^5b}$ $\ \frac{9b}{8a^2}$ (§6.2)

88. $\dfrac{x^2 - 9y^2}{x^2 - 6xy + 9y^2}$ $\ \frac{x + 3y}{x - 3y}$ (§6.2)

89. $\dfrac{2a^3 - 5a^2b - 3ab^2}{a^4 - 2a^3b - 3a^2b^2}$ $\ \frac{2a + b}{a(a + b)}$ (§6.2)

90. $\dfrac{27x^3 - y^3}{18x^2 + 6xy + 2y^2}$ $\ \frac{3x - y}{2}$ (§6.2)

In Exercises 91–100, perform the operations. Express your answer in simplest terms.

91. $\dfrac{3xy^2}{5a^3b} \div \dfrac{21x^3y}{25ab^3}$ $\ \frac{5yb^2}{7a^2x^2}$ (§§6.3–6.4)

92. $\dfrac{2x^2 - xy - y^2}{x + y} \cdot \dfrac{x^2 - y^2}{x - y}$ $\ (2x - y)(x - y)$ (§§6.3–6.4)

93. $\dfrac{3x}{2y} + \dfrac{2y}{3x}$ $\ \frac{9x^2 + 4y^2}{6xy}$ (§§6.3–6.4)

94. $\dfrac{6}{x - 2} - \dfrac{3x}{x - 2}$ $\ -3$ (§§6.3–6.4)

95. $\left(\dfrac{2x^3 - 2x^2 - 24x}{x + 2}\right)\left(\dfrac{x + 2}{4x^2 + 12x}\right)$ $\ \frac{x - 4}{2}$ (§§6.3–6.4)

96. $\dfrac{9a^2 - 6ab + b^2}{3a + b} \div \dfrac{3a^2 - 4ab + b^2}{3a^2 - 2ab - b^2}$ $\ 3a - b$ (§§6.3–6.4)

97. $\dfrac{2}{x - 5} - \dfrac{3}{x - 2}$ $\ \frac{-x + 11}{(x - 5)(x - 2)}$ (§§6.3–6.4)

98. $\dfrac{3}{2x^2 - 5xy - 3y^2} - \dfrac{5}{x^2 - 2xy - 3y^2}$ $\ $ See Answer Section

99. $\dfrac{x^3 + x^2y}{2x^2 + xy} \div \left[\left(\dfrac{x^2 + 2xy - 3y^2}{2x^2 - xy - y^2}\right)(x + y)\right]$ $\ $ See Answer Section

100. $\dfrac{x}{x - 5y} - \dfrac{2y}{x + 5y} - \dfrac{20y^2}{x^2 - 25y^2}$ $\ \frac{x - 2y}{x - 5y}$ (§§6.3–6.4)

In Exercises 101–102, write as a simple fraction reduced to lowest terms.

101. $\dfrac{x - \dfrac{2}{x}}{\dfrac{1}{2} - x}$ $\ \frac{2x^2 - 4}{x - 2x^2}$ (§6.5)

102. $\dfrac{\dfrac{x}{x + y} - \dfrac{4y}{x + 4y}}{\dfrac{x - 2y}{x + y} + 1}$ $\ \frac{(x - 2y)(x + 2y)}{(2x - y)(x + 4y)}$ (§6.5)

In Exercises 103–104, given $f(x) = \dfrac{x + 4}{2x + 1}$ and $g(x) = \dfrac{3}{x + 1}$, find and simplify the following.

103. $f\left(\frac{2}{3}\right)$ $\ 2$ (§6.5)

104. $g(x + h) - g(x)$ $\ \dfrac{-3h}{(x + h + 1)(x + 1)}$ (§6.5)

In Exercises 105–110, solve the equations or inequalities.

105. $\dfrac{2}{x} - \dfrac{1}{2} = 2 - \dfrac{1}{x}$ $x = \frac{6}{5}$ (§6.6)

106. $\dfrac{4}{x+4} - 2 = 0$ $x = -2$ (§6.6)

107. $\dfrac{x}{2} - \dfrac{x+1}{3} > \dfrac{2}{3}$ $x > 6$ (§6.6)

108. $\dfrac{3}{x-2} + 3 = \dfrac{x+1}{x-2}$ No solution (§6.6)

109. $\dfrac{3}{x-2} + \dfrac{5}{x+1} = \dfrac{1}{x^2 - x - 2}$ $x = 1$ (§6.6)

110. $\dfrac{x+3}{4} - \dfrac{2x+1}{3} \le \dfrac{1}{2}$ $x \ge -\frac{1}{5}$ (§6.6)

In Exercises 111–114, solve for the given variable.

111. $2a + 3b = 5b - 4a$ (for a) $a = \frac{b}{3}$ (§6.7)

112. $3xy - 2y = 5x + 3y$ (for y) $y = \dfrac{5x}{3x-5}$ (§6.7)

113. $\dfrac{x-y}{y} = x$ (for y) $y = \dfrac{x}{x+1}$ (§6.7)

114. $\dfrac{2x+3}{x-2} = y$ (for x) $x = \dfrac{2y+3}{y-2}$ (§6.7)

In Exercises 115–119, solve algebraically.

115. The ratio of foreign-made cars to American-made cars in a town is 5 to 6. If there are 1,200 American-made cars in the town, how many cars are there altogether? Let x = total # of cars; $\dfrac{5}{6} = \dfrac{x - 1,200}{1,200}$; 2,200 cars (§6.8)

116. How much of a 20% solution of alcohol should be mixed with 8 liters of a 35% solution of alcohol to get a solution that is 30% alcohol? Let x = amount of 20% solution; $0.20x + 0.35(8) = 0.30(8 + x)$; 4 liters (§6.8)

117. Carmen can paint a room in 4 hours, and Judy can paint the same room in $4\frac{1}{2}$ hours. How long would it take for them to paint the room working together? Let t = time they would take to work together; $\dfrac{t}{4} + \dfrac{t}{4\frac{1}{2}} = 1$; $2\frac{2}{17}$ hours (§6.8)

118. Megan takes 5 hours to complete a particular job. Suppose she starts the job at 8 A.M. and two hours later Jason shows up and helps her finish at 11 A.M. How long would it have taken Jason to do the job by himself? Let x = # hours it takes Jason to complete the job alone; $\dfrac{2}{5} + \dfrac{1}{5} + \dfrac{1}{x} = 1$; $2\frac{1}{2}$ hours (§6.8)

119. General admission tickets at a theater sold for $3.50, and reserved seats sold for $4.25. If 505 tickets were sold for a performance that took in $1,861.25, how many of each type of ticket were sold? Let x = number of general admission tickets sold; $3.50x + 4.25(505 - x) = 1,861.25$; 380 general admission tickets, 125 reserved seat tickets (§6.8)

120. A company finds that the cost C per unit of producing x units is given by the function

$$C = C(x) = 325 + \dfrac{750}{x^2}$$

(a) Compute $C(5)$, $C(10)$, $C(50)$, $C(100)$, $C(300)$, $C(500)$, and $C(1,000)$, and explain what each of these values means. $C(5) = 355$; $C(10) = 332.5$; $C(50) = 325.3$; $C(100) = 325.08$; $C(300) = 325.0083$; $C(500) = 325.003$; $C(1,000) = 325.00075$ (§6.1)

(b) Use the points found in part **(a)** to sketch a rough graph of this function. Describe the behavior of this cost function. What is happening to the cost per item as the company manufactures more and more items? See Answer Section

CHAPTERS 4–6 CUMULATIVE PRACTICE TEST

1. Perform the operations and simplify:

(a) $(2x^2 - 3xy + 4y^2) - (5x^2 - 2xy + y^2)$ $-3x^2 - xy + 3y^2$ **(b)** $(3a - 2b)(5a + 3b)$ $15a^2 - ab - 6b^2$

(c) $(2x^2 - y)(2x^2 + y)$ $4x^4 - y^2$ **(d)** $(3y - 2z)^2$ $9y^2 - 12yz + 4z^2$

(e) $(x + y - 3)^2$ $x^2 + 2xy + y^2 - 6x - 6y + 9$

2. Factor the following completely:

(a) $a^2 - 9a + 14$ $\;(a-7)(a-2)$

(b) $6r^2 - 5rs - 6s^2$ $\;(3r+2s)(2r-3s)$

(c) $10a^4 - 9a^2y - 9y^2$ $\;(5a^2+3y)(2a^2-3y)$

(d) $4a^4 - 12a^2b^2 + 9b^4$ $\;(2a^2-3b^2)^2$

(e) $(a + 2b)^2 - 25$ $\;(a+2b-5)(a+2b+5)$

(f) $125x^3 - 1$ $\;(5x-1)(25x^2+5x+1)$

3. If $f(x) = (x - 3)(x + 9)$, solve the polynomial equation $f(x) = 0$. $\;x=-9,3$

4. Solve the following:

(a) $2x^2 + 3x - 2 = 0$ $\;x=-2,\frac{1}{2}$

(b) $(x - 3)(x + 4) = 8$ $\;x=-5,4$

5. Find the quotient: $(2x^3 + 6x + 5) \div (x + 2)$ $\;2x^2-4x+14,\,R-23$

6. Find the domain of the rational function $f(x) = \dfrac{2x}{(3x - 2)(x + 1)}$. $\;\left\{x \mid x \neq -1, \frac{2}{3}\right\}$

7. Perform the indicated operations and express your answer in simplest form.

(a) $\left(\dfrac{25x^2 - 9y^2}{x - y}\right)\left(\dfrac{2x^2 + xy - 3y^2}{10x^2 + 9xy - 9y^2}\right)$ $\;5x+3y$

(b) $\dfrac{2y}{2x - y} + \dfrac{4x}{y - 2x}$ $\;-2$

(c) $\dfrac{3x}{x^2 - 10x + 21} - \dfrac{2}{x^2 - 8x + 15}$ $\;\dfrac{(3x-14)(x-1)}{(x-5)(x-3)(x-7)}$

8. Write as a simple fraction reduced to lowest terms.

$$\dfrac{\dfrac{1}{x} - 2}{3 + \dfrac{1}{x + 1}} \quad \dfrac{-2x^2 - x + 1}{3x^2 + 4x}$$

9. If $f(x) = \dfrac{x}{2x + 1}$, find and simplify $f(x + h) - f(x)$. $\;\dfrac{h}{(2x+2h+1)(2x+1)}$

10. Solve the following:

(a) $\dfrac{2}{x + 1} + \dfrac{3}{2x} = \dfrac{6}{x^2 + x}$ $\;x=\frac{9}{7}$

(b) $\dfrac{x}{3} - \dfrac{x + 2}{7} < 4$ $\;x<\frac{45}{2}$

(c) $\dfrac{5}{x + 3} + 2 = \dfrac{x + 8}{x + 3}$ $\;$No solution

11. Solve for a:

$$y = \dfrac{a}{a + 1} \quad a=\frac{y}{1-y}$$

12. Carol can process 80 forms in 6 hours, whereas Joe can process the same 80 forms in $5\frac{1}{2}$ hours. How long would it take them to process the same forms if they work together? $\;$Let t = time it takes to process 80 forms together; $\dfrac{t}{6} + \dfrac{t}{5\frac{1}{2}} = 1;\, 2\frac{20}{23}$ hours

13. Sketch the graph of the equation $y = -\dfrac{3}{4}x + 6$ using the intercept method. $\;$See Answer Section

14. Find the slope of a line passing through the points $(2, -3)$ and $(3, -4)$. $\;m=-1$

15. Find the equations of the line passing through the point $(2, -3)$ and

(a) Parallel to the line $3y - 2x = 4$ $\;y=\frac{2}{3}x-\frac{13}{3}$

(b) Perpendicular to the line $3y - 2x = 4$ $\;y=-\frac{3}{2}x$

16. There is a perfect linear relationship between the scores on two tests, test A and test B. Anyone scoring 40 on test A will score 30 on test B; anyone scoring 60 on test A will score 25 on test B. What score on test B should a person get if he or she scored 48 on test A? $B = 28$

17. Solve the following system of equations:

$$\begin{cases} 5x + 2y = 4 \\ 2x - 3y = 13 \end{cases} \quad x = 2, y = -3$$

18. Sketch a graph of $3x + 6y > 18$ on a rectangular coordinate system. See Answer Section

7

Exponents and Radicals

© Keith Levit Photography/Index Stock Imagery

Although we have dealt with exponents in previous chapters, in this chapter we will engage in a more formal treatment of exponents. Beginning with the definition of *natural number* exponents, we will discuss the rules for multiplying and dividing simple expressions. Then we will examine what happens when we allow exponents to take on integer and rational values.

In the latter half of this chapter we will discuss radical notation and examine radical functions. For example, the distance traveled by a car *after* applying brakes is related to the velocity of the car before brakes were applied in the following way:

$$v = v(s) = 2\sqrt{5s}$$

where v is the speed of the car in miles per hour, and s is the distance the car travels, in feet, after the brakes are applied. We will further study this particular function in Section 7.7.

7.1 Natural Number and Integer Exponents

Natural Number Exponents

First, recall the definition of exponential notation:

$$x^n = x \cdot x \cdot x \cdot \cdots \cdot x, \text{ where the factor } x \text{ occurs } n \text{ times in the product.}$$

Observe that for this definition to make sense, n must be a natural number. (How can you have -3 factors of x?) With this definition of exponents, and the suitable properties of the real number system, we will state the first three rules of natural number exponents:

Rules of Natural Number Exponents	1. $x^n \cdot x^m = x^{n+m}$ 2. $(x^n)^m = x^{nm}$ 3. $(xy)^n = x^n y^n$

As we stated in Chapter 1, rule 1 is a matter of counting x's:

$$x^n \cdot x^m = \underbrace{(x \cdot x \cdot x \cdot \cdots \cdot x)}_{n \text{ times}}\underbrace{(x \cdot x \cdot x \cdot \cdots \cdot x)}_{m \text{ times}} = \underbrace{x \cdot x \cdot x \cdot x \cdot \cdots \cdot x}_{n + m \text{ times}} = x^{n+m}$$

Rule 2 is derived by applying rule 1:

$$(x^n)^m = \underbrace{x^n \cdot x^n \cdot \cdots \cdot x^n}_{x^n \text{ occurs } m \text{ times in the product.}}$$

$$= x^{n+n+\cdots+n} \quad \textit{By rule 1 the exponent } n \textit{ is added } m \textit{ times.}$$

$$= x^{nm}$$

Rule 1 and rule 2 are often confused (when do I add exponents and when do I multiply?). Keep the differences in mind: Rule 1 states that when *two powers of the same base are to be multiplied*, the exponents are *added*; rule 2 states that when *a power is raised to a power*, the exponents are *multiplied*.

Rule 3 is derived from the associative and commutative properties of multiplication:

$$(xy)^n = \underbrace{(xy) \cdot (xy) \cdot \cdots \cdot (xy)}_{n \text{ times}}$$

Reorder and regroup by the associative and commutative properties of multiplication.

$$= (\underbrace{x \cdot x \cdot \cdots \cdot x}_{n \text{ times}})(\underbrace{y \cdot y \cdot \cdots \cdot y}_{n \text{ times}})$$

$$= x^n y^n$$

EXAMPLE 1

Perform the indicated operations and simplify.

(a) $x^7 \cdot x^6$ (b) $3^8 \cdot 3^2$ (c) $(x^7)^6$

(d) $(3xy)^4$ (e) $(-3x^3)^2(-2x^2)^5$ (f) $(x^2 + y^3)^2$

Solution

(a) $x^7 \cdot x^6$ *Apply rule 1.*

$$= x^{7+6} = \boxed{x^{13}}$$

(b) $3^8 \cdot 3^2$ *Multiply using rule 1.*

$$= 3^{8+2} = \boxed{3^{10}}$$ *Note that the answer is not 9^{10}.*

(c) $(x^7)^6$ *Apply rule 2.*

$$= x^{7 \cdot 6} = \boxed{x^{42}}$$

(d) $(3xy)^4$ *Apply rule 3:*

$$= 3^4 x^4 y^4$$ *Each factor is raised to the 4th power.*

$$= \boxed{81x^4 y^4}$$

(e) $(-3x^3)^2(-2x^2)^5$ *First apply rule 3.*

$$= (-3)^2(x^3)^2(-2)^5(x^2)^5$$ *Note that the sign is part of the coefficient and is raised to the given power. Then use rule 2.*

$$= 9x^6(-32)x^{10}$$ *Multiply using rule 1.*

$$= \boxed{-288x^{16}}$$

(f) $(x^2 + y^3)^2 = (x^2 + y^3)(x^2 + y^3)$ *We cannot apply rule 2 for exponents in this case since rule 2 applies only to factors and not terms. We use polynomial multiplication instead.*

$$= \boxed{x^4 + 2x^2 y^3 + y^6}$$

Until now we have concentrated on developing rules for exponential expressions involving multiplication. Let's now examine division. We consider the expression $\dfrac{x^n}{x^m}$ (we assume $x \neq 0$):

Case (i): $n > m$

$$\frac{x^n}{x^m} = \frac{\overbrace{x \cdot \cdots \cdot x}^{n \text{ times}}}{\underbrace{x \cdot \cdots \cdot x}_{m \text{ times}}} = \frac{\overbrace{\cancel{x} \cdot \cdots \cdot \cancel{x}}^{m \text{ factors reduced}} \cdot \overbrace{x \cdot \cdots \cdot x}^{n - m \text{ factors left}}}{\underbrace{\cancel{x} \cdot \cdots \cdot \cancel{x}}_{m \text{ factors reduced}}} = \underbrace{x \cdot x \cdot \cdots \cdot x}_{n - m \text{ factors}} = x^{n-m}$$

Case (ii): $n < m$

$$\frac{x^n}{x^m} = \frac{\overbrace{x \cdot \cdots \cdot x}^{n\ times}}{\underbrace{x \cdot \cdots \cdot x}_{m\ times}} = \frac{\overbrace{\overset{n\ factors}{reduced}}{\cancel{x} \cdot \cdots \cdot \cancel{x}}}{\underbrace{\cancel{x} \cdot \cdots \cdot \cancel{x}}_{\substack{n\ factors \\ reduced}} \cdot \underbrace{x \cdot \cdots \cdot x}_{\substack{m-n \\ factors\ left}}} = \frac{1}{\underbrace{x \cdot x \cdot \cdots \cdot x}_{m-n\ factors}} = \frac{1}{x^{m-n}}$$

Case (iii): $n = m$

Since $n = m$, we have $x^n = x^m$.

Hence, $\dfrac{x^n}{x^m}$ is a fraction with the numerator identical to the denominator. Thus,

$$\frac{x^n}{x^m} = \frac{x^n}{x^n} = 1$$

We summarize this discussion as follows:

Rule 4 of Natural Number Exponents	If $x \ne 0$, $$\frac{x^n}{x^m} = \begin{cases} x^{n-m} & \text{if } n > m \\ \dfrac{1}{x^{m-n}} & \text{if } n < m \\ 1 & \text{if } n = m \end{cases}$$

We had to split rule 4 into three cases to ensure natural number values for the exponents. Shortly, when we allow integer values for exponents, we will consolidate these three cases into one.

The next and final rule for natural number exponents is a result of the associative and commutative properties of rational multiplication.

We consider the expression $\left(\dfrac{x}{y}\right)^n$.

$$\left(\frac{x}{y}\right)^n = \overbrace{\left(\frac{x}{y}\right)\left(\frac{x}{y}\right) \cdot \cdots \cdot \left(\frac{x}{y}\right)}^{n\ times} = \frac{\overbrace{x \cdot x \cdot \cdots \cdot x}^{n\ times}}{\underbrace{y \cdot y \cdot \cdots \cdot y}_{n\ times}} = \frac{x^n}{y^n}$$

Thus, we have:

Rule 5 of Natural Number Exponents	If $y \ne 0$, $$\left(\frac{x}{y}\right)^n = \frac{x^n}{y^n}$$

For convenience, we assume that all variables are nonzero real numbers.

EXAMPLE 2	Perform the indicated operations and simplify.

(a) $\dfrac{x^7}{x^4}$ **(b)** $\dfrac{x^4}{x^7}$ **(c)** $\left(\dfrac{r^5}{s^2}\right)^4$

Solution   These expressions require straightforward applications of rules 4 and 5.

(a) $\dfrac{x^7}{x^4}$ *Divide using rule 4, case* (i).

$$= x^{7-4} = \boxed{x^3}$$

(b) $\dfrac{x^4}{x^7}$ *Divide using rule 4, case* (ii).

$$= \dfrac{1}{x^{7-4}} = \boxed{\dfrac{1}{x^3}}$$

(c) $\left(\dfrac{r^5}{s^2}\right)^4 = \dfrac{(r^5)^4}{(s^2)^4}$ *Apply rule 5, then raise to a power using rule 2.*

$$= \boxed{\dfrac{r^{20}}{s^8}}$$

Let's examine problems requiring us to combine the use of several of the rules of natural number exponents.

EXAMPLE 3

Perform the indicated operations and simplify.

(a) $\left(\dfrac{a^2 a^5}{a^6}\right)^4$ **(b)** $\dfrac{(3a^2)^4(2a)^2}{(-2a^3)^5}$

Solution   **(a)**   We will show two ways to approach this problem.

First applying rule 1 inside parentheses:

$$\left(\dfrac{a^2 a^5}{a^6}\right)^4 = \left(\dfrac{a^7}{a^6}\right)^4 \quad \textit{Then rule 4}$$

$$= (a)^4$$

$$= \boxed{a^4}$$

Applying rule 5 first:

$$\left(\dfrac{a^2 a^5}{a^6}\right)^4 = \dfrac{(a^2 a^5)^4}{(a^6)^4} \quad \textit{Then rule 3}$$

$$= \dfrac{(a^2)^4(a^5)^4}{(a^6)^4} \quad \textit{Then rule 2}$$

$$= \dfrac{a^8 a^{20}}{a^{24}} \quad \textit{Then rule 1}$$

$$= \dfrac{a^{28}}{a^{24}} \quad \textit{Then rule 4}$$

$$= \boxed{a^4}$$

Experience will teach you which approach is more efficient.

(b) $\dfrac{(3a^2)^4(2a)^2}{(-2a^3)^5} = \dfrac{3^4(a^2)^4 2^2 a^2}{(-2)^5(a^3)^5}$ *Apply rule 3, then rule 2*

$$= \dfrac{81a^8(4)a^2}{-32a^{15}} \quad \textit{Then rule 1}$$

$$= \dfrac{81 \cdot 4a^{10}}{-32a^{15}} \quad \textit{Then rule 4}$$

$$= \dfrac{(81)(4)}{-32a^5} \quad \textit{Reduce.}$$

$$= \boxed{-\dfrac{81}{8a^5}}$$

Integer Exponents

We originally defined x^n as x taken as a factor n times. Based on that definition, it does not make sense for n to be negative or 0. Hence, the exponents we discussed were natural numbers. Now we would like to allow exponents to take on any *integer* values, and determine what it means when exponents have nonpositive integer values. This is called extending the definition of exponents to the integers.

In previous discussions, our understanding of the definition of a positive integer exponent led quite naturally to the five rules for exponents. To define negative and zero exponents, we will work the other way around. Since we want the exponent rules developed previously to continue to apply, we assume that the exponent rules developed for natural number exponents will still be valid for integer exponents. We then will use these rules to show us how to define integer exponents.

Let's assume x^0 exists and $x \neq 0$; see what happens when we find the product $x^0 \cdot x^n$ by applying the first rule of exponents:

$$x^0 \cdot x^n = x^{0+n} \qquad \text{Rule 1}$$
$$= x^n$$

Hence, $x^0 \cdot x^n = x^n$. Verbally stated, multiplying an expression by x^0 does not change the expression. Thus, x^0 has the same property as 1, the multiplicative identity. Since 1 is the only number that has this property, it would be convenient to have $x^0 = 1$. We therefore make the following definition:

DEFINITION

A zero exponent is defined as follows:

$$x^0 = 1 \qquad (x \neq 0)$$

Note that $\mathbf{0^0}$ *is undefined.*

For example (assume all variables are nonzero and real numbers):

$$5^0 = 1$$
$$(xy^3)^0 = 1$$
$$(-165)^0 = 1 \qquad \textit{Even if the base is negative, raising to the zero power still yields } +1.$$
$$\left(\frac{x^2y^5z^2}{xy^4z^9}\right)^0 = 1 \qquad \textit{Look carefully before simplifying an expression.}$$

Now let's examine x^{-n} when n is a natural number and $x \neq 0$. What does it mean to have a negative exponent? We assume x^{-n} exists and will examine what happens when we find the product $x^n \cdot x^{-n}$ by applying rule 1 again:

$$x^{-n} \cdot x^n = x^{-n+n} \qquad \text{Rule 1}$$
$$= x^0$$
$$= 1 \qquad \textit{Definition of zero exponent}$$

Hence, $x^{-n} \cdot x^n = 1$. *Dividing both sides of the equation above by x^n, we get*

$$x^{-n} = \frac{1}{x^n}$$

We therefore make the following definition:

DEFINITION

Negative exponents are defined as follows:

If n is a natural number and $x \neq 0$, then

$$x^{-n} = \frac{1}{x^n}$$

Verbally stated, an expression with a negative exponent is the reciprocal of the same expression with the exponent made positive. Thus,

$$x^{-3} = \frac{1}{x^3} \qquad 5^{-4} = \frac{1}{5^4} \qquad 10^{-5} = \frac{1}{10^5}$$

Let's examine what happens when there is a negative exponent in the denominator of a fraction:

$$\frac{1}{x^{-n}} = \frac{1}{\frac{1}{x^n}} \qquad \textit{Definition of negative exponents}$$

$$= 1 \cdot \frac{x^n}{1} \qquad \textit{Rule for dividing fractions}$$

$$= x^n$$

Hence, $\dfrac{1}{x^{-n}} = x^n$. Thus we have

$$x^{-n} = \frac{1}{x^n} \quad \text{and} \quad \frac{1}{x^{-n}} = x^n$$

In words, this says that x^{-n} is the reciprocal of x^n, and x^n is the reciprocal of x^{-n}. For example,

$$\frac{1}{x^{-4}} = \frac{x^4}{1} = x^4 \qquad\qquad \frac{1}{2^{-2}} = \frac{2^2}{1} = 4$$

$$x^{-2} = \frac{1}{x^2} \qquad\qquad \frac{1}{(-2)^{-3}} = (-2)^{+3} = -8$$

Do not confuse the sign of an exponent with the sign of the base. For example, -2 is a number 2 units to the left of 0 on the number line, whereas $2^{-1} = \frac{1}{2}$ is a positive number, one-half unit to the right of 0, as shown in the figure:

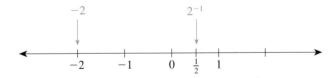

EXAMPLE 4

Using a calculator, evaluate 3^{-6}.

(a) Find the answer rounded to six decimal places.

(b) Find the answer exactly.

Solution

(a) To evaluate 3^{-6} as a decimal, we would press the following sequence of keys: ⟦3⟧ ⟦y^x⟧ ⟦6⟧ ⟦+/−⟧ ⟦=⟧ to get ⟦0.001372⟧ rounded to six decimal places.

(b) To find the answer exactly, we would have to rewrite 3^{-6} as $\dfrac{1}{3^6}$. We can use the calculator to evaluate 3^6 as follows: ⟦3⟧ ⟦y^x⟧ ⟦6⟧ ⟦=⟧ to get ⟦729⟧. The exact answer is then

$$\boxed{\dfrac{1}{729}}$$

It is important to note that our definitions of zero and negative exponents are consistent with the five exponent rules we have been using, so we are free to use the exponent rules for expressions involving both positive and negative integer exponents.

Since we need not be concerned whether the exponents are integers, we can now consolidate rule 4 for natural number exponents.

Rule 4 for Integer Exponents	$$\frac{x^n}{x^m} = x^{n-m} \qquad (x \neq 0)$$

For example,

According to natural number exponent rule 4:	**According to integer exponent rule 4:**
$$\frac{x^3}{x^3} = 1$$	$$\frac{x^3}{x^3} = x^{3-3} = x^0 = 1$$
$$\frac{x^2}{x^6} = \frac{1}{x^{6-2}} = \frac{1}{x^4}$$	$$\frac{x^2}{x^6} = x^{2-6} = x^{-4}$$
	$$= \frac{1}{x^4} \qquad \textit{By definition of negative exponents}$$

Definitions and Rules for Integer Exponents

Rule 1: $x^n x^m = x^{n+m}$

Rule 2: $(x^n)^m = x^{nm}$

Rule 3: $(xy)^n = x^n y^n$

Rule 4: $\dfrac{x^n}{x^m} = x^{n-m} \qquad (x \neq 0)$

Rule 5: $\left(\dfrac{x}{y}\right)^n = \dfrac{x^n}{y^n} \qquad (y \neq 0)$

Definition: $a^0 = 1 \qquad (a \neq 0)$

Definition: $a^{-n} = \dfrac{1}{a^n} \qquad (a \neq 0)$

The only difference between the rules for integer and natural number exponents is the consolidation of rule 4.

Let's now examine some expressions involving integer exponents.

EXAMPLE 5	Perform the indicated operations and simplify. Express your answers with positive exponents only.

(a) $y^{-4} y^8 y^{-6}$ **(b)** $\dfrac{x^{-6}}{x^{-8}}$ **(c)** $(x^{-4})^{-6}$

(d) $(a^{-2} b^3 c^{-2})^2$ **(e)** $\left(\dfrac{2}{3}\right)^{-1}$ **(f)** $\dfrac{c^{-4} d^6}{c^5 d^{-3}}$ **(g)** $\left(\dfrac{r^{-3}}{r^2 r^{-2}}\right)^{-2}$

Solution We *could* simplify all these expressions by using the definition of negative exponents and changing all negative exponents to positive exponents by substituting reciprocals.

For example:

(a) $y^{-4}y^8y^{-6} = \left(\dfrac{1}{y^4}\right)(y^8)\left(\dfrac{1}{y^6}\right)$ *Definition of negative exponents*

$= \dfrac{1}{y^4} \cdot \dfrac{y^8}{1} \cdot \dfrac{1}{y^6}$

$= \dfrac{y^8}{y^4 y^6}$ *Apply rule 1 for natural number exponents.*

$= \dfrac{y^8}{y^{10}}$ *Apply rule 4 for natural number exponents.*

$= \boxed{\dfrac{1}{y^2}}$

A more efficient method would be to apply the rules for integer exponents first, and then use the definition, if needed, to express the final answer with positive exponents.

Alternative Solution for Part (a):

$y^{-4}y^8y^{-6}$ *Apply rule 1 for integer exponents.*

$= y^{-4+8+(-6)}$

$= y^{-2}$ *Apply the definition of negative exponents.*

$= \boxed{\dfrac{1}{y^2}}$

When simplifying an expression with exponents it is easier, in general, to use the exponent rules rather than using the definition, if you have a choice.

(b) $\dfrac{x^{-6}}{x^{-8}} = x^{-6-(-8)}$ *We have applied rule 4 for integer exponents. Note: Numerator exponent minus denominator exponent*

$= x^{-6+8}$ *Remember that you are subtracting a negative exponent.*

$= \boxed{x^2}$

(c) $(x^{-4})^{-6} = x^{(-6)(-4)} = \boxed{x^{24}}$ *Rule 2*

(d) $(a^{-2}b^3c^{-2})^2$ *This is a product raised to a power: Apply rule 3.*

$= (a^{-2})^2(b^3)^2(c^{-2})^2$ *This is a power of a power: Apply rule 2.*

$= a^{-4}b^6c^{-4}$ *Apply the definition of negative exponents.*

$= \left(\dfrac{1}{a^4}\right)(b^6)\left(\dfrac{1}{c^4}\right)$

$= \boxed{\dfrac{b^6}{a^4c^4}}$

In general, x^{-n} is the reciprocal of x^n.

(e) $\left(\dfrac{2}{3}\right)^{-1}$ *Since x^{-1} is the reciprocal of x (remember that x^{-n} is the reciprocal of x^n)*

$= \boxed{\dfrac{3}{2}}$

(f) $\dfrac{c^{-4}d^6}{c^5 d^{-3}} = c^{-4-5}d^{6-(-3)}$ *Rule 4*

$$= c^{-9}d^9 = \boxed{\dfrac{d^9}{c^9}}$$

(g) $\left(\dfrac{r^{-3}}{r^2 \cdot r^{-2}}\right)^{-2} = \left(\dfrac{r^{-3}}{r^0}\right)^{-2}$ *Apply rule 1 in the denominator*

$$= \left(\dfrac{r^{-3}}{1}\right)^{-2} \quad \text{\textit{Definition of zero exponent}}$$

$$= (r^{-3})^{-2} \quad \text{\textit{Apply rule 2.}}$$

$$= \boxed{r^6}$$

EXAMPLE 6

Perform the indicated operations and simplify. Express your answers with positive exponents only.

(a) $(5x^{-2}y^{-1})^{-2}$ **(b)** $(5x^{-2} + y^{-1})^{-2}$

(c) $\dfrac{a^{-2}b^{-2}}{a^{-1}b^{-1}}$ **(d)** $\dfrac{a^{-2} - b^{-2}}{a^{-1} + b^{-1}}$

Solution

(a) $(5x^{-2}y^{-1})^{-2}$ *A product raised to a power: Apply rule 3.*

$$= 5^{-2}(x^{-2})^{-2}(y^{-1})^{-2} \quad \text{\textit{Now apply rule 2 to x and y.}}$$

$$= 5^{-2}(x^4)(y^2) \quad \text{\textit{Apply the definition of negative exponents to } 5^{-2}.}$$

$$= \left(\dfrac{1}{5^2}\right)(x^4 y^2) = \boxed{\dfrac{x^4 y^2}{25}}$$

Note the differences between parts (a) and (b).

(b) Do not try to apply rule 3 in this problem; x^{-2} and y^{-1} are *terms*, not factors. Rule 3 applies to *factors only*.

$$(5x^{-2} + y^{-1})^{-2} = \left(\dfrac{5}{x^2} + \dfrac{1}{y}\right)^{-2} \quad \begin{array}{l}\text{\textit{By the definition of negative exponents. Note that}}\\\text{\textit{5 remains in the numerator of the first fraction.}}\\\text{\textit{Now add fractions in parentheses; the LCD is }} x^2 y.\end{array}$$

$$= \left(\dfrac{5y + x^2}{x^2 y}\right)^{-2} \quad \begin{array}{l}\text{\textit{Now we can change the sign of the exponent}}\\\text{\textit{outside the parentheses and rewrite the entire}}\\\text{\textit{fraction as its reciprocal.}}\end{array}$$

$$= \left(\dfrac{x^2 y}{5y + x^2}\right)^2$$

$$= \dfrac{(x^2 y)^2}{(5y + x^2)^2} = \boxed{\dfrac{x^4 y^2}{(5y + x^2)^2}}$$

(c) $\dfrac{a^{-2}b^{-2}}{a^{-1}b^{-1}}$ *Apply rule 4.*

$$= a^{-2-(-1)}b^{-2-(-1)} = a^{-1}b^{-1} = \boxed{\dfrac{1}{ab}}$$

Note the differences between parts (c) and (d).

(d) $\dfrac{a^{-2} - b^{-2}}{a^{-1} + b^{-1}}$ *We rewrite each* term *in the numerator and denominator using the definition of negative exponents.*

$$= \dfrac{\dfrac{1}{a^2} - \dfrac{1}{b^2}}{\dfrac{1}{a} + \dfrac{1}{b}} \quad \begin{array}{l}\text{\textit{A complex fraction. Multiply the numerator and denominator of}}\\\text{\textit{the complex fraction by }} a^2 b^2. \text{\textit{ Why?}}\end{array}$$

$$= \frac{\left(\dfrac{1}{a^2} - \dfrac{1}{b^2}\right)a^2b^2}{\left(\dfrac{1}{a} + \dfrac{1}{b}\right)a^2b^2}$$

$$= \frac{b^2 - a^2}{ab^2 + a^2b} \qquad \textit{Factor and reduce.}$$

$$= \frac{(b - a)(b + a)}{ab(b + a)} = \boxed{\frac{b - a}{ab}}$$

EXAMPLE 7

Perform the indicated operations and simplify. Express your answer with positive exponents only.

$$\frac{9^{-3} \cdot 3^{-4}}{3^{-2} \cdot 9^2}$$

Solution

$$\frac{9^{-3} \cdot 3^{-4}}{3^{-2} \cdot 9^2} \qquad \textit{If we notice that everything can be expressed as a power of 3, we can proceed as follows.}$$

$$= \frac{(3^2)^{-3} \cdot 3^{-4}}{3^{-2}(3^2)^2}$$

$$= \frac{3^{-6} \cdot 3^{-4}}{3^{-2} \cdot 3^4}$$

$$= \frac{3^{-10}}{3^2}$$

$$= 3^{-10-2}$$

$$= 3^{-12}$$

$$= \boxed{\frac{1}{3^{12}}}$$

EXERCISES 7.1

In Exercises 1–46, perform the indicated operations and simplify. Assume that all variables represent nonzero real numbers. Express your answers with positive exponents only.

1. $(x^2x^5)(x^3x)$ x^{11}

2. $(a^2b^5)(a^4b^7)$ a^6b^{12}

3. $(-2a^2b^3)(3a^5b^7)$ $-6a^7b^{10}$

4. $(-4r^3s)(-2rst^3)$ $8r^4s^2t^3$

5. $(a^2)^5$ a^{10}

6. $(y^3)^8$ y^{24}

7. $\left(\dfrac{2}{3}\right)^2$ $\dfrac{4}{9}$

8. $\left(\dfrac{3}{2}\right)^3$ $\dfrac{27}{8}$

9. $(x + y^3)^2$ $x^2 + 2xy^3 + y^6$

10. $(a^2b)^2$ a^4b^2

11. $(xy^3)^2$ x^2y^6

12. $(a^2 + b)^2$ $a^4 + 2a^2b + b^2$

13. $(2^3 \cdot 3^2)^2$ $2^6 \cdot 3^4$

14. $(5^2 \cdot 2^3)^2$ $5^4 2^6$

15. $(x^4y^3)^5(x^3y^2)^2$ $x^{26}y^{19}$

16. $(a^2a^3)^5(a^3a)^6$ a^{49}

17. $(-2a^2)^3(ab^2)^4$ $-8a^{10}b^8$

18. $(r^2s^3)^2(-rs^4)^3$ $-r^7s^{18}$

19. $(r^2st)^3(-2rs^2t)^4$ $16r^{10}s^{11}t^7$

20. $(-2x^2y)^2(2xy^3)^3$ $32x^7y^{11}$

21. $\dfrac{x^5}{x^2}$ $\;x^3$

22. $\dfrac{x^2}{x^5}$ $\;\dfrac{1}{x^3}$

23. $\dfrac{x^3y^2}{xy^4}$ $\;\dfrac{x^2}{y^2}$

24. $\dfrac{a^7b^4}{a^7b^4}$ $\;1$

25. $\dfrac{5^4\cdot 2^2}{25^2\cdot 4^2}$ $\;\dfrac{1}{4}$

26. $\dfrac{2^2\cdot 3^4}{9^2\cdot 4^3}$ $\;\dfrac{1}{16}$

27. $\dfrac{a^5b^9c}{a^4bc^5}$ $\;\dfrac{ab^8}{c^4}$

28. $\dfrac{2^2x^2y^3}{2^4xy^3}$ $\;\dfrac{x}{4}$

29. $\dfrac{(-3)^2xy^4}{-3^2xy^5}$ $\;-\dfrac{1}{y}$

30. $\dfrac{(-2)^2xy^5}{-2^2xy^5}$ $\;-1$

31. $\dfrac{3^2(-2)^3}{(-9^2)(-4)^2}$ $\;\dfrac{1}{18}$

32. $\dfrac{(-9)^2(-4)^3}{-3^3\cdot 2^4}$ $\;12$

33. $\left(\dfrac{y^5}{y^8}\right)^3$ $\;\dfrac{1}{y^9}$

34. $\left(\dfrac{x^5}{x^4}\right)^5$ $\;x^5$

35. $\left(\dfrac{y^2y^7}{y^4}\right)^3$ $\;y^{15}$

36. $\left(\dfrac{x^5}{x^3x^9}\right)^4$ $\;\dfrac{1}{x^{28}}$

37. $\dfrac{(3r^2s)^3(-2rs^2)^4}{(-18rs)^2}$ $\;\dfrac{4r^8s^9}{3}$

38. $\dfrac{(2xy^2)^3(-2xy)^2}{(-2x)^3}$ $\;-4x^2y^8$

39. $\left(\dfrac{2x^2y^3}{xy^4}\right)^2\!\left(\dfrac{3xy^2}{6}\right)^3$ $\;\dfrac{x^5y^4}{2}$

40. $\left(\dfrac{x^2x^3}{x}\right)^4\!\left(\dfrac{x^2x^5}{x^3}\right)^2$ $\;x^{24}$

41. $\left(\dfrac{(6ab^2)^2}{-3ab}\right)^3$ $\;-1{,}728a^3b^9$

42. $\left(\dfrac{(-5b^2c)^2}{-10b^2c^3}\right)^3$ $\;-\dfrac{125b^6}{8c^3}$

43. $\left(\dfrac{-2a^2b^3}{ab}\right)^3(-3xy^2)^3$ $\;216a^3b^6x^3y^6$

44. $\left(\dfrac{(3ab^2)^3(5a^2b)}{9ab}\right)^3$ $\;3{,}375a^{12}b^{18}$

45. $[(r^3s^2)^3(rs^2)^4]^2$ $\;r^{26}s^{28}$

46. $[(-3rs^2)^2(-2st^2)^3]^3$ $\;-2^9\cdot 3^6r^6s^{21}t^{18}$

In Exercises 47–50, use a calculator to evaluate the following to six decimal places.

47. 5^{-8} $\;0.000003$

48. -3^{-5} $\;-0.004115$

49. $-2\cdot 5^{-3}+8$ $\;7.984000$

50. $4\cdot 2^{-6}-4$ $\;-3.937500$

In Exercises 51–104, perform the indicated operations and simplify. Express your answers with positive exponents only. (Assume that all variables represent nonzero real numbers.)

51. $x^{-2}x^4x^{-3}$ $\;\dfrac{1}{x}$

52. $a^2a^{-3}b^2$ $\;\dfrac{b^2}{a}$

53. $(x^5y^{-4})(x^{-3}y^2x^0)$ $\;\dfrac{x^2}{y^2}$

54. $(a^{-1}b^2)(a^3b^{-4})$ $\;\dfrac{a^2}{b^2}$

55. $(3^{-2})^{-3}$ $\;3^6$

56. $(2^{-3})^{-2}$ $\;2^6$

57. $(a^{-2}b^{-3})^2$ $\;\dfrac{1}{a^4b^6}$

58. $(x^{-5}y^{-3}x^0)^{-2}$ $\;x^{10}y^6$

59. $(r^{-3}s^2)^{-4}(r^2)^{-3}$ $\;\dfrac{r^6}{s^8}$

60. $(a^{-2}b^3)^{-5}(a^3)^{-2}$ $\;\dfrac{a^4}{b^{15}}$

61. -2^{-2} $\;-\dfrac{1}{4}$

62. $(-2)^{-2}$ $\;\dfrac{1}{4}$

63. $(-3)^{-2}$ $\;\dfrac{1}{9}$

64. -3^{-2} $\;-\dfrac{1}{9}$

65. $(2^{-2})^{-3}(3^{-3})^2$ $\;\dfrac{2^6}{3^6}$

66. $(3^{-2})^4(2^{-4})^{-3}$ $\;\dfrac{2^{12}}{3^8}$

67. $(3^{-2}s^3)^4(9s^{-3})^{-2}$ $\;\dfrac{s^{18}}{3^{12}}$

68. $(2r^2s)^{-2}(4r)^3$ $\;\dfrac{16}{rs^2}$

69. $\dfrac{x^4}{x^{-2}}$ $\;x^6$

70. $\dfrac{x^{-2}}{x^4}$ $\;\dfrac{1}{x^6}$

71. $\dfrac{x^{-3}y^2}{x^{-5}y^0}$ $\;y^2x^2$

72. $\dfrac{a^{-2}b^2}{a^{-3}b^4a^0}$ $\;\dfrac{a}{b^2}$

73. $\left(\dfrac{1}{2}\right)^{-1}$ $\;2$

74. $\left(-\dfrac{3}{4}\right)^{-1}$ $\;-\dfrac{4}{3}$

75. $\left(-\dfrac{3}{5}\right)^{-3}$ $\;-\dfrac{125}{27}$

76. $\left(\dfrac{5}{8}\right)^{-3}$ $\;\dfrac{512}{125}$

77. $\dfrac{x^{-1}xy^{-2}}{x^4y^{-3}y}$ $\;\dfrac{1}{x^4}$

78. $\dfrac{a^{-4}b^{-3}}{a^{-2}b^{-4}}$ $\;\dfrac{b}{a^2}$

79. $\dfrac{(a^{-2}b^2)^{-3}}{ab^{-2}}$ $\;\dfrac{a^5}{b^4}$

80. $\dfrac{(2r^{-4}s^{-2})^{-3}}{2r^2s^3}$ $\;\dfrac{r^{10}s^3}{16}$

81. $\left(\dfrac{x^{-1}x^{-3}}{x^{-2}}\right)^{-3}$ $\;x^6$

82. $\left(\dfrac{a^{-2}a^3}{a^{-1}}\right)^{-2}$ $\;\dfrac{1}{a^4}$

83. $\dfrac{2^{-2}\cdot 3^2}{6^{-2}}$ $\;81$

84. $\dfrac{3^{-5}\cdot 9^{-1}}{27^2}$ $\;\dfrac{1}{3^{13}}$

85. $\dfrac{(3x)^{-2}(2xy^{-1})^0}{(2x^{-2}y^3)^{-2}}$ $\;\dfrac{4y^6}{9x^6}$

86. $\dfrac{(2xy^2)^0(3x^2y^{-1})^{-1}}{(3x^{-1}y)^{-1}}$ $\;\dfrac{y^2}{x^3}$

87. $\left(\dfrac{x^{-3}y^{-4}}{x^{-5}y^{-7}}\right)^{-3}$ $\;\dfrac{1}{x^6y^9}$

88. $\left(\dfrac{a^{-2}b^3}{a^3b^{-4}}\right)^{-2}$ $\;\dfrac{a^{10}}{b^{14}}$

89. $\dfrac{(-3a^{-4}b^{-2})(-4ab^{-3})^{-1}}{(12ab^2)^{-1}}$ $\dfrac{9b^3}{a^4}$

90. $\dfrac{(-2a^2b^{-1})^{-2}(-5ab^2)^{-1}}{(50a^{-2}b)^{-1}}$ $\dfrac{-5b}{2a^7}$

91. $(x^2y^{-3})^{-1}$ $\dfrac{y^3}{x^2}$

92. $(a^3b^{-2})^{-2}$ $\dfrac{b^4}{a^6}$

93. $(x^2 + y^{-3})^{-1}$ $\dfrac{y^3}{x^2y^3 + 1}$

94. $(a^3 + b^{-2})^{-2}$ $\dfrac{b^4}{a^6b^4 + 2a^3b^2 + 1}$

95. $(x^{-2} + y^{-2})^{-2}$ $\dfrac{x^4y^4}{(x^2 + y^2)^2}$

96. $(x^{-2} + y^{-2})(x^{-2} - y^{-2})$ $\dfrac{y^4 - x^4}{x^4y^4}$

97. $\left(\dfrac{x^{-4}y^{-7}z^{-6}}{x^{-24}y^{-16}}\right)^0$ 1

98. $\left(\dfrac{x^{-2} + y^{-5}}{x^{-6} - y^{-4}}\right)^0$ 1

99. $\dfrac{x^{-1} + y^{-1}}{xy^{-1}}$ $\dfrac{x + y}{x^2}$

100. $\dfrac{a^{-1} - b^{-1}}{(ab)^{-1}}$ $b - a$

101. $\dfrac{r^{-2} + s^{-1}}{r^{-1} + s^{-2}}$ $\dfrac{s(s + r^2)}{r(s^2 + r)}$

102. $\dfrac{c^{-2} - d^{-3}}{c^{-3} - d^{-2}}$ $\dfrac{cd^3 - c^3}{d^3 - c^3d}$

103. $\dfrac{2a^{-1} + b^{-2}}{a^{-2} + b}$ $\dfrac{a(2b^2 + a)}{b^2(1 + a^2b)}$

104. $\dfrac{3x^{-2} - y^{-3}}{x + y^{-1}}$ $\dfrac{3y^3 - x^2}{x^3y^3 + x^2y^2}$

? Questions for Thought

105. Discuss what is *wrong* (if anything) in each of the following.

(a) $x^2x^3 \overset{?}{=} x^6$

(b) $(x^2)^3 \overset{?}{=} x^5$

(c) $(x^2y^2)^3 \overset{?}{=} x^6y^6$

(d) $(x^2 + y^2)^3 \overset{?}{=} x^6 + y^6$

(e) $\dfrac{6x^6}{2x^2} \overset{?}{=} 3x^3$

(a) This is an incorrect application of the first rule of exponents: We should add exponents, not multiply. (b) This is an incorrect application of the second rule of exponents: We should multiply exponents, not add. (c) This is a correct application of exponential rules 2 and 3. (d) This is an incorrect application of exponential rules: The exponential rules apply only to products and quotients, not to sums. (e) This is an incorrect application of the fourth rule of exponents: We should subtract exponents, not divide.

106. Fill in the operation(s) that makes the following true for all *n:*   The operations of multiplication and division make the statement true.

$$(x \text{ ? } y)^n = x^n \text{ ? } y^n$$

107. Why do we have to separate rule 4 of natural number exponents into cases?   In order to ensure that the exponents are still natural numbers

108. Describe what is *wrong* (if anything) with the following: See Additional Answers

(a) $8^{-1} \overset{?}{=} -8$

(b) $8^{-1} \overset{?}{=} \dfrac{1}{8}$

(c) $(-2)^{-3} \overset{?}{=} +8$

(d) $(-2)^{-3} \overset{?}{=} +6$

(e) $(-2)^{-3} \overset{?}{=} \dfrac{1}{-8}$

(f) $\dfrac{x^{16}}{x^{-5}} \overset{?}{=} x^{11}$

(g) $xy^{-1} \overset{?}{=} \dfrac{1}{xy}$

(h) $xy^{-1} \overset{?}{=} \dfrac{x}{y}$

109. Suppose that we allow exponents to be fractions and that the rules of integer exponents still hold. Notice that if we square $9^{1/2}$, we get

$$(9^{1/2})^2 = 9^{2/2} = 9^1 = 9$$

Hence,

$$(9^{1/2})^2 = 9$$

What does this imply about the value of $9^{1/2}$? $9^{1/2}$ is the number which when squared yields 9. (This is also called the square root of 9.)

110. Consider the equation $y = 2^x$.   See Additional Answers

(a) Construct a table of values for y for $x = -3, -2, -1, 0, 1, 2,$ and 3.

(b) Plot the points (x, y) on the rectangular coordinate system. You should arrive at the graph shown in Figure 7.1. Check that your graph agrees with the figure. The values agree with the given graph.

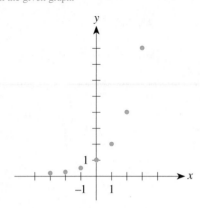

Figure 7.1

(c) Use the figure to discuss how changes in the value of x affect the value of y. (If x changes by 1 unit, how will this affect y? Will y always change by the same amount?) Is y growing quickly or slowly? Is y ever negative? See Additional Answers

 Mini-Review

111. *Solve for x.* Express your answer using interval notation. $-2 \leq 5 - 2x < 11$ $\left(-3, \dfrac{7}{2}\right]$

112. The length of a rectangular area is 5 m less than twice its width. If the rectangular area is to be carpeted at a cost of $14 per sq m, express the cost of carpeting this area in terms of the width, w. $28w^2 - 70w$ dollars

113. Write an equation of the line passing through the points $(0, -4)$ and $(6, 0)$. $y = \frac{2}{3}x - 4$

114. Given $f(x) = 2x^2 - 5x - 1$, compute $f(3) + f(2)$ and $f(5)$. $f(3) + f(2) = -1; f(5) = 24$

 7.2 **Scientific Notation**

There are many occasions (especially in science-related fields) when we may come across either very large or very small numbers. For example, we may read statements such as:

The mass of the earth is approximately 5,980,000,000,000,000,000,000,000 kg.

The wavelength of yellow-green light is approximately 0.0000006 m.

Writing numbers like this in their decimal form can be quite messy and is very prone to errors. Using our knowledge of integer exponents, we can describe an alternative form for writing such numbers that makes them easier to work with. This other form is called *scientific notation.*

Scientific notation is a way of concisely expressing very large or very small numbers. Before we discuss how to write numbers using scientific notation, we will examine powers of 10 with integer exponents:

$$10^3 = 10 \cdot 10 \cdot 10 = 1,000$$
$$10^2 = 10 \cdot 10 = 100$$
$$10^1 = 10$$
$$10^0 = 1$$

$$10^{-1} = \frac{1}{10} = 0.1$$

$$10^{-2} = \frac{1}{10^2} = \frac{1}{100} = 0.01$$

$$10^{-3} = \frac{1}{10^3} = \frac{1}{1,000} = 0.001$$

You can see from these examples that the exponent of 10 indicates the direction and number of places to move the decimal point from the 1. We start with $10^0 = 1.0$ and move the decimal point to the right if the exponent is positive, to the left if the exponent is negative. The number of places we move is equal to the absolute value of the exponent. For example,

$$10^{+4} = 1{,}0000. = 10,000$$
*Start here, move **4** places right.*

$$10^{-6} = 0.000001$$
*Start here, move **6** places left.*

We already know how to multiply and divide powers of 10 by applying the rules for integer exponents.

EXAMPLE 1

Evaluate each of the following:

(a) $10^{-3} \cdot 10^7 \cdot 10^{-6}$ (b) $\dfrac{10^{-4} \cdot 10^{-2}}{10^{-5}}$

Solution   (a) $10^{-3} \cdot 10^7 \cdot 10^{-6} = 10^{-3+7+(-6)} = 10^{-2} = \boxed{0.01}$

(b) $\dfrac{10^{-4} \cdot 10^{-2}}{10^{-5}} = \dfrac{10^{-6}}{10^{-5}} = 10^{-6-(-5)} = 10^{-6+5} = 10^{-1} = \boxed{0.1}$

Scientific Notation

Standard notation is the decimal notation we normally use, as in writing 12.86, 7.954, or 0.0072. *Scientific notation* has the following form:

$a \times 10^n$ where $1 \le a < 10$ and n is an integer

For example, 1.645×10^4, 6.8×10^{-3}, and 8.796×10^4 are numbers written in scientific notation.

The following are *not* in scientific notation.

62×10^1 **62 is not between 1 and 10. This should be written as 6.2×10^2.**

0.064 **0.064 is not between 1 and 10. This should be written as 6.4×10^{-2}.**

Converting from scientific to standard notation is straightforward; it is merely a matter of moving the decimal point right or left.

EXAMPLE 2

Express the following in standard notation:

(a) 1.642×10^5 (b) 7.3×10^{-4}

Solution   (a) $1.642 \times 10^5 = 1.642 \times 100,000 = \boxed{164,200}$

The middle step may be bypassed by observing that the number of places the decimal point is moved is given by the absolute value of the exponent of 10; the direction the decimal point is moved is determined by the sign of the exponent.

$$1.642 \times 10^5 = 164,200 \qquad \textit{Move 5 places right.}$$

(b) $7.3 \times 10^{-4} = 7.3 \times 0.0001 = \boxed{0.00073}$

Converting from standard to scientific notation takes a bit more thought, but again is merely a matter of shifting decimal points right or left. For example, suppose we want to convert 6,748 to scientific notation. We know it must take on the following form:

$$6,748 = 6.748 \times 10^?$$

All that remains is to fill in the question mark.

Since we moved the decimal point three places left to put 6,748 into scientific notation form, we must make up for this by multiplying the expression by 10^{+3}:

$$6,748 \qquad = \qquad 6.748 \qquad \times \qquad 10^3$$

<table>
<tr><td>We want to move the
decimal point 3 places
to the left.</td><td>We put the decimal point
where we want it.</td><td>This power of 10 compensates
by putting the decimal point
back where it really belongs,
3 places to the right.</td></tr>
</table>

In other words, the power of 10 restores the decimal point to its original position. The net effect is that we have *not* changed the value of the number; we have changed only its appearance—which was the whole idea.

$$\text{Answer:} \quad 6,748 = 6.748 \times 10^3$$

EXAMPLE 3

Express the following in scientific notation:

(a) 78,964   (b) 0.00751   (c) 62

Solution   (a) $78,964 = 7.8964 \times 10^?$ *Since we moved the decimal point 4 places left we must multiply by 10^{+4} to compensate for it.*

$$= \boxed{7.8964 \times 10^4}$$

(b) $0.00751 = 7.51 \times 10^?$ *Since we moved the decimal point 3 places right we must multiply by 10^{-3}.*

$$= \boxed{7.51 \times 10^{-3}}$$

(c) $62 = \boxed{6.2 \times 10^1}$

Check: Always check your answers.

(a) $7.8964 \times 10^4 \stackrel{\checkmark}{=} 78,964$ (b) $7.51 \times 10^{-3} \stackrel{\checkmark}{=} 0.00751$

(c) $6.2 \times 10^1 \stackrel{\checkmark}{=} 62$

Computations can often be simplified using scientific notation. We would first convert all numbers from standard to scientific notation and then perform the computations.

EXAMPLE 4

Over a 15-year period, a country with a population of 100 million taxpayers had a national debt accumulating at 43 billion dollars per year. If each taxpayer were to pay an equal amount, how much would each have to pay if they were to pay off the entire accumulated 15-year debt?

Solution

If each taxpayer were to pay an equal amount, A, then each would pay

$$A = \frac{\text{Total debt}}{\text{Number of taxpayers}} = \frac{43 \text{ billion/year} \times 15 \text{ years}}{100 \text{ million}}$$

$$= \frac{43{,}000{,}000{,}000 \times 15}{100{,}000{,}000}$$

$$= \frac{(4.3 \times 10^{10})(1.5 \times 10^{1})}{1 \times 10^{8}} \qquad \textit{Separate out powers of } 10.$$

$$= \frac{(4.3)(1.5)}{1} \times \frac{(10^{10})(10^{1})}{10^{8}}$$

$$= 6.45 \times 10^{3}$$

$$= \boxed{\$6{,}450 \text{ per taxpayer}}$$

Most scientific calculators have an **EE** *key, which will allow you to enter* 4.3×10^{10} *by typing* **4.3** **EE** **10** *.*

If you use a calculator for this example, you may find it quicker to enter 43 billion using scientific notation rather than standard notation. We first change 43 billion to 4.3×10^{10}, and we enter this number in the calculator as follows:

4.3 **×** **10** **y^x** **10** **=** .

Another advantage to using scientific notation is that it makes it easier for you to estimate your answer. In Example 4, after converting the numbers to scientific notation, we could have mentally estimated the answer to be around $4 \times 1.5 \times 10^{3}$, or 6,000 dollars.

EXAMPLE 5

Solution

The sun is approximately 93 million miles from Earth. How long does it take light, which travels at approximately 186,000 miles per second, to reach us from the sun?

To compute the time it takes light to reach us from the sun, we will use the familiar relationship $d = rt$ in the form

$$t = \frac{d}{r}$$

$$= \frac{93{,}000{,}000}{186{,}000}$$

This computation will be much easier to do with the numbers in scientific notation:

$$t = \frac{9.3 \times 10^{7}}{1.86 \times 10^{5}} = 5 \times 10^{2} = 500 \text{ seconds or } \boxed{8 \text{ min } 20 \text{ sec}}$$

EXERCISES 7.2

In Exercises 1–10, perform the indicated operations. Express your answers in standard form.

1. $10^{-4} \cdot 10^{7}$ 1,000

2. $10^{-6} \cdot 10^{5} \cdot 10^{-3}$ 0.0001

3. $\dfrac{10^{4}}{10^{-5}}$ 1,000,000,000

4. $(10^{-2})^{-3}$ 1,000,000

5. $\dfrac{10^{-4} \cdot 10^{2}}{10^{-3}}$ 10

6. $\dfrac{10^{4} \cdot 10^{-5}}{10^{-7} \cdot 10^{-2}}$ 100,000,000

7. $\dfrac{10^{-4} \cdot 10^{2} \cdot 10^{-3}}{10^{4} \cdot 10^{-3}}$ 0.000001

8. $\dfrac{10^{-5} \cdot 10^{0} \cdot 10^{2}}{10^{3} \cdot 10^{-2}}$ 0.0001

9. $\dfrac{10^{-4} \cdot 10^{-5} \cdot 10^{7}}{10^{-6} \cdot 10^{-2} \cdot 10^{0}}$ 1,000,000

10. $\dfrac{10^{5} \cdot 10^{-5} \cdot 10^{4}}{10^{2} \cdot 10^{-2} \cdot 10^{-3}}$ 10,000,000

In Exercises 11–18, convert to standard form.

11. 1.62×10^{1} 16.2

12. 8.654×10^{-2} 0.08654

13. 7.6×10^{8} 760,000,000

14. 9.37×10^{-5} 0.0000937

15. 8.51×10^{-7} 0.000000851

16. 3.4×10^{0} 3.4

17. 6.0×10^{3} 6,000

18. 7.924×10^{-5} 0.00007924

In Exercises 19–30, convert to scientific notation.

19. 824 8.24×10^{2}

20. 72 7.2×10^{1}

21. 5 5.0×10^{0}

22. 0.06 6.0×10^{-2}

23. 0.0093 9.3×10^{-3}

24. 65,789 6.5789×10^{4}

25. 827,546,000 8.27546×10^{8}

26. 70,000,000,000 7.0×10^{10}

27. 0.00000072 7.2×10^{-7}

28. 685.4 6.854×10^{2}

29. 79.32 7.932×10^{1}

30. 0.0000573 5.73×10^{-5}

In Exercises 31–34, perform the computations by first converting the numbers to scientific notation. Express your answers in standard notation.

31. $\dfrac{(6,000)(0.007)}{(0.021)(12,000)}$ 0.166 or 0.17

32. $\dfrac{(720,000)(0.005)}{(0.8)(0.0003)}$ 15,000,000

33. $\dfrac{(120)(0.005)}{(10,000)(60)}$ 0.000001

34. $\dfrac{(200)(25)}{(0.00004)(0.005)}$ 25,000,000,000

35. An adult has an average of about 5 million red blood cells and 7,500 white blood cells per cubic millimeter of blood. Express the number of red and white blood cells per cubic millimeter of blood using scientific notation. Red: 5×10^{6}; White: 7.5×10^{3}

36. In Australia at the end of June 2000, credit card debt reached 14.5 billion dollars. There were 9 million credit card accounts at that time in Australia. Express the debt and number of cards using scientific notation, and find the average debt per credit card. Debt: $\$1.45 \times 10^{10}$; cards: 9×10^{6}, \$1611.11

37. Use the information in Exercise 35 to write the ratio of red to white blood cells per cubic millimeter of blood. $\frac{2,000}{3}$

38. At the end of 1997, of the 100 million households in the United States that carry credit, 58% had average balances of more that \$7,000, costing these households an average of \$1,200 per year in interest and fees. Express the number of households having average balances over \$7,000 using scientific notation. Use scientific notation to compute the total amount of money spent on credit interest and fees in 1997 in the United States. 5.8×10^{7} households; $\$6.96 \times 10^{10}$

39. Pluto is the planet in our solar system that is farthest from the sun, at a distance of 3,670,000,000 miles. How long does it take light from the sun to reach Pluto? (Light travels approximately 186,000 miles per second.)   5 hr 28 min 51 sec

40. If it takes light 4 years to reach the Earth from Proxima Centauri (the star closest to our sun), how far, in miles, is Proxima Centauri from Earth? 2.3462784×10^{13} miles

41. Scientists estimate that the sun converts about 700 million tons of hydrogen to helium every second. How much hydrogen does the sun use in 1 day? In 1 year? 6.048×10^{13} tons in one day, 2.20752×10^{16} tons in one year

42. The sun contains about 1.49×10^{27} tons of hydrogen. Use the result of Exercise 41 to compute how many years it will take to use up the sun's supply of hydrogen. 6.750×10^{10} years

43. Two common units of measurements in physics and chemistry are the *angstrom* unit (written Å), which is 10^{-8} cm, and the *micron* unit (written μm), which is 10^{-4} cm. How many angstroms are there in 1 μm? 10^4 Å in 1 μm

44. The mass of the earth is approximately 5.98×10^{24} kg and the mass of the sun is approximately 1.99×10^{30} kg. How many times greater is the mass of the sun than that of the earth? 3.328×10^5 times greater

45. An atom of oxygen has an atomic radius of 0.66 Å. Use scientific notation to write the number in centimeters. (See Exercise 43.) 0.66×10^{-8} cm $= 6.6 \times 10^{-9}$ cm

46. A colloidal suspension may contain particles that are 17 thousandths of a micron wide. Use scientific notation to write this number in centimeters. (See Exercise 43.) 1.7×10^{-6} cm

47. One light-year (the distance that light travels in 1 year) is approximately equal to 5.86×10^{12} miles. If 1 mile is 1.6 km, what is the distance of 1 light-year in kilometers?

$\dfrac{1 \text{ mile}}{1.6 \text{ km}} = \dfrac{5.86 \times 10^{12} \text{ miles}}{x \text{ km}}$;

9.376×10^{12} km

48. Another unit of measurement used by astronomers is the *parsec*. One parsec is equal to 3.08×10^{13} km. How many light-years are there in 1 parsec? (See Exercise 47.)   3.285 light years

49. The diameter of a typical cheek cell is 60 μm. Use scientific notation to write this number in centimeters. (See Exercise 43.) 60×10^{-4} cm $= 6.0 \times 10^{-3}$ cm

50. If one atom of hydrogen has a mass of 1.67×10^{-24} g, what is the mass of 75,000 atoms of hydrogen? 1.2525×10^{-19} g

51. If one atom of an element has a mass of 10^{-23} g, how many atoms are there in 1 g of this element? 10^{23} atoms

52. If one atom of an element has a mass of 10^{-23} g, how many atoms are there in 1 kg of this element? 10^{26} atoms

53. If 40,000 cheek cells are lined up in a row, how far would they stretch, in centimeters? (Use the result of Exercise 49.)   240 cm

54. If the mass of an atom of iron is 9.3×10^{-23} g, how many atoms are there in 1 kg of iron? 1.075×10^{25} atoms

55. Another unit of measure used by astronomers is the *astronomical unit,* where 1 *AU* is approximately 93 million miles. How many *AU* are in a light-year? (See Exercise 47.)   Approximately 63,011 *AU*

56. A *parsec* is approximately 1.91×10^{13} miles. If an *AU* is approximately 93 million miles, how many *AU* are in a parsec? (See Exercises 48 and 55.) 2.054×10^5 *AU*

Questions for Thought

57. Explain how you would multiply 2.58×10^{-3} by 4.22×10^4 without converting
either number to standard notation.   Multiply 2.58 by 4.22 to get 10.8876. Then multiply 10^{-3} by 10^4 to get 10^1. We now have
10.8876×10^1. It is not quite in scientific notation form yet—we rewrite it as 1.08876×10^2.

58. In scientific notation, what is the sign of the exponent of 10 of a positive number
that is greater than 1? Less than 1?   The sign of the exponent of 10 of a positive number in scientific notation form is nonnegative if the number is greater than 1 and negative if the number is less than 1.

◆ **Mini-Review**

59. *Simplify:* $(3x - 2)(x - 8) - (x - 4)^2$ $2x^2 - 18x$

60. Given $f(x) = \dfrac{2x}{x - 1}$, find $f(x + 1)$. $\dfrac{2x + 2}{x}$

61. Evaluate $(x - y)^3 - x^2$ for $x = -3$ and $y = -1$. -17

62. Lenore leaves home at 10:00 A.M. driving at 40 mph. One hour later Haleema
leaves the same location and drives in the opposite direction at 46 mph. At
approximately what time will they be 100 miles apart?   Approx. 11:42 A.M.

7.3 | **Rational Exponents and Radical Notation**

Our original definition of exponents required natural number exponents. We then let the
exponent rules tell us how to extend the definition of exponents to include integer exponents. Next we will extend the definition even further to define rational number exponents. We will approach this task in the same way we did in Section 7.1, except that we
will primarily use rule 2 for exponents to extend our definition to include rational number exponents. Remember that a rational number is a number of the form $\frac{p}{q}$, where p and
q are integers and $q \neq 0$.

We will start with the simplest case: $a^{1/2}$. How do we define $a^{1/2}$ in general? Let's
examine $9^{1/2}$. First observe that if we apply rule 2 and square $9^{1/2}$ we get

$$(9^{1/2})^2 = 9^{2/2} = 9^1 = 9$$

So $9^{1/2}$ is a number that, *when squared,* yields 9. There are two possible answers:

$$3 \quad \text{since} \quad 3^2 = 9 \quad \text{and} \quad -3 \quad \text{since} \quad (-3)^2 = 9$$

Therefore, we create the following definition to avoid ambiguity:

DEFINITION	$a^{1/2}$ (called the ***principal square root*** of a) is the *nonnegative* quantity that, when squared, yields a.

Thus, $9^{1/2} = 3$.

EXAMPLE 1   Evaluate the following:

(a) $16^{1/2}$ **(b)** $-16^{1/2}$ **(c)** $(-16)^{1/2}$

Solution   **(a)** $16^{1/2} = \boxed{4}$ *Since $4^2 = 16$ and 4 is nonnegative*

(b) $-16^{1/2} = \boxed{-4}$ *Note: We want the negative of $16^{1/2}$.*

(c) $(-16)^{1/2}$ is $\boxed{\text{not a real number}}$.   *Since no real number squared will yield a negative number*

Can you find a real number whose square is -16?

We arrive at the definition of $a^{1/3}$ in the same way we did for $a^{1/2}$. For example, if we cube $8^{1/3}$, we get

$$(8^{1/3})^3 = 8^{3/3} = 8^1 = 8$$

Thus, $8^{1/3}$ is the number that, *when cubed,* yields 8.

Since $2^3 = 8$, we have $8^{1/3} = 2$.

DEFINITION $a^{1/3}$ (called the ***cube root*** of a) is the quantity that, when cubed, yields a.

Hence,

$$27^{1/3} = 3 \quad \text{since } 3^3 = 27$$
$$(-125)^{1/3} = -5 \quad \text{since } (-5)^3 = -125$$

Let's examine one more root before we generalize our findings. We will look at $a^{1/4}$. Let's raise $16^{1/4}$ to the fourth power:

$$(16^{1/4})^4 = 16^{4/4} = 16^1 = 16$$

Thus, $16^{1/4}$ is a number that, when raised to the fourth power, yields 16. As with the square root, we have two possible answers:

$$2 \quad \text{since } 2^4 = 16 \quad \text{and} \quad -2 \quad \text{since } (-2)^4 = 16$$

Again, to avoid ambiguity, we define $a^{1/4}$ as follows:

DEFINITION $a^{1/4}$ (called the ***principal fourth root*** of a) is the *nonnegative* quantity that, when raised to the fourth power, yields a.

Looking at the discussion preceding each of the definitions above, you will notice that there is a possibility for ambiguity when the root is even. That is, there are two possible answers. To eliminate this ambiguity, we inserted the word *nonnegative* in our definitions of the square root and fourth root. On the other hand, for odd roots (such as cube roots), there is only one real-number answer, so we need not be concerned with ambiguity. This leads us to the following general definition.

DEFINITION $a^{1/n}$ (*called the **principal nth root of a***) is the real number (positive when n is even) that, when raised to the nth power, yields a. We can write this symbolically as:

$$a^{1/n} = \begin{cases} b & \text{if } n \text{ is odd and } b^n = a \\ |b| & \text{if } n \text{ is even, } a \geq 0, \text{ and } b^n = a \end{cases}$$

Earlier we pointed out that the square root of -16 is not a real number, since no real number squared will yield a negative number. In general, raising any real number to an even power will always yield a nonnegative number. Therefore, the even root of a negative number is not a real number.

We can summarize the various types of roots as indicated in the box.

	n is even	**n is odd**
$a > 0$	$a^{1/n}$ is the positive nth root of a	$a^{1/n}$ is the nth root of a
$a < 0$	$a^{1/n}$ is not a real number	$a^{1/n}$ is the nth root of a
$a = 0$	$0^{1/n} = 0$	$0^{1/n} = 0$

EXAMPLE 2

Evaluate the following:

(a) $(-32)^{1/5}$ **(b)** $(-64)^{1/6}$ **(c)** $(-8)^{1/3}$ **(d)** $\left(\dfrac{1}{81}\right)^{1/4}$

Solution

(a) $(-32)^{1/5} = \boxed{-2}$ *Since $(-2)^5 = -32$*

(b) $(-64)^{1/6}$ $\boxed{\text{is not a real number.}}$ *What number raised to the sixth power will yield -64?*

(c) $(-8)^{1/3} = \boxed{-2}$ *Since $(-2)^3 = -8$*

(d) $\left(\dfrac{1}{81}\right)^{1/4} = \boxed{\dfrac{1}{3}}$ *Since $\left(\dfrac{1}{3}\right)^4 = \dfrac{1}{81}$*

We will often have occasion to use an alternate notation for fractional exponents, called **radical notation:**

$$x^{1/n} \quad \text{is also written as} \quad \sqrt[n]{x}$$

For example:

$$x^{1/5} = \sqrt[5]{x}$$
$$x^{1/7} = \sqrt[7]{x}$$

In particular,

$$x^{1/2} \quad \text{is written as} \quad \sqrt{x}$$

Thus far we have defined $a^{1/n}$ where n is a natural number. With some help from rule 2 for exponents, we can define the expression $a^{m/n}$ where n and m are natural numbers.

DEFINITION

If $a^{1/n}$ is a real number, then

$$a^{m/n} = (a^{1/n})^m$$

That is, $a^{m/n}$ is the nth root of a raised to the mth power.

We define $a^{-m/n}$ as follows:

DEFINITION

$$a^{-m/n} = \frac{1}{a^{m/n}} \qquad (a \neq 0).$$

Now that we have defined rational exponents, we assert that the *rules for integer exponents hold for rational exponents as well, provided the root is a real number* (that is, provided we avoid even roots of negative numbers). For example, to use the rule $(a^r)^s = a^{rs}$, where r and s are rational, it is necessary that both a^r and a^s be defined.

Since $a^{\frac{m}{n}} = a^{m \cdot \frac{1}{n}} = (a^m)^{1/n}$, we find that $(a^{1/n})^m = (a^m)^{1/n}$. Hence, we can interpret $a^{m/n}$ in two ways, as indicated in the next box.

If $a^{1/n}$ is a real number, then

$$a^{m/n} = (a^{1/n})^m = (a^m)^{1/n}$$

EXAMPLE 3

Evaluate each of the following:

(a) $27^{2/3}$ **(b)** $16^{3/4}$ **(c)** $(-64)^{2/3}$ **(d)** $-64^{2/3}$

(e) $36^{-1/2}$ **(f)** $(-125)^{-1/3}$ **(g)** $\left(\dfrac{1}{125}\right)^{-1/3}$

Solution

In general, you will find it easier to find the root *before* raising to a power.

(a) $27^{2/3} = (27^{1/3})^2 = 3^2 = \boxed{9}$

(b) $16^{3/4} = (16^{1/4})^3 = 2^3 = \boxed{8}$

We could have done this same problem in the following way:

$$16^{3/4} = (16^3)^{1/4} = (4{,}096)^{1/4} = 8$$

But this approach requires more multiplication and finding a difficult root.

(c) $(-64)^{2/3} = [(-64)^{1/3}]^2 = [-4]^2 = \boxed{+16}$

(d) $-64^{2/3} = -[(64)^{1/3}]^2 = -[4]^2 = \boxed{-16}$ *Note the difference between this problem and part* (c).

(e) $36^{-1/2} = \dfrac{1}{36^{1/2}}$ *Begin by changing the negative exponent into a positive exponent by using the definition of a negative rational exponent.*

$$= \boxed{\dfrac{1}{6}}$$

(f) $(-125)^{-1/3} = \dfrac{1}{(-125)^{1/3}}$ *Definition of negative rational exponent*

$$= \dfrac{1}{-5} = \boxed{-\dfrac{1}{5}}$$

Note that *the sign of the exponent has no effect on the sign of the base.*

(g) We will approach this problem a bit differently, using the fact that an exponent of -1 turns an expression into its reciprocal.

$$\left(\dfrac{1}{125}\right)^{-1/3} = \left[\left(\dfrac{1}{125}\right)^{-1}\right]^{1/3}$$

$$= \left[\dfrac{125}{1}\right]^{1/3}$$ *Since the reciprocal of $\dfrac{1}{125}$ is $\dfrac{125}{1}$*

$$= 125^{1/3} = \boxed{5}$$

As we stated earlier, the rules for integer exponents are also valid for rational exponents. Thus, we simplify expressions involving rational exponents by following the same procedures we used for integer exponents. The main difficulty in simplifying rational exponent expressions is the fractional arithmetic involved.

One point of confusion that frequently arises involves the difference between negative exponents and fractional exponents. *Negative exponents involve **reciprocals** of the base.* For example,

$$x^{-4} = \frac{1}{x^4} \quad \text{or} \quad 16^{-4} = \frac{1}{16^4} = \frac{1}{65,536}$$

On the other hand, *fractional exponents yield **roots** of the base.* For example,

$$x^{1/4} = \text{the fourth root of } x \quad \text{or} \quad 16^{1/4} = 2$$

As with integer exponents, when we are asked to simplify an expression, the bases and exponents should appear as few times as possible. Unless otherwise noted, we assume that all variables represent positive real numbers.

EXAMPLE 4

Perform the operations and simplify. Express your answer with positive exponents only.

(a) $x^{1/2}x^{2/3}x^{3/4}$ **(b)** $\dfrac{x^{2/5}}{x^{3/4}}$ **(c)** $(y^{2/3}y^{-1/2})^2$

(d) $\dfrac{a^{1/2}a^{-2/3}}{a^{1/4}}$ **(e)** $\left(\dfrac{x^{-1/2}y^{-1/4}}{x^{1/4}}\right)^{-4}$

Solution We apply the rules of rational exponents.

(a) $x^{1/2}x^{2/3}x^{3/4}$ *Since all factors have the same base, we apply rule 1 and add the exponents.*

$= x^{1/2+2/3+3/4}$

$= x^{6/12+8/12+9/12}$ *Leave the exponent as an improper fraction (reduced).*

$= \boxed{x^{23/12}}$

(b) $\dfrac{x^{2/5}}{x^{3/4}}$ *Use rule 4: Subtract the exponents.*

$= x^{2/5-3/4}$

$= x^{8/20-15/20}$

$= x^{-7/20}$ *Use the definition of negative exponents.*

$= \boxed{\dfrac{1}{x^{7/20}}}$

(c) $(y^{2/3}y^{-1/2})^2$ *Make sure you can follow the arithmetic. Apply rule 1.*

$= (y^{2/3+(-1/2)})^2$

$= (y^{1/6})^2$ *Next rule 2: Multiply exponents.*

$= y^{2/6}$ *Reduce the exponent.*

$= \boxed{y^{1/3}}$

(d) $\dfrac{a^{1/2}a^{-2/3}}{a^{1/4}}$

$= \dfrac{a^{1/2+(-2/3)}}{a^{1/4}}$ *Rule 1*

$= \dfrac{a^{-1/6}}{a^{1/4}}$ *Apply rule 4.*

$$= a^{-1/6 - 1/4}$$

$$= a^{-5/12}$$

$$= \boxed{\dfrac{1}{a^{5/12}}}$$

(e) This example is a lot easier if we bring in the outside exponent by rules 5 and 3.

$$\left(\dfrac{x^{-1/2} y^{-1/4}}{x^{1/4}} \right)^{-4} \qquad \textit{First rule 5}$$

$$= \dfrac{(x^{-1/2} y^{-1/4})^{-4}}{(x^{1/4})^{-4}} \qquad \textit{Next rule 3}$$

$$= \dfrac{(x^{-1/2})^{-4}(y^{-1/4})^{-4}}{(x^{1/4})^{-4}} \qquad \textit{Now rule 2}$$

$$= \dfrac{x^2 y^1}{x^{-1}}$$

$$= x^{2-(-1)} y = \boxed{x^3 y}$$

Again, keep in mind that rules 3 and 5 for exponents apply to *factors,* not terms.

EXAMPLE 5

Perform the indicated operations and simplify the following:

(a) $(x^{1/2} + 2x^{1/2})x^{-1/3}$ **(b)** $(5a^{1/2} + 3b^{1/2})^2$

Solution

(a) $(x^{1/2} + 2x^{1/2})x^{-1/3}$ *Combine like terms in parentheses.*

$$= (3x^{1/2})x^{-1/3} \qquad \textit{Then apply rule 1.}$$

$$= 3x^{1/2 + (-1/3)}$$

$$= \boxed{3x^{1/6}}$$

(b) For squaring a binomial, we can use the perfect square special product.

$$(5a^{1/2} + 3b^{1/2})^2 = (5a^{1/2})^2 + 2(5a^{1/2})(3b^{1/2}) + (3b^{1/2})^2$$

$$= 5^2 a^{2/2} + 30a^{1/2} b^{1/2} + 3^2 b^{2/2}$$

$$= \boxed{25a + 30a^{1/2} b^{1/2} + 9b}$$

Radicals

In Chapter 1, we stated that the real numbers consist of rational and irrational numbers. Recall that irrational numbers are defined as real numbers that cannot be expressed as a quotient of two integers. Some of the examples of irrational numbers given included expressions such as $\sqrt{2}$, $\sqrt{3}$, and $\sqrt{7}$, which are also examples of square root expressions. Here we will examine the more general radical expression, $\sqrt[n]{a}$.

Recall that radicals are an alternative way of writing an expression with fractional exponents. That is, we have the following definition:

DEFINITION

$\sqrt[n]{a} = a^{1/n}$ where n is a positive integer.

$\sqrt[n]{a}$ is called the principal ***nth root of a.***

In $\sqrt[n]{a}$, n is called the *index* of the radical, the symbol $\sqrt{}$ is called the *radical* or radical sign, and the expression, a, under the radical is called the *radicand.* Thus,

$\sqrt[4]{3} = 3^{1/4}$ is called the *fourth root* of 3.

$\sqrt[5]{9} = 9^{1/5}$ is called the *fifth root* of 9.

$\sqrt[3]{x} = x^{1/3}$ is usually called the *cube root* of x.

$\sqrt{y} = y^{1/2}$ is usually called the *square root* of y.

Note that we usually drop the index for square roots and write $\sqrt{a}$ rather than $\sqrt[2]{a}$. Thus, $\sqrt[n]{a}$ is that quantity (nonnegative when n is even) that, when raised to the nth power, yields a.

Symbolically, we define the nth root for n a positive odd integer as follows:

For $a \in R$ and n a positive *odd* integer,
$$\sqrt[n]{a} = b \quad \text{if} \quad b^n = a.$$

Thus,

$$\sqrt[3]{64} = 4 \quad \text{since} \quad 4^3 = 64.$$
$$\sqrt[5]{-32} = -2 \quad \text{since} \quad (-2)^5 = -32.$$

When the index of the radical is *even,* we require the root to be *positive:*

For $a \in R$, and n a positive *even* integer.
$$\sqrt[n]{a} = b \quad \text{if } b^n = a \text{ and } b \geq 0$$

In either case, where n is odd or even, if $\sqrt[n]{a}$ is real, $\sqrt[n]{a}$ is called the *principal nth root* of a. Keep in mind that when n is even, the principal nth root cannot be negative.

$\sqrt{9} = 3$ since $3^2 = 9$ *Note that even though $(-3)^2 = 9$, we are interested only in the principal or positive square root.*

$\sqrt[4]{16} = 2$ since $2^4 = 16$

$\sqrt[6]{-64}$ is not a real number since no real number when raised to the sixth power will yield a negative number.

As with rational exponents, we summarize the various types of roots as follows:

	n is even	**n is odd**
$a > 0$	$\sqrt[n]{a}$ is the positive nth root of a	$\sqrt[n]{a}$ is the nth root of a
$a < 0$	$\sqrt[n]{a}$ is not a real number	$\sqrt[n]{a}$ is the nth root of a
$a = 0$	$\sqrt[n]{0} = 0$	$\sqrt[n]{0} = 0$

EXAMPLE 6

Evaluate the following.

(a) $\sqrt[4]{81}$ (b) $\sqrt[5]{-243}$ (c) $\sqrt{-81}$ (d) $\sqrt[3]{0}$

Solution

(a) $\sqrt[4]{81} = \boxed{3}$ *Because $(3)^4 = 81$ (principal root only)*

(b) $\sqrt[5]{-243} = \boxed{-3}$ *Because $(-3)^5 = -243$*

(c) $\sqrt{-81}$ $\boxed{\text{is not a real number}}$. *Since the radicand is negative* **and** *the index is even*

(d) $\sqrt[3]{0} = \boxed{0}$

By defining $\sqrt[n]{a} = a^{1/n}$, we can rewrite $a^{m/n}$ in terms of radicals, assuming $a^{1/n}$ is a real number.

Using exponent rule 2, we have

$$a^{m/n} = (a^m)^{1/n} = \sqrt[n]{a^m}$$

Or, again by exponent rule 2, we can equivalently write

$$a^{m/n} = (a^{1/n})^m = \left(\sqrt[n]{a}\right)^m$$

Hence, we have the property stated in the box.

If $\sqrt[n]{a}$ is a real number and m and n are positive integers, then
$$\sqrt[n]{a^m} = \left(\sqrt[n]{a}\right)^m$$

EXAMPLE 7

Rewrite the following using radical notation.

(a) $a^{1/6}$ **(b)** $b^{2/3}$ **(c)** $(8)^{4/9}$ **(d)** $5x^{1/3}$ **(e)** $x^{-3/4}$

Solution

(a) $a^{1/6} = \boxed{\sqrt[6]{a}}$

(b) $b^{2/3} = \boxed{\sqrt[3]{b^2} \quad \text{or} \quad \left(\sqrt[3]{b}\right)^2}$

(c) $(8)^{4/9} = \boxed{\sqrt[9]{8^4} \quad \text{or} \quad \left(\sqrt[9]{8}\right)^4}$

(d) $5x^{1/3} = \boxed{5\sqrt[3]{x}}$ *Note that the exponent applies only to x and not to 5.*

(e) $x^{-3/4}$ *Change to positive exponents first.*

$$= \frac{1}{x^{3/4}}$$

$$= \boxed{\frac{1}{\sqrt[4]{x^3}} \quad \text{or} \quad \frac{1}{\left(\sqrt[4]{x}\right)^3}}$$

EXAMPLE 8

Express in radical form.

(a) $x^{5/8}$ **(b)** $(x^2 + y^2)^{3/2}$

Solution

(a) $x^{5/8} = \boxed{\left(\sqrt[8]{x}\right)^5}$ or $\boxed{\sqrt[8]{x^5}}$

Note: It is standard practice to leave a radical in the form $\sqrt[8]{x^5}$ rather than the form $\left(\sqrt[8]{x}\right)^5$.

(b) $(x^2 + y^2)^{3/2} = \boxed{\sqrt{(x^2 + y^2)^3}}$

As with rational exponents, depending on the problem, you may find one form more convenient to use than another. Using the form $\left(\sqrt[n]{a}\right)^m$, or finding the root first, is most useful when evaluating a number. For example, to evaluate $27^{2/3}$, we *could* interpret this as $\sqrt[3]{27^2}$. Then

$$\sqrt[3]{27^2} = \sqrt[3]{729} = 9$$

which requires quite a bit of multiplication and being able to figure out that the cube root of 729 is 9.

On the other hand, if we interpret $27^{2/3}$ as $\left(\sqrt[3]{27}\right)^2$, then

$$27^{2/3} = \left(\sqrt[3]{27}\right)^2 = (3)^2 = 9$$

which is obviously much less work.

When we evaluate numerical expressions it is preferable to use the form that finds the root first. The form $\sqrt[n]{a^m}$ will be most useful when we simplify radical expressions with variables, as we will discuss in the next section.

Let's examine $\sqrt[n]{a^n}$. It is tempting to say that $\sqrt[n]{a^n} = a$. However, if a is negative and n is even, this causes a problem. For example,

$$\sqrt[4]{(-2)^4} = \sqrt[4]{+16} = 2 \qquad \textit{Note that the answer is not } -2.$$

Regardless of the sign of a, if n is even, a^n is always nonnegative and therefore its root always exists. In addition, by definition, an even root is always nonnegative. We therefore have the following.

> For $a \in R$ and n an even positive integer,
> $$\sqrt[n]{a^n} = |a|$$

On the other hand, if n is an odd positive integer, we *do* find the following:

> For $a \in R$ and n an odd positive integer,
> $$\sqrt[n]{a^n} = a$$

EXAMPLE 9

Evaluate the following:

(a) $\sqrt[3]{5^3}$ **(b)** $\sqrt{(-2)^2}$ **(c)** $\sqrt[7]{(-8)^7}$ **(d)** $\sqrt[4]{(-3)^4}$

Solution

(a) $\sqrt[3]{5^3} = \boxed{5}$

(b) $\sqrt{(-2)^2} = |-2| = \boxed{2}$ *Note that the answer is not* -2.

(c) $\sqrt[7]{(-8)^7} = \boxed{-8}$

(d) $\sqrt[4]{(-3)^4} = |-3| = \boxed{3}$ *Note:* $\sqrt[4]{(-3)^4} = \sqrt[4]{81} = 3$

Also note that $\left(\sqrt[4]{-3}\right)^4$ *is not equal to* $\sqrt[4]{(-3)^4}$, *since* $\sqrt[4]{-3}$ *is not a real number.*

Until now we have restricted our discussion to numbers that are perfect nth powers (the nth root is a rational number). Consider the two numbers $\sqrt{9}$ and $\sqrt{21}$. They are numerically different, but conceptually, they are the same: $\sqrt{9}$ is the positive number that when squared yields 9; $\sqrt{21}$ is the positive number that when squared yields 21. The only difference is that $\sqrt{9}$ turns out to be a nice rational number but $\sqrt{21}$ does not.

Keep in mind that decimal representations of irrational numbers must be approximations since the decimal neither terminates nor repeats. When we write approximations, we usually use the symbol "≈." Thus, we can state

$\sqrt{21} \approx 4.58$ if we want an approximate value of $\sqrt{21}$ rounded to the nearest hundredth.

$\sqrt{21} \approx 4.583$ if we want the value rounded to the nearest thousandth.

On occasion we need to do computations involving roots that are irrational numbers. In these situations, when we need to approximate the value of a radical expression, we are expected to use a calculator.

EXERCISES 7.3

In Exercises 1–46, evaluate the expression, if possible.

1. $8^{1/3}$
2. $16^{1/4}$
3. $(-32)^{1/5}$
4. $(-64)^{1/3}$
5. $-100^{1/2}$
6. $(-100)^{1/2}$
7. $\sqrt[3]{64}$
8. $\sqrt[6]{64}$
9. $\sqrt[4]{81}$
10. $\sqrt{-81}$
11. $\sqrt[8]{-1}$
12. $\sqrt[3]{-125}$
13. $-\sqrt[9]{-1}$
14. $\sqrt[10]{0}$
15. $\sqrt[3]{-343}$
16. $\sqrt[6]{-78}$
17. $\sqrt[4]{-1,296}$
18. $-\sqrt[5]{-243}$
19. $\sqrt[8]{256}$
20. $\sqrt[3]{-729}$
21. $\sqrt[7]{78,125}$
22. $-\sqrt[3]{343}$
23. $(-32)^{3/5}$
24. $(-243)^{3/5}$
25. $32^{-1/5}$
26. $27^{-1/3}$
27. $(-32)^{-1/5}$
28. $(-27)^{-1/3}$
29. $-(81)^{-1/2}$
30. $-(16)^{-1/4}$
31. $(-64)^{-2/3}$
32. $(-81)^{-3/4}$
33. $\sqrt{(-16)^2}$
34. $\sqrt{-16^2}$
35. $\left(\sqrt{-16}\right)^2$
36. $\sqrt{(-3)^4}$
37. $-\left(\sqrt{16}\right)^2$
38. $\left(\sqrt{-3}\right)^4$
39. $\sqrt[n]{3^{2n}}$
40. $\sqrt[n]{2^n}$
41. $\left(\frac{64}{27}\right)^{1/3}$
42. $\left(\frac{16}{81}\right)^{1/4}$
43. $\left(\frac{81}{16}\right)^{-1/4}$
44. $\left(\frac{27}{64}\right)^{-1/3}$
45. $\left(-\frac{1}{32}\right)^{-4/5}$
46. $\left(-\frac{1}{243}\right)^{-4/5}$

In Exercises 47–70, perform the operations and simplify. Express your answers with positive exponents only. Assume that all variables represent positive real numbers.

47. $x^{1/2}x^{2/3}$
48. $x^{1/3}x^{-2/5}$
49. $(a^{-1/2})^{-3/4}$
50. $(s^{-1/3})^{2/5}$
51. $(2^{-1} \cdot 4^{1/2})^{-2}$
52. $(3^2 \cdot 27^{-1/3})^{-2}$
53. $(r^{1/2}r^{-2/3}s^{1/2})^{-2}$
54. $(a^{1/2}b^{-1/3})^{-1/2}$
55. $(r^{-1}s^{1/2})^{-2}(r^{-1/2}s^{1/3})^2$
56. $(a^{1/3}b^{-2})^{-2}(a^{2/3}b^{-2})^3$
57. $\frac{x^{-1/2}}{x^{-1/3}}$
58. $\frac{x^{1/2}}{x^{-1/4}}$
59. $\frac{a^{-1/2}b^{1/3}}{a^{1/4}b^{1/5}}$
60. $\frac{x^{1/2}y^{-1/3}}{x^{1/3}y^{1/5}}$
61. $\left(\frac{x^{1/2}x^{-1}}{x^{1/3}}\right)^{-6}$
62. $\left(\frac{x^{-1/3}x^{1/5}}{x}\right)^{-15}$
63. $\frac{(4^{-1/2} \cdot 16^{3/4})^{-2}(64^{5/6})}{(-64)^{1/3}}$
64. $\left(\frac{9^{-1/2} \cdot 27^{2/3} \cdot 81^{1/4}}{(-3)^{-1} \cdot 81^{3/4}}\right)^{-2}$
65. $\frac{(x^{1/2}y^{1/3})^{-2}(x^{1/3}y^{1/4})^{-12}}{xy^{1/4}}$
66. $\frac{(a^{1/2}b^{1/3})^{-6}(a^{1/3}b^{-1/2})^2}{a^{-1/3}b}$

67. $(x^{1/2} + x)x^{1/2}$

68. $(x^{1/2} + x^{3/2})x^{1/2}$

69. $(x^{1/2} - 2x^{-1/2})^2$

70. $4x(x^{1/2} + 1)^2$

In Exercises 71–84, change to exponential form.

71. $\sqrt[3]{xy}$

72. $\sqrt[4]{x^3y^2}$

73. $\sqrt{x^2 + y^2}$

74. $\sqrt[3]{x^3 - y^3}$

75. $\sqrt[5]{5a^2b^3}$

76. $3\sqrt[5]{2x^2y^3}$

77. $2\sqrt[3]{3xyz^4}$

78. $2\sqrt[3]{ab}$

79. $5\sqrt[3]{(x - y)^2}$

80. $3\sqrt[7]{(x - y)^2(x + y)^3}$

81. $\sqrt[n]{x^n - y^n}$

82. $\sqrt[n]{x^{2n}y^{3n}}$

83. $\sqrt[n]{x^{5n+1}y^{2n-1}}$

84. $\sqrt[n]{x^{2n-3}y^{4n}}$

In Exercises 85–98, change to radical form.

85. $x^{1/3}$

86. $a^{3/5}$

87. $mn^{1/3}$

88. $(mn)^{1/3}$

89. $(-a)^{2/3}$

90. $(-x)^{3/4}$

91. $-a^{2/3}$

92. $-x^{3/4}$

93. $(a^2b)^{1/3}$

94. $(x^3y^2)^{3/5}$

95. $(x^2 + y^2)^{1/2}$

96. $(x^3 - y^3)^{1/3}$

97. $(x^n - y^n)^{1/2}$

98. $(x^2 - y)^{1/n}$

In Exercises 99–102, use your calculator to evaluate the following to four decimal places.

99. $36^{-1/2}$

100. $64^{-2/3}$

101. $3 - 18^{-3/4}$

102. $12 - 28^{2/5}$

? Questions for Thought

103. Explain what is *wrong* (if anything) with each of the following:

(a) $8^{1/3} \stackrel{?}{=} \dfrac{1}{8^3}$ (b) $8^{-3} \stackrel{?}{=} -512$ (c) $8^{-1/3} \stackrel{?}{=} -2$

104. What is the difference between a negative exponent and a fractional exponent?

105. What are the differences among the rules of exponents for rational, integer, and natural number exponents?

106. Discuss the similarities and differences among 9^{-2}, $9^{1/2}$, and $9^{-1/2}$.

107. Prove the following using the rules of rational exponents, assuming all roots are real numbers:

$$\sqrt[m]{\sqrt[n]{a}} = \sqrt[mn]{a}$$

[*Hint:* Change the radicals to rational exponents.]

108. Given the result of Exercise 107, rewrite the following as a single radical:

(a) $\sqrt[3]{\sqrt[4]{6}}$ (b) $\sqrt{\sqrt[3]{x^2y}}$ (c) $\sqrt[4]{\sqrt[3]{\sqrt{x}}}$

109. Explain why the following is *wrong:*

$$\sqrt{(-4)^2} \stackrel{?}{=} (\sqrt{-4})^2$$

Mini-Review

110. *Solve:* $2x^2 = 3x$

111. Use the following graph of $y = f(x)$ to find

  (a) $f(4)$

  (b) $f(-3)$

  (c) $f(0)$

  (d) $f(2)$

  (e) x such that $f(x) = 0$

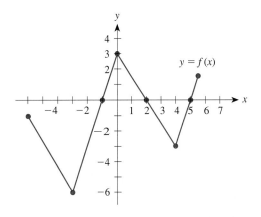

112. *Combine:* $\dfrac{x}{3} - \dfrac{x}{2} + 4$

113. *Solve:* $\dfrac{x}{3} - \dfrac{x}{2} = 4$

7.4 Simplifying Radical Expressions

Now that we have defined radicals, our next step is to determine what constitutes simplified form. Along with the definition of a radical, the following three properties of radicals will provide us with much of what we will need to simplify radicals:

Properties of Radicals

If $\sqrt[n]{a}$ and $\sqrt[n]{b}$ are real numbers, then

1. $\sqrt[n]{ab} = \sqrt[n]{a}\sqrt[n]{b}$

2. $\sqrt[n]{\dfrac{a}{b}} = \dfrac{\sqrt[n]{a}}{\sqrt[n]{b}}$ $(b \neq 0)$

3. $\sqrt[np]{a^{mp}} = \sqrt[n]{a^m}$ $(a \geq 0)$

Properties 1 and 2 are actually forms of rules 3 and 5 of rational exponents, whereas property 3 is a result of reducing a fractional exponent.

Property 1: $\sqrt[n]{ab} = (ab)^{1/n} = a^{1/n}b^{1/n} = \sqrt[n]{a}\sqrt[n]{b}$

Property 2: $\sqrt[n]{\dfrac{a}{b}} = \left(\dfrac{a}{b}\right)^{1/n} = \dfrac{a^{1/n}}{b^{1/n}} = \dfrac{\sqrt[n]{a}}{\sqrt[n]{b}}$

Property 3: $\sqrt[np]{a^{mp}} = a^{(mp)/(np)} = a^{m/n} = \sqrt[n]{a^m}$

Our goal in this section is to write expressions in simplest radical form. We *define* simplest radical form in the accompanying box.

> An expression is in **simplest radical form** if:
>
> 1. All factors of the radicand have exponents less than the index.
>   For example, $\sqrt[3]{y^5}$ violates this condition.
>
> 2. There are no fractions under the radical.
>   For example, $\sqrt{\dfrac{3}{5}}$ violates this condition.
>
> 3. There are no radicals in the denominator of a fraction.
>   For example, $\dfrac{4}{\sqrt[3]{x}}$ violates this condition.
>
> 4. The greatest common factor of the index and the exponents of all the radicand factors is 1.
>   For example, $\sqrt[8]{x^4}$ violates this condition.

To simplify matters, from this point on, all variables will represent positive real numbers.

Given the definition of a radical and the three properties of radicals, we can simplify many radical expressions according to the criteria given in the box. Here are a few examples of how we would use the properties to simplify radicals:

1. $\sqrt{18} = \sqrt{9 \cdot 2} = \sqrt{3^2 \cdot 2}$ *Factor out the greatest perfect square. Then apply property 1.*

$= \sqrt{3^2} \sqrt{2}$

$= 3\sqrt{2}$ *Since $\sqrt{3^2} = 3$*

2. $\sqrt[5]{64} = \sqrt[5]{2^6}$ *Factor out the greatest perfect fifth power.*

$= \sqrt[5]{2^5 \cdot 2}$ *Then apply property 1.*

$= \sqrt[5]{2^5} \sqrt[5]{2}$

$= 2\sqrt[5]{2}$ *Since $\sqrt[5]{2^5} = 2$*

3. $\sqrt[4]{\dfrac{81}{16}} = \dfrac{\sqrt[4]{81}}{\sqrt[4]{16}}$ *Property 2*

$= \dfrac{3}{2}$

4. $\sqrt[8]{x^6}$ *Factor the greatest common factor from the index and the exponent of the radicand. Then apply property 3.*

$= \sqrt[4 \cdot 2]{x^{3 \cdot 2}}$

$= \sqrt[4]{x^3}$

An alternative way to simplify this last expression is to first convert the radical expression to rational exponents:

$$\sqrt[8]{x^6} = x^{6/8} = x^{3/4} = \sqrt[4]{x^3}$$

EXAMPLE 1

Express the following in simplest radical form. Assume all variables are positive numbers.

(a) $\sqrt{25x^8y^6}$ (b) $\sqrt[3]{\dfrac{27}{y^{18}}}$

Solution   (a) $\sqrt{25x^8y^6}$

$= \sqrt{25} \sqrt{x^8} \sqrt{y^6}$ *Apply property 1*

$= 5x^4 \cdot y^3$ *Since $5^2 = 25$, $(x^4)^2 = x^8$, and $(y^3)^2 = y^6$*

$= \boxed{5x^4y^3}$

(b) $\sqrt[3]{\dfrac{27}{y^{18}}} = \dfrac{\sqrt[3]{27}}{\sqrt[3]{y^{18}}}$ *Property 2*

$= \boxed{\dfrac{3}{y^6}}$ *Since $3^3 = 27$ and $(y^6)^3 = y^{18}$*

EXAMPLE 2

Express the following in simplest radical form. Assume all variables are positive numbers.

(a) $\sqrt{x^{10}}$ (b) $\sqrt[3]{y^{18}}$ (c) $\sqrt[4]{x^{12}}$ (d) $\sqrt[7]{x^{21}}$

Solution   (a) $\sqrt{x^{10}} = \boxed{x^5}$ *Since $(x^5)^2 = x^{10}$*

(b) $\sqrt[3]{y^{18}} = \boxed{y^6}$ *Since $(y^6)^3 = y^{18}$*

(c) $\sqrt[4]{x^{12}} = \boxed{x^3}$ *Since $(x^3)^4 = x^{12}$*

(d) $\sqrt[7]{x^{21}} = \boxed{x^3}$ *Since $(x^3)^7 = x^{21}$*

For a power to be a perfect *n*th power, the exponent must be divisible by *n*. In all cases of Example 2, the factors were perfect *n*th powers, where *n* was the index of the radical. Let's examine how to simplify radicals with radicand factors that are not perfect *n*th powers. In these cases we try to factor the greatest perfect *n*th power from the expression.

EXAMPLE 3

Express the following in simplest radical form:

(a) $\sqrt{x^7}$ **(b)** $\sqrt[6]{x^{17}}$ **(c)** $\sqrt[3]{24}$ **(d)** $\sqrt[4]{64a^{11}c^{24}}$

Solution   **(a)** $\sqrt{x^7}$ *Factor the greatest perfect square factor of x^7, which is x^6, from the radicand.*

$= \sqrt{x^6 \cdot x}$

$= \sqrt{x^6}\,\sqrt{x}$

$= \boxed{x^3\sqrt{x}}$ *$\sqrt{x^6} = x^3$ because $(x^3)^2 = x^6$.*

(b) $\sqrt[6]{x^{17}}$ *The greatest perfect sixth-power factor of x^{17} is x^{12}. Thus, we factor x^{12} from x^{17}.*

$= \sqrt[6]{x^{12} \cdot x^5}$

$= \sqrt[6]{x^{12}}\,\sqrt[6]{x^5}$

$= \boxed{x^2\sqrt[6]{x^5}}$ *$\sqrt[6]{x^{12}} = x^2$ because $(x^2)^6 = x^{12}$.*

Note that the exponent of *x* under the radical is less than the index, thereby satisfying criterion 1 for simplifying radicals.

(c) $\sqrt[3]{24} = \sqrt[3]{8 \cdot 3}$

$\qquad = \sqrt[3]{8}\sqrt[3]{3} = \boxed{2\sqrt[3]{3}}$

(d) $\sqrt[4]{64a^{11}c^{24}}$ *Factor the greatest perfect fourth power from each factor.*

$= \sqrt[4]{(16 \cdot 4)(a^8 \cdot a^3)(c^{24})}$

$= \sqrt[4]{(16a^8c^{24})(4a^3)}$

$= \sqrt[4]{16}\,\sqrt[4]{a^8}\,\sqrt[4]{c^{24}}\,\sqrt[4]{4a^3}$ *By property 1*

$= \boxed{2a^2c^6\sqrt[4]{4a^3}}$

Again, note that all exponents of factors under the radical are less than the index.

EXAMPLE 4

Express in simplest radical form.

(a) $\sqrt[5]{(x^2 + 2)^5}$ **(b)** $\sqrt[3]{x^3 + y^3}$

Solution (a) $\sqrt[5]{(x^2 + 2)^5} = x^2 + 2$ *Remember that $\sqrt[5]{a^5} = a$.*

(b) $\sqrt[3]{x^3 + y^3}$ $\boxed{\text{cannot be simplified.}}$ *Remember, $x^3 + y^3$ is not the same as $(x + y)^3$.*

The properties of radicals can also be used to simplify products and quotients of radicals. When the first two properties of radicals are read from right to left, they state that *if the indices are the same,* the product (quotient) of radicals is the radical of the product (quotient).

EXAMPLE 5

Perform the operations and simplify.

(a) $\sqrt{3a^2b}\,\sqrt{6ab^3}$ (b) $\left(ab\sqrt[3]{2a^2b}\right)\left(2a\sqrt[3]{4ab}\right)$

Solution (a) We could simplify each radical first, but after we multiply, it may be necessary to simplify again. It is often better to multiply radicands first.

$$\sqrt{3a^2b}\sqrt{6ab^3} = \sqrt{(3a^2b)(6ab^3)}$$
$$= \sqrt{18a^3b^4} \quad \textit{Then simplify.}$$
$$= \sqrt{9a^2b^4}\sqrt{2a}$$
$$= \boxed{3ab^2\sqrt{2a}}$$

(b) Since this is all multiplication, we use the associative and commutative properties to reorder and regroup variables, and then use radical property 1 to multiply the radicals.

$$\left(ab\sqrt[3]{2a^2b}\right)\left(2a\sqrt[3]{4ab}\right) = (ab)(2a)\sqrt[3]{(2a^2b)(4ab)}$$
$$= 2a^2b\,\sqrt[3]{8a^3b^2} \quad \textit{Then simplify the radical.}$$
$$= 2a^2b\left(\sqrt[3]{8a^3}\,\sqrt[3]{b^2}\right)$$
$$= 2a^2b\left(2a\sqrt[3]{b^2}\right) \quad \textit{Multiply the expressions outside the radical.}$$
$$= \boxed{4a^3b\sqrt[3]{b^2}}$$

Next we examine radicals of quotients.

EXAMPLE 6

Express in simplest radical form: $\sqrt{\dfrac{32s^9}{4s^{17}}}$

Solution $\sqrt{\dfrac{32s^9}{4s^{17}}}$ *Do not apply property 2 without thinking—try to simplify the fraction first.*

$$= \sqrt{\frac{8}{s^8}} \quad \textit{Then apply property 2.}$$

$$= \frac{\sqrt{8}}{\sqrt{s^8}} \quad \textit{We simplify the numerator and denominator.}$$

$$= \boxed{\frac{2\sqrt{2}}{s^4}}$$

Rationalizing the Denominator

A radical in the denominator of a fraction violates criterion 3 for simplifying radicals. In such a case we may have to **rationalize,** or eliminate the radical, in the denominator. To do this, we apply the Fundamental Principle of Fractions, as shown in Example 7.

EXAMPLE 7

Simplify: $\sqrt{\dfrac{2}{5}}$

Solution

$\sqrt{\dfrac{2}{5}}$ *This violates criterion 2 of simplified form for radicals. We apply property 2.*

$= \dfrac{\sqrt{2}}{\sqrt{5}}$ *This violates criterion 3 (a radical in the denominator).*

We apply the Fundamental Principle of Fractions and multiply the numerator and denominator by $\sqrt{5}$. Why $\sqrt{5}$? We choose $\sqrt{5}$ because the product of $\sqrt{5}$ and the denominator will yield the square root of a perfect square, which yields a rational expression. Thus,

$$\frac{\sqrt{2}}{\sqrt{5}} = \frac{\sqrt{2} \cdot \sqrt{5}}{\sqrt{5} \cdot \sqrt{5}} = \frac{\sqrt{2 \cdot 5}}{\sqrt{5^2}} = \boxed{\frac{\sqrt{10}}{5}}$$

In rationalizing a denominator, what we try to do is to multiply the numerator *and* the denominator of the fraction by the expression that will make the denominator the *n*th root of a perfect *n*th power.

EXAMPLE 8

Simplify: $\sqrt[3]{\dfrac{1}{y}}$

Solution

$$\sqrt[3]{\frac{1}{y}} = \frac{\sqrt[3]{1}}{\sqrt[3]{y}} = \frac{1}{\sqrt[3]{y}}$$

We multiply the numerator and denominator by $\sqrt[3]{y^2}$. Why $\sqrt[3]{y^2}$? Because multiplying $\sqrt[3]{y^2}$ by the original denominator, $\sqrt[3]{y}$, will yield $\sqrt[3]{y^3}$, the cube root of a perfect cube. Thus,

$$\sqrt[3]{\frac{1}{y}} = \frac{1}{\sqrt[3]{y}} = \frac{1 \cdot \sqrt[3]{y^2}}{\sqrt[3]{y} \cdot \sqrt[3]{y^2}}$$

$$= \frac{\sqrt[3]{y^2}}{\sqrt[3]{y \cdot y^2}}$$

$$= \frac{\sqrt[3]{y^2}}{\sqrt[3]{y^3}} \qquad \textit{Since } \sqrt[3]{y^3} = y$$

$$= \boxed{\frac{\sqrt[3]{y^2}}{y}} \qquad \textit{Notice that we no longer have a radical in the denominator.}$$

EXAMPLE 9

Express the following in simplest radical form:

(a) $\sqrt{\dfrac{3}{2x^2}}$ (b) $\sqrt[3]{\dfrac{3}{2x^2}}$

Solution (a) $\sqrt{\dfrac{3}{2x^2}}$ *Apply property 2.*

$$= \dfrac{\sqrt{3}}{\sqrt{2x^2}}$$ *Simplify the denominator by property 1:*
$\sqrt{2x^2} = x\sqrt{2}$

$$= \dfrac{\sqrt{3}}{x\sqrt{2}}$$ *To make the denominator contain the square root of a perfect square, multiply the numerator and denominator by $\sqrt{2}$.*

$$= \dfrac{\sqrt{3} \cdot \sqrt{2}}{x\sqrt{2} \cdot \sqrt{2}}$$ *Note $\sqrt{2}\sqrt{2} = 2$.*

$$= \boxed{\dfrac{\sqrt{6}}{2x}}$$

(b) $\sqrt[3]{\dfrac{3}{2x^2}}$ *Apply property 2.*

$$= \dfrac{\sqrt[3]{3}}{\sqrt[3]{2x^2}}$$ *To change $\sqrt[3]{2x^2}$ into the cube root of a perfect cube, we can multiply the numerator and denominator by $\sqrt[3]{4x}$.*

$$= \dfrac{\sqrt[3]{3} \cdot \sqrt[3]{4x}}{\sqrt[3]{2x^2} \cdot \sqrt[3]{4x}}$$ *Then $\sqrt[3]{2x^2}\,\sqrt[3]{4x} = \sqrt[3]{8x^3}$.*

$$= \dfrac{\sqrt[3]{12x}}{\sqrt[3]{8x^3}}$$ *Now the radicand in the denominator is a perfect cube.*

$$= \boxed{\dfrac{\sqrt[3]{12x}}{2x}}$$

EXAMPLE 10

Express the following in simplest radical form. $\dfrac{2x\sqrt{24xy^3}}{5x^2y\sqrt{12xy}}$

Solution We could simplify each radical first, but after simplifying, multiplying, and reducing, we may have to simplify again. Since this expression is a quotient, we can use property 2 to collect all radicands under one radical.

$$\dfrac{2x\sqrt{24xy^3}}{5x^2y\sqrt{12xy}}$$

$$= \dfrac{2x}{5x^2y}\sqrt{\dfrac{24xy^3}{12xy}}$$ *Then reduce fractions inside and outside the radicals.*

$$= \dfrac{2}{5xy}\sqrt{2y^2}$$ *Now simplify the radical expression.*

$$= \dfrac{2}{5xy}\left(y\sqrt{2}\right)$$ *Then reduce again.*

$$= \boxed{\dfrac{2\sqrt{2}}{5x}}$$

The last criterion for simplifying radical expressions requires us to factor the greatest common factor of the index and all the exponents of factors of the radicand; this requires property 3 for radicals:

$$\sqrt[np]{a^{mp}} = \sqrt[n]{a^{m}}$$ *We can factor out the greatest common factor of the index and the exponent of the radicand.*

For example, $\sqrt[3]{x}$ and $\sqrt[6]{x^2}$ are identical (assuming $x \geq 0$). If we convert to exponential notation, we can see that

$$\sqrt[6]{x^2} = x^{2/6} = x^{1/3} = \sqrt[3]{x}$$

Thus, we can find $\sqrt[6]{27^2}$ by the following:

$$\sqrt[6]{27^2} = \sqrt[6]{729} = 3$$

Or we can recognize that $\sqrt[6]{27^2} = \sqrt[3]{27} = 3$, which takes less effort to evaluate.

We can use property 3 to "reduce the index with the exponent of the radicand" or we can convert to fractional exponents, reduce the fractional exponent, and then rewrite the answer in radical form.

Property 3 is also used to find products and quotients of radicals with different indices. (Keep in mind that properties 1 and 2 require that the indices be the same.)

EXAMPLE 11   Simplify: $\sqrt[3]{x^2}\, \sqrt[5]{x^4}$

Solution $\sqrt[3]{x^2}\, \sqrt[5]{x^4}$ *The indices are not the same; thus, we cannot use property 1.*

The least common multiple of the indices, 5 and 3, is 15. We change the radicals using property 3 as follows:

$$\sqrt[3]{x^2}\, \sqrt[5]{x^4} = \sqrt[3 \cdot 5]{x^{2 \cdot 5}}\, \sqrt[5 \cdot 3]{x^{4 \cdot 3}}$$

$$= \sqrt[15]{x^{10}}\, \sqrt[15]{x^{12}}$$ *Since the indices are the same we can now apply property 1.*

$$= \sqrt[15]{x^{22}}$$

$$= \sqrt[15]{x^{15} \cdot x^{7}}$$ *Don't forget to simplify the radical.*

$$= \boxed{x\, \sqrt[15]{x^{7}}}$$

Actually, it is more instructive to change radicals into rational exponents and to simplify the problem as follows:

$$\sqrt[3]{x^2}\, \sqrt[5]{x^4} = x^{2/3}x^{4/5}$$

$$= x^{2/3 + 4/5}$$

$$= x^{10/15 + 12/15}$$

$$= x^{22/15}$$ *Change back to radical notation.*

$$= \sqrt[15]{x^{22}} = \boxed{x\, \sqrt[15]{x^{7}}}$$

EXERCISES 7.4

Simplify the following. Assume that all variables represent positive real numbers.

1. $\sqrt{56}$ **2.** $\sqrt{45}$ **3.** $\sqrt{48}$ **4.** $\sqrt{54}$

5. $\sqrt[5]{64}$ **6.** $\sqrt[4]{243}$ **7.** $\sqrt{8}\, \sqrt{18}$ **8.** $\sqrt{12}\, \sqrt{75}$

9. $\sqrt{64x^8}$ **10.** $\sqrt[4]{64x^8}$ **11.** $\sqrt[4]{81x^{12}}$ **12.** $\sqrt{81x^{12}}$

13. $\sqrt{128x^{60}}$

14. $\sqrt[3]{128x^{60}}$

15. $\sqrt[4]{128x^{60}}$

16. $\sqrt[6]{128x^{60}}$

17. $\sqrt[5]{128x^{60}}$

18. $\sqrt[7]{128x^{60}}$

19. $\sqrt[3]{x^3y^6}$

20. $\sqrt{x^4y^8}$

21. $\sqrt{32a^2b^4}$

22. $\sqrt{54x^6y^{12}}$

23. $\sqrt[5]{a^{35}b^{75}}$

24. $\sqrt[6]{x^{36}y^{72}}$

25. $\sqrt{x^3y}\,\sqrt{xy^3}$

26. $\sqrt{6ab}\,\sqrt{6a^5b^5}$

27. $\sqrt{\dfrac{1}{2}}$

28. $\sqrt{\dfrac{2}{7}}$

29. $\dfrac{\sqrt{x}}{\sqrt{5}}$

30. $\dfrac{\sqrt{5}}{\sqrt{a}}$

31. $\sqrt{\dfrac{45}{4}}$

32. $\sqrt{\dfrac{8}{9}}$

33. $\dfrac{1}{\sqrt{75}}$

34. $\sqrt{\dfrac{3}{8}}$

35. $\sqrt{64x^5y^8}$

36. $\sqrt{12a^3b^9}$

37. $\sqrt[3]{81x^8y^7}$

38. $\sqrt[3]{24a^5b^2}$

39. $\sqrt[4]{54y^2}\,\sqrt[4]{48y^4}$

40. $\sqrt[3]{20ab^5}\,\sqrt[3]{50b^4}$

41. $\sqrt[6]{(x+y^2)^6}$

42. $\sqrt[6]{(a^2+b)^6}$

43. $\sqrt[4]{x^4-y^4}$

44. $\sqrt[3]{x^3-y^3}$

45. $\left(2s\sqrt{6t}\right)\left(5t\sqrt{3s}\right)$

46. $\left(4a\sqrt{10b}\right)\left(3b\sqrt{2a}\right)$

47. $\left(3a\sqrt[3]{2b^4}\right)\left(2a^2\sqrt[3]{4b^2}\right)$

48. $\left(3a\sqrt[3]{5b^2a}\right)\left(2b\sqrt[3]{25a^2b}\right)$

49. $\sqrt[3]{\dfrac{x^3y^6}{8}}$

50. $\sqrt[5]{\dfrac{r^{10}s^{15}}{32}}$

51. $\sqrt[4]{\dfrac{32x^9}{y^{12}}}$

52. $\sqrt[3]{\dfrac{32x^9}{y^{12}}}$

53. $\dfrac{\sqrt{54xy}}{\sqrt{2xy}}$

54. $\dfrac{\sqrt{6x^2y}}{\sqrt{3xy}}$

55. $\dfrac{\sqrt[4]{x^2y^{17}}}{\sqrt[4]{x^{14}y}}$

56. $\dfrac{\sqrt[3]{a^{13}b^2}}{\sqrt[3]{a^4b^5}}$

57. $\sqrt{\dfrac{3xy}{5x^2y}}$

58. $\sqrt{\dfrac{5ab}{3a^2b}}$

59. $\sqrt{\dfrac{3x^2y}{x^3y^4}}$

60. $\sqrt{\dfrac{2a^2b^3}{a^5b^{18}}}$

61. $\sqrt[3]{\dfrac{3}{2}}$

62. $\sqrt[3]{\dfrac{2}{3}}$

63. $\sqrt[3]{\dfrac{9}{4}}$

64. $\sqrt[3]{\dfrac{4}{9}}$

65. $\sqrt[4]{\dfrac{9}{4}}$

66. $\sqrt[4]{\dfrac{4}{9}}$

67. $\sqrt[3]{\dfrac{81x^2y^4}{2x^3y}}$

68. $\sqrt[3]{\dfrac{64a^2b^5}{9a^4b^2}}$

69. $\dfrac{3a^2\sqrt{a^2x^5}}{9a^5\sqrt{a^6x}}$

70. $\dfrac{5x^2\sqrt{a^3b^2}}{4y^2\sqrt{a^7b^3}}$

71. $\dfrac{-3r^2s\sqrt{32r^2s^5}}{2r\sqrt{2r^5}}$

72. $\dfrac{-7a^2b\sqrt{81a^2b}}{14a\sqrt{9a^7}}$

73. $\sqrt[12]{a^6}$

74. $\sqrt[16]{x^4}$

75. $\left(\sqrt[3]{x}\right)\left(\sqrt[4]{x^3}\right)$

76. $\left(\sqrt[3]{x^2}\right)\left(\sqrt{x^3}\right)$

77. $\dfrac{\sqrt[3]{a^2}}{\sqrt{a}}$

78. $\dfrac{\sqrt[3]{a}}{\sqrt[4]{a^2}}$

79. $\sqrt[n]{x^{5n}y^{3n}}$

80. $\sqrt[n]{x^{2n}y^{4n}}$

Question for Thought

81. Discuss what is *wrong* (if anything) with each of the following:

    (a) $\left(\sqrt{-2}\right)^2 \overset{?}{=} \sqrt{(-2)^2} \overset{?}{=} \sqrt{4} \overset{?}{=} 2$

    (b) $2 \overset{?}{=} \sqrt[6]{64} \overset{?}{=} \sqrt[6]{(-8)^2} \overset{?}{=} \sqrt[3]{-8} \overset{?}{=} -2$

    (c) $\sqrt{x^2-y^2} \overset{?}{=} x-y$

    (d) $(a^5+b^5)^{1/5} \overset{?}{=} a+b$

Mini-Review

82. *Factor as completely as possible:* $2x^2+7x-15$.

83. *Divide:* $\dfrac{x^3-2x+3}{x+4}$

7.5

Adding and Subtracting Radical Expressions

We combine terms with the radical factors in the same way we combined terms with variable factors—through the use of the distributive property. Just as we can combine

$$3x + 4x = (3 + 4)x \qquad \textit{Distributive property}$$
$$= 7x$$

we can also combine

$$3\sqrt{2} + 4\sqrt{2} = (3 + 4)\sqrt{2} \qquad \textit{Distributive property}$$
$$= 7\sqrt{2}$$

EXAMPLE 1

Simplify.

(a) $7\sqrt{3} - 4\sqrt{3} + 6\sqrt{3}$ **(b)** $5\sqrt{2} - 8\sqrt{3} - \left(\sqrt{2} - 7\sqrt{3}\right)$

Solution

(a) $7\sqrt{3} - 4\sqrt{3} + 6\sqrt{3}$

$\qquad = (7 - 4 + 6)\sqrt{3} \qquad \textit{Apply the distributive property.}$

$\qquad = \boxed{9\sqrt{3}}$

(b) $5\sqrt{2} - 8\sqrt{3} - \left(\sqrt{2} - 7\sqrt{3}\right) \qquad \textit{First remove parentheses.}$

$\qquad = 5\sqrt{2} - 8\sqrt{3} - \sqrt{2} + 7\sqrt{3}$

$\qquad = (5 - 1)\sqrt{2} + (-8 + 7)\sqrt{3}$

$\qquad = \boxed{4\sqrt{2} - \sqrt{3}}$

We may encounter an expression in which, at first glance, it may seem that the radicals cannot be combined, as in the following case:

$$\text{Simplify:} \quad \sqrt{18x^3} - x\sqrt{32x}$$

However, if we simplify each radical term first, we may find that we *can* combine the two radicals:

$$\sqrt{18x^3} - x\sqrt{32x} = \sqrt{9x^2}\sqrt{2x} - x\sqrt{16}\sqrt{2x}$$
$$= 3x\sqrt{2x} - 4x\sqrt{2x}$$
$$= (3 - 4)x\sqrt{2x}$$
$$= -x\sqrt{2x}$$

Thus, our first step should be to simplify each radical expression.

EXAMPLE 2

Simplify each of the following.

(a) $\sqrt{27} - \sqrt{81} + \sqrt{12}$ **(b)** $5\sqrt{4x^3} + 7x\sqrt{8x} - 2x\sqrt{9x} + \sqrt{2x}$

(c) $3x\sqrt[3]{24x^2} + 4x\sqrt[3]{54x^5} - 2\sqrt[3]{81x^5}$

Solution

(a) $\sqrt{27} - \sqrt{81} + \sqrt{12} \qquad \textit{Simplify each term first.}$

$\qquad = 3\sqrt{3} - 9 + 2\sqrt{3} \qquad \textit{Then combine where possible.}$

$\qquad = \boxed{5\sqrt{3} - 9}$

Note that we cannot combine $5\sqrt{3} - 9$, just as we cannot combine $5x - 9$.

(b) $5\sqrt{4x^3} + 7x\sqrt{8x} - 2x\sqrt{9x} + \sqrt{2x}$

$\qquad = 5\left(\sqrt{4x^2}\sqrt{x}\right) + 7x\left(\sqrt{4}\sqrt{2x}\right) - 2x\left(\sqrt{9}\sqrt{x}\right) + \sqrt{2x}$

$\qquad = 5\left(2x\sqrt{x}\right) + 7x\left(2\sqrt{2x}\right) - 2x\left(3\sqrt{x}\right) + \sqrt{2x}$

$\qquad = 10x\sqrt{x} + 14x\sqrt{2x} - 6x\sqrt{x} + \sqrt{2x}$

$\qquad = \boxed{4x\sqrt{x} + 14x\sqrt{2x} + \sqrt{2x}}$

(c) $3x\sqrt[3]{24x^2} + 4x\sqrt[3]{54x^5} - 2\sqrt[3]{81x^5}$ *Simplify each radical.*

$\qquad = 3x\left(\sqrt[3]{8}\,\sqrt[3]{3x^2}\right) + 4x\left(\sqrt[3]{27x^3}\,\sqrt[3]{2x^2}\right) - 2\left(\sqrt[3]{27x^3}\,\sqrt[3]{3x^2}\right)$

$\qquad = 3x\left(2\sqrt[3]{3x^2}\right) + 4x\left(3x\sqrt[3]{2x^2}\right) - 2\left(3x\sqrt[3]{3x^2}\right)$

$\qquad = 6x\sqrt[3]{3x^2} + 12x^2\sqrt[3]{2x^2} - 6x\sqrt[3]{3x^2}$ *Then combine where possible.*

$\qquad = \boxed{12x^2\sqrt[3]{2x^2}}$

EXAMPLE 3

Perform the indicated operations and simplify.

(a) $\sqrt{3} + \dfrac{6}{\sqrt{3}}$ **(b)** $5\sqrt{\dfrac{x}{y}} - \dfrac{\sqrt{xy}}{y}$

Solution **(a)** $\sqrt{3} + \dfrac{6}{\sqrt{3}}$ *Rationalize the denominator.*

$\qquad = \sqrt{3} + \dfrac{6\sqrt{3}}{\sqrt{3}\sqrt{3}} = \sqrt{3} + \dfrac{6\sqrt{3}}{3}$ *Reduce.*

$\qquad\qquad = \sqrt{3} + 2\sqrt{3}$ *Combine.*

$\qquad\qquad = \boxed{3\sqrt{3}}$

(b) $5\sqrt{\dfrac{x}{y}} - \dfrac{\sqrt{xy}}{y}$ *Apply property 2.*

$\qquad = \dfrac{5}{1} \cdot \dfrac{\sqrt{x}}{\sqrt{y}} - \dfrac{\sqrt{xy}}{y}$

$\qquad = \dfrac{5\sqrt{x}}{\sqrt{y}} - \dfrac{\sqrt{xy}}{y}$ *Rationalize the first denominator.*

$\qquad = \dfrac{5\sqrt{x}\sqrt{y}}{\sqrt{y}\sqrt{y}} - \dfrac{\sqrt{xy}}{y}$

$\qquad = \dfrac{5\sqrt{xy}}{y} - \dfrac{\sqrt{xy}}{y}$ *Combine fractions.*

$\qquad = \dfrac{5\sqrt{xy} - \sqrt{xy}}{y} = \boxed{\dfrac{4\sqrt{xy}}{y}}$

EXERCISES 7.5

Perform the indicated operations. Express your answers in simplest radical form. Assume that all variables represent positive real numbers.

1. $5\sqrt{3} - \sqrt{3}$

2. $8\sqrt{7} - 2\sqrt{7}$

3. $2\sqrt{5} - 4\sqrt{5} - \sqrt{5}$

4. $9\sqrt{2} - \sqrt{2} - 12\sqrt{2}$

5. $8\sqrt{3} - \left(4\sqrt{3} - 2\sqrt{6}\right)$

6. $6\sqrt{5} - \left(3\sqrt{7} - \sqrt{5}\right)$

7. $2\sqrt{3} - 2\sqrt{5} - \left(\sqrt{3} - \sqrt{5}\right)$

8. $\sqrt{6} - 2\sqrt{2} - \left(\sqrt{2} - \sqrt{6}\right)$

9. $2\sqrt{x} - 5\sqrt{x} + 3\sqrt{x}$

10. $5\sqrt{ab} - 7\sqrt{ab} + 3\sqrt{a}$

11. $5a\sqrt{b} - 3a^3\sqrt{b} + 2a\sqrt{b}$

12. $3xy - 2\sqrt{y} - 5x\sqrt{y}$

13. $3x\sqrt[3]{x^2} - 2\sqrt[3]{x^2} + 6\sqrt[3]{x^2}$

14. $6 - 4x\sqrt[3]{x^2} + 3\sqrt[3]{x^2} - 8$

15. $\left(7 - 3\sqrt[3]{a}\right) - \left(6 - \sqrt[3]{a}\right)$

16. $\left(2 - 5\sqrt[3]{x}\right) - \left(6 + \sqrt[3]{x}\right)$

17. $\sqrt{12} - \sqrt{27}$

18. $\sqrt{18} - \sqrt{8} + \sqrt{32}$

19. $\sqrt{24} - \sqrt{27} + \sqrt{54}$

20. $\sqrt{12} + \sqrt{18} + \sqrt{24}$

21. $6\sqrt{3} - 4\sqrt{81}$

22. $5\sqrt{2} - 6\sqrt{16}$

23. $3\sqrt{24} - 5\sqrt{48} - \sqrt{6}$

24. $3\sqrt{8} - 5\sqrt{32} + 2\sqrt{27}$

25. $3\sqrt[3]{24} - 5\sqrt[3]{48} - \sqrt[3]{6}$

26. $3\sqrt[3]{8} - 5\sqrt[3]{32} + 2\sqrt[3]{27}$

27. $2a\sqrt{ab^2} - 3b\sqrt{a^2b} - ab\sqrt{ab}$

28. $x\sqrt{x^3y} + y\sqrt{xy^3}$

29. $\sqrt[3]{x^4} - x\sqrt[3]{x}$

30. $3\sqrt[3]{16x} - 5\sqrt[3]{2x} - 3x\sqrt[3]{2}$

31. $\sqrt{20x^9y^8} + 2xy\sqrt{5x^7y^6}$

32. $2b^2\sqrt{48a^7b^6} - 5a\sqrt{27a^5b^{10}}$

33. $5\sqrt[3]{9x^5} - 3x\sqrt[3]{x^2} + 2x\sqrt[3]{72x^2}$

34. $3a\sqrt[3]{ab^4} - 5b\sqrt[3]{8a^4b} - ab\sqrt[3]{2ab}$

35. $4\sqrt[4]{16x} - 7\sqrt[4]{x^5} + x\sqrt[4]{81x}$

36. $5\sqrt[4]{32x^5} - 3x\sqrt[4]{2x} + 7\sqrt[4]{x^5}$

37. $\dfrac{1}{\sqrt{5}} + 2$

38. $\dfrac{1}{\sqrt{3}} - 3$

39. $\dfrac{12}{\sqrt{6}} - 2\sqrt{6}$

40. $\dfrac{15}{\sqrt{3}} - 3\sqrt{3}$

41. $\sqrt{\dfrac{1}{2}} + \sqrt{2}$

42. $\sqrt{\dfrac{2}{7}} - \sqrt{7}$

43. $\sqrt{\dfrac{5}{2}} + \sqrt{\dfrac{2}{5}}$

44. $\sqrt{\dfrac{3}{5}} - \sqrt{\dfrac{5}{3}}$

45. $\sqrt{\dfrac{1}{7}} - 3\sqrt{\dfrac{1}{5}}$

46. $2\sqrt{\dfrac{1}{5}} + \sqrt{\dfrac{2}{3}}$

47. $\dfrac{1}{\sqrt[3]{2}} - 6\sqrt[3]{4}$

48. $\dfrac{1}{\sqrt[3]{5}} - 4\sqrt[3]{25}$

49. $\sqrt{\dfrac{1}{x}} + \sqrt{\dfrac{1}{y}}$

50. $\sqrt{\dfrac{1}{x}} - \sqrt{y}$

51. $\dfrac{1}{\sqrt[3]{9}} - \dfrac{3}{\sqrt[3]{3}}$

52. $\dfrac{5}{\sqrt[3]{25}} - \dfrac{15}{\sqrt[3]{5}}$

53. $3\sqrt{\dfrac{2}{49}} + 3\sqrt{7}$

54. $2\sqrt{2} - 3\sqrt{\dfrac{5}{4}}$

55. $3\sqrt{10} - \dfrac{4}{\sqrt{10}} + \dfrac{2}{\sqrt{10}}$

56. $2\sqrt{30} - \dfrac{5}{\sqrt{30}} + \dfrac{1}{\sqrt{30}}$

57. $6\sqrt[3]{25} - \dfrac{15}{\sqrt[3]{5}} + 5\sqrt[3]{\dfrac{1}{5}}$

58. $2\sqrt[3]{49} - \dfrac{14}{\sqrt[3]{7}} + 5\sqrt[3]{\dfrac{1}{7}}$

59. $\dfrac{1}{2\sqrt{x-1}} + \sqrt{x-1}$

60. $3\sqrt{x^2-2} - \dfrac{2}{\sqrt{x^2-2}}$

 Mini-Review

61. Use the accompanying graph of $y = f(x)$ to find

    (a) the domain

    (b) the range

    (c) where $f(x)$ is increasing

    (d) where $f(x)$ is decreasing

62. *Factor as completely as possible:* $3x^3 - 75x$

63. Sketch the graph of the equation $y = \dfrac{2x-3}{5}$. Label the intercepts.

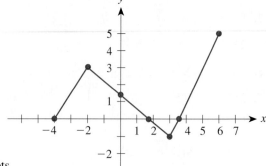

7.6 Multiplying and Dividing Radical Expressions

As with multiplying polynomials, we use the distributive property to multiply expressions with more than one radical term. For example,

$$\sqrt{2}\left(\sqrt{2} - \sqrt{3}\right) = \sqrt{2}\sqrt{2} - \sqrt{2}\sqrt{3} \qquad \textit{Distributive property}$$

$$= 2 - \sqrt{6}$$

EXAMPLE 1

Perform the indicated operations.

(a) $\left(2\sqrt{3} - 5\sqrt{2}\right)\left(2\sqrt{3} - \sqrt{2}\right)$ **(b)** $\left(2\sqrt{x} - 3\sqrt{y}\right)\left(2\sqrt{x} + 3\sqrt{y}\right)$

Solution

We apply what we have learned about multiplying polynomials. Each term in the first set of parentheses multiplies each term in the second set.

(a) $\left(2\sqrt{3} - 5\sqrt{2}\right)\left(2\sqrt{3} - \sqrt{2}\right)$

$$= \left(2\sqrt{3}\right)\left(2\sqrt{3}\right) - \left(2\sqrt{3}\right)\left(\sqrt{2}\right) - \left(5\sqrt{2}\right)\left(2\sqrt{3}\right) + \left(5\sqrt{2}\right)\left(\sqrt{2}\right)$$

$$= 4 \cdot 3 - 2\sqrt{6} - 10\sqrt{6} + 5 \cdot 2$$

$$= 12 - 12\sqrt{6} + 10 = \boxed{22 - 12\sqrt{6}}$$

(b) $\left(2\sqrt{x} - 3\sqrt{y}\right)\left(2\sqrt{x} + 3\sqrt{y}\right)$ *This is a difference of squares.*

$$= \left(2\sqrt{x}\right)^2 - \left(3\sqrt{y}\right)^2 \qquad \textit{Square each term.}$$

$$= 2^2\left(\sqrt{x}\right)^2 - 3^2\left(\sqrt{y}\right)^2$$

$$= \boxed{4x - 9y}$$

EXAMPLE 2

Perform the indicated operations.

(a) $\left(\sqrt{x+y}\right)^2 - \left(\sqrt{x} + \sqrt{y}\right)^2$ **(b)** $\left(\sqrt[3]{x} - \sqrt[3]{y}\right)\left(\sqrt[3]{x^2} + \sqrt[3]{xy} + \sqrt[3]{y^2}\right)$

Solution   **(a)** $\left(\sqrt{x+y}\right)^2 - \left(\sqrt{x} + \sqrt{y}\right)^2$

$= x + y - \left[\left(\sqrt{x}\right)^2 + 2\sqrt{x}\sqrt{y} + \left(\sqrt{y}\right)^2\right]$ *Note the difference in the way we handle* $\left(\sqrt{x+y}\right)^2$ *and* $\left(\sqrt{x} + \sqrt{y}\right)^2$.

$= x + y - \left[x + 2\sqrt{xy} + y\right]$

$= x + y - x - 2\sqrt{xy} - y = \boxed{-2\sqrt{xy}}$

(b) $\left(\sqrt[3]{x} - \sqrt[3]{y}\right)\left(\sqrt[3]{x^2} + \sqrt[3]{xy} + \sqrt[3]{y^2}\right)$

$= \sqrt[3]{x}\sqrt[3]{x^2} + \sqrt[3]{x}\sqrt[3]{xy} + \sqrt[3]{x}\sqrt[3]{y^2} - \sqrt[3]{y}\sqrt[3]{x^2} - \sqrt[3]{y}\sqrt[3]{xy} - \sqrt[3]{y}\sqrt[3]{y^2}$

$= \sqrt[3]{x^3} + \sqrt[3]{x^2y} + \sqrt[3]{xy^2} - \sqrt[3]{x^2y} - \sqrt[3]{xy^2} - \sqrt[3]{y^3}$

$= \sqrt[3]{x^3} - \sqrt[3]{y^3}$

$= \boxed{x - y}$

Part **(b)** of Examples 1 and 2 illustrate that any expression that is a difference can be factored if we lift our restriction that factors be polynomials with integer coefficients. For example, $3a - 2b$ can be factored into

$$\left(\sqrt{3a} - \sqrt{2b}\right)\left(\sqrt{3a} + \sqrt{2b}\right)$$

as a *difference of squares.* Multiply this out and check to verify that it equals $3a - 2b$.

The last operation to cover is division of radical expressions. We begin with division by a single term.

EXAMPLE 3

Simplify: $\dfrac{\sqrt{24} - 8}{10}$

Solution $\dfrac{\sqrt{24} - 8}{10}$ *Simplify the radical.*

$= \dfrac{\sqrt{4}\sqrt{6} - 8}{10}$

$= \dfrac{2\sqrt{6} - 8}{10}$ *Remember, we can only reduce common factors. Factor first. Then reduce.*

$= \dfrac{2\left(\sqrt{6} - 4\right)}{10}$

$= \boxed{\dfrac{\sqrt{6} - 4}{5}}$

In Section 7.4 we discussed rationalizing the denominator of a fraction with a single radical term in the denominator. In this section we will discuss how to rationalize the denominator of a fraction with more than one term in the denominator.

We cannot rationalize the denominator of $\dfrac{2}{3 + \sqrt{5}}$ as easily as we rationalized denominators consisting of only a single term. What shall we multiply the denominator by: $\sqrt{5}$ or $3 + \sqrt{5}$? Let's try $\sqrt{5}$ and see what happens:

$$\frac{2}{3 + \sqrt{5}} = \frac{2 \cdot \sqrt{5}}{\left(3 + \sqrt{5}\right) \cdot \sqrt{5}}$$

$$= \frac{2\sqrt{5}}{3\sqrt{5} + \sqrt{5}\sqrt{5}} = \frac{2\sqrt{5}}{3\sqrt{5} + 5}$$

Notice that we still have a radical remaining in the denominator. If we were to try $3 + \sqrt{5}$, we would again find that a radical would still remain in the denominator.

The denominator can be rationalized by another method, however. We exploit the difference of squares and multiply the numerator and denominator by a ***conjugate*** of the denominator (recall from Section 5.3 that a conjugate of $a + b$ is $a - b$):

Recall the difference of squares:
$(a + b)(a - b) = a^2 - b^2$

$$\frac{2}{3 + \sqrt{5}} = \frac{2\left(3 - \sqrt{5}\right)}{\left(3 + \sqrt{5}\right)\left(3 - \sqrt{5}\right)}$$ *Multiply numerator and denominator by $3 - \sqrt{5}$.*

$$= \frac{2\left(3 - \sqrt{5}\right)}{(3)^2 - \left(\sqrt{5}\right)^2}$$ *The denominator is the difference of squares.*

$$= \frac{2\left(3 - \sqrt{5}\right)}{9 - 5}$$

$$= \frac{2\left(3 - \sqrt{5}\right)}{4}$$ *Then reduce.*

$$= \frac{3 - \sqrt{5}}{2}$$

EXAMPLE 4 Simplify.

(a) $\dfrac{\sqrt{a} + \sqrt{b}}{\sqrt{a} - \sqrt{b}}$ **(b)** $\dfrac{8\sqrt{3}}{\sqrt{7} - \sqrt{3}}$

Solution **(a)** $\dfrac{\sqrt{a} + \sqrt{b}}{\sqrt{a} - \sqrt{b}}$ *Multiply the numerator and denominator by $\sqrt{a} + \sqrt{b}$, a conjugate of the denominator.*

$$= \frac{\left(\sqrt{a} + \sqrt{b}\right)\left(\sqrt{a} + \sqrt{b}\right)}{\left(\sqrt{a} - \sqrt{b}\right)\left(\sqrt{a} + \sqrt{b}\right)}$$

$$= \frac{a + \sqrt{ab} + \sqrt{ab} + b}{a - b} = \boxed{\frac{a + 2\sqrt{ab} + b}{a - b}}$$

(b) $\dfrac{8\sqrt{3}}{\sqrt{7} - \sqrt{3}}$ *Multiply the numerator and denominator by $\sqrt{7} + \sqrt{3}$, a conjugate of the denominator.*

$$= \frac{8\sqrt{3}\left(\sqrt{7} + \sqrt{3}\right)}{\left(\sqrt{7} - \sqrt{3}\right)\left(\sqrt{7} + \sqrt{3}\right)}$$

$$= \frac{8\sqrt{3}\left(\sqrt{7} + \sqrt{3}\right)}{\left(\sqrt{7}\right)^2 - \left(\sqrt{3}\right)^2}$$

$$= \frac{8\sqrt{3}\left(\sqrt{7} + \sqrt{3}\right)}{7 - 3} = \frac{8\sqrt{3}\left(\sqrt{7} + \sqrt{3}\right)}{4}$$ *Now reduce.*

$$= 2\sqrt{3}\left(\sqrt{7} + \sqrt{3}\right) = \boxed{2\sqrt{21} + 6}$$

EXAMPLE 5

Solution

Try finding the value of the original expression by using a calculator.

Perform the operations and simplify: $\dfrac{21}{3 - \sqrt{2}} - \dfrac{6}{\sqrt{2}}$

First we rationalize the denominator of each fraction:

$$\dfrac{21}{3 - \sqrt{2}} - \dfrac{6}{\sqrt{2}} = \dfrac{21(3 + \sqrt{2})}{(3 - \sqrt{2})(3 + \sqrt{2})} - \dfrac{6\sqrt{2}}{\sqrt{2}\sqrt{2}}$$

$$= \dfrac{21(3 + \sqrt{2})}{9 - 2} - \dfrac{6\sqrt{2}}{2}$$

$$= \dfrac{21(3 + \sqrt{2})}{7} - \dfrac{6\sqrt{2}}{2} \qquad \textit{Reduce.}$$

$$= 3(3 + \sqrt{2}) - 3\sqrt{2} = 9 + 3\sqrt{2} - 3\sqrt{2} = \boxed{9}$$

EXERCISES 7.6

Perform the operations. Express your answer in simplest radical form.

1. $5\left(\sqrt{5} - \sqrt{3}\right)$

2. $3\left(\sqrt{7} + 3\right)$

3. $2\left(\sqrt{3} - \sqrt{5}\right) - 4\left(\sqrt{3} + \sqrt{5}\right)$

4. $4\left(\sqrt{6} - \sqrt{2}\right) - 3\left(\sqrt{2} - \sqrt{6}\right)$

5. $\sqrt{a}\left(\sqrt{a} + \sqrt{b}\right)$

6. $\sqrt{x}\left(\sqrt{x} - \sqrt{y}\right)$

7. $\sqrt{2}\left(\sqrt{5} + \sqrt{2}\right)$

8. $\sqrt{3}\left(\sqrt{2} - \sqrt{3}\right)$

9. $3\sqrt{5}\left(2\sqrt{3} - 4\sqrt{5}\right)$

10. $5\sqrt{2}\left(7\sqrt{3} - 6\sqrt{2}\right)$

11. $\sqrt{2}\left(\sqrt{3} + \sqrt{2}\right) - 3\left(2\sqrt{6} - 4\right)$

12. $2\sqrt{5}\left(\sqrt{3} - \sqrt{5}\right) + 2\left(\sqrt{10} - 3\right)$

13. $\left(\sqrt{5} - 2\right)\left(\sqrt{3} + 1\right)$

14. $\left(\sqrt{7} - 2\right)\left(\sqrt{5} + 3\right)$

15. $\left(\sqrt{5} - \sqrt{3}\right)\left(\sqrt{5} + \sqrt{3}\right)$

16. $\left(\sqrt{2} + \sqrt{5}\right)^2$

17. $\left(\sqrt{5} - \sqrt{3}\right)^2$

18. $\left(\sqrt{2} + \sqrt{5}\right)\left(\sqrt{2} - \sqrt{5}\right)$

19. $\left(2\sqrt{7} - 5\right)\left(2\sqrt{7} + 5\right)$

20. $\left(2\sqrt{5} - 7\right)\left(2\sqrt{5} + 7\right)$

21. $\left(5\sqrt{2} - 3\sqrt{5}\right)\left(5\sqrt{2} + 3\sqrt{5}\right)$

22. $\left(3\sqrt{5} - 2\sqrt{3}\right)\left(3\sqrt{5} + 2\sqrt{3}\right)$

23. $\left(2\sqrt{a} - \sqrt{b}\right)^2$

24. $\left(5\sqrt{x} - \sqrt{y}\right)^2$

25. $\left(3\sqrt{x} - 2\sqrt{y}\right)\left(3\sqrt{x} + 2\sqrt{y}\right)$

26. $\left(5\sqrt{a} - 3\sqrt{b}\right)\left(5\sqrt{a} + 3\sqrt{b}\right)$

27. $\left(\sqrt{x} - 3\right)^2$

28. $\left(\sqrt{a} + 5\right)^2$

29. $\left(\sqrt{x} - 3\right)^2$

30. $\left(\sqrt{a} + 5\right)^2$

31. $\left(\sqrt{x + 1}\right)^2 - \left(\sqrt{x} + 1\right)^2$

32. $\left(\sqrt{a} - 2\right)^2 - \left(\sqrt{a - 2}\right)^2$

33. $\left(\sqrt[3]{2} - \sqrt[3]{3}\right)\left(\sqrt[3]{4} + \sqrt[3]{6} + \sqrt[3]{9}\right)$

34. $\left(\sqrt[3]{4} + \sqrt[3]{5}\right)\left(\sqrt[3]{16} - \sqrt[3]{20} + \sqrt[3]{25}\right)$

35. $\dfrac{4\sqrt{2} - 6\sqrt{3}}{2}$

36. $\dfrac{12\sqrt{3} - 8\sqrt{2}}{4}$

37. $\dfrac{5\sqrt{8} - 2\sqrt{7}}{8}$

38. $\dfrac{2\sqrt{27} + 6\sqrt{5}}{3}$

39. $\dfrac{3\sqrt{50} + 5\sqrt{5}}{5}$

40. $\dfrac{2\sqrt{12} + 5\sqrt{8}}{12}$

41. $\dfrac{4 + \sqrt{28}}{4}$

42. $\dfrac{20 + \sqrt{60}}{4}$

43. $\dfrac{1}{\sqrt{2} - 3}$

44. $\dfrac{1}{2 - \sqrt{3}}$

45. $\dfrac{10}{\sqrt{5} + 1}$

46. $\dfrac{15}{\sqrt{6} - 1}$

47. $\dfrac{2}{\sqrt{3} - \sqrt{a}}$

48. $\dfrac{3}{\sqrt{x} - \sqrt{2}}$

49. $\dfrac{\sqrt{x}}{\sqrt{x} - \sqrt{y}}$

50. $\dfrac{\sqrt{x}}{\sqrt{x} + \sqrt{y}}$

51. $\dfrac{\sqrt{2}}{\sqrt{5} - \sqrt{2}}$

52. $\dfrac{\sqrt{5}}{\sqrt{7} + \sqrt{3}}$

53. $\dfrac{2\sqrt{2}}{2\sqrt{5} - \sqrt{2}}$

54. $\dfrac{3\sqrt{5}}{4\sqrt{3} + \sqrt{5}}$

55. $\dfrac{2\sqrt{5} - \sqrt{2}}{2\sqrt{2}}$

56. $\dfrac{4\sqrt{3} + \sqrt{5}}{3\sqrt{5}}$

57. $\dfrac{\sqrt{3} + \sqrt{2}}{\sqrt{3} - \sqrt{2}}$

58. $\dfrac{\sqrt{x} + \sqrt{y}}{\sqrt{x} - \sqrt{y}}$

59. $\dfrac{3\sqrt{5} - 2\sqrt{2}}{2\sqrt{5} - 3\sqrt{2}}$

60. $\dfrac{2\sqrt{7} + 3\sqrt{2}}{3\sqrt{7} + 2\sqrt{2}}$

61. $\dfrac{x - y}{\sqrt{x} - \sqrt{y}}$

62. $\dfrac{a^2 - b^2}{\sqrt{a} + \sqrt{b}}$

63. $\dfrac{x^2 - x - 2}{\sqrt{x} - \sqrt{2}}$

64. $\dfrac{x^2 - 3x - 4}{\sqrt{x} + 2}$

65. $\dfrac{12}{\sqrt{6} - 2} - \dfrac{36}{\sqrt{6}}$

66. $\dfrac{15}{4 + \sqrt{11}} + \dfrac{33}{\sqrt{11}}$

67. $\dfrac{20}{\sqrt{7} + \sqrt{3}} + \dfrac{28}{\sqrt{7}}$

68. $\dfrac{30}{\sqrt{10} - \sqrt{5}} - \dfrac{15}{\sqrt{5}}$

One solution to the quadratic equation $Ax^2 + Bx + C = 0$ is given by the quadratic

formula, $x = \dfrac{-B + \sqrt{B^2 - 4AC}}{2A}$. In Exercises 69–74, use the quadratic formula to

find and simplify x for the given values of A, B and C.

69. $A = 1$, $B = 3$, and $C = 2$

70. $A = 1$, $B = 5$, and $C = 4$

71. $A = 2$, $B = 6$, and $C = 3$

72. $A = 2$, $B = 6$, and $C = 1$

73. $A = 2$, $B = 6$, and $C = -3$

74. $A = 4$, $B = 2$, and $C = -3$

Mini-Review

75. Compute the value of the expressions given in Exercises 65 and 66 using a calculator. How do your answers compare to the answers found as a result of simplifying the expressions?

76. *Multiply and simplify:* $(2x - 5)^2 - (x - 10)^2$

77. *Solve:* $(2x - 1)(2x + 3) = 12$

78. *Given* $f(x) = \dfrac{2.4x^2 - 3.8}{5.1x - 7}$, *compute $f(-1.6)$ to the nearest tenth.*

79. *Factor as completely as possible:* $12x^2 - 46x + 40$

7.7 Radical Functions and Equations

When analyzing the scene of a car accident, a police officer can sometimes estimate the speed of a vehicle (before the brakes were applied) by measuring the length of the skid marks and using the formula

$$v = v(s) = 2\sqrt{5s}$$

where v is the vehicle speed in miles per hour, and s is the length of the skid marks in feet. Let's examine this function by constructing a table and approximating values of $v(s)$ for a few values of s:

s	$v = v(s) = 2\sqrt{5s}$
0	$v(0) = 2\sqrt{5(0)} = 0$
10	$v(10) = 2\sqrt{5(10)} \approx 14.1$
50	$v(50) = 2\sqrt{5(50)} \approx 31.6$
100	$v(100) = 2\sqrt{5(100)} \approx 44.7$
200	$v(200) = 2\sqrt{5(200)} \approx 63.25$

We can graph this function, letting s represent the horizontal axis and v represent the vertical axis. The graph of this function appears in Figure 7.2. This is the general shape of the square root function, which we will explore in more detail in Chapter 9.

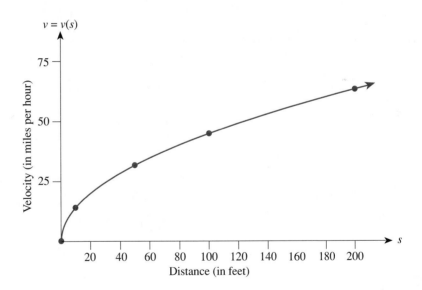

Figure 7.2
The graph of $v(s) = 2\sqrt{5s}$

The function $v(s) = 2\sqrt{5s}$ is an example of a radical function. Radical functions are functions that contain a radical expression involving a variable.

Suppose the driver of this car is involved in an accident and claims that he was driving below the legal speed limit of 40 mph for this area. If his vehicle left skid marks of 175 ft at the accident site, is his claim plausible?

To solve this problem, we could estimate the answer by using the graph, but because the formula is already solved for v in terms of s, we can simply substitute $s = 175$ into the equation $v = 2\sqrt{5s}$ to get $v = 2\sqrt{5(175)} \approx 59.2$. This means that he was driving at approximately 59 mph, which is well above the speed limit for the area.

Suppose, however, we were given the velocity of the car and were asked to find the length of the skid marks; that is, find s given v. For example, we may want to know the length of the skid marks if the car was traveling at 50 mph before the brakes were applied. We could get a rough estimate of v using the graph above, by guessing at values using a table, or by substituting $v = 50$ in the equation $v = 2\sqrt{5s}$ and solving for s.

An equation that contains a radical expression involving a variable is called a ***radical equation.*** In this section we will begin to discuss methods for solving equations involving radical expressions; our discussion will continue in Chapter 8.

In Chapter 6, when we solved equations containing fractions, our first step was to transform the equation to an equivalent one without fractions. Our strategy for treating radicals in equations will be similar. That is, our first step in solving an equation with radicals will be to transform the equation to one that does not contain radicals.

We utilize the following:

$$\text{If } a = b \text{ then } a^2 = b^2.$$

In words, this says that squares of equal quantities are equal to each other. We can apply this property to "get rid of the radical." For example, to solve the equation

$$\sqrt{x} = 5$$

we would square both sides of the equation to get

$$\left(\sqrt{x}\right)^2 = 5^2$$

Remember by definition of $\sqrt{x}$*, we have* $\left(\sqrt{x}\right)^2 = x$

or

$$x = 25$$

Simply squaring both sides of an equation containing a radical does not necessarily eliminate the radical, however. For example, in solving

$$\sqrt{x} + 6 = 9$$

if we simply square both sides we would have

$$\left(\sqrt{x} + 6\right)^2 = 9^2$$
$$x + 12\sqrt{x} + 36 = 81 \qquad \text{\textit{Remember that when squaring a binomial, we will have a middle term.}}$$

Note that a radical still remains in the equation. Therefore, we should *isolate the radical before squaring both sides of the equation.* Thus, we would solve the equation as follows:

$$\sqrt{x} + 6 = 9 \qquad \textit{First isolate } \sqrt{x}.$$
$$\sqrt{x} = 3 \qquad \textit{Then \textit{square both sides of the equation.}}$$
$$\left(\sqrt{x}\right)^2 = 3^2$$
$$x = 9$$

EXAMPLE 1

Solve for x.

(a) $\sqrt{2x - 7} = 9$ **(b)** $\sqrt{2x} - 7 = 9$ **(c)** $2\sqrt{x - 7} = 9$

Solution   **(a)**
$$\sqrt{2x - 7} = 9 \qquad \textit{Since the radical is already isolated, we square both sides of the equation.}$$
$$\left(\sqrt{2x - 7}\right)^2 = 9^2$$
$$2x - 7 = 81$$
$$2x = 88$$
$$\boxed{x = 44}$$

Check $x = 44$:

$$\sqrt{2x - 7} = 9$$

$$\sqrt{2(44) - 7} \overset{?}{=} 9$$

$$\sqrt{88 - 7} \overset{?}{=} 9$$

$$\sqrt{81} \overset{\checkmark}{=} 9$$

(b) $\sqrt{2x} - 7 = 9$ *First isolate the radical.*

$\sqrt{2x} = 16$ *Then square both sides of the equation.*

$\left(\sqrt{2x}\right)^2 = 16^2$

$2x = 256$

$$\boxed{x = 128}$$

Check $x = 128$:

$$\sqrt{2x} - 7 = 9$$

$$\sqrt{2(128)} - 7 \overset{?}{=} 9$$

$$\sqrt{256} - 7 \overset{?}{=} 9$$

$$16 - 7 \overset{\checkmark}{=} 9$$

Note the similarities and differences between the two equations in parts **(a)** and **(b)** and how they were solved.

(c) $2\sqrt{x - 7} = 9$ *Isolate the radical; divide both sides by 2.*

$\sqrt{x - 7} = \dfrac{9}{2}$ *Square both sides of the equation.*

$\left(\sqrt{x - 7}\right)^2 = \left(\dfrac{9}{2}\right)^2$

$x - 7 = \dfrac{81}{4}$ *Add 7 to both sides of the equation.*

$x = \dfrac{81}{4} + 7 = \boxed{\dfrac{109}{4}}$

You should check to see that this is a valid solution.

Based on our original properties of equality (Chapter 1), we were able to state that certain operations performed on both sides of an equation yield ***equivalent equations.*** By *equivalent* we mean that the transformed equation and the original equation have identical solutions.

We cannot make the same statement for the property of squaring used in Example 1. Note that it is written "if $a = b$ then $a^2 = b^2$." For equations, this means that solutions for the equation $a = b$ are also solutions for $a^2 = b^2$, but all solutions for $a^2 = b^2$ are not necessarily solutions for $a = b$.

For example, let's suppose we have the equation

$$x = -2$$

If we square both sides, we get

$$x^2 = (-2)^2$$

$$x^2 = 4$$

We had only one solution to the first equation, $x = -2$, but you can check to see that the transformed equation $x^2 = 4$ has two solutions: $x = -2$ and $x = 2$.

By squaring both sides of the equation, we picked up an extra number that is a solution to the transformed equation, but *not* a solution to the original equation. This "extra" solution is called an **extraneous solution,** or an **extraneous root.** Because of this possibility, we have to *check all solutions in the original equation* to ensure that our solution is valid. (In the next chapter we will encounter some radical equations that seem to yield two solutions and we will have to check to see whether both are valid.)

EXAMPLE 2

Solve for a.

(a) $\sqrt{3a + 1} + 13 = 8$ **(b)** $\sqrt{4a - 1} - \sqrt{2a + 3} = 0$

Solution

(a) $\sqrt{3a + 1} + 13 = 8$ *First isolate the radical $\sqrt{3a + 1}$. Subtract 13 from both sides of the equation.*

$$\sqrt{3a + 1} = -5$$ *Does this make sense?*

Let's continue and see what we get:

$$\left(\sqrt{3a + 1}\right)^2 = (-5)^2$$
$$3a + 1 = 25$$
$$3a = 24$$
$$a = 8$$

Check $a = 8$:

$$\sqrt{3a + 1} + 13 = 8$$
$$\sqrt{3(8) + 1} + 13 \overset{?}{=} 8$$
$$\sqrt{24 + 1} + 13 \overset{?}{=} 8$$
$$\sqrt{25} + 13 \overset{?}{=} 8$$
$$5 + 13 \neq 8$$

The only solution is extraneous.

Therefore, there is $\boxed{\text{no solution}}$.

We could have stopped at the point at which we had $\sqrt{3a + 1} = -5$ and concluded then that there is no solution to this problem. Why?

(b) $\sqrt{4a - 1} - \sqrt{2a + 3} = 0$ *Isolate one of the radicals.*

$$\sqrt{4a - 1} = \sqrt{2a + 3}$$ *Then square both sides of the equation.*
$$\left(\sqrt{4a - 1}\right)^2 = \left(\sqrt{2a + 3}\right)^2$$
$$4a - 1 = 2a + 3$$ *Isolate a.*
$$2a = 4$$
$$a = 2$$

Check $a = 2$:

$$\sqrt{4a - 1} - \sqrt{2a + 3} = 0$$
$$\sqrt{4(2) - 1} - \sqrt{2(2) + 3} \overset{?}{=} 0$$
$$\sqrt{7} - \sqrt{7} \overset{\checkmark}{=} 0$$

The solution is $\boxed{a = 2}$.

We can generalize the squaring property to other powers:

$$\boxed{\text{If } a = b \text{ then } a^n = b^n.}$$

This property allows us to solve radical equations of a higher index, as shown in the next example.

EXAMPLE 3

Solve the equation $\sqrt[3]{x} = 5$

Solution

$$\sqrt[3]{x} = 5 \qquad \textit{Cube both sides of the equation.}$$
$$\left(\sqrt[3]{x}\right)^3 = (5)^3$$
$$x = 125$$

Check $x = 125$: $\sqrt[3]{x} = 5$
$$\sqrt[3]{125} \overset{\checkmark}{=} 5$$

The solution is $\boxed{x = 125}$.

EXAMPLE 4

If $f(x) = \sqrt[4]{x - 7} + 2$, find the solution(s) to $f(x) = 6$.

Solution

Since $f(x) = \sqrt[4]{x - 7} + 2$, we set the expression $\sqrt[4]{x - 7} + 2$ equal to 6.

$$\sqrt[4]{x - 7} + 2 = 6 \qquad \textit{This is the statement } f(x) = 6.\textit{ Isolate the radical.}$$
$$\sqrt[4]{x - 7} = 4 \qquad \textit{Raise each side to the fourth power.}$$
$$\left(\sqrt[4]{x - 7}\right)^4 = (4)^4$$
$$x - 7 = 256$$
$$x = 263$$

Check $x = 263$:   Here we want to check that $f(236) = 6$:

$$f(x) = \sqrt[4]{x - 7} + 2$$
$$f(263) = \sqrt[4]{263 - 7} + 2 \overset{?}{=} 6$$
$$= \sqrt[4]{256} + 2 \overset{?}{=} 6$$
$$= 4 + 2 \overset{\checkmark}{=} 6$$

The solution is $\boxed{x = 263}$.

Since we can rewrite radicals using rational exponents, the same principles can be applied to equations with rational exponents. Note that in raising an expression such as $a^{1/6}$ to the sixth power, we use the second rule of the exponents as follows:

$$(a^{1/6})^6 = a^{(1/6)(6)} = a^1 = a$$

EXAMPLE 5

Solve each of the following equations:

(a) $x^{1/4} = 2$ (b) $x^{1/3} - 2 = 5$
(c) $(x - 2)^{1/3} = 5$ (d) $a^{-1/3} = 4$

Solution

(a) $x^{1/4} = 2$ *Raise each side to the fourth power.*
$$(x^{1/4})^4 = 2^4$$
$$x = 2^4 \qquad \textit{Since } (x^{1/4})^4 = x^{(1/4)4} = x^1 = x$$
$$x = 16$$

Check $x = 16$: $x^{1/4} = 2$

$$16^{1/4} \overset{\checkmark}{=} 2$$

Thus, $\boxed{x = 16}$.

(b) $x^{1/3} - 2 = 5$ *Isolate $x^{1/3}$.*

$$x^{1/3} = 7 \qquad \textit{Cube both sides.}$$

$$(x^{1/3})^3 = 7^3$$

$$\boxed{x = 343}$$

The check is left to the student.

(c) $(x - 2)^{1/3} = 5$ *Cube both sides.*

$$[(x - 2)^{1/3}]^3 = 5^3$$

$$x - 2 = 125$$

$$\boxed{x = 127}$$

The check is left to the student.

Note the differences between parts **(b)** and **(c)**.

(d) $a^{-1/3} = 4$ *We raise both sides to the -3 power to change $a^{-1/3}$ to $a^1 = a$.*

$$(a^{-1/3})^{-3} = 4^{-3}$$

$$a^1 = 4^{-3} \qquad \textit{Rewrite 4^{-3} using positive exponents.}$$

$$a = \frac{1}{4^3} = \frac{1}{64}$$

Check $a = \dfrac{1}{64}$: $a^{-1/3} = 4$

$$\left(\frac{1}{64}\right)^{-1/3} \overset{?}{=} 4$$

$$64^{1/3} \overset{\checkmark}{=} 4$$

The solution is $\boxed{a = \dfrac{1}{64}}$.

Let's return to the problem we discussed at the beginning of this section.

EXAMPLE 6

The distance traveled by a car *after* applying brakes is related to the velocity of the car before brakes were applied in the following way:

$$v = v(s) = 2\sqrt{5s}$$

where v is the speed of the vehicle in miles per hour, and s is the distance the car travels, in feet, after the brakes are applied. What is the length of the skid marks made by a car that was traveling at 50 mph before the brakes were applied?

Solution

We are given the function $v(s) = 2\sqrt{5s}$, and we need to find s when $v(s) = 50$. We substitute $v(s) = 50$ into the equation and solve for s:

$$v(s) = 2\sqrt{5s} \qquad \textit{Since $v(s) = 50$, we have}$$

$$50 = 2\sqrt{5s} \qquad \textit{Solve for s. Isolate the radical (divide both sides by 2).}$$

$$25 = \sqrt{5s} \quad \text{\textit{Square each side.}}$$
$$25^2 = \left(\sqrt{5s}\right)^2$$
$$625 = 5s \quad \text{\textit{Divide each side by 5 to get}}$$
$$\frac{625}{5} = 125 = s$$

Hence the length of the skid marks is $\boxed{125 \text{ ft}}$.

EXERCISES 7.7

In Exercises 1–36, solve for the given variable. Round your answer to the nearest hundredth where necessary.

1. $\sqrt{a} = 7$

2. $\sqrt{x - 3} = 5$

3. $6 = \sqrt{a} + 5$

4. $6 = \sqrt{a + 5}$

5. $\sqrt{3x - 1} = 5$

6. $\sqrt{3x} - 1 = 5$

7. $\sqrt{y + 5} - 4 = 7$

8. $\sqrt{s + 7} + 6 = 3$

9. $2 = 5 - \sqrt{3x - 1}$

10. $-4 = 7 - \sqrt{5 - 2x}$

11. $3\sqrt{x} = 12$

12. $5\sqrt{y} = 20$

13. $6\sqrt{a} + 3.4 = 21.3$

14. $7\sqrt{s} + 4 = 8.8$

15. $7 + \sqrt{x - 1} = 3$

16. $5 + \sqrt{x + 2} = 10$

17. $\sqrt{2x - 3} - \sqrt{x + 5} = 0$

18. $\sqrt{3x - 2} - \sqrt{x + 6} = 0$

19. $\sqrt{5 - 4x} - \sqrt{6 - 3x} = 0$

20. $6 = -4 + 5\sqrt{2a + 1}$

21. $4 = \sqrt[3]{x}$

22. $\sqrt[3]{y} + 7 = 5$

23. $\sqrt[3]{s} = -3$

24. $\sqrt[3]{x - 1} - 2 = 5$

25. $-3 = \sqrt[3]{y + 5}$

26. $\sqrt[3]{4 - 3y} + 2 = 1$

27. $x^{1/3} = 5$

28. $y^{1/3} = -3$

29. $x^{1/3} + 7 = 5$

30. $y^{1/3} - 6 = 2$

31. $(x + 7)^{1/3} = 5$

32. $(b - 7)^{1/3} = 4$

33. $x^{-1/4} = 4$

34. $x^{-1/3} = 8$

35. $(x + 3)^{1/3} + 4 = 2$

36. $(2x - 1)^{1/3} - 2 = 5$

In Exercises 37–48, find the solution accurate to two decimal places.

37. If $f(x) = \sqrt{x}$, solve $f(x) = 7.8$.

38. If $g(x) = \sqrt{x + 4}$, solve $g(x) = 3$.

39. If $h(x) = \sqrt{x - 4.3}$, solve $h(x) = 9.38$.

40. If $f(x) = \sqrt{x} + 5$, solve $f(x) = 6$.

41. If $g(x) = (3x - 1)^{1/2}$, solve $g(x) = 5.24$.

42. If $h(x) = (3x)^{1/2} - 1$, solve $h(x) = 5.9$.

43. If $f(x) = 6\sqrt{x} + 3.4$, solve $f(x) = 21$.

44. If $g(x) = 5\sqrt{x} - 4$, solve $g(x) = 8.1$.

45. If $f(x) = 7 - \sqrt{x - 2}$, solve $f(x) = 4$.

46. If $h(x) = 2\sqrt{x - 4} - \sqrt{3x - 1}$, solve $h(x) = 0$.

47. If $f(x) = x^{1/3} - 3$, solve $f(x) = 5$.

48. If $f(x) = (x - 3)^{1/3}$, solve $f(x) = 5$.

In Exercises 49–58, solve for the given variable accurate to two decimal places, by any method.

49. $\sqrt{3a} = 5$

50. $\sqrt{x - 7} = 9.1$

51. $8 = \sqrt{3x - 2}$

52. $\sqrt{3x} - 1.25 = 5.4$

53. $\sqrt{y+4} + 8 = 15.6$ **54.** $2 = 5 - \sqrt{2x+3}$ **55.** $3.5\sqrt{2x} = 8.4$ **56.** $6\sqrt{3x} + 3 = 21$

57. $\sqrt{7-9x} - \sqrt{13-6x} = 0$ **58.** $(a - 3.2)^{1/3} + 4 = 5.9$

59. In the example given at the beginning of this section, we used the function $v(s) = 2\sqrt{5s}$ to estimate the speed of a car, given the length of the skid marks made by the car. Determine the length of the skid marks made by a car braking at 55 mph.

60. Using the formula given in Exercise 59, determine the speed of a car, given that the length of the skid marks is 30 ft.

61. A pendulum consists of a weight on a string. The *period, T,* of a pendulum is the time it takes to complete one full cycle (swinging out and returning back) and is given by the formula

$$T = 2\pi\sqrt{\frac{L}{980}}$$

where L is the length of the pendulum in centimeters. Determine the length of the pendulum if the period T is 8 seconds.

62. The time t, in seconds, it takes for an object to fall a distance d, in feet, is given by

$$t = \sqrt{\frac{d}{16}}$$

How far would an object fall in 15 seconds?

63. The volume V of a sphere is related to the radius r of the sphere in the following way:

$$r = \sqrt[3]{\frac{3V}{4\pi}}$$

To the nearest tenth of an inch, find the radius of a sphere with volume 100 cubic inches.

64. The surface area S of a sphere is related to the radius r of the sphere in the following way:

$$r = \sqrt{\frac{S}{4\pi}}$$

To the nearest tenth of an inch, find the radius of a sphere with surface area 100 square inches.

? Question for Thought

65. What is the difference between squaring the expression $\sqrt{x+4}$ and squaring the expression $\sqrt{x} + 4$?

Mini-Review

66. *Solve:* $(x - 3)^2 = 25$

67. *Solve:* $\dfrac{x-3}{4} - \dfrac{x+2}{5} = \dfrac{x-5}{2}$

68. *Combine and simplify:* $\dfrac{x-3}{4} - \dfrac{x+2}{5} + \dfrac{x-5}{2}$

7.8   Complex Numbers

In Chapter 1, we pointed out the need to extend number systems to solve various types of equations. For example, the set of rational numbers contains the solution to the equation $3x + 7 = 2$. However, for the equation $x^2 = 3$, we have to "go beyond" the rationals and define the real numbers to obtain the solutions $x = +\sqrt{3}$ and $x = -\sqrt{3}$.

It seems as though our system is complete, but we are still unable to solve all polynomial equations. For example, the equation $x^2 = -5$ has no real-number solution because no real number will yield a negative number when squared. Thus, we again define a system even larger than the real numbers to have the solutions to such equations. We will begin by defining the quantity i.

DEFINITION

The *imaginary unit, i,* is defined by

$$i = \sqrt{-1}$$

Given this definition, we have:

$$i$$
$$i^2 = \left(\sqrt{-1}\right)^2 = -1$$
$$i^3 = i^2 \cdot i = (-1)i = -i$$
$$i^4 = i^2 \cdot i^2 = (-1)(-1) = +1$$
$$i^5 = i^4 i = (1)i = i$$
$$i^6 = i^4 i^2 = (1)(-1) = -1$$

Notice that this cycle repeats itself after i^4, so that any power of i can be written as i, -1, $-i$, or 1.

EXAMPLE 1

Simplify each of the following:

(a) i^{39} **(b)** i^{26}

Solution   Because $i^4 = 1$, it would be most convenient to factor the largest perfect fourth power of i.

(a) $i^{39} = i^{36} \cdot i^3$ *36 is the largest multiple of 4 less than 39; hence i^{36} is the greatest perfect fourth-power factor of i^{39}. We rewrite i^{36} as a perfect fourth power.*
$\qquad = (i^4)^9 i^3$
$\qquad = (1)^9 i^3$ *Since $i^4 = 1$*
$\qquad = i^3$
$\qquad = -i$

Hence, $i^{39} = \boxed{-i}$.

Actually, we find that $i^s = i^r$, where r is the remainder when s is divided by 4, and then rewrite the expression as i, $-i$, 1, or -1.

(b) $i^{26} = (i^4)^6 i^2$ *Note:* $\begin{array}{r} 6 \\ 4\overline{)26} \\ -24 \\ \hline 2 \end{array}$

$\qquad = 1^6 i^2$ *We could have divided 4 into 26 to find a remainder of 2; hence, $i^{26} = i^2$.*
$\qquad = i^2$

$\qquad = \boxed{-1}$

We note the following:

$$(i\sqrt{a})^2 = i^2(\sqrt{a})^2 = (-1)a = -a$$

Thus, we see that $i\sqrt{a}$ is a number whose square is $-a$. Therefore, for $a > 0$ we have the following:

If a is a real number and $a > 0$, then

$$\sqrt{-a} = i\sqrt{a}$$

Using i, we can now translate square roots with negative radicands as follows:

$$\sqrt{-4} = i\sqrt{4} = i(2) = 2i$$

$$\sqrt{-\frac{1}{16}} = i\sqrt{\frac{1}{16}} = i\left(\frac{1}{4}\right) = \frac{1}{4}i$$

We will use i, the imaginary unit, to define a new type of number—a *complex number.*

DEFINITION

A *complex number* is a number that can be written in the form $a + bi$, where a and b are real numbers, and i is the imaginary unit. a is called the *real part* of $a + bi$, and b is called the *imaginary part.*

For example, in the complex number $3 + 4i$, 3 is the real part and 4 is the imaginary part.

If a is a real number, then we can rewrite it as a complex number in the following way:

$$a = a + 0i$$

Since we can put any real number into complex form (with b equal to 0), we conclude that all real numbers are complex numbers. If we designate the set of complex numbers as C, then $R \subset C$.

If a nonzero complex number does not have a real part (real part is 0), then we say that the number is *pure imaginary.* For example, $3i$, $-4i$, and $\frac{1}{2}i$ are pure imaginary numbers.

EXAMPLE 2

Put the following in complex number form:

(a) $5 + \sqrt{-16}$ **(b)** 0 **(c)** -5 **(d)** $\dfrac{6 - \sqrt{-3}}{2}$

Solution

(a) Since $\sqrt{-16} = i\sqrt{16} = i(4) = 4i$, we have

$$5 + \sqrt{-16} = \boxed{5 + 4i}$$

(b) $0 = \boxed{0 + 0i}$

(c) $-5 = \boxed{-5 + 0i}$

(d) $\dfrac{6 - \sqrt{-3}}{2} = \dfrac{6 - i\sqrt{3}}{2}$ *We place i before the radical because it might be confused with $\sqrt{3}i$.*

$$= \dfrac{6}{2} - \dfrac{\sqrt{3}}{2}i$$

$$= \boxed{3 - \dfrac{\sqrt{3}}{2}i}$$ *The real part is 3; the imaginary part is $-\sqrt{3}/2$.*

Now that we have defined complex numbers, our next step will be to examine operations on complex numbers. First we will define the following:

DEFINITION

$$a + bi = c + di \quad \text{if and only if} \quad a = c \text{ and } b = d.$$

Verbally stated, two complex numbers are equal if and only if their real parts are identical *and* their imaginary parts are identical.

Addition and Subtraction of Complex Numbers

Addition and subtraction of complex numbers are relatively straightforward:

Addition and Subtraction of Complex Numbers

$(a + bi) + (c + di) = (a + c) + (b + d)i$ Addition

$(a + bi) - (c + di) = (a - c) + (b - d)i$ Subtraction

The real part of the sum (difference) is the sum (difference) of the real parts and the imaginary part of the sum (difference) is the sum (difference) of the imaginary parts.

EXAMPLE 3

Perform the indicated operations.

(a) $(3 + 4i) + (5 - 6i)$ **(b)** $(6 + 9i) - (3 - 2i)$

Solution

We could either use the definition of addition and subtraction given in the box, or treat i as a variable and combine "like" terms.

(a) $(3 + 4i) + (5 - 6i) = 3 + 4i + 5 - 6i = \boxed{8 - 2i}$ *By treating i as a variable and combining terms*

Or, alternatively, by using the definition, we have

$$(3 + 4i) + (5 - 6i) = (3 + 5) + (4 - 6)i = \boxed{8 - 2i}$$

(b) $(6 + 9i) - (3 - 2i)$ *Use the definition.*

$$= (6 - 3) + [9 - (-2)]i$$

$$= \boxed{3 + 11i}$$

Products of Complex Numbers

We can treat a complex number as if it were a binomial and multiply two complex numbers using binomial multiplication:

$$(a + bi)(c + di) = ac + adi + bci + bdi^2 \qquad \textit{Binomial multiplication}$$
$$= ac + adi + bci + bd(-1) \qquad \textit{Since } i^2 = -1$$
$$= ac + adi + bci - bd \qquad \textit{Rearrange terms so that the real}$$
$$\qquad\qquad\qquad\qquad\qquad\qquad \textit{parts are together and the imagi-}$$
$$= ac - bd + (ad + bc)i \qquad \textit{nary parts are together, then factor i}$$
$$\qquad\qquad\qquad\qquad\qquad\qquad \textit{from adi + bci.}$$

Thus, we have the rule for multiplying complex numbers:

Multiplication of Complex Numbers	$(a + bi)(c + di) = (ac - bd) + (ad + bc)i$

It is probably not very useful to memorize this rule as a special product at this point. It is better to multiply two complex numbers as binomials, substitute -1 for i^2, and finally combine the real parts and imaginary parts to conform to complex number form.

EXAMPLE 4

Perform the indicated operations.

(a) $(3 + 2i)(4 - 5i)$ **(b)** $(3 - 7i)^2$
(c) $(5 - 2i)(5 + 2i)$ **(d)** $(3 - i)^2 - 6(3 - i)$

Solution

(a) $(3 + 2i)(4 - 5i) = 12 - 15i + 8i - 10i^2$ *Binomial multiplication*

$$= 12 - 15i + 8i + 10 \qquad \textit{Since } i^2 = -1,$$
$$\qquad\qquad\qquad\qquad\qquad\qquad -10i^2 = -10(-1) = +10.$$

$$= \boxed{22 - 7i}$$

(b) $(3 - 7i)^2 = (3)^2 - 2(3)(7i) + (7i)^2$ *This is a special product: a perfect square.*

$$= 9 - 42i + 49i^2 \qquad \textit{Since } i^2 = -1, 49i^2 = 49(-1) = -49.$$
$$= 9 - 42i - 49$$

$$= \boxed{-40 - 42i}$$

(c) $(5 - 2i)(5 + 2i) = 5^2 - (2i)^2$ *This is a special product: the difference of squares.*

$$= 25 - 4i^2$$
$$= 25 + 4 \qquad \textit{Since } i^2 = -1$$

$$= \boxed{29} \qquad \textit{Note that the result is a real number.}$$

(d) $(3 - i)^2 - 6(3 - i)$ *Powers first: square $(3 - i)$.*

$$= 9 - 6i + i^2 - 6(3 - i) \qquad \textit{Then multiply.}$$
$$= 9 - 6i + i^2 - 18 + 6i \qquad i^2 = -1$$
$$= 9 - 6i - 1 - 18 + 6i$$

$$= \boxed{-10}$$

EXAMPLE 5

Show that $2 + 3i$ is a solution to the equation $x^2 - 4x + 13 = 0$.

Solution

To determine whether $2 + 3i$ is a solution to the equation, we substitute the value $2 + 3i$ for x in the equation and check to see whether, after simplifying, the right and left sides of the equation agree:

$$x^2 - 4x + 13 = 0$$
$$(2 + 3i)^2 - 4(2 + 3i) + 13 \overset{?}{=} 0$$
$$4 + 12i + 9i^2 - 8 - 12i + 13 \overset{?}{=} 0$$
$$4 + 12i - 9 - 8 - 12i + 13 \overset{\checkmark}{=} 0$$

Thus, ┌───┐
      $2 + 3i$ does satisfy the equation $x^2 - 4x + 13 = 0$.
      └───┘

We now define the *conjugate* of a complex number.

DEFINITION	The ***complex conjugate*** of $a + bi$ is $a - bi$.

By this definition, the complex conjugate of $a - bi$ is $a + bi$.

Observe, as in part **(c)** of Example 4, that the product of complex conjugates will yield a real number. In general,

$$(a + bi)(a - bi) = a^2 - (bi)^2 \quad \textit{Difference of squares}$$
$$= a^2 - b^2i^2 \quad \textit{Since } i^2 = -1, -b^2i^2 = -b^2(-1) = +b^2.$$
$$= a^2 + b^2 \quad \textit{Note that } a^2 + b^2 \textit{ is a real number.}$$

We will use this result to help us find certain quotients of complex numbers.

Quotients of Complex Numbers

Our goal is to express a quotient of complex numbers in the form $a + bi$.

If we have a quotient of a complex number and a real number, where the real number is the divisor, then expressing the quotient in the form $a + bi$ is similar to dividing a polynomial by a monomial. For example,

$$\frac{6 - 5i}{2} = \frac{6}{2} - \frac{5}{2}i = 3 - \frac{5}{2}i$$

Keeping in mind that $i = \sqrt{-1}$ is a radical expression, we will find the rationalizing techniques we used with quotients of radicals useful when working with quotients of complex numbers when the divisor has an imaginary part.

EXAMPLE 6 Express the following in the form $a + bi$:

$$\frac{4 + 3i}{2i}$$

Solution $\dfrac{4 + 3i}{2i}$ *We will treat this expression in the same manner as we would if we were rationalizing a denominator.*

$$= \frac{(4 + 3i) \cdot i}{2i \cdot i} \quad \textit{Multiply the numerator and denominator by } i.$$

$$= \frac{4i + 3i^2}{2i^2}$$

$$= \frac{4i + 3(-1)}{2(-1)} \quad \textit{Since } i^2 = -1.$$

$$= \frac{-3 + 4i}{-2} = \boxed{\frac{3}{2} - 2i}$$

EXAMPLE 7

Express the following in the form $a + bi$: $\dfrac{7 - 4i}{2 + i}$

Solution

Our goal is to get a real number in the denominator so that we may express the result in the form $a + bi$. We proceed as follows:

$\dfrac{7 - 4i}{2 + i}$ *Multiply the numerator and the denominator by the conjugate of the denominator, which is $2 - i$.*

$= \dfrac{(7 - 4i)(2 - i)}{(2 + i)(2 - i)}$

$= \dfrac{14 - 7i - 8i + 4i^2}{2^2 - i^2}$ *Recall that $i^2 = -1$.*

$= \dfrac{14 - 7i - 8i - 4}{4 + 1}$ *Note that the denominator is now a real number. Simplify the numerator and denominator.*

$= \dfrac{10 - 15i}{5}$ *Then rewrite in complex number form.*

$= \dfrac{10}{5} - \dfrac{15}{5}i$ *Simplify.*

$= \boxed{2 - 3i}$

We have defined complex numbers and their operations. Complex numbers, together with the operations defined, obey the same properties as the real numbers (associative, commutative, distributive, inverses, etc.) and form a system called the **complex number system.**

We have extended the natural numbers to the whole numbers, the whole numbers to the integers, the integers to the reals, and finally, the reals to the complex numbers, C (see Figure 7.3):

$$N \subset W \subset Z \subset Q \subset R \subset C$$

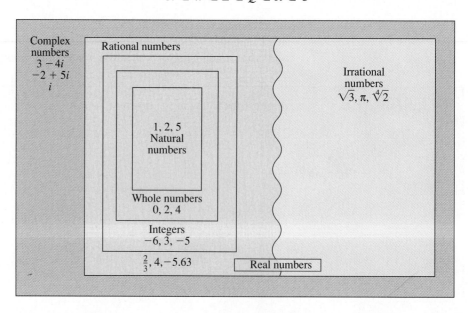

Figure 7.3
The complex number system

It turns out that any polynomial equation has all its solutions in C. In the next chapter we will see that any second-degree equation has its solutions in C, the complex number system.

EXERCISES 7.8

In Exercises 1–12, express as i, −i, −1, or +1:

1. i^3

2. i^7

3. $-i^{21}$

4. $-i^{13}$

5. $(-i)^{29}$

6. $(-i)^{56}$

7. i^{72}

8. i^{35}

9. i^{67}

10. i^{103}

11. $-i^{16}$

12. $-i^{24}$

In Exercises 13–22, express in the form a + bi:

13. $\sqrt{-5}$

14. $\sqrt{-8}$

15. $3 - \sqrt{-2}$

16. $5 - \sqrt{-64}$

17. $2\sqrt{-4} - \sqrt{28}$

18. $\sqrt{-6} - \sqrt{64}$

19. $\dfrac{3 - \sqrt{-2}}{5}$

20. $\dfrac{5 - \sqrt{-3}}{2}$

21. $\dfrac{6 - \sqrt{-3}}{3}$

22. $\dfrac{8 - 4\sqrt{-2}}{2}$

In Exercises 23–70, perform the indicated operations and express your answers in the form a + bi.

23. $(3 + 2i) - (2 + 3i)$

24. $(5 - 4i) + (3 + 6i)$

25. $(2 - i) + (2 + i)$

26. $(5 - 3i) - (2 - 4i)$

27. $5(3 - 7i)$

28. $6(2 + 5i)$

29. $(2 - i)(-2)$

30. $(3 - 8i)(-4)$

31. $(5i)(3i)$

32. $(2i)(8i)$

33. $(-3i)^2$

34. $(-7i)^2$

35. $2i(3 + 2i)$

36. $3i(5 - 4i)$

37. $-2i(-3 + 5i)$

38. $-5i(-2 - 2i)$

39. $(3 + 2i)(0)$

40. $(0)(7 + 2i)$

41. $(7 - 4i)(3 + i)$

42. $(5 - 2i)(5 + 3i)$

43. $(3 + i)(3 - i)$

44. $(2 + i)(2 - i)$

45. $(2 - 7i)(2 + 7i)$

46. $(3 + 7i)(3 - 7i)$

47. $(i + 1)^2$

48. $(5 + i)^2$

49. $(2 - i)^2 - 4(2 - i)$

50. $(3 - 4i)^2 + (3 - 4i)$

51. $(1 + i)^2 - 2(1 + i) + 1$

52. $(5 - 2i)^2 + (5 - 2i) + 6$

53. $\dfrac{6 - 5i}{3}$

54. $\dfrac{-4 + 9i}{3}$

55. $\dfrac{3 - i}{i}$

56. $\dfrac{8 - i}{i}$

57. $\dfrac{5 - 2i}{2i}$

58. $\dfrac{3 - 5i}{3i}$

59. $\dfrac{2i}{5 - 2i}$

60. $\dfrac{3i}{3 - 5i}$

61. $\dfrac{2}{5 - 2i}$

62. $\dfrac{3}{3 - 5i}$

63. $\dfrac{2 + i}{2 - i}$

64. $\dfrac{3 + i}{3 - i}$

65. $\dfrac{2 - 5i}{2 + 5i}$

66. $\dfrac{2 - 3i}{2 + 3i}$

67. $\dfrac{3 - 7i}{5 + 2i}$

68. $\dfrac{4 + 5i}{3 + 7i}$

69. $\dfrac{5 - \sqrt{-4}}{3 + \sqrt{-25}}$

70. $\dfrac{3 - \sqrt{-9}}{3 + \sqrt{-16}}$

One solution to the quadratic equation $Ax^2 + Bx + C = 0$ is given by the quadratic

formula, $x = \dfrac{-B + \sqrt{B^2 - 4AC}}{2A}$. In Exercises 71–74, use the quadratic formula to

find and simplify x for the given values of A, B, and C.

71. $A = 1$, $B = 4$, and $C = 5$

72. $A = 2$, $B = 5$, and $C = 4$

73. $A = 2$, $B = 4$, and $C = 5$

74. $A = 1$, $B = 4$, and $C = 7$

75. Show that $2i$ is a solution for $x^2 + 4 = 0$.

76. Show that $3i$ is a solution for $x^2 + 9 = 0$.

77. Show that $2 - i$ is a solution for $x^2 - 4x + 5 = 0$.

78. Show that $2 + i$ is a solution for $x^2 - 4x + 5 = 0$.

? Questions for Thought

79. Are the irrational numbers closed under multiplication? (Does the product of two irrational numbers always yield an irrational number?) If not, give an example.

80. The number i allows us to represent even roots of negative numbers other than square roots. Show how we can represent $\sqrt[6]{-1}$ using i. [*Hint:* $\sqrt[6]{x} = \sqrt{\sqrt[3]{x}}$]. How about $\sqrt[10]{-1}$? $\sqrt[14]{-1}$?

81. Express the following as i, $-i$, -1, or $+1$: i^{4n}, i^{4n+1}, i^{4n+2}, i^{4n+3}, (where n is an integer).

◆ Mini-Review

82. *Reduce to lowest terms:* $\dfrac{16x - 4x^2}{x^2 - 16}$

83. *Perform the indicated operations and simplify:* $\dfrac{x + 3}{x - 6} \div (x^2 - 3x - 18)$

84. *Solve for x:* $ax + 5 = 6x + 12$

85. Find the domain of $f(x) = \dfrac{x}{x^2 - 25}$.

CHAPTER 7 SUMMARY

After having completed this chapter you should:

1. Know the five rules of exponents and be able to use them to simplify expressions (Section 7.1).

For example:

$$\frac{(2x^3y^2)^5}{8(x^2y^3)^4} = \frac{2^5(x^3)^5(y^2)^5}{8(x^2)^4(y^3)^4} \qquad \textit{By rule 3}$$

$$= \frac{32x^{15}y^{10}}{8x^8y^{12}} \qquad \textit{By rule 2}$$

$$= \frac{4x^7}{y^2}$$

2. Understand the definition of zero and negative exponents (Section 7.1).

For example:

(a) $18^0 = 1$

(b) $4^{-3} = \dfrac{1}{4^3} = \dfrac{1}{4 \cdot 4 \cdot 4} = \dfrac{1}{64}$

3. Be able to use the exponent rules to simplify expressions with integer exponents (Section 7.1).

   For example: Simplify and express with positive exponents only.

   (a) $\dfrac{(x^{-3}y^8)^{-2}}{(x^{-1}y^{-3})^4} = \dfrac{(x^{-3})^{-2}(y^8)^{-2}}{(x^{-1})^4(y^{-3})^4}$ *By rule 3*

   $= \dfrac{x^6 y^{-16}}{x^{-4}y^{-12}}$

   $= x^{6-(-4)}y^{-16-(-12)}$ *By rule 4*

   $= x^{10}y^{-4}$

   $= x^{10}\left(\dfrac{1}{y^4}\right)$ *By definition of negative exponents*

   $= \dfrac{x^{10}}{y^4}$

   (b) $(x^{-2} + y^{-3})^{-1} = \left(\dfrac{1}{x^2} + \dfrac{1}{y^3}\right)^{-1}$ *Add fractions.*

   $= \left(\dfrac{y^3 + x^2}{x^2 y^3}\right)^{-1}$ *Take the reciprocal.*

   $= \dfrac{x^2 y^3}{y^3 + x^2}$

4. Be able to write and compute with numbers in scientific notation (Section 7.2).

   For example:

   A *nanometer* (nm) is a unit of measurement used in microscopy, where 1 nm = 10^{-7} cm. If a microorganism is 1.5×10^{-4} cm long, what is its length in nanometers?

   Since 1 nm = 10^{-7} cm, the length of the microorganism in nm is

   $$\dfrac{1.5 \times 10^{-4}}{10^{-7}} = 1.5 \times 10^3 \text{ nm} \quad \text{or} \quad 1,500 \text{ nm}$$

5. Understand and be able to evaluate expressions involving rational exponents (Section 7.3).

   For example:
   (a) $16^{1/2} = 4$ because $4^2 = 16$
   (b) $(-32)^{1/5} = -2$ because $(-2)^5 = -32$

   (c) $27^{-2/3} = \dfrac{1}{27^{2/3}} = \dfrac{1}{(27^{1/3})^2} = \dfrac{1}{(3)^2} = \dfrac{1}{9}$

6. Be able to simplify expressions involving rational exponents (Section 7.3).

   For example:

   (a) $\dfrac{(4x^2)^{1/2}(x^{-2/3})}{(2x)^2(x^5)^{-1/2}} = \dfrac{4^{1/2}x \cdot x^{-2/3}}{2^2 x^2 \cdot x^{-5/2}} = \dfrac{2x^{1+(-2/3)}}{4x^{2+(-5/2)}}$

   $= \dfrac{2x^{1/3}}{4x^{-1/2}} = \dfrac{x^{5/6}}{2}$

   (b) $(x^{1/3} + x^{2/5})^2 = (x^{1/3} + x^{2/5})(x^{1/3} + x^{2/5})$

   $= x^{1/3}x^{1/3} + 2x^{1/3}x^{2/5} + x^{2/5}x^{2/5}$

   $= x^{2/3} + 2x^{1/3+2/5} + x^{4/5}$

   $= x^{2/3} + 2x^{11/15} + x^{4/5}$

7. Convert expressions from radical form to exponential form, and vice versa (Section 7.3).

For example:

(a) $2a^{5/4} = 2\sqrt[4]{a^5}$ or $2\left(\sqrt[4]{a}\right)^5$

(b) $\sqrt[5]{x^3y} = (x^3y)^{1/5}$ or $x^{3/5}y^{1/5}$

8. Evaluate numerical expressions given in radical or exponential form (Section 7.3).

For example:

(a) $\sqrt{64} = 8$ because $8^2 = 64$

(b) $\sqrt[3]{-64} = -4$ because $(-4)^3 = -64$

(c) $\sqrt[6]{64} = 2$ because $2^6 = 64$

9. Write radical expressions in simplest radical form (Section 7.4).

For example:

(a) $\sqrt{24x^5y^6} = \sqrt{(4x^4y^6)(6x)} = \sqrt{4x^4y^6}\sqrt{6x} = 2x^2y^3\sqrt{6x}$

(b) $\sqrt[3]{24x^5y^6} = \sqrt[3]{(8x^3y^6)(3x^2)} = \sqrt[3]{8x^3y^6}\sqrt[3]{3x^2} = 2xy^2\sqrt[3]{3x^2}$

(c) $\left(a\sqrt{3a^2}\right)\left(b\sqrt{9ab}\right) = ab\left(\sqrt{27a^3b}\right) = ab\left(\sqrt{9a^2}\sqrt{3ab}\right) = ab\left(3a\sqrt{3ab}\right) = 3a^2b\sqrt{3ab}$

(d) $\sqrt{\dfrac{3}{x}} = \dfrac{\sqrt{3}}{\sqrt{x}} = \dfrac{\sqrt{3}\sqrt{x}}{\sqrt{x}\sqrt{x}} = \dfrac{\sqrt{3x}}{\sqrt{x^2}} = \dfrac{\sqrt{3x}}{x}$

(e) $\dfrac{\sqrt[3]{3}}{\sqrt[3]{x^2y}} = \dfrac{\sqrt[3]{3}\sqrt[3]{xy^2}}{\sqrt[3]{x^2y}\sqrt[3]{xy^2}} = \dfrac{\sqrt[3]{3xy^2}}{\sqrt[3]{x^3y^3}} = \dfrac{\sqrt[3]{3xy^2}}{xy}$

(f) $\sqrt[3]{3}\sqrt{3} = \sqrt[6]{3^2}\sqrt[6]{3^3} = \sqrt[6]{3^5} = \sqrt[6]{243}$

10. Combine radicals (Section 7.5).

For example:

(a) $2\sqrt{75} - \sqrt{12} = 2\left(\sqrt{25}\sqrt{3}\right) - \sqrt{4}\sqrt{3} = 10\sqrt{3} - 2\sqrt{3} = 8\sqrt{3}$

(b) $\sqrt{\dfrac{3}{2}} - 5\sqrt{6} = \dfrac{\sqrt{3}}{\sqrt{2}} - 5\sqrt{6} = \dfrac{\sqrt{3}\sqrt{2}}{\sqrt{2}\sqrt{2}} - 5\sqrt{6} = \dfrac{\sqrt{6}}{2} - 5\sqrt{6}$

$$= \dfrac{\sqrt{6}}{2} - \dfrac{10\sqrt{6}}{2} = \dfrac{-9\sqrt{6}}{2}$$

11. Find products and quotients of radicals (Section 7.6).

For example:

(a) $\left(2\sqrt{3} - 4\right)\left(3\sqrt{2} + 5\right) = \left(2\sqrt{3}\right)\left(3\sqrt{2}\right) + 5\left(2\sqrt{3}\right) - 4\left(3\sqrt{2}\right) - 4(5)$

$$= 6\sqrt{6} + 10\sqrt{3} - 12\sqrt{2} - 20$$

(b) $\dfrac{18}{\sqrt{6} - \sqrt{3}} = \dfrac{18\left(\sqrt{6} + \sqrt{3}\right)}{\left(\sqrt{6} - \sqrt{3}\right)\left(\sqrt{6} + \sqrt{3}\right)}$

$$= \dfrac{18\left(\sqrt{6} + \sqrt{3}\right)}{6 - 3}$$

$$= \dfrac{\overset{6}{\cancel{18}}\left(\sqrt{6} + \sqrt{3}\right)}{\cancel{3}}$$

$$= 6\left(\sqrt{6} + \sqrt{3}\right)$$

12. Solve radical equations (Section 7.7).

For example:

Solve for x: $\sqrt{5x - 2} = 4$ *Square both sides.*

$$\left(\sqrt{5x - 2}\right)^2 = 4^2$$

$$5x - 2 = 16$$

$$5x = 18$$

$$x = \frac{18}{5}$$

Check $\sqrt{5\left(\dfrac{18}{5}\right) - 2} \overset{?}{=} 4$

$$\sqrt{18 - 2} \overset{?}{=} 4$$

$$\sqrt{16} \overset{\checkmark}{=} 4$$

13. Add, subtract, multiply, and divide complex numbers (Section 7.8).

For example:

(a) $(5 + 3i) + (2 - 6i) = 7 - 3i$

(b) $(5 + 3i)(2 - 6i) = 10 - 30i + 6i - 18i^2$

$$= 10 - 24i - 18(-1)$$

$$= 10 - 24i + 18$$

$$= 28 - 24i$$

(c) $\dfrac{5 + 3i}{2 - 6i} = \dfrac{(5 + 3i)(2 + 6i)}{(2 - 6i)(2 + 6i)}$

$$= \frac{10 + 36i - 18}{4 + 36}$$

$$= \frac{-8 + 36i}{40} = \frac{-8}{40} + \frac{36}{40}i = -\frac{1}{5} + \frac{9}{10}i$$

CHAPTER 7 REVIEW EXERCISES

In Exercises 1–46, perform the operations and simplify; express your answers with positive exponents only. Assume all variables represent nonzero real numbers.

1. $(x^2x^5)(x^4x)$

2. $(x^5y^2)(x^6y^7)$

3. $(-3x^2y)(-2xy^4)(-x)$

4. $(-3xy^5)(-2xy)(-5x)$

5. $(a^3)^4$

6. $(b^5)^4$

7. $(a^2b^3)^7$

8. $(r^2s^3)^8$

9. $(a^2b^3)^2(a^2b)^3$

10. $(x^2y^5)^2(xy^4)^3$

11. $(a^2bc^2)^2(ab^2c)^3$

12. $(-2xy^2)(-3x^2y)^3(-3x)$

13. $\dfrac{a^5}{a^6}$

14. $\dfrac{x^7}{x^2}$

15. $\dfrac{x^2x^5}{x^4x^3}$

16. $\dfrac{y^5y^7}{y^8y^2}$

17. $\dfrac{(x^3y^2)^3}{(x^5y^4)^5}$

18. $\dfrac{(a^2b^3)^2}{(a^2b^4)^3}$

19. $\left(\dfrac{a^2b}{ab}\right)^4$

20. $\left(\dfrac{ab^2}{cb}\right)^4$

21. $\dfrac{(2ax)^2(3ax)^2}{(-2x)^2}$

22. $\dfrac{(-5xy)^2(-3x^2)^3}{(-15x)^2}$

23. $\left(\dfrac{-3xy}{x^2}\right)^2\left(\dfrac{-2xy^2}{x}\right)^3$

24. $\left(\dfrac{-3ab^2}{a^2b}\right)^2\left(\dfrac{-2ab}{5a^2}\right)^3$

25. $a^{-3}a^{-4}a^5$

26. $x^{-5}x^{-4}x^0$

27. $(x^{-2}y^5)^{-4}$

28. $(x^{-2}x^3)^{-4}$

29. $(-3)^{-4}(-2)^{-1}$

30. $(-2)^{-2}(-2)^2$

31. $\left(\dfrac{3x^{-5}y^2z^{-4}}{2x^{-7}y^{-4}}\right)^0$

32. $(-156{,}794)^0$

33. $\dfrac{x^{-3}x^{-6}}{x^{-5}x^0}$

34. $\dfrac{x^{-2}y^{-3}}{x^2y^{-4}}$

35. $\left(\dfrac{x^{-2}y^{-3}}{y^{-3}x^2}\right)^{-2}$

36. $\left(\dfrac{a^{-2}b^{-3}}{a^2b^3}\right)^{-2}$

37. $\left(\dfrac{r^{-2}s^{-3}r^{-2}}{s^{-4}}\right)^{-2}\left(\dfrac{r^{-1}}{s^{-1}}\right)^{-3}$ **38.** $\left(\dfrac{2r^{-1}s^{-2}}{r^{-3}s^2}\right)^{-2}\left(\dfrac{-3r^{-2}s^{-2}}{4r^{-3}s}\right)^{-1}$ **39.** $\left(\dfrac{2}{5}\right)^{-2}$ **40.** $\left(\dfrac{3}{4}\right)^{-4}$

41. $\dfrac{(2x^2y^{-1}z)^{-2}}{(3xy^2)^{-3}}$ **42.** $\dfrac{(3x^{-2}y^2z^4)^{-2}}{(2x^{-1}y)^{-3}}$ **43.** $(x^{-1}+y^{-1})(x-y)$ **44.** $(x^{-1}+2y^{-2})^2$

45. $\dfrac{x^{-1}+y^{-3}}{x^{-1}y^2}$ **46.** $\dfrac{a^{-2}+b^{-1}}{a^{-1}b^{-2}}$

In Exercises 47–50, convert to standard form.

47. 2.83×10^4 **48.** 6.29×10^0 **49.** 7.96×10^{-5} **50.** 8.264×10^{-7}

In Exercises 51–54, convert to scientific notation.

51. 92.59 **52.** 0.00578 **53.** 625,897 **54.** 0.0000073

In Exercise 55, perform the computation by first converting the numbers to scientific notation. Express your answer in standard notation.

55. $\dfrac{(0.0014)(9,000)}{(20,000)(63,000)}$

56. A *nanometer* (nm) and a *micron* (μm) are two units of measurement used in microscopy, where 1 nm = 10^{-9} m, and 1 μm = 10^{-6} m. How many nm are in a μm?

In Exercises 57–70, perform the operations and simplify; express your answers with positive exponents only. Assume all variables represent positive real numbers only.

57. $x^{1/2}x^{1/3}$ **58.** $y^{1/3}y^{-1/2}$

59. $(x^{-1/2}x^{1/3})^{-6}$ **60.** $(y^{-1/3}y^{-1/2}y^{2/3})^{-1/2}$

61. $\dfrac{x^{1/2}x^{1/3}}{x^{2/5}}$ **62.** $\dfrac{r^{-1/2}s^{-1/3}}{r^{1/3}s^{-2/3}}$

63. $\left(\dfrac{a^{1/2}a^{-1/3}}{a^{1/2}b^{1/5}}\right)^{-15}$ **64.** $\left(\dfrac{x^{-1/2}x^{1/3}}{x^{-1/3}y^{1/2}}\right)^{-6}$

65. $\dfrac{(x^{-1/2}y^{1/2})^{-2}}{(x^{-1/3}y^{-1/3})^{-1/2}}$ **66.** $\dfrac{(a^{1/2}a^{1/3}a^{-2/3})^{-2}}{a^{1/3}a^{-1/2}}$

67. $\left(\dfrac{4^{-1/2}\cdot 16^{-3/4}}{8^{1/3}}\right)^{-2}$ **68.** $(a^{1/2}-b^{1/2})(a^{-2/3})$

69. $(a^{1/3}+b^{1/3})(a^{1/3}-b^{1/3})$ **70.** $(a^{-1/2}+2b^{1/2})^2$

In Exercises 71–78, write the expression in radical form. Assume that all variables represent positive real numbers only.

71. $x^{1/2}$ **72.** $x^{1/3}$ **73.** $xy^{1/2}$ **74.** $(xy)^{1/2}$

75. $m^{2/3}$ **76.** $m^{3/2}$ **77.** $(5x)^{3/4}$ **78.** $5x^{3/4}$

In Exercises 79–84, write the expression in exponential form.

79. $\sqrt[3]{a}$ **80.** $\sqrt[5]{t}$ **81.** $-\sqrt[5]{n^4}$ **82.** $\left(\sqrt[3]{t}\right)^5$

83. $\dfrac{1}{\sqrt[5]{t^7}}$ **84.** $\dfrac{1}{\sqrt[7]{t^5}}$

In Exercises 85–108, express in simplest radical form.

85. $\sqrt{54}$ **86.** $\sqrt[3]{54}$ **87.** $\sqrt{x^{60}}$ **88.** $\sqrt[4]{x^{60}}$

89. $\sqrt[3]{48x^4y^8}$ **90.** $\sqrt{28x^9y^{13}}$ **91.** $\sqrt{75xy}\,\sqrt{3x}$ **92.** $\sqrt{48a^2b}\,\sqrt{12b}$

93. $\left(x\sqrt{xy}\right)\left(2x^2y\sqrt{xy^2}\right)$ **94.** $\left(2x^3\sqrt{x^2}\right)\left(3x^2\sqrt{x^2y^4}\right)$ **95.** $\dfrac{\sqrt{28}}{\sqrt{63}}$ **96.** $\dfrac{\sqrt{12}}{\sqrt{27}}$

97. $\dfrac{y}{x\sqrt{y}}$ **98.** $\dfrac{y}{\sqrt{xy}}$ **99.** $\sqrt{\dfrac{48a^2b}{3a^5b^2}}$ **100.** $\sqrt{\dfrac{81x^3y^2}{2x^4}}$

101. $\sqrt{\dfrac{5}{a}}$ **102.** $\sqrt[3]{\dfrac{5}{a}}$ **103.** $\dfrac{4}{\sqrt[3]{2a}}$ **104.** $\dfrac{4}{\sqrt[3]{2a^2}}$

105. $\sqrt[6]{x^4}$ **106.** $\sqrt[16]{x^6}$ **107.** $\sqrt{2}\sqrt[3]{2}$ **108.** $\dfrac{\sqrt{2}}{\sqrt[3]{2}}$

In Exercises 109–126, perform the indicated operations and simplify as completely as possible.

109. $4\sqrt{2} + \sqrt{2} - 5\sqrt{2}$ **110.** $7\sqrt{54} + 6\sqrt{24}$

111. $6\sqrt{12} - 4\sqrt{27}$ **112.** $2t\sqrt{s^5t^2} - 3s\sqrt{s^2t^5}$

113. $\sqrt{\dfrac{3}{2}} + \sqrt{\dfrac{5}{3}}$ **114.** $\sqrt{\dfrac{5}{7}} + \sqrt{\dfrac{2}{3}}$

115. $\sqrt{3}\left(\sqrt{6} - \sqrt{2}\right) + \sqrt{2}\left(\sqrt{3} - 3\right)$ **116.** $\sqrt{5}\left(\sqrt{2} - 2\right) + \sqrt{10}\left(\sqrt{2} - 2\right)$

117. $\left(3\sqrt{7} - \sqrt{3}\right)\left(\sqrt{7} - 2\sqrt{3}\right)$ **118.** $\left(2\sqrt{x} - \sqrt{5}\right)\left(3\sqrt{x} + \sqrt{4}\right)$

119. $\left(\sqrt{x} - 5\right)^2$ **120.** $\left(\sqrt{7} - \sqrt{3}\right)^2$

121. $\left(\sqrt{a+7}\right)^2 - \left(\sqrt{a} + 7\right)^2$ **122.** $\left(\sqrt{b} - 3\right)^2 - \left(\sqrt{b-3}\right)^2$

123. $\dfrac{12}{\sqrt{5} + \sqrt{3}}$ **124.** $\dfrac{m - n^2}{\sqrt{m} - n}$

125. $\dfrac{8x - 20y}{\sqrt{2x} - \sqrt{5y}}$ **126.** $\dfrac{15}{\sqrt{7} + \sqrt{2}} - \dfrac{21}{\sqrt{7}}$

In Exercises 127–136, solve the equations.

127. $\sqrt{2x - 5} = 7$ **128.** $\sqrt{3x + 2} = 4$ **129.** $\sqrt{2x - 5} = 7$ **130.** $\sqrt{3x + 2} = 4$

131. $\sqrt[4]{x - 4} = 3$ **132.** $\sqrt[3]{x + 3} = 2$ **133.** $x^{1/4} + 1 = 3$ **134.** $(x + 1)^{1/4} = 3$

135. If $f(x) = \sqrt{3x}$, solve $f(x) = 7.25$ accurate to two decimal places.

136. If $g(x) = \sqrt{x - 2.1}$, solve $g(x) = 5.8$ accurate to two decimal places.

137. The pressure, p (in pounds per square inch or psi), of the stream from a fire hose is related to the number of gallons per minute, G, discharged from the hose, and the diameter, d, of the hose in inches according to the formula $G = 34.2d^2\sqrt{p}$. Find the pressure of the water coming out of a hose that is 3 in. in diameter and that is discharging 450 gallons per minute. Round your answer to the nearest tenth.

138. Use the formula given in Exercise 137 to find the pressure of the water coming out of a hose that is 4.5 in. in diameter and that is discharging 900 gallons per minute. Round your answer to the nearest tenth.

In Exercises 139–142, express as i, $-i$, 1, or -1.

139. i^{11} **140.** i^{29} **141.** $-i^{14}$ **142.** $(-i)^{14}$

In Exercises 143–154, perform the indicated operations with complex numbers and express your answers in complex number form.

143. $(5 + i) + (4 - 2i)$ **144.** $(7 - 2i) - (6 + 3i)$ **145.** $(7 - 2i)(2 - 3i)$ **146.** $(5 - 4i)(5 + 4i)$

147. $2i(3i - 4)$ **148.** $5i(i - 2)$ **149.** $(3 - 2i)^2$ **150.** $(2 - 3i)^2$

151. $(6 - i)^2 - 12(6 - i)$ **152.** $(7 - i)^2 - 2(7 - 4i)$ **153.** $\dfrac{4 - 3i}{3 + i}$ **154.** $\dfrac{2 + 3i}{-3 + 2i}$

155. Show that $1 - 2i$ is a solution for $x^2 - 2x + 5 = 0$.

156. Show that $1 + 2i$ is a solution for $x^2 - 2x + 5 = 0$.

CHAPTER 7 PRACTICE TEST ⊙

Perform the indicated operations and express your answers in simplest form with positive exponents only. Assume that all variables represent positive real numbers.

1. $(a^2b^3)(ab^4)^2$

2. $(-2x^2y)^3(-5x)^2$

3. $(-3ab^{-2})^{-1}(-2x^{-1}y)^2$

4. $\dfrac{(-2x^2y)^3}{(-6xy^4)^2}$

5. $\left(\dfrac{5r^{-1}s^{-3}}{3rs^2}\right)^{-2}$

6. $\dfrac{27^{2/3} \cdot 3^{-4}}{9^{-1/2}}$

7. $\left(\dfrac{x^{1/4}x^{-2/3}}{x^{-1}}\right)^4$

8. $\dfrac{x^{-3} + x^{-1}}{yx^{-2}}$

9. $x^{1/3}(x^{2/3} - x)$

10. Evaluate the following:

    (a) $(-125)^{1/3}$

    (b) $(-128)^{-3/7}$

11. Express $3a^{2/3}$ in radical form.

12. Compute the following by first converting the numbers to scientific notation. Express your answer in standard notation.

$$\frac{(64)(28)}{(8,000)(70,000)}$$

13. Light travels at approximately 186,000 miles per second. How far does light travel in 1 hour? (Use scientific notation to compute your answer.)

Perform the indicated operations and express the following in simplest radical form. Assume that all variables represent positive real numbers.

14. $\sqrt[3]{27x^6y^9}$

15. $\sqrt[3]{4x^2y^2}\sqrt[3]{2x}$

16. $\left(2x^2\sqrt{x}\right)\left(3x\sqrt{xy^2}\right)$

17. $\sqrt{\dfrac{5}{7}}$

18. $\sqrt[3]{\dfrac{8}{9}}$

19. $\dfrac{\left(xy\sqrt{2xy}\right)\left(3x\sqrt{y}\right)}{\sqrt{4x^3}}$

20. $\sqrt[4]{5}\sqrt{5}$

21. $\sqrt{50} - 3\sqrt{8} + 2\sqrt{18}$

22. $\left(\sqrt{x} - 3\right)^2$

23. $\dfrac{\sqrt{6}}{\sqrt{6} - 2}$

Solve the following equations.

24. $\sqrt{x - 3} + 4 = 8$

25. $\sqrt{x} - 3 = 8$

26. If $f(x) = \sqrt{5x}$, solve $f(x) = 9.3$ accurate to two decimal places.

Express the following in the form $a + bi$:

27. i^{51}

28. $(3i + 1)(i - 2)$

29. $\dfrac{2i + 3}{i - 2}$

8

Quadratic Functions and Equations

© Royalty-Free/CORBIS

In Section 5.1, we examined functions that were polynomials. In this chapter we turn our attention to polynomial functions of the *second degree*—that is, polynomial functions in which the highest degree of the variable is 2. These functions are usually called *quadratic functions* and have the form

$$f(x) = Ax^2 + Bx + C \qquad (A \neq 0)$$

where A is the coefficient of the second-degree term, B is the coefficient of the first-degree term, and C is the constant term.

In the first section of this chapter, we will discuss a few properties of quadratic functions and examine some examples of quadratic models. In subsequent sections, we will examine algebraic methods of solving quadratic equations and the graphs of quadratic functions. We will then apply methods of solving quadratic equations to more complex equations as well as to inequalities.

8.1 Quadratic Functions as Mathematical Models

In Chapter 5 we looked at the function $h(t) = 10 + 50t - 4.9t^2$, which described the height of an object, $h(t)$, above the ground (in meters) as a function of the time, t (in seconds). In that example we looked at a table of values and noted that there is a single maximum value of $f(x)$, which occurs at the turning point of the graph. We graphed the function as shown in Figure 8.1.

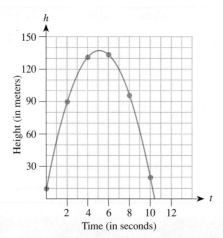

Figure 8.1
The graph of
$h = h(t) = 10 + 50t - 4.9t^2$

In general, the graph of a quadratic function $y = f(x) = Ax^2 + Bx + C$ (where $A \neq 0$, and A, B, and C are real numbers) has an umbrella shape that is called a *parabola.* Figure 8.2 shows the graphs of two quadratic functions.

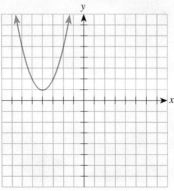
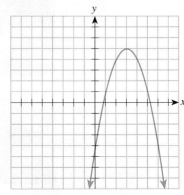

Figure 8.2   **(a)** The graph of $f(x) = x^2 + 8x + 17$ **(b)** The graph of $f(x) = -x^2 + 6x - 4$

Depending on the value of *A* (as we will see in Section 8.6), this umbrella shape opens either up or down. Note that there is only a single turning point, which is called the ***vertex*** of the parabola. If the parabola opens downward, the vertex is the highest point and the corresponding *y*-value is called the *maximum* value of *f*(*x*). If the parabola opens upward, the vertex is the lowest point and the corresponding *y*-value is called the *minimum* value of *f*(*x*).

Although the parabola may be stretched (appear narrower or wider) or shifted (moved up, down, right, or left), a quadratic function will always have this shape.

If we look at either a table of values or the graph of a quadratic function, we note that each *y*-value is assigned to *two* *x*-values except at the vertex: *At the vertex the y-value is assigned to only one x-value.* Figures 8.3a and 8.3b show an example of a table of values and the graph of the quadratic function $y = f(x) = x^2 - 2x - 2$.

x	$f(x) = y$
-2	6
-1	1
0	-2
1	-3
2	-2
3	1
4	6

(a) $y = f(x) = x^2 - 2x - 2$

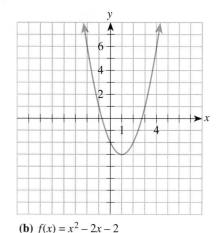

(b) $f(x) = x^2 - 2x - 2$

Figure 8.3

In Section 8.6 we will examine the graphs of quadratic functions in more detail. For now, we will use the information we have discussed thus far to further our understanding of the general behavior of quadratic models.

EXAMPLE 1

The profit in dollars, *P*(*x*), made on a concert is related to the price of a ticket in dollars, *x*, in the following way:

$$P(x) = 10{,}000(-x^2 + 11x - 15)$$

(a) Use this function to find the profit if the ticket price is $0, $2, $4, $6, $8, and $10 per ticket, and describe what happens to the profit as the ticket price increases.

(b) Estimate the maximum profit that can be made for this concert and what ticket price will produce that maximum profit.

Solution

(a) We are looking for the value of *P*(*x*) when *x* = $0, $2, $4, $6, $8, and $10. Using function notation, this means we are looking for *P*(0), *P*(2), *P*(4), *P*(6), *P*(8), and *P*(10).

To find P(a), substitute x = a in $P(x) = \mathbf{10{,}000(-x^2 + 11x - 15)}$

To find P(0), substitute x = 0. $P(0) = 10{,}000(-0^2 + 11 \cdot 0 - 15) = -150{,}000$

To find P(2), substitute x = 2. $P(2) = 10{,}000(-2^2 + 11 \cdot 2 - 15) = 30{,}000$

To find P(4), substitute x = 4. $P(4) = 10{,}000(-4^2 + 11 \cdot 4 - 15) = 130{,}000$

To find P(6), substitute x = 6. $P(6) = 10{,}000(-6^2 + 11 \cdot 6 - 15) = 150{,}000$

To find P(8), substitute x = 8. $P(8) = 10{,}000(-8^2 + 11 \cdot 8 - 15) = 90{,}000$

To find P(10), substitute x = 10. $P(10) = 10{,}000(-10^2 + 11 \cdot 10 - 15) = -50{,}000$

From the values of $P(x)$ above, we can see that the profit function can be negative: We interpret this as a loss. When ticket prices are $0 (a free concert), $150,000 is lost ($P(0) = -150,000$). We see that a profit of $30,000 is realized when the ticket price is $2 ($P(2) = 30,000$); therefore, the concert begins to earn a profit somewhere between $0 and $2. As the ticket price increases, the profit rises for a while; when the ticket price exceeds $6 per ticket, the profit begins to decrease. The profit seems to reach a maximum of about $150,000 at a ticket price of $6.

As the ticket price increases beyond $6, the profit function decreases, or there is less profit. Perhaps fewer people go to the concert because the ticket prices are too expensive. Profits continue to decrease until there is a loss again, which occurs when the price of a ticket is between $8 and $10.

We can represent the data as ordered pairs $(x, P(x))$, letting P represent the dependent variable (and therefore the vertical axis), and plot the data points as shown in Figure 8.4. $P(x)$ is a quadratic function, and since we have stated that the graph of a quadratic function will give us a parabola, we can "fit" the parabola around the data points to get the rough graph of the function $P(x)$ shown in Figure 8.4.

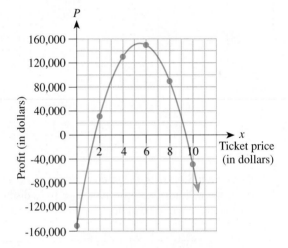

Figure 8.4
The graph of $P(x) =$
$10,000 \left(-x^2 + 11x - 15\right)$

(b) We can see by the graph in part **(a)** that the maximum, or highest, profit occurs somewhere around a ticket price of $6. We are not sure how close until we examine other values around $6. We check the value of $P(x)$ for $x = 5$ to get

$$P(5) = 10,000(-5^2 + 11 \cdot 5 - 15) = 150,000$$

which is the same table value as for $x = 6$. Hence, the maximum value must lie between $x = 5$ and $x = 6$. We could improve this estimate by computing other ticket prices between $5 and $6; however, later in this chapter we will discuss methods for finding the exact values. Hence, the estimated maximum profit is around $150,000, which occurs for a ticket price between $5 and $6.

EXAMPLE 2

Raju wants to fence in a rectangular garden against his house. Since the house is to serve as one of the sides, he needs to fence in only three sides (see Figure 8.5). Suppose he uses 100 linear feet of fencing.

(a) Express the area of the garden as a function of x (the length of side given in Figure 8.5).

(b) Use the area function found in part **(a)** to find the area of the garden for $x = 5, 10, 20, 30, 40,$ and 45 ft, and describe what happens to the area of the garden as the length x increases.

(c) Estimate the maximum area of the garden and the dimensions of the garden that would give the maximum possible area.

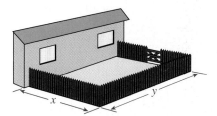

Figure 8.5

Solution   (a) We want to find the area of the garden in the diagram with given side x. Hence, we first write the formula for the area of the garden:

$$A = (\text{Length})(\text{Width})$$

Let's apply this formula to the rectangular garden in Figure 8.5:

$$A = xy$$

Now we have expressed the area of the garden as a function of two variables, but we are asked to express the area in terms of one variable, x. Let's see whether we can find a relationship between the sides x and y of the rectangle.

We know that the length of the fence used is 100 ft; hence, from the figure,

$$2x + y = 100 \qquad \textit{Solve for y.}$$
$$y = 100 - 2x$$

We have just expressed the third side, y, in terms of x. Now we can substitute into the area equation:

$$A = x \cdot y \qquad \textit{Substitute 100} - 2x \textit{ for y.}$$
$$A = A(x) = x(100 - 2x) = 100x - 2x^2$$

This expresses the area, A, as a function of x. We note that this is a quadratic function.

(b) We make a table of values for the six values of x: 5, 10, 20, 30, 40, and 45:

X	Y1
5	450
10	800
20	1200
30	1200
40	800
45	450

Y1☰100X-2X²

A table of areas on a graphing calculator.

x	$A(x) = 100x - 2x^2$	$= $ *Area of the garden*
5	$A(5) = 100(5) - 2(5)^2$	$= 450$
10	$A(10) = 100(10) - 2(10)^2$	$= 800$
20	$A(20) = 100(20) - 2(20)^2$	$= 1{,}200$
30	$A(30) = 100(30) - 2(30)^2$	$= 1{,}200$
40	$A(40) = 100(40) - 2(40)^2$	$= 800$
45	$A(45) = 100(45) - 2(45)^2$	$= 450$

From the table, we can see that the area $A(x)$ is 450 sq ft when the length x is 5 ft. As the length x increases, the area $A(x)$ increases until $A(x)$ reaches 1,200 sq ft at $x = 20$. The area is also 1,200 sq ft at $x = 30$, but then the area $A(x)$ decreases for larger values of x until it moves down to 450 sq ft again at $x = 45$.

Looking at the table, we may get the (incorrect) impression that the largest value of $A(x)$ is 1,200, occurring at $x = 20$ and $x = 30$. However, let's plot the data points (see Figure 8.6), letting A represent the vertical axis.

Since we know that $A(x)$ is a quadratic function and that the graph of a quadratic function will give us a parabola, we again "fit" the parabola around the data points to get the rough graph of the function $A(x)$ shown in Figure 8.6.

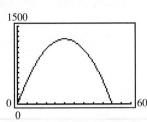

The graph of A(x) in the [0, 60] by [0, 1500] window in a graphing calculator.

Figure 8.6
The graph of
$A = A(x) = 100x - 2x^2$

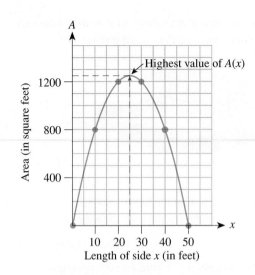

Later in this chapter we will be able to verify the maximum value for A(x). (See Exercise 53 in Section 8.6.)

(c) We can see from the table and the graph in part **(b)** that the maximum area, or highest value of $A(x)$, occurs somewhere between $x = 20$ and $x = 30$.
We check the value $A(25)$ to get

$$A(25) = 100(25) - 2(25)^2 = 1,250$$

If we check the values $A(24)$ and $A(26)$, we will find that these values will be smaller than $A(25)$. The maximum value of the area $A(x)$ is in fact 1,250 sq ft, occurring at $x = 25$ ft.
When $x = 25$, $y = 100 - 2x = 100 - 2(25) = 50$ ft. Hence we can estimate that the dimensions for the maximum area are 25 by 50 ft, where the side opposite the house is 50 ft.

EXERCISES 8.1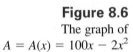

1. A company finds that its daily profit in dollars, $P(x)$, is related to the number of items it produces daily, x, in the following way:

$$P(x) = -3.5x^2 + 600.4x$$

(a) Use this function to find the profit if the number of items produced daily is 30, 50, 70, 100, and 120, and describe what happens to the profit as the company increases its daily production of items.

(b) Estimate the maximum profit that can be made by this company and what the daily production would be for that maximum profit.

2. The daily production cost, in dollars, for a desk manufacturer is

$$C(x) = 0.3x^2 - 5.3x + 546.8$$

where x is the number of desks produced daily.
(a) Use this function to find the daily production cost if the number of desks produced daily is 4, 6, 8, 10, and 12, and describe what happens to the daily production cost as the company increases its daily production of desks.

(b) Estimate how many desks the manufacturer should produce to minimize the daily production cost. Estimate the minimum production cost.

3. Harry fires a rocket upward, which travels according to the equation

$$s(t) = -16t^2 + 725t$$

where $s(t)$ is the height (in feet) of the rocket above the ground t seconds after the rocket is fired.

(a) Use this function to find the height of the rocket at the end of 5, 10, 20, 30, 40, and 45 seconds.

(b) Estimate how many seconds it takes for the rocket to reach its maximum height. Estimate the maximum height of the rocket.

(c) Estimate how long it takes for the rocket to hit the ground.

4. A manufacturer finds that the revenue, $R(x)$, earned on the production and sale of n items is given by the function

$$R(n) = 1.82n - 0.0001n^2$$

(a) Use this function to find the revenue if the number of items produced and sold is 3,000, 6,000, 9,000, 12,000, and 15,000. Describe what happens to the revenue as the company increases its production and sales.

(b) Estimate how many items should be produced and sold to maximize the revenue. Estimate the maximum revenue.

5. A rectangular area is to be fenced in with 200 ft of fencing (see accompanying figure).

(a) Express the area of the rectangle as a function of x (the length of side x in the accompanying figure).

(b) Use the area function found in part **(a)** to find the area of the rectangle for $x = 10, 40, 60, 70,$ and 90 ft, and describe what happens to the area of the rectangle as the length x increases.

(c) Estimate the maximum area of the rectangle and the dimensions of the rectangle that would give the maximum possible area.

6. For a fixed perimeter of 100 ft, estimate what dimensions will yield a rectangle with the maximum area.

7. Estimate which two numbers whose sum is 225 will yield the maximum product.

8. Estimate which two numbers whose difference is 100 and will yield a minimum product.

9. Two rectangular pens are to be fenced in with 1,000 ft of fencing as illustrated in the accompanying figure.

(a) Express the total area of the pens as a function of x (the length of side x in the accompanying figure).

(b) Use the area function found in part **(a)** to find the total area of the pens for $x = 50, 100, 120, 150,$ and 200 ft, and describe what happens to the total area of the pens as the length x increases.

(c) Estimate the total maximum area of the pens and the values of x and y that would give the maximum possible area.

10. A rectangular pen is to be fenced in against a house with 1,000 ft of fencing as illustrated in the accompanying figure.

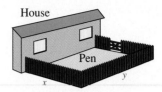

(a) Express the area of the pen as a function of x (the length of side x in the accompanying figure).

(b) Use the area function found in part (a) to find the area of the pen for $x = 50$, 100, 200, 300, and 400 ft, and describe what happens to the area of the pen as the length x increases.

(c) Estimate the maximum area of the pen and the dimensions of the rectangle that would give the maximum possible area.

11. In economics, the demand function for a given item indicates how the price per unit, p, is related to the number of units, x, that are sold. Suppose a company finds that the demand function for one of the items it produces is

$$p = 10 - \frac{x}{6}$$

where p is in dollars.

(a) How many items would be sold if the price were $6 per unit?

(b) What should the price per unit be if 30 units were to be sold?

(c) Sketch a graph of this function.

(d) The revenue function, $R(x)$, is found by multiplying the price per unit, p, by the number of items sold, x (that is, $R = xp$). Find the revenue function corresponding to this demand function.

(e) Use the revenue function found in part (d) to find the revenue if the number of items sold is 5, 15, 25, 35, and 40; describe what happens to the revenue as the company increases its sale of items.

(f) Estimate how many items should be sold to maximize the revenue. Estimate the maximum revenue.

12. In economics, the demand function for a given item indicates how the price per unit, p, is related to the number of units, x, that are sold. Suppose a company finds that the demand function for one of the items it produces is

$$p = 60 - \frac{x}{4}$$

where p is in dollars.

(a) How many items would be sold if the price were $30 per unit?

(b) What should the price per unit be if 80 units are to be sold?

(c) Sketch a graph of this function.

(d) The revenue function, $R(x)$, is found by multiplying the price per unit, p, by the number of items sold, x; that is, $R = xp$. Find the revenue function corresponding to this demand function.

(e) Use the revenue function found in part (d) to find the revenue if the number of items sold is 50, 75, 100, 125, and 150; and describe what happens to the revenue as the company increases its sale of items.

(f) Estimate how many items should be sold to maximize the revenue. Estimate the maximum revenue.

Mini-Review

13. *Evaluate:* $3x^3y^2 - x^2 + y^3$ for $x = -2$ and $y = -3$

14. *Solve for x:* $|3x - 4| < 6$

15. *Solve for x:* $|5 - 2x| \geq 6$

16. A manufacturer produces two types of telephones. The more expensive model requires 1 hour to manufacture and 30 minutes to assemble. The less expensive model requires 45 minutes to manufacture and 15 minutes to assemble. If the company allocates 150 hours for manufacture and 60 hours for assembly, how many of each type can be produced?

8.2 Solving Quadratic Equations: The Factoring and Square Root Methods

A *second-degree equation* is a polynomial equation in which the highest degree of the variable is 2. In particular, a second-degree equation in one unknown is called a *quadratic equation.* We define the *standard form* of a quadratic equation as

$$Ax^2 + Bx + C = 0 \qquad (A \neq 0)$$

where A is the coefficient of the second-degree term, x^2; B is the coefficient of the first-degree term, x; and C is a numerical constant.

As with all other equations, the solutions of quadratic equations are values that, when substituted for the variable, will make the equation a true statement. The solutions to the equation $Ax^2 + Bx + C = 0$ are also called the *roots* of the equation $Ax^2 + Bx + C = 0$.

The Factoring Method

We begin by recalling the zero-product rule from Section 5.6:

The Zero-Product Rule	If $a \cdot b = 0$, then $a = 0$ or $b = 0$.

In words, if the product of two quantities is 0, then one or both of the quantities is 0.

The next example reviews the factoring method discussed in Section 5.6. You may want to review some of the simpler quadratic equations covered there before continuing in this section.

EXAMPLE 1

Solve each of the following equations.

(a) $2x^2 + 9x - 5 = 0$

(b) $(3x + 1)(x - 1) = (5x - 3)(2x - 3)$

(c) $\dfrac{x}{x - 5} - \dfrac{3}{x + 1} = \dfrac{30}{x^2 - 4x - 5}$

Solution   **(a)** $2x^2 + 9x - 5 = 0$ *Factor the left side.*

$(2x - 1)(x + 5) = 0$ *Since the product is 0, set each factor equal to 0.*

$2x - 1 = 0$ or $x + 5 = 0$ *Then solve each first-degree equation.*

$$\boxed{x = \frac{1}{2}}$$ or $\boxed{x = -5}$

(b) $(3x + 1)(x - 1) = (5x - 3)(2x - 3)$

To determine what type of equation we have, we must first multiply out each side.

$3x^2 - 2x - 1 = 10x^2 - 21x + 9$ *Then put in standard form.*

$0 = 7x^2 - 19x + 10$ *Factor the right side of the equation.*

$0 = (7x - 5)(x - 2)$ *Then set each factor equal to 0.*

$7x - 5 = 0$ or $x - 2 = 0$

$7x = 5$ or $x = 2$

$$\boxed{x = \frac{5}{7}}$$ or $\boxed{x = 2}$

(c) Since this equation contains fractions, we begin by clearing the denominators. First we find the LCD of the fractions, which is $(x - 5)(x + 1)$.

$$\frac{x}{x - 5} - \frac{3}{x + 1} = \frac{30}{x^2 - 4x - 5}$$ *Multiply both sides of the equation by $(x - 5)(x + 1)$.*

$$(x - 5)(x + 1)\left(\frac{x}{x - 5} - \frac{3}{x + 1}\right) = \frac{30}{(x - 5)(x + 1)}(x - 5)(x + 1)$$

$$(x - 5)(x + 1)\frac{x}{x - 5} - (x - 5)(x + 1)\frac{3}{x + 1} = \frac{30}{(x - 5)(x + 1)}(x - 5)(x + 1)$$

$x(x + 1) - 3(x - 5) = 30$

$x^2 + x - 3x + 15 = 30$ *Simplify.*

$x^2 - 2x + 15 = 30$ *Put in standard form.*

$x^2 - 2x - 15 = 0$ *Factor and solve each first-degree equation.*

$(x - 5)(x + 3) = 0$

$x - 5 = 0$ or $x + 3 = 0$

$x = 5$ or $x = -3$

Remember, you must check the solutions to rational equations to determine whether the solutions are valid.

As we pointed out in Chapter 6, we can observe at the outset that neither 5 nor -1 is an admissible value since each makes a denominator of one of the fractions 0 and therefore the fraction undefined. [Also, multiplying both sides of the equation by $(x - 5)(x + 1)$ is not valid if x is 5 or -1, since multiplying both sides by 0 does not yield an equivalent equation.] Since 5 is one of the proposed solutions, we must eliminate this value, and therefore -3 is the only solution.

Throughout the chapter, any checks not done are left to the student.

Hence, the solution is $\boxed{x = -3}$.

EXAMPLE 2   The sum of a number and its reciprocal is $\frac{29}{10}$. Find the numbers.

Solution   Let $x =$ the number.

The sum of a number and its reciprocal is $\dfrac{29}{10}$

$$x + \frac{1}{x} = \frac{29}{10}$$

The equation is

$$x + \frac{1}{x} = \frac{29}{10}$$ *Multiply both sides of the equation by the LCD, 10x.*

$$10x\left(x + \frac{1}{x}\right) = \left(\frac{29}{10}\right)10x$$ *Distribute 10x and reduce.*

$$10x(x) + 10x\left(\frac{1}{x}\right) = \frac{29}{10} \cdot \frac{10x}{1}$$

$$10x^2 + 10 = 29x$$ *Now we have a quadratic equation and we solve for x.*

$$10x^2 - 29x + 10 = 0$$

$$(5x - 2)(2x - 5) = 0$$

$$5x - 2 = 0 \quad \text{or} \quad 2x - 5 = 0$$

$$5x = 2 \qquad\qquad 2x = 5$$

$$x = \frac{2}{5} \quad \text{or} \quad x = \frac{5}{2}$$

Thus, the answers are $\boxed{\dfrac{2}{5} \text{ and } \dfrac{5}{2}}$. Notice that the two answers are reciprocals of each other.

Often we find that the relationship between two quantities can be expressed as a second-degree equation in two unknowns.

EXAMPLE 3

The profit in dollars, $P(x)$, on each television set made daily by AAA Television Company is related to the number of television sets, x, produced at the AAA factory as follows:

$$P(x) = -\frac{x^2}{4} + 45x - 1{,}625$$

(a) What is the company's profit for each TV set if it produces 90 TV sets?

(b) How many sets must be produced for the company to make a profit of \$175 per set?

Solution **(a)** Since profit per set is

$$P(x) = -\frac{x^2}{4} + 45x - 1{,}625$$

where x is the number of sets produced daily, we simply substitute the daily number of sets produced for x and find $P(90)$:

$$P(90) = -\frac{(90)^2}{4} + 45(90) - 1{,}625$$

$$= -\frac{8{,}100}{4} + 4{,}050 - 1{,}625$$

$$= -2{,}025 + 4{,}050 - 1{,}625$$

$$= \boxed{\$400 \text{ profit per TV set}}$$

(b) To find the number of sets to be produced to make a profit of $175 per set, we set $P(x) = 175$ and find x by solving the quadratic equation:

$$175 = -\frac{x^2}{4} + 45x - 1{,}625 \qquad \textit{Multiply both sides of the equation by 4.}$$

$$700 = -x^2 + 180x - 6{,}500 \qquad \textit{Put in standard form.}$$

$$0 = x^2 - 180x + 7{,}200 \qquad \textit{Solve by factoring.}$$

$$0 = (x - 60)(x - 120)$$

$$x = 60 \qquad \text{or} \qquad x = 120$$

Thus, for the company to make a profit of $175 per set, it must produce either 60 TV sets or 120 TV sets daily .

EXAMPLE 4

The rate of a stream is 4 mph. Meri rowed her boat downstream (with the current) a distance of 9 miles and then back upstream (against the current) to her starting point. If the complete trip took 10 hours, what was her rate in still water?

Solution   First we must realize that when Meri is traveling downstream, the net rate (the rate her boat is traveling relative to the land) is equal to the rate she can row in still water plus the rate of the stream. That is,

$$\text{Net rate downstream} = r_{\text{rowing}} + r_{\text{stream}}$$

On the other hand, the net rate upstream (the rate the boat travels upstream relative to the land) is the difference of the rate rowing in still water and the rate of the stream:

$$\text{Net rate upstream} = r_{\text{rowing}} - r_{\text{stream}}$$

Let's call Meri's rate rowing r. Then, since the rate of the stream is 4 mph we have

$$\text{Net rate downstream} = r + 4$$

$$\text{Net rate upstream} = r - 4$$

Since distance = (rate)(time), the *time* it takes for her to go downstream is

$$\text{Time downstream} = \frac{\text{Distance}}{\text{Net rate downstream}} = \frac{9}{r + 4}$$

The *time* it takes for her to go upstream is

$$\text{Time upstream} = \frac{\text{Distance}}{\text{Net rate upstream}} = \frac{9}{r - 4}$$

Since the total time for the round trip is 10 hours, we have

$$10 = \frac{9}{r + 4} + \frac{9}{r - 4}$$

We solve for r. First multiply both sides of the equation by $(r + 4)(r - 4)$:

$$(r + 4)(r - 4) \cdot 10 = \left(\frac{9}{r + 4} + \frac{9}{r - 4} \right)(r + 4)(r - 4)$$

$$10(r + 4)(r - 4) = 9(r - 4) + 9(r + 4)$$

$$10(r^2 - 16) = 9r - 36 + 9r + 36$$

$$10r^2 - 160 = 18r$$

$$10r^2 - 18r - 160 = 0$$

$$2(5r^2 - 9r - 80) = 0 \qquad \textit{Divide both sides by 2.}$$

$$5r^2 - 9r - 80 = 0$$

$$(5r + 16)(r - 5) = 0$$

$$r = -\frac{16}{5} \quad \text{or} \quad r = 5 \qquad \textit{We eliminate the negative value.}$$

Thus, her rate in still water is $\boxed{5 \text{ mph}}$.

Check:

Going downstream takes her $\dfrac{9}{r + 4} = \dfrac{9}{5 + 4} = \dfrac{9}{9} = 1$ hour.

Going upstream takes her $\dfrac{9}{r - 4} = \dfrac{9}{5 - 4} = \dfrac{9}{1} = 9$ hours.

Total time $= 1 + 9 \overset{\checkmark}{=} 10$ hours.

The Square Root Method

The solutions to the equation $x^2 = d$ are numbers that, when squared, yield d. In Chapter 7 we defined $\sqrt{d}$ as the nonnegative expression that, when squared, yields d. However, since we have no information as to whether x is positive or negative, we must take into account the negative square root as well. For example, in solving $x^2 = 4$, since $(+2)^2 = 4$ and $(-2)^2 = (-2)(-2) = 4$, both $+2$ *and* -2 are solutions. Hence, if $x^2 = d$, then $x = +\sqrt{d}$ or $x = -\sqrt{d}$.

A shorter way to write a quantity and its opposite would be to write "$\pm$" in front of the quantity to symbolize both answers. Thus, the answers to $x^2 = 4$ can be written as $x = \pm 2$ rather than $x = +2$ or $x = -2$.

THEOREM 11.1	If $x^2 = d$, then $x = \pm\sqrt{d}$.

The square root method of solving quadratic equations is used mainly when there is no x term in the standard form of the quadratic equation—that is, when $B = 0$ in $Ax^2 + Bx + C = 0$.

The square root method requires us to isolate x^2 on one side of the equation and then apply the theorem. When we use this theorem we will say that we are *taking square roots*. For example:

Solve for x if $x^2 + 5 = 8$.

We proceed as follows:

$$x^2 + 5 = 8 \qquad \textit{Isolate } x^2.$$

$$x^2 = 3 \qquad \textit{Take square roots.}$$

$$x = \pm\sqrt{3} \qquad \textit{This means that } x = +\sqrt{3} \textit{ or } x = -\sqrt{3}.$$

EXAMPLE 5	Solve each of the following equations.

(a) $x^2 + 2 = 5 - 2x^2$ **(b)** $2x^2 + 15 = 7$

Solution **(a)** $x^2 + 2 = 5 - 2x^2$ *First isolate x^2 on one side of the equation.*

$$3x^2 = 3$$

$$x^2 = 1 \qquad \textit{Then take the square roots.}$$

$$x = \pm\sqrt{1} \qquad \textit{Do not forget the negative root.}$$

$$\boxed{x = \pm 1}$$

(b) $2x^2 + 15 = 7 \qquad \textit{Isolate } x^2.$

$$2x^2 = -8$$

$$x^2 = \frac{-8}{2} = -4 \qquad \textit{Take the square roots.}$$

$$x = \pm\sqrt{-4}$$

This means there are no *real* solutions, but if we use complex numbers, then

$$x = \pm\sqrt{-4} = \pm i\sqrt{4} = \boxed{\pm 2i}$$

Let's check the solutions:

Check: $x = +2i$:	**Check:** $x = -2i$:
$2x^2 + 15 = 7$	$2x^2 + 15 = 7$
$2(2i)^2 + 15 \overset{?}{=} 7$	$2(-2i)^2 + 15 \overset{?}{=} 7$
$2(4i^2) + 15 \overset{?}{=} 7$	$2(4i)^2 + 15 \overset{?}{=} 7$
$8i^2 + 15 \overset{?}{=} 7$	$8i^2 + 15 \overset{?}{=} 7$
$-8 + 15 \overset{\checkmark}{=} 7 \quad \textit{Since } i^2 = -1$	$-8 + 15 \overset{\checkmark}{=} 7$

EXAMPLE 6

If $f(x) = 3x^2 - 2$, find the value of x for which $f(x) = 6$.

Solution

Since $f(x) = 3x^2 - 2$, we are in effect being asked to solve $3x^2 - 2 = 6$.

$$f(x) = 6 \qquad \textit{Substitute } f(x) = 3x^2 - 2 \text{ to get}$$

$$3x^2 - 2 = 6 \qquad \textit{Isolate } x^2 \text{ (add +2 to both sides of the equation).}$$

$$3x^2 = 8 \qquad \textit{Divide both sides of the equation by 3.}$$

$$x^2 = \frac{8}{3} \qquad \textit{Take square roots.}$$

$$x = \pm\sqrt{\frac{8}{3}} \qquad \textit{Simplify your answers: rationalize the denominator.}$$

If you are asked to estimate $\sqrt{\dfrac{8}{3}}$ using a calculator, you do not need to simplify first.

$$x = \frac{\pm\sqrt{8}}{\sqrt{3}} = \frac{\pm\sqrt{8}\sqrt{3}}{\sqrt{3}\sqrt{3}} = \frac{\pm\sqrt{24}}{3} = \frac{\pm 2\sqrt{6}}{3}$$

Recall that this means the solutions are $\boxed{\dfrac{2\sqrt{6}}{3} \text{ and } \dfrac{-2\sqrt{6}}{3}}$.

We can solve literal quadratic equations by the same methods.

EXAMPLE 7

Solve the following explicitly for x: $\quad 9x^2 + a = b$

Solution

$$9x^2 + a = b \qquad \textit{Isolate } x^2.$$

$$9x^2 = b - a$$

$$\frac{9x^2}{9} = \frac{b - a}{9}$$

$$x^2 = \frac{b - a}{9} \qquad \textit{Take the square roots.}$$

$$x = \pm \sqrt{\frac{b - a}{9}} \qquad \textit{Simplify.}$$

$$\boxed{x = \pm \frac{\sqrt{b - a}}{3}}$$

We mentioned that the square root method is used mainly when the standard form of a quadratic equation has no first-degree term. The square root method can also be used immediately if we have the square of a binomial equal to a constant. For example, consider solving the following equation for *x:*

$$(x - 5)^2 = 7$$

If we tried to solve this equation by first putting it in standard form, we would square $x - 5$ to get $x^2 - 10x + 25 = 7$, which becomes

$$x^2 - 10x + 18 = 0$$

Note that $x^2 - 10x + 18$ does not factor (with integer coefficients), so we do not use the factoring method. In addition, we cannot use the square root method on the equation in standard form since there is a first-degree term.

We can, however, take advantage of the given form of the equation and "take square roots" as our *first* step in solving the equation.

If $(x - 5)^2 = 7$ then $x - 5$ must be either $\sqrt{7}$ or $-\sqrt{7}$.

$$(x - 5)^2 = 7 \qquad \textit{Take square roots.}$$
$$x - 5 = \pm\sqrt{7} \qquad \textit{Now isolate x by adding 5 to both sides of the equation.}$$
$$x = 5 \pm\sqrt{7}$$

The two solutions are $5 + \sqrt{7}$ and $5 - \sqrt{7}$.

EXAMPLE 8   Solve for *x* if $(x - 3)^2 = 4$.

Solution
$$(x - 3)^2 = 4 \qquad \textit{Take square roots.}$$
$$(x - 3) = \pm\sqrt{4}$$
$$x - 3 = \pm 2 \qquad \textit{Then add 3 to both sides of the equation.}$$
$$x = \pm 2 + 3$$

Then,

$$x = 2 + 3 = 5 \quad \text{or} \quad x = -2 + 3 = 1$$

Hence, $\boxed{x = 5 \quad \text{or} \quad x = 1}$.

The Pythagorean Theorem

A *right triangle* is a triangle with a right (90°) angle. The sides forming the right angle in a right triangle are called the *legs* of the right triangle. The side opposite the right angle is called the *hypotenuse* (see Figure 8.7).

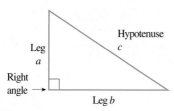

Figure 8.7
Right triangle

| The Pythagorean Theorem | Given a right triangle with legs of length a and b and hypotenuse c, we have the following relationship: |

$$a^2 + b^2 = c^2$$

In other words, the sum of the squares of the legs of a right triangle is equal to the square of the hypotenuse. This is called the ***Pythagorean theorem.*** The converse is also true; that is, if the sum of the squares of two sides of a triangle is equal to the square of the third side, then the triangle is a right triangle.

EXAMPLE 9

Given a right triangle with one leg equal to 9 in. and the hypotenuse equal to 15 in., find the other leg.

Solution

Let $a = 9$ in. and $c = 15$ in. (the hypotenuse) and use the Pythagorean theorem to find the other leg (see Figure 8.8).

$$a^2 + b^2 = c^2$$
$$9^2 + b^2 = 15^2$$
$$81 + b^2 = 225$$
$$b^2 = 144$$
$$b = \pm\sqrt{144} = \pm 12 \qquad \textit{Eliminate the negative answer.}$$

Figure 8.8

Thus, the other leg is $\boxed{12 \text{ in.}}$

EXAMPLE 10

Find the side of a square with diagonal 10 in.

Solution

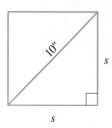

We draw a diagram and label the diagonal as shown in Figure 8.9.

A square has all right angles and so its diagonal will cut the square into two right triangles. Since a square has equal sides, the legs of the right triangles will be equal. The hypotenuse is 10 in. since the hypotenuse of the right triangle is the diagonal of the square.

Using the Pythagorean theorem with each leg equal to s and hypotenuse equal to 10 in., we have

$$s^2 + s^2 = 10^2$$
$$2s^2 = 100$$
$$s^2 = \frac{100}{2} = 50$$
$$s = \pm\sqrt{50} = \pm 5\sqrt{2}$$

Figure 8.9

The length of the sides of the square is $\boxed{5\sqrt{2} \text{ in.} \approx 7.07 \text{ in.}}$

EXAMPLE 11   Two cars start at the same place: one car travels due north at a rate of 40 mph, and the other car travels due east at a rate of 50 mph. How long will it take them to be 100 miles apart?

Solution   Let's analyze this carefully.

THINKING OUT LOUD

What do we need to find?

How much time it will take for the cars to be 100 miles apart.

What information are we given?

We are given that one car travels due north at 40 mph, while the other car travels due east at 50 mph.

How is the given information related to what we need to find?

We first draw a diagram and note that the cars are traveling at right angles to each other, and hence we have a right triangle.

How do we label the diagram? We are not given the lengths of the sides.

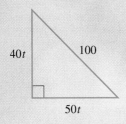

We want to know when the cars are 100 miles apart, so we can label the hypotenuse 100 miles. We are not given the distances the cars traveled, but we are given that the car traveling north is traveling at 40 mph. If we let t = number of hours traveled since they started, then using $d = rt$, the car traveling north would have traveled $40t$ miles. Likewise the car traveling east would have traveled $50t$ miles. We can now label the other two sides of the right triangle as given in the accompanying figure.

Hence, by the Pythagorean theorem, we get

$$(40t)^2 + (50t)^2 = 100^2 \qquad \textit{Simplify. Note we square each entire term: 40t and 50t.}$$
$$1{,}600t^2 + 2{,}500t^2 = 10{,}000$$
$$4{,}100t^2 = 10{,}000 \qquad \textit{Solve for t.}$$
$$t^2 = \frac{10{,}000}{4{,}100} = \frac{100}{41}$$
$$t = \pm\sqrt{\frac{100}{41}} = \pm\frac{10\sqrt{41}}{41}$$

Hence, the time it takes for the cars to be 100 miles apart is

$$t = \frac{10\sqrt{41}}{41} \approx \boxed{1.56 \text{ hours}}.$$

EXERCISES 8.2

In Exercises 1–58, *solve the equation using the factoring or square root method where appropriate.*

1. $(x + 2)(x - 3) = 0$

2. $(a - 5)(a + 7) = 0$

3. $x^2 = 25$

4. $a^2 = 81$

5. $x(x - 4) = 0$

6. $y(y + 3) = 0$

7. $12 = x(x - 4)$

8. $15 = y(y + 2)$

9. $5y(y - 7) = 0$

10. $3x(x + 2) = 0$

11. $x^2 - 16 = 0$

12. $a^2 - 225 = 0$

13. $0 = 9c^2 - 16$

14. $0 = 25x^2 - 4$

15. $8a^2 - 18 = 0$

16. $2x^2 = 32$

17. $0 = x^2 - x - 6$

18. $0 = a^2 + 2a - 8$

19. $2y^2 - 3y + 1 = 0$

20. $5a^2 - 14a - 3 = 0$

21. If $f(x) = 4x^2 - 24$, solve the equation $f(x) = 0$.

22. If $g(x) = 5x^2$, solve the equation $g(x) = 65$.

23. If $g(x) = 7x^2$, find the value(s) of x for which $g(x) = 19$.

24. If $f(x) = 6x^2 - 25$, find the value(s) of x for which $f(x) = 0$.

25. $10 = x^2 - 3x$

26. $1 - 2x = 3x^2$

27. $6a^2 + 3a - 1 = 4a$

28. $9a^2 + 4 = 12a^2$

29. $3y^2 - 5y + 8 = 9y^2 - 10y + 2$

30. $5y^2 + 15y + 4 = 2y + 10$

31. $0 = 3x^2 + 5$

32. $0 = 2x^2 + 7$

33. $(2s - 3)(3s + 1) = 7$

34. $x(x - 2) = -1$

35. $(x + 2)^2 = 25$

36. $(y - 15)^2 = 36$

37. $(x - 2)(3x - 1) = (2x - 3)(x + 1)$

38. $(x + 5)(2x - 3) = (x - 1)(3x + 1)$

39. $(x + 3)^2 = x(x + 5)$

40. $(2x + 1)^2 = 4x(x - 3)$

41. $x^2 + 3 = 1$

42. $x^2 - 7 = -12$

43. $8 = (x - 8)^2$

44. $12 = (x + 12)^2$

45. $(y - 5)^2 = -16$

46. $(y + 3)^2 = -4$

47. $\dfrac{2}{x - 1} + x = 4$

48. $\dfrac{6}{x - 3} + x = 8$

49. $2a - 5 = \dfrac{3(a + 2)}{a + 4}$

50. $3a - 4 = \dfrac{a + 2}{a}$

51. $\dfrac{x}{x + 2} - \dfrac{3}{x} = \dfrac{x + 1}{x}$

52. $\dfrac{x}{x - 1} + \dfrac{4}{x} = \dfrac{x + 2}{x}$

53. $\dfrac{1}{x} + x = 2$

54. $2x - \dfrac{5}{x} = 9$

55. $\dfrac{3}{x - 2} + \dfrac{7}{x + 2} = \dfrac{x + 1}{x - 2}$

56. $\dfrac{3}{a + 4} + \dfrac{2}{a - 2} = \dfrac{4a + 5}{a + 4}$

57. $\dfrac{2}{x-1} + \dfrac{3x}{x+2} = \dfrac{2(5x+9)}{x^2+x-2}$

58. $\dfrac{3}{y-1} + \dfrac{2}{y+1} = \dfrac{2y+3}{3(y-1)}$

In Exercises 59–68, solve explicitly for the given variable using the factoring or square root method.

59. $8a^2 + 3b = 5b$ for a

60. $5r^2 - 8a = 2a - 3r^2$ for r

61. $5x^2 + 7y^2 = 9$ for y

62. $14a^2 + 3b^2 = c$ for b

63. $V = \dfrac{2}{3}\pi r^2$ for $r > 0$

64. $K = \dfrac{2gm}{s^2}$ for $s > 0$

65. $a^2 - 4b^2 = 0$ for a

66. $x^2 - 9y^2 = 0$ for x

67. $x^2 - xy - 6y^2 = 0$ for x

68. $a^2 + 3ab - 10b^2 = 0$ for a

Solve each of the following problems algebraically.

69. Five less than the square of a positive number is 1 more than 5 times the number. Find the number.

70. Seven more than the square of a positive number is 1 less than 6 times the number. Find the number(s).

71. The sum of the squares of two positive numbers is 68. If one of the numbers is 8, find the other number.

72. The sum of the squares of two negative numbers is 30. If one of the numbers is -5, find the other number.

73. Find the numbers such that the square of the sum of the number and 6 is 169.

74. Find the numbers such that the square of the sum of the number and 3 is 100.

75. The sum of a number and its reciprocal is $\frac{13}{6}$. Find the number(s).

76. The sum of a number and twice its reciprocal is 3. Find the number(s).

77. The profit in dollars (P) made on a concert is related to the price in dollars (d) of a ticket in the following way:

$$P(d) = 10{,}000(-d^2 + 12d - 35)$$

(a) How much profit is made by selling tickets for $5?

(b) What must the price of a ticket be to make a profit of $10,000?

78. The daily cost in dollars (C) of producing items in a factory is related to the number of items (x) produced in the following way:

$$C(x) = -\dfrac{x^2}{10} + 100x - 24{,}000$$

(a) How much does it cost to produce 450 items?

(b) How many items must be produced so that the daily cost is $1,000?

79. A man jumps off a diving board into a pool. The equation

$$h(t) = -16t^2 + 40$$

gives the height h (in feet) the man is above the pool t seconds after he jumps.

(a) How high above the pool is the diver after the first second?

(b) How long does it take for him to hit the water?

(c) How high is the diving board? [*Hint:* Let $t = 0$.]

80. Alex throws a ball straight up into the air. The equation

$$h(t) = -16t^2 + 80t + 44$$

gives the height h (in feet) the ball is above the ground t seconds after he throws it.

(a) How high is the ball at $t = 2$ seconds?

(b) How long does it take for the ball to hit the ground?

81. Find the dimensions of a rectangle whose area is 80 sq ft and whose length is 2 ft more than its width.

82. Find the dimensions of a rectangle whose area is 108 sq ft and whose length is 3 ft more than its width.

83. Find the dimensions of a square with area 60 sq in.

84. Find the dimensions of a square with area 75 sq in.

85. The product of two numbers is 120. What are the numbers if one number is 2 less than the other?

86. The sum of two numbers is 20. Find the numbers if their product is 96.

87. The product of two numbers is 85. What are the numbers if one number is 2 more than 3 times the other?

88. The sum of two numbers is 25. Find the numbers if their product is 144.

89. In a right triangle, a leg is 4″ and the hypotenuse is 7″. Find the length of the other leg.

90. A leg of a right triangle is 7″ and the hypotenuse is 15″. Find the length of the other leg.

91. Find the lengths of the sides of the right triangle shown in the accompanying figure.

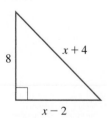

92. The lengths of the sides of the right triangle are shown in the figure below. Find x.

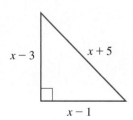

93. Find the length of the diagonals of a 7″ by 12″ rectangle.

94. Find the length of the diagonals of a 4″ by 5″ rectangle.

95. Harold leans a 30′ ladder against a building. If the base of the ladder is 8′ from the building, how high up the building does the ladder reach? (See the accompanying figure on page 475.)

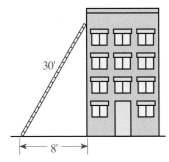

96. Harold leans a 30′ ladder against a building. If the base of the ladder is 4′ from the building, how high up the building does the ladder reach?

97. Find the area of a square if its diagonal is 8″.

98. Find the area of a square if its diagonal is 9″.

99. Find the area of a rectangle with diagonal 12″ and width 5″.

100. Find the area of a rectangle with diagonal 15″ and width 3″.

101. Two cars start at the same time and place: one car travels due west at a rate of 50 mph, and the other car travels due south at a rate of 60 mph. How long will it take them to be 160 miles apart?

102. Two cars start at the same time and place: one car travels due west at a rate of 35 mph, and the other car travels due south at a rate of 45 mph. How long will it take them to be 90 miles apart?

103. At 3 P.M., two trains leave the same station: one train travels due north at a rate of 90 mph, and the other travels due west at a rate of 110 mph. At what time will the trains be 500 miles apart?

104. At 7 P.M., two trains leave the same location: one train travels due south at a rate of 80 mph, while the other travels due east at a rate of 60 mph. At what time will the two cars be 200 miles apart?

105. Find the value(s) of n so that the line passing through the points $(1, n)$ and $(3, n^2)$ is parallel to the line passing through the points $(-6, 0)$ and $(-5, 6)$.

106. Find the value(s) of a so that the line passing through the points $(-1, a)$ and $(2, a^2)$ is perpendicular to the line passing through the points $(0, 2)$ and $(2, 1)$.

107. A rectangular picture has a 1-in. (uniform-width) frame as its border. If the length of the picture is twice its width and the total area of the picture and frame is 60 sq in., find the dimensions of the picture.

108. A 7″ by 10″ rectangular picture has a frame of uniform width. If the area of the frame is 60 sq in., how wide is the frame?

109. A boat travels upstream for 20 km against a current of 5 kph, then it travels downstream for 10 km with the same current. If the 30-km trip took $1\frac{1}{3}$ hours, how fast was the boat traveling in still water?

110. An airplane heads into the wind to get to its destination 200 miles away and returns traveling with the wind to its starting point. If the wind currents were at a constant 20 mph and the total round trip took $2\frac{1}{4}$ hours, find the speed of the airplane.

? Questions for Thought

111. On what property of real numbers is the factoring method based?

112. Discuss what is **wrong** (and why) with the solutions to the following equations:

    (a) *Problem:* Solve $(x - 3)(x - 4) = 7$.

        Solution: $x - 3 = 7$ or $x - 4 = 7$

                    $x = 10$ or $x = 11$

    (b) *Problem:* Solve $3x(x - 2) = 0$.

        Solution: $x = 3$, $x = 0$,   or $x = 2$

    (c) *Problem:* Solve $(x - 4) + (x + 3) = 0$.

        Solution: $x = 4$ or $x = -3$

113. Discuss the differences between the solutions of

$$x^2 = 36 \quad \text{and} \quad x = \sqrt{36}$$

114. Solve each of the following by the factoring method *and* the square root method.

    (a) $x^2 = 9$ **(b)** $9x^2 - 16 = 0$

    (c) $7x^2 - 5 = 3x^2 + 4$

115. Look at the solution of the following problem. See whether you can justify each step.

    Problem: $x^2 + 4x - 7 = 0$

    Solution: 1. $x^2 + 4x - 7 = 0$

                   2. $x^2 + 4x = 7$

                   3. $x^2 + 4x + 4 = 7 + 4$

                   4. $x^2 + 4x + 4 = 11$

                   5. $(x + 2)^2 = 11$

                   6. $x + 2 = \pm\sqrt{11}$

                   7. $x = -2 \pm \sqrt{11}$

116. Check that $2 + \sqrt{3}$ is a solution to the equation $x^2 - 4x + 1 = 0$ by substituting $2 + \sqrt{3}$ for x in the equation. Check to see whether $2 - \sqrt{3}$ is also a solution of the same equation.

8.3   Solving Quadratic Equations: Completing the Square

Thus far we have discussed the factoring and square root methods for solving quadratic equations algebraically. However, those methods cannot be applied to an equation such as $x^2 + 3x - 5 = 0$ because the expression $x^2 + 3x - 5$ is not factorable using integer coefficients and is not in the form appropriate to the square root method (there *is* a first-degree term).

Our interest in this section is to find a method for solving quadratic equations that can be applied to all cases.

If we can take any equation and put it in the form $(x + p)^2 = d$, where p and d are constants, then all that remains is to apply the square root method—that is, take square roots, solve for x, and simplify the answer. But can all quadratic equations be put in the form $(x + p)^2 = d$, where p and d are constants?

To answer this, we first examine the squares of binomials of the form $(x + p)$, where p is a constant. When the squares are multiplied out, we call them **perfect squares.**

First we will square binomials of the form $x + p$ and look at the relationship between the x coefficient and the numerical term. Observe the following:

	Coefficient of the x term	The numerical term
$(x - 3)^2 = x^2 - 6x + 9$	-6	$+9$
$(x + 5)^2 = x^2 + 10x + 25$	$+10$	$+25$
$(x + 4)^2 = x^2 + 8x + 16$	$+8$	$+16$

Let's examine what happens when we square $(x + p)$:

Square of second term in the binomial

$$(x + p)^2 = x^2 + 2px + p^2$$

Square of first term in the binomial — — *Twice the product of the terms in the binomial*

Given that p is a constant and x is a variable, the middle (first-degree) term *coefficient* will be $2p$, and the numerical term will be p^2. What is the relationship between the middle-term coefficient, $2p$, and the numerical term, p^2?

If you take half of $2p$ and square it, you will get p^2:

$$\left[\frac{1}{2}(2p)\right]^2 = \left(\frac{2p}{2}\right)^2 = p^2$$

The square of half of $2p$ is p^2

Thus, if you square the binomial $(x + p)$, *the square of one-half the middle-term coefficient will yield the numerical term.*

We now return to the examples above:

$(x - 3)^2 = x^2 - 6x + 9$

1. Take half of -6 to get -3.
2. Square -3 to get the numerical term, $+9$.

$(x + 5)^2 = x^2 + 10x + 25$

1. Take half of 10 to get 5.
2. Square 5 to get the numerical term, 25.

Now we will demonstrate how we can take any second-degree equation and put it in the form $(x + p)^2 = d$, where p and d are constants. Suppose we have an equation such as

$$x^2 + 6x - 8 = 0$$

To make it clearer how to make the left side into a perfect square, we first add 8 to both sides of the equation:

$$x^2 + 6x - 8 = 0 \qquad \text{\textit{Add 8 to both sides of the equation.}}$$
$$x^2 + 6x \quad\;\; = 8$$

Now what is missing to make the left side a perfect square? Take half the middle-term coefficient, 6, and square it:

$$\left[\frac{1}{2}(6)\right]^2 = (3)^2 = 9$$

Thus, 9 must be added to make the left side a perfect square. But since we are dealing with equations, we must add 9 to *both sides* of the equation:

$$x^2 + 6x + 9 = 8 + 9$$
$$x^2 + 6x + 9 = 17 \qquad \text{\textit{Write the left side as a perfect square}}$$
$$(x + 3)^2 = 17 \qquad\quad \text{\textit{in factored form.}}$$

The only difference between the equation $x^2 + 6x = 8$ and the perfect square version $x^2 + 6x + 9 = 17$ is that 9 was added to both sides. Why 9? To make the left side a perfect square so that it could be written in factored form. This process of adding a number to make a perfect square is called **completing the square.**

Now we can solve this equation by the square root method, as demonstrated in Section 8.2.

$$(x + 3)^2 = 17 \qquad \textit{Take square roots.}$$

$$x + 3 = \pm\sqrt{17} \qquad \textit{Then isolate x by adding -3 to both sides of the equation.}$$

$$x = -3 \pm\sqrt{17}$$

Similarly, *any* quadratic equation can be put in the form $(x + p)^2 = d$, where p and d are constants.

EXAMPLE 1

Solve the following by completing the square: $x^2 + 8x - 4 = 0$

Solution

1. Add $+4$ to both sides of the equation. $x^2 + 8x \qquad = 4$

2. Take one-half the middle-term coefficient and square it: $[\frac{1}{2}(8)]^2 = 4^2 = 16$.

3. Add result of step 2 to both sides of the equation. $x^2 + 8x \boxed{+ 16} = 4 \boxed{+ 16}$

   $$x^2 + 8x + 16 = 20$$

4. Factor the left side as a perfect square. $(x + 4)^2 = 20$

5. Take square roots. $x + 4 = \pm\sqrt{20}$

6. Isolate x by adding -4 to both sides of the equation. $x = -4 \pm \sqrt{20}$

7. Simplify the radical. $x = -4 \pm 2\sqrt{5}$

The solutions are $\boxed{-4 + 2\sqrt{5} \text{ and } -4 - 2\sqrt{5}}$.

EXAMPLE 2

Solve $2x^2 - 6x + 5 = 0$ by completing the square.

Solution

Keep in mind that the number we add to a quadratic expression to make it a perfect square is found by exploiting the relationship between coefficients in $x^2 + 2px + p^2$, when x^2 has a coefficient of 1. However, in this example, the coefficient of x^2 is not 1. Thus, the relationship between the x-term coefficient and the numerical coefficient (being the square of half the x-term coefficient) does not apply yet. We must first divide both sides of the equation by 2, so that the leading (highest-degree) term has a coefficient of 1.

1. Divide both sides of the equation by the leading-term coefficient, 2. $\dfrac{2x^2}{2} - \dfrac{6x}{2} + \dfrac{5}{2} = \dfrac{0}{2}$

   $$x^2 - 3x + \frac{5}{2} = 0$$

2. Add $-\frac{5}{2}$ to both sides of equation. $x^2 - 3x \qquad = -\dfrac{5}{2}$

3. Take half the middle-term coefficient, square it, and add the result to both sides of the equation.

$$\left[\frac{1}{2}(-3)\right]^2 = \left(\frac{-3}{2}\right)^2 = \frac{9}{4} \qquad x^2 - 3x \;\boxed{+ \frac{9}{4}}\; = -\frac{5}{2} \;\boxed{+ \frac{9}{4}}$$

4. Factor the left side as a perfect square and simplify the right side.

$$\left(x - \frac{3}{2}\right)^2 = -\frac{10}{4} + \frac{9}{4} = -\frac{1}{4}$$

5. Take square roots.

$$x - \frac{3}{2} = \pm\sqrt{-\frac{1}{4}}$$

This equation has *no real solutions,* but if we use complex numbers,

6. Isolate x by adding $\frac{3}{2}$ to both sides.

$$x = \frac{3}{2} \pm \sqrt{-\frac{1}{4}}$$

7. Simplify.

$$\boxed{x = \frac{3}{2} \pm \frac{i}{2}}$$

In general, we solve quadratic equations by completing the square as indicated in the following example:

Solve: $3x^2 - 5 = -12x$

1. Simplify to standard quadratic form.

$$3x^2 + 12x - 5 = 0$$

2. Divide both sides of the equation by the leading coefficient (if not 1).

$$\frac{3x^2}{3} + \frac{12x}{3} - \frac{5}{3} = \frac{0}{3}$$

$$x^2 + 4x - \frac{5}{3} = 0$$

3. Put the numerical term on the other side of the equation by adding its opposite to both sides of the equation.

$$x^2 + 4x \;= \frac{5}{3}$$

4. Take half the middle-term coefficient, square it, and add this result to both sides of the equation.

$$\left[\frac{1}{2}(4)\right]^2 = 2^2 = 4 \qquad\qquad x^2 + 4x \;\boxed{+ 4}\; = \frac{5}{3} \;\boxed{+ 4}$$

*Add **4** to both sides of the equation.*

5. Factor the left side as a perfect square and simplify the right side.

$$(x + 2)^2 = \frac{5}{3} + \frac{12}{3} = \frac{17}{3}$$

6. Take square roots.

$$x + 2 = \pm\sqrt{\frac{17}{3}}$$

7. Isolate x.

$$x = -2 \pm \sqrt{\frac{17}{3}}$$

8. Express your answer in simplest radical form.

$$x = -2 \pm \frac{\sqrt{17}}{\sqrt{3}} \qquad \textit{Rationalize the denominator.}$$

$$= -2 \pm \frac{\sqrt{17}\sqrt{3}}{\sqrt{3}\sqrt{3}} = \boxed{-2 \pm \frac{\sqrt{51}}{3} = \frac{-6 \pm \sqrt{51}}{3}}$$

EXAMPLE 3

Solve the following by completing the square: $\quad 2x^2 + 3x + 5 = 7$

Solution

$2x^2 + 3x + 5 = 7$ *Put in standard form.*

$2x^2 + 3x - 2 = 0$ *Divide both sides by the leading-term coefficient, 2.*

$x^2 + \dfrac{3}{2}x - 1 = 0$ *Isolate the constant term on the right-hand side.*

$x^2 + \dfrac{3}{2}x \quad\quad = 1$ *Take half the middle term coefficient, square it and add it to both sides of the equation.*

$\left[\dfrac{1}{2}\left(\dfrac{3}{2}\right)\right]^2 = \dfrac{9}{16}$

$x^2 + \dfrac{3}{2}x + \dfrac{9}{16} = 1 + \dfrac{9}{16}$ *Write the left-hand side as a perfect square and simplify the right-hand side*

$\left(x + \dfrac{3}{4}\right)^2 = \dfrac{25}{16}$ *Apply the square root theorem*

$x + \dfrac{3}{4} = \pm\sqrt{\dfrac{25}{16}}$ *Isolate x*

$x = -\dfrac{3}{4} \pm \sqrt{\dfrac{25}{16}} = -\dfrac{3}{4} \pm \dfrac{5}{4} = \dfrac{-3 \pm 5}{4}$

Hence, the two solutions are $x = \dfrac{-3 - 5}{4} = \boxed{-2}$ and $x = \dfrac{-3 + 5}{4} = \boxed{\dfrac{1}{2}}$. The fact that the solutions are rational numbers $\left(-2 \text{ and } \frac{1}{2}\right)$ indicates that we could have solved this equation by factoring (Try it).

EXERCISES 8.3

Solve the following by completing the square only.

1. $x^2 - 6x - 1 = 0$

2. $s^2 - 2s - 15 = 0$

3. $0 = c^2 - 2c - 5$

4. $0 = a^2 - 2a - 4$

5. $y^2 + 5y - 2 = 0$

6. $x^2 + 3x - 2 = 0$

7. $2x^2 + 3x - 1 = x^2 - 2$

8. $y^2 - 3y - 2 = 2y - 3$

9. $(a - 2)(a + 1) = 2$

10. $(2b + 3)(b - 4) = 3b$

11. $10 = 5a^2 + 10a + 20$

12. $y + 1 = (y + 2)(y + 3)$

13. $3x^2 + 3x = x^2 - 5x + 4$

14. $3x^2 - 7x - 4 = 2x^2 - 3x + 1$

15. $(a - 2)(a + 1) = 6$

16. $(c - 3)(2c + 1) = c(c - 1)$

17. $2x - 7 = x^2 - 3x + 4$

18. $2t - 4 = t^2 - 3$

19. $5x^2 + 10x - 14 = 20$

20. $2a^2 - 4a - 14 = 0$

21. $2t^2 + 3t - 4 = 2t - 1$

22. $3s^2 + 12s + 4 = 0$

23. $(2x + 5)(x - 3) = (x + 4)(x - 1)$

24. $(3x - 2)(x - 1) = (2x - 1)(x - 4)$

25. $\dfrac{2x}{2x - 3} = \dfrac{3x - 1}{x + 1}$

26. $\dfrac{x + 1}{2x + 3} = \dfrac{3x + 2}{x + 2}$

27. $a^2 - a + 1 = 0$

28. $a^2 + a + 1 = 0$

29. $0 = 2y^2 + 2y + 5$

30. $0 = 2y^2 + 2y - 5$

31. $5n^2 - 3n = 2n^2 - 6$

32. $(3n - 1)(n - 2) = n + 8$

33. $(3t + 5)(t + 1) = (t + 4)(t + 2)$

34. $(2z - 1)(z + 3) = (3z - 2)(2z + 1)$

35. $\dfrac{3}{x + 2} - \dfrac{2}{x - 1} = 5$

36. $\dfrac{2y + 1}{y + 1} - \dfrac{2}{y + 4} = 7$

? Questions for Thought

37. Verbally state the relationship between the middle-term coefficient and the numerical term of a perfect square.

38. Solve the equation $3x^2 + 11x = 4$ by completing the square. What does the fact that the answers are rational tell you? Solve the same equation by factoring. Which method was easier?

39. Solve for x in the equation $x^2 + rx + q = 0$ by completing the square; that is, solve for x in terms of r and q.

40. Solve for x in the equation $Ax^2 + Bx + C = 0$ (where $A > 0$) by completing the square.

Mini-Review

41. *Solve for x and y:* $\begin{cases} 0.3x + 0.4y = 15 \\ 0.5x + 0.6y = 24 \end{cases}$

42. *Simplify:* $(3x^2)^3 (x^{-1}y)^4$

43. Evaluate $5n^0 + n^{-1} + 6n^{-2}$ for $n = 3$.

44. Write an equation of the line that crosses the x-axis at 4 and is parallel to the line whose equation is $7x - 2y = 12$.

8.4 Solving Quadratic Equations: The Quadratic Formula

Completing the square is a useful algebraic technique that will be needed elsewhere in intermediate algebra, as well as in precalculus and calculus. It is the most powerful of the methods for solving quadratic equations covered so far because, unlike the previous methods, completing the square works for *all* quadratic equations. It can be, however, a tedious method.

What we would like is a method that works for all quadratic equations without the effort required in completing the square. Algebraically, we can derive a formula that will allow us to produce the solutions to the "general" quadratic equation.

To derive the formula, we start with the general equation $Ax^2 + Bx + C = 0$ and solve it for x by the method of completing the square. (We start by assuming $A > 0$.)

Start with the equation $Ax^2 + Bx + C = 0$.

1. Divide both sides of the equation by A.

$$\frac{Ax^2}{A} + \frac{Bx}{A} + \frac{C}{A} = \frac{0}{A}$$

$$x^2 + \frac{Bx}{A} + \frac{C}{A} = 0$$

2. Subtract $\dfrac{C}{A}$ from both sides of the equation.

$$x^2 + \frac{B}{A}x \quad = -\frac{C}{A}$$

3. Take half the middle-term coefficient, square it,

$$\left[\frac{1}{2} \cdot \frac{B}{A}\right]^2 = \left(\frac{B}{2A}\right)^2 = \frac{B^2}{4A^2}$$

and add the result to both sides of the equation.

$$x^2 + \frac{B}{A}x + \frac{B^2}{4A^2} = \frac{B^2}{4A^2} - \frac{C}{A}$$

4. Factor the left side as a perfect square and simplify the right side.

$$\left(x + \frac{B}{2A}\right)^2 = \frac{B^2}{4A^2} - \frac{4AC}{4A^2}$$

$$\left(x + \frac{B}{2A}\right)^2 = \frac{B^2 - 4AC}{4A^2}$$

5. Take square roots.

$$x + \frac{B}{2A} = \pm\sqrt{\frac{B^2 - 4AC}{4A^2}}$$

6. Isolate x.

$$x = -\frac{B}{2A} \pm \sqrt{\frac{B^2 - 4AC}{4A^2}}$$

7. Simplify the solution.

$$x = \frac{-B}{2A} \pm \frac{\sqrt{B^2 - 4AC}}{\sqrt{4A^2}}$$

Note: Since $A > 0$, $\sqrt{4A^2} = 2A$.

$$x = \frac{-B}{2A} \pm \frac{\sqrt{B^2 - 4AC}}{2A}$$

$$x = \frac{-B \pm \sqrt{B^2 - 4AC}}{2A}$$

This solution is known as the ***quadratic formula.*** A similar proof applies where $A < 0$.

The Quadratic Formula

The solutions to the equation $Ax^2 + Bx + C = 0$, $A \neq 0$, are given by

$$x = \frac{-B \pm \sqrt{B^2 - 4AC}}{2A}$$

As long as we can put an equation in standard quadratic form, we can identify A (the coefficient of x^2), B (the coefficient of x), and C (the constant). Once we have identified A, B, and C, we substitute the numbers into the quadratic formula and find the solutions to the equation.

EXAMPLE 1

Solve each of the following equations by using the quadratic formula.

(a) $y^2 - 4y - 3 = 0$ **(b)** $\dfrac{2}{z + 2} + \dfrac{1}{z} = 1$ **(c)** $(x + 4)(x - 1) = -8$

Solution

(a) The equation $y^2 - 4y - 3 = 0$ is already in standard form, so we can identify A, B, and C: $A = 1$ (the coefficient of y^2), $B = -4$ (the coefficient of y), and $C = -3$ (the constant).

We now find y using the quadratic formula:

$$y = \frac{-B \pm \sqrt{B^2 - 4AC}}{2A} = \frac{-(-4) \pm \sqrt{(-4)^2 - 4(1)(-3)}}{2(1)}$$

$$= \frac{4 \pm \sqrt{16 + 12}}{2}$$

$$= \frac{4 \pm \sqrt{28}}{2} = \frac{4 \pm 2\sqrt{7}}{2} = \frac{2(2 \pm \sqrt{7})}{2}$$

$$= \frac{\cancel{2}(2 \pm \sqrt{7})}{\cancel{2}} = \boxed{2 \pm \sqrt{7}}$$

Notice that we must factor the numerator first before reducing.

We will check these solutions:

Check: $y = 2 + \sqrt{7}$:

$$y^2 - 4y - 3 = 0$$
$$(2 + \sqrt{7})^2 - 4(2 + \sqrt{7}) - 3 \overset{?}{=} 0$$
$$4 + 4\sqrt{7} + 7 - 4(2 + \sqrt{7}) - 3 \overset{?}{=} 0$$
$$4 + 4\sqrt{7} + 7 - 8 - 4\sqrt{7} - 3 \overset{\checkmark}{=} 0$$

Check: $y = 2 - \sqrt{7}$:

$$y^2 - 4y - 3 = 0$$
$$(2 - \sqrt{7})^2 - 4(2 - \sqrt{7}) - 3 \overset{?}{=} 0$$
$$4 - 4\sqrt{7} + 7 - 4(2 - \sqrt{7}) - 3 \overset{?}{=} 0$$
$$4 - 4\sqrt{7} + 7 - 8 + 4\sqrt{7} - 3 \overset{\checkmark}{=} 0$$

(b) $\dfrac{2}{z + 2} + \dfrac{1}{z} = 1$ *First put the equation in standard form.*
Multiply both sides by $z(z + 2)$.

$$z(z + 2)\left(\frac{2}{z + 2} + \frac{1}{z}\right) = 1 \cdot z(z + 2)$$

$$2z + (z + 2) = z(z + 2)$$
$$3z + 2 = z^2 + 2z$$
$$0 = z^2 - z - 2$$

Now identify A, B, and C:

$$A = 1, \quad B = -1, \quad C = -2$$

Solve for z using the quadratic formula:

$$z = \frac{-B \pm \sqrt{B^2 - 4AC}}{2A} = \frac{-(-1) \pm \sqrt{(-1)^2 - 4(1)(-2)}}{2(1)}$$

$$= \frac{1 \pm \sqrt{1 + 8}}{2}$$

$$= \frac{1 \pm \sqrt{9}}{2} = \frac{1 \pm 3}{2}$$

Thus,

$$z = \frac{1 + 3}{2} = \frac{4}{2} = 2 \quad \text{or} \quad z = \frac{1 - 3}{2} = \frac{-2}{2} = -1$$

The solutions are $\boxed{2 \text{ and } -1}$.   *Since the solutions are rational, we could have solved the*
equation by the factoring method. Try it for yourself.

You should check these solutions to see that both are valid.

(c) $(x + 4)(x - 1) = -8$ *We put the equation in standard form.*

$$x^2 + 3x - 4 = -8$$

$$x^2 + 3x + 4 = 0$$ *Now we identify A, B, and C.*

$$A = 1, \quad B = 3, \quad C = 4$$

Solve for x by the quadratic formula:

$$x = \frac{-B \pm \sqrt{B^2 - 4AC}}{2A} = \frac{-3 \pm \sqrt{(3)^2 - 4(1)(4)}}{2(1)}$$

$$= \frac{-3 \pm \sqrt{9 - 16}}{2}$$

$$= \frac{-3 \pm \sqrt{-7}}{2}$$

Hence, there are *no real solutions*. The complex number solutions are

$$\boxed{\dfrac{-3 + i\sqrt{7}}{2} \quad \text{and} \quad \dfrac{-3 - i\sqrt{7}}{2}}$$

Here are a few things to be careful about when you use the quadratic formula:

1. If B is a negative number, remember that $-B$ will be positive [see parts **(a)** and **(b)** of Example 1].

2. If C is negative (and A is positive), then you will end up *adding* the expressions under the radical [see parts **(a)** and **(b)** of Example 1].

3. Do not forget that $2A$ is in the denominator of the *entire expression:*

$$\frac{-B \pm \sqrt{B^2 - 4AC}}{2A}$$

The Discriminant

Let us further examine the quadratic formula; in particular, let's look at the radical portion of the quadratic formula: $\sqrt{B^2 - 4AC}$. Note that if we solve a quadratic equation and the quantity $B^2 - 4AC$ is negative [that is, if $B^2 - 4AC < 0$, as in part **(c)** of Example 1], then we have the square root of a negative number. As we discussed in Chapter 7, the square root of a negative number is not a real number. If $B^2 - 4AC < 0$, then the two solutions or roots

$$x = \frac{-B + \sqrt{B^2 - 4AC}}{2A} \qquad \text{and} \qquad x = \frac{-B - \sqrt{B^2 - 4AC}}{2A}$$

are not real.

On the other hand, if $B^2 - 4AC = 0$, then $\sqrt{B^2 - 4AC} = \sqrt{0} = 0$. Thus, the two roots

$$x = \frac{-B + \sqrt{0}}{2A} = \frac{-B}{2A} \qquad \text{and} \qquad x = \frac{-B - \sqrt{0}}{2A} = \frac{-B}{2A}$$

are equal. They are also real, since the quotient of two real numbers yields a real number.

Finally, if $B^2 - 4AC$ is positive (that is, $B^2 - 4AC > 0$), then the two roots are real and distinct (unequal):

$$x = \frac{-B + \sqrt{B^2 - 4AC}}{2A} \qquad \text{and} \qquad x = \frac{-B - \sqrt{B^2 - 4AC}}{2A}$$

We call $B^2 - 4AC$ the ***discriminant*** of the equation $Ax^2 + Bx + C = 0$. The above discussion is summarized in the box.

In Section 8.6 we will see how the discriminant relates to the graph of a quadratic function.

For the equation $Ax^2 + Bx + C = 0$ $(A \neq 0)$:

If $B^2 - 4AC < 0$, the roots are not real.

If $B^2 - 4AC = 0$, the roots are real and equal.

If $B^2 - 4AC > 0$, the roots are real and distinct.

EXAMPLE 2

Without solving the given equation, determine the nature of the roots of

$$3x^2 - 2x + 5 = 0$$

Solution   Identify A, B, and C:

$$A = 3, \quad B = -2, \quad C = 5$$

The discriminant is $B^2 - 4AC = (-2)^2 - 4(3)(5) = 4 - 60 = -56 < 0$.

Thus, $\boxed{\text{the roots are not real}}$.

Choosing a Method for Solving Quadratic Equations

In this chapter we have discussed a variety of methods for solving quadratic equations. The methods of completing the square and using the quadratic formula will work for all quadratic equations, but both methods can be a bit messy computationally. The factoring and square root methods are often easiest, but do not always work. In general, we recommend the following:

Unless the equation is in the form $(x + a)^2 = b$ or $(ax + b)(cx + d) = 0$, put the equation in the standard form $Ax^2 + Bx + C = 0$.

If there is no x term (that is, if $B = 0$), then use the square root method.

If there is an x term, try the factoring method.

If the quadratic expression does not factor easily, then use either the quadratic formula or completing the square.

EXERCISES 8.4

In Exercises 1–12, solve using the quadratic formula only.

1. $x^2 - 4x - 5 = 0$

2. $a^2 + 4a - 5 = 0$

3. $2a^2 - 3a - 4 = 0$

4. $5x^2 + x - 2 = 0$

5. $(3y - 1)(2y - 3) = y$

6. $(5s + 2)(s - 1) = 2s + 1$

7. $y^2 - 3y + 4 = 2y^2 + 4y - 3$

8. $2a^2 - 4a - 9 = 0$

9. $3a^2 + a + 2 = 0$

10. $2y^2 - 3y = -4$

11. $(s - 3)(s + 4) = (2s - 1)(s + 2)$

12. $\dfrac{2x}{x + 1} + \dfrac{1}{x} = 1$

In Exercises 13–20, use the discriminant to determine the nature of the roots.

13. $3a^2 - 2a + 5 = 0$

14. $(2z - 3)(z + 4) = z - 2$

15. $(3y + 5)(2y - 8) = (y - 4)(y + 1)$

16. $\dfrac{3a - 2}{2a + 1} = \dfrac{1}{a - 2}$

17. $2a^2 + 4a = 0$

18. $(5s + 3)(2s + 1) = s^2 - s - 1$

19. $(2y + 3)(y - 1) = y + 5$

20. $\dfrac{2}{2x + 1} + \dfrac{3x}{x - 2} = 4$

In Exercises 21–60, solve by any method. Answer to the nearest hundredth where necessary.

21. $a^2 - 3a - 4 = 0$

22. $x^2 - 5x + 4 = 0$

23. $8y^2 = 3$

24. $12y^2 = 5$

25. $(5x - 4)(2x - 3) = 0$

26. $(2z - 3)(3z + 1) = 0$

27. $(5x - 4) + (2x - 3) = 0$

28. $(2z - 3) + (3z + 1) = 0$

29. $(5x - 4)(2x - 3) = 17$

30. $(2z - 3)(3z + 1) = 17$

31. $(a - 1)(a + 2) = -2$

32. $(5x - 1)(x + 2) = 2x^2 - 3x + 10$

33. $2.4x^2 - 12.72x + 3.6 = 0$

34. $2.3x^2 - 13.11x + 8.05 = 0$

35. $x^2 - 3x + 5 = 0$

36. $3z^2 - 5z - 4 = -4z - 3$

37. $2s^2 - 5s - 12 = -5s$

38. $5a^2 - 2a + 4 = 8 - 2a$

39. $3x^2 - 2x + 9 = 2x^2 - 3x - 1$

40. $(3z - 2)(2z + 1) = 2z - 3$

41. $x^2 + 3x - 8 = x^2 - x + 11$

42. $t^2 - 6t + 1 = t^2 - 5t + 3$

43. $3x^2 + 2.7x = 14.58$

44. $x^2 + 1.1x - 2.1 = 3.2x + 2.8$

45. $3a^2 - 4a + 2 = 0$

46. $x^2 - 2x + 4 = 0$

47. $(x - 4)(2x + 3) = x^2 - 4$

48. $(a - 5)(a + 2) = (2a - 3)(a - 1)$

49. $(a + 1)(a - 3) = (3a + 1)(a - 2)$

50. $(2y + 3)(y + 5) = (y - 1)(y + 4)$

51. $(z + 3)(z - 1) = (z - 2)^2$

52. $(w - 3)^2 = (w + 4)(w - 2)$

53. $\dfrac{y}{y - 2} = \dfrac{y - 3}{y}$

54. $\dfrac{x - 4}{3} = \dfrac{2}{x + 5}$

55. $0.001x^2 - 2x = 0.1$

56. $0.35x - \dfrac{4}{x} = 0.8$

57. $\dfrac{3a}{a + 1} + \dfrac{2}{a - 2} = 5$

58. $\dfrac{3x}{2} + \dfrac{2}{x + 1} = 4$

59. $\dfrac{3}{a + 2} - \dfrac{5}{a - 2} = 2$

60. $\dfrac{2}{a + 4} - \dfrac{3}{a + 1} = 4$

61. The product of two numbers is 40. Find the numbers if the difference between the two numbers is 3.

62. The product of two numbers is 72. Find the numbers if the difference between the two numbers is 1.

63. In a right triangle, the legs are 3″ and 8″. Find the hypotenuse.

64. In a right triangle, the legs are 5″ and 7″. Find the hypotenuse.

65. Find the length of the diagonals of a square with a side equal to 5″.

66. Find the length of the diagonals of a square with a side equal to 12″.

67. Find the length of the side of a square with the diagonal equal to 8″.

68. Find the length of a rectangle if the width is 3″ and the diagonal is 18″.

69. Find the value(s) of c so that the line passing through the points $(0, 1)$ and (c, c) is perpendicular to the line passing through the points $(0, 2)$ and (c, c).

70. Find the value(s) of t so that the line passing through the points $(-2, t)$ and $(-t, 1)$ is parallel to the line passing through the points $(-4, t)$ and $(-t, 3)$.

71. A company determines that the profit, P, earned on the sale of w items is given by the formula

$$P(w) = 20w - w^2$$

Determine the number of items sold if the company loses $2,400.

72. If a line segment (called an *edge*) is drawn from each vertex of a polygon of n sides to every other vertex of the polygon, we obtain a *complete graph* on n vertices. A complete graph on 5 vertices is illustrated in the accompanying figure. The number, E, of edges in a complete graph of n vertices is given by the formula $E = \dfrac{n(n-1)}{2}$.

A complete graph on 5 vertices

   (a) How many edges are there in a complete graph on 8 vertices?

   (b) If a complete graph on a polygon has 15 edges, how many vertices are there?

73. A circular garden is surrounded by a path of uniform width. If the path has area 44π sq ft and the radius of the garden is 10 ft, find the width of the path.
[*Hint:* See Section 5.2, Example 6.]

74. A circular garden is surrounded by a path of uniform width. If the path has area 57π sq ft and the radius of the garden is 8 ft, find the width of the path.

75. A discount wallpaper store has a policy of setting a fixed price on all rolls of wallpaper in the store. During a certain week $10,000 worth of wallpaper was sold. The next week the price was reduced by $2 per roll and another $10,000 dollars worth of wallpaper was sold. If a total of 2,250 rolls were sold during the two weeks, what was the price per roll during each week?

76. The Books-R-Us discount bookstore charges the same price for all the books in the store. During a certain week $1,000 worth of books were sold. The next week the price was reduced by $2 per book and another $1,000 worth of books were sold. If a total of 1,500 books were sold during the two weeks, what was the price per book during each week?

? Questions for Thought

77. Compare and contrast the various methods for solving quadratic equations.

78. Discuss what is **wrong** (if anything) with the solutions to the following problems:

   (a) *Solve:* $x^2 - 3x + 1 = 0$.

   Solution: $x \overset{?}{=} \dfrac{-3 \pm \sqrt{9 - 4(1)}}{2} \overset{?}{=} \dfrac{-3 \pm \sqrt{5}}{2}$

(b) *Solve:* $x^2 - 5x - 3 = 0$.

   Solution: $x \overset{?}{=} \dfrac{5 \pm \sqrt{25 - 12}}{2} \overset{?}{=} \dfrac{5 \pm \sqrt{13}}{2}$

(c) *Solve:* $x^2 - 3x + 2 = 0$.

   Solution: $x \overset{?}{=} -3 \pm \dfrac{\sqrt{9 - 8}}{2} \overset{?}{=} -3 \pm \dfrac{\sqrt{1}}{2} \overset{?}{=} -3 \pm \dfrac{1}{2}$

(d) *Solve:* $x^2 - 6x - 3 = 0$.

   Solution: $x \overset{?}{=} \dfrac{6 \pm \sqrt{36 + 12}}{2} \overset{?}{=} \dfrac{6 \pm \sqrt{48}}{2} \overset{?}{=} \dfrac{6 \pm 4\sqrt{3}}{2}$

$$\overset{?}{=} \dfrac{6 \overset{2}{\pm} 4\sqrt{3}}{\cancel{2}} \overset{?}{=} 6 \pm 2\sqrt{3}$$

79. Write an equation with the following roots:

   (a) 3   and   4   **(b)** -4 and $+5$ **(c)** $\dfrac{3}{5}$ and -2 **(d)** $3i$ and $-3i$

Mini-Review

80. Given $f(x) = \dfrac{2x - 1}{1 - x}$, find and simplify $f\left(\dfrac{2}{3}\right)$.

81. *Evaluate:* $(-27)^{-2/3}$

82. *Combine and simplify:* $\dfrac{7}{x^2 + x - 12} - \dfrac{4}{x^2 + 4x} + \dfrac{3}{x^2 - 3x}$

83. A person invests a certain amount of money, part at 8% and the remainder at 10%, yielding a total of $730 in annual interest. Had the amounts invested been reversed, the annual interest would have been $710. How much was invested altogether?

 8.5   # Equations Reducible to Quadratic Form (and More Radical Equations)

Radical Equations

In Chapter 7, we solved radical equations by first isolating the radical and then squaring both sides of the equation to eliminate the radical where possible.

   In this chapter we use the same techniques to solve radical equations, except, unlike the radical equations we covered in Chapter 7, the radical equations we will solve in this section may give rise to quadratic equations.

EXAMPLE 1   Solve for x. $x - \sqrt{2x} = 0$

   Solution   Keep in mind that simply squaring both sides of an equation containing a radical does not necessarily eliminate the radical. We should isolate the radical first.

$$x - \sqrt{2x} = 0 \qquad \textit{Isolate the radical.}$$

$$x = \sqrt{2x} \qquad \textit{Square both sides of the equation.}$$

$$(x)^2 = \left(\sqrt{2x}\right)^2$$

$$x^2 = 2x \qquad \textit{Now we have a quadratic equation that we can}$$
$$\textit{solve by factoring.}$$

$$x^2 - 2x = 0$$

$$x(x - 2) = 0$$

$$x = 0 \quad \text{or} \quad x - 2 = 0$$

$$x = 2$$

As usual, we check the solutions to all radical equations to ensure that all our solutions are valid.

Check: $x = 0$: $\qquad x - \sqrt{2x} = 0$ $\qquad$ **Check:** $\quad x = 2$: $\qquad x - \sqrt{2x} = 0$

$$0 - \sqrt{2(0)} \overset{\checkmark}{=} 0 \qquad\qquad\qquad 2 - \sqrt{2(2)} \overset{?}{=} 0$$

$$2 - \sqrt{4} \overset{\checkmark}{=} 0$$

Therefore, the solutions are $\boxed{x = 0 \quad \text{and} \quad x = 2}$.

EXAMPLE 2

Solution

Solve for z: $\quad \sqrt{z + 3} + 2z = 0$

$$\sqrt{z + 3} + 2z = 0 \qquad \textit{First, isolate the radical.}$$

$$\sqrt{z + 3} = -2z \qquad \textit{Next, square both sides to eliminate the radical.}$$

$$\left(\sqrt{z + 3}\right)^2 = (-2z)^2$$

$$z + 3 = 4z^2 \qquad \textit{Now we have a quadratic equation to solve. Put it in}$$
$$\textit{standard form.}$$

$$0 = 4z^2 - z - 3 \qquad \textit{Solve for z.}$$

$$0 = (4z + 3)(z - 1)$$

$$0 = 4z + 3 \quad \text{or} \quad 0 = z - 1$$

$$z = -\frac{3}{4} \qquad \text{or} \quad z = 1$$

Now we must check both solutions in the original equation.

Check: $\quad z = 1$: $\qquad \sqrt{z + 3} + 2z = 0$

$$\sqrt{1 + 3} + 2(1) \overset{?}{=} 0$$

$$\sqrt{4} + 2 \overset{?}{=} 0$$

$$2 + 2 \neq 0$$

Thus, $z = 1$ is an extraneous root.

Check: $\quad z = -\dfrac{3}{4}$: $\qquad \sqrt{z + 3} + 2z \overset{?}{=} 0$

$$\sqrt{\frac{-3}{4} + 3} + 2\left(\frac{-3}{4}\right) \overset{?}{=} 0$$

$$\sqrt{\frac{9}{4}} + \frac{-3}{2} \overset{?}{=} 0$$

$$\frac{3}{2} + \frac{-3}{2} \overset{\checkmark}{=} 0$$

Hence, the *only* solution is $\boxed{z = -\dfrac{3}{4}}$.

When we are confronted with two radicals in an equation, the algebra is more complicated. For example, the process of squaring $\sqrt{5a} - \sqrt{2a - 1}$ can be messy. It is usually less trouble to *first isolate the more complicated radical and then square both sides of the equation,* as demonstrated in the next example.

EXAMPLE 3

Solve for a: $\sqrt{5a} - \sqrt{2a - 1} = 2$

Solution

$$\sqrt{5a} - \sqrt{2a - 1} = 2$$ *Isolate $\sqrt{2a - 1}$.*

$$\sqrt{5a} - 2 = \sqrt{2a - 1}$$ *Square both sides of the equation.*

$$\left(\sqrt{5a} - 2\right)^2 = \left(\sqrt{2a - 1}\right)^2$$

$$5a - 4\sqrt{5a} + 4 = 2a - 1$$ *Note how each side is squared differently.*
Isolate $4\sqrt{5a}$.

$$3a + 5 = 4\sqrt{5a}$$ *Square both sides of the equation again.*

$$(3a + 5)^2 = \left(4\sqrt{5a}\right)^2$$

$$9a^2 + 30a + 25 = 16(5a)$$ *Simplify.*

$$9a^2 + 30a + 25 = 80a$$

$$9a^2 - 50a + 25 = 0$$ *Solve for a by factoring.*

$$(9a - 5)(a - 5) = 0$$

$$a = \frac{5}{9} \quad \text{or} \quad a = 5$$

Check for extraneous solutions.

Check: $a = \dfrac{5}{9}$:

$$\sqrt{5a} - \sqrt{2a - 1} = 2$$

$$\sqrt{5\left(\frac{5}{9}\right)} - \sqrt{2\left(\frac{5}{9}\right) - 1} \overset{?}{=} 2$$

$$\sqrt{\frac{25}{9}} - \sqrt{\frac{10}{9} - 1} \overset{?}{=} 2$$

$$\sqrt{\frac{25}{9}} - \sqrt{\frac{1}{9}} \overset{?}{=} 2$$

$$\frac{5}{3} - \frac{1}{3} \neq 2$$

Check: $a = 5$:

$$\sqrt{5a} - \sqrt{2a - 1} = 2$$

$$\sqrt{5 \cdot 5} - \sqrt{2 \cdot 5 - 1} \overset{?}{=} 2$$

$$\sqrt{25} - \sqrt{9} \overset{?}{=} 2$$

$$5 - 3 \overset{\checkmark}{=} 2$$

Thus, $\dfrac{5}{9}$ is extraneous.

The solution is $\boxed{a = 5}$.

EXAMPLE 4

Points A and B are on opposite sides of a straight river that is 1 mile wide. Point D is 6 miles down the river on the same side as B. A cable company needs to lay cable from point A to point D. The company will lay the cable underwater from A to some point P on the opposite shore, x miles from B, and then the cable will be run above ground from P to D. (See figure on page 491.)

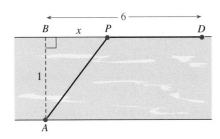

(a) If it costs $2,500 per mile for special underwater cable, but $2000 per mile for standard above-ground cable, express the total cost of cable D for the project as a function of x.

(b) Find x, the distance from B to P if the total cost of cable for this project is $13,500.

Solution   Let's analyze this problem.

THINKING OUT LOUD

What do we need to find?

We need to find C, the total cost of the cable, in terms of x.

How is C related to the information we are given?

We are given the diagram indicating where the cable is to be laid, along with the costs of each type of cable.

How are the different costs and the lengths related?

In general, the cost of cable = length of cable (in miles) $\times$ cost per mile of cable. So we have to find the lengths of each type of cable before we can find the total cost.

How do we find the length and cost of the underwater cable?

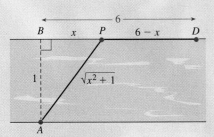

We observe that the width of the river is 1 mile, and the distance along the bank from B to P is x. We can see that side 1 and side x form the legs of a right triangle, and the length of the cable underwater is the hypotenuse of this right triangle. Using the Pythagorean theorem, this distance is $\sqrt{x^2 + 1}$. Hence, the cost of the underwater cable is $2,500\sqrt{x^2 + 1}$ dollars. (See figure.)

How do we find the length and cost of the above-ground cable?

The cable is above ground from point P to point D, which is a distance of $6 - x$ miles. Hence, the cost of the above-ground cable is $2,000(6 - x)$ dollars.

We write the equation expressing the costs of cable as follows:

$$C = 2,500\sqrt{x^2 + 1} + 2,000(6 - x)$$

(b) Now that we have found a model relating the cost of cable to the distance x, we are asked to find x, given that the cost of cable, C, is $13,500:

$$C = 2{,}500\sqrt{x^2 + 1} + 2{,}000(6 - x)$$ *Substitute C = 13,500.*

$$13{,}500 = 2{,}500\sqrt{x^2 + 1} + 2{,}000(6 - x)$$ *Divide both sides by 500.*

$$27 = 5\sqrt{x^2 + 1} + 4(6 - x)$$ *Simplify the right-hand side and isolate the radical.*

$$3 + 4x = 5\sqrt{x^2 + 1}$$ *Square each side.*

$$(3 + 4x)^2 = \left(5\sqrt{x^2 + 1}\right)^2$$ *Simplify each side.*

$$9 + 24x + 16x^2 = 25(x^2 + 1)$$ *Put into standard form.*

$$0 = 9x^2 - 24x + 16$$ *Solve for x by factoring.*

$$0 = (3x - 4)^2$$ *There is only one solution here.*

$$3x - 4 = 0$$

$$x = 4/3 = \boxed{1\frac{1}{3} \text{ miles.}}$$

We follow the same principles in solving literal equations, as demonstrated in Example 5.

EXAMPLE 5

Solve for a: $\quad \sqrt{ab + c} - b = d$

Solution

$$\sqrt{ab + c} - b = d$$ *Isolate the radical.*

$$\sqrt{ab + c} = d + b$$ *Square both sides of the equation.*

$$\left(\sqrt{ab + c}\right)^2 = (d + b)^2$$

$$ab + c = d^2 + 2bd + b^2$$ *Isolate a: Subtract c from both sides of the equation.*

$$ab = d^2 + 2bd + b^2 - c$$ *Divide both sides of the equation by b.*

$$\boxed{a = \frac{d^2 + 2bd + b^2 - c}{b}}$$

Miscellaneous Equations

As we saw in Sections 5.6 and 8.2, we can apply the factoring method to any higher-degree equation that can be factored into first-degree factors. For example:

$$x^4 - 13x^2 + 36 = 0$$ *We factor the left-hand expression.*

$$(x^2 - 4)(x^2 - 9) = 0$$ *This can be factored further.*

$$(x - 2)(x + 2)(x - 3)(x + 3) = 0$$ *Then set each factor equal to 0.*

$$x - 2 = 0 \quad \text{or} \quad x + 2 = 0 \quad \text{or} \quad x - 3 = 0 \quad \text{or} \quad x + 3 = 0$$ *Solve each first-degree equation.*

$$x = 2 \quad \text{or} \quad x = -2 \quad \text{or} \quad x = 3 \quad \text{or} \quad x = -3$$ *Check these solutions in the original equation.*

By applying our knowledge of solving quadratic equations, we can also solve equations that have second-degree factors that cannot be factored using integer coefficients. For example:

$$x^3 - 5x = 0$$ *We factor the left side.*

$$x(x^2 - 5) = 0$$ *Set each factor equal to 0.*

$$x = 0 \quad \text{or} \quad x^2 - 5 = 0$$

The first equation is a simple linear equation, $x = 0$. The second equation is a quadratic equation that factors no further with integer coefficients, but we can solve it using the square root method, as follows:

$$x^2 - 5 = 0$$
$$x^2 = 5$$
$$x = \pm\sqrt{5}$$

Therefore, the solutions are $x = 0$, $-\sqrt{5}$, and $+\sqrt{5}$.

EXAMPLE 6

Solution

Solve for y: $y^4 - 15y^2 = 16$

$$y^4 - 15y^2 = 16$$ *Rewrite the equation so that one side of the equation is 0.*

$$y^4 - 15y^2 - 16 = 0$$ *Factor.*
$$(y^2 - 16)(y^2 + 1) = 0$$
$$(y - 4)(y + 4)(y^2 + 1) = 0$$ *Set each factor equal to 0, and solve each equation.*
$$y - 4 = 0 \quad \text{or} \quad y + 4 = 0 \quad \text{or} \quad y^2 + 1 = 0$$
$$y = 4 \quad \text{or} \quad y = -4 \quad \text{or} \quad y^2 = -1 \quad \text{\textit{This equation has no real solutions.}}$$

Note that two of the solutions are real and two are not. If we use complex numbers, we solve $y^2 = -1$ to get:

$$y = \pm\sqrt{-1} = \pm i$$

The solutions are $\boxed{-4, 4, -i, \text{ and } i}$.

The student should check to see that the answers are correct.

Let's reexamine how we solved the equation $x^4 - 13x^2 + 36 = 0$ previously. To begin with, this is a fourth-degree equation, and at this point our only method of solving an equation of degree greater than 2 is by the factoring method. Although $x^4 - 13x^2 + 36$ is not a quadratic expression, we initially factored it as though it were. That is,

We factored $x^4 - 13x^2 + 36$ into $(x^2 - 9)(x^2 - 4)$ as though we were factoring the quadratic $u^2 - 13u + 36$ into $(u - 9)(u - 4)$, where u would equal x^2.

Thus, although $x^4 - 13x^2 + 36 = 0$ is not a quadratic equation, it does have the form of a quadratic equation, which is why we call it ***an equation in quadratic form.***

Having examined how we originally solved the equation, now let's approach it in a slightly different way.

Suppose we let $u = x^2$. [Then $u^2 = (x^2)^2 = x^4$.] Now let's substitute u for x^2 and u^2 for x^4 in the original equation. Then

$$x^4 - 13x^2 + 36 = 0 \quad \text{becomes} \quad u^2 - 13u + 36 = 0$$

Now let's solve for u:

$$u^2 - 13u + 36 = 0$$
$$(u - 9)(u - 4) = 0$$
$$u = 9 \quad \text{or} \quad u = 4$$

Since we were asked originally to solve for *x,* we substitute x^2 back for *u:*

$u = 9$	or	$u = 4$	*Substitute x^2 for u.*
$x^2 = 9$	or	$x^2 = 4$	*Now we solve for x. Take square roots.*
$x = \pm\sqrt{9}$	or	$x = \pm\sqrt{4}$	
$x = \pm 3$	or	$x = \pm 2$	*Note that we get the same solutions.*

We used this example to illustrate the method of **substitution of variables.** We are substituting a single variable for a more complex expression in the equation to help us determine whether the equation is in quadratic form.

You may be able to factor $x^4 - 13x^2 + 36$ by recognizing that it is in quadratic form, and therefore the substitution of variables method may seem to be an unnecessary complication. However, the method can help us in solving more complicated equations, such as in Example 7.

EXAMPLE 7

Solve the following: $2a^{1/2} + 3a^{1/4} - 2 = 0$.

Solution

The given equation is not a quadratic equation (for that matter, it is not even a polynomial equation), but it is an equation in quadratic form. Let's examine

$$2a^{1/2} + 3a^{1/4} - 2 = 0$$

Is it obvious that we can solve this equation by factoring? It is probably not obvious that

$$2a^{1/2} + 3a^{1/4} - 2 \quad \text{factors into} \quad (2a^{1/4} - 1)(a^{1/4} + 2)$$

Let's see what happens when we make the appropriate substitution of variables. Let $u = a^{1/4}$. [Then $u^2 = (a^{1/4})^2 = a^{2/4} = a^{1/2}$.] So

$$2a^{1/2} + 3a^{1/4} - 2 = 0 \quad \text{becomes} \quad 2u^2 + 3u - 2 = 0$$

We can solve for *u* by factoring:

$$(2u - 1)(u + 2) = 0$$
$$2u - 1 = 0 \quad \text{or} \quad u + 2 = 0$$

$$\text{Hence, } u = \frac{1}{2} \quad \text{or} \quad u = -2.$$

We found the values of *u,* but we are looking for the values of *a,* so we substitute $a^{1/4}$ back in for *u:*

$$a^{1/4} = \frac{1}{2} \quad \text{or} \quad a^{1/4} = -2 \qquad \text{\textit{Since } } a^{1/4} = u$$

We can discard the equation on the right, as $a^{1/4} = -2$ has no solution. The remaining equation above is similar to the equations we solved in Chapter 7. Raise both sides of the equation to the fourth power:

$$(a^{1/4})^4 = \left(\frac{1}{2}\right)^4$$

$$a = \frac{1}{16}$$

Now we must check for extraneous roots:

Check: $a = \dfrac{1}{16}$: $\qquad\qquad 2a^{1/2} + 3a^{1/4} - 2 = 0$

$$2\left(\frac{1}{16}\right)^{1/2} + 3\left(\frac{1}{16}\right)^{1/4} - 2 \stackrel{?}{=} 0$$

$$2\left(\frac{1}{4}\right) + 3\left(\frac{1}{2}\right) - 2 \overset{?}{=} 0$$

$$\frac{1}{2} + \frac{3}{2} - 2 \overset{?}{=} 0$$

$$2 - 2 \overset{\checkmark}{=} 0$$

The solution is $\boxed{a = \dfrac{1}{16}}$.

How do we know an equation is in quadratic form? How do we know what or where to substitute? To begin with, we observe that an equation in standard quadratic form should look like

$$Ax^2 + Bx + C = 0$$

But x can represent any expression. For example,

$$5(3a + 2)^2 + 7(3a + 2) + 8 = 0 \quad \text{and} \quad 5(y^{1/3})^2 + 7y^{1/3} + 8 = 0$$

are expressions in quadratic form. The important condition is that, ignoring the coefficients, the expression with the higher power should be the square of the expression with the lower power. (The reverse is true if the exponents are negative.)

The following are more examples of equations in quadratic form:

$$a^6 - 7a^3 - 8 = 0 \qquad \textit{a^6 is the square of a^3.}$$
$$2x^{2/3} + x^{1/3} - 15 = 0 \qquad \textit{$x^{2/3}$ is the square of $x^{1/3}$.}$$
$$a^{-4} - 6a^{-2} - 7 = 0 \qquad \textit{a^{-4} is the square of a^{-2}.}$$

We let u equal the middle-term literal expression. Note that we always find u^2 to make sure it checks out. For example, suppose in $2x^{2/3} - 7x^{1/3} + 3 = 0$, we let $u = x^{2/3}$. Then $u^2 = (x^{2/3})^2 = x^{4/3}$. Since $x^{4/3}$ is not in the equation, we reason that we chose the wrong variable for substituting u. Therefore, if the equation *is* in quadratic form, then u *must be* $x^{1/3}$.

EXERCISES 8.5

In Exercises 1–20, *solve the equation.*

1. $\sqrt{x + 3} = 2x$

2. $\sqrt{a - 2} = a - 4$

3. $\sqrt{x + 5} = 7 - x$

4. $\sqrt{y - 2} = 22 - y$

5. $\sqrt{5a - 1} + 5 = a$

6. $\sqrt{7a + 4} - 2 = a$

7. $\sqrt{a + 1} + a = 11$

8. $\sqrt{3a + 1} + a = 9$

9. $\sqrt{3x + 1} + 3 = x$

10. $5a - \sqrt{2a - 3} = 4a + 9$

11. $5a - 2\sqrt{a + 3} = 2a - 1$

12. $6a - 3\sqrt{a + 5} = 4a - 1$

13. $\sqrt{y + 3} = 1 + \sqrt{y}$

14. $\sqrt{3x + 4} = 2 + \sqrt{x}$

15. $\sqrt{a + 7} = 1 + \sqrt{2a}$

16. $\sqrt{2r - 1} = \sqrt{5r} - 2$

17. $\sqrt{7s + 1} - 2\sqrt{s} = 2$

18. $\sqrt{3s + 4} - \sqrt{s} = 2$

19. $\sqrt{7 - a} - \sqrt{3 + a} = 2$

20. $\sqrt{3 - 2a} - \sqrt{3 - 3a} = -1$

21. Points A and B are on opposite sides of a straight river that is 2 miles wide. Point C is 9 miles down the river on the same side as B. A person swims from A to some point P between B and C and then runs from P to C.

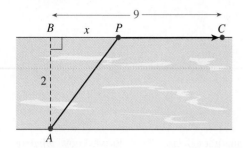

(a) Express the total distance the person covers as a function of x.

(b) If the person can swim at 2 mph and can run at 6 mph, express the total time, T, it would take to get from A to C as a function of x.

22. Points A and B are on opposite sides of a straight river that is 1 mile wide. Point D is 6 miles down the river on the same side as B. A cable company needs to lay cable from point A to point D. The company will lay the cable underwater from A to some point P on the opposite shore, x miles from B, and then the cable will be run above ground from P to D.

(a) Express the total distance covered as a function of x.

(b) It costs \$3,600 per mile to lay the cable underwater, but \$2,500 per mile above ground. Express the total cost of laying the cable as a function of x.

23. A 6-ft long piece of wire is cut into two pieces, x feet from one end. A square is formed with each piece.

(a) In terms of x, what is the length of the side of each square?

(b) Express the total area of each square as a function of x.

24. A 6-ft long piece of wire is cut into two pieces, x feet from one end. A square is formed with each piece. If area of both squares totals $1\frac{1}{8}$ sq ft, what is x?

25. Two poles, 20 and 30 ft high, are to be anchored to the ground with guy wires from the top of each to a point directly between the two poles, which are 50 ft apart. (See accompanying figure.) Express the total length of the wire, L, needed as a function of x.

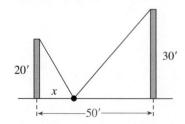

26. Two poles, 20 and 30 ft high, are to be anchored to the ground with guy wires from the top of each to a point directly between the two poles, which are 50 ft apart. If the total length of the wire is 75 ft, find x.

In Exercises 27–34, solve for the given variable.

27. $\sqrt{x} + a = b$ for x

28. $\sqrt{3a} - x = b$ for a

29. $\dfrac{\sqrt{\pi L}}{g} = T$ for L

30. $K = \sqrt{\dfrac{2gs}{l}}$ for s

31. $\sqrt{5x + b} = 6 + b$ for x

32. $\sqrt{3x + y} = 5 - y$ for x

33. $t = \dfrac{\overline{X} - a}{\dfrac{s}{\sqrt{n}}}$ for n

34. $s_e = s_y\sqrt{1 - r^2}$ for r

In Exercises 35–74, solve the equation.

35. $x^3 - 2x^2 - 15x = 0$

36. $x^3 + x^2 - 20x = 0$

37. $6a^3 - a^2 - 2a = 0$

38. $2s^4 - 7s^3 + 3s^2 = 0$

39. $y^4 - 17y^2 + 16 = 0$

40. $2t^4 - 34t^2 = -32$

41. $3a^4 + 24 = 18a^2$

42. $a^4 + 45 = 14a^2$

43. $b^4 + 112 = 23b^2$

44. $b^4 + 75 = 28b^2$

45. $9 - \dfrac{8}{x^2} = x^2$

46. $3 - \dfrac{1}{x^2} = 2x^2$

47. $x^3 + x^2 - x - 1 = 0$

48. $a^3 - 2a^2 - 4a + 8 = 0$

49. $x^{2/3} - 4 = 0$

50. $a^{2/3} = 9$

51. $x + x^{1/2} - 6 = 0$ [*Hint:*   Let $u = x^{1/2}$.]

52. $x - 5x^{1/2} + 6 = 0$

53. $y^{2/3} - 4y^{1/3} - 5 = 0$

54. $b^{2/3} - 4b^{1/3} = -3$

55. $x^{1/2} + 8x^{1/4} + 7 = 0$

56. $x^{1/2} + 3x^{1/4} - 10 = 0$

57. $x^{-2} - 5x^{-1} + 6 = 0$ [*Hint:*   Let $u = x^{-1}$.]

58. $x^{-2} - 2x^{-1} - 15 = 0$

59. $6x^{-2} + x^{-1} - 1 = 0$

60. $15x^{-2} + 7x^{-1} - 2 = 0$

61. $x^{-4} - 13x^{-2} + 36 = 0$

62. $x^{-4} - 3x^{-2} = 4$ [*Hint:*   Let $u = x^{-2}$.]

63. $\sqrt{a} - \sqrt[4]{a} - 6 = 0$
[*Hint:*   Change radicals to rational exponents.]

64. $\sqrt{b} + 2\sqrt[4]{b} = 3$

65. $\sqrt{x} - 4\sqrt[4]{x} = 5$

66. $\sqrt{x} - 2\sqrt[4]{x} = 8$

67. $(a + 4)^2 + 6(a + 4) + 9 = 0$

68. $(a - 1)^2 - 3(a - 1) = 10$

69. $2(3x + 1)^2 - 5(3x + 1) - 3 = 0$

70. $3(2x - 1)^2 - 1 = 0$

71. $(x^2 + x)^2 - 4 = 0$

72. $(x^2 - 4x)^2 - 25 = 0$

73. $\left(a - \dfrac{10}{a}\right)^2 - 12\left(a - \dfrac{10}{a}\right) + 27 = 0$

74. $\left(y + \dfrac{12}{y}\right)^2 - 15\left(y + \dfrac{12}{y}\right) + 56 = 0$

75. Given $s_e = s_y\sqrt{1 - r^2}$, if $s_e = 1.4$ and $s_y = 2.2$, compute r to two decimal places.

76. Given $t = \dfrac{\overline{X} - a}{\dfrac{s}{\sqrt{n}}}$, compute the nearest whole number n for $t = 2.2$, $s = 3.4$, $\overline{X} = 68$, and $a = 66$.

? Question for Thought

77. Discuss what is ***wrong*** (if anything) with solving the equation $\sqrt{3x - 2} - \sqrt{x} = 2$ in the following way:

$$\sqrt{3x - 2} - \sqrt{x} = 2$$
$$\left(\sqrt{3x - 2} - \sqrt{x}\right)^2 = 2^2$$
$$3x - 2 - x = 4$$

$$2x - 2 = 4$$
$$2x = 6$$
$$x = 3$$

 Mini-Review

78. *Solve:* $(x - 2)^2 = (x + 1)^2 - 15$

79. Find the slope of the line whose equation is $2x + 5y - 8 = 0$.

80. *Solve for a:* $\dfrac{3}{a} = \dfrac{r}{1 - a}$

81. In a Senate race, the two candidates received votes in the ratio of 8 to 5. If the winner received 875,400 votes, how many votes were cast altogether?

 8.6

Graphing Quadratic Functions

In Section 8.1 we saw that the graph of a quadratic function has an "umbrella" shape called a parabola. In this section we will examine quadratic functions in more detail. We begin with the basic parabola $y = f(x) = Ax^2$, where A is any real nonzero number.

The Graph of $y = f(x) = Ax^2$

We begin by examining the graph of $y = f(x) = x^2$. First we let x take on various values and then find y:

Table of values for $y = f(x) = x^2$

x	-3	-2	-1	0	1	2	3
$y = f(x)$	9	4	1	0	1	4	9

Sometimes it is more convenient to write tables horizontally.

We then plot the points (see Figure 8.10). This shape is called a *parabola.* Know this picture well.

The parabola shown in Figure 8.10 is the simplest parabola, and it is important that we examine and understand its basic properties. The lowest point on this parabola is called the **vertex.** Notice that for each point on the parabola to the right of the vertex, there is a corresponding point to the left of the vertex. The right side, as a matter of fact, is the mirror image of the left side (put a mirror on the y-axis). We say that "the parabola is symmetric about the y-axis" (see Figure 8.11).

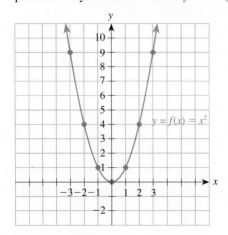

Figure 8.10

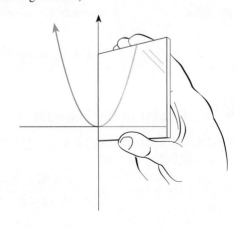

Figure 8.11

The **axis of symmetry** is the line that divides the parabola into two identical (mirror-image) parts: It is where we would place the mirror to get a full picture of the parabola. In this case, the axis of symmetry is the y-axis, which has the equation $x = 0$.

Now let's look at the graphs of $y = 2x^2$ and $y = \frac{1}{3}x^2$. In Figures 8.12 and 8.13, we plot some of the points given in the accompanying tables.

Table of values for $y = f(x) = 2x^2$

x	-3	-2	-1	0	1	2	3
y	18	8	2	0	2	8	18

Table of values for $y = f(x) = \frac{1}{3}x^2$

x	-3	-2	-1	0	1	2	3
y	3	$\frac{4}{3}$	$\frac{1}{3}$	0	$\frac{1}{3}$	$\frac{4}{3}$	3

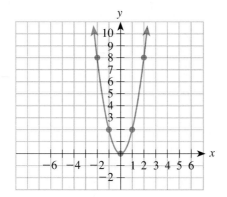

Figure 8.12
Graph of $y = 2x^2$

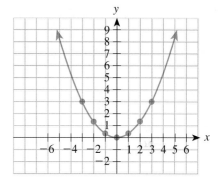

Figure 8.13
Graph of $y = \frac{1}{3}x^2$

Notice that the graphs of $y = 2x^2$ (Figure 8.12) and $y = \frac{1}{3}x^2$ (Figure 8.13) have the same general shape, the same axis of symmetry ($x = 0$) and vertex $(0, 0)$, and they both open upward. However,

$y = 2x^2$ is narrower than $y = x^2$ and $y = \frac{1}{3}x^2$ is wider than $y = x^2$.

We find that A, the coefficient of x^2, "stretches" the parabola. As Figure 8.14 illustrates, the larger the positive coefficient, the more stretched, or narrower, the parabola becomes. The smaller the positive coefficient, the wider the parabola becomes. Notice that the parabola stays at or above the x-axis, that is, y is always nonnegative.

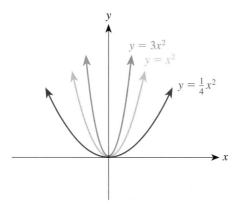

Figure 8.14

Now let's examine what happens when the coefficient of x^2 is negative. Let's look at $y = f(x) = -x^2$. The points given in the accompanying table are plotted in Figure 8.15.

Table of values for $y = f(x) = -x^2$

x	-3	-2	-1	0	1	2	3
$y = f(x)$	-9	-4	-1	0	-1	-4	-9

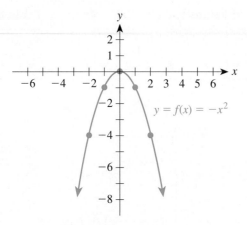

Figure 8.15

The highest point of this parabola is also called the *vertex*. Notice that the y-values for $y = -x^2$ are the opposite of the y-values for $y = x^2$. All y-values [except at the point $(0, 0)$] are negative. Thus, we have the same parabola as $y = x^2$, but now it is upside down. The vertex (now the highest point) is still $(0, 0)$ and the axis of symmetry is still the y-axis. In this case, we say the parabola *opens downward.*

A negative coefficient of x^2 still stretches the parabola, but now it is upside down. Figure 8.16 illustrates what happens to the parabola as $A < 0$ varies. Compare this figure with Figure 8.14, in which $A > 0$ varies.

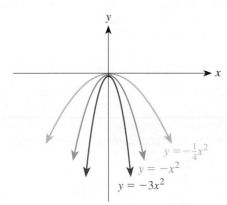

Figure 8.16

Hence, we have the results summarized in the box.

$y = Ax^2$ is a parabola with vertex $(0, 0)$, and the equation of the axis of symmetry is $x = 0$ (the y-axis).

If $A > 0$, the parabola opens up.

If $A < 0$, the parabola opens downward.

EXAMPLE 1

Sketch the graph of $y = f(x) = -5x^2$.

Solution

For this equation, we have $A = -5$. Thus, the parabola opens downward (since $A < 0$). Plot a few points (see Figure 8.17).

Axis of symmetry: $x = 0$ (the *y*-axis)

Vertex at (0, 0)

Stretched by a factor of 5

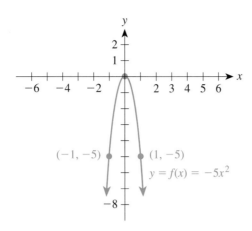

Figure 8.17

The Graph of $y = f(x) = Ax^2 + k$

The next case is the form $y = Ax^2 + k$. Basically, we are adding a constant to the parabola $y = f(x) = Ax^2$. For example, let's examine the table and graph of the equation $y = f(x) = x^2 + 3$.

$$y = f(x) = x^2 + 3$$

x	-3	-2	-1	0	1	2	3
$y = f(x)$	12	7	4	3	4	7	12

As we may expect, comparing this table with the table of $y = x^2$ on page 498, we find that for each *x*, the *y*-value of $y = f(x) = x^2 + 3$ is 3 more than the *y*-value of $y = x^2$. Since we are simply adding 3 to each *y*-value of the "basic" parabola, we end up with the same shape, but the parabola is shifted up 3 units (see Figure 8.18).

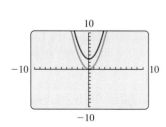

The graphs of $f(x) = x^2 + 3$ and $y = x^2$, on a graphing calculator.

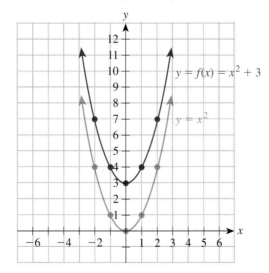

Figure 8.18

Notice that the axis of symmetry is still $x = 0$, but now the vertex is shifted up 3 units to (0, 3). We can generalize this as follows:

The graph of $y = Ax^2 + k$ is a parabola with the same shape as the parabola $y = Ax^2$, but shifted $|k|$ units up if $k > 0$ or down if $k < 0$. The vertex is $(0, k)$ and the axis of symmetry is $x = 0$ (the y-axis).

EXAMPLE 2

Sketch a graph of $y = f(x) = 9x^2 - 4$.

Solution

For this parabola, we have $A = 9$ and $k = -4$ (see Figure 8.19).

Graph has shape of $y = 9x^2$, but moved *down* 4 units ($k < 0$)

Axis of symmetry is $x = 0$

Vertex is $(0, -4)$

Parabola opens up ($A > 0$)

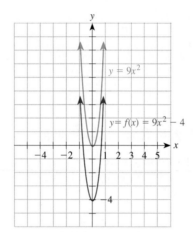

Figure 8.19

Thus far, the important information about a parabola includes its direction (opening up or down), its vertex, its axis of symmetry, and its stretching factor. We should also note where the parabola intersects the coordinate axes, that is, the x- and y-intercepts of the parabola.

The *x*- and *y*-Intercepts of a Parabola

From graphing linear or first-degree equations, we already know that the x- (or y-) intercept is the x (or y) value where the graph of the figure crosses the x- (or y-) axis. Thus, the x-intercept is the value of x when $y = 0$ (why?), and the y-intercept is the value of y when $x = 0$ (why?).

Locating the y-intercept is a straightforward process. For example, to find the y-intercept of $y = f(x) = 9x^2 - 4$ from Example 2, we set $x = 0$ and solve for y:

$$y = 9x^2 - 4 = 9(0)^2 - 4 = -4$$

Thus, the y-intercept is -4, and the graph crosses the y-axis at $y = -4$, as illustrated in Figure 8.20.

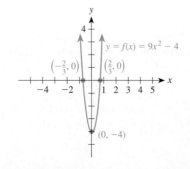

Figure 8.20

Graph of $y = f(x) = 9x^2 - 4$

of Example 2

For the equations we have had thus far, those of the form $y = Ax^2 + k$, the y-intercept is $y = k$ (set $x = 0$ in $y = Ax^2 + k$). This corresponds to the vertex since the vertex lies on the axis of symmetry, which is the y-axis. (This will *not* be the case for more general parabolas, as we shall soon see.)

To locate the x-intercepts of $y = 9x^2 - 4$, we set $y = 0$ and solve for x:

$$y = 9x^2 - 4 \quad \text{\textit{Set y = 0.}}$$
$$0 = 9x^2 - 4 \quad \text{\textit{Solve for x.}}$$
$$0 = (3x + 2)(3x - 2)$$
$$x = -\frac{2}{3} \quad \text{or} \quad x = +\frac{2}{3}$$

Thus, the x-intercepts are $-\frac{2}{3}$ and $+\frac{2}{3}$, and the graph crosses the x-axis at $x = +\frac{2}{3}$ and $x = -\frac{2}{3}$, as can be seen in Figure 8.20.

Using function notation, finding the y-intercept of $y = f(x)$ means finding $f(0)$; finding the x-intercept means finding x such that $f(x) = 0$.

The Graph of $y = f(x) = A(x - h)^2$

Let's examine a table of values for $y = f(x) = (x - 2)^2$, and put it next to a table for $y = x^2$.

x	$y = (x - 2)^2$	$y = x^2$
-3	25	9
-2	16	4
-1	9	1
0	4	0
1	1	1
2	0	4
3	1	9
4	4	16

You may be able to discern the identical pattern of y-values of the two functions as x grows from -3 to 4 in the table: The same y-values in the table of $y = (x - 2)^2$ appear as y-values in the table for $y = x^2$, but for x-values that are 2 units greater. We graph the 2 functions together to help clarify the pattern (see Figure 8.21).

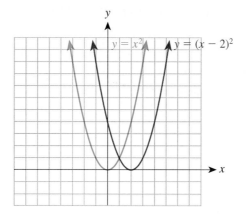

Figure 8.21
The graphs of $y = x^2$ and $y = (x - 2)^2$

Note that the graph of $y = (x - 2)^2$ has exactly the same shape as the graph of $y = x^2$, only it is shifted *horizontally* 2 units to the *right*.

Could you guess what the graph of $f(x) = (x + 5)^2$ would look like? How is the graph of $f(x) = (x - h)^2$ affected when h varies?

Figure 8.22 shows the graphs of the functions $f(x) = x^2$, $f(x) = (x - 4)^2$, and $f(x) = (x + 3)^2$. Note that the graph of $f(x) = (x - 4)^2$ has exactly the same shape as $f(x) = x^2$, only it is shifted 4 units *horizontally to the right*. On the other hand, the graph of $f(x) = (x + 3)^2$ has exactly the same shape as $f(x) = x^2$, only it is shifted 3 units *horizontally to the left.*

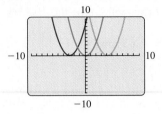

The graph of $y = x^2$, $y = (x + 3)^2$ and $y = (x - 4)^2$ on a graphing calculator.

Figure 8.22
The graphs of $y = x^2$,
$y = (x + 3)^2$,
and $y = (x - 4)^2$

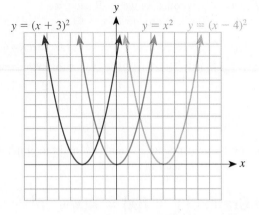

We can generalize this as follows:

The graph of $y = A(x - h)^2$ has exactly the same shape as the graph of $y = Ax^2$, only it is shifted *horizontally* $|h|$ units: to the right if h is positive, and to the left if h is negative. [Keep in mind that h is always *subtracted* in the form $f(x) = A(x - h)^2$.] Note that the vertex as well as the axis of symmetry is shifted horizontally $|h|$ units.

The Graph of $y = f(x) = A(x - h)^2 + k$

Can you guess how the graph of $f(x) = 3(x - 2)^2 + 5$ compares with the graph of $f(x) = 3x^2$? Can you guess the vertex and axis of symmetry? In general, can you determine how the graph of $f(x) = A(x - h)^2 + k$ compares with the graph of $f(x) = Ax^2$?

We call the form $y = f(x) = A(x - h)^2 + k$ the ***standard form of the parabola.***

Figure 8.23 illustrates what happens when we shift a parabola that is symmetric with respect to the y-axis, with vertex $(0, 0)$, *horizontally h units* and *vertically k units.* We arrive at a parabola with the following properties: Its vertex is (h, k) and its axis of symmetry is the vertical line $x = h$.

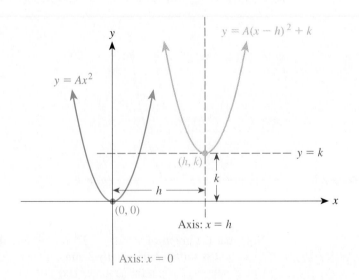

Figure 8.23
The graphs of $y = Ax^2$ and
$y = A(x - h)^2 + k$

Horizontal and vertical shifting of the figure will produce a figure with axis of symmetry parallel to its original axis. In the case illustrated in Figure 8.23, the new axis of the parabola will be parallel to the y-axis.

The Standard Form for the Equation of a Parabola

The equation

$$y = f(x) = A(x - h)^2 + k$$

is the standard form of a parabola with vertex (h, k) and axis of symmetry $x = h$.

Note that for the standard form shown in the box, h is subtracted and k is added.

Compare the form of the parabola with vertex at $(0, 0)$ and symmetric with respect to the y-axis with the form of the parabola with vertex at (h, k) and symmetric with respect to a line parallel to the y-axis.

To change the graph of the parabola $y = Ax^2$ into $y = A(x - h)^2 + k$:

Shift the vertex from $(0, 0)$ to (h, k)

Shift the axis of symmetry from $x = 0$ to $x = h$

This suggests that if we have the parabola in the form $y = A(x - h)^2 + k$, then we can identify h and k, *draw a new set of coordinate axes through the point* (h, k), *and graph the equation* $y = Ax^2$ *on the new set of axes.*

EXAMPLE 3

Identify the vertex and axis of symmetry, and sketch the graph of the following parabolas:

(a) $y = 2(x - 3)^2 + 1$ **(b)** $y = -(x + 2)^2 - 4$

Solution

(a) For the graph of $y = 2(x - 3)^2 + 1$, we compare it with the standard form given in the box:

$$y = A(x - h)^2 + k$$
$$y = 2(x - 3)^2 + 1 \qquad \textit{Now we can identify } h = 3, k = 1, \textit{ and } A = 2.$$

Hence, the vertex, (h, k), is $(3, 1)$. We draw a new set of coordinate axes centered at $(3, 1)$ (and parallel to the x- and y-axes). The axis of symmetry is the new vertical axis, which has the equation $x = 3$. (Note that since the axis of symmetry is parallel to and 3 units to the right of the original y-axis, it *does* have the equation $x = 3$.)

Since $A = 2$, the parabola is narrower than $y = x^2$. We plot a few more points and sketch the graph given in Figure 8.24.

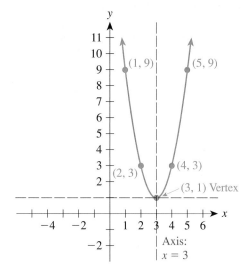

Figure 8.24
The graph of
$y = 2(x - 3)^2 + 1$

(b) For the graph of $y = -(x + 2)^2 - 4$, we compare it with the standard form:

$$y = A(x - h)^2 + k$$

$$y = -(x + 2)^2 - 4$$ *Note that the expression within parentheses is a sum, not a difference. Hence we rewrite $(x + 2)$ as $[x - (-2)]$ to get*

$$y = -[x - (-2)]^2 - 4$$ *Now we can identify $h = -2, k = -4,$ and $A = -1$.*

The vertex is $(-2, -4)$. We draw a new set of coordinate axes centered at $(-2, -4)$ and parallel to the x- and y-axes. The axis of symmetry is $x = -2$. Since the coefficient A is negative, the parabola opens downward. The parabola has exactly the same shape as $y = -x^2$. We plot a few more points and sketch the graph given in Figure 8.25.

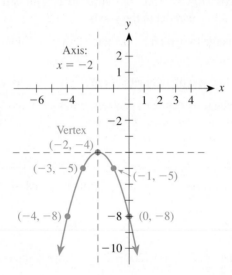

Figure 8.25
The graph of
$y = -(x + 2)^2 - 4$

We see that when the equation of the parabola is given in standard form, we can easily identify the vertex and axis of symmetry, and sketch its graph. If the equation of a parabola is not in standard form (as, for example, $y = x^2 - 6x + 2$), then we use the procedure of completing the square to put the equation in standard form, as demonstrated in the next example.

EXAMPLE 4   Put the equation of the parabola $y = x^2 - 6x + 2$ into standard form and identify its vertex and axis of symmetry.

Solution   We want to put the equation $y = x^2 - 6x + 2$ into standard form. Since standard form requires the perfect square $(x - h)^2$, we use the method of completing the square as follows:

$$y = x^2 - 6x + 2$$ *First group together the terms containing powers of x.*

$$y = (x^2 - 6x \qquad) + 2$$ *Next, complete the square for the expression in parentheses, $x^2 - 6x$. Take half of the x-term coefficient and square it to get*

$$\left[\frac{1}{2}(-6)\right]^2 = 9$$

At this point we note that when we solved quadratic equations by completing the square, we added the number we found to both sides of the equation. However, standard form for the parabola requires y to be isolated on one side of the equation, so we take a slightly different approach: We add *and subtract* the same number on the same side of the equation as follows:

$$y = (x^2 - 6x \qquad) + 2$$ *To complete the square add and subtract 9 within the parentheses.*

$$y = (x^2 - 6x + 9 - 9) + 2$$ *Regroup the expression, leaving a perfect square within parentheses.*

$$y = (x^2 - 6x + 9) - 9 + 2$$ *Rewrite the quadratic expression in factored form and combine the constants outside parentheses.*

$$y = (x - 3)^2 - 7$$

Comparing this equation with the standard form of the parabola, we get

$$h = 3, \quad k = -7, \quad \text{and} \quad A = 1$$

Thus the vertex is $(3, -7)$ and the axis of symmetry is $x = 3$.

EXAMPLE 5

Identify the vertex and axis of symmetry, and sketch the graph of the parabola $y = 3x^2 + 12x + 13$.

Solution

We start by putting the equation $y = 3x^2 + 12x + 13$ into standard form:

$$y = 3x^2 + 12x + 13$$ *First group together the terms containing powers of x.*

$$y = (3x^2 + 12x \quad) + 13$$

The x^2 term has a coefficient of 3, but completing the square requires the coefficient to be 1. We can handle this by factoring out 3 from the terms containing powers of x as follows:

$$y = 3(x^2 + 4x \quad) + 13$$ *Complete the square for the expression $x^2 + 4x$:*

$$\left[\frac{1}{2}(4)\right]^2 = 4$$

Hence, we add and subtract 4 within parentheses.

$$y = 3(x^2 + 4x + 4 - 4) + 13$$ *Regroup the expression, leaving a perfect square within parentheses.*

$$y = 3(x^2 + 4x + 4) + 3(-4) + 13$$ *Notice that we multiplied -4 by the factor of 3. Rewrite the quadratic expression in factored form.*

$$y = 3(x + 2)^2 - 12 + 13$$

$$y = 3(x + 2)^2 + 1$$

Comparing this equation with the standard form of the parabola, we get $h = -2$, $k = 1$, and $A = 3$. Thus, the vertex is $(-2, 1)$ and the axis of symmetry is $x = -2$. We plot a few additional points and sketch the graph in Figure 8.26.

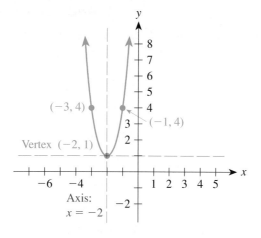

Figure 8.26
$y = 3(x + 2)^2 + 1$

As we know from graphing linear equations, the *x*- (or *y*-) intercept is the x (or *y*-) value at which the graph crosses the *x*- (or *y*-) axis. Thus, the *x*-intercept is the value of x when $y = 0$, and the *y*-intercept is the value of y when $x = 0$. The intercepts are important points and provide us with information we can use to get a more accurate sketch of the graph.

EXAMPLE 6   Sketch a graph of the parabola $y = -2x^2 + 2x + 12$. Identify the vertex, axis of symmetry, and intercepts.

Solution   We start by finding the vertex and the axis of symmetry. First we put the equation $y = -2x^2 + 2x + 12$ into the standard form of a parabola:

$y = -2x^2 + 2x + 12$	*First group together the terms containing powers of x.*
$y = (-2x^2 + 2x \quad) + 12$	*Next factor −2 from the expression in parentheses.*
$y = -2(x^2 - x \quad) + 12$	*Note the change of sign of the x term. Complete the square for $x^2 - x$:*

$$\left[\frac{1}{2}(-1)\right]^2 = \frac{1}{4}$$

Hence we add and subtract $\frac{1}{4}$ within parentheses.

$y = -2\left(x^2 - x + \dfrac{1}{4} - \dfrac{1}{4}\right) + 12$	*Regroup the expression, leaving a perfect square within parentheses.*
$y = -2\left(x^2 - x + \dfrac{1}{4}\right) - 2\left(-\dfrac{1}{4}\right) + 12$	*Rewrite the quadratic expression in factored form, perform operations, and combine the constants outside parentheses.*
$y = -2\left(x - \dfrac{1}{2}\right)^2 + \dfrac{1}{2} + 12$	
$y = -2\left(x - \dfrac{1}{2}\right)^2 + \dfrac{25}{2}$	

Comparing this equation with the standard form of the parabola, we get $h = \frac{1}{2}$, $k = \frac{25}{2}$, and $A = -2$. Thus, the vertex is $\left(\frac{1}{2}, \frac{25}{2}\right)$ and the axis of symmetry is $x = \frac{1}{2}$. The parabola opens downward since $A < 0$.

Next we find the intercepts:

$y = -2x^2 + 2x + 12$	*To find the y-intercept, let $x = 0$ and solve for y.*
$y = -2(0)^2 + 2(0) + 12 = 12$	*Hence, the y-intercept is 12.*
$y = -2x^2 + 2x + 12$	*To find the x-intercepts, let $y = 0$ and solve for x.*
$0 = -2x^2 + 2x + 12$	*We can solve this equation by factoring.*
$0 = -2(x + 2)(x - 3)$	
$x = -2 \quad \text{or} \quad x = 3$	*Hence, the x-intercepts are −2 and 3.*

We plot the intercepts and vertex, draw in the axis of symmetry, and sketch the graph shown in Figure 8.27.

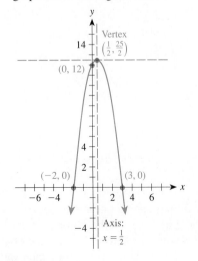

Figure 8.27

We see that we can take any equation of the form $f(x) = Ax^2 + Bx + C$ and put it into the standard form of the parabola, $y = A(x - h)^2 + k$.

A Formula for the Vertex

Let's think about the vertex of the general parabola $y = f(x) = Ax^2 + Bx + C$ from a slightly different point of view. Recall that the axis of symmetry divides the parabola into two identical mirror-image parts. In particular, if there are x-intercepts, *the axis of symmetry must cut through the midpoint between the x-intercepts.*

For example, the x-intercepts of $y = f(x) = x^2 - 4x$ can easily be found by factoring: $0 = x^2 - 4x = x(x - 4) \Rightarrow x = 0$ or $x = 4$, and so the axis of symmetry must be the vertical line $x = 2$ (see Figure 8.28). Since the vertex falls on the axis of symmetry, we know that the x-coordinate of the vertex is $x = 2$, and we can find the y-coordinate by substituting $x = 2$ into $y = x^2 - 4x$, obtaining $y = -4$. Thus, the vertex is $(2, -4)$, as indicated in Figure 8.28.

We can use this same approach to find the x-coordinate of the vertex of any parabola of the form $y = f(x) = Ax^2 + Bx$. We find the x-intercepts by setting $y = 0$ and solving for x. See Figure 8.29.

$$0 = Ax^2 + Bx = x(Ax + B) \Rightarrow x = 0 \quad \text{or} \quad x = -\frac{B}{A}$$

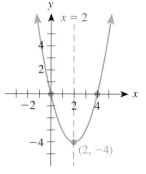

Figure 8.28

Note that the axis of symmetry is the vertical line midway between the x-intercepts.

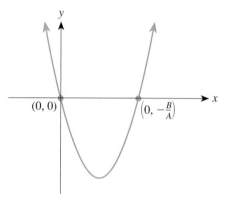

Figure 8.29

The graph of $f(x) = Ax^2 + Bx$ with $A > 0$ has x-intercepts 0 and $-\dfrac{B}{A}$.

Since the x-intercepts are $x = 0$ and $x = -\dfrac{B}{A}$, the axis of symmetry will again be the vertical line that passes through the midpoint between them. See Figure 8.30.

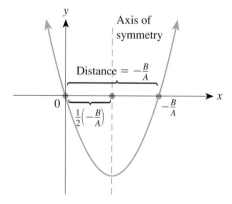

Figure 8.30

The midpoint between x-intercepts has x-coordinate $= \dfrac{1}{2}\left(-\dfrac{B}{A}\right)$.

As we can see by Figure 8.30, the distance between points $(0, 0)$ and $\left(0, -\dfrac{B}{A}\right)$ is $-\dfrac{B}{A}$. Hence, the x-coordinate of the midpoint must be half this distance:

$$x = \left(\frac{1}{2}\right)\left(-\frac{B}{A}\right) = -\frac{B}{2A}$$

Knowing that the axis of symmetry is vertical and passes through the point $\left(-\dfrac{B}{2A}, 0\right)$, can you find its equation?

How are the graphs of $y = Ax^2 + Bx$ and $y = Ax^2 + Bx + C$ related?

Thus, we have found that the x-coordinate of the vertex of any parabola of the form $y = Ax^2 + Bx$ is $x = -\dfrac{B}{2A}$.

Our previous discussion of vertical shift tells us that the graph of the general parabola $y = f(x) = Ax^2 + Bx + C$ is obtained by shifting the parabola $y = Ax^2 + Bx$ up or down C units. This shifts the y-coordinate of the vertex vertically up or down *but has no effect on the x-coordinate of the vertex,* and so the x-coordinate of the vertex of $y = Ax^2 + Bx + C$ is still given by $x = -\dfrac{B}{2A}$. Knowing that the x-coordinate of the vertex is $-\dfrac{B}{2A}$ allows us to find the vertex without the necessity of completing the square. We summarize this discussion as follows:

Finding the Vertex of a Parabola

An equation of the form $y = f(x) = Ax^2 + Bx + C$, $A \neq 0$, is a parabola with axis of symmetry $x = -\dfrac{B}{2A}$. The x-coordinate of the vertex is $-\dfrac{B}{2A}$.

EXAMPLE 7

Graph the function $y = f(x) = -x^2 + 6x - 7$. Label the intercepts, vertex, and the axis of symmetry.

Solution

This is the quadratic function $f(x) = Ax^2 + Bx + C$ with $A = -1$, $B = 6$, and $C = -7$. Hence, the x-coordinate of the vertex is

$$x = -\frac{B}{2A} = -\frac{6}{2(-1)} = 3$$

and the axis of symmetry is $x = 3$. Since the x-coordinate of the vertex is 3, the y-coordinate of the vertex is

$$f(3) = -(3)^2 + 6(3) - 7 = 2$$

Hence, the coordinates of the vertex are $(3, 2)$.

The y-intercept is -7 since $f(0) = -(0)^2 + 6(0) - 7 = -7$.
The x-intercepts are found by using the quadratic formula:

$$x = \frac{-6 \pm \sqrt{6^2 - 4(-1)(-7)}}{2(-1)} = \frac{-6 \pm \sqrt{8}}{-2} = \frac{-6 \pm 2\sqrt{2}}{-2} = 3 \pm \sqrt{2}$$

Since $\sqrt{2} \approx 1.41$, the x-intercepts are

$$x = 3 + \sqrt{2} \approx 4.41 \quad \text{and} \quad x = 3 - \sqrt{2} \approx 1.59.$$

The graph appears in Figure 8.31.

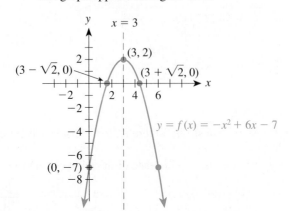

Figure 8.31

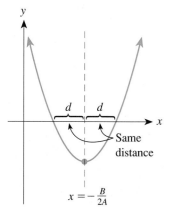

Figure 8.32

If a parabola has *x*-intercepts, they are equidistant from the axis of symmetry.

What does the fact that the roots of $Ax^2 + Bx + C = 0$ are imaginary say about the x-intercepts of the graph of $f(x) = Ax^2 + Bx + C$?

Note that we did not bother to develop a formula to find the *y*-coordinate of the vertex, since it is much easier to substitute the *x*-value in the equation to find the *y*-coordinate.

Let's look at Figure 8.32 and note a few things about the quadratic formula and its relationship to the axis of symmetry, vertex, and *x*-intercepts of the graph of $f(x) = Ax^2 + Bx + C$. The two solutions to the equation $Ax^2 + Bx + C = 0$ are the *x*-intercepts of the graph of $f(x) = Ax^2 + Bx + C$. The two solutions are found by applying the quadratic formula, which tells us to add and subtract the same quantity, $\frac{\sqrt{B^2 - 4AC}}{2A}$, to and from $-\frac{B}{2A}$, which is the *x*-coordinate of the vertex.

When solving a quadratic equation, we know from the quadratic formula that if the expression under the radical, $B^2 - 4AC$, is 0, then there is only one solution to $Ax^2 + Bx + C = 0$. This means the parabola $f(x) = Ax^2 + Bx + C$ has only one *x*-intercept and therefore its vertex is *on* the *x*-axis. If $B^2 - 4AC < 0$, then there are two solutions, but the solutions are imaginary; hence, the graph of $f(x) = Ax^2 + Bx + C$ has no *x*-intercepts. On the other hand, if $B^2 - 4AC > 0$, then there are two real solutions and, therefore, two *x*-intercepts for the graph of $f(x) = Ax^2 + Bx + C$.

Recall that the expression $B^2 - 4AC$ is called the ***discriminant*** of the equation $Ax^2 + Bx + C = 0$. In Figure 8.33, we consider the three possibilities for the discriminant and illustrate two possible graphs of $y = Ax^2 + Bx + C$ for each of these cases.

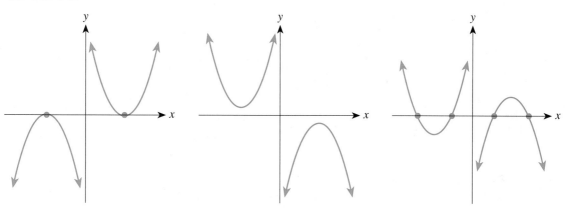

$B^2 - 4AC = 0$ means one *x*-intercept. $B^2 - 4AC < 0$ means no *x*-intercepts. $B^2 - 4AC > 0$ means two *x*-intercepts.

Figure 8.33

Let's return to the first example discussed at the beginning of this chapter on page 457, and see how we can apply what we know about the graphs of quadratic functions.

EXAMPLE 8

The profit in dollars, $P(x)$, made on a concert is related to the price of a ticket in dollars, *x,* in the following way:

$$P(x) = 10{,}000(-x^2 + 11x - 15)$$

How much should a ticket cost in order to achieve the maximum profit? What is the maximum profit?

Solution

At the beginning of this chapter we found that we could get a rough estimate of the maximum value of a quadratic function by knowing the shape of its graph and plotting points or making a table of values. Now we can find this value exactly with less effort.

We note that $P(x)$ is quadratic. Therefore, if we let the horizontal axis be the price of a ticket, *x,* and the vertical axis be the profit, $P(x)$, we would have a parabola that opens downward (why?). Our ordered pairs are of the form $(x, P(x))$.

Since the parabola opens downward, the vertex is the highest point. This means that the vertex is the point that yields the highest value of $P(x)$, the profit. Hence we are looking for the coordinates of the vertex. To find the vertex we identify *A* and *B,*

In this example, is it really necessary to multiply out P(x) to get the x-coordinate of the vertex?

the coefficients of x^2 and x, respectively, by first multiplying out $P(x)$ to get $P(x) = -10{,}000x^2 + 110{,}000x - 150{,}000$. We see that $A = -10{,}000$ and $B = 110{,}000$. The x-coordinate of the vertex is

$$x = -\frac{B}{2A} = -\frac{110{,}000}{2(-10{,}000)} = 5.5$$

Thus, the ticket price must be \$5.50 to maximize the profit. Since the maximum value of $P(x)$ occurs when $x = 5.5$, the maximum profit is $P(5.5)$, which is

$$P(5.5) = 10{,}000[-(5.5)^2 + 11(5.5) - 15] = 10{,}000(15.25) = 152{,}500$$

Hence, the highest $P(x)$ is 152,500, which occurs when $x = 5.5$. This means that the maximum profit is \$152,500 when the price of a ticket is \$5.50. Figure 8.34 depicts the graph of this equation.

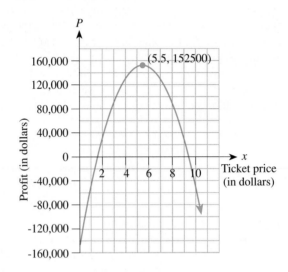

Figure 8.34
The graph of $P(x) = 10{,}000\,(-x^2 + 11x - 15)$

EXAMPLE 9

A printer manufacturing company determines that it can sell 3,000 printers at a price of \$350 and that for each \$25 increase in the price, 100 fewer printers will be sold. Let x represent the number of \$25 increases in price.

(a) Express the total revenue from printer sales as a function of x.

(b) Determine how many price increases are needed to maximize the revenue.

(c) What price will produce the maximum revenue?

(d) Determine the maximum revenue.

Solution

(a) The revenue is computed by multiplying the number of printers sold by the price per printer:

Revenue = (Number of printers)(Price per printer)

Before trying to write out a formula, it would be helpful to first set up a table with some numerical values for x and see whether we can identify a pattern. Let $x = $ the number of \$25 increases.

x	Price per printer	Number of printers sold	Revenue
0	350	3,000	350(3,000)
1	350 + 25	3,000 − 100	(350 + 25)(3,000 − 100)
2	350 + 2(25)	3,000 − 2(100)	(350 + 2(25))(3,000 − 2(100))
3	350 + 3(25)	3,000 − 3(100)	(350 + 3(25))(3,000 − 3(100))
⋮	⋮	⋮	⋮
x	350 + x(25)	3,000 − x(100)	(350 + x(25))(3,000 − x(100))

Thus, we have

$$R(x) = (350 + 25x)(3{,}000 - 100x) = -2{,}500x^2 + 40{,}000x + 1{,}050{,}000$$

The revenue function

$$R(x) = -2{,}500x^2 + 40{,}000x + 1{,}050{,}000$$

represents the relationship between $R(x)$, the total revenue from printer sales, and x, the *number of* $25 *increases in price (from the original price of* $350). Figure 8.35 shows the graph of this function.

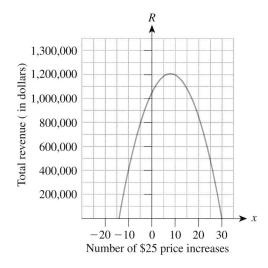

Figure 8.35
The graph of $R = R(x) = -2{,}500x^2 + 40{,}000x + 1{,}050{,}000$

(b) By Figure 8.35, we see that the maximum revenue occurs at the vertex. Keeping in mind that x is the number of $25 increases, the x-coordinate of the vertex will give us the number of increases needed for the maximum revenue.

The x-coordinate of the vertex is

$$x = -\frac{B}{2A} = -\frac{40{,}000}{2(-2{,}500)} = 8$$

Hence, the maximum revenue occurs when $x = 8$, or when there are eight $25 price increases.

(c) From part **(b)** above, the maximum revenue occurs at $x = 8$, when the price is increased by $25 eight times: this comes to $(8 \times 25 =)$ $200. Since the original price is $350, the *price* that yields the highest revenue is $350 + $200 = $550.

(d) Since the vertex occurs at $x = 8$, the maximum revenue is

$$R(8) = -2{,}500(8)^2 + 40{,}000(8) + 1{,}050{,}000 = \$1{,}210{,}000.$$

EXERCISES 8.6

In Exercises 1–16, sketch a graph of the parabola.

1. $y = 3x^2$

2. $y = 5x^2$

3. $y = \frac{1}{2}x^2$

4. $y = \frac{1}{4}x^2$

5. $y = -4x^2$

6. $y = -\frac{1}{3}x^2$

7. $6x^2 - y = 0$

8. $-2x^2 + 10y = 0$

9. $y = x^2 + 4$

10. $y = 2x^2 - 3$

11. $y = -2x^2 + 1$

12. $y = 3x^2 - 2$

13. $y = (x - 4)^2$

14. $y = (x - 3)^2$

15. $y = -(x + 2)^2$

16. $y = -(x + 4)^2$

In Exercises 17–22, sketch a graph and identify the vertex and axis of symmetry.

17. $y = (x - 3)^2 + 4$

18. $y = -(x - 4)^2 + 1$

19. $y = -3(x - 1)^2 - 8$

20. $y = 2(x + 1)^2 - 3$

21. $y = 2(x + 4)^2 - 2$

22. $y = -4(x + 1)^2 + 9$

In Exercises 23–28, put the equation into the standard form of a parabola and identify its vertex and axis of symmetry.

23. $y = f(x) = 2x^2 - 12x + 19$

24. $y = f(x) = 3x^2 - 6x + 1$

25. $y = f(x) = -x^2 + 4x - 1$

26. $y = f(x) = -x^2 - 4x$

27. $y = f(x) = -3x^2 + 30x - 70$

28. $y = f(x) = 2x^2 + 8x + 2$

In Exercises 29–44, sketch a graph of the quadratic function. Label the vertex, axis of symmetry, and the intercepts.

29. $f(x) = x^2 - 4$

30. $f(x) = x^2 - 9$

31. $f(x) = 2x^2 + 8$

32. $f(x) = 2x^2 - 8$

33. $f(x) = -x^2 - 10x - 25$

34. $f(x) = x^2 + 8x + 16$

35. $f(x) = x^2 - 2x - 35$

36. $f(x) = x^2 - 8x + 15$

37. $f(x) = 2x^2 - 4x + 4$

38. $f(x) = 3x^2 + 6x - 24$

39. $f(x) = -x^2 - 3x - 4$

40. $f(x) = -4x^2 - 2x - 8$

41. $f(x) = 2x^2 - 3x + 2$

42. $f(x) = 2x^2 + 4x + 1$

43. $f(x) = -2x^2 + 4x - 1$

44. $f(x) = -2x^2 + 4x + 1$

45. The bagel factory finds that its profit is related to the number of bagels produced as follows:

$$P(x) = -x^2 + 70x$$

where x is the number of hundreds of bagels produced daily and $P(x)$ is the daily profit in dollars. How many bagels must be produced daily to maximize the profit? What would the maximum profit be?

46. The profit, $P(x)$, made on a concert is related to the price (x) of a ticket in the following way:

$$P(x) = 10{,}000(-x^2 + 12x - 35)$$

where $P(x)$ is the profit in dollars and x is the price of a ticket. What ticket price would produce the maximum profit?

47. In the United States, the number of marriages per year from 1975 to 1988 can be estimated by the equation

$$M(t) = -4.185t^2 + 703.65t - 27{,}132.38$$

where $M(t)$ is the number of marriages (in thousands), and t is the last two digits of the year. In what year (between 1975 and 1988) was the number of marriages at a maximum?

48. In the United States, the number of first-year students enrolled in medical and osteopathy school from 1988 to 1996 can be estimated by the equation

$$S(t) = -0.929t^2 + 174.734t - 8{,}090.367$$

where $S(t)$ is the number of students (in thousands), and t is the last two digits of the year. In what year (between 1988 and 1996) was the number of students at a maximum?

49. The daily profit earned by the Barrie factory is related to the number of cases of candy canes produced in the following way:

$$P(x) = -x^2 + 112x - 535$$

where $P(x)$ is the daily profit in dollars and x is the number of cases of candy canes made daily. Find the number of cases of candy canes to be made daily to maximize the daily profit. What is the maximum profit?

50. The number of portable widgets produced weekly by Widgets, Inc., is related to the weekly profit in the following way:

$$P(x) = -2x^2 + 88x - 384$$

where $P(x)$ is the weekly profit in hundreds of dollars and x is the number of widgets produced weekly. How many widgets must be produced weekly for the maximum weekly profit? What is the maximum weekly profit?

51. Jerry launches a rocket upward and the rocket travels according to the equation

$$y = -16t^2 + 400t$$

where y is the height of the rocket off the ground (in feet) at t seconds after he launches it. How many seconds does it take for the rocket to reach maximum height? What is the maximum height of the rocket?

52. Carol stands on the roof of a building and launches a rocket upward. The rocket travels according to the equation

$$y = -16t^2 + 400t + 50$$

where y is the height of the rocket off the *ground* in feet at t seconds after it is launched.

(a) How far is Carol above the ground when she launches the rocket?

(b) How high does the rocket travel relative to the ground?

(c) After how many seconds does the rocket hit the ground?

53. Raju wants to fence in a rectangular garden against his house. Since the house is to serve as one of the sides, he needs to fence in only three sides. If he uses 100 linear feet of fencing, what are the dimensions of the garden that would give him the maximum possible area?

54. Susan wanted to fence in a rectangular vegetable garden against her house. She needed to fence in only three sides since the house protected the fourth side. If she used 50 linear feet of fencing, what are the dimensions of the garden that would give her the maximum possible area?

55. In economics, the demand function for a given item indicates how the price per unit, p, is related to the number of units, x, that are sold. Suppose a company finds that the demand function for one of the items it produces is

$$p = 10 - \frac{x}{6}$$

where p is in dollars. The revenue function, $R(x)$, is found by multiplying the price per unit, p, by the number of items sold, x (that is, $R = xp$). Find the revenue function corresponding to this demand function, and determine the maximum revenue and how many items should be sold to maximize the revenue.

56. A computer company determines that it can sell 5,000 computers at a price of $800 and that for each $50 increase in price, 60 fewer computers will be sold.

(a) Let x represent the number of $50 increases in price. Express the total revenue from computer sales as a function of x.

(b) If $R(x)$ represents the revenue function found in part **(a)**, use this function to determine how many price increases are needed to maximize the revenue. What computer price will produce the maximum revenue? What is the maximum revenue?

57. The managers of Games Inc. determined that they can sell 200,000 puzzler cubes at a price of $3.00 and that for each 25¢ increase in price, 7,250 fewer puzzler cubes will be sold. Let x represent the number of 25¢ increases in price.

    (a) Express the total revenue from puzzler cube sales as a function of x.

    (b) If $R(x)$ represents the revenue function found in part (a), use $R(x)$ to find how many price increases are needed to maximize the revenue. What price will produce the maximum revenue? Determine the maximum revenue.

58. A shoe manufacturer can sell 3,500 pair of shoes for $24 each. It is determined that for each 50¢ reduction in price, 150 more pairs of shoes can be sold.

    (a) Express the revenue, R, as a function of n, where n is the number of 50¢ reductions.

    (b) If $R(n)$ represents the revenue function found in part (a), use $R(n)$ to find how many price reductions are needed to maximize the revenue. What price will produce the maximum revenue? Determine the maximum revenue.

? Questions for Thought

59. In the equation $y = Ax^2 + Bx + C$, what can you say in general about the parabola if $A > 0$? If $A < 0$?

60. Given the equation $y = Ax^2 + Bx + C$, what can be said of a parabola if $B^2 - 4AC > 0$? If $B^2 - 4AC < 0$? If $B^2 - 4AC = 0$?

61. Find the formula for the y-coordinate of the vertex of the parabola $y = Ax^2 + Bx + C$ by substituting the x-value $-\dfrac{B}{2A}$ in the equation. Simplify.

62. Consider the equation $y = 2(x - 3)^2 + 5$. Substitute various values for x and look at the numerical values of y. Give an algebraic explanation as to what the smallest value of y must be. [*Hint: What is the smallest possible value of* $(x - 3)^2$?] For what value of x does the smallest value of y occur?

63. Consider the equation $y = A(x - h)^2 + k$, where $A > 0$. Give an algebraic explanation as to what the smallest value of y must be. [*Hint: What is the smallest possible value of* $(x - h)^2$?] For what value of x does the smallest value of y occur?

64. Use the method of completing the square to put the general equation $y = Ax^2 + Bx + C$ $(A \neq 0)$ into the standard form of a parabola, $y = A(x - h)^2 + k$. Show that the vertex of the parabola is $\left(-\dfrac{B}{2A}, \dfrac{4AC - B^2}{4A}\right)$.

◇ Mini-Review

65. *Solve:* $\dfrac{3}{2x - 4} = \dfrac{x - 2}{x + 1}$

66. *Perform the indicated operations and simplify:* $\dfrac{x - 3}{x + 3} \div (x^2 - 9)$

67. A bottle contains 60 oz of a 5% iodine solution. How much of the solution must be removed and replaced with water to dilute it to a 4% iodine solution?

68. Sketch the graph of the line passing through the point $(0, -2)$ with slope $-\frac{2}{3}$.

8.7 Quadratic and Rational Inequalities

In this section we will solve quadratic inequalities such as $x^2 + x - 12 > 0$. Recall that in solving quadratic equations, we found it convenient to work with them in standard form. We will find similarly that quadratic inequalities will be easier to work with if they are in standard form.

We say that a quadratic inequality is in **_standard form_** if it is in the form $Ax^2 + Bx + C > 0$ or $Ax^2 + Bx + C < 0$ with $A \neq 0$. (Note that $<$ and $>$ can be replaced, respectively, by $\leq$ and $\geq$.)

To solve the inequality $x^2 + x - 12 > 0$ we must realize that we are looking for values of x that make $x^2 + x - 12$ _positive_. (Remember that $u > 0$ means u is positive.) On the other hand, to solve the inequality $x^2 + x - 12 < 0$ means that we are seeking the values of x that make the expression negative. Hence, we are concerned only with the _sign_ of the expression $x^2 + x - 12$.

When is the expression $x^2 + x - 12$ positive and when is it negative? The easiest way to determine this is to factor the expression and _examine the sign of each factor_ as x takes on various values on the number line. Then use the multiplication rules for signed numbers to determine the sign of the product to find the solution.

We begin by factoring $x^2 + x - 12$ into $(x + 4)(x - 3)$ and for convenience we will use this factored form.

Looking at $(x + 4)(x - 3)$, we can tell at a glance that $x = -4$ and $x = +3$ make $(x + 4)(x - 3)$ equal to 0. What happens when x takes on values around -4 and $+3$? Let's examine the signs of each factor of $(x + 4)(x - 3)$ using the number line.

First we examine the sign of $x + 4$ as x takes on various values on the number line. Notice that $x + 4$ is negative for $x < -4$ (for example, if $x = -5$, then we have $x + 4 = -5 + 4 = -1$), and $x + 4$ is positive for $x > -4$.

We do the same for the factor $x - 3$. Notice that $x - 3$ is negative for $x < 3$ and positive for $x > 3$.

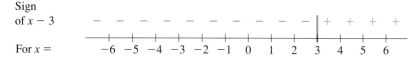

If we put these two pictures together, we can determine the sign of the product by looking at the signs of the factors above the number line:

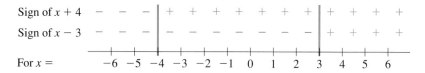

By using the multiplication rules, at a glance we arrive at the signs of the product. The sign of the product is positive when $x < -4$ or $x > 3$, and negative when $-4 < x < 3$.

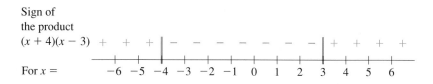

From the number line above, we find that $(x + 4)(x - 3) > 0$ or, equivalently, that $x^2 + x - 12 > 0$ (positive), when $x < -4$ or $x > 3$.

The x-values where the quadratic expression $x^2 + x - 12$ is exactly 0 are the x-values where the factors $x + 4$ and $x - 3$ are 0: at $x = -4$ and $x = +3$. These are called the ***cutpoints*** for the inequality $(x + 4)(x - 3) > 0$. The cutpoints serve as endpoints for the solution set(s) of the quadratic inequality.

Let's demonstrate with another example.

EXAMPLE 1

Solve for x. $2x^2 - 9x - 5 < 0$

Solution

Keep in mind that we are looking for the values of x that make the expression $2x^2 - 9x - 5$ negative.

$2x^2 - 9x - 5 < 0$ *This inequality is asking: When is $2x^2 - 9x - 5$ negative? Factor.*

$(2x + 1)(x - 5) < 0$ *Find the cutpoints.*

$$2x + 1 = 0 \Rightarrow x = -\frac{1}{2} \quad \text{and} \quad x - 5 = 0 \Rightarrow x = 5$$

Now we look at the sign of each factor as x takes on different values around the cutpoints.

$2x + 1$ is negative when $x < -\frac{1}{2}$, and positive when $x > -\frac{1}{2}$:

$x - 5$ is negative when $x < 5$, and positive when $x > 5$:

Therefore, the sign of the product is as shown here:

Sign of the product
$(2x + 1)(x - 5)$

As we can see, $(2x + 1)(x - 5)$ is negative when $-\frac{1}{2} < x < 5$ or

$$(2x + 1)(x - 5) < 0 \qquad \text{when} \qquad -\frac{1}{2} < x < 5$$

Therefore, the solution to $2x^2 - 9x - 5 < 0$ is

$$-\frac{1}{2} < x < 5 \quad \text{or} \quad \left(-\frac{1}{2}, 5\right), \text{ using interval notation}$$

We can graph the solution set to the inequality $2x^2 - 9x - 5 < 0$ as shown in Figure 8.36.

Figure 8.36
Solution set for
Example 1

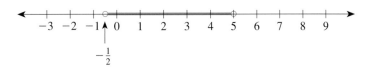

Note that we exclude the points $-\frac{1}{2}$ and 5.

Looking at the number line representation of the signs of $(2x + 1)(x - 5)$ in Example 1, we can observe that the cutpoints break up the number line into three distinct intervals. Most important, note that the sign of the product does not change within an interval (excluding the cutpoints). If the expression is positive (or negative) for one value of x within an interval, it is positive (or negative) for *all* values within that interval.

Hence, if we want to know the sign of the product in an interval, *we need only check one value within the interval.* For example, we found that the cutpoints for $(2x + 1)(x - 5)$ are $-\frac{1}{2}$ and 5. This breaks up the number line into three intervals, as shown in Figure 8.37.

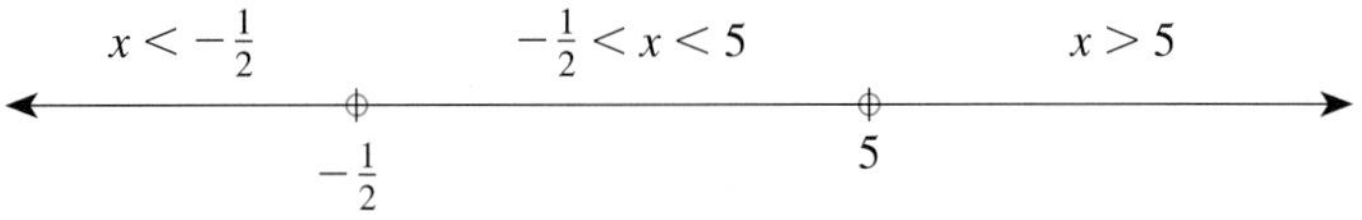

Figure 8.37

Choose any value in the leftmost interval; that is, choose any number less than $-\frac{1}{2}$, say $x = -6$.

When $x = -6$, the sign of $(2x + 1)(x - 5) = [2(-6) + 1][-6 - 5]$ is positive and therefore the sign of $(2x + 1)(x - 5)$ is positive for *all* values of x in the interval $x < -\frac{1}{2}$.

Choose any value in the middle interval, $-\frac{1}{2} < x < 5$, say $x = 0$.

When $x = 0$, the sign of $(2x + 1)(x - 5) = (2 \cdot 0 + 1)(0 - 5)$ is negative and therefore the sign of $(2x + 1)(x - 5)$ is negative for *all* values of x in the interval $-\frac{1}{2} < x < 5$.

Finally, choose any value in the rightmost interval, $x > 5$, say $x = 10$.

When $x = 10$, the sign of $(2x + 1)(x - 5) = (2 \cdot 10 + 1)(10 - 5)$ is positive and therefore the sign of $(2x + 1)(x - 5)$ is positive for *all* values of x in the interval $x > 5$.

Note: The results agree with the solution we found to Example 1. (See Figure 8.36.)

The values we choose in the intervals to check the sign of the product are called the ***test values*** for the interval. We demonstrate the use of this method in the following example.

EXAMPLE 2

Solve for x and sketch a graph of the solution set. $x^2 - 7x + 6 > 0$

Solution

$x^2 - 7x + 6 > 0$ *This inequality is asking: When is $x^2 - 7x + 6$ positive? First factor the left side.*

$(x - 1)(x - 6) > 0$ *Find the cutpoints.*

$$x - 1 = 0 \Rightarrow x = 1 \quad \text{and} \quad x - 6 = 0 \Rightarrow x = 6$$

Therefore, the cutpoints break up the number line into three intervals as shown below. Now pick a test value in each interval:

Test value:

For $x = 0$,	For $x = 4$,	For $x = 10$,
$(x - 1)(x - 6) =$	$(x - 1(x - 6) =$	$(x - 1)(x - 6) =$
$(0 - 1)(0 - 6)$	$(4 - 1)(4 - 6)$	$(10 - 1)(10 - 6)$
is positive. Hence,	is negative. Hence,	is positive. Hence,
$(x - 1)(x - 6)$ is	$(x - 1)(x - 6)$	$(x - 1)(x - 6)$ is
positive for $x < 1$.	is negative for	positive for $x > 6$.
	$1 < x < 6$.	

The original problem asks for what values of x is $x^2 - 7x + 6 > 0$? The solution is $\boxed{x < 1 \quad \text{or} \quad x > 6 \quad \text{or} \quad (-\infty, 1) \cup (6, \infty), \text{ using interval notation}}$.

We can graph the solution to $x^2 - 7x + 6 > 0$ on the number line as shown in Figure 8.38.

Figure 8.38
Solution set for
Example 2

The general procedure for solving quadratic inequalities is summarized in the box.

Procedure for Solving Quadratic Inequalities

1. *Put the inequality into standard form* (one side of the inequality is 0).
2. *Factor* the quadratic expression.
3. *Find the cutpoints* by setting each factor equal to 0.
4. The cutpoints divide the number line into intervals. Draw a number line, clearly indicating the cutpoints.
5. *Pick test values* within each interval determined by the cutpoints and check the sign of the product for each test value.
6. *Choose the interval(s)* that satisfy the sign requirements of the inequality in step 1.

EXAMPLE 3

Solve for x and graph the solution on the number line:

$$x^2 - 2x \geq 15$$

Solution

$$x^2 - 2x \geq 15 \qquad \textit{First rewrite the inequality so that one side of the inequality is 0.}$$

$$x^2 - 2x - 15 \geq 0 \qquad \textit{Then factor.}$$

$$(x - 5)(x + 3) \geq 0$$

Find the cutpoints:

$$x - 5 = 0 \Rightarrow x = 5 \qquad x + 3 = 0 \Rightarrow x = -3 \qquad \textit{Cutpoints are 5 and } -3.$$

The intervals are $x < -3$, $\quad -3 < x < 5$, $\quad$ and $\quad x > 5$.

Sign of
$(x - 5)(x + 3)$

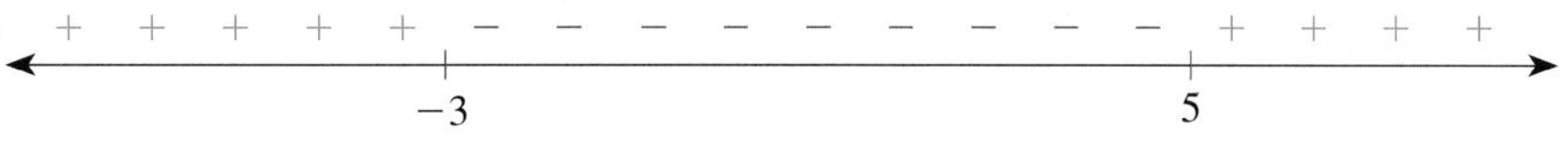

Test value:

For $x = -10$,   For $x = 0$,   For $x = 10$,
$(x - 5)(x + 3) =$ $(x - 5)(x + 3) =$ $(x - 5)(x + 3) =$
$(-10 - 5)(-10 + 3)$ $(0 - 5)(0 + 3)$ $(10 - 5)(10 + 3)$
is positive.   is negative.   is positive.

We see that the product $(x - 5)(x + 3)$ is positive for $x < -3$ or $x > 5$, and $(x - 5)(x + 3)$ is 0 for $x = -3$ and $x = 5$. Thus, $(x - 5)(x + 3) \geq 0$ (positive or equal to 0) for

$$\boxed{x \leq -3 \quad \text{or} \quad x \geq 5}$$

We graph the *solution* in Figure 8.39.

Figure 8.39
Solution set for
Example 3

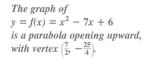

Notice that the endpoints are included. (Why?)

In Example 2 we solved the inequality $x^2 - 7x + 6 > 0$, which meant that we were looking for the values of x that make $x^2 - 7x + 6$ positive, and found the solution to be $\{x \mid x < 1 \text{ or } x > 6\}$. We now look at this question graphically.

Let's consider the function $y = f(x) = x^2 - 7x + 6$ and examine how the graph of the function $y = f(x)$ relates to finding the solutions to the inequality $x^2 - 7x + 6 > 0$, as well as solutions to the inequality $x^2 - 7x + 6 < 0$. See Figure 8.40.

The graph of
$y = f(x) = x^2 - 7x + 6$
is a parabola opening upward,
with vertex $\left(\frac{7}{2}, -\frac{25}{4}\right)$.

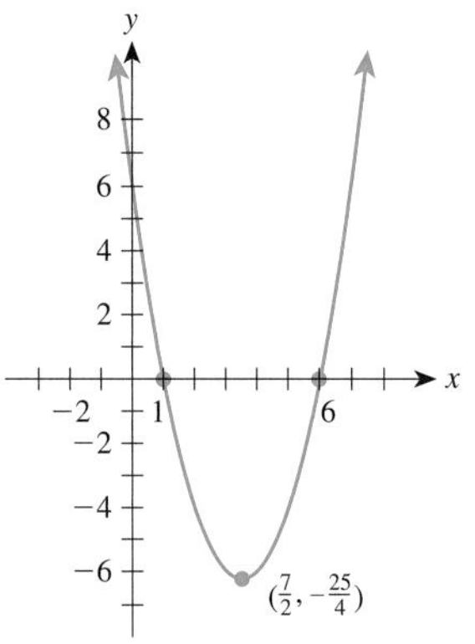

Figure 8.40
The graph of
$y = f(x) = x^2 - 7x + 6$

We can also say $f(1) = 0$ and
$f(6) = 0$.

We can see in Figure 8.40 that the graph of $y = f(x) = x^2 - 7x + 6$ crosses the x-axis at $x = 1$ and $x = 6$ or, equivalently, $y = f(x) = 0$ when $x = 1$ or $x = 6$.

We know that when the graph of any function is below the x-axis, y *is negative*, that is, $y < 0$. For this function, we can see that the graph dips below the x-axis when

x is between 1 and 6; see Figure 8.41. (Notice that we talk about the behavior of y as x changes.)

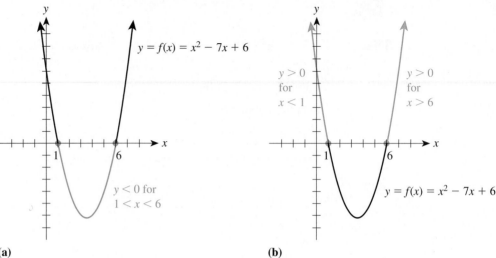

Figure 8.41 **(a)**   **(b)**

Hence,

The graph of $y = f(x)$ dips below the x-axis when $1 < x < 6$,   *which means that*

y is negative or $y < 0$ when $1 < x < 6$.   *Since*
$y = x^2 - 7x + 6,$
we have

$$x^2 - 7x + 6 < 0 \text{ when } 1 < x < 6.$$

Thus, $1 < x < 6$ is the solution to the inequality $x^2 - 7x + 6 < 0$.

A similar analysis of the graph in Figure 8.41**(b)** tells us that $x^2 - 7x + 6$ is greater than 0 for $x < 1$ and for $x > 6$.

In Figure 8.42 we have indicated on the x-axis those values of y for which the function $y = f(x)$ is positive (above the x-axis) or negative (below the x-axis). The result is the same as if we had done a sign analysis.

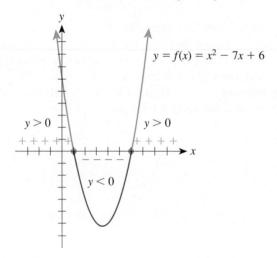

Figure 8.42

This gives us another method for solving the inequality $f(x) > 0$ or $f(x) < 0$.

To solve the inequality $f(x) > 0$, we graph the function $y = f(x)$ and determine when (for which intervals of x) the graph is *above* the x-axis.

To solve the inequality $f(x) < 0$, we graph the function $y = f(x)$ and determine when (for which intervals of x) the graph is *below* the x-axis.

EXAMPLE 4

Use the graph of $y = f(x) = -x^2 - 5x - 2$ given in Figure 8.43 to estimate the solution to the inequality $-x^2 - 5x - 2 < 0$.

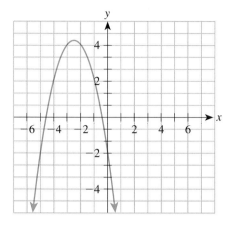

Figure 8.43
The graph of
$f(x) = -x^2 - 5x - 2$

Solution

We are being asked to use the graph to estimate when $-x^2 - 5x - 2$ is less than 0. If we let $y = -x^2 - 5x - 2$, then we can rephrase this question as "When is y less than 0?" Having the graph of $y = -x^2 - 5x - 2$ allows us to interpret this question graphically as "For what values of x does the graph lie *below* the x-axis?"

We estimate that the graph crosses the x-axis at $x \approx -0.5$ and $x \approx -4.5$. We can see by the graph (and what we know about the graph of a quadratic function) that $y < 0$ for $x < -4.5$ and for $x > -0.5$. Hence $-x^2 - 5x - 2 < 0$ when

$x < -4.5$ or $x > -0.5$ or   using interval notation, $(-\infty, -4.5) \cup (-0.5, \infty)$

Solving the actual equation, the cutpoints accurate to one decimal place are −4.6 and −0.4.

We can apply the sign analyses to solve *rational* inequalities such as

$$\frac{a - 7}{a + 2} < 0 \quad \text{or} \quad \frac{6}{a - 3} \leq 3$$

Note that these inequalities are different from those covered in Chapter 6; the rational inequalities in Chapter 6 did not have variables in the denominator.

In attempting to solve $\dfrac{a - 7}{a + 2} < 0$, your first inclination may be to "eliminate the denominator" by multiplying both sides of the inequality by $a + 2$. With *equations* this approach is appropriate, provided we keep in mind $a \neq -2$. With inequalities, however, we need to know whether this multiplier is positive or negative to determine whether the inequality symbol should be reversed. Is $a + 2$ positive or negative? We could reason out an answer on a case-by-case basis, but an easier approach is to use the same method we used for quadratic inequalities, as illustrated in the next example.

Recall that when we multiply both sides of an inequality by a negative number, we change (reverse) the direction of the inequality symbol. If we multiply both sides by a positive number, we leave the inequality symbol alone. What should we do with the inequality symbol when the multiplier is $(a + 2)$?

EXAMPLE 5

Solution

Solve for a.

(a) $\dfrac{a - 7}{a + 2} < 0$ **(b)** $\dfrac{6}{a - 3} \leq 3$

(a) We can analyze a quotient in the same way we analyzed a product: If the signs are the same, the quotient is positive; if the signs are opposite, the quotient is negative.

First we find the cutpoints. In the case of rational expressions, the cutpoints are the points at which the rational expression is 0 or undefined. Note that *the rational expression is 0 when the numerator is 0 and undefined when the denominator is 0*. Thus, the cutpoints are as follows:

The value of a for which $a - 7 = 0 \Rightarrow a = 7$

The value of a for which $a + 2 = 0 \Rightarrow a = -2$

The intervals are $a < -2$, $-2 < a < 7$, and $a > 7$

Test value:

For $a = -5$,	For $a = 0$,	For $a = 10$,
$\dfrac{a-7}{a+2} = \dfrac{-5-7}{-5+2}$	$\dfrac{a-7}{a+2} = \dfrac{0-7}{0+2}$	$\dfrac{a-7}{a+2} = \dfrac{10-7}{10+2}$
is positive.	is negative.	is positive.

Sign of
$\dfrac{a-7}{a+2}$

Since $\dfrac{a-7}{a+2}$ is negative (or less than 0) when $-2 < a < 7$, the solution is

$$\boxed{-2 < a < 7 \quad \text{or} \quad (-2, 7), \text{ using interval notation}}$$

(b) Our method of solving quadratic and rational inequalities is a matter of determining when the expression is positive or negative. Therefore, we must take the original inequality and transform it into the form $R \le 0$, where R is a rational expression.

$$\frac{6}{a-3} \le 3 \qquad \textit{Subtract 3 from both sides of the inequality.}$$

$$\frac{6}{a-3} - 3 \le 0$$

Then we combine the left side into one fraction.

$$\frac{6}{a-3} - \frac{3(a-3)}{a-3} \le 0 \qquad \textit{Find the LCD and change to equivalent fractions.}$$
$$\textit{Add numerators.}$$

$$\frac{6 - 3(a-3)}{a-3} \le 0$$

$$\frac{15 - 3a}{a-3} \le 0$$

The cutpoints are as follows:

The value of a for which $15 - 3a = 0 \Rightarrow a = 5$

The value of a where $a - 3 = 0 \Rightarrow a = 3$

The intervals are $a < 3$, $3 < a < 5$, and $a > 5$.

Given the test value, you should be able to determine the sign of the expression mentally.

Test value:

For $a = 0$,	For $a = 4$,	For $a = 10$,
$\dfrac{15-3a}{a-3}$ is negative.	$\dfrac{15-3a}{a-3}$ is positive.	$\dfrac{15-3a}{a-3}$ is negative.

Thus, $\dfrac{15 - 3a}{a - 3} < 0$ when $a < 3$ or $a > 5$.

Our problem asks for a solution in which the rational expression is less than *or equal to* 0. Since $\dfrac{15 - 3a}{a - 3} = 0$ when $a = 5$, we include the point 5 in the solution.

Therefore, the solution is $\boxed{a < 3 \quad \text{or} \quad a \geq 5}$.

Note that we cannot include the point $a = 3$ since this value makes the fraction undefined.

EXERCISES 8.7

In Exercises 1–8, solve the inequalities and sketch a graph of the solution set.

1. $(x + 4)(x - 2) < 0$ **2.** $(x - 7)(x + 2) > 0$ **3.** $(x + 4)(x - 2) > 0$ **4.** $(x - 7)(x + 2) < 0$

5. $(x + 2)(x - 5) \leq 0$ **6.** $(2x - 1)(x + 3) \geq 0$ **7.** $(x - 3)(2x - 1) \geq 0$ **8.** $(5x - 8)(x - 2) < 0$

In Exercises 9–36, solve the inequalities. Express your answer using interval notation.

9. $a^2 - a - 20 < 0$ **10.** $x^2 - 3x - 18 > 0$ **11.** $x^2 + x - 12 \geq 0$ **12.** $y^2 + 7y - 8 \leq 0$

13. $2a^2 - 9a \leq 5$ **14.** $6x^2 + 7x > -2$ **15.** $6y^2 - y > 1$ **16.** $5a^2 < 7 - 34a$

17. $3x^2 \leq 10 - 13x$ **18.** $2y^2 \geq 5 - 9y$ **19.** $x^2 + 2x + 1 \geq 0$ **20.** $y^2 - 4y + 4 < 0$

21. $x^2 - 6x + 9 < 0$ **22.** $a^2 - 10a + 25 \geq 0$ **23.** $2x^2 - 13x > -15$ **24.** $10x^2 < -2 - 9x$

25. $3y^2 \geq 5y + 2$ **26.** $2x^2 \leq 12 - 5x$ **27.** $x^2 + 2x \leq -1$ **28.** $x^2 - 6x \leq -9$

29. $\dfrac{x - 2}{x + 1} < 0$ **30.** $\dfrac{x + 7}{x + 1} > 0$ **31.** $\dfrac{y + 3}{y - 5} \geq 0$ **32.** $\dfrac{y - 6}{y - 2} \leq 0$

33. $\dfrac{a - 6}{a + 4} < 0$ **34.** $\dfrac{z - 5}{z + 2} \geq 0$ **35.** $\dfrac{5}{y - 4} > 0$ **36.** $\dfrac{7}{y + 2} < 0$

In Exercises 37–50, solve the inequalities and sketch a graph of the solution set.

37. $\dfrac{3}{y - 1} < 1$ **38.** $\dfrac{4}{x + 5} > 1$ **39.** $\dfrac{y + 1}{y + 2} > 3$ **40.** $\dfrac{y + 4}{y - 1} < 2$

41. $\dfrac{2y + 3}{y - 1} \leq 2$ **42.** $\dfrac{x + 3}{2x + 1} \geq 4$ **43.** $\dfrac{y - 3}{y + 1} > -2$ **44.** $\dfrac{a + 3}{a - 5} < -3$

45. $\dfrac{x}{x - 1} \leq \dfrac{3}{x - 1}$ **46.** $\dfrac{a}{2a + 1} > \dfrac{3a + 1}{2a + 1}$ **47.** $\dfrac{x - 2}{x + 3} > \dfrac{x + 4}{x + 3}$ **48.** $\dfrac{y + 1}{y - 1} < \dfrac{y - 1}{y - 1}$

49. $\dfrac{1}{x - 2} + \dfrac{2}{x + 3} \leq \dfrac{3}{x + 3}$ **50.** $\dfrac{2}{x + 1} + \dfrac{3}{x + 2} \geq \dfrac{5}{x + 2}$

51. Use the graph of the function $y = f(x) = -x^2 + 5x - 3$ given in the accompanying figure to estimate the solution to the inequality $-x^2 + 5x - 3 \le 0$.

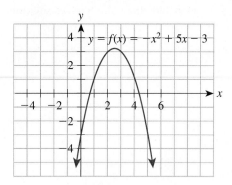

52. Use the graph of the function $y = f(x) = x^2 - 5x - 3$ given in the accompanying figure to estimate the solution to the inequality $x^2 - 5x - 3 \ge 0$.

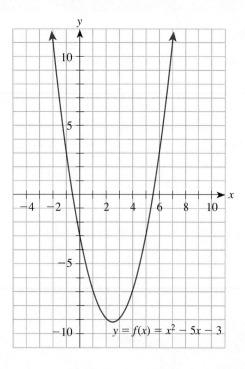

? Questions for Thought

53. For what values of x would $\dfrac{3}{x - 2} = 0$?

54. Discuss what is ***wrong*** (if anything) with the following:

$$\frac{x - 4}{x + 3} < 2$$

$$x - 4 \overset{?}{<} 2(x + 3)$$

$$x - 4 \overset{?}{<} 2x + 6$$

$$-10 \overset{?}{<} x$$

55. Sketch a graph of the function $f(x) = 2x^2 + 5x - 3$, and explain how to use the graph to solve the inequality $2x^2 + 5x - 3 < 0$.

Mini-Review

56. Given $f(x) = \dfrac{2x - 3}{3x - 2}$, find and simplify $f\left(\dfrac{4}{x}\right)$.

57. Suppose that a manufacturer determines that there is a linear relationship between P, the profit earned, and x, the number of items produced. The profit is $600 on 50 items and $750 on 65 items.

   (a) Write an equation relating x and P.

   (b) What would the expected profit be if 90 items were produced?

58. *Simplify:* $(x^{-2}y)^{-1}$ **59.** *Simplify:* $(x^{-2} + y)^{-1}$

8.8 The Distance Formula: Circles

The Distance Formula

In Section 8.2 we discussed the Pythagorean theorem, which gives us the relationship between the sides of a right triangle. We will exploit this relationship to derive a method, or formula, for finding the distance between two points in a Cartesian plane.

Let's begin by first noting that *on the x-axis*, the distance between two points with x-coordinates x_1 and x_2 is $|x_2 - x_1|$ (see Figure 8.44(**a**)). This is also true for any two points on a line *parallel to the x-axis*.

On the y-axis, the distance between two points with y-coordinates y_1 and y_2 is $|y_2 - y_1|$ (see Figure 8.44(**b**)). Again, the same is true for any two points on a line *parallel to the y-axis*.

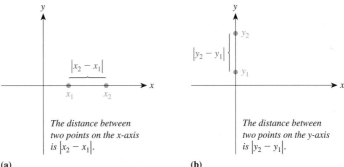

Figure 8.44   (a)   (b)

For example, the distance between points $(4, 0)$ and $(-2, 0)$ (see Figure 8.45(**a**)) is

$$|x_2 - x_1| = |4 - (-2)| = |6| = 6 \quad \text{or} \quad |-2 - (4)| = |-6| = 6$$

The distance between $(4, 3)$ and $(4, -5)$ (see Figure 8.45(**b**)) is

$$|y_2 - y_1| = |3 - (-5)| = |8| = 8 \quad \text{or} \quad |-5 - 3| = |-8| = 8$$

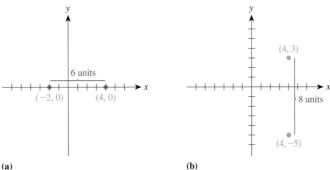

Figure 8.45   (a)   (b)

Now let's take any two points, P_1 with coordinates (x_1, y_1) and P_2 with coordinates (x_2, y_2), and see whether we can find the distance, d, between the two points (see Figure 8.46).

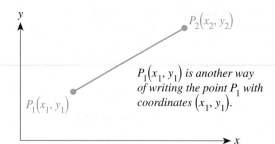

Figure 8.46

First we draw a line through P_2 that is perpendicular to the x-axis (and therefore parallel to the y-axis). Then we draw another line through P_1 that is perpendicular to the y-axis (and therefore parallel to the x-axis). The point at which the lines intersect we will label Q (see Figure 8.47).

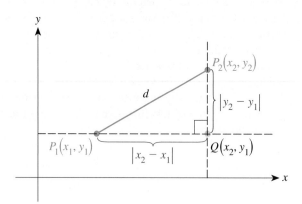

Figure 8.47

We note the following:

Q has coordinates (x_2, y_1). Why?

Triangle $P_1 Q P_2$ is a right triangle with right angle $P_1 Q P_2$.

Because our newly drawn lines are parallel to the x- and y-axes,

The distance between points P_1 and Q is $|x_2 - x_1|$.
The distance between points P_2 and Q is $|y_2 - y_1|$.

Applying the Pythagorean theorem to triangle $P_1 Q P_2$, we have

$$\left(|x_2 - x_1|\right)^2 + \left(|y_2 - y_1|\right)^2 = d^2$$

We can drop the absolute values, since the squares of expressions are automatically nonnegative, to get

$$(x_2 - x_1)^2 + (y_2 - y_1)^2 = d^2 \quad \textit{Then take square roots.}$$
$$\sqrt{(x_2 - x_1)^2 + (y_2 - y_1)^2} = d \quad \textit{We are interested only in the positive root since distances are always nonnegative.}$$

We thus obtain the result stated in the box.

The Distance Formula

The distance, d, between points $P_1(x_1, y_1)$ and $P_2(x_2, y_2)$ is

$$d = \sqrt{(x_2 - x_1)^2 + (y_2 - y_1)^2}$$

EXAMPLE 1

Find the distance between the following pairs of points:

(a) (2, 4) and (5, 8)   **(b)** (4, 2) and (−2, 4)

Solution   In Figure 8.48 we plot the points and show the distances we want to find. Since the differences are squared, it does not matter which point you designate (x_1, y_1) and (x_2, y_2), as long as you remember to subtract x-coordinate from x-coordinate and y-coordinate from y-coordinate.

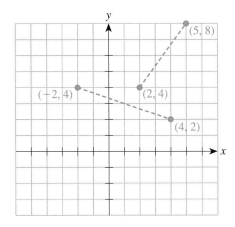

Figure 8.48

(a) Let $(x_1, y_1) = (2, 4)$ and $(x_2, y_2) = (5, 8)$. Then

$$d = \sqrt{(5 - 2)^2 + (8 - 4)^2}$$
$$= \sqrt{3^2 + 4^2}$$
$$= \sqrt{9 + 16}$$
$$= \sqrt{25}$$
$$= \boxed{5}$$

(b) Let $(x_1, y_1) = (-2, 4)$ and $(x_2, y_2) = (4, 2)$. Then

$$d = \sqrt{[4 - (-2)]^2 + (2 - 4)^2}$$
$$= \sqrt{6^2 + (-2)^2}$$
$$= \sqrt{36 + 4}$$
$$= \sqrt{40}$$
$$= \boxed{2\sqrt{10}}$$ *Always simplify radicals in answers.*

EXAMPLE 2

Show that the points $P(1, 1)$, $Q(3, 4)$, and $R(6, 2)$ form the vertices of a right triangle.

Solution   If we join together the three points, we have a triangle. To determine whether it is a right triangle, we refer to the converse of the Pythagorean theorem in Section 8.2, which states that if the sum of the squares of the two sides of a triangle is equal to the square of the third side, then the triangle is a right triangle. Thus, we must show that the sum of the squares of two sides of our triangle is equal to the square of the third side (see Figure 8.49).

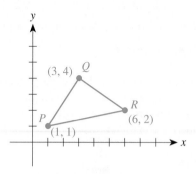

Figure 8.49

The lengths of the sides are computed as follows:
The distance from P to Q is

$$\sqrt{(1-3)^2 + (1-4)^2} = \sqrt{(-2)^2 + (-3)^2} = \sqrt{4+9} = \sqrt{13}$$

The distance from Q to R is

$$\sqrt{(3-6)^2 + (4-2)^2} = \sqrt{(-3)^2 + (2)^2} = \sqrt{9+4} = \sqrt{13}$$

The distance from P to R is

$$\sqrt{(1-6)^2 + (1-2)^2} = \sqrt{(-5)^2 + (-1)^2} = \sqrt{25+1} = \sqrt{26}$$

If we square the distance from P to Q, we get $(\sqrt{13})^2 = 13$.

If we square the distance from Q to R, we get $(\sqrt{13})^2 = 13$.

As we can see, they sum to the square of the distance from P to R: $(\sqrt{26})^2 = 26$.

Hence, the sum of the squares of two sides is equal to the square of the third side and therefore we do have a right triangle.

Alternatively, we can solve this problem by showing that the slope of $\overline{PQ}$ is the negative reciprocal of the slope of $\overline{QR}$ and hence $\overline{PQ}$ is perpendicular to $\overline{QR}$. Either method can be used.

Another formula that we will find useful is the midpoint formula, given in the box.

The Midpoint Formula

The *midpoint* of the line segment joining the two points $P(x_1, y_1)$ and $Q(x_2, y_2)$ has coordinates

$$\left(\frac{x_1 + x_2}{2}, \frac{y_1 + y_2}{2} \right)$$

Hence, the point exactly midway between (2, 5) and (6, 1) has

$$x\text{-coordinate:} \quad \frac{2+6}{2} = \frac{8}{2} = 4 \quad \text{and} \quad y\text{-coordinate:} \quad \frac{5+1}{2} = \frac{6}{2} = 3$$

Thus, the midpoint is (4, 3) (see Figure 8.50). Note that the x-coordinate of the midpoint, $\dfrac{x_1 + x_2}{2}$, is the *average* of the x-coordinates of P and Q, and the y-coordinate of the midpoint, $\dfrac{y_1 + y_2}{2}$, is the *average* of the y-coordinates of P and Q. Do not confuse the midpoint formula with the distance formula, which uses *differences*.

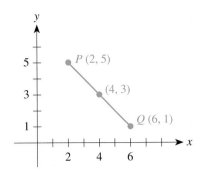

Figure 8.50

EXAMPLE 3	Find the midpoint of the following pair of points: $(6, -5)$ and $(3, 7)$.
Solution	By the midpoint formula, the midpoint is

$$\left(\frac{6 + 3}{2}, \frac{-5 + 7}{2}\right) = \boxed{\left(\frac{9}{2}, 1\right) \quad \text{or} \quad (4.5, 1)}$$

The Circle

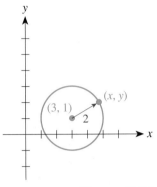

Figure 8.51
Circle with radius 2
and center (3, 1)

A *circle* is defined to be the set of all points whose distance from a fixed point is constant. The fixed point is called the center, *C*, and the constant distance from the center to the circle is called the radius, *r*.

Let's put a circle of radius 2 on a Cartesian plane and center it at the point $(3, 1)$. Pick any point on the circle and call it (x, y) (see Figure 8.51).

Note that by definition of the circle, 2 is the distance from the center, $(3, 1)$, to any point, (x, y), on the circle. Hence, by the distance formula, we have

$$\sqrt{(x - 3)^2 + (y - 1)^2} = 2$$

For convenience, we eliminate the radical by squaring both sides of the equation to get

$$(x - 3)^2 + (y - 1)^2 = 2^2 \quad \textit{This is the equation of the circle with center (3, 1) and radius 2.}$$

In general, we have the result given in the box.

Standard Form of the Equation of a Circle	The equation

$$(x - h)^2 + (y - k)^2 = r^2$$

is the standard form of the equation of a circle with center (h, k) and radius r.

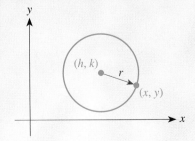

This equation is called the ***standard form of the equation of a circle.***

EXAMPLE 4

Find the equation of the circle with

(a) Center $(3, 5)$ and radius 6

(b) Center $(-2, 4)$ and radius 3

(c) Center $(0, 0)$ and radius 5

Solution

(a) Looking at the standard form, we note that since the center is $(3, 5)$, we have $h = 3$ and $k = 5$. The radius is 6; thus, $r = 6$. Therefore, the equation of the circle is

$$(x - h)^2 + (y - k)^2 = r^2$$
$$(x - 3)^2 + (y - 5)^2 = 6^2$$

$$\boxed{(x - 3)^2 + (y - 5)^2 = 36}$$ *Which, when multiplied out, is*

$$x^2 - 6x + 9 + y^2 - 10y + 25 = 36 \quad \text{or} \quad \boxed{x^2 + y^2 - 6x - 10y - 2 = 0}$$

(b) The circle with center $(-2, 4)$ and radius 3 has $h = -2$, $k = 4$, and $r = 3$. The equation is

$$(x - h)^2 + (y - k)^2 = r^2$$

$$[x - (-2)]^2 + (y - 4)^2 = 3^2 \quad \text{or} \quad \boxed{(x + 2)^2 + (y - 4)^2 = 9}$$

Note that the x-coordinate of the center is negative, which yields the expression $(x + 2)^2$.

(c) The circle with center $(0, 0)$ and radius 5 has $h = 0$, $k = 0$, and $r = 5$. Hence, its equation is

$$(x - 0)^2 + (y - 0)^2 = (5)^2$$

$$\boxed{x^2 + y^2 = 25}$$

Given an equation of a circle in *standard form,* we can easily identify the center and radius, as indicated in the next example.

EXAMPLE 5

Find the center and radius of the following circles and sketch their graphs:

(a) $(x - 5)^2 + (y - 3)^2 = 16$ (b) $(x + 4)^2 + (y - 2)^2 = 9$

Solution

(a) We compare the given equation with the standard form of the equation of a circle:

$$(x - h)^2 + (y - k)^2 = r^2$$
$$(x - 5)^2 + (y - 3)^2 = 16$$

We can identify $h = 5$, $k = 3$, and $r^2 = 16$, which means $r = 4$.

$$\boxed{\text{The center is } (5, 3) \text{ and the radius is } r = 4.}$$

See Figure 8.52.

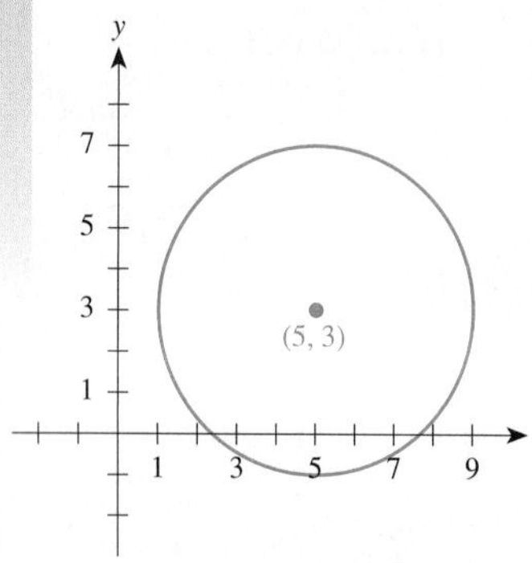

Figure 8.52
Circle with radius 4
and center $(5, 3)$

(b) Note that the left-hand side of $(x + 4)^2 + (y - 2)^2 = 9$ is not quite in standard form. In the standard form, the squared expressions are written as *differences.*

$$(x - h)^2 + (y - k)^2 = r^2$$
$$(x + 4)^2 + (y - 2)^2 = 9 \quad \textit{Which can be written as}$$
$$[x - (-4)]^2 + (y - 2)^2 = 9$$

Hence, $h = -4$, $k = 2$, and $r^2 = 9$, or $r = 3$.

Thus, the center is $(-4, 2)$ and the radius is 3.

See Figure 8.53.

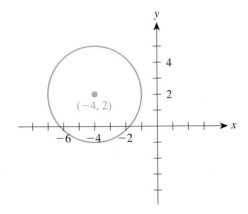

Figure 8.53
Circle with center
$(-4, 2)$ and radius 3

We can see that if we are given an equation of a circle in standard form, then finding the center and radius of the circle is quite straightforward. However, what if the equation is not in standard form, such as $x^2 + y^2 - 14x - 6y - 42 = 0$? How do we find the radius and center?

We want to change the equation into the form $(x - h)^2 + (y - k)^2 = r^2$. Note that the left side of the equation consists of perfect squares. This suggests that we use the technique of completing the square. (You may want to review completing the square in Section 8.3.)

Let's look at an example.

EXAMPLE 6

Find the center and radius of the following circle:

$$x^2 + y^2 - 14x - 6y - 42 = 0$$

Solution

We want to change the equation into standard form. Since standard form consists of perfect squares, we complete the square *for each variable*.

$$x^2 + y^2 - 14x - 6y - 42 = 0 \qquad \text{\textit{Add 42 to both sides of the equation.}}$$

$$x^2 + y^2 - 14x - 6y = 42 \qquad \text{\textit{Separate the terms containing x from the terms containing y.}}$$

$$(x^2 - 14x \quad) + (y^2 - 6y \quad) = 42 \qquad \text{\textit{Complete the square for each quadratic expression: Take half the middle-term coefficient and square it.}}$$

$$\left[\tfrac{1}{2}(-14)\right]^2 = (-7)^2 = 49$$

$$\left[\tfrac{1}{2}(-6)\right]^2 = (-3)^2 = 9$$

Add both numbers, 9 and 49, to both sides of the equation.

$$(x^2 - 14x + \boxed{49}) + (y^2 - 6y + \boxed{9}) = 42 + \boxed{49} + \boxed{9} \qquad \text{\textit{Rewrite quadratic expressions in factored form and simplify the right side.}}$$

$$(x - 7)^2 + (y - 3)^2 = 100$$

Thus, we have a circle with   center $(7, 3)$ and radius $\sqrt{100} = 10$.

EXAMPLE 7

Find the center and radius of the following circles:

(a) $x^2 + y^2 + 6x - 4y = 12$ **(b)** $2x^2 + 2y^2 + 20y = 10$

Solution   **(a)**

$$x^2 + y^2 + 6x - 4y = 12$$

Separate the terms containing x from the terms containing y.

$$(x^2 + 6x \quad) + (y^2 - 4y \quad) = 12$$

Complete the square for each quadratic expression:

$$\left[\frac{1}{2}(6)\right]^2 = 3^2 = 9$$

$$\left[\frac{1}{2}(-4)\right]^2 = (-2)^2 = 4$$

Add both numbers to both sides of the equation.

$$(x^2 + 6x + \boxed{9}) + (y^2 - 4y + \boxed{4}) = 12 + \boxed{9} + \boxed{4}$$

Rewrite quadratic expressions in factored form.

$$(x + 3)^2 + (y - 2)^2 = 25$$
$$\uparrow$$

Note that this sign is positive, so the x-coordinate of the center is negative.

Thus, we have a circle with $\boxed{\text{center } (-3, 2) \text{ and radius } \sqrt{25} = 5}$.

(b) Since the standard form of the equation of a circle has a coefficient of 1 in the square terms, divide both sides of the equation by 2.

$$2x^2 + 2y^2 + 20y = 10$$

Divide both sides of the equation by 2.

$$\frac{2(x^2 + y^2 + 10y)}{2} = \frac{10}{2}$$

$$x^2 + y^2 + 10y = 5$$

We now put the equation in standard form. Separate terms containing x from terms containing y.

$$x^2 + (y^2 + 10y \quad) = 5$$

We have only to complete the square for the quadratic expression in y:

$$\left[\frac{1}{2}(10)\right]^2 = 5^2 = 25$$

Add 25 to both sides.

$$x^2 + (y^2 + 10y + \boxed{25}) = 5 + \boxed{25}$$

Rewrite quadratic expressions in factored form.

$$(x - 0)^2 + (y + 5)^2 = 30$$

Thus, we have a circle with $\boxed{\text{center } (0, -5) \text{ and radius } \sqrt{30}}$.

Observe that with completing the square, we can take any equation of the form $Ax^2 + By^2 + Cx + Dy + E = 0$ and put it in the standard form of a circle *if the coefficients of the squared terms, A and B, are 1 (or can be made to equal 1).*

EXAMPLE 8

Find the equation of a circle whose diameter has endpoints $(2, 3)$ and $(-4, 7)$.

Solution

We draw a diagram (Figure 8.54) so that we may visualize what is given and what needs to be found. Let's analyze this problem carefully to develop a strategy for the solution.

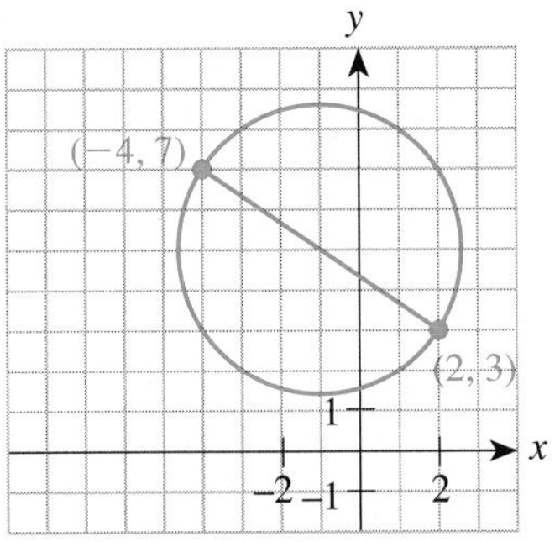

Figure 8.54
The circle with diameter
having endpoints (2, 3)
and (−4, 7)

THINKING OUT LOUD

What do we need to find?	The equation of the circle with the given diameter
What is needed to find an equation of a circle?	The center and the radius
What information is given in the problem?	The diameter's endpoints
How can I restate the problem in simpler terms?	Find the center and radius of the circle given the endpoints of a diameter.
What additional knowledge or information do I need to solve this simpler problem?	That the center of the circle is the midpoint of the diameter (requiring the midpoint formula), and that the radius of a circle is the distance from the center of the circle to any point of the circle (requiring the distance formula)

To locate the center, we find the midpoint between (2, 3) and (−4, 7), which is

$$\left(\frac{2 + (-4)}{2}, \frac{3 + 7}{2}\right) = (-1, 5)$$

Therefore, the center of the circle is (−1, 5).

To find the radius we use the distance formula to find the distance between the center, (−1, 5), and one of the endpoints of the diameter; we use the endpoint (2, 3):

$$r = \sqrt{(-1 - 2)^2 + (5 - 3)^2} = \sqrt{9 + 4} = \sqrt{13} \quad \textit{Hence the radius is } \sqrt{13}.$$

Can you find the radius using only the endpoints of the diameter?

Having found that the center is (−1, 5), and the radius is $\sqrt{13}$, the equation of the circle is

$$(x + 1)^2 + (y - 5)^2 = \left(\sqrt{13}\right)^2 \quad \text{or} \quad \boxed{(x + 1)^2 + (y - 5)^2 = 13}$$

EXERCISES 8.8

In Exercises 1–12, find the distance between the two given points, P and Q.

1. $P(3, 5)$ and $Q(6, 9)$

2. $P(0, 2)$ and $Q(12, 7)$

3. $P(6, 3)$ and $Q(3, 6)$

4. $P(5, 7)$ and $Q(7, 5)$

5. $P(6, -9)$ and $Q(-6, 9)$

6. $P(-5, 4)$ and $Q(5, -4)$

7. $P(-8, -3)$ and $Q(-7, -3)$

8. $P(-6, 2)$ and $Q(-6, -4)$

9. $P\left(\dfrac{1}{2}, 0\right)$ and $Q\left(\dfrac{1}{3}, 2\right)$

10. $P\left(\dfrac{1}{4}, 2\right)$ and $Q\left(\dfrac{1}{3}, 0\right)$

11. $P(1.7, 1.2)$ and $Q(1.4, 0.8)$

12. $P(1.4, 3.8)$ and $Q(0.9, 2.6)$

In Exercises 13–20, find the midpoint between the two given points, P and Q.

13. $P(0, 5)$ and $Q(0, 7)$

14. $P(7, 0)$ and $Q(9, 0)$

15. $P(-3, 1)$ and $Q(3, 1)$

16. $P(6, -2)$ and $Q(-6, -2)$

17. $P(-3, 4)$ and $Q(3, -4)$

18. $P(5, -2)$ and $Q(-5, 2)$

19. $P\left(\dfrac{2}{5}, \dfrac{3}{4}\right)$ and $Q\left(\dfrac{1}{3}, 2\right)$

20. $P\left(\dfrac{2}{3}, \dfrac{3}{5}\right)$ and $Q\left(3, \dfrac{1}{2}\right)$

In Exercises 21–24, determine whether the given points are vertices of right triangles.

21. $P(5, 2)$, $Q(8, 2)$, $R(8, 6)$

22. $P(3, 5)$, $Q(7, 5)$, $R(6, 9)$

23. $P(4, 2)$, $Q(-1, 5)$, $R(3, 9)$

24. $P(0, 8)$, $Q(2, 11)$, $R(5, 9)$

In Exercises 25–28, plot the points P, Q, R, and S. If the diagonals of a quadrilateral bisect each other, then the quadrilateral is a parallelogram. Knowing this, determine whether the quadrilateral formed by P, Q, R, and S is a parallelogram.

25. $P(5, 3)$, $Q(7, 4)$, $R(9, 7)$, $S(7, 6)$

26. $P(3, 9)$, $Q(0, 7)$, $R(7, 1)$, $S(10, 3)$

27. $P(-2, 3)$, $Q(5, -4)$, $R(-6, 5)$, $S(3, -4)$

28. $P(3, -5)$, $Q(6, -2)$, $R(9, -3)$, $S(6, -6)$

In Exercises 29–38, give the equation of the circle, given the center C and radius r.

29. $C(0, 0)$, $r = 1$

30. $C(0, 0)$, $r = 5$

31. $C(1, 0)$, $r = 7$

32. $C(0, 1)$, $r = 1$

33. $C(2, 5)$, $r = 6$

34. $C(5, 2)$, $r = 1$

35. $C(6, -2)$, $r = 5$

36. $C(-6, 2)$, $r = 4$

37. $C(-3, -2)$, $r = 1$

38. $C(-2, -3)$, $r = 1$

In Exercises 39–48, find the center and radius of the circle.

39. $x^2 + y^2 = 16$

40. $x^2 + y^2 = 36$

41. $x^2 + y^2 = 24$

42. $x^2 + y^2 = 98$

43. $(x - 3)^2 + y^2 = 16$

44. $(x - 4)^2 + y^2 = 25$

45. $(x - 2)^2 + (y - 1)^2 = 1$

46. $(x - 5)^2 + (y - 3)^2 = 36$

47. $(x + 1)^2 + (y - 3)^2 = 25$

48. $(x - 2)^2 + (y + 7)^2 = 49$

In Exercises 49–58, find the center and radius, and graph the given equation of the circle.

49. $(x + 2)^2 + (y + 3)^2 = 32$

50. $(x + 5)^2 + (y + 3)^2 = 27$

51. $(x + 7)^2 + (y + 1)^2 = 2$

52. $(y + 5)^2 + (x + 2)^2 = 8$

53. $x^2 + y^2 - 2x = 15$

54. $x^2 + y^2 - 6y = 7$

55. $x^2 + y^2 - 4x - 2y = 20$

56. $x^2 + y^2 - 2y = 6 + 6x$

57. $x^2 + y^2 - 2x = 20 + 4y$

58. $x^2 + y^2 - 2x - 6y = 6$

In Exercises 59–66, find the center and radius of the circle.

59. $x^2 + 10y = 71 - y^2 + 4x$

60. $x^2 + y^2 = 6x - 14y - 32$

61. $x^2 + y^2 = 2y - 6x - 2$

62. $2x - 6y = 2 + x^2 + y^2$

63. $2x^2 + 2y^2 - 4x + 4y = 22$

64. $3x^2 + 3y^2 + 18x - 6y = 45$

65. $x^2 + y^2 - x + 2y = \dfrac{59}{4}$

66. $x^2 + y^2 - 2x - 3y = \dfrac{75}{4}$

In Exercises 67–70, graph the given equation.

67. $x^2 + y^2 = 16$

68. $x^2 = 36 - y^2$

69. $x + y = 4$

70. $y = 6 - x$

71. Find the equation of the circle with a diameter having endpoints $(-2, 8)$ and $(4, -5)$.

72. Find the equation of the circle with a diameter having endpoints $(-3, 5)$ and $(-4, 0)$.

73. Find the equation of the circle passing through the point $(2, 6)$ with center $(3, -5)$.

74. Find the equation of the circle passing through the point $(3, -2)$ with center $(2, 5)$.

75. Find the circumference of the circle passing through the point $(-3, 4)$ with center $(5, 2)$.

76. Find the area of the circle passing through the point $(3, -2)$ with center $(5, 2)$.

77. Find the equation of the circle tangent to the x-axis with center $(3, -2)$.

78. Find the equation of the circle tangent to the y-axis with center $(3, -2)$.

79. Find the equation of the circle tangent to the x-axis at $(3, 0)$ and tangent to the y-axis at $(0, -3)$.

? Questions for Thought

80. Explain why we can drop the absolute value symbols in the equation

$$(|x_2 - x_1|)^2 + (|y_2 - y_1|)^2 = d^2$$

81. Show that the point with coordinates $\left(\dfrac{x_1 + x_2}{2}, \dfrac{y_1 + y_2}{2} \right)$ is the same distance

from (x_1, y_1) as it is from (x_2, y_2).

82. Describe the graph of the following:

(a) $x^2 + y^2 = 0$

(b) $x^2 + y^2 = -4$

Mini-Review

83. Use scientific notation to determine whether $\dfrac{83{,}700}{0.0042}$ is closest to 10^5, 10^6, 10^7, or 10^8.

84. *Simplify:* $\dfrac{x^{-1} + y^{-2}}{xy^{-1}}$

85. *Express in radical form:* $2x^{1/2} - (5x)^{2/3}$

86. *Evaluate:* $\left(-\dfrac{1}{8} \right)^{-4/3}$

CHAPTER 8 SUMMARY

After having completed this chapter you should be able to:

1. Create or use a mathematical model that describes a quadratic relationship between variables in an application, and create a table or rough graph to estimate certain values, including the maximum or minimum value (Section 8.1).

   For example: A company finds that if it spends x dollars on advertising, it earns a profit, $P(x)$, given by the function

   $$P(x) = 250 + 30x - 0.04x^2$$

   Compute the profit for advertising expenditures of $100, $300, $500, and $700, and use this information to estimate the amount to be spent on advertising that would maximize the profit.

   Solution:

   We are looking for the value of $P(x)$ when $x = 100$, 300, 500, and 700. Using function notation, this means we are looking for $P(100)$, $P(300)$, $P(500)$, and $P(700)$.

To find P(a), substitute x = a in	$P(x) = 250 + 30x - 0.04x^2$
To find P(100), substitute x = 100.	$P(100) = 250 + 30(100) - 0.04(100)^2 = 2{,}850$
To find P(300), substitute x = 300.	$P(300) = 250 + 30(300) - 0.04(300)^2 = 5{,}650$
To find P(500), substitute x = 500.	$P(500) = 250 + 30(500) - 0.04(500)^2 = 5{,}250$
To find P(700), substitute x = 700.	$P(700) = 250 + 30(700) - 0.04(700)^2 = 1{,}650$

   We can represent the data as ordered pairs, $(x, P(x))$, letting P represent the dependent variable (and therefore the vertical axis), and plot the data points as shown in Figure 8.55. $P(x)$ is a quadratic function, and, since we know that the graph of a quadratic function will give us a parabola, we can "fit" the parabola around the data points to get the rough graph of the function $P(x)$ shown in Figure 8.55.

   We can see from this graph that the maximum, or highest point, occurs somewhere between $x = 350$ and 400. This means that an advertising expenditure of $350 to $400 will produce a maximum profit. Note also that the highest value of $P(x)$ is above 5,500. To get a better estimate of this value, we can compute $P(375)$ (halfway between 350 and 400) to get $P(375) = 5{,}875$. Hence the maximum profit seems to lie between $5,500 and $6,000.

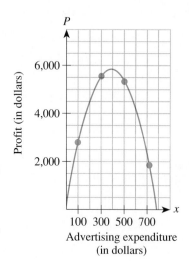

Figure 8.55
The graph of
$P(x) = 250 + 30x - 0.04x^2$

Profit (in dollars)

Advertising expenditure
(in dollars)

2. Solve quadratic equations by the square root method or factoring method (Section 8.2).

   For example: Solve the following:

   (a) $10x^2 + 13x = 3$ **(b)** $\dfrac{2x + 3}{x + 3} = \dfrac{x - 2}{x - 1}$

   Solution:

   (a) $10x^2 + 13x = 3$ *Put in standard form.*

   $10x^2 + 13x - 3 = 0$ *Factor.*

   $(5x - 1)(2x + 3) = 0$ *Set each factor equal to 0.*

   $5x - 1 = 0$ or $2x + 3 = 0$ *Solve each equation.*

   $$x = \frac{1}{5} \quad \text{or} \quad x = -\frac{3}{2}$$

   (b) $\dfrac{2x + 3}{x + 3} = \dfrac{x - 2}{x - 1}$ *Eliminate denominators [multiply both sides by $(x - 1)(x + 3)$].*

   $(2x + 3)(x - 1) = (x + 3)(x - 2)$ *Simplify each side.*

   $2x^2 + x - 3 = x^2 + x - 6$

   $x^2 = -3$ *Take square roots.*

   $x = \pm\sqrt{-3}$ *There are no real solutions, but we do get two complex solutions.*

   $x = \pm i\sqrt{3}$

   The solutions are $-i\sqrt{3}$ and $+i\sqrt{3}$.

   Check to ensure that the denominators are not 0.

3. Solve a quadratic equation by completing the square (Section 8.3).

   For example: Solve for y. $2y^2 + 2y - 3 = 0$

   Solution:

   $2y^2 + 2y - 3 = 0$ *Divide both sides of the equation by 2.*

   $y^2 + y - \dfrac{3}{2} = 0$ *Next, add $\dfrac{3}{2}$ to both sides of the equation.*

   $y^2 + y = \dfrac{3}{2}$ *Take $\dfrac{1}{2}$ of the middle-term coefficient and square it: $\left[\dfrac{1}{2}(1)\right]^2 = \dfrac{1}{4}$.*

   Add $\dfrac{1}{4}$ to both sides of the equation.

   $y^2 + y + \dfrac{1}{4} = \dfrac{3}{2} + \dfrac{1}{4}$ *Write the left side as a perfect square and simplify the right side.*

   $\left(y + \dfrac{1}{2}\right)^2 = \dfrac{7}{4}$ *Take square roots.*

   $y + \dfrac{1}{2} = \pm\sqrt{\dfrac{7}{4}}$ *Isolate y.*

   $y = -\dfrac{1}{2} \pm \sqrt{\dfrac{7}{4}}$ *Simplify the answer.*

   $y = \dfrac{-1 \pm \sqrt{7}}{2}$

4. Solve a quadratic equation by using the quadratic formula (Section 8.4).

For example: $3t^2 - 2t - 2 = 0$

Solution: $A = 3$, $B = -2$, $C = -2$

$$t = \frac{-(-2) \pm \sqrt{(-2)^2 - 4(3)(-2)}}{2(3)}$$

$$t = \frac{2 \pm \sqrt{4 + 24}}{6} = \frac{2 \pm \sqrt{28}}{6} = \frac{2 \pm \sqrt{4 \cdot 7}}{6}$$

$$= \frac{2 \pm 2\sqrt{7}}{6}$$

$$= \frac{\overset{1}{\cancel{2}}(1 \pm \sqrt{7})}{\underset{3}{\cancel{6}}}$$

The solutions are $\dfrac{1 \pm \sqrt{7}}{3}$, which are approximately -0.55 and 1.22 (accurate to two decimal places).

5. Determine the nature of the roots of a quadratic equation without directly solving the equation (Section 8.4).

For example: Determine the nature of the roots of

$$3x^2 - 2x + 5 = 0$$

Solution: $A = 3$, $B = -2$, $C = 5$

The discriminant, $B^2 - 4AC = (-2)^2 - 4(3)(5) = 4 - 60 = -56 < 0$

Therefore, the roots are not real.

6. Solve radical equations (Section 8.5).

For example: Solve for x. $\sqrt{2x + 1} + 7 = x$

Solution:

$\sqrt{2x + 1} + 7 = x$	*Isolate the radical.*
$\sqrt{2x + 1} = x - 7$	*Square both sides of the equation.*
$\left(\sqrt{2x + 1}\right)^2 = (x - 7)^2$	*Simplify.*
$2x + 1 = x^2 - 14x + 49$	*Solve the quadratic equation; put into standard form.*
$x^2 - 16x + 48 = 0$	*Solve by factoring.*
$(x - 4)(x - 12) = 0$	
$x = 4 \quad \text{and} \quad x = 12$	*These are two possible solutions to the original radical equation.*

If you check these values in the original equation, you will find that $x = 4$ is extraneous. The solution is $x = 12$.

7. Solve equations in quadratic form (Section 8.5).

For example: Solve for x. $x^{2/3} + 2x^{1/3} = 15$

Solution: Let $u = x^{1/3}$. [Then $u^2 = (x^{1/3})^2 = x^{2/3}$.] Substitute u for $x^{1/3}$ and

$$x^{2/3} + 2x^{1/3} = 15$$

becomes $\quad u^2 + 2u = 15$

$$u^2 + 2u - 15 = 0 \qquad \textit{Solve for u.}$$

$$(u + 5)(u - 3) = 0$$

$$u = -5 \quad \text{or} \quad u = 3$$

Since $x^{1/3} = u$,

$$x^{1/3} = -5 \quad \text{or} \quad x^{1/3} = 3 \qquad \textit{Solve for x. Cube both sides of the equation.}$$
$$(x^{1/3})^3 = (-5)^3 \quad \text{or} \quad (x^{1/3})^3 = (3)^3$$
$$x = -125 \quad \text{or} \quad x = 27$$

8. Graph a quadratic function and identify its vertex, axis of symmetry, and intercepts (if they exist) (Section 8.6).

   For example: Graph the following parabola and identify its vertex, axis of symmetry, and intercepts (if they exist).

   $$y = 2x^2 + 4x - 6$$

   Solution: The parabola is in the form $y = Ax^2 + Bx + C$, with $A = 2$, $B = 4$, and $C = -6$.

   The axis of symmetry is $x = -\dfrac{B}{2A} = -\dfrac{4}{2(2)} = -\dfrac{4}{4} = -1$

   The x component of the vertex is -1. The y component of the vertex is

   $$y = 2(-1)^2 + 4(-1) - 6 = -8$$

   The vertex is $(-1, -8)$.

   The y-intercept occurs at $x = 0$: $y = 2(0)^2 + 4(0) - 6 = -6$

   The x-intercepts occur at $y = 0$: $0 = 2x^2 + 4x - 6$

   $$0 = 2(x - 1)(x + 3)$$

   The x-intercepts are 1 and -3.

   The graph of the parabola is shown in Figure 8.56.

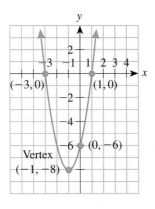

Figure 8.56
Graph of $y = 2x^2 + 4x - 6$

9. Solve verbal problems involving quadratic equations (Sections 8.1–8.8).

10. Solve quadratic and rational inequalities (Section 8.7).

    For example: Solve for x. $x^2 - 9x - 10 \le 0$

    Solution: We will use a sign analysis:

    $$x^2 - 9x - 10 \le 0$$
    $$(x - 10)(x + 1) \le 0 \qquad \textit{Factor.}$$
    $$x - 10 = 0 \quad x + 1 = 0 \qquad \textit{Find cutpoints.}$$
    $$x = 10 \qquad x = -1$$

The intervals are

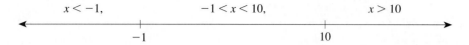

Test value:

For $x = -5$,	For $x = 0$,	For $x = 20$,
$(x - 10)(x + 1) =$	$(x - 10)(x + 1) =$	$(x - 10)(x + 1) =$
$(-5 - 10)(-5 + 1)$	$(0 - 10)(0 + 1)$	$(20 - 10)(20 + 1)$
is positive. Hence,	is negative. Hence,	is positive. Hence,
$(x - 10)(x + 1)$	$(x - 10)(x + 1)$	$(x - 10)(x + 1)$
is positive for	is negative for	is positive for
$x < -1$.	$-1 < x < 10$.	$x > 10$.

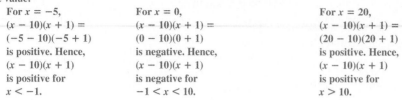

$(x - 10)(x + 1)$ is negative when $-1 < x < 10$.

$(x - 10)(x + 1)$ is 0 when $x = -1$ and $x = 10$.

Therefore, $x^2 - 9x - 10 \le 0$ when $-1 \le x \le 10$. *Note that the cutpoints are included.*

11. Find the distance and midpoint between two points (Section 8.8).

For example: Find the distance between the points $(5, 3)$ and $(-4, 7)$, and find the midpoint between the two points.

Solution: Let $(x_2, y_2) = (5, 3)$ and $(x_1, y_1) = (-4, 7)$. Then the distance between the two points is

$$d = \sqrt{(x_2 - x_1)^2 + (y_2 - y_1)^2} = \sqrt{[5 - (-4)]^2 + (3 - 7)^2}$$
$$= \sqrt{(9)^2 + (-4)^2}$$
$$= \sqrt{81 + 16}$$
$$= \sqrt{97}$$

The midpoint between the two points is found by

$$M = \left(\frac{x_1 + x_2}{2}, \frac{y_1 + y_2}{2}\right) = \left(\frac{5 + (-4)}{2}, \frac{3 + 7}{2}\right)$$
$$= \left(\frac{1}{2}, 5\right)$$

12. Determine the equation of a circle given its radius and center (Section 8.8).

For example: Determine the equation of the circle with center $(5, -2)$ and radius 3.

Solution: The equation of a circle is $(x - h)^2 + (y - k)^2 = r^2$. Substitute $h = 5, k = -2$, and $r = 3$ to get

$$(x - 5)^2 + [y - (-2)]^2 = 3^2$$
$$(x - 5)^2 + (y + 2)^2 = 9$$

13. Graph the equation of a circle and identify its center and radius (Section 8.8).

For example: Graph the following circles and identify the center and radius.

(a) $(x - 3)^2 + (y + 2)^2 = 36$ **(b)** $x^2 + y^2 - 10x + 2y = -16$

Solution:

(a) Since the standard form of a circle is

$$(x - h)^2 + (y - k)^2 = r^2$$

then $h = 3$, $k = -2$, and $r^2 = 36 \Rightarrow r = 6$.

Hence, the center is $(3, -2)$ and the radius is 6.

A graph of the circle is given in Figure 8.57.

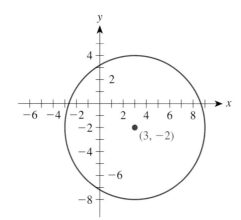

Figure 8.57
Graph of
$(x - 3)^2 + (y + 2)^2 = 36$

(b) We use the method of completing the square to put the equation in standard form:

$$x^2 + y^2 - 10x + 2y = -16$$
$$(x^2 - 10x \quad) + (y^2 + 2y \quad) = -16 \qquad \textit{Complete the squares:}$$

$$\left[\frac{1}{2}(-10)\right]^2 = 25 \qquad \left[\frac{1}{2}(2)\right]^2 = 1^2 = 1$$

Hence, we add $+25$ and $+1$ to both sides of the equation:

$$(x^2 - 10x + 25) + (y^2 + 2y + 1) = -16 + 25 + 1$$
$$(x - 5)^2 + (y + 1)^2 = 10 \qquad h = 5, k = -1, \textit{and } r^2 = 10$$

Therefore, the center is $(5, -1)$ and the radius is $\sqrt{10} \approx 3.2$.

A graph of the circle is given in Figure 8.58.

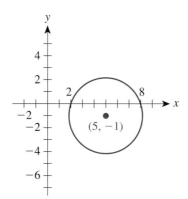

Figure 8.58
Graph of
$x^2 + y^2 - 10x + 2y = -16$

CHAPTER 8 REVIEW EXERCISES

1. Suppose that a motor vehicle consumer ratings service finds that the distance, *d,* in miles that a certain car can travel on one (15-gallon) tank of gasoline depends on its speed, *v,* in miles per hour, according to the function

$$d = d(v) = 18v - \left(\frac{v}{2.3}\right)^2 \qquad \text{for } 10 \le v \le 90$$

(a) To the nearest tenth of a mile, determine the distance the car can travel on a tank of gas at a constant speed of 30 mph, 40 mph, 50 mph, and 60 mph.

(b) Estimate the constant speed that must be maintained for the car to travel 400 miles on one tank (to the nearest mile per hour).

2. In economics, the demand function for a given item indicates how the price per unit, p, is related to the number of units, x, that are sold. Suppose a company finds that the demand function for one of the items it produces is

$$p = 80 - \frac{x}{6}$$

where p is in dollars.

(a) How many items would be sold if the price were \$40 per unit?

(b) What should the price per unit be if 200 units are to be sold?

(c) Sketch a graph of this function.

(d) The revenue function, $R(x)$, is found by multiplying the price per unit, p, by the number of items sold, x (that is, $R = xp$). Find the revenue function corresponding to this demand function.

(e) Use the revenue function found in part (d) to find the revenue if the number of items sold is 150, 200, 250, 300, and 350; and describe what happens to the revenue as the company increases its sale of items.

(f) Estimate how many items should be sold to maximize the revenue. Estimate the maximum revenue.

Solve Exercises 3–18 by factoring or the square root method.

3. $2y^2 - y - 1 = 0$ 4. $3a^2 - 13a - 10 = 0$

5. $3x^2 - 17x = 28$ 6. $10a^2 - 3a = 1$

7. $81 = a^2$ 8. $0 = y^2 - 65$

9. $z^2 + 7 = 2$ 10. $3r^2 + 5 = r^2$

11. $4x^2 + 36 = 24x$ 12. $5a^2 - 5a = 10$

13. $(a + 7)(a + 3) = (3a + 1)(a + 1)$ 14. $(y + 2)(y + 1) = (3y + 1)(y - 1)$

15. $x - 2 = \dfrac{1}{x + 2}$ 16. $a - 5 = \dfrac{1}{a - 5}$

17. $\dfrac{2}{x - 2} - \dfrac{5}{x + 2} = 1$ 18. $\dfrac{3}{x + 4} + \dfrac{5}{x + 2} = 6$

Solve Exercises 19–26 by completing the square.

19. $x^2 + 2x - 4 = 0$ 20. $x^2 - 2x - 4 = 0$

21. $2y^2 + 4y - 3 = 0$ 22. $2y^2 + 4y + 3 = 0$

23. $3a^2 + 6a - 5 = 0$ 24. $3a^2 - 6a + 5 = 0$

25. $\dfrac{1}{a - 5} + \dfrac{3}{a + 2} = 4$ 26. $\dfrac{3}{a - 1} - \dfrac{1}{a + 2} = 2$

Solve Exercises 27–48 by any method.

27. $6a^2 - 13a = 5$ 28. $5x^2 - 18x = 8$

29. $3.2a^2 - 5.8a + 4 = 9.6$ 30. $2y^2 + 5y = 7$

31. $5a^2 - 3a = 3 - 3a + 2a^2$ 32. $5a^2 + 6a - 4 = 3a^2 + 10a + 26$

33. $(x - 4)(x + 1) = x - 2$

34. $(y + 5)(y - 7) = 3$

35. $(t + 3)(t - 4) = t(t + 2)$

36. $(u - 5)(u + 1) = (u - 3)(u + 5)$

37. $8x^2 = 12$

38. $3x^2 - 14 = 5$

39. $3x^2 - 2x + 5 = 7x^2 - 2x + 5$

40. $7.4x^2 - 3.8x = -5.4$

41. $(x + 2)(x - 4) = 2x - 10$

42. $2.8x^2 - 1.4x - 8.7 = 1.3$

43. $\dfrac{1}{z + 2} = z - 4$

44. $\dfrac{3}{z - 2} = z$

45. $\dfrac{1}{x + 4} - \dfrac{3}{x + 2} = 5$

46. $\dfrac{2}{x - 1} - \dfrac{3}{x + 4} = 2$

47. $\dfrac{3}{x - 4} + \dfrac{2x}{x - 5} = \dfrac{3}{x - 5}$

48. $\dfrac{2}{x - 6} - \dfrac{x}{x + 6} = \dfrac{3}{x^2 - 36}$

In Exercises 49–52, solve for the given variable.

49. $A = \pi r^2 h$ for $r > 0$

50. $l = \dfrac{gt^2}{2}$ for $t > 0$

51. $2x^2 + xy - 3y^2 = 0$ for x

52. $5x^2 + 4xy - y^2 = 0$ for y

In Exercises 53–60, solve the equation.

53. $\sqrt{2a + 3} = a$

54. $\sqrt{5a - 4} = a$

55. $\sqrt{3a + 1} + 1 = a$

56. $\sqrt{7y + 4} - 2 = y$

57. $\sqrt{2x + 1} - \sqrt{x - 3} = 4$

58. $\sqrt{3x + 1} - \sqrt{x - 1} = 2$

59. $\sqrt{3x + 4} - \sqrt{x - 3} = 3$

60. $\sqrt{2x - 1} - \sqrt{9 - x} = 1$

In Exercises 61–64, solve for y.

61. $\sqrt{3y + z} = x$

62. $\sqrt{5y - z} = x$

63. $\sqrt{3y} + z = x$

64. $\sqrt{5y} - z = x$

In Exercises 65–82, solve the equation.

65. $x^3 - 2x^2 - 15x = 0$

66. $x^4 - 2x^3 = 35x^2$

67. $4x^3 - 10x^2 - 6x = 0$

68. $9x^3 + 3x^2 - 6x = 0$

69. $a^4 - 17a^2 = -16$

70. $y^4 - 18y^2 = -81$

71. $y^4 - 3y^2 = 4$

72. $a^4 - 5a^2 = 36$

73. $z^4 = 6z^2 - 5$

74. $z^4 = 11z^2 - 18$

75. $a^{1/2} - a^{1/4} - 6 = 0$

76. $a^{2/3} - a^{1/3} - 6 = 0$

77. $2x^{2/3} = 5x^{1/3} + 3$

78. $2x^{1/2} = 5x^{1/4} + 3$

79. $\sqrt{x} + 2\sqrt[4]{x} - 35 = 0$

80. $\sqrt{x} - 5\sqrt[4]{x} + 6 = 0$

81. $3x^{-2} + x^{-1} - 2 = 0$

82. $5x^{-2} - 2x^{-1} - 3 = 0$

In Exercises 83–86, sketch a graph and identify the vertex and axis of symmetry.

83. $y = (x - 2)^2 + 1$

84. $y = -(x - 4)^2 + 5$

85. $y = -3(x - 2)^2 - 4$

86. $y = 2(x + 3)^2 - 6$

In Exercises 87–90, put the equation into the standard form of a parabola and identify its vertex and axis of symmetry.

87. $y = 2x^2 - 12x + 4$

88. $y = 2x^2 - 6x + 1$

89. $y = -x^2 + 4x - 12$

90. $y = -x^2 - 6x$

In Exercises 91–100, graph the parabola and identify its vertex, axis of symmetry, and intercepts (if they exist).

91. $y = 7x^2$

92. $y = -6x^2$

93 $y = -7x^2 + 3$

94. $y = \dfrac{1}{6}x^2 - 3$

95. $y = x^2 - 6x$

96. $y = x^2 - 9x$

97. $y = x^2 - 2x - 8$

98. $y = x^2 - 12x + 35$

99. $y = -x^2 + 2x - 5$

100. $y = -x^2 + 4x - 7$

In Exercises 101–104, solve the inequality.

101. $(x - 2)(x + 1) > 0$

102. $(a + 5)(a - 1) \geq 0$

103. $(3x + 1)(x - 2) \leq 0$

104. $(y - 6)(2y + 1) < 0$

In Exercises 105–114, solve the inequality. Express your answer using internal notation and graph the solution set on the number line.

105. $y^2 - 5y + 4 > 0$

106. $a^2 - 13a + 36 \leq 0$

107. $a^2 < 81$

108. $a^2 > 81$

109. $5s^2 - 18s \geq 8$

110. $2a^2 \leq 4a + 30$

111. $\dfrac{x - 3}{x + 2} < 0$

112. $\dfrac{x + 4}{x - 2} > 0$

113. $\dfrac{x - 3}{x + 2} \geq 0$

114. $\dfrac{x + 4}{x - 2} \leq 0$

In Exercises 115–118, solve the inequality.

115. $\dfrac{2x + 1}{x - 3} < 2$

116. $\dfrac{3}{x - 2} \leq 4$

117. $\dfrac{5}{x + 4} \geq 4$

118. $\dfrac{2 - x}{5 + x} < 2$

119. Use the graph of the function $y = f(x) = -x^2 - 3x + 2$ shown here to estimate the solution to the inequality $-x^2 - 3x + 2 < 0$.

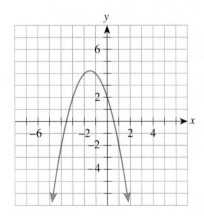

120. Use the graph of the function $y = f(x) = -x^2 - 3x + 2$ given in the figure to estimate the solution to the inequality $-x^2 - 3x + 2 > 0$.

In Exercises 121–128, find the distance between P and Q, and the midpoint between P and Q.

121. $P(0,0)$, $Q(2, 6)$

122. $P(4, 8)$, $Q(2, 12)$

123. $P(2, 5)$, $Q(-2, 5)$

124. $P(3, 8)$, $Q(3, -8)$

125. $P(6, -4)$, $Q(4, -6)$

126. $P(-7, -3)$, $Q(7, 3)$

127. $P(-2, -5)$, $Q(2, 5)$

128. $P(-2, 5)$, $Q(2, -5)$

In Exercises 129–130, determine whether the points P, Q, and R form the vertices of a right triangle.

129. $P(3, 6)$, $Q(5, 9)$, $R(8, 7)$

130. $P(4, 2)$, $Q(6, 5)$, $R(8, 3)$

In Exercises 131–136, graph the circle and indicate its center and radius.

131. $x^2 + y^2 = 100$

132. $x^2 + y^2 = 28$

133. $x^2 + y^2 - 4x - 14y = -52$

134. $x^2 + y^2 - 2x = 15$

135. $x^2 + y^2 - 6x + 4y = 68$

136. $x^2 + y^2 + 4x - 6y = 14$

137. Find the equation of the circle whose diameter has endpoints $(-2, 4)$ and $(6, 8)$.

138. Find the equation of the circle that passes through the point $(3, 6)$ and has the center $(2, -4)$.

139. The profit, *P*, made on a concert is related to the price, *x*, of a ticket in the following way:

$$P(x) = 10,000(-x^2 + 12x - 35)$$

where *P* is the profit in dollars and *x* is the price of a ticket. What ticket price would produce the maximum profit?

140. Jake jumps off a diving board into a pool. The equation $S(t) = -2t^2 + 5t + 5$ gives the distance (in feet) Jake is above the pool at *t* seconds after he jumps.

   (a) What is the maximum height Jake reaches when he dives?

   (b) How high is the diving board?

141. The sum of the square of a number and 4 is 36. What are the numbers?

142. The square of the sum of a number and 4 is 36. What are the numbers?

143. The sum of a number and its reciprocal is $\frac{53}{14}$. Find the number(s).

144. The difference between a number and its reciprocal is $\frac{40}{21}$. Find the number(s).

145. Find the dimensions of a rectangle whose length is twice its width if its area is 50 sq ft.

146. Find the dimensions of a rectangle whose length is 2 more than 3 times its width if its area is 85 sq in.

147. A 5″ by 8″ rectangular picture has a frame of uniform width. If the area of the frame is 114 sq in., what is the width of the frame?

148. A 20′ by 30′ rectangular pool is to be bordered by a cement walkway of uniform width. If the area of the walkway is 216 sq ft, how wide is the walkway?

149. Find the hypotenuse of a right triangle with legs 5″ and 15″.

150. In a right triangle, the hypotenuse is 30 ft and one leg is 15 ft. Find the length of the other leg.

151. Find the length of the diagonals of a 5″ by 4″ rectangle.

152. Find the length of the side of a square with a 20″ diagonal.

153. An airplane heads into the wind to get to its destination 300 miles away, and returns traveling with the wind to its starting point. If the airplane's speed is 200 mph, what was the rate of the wind if it took $3\frac{1}{8}$ hours for the round trip?

154. A shoe manufacturer can sell 3,500 pairs of shoes for $24 each. It is determined that for each 50¢ reduction in price, 150 more pairs of shoes can be sold.

(a) Express the revenue, R, as a function of n, where n is the number of 50¢ reductions.

(b) Determine how many price reductions are needed to maximize the revenue. What price will produce the maximum revenue? Estimate the maximum revenue.

155. To encourage the use of mass transit, a local transit authority projects that 36,000 people will use buses if the fare is $4, and that for each $0.25 decrease in fare, 3,000 additional passengers will take the bus rather than drive.

(a) Express the total revenue from bus fares, $R(x)$, as a function of the number of $0.25 fare decreases.

(b) Using $R(x)$ found in part **(a)**, predict the revenue if the fare is $3.50.

CHAPTER 8 PRACTICE TEST

1. Solve the following by either the factoring method or the square root method.

(a) $(3z - 1)(z - 4) = z^2 - 8z + 7$

(b) $3 + \dfrac{5}{x^2} = 4$

2. Solve the following by any method:

(a) $y^2 + 4y - 1 = 0$

(b) $(a - 2)(a + 1) = 3a^2 - 4$

(c) $\dfrac{3}{x - 2} + \dfrac{3x}{x + 2} = \dfrac{66}{x^2 - 4}$

3. Use the discriminant to determine the nature of the roots of $2x^2 - 3x + 5 = 0$.

4. Solve for $r > 0$: $V = 2\pi r^2 h$

5. Solve the following radical equation: $\sqrt{3x} = 2 + \sqrt{x + 4}$

6. Solve the following: $x^{1/2} - x^{1/4} - 20 = 0$

7. Solve the following and graph the solution set:

(a) $x^2 - 5x \geq 36$

(b) $\dfrac{3x - 2}{x - 6} < 0$

8. Find the length of the diagonal of a square with side equal to 9 ft.

9. A boat travels upstream for 15 miles against a current of 3 mph, then it travels downstream for 12 miles with the same current. If the 27-mile trip took $2\frac{1}{4}$ hours, how fast was the boat traveling in still water?

10. Sketch a graph of the following. Label the x- and y-intercepts, vertex, and axis of symmetry.

(a) $y = 2x^2 + 3x + 1$

(b) $y = -2x^2 + x - 3$

11. The Chico Company found that the daily cost of producing widgets in a factory is related to the number of widgets produced in the following way:

$$C(x) = \dfrac{-x^2}{10} + 100x - 24,000$$

where C is the cost in dollars and x is the number of widgets. What number of widgets would produce the maximum cost? What would this cost be?

12. For $P(2, 5)$ and $Q(-4, 7)$, find the distance between P and Q, and the midpoint between P and Q.

13. Determine the center and radius of the circle $x^2 + y^2 - 8x + 12y = -2$.

9

More on Functions

© Dynamic Graphics Group/Creatas/Alamy

In this chapter we will discuss some additional ideas related to functions. In particular, we will describe various types of functions and how to make new functions from existing ones. Before we proceed, however, let's step back and review some of the basic notions regarding functions.

9.1 More on Function Notation: Split Functions

EXAMPLE 1

Given the function $f(x) = \sqrt{4 - 2x}$, find the following:

(a) $f(1)$ **(b)** $f(-6)$ **(c)** $f(3)$ **(d)** the domain of $f(x)$

Solution

(a) Since $f(x) = \sqrt{4 - 2x}$, we find $f(1)$ by substituting $x = 1$ in $\sqrt{4 - 2x}$.

$$f(x) = \sqrt{4 - 2x}$$

$$f(1) = \sqrt{4 - 2(1)} = \sqrt{2}$$

Thus, $\boxed{f(1) = \sqrt{2}}$.

(b) $f(x) = \sqrt{4 - 2x}$ *To find $f(-6)$, substitute $x = -6$ in $\sqrt{4-2x}$.*

$$f(-6) = \sqrt{4 - 2(-6)} = \sqrt{16} = 4$$

Thus, $\boxed{f(-6) = 4}$.

(c) $f(x) = \sqrt{4 - 2x}$ *To find $f(3)$, substitute $x = 3$ in $\sqrt{4-2x}$.*

$$f(3) = \sqrt{4 - 2(3)} = \sqrt{-2}$$

Recall that according to our ground rules, the (natural) domain of a function is the set of real numbers that make $f(x)$ a real number. Since $\sqrt{-2}$ is not a real number, we say that $\boxed{3 \text{ is not in the domain of } f(x).}$

(d) The domain of a function is the set of real numbers that make $f(x)$ a real number. For the square root function, we need to find the values of x that make the radicand (in this case, $4 - 2x$) nonnegative. (Remember that the square root of the negative number is not a real number.) Since nonnegative means greater than or equal to 0, finding the domain of $f(x)$ can be restated algebraically as solving the inequality $4 - 2x \geq 0$.

$$4 - 2x \geq 0$$

$$-2x \geq -4 \qquad \text{\textit{When we divide by a negative number, we must reverse}}$$
$$\downarrow \qquad\qquad \text{\textit{the direction of the inequality.}}$$

$$\frac{-2x}{-2} \leq \frac{-4}{-2}$$

$$x \leq 2$$

Hence, the domain is $\boxed{\{x \mid x \leq 2\}}$.

From the accompanying table of values we can sketch the graph of $f(x) = \sqrt{4 - 2x}$ shown in Figure 9.1.

x	$f(x) = \sqrt{4 - 2x}$
3	$\sqrt{-2}$ (not real)
2	0
1	$\sqrt{2} \approx 1.4$
0	2
−2	$2\sqrt{2} \approx 2.8$
−6	4

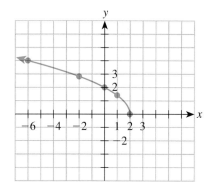

Figure 9.1
The graph of
$y = f(x) = \sqrt{4 - 2x}$

Note that we can determine the domain and range of this function from its graph: The domain is $\{x \mid x \leqslant 2\}$, and the range is $\{y \mid y \geqslant 0\}$.

EXAMPLE 2

Given $f(x) = 3x^2 - 2x + 7$, find the following:

(a) $f(2)$ **(b)** $f(x + 2)$ **(c)** $f(x) + f(2)$ **(d)** $f(x + 2) - f(x)$

Solution

(a) $f(x) = 3x^2 - 2x + 7$ *To find f(2), substitute x = 2 in $3x^2 - 2x + 7$.*

$f(2) = 3(2)^2 - 2(2) + 7$

$= 15$

Remember, $f(x)$ can be thought of as $f(\ ) = 3(\ )^2 - 2(\ ) + 7$.

Thus, $\boxed{f(2) = 15}$.

(b) $f(x) = 3x^2 - 2x + 7$ *To find f(x + 2), substitute $x + 2$ for x in $3x^2 - 2x + 7$.*

$f(x + 2) = 3(x + 2)^2 - 2(x + 2) + 7$

$= 3(x^2 + 4x + 4) - 2(x + 2) + 7$

$= 3x^2 + 12x + 12 - 2x - 4 + 7$

$= 3x^2 + 10x + 15$

Thus, $\boxed{f(x + 2) = 3x^2 + 10x + 15}$.

(c) $f(x) + f(2)$ is the sum of $f(x)$ and $f(2)$. We know what $f(x)$ is, and we already evaluated $f(2)$ in part **(a)** above:

$$f(x) + f(2) = \underbrace{3x^2 - 2x + 7}_{f(x)} + \underbrace{15}_{f(2)}$$

$$= 3x^2 - 2x + 22$$

Thus, $\boxed{f(x) + f(2) = 3x^2 - 2x + 22}$.   *Compare this answer with that of part (b) and note that $f(x + 2) \neq f(x) + f(2)$.*

The parentheses around $f(x)$ are necessary because we must subtract all of $f(x)$.

(d) $f(x + 2) - f(x)$ is the difference between two expressions; we evaluated $f(x + 2)$ in part **(b)** above:

$$f(x + 2) - f(x) = \underbrace{3x^2 + 10x + 15}_{f(x + 2)} - \underbrace{(3x^2 - 2x + 7)}_{f(x)}$$ *Note the parentheses around $f(x)$.*

$$= 3x^2 + 10x + 15 - 3x^2 + 2x - 7$$

$$= 12x + 8$$

Thus, $\boxed{f(x + 2) - f(x) = 12x + 8}$.

EXAMPLE 3

Given $f(x) = 2x^2 - 9$, find $\dfrac{f(x+h)-f(x)}{h}$.

Solution   We recognize that we need to compute the difference between the two expressions, $f(x+h)$ and $f(x)$, and then divide this difference by the variable h. Since we are given $f(x)$, our first step is to find $f(x+h)$:

$$f(x) = 2x^2 - 9 \qquad \text{\textit{To find }} f(x+h), \text{ \textit{substitute} } x+h \text{ \textit{for} } x \text{ \textit{in} } 2x^2 - 9.$$
$$f(x+h) = 2(x+h)^2 - 9$$
$$= 2(x^2 + 2xh + h^2) - 9$$
$$= 2x^2 + 4xh + 2h^2 - 9$$

Now that we have $f(x+h)$, we can find $\dfrac{f(x+h)-f(x)}{h}$.

$$\frac{f(x+h)-f(x)}{h} = \frac{\overbrace{2x^2 + 4xh + 2h^2 - 9}^{f(x+h)} - \overbrace{(2x^2 - 9)}^{f(x)}}{h} \qquad \begin{array}{l}\textit{Note the parentheses}\\\textit{around } f(x).\end{array}$$

$$= \frac{2x^2 + 4xh + 2h^2 - 9 - 2x^2 + 9}{h} \qquad \begin{array}{l}\textit{Simplify the}\\\textit{numerator.}\end{array}$$

$$= \frac{4xh + 2h^2}{h} \qquad \textit{Factor the numerator.}$$

$$= \frac{h(4x + 2h)}{\not h} \qquad \textit{Reduce.}$$

$$= 4x + 2h$$

Thus, $\boxed{\dfrac{f(x+h)-f(x)}{h} = 4x + 2h}$.

Split Functions

The functions we have examined thus far have all had a single rule of assignment for all values in the domain, but this does not necessarily have to be the case. For example, consider the following function:

$$y = f(x) = \begin{cases} 4x - 1 & \text{if } 0 \le x < 3 \\ x^2 - 2x + 3 & \text{if } x \ge 6 \end{cases}$$

This notation means that the rule we use to determine y depends on the input value for x. If the input value x is greater than or equal to 0 and less than 3, then we are to use the top rule, $f(x) = 4x - 1$. If the input value is greater than or equal to 6, then we are to use the bottom rule, $f(x) = x^2 - 2x + 3$.

In particular, we would compute values of this function as follows:

$f(2) = 4(2) - 1 = 7$ *We use the top rule because 2 is a number greater than or equal to 0 and less than 3. This means that when $x = 2$, $y = 7$.*

$f(7) = 7^2 - 2(7) + 3 = 38$ *We use the bottom rule because 7 is a number greater than or equal to 6. This means that when $x = 7$, $y = 38$.*

$f(4)$ is undefined   *because there is no rule of assignment for $x = 4$. In other words, $x = 4$ is not in the domain of this function.*

In fact, the domain of this function is $\{x \mid 0 \le x < 3 \text{ or } x \ge 6\}$.

EXAMPLE 4

Given the function

$$f(x) = \begin{cases} x + 2 & \text{if } x < -1 \\ x^2 - 7 & \text{if } -1 \le x \le 4 \end{cases}$$

(a) Find $f(-4)$.   **(b)** Find $f(3)$.   **(c)** Find $f(-1)$.   **(d)** Find $f(6)$.

(e) Sketch the graph of $f(x)$.

Solution

(a) Remember—which rule we use depends on the value of x. To find $f(-4)$ we first note that $x = -4$ satisfies the top condition for x ($x < -1$), and therefore we use the top rule of $f(x)$ to compute $f(-4)$:

$$f(-4) = (-4) + 2 = -2 \qquad \textit{We use the top rule, } x + 2\textit{, to compute } f(-4)\textit{, since } -4 < -1.$$

(b) To find $f(3)$ we first note that $x = 3$ satisfies the bottom condition $-1 \le x \le 4$, and therefore we use the bottom rule of $f(x)$ to compute $f(3)$:

$$f(3) = (3)^2 - 7 = 2 \qquad \textit{We use the bottom rule, } x^2 - 7\textit{, to compute } f(3) \textit{ since } -1 \le 3 \le 4.$$

(c) $f(-1) = (-1)^2 - 7 = -6$ *We use the bottom rule, $x^2 - 7$, to compute $f(-1)$ since*
$-1 \le -1 \le 4.$

(d) $f(6)$ is undefined. This function is defined only for values for x that are less than or equal to 4. Alternatively, we can say that 6 is not in the domain of the function.

(e) To graph the function $f(x)$, notice that we are actually piecing together portions of two functions, $y = x + 2$ and $y = x^2 - 7$. The first rule gives the line $y = x + 2$, but it is only for values of x less than -1. Hence, we "cut off" the portion of this line for values of x greater than (or equal to) -1. (We cut on the line $x = -1$, as shown in Figure 9.2a.)

The second rule gives the parabola $y = x^2 - 7$, but defined only for values of x between (and including) -1 and 4. Hence, we "cut off" the portions of the line for values of x less than -1 and greater than 4. (We cut on the lines $x = -1$ and $x = 4$, as shown in Figure 9.2b.) We put the two pieces together as shown in Figure 9.3.

Split functions are also called piecewise functions.

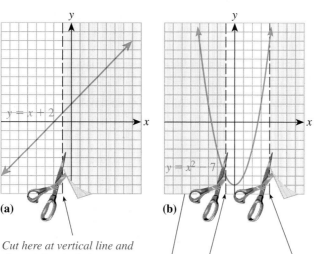

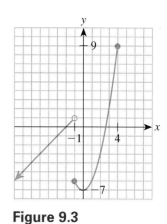

Figure 9.2   (a)   (b)   **Figure 9.3**

Cut here at vertical line and discard the shaded section.

Piece together the remaining portions to get the graph shown in Figure 9.3.

Cut here at vertical line
$x = -1$

Cut here at vertical line
$x = 4$

and

Note that we put an empty circle at the right end of the line (the portion of the graph that is $y = x + 2$). This occurs where $x = -1$, and indicates that the point where $x = -1$ on the *line* portion of the graph is excluded. Remember, the graph of $f(x)$ is a line only for values of x strictly less than -1.

We put a filled-in circle on the left end of the parabola (the portion of the graph that is $y = x^2 - 7$). This occurs where $x = -1$, and indicates that the point where $x = -1$ on the *parabola* portion of the graph is included. The filled-in circle at the right end of the parabola (at $x = 4$) indicates that this point on the parabola is included as well.

EXAMPLE 5

Because of the pressure to complete a project, an employer offers to pay its employees $12 for each hour of regular time per week, and triple-time for each hour of overtime. The first 35 hours worked per week are considered regular time, and all hours in excess of 35 hours per week are considered overtime.

(a) Express an employee's weekly income, I, from this project as a function of h, the number of hours worked.

(b) Use this function to compute the weekly income if the employee works 35 hours, 42 hours, or 48 hours per week on this project.

Solution

(a) Let's examine how we would compute an employee's weekly income. If the employee works 35 hours or less, we simply multiply the number of hours worked by $12 per hour. If the employee works more than 35 hours, he or she would earn $12 per hour for the first 35 hours plus $36 per hour (triple-time) for all hours in excess of 35 per week.

For instance, if an employee works for 38 hours per week, he or she will earn $12 per hour for the first 35 hours and $36 per hour for the additional 3 hours. The weekly income would be $I = 12(35) + 36(38 - 35) = 420 + 36(3) = \$528$.

Similarly, if the employee works h hours per week, the formula we use to compute the salary depends on *whether or not h is greater than* 35. We need two different rules to determine the salary, depending on how many hours are worked. This suggests that we can use a split function to define I as a function of h, the number of hours the employee works per week:

$$I = I(h) = \begin{cases} 12h & \text{if } 0 \le h \le 35 \\ 36(h - 35) + 420 & \text{if } h > 35 \end{cases}$$

In the second part of this split function, we multiply $36 per hour by $(h - 35)$, which is the number of hours *above* 35 that the employee works, and add to that $420 (= 35 \times 12)$, which is the income earned for the first 35 hours.

(b) To compute the weekly income for an employee who works 35, 42, or 48 hours, we must substitute the number of hours into the applicable rule of the split function.

If the employee works 35 hours, we use the first rule to get

$$I(35) = 12(35) = 420$$

If the employee works 42 hours, we use the second rule to get

$$I(42) = 36(42 - 35) + 420 = 36(7) + 420 = 252 + 420 = 672$$

If the employee works 48 hours, we use the second rule to get

$$I(48) = 36(48 - 35) + 420 = 36(13) + 420 = 468 + 420 = 888$$

Therefore, we see that if the employee works 35 hours, he or she earns $420; 42 hours, $672; and 48 hours, $888.

EXERCISES 9.1

In Exercises 1–10, evaluate f(–2), f(0), and f(2).

1. $f(x) = 7x - 2$

2. $f(x) = 2x^2 - 5x + 3$

3. $f(x) = \sqrt{3x - 2}$

4. $f(x) = \sqrt[3]{3x - 2}$

5. $f(x) = \sqrt{2 - x}$

6. $f(x) = \sqrt{1 - x}$

7. $f(x) = \dfrac{2}{x}$

8. $f(x) = \dfrac{x + 2}{x}$

9. $f(x) = \dfrac{x + 2}{x - 2}$

10. $f(x) = \dfrac{3x + 1}{x + 2}$

In Exercises 11–32, determine the domains of the functions.

11. $f(x) = 5x - 3$

12. $g(x) = 3 - 2x$

13. $h(x) = 2x^2 - x + 1$

14. $g(x) = x^2 - 3x + 4$

15. $s(x) = 3x^3 - 1$

16. $f(x) = 2 - 5x^4$

17. $f(x) = \sqrt{2x - 1}$

18. $f(x) = \sqrt{5 - x}$

19. $r(x) = \sqrt{2 - 3x}$

20. $r(x) = \sqrt{2x - 3}$

21. $h(x) = \dfrac{3}{x}$

22. $k(x) = \dfrac{2}{x - 5}$

23. $k(x) = \dfrac{x}{x - 4}$

24. $h(x) = \dfrac{x - 1}{x + 2}$

25. $f(x) = \dfrac{5}{(x - 1)(x + 2)}$

26. $r(x) = \dfrac{x}{(x + 1)(x + 5)}$

27. $h(x) = \dfrac{x}{x^2 - x - 12}$

28. $f(x) = \dfrac{x - 5}{x^2 + x - 12}$

29. $s(x) = \dfrac{2}{\sqrt{x}}$

30. $g(x) = \dfrac{3}{\sqrt{x - 2}}$

31. $h(x) = \dfrac{x + 2}{\sqrt{2x - 1}}$

32. $f(x) = \dfrac{x - 4}{\sqrt{4 - 3x}}$

33. Given $f(x) = 2x - 3$, find the following.

(a) $f(x) + f(5)$

(b) $f(x + 5)$

(c) $f(2x)$

(d) $2f(x)$

(e) $f(x + 5) - f(x)$

(f) $\dfrac{f(x + 5) - f(x)}{5}$

(g) $f(x + h)$

(h) $f(x) + f(h)$

(i) $\dfrac{f(x + h) - f(x)}{h}$

34. Given $g(x) = 3x^2 - 2x + 1$, find the following.

(a) $g(x) + g(4)$

(b) $g(x + 4)$

(c) $g(4x)$

(d) $4g(x)$

(e) $g(x + 4) - g(x)$

(f) $\dfrac{g(x + 4) - g(x)}{4}$

(g) $g(x + h)$

(h) $g(x) + g(h)$

(i) $\dfrac{g(x + h) - g(x)}{h}$

35. Given $f(x) = \dfrac{5}{x - 1}$, find the following.

(a) $f(x) + f(3)$

(b) $f(x + 3)$

(c) $f(3x)$

(d) $3f(x)$

(e) $f(x + 3) - f(x)$

(f) $\dfrac{f(x + 3) - f(x)}{3}$

(g) $f(x + h)$

(h) $f(x) + f(h)$

(i) $\dfrac{f(x + h) - f(x)}{h}$

36. Given $f(x) = \dfrac{2x}{x-3}$, find the following.

(a) $f(x) + f(6)$ (b) $f(x + 6)$ (c) $f(6x)$ (d) $6f(x)$

(e) $f(x + 6) - f(x)$ (f) $\dfrac{f(x+6) - f(x)}{6}$ (g) $f(x + h)$ (h) $f(x) + f(h)$

(i) $\dfrac{f(x+h) - f(x)}{h}$

In Exercises 37–44, sketch the graph of the given function and find the requested values.

37. $f(x) = \begin{cases} x - 4 & \text{if } x < 2 \\ x + 1 & \text{if } x > 2 \end{cases}$

Find $f(0), f(2),$ and $f(3)$.

38. $g(x) = \begin{cases} x - 3 & \text{if } x \le 2 \\ 5 - x & \text{if } x > 2 \end{cases}$

Find $g(-3), g(2),$ and $g(8)$.

39. $h(x) = \begin{cases} 2x + 3 & \text{if } -4 \le x < 2 \\ 6 - x & \text{if } x \ge 2 \end{cases}$

Find $h(-1), h(2),$ and $h(5)$.

40. $f(x) = \begin{cases} 5 - 3x & \text{if } 0 < x \le 3 \\ x - 4 & \text{if } x \ge 4 \end{cases}$

Find $f(2), f(4),$ and $f(10)$.

41. $g(x) = \begin{cases} x^2 - 1 & \text{if } -3 \le x \le 3 \\ 14 - 2x & \text{if } x > 3 \end{cases}$

Find $g(-3), g(3),$ and $g(6)$.

42. $f(x) = \begin{cases} x^2 + 3 & \text{if } x \le 1 \\ 2x + 1 & \text{if } x > 1 \end{cases}$

Find $f(-4), f(1),$ and $f(3)$.

43. $f(x) = \begin{cases} 9 - x^2 & \text{if } -2 \le x < 2 \\ x + 1 & \text{if } x \ge 2 \end{cases}$

Find $f(-3), f(1),$ and $f(5)$.

44. $g(x) = \begin{cases} 1 - x^2 & \text{if } -4 < x \le 1 \\ x - 4 & \text{if } x > 1 \end{cases}$

Find $g(-4), g(1),$ and $g(7)$.

45. A local gas company charges its customers for the monthly number of *therms* used by the household. The company charges $0.61 per therm for the first 200 therms used, and $0.59 per therm for any additional therms used.

(a) Express the total monthly gas usage charge, *C,* as a function of *t,* the number of therms used monthly.

(b) Use this function to compute the monthly charge if the customer uses 150 therms, 200 therms, and 300 therms.

46. A company pays its employees $15 an hour for regular time, and one and a half times the hourly rate for each hour of overtime. The first 40 hours worked per week are considered to be regular time, and all hours in excess of 40 hours per week are considered overtime.

(a) Express an employee's weekly income, *I,* for this company as a function of *h,* the number of hours worked per week.

(b) Use this function to compute the weekly income if the employee works 40 hours, 45 hours, or 50 hours per week.

47. In addition to a $2.50 flat monthly service fee, a local electric company charges its customers for the monthly number of kilowatt-hours (KWH) used by the household. The company charges $0.125 per KWH for the first 560 KWH used, and $0.118 per KWH for any additional KWH used.

(a) Express the total monthly electric usage charge, *C,* as a function of *k,* the number of KWH used monthly.

(b) Use this function to compute the monthly charge if the customer uses 300 KWH, 600 KWH, and 1,000 KWH.

48. For a particular long-distance phone company, the cost to call Spain from the United States is $2.95 for the first 3 minutes and $0.67 for each additional minute. The minimum charge is $2.95 and, except for the first 3 minutes, rates are computed on fractions of a minute.

(a) Express the charge, C, for a call (from the United States to Spain) as a function of m, the number of minutes.

(b) Use this function to compute the charges for a call that lasts 2 minutes, 15 minutes, and 1 hour.

Question for Thought

49. Find the domain of the function $g(x) = \sqrt{3x - x^2}$. (Hint: Recall how to solve a quadratic inequality.)

Mini-Review

50. *Solve for x:* $\dfrac{x}{x + 1} - \dfrac{3}{x - 1} = 1$

51. *Combine and simplify:* $\dfrac{x}{x + 1} - \dfrac{3}{x - 1} + 1$

52. Refer to the accompanying graph of $y = f(x)$.

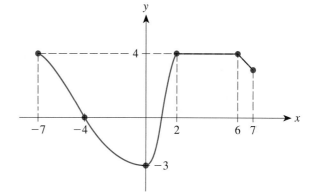

   (a) Determine $f(-4)$.

   (b) Find $f(0)$.

   (c) For what values of x is $f(x) = 4$?

   (d) Which is greater, $f(6)$ or $f(7)$? Explain.

53. Marla makes a round trip by car totaling 120 miles. Her return trip took 15 minutes less than her trip going because she drove 12 mph faster returning than she did going. Find her rate and time in each direction.

9.2   Composition and the Algebra of Functions

Just as we can add, subtract, multiply, and divide two real numbers to create another real number, we now define what it means to perform arithmetic operations on functions. This is often called the *algebra of functions*.

The Algebra of Functions

We can define operations with functions as long as the domains have elements in common. If $f(x)$ and $g(x)$ are functions with some domain elements in common, then we have the definitions listed in the box.

$$(f + g)(x) = f(x) + g(x) \qquad \textit{Sum of two functions}$$

$$(f - g)(x) = f(x) - g(x) \qquad \textit{Difference of two functions}$$

$$(fg)(x) = f(x) \cdot g(x) \qquad \textit{Product of two functions}$$

$$\left(\frac{f}{g}\right)(x) = \frac{f(x)}{g(x)}; \ \ g(x) \neq 0 \qquad \textit{Quotient of two functions}$$

Thus, if $f(x)$ and $g(x)$ are functions, their sum, product, difference, and quotient are functions, provided $f(x)$ and $g(x)$ exist. In the case of the quotient, we must eliminate any value of x that makes $g(x) = 0$ and therefore the quotient $\dfrac{f(x)}{g(x)}$ undefined.

EXAMPLE 1

Given $f(x) = x^2 - 16$ and $g(x) = x - 3$, express each of the following as a function of x:

(a) $(f + g)(x)$ **(b)** $\left(\dfrac{f}{g}\right)(x)$

Solution

(a) $(f + g)(x) = f(x) + g(x)$ *By definition of $(f + g)$*

$\qquad\qquad = (x^2 - 16) + (x - 3)$

$\qquad\qquad = x^2 + x - 19$

> Thus, $(f + g)(x)$ is the function $x^2 + x - 19$

(b) $\left(\dfrac{f}{g}\right)(x) = \dfrac{f(x)}{g(x)}$ *By definition of $\left(\dfrac{f}{g}\right)(x)$*

$\qquad\qquad = \dfrac{x^2 - 16}{x - 3}$

> Thus, $\left(\dfrac{f}{g}\right)(x)$ is the function $\dfrac{x^2 - 16}{x - 3}$ provided $x \neq 3$. (Why?)

Another way of producing new functions from old functions is described in the next subsection.

Composition of Functions

EXAMPLE 2

An oil rig is leaking oil into the ocean, creating a circular slick of oil on the ocean surface (see Figure 9.4). Suppose that the radius is growing at a constant rate of 4 m/min. How large is the oil slick 30 minutes after the start of the leak?

Figure 9.4

Solution

We want to find the area of the oil slick. Since it is circular, we use the formula for the area of a circle, $A = A(r) = \pi r^2$, where A is the area of the circle and r is the radius. We sometimes write $A(r)$ to emphasize that the area is a function of the radius in this formula, and to remind us that to find the area, we must first find the radius.

Since the radius is growing at the rate of 4 m/min, we use distance equals rate $\times$ time to find the length of the radius t minutes after the start of the spill. For example, after 5 minutes, the radius is $5 \times 4 = 20$ m; after 10 minutes, the radius is $10 \times 4 = 40$ m. Hence, $r = r(t) = 4t$, where $r(t)$ is the radius and t is the time in minutes since the leak began. [Again, we use function notation $r(t)$ to emphasize that the radius is a function of time.] We want the radius when $t = 30$ minutes, so we are looking for $r(30)$. We have $r(30) = 4 \times 30 = 120$ m.

At 30 minutes the radius is 120 m, and the area of the circular slick when $r = 120$ is

$$A(r) = \pi r^2 \qquad \textit{Substitute } r = 120 \textit{ to get}$$

$$A(120) = \pi 120^2 = 14,400\pi \qquad \textit{or 45,239 sq m (rounded to the nearest sq m)}$$

EXAMPLE 3

If $f(x) = 2x^3 - x + 5$ and $g(x) = x + 2$, find the following:

(a) $f[g(2)]$ **(b)** $g[f(2)]$

Solution

(a) $f[g(2)]$ means to evaluate $f(x)$ when $x = g(2)$.

$$f(x) = 2x^3 - x + 5 \qquad \textit{Substitute } g(2) \textit{ for } x \textit{ in } f(x).$$

$$f[g(2)] = 2[g(2)]^3 - g(2) + 5 \qquad \textit{Find } g(2) \textit{ by substituting 2 for } x \textit{ in } g(x): \\ g(2) = 2 + 2 = 4. \textit{ Now substitute } \\ g(2) = 4.$$

$$= 2(4)^3 - 4 + 5$$

$$= 129$$

Hence, $\boxed{f[g(2)] = 129}$.

(b) $g[f(2)]$ means to evaluate $g(x)$ when $x = f(2)$.

$$g(x) = x + 2 \qquad \textit{Substitute } f(2) \textit{ for } x \textit{ in } g(x).$$

$$g[f(2)] = f(2) + 2 \qquad \textit{Find } f(2) \textit{ by substituting 2 for } x \textit{ in } f(x): \\ f(2) = 2(2)^3 - 2 + 5 = 19. \textit{ Now substitute } \\ f(2) = 19.$$

$$= 19 + 2 = 21$$

Thus, $\boxed{g[f(2)] = 21}$. $\textit{Note that } f[g(2)] \neq g[f(2)].$

To compute the requested values in Example 3, we had to substitute values into two functions—one function dependent on the other. Once we obtained a value from one function, we used that value in the other function to get the final answer. Mathematically, we describe this process as "evaluating a function within another function."

In Example 2, we had the function $A(r) = \pi r^2$ showing that the area, $A(r)$, is dependent on r, and we also used $r(t) = 4t$ to express the radius as a function of t. In Example 2 we first computed $r(30) = 120$ (that is, the radius is 120 when $t = 30$), and then we computed $A(120)$, which is the area when $r = 120$. In short, we computed $A[r(30)]$.

Suppose for Example 2 we had to compute the area of several values of t. Perhaps we wanted to find the area of the oil slick after 30, 60, and 120 min, etc. If we had a single formula to express the area of the oil slick in terms of the time, then we could reduce the computational work.

EXAMPLE 4

An oil rig is leaking oil into the ocean, creating a circular slick of oil on the ocean surface (see Figure 9.4). Suppose that the radius is growing at a constant rate of 4 m/min. Express the area of the oil slick as a function of t, where t is the number of minutes since the leak began.

Solution   The oil slick is circular, so we use the formula for the area of a circle, $A = A(r) = \pi r^2$, where $A(r)$ is the area of the circle and r is the radius. Since the radius is growing at the rate of 4 m/min, the length of the radius given t is expressed as $r = r(t) = 4t$, where $r(t)$ is the radius and t is the time in minutes:

$$A(r) = \pi r^2 \qquad \text{\textit{Substitute } r = r(t) = 4t.}$$

$$A[r(t)] = \pi (4t)^2 \qquad \text{\textit{Simplify the right side.}}$$

$$A[r(t)] = \pi (16t^2) = 16\pi t^2$$

We write this as $\boxed{a(t) = 16\pi t^2}$ to show that the area is expressed as a function of t. (We cannot use A, as we have already used it to express area as a function of r.)

We now have a direct formula for finding the area of the oil slick given the time, without having to compute the radius. For example, the area of the slick when $t = 30$ minutes is $a(30) = 16\pi(30)^2 = 14{,}400\pi$ sq m, which agrees with the answer we obtained in Example 2.

EXAMPLE 5   If $f(x) = x^2 - 3x + 8$ and $g(x) = x - 7$, find the following:

(a) $f[g(x)]$ **(b)** $g[f(x)]$

Solution   **(a)** $f[g(x)]$ means to evaluate $f(x)$ when the input is $g(x)$.

$$f(x) = x^2 - 3x + 8 \qquad \text{\textit{Substitute } g(x) \text{ \textit{for} } x \text{ \textit{in} } f(x).}$$

$$f[g(x)] = [g(x)]^2 - 3g(x) + 8 \qquad \text{\textit{Since } g(x) = x - 7, \text{ we have}}$$

$$= (x - 7)^2 - 3(x - 7) + 8 \qquad \text{\textit{Simplify.}}$$

$$= x^2 - 14x + 49 - 3x + 21 + 8$$

$$= x^2 - 17x + 78$$

Thus, $\boxed{f[g(x)] = x^2 - 17x + 78}$.

(b) $g[f(x)]$ means to evaluate $g(x)$ when the input is $f(x)$.

$$g(x) = x - 7 \qquad \text{\textit{Substitute } f(x) \text{ \textit{for} } x \text{ \textit{in} } g(x).}$$

$$g[f(x)] = f(x) - 7 \qquad \text{\textit{Since } f(x) = x^2 - 3x + 8, \text{ we have}}$$

$$= (x^2 - 3x + 8) - 7$$

$$= x^2 - 3x + 1$$

Note that $f[g(x)] \neq g[f(x)]$.   Thus, $\boxed{g[f(x)] = x^2 - 3x + 1}$.

EXAMPLE 6

If $f(x) = \dfrac{x + 2}{x - 1}$ and $g(x) = \dfrac{1}{x}$, find $f[g(x)]$ and its domain.

Solution $f[g(x)]$ means to evaluate $f(x)$ when the input is $g(x)$.

$$f(x) = \frac{x + 2}{x - 1} \qquad \text{\textit{Substitute } g(x) \text{ \textit{for} } x \text{ \textit{in} } f(x).}$$

$$f[g(x)] = \frac{g(x) + 2}{g(x) - 1} \qquad \text{\textit{Since } g(x) = \frac{1}{x}, \text{ we have}}$$

$$= \frac{\dfrac{1}{x} + 2}{\dfrac{1}{x} - 1}$$

Simplify the complex fraction:
Multiply numerator and denominator by x.

$$= \frac{\left(\dfrac{1}{x} + 2\right) \cdot x}{\left(\dfrac{1}{x} - 1\right) \cdot x}$$

$$= \frac{1 + 2x}{1 - x}$$

From which we can see that x ≠ 1

To find the domain of $f[g(x)]$, we must determine all values of x for which $f[g(x)]$ is defined. In particular, for the function $f[g(x)]$ to be defined, the function $g(x)$ must be defined. But $g(x)$ is not defined for $x = 0$ [0 is not in the domain of $g(x)$]. Unfortunately, when we simplify $f[g(x)]$, the restrictions on the domain of $g(x)$ may be hidden. Even though the expression $\dfrac{1 + 2x}{1 - x}$ is defined when $x = 0$, since 0 is not in the domain of $g(x)$, it must be excluded from the domain of $f[g(x)]$ as well. Hence, we must explicitly state that

$$f[g(x)] = \frac{1 + 2x}{1 - x}, \qquad \text{for } x \neq 0,\, x \neq 1$$

In general, when we find an expression for $f[g(x)]$, we must explicitly state any domain restrictions for the "inner function" $g(x)$.

Examples 2–5 illustrate an operation on functions called ***composition of functions***. While $f[g(x)]$ is the composition of f with g, $g[f(x)]$ is the composition of g with f. As we have seen, in general, $g[f(x)] \neq f[g(x)]$.

EXERCISES 9.2

In Exercises 1–12, given $f(x) = x^2 - 2x - 3$ and $g(x) = x - 1$, find the following.

1. $(f + g)(0)$ **2.** $(f - g)(0)$ **3.** $(g - f)(2)$ **4.** $(f - g)(2)$

5. $\left(\dfrac{f}{g}\right)(3)$ **6.** $\left(\dfrac{g}{f}\right)(3)$ **7.** $\left(\dfrac{f}{g}\right)(1)$ **8.** $\left(\dfrac{g}{f}\right)(1)$

9. $(f + g)(x)$ **10.** $(f - g)(x)$ **11.** $\left(\dfrac{g}{f}\right)(x)$ **12.** $\left(\dfrac{f}{g}\right)(x)$

In Exercises 13–20, given $h(x) = \dfrac{1}{2x - 1}$ and $g(x) = 2x + 1$, find the following.

13. $(h + g)(2)$ **14.** $(h - g)(2)$ **15.** $(g - h)\left(\dfrac{1}{2}\right)$ **16.** $(h - g)\left(\dfrac{1}{2}\right)$

17. $\left(\dfrac{h}{g}\right)(3)$ **18.** $\left(\dfrac{g}{h}\right)(3)$ **19.** $\left(\dfrac{h}{g}\right)(x)$ **20.** $\left(\dfrac{g}{h}\right)(x)$

In Exercises 21–36, given $f(x) = x^2 - 4$, $g(x) = \sqrt{x + 1}$, and $h(x) = \dfrac{1}{x}$, find (and simplify) the following.

21. $f[g(3)]$

22. $f[g(0)]$

23. $g[f(3)]$

24. $g[f(2)]$

25. $f[g(x)]$

26. $f[g(a)]$

27. $g[f(x)]$

28. $g[f(a)]$

29. $g[h(3)]$

30. $h[g(3)]$

31. $f\left[g\left(\dfrac{1}{2}\right)\right]$

32. $h\left[f\left(\dfrac{1}{2}\right)\right]$

33. $g[h(x)]$

34. $h[g(x)]$

35. $g[f(1)]$

36. $g[h(0)]$

In Exercises 37–52, given $f(x) = x^2 - 3x + 5$, $g(x) = 3x - 1$, and $h(x) = \dfrac{x - 1}{4 - 2x}$, find and simplify the following.

37. $f[g(2)]$

38. $g[f(2)]$

39. $f[g(-1)]$

40. $g[f(-1)]$

41. $f\left[h\left(\dfrac{1}{3}\right)\right]$

42. $h\left[f\left(\dfrac{1}{3}\right)\right]$

43. $g[h(2)]$

44. $h[g(2)]$

45. $g\left[f\left(\dfrac{1}{2}\right)\right]$

46. $f\left[g\left(\dfrac{1}{2}\right)\right]$

47. $f[g(x)]$

48. $g[f(x)]$

49. $g[h(x)]$

50. $h[g(x)]$

51. $g[g(x)]$

52. $h[h(x)]$

53. The radius, r, of a circle is growing at a constant rate of 3 in./sec.

    (a) What is the area of the circle after 2 seconds (starting at $r = 0$)?

    (b) What is the area of the circle after t seconds (starting at $r = 0$)?

54. The area, A, of a circle is growing at a constant rate of 10 sq in./sec.

    (a) What is the *radius* of the circle after 2 seconds (starting at $A = 0$)?

    (b) Express the *radius* as a function of the time, t, in seconds.

55. The radius, r, of a snowball is growing at a constant rate of 3 in./min.

    (a) What is the volume of the snowball after 5 minutes (starting at $r = 0$)? [The volume, V, of a snowball is given by $V = \frac{4}{3}\pi r^3$, where r is the radius.]

    (b) Express the volume of the snowball as a function of the time, t, in minutes.

56. The volume of a snowball is growing at a constant rate of 8 cu in./min. Express the radius of the snowball as a function of the time, t, in minutes. [See Exercise 55.]

57. A farmer wants to enclose a rectangular field with fencing material; the fencing material costs \$8 per foot. If the perimeter of the field is P ft, express the cost of the fencing as a function of the perimeter.

58. A farmer wants to enclose a square field with fencing material; the fencing material costs \$8 per foot. If the *area* of the field is 200 sq ft, how much would it cost to enclose the field with fencing?

59. A farmer wants to enclose a square field with fencing material; the fencing material costs x dollars per foot. If the *area* of the field is 1,000 sq ft, express the cost of enclosing the field with fencing as a function of x.

60. A gardener wants to enclose a circular garden with fencing material; the fencing material costs \$3 per foot. If the *area* of the garden is A sq ft, express the cost of enclosing the garden with fencing as a function of A.

Question for Thought

61. Consider the functions $f(x) = x^2 + 5$ and $g(x) = \sqrt{x}$.

(a) Compute $f[g(5)]$ by first computing $g(5)$. Compute $f[g(-2)]$ by first computing $g(-2)$.

(b) Find (and simplify) $f[g(x)]$.

(c) Use the result of part **(b)** to compute $f[g(5)]$ and $f[g(-2)]$.

(d) Is there a difference between $f[g(5)]$ and $f[g(-2)]$ computed in part **(a)** and part **(b)**? If so, explain.

 Mini-Review

62. Find the dimensions of a rectangle if its area is 210 sq in. and the perimeter is 59 in.

63. *Simplify:* $\left(\dfrac{16x^2y^8}{x^{-1/2}}\right)^{1/4}$

64. *Reduce to lowest terms:* $\dfrac{2x^2 - 5x - 12}{x^2 - 7x + 12}$

65. *Simplify:* $\sqrt{32x^8y^5}$

9.3 Types of Functions

Most of our previous work with graphs in this textbook has involved functions, although we often used y rather than $f(x)$ as the dependent variable. In this section we will present a few of the basic types of functions.

Linear Functions

DEFINITION   A *linear function* is a function of the form

$$f(x) = mx + b$$

We worked with linear functions of the form $y = mx + b$ in Chapter 4. As the name suggests, the graph of a linear function is a straight line. For the function $f(x) = mx + b$, m is the slope and b is the y-intercept. Thus, all nonvertical lines can be represented by linear functions.

EXAMPLE 1   Graph the linear functions:

(a) $f(x) = 3x - 4$ (b) $f(x) = 5$

Solution   (a) It is helpful to write $y = f(x) = 3x - 4$ to remind us that $f(x)$ is just another name for the dependent variable, y. We graph the function $f(x) = 3x - 4$ as we would $y = 3x - 4$. The slope is 3 and the y-intercept is -4. The function is graphed in Figure 9.5.

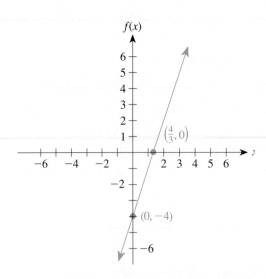

Figure 9.5

(b) We are graphing $y = f(x) = 5$. The graph of the function is shown in Figure 9.6.

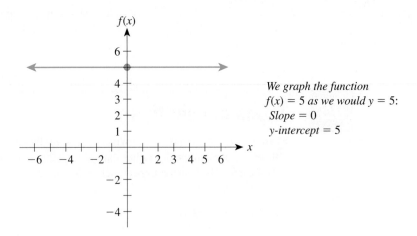

We graph the function
$f(x) = 5$ *as we would* $y = 5$:
Slope $= 0$
y-intercept $= 5$

Figure 9.6

The function $f(x)$ in part **(b)** of Example 1 is a special case of a linear function. In general, the function $f(x) = k$, where k is a constant, is called a ***constant function.***

Quadratic Functions

DEFINITION A ***quadratic function*** is a function of the form

$$f(x) = Ax^2 + Bx + C \quad \text{where } A \neq 0$$

We found in Section 8.6 that the graph of the equation $y = Ax^2 + Bx + C$ (where $A \neq 0$) is a parabola. Hence,

> Thee graph of the function $f(x) = Ax^2 + Bx + C$ (where $A \neq 0$) is a parabola that opens either upward (if $A > 0$) or downward (if $A < 0$).

EXAMPLE 2

Sketch a graph of $f(x) = x^2 - 4x$.

Solution

The x-coordinate of the vertex is $-\dfrac{B}{2A} = -\dfrac{-4}{2(1)} = 2$, as we derived in Section 8.6.

Next we need to find the second coordinate when $x = 2$. Instead of saying "the y-coordinate when $x = 2$" or "the $f(x)$-coordinate when $x = 2$," in the language of functions we are looking for $f(2)$—that is, $f(2)$ is the value of $f(x)$ when $x = 2$. Hence,

$$f(2) = 2^2 - 4(2) = 4 - 8 = -4 \qquad \text{"}f(2) = -4\text{" means } y = -4 \text{ when } x = 2.$$

The vertex is at $(2, -4)$.

The y-intercept is $f(0) = 0^2 - 4(0) = 0$.

The x-intercept is the value of x when $f(x) = 0$:

$$0 = f(x) = x^2 - 4x$$
$$0 = x(x - 4) \Rightarrow x = 0 \quad \text{or} \quad x = 4$$

The x-intercepts are 0 and 4.

The graph of the parabola is shown in Figure 9.7.

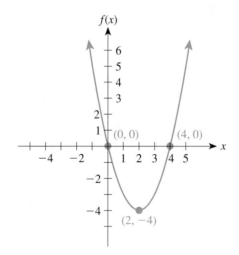

Figure 9.7
Graph of $f(x) = x^2 - 4x$

Polynomial Functions

The linear and quadratic functions are special cases of polynomial functions.

DEFINITION

Polynomial functions are functions of the form

$$f(x) = a_n x^n + a_{n-1} x^{n-1} + \cdots + a_2 x^2 + a_1 x + a_0 \quad (a_n \neq 0)$$

where n is a nonnegative integer.

For example, the polynomial functions $f(x) = x^3$ and $g(x) = x^4 + 2x^3 + x - 3$ are graphed in Figure 9.8(**a**) and (**b**).

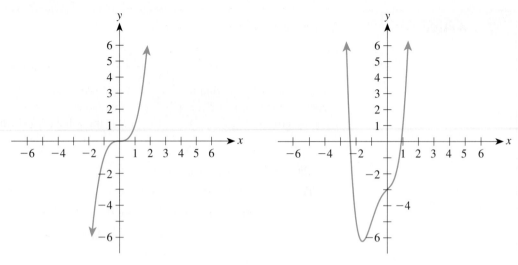

Figure 9.8   **(a)** The graph of $y = f(x) = x^3$ **(b)** The graph of $y = g(x) = x^4 + 2x^3 + x - 3$

The Square Root Function

DEFINITION   The *square root function* is a function of the form

$$f(x) = \sqrt{x}$$

For example, if we compute a table of values of $\sqrt{x}$, we find:

x	$f(x) = \sqrt{x}$
-2	$\sqrt{-2}$ (not a real number)
-1	$\sqrt{-1}$ (not a real number)
0	0
1	1
2	$\sqrt{2}$
4	2
9	3

Note that when x is negative, $f(x)$ is not a real number. Therefore, the domain is $\{x \mid x \geq 0\}$, which is the set of nonnegative real numbers. Since $\sqrt{}$ means the non-negative root, $f(x)$ is always nonnegative. Hence, $f(x) \geq 0$, or the range is the set of all nonnegative real numbers. The graph of $f(x) = \sqrt{x}$ is shown in Figure 9.9.

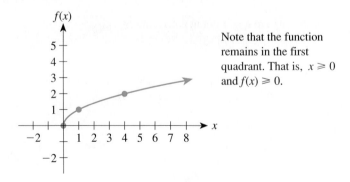

Note that the function remains in the first quadrant. That is, $x \geq 0$ and $f(x) \geq 0$.

Figure 9.9
Graph of $f(x) = \sqrt{x}$

EXAMPLE 3

Sketch a graph of the following functions:

(a) $f(x) = \sqrt{x-3}$ **(b)** $f(x) = \sqrt{x} - 3$

Solution

(a) The domain of $f(x)$ is $\{x \mid x \geq 3\}$ (so that $\sqrt{x-3}$ will be a real number). We use the values of x and $f(x)$ shown in the accompanying table to sketch a graph of $f(x) = \sqrt{x-3}$ (see Figure 9.10).

x	$f(x) = \sqrt{x-3}$
3	0
4	1
5	$\sqrt{2}$
7	2
12	3

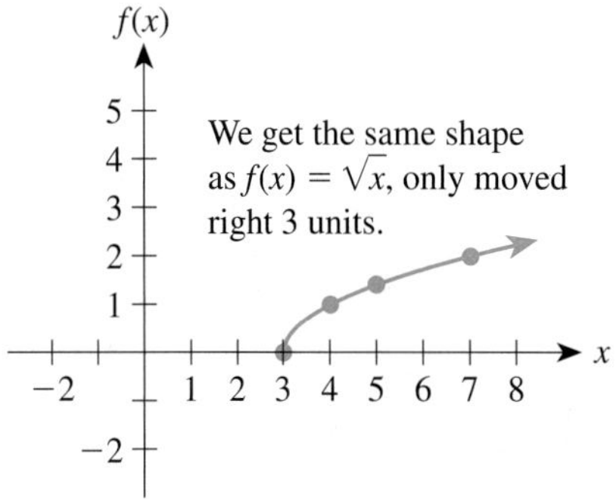

f(x)
We get the same shape
as f(x) = √x, only moved
right 3 units.
x
5
4
3
2
1
−2
1 2 3 4 5 6 7 8
−2

Figure 9.10
Graph of $f(x) = \sqrt{x-3}$

From the figure we can see that the domain must be $\{x \mid x \geq 3\}$; the range is $\{y \mid y \geq 0\}$.

(b) From the accompanying table of values, we see that the graph of $f(x)$ can take on negative values. We sketch a graph of $f(x)$ as shown in Figure 9.11.

x	$f(x) = \sqrt{x} - 3$
0	-3
1	-2
2	$\sqrt{2} - 3 \approx -1.59$
4	-1
9	0
16	1

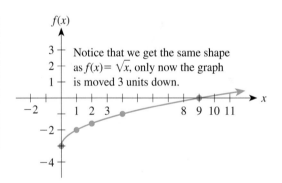

f(x)
Notice that we get the same shape
as f(x)= √x, only now the graph
is moved 3 units down.
x
3
2
1
−2
1 2 3
8 9 10 11
−2
−4

Figure 9.11
Graph of $f(x) = \sqrt{x} - 3$

The domain is $\{x \mid x \geq 0\}$ and the range is $\{y \mid y \geq -3\}$.

The Absolute Value Function

DEFINITION The ***absolute value function*** is a function of the form

$$f(x) = |x|$$

In Chapter 1 we defined the absolute value of x algebraically as:

$$|x| = \begin{cases} x & \text{if } x \geq 0 \\ -x & \text{if } x < 0 \end{cases}$$

If we make a table of values, we obtain the following:

Consider the algebraic definition of |x| above: We use the graph of y = x for all points on and right of the y-axis (x ≥ 0), and the graph of y = −x for all points left of the y-axis (x < 0).

x	$f(x) = |x|$
−2	2
−1	1
0	0
1	1
2	2

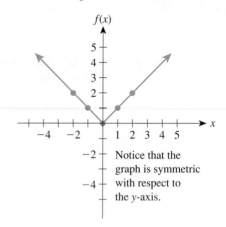

Notice that the graph is symmetric with respect to the y-axis.

Figure 9.12
Graph of $f(x) = |x|$

We graph the function as shown in Figure 9.12. The domain is the set of all real numbers. Since $f(x)$ is always nonnegative, the range is $\{y \mid y \geq 0\}$.

EXAMPLE 4

Sketch a graph of each of the following:

(a) $f(x) = |x + 2|$ **(b)** $f(x) = |x| + 2$

(a) We construct the following table:

x	$f(x) = |x + 2|$
−6	4
−4	2
−2	0
0	2
2	4

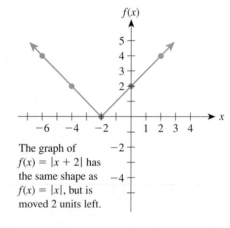

The graph of $f(x) = |x + 2|$ has the same shape as $f(x) = |x|$, but is moved 2 units left.

Figure 9.13
Graph of $f(x) = |x + 2|$

The graph is sketched in Figure 9.13. From the figure we see that the domain is the set of all real numbers, and the range is $\{y \mid y \geq 0\}$.

(b) We make a table:

x	$f(x) = |x| + 2$
−2	4
−1	3
0	2
1	3
2	4

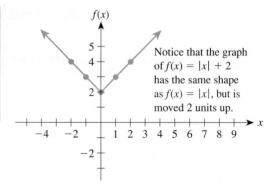

Notice that the graph of $f(x) = |x| + 2$ has the same shape as $f(x) = |x|$, but is moved 2 units up.

Figure 9.14
Graph of $f(x) = |x| + 2$

We sketch the graph in Figure 9.14. The domain is still the set of all real numbers, but the range is now $\{y \mid y \geq 2\}$.

For completeness, we mention two additional types of functions that will be studied in the next chapter: the exponential function and the logarithmic function.

EXERCISES 9.3

In Exercises 1–16, determine the type of function.

1. $f(x) = 5x + 2$

2. $f(x) = 3x^2 + 2$

3. $f(x) = x^2 - 4x + 1$

4. $f(x) = 5x - 4$

5. $f(x) = 3x^3 - 2x + 4$

6. $f(x) = \sqrt{3x + 5}$

7. $f(x) = 2x^2 + 1$

8. $f(x) = 4x^4 - 2x^3 + 3x$

9. $f(x) = \sqrt{2x - 5}$

10. $f(x) = 3x^2 - 2$

11. $f(x) = |x| - 1$

12. $f(x) = |x - 2|$

13. $f(x) = 2x - 1$

14. $f(x) = 3x^2 + 2x$

15. $f(x) = 2x^2 - 1$

16. $f(x) = \sqrt{2x + 5}$

In Exercises 17–24, evaluate $f(-2)$, $f(-1)$, $f(0)$, $f(1)$, and $f(2)$ for the functions.

17. $f(x) = |4x - 1|$

18. $f(x) = |3x - 2|$

19. $f(x) = |4x| - 1$

20. $f(x) = |3x| - 2$

21 $f(x) = |4x| + 1$

22. $f(x) = |3x| + 2$

23. $f(x) = |4x + 1|$

24. $f(x) = |3x + 2|$

In Exercises 25–58, determine the type of function and graph it.

25. $f(x) = 4 - 3x$

26. $f(x) = 5 - 4x$

27. $f(x) = x^2 - 9$

28. $f(x) = 4 - x^2$

29. $f(x) = x^3$

30. $f(x) = x^4$

31. $f(x) = 2x^3$

32. $f(x) = 2x^4$

33. $f(x) = \sqrt{x - 5}$

34. $f(x) = \sqrt{x + 4}$

35. $f(x) = \sqrt{x + 5}$

36. $f(x) = \sqrt{x - 4}$

37. $f(x) = \sqrt{x} + 5$

38. $f(x) = \sqrt{x} - 4$

39. $f(x) = \sqrt{x} - 5$

40. $f(x) = \sqrt{x} + 4$

41. $f(x) = x^2 - 4x + 1$

42. $f(x) = x^2 + 4x - 1$

43. $f(x) = \sqrt{3x - 2}$

44. $f(x) = \sqrt{2x - 5}$

45. $f(x) = 8 - 2x - x^2$

46. $f(x) = 4 + 3x - x^2$

47. $f(x) = \sqrt{8 - 2x}$

48. $f(x) = \sqrt{9 - 3x}$

49. $f(x) = \sqrt{6 - 4x}$

50. $f(x) = \sqrt{8 - 6x}$

51. $f(x) = |x + 5|$

52. $f(x) = |x + 4|$

53. $f(x) = |x| + 5$

54. $f(x) = |x| + 4$

55. $f(x) = x + 5$

56. $f(x) = x + 4$

57. $f(x) = |5 - x|$

58. $f(x) = |4 - x|$

? Questions for Thought

59. Graph the following absolute value functions on the same set of coordinate axes:

$$y = |x| \qquad y = |x| + 2 \qquad y = |x| - 3$$

(a) What can you conclude about the effect on the graph of $y = |x|$ of adding a constant to $|x|$?

(b) Predict what the graph of $y = |x| + 4$ looks like and check your prediction by graphing the function.

60. Graph the following square root functions on the same set of coordinate axes:

$$y = \sqrt{x} \qquad y = \sqrt{x} - 2 \qquad y = \sqrt{x} + 3$$

(a) What can you conclude about the effect on the graph of $y = \sqrt{x}$ of adding a constant to $\sqrt{x}$?

(b) Predict what the graph of $y = \sqrt{x} - 5$ looks like and check your prediction by graphing the function.

61. Based on your conclusions to Exercises 59 and 60, describe how the graph of $y = f(x) + 3$ would be similar to or different from the graph of $y = f(x)$. How would you construct the graph of $y = f(x) + k$ from the graph of $y = f(x)$?

62. Graph the following absolute value functions on the same set of coordinate axes:

$$y = |x| \qquad y = |x + 2| \qquad y = |x - 3|$$

  (a) What can you conclude about the effect on the graph of $y = |x|$ of adding a constant to x before finding its absolute value?

  (b) Predict what the graph of $y = |x + 4|$ looks like and check your prediction by graphing the function.

63. Graph the following square root functions on the same set of coordinate axes:

$$y = \sqrt{x} \qquad y = \sqrt{x - 2} \qquad y = \sqrt{x + 3}$$

  (a) What can you conclude about the effect on the graph of $y = \sqrt{x}$ of adding a constant to x before taking the square root?

  (b) Predict what the graph of $y = \sqrt{x - 5}$ looks like and check your prediction by graphing the function.

64. Based on your conclusions to Exercises 62 and 63, describe how the graph of $y = f(x + 3)$ would be similar to or different from the graph of $y = f(x)$. How would you construct the graph of $y = f(x + h)$ from the graph of $y = f(x)$?

 Mini-Review

65. An electronic mail sorter processes two batches of 500 letters each. Set at high speed, it sorts the second batch in 10 minutes less than it took to sort the first batch at medium speed. If high speed sorts 150 letters more per hour than medium speed, find the rates at medium and high speeds, and the time needed to sort each batch.

66. *Combine and simplify:* $\dfrac{9}{\sqrt{7} + 2} - \dfrac{21}{\sqrt{7}}$

67. Given $f(x) = 3x^2 - 2x + 5$, find

  (a) $f(x - 2)$ (b) $f(x) - f(2)$

9.4   Inverse Functions

So far in this chapter we have built new functions from existing functions by arithmetic operations and by an operation called composition. In this section we will create new functions from existing ones in a different manner.

We defined a function as a correspondence between two sets, called the domain and the range, such that to each element of the domain there is assigned exactly one element of the range. If we wrote the function as a set of ordered pairs, this would mean that no two ordered pairs in the set can have the same x-coordinate but different y-coordinates.

Let's examine what happens when we reverse the assignment of a given function. For example, suppose we have a function F with domain $A = \{1, 2, 3\}$ and range $B = \{5, 8\}$ defined as follows:

F

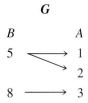

Using ordered pair notation, $F = \{(1, 5), (2, 5), (3, 8)\}$. F is a function, since each *x*-value is assigned exactly one *y* value.

If we reverse this assignment we get the following:

G

$$
\begin{array}{ccc}
B & & A \\
5 & \begin{array}{c}\searrow\end{array} & 1 \\
 & & 2 \\
8 & \longrightarrow & 3
\end{array}
$$

Let's call this assignment (or relation) *G*, which can be written using ordered pair notation as $G = \{(5, 1), (5, 2), (8, 3)\}$. Notice that *G* is the relation obtained by interchanging the *x*- and *y*-coordinates of all ordered pairs in *F*. However, *G* is *not* a function because the *x*-value 5 is assigned to two different *y*-values: 1 and 2.

On the other hand, the relation $H = \{(-2, 2), (1, 3), (4, 0), (5, 8)\}$ is a function. We can form a new relation, *I*, by interchanging the *x*- and *y*-coordinates of the ordered pairs in *H*, and get $I = \{(2, -2), (3, 1), (0, 4), (8, 5)\}$.

We can see that *I* is also a function. Observe that the domains and ranges of *H* and *I* are interchanged as well: Domain of $H = \{-2, 1, 4, 5\}$ = Range of *I*; and Range of $H = \{2, 3, 0, 8\}$ = Domain of *I*.

We see that if a function *F* is defined by a set of ordered pairs, then we can form a new relation *G* by interchanging the *x*- and *y*-coordinates of the ordered pairs; however, the relation *G* is not always a function.

DEFINITION   Suppose *F* is a function and *G* is a relation obtained by interchanging the *x*- and *y*-coordinates of all ordered pairs in *F*. If *G* is also a function, then *G* is called the *inverse of F* and is denoted by F^{-1}.

Let's note a few things about this definition. First, by interchanging the *x*- and *y*-coordinates of the ordered pairs of a function, we are reversing the original assignment. Second, the range and the domain of a function and its inverse are also interchanged: The domain of *F* is the range of F^{-1}; the range of *F* is the domain of F^{-1}. Third, as we showed in our first example, we can always form a new *relation* by reversing the assignment of *F*, but this new relation may not be a function. If this reverse assignment is not a function, we say that *the function has no inverse*.

A *word about notation:* Although the notation we use is the same, F^{-1} is *not* the reciprocal of *F*. With functions, the exponent "-1" symbolizes the inverse of the function. The use of the notation F^{-1} to mean the inverse of F rather than the reciprocal of *F* is usually clear in the context of the given example or discussion.

EXAMPLE 1   Find the inverse of each of the following functions, if possible:

(a) $F = \{(1, 3), (5, 2), (2, 4)\}$ **(b)** $G = \{(3, 5), (6, 8), (7, 8), (4, 5)\}$

Solution   **(a)** F^{-1} is found by interchanging the *x*- and *y*-coordinates of the ordered pairs of *F*.

$$\boxed{F^{-1} = \{(3, 1), (2, 5), (4, 2)\}}$$ *Note that F^{-1} is a function.*

(b) $G = \{(3, 5), (6, 8), (7, 8), (4, 5)\}$. To find the inverse, we interchange the *x*- and *y*-coordinates of the ordered pairs of *G* to get the set

$$\{(5, 3), (8, 6), (8, 7), (5, 4)\}$$

Since this set is not a function, we say that $\boxed{G \text{ has no inverse}}$.

To find the inverse of a function that is written in the form of an equation such as $y = 2x + 3$, we *interchange the independent variable with the dependent variable* as demonstrated in Example 2.

EXAMPLE 2

Find the inverse of the function defined by $y = 2x + 3$.

Solution

The function $y = 2x + 3$ is defined by all ordered pairs (x, y) such that *y* is 3 more than twice *x*. For example, the ordered pairs $(1, 5)$, $(0, 3)$, $(-1, 1)$, and $(-2, -1)$ satisfy $y = 2x + 3$.

To find the inverse, we interchange the independent variable with the dependent variable. We can do this with the equation

If *F* is represented by $y = 2x + 3$,

then F^{-1} is represented by $x = 2y + 3$.

Since we prefer to have our function expressed in a form that is solved explicitly for the dependent variable, we now solve for *y*:

$$x = 2y + 3$$

$$x - 3 = 2y$$

$$\frac{x - 3}{2} = y$$

Hence, F^{-1} is represented by

$$\boxed{y = \frac{x - 3}{2}}$$

In Example 2, note that $(5, 1)$, $(3, 0)$, $(1, -1)$, and $(-1, -2)$ are members of F^{-1}, as we would expect. Let's graph this last function, $y = 2x + 3$, and graph its inverse, $y = \dfrac{x - 3}{2}$ (see Figure 9.15).

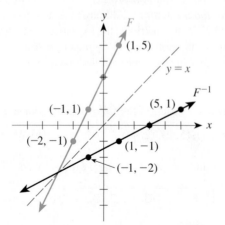

Figure 9.15

We find that by interchanging the x and y variables of the function F, we arrive at a picture that is the mirror image of the graph of F, with the mirror placed on the line $y = x$; this reflected image is F^{-1}. We call such a graph the **reflection** of F about the line $y = x$. We offer another example in Figure 9.16.

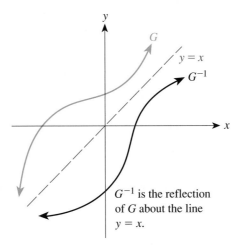

G^{-1} is the reflection of G about the line $y = x$.

Figure 9.16

As we saw in part **(b)** of Example 1, not all functions have inverses—that is, the relation we form by reversing an assignment may not necessarily be a function. We need an additional restriction on functions to guarantee that interchanging the roles of x and y will produce another *function*.

DEFINITION

A *one-to-one function* is a function in which at most one x-value is associated with a y-value.

For a relation to be a function, each x must be assigned a unique y, and for a function to be one-to-one, each y must be assigned to a unique x. This means that a one-to-one function is a one-to-one correspondence between elements of the domain and elements of the range. The relationship among relations, functions, and one-to-one functions is depicted in Figure 9.17.

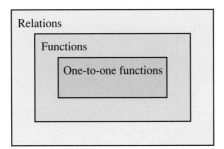

Figure 9.17

EXAMPLE 3

State whether each of the following relations is a function, a one-to-one function, or neither.

(a) $\{(2, 3), (2, 4), (5, 1)\}$ **(b)** $\{(1, 3), (5, 7), (2, 7)\}$

(c) $\{(6, 2), (5, 8), (3, 1)\}$ **(d)** $x + 2y = 4$ **(e)** $y = x^2$

Solution

(a) $\{(2, 3), (2, 4), (5, 1)\}$ is $\boxed{\text{not a function}}$ since $x = 2$ is assigned two values of y: $y = 3$ and $y = 4$.

(b) $\{(1, 3), (5, 7), (2, 7)\}$ $\boxed{\text{is a function}}$ since each x is assigned a unique y. It is not one-to-one since $y = 7$ is associated with two values of x: $x = 5$ and $x = 2$.

(c) $\{(6, 2), (5, 8), (3, 1)\}$ is a one-to-one function .

Each x component is assigned a unique y component *and* each y component is associated with only one x component.

(d) $x + 2y = 4$ is a one-to-one function .

If you substitute any value for x you will get one value for y, and if you substitute any value for y you will get one value for x.

(e) $y = x^2$ is a function as we discussed in Section 8.1. If we let $x = 3$, then $y = 3^2 = 9$; if we let $x = -3$, then $y = (-3)^2 = 9$. Hence, $(3, 9)$ *and* $(-3, 9)$ are members of this function. Since we have one y value, $y = 9$, assigned to two x-values, $x = -3$ and $x = 3$, the function is *not* a one-to-one function .

Just as we used a vertical line to tell whether a relation is a function, we can use a horizontal line to tell whether a function is one-to-one.

The Horizontal Line Test

If a horizontal line intersects the graph of a function at more than one point, the function is *not* a one-to-one function.

For example, the function graphed in Figure 9.18, $y = x^2 + 2$, is a function, as we can tell by the vertical line test, but it is not a one-to-one function, since a horizontal line intersects the graph at more than one point.

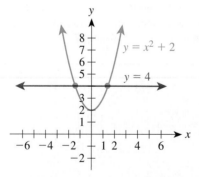

Figure 9.18
Graph of $y = x^2 + 2$

The horizontal line, $y = 4$, intersects the graph of $y = x^2 + 2$ at more than one point. Thus, these two points on the graph of $y = x^2 + 2$ will have the same y value but different x values.

Now we are in the position to determine whether a function has an inverse.

THEOREM

If $f(x)$ is a one-to-one function, then it has an inverse.

Hence, since $\{(6, 2) (5, 8), (3, 1)\}$ of Example 3(**c**) is one-to-one, by this theorem it has an inverse. Similarly, the equation $x + 2y = 4$ of Example 3(**d**) is a one-to-one function and therefore has an inverse.

EXAMPLE 4

Find the inverse of $f(x)$ given the following graph:

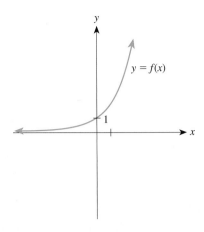

Solution   The graph of the inverse function is found by drawing the reflected graph of the function about the line $y = x$, as shown in the figure below.

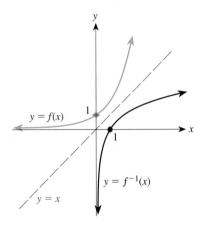

The dashed line represents $y = x$.

The black curve represents the inverse function $f^{-1}(x)$.

If a function $f(x)$ is one-to-one, we can define its inverse, designated $f^{-1}(x)$, by interchanging the roles of the dependent and independent variables. However, keep in mind that we generally write our function solved explicitly for the dependent variable; so after we interchange the roles of the variables, we solve the equation explicitly for the dependent variable.

EXAMPLE 5

Find the inverse of each of the following. Indicate the domain of the function and the domain of its inverse.

(a) $g(x) = \dfrac{x+1}{x-2}$ **(b)** $h(x) = x^3 - 3$

Solution   **(a)** First note that ⟦ the domain of $g(x)$ is $\{x \mid x \neq 2\}$ ⟧.

Let $g(x) = y$. Then

$$y = \frac{x+1}{x-2} \qquad \text{\textit{Then interchange x with y.}}$$

$$x = \frac{y+1}{y-2}$$ *Solve explicitly for y: Multiply both sides of the equation by y − 2.*

$$x(y-2) = y+1$$ *Use the distributive property.*

$$xy - 2x = y + 1$$ *Collect all y terms on one side and non-y terms on the other side.*

$$xy - y = 2x + 1$$ *Factor y from the left side.*

$$y(x-1) = 2x + 1$$ *Divide both sides of the equation by x − 1.*

$$y = \frac{2x+1}{x-1}$$

The inverse is defined by $y = \dfrac{2x+1}{x-1}$ or $\boxed{g^{-1}(x) = \dfrac{2x+1}{x-1}}$.

The domain of g^{-1} is $\{x \mid x \neq 1\}$.

(b) $h(x) = x^3 - 3$. First note that $\boxed{\text{the domain of } h(x) \text{ is the set of all real numbers}}$.

Let $y = h(x) = x^3 - 3$. The inverse is found by interchanging the x- and y-variables:

$$x = y^3 - 3$$ *Now solve for y. First isolate y^3.*

$$x + 3 = y^3$$ *Take cube roots.*

$$\sqrt[3]{x+3} = y$$

Therefore, the inverse function is defined by $y = \sqrt[3]{x+3}$ or, simply,

$$\boxed{h^{-1}(x) = \sqrt[3]{x+3}}$$

$$\boxed{\text{The domain of } h^{-1}(x) \text{ is the set of all real numbers.}}$$

The graph of the function $h(x) = x^3 - 3$ and its inverse appears in Figure 9.19.

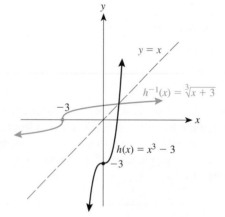

Figure 9.19
The graph of $h(x) = x^3 - 3$
and its inverse,
$h^{-1}(x) = \sqrt[3]{x+3}$

EXERCISES 9.4

In Exercises 1–6, find the inverses of the given functions.

1. $\{(1, 3), (2, 5)\}$

2. $\{(6, 3), (2, 4)\}$

3. $\{(3, 2), (-3, -1), (-1, 3)\}$

4. $\{(5, -2), (-2, 5), (6, -5)\}$

5. $\{(3, 5), (2, -5), (1, 5), (4, 6)\}$

6. $\{(2, 5), (3, 2), (4, 2), (1, 3)\}$

In Exercises 7–24, determine whether the relations are functions, one-to-one functions, or neither.

7. $\{(3, 1), (4, 1)\}$

8. $\{(5, 2), (6, 2)\}$

9. $\{(3, 2), (2, 3)\}$

10. $\{(5, -6), (-6, 5)\}$

11. $\{(3, 7), (3, 8)\}$

12. $\{(4, 9), (4, 1)\}$

13. $\{(-2, 3), (3, -2), (5, -2)\}$

14. $\{(6, 4), (6, -3), (5, 2)\}$

15. $\{(6, -1), (-1, 6), (3, 4)\}$

16. $\{(5, -2), (3, -1), (-2, 5)\}$

17. $y = 3x - 4$

18. $y = 2x + 5$

19. $y = 2x^2$

20. $y = x^2 - 4$

21. $y^2 = x$

22. $y^2 = x + 2$

23. $y = \sqrt{x - 3}$

24. $y = \sqrt{x + 5}$

In Exercises 25–48, find the domain of the function, the inverse of the function (if the inverse is a function), and the domain of the inverse. If there is no inverse, state so.

25. $\{(3, -2), (2, -3)\}$

26. $\{(5, 4), (4, 5)\}$

27. $\{(6, -3), (2, -4), (-3, 6)\}$

28. $\{(5, 1), (2, -3), (-3, 2)\}$

29. $\{(2, 3), (3, 3), (4, 2)\}$

30. $\{(3, 4), (2, 5), (7, 4)\}$

31. $f(x) = 3x + 4$

32. $f(x) = 5x - 3$

33. $g(x) = 2x - 3$

34. $g(x) = 3x - 2$

35. $h(x) = 4 - 5x$

36. $h(x) = 5 - 4x$

37. $f(x) = x^2 + 2$

38. $f(x) = x^2 - 4$

39. $f(x) = x^3 + 4$

40. $f(x) = x^3 - 6$

41. $g(x) = \dfrac{1}{x}$

42. $g(x) = \dfrac{5}{x}$

43. $g(x) = \dfrac{2}{x + 3}$

44. $g(x) = \dfrac{4}{x - 1}$

45. $g(x) = \dfrac{x - 1}{x}$

46. $g(x) = \dfrac{x + 2}{x}$

47. $h(x) = \dfrac{x + 2}{x - 1}$

48. $h(x) = \dfrac{x - 2}{x + 1}$

In Exercises 49–52, sketch a graph of the inverse of the graph of the function:

49.

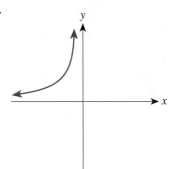

50.

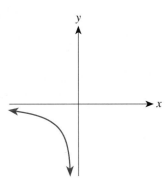

51.

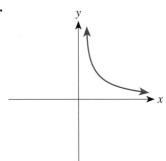

52.

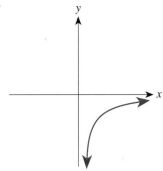

Questions for Thought

53. Given $f(x) = 2x - 3$.

    (a) Find $f^{-1}(x)$.   **(b)** Find $f[f^{-1}(2)]$ and $f^{-1}[f(2)]$.

    (c) Find $f[f^{-1}(x)]$ and $f^{-1}[f(x)]$.

54. **(a)** Graph the function $y = 2x - 3$.

    (b) Graph its inverse on the same set of coordinate axes.

    (c) Compare the graphs of parts **(a)** and **(b)**, and discuss the relationship between them.

◆ Mini-review

55. *Multiply and simplify:* $(\sqrt{x} - 5)^2$

56. *Solve:* $3x^2 - 5x - 7 = x^2 - 2x - 1$

57. Jean invests \$2,800 at 6.2% interest per year. At what rate must she invest another \$1,800 so that her total annual interest from both investments is \$300? (Round to the nearest tenth.)

 # Variation

When we say "varies directly," it is usually the case that as one quantity increases, so does the other.

If Elaine is working for a fixed hourly wage of \$8 per hour, we say that her salary, S, *varies directly* as the number of hours she works, h. We are saying that S is a particular function of h. Specifically,

$$S = 8h$$

When we say "varies inversely," it is usually the case that as one quantity increases, the other decreases.

Similarly, if we were to take an automobile trip of 500 miles, we say that the time of the trip, t, *varies inversely* as our speed, r. We are saying that t is a particular function of r. In fact, since $d = rt$, we have $t = \dfrac{d}{r}$, and therefore, for this case we have

$$t = \frac{500}{r}$$

In this section we will use polynomial, rational, and radical functions to construct models of such "variation" situations. We begin with the following definition.

DEFINITION   *y varies directly* as x means that there is a constant, $k \neq 0$, such that $y = kx$.

We can also say that y is *directly proportional to* x. The quantity k is called the **constant of proportionality**.

For example, the relationship between the circumference of a circle and its radius is

$$C = 2\pi r$$

We could say that the circumference, C, varies directly as r. The constant of proportionality is 2π.

Suppose we know that y varies directly as x, and $y = 15$ when $x = 3$. We can write

$$y = kx \qquad \textit{Since y varies directly as x}$$

Then we can find k by substituting $y = 15$ and $x = 3$. Hence,

$$y = kx$$
$$15 = k(3)$$
$$5 = k$$

Thus, we have the exact relationship between x and y:

$$y = 5x$$

Now if we want to find y given another value for x, say $x = 7$, we can substitute 7 for x in the equation $y = 5x$ to get

$$y = 5x$$
$$y = 5 \cdot 7$$
$$y = 35$$

EXAMPLE 1

If r is directly proportional to s, and $r = 8$ when $s = 3$, find r when $s = 10$.

Solution   Since r varies directly as s, we have

$$r = ks \qquad \textit{Now we can find k by substituting in the given values of r and s.}$$

$$8 = k(3)$$

$$\frac{8}{3} = k$$

Thus, the variation equation is $\boxed{r = \frac{8}{3}s}$.

To find r when $s = 10$, we substitute in 10 for s to get

$$r = \frac{8}{3}(10)$$

$$r = \frac{80}{3}$$

Hence, $\boxed{r = \dfrac{80}{3}}$ when $s = 10$.

An alternative approach is to realize what we mean when we say that r is directly proportional to s, or that $r = ks$. First, we see that if $r = ks$, then

$$\frac{r}{s} = k$$

and since k is a constant, the ratio $\dfrac{r}{s}$ will never change. Thus, given $r = 8$ when $s = 3$, we can find r when $s = 10$ by setting up the following proportion:

$$\frac{r_1}{s_1} = \frac{r_2}{s_2} \qquad \textit{Let } r_1 = 8,\ s_1 = 3,\ \textit{and } s_2 = 10.$$

$$\frac{8}{3} = \frac{r_2}{10}$$

Solving for r_2, we get

$$r_2 = \frac{80}{3} \qquad \textit{Note that this is the same answer obtained above.}$$

Although both approaches result in the same answer, the first has the advantage that it yields the formula $r = \frac{8}{3}s$, which can be used for additional values of r or s.

| DEFINITION | *y varies directly as the nth power of x* means that there is a constant, $k \neq 0$, such that $y = kx^n$. |

EXAMPLE 2

The cost of constructing a carton in the shape of a cube is directly proportional to the volume. If it costs $10.80 to construct a box that is 3 ft by 3 ft by 3 ft, how much would it cost to construct a box that is 2 ft by 2 ft by 2 ft?

Solution

The cost, C, is directly proportional to the volume, V. So we have

$$C = kV$$

Recall that the volume of a rectangular box is given by $V = lwh$, where l is the length, w is the width, and h is the height. In the case of the cube, it gives $V = s \cdot s \cdot s = s^3$:

$$V = s^3, \text{ where } s \text{ is the length of an edge.}$$

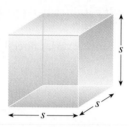

Then

$$C = ks^3 \qquad \textit{Since C = 10.80 when s = 3, we have}$$
$$10.80 = k(3)^3 \qquad \textit{Solve for k.}$$
$$10.80 = 27k$$
$$\frac{10.80}{27} = k$$
$$0.4 = k \qquad \textit{Hence, the cost equation is}$$
$$C = 0.4s^3$$

The cost of constructing a box that is 2 ft by 2 ft by 2 ft is

$$C = 0.4(2)^3$$
$$= 0.4(8)$$
$$= \boxed{\$3.20}$$

| DEFINITION | *y varies inversely as x* means that there is a constant, $k \neq 0$, such that $y = \dfrac{k}{x}$. |

We also say that y is *inversely proportional to x*.

EXAMPLE 3

If y varies inversely as x, and $y = 5$ when $x = 4$, find x when $y = 7$.

Solution

We are given that y varies inversely as x; hence,

$$y = \frac{k}{x} \qquad \textit{Since y = 5 when x = 4, we have}$$

$$5 = \frac{k}{4}$$

$$20 = k$$

Thus, the variation equation becomes

$$y = \frac{20}{x}$$

and we find x when $y = 7$ by substituting 7 for y to get

$$7 = \frac{20}{x}$$

Hence, $\boxed{x = \frac{20}{7}}$.

The dependent variable may be related to more than one independent variable.

DEFINITION   *z varies jointly as x and y* means that there is a constant, $k \neq 0$, such that $z = kxy$.

EXAMPLE 4   If z varies jointly as x and y, and $z = 24$ when $x = 2$ and $y = 4$, find z when $x = 2$ and $y = 5$.

Solution   Since z varies jointly as x and y, we have

$$z = kxy \qquad \textit{Since z = 24 when x = 2 and y = 4, we have}$$

$$24 = k(2)(4)$$

$$3 = k$$

Thus, the variation equation is

$$z = 3xy \qquad \textit{Now find z when x = 2 and y = 5.}$$

$$z = 3(2)(5)$$

$$\boxed{z = 30}$$

A type of question that is often relevant when certain quantities are related by a variation equation is one such as "What happens to one of the quantities when the other quantity is doubled? Tripled? Halved?" Questions of this type become fairly straightforward to answer by using the variation equation.

EXAMPLE 5   What happens to the volume of a cube when the length of one of its sides is doubled? Tripled? Halved?

Solution   Before we examine this question algebraically, let's look at some numerical evidence. We begin with the formula for the volume of a cube, which appears together with a typical cube in Figure 9.20.

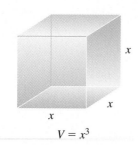

Figure 9.20
The volume, V, of a cube is given by $V = x^3$.

$V = x^3$

We pick an arbitrary value of x, say $x = 5$, and compute x^3, the volume of a cube with side of length x, and then change the x-value by doubling it a few times. This is illustrated in Table 9.1.

Table 9.1

x	$V(x) = x^3$	= Volume of the cube
5	$V(5) = (5)^3 = 125$	
10	$V(10) = (10)^3 = 1{,}000$	
20	$V(20) = (20)^3 = 8{,}000$	

Examining the data in Table 9.1, we can see that each time we double the length of the side, the volume gets multiplied by 8.

Table 9.2 contains the results of carrying out a similar process, in which we triple the length of the side of the cube each time. The data suggest that each time we triple the length of the side of the cube, the volume of the cube gets multiplied by 27.

Table 9.2

x	$V(x) = x^3$	= Volume of the cube
10	$V(10) = (10)^3 = 1{,}000$	
30	$V(30) = (30)^3 = 27{,}000$	
90	$V(90) = (90)^3 = 729{,}000$	

Table 9.3 contains the results of carrying out a similar process, in which we divide the length of the side of the cube by 2 each time. The data suggest that each time we halve the length of the side of the cube, the volume of the cube gets divided by 8.

To see more clearly what is happening to the volume each time, it is helpful to divide the volume at each step by the volume at the previous step. When we double the length of a side, the result of this quotient is 8. When we triple the length of a side, the result of this quotient is 27. When we halve the length of a side, the result of this quotient is $\frac{1}{8}$.

Table 9.3

x	$V(x) = x^3$	= Volume of the cube
40	$V(40) = (40)^3 = 64{,}000$	
20	$V(20) = (20)^3 = 8{,}000$	
10	$V(10) = (10)^3 = 1{,}000$	

Alternatively, we can say that multiplying the length of the side of the cube by $\frac{1}{2}$ causes the volume to be multiplied by $\frac{1}{8}$.

In general, it appears that when we multiply the length of a side of the cube by a constant, k, the volume gets multiplied by k^3.

Let's now examine this situation algebraically. If we start with a cube with side of length x, its volume is $V = x^3$. If we now double the length of the side to $2x$, the volume becomes $V = (2x)^3 = 8x^3$, which is 8 times the original volume.

Similarly, if we triple the length of the side of the cube to $3x$, the volume becomes $V = (3x)^3 = 27x^3$, which is 27 times the original volume. And if we halve the length of the side to $\dfrac{x}{2}$, the volume becomes $V = \left(\dfrac{x}{2}\right)^3 = \dfrac{x^3}{8} = \dfrac{1}{8}x^3$, which is one-eighth the original volume. These results agree with the results indicated by the numerical data we accumulated previously.

Again, in general, if we multiply the length of the side by k to become kx, the volume becomes $V = (kx)^3 = k^3x^3$, which is k^3 times the original volume.

EXAMPLE 6

The volume, V, of a gas varies directly as its temperature, T, and inversely as its pressure, P. A volume of 27.6 m³ of a gas yields a pressure of 43.5 kg/m² at a temperature of 230 K (degrees Kelvin is a scale for measuring temperature). To the nearest tenth m³, what will be the volume of the same gas if the pressure is decreased to 14 kg/m² and the temperature is increased to 260 K?

Solution   We translate the information that V varies directly as T and inversely as P:

$$V = \frac{kT}{P} \qquad \textit{Note that we need only one constant, k.}$$

To find k, we let $V = 27.6$, $P = 43.5$, and $T = 230$:

$$27.6 = \frac{k(230)}{43.5} \qquad \textit{Solve for k.}$$

$$\frac{(43.5)(27.6)}{230} = k$$

$$5.22 = k$$

The variation equation is

$$V = \frac{5.22T}{P}$$

Now we find V when $P = 14$ and $T = 260$:

$$V = \frac{(5.22)(260)}{14} = 96.9 \qquad \text{(to the nearest tenth)}$$

Thus, $V = \boxed{96.9 \text{ m}^3}$.

Alternatively, we could have solved for k in $V = \dfrac{kT}{P}$ to get

$$\frac{PV}{T} = k$$

and set up the proportion

$$\frac{P_1V_1}{T_1} = \frac{P_2V_2}{T_2}$$

Letting $P_1 = 43.5$, $V_1 = 27.6$, $T_1 = 230$, $P_2 = 14$, and $T_2 = 260$, we get

$$\frac{(43.5)(27.6)}{230} = \frac{(14)V_2}{260} \qquad \textit{Solving for V₂, we get}$$

$$\frac{(43.5)(27.6)(260)}{(14)(230)} = V_2$$

$$96.9 = V_2 \qquad \textit{To the nearest tenth}$$

This method gives the same answer: 96.9 m³.

EXERCISES 9.5 ●

Solve each of the following problems algebraically.

1. If y varies directly as x, and $y = 8$ when $x = 4$, find y when $x = 3$.

2. If y is directly proportional to x, and $y = 36$ when $x = 12$, find y when $x = 14$.

3. If y varies directly as x, and $y = 25$ when $x = 15$, find y when $x = 8$.

4. If y varies directly as x, and $y = 14$ when $x = 21$, find y when $x = 9$.

5. If y is directly proportional to x, and $y = 22$ when $x = 3$, find x when $y = 5$.

6. If y is directly proportional to x, and $y = 30$ when $x = 7$, find x when $y = 3$.

7. If a varies directly as the square of b, and $a = 4$ when $b = 3$, find a when $b = 9$.

8. If r varies directly as the cube of s, and $r = 8$ when $s = 2$, find r when $s = 5$.

9. If r varies directly as the fourth power of s, and $r = 12$ when $s = 2$, find r when $s = 3$.

10. If r varies directly as the fifth power of s, and $r = 16$ when $s = 2$, find r when $s = 3$.

11. If y varies inversely as x, and $y = 20$ when $x = 4$, find y when $x = 8$.

12. If y is inversely proportional to x, and $y = 24$ when $x = 6$, find y when $x = 8$.

13. If y is inversely proportional to x, and $y = 21$ when $x = 12$, find x when $y = 9$.

14. If y varies inversely as x, and $y = 24$ when $x = 8$, find x when $y = 10$.

15. If y is inversely proportional to x, and $y = 25$ when $x = 10$, find x when $y = 12$.

16. If y varies inversely as x, and $y = 35$ when $x = 14$, find x when $y = 15$.

17. At a constant pressure, the volume of a gas is directly proportional to the temperature. If the volume of a gas is 250 m³ when the temperature is 30 K, what is the volume of the gas when the temperature is 40 K?

18. At a constant temperature, the volume of a gas is inversely proportional to the pressure. If the volume of a gas is 300 cm³ when the pressure is 5 kg/m², find the volume when the pressure is 12 kg/m².

19. The volume of a sphere is directly proportional to the cube of its radius. If the volume of a sphere is 36π cm³ when its radius is 3 cm, find its volume when the radius is 4 cm.

20. The volume of a sphere is directly proportional to the cube of its radius. If the volume of a sphere is 36π cm³ when its radius is 3 cm, find its radius when the volume is $\dfrac{32\pi}{3}$ cm³.

21. If a is inversely proportional to the cube of b, and $a = 6$ when $b = 2$, find a when $b = 16$.

22. If r varies inversely as the square of s, and $r = 16$ when $s = 2$, find r when $s = 12$.

23. If a varies inversely as the square root of b, and $a = 16$ when $b = 4$, find a when $b = 9$.

24. If x varies inversely as the cube root of y, and $x = 27$ when $y = 27$, find x when $y = 8$.

25. The distance an object falls is directly proportional to the square of the length of time it falls. If an object falls 256 ft in 4 seconds, how long does it take for it to fall 800 ft?

26. The distance an object falls is directly proportional to the square of the length of time it falls. If an object falls 144 ft in 3 seconds, how far does it fall in 8 seconds?

27. The intensity of illumination on a surface, E, in foot-candles, varies inversely as the square of the distance, d, in feet, of the light source from the surface. If the illumination from a source is 25 foot-candles when d is 4 ft, find the illumination when the distance is 8 ft.

28. The intensity of illumination on a surface, E, in foot-candles, varies inversely as the square of the distance, d, in feet, of the light source from the surface. If the illumination from a source is 25 foot-candles when d is 8 ft, find the distance when the illumination is 4 foot-candles.

29. If z varies jointly as x and y, and $z = 12$ when $x = 2$ and $y = 4$, find z when $x = 5$ and $y = 2$.

30. If z varies jointly as x and y, and $z = 24$ when $x = 3$ and $y = 4$, find z when $x = 3$ and $y = 2$.

31. If z varies jointly as x and y, and $z = 20$ when $x = 3$ and $y = 4$, find z when $x = 2$ and $y = 5$.

32. If z varies jointly as x and y, and $z = 32$ when $x = 3$ and $y = 4$, find z when $x = 3$ and $y = 5$.

33. If a varies jointly as c and d, and $a = 20$ when $c = 2$ and $d = 4$, find d when $a = 25$ and $c = 8$.

34. If a varies jointly as c and d, and $a = 15$ when $c = 4$ and $d = 5$, find c when $a = 25$ and $d = 2$.

35. If z varies jointly as x and the square of y, and $z = 20$ when $x = 4$ and $y = 2$, find z when $x = 2$ and $y = 4$.

36. If z varies jointly as x and the square of y, and $z = 40$ when $x = 5$ and $y = 4$, find z when $x = 4$ and $y = 5$.

37. If z varies directly as x and inversely as y, and $z = 16$ when $x = 3$ and $y = 2$, find z when $x = 5$ and $y = 3$.

38. If z varies directly as x and inversely as y, and $z = 12$ when $x = 2$ and $y = 5$, find z when $x = 3$ and $y = 4$.

39. If z varies directly as the square of x and inversely as y, and $z = 20$ when $x = 2$ and $y = 4$, find z when $x = 4$ and $y = 2$.

40. If z varies directly as the square of x and inversely as y, and $z = 50$ when $x = 5$ and $y = 2$, find z when $x = 4$ and $y = 5$.

41. If z varies directly as x and inversely as the square of y, and $z = 32$ when $x = 4$ and $y = 2$, find z when $x = 3$ and $y = 3$.

42. If z varies directly as x and inversely as the square of y, and $z = 45$ when $x = 5$ and $y = 3$, find z when $x = 3$ and $y = 5$.

43. The volume of a cone varies jointly as its height and the square of its radius. If its volume is 4π m^3 when its height is 3 m and its radius is 2 m, find its volume when its height is 2 m and its radius is 3 m.

44. The volume of a cylinder varies jointly as the height and the square of its radius. If the volume of a cylinder is 36π cm^3 when the height is 4 cm and its radius is 3 cm, find its volume when its height is 5 cm and its radius is 4 cm.

45. The resistance (R) of a wire is directly proportional to the length (l) of the wire and inversely proportional to the square of the diameter (d) of the wire. If $R = 12$ ohms when the length is 80 ft and the diameter is 0.01 in., find R when the length is 100 ft and the diameter is 0.02 in.

46. The volume (V) of a gas is directly proportional to its temperature (T) and inversely proportional to the pressure (P). The volume of the gas is 20 m^3 when the temperature is 100 K and the pressure is 15 kg/m^2. What is the volume when the temperature is 150 K and the pressure is 20 kg/m^2?

47. The circumference, C, of a circle varies directly as its radius, r, according to the formula $C = 2\pi r$. Describe what happens to the circumference of a circle if its radius is doubled, tripled, and halved.

48. The area, A, of a circle varies directly as the square of its radius, r, according to the formula $A = \pi r^2$. Describe what happens to the area of a circle if its radius is doubled, tripled, and halved.

49. Suppose that y varies inversely as x. What happens to y if x is multiplied by a factor of 4?

50. Suppose that y varies inversely as the square of x. What happens to y if x is multiplied by a factor of 4?

51. Suppose that z varies jointly as x and y. What happens to z if x is multiplied by a factor of 4 and y is multiplied by a factor of 5?

52. Suppose that z varies jointly as x and y. What happens to z if x is doubled and y is tripled?

53. Suppose that s varies directly as r and inversely as t. What happens to s if r is doubled and t is halved?

54. Suppose that s varies directly as r and inversely as t. What happens to s if r is halved and t is doubled?

 ## Mini-review

55. *Write in exponential form:* $\sqrt{5x} + 5\sqrt[3]{x}$

56. Sketch the graph of the solution set of the inequality $5x - 4 < 2y$.

57. Anne makes two investments. She invests $5,000 more at 7% than she invests at 6.5%. If the total yearly interest from both investments is $1,430, how much is invested at each rate?

58. If $f(x) = \dfrac{2x - 3}{x^2 + 4}$, find and simplify $f\left(\dfrac{5}{x}\right)$.

CHAPTER 9 SUMMARY

After having completed this chapter you should be able to:

1. Evaluate and perform operations with functions and determine their domains (Section 9.1).

   For example:

   (a) If $g(x) = \dfrac{x}{x - 5}$, then we must exclude any value of x that makes the denominator 0; hence, the domain of $g(x)$ is $\{x \mid x \neq 5\}$.

(b) $g\left(\dfrac{1}{2}\right) = \dfrac{\dfrac{1}{2}}{\dfrac{1}{2} - 5} = \dfrac{\dfrac{1}{2}}{\dfrac{-9}{2}} = -\dfrac{1}{9}$

(c) $g(x + 5) - g(x) = \dfrac{x + 5}{(x + 5) - 5} - \dfrac{x}{x - 5} = \dfrac{x + 5}{x} - \dfrac{x}{x - 5}$

The LCD is $x(x - 5)$.

$$= \dfrac{(x + 5)(x - 5)}{x(x - 5)} - \dfrac{x \cdot x}{x(x - 5)}$$

$$= \dfrac{(x + 5)(x - 5) - x^2}{x(x - 5)} = \dfrac{-25}{x(x - 5)}$$

2. Evaluate a split function (Section 9.1).

   For example: If

   $$y = f(x) = \begin{cases} 3x + 5 & \text{if } x < 2 \\ 1 - x^2 & \text{if } 2 \le x < 5 \\ 7 & \text{if } x \ge 5 \end{cases}$$

   Then

   $f(12) = 7$ *Use the third rule, since $12 \ge 5$.*

   $f(-4) = 3(-4) + 5 = -7$ *Use the first rule, since $-4 < 2$.*

   $f(2) = 1 - (2)^2 = -3$ *Use the second rule, since $2 \le 2 < 5$.*

3. Find sums, products, differences, and quotients of functions (Section 9.2).

   For example: If $f(x) = 2x^2 - 3$ and $g(x) = x - 4$, then

   (a) $(f + g)(3) = f(3) + g(3) = [2(3)^2 - 3] + [3 - 4]$

   $= 2 \cdot 9 - 3 + 3 - 4 = 14$

   (b) $\left(\dfrac{f}{g}\right)(x) = \dfrac{f(x)}{g(x)} = \dfrac{2x^2 - 3}{x - 4}$ $(x \ne 4)$

4. Evaluate and simplify composite functions (Section 9.2).

   For example:

   (a) If $f(x) = 2x - 1$ and $g(x) = x^2 + 3$, then

   $f[g(4)] = 2[g(4)] - 1$ *Since $g(4) = 4^2 + 3 = 19$, we have*

   $= 2(19) - 1 = 37$

   $g[f(4)] = [f(4)]^2 + 3$ *Since $f(4) = 2(4) - 1 = 7$, we have*

   $= (7)^2 + 3 = 52$

   $g[f(x)]^2 = [f(x)]^2 + 3$ *Since $f(x) = 2x - 1$, we have*

   $= (2x - 1)^2 + 3$

   $= 4x^2 - 4x + 1 + 3$

   $= 4x^2 - 4x + 4$

   (b) To express the area of a circle as a function of its diameter, d, first note that the area of a circle is given by $A = A(r) = \pi r^2$, where $A(r)$ is the area of the circle and r is the length of its radius. So we have the area of the circle as a function of its radius. We know that a radius is half a diameter, so we can write $r = \dfrac{d}{2}$, where r is the radius and d is the diameter. Hence

$$A = \pi r^2 \qquad \textit{Since } r = \frac{d}{2}, \textit{ we substitute } \frac{d}{2} \textit{ for r in A to get}$$

$$= \pi \left(\frac{d}{2}\right)^2 \qquad \textit{Simplify.}$$

$$= \pi \left(\frac{d^2}{4}\right)$$

$$= \frac{\pi d^2}{4}$$

Thus,

$$A = \frac{\pi d^2}{4}$$

5. Graph and classify the graph as one of the basic types of functions (Section 9.3).

For example:

(a) $f(x) = \sqrt{3x - 1}$ is a square root function that has the graph shown in Figure 9.21.

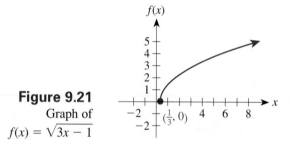

Figure 9.21
Graph of
$f(x) = \sqrt{3x - 1}$

(b) $f(x) = 3x^2 - 4x + 1$ is a quadratic function that has the graph shown in Figure 9.22.

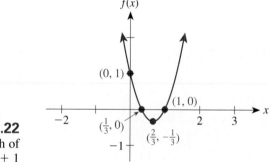

Figure 9.22
Graph of
$f(x) = 3x^2 - 4x + 1$

6. Identify a one-to-one function (Section 9.4).

For example:

(a) $\{(3, 2), (5, 2)\}$ is a function, but it is not a one-to-one function, since the y value, 2, is assigned to more than one x-value, $x = 3$ and $x = 5$.

(b) $\{(3, 2), (4, 5)\}$ is a one-to-one function, since each x-value is assigned a unique y value and each y value is assigned a unique x-value.

(c) $y = 3x^2 - 1$ is a function for which

$$x = -2 \Rightarrow y = 3(-2)^2 - 1 = 3 \cdot 4 - 1 = 11$$

$$x = 2 \Rightarrow y = 3(2)^2 - 1 = 3 \cdot 4 - 1 = 11$$

Since $y = 11$ corresponds to two different values of x, $x = 2$ and $x = -2$, the function is not one-to-one.

7. Find the inverse of a function (Section 9.4).

For example:

(a) Given $f(x) = 3x - 4$. To find $f^{-1}(x)$, let $y = f(x)$. Then

$$y = 3x - 4 \qquad \text{\textit{Interchange x and y variables.}}$$

$$x = 3y - 4 \qquad \text{\textit{Solve explicitly for y.}}$$

$$\frac{x + 4}{3} = y$$

Hence,

$$f^{-1}(x) = \frac{x + 4}{3}$$

(b) Sketch the graph of the inverse of a function given the graph of the function $f(x)$. See Figure 9.23.

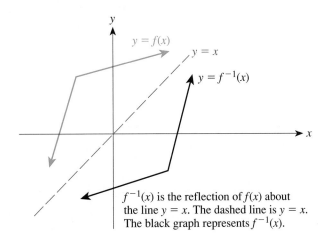

Figure 9.23

$f^{-1}(x)$ is the reflection of $f(x)$ about the line $y = x$. The dashed line is $y = x$. The black graph represents $f^{-1}(x)$.

8. Solve variation problems (Section 9.5).

For example: Suppose z varies directly as x and inversely as y. If $z = 15$ when $x = 3$ and $y = 2$, find z when $x = 3$ and $y = 4$.

Let $z = \dfrac{kx}{y}$. To find k, we use the information that $z = 15$ when $x = 3$ and $y = 2$:

$$15 = \frac{k(3)}{2}$$

$$30 = 3k$$

$$10 = k$$

Thus,

$$z = \frac{10x}{y}$$

Now we find z when $x = 3$ and $y = 4$:

$$z = \frac{10(3)}{4} = \frac{30}{4} = \frac{15}{2}$$

CHAPTER 9 REVIEW EXERCISES

In Exercises 1–6, evaluate $f(-3)$, $f(1)$, and $f(3)$.

1. $f(x) = 5x + 4$ **2.** $f(x) = x^2 + 3x + 1$ **3.** $f(x) = \sqrt{x - 1}$ **4.** $f(x) = \sqrt{2 - 3x}$

5. $f(x) = \dfrac{3}{\sqrt{1 - x}}$ **6.** $f(x) = x - 2\sqrt{3 - x}$

In Exercises 7–15, determine the domain of the following functions.

7. $f(x) = 2x + 1$ **8.** $g(x) = 3x$ **9.** $h(x) = x^2 - x$ **10.** $g(x) = \sqrt{4 - 2x}$

11. $s(x) = \sqrt{3 - 5x}$ **12.** $f(x) = \dfrac{5}{x - 2}$ **13.** $f(x) = \dfrac{2}{\sqrt{2x - 1}}$ **14.** $g(x) = \dfrac{x}{\sqrt{x - 5}}$

15. $h(x) = \dfrac{x + 3}{\sqrt{2x - 3}}$

In Exercises 16–32, $f(x) = 5x + 2$ and $g(x) = \sqrt{3 - 2x}$. Find the following.

16. $f(x + 2)$ **17.** $f(x + 3)$ **18.** $f(x) + 2$ **19.** $f(x) + 3$

20. $f(x) + f(2)$ **21.** $f(x) + f(3)$ **22.** $g(x + 2)$ **23.** $g(x + 3)$

24. $g(2x)$ **25.** $g(3x)$ **26.** $2g(x)$ **27.** $3g(x)$

28. $f(x + h) - f(x)$ **29.** $g(x + h) - g(x)$ **30.** $f[g(x)]$ **31.** $g[f(x)]$

32. $\dfrac{f(x + h) - f(x)}{h}$

In Exercises 33–38, given $f(x) = x^2 - 3$ and $g(x) = x - 8$, find the following.

33. $(f + g)(3)$ **34.** $(f - g)(2)$ **35.** $\left(\dfrac{f}{g}\right)(3)$ **36.** $\left(\dfrac{g}{f}\right)(3)$

37. $\left(\dfrac{f}{g}\right)(8)$ **38.** $\left(\dfrac{g}{f}\right)(8)$

In Exercises 39–50, given $f(x) = x^2 - x - 5$, $g(x) = x + 1$, and $h(x) = \dfrac{x + 1}{2x}$, find the following.

39. $f[g(3)]$ **40.** $g[f(3)]$ **41.** $f[g(-1)]$ **42.** $g[f(-1)]$

43. $f\left[h\left(\dfrac{1}{2}\right)\right]$ **44.** $h\left[f\left(\dfrac{1}{2}\right)\right]$ **45.** $g[h(-1)]$ **46.** $h[g(-1)]$

47. $g[f(x)]$ **48.** $f[g(x)]$ **49.** $g[h(x)]$ **50.** $h[g(x)]$

51. The area, A, of a circle is growing at a constant rate of 8 sq in./sec.

     (a) What is the *radius* of the circle after 4 seconds (starting at $A = 0$)?

     (b) Express the *radius* as a function of the time, t, in seconds.

52. The side, s, of a square is growing at a constant rate of 4 in./min.

     (a) What is the area of the square after 5 minutes (starting at $s = 0$)?

     (b) Express the area as a function of the time, t, in minutes.

53. A gardener wants to enclose a circular garden with fencing material; the fencing material costs x dollars per foot. If the *area* of the garden is 200π sq ft, express the cost of enclosing the garden with fencing as a function of x.

54. **(a)** Find the area of a square if its diagonal is 5 ft.

(b) Express the area of a square in terms of its diagonal.

In Exercises 55–62, graph the function and identify its type.

55. $f(x) = 2x - 3$ **56.** $f(x) = 3x - 1$ **57.** $f(x) = \sqrt{2x - 3}$ **58.** $f(x) = |3x - 1|$

59. $f(x) = 4x - 2x^2$ **60.** $f(x) = \sqrt{3x - 1}$ **61.** $f(x) = |2x| - 3$ **62.** $f(x) = 6x - x^2$

In Exercises 63–66, determine whether the relation is a function, a one-to-one function, or neither.

63. $\{(2, 0), (3, 0)\}$ **64.** $\{(5, 2), (5, 4)\}$ **65.** $y = 3x$ **66.** $2y = x + 1$

In Exercises 67–70, determine whether the graph of the relation is a function, a one-to-one function, or neither.

67.

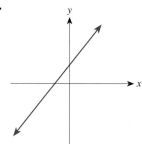

68.

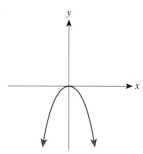

69.

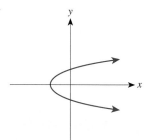

70.

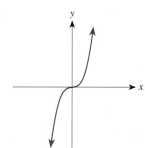

In Exercises 71–78, find the inverse of the function.

71. $\{(2, 3), (3, 4)\}$

72. $\{(5, -8), (3, 7), (2, 6)\}$

73. $y = 3x + 8$

74. $y = 5x - 1$

75. $y = x^3$

76. $y = x^5$

77. $y = \dfrac{3}{x + 1}$

78. $y = \dfrac{2x}{x + 3}$

In Exercises 79–80, sketch the inverse of the function given on the graph.

79.

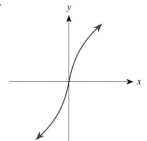

80.

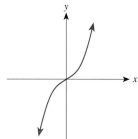

81. If x varies directly as y, and $x = 8$ when $y = 3$, find x when $y = 2$.

82. If x varies inversely as y, and $x = 8$ when $y = 3$, find x when $y = 2$.

83. If r is inversely proportional to the square of s, and $s = 2$ when $r = 8$, find s when $r = 4$.

84. If r is directly proportional to the cube of s, and $r = 16$ when $s = 2$, find s when $r = 4$.

85. If z varies jointly as x and y, and $z = 16$ when $y = 2$ and $x = 7$, find z when $x = 3$ and $y = 4$.

86. If z is directly proportional to x and inversely proportional to y, and $z = 12$ when $x = 2$ and $y = 4$, find z when $x = 3$ and $y = 2$.

87. The volume (V) of a sphere is directly proportional to the cube of its radius. If the volume is $\dfrac{500\pi}{3}$ in.^{3} when its radius is 5 in., find its volume when the radius is 6 in.

88. The weight of an object on or above the surface of the earth is inversely proportional to the square of its distance from the center of the earth. A man weighs 175 lb on the surface of the earth. Assuming the radius of the earth is 4,000 miles, how much would the man weigh 200 miles above the earth's surface?

89. The volume (V) of a gas is directly proportional to its temperature (T) and inversely proportional to its pressure (P). If $V = 80$ m^3 when $T = 20$ K and $P = 30$ kg/m^2, find V when $T = 10$ K and $P = 20$ kg/m^2.

90. The production cost of manufacturing widgets varies jointly as the number of widget machines running and the number of hours the machines are in operation. When 5 widget machines are running 8 hours, the production cost is $3,800. What is the production cost for 6 widget machines running 20 hours?

CHAPTER 9 PRACTICE TEST ●

1. Determine the domain of each of the following:

   (a) $f(x) = 3x^2 - 5$ **(b)** $g(x) = \dfrac{x}{2x - 3}$ **(c)** $h(x) = \sqrt{8 - 3x}$

2. Given $f(x) = 3x - 5$ and $g(x) = \dfrac{3}{x + 1}$, find the following:

   (a) $g(-1)$ **(b)** $f(x + 2)$ **(c)** $g(x + 3) - g(x)$

3. Given $f(x) = x^2 - 3$ and $h(x) = \sqrt{x - 2}$, find the following:

   (a) $\left(\dfrac{f}{h}\right)(3)$ **(b)** $f[h(6)]$ **(c)** $h[f(6)]$ **(d)** $f[h(x)]$

   (e) $h[f(x)]$

4. (a) Find the perimeter of a square if its diagonal is 8 ft.

   (b) Express the perimeter of a square as a function of its diagonal.

5. Sketch a graph of $f(x) = |2x - 1|$.

6. Find the inverse of the function $\{(2, -3), (5, 8), (3, 8)\}$.

7. Determine whether the following function is a one-to-one function.

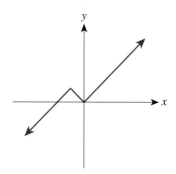

8. Find the inverse of each of the following functions:

(a) $f(x) = 2x - 4$ 　　　　　　　　　　(b) $f(x) = \dfrac{x}{x + 3}$

9. If y varies inversely as the square of x, and $y = 24$ when $x = 2$, find y when $x = 6$.

10. If z varies directly as the square of x and inversely as y, and $z = 120$ when $x = 2$ and $y = 3$, find z when $x = 3$ and $y = 2$.

CHAPTERS 7–9 CUMULATIVE REVIEW

In Exercises 1–12, perform the indicated operations and simplify. Express your answer with positive exponents only.

1. $(-2x^2y)(-3xy^2)^3$

2. $\dfrac{8x^4y^5}{16x^6y}$

3. $\left(\dfrac{2xy^2}{4x^2y^3}\right)^2$

4. $\dfrac{(-2ab^2)^2(-3a)^2}{-6a^2b}$

5. $(x^{-5}y^{-2})(x^{-7}y)$

6. $(2x^{-3}y)^{-2}$

7. $\dfrac{r^{-3}s^{-2}}{r^{-2}s^0}$

8. $\left(-\dfrac{2}{3}\right)^{-3}$

9. $(-2r^{-3}s)^{-1}(-4r^{-2}s^{-3})^2$

10. $\left(\dfrac{5x^{-1}y}{(-5x^{-2}y^2)^2}\right)^{-1}$

11. $(x^{-1}-y^{-1})(x+y)$

12. $\dfrac{a^{-1}+b^{-2}}{a^{-1}+1}$

In Exercises 13–16, express the following using scientific notation.

13. 56,429.32

14. 0.0000752

15. How far does light travel in 1 day if it travels 186,000 miles per second?

16. Perform the following computations and express your answer using scientific notation:

$$\dfrac{(7.2 \times 10^{-3})(8.0 \times 10^5)}{1.2 \times 10^{-4}}$$

In Exercises 17–18, evaluate.

17. $(-1{,}000)^{1/3}$

18. $\left(\dfrac{27}{64}\right)^{-2/3}$

In Exercises 19–22, perform the indicated operations and express your answers with positive exponents only.

19. $a^{2/3}a^{-1/2}$

20. $\dfrac{x^{2/5}}{x^{1/3}}$

21. $[(81)^{-1/3}3^2]^{-3}$

22. $\left(\dfrac{x^{-1/3}x^{2/3}}{x^{1/4}}\right)^{-6}$

In Exercises 23–24, change to radical notation.

23. $x^{3/4}$

24. $2x^{1/2}$

In Exercises 25–32, express in simplest radical form. Assume that all variables represent positive real numbers only.

25. $\sqrt{16a^4b^8}$

26. $\sqrt{8a^3b^{10}}$

27. $\left(3x\sqrt{x^2y}\right)\left(2x\sqrt{8xy^3}\right)$

28. $\sqrt[3]{81x^4y^5}$

29. $\dfrac{\sqrt{48}}{\sqrt{3}}$

30. $\sqrt{\dfrac{3}{xy^2}}$

31. $\sqrt[8]{x^4}$

32. $\sqrt[3]{\dfrac{2}{5x}}$

In Exercises 33–42, perform the indicated operations and simplify as completely as possible.

33. $2\sqrt{5}-3\sqrt{5}+8\sqrt{5}$

34. $2\sqrt{27}-\sqrt{75}-\sqrt{3}$

35. $3a\sqrt[3]{a^4}-a^2\sqrt[3]{a}$

36. $\sqrt{\dfrac{2}{3}}-\sqrt{\dfrac{3}{2}}$

37. $\sqrt{2}\left(\sqrt{2}-1\right)+2\sqrt{2}$

38. $\left(\sqrt{5}-\sqrt{3}\right)^2$

39. $\left(\sqrt{5}-\sqrt{3}\right)\left(\sqrt{5}+\sqrt{3}\right)$

40. $\left(\sqrt{x+4}\right)^2-\left(\sqrt{x}+4\right)^2$

41. $\dfrac{5}{\sqrt{3}+\sqrt{2}}-\dfrac{3}{\sqrt{3}}$

42. $\dfrac{2\sqrt{3}}{\sqrt{5}+\sqrt{3}}$

In Exercises 43–46, solve the equation.

43. $\sqrt{3x + 2} = 7$ **44.** $\sqrt{3x} + 2 = 7$ **45.** $\sqrt[3]{x - 1} = -2$ **46.** $x^{1/5} + 5 = 4$

In Exercises 47–48, express as i, −1, −i, or 1.

47. i^{35} **48.** i^{-2}

In Exercises 49–50, express in the form a + bi.

49. $(3 - 2i)(5 + i)$ **50.** $\dfrac{1 + 3i}{1 + 2i}$

In Exercises 51–54, solve by the factoring or square root method.

51. $a^2 - 2a - 15 = 0$ **52.** $a^2 - 6a + 9 = a + 3$

53. $2x^2 - 3x - 4 = 9 - 3(x - 2)$ **54.** $(x - 5)^2 = 28$

In Exercises 55–56, solve by completing the square.

55. $y^2 + 6y - 1 = 0$ **56.** $3x^2 + 6x = 2$

In Exercises 57–64, solve by any method.

57. $3a^2 - 2a - 2 = 0$ **58.** $(x - 2)(x + 3) = 2x^2 - 3x + 1$

59. $\dfrac{1}{x + 2} = x + 2$ **60.** $(y - 2)(y + 3) = 6$

61. $\dfrac{2}{x + 3} - \dfrac{3}{x} = -\dfrac{2}{3}$ **62.** $(b - 2)(b - 3) = (b - 4)(b + 1)$

63. $\dfrac{3}{x - 3} + \dfrac{2x}{x + 3} = \dfrac{5}{x - 3}$ **64.** $\dfrac{3}{x + 1} + \dfrac{2}{x - 5} = 5$

In Exercises 65–76, sketch a graph of the equation; identify its intercept(s), axis of symmetry, and vertex where appropriate.

65. $y = 2x^2$ **66.** $y = -5x^2$ **67.** $y = 2x$ **68.** $y = -5x$

69. $y = -\dfrac{1}{3}x^2$ **70.** $y = \dfrac{1}{4}x^2$ **71.** $y = (x - 2)^2 + 4$ **72.** $y = -(x + 4)^2 - 3$

73. $y = -(x - 3)^2 + 5$ **74.** $y = -(x + 4) - 3$ **75.** $y = 2x^2 - 4x + 5$ **76.** $y = 3x^2 + 12x + 9$

In Exercises 77–82, solve the inequality and sketch the solution set on the real number line.

77. $(x - 7)(2x + 1) \leq 0$ **78.** $(x + 2)(x - 8) < 0$ **79.** $2x^2 - x > 6$ **80.** $\dfrac{x - 6}{2x + 1} > 0$

81. $\dfrac{x - 1}{x + 3} \leq 2$ **82.** $\dfrac{x}{x - 2} \leq \dfrac{x + 9}{x + 1}$

83. The profit, P, made on a 15-minute sideshow is related to the price (x) of a ticket in the following way:

$$P(x) = -200x^2 + 500x$$

where P is the profit in dollars and x is the price of a ticket in dollars. What ticket price would produce the maximum profit? What is the maximum profit?

84. Write an equation of the circle with center $(-3, 5)$ and radius 4.

85. Write an equation of the circle with center $(0, 2)$ that passes through the point $(1, -3)$.

86. Given the points $A(2, -4)$ and $B(3, 5)$, find the distance between A and B; find the coordinates of the midpoint between A and B.

87. Write an equation of the circle with a diameter having endpoints $(-2, 1)$ and $(4, 5)$.

88. Identify the center and radius of the circle with equation $x^2 + (y + 2)^2 = 10$.

In Exercises 89–98, use $f(x) = 2x^2 - 4x$, $g(x) = 3x - 6$, $h(x) = \dfrac{2x}{x + 1}$, and $r(x) = \dfrac{1}{x}$ to compute each of the following.

89. $f(x) + g(x)$ **90.** $f(x) \cdot g(x)$ **91.** $f[g(x)]$ **92.** $g[f(x)]$

93. $h(x) \cdot r(x)$ **94.** $h(x) - r(x)$ **95.** $\left(\dfrac{f}{g}\right)(x)$ **96.** $\left(\dfrac{g}{r}\right)(x)$

97. $h[r(x)]$ **98.** $r[h(x)]$

In Exercises 99–100, sketch the graph of the given function and find the required values.

99. $f(x) = \begin{cases} x - 1 & \text{if } x \le 1 \\ 2x + 1 & \text{if } x > 1 \end{cases}$

Find $f(-2), f(1),$ and $f(3)$.

100. $g(x) = \begin{cases} x^2 - 3 & \text{if } x \le 2 \\ 5 - x & \text{if } x > 2 \end{cases}$

Find $g(0), g(2),$ and $g(3)$.

In Exercises 101–104, find the inverse of the given function, if it exists.

101. $y = x^3 - 1$ **102.** $y = \dfrac{2x - 5}{3}$ **103.** $y = \dfrac{x + 1}{x}$ **104.** $y = x^4$

In Exercises 105–110, sketch the graph of the given equation. Be sure to label intercepts.

105. $y = \sqrt{x + 4}$ **106.** $y = \sqrt{x} + 4$ **107.** $y = x^2 - 4x - 5$ **108.** $y = |x - 3|$

109. $(x - 2)^2 + (y + 3)^2 = 4$ **110.** $x^2 + 6x + y^2 - 10y + 33 = 0$

111. If y varies directly as x, and $y = 8$ when $x = 6$, find y when $x = 20$.

112. If y varies inversely as x, and $y = 8$ when $x = 6$, find y when $x = 20$.

113. If x varies jointly as y and z, and $x = 10$ when $y = 4$ and $z = 15$, find x when $y = 6$ and $z = 20$.

114. If x varies directly as y and inversely as the square of z, and $x = 2$ when $y = 12$ and $z = 6$, find x when $y = 20$ and $z = 5$.

115. The area, A, of a rectangle varies jointly with its base, b, and height, h. What happens to the area of a rectangle if the base is doubled and the height is tripled?

116. The amount of illumination, I, reaching a surface from a light source is inversely proportional to the square of the distance, d, between the light source and the surface. What happens to the amount of illumination reaching a surface if the surface is moved to one-half its original distance from the light source?

CHAPTERS 7–9 CUMULATIVE PRACTICE TEST

1. Perform the indicated operations and simplify. Express your answers with positive exponents. Assume all variables represent positive real numbers.

   (a) $(3x^2y)^2(-2xy^3)^3$

   (b) $(-2x^{-1}y^3)(-3x^2y^{-3})^2$

   (c) $\left(\dfrac{3x^{-2}y^{-1}}{9xy^{-3}}\right)^{-2}$

   (d) $\dfrac{a^{-1}}{a^{-1}+b^{-1}}$

2. Express 0.000034 using scientific notation.

3. Perform the operations using scientific notation. Express your answer in standard form.

$$\frac{(150,000)(0.00028)}{(0.07)(0.0002)}$$

4. Evaluate $(-128)^{-3/7}$

5. Perform the operations and simplify. Express your answers using positive exponents only. Assume that all variables represent positive real numbers.

   (a) $(x^{1/3}x^{-1/5})^5$

   (b) $\dfrac{x^{-2/3}y^{-3/4}}{x^{1/2}}$

6. Express the following in simplest form. Assume that all variables represent positive real numbers.

   (a) $\sqrt{24x^2y^5}$

   (b) $\left(a^2\sqrt{2ab^2}\right)\left(b\sqrt{4a^2b^3}\right)$

   (c) $\sqrt[3]{\dfrac{5}{a}}$

7. Perform the operations and simplify as completely as possible.

   (a) $5\sqrt{8}-5\sqrt{2}-\sqrt{50}$

   (b) $\sqrt{\dfrac{x}{y}}-\sqrt{\dfrac{x}{2}}$

   (c) $(2-\sqrt{3})(\sqrt{3}-2)$

   (d) $\dfrac{5\sqrt{2}}{\sqrt{5}-\sqrt{2}}$

8. *Solve for x:* $\sqrt{2x}+3=4$

9. *Express the following in the form $a+bi$:* $\dfrac{2-i}{3-i}$

10. *Solve the following equations by any method.*

    (a) $2x^2+x=3$

    (b) $\dfrac{x}{3}=\dfrac{4}{x}$

    (c) $2x^2+4x-3=0$

    (d) $\dfrac{x}{x+1}+\dfrac{2}{x-3}=4$

11. Sketch a graph of the parabola; identify its intercept(s), axis of symmetry, and vertex.

    (a) $y=5x^2$

    (b) $y=(x-4)^2+1$

    (c) $y=-(x-1)^2$

    (d) $y=3x^2-6x+4$

12. Solve the inequality and sketch the solution set on the real number line.

    (a) $(2x-1)(x+1)\le 0$

    (b) $3x^2-11x<4$

    (c) $\dfrac{x-3}{x+5}\ge 0$

    (d) $\dfrac{x}{x-2}\le 3$

13. Write an equation of the circle with center $(-2,0)$ that passes through the point $(5,-1)$.

14. Given the points $A(3, 8)$ and $B(-2, 6)$, find the distance between A and B; find the coordinates of the midpoint between A and B.

15. Sketch the graph of the given function and find the required values.

$$f(x) = \begin{cases} x + 1 & \text{if} \quad x \le 2 \\ x^2 - 2 & \text{if} \quad x > 2 \end{cases}$$

Find $f(0)$, $f(2)$, and $f(3)$.

16. For $f(x) = 3x^2 - 4x + 3$ and $g(x) = 3x - 1$, find

(a) $f(x) \cdot g(x)$ **(b)** $f[g(x)]$

17. Sketch the graphs of the following equations. Be sure to label the intercepts.

(a) $y = \sqrt{x} - 2$ **(b)** $y = x^2 - 6x$

(c) $x^2 - 4x + y^2 = 0$

18. Suppose that y varies directly with the square of x and inversely as the cube root of t. If $y = 4$ when $x = 3$ and $t = 8$, find y when $x = 5$ and $t = 1$.

10

Exponential and Logarithmic Functions

© Royalty-Free/CORBIS

10.1 Exponential Functions

Throughout this textbook we have discussed a variety of equations and how they define various relations and functions. We now turn our attention to a new type of relationship and its corresponding equation. We begin by considering the following example.

EXAMPLE 1

A bacteria culture initially has 1,000 bacteria of a type whose number doubles every 12 hours.

(a) How many bacteria are there after 12 hours?

(b) How many bacteria are there after 24 hours?

(c) How many bacteria are there after 36 hours?

(d) How many bacteria are there after t hours?

(e) How many bacteria are there after 15 hours?

Solution

(a) Since the number of bacteria doubles every 12 hours, after 12 hours there will be twice as many as there were initially:

$$1{,}000(2) = \boxed{2{,}000 \text{ bacteria}}$$

(b) After 24 hours the number of bacteria has doubled *two* times $\left(\frac{24}{12} = 2\right)$:

$$[1{,}000(2)](2) = 1{,}000(2)^2 = 1{,}000(4) = \boxed{4{,}000 \text{ bacteria}}$$

(c) After 36 hours the number of bacteria has doubled *three* times $\left(\frac{36}{12} = 3\right)$:

$$[1{,}000(2)(2)](2) = 1{,}000(2)^3 = 1{,}000(8) = \boxed{8{,}000 \text{ bacteria}}$$

(d) Note that in parts **(a)**, **(b)**, and **(c)**, the exponent of 2 is the number of times the population has doubled. Since the number of bacteria doubles every 12 hours, in t hours the number will double $\dfrac{t}{12}$ times.

Note how useful function notation is here: It allows us to emphasize how N is dependent on t.

If we let $N(t) =$ the number of bacteria present after t hours, then

$$\boxed{N(t) = 1{,}000 \cdot 2^{t/12}}$$

(e) We use the answer obtained in part **(d)** with $t = 15$:

$$N(t) = 1{,}000 \cdot 2^{t/12} \qquad \textit{Substitute } t = 15.$$
$$N(15) = 1{,}000 \cdot 2^{15/12}$$
$$= 1{,}000 \cdot 2^{1.25} \qquad \textit{Using a calculator, we get}$$
$$= \boxed{2{,}378 \text{ bacteria}}$$

rounded to the nearest whole number.

Unlike the functions or equations we have discussed previously in this textbook, the mathematical model we constructed in Example 1 required the use of a variable in an exponent. Functions or equations that contain variables in their exponents are called *exponential functions or equations*.

DEFINITION

A function of the form $y = f(x) = b^x$, where $b > 0$ and $b \neq 1$, is called an *exponential function*.

A question that should immediately come to mind is the following: For what values of x does this function make sense? As an example, for what values of x is $y = 2^x$ defined?

Based on our work in Chapter 7, we know what 2^x means when x is a rational number. In other words, we know what 2^x means when x takes on values such as $x = 5$, $x = -3$, and $x = \frac{2}{7}$.

$$2^5 = 2 \cdot 2 \cdot 2 \cdot 2 \cdot 2 = 32$$

$$2^{-3} = \frac{1}{2^3} = \frac{1}{8}$$

$$2^{2/7} = \left(\sqrt[7]{2}\right)^2$$

But what about irrational values of x? What do we mean by $2^{\sqrt{3}}$?

Although a formal definition of b^x for x irrational is beyond the scope of this book, we can make the following remarks. We know that we can get arbitrarily accurate approximations to $\sqrt{3}$. For example, we find that

$$1.732 < \sqrt{3} < 1.733$$

and so it seems reasonable that

$$2^{1.732} < 2^{\sqrt{3}} < 2^{1.733}$$

Since 1.732 and 1.733 are rational numbers, $2^{1.732}$ and $2^{1.733}$ are well defined (not easy to compute, perhaps, but well defined nonetheless).

In this way, we can get better and better approximations to what we expect $2^{\sqrt{3}}$ to be. Consequently, we will make the following assumption.

> The exponential function is defined for all real numbers x, and all the properties of exponents developed in Chapter 7 extend to real exponents as well.

Let's examine the function $y = f(x) = 2^x$. Since we have not worked with this type of function before, we construct a brief table of values as shown here. If we plot these points and connect them with a smooth curve, we obtain the graph shown in Figure 10.1.

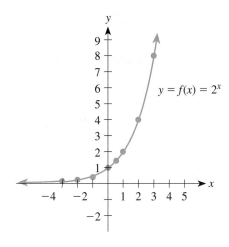

x	$y = f(x) = 2^x$
-3	$2^{-3} = \dfrac{1}{2^3} = \dfrac{1}{8}$
-2	$2^{-2} = \dfrac{1}{2^2} = \dfrac{1}{4}$
-1	$2^{-1} = \dfrac{1}{2}$
0	$2^0 = 1$
$\frac{1}{2}$	$2^{1/2} = \sqrt{2} \approx 1.414$
1	$2^1 = 2$
2	$2^2 = 4$
3	$2^3 = 8$

Figure 10.1
The graph of $y = f(x) = 2x$

Let's examine the graph of this function more carefully. Notice that as x takes on smaller and smaller values (as we move left on the graph), the y values get closer and closer to 0 (the curve $y = 2^x$ gets closer to the x-axis), but the graph never touches the x-axis. The graph of $y = 2^x$ will never intersect the x-axis because there is no value of x for which $2^x = 0$. If the graph of a function approaches a line for extreme values of x and never touches it, we call that line an ***asymptote*** for the graph. Thus, the x-axis is an asymptote for the graph of $y = 2^x$.

Looking at the graph of $y = 2^x$, we note that as x increases (as we move right), y also increases, but very rapidly as compared with x. The graph of the equation $y = 2^x$ exhibits what is called **exponential growth**.

Figure 10.2 shows how the graphs of $y = 3^x$ and $y = 5^x$ compare with the graph of $y = 2^x$.

Construct a table for each of these functions and verify their graphs.

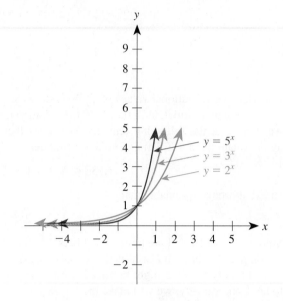

Figure 10.2
The graphs of $y = 2^x$,
$y = 3^x$, and $y = 5^x$

Note that they all have the same basic shape, but the larger the base, the more rapidly the graph rises. Hence, the graph of $y = 3^x$ rises more quickly than the graph of $y = 2^x$, and the graph of $y = 5^x$ rises more quickly than the graph of $y = 3^x$. [Also note that they all intersect the y-axis at $(0, 1)$.]

Again we note that y increases very rapidly as compared with x, and generalize that the graph of the equation $y = b^x$ for $b > 1$ exhibits exponential growth.

So far we have discussed the graph of $y = b^x$ when $b > 1$. Next, we examine the graph of $y = f(x) = b^x$ when $0 < b < 1$.

EXAMPLE 2

Sketch the graph of $y = f(x) = \left(\dfrac{1}{2}\right)^x$.

Solution   Again, we begin by constructing a table of values.

x	-3	-2	-1	0	1	2	3
$y = f(x) = \left(\dfrac{1}{2}\right)^x$	8	4	2	1	$\dfrac{1}{2}$	$\dfrac{1}{4}$	$\dfrac{1}{8}$

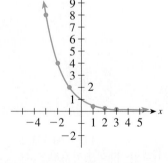

Figure 10.3

Graph of $y = \left(\dfrac{1}{2}\right)^x$

We plot these points and connect them with a smooth curve to obtain the graph shown in Figure 10.3. The graph again approaches, but does not touch, the x-axis (this time on the right side).

The equation $y = f(x) = \left(\dfrac{1}{3}\right)^x$ will have a graph very similar to the one obtained in Example 2. (*Try it!*) However, because the base is smaller, it will *fall* more sharply.

Notice that as x increases, y decreases very rapidly as compared with x. In general, the graph of the equation $y = f(x) = b^x$ when $0 < b < 1$ exhibits what is called ***exponential decay***.

Based on our previous discussion, we make the following observations about the exponential function $f(x) = b^x$:

1. The domain (the set of possible x values) for an exponential function is R, the set of all real numbers.

2. The range (the set of possible y values) for an exponential function is $\{y \mid y > 0\}$. That is, the range of an exponential function is the set of all positive real numbers.

3. The graph of an exponential function always passes through the point $(0, 1)$ (because $b^0 = 1$ for all $b \neq 0$).

4. The graph of an exponential function rises if $b > 1$ and falls if $0 < b < 1$ (see Figure 10.4). Remember that we always describe a graph as we move from left to right, that is, for increasing values of x.

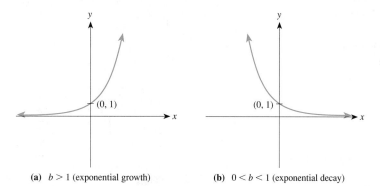

Figure 10.4
Graphs of $y = f(x) = b^x$

(a) $b > 1$ (exponential growth)   (b) $0 < b < 1$ (exponential decay)

5. The graph of $y = b^x$ also satisfies the *horizontal line test*. That is, a horizontal line crosses the graph of an exponential function at most once. Based on our discussion in Section 9.4, this means that the exponential function is one-to-one and so has an inverse.

Higher-level courses discuss how to derive such a mathematical model from the given data.

Exponential functions can be used to model real-life data. The table and graph below show the population of the United States from 1900 to 2000. See Figure 10.5. The graph shows how an exponential function can accurately model population growth. The function is $P(t) = 78.8(1.01315)^x$, where t the number of years after 1900, and $P(t)$ is the population in millions. For example, based on the table below, the population of the United States in 1980 was 226.542, or 226,542,000. Using the function $P(t)$, we can estimate the population in 1980 by $P(80) = 78.8(1.01315)^{80} = 224.092$, or 224,092,000 (a reasonably good approximation).

Year	1900	1920	1940	1960	1980	2000
Population (millions)	75.955	105.711	131.669	179.323	226.542	281.422

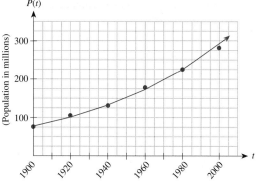

Figure 10.5

EXAMPLE 3

In a certain medium, a population of a strain of bacteria doubles every 5 hours. If we start with 1,000 bacteria, the population $P(t)$ at t hours can be described by the model

$$P(t) = (1,000)2^{t/5}$$

The graph of this model appears in Figure 10.6.

(a) Use the formula for $P(t)$ to determine the size of the population at $t = 24$ hours, and check your value with the graph.

(b) Use the graph to estimate at which hour the population reaches 20,000.

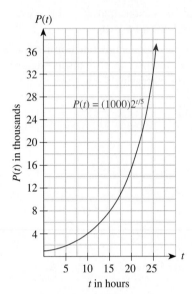

Figure 10.6
The graph of
$P(t) = (1,000)2^{t/5}$

Solution

(a) To find the number of bacteria present at 24 hours, we substitute $t = 24$ in the given function; that is, we want $P(24)$:

$$P(t) = (1,000)2^{t/5} \qquad \textit{Substitute } t = 24 \textit{ to get}$$
$$P(24) = (1,000)2^{24/5} \qquad \textit{Use a calculator for the computation.}$$
$$P(24) = 27,857.61803$$

Rounded, our answer is 27,858 bacteria in 24 hours.

We check our answer on the graph in Figure 10.7**(a)** and note that when $t = 24$, $P(24)$ is around 28,000.

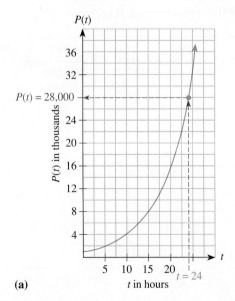

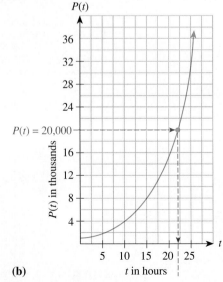

Figure 10.7   **(a)**   **(b)**

(b) Algebraically, to find the time at which the population of bacteria reaches 20,000, we would have to solve the equation $P(t) = 20{,}000$ for t. In other words, we would have to solve $20{,}000 = (1{,}000)2^{t/5}$ for t. At this point we have not discussed solving equations with the variable in the exponent, so we are restricted to estimating the solution by using the graph.

Graphically, finding the time at which the population of bacteria reaches 20,000 means we are given that $P(t)$, the vertical coordinate, is 20,000, and we use the graph to find t, the horizontal coordinate, when $P(t) = 20{,}000$.

On the graph in Figure 10.7**(b)** we start on the y-axis at $y = 20{,}000$, trace a horizontal line until we touch the graph of $y = P(t)$, and then draw a vertical line at that intersection to the t-axis to get the estimate $t = 22$. Hence, it takes about 22 hours for the population to reach 20,000.

Use a calculator to compute $P(22)$. How close is $P(22)$ to 20,000? We will return to this example in Section 10.5 to get a more accurate answer.

EXAMPLE 4

(a) Sketch the graph of $f(x) = 2^x + 1$.

(b) Use the graph to determine the value of x for which $f(x) = 9$ and verify this solution algebraically.

Solution   **(a)** Again, we begin by constructing a table of values.

x	-3	-2	-1	0	1	2	3
$f(x) = 2^x + 1$	$1\frac{1}{8}$	$1\frac{1}{4}$	$1\frac{1}{2}$	2	3	5	9

We plot these points and connect them with a smooth curve to obtain the graph shown in Figure 10.8**(a)**. Notice that as x gets smaller, the graph of $f(x)$ gets closer to but never touches $y = 1$: The horizontal asymptote is $y = 1$.

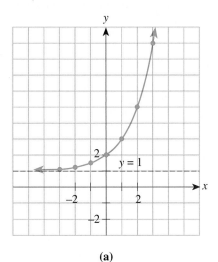

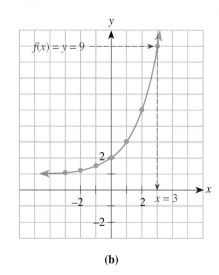

Figure 10.8
The graph of $f(x) = 2^x + 1$

(a)   (b)

Is there an x such that $2^x = 0$? That is, to what power should 2 be raised in order to get 0?

Looking at the table on page 602, we see that as x gets smaller (more negative), 2^x gets closer and closer to 0. Hence, as x gets smaller, $2^x + 1$ gets closer and closer to $0 + 1 = 1$. We can see that $2^x + 1$ can never be 1 since 2^x is never 0.

(b) Finding the value of x for which $f(x) = 9$ means looking for the value of x at the point on the graph where $y = 9$. On the graph of $f(x) = 2^x + 1$ in Figure 10.8b, we start on the y-axis at $y = 9$, trace a horizontal line until we touch the graph of $y = f(x)$, and then draw a vertical line at that intersection to the x-axis to get the estimate $x = 3$. We can verify this solution by noting that if $f(x) = 2^x + 1$, then $f(3) = 2^3 + 1$, which is 9.

An *exponential equation* is an equation in which an exponent contains a variable. In Chapter 9 we discussed one-to-one functions and noted that by definition, each y value is assigned to only one x value. Symbolically, we can state this as follows: If $f(r) = f(s)$, then $r = s$ (if the y values are the same, the x values must be the same). In particular, since the exponential function $f(x) = b^x$ is a one-to-one function, we can write this general statement for exponential functions as follows: If $b^r = b^s$, then $r = s$. This gives a procedure for solving certain types of exponential equations, as demonstrated in the next example.

EXAMPLE 5

Solve each of the following exponential equations.

(a) $2^{3x-1} = 2^x$ **(b)** $3^{5x+1} = 9^{2x}$ **(c)** $\dfrac{4^{x^2}}{4^{3x}} = \dfrac{1}{16}$

Solution

The fact that the exponential function is one-to-one means that if $b^r = b^t$, then $r = t$.

(a) $2^{3x-1} = 2^x$ *Since the bases are equal, the exponents must be equal.*

$$3x - 1 = x$$

$$2x = 1$$

$$\boxed{x = \frac{1}{2}}$$

Exponential equations are discussed further in Section 10.5.

Check: $x = \dfrac{1}{2}$:

$$2^{3x-1} = 2^x$$
$$2^{3(1/2)-1} \stackrel{?}{=} 2^{1/2}$$
$$2^{(3/2)-1} \stackrel{?}{=} 2^{1/2}$$
$$2^{1/2} \stackrel{\checkmark}{=} 2^{1/2}$$

(b) We try to express both sides of the equation in terms of the same base.

$$3^{5x+1} = 9^{2x}$$ *Write 9 as 3^2.*

$$3^{5x+1} = (3^2)^{2x}$$ *Use exponent rule 2 (raising a power to a power).*

$$3^{5x+1} = 3^{4x}$$ *Since the bases are equal, the exponents must be equal.*

$$5x + 1 = 4x$$

$$\boxed{x = -1}$$

Check: $x = -1$:

$$3^{5x+1} = 9^{2x}$$
$$3^{5(-1)+1} \stackrel{?}{=} 9^{2(-1)}$$
$$3^{-4} \stackrel{?}{=} 9^{-2}$$
$$\frac{1}{81} \stackrel{\checkmark}{=} \frac{1}{81}$$

(c) We express both sides in terms of the base 4.

$$\frac{4^{x^2}}{4^{3x}} = \frac{1}{16}$$ *Use exponent rule 3 (the quotient of powers).*

$$4^{x^2-3x} = \frac{1}{4^2}$$

$$4^{x^2-3x} = 4^{-2}$$

$$x^2 - 3x = -2$$ *This is a quadratic equation.*

$$x^2 - 3x + 2 = 0$$ *Factor and solve.*

$$(x - 2)(x - 1) = 0$$

$$\boxed{x = 2 \text{ or } x = 1}$$

The check is left to the reader.

In Example 5 we were able to rewrite the equations so that each side of the equation was an exponential expression with the same base; this allowed us to set the exponents equal to each other and to solve a simpler equation. In Section 10.5, we will discuss other algebraic techniques for solving exponential equations.

EXAMPLE 6

Suppose $5,000 is deposited in a bank that pays interest at an effective rate of 7.5% simple yearly interest. Make a table to determine how much money would be in the bank at the end of each of the first 4 years. Use this table to develop a formula to determine the amount in the bank at the end of year t, and use this formula to determine the amount in the bank at the end of 14 years.

Solution

In Section 1.3, we showed that we can find the amount in the bank at the end of 1 year by multiplying the amount in the bank at the beginning of the year, P, by 1.075 to get $1.075P$ (we can view this as 107.5% of the principal). Hence we found that we would have $1.075(5,000) = \$5,375$ in the bank at the end of the first year.

To find out how much we have in the bank at the end of the *second* year, we do the same thing; that is, we multiply the amount in the bank at the *beginning of the second year*, which is now $5,375 = 1.075(5,000)$, by 1.075. Therefore, the amount in the bank at the end of the second year is $(1.075)[1.075(5,000)]$, which we write as $(1.075)^2(5,000)$.

From the table below, we can see a pattern emerging between t (the year) and the amount in the bank at the end of year t, which we will call $A(t)$.

$$A(t) = 1.075 \cdot (\textit{Amount in the bank at the beginning of the year})$$
$$A(1) = (1.075)(5,000) \qquad\qquad = (1.075)^1(5,000) = \$5,375$$
$$A(2) = (1.075)[(1.075)(5,000)] \ = (1.075)^2(5,000) = \$5,778.13$$
$$A(3) = (1.075)[(1.075)^2(5,000)] = (1.075)^3(5,000) = \$6,211.48$$
$$A(4) = (1.075)[(1.075)^3(5,000)] = (1.075)^4(5,000) = \$6,677.35$$

By the table we can see that the pattern gives us the formula

$$A(t) = 1.075^t(5,000)$$

where $A(t)$ is the amount in the bank at the end of year t.

At the end of the 14th year, the amount in the bank is

$$A(14) = 1.075^{14}(5,000) = \boxed{\$13,762.22}$$

EXERCISES 10.1

In Exercises 1–10, sketch the graph of the given function.

1. $f(x) = 4^x$

2. $f(x) = \left(\dfrac{1}{4}\right)^x$

3. $f(x) = \left(\dfrac{1}{5}\right)^x$

4. $y = 5^x$

5. $f(x) = 3^{-x}$

6. $y = \left(\dfrac{1}{3}\right)^x$

7. $y = 2^{x+1}$

8. $f(x) = 3^x - 1$

9. $f(x) = 2^x + 1$ **10.** $f(x) = 3^{x-1}$

11. The table below shows the number of bacteria (in millions) present in a certain culture at 2-hour intervals, from time $t = 0$ (the initial time) to time $t = 10$ hours. The exponential function $P(t) = 4.28(2.0263)^t$ is a good model of these data, where $P(t)$ is the number in millions at time t.

Hour	0	2	4	6	8	10
Population (millions)	4.0	18.5	73.8	306.0	1,218.1	4,800.0

    (a) Compute the values for each of the hours in the table using the function $P(t) = 4.28(2.0263)^t$. How do these values compare with the actual values in the table?

    (b) Plot the table values on a coordinate system and sketch the function $P(t)$ on the same set of coordinate axes.

12. Moore's Law: In 1965, Gordon Moore published a paper in which he observed that the number of transistors on integrated circuits doubled every few years: he predicted that this trend would continue. The table below shows the number of transistors (in the thousands) per circuit, by the year in which it was introduced.

Year	1972	1978	1985	1993	1999	2002
Transistors (thousands)	2.5	29	275	3,100	24,000	220,000

    (a) Compute the values for each of the years in the table using the function $N(t) = 2.6376(2)^{0.51324t}$, where t is the number of years after 1972 ($t = 0$ for 1972, $t = 6$ for 1978, etc.). How do these values compare to the actual values in the table?

    (b) Plot the table values on a coordinate system and sketch the function $N(t)$ on the same set of coordinate axes.

In Exercises 13–26, *solve the given exponential equation.*

13. $2^x = 2^{3x-2}$ **14.** $3^{5x-3} = 3^{x+5}$ **15.** $5^x = 25^{x-1}$ **16.** $9^x = 3^{x+3}$

17. $4^{1-x} = 16$ **18.** $4^{1-x} = 8$ **19.** $8^x = 4^{x+1}$ **20.** $27^{x-1} = 9^{3x}$

21. $\dfrac{9^{x^2}}{9^x} = 81$ **22.** $\dfrac{4^{x^2}}{4^{2x}} = 64$ **23.** $4^{\sqrt{x}} = 4^{2x-3}$ **24.** $25^{\sqrt{x}} = 5^x$

25. $16^{x^2-1} = 8^{x-1}$ **26.** $16^{x-1} = 8^{x^2-1}$

27. The number of bacteria in a culture that initially contains 2,500 bacteria doubles every 10 hours. How many bacteria are there after 10 hours? After 20 hours? After 50 hours? After t hours?

28. The population of Centerville is 8,000 and triples every 25 years. What is the population in 25 years? In 75 years? In 200 years? In Y years?

29. In 2004, the population of Capitol City was 20,000. As a result of industrial development, the population will double every 14 years. What will the population be in 2018? In 2032? In year Y?

30. In 2005, the rabbit population in a certain area was estimated at 12,000, and is assumed to quadruple (grow by a factor of 4) every 3 years unless controlled by some outside agent. Estimate what the uncontrolled population will be in 2008. In 2017. In year Y.

31. An antibacterial substance is introduced into a bacterial culture that contains 10,000 bacteria. The substance destroys bacteria so that there are half as many bacteria as there were 1 hour ago. What is the number of bacteria present 2 hours after the substance is introduced? After 3 hours? After t hours?

32. The oil reserves of a certain oil field that contains 1 million barrels of oil are being depleted at the rate of $\frac{1}{10}$ every 5 years. How many barrels are left after 5 years? After 10 years? After 30 years?

33. A car loses $\frac{1}{6}$ of its value each year. If the car is worth $16,000 today, how much will it be worth 4 years from now?

34. A boat is now worth $14,000, and loses 10% of its value each year. What would it be worth 5 years from now?

35. A machine depreciates in value by $\frac{1}{5}$ each year. If the machine is now worth $24,000, how much will it be worth 3 years from now?

36. An antique coin is now worth $8,000. If it appreciates in value at the rate of 10% per year, how much will it be worth 5 years from now?

37. The population of Davisville is increasing at a rate of 5% per year. If the population is currently 100,000, what will the population be 4 years from now?

38. The population of Sivadville is increasing at a rate of 10% per year. If the population is currently 100,000, what will the population be 4 years from now?

39. Roberta invests $10,000 in a bank that pays 8% simple annual interest. How much will Roberta have in the bank at the end of 5 years?

40. How much would Roberta of Exercise 39 have in the bank at the end of 5 years if the bank paid 6% simple annual interest?

41. Beverly invests $5,000 in a bond that pays 10% simple annual interest. How much is her bond worth at the end of 5 years?

42. How much would the bond of Exercise 41 be worth at the end of 10 years?

43. If $10,000 is invested at 6% interest compounded monthly, the amount present $A(t)$ at the end of t years can be found by the model

$$A(t) = (10{,}000)1.005^{12t}$$

The graph of this model appears below.

(a) Use the formula for $A(t)$ to determine how much is in the bank at $t = 10$ years, and check your value with the graph.

(b) Use the graph to estimate at what year the amount will reach $30,000.

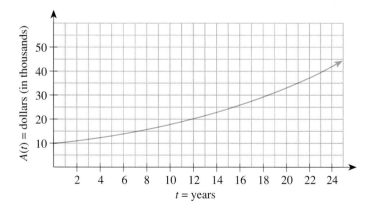

Question for Thought

44. Sketch the graphs of the equations $y = 2^x$ and $y = x^2$ on the same set of coordinate axes for $x \geq 0$. Compare the two graphs and discuss their differences. (For each graph, how do changes in the value of x affect the value of y? Does y increase as x increases, or does y decrease? If x changes by 1 unit, how will this affect y? Will y always change by the same amount?) For $x > 4$, which "rises" faster?

Mini-Review

45. *Evaluate:* $(2^{-3} + 3^{-2})^{-1}$

46. *Combine and simplify:* $\dfrac{3}{2x} + \dfrac{6}{x + 2}$

47. Sketch the graph of the equation $\dfrac{x}{2} - \dfrac{y}{3} = 4$. Label the intercepts.

48. *Evaluate:* $\left(\sqrt{3}\right)^6 + \left(\sqrt[3]{5}\right)^6$

10.2   Logarithms and Logarithmic Functions

In the previous section, we discussed exponential functions—that is, functions of the form

$$y = f(x) = b^x \qquad (b > 0, b \neq 1)$$

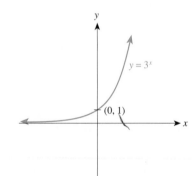

We pointed out that the graphs of exponential functions satisfy the horizontal line test and so an exponential function has an inverse. For example, if we sketch the graph of $y = f(x) = 3^x$, we can see that each y value in the range is associated with exactly one x value in the domain. That is, each horizontal line crosses the graph at most once, and so the graph of $y = 3^x$ satisfies the horizontal line test and is therefore a one-to-one function. Thus, the function $y = f(x) = 3^x$ has an inverse.

Let's look at the functions $y = x^3$ and $y = 3^x$ and their inverse functions. Keep in mind that we let x represent the *independent* variable and y the *dependent* variable, so that whenever possible we want to write y explicitly as a function of x.

Recall that to obtain the inverse of a given function, we interchange x and y and then solve for y.

Function	Inverse function		
$y = x^3$	$x = y^3$	or	$y = \sqrt[3]{x}$
$y = 3^x$	$x = 3^y$	or	$y = ?$

Notice that for the inverse function of the exponential function $y = 3^x$, we have no way of writing y explicitly as a function of x. (Actually, if you think about it, if we did not have radicals and fractional exponents, then to take the function $x = y^3$ and solve for y, we would have had to "invent" radical notation so that we could write $y = \sqrt[3]{x}$.) To write the function $x = 3^y$, which is the inverse of the exponential function, explicitly as a function of x we must invent a new notation.

DEFINITION   We write $y - \log_b x$ to mean $x = b^y$, where $b > 0$, $x > 0$, and $b \neq 1$. $y = \log_b x$ is read "$y = \log$ base b of x."

In words, $y = \log_b x$ means that "y is the exponent of b that gives x."

Remember

$y = \log_b x$ and $x = b^y$ are alternative ways of expressing the same relationship. *A logarithm is just an exponent.*

We will often use the following terminology:

When an expression is written in the form $x = b^y$, it is said to be in *exponential form*.

When an expression is written in the form $y = \log_b x$, it is said to be in *logarithmic form*.

Thus, $5^2 = 25$ and $\log_5 25 = 2$ are the exponential and logarithmic forms, respectively, of the same relationship.

EXAMPLE 1

Write each of the following in exponential form.

(a) $\log_{10} 100 = 2$ **(b)** $\log_4 \dfrac{1}{64} = -3$ **(c)** $\log_7 \sqrt{7} = \dfrac{1}{2}$

(d) $\log_9 27 = \dfrac{3}{2}$

Solution

We use the fact that $\log_b x = y$ means that $b^y = x$.

(a) $\log_{10} 100 = 2$ means $\boxed{10^2 = 100}$.

(b) $\log_4 \dfrac{1}{64} = -3$ means $\boxed{4^{-3} = \dfrac{1}{64}}$.

(c) $\log_7 \sqrt{7} = \dfrac{1}{2}$ means $\boxed{7^{1/2} = \sqrt{7}}$.

(d) $\log_9 27 = \dfrac{3}{2}$ means $\boxed{9^{3/2} = 27}$.

Keep in mind that all we did in Example 1 was to *translate* from logarithmic form to exponential form. You should verify for yourself that we were translating *true* statements.

EXAMPLE 2

Write each of the following in logarithmic form.

(a) $2^4 = 16$ **(b)** $4^2 = 16$ **(c)** $3^{-4} = \dfrac{1}{81}$ **(d)** $100^{1/2} = 10$

Solution

We use the fact that $b^y = x$ means that $\log_b x = y$.

(a) $2^4 = 16$ means $\boxed{\log_2 16 = 4}$.

(b) $4^2 = 16$ means $\boxed{\log_4 16 = 2}$.   *Notice the difference between parts* (a) *and* (b).

(c) $3^{-4} = \dfrac{1}{81}$ means $\boxed{\log_3 \dfrac{1}{81} = -4}$.

(d) $100^{1/2} = 10$ means $\boxed{\log_{100} 10 = \dfrac{1}{2}}$.

Being able to translate from logarithmic form to exponential form often makes it easier to evaluate logarithmic expressions. For example, evaluating

$$\log_5 5^3 = ?$$

is the same as asking (in exponential form)

$$5^? = 5^3$$

and so the answer is clearly 3. Remember that since the exponential function is one-to-one, $b^t = b^s$ implies that $t = s$.

Thus, we have $\log_5 5^3 = 3$.

Similarly, $\log_6 6^{1/4} = ?$ is asking $6^? = 6^{1/4}$ and so that answer is $\frac{1}{4}$.

Thus, $\log_6 6^{1/4} = \dfrac{1}{4}$.

In general, we have

$$\boxed{\log_b b^r = r}$$

In words, this says that "the exponent of b that gives b^r is r."

EXAMPLE 3

Find each of the following:

(a) $\log_2 32$ **(b)** $\log_5 \dfrac{1}{25}$ **(c)** $\log_6 \sqrt[3]{6}$ **(d)** $\log_8 16$

(e) $\log_5 5$ **(f)** $\log_5 1$ **(g)** $\log_3(-8)$

Solution   We will try to make use of the fact just stated that $\log_b b^r = r$. Consequently, we try to express the number whose logarithm we are trying to find as a power of the base b.

(a) $\log_2 32 = \log_2 2^5 = \boxed{5}$

(b) $\log_5 \dfrac{1}{25} = \log_5 \dfrac{1}{5^2} = \log_5 5^{-2} = \boxed{-2}$

(c) Since a logarithm is an exponent, when we are working with logarithms we generally prefer to write radical expressions in exponential form:

$$\log_6 \sqrt[3]{6} = \log_6 6^{1/3} = \boxed{\dfrac{1}{3}}$$

(d) For $\log_8 16$, it is not quite as obvious how to express 16 in terms of the base 8. Let's call the answer t, translate the required logarithm into exponential form, and solve the resulting exponential equation.

$\log_8 16 = t$ means

$$8^t = 16 \qquad \textit{Let's write both 8 and 16 as powers of the same base, 2.}$$
$$(2^3)^t = 2^4$$
$$2^{3t} = 2^4 \qquad \textit{Since the exponential function is one-to-one, if } b^r = b^t \textit{ then } r = t.$$
$$3t = 4$$
$$t = \dfrac{4}{3}$$

Thus, $\log_8 16 = \boxed{\dfrac{4}{3}}$.

(e) $\log_5 5 = \log_5 5^1 = \boxed{1}$

(f) $\log_5 1 = \log_5 5^0 = \boxed{0}$

(g) If we translate the expression $\log_3(-8)$ into exponential form, we get $3^? = -8$. Since there is no exponent of 3 that gives a negative number, $\log_3(-8)$ $\boxed{\text{does not exist}}$.

Note that since $b^0 = 1$ for all $b \neq 0$, we have

$$\boxed{\log_b 1 = 0}$$

If we use $y = \log_b x$ to replace y in the equation $x = b^y$, we obtain the relationship

$$\boxed{b^{\log_b x} = x}$$

Thus, for example, $5^{\log_5 8} = 8$.

EXAMPLE 4	Evaluate $\log_2(\log_4 16)$.

Solution We first find $\log_4 16 = \log_4 4^2 = 2$. Therefore, our original expression becomes

$$\log_2(\log_4 16) = \log_2 2 = \boxed{1}$$

The same idea of translating a logarithmic statement into exponential form allows us to solve some logarithmic equations.

EXAMPLE 5

Solve each of the following equations for t.

(a) $\log_7 t = 3$ **(b)** $\log_9 \frac{1}{3} = t$ **(c)** $\log_t 125 = 3$

Solution **(a)** $\log_7 t = 3$ means $7^3 = t$, so $\boxed{343 = t}$

(b) $\log_9 \frac{1}{3} = t$ means

$$9^t = \frac{1}{3}$$
$$(3^2)^t = 3^{-1}$$
$$3^{2t} = 3^{-1}$$
$$2t = -1$$
$$\boxed{t = -\frac{1}{2}}$$

(c) $\log_t 125 = 3$ means

$$t^3 = 125$$
$$t = \sqrt[3]{125}$$
$$\boxed{t = 5}$$

The checks are left to the student.

EXAMPLE 6

On the same set of coordinate axes, sketch the graphs of each pair of equations.

(a) $f(x) = 3^x$ and $g(x) = \log_3 x$ **(b)** $f(x) = \left(\dfrac{1}{2}\right)^x$ and $g(x) = \log_{1/2} x$

Solution

Using the properties of inverse functions that we developed in Chapter 9, we note the following:

1. Since $f(x) = b^x$ has as its domain R, the set of all real numbers, and its range is $\{y \mid y > 0\}$, the domain of the inverse function $y = \log_b x$ is $\{x \mid x > 0\}$ and the range of the inverse function is R, the set of all real numbers.

2. Since the graphs of all exponential functions pass through the point $(0, 1)$, the graphs of all logarithmic functions pass through the point $(1, 0)$.

3. We can draw the graph of an inverse function by reflecting the graph of the original function about the line $y = x$.

Based on these observations, we obtain the graphs appearing in Figures 10.9(a) and 10.9(b).

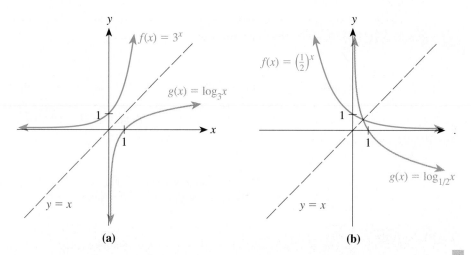

Figure 10.9   **(a)**   **(b)**

Note that since $y = 3^x$ and $y = \log_3 x$ are inverses of each other, we interchange the domain and range: The domain of $y = 3^x$ (all real numbers) is the range of $y = \log_3 x$, and the range of $y = 3^x$ (all positive real numbers) is the domain of $y = \log_3 x$.

You have probably noticed that there is a **LOG** key on your calculator, but the key gives no indication of the base of the logarithm. When no base is indicated with a logarithm, it is assumed that the base is 10. That is,

$$\log_{10} x = \log x$$

To find $\log_{10} 100$, we are looking for the power of 10 that gives 100, or $10^? = 100$. In this case, the answer is easily found to be 2. To find $\log_{10} 842$, however, requires that we solve $10^? = 842$. Since this solution is not easily obtainable, we use the calculator to find $\log_{10} 842$ by pressing the keys **842** **LOG** to get the answer **2.9253121**. Note that this answer is an approximation: If we raise 10 to the power 2.9253121, we get (the approximation) 842.00002. We will cover log base 10 in more detail in Section 10.4.

Let's end this section by summarizing some of the basic facts about logarithms that we have developed in this section.

In the next section we will derive some important properties of logarithms.

Summary

1. The functions $f(x) = \log_b x$ and $g(x) = b^x$ are inverse functions. In other words, $y = \log_b x$ is equivalent to $x = b^y$, where $b > 0$ and $b \neq 1$.

2. The domain of the function $f(x) = \log_b x$ is $\{x \mid x > 0\}$ and the range is all real numbers.

3. $\log_b b^x = x$

4. $\log_b 1 = 0$

5. $b^{\log_b x} = x$

6. The graphs of $f(x) = \log_b x$ are shown below.

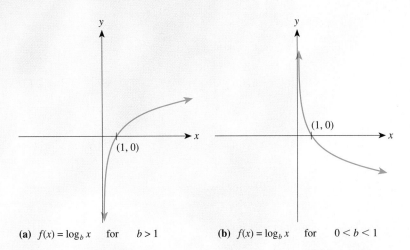

(a) $f(x) = \log_b x$ for $b > 1$ **(b)** $f(x) = \log_b x$ for $0 < b < 1$

EXERCISES 10.2

In Exercises 1–38, write each logarithmic statement in exponential form and each exponential statement in logarithmic form.

1. $\log_7 49 = 2$ **2.** $\log_4 64 = 3$ **3.** $3^4 = 81$ **4.** $2^6 = 64$

5. $\log_{10} 10{,}000 = 4$ **6.** $\log_2 32 = 5$ **7.** $10^3 = 1{,}000$ **8.** $4^5 = 1{,}024$

9. $\log_9 81 = 2$ **10.** $\log_5 125 = 3$ **11.** $\log_{81} 9 = \dfrac{1}{2}$ **12.** $\log_{125} 5 = \dfrac{1}{3}$

13. $6^{-2} = \dfrac{1}{36}$ **14.** $10^{-3} = 0.001$ **15.** $\log_3 \dfrac{1}{3} = -1$ **16.** $8^{1/3} = 2$

17. $25^{1/2} = 5$ **18.** $\log_4 \dfrac{1}{16} = -2$ **19.** $\log_8 8 = 1$ **20.** $11^1 = 11$

21. $8^0 = 1$ **22.** $\log_{11} 11 = 1$ **23.** $\log_{16} 8 = \dfrac{3}{4}$ **24.** $\log_{27} 9 = \dfrac{2}{3}$

25. $27^{-2/3} = \dfrac{1}{9}$ **26.** $16^{-3/4} = \dfrac{1}{8}$ **27.** $\log_8 \dfrac{1}{2} = -\dfrac{1}{3}$ **28.** $\log_4 \dfrac{1}{2} = -\dfrac{1}{2}$

29. $\log_{1/2} 4 = -2$ **30.** $\log_{1/3} 27 = -3$ **31.** $3^0 = 1$ **32.** $\log_3 1 = 0$

33. $\log_7 1 = 0$ **34.** $7^0 = 1$ **35.** $\log_6 \sqrt{6} = \dfrac{1}{2}$ **36.** $\log_2 \sqrt[3]{2} = \dfrac{1}{3}$

37. $6^{1/2} = \sqrt{6}$ **38.** $5^{1/4} = \sqrt[4]{5}$

In Exercises 39–64, evaluate the given logarithm.

39. $\log_2 8$

40. $\log_3 81$

41. $\log_9 81$

42. $\log_{10} 1,000$

43. $\log_4 \dfrac{1}{4}$

44. $\log_6 \dfrac{1}{36}$

45. $\log_5 \dfrac{1}{125}$

46. $\log_2 \dfrac{1}{16}$

47. $\log_4 \dfrac{1}{2}$

48. $\log_9 \dfrac{1}{3}$

49. $\log_8 4$

50. $\log_4 8$

51. $\log_9(-27)$

52. $\log_{27} 0$

53. $\log_4 \dfrac{1}{8}$

54. $\log_8 \dfrac{1}{4}$

55. $\log_6 \sqrt{6}$

56. $\log_{10} \sqrt[8]{10}$

57. $\log_5 \sqrt[3]{25}$

58. $\log_3 \sqrt{243}$

59. $\log_5(\log_3 243)$

60. $\log_4(\log_5 625)$

61. $\log_8(\log_7 7)$

62. $\log_2(\log_9 9)$

63. $5^{\log_5 7}$

64. $3^{\log_3 6}$

In Exercises 65–80, solve the given equation for x, y, or b.

65. $\log_5 x = 3$

66. $\log_2 x = 5$

67. $y = \log_{10} 1,000$

68. $y = \log_4 \dfrac{1}{64}$

69. $\log_b 64 = 3$

70. $\log_b 64 = 2$

71. $\log_6 x = -2$

72. $\log_3 x = -4$

73. $\log_4 x = \dfrac{3}{2}$

74. $\log_{27} x = \dfrac{2}{3}$

75. $y = \log_8 32$

76. $y = \log_{32} \dfrac{1}{8}$

77. $\log_b \dfrac{1}{8} = -3$

78. $\log_b 4 = -2$

79. $\log_5 x = 0$

80. $\log_b 1 = 0$

In Exercises 81–84, use a calculator to find the following.

81. $\log_{10} 23,596$

82. $\log_{10} 32$

83. $\log_{10} 0.000925$

84. $\log_{10} 628$

85. On the same set of coordinate axes, sketch the graphs of $f(x) = \log_2 x$ and $g(x) = \log_5 x$.

86. On the same set of coordinate axes, sketch the graphs of $f(x) = \log_{1/2} x$ and $g(x) = \log_{1/5} x$.

Questions for Thought

87. Using what you know about the graph of $y = \log_3 x$, sketch the graph of $y = \log_3(x - 1)$.

88. Repeat Exercise 87 for $y = \log_3(x + 1)$.

89. Repeat Exercise 87 for $y = \log_3 x - 1$.

90. Repeat Exercise 87 for $y = \log_3 x + 1$.

Mini-Review

91. Given $f(x) = \dfrac{\sqrt{x + 2}}{\sqrt{x + 2}}$, express $f(7)$ in simplest radical form.

92. *Multiply and simplify:* $(x - y)(x^2 + xy + y^2)$

93. *Divide:* $\dfrac{x^3 + 27}{x + 3}$

94. Use the discriminant to determine the nature of the roots of the equation $8x - x^2 = 5$.

10.3 Properties of Logarithms

Historically, logarithms were developed to simplify complex numerical computations. The availability of inexpensive hand-held calculators has made this use of logarithms virtually obsolete. Nevertheless, the properties of logarithms, which we discuss in this section, serve as the basis for using logarithms for both numerical and nonnumerical purposes.

As we have already emphasized in the previous section, a logarithm is just an exponent:

$$x = b^y \text{ is equivalent to } y = \log_b x$$

Consequently, it seems only natural to expect that logarithms will "inherit" the properties of exponents. The following box contains the three basic properties of logarithms. Again, we assume $b > 0$, $u > 0$, and $v > 0$.

Properties of Logarithms		
Product rule	**1.**	$\log_b(uv) = \log_b u + \log_b v$
Quotient rule	**2.**	$\log_b\left(\dfrac{u}{v}\right) = \log_b u - \log_b v$
Power rule	**3.**	$\log_b u^r = r \log_b u$

In words, these properties can be remembered as follows:

1. The product rule says that *"the log of a product is equal to the sum of the logs."*
2. The quotient rule says that *"the log of a quotient is equal to the difference of the logs."*
3. The power rule says that *"the log of a power is the exponent times the log."*

We will prove the product rule for logarithms and leave the proofs of the other two properties as exercises for the student.

Proof of the Product Rule: Rule 1 for exponents (the "product rule" for exponents) says that

$$b^m \cdot b^n = b^{m+n}$$

Let $u = b^m$ and $v = b^n$, and let's write these exponential statements in logarithmic form:

$$u = b^m \quad \text{is equivalent to} \quad \log_b u = m$$
$$v = b^n \quad \text{is equivalent to} \quad \log_b v = n$$

Thus, we have

$$u \cdot v = b^m \cdot b^n = b^{m+n}$$

Hence,

$$uv = b^{m+n} \quad \text{which is equivalent to} \quad \log_b(uv) = m + n$$

If we now substitute $m = \log_b u$ and $n = \log_b v$ into the last equation, we get

$$\log_b(uv) = \log_b u + \log_b v$$

Use a calculator to verify that $\log 10 = \log 2 + \log 5$.

which is exactly what the product rule states.

Just as the product rule for logarithms is the logarithmic form of exponent rule 1, so too the quotient and power rules for logarithms are the logarithmic forms of exponent rules 4 and 2, respectively.

These three properties of logarithms enable us to take the logarithm of a complicated expression that involves products, quotients, powers, and roots, and to write it as a sum of logarithms of much simpler expressions.

Several examples will illustrate this process.

EXAMPLE 1   Express as a sum of simpler logarithms. $\log_b(x^2y)$

Solution   We begin by using the product rule:

$$\log_b(x^2y) = \log_b x^2 + \log_b y \qquad \textit{Now use the power rule.}$$

$$= \boxed{2\log_b x + \log_b y}$$

EXAMPLE 2   Express as a sum of simpler logarithms:

$$\log_b \frac{\sqrt{x}}{z^5}$$

Solution   As we have mentioned before, since logarithms are exponents, when we are working with logarithms we generally write radical expressions in exponential form. Thus, we write $\sqrt{x} = x^{1/2}$.

$$\log_b \frac{\sqrt{x}}{z^5} = \log_b \frac{x^{1/2}}{z^5} \qquad \textit{Using the quotient rule, we get}$$

$$= \log_b x^{1/2} - \log_b z^5 \qquad \textit{Now use the power rule.}$$

$$= \boxed{\frac{1}{2}\log_b x - 5\log_b z}$$

As the expressions get more complex, we must be very careful using the properties of logarithms—particularly the quotient rule.

EXAMPLE 3   Express as a sum of simpler logarithms:

$$\log_8 \frac{\sqrt[3]{x^3y^5}}{8z^4}$$

Solution   As we did in Example 2, we first write the radical as a fractional exponent.

$$\log_8 \frac{\sqrt[3]{x^3y^5}}{8z^4} = \log_8 \frac{(x^3y^5)^{1/3}}{8z^4} \qquad \textit{Using the quotient rule, we get}$$

$$= \log_8(x^3y^5)^{1/3} - \log_8(8z^4) \qquad \textit{Using the power rule, we get}$$

$$= \frac{1}{3}\log_8(x^3y^5) - \log_8(8z^4) \qquad \textit{Using the product rule, we get}$$

$$= \frac{1}{3}(\log_8 x^3 + \log_8 y^5) - (\log_8 8 + \log_8 z^4)$$

Note that both sets of parentheses are essential. Now use the power rule again.

$$= \frac{1}{3}(3\log_8 x + 5\log_8 y) - (\log_8 8 + 4\log_8 z)$$

Now we remove parentheses; also, $\log_8 8 = 1$.

$$\boxed{= \log_8 x + \frac{5}{3}\log_8 y - 1 - 4\log_8 z}$$

It is also important to point out that an expression such as $\log_b(x^3 + y^5)$ *cannot* be expressed as the sum of simpler logs. The properties of logarithms *do not* apply to the logarithm of a sum. Thus,

$$\log_b(x^3 + y^5) \neq 3 \log_b x + 5 \log_b y$$

Similarly,

$$\frac{\log_b x}{\log_b y} \neq \log_b x - \log_b y$$

because the quotient rule applies to the log of a quotient, *not* to a quotient of logs.

Correct	*Incorrect*
$\log_b(uv) = \log_b u + \log_b v$	$\log_b(u + v) = \log_b u + \log_b v$
$\log_b\left(\dfrac{u}{v}\right) = \log_b u - \log_b v$	$\dfrac{\log_b u}{\log_b v} = \log_b u - \log_b v$
$\log_b(u^r) = r \log_b u$	$(\log_b u)^r = r \log_b u$

EXAMPLE 4

Write as a single logarithm. $3 \log_b x + \dfrac{1}{2} \log_b y - 6 \log_b z$

Solution We are going to use the properties of logarithms in the reverse direction, by first noting that the power rule allows us to write the coefficients of the logarithms as exponents of the variables.

$$3 \log_b x + \frac{1}{2} \log_b y - 6 \log_b z = \log_b x^3 + \log_b y^{1/2} - \log_b z^6 \qquad \text{\textit{Now use the product rule.}}$$

$$= \log_b(x^3 y^{1/2}) - \log_b z^6 \qquad \text{\textit{Now use the quotient rule.}}$$

$$= \boxed{\log_b \frac{x^3 y^{1/2}}{z^6}}$$

EXAMPLE 5 Given $\log_b 2 = 1.2$ and $\log_b 3 = 1.38$, find each of the following:

(a) $\log_b 6$ **(b)** $\log_b 81$ **(c)** $\log_b \dfrac{3}{2}$ **(d)** $\log_b 48$

Solution **(a)** $\log_b 6$ 　　　　　　　　*First we rewrite 6 as a product of 2 and 3:* $6 = 2 \cdot 3$.

$= \log_b(2 \cdot 3)$ 　　　　*Then use the product rule.*

$= \log_b 2 + \log_b 3$ 　　*Since* $\log_b 2 = 1.2$ *and* $\log_b 3 = 1.38$,

$= 1.2 + 1.38$

$= \boxed{2.58}$

(b) $\log_b 81$ 　　　　　　*Rewrite 81 as* 3^4.

$= \log_b 3^4$ 　　　　*Then use the power rule.*

$= 4 \log_b 3$ 　　*Since* $\log_b 3 = 1.38$,

$$= 4(1.38)$$

$$= \boxed{5.52}$$

(c) $\log_b \dfrac{3}{2}$ 　　　　　　　　　　*Use the quotient rule.*

$$= \log_b 3 - \log_b 2$$

$$= 1.38 - 1.2$$

$$= \boxed{0.18}$$

(d) $\log_b 48$ 　　　　　　*Rewrite 48 as a product of factors of 2 and 3:* $48 = 2^4 \cdot 3$.

$$= \log_b(2^4 \cdot 3)$$ 　　　*Use the product rule.*

$$= \log_b 2^4 + \log_b 3$$ 　*Use the power rule.*

$$= 4 \log_b 2 + \log_b 3$$

$$= 4(1.2) + 1.38$$

$$= \boxed{6.18}$$

EXERCISES 10.3

In Exercises 1–26, use the properties of logarithms to express the given logarithm as a sum of simpler ones (wherever possible). Assume that all variables represent positive quantities.

1. $\log_5(xyz)$ 　　　　**2.** $\log_2(rst)$ 　　　　**3.** $\log_7 \dfrac{2}{3}$ 　　　　**4.** $\log_6 \dfrac{4}{7}$

5. $\log_3 x^3$ 　　　　**6.** $\log_4 y^6$ 　　　　**7.** $\log_b a^{2/3}$ 　　　　**8.** $\log_b t^{3/4}$

9. $\log_b b^8$ 　　　　**10.** $\log_c c^5$ 　　　　**11.** $\log_s s^{-1/4}$ 　　　　**12.** $\log_n n^{-5/6}$

13. $\log_b(x^2 y^3)$ 　　　**14.** $\log_b(uv^2 w^7)$ 　　　**15.** $\log_b \dfrac{m^4}{n^2}$ 　　　**16.** $\log_b \dfrac{r^3 s}{t^5}$

17. $\log_b \sqrt{xy}$ 　　　**18.** $\log_b \sqrt[3]{mn}$ 　　　**19.** $\log_2 \sqrt[5]{\dfrac{x^2 y}{z^3}}$ 　　　**20.** $\log_7 \sqrt[4]{\dfrac{x^3 y^5}{z}}$

21. $\log_b(xy + z^2)$ 　　**22.** $\log_b(m^2 - n^3)$ 　　**23.** $\log_b \dfrac{x^2}{yz}$ 　　**24.** $\log_b \dfrac{u^3}{4v^2}$

25. $\log_6 \sqrt{\dfrac{6m^2 n}{p^5 q}}$ 　　　**26.** $\log_3 \sqrt{\dfrac{27 r^3 s}{t^5}}$

In Exercises 27–40, write the given expressions as a single logarithm.

27. $\log_b x + \log_b y$ 　　　　　　　　**28.** $\log_b x - \log_b y$

29. $2 \log_b m - 3 \log_b n$ 　　　　　　**30.** $4 \log_b u + 7 \log_b v$

31. $4 \log_b 2 + \log_b 5$ 　　　　　　　　**32.** $2 \log_b 3 + \log_b 4$

33. $\dfrac{1}{3} \log_b x + \dfrac{1}{4} \log_b y - \dfrac{1}{5} \log_b z$ 　　　**34.** $\log_b \dfrac{x}{3} + \log_b \dfrac{y}{4} - \log_b \dfrac{z}{5}$

35. $\dfrac{1}{2}(\log_b x + \log_b y) - 2 \log_b z$ 　　　**36.** $\dfrac{1}{3}(\log_b m + \log_b n) - 6 \log_b p$

37. $2 \log_b x - (\log_b y + 3 \log_b z)$

38. $5 \log_p u - \left(\dfrac{1}{2} \log_p v + \log_p w \right)$

39. $\dfrac{2}{3} \log_p x + \dfrac{4}{3} \log_p y - \dfrac{3}{7} \log_p z$

40. $\dfrac{1}{4} \log_n x - \log_n y - \log_n z$

In Exercises 41–52, evaluate the logarithm given $\log_b 2 = 1.2$, $\log_b 3 = 1.42$, and $\log_b 5 = 2.1$.

41. $\log_b 10$

42. $\log_b 15$

43. $\log_b \dfrac{2}{5}$

44. $\log_b \dfrac{3}{5}$

45. $\log_b \dfrac{1}{3}$

46. $\log_b \dfrac{1}{5}$

47. $\log_b 32$

48. $\log_b 125$

49. $\log_b 100$

50. $\log_b 1{,}000$

51. $\log_b \sqrt{20}$

52. $\log_b \sqrt{18}$

In Exercises 53–58, let $\log_b x = A$, $\log_b y = B$, and $\log_b z = C$. Express each logarithm in terms of A, B, and C.

53. $\log_b \sqrt[3]{x^2}$

54. $\log_b \dfrac{1}{\sqrt[5]{x}}$

55. $\log_b \dfrac{x^3 y^2}{z}$

56. $\log_b \dfrac{z}{x^3 y^2}$

57. $\log_b \sqrt{\dfrac{x^5 y}{z^3}}$

58. $\log_b \dfrac{\sqrt{x}}{\sqrt[3]{yz}}$

 Questions for Thought

59. Describe what is *wrong* (if anything) with each of the following:

(a) $\log_b(x^3 + y^4) \overset{?}{=} 3 \log_b x + 4 \log_b y$

(b) $\dfrac{\log_b x^3}{\log_b y^2} \overset{?}{=} 3 \log_b x - 2 \log_b y$

60. Use the outline for the proof of the product rule for logarithms to prove the quotient and power rules for logarithms.

61. Using a calculator, find $\log_{10} 3$ and $\log_{10} 5$. Using the properties of logarithms and the values you found above, determine the value of

(a) $\log_{10} 15$
(b) $\log_{10}\left(\dfrac{3}{5} \right)$
(c) $\log_{10} 45$
(d) $\log_{10}\left(\dfrac{9}{50} \right)$

Check your answers with a calculator.

Mini-Review

62. Find the axis of symmetry and the vertex of the graph of the equation $y = x^2 - 3x + 7$.

63. *Multiply and simplify:* $(2\sqrt{x} - 5)(3\sqrt{x} + 4)$

64. Given $y = f(x) = 5x - 3$, find $f^{-1}(x)$, and sketch the graphs of $f(x)$ and $f^{-1}(x)$ on the same coordinate system.

65. In physiology, a jogger's heart rate, N, in beats per minute is related linearly to the jogger's speed, s. A jogger's heartbeat is 80 beats per minute at a speed of 15 ft/sec and 85 beats per minute at a speed of 18 ft/sec.

(a) Write an equation expressing N in terms of s.

(b) Using the equation obtained in part (a), predict the jogger's heart rate at a speed of 30 ft/sec.

Common Logarithms, Natural Logarithms, and Change of Base

As we mentioned before, logarithms were invented to make complex numerical calculations easier to do. While calculators have made this use of logarithms obsolete, logarithms do still have widespread applications in the physical and social sciences. The material we develop in this section will enable us to do some actual computations with logarithms, which in turn gives us an opportunity to practice using the properties of logarithms we developed in the previous section.

Our development of the properties of logarithms in the previous section was independent of the choice of the base b (as long as $b > 0$ and $b \neq 1$). Since we use the decimal number system (that is, a number system using the base 10), one of the most frequently used bases for logarithms is 10.

DEFINITION

Logarithms with base 10 are called ***common logarithms.*** In other words, $\log_{10} x$ is called a common logarithm.

Normally we write

$$\log_{10} x \quad \text{as} \quad \log x$$

That is, the base 10 is understood. (This is similar to the convention that the index 2 is understood when we write $\sqrt{x}$.)

As we have seen on numerous occasions,

$$\log_b b^r = r \quad \text{which is just the logarithmic form of } b^r = b^r$$

Thus,

$$\log_{10} 1{,}000 = \log_{10} 10^3 = 3$$
$$\log_{10} 100 = \log_{10} 10^2 = 2$$
$$\log_{10} 10 = \log_{10} 10^1 = 1$$
$$\log_{10} 1 = \log_{10} 10^0 = 0$$
$$\log_{10} 0.1 = \log_{10} 10^{-1} = -1$$
$$\log_{10} 0.01 = \log_{10} 10^{-2} = -2$$
$$\log_{10} 0.001 = \log_{10} 10^{-3} = -3$$

For numbers not easily expressible as powers of 10, we will use the $\boxed{\text{LOG}}$ key on a calculator. For example, to find log 495, we press the sequence of keys

$$\boxed{\text{495}} \ \boxed{\text{LOG}}$$

to get the answer $\boxed{\text{2.6946052}}$ (to seven decimal places). This seven-place value is an approximation since the actual value of log 495 is irrational. We should actually write $\log 495 \approx 2.6946052$; however, it is customarily accepted to write instead $\log 495 = 2.6946052$.

The logarithmic statement $\log 495 = 2.6946052$ is equivalent to the exponential statement $10^{2.6946052} = 495$. Check this exponential statement with your calculator.

Let's examine this further to see whether the answer makes sense. Since $100 < 495 < 1{,}000$, we would expect it to be the case that the exponent of 10 that gives 495 lies between the exponent of 10 that gives 100 and the exponent of 10 that gives 1,000, which is another way of saying $2 < \log 495 < 3$.

EXAMPLE 1

Estimate the following between two consecutive integers, then check your estimate with a calculator:

(a) log 28,300   **(b)** log 0.00749

Solution   **(a)** Since log 28,300 is the exponent of 10 that yields 28,300, we look for powers of 10 that are closest to 28,300. Since 28,300 lies between 10,000 and 100,000, and log 10,000 = 4 and log 100,000 = 5, we can estimate log 28,300 to be

$$\boxed{\text{between 4 and 5}}.$$

To determine the value to four decimal places using a calculator, enter the following sequence of keys:

$$\boxed{\texttt{28300}}\ \boxed{\texttt{LOG}}$$

which yields $\boxed{\texttt{4.4518}}$ to four decimal places.

(b) Since log 0.00749 is the exponent of 10 that yields 0.00749, we look for powers of 10 that are closest to 0.00749. Since 0.00749 lies between 0.001 and 0.01, and log 0.001 = −3 and log 0.01 = −2, we can estimate log 0.00749 to be

Note $10^{-3} = 0.001$ and $10^{-2} = 0.01$.

$$\boxed{\text{between } -3 \text{ and } -2}.$$

Using a calculator, we arrive at the value $\boxed{-2.1255}$ to four places.

EXAMPLE 2

Find N if log $N = 3.93$.

Solution

The process demonstrated in Example 2 is sometimes called finding the antilogarithm of 3.93. Hence, we could have worded Example 2 as "Find antilog 3.93."

To help us determine what we are trying to find, we change log $N = 3.93$ into exponential form to get $10^{3.93} = N$. Hence, we want to evaluate a power of 10. Since 3.93 is close to 4, we estimate that our answer will be close to $10^4 = 10,000$.

To compute $10^{3.93}$ we enter the sequence of keys

$$\boxed{\texttt{10}}\ \boxed{\texttt{y}^x}\ \boxed{\texttt{3.93}}\ \boxed{\texttt{=}}$$

to get $\boxed{\texttt{8511.3804}}$ to four decimal places.

The graphs of $f(x) = 10^x$ and $g(x) = \log x$ are shown in Figure 10.10. As expected, because $f(x) = 10^x$ and $g(x) = \log x$ are inverses of each other, their graphs are mirror images of each other (reflected about the line $y = x$).

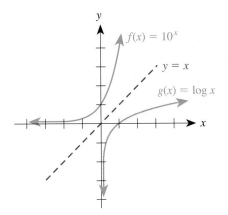

Figure 10.10
The graphs of $f(x) = 10^x$, $y = x$, and $g(x) = \log x$

e: The Natural Log Base

There is a particular number, designated by the letter e, that plays a prominent role in mathematics and the sciences. The number e is an irrational number and its value (correct to seven decimal places) is $e = 2.7182818$. (We will briefly discuss how e is defined in the next section.)

The number e comes up so frequently that it is called the *natural base*, and $\log_e$ is called the *natural logarithm*. The natural logarithm is also given a special notation:

$$\log_e x \text{ is written } \ln x$$

Your calculator should have an LN key. To become familiar with e and ln, let's evaluate a few expressions using the calculator.

EXAMPLE 3

Using a calculator, evaluate

(a) ln 78   **(b)** ln 0.007   **(c)** ln(−6)   **(d)** $e^{-4.5}$

Solution

(a) Remember that $\ln 78 = \log_e 78$, which is the exponent of e that yields 78. Since e is about 3, we can estimate the exponent to be somewhere near 4 (since $3^4 = 81$). To find this value with a calculator, we enter the following sequence of keys:

$$\boxed{78}\ \boxed{\text{LN}}$$

which yields $\boxed{\text{4.3567088}}$ to seven decimal places.

(b) Using the calculator, we arrive at the value $\ln(0.007) = \boxed{\text{−4.9618451}}$ (to seven decimal places).

(c) To find the value of $\ln(-6)$ with a calculator, we enter $\boxed{6}$ $\boxed{\text{+/−}}$ $\boxed{\text{LN}}$. When you press the LN key, rather than returning a numerical answer, the calculator display will indicate that an error has been made.

The calculator is letting you know that you have made a domain error: You are trying to evaluate a function for a value that is not in its domain. In this case, you are trying to take the ln of a negative number, and you cannot take the ln (or *any* logarithm) of any nonpositive number.

Remember that $\ln(-6) = t$ means $\log_e(-6) = t$, or, in exponential form, $e^t = -6$. Since e is a positive number, no matter what the value of t may be, e^t will never be negative. Hence, ln (-6) is undefined.

(d) Recall that $y = e^x$ and $y = \ln x$ are inverses of each other. We use the $\boxed{\text{2nd}}$ $\boxed{\text{LN}}$ keys to compute $e^{-4.5}$ as follows.

$$\boxed{4.5}\ \boxed{\text{+/−}}\ \boxed{\text{2nd}}\ \boxed{\text{LN}}$$

to get $\boxed{\text{0.0111090}}$ to seven decimal places.

If we want to display e to as many places as the calculator will allow, we use the $\boxed{\text{2nd}}$ and $\boxed{\text{LN}}$ keys to evaluate e^1 in the same way as in part **(d)** of Example 3. Press $\boxed{1}$ $\boxed{\text{2nd}}$ $\boxed{\text{LN}}$ to get $\boxed{\text{2.7182818}}$ to seven decimal places.

Let's examine the graphs of $f(x) = e^x$ and $g(x) = \ln x$ in Figure 10.11. We observe that as with the graphs of $f(x) = 10^x$ and $g(x) = \log x$, the graph of $g(x) = \ln x$ is the reflection of the graph of $f(x) = e^x$ about the line $y = x$, since $g(x) = \ln x$ and $f(x) = e^x$ are inverses of each other.

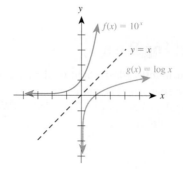

Figure 10.11
The graphs of $y = e^x$,
$y = x$, and $y = \ln x$

Change of Base

Calculators have the capability of computing both common logs and natural logs. How do we compute logarithms of numbers with other bases? To compute $\log_8 15$, we find that we need only have available the log values of one base, as demonstrated below.

Why can't we find $\log_8 15$ as we found $\log_8 16$ in Example 3(d) of Section 10.2?

$$\text{Let}\quad \log_8 15 = y \qquad \textit{Write in exponential form.}$$

$$8^y = 15 \qquad \textit{We can take the log of both sides of the equation. (We will discuss this step in more detail in Section 10.5.)}$$

$$\log 8^y = \log 15 \qquad \textit{Use the power rule for logs.}$$

$$y \log 8 = \log 15 \qquad \textit{Divide both sides by } \log 8.$$

$$y = \frac{\log 15}{\log 8} \qquad \textit{But } y = \log_8 15; \textit{ therefore, we have}$$

$$\log_8 15 = \frac{\log 15}{\log 8} \qquad \textit{Using a calculator, we get}$$

$$= 1.3022969$$

rounded to seven decimal places. Check this answer by raising 8 to the power 1.3022969.

We can apply the technique shown above to derive the following formula for change of base in a logarithmic computation:

Change of Base Formula	$$\log_a x = \frac{\log_b x}{\log_b a}$$

EXAMPLE 4

Convert $\log_3 x$ into a logarithm using base 9.

Solution

According to the change of base formula,

$$\log_3 x = \frac{\log_9 x}{\log_9 3} \qquad \textit{Since } \log_9 3 = \frac{1}{2}$$

$$= \frac{\log_9 x}{\dfrac{1}{2}}$$

$$= \boxed{2 \log_9 x}$$

We leave the proof of the change of base formula as an exercise (see Exercise 39).

EXAMPLE 5

Compute $\log_7 165$ to three decimal places.

Solution

Since the calculator does not have a log base 7 key, we use the change of base formula to express $\log_7 165$ in terms of logarithms with bases for which we do have keys—such as the natural log or log base 10. By the change of base formula:

$$\log_7 165 = \frac{\log 165}{\log 7}$$

$$= \boxed{2.624}$$

to three decimal places. Equivalently, we could have used the change of base formula with ln to get

$$\log_7 165 = \frac{\ln 165}{\ln 7}$$

$$= \boxed{2.624}$$

to three decimal places.

Be careful! As we mentioned in Section 10.3, $\frac{\log 165}{\log 7}$ and $\log\left(\frac{165}{7}\right)$ are *not* the same. We cannot apply the quotient rule for logarithms to the expression $\frac{\log 165}{\log 7}$. We note that $\log\left(\frac{165}{7}\right) = \log 165 - \log 7 = 2.2174839 - 0.845098 = 1.3723859$. We can see that this is not the same answer we obtained in Example 6.

EXERCISES 10.4

In Exercises 1–10, use your calculator to find the given logarithm accurate to four decimal places.

1. log 584

2. log 0.0584

3. log 0.00371

4. log 37,100

5. log 280,000

6. log 3,600,000

7. log 0.0000553

8. log 0.000939

9. log 8

10. log 0.6

In Exercises 11–20, use your calculator to find the following.

11. ln 0.941

12. log 0.941

13. $e^{0.941}$

14. ln 375

15. $e^{1.24}$

16. ln 0.0045

17. ln 4.5

18. $e^{-4.5}$

19. ln(−23)

20. ln 0.0023

In Exercises 21–24, find x, accurate to four decimal places.

21. log x = 0.941

22. log x = 4.85

23. ln x = 4

24. ln x = −3.2

In Exercises 25–28, use the change of base formula to rewrite the following as logarithms using base 10.

25. $\log_5 x$

26. $\log_6 4$

27. $\log_7 8$

28. $\log_5 10$

29. Rewrite $\log_5 8$ as a logarithm using base 25.

30. Rewrite $\log_4 5$ as a logarithm using base 2.

In Exercises 31–38, use your calculator to evaluate the following.

31. $\log_5 87$

32. $\log_5 0.09$

33. $\log_4 265$

34. $\log_2 93$

35. $\log_3 821$

36. $\log_8 0.0014$

37. $\log_7 52$

38. $\log_{52} 7$

39. Follow this outline to derive the change of base formula.

(a) We want to express $\log_a x$ in terms of a logarithm of another base. Start by letting $y = \log_a x$.

(b) Rewrite $y = \log_a x$ as an equivalent equation in exponential form.

(c) Take $\log_b$ of both sides of the exponential equation found in part (b).

(d) To remove y from the exponent, apply the power rule.

(e) Solve for y.

In Exercises 40–45, graph the following functions.

40. $f(x) = \log_4 x$ **41.** $y = \log_8 x$ **42.** $f(x) = 2 + \log_6 x$ **43.** $f(x) = -4 + \log_3 x$

44. $y = \log_6(x + 2)$ **45.** $y = \log_3(x - 4)$

 Mini-Review

46. Find the length of the line segment joining the points $(2, 4)$, and $(5, -6)$.

47. Find the midpoint of the line segment joining the points $(2, 4)$, and $(5, -6)$.

48. Sketch the graph of $y = x^2 - 4x - 5$. Label the intercepts.

49. *Solve for x:* $\sqrt{7x + 4} - 2 = x$

 Exponential and Logarithmic Equations

Let's consider the following problem:

> Lab technicians are trying to develop a fast lab test for doctors to use in their office to determine whether patients have a certain bacterial infection. They found that when this strain of infectious bacteria is placed in a certain medium, the population doubles every hour. They estimate that an average throat culture of an infected person usually contains about 1,000 bacteria, and that in the medium, the culture must grow to at least 2 million for it to be clearly seen with the naked eye. If a doctor takes a throat culture from a patient and puts it in the medium, how long will the doctor and patient have to wait before the bacteria can be clearly seen?

We know that we start with 1,000 bacteria and the bacteria count doubles every hour, so after the first hour there are $1,000(2)$ bacteria; after the second hour there are $1,000(2)(2) = 1,000(2^2)$ bacteria; after the third hour there are $1,000(2^2)(2) = 1,000(2^3)$ bacteria, and so on. The model is

$$N(t) = 1,000(2^t)$$

where $N(t)$ is the number of bacteria present after t hours.

The question is: How long will it take before we can see the bacteria with the naked eye, or how long will it take before there are 2,000,000 bacteria? To answer this question, we need to solve the exponential equation $2,000,000 = 1,000(2^t)$ for t.

In this section we discuss techniques for algebraically solving both exponential and logarithmic equations. We begin by discussing how to solve logarithmic equations and will return to this example later in this section when we discuss methods for solving exponential equations.

In Section 10.2 we solved simple logarithmic equations using the definition of a logarithm. We can use the properties of logarithms to solve more complicated logarithmic equations. For example, to solve the logarithmic equation

$$\log_3 x - \log_3 9 = 2$$

we proceed as follows:

$\log_3 x - \log_3 9 = 2$ *Use the quotient rule to write the left side as a single logarithm.*

$\log_3\left(\dfrac{x}{9}\right) = 2$ *Write this equation in exponential form.*

$3^2 = \dfrac{x}{9}$

$9 = \dfrac{x}{9}$

$\boxed{81 = x}$

Check: $x = 81$:
$$\log_3 x - \log_3 9 = 2$$
$$\log_3 81 - \log_3 9 \stackrel{?}{=} 2$$
$$4 - 2 \stackrel{\checkmark}{=} 2$$

Looking at this solution, we can see that our basic strategy in solving a logarithmic equation algebraically is to *rewrite the equation so that it involves a single logarithm, and then translate it into exponential form.*

EXAMPLE 1

Solve the following equation:
$$\log_6 x + \log_6(x + 1) = 1$$

Solution

$\log_6 x + \log_6(x + 1) = 1$	*Use the product rule.*
$\log_6 x(x + 1) = 1$	*Write this equation in exponential form.*
$x(x + 1) = 6^1$	
$x^2 + x = 6$	
$x^2 + x - 6 = 0$	*This is a quadratic equation.*
$(x + 3)(x - 2) = 0$	
$x = -3 \ \text{ or } \ x = 2$	

Check: $x = -3$:
$$\log_6 x + \log_6(x + 1) = 1$$
$$\log_6(-3) + \log_6(-3 + 1) \stackrel{?}{=} 1$$

Since the domain for the logarithm function is restricted to positive numbers, $x = -3$ is not an allowable value.

Check: $x = 2$:
$$\log_6 x + \log_6(x + 1) = 1$$
$$\log_6 2 + \log_6(2 + 1) \stackrel{?}{=} 1 \qquad \textit{Use the product rule.}$$
$$\log_6(2 \cdot 3) \stackrel{?}{=} 1$$
$$\log_6 6 \stackrel{\checkmark}{=} 1$$

Thus, the solution is $\boxed{x = 2}$.

If logarithmic expressions appear on both sides of the equation, we first use the properties of equality to collect all log terms on one side of the equation and all terms without logs on the other side. Then we try to rewrite the equation as a single logarithm, as illustrated in the next example.

EXAMPLE 2

Solve the following equation for x:
$$\log x = 2 + \log(x - 1)$$

Solution

$\log x = 2 + \log(x - 1)$	*Collect all log terms on one side and "nonlog" terms on the other side.*
$\log x - \log(x - 1) = 2$	*Use the quotient rule.*
$\log\left(\dfrac{x}{x - 1}\right) = 2$	*Write this equation in exponential form.*
$\dfrac{x}{x - 1} = 10^2 = 100$	*Remember that $\log\left(\dfrac{x}{x + 1}\right)$ has base 10.*
	Solve the equation.
$x = 100(x - 1)$	

$$x = 100x - 100$$

$$99x = 100$$

$$x = \boxed{\frac{100}{99}}$$

Check to verify that this solution is valid.

The log function is a one-to-one function. Thus,

$$\boxed{\text{If } \log_b r = \log_b s, \text{ then } r = s.}$$

We can use this fact to solve equations that consist entirely of log terms, as shown in the next example.

EXAMPLE 3

Solution

Solve for a: $2 \log_p a = \log_p 3$

As with the previous examples, we could start by collecting all log terms on one side of the equation and nonlog terms on the other side, and then use the log properties. Instead, we will just use the power rule on the left side of the equation.

$$\log_p a^2 = \log_p 3 \qquad \textit{Now we have it in the form } \log_b r = \log_b s, \textit{ and because the log function is one-to-one, we must have}$$

$$a^2 = 3 \qquad \textit{Now solve for } a.$$

$$a = \pm\sqrt{3}$$

We must check the answer in the original equation.

Check: When we check $a = -\sqrt{3}$, we get

$$2 \log_p(-\sqrt{3}) = \log_p 3$$

Since the domain of the log function is restricted to positive numbers, we eliminate $a = -\sqrt{3}$. The student should check that the value $\boxed{a = \sqrt{3}}$ is a valid solution.

We have mentioned that the exponential function is a one-to-one function. Hence, if $b^r = b^t$, then we can conclude that $r = t$. This allowed us to solve certain types of exponential equations. We can solve other types of exponential equations using the fact that log is a well-defined function. That is,

If r and s are positive and $r = s$, then

$$\log_b r = \log_b s$$

When we use this fact, we will say that we are "taking the log of both sides of the equation."

EXAMPLE 4

Solution

Solve for x: $3^{2x} = 5$

Note that in this equation, the variable for which we are solving appears in the exponent. We cannot rewrite 5 and 3 as powers of the same base as we did in Section 10.1. Let's take the log of both sides and see what happens. We will take the log base 10.

$$3^{2x} = 5 \qquad \textit{Take the log of both sides of the equation.}$$

$$\log 3^{2x} = \log 5 \qquad \textit{Use the power rule on the left side.}$$

$$2x(\log 3) = \log 5 \qquad \textit{Now we can solve for x. Keep in mind that both log 3 and log 5 are constants.}$$

$$x = \boxed{\dfrac{\log 5}{2 \log 3}} \qquad \textit{Which we evaluate as}$$

Check this solution with a calculator.

$$x \approx \boxed{0.732} \text{ to three decimal places.}$$

Let's return now to the model described in Example 3 of Section 10.1.

EXAMPLE 5

If $P(t) = 1{,}000 \cdot 2^{t/5}$, then solve the equation $P(t) = 20{,}000$.

Solution

Solving the equation $P(t) = 20{,}000$ means solving the equation $1{,}000 \cdot 2^{t/5} = 20{,}000$ for t. In Example 3 of Section 10.1, the only way we could solve this equation was to graph the model $P(t) = 1{,}000 \cdot 2^{t/5}$ and estimate t on the graph when $P(t) = 20{,}000$. Now we can solve it algebraically:

$$1{,}000 \cdot 2^{t/5} = 20{,}000 \qquad \textit{Divide both sides by 1,000.}$$

$$2^{t/5} = 20 \qquad \textit{Take the log of both sides of the equation.}$$

$$\log(2^{t/5}) = \log(20) \qquad \textit{Use the power rule on the left side.}$$

$$\frac{t}{5} \log(2) = \log(20) \qquad \textit{Solve for t. Multiply both sides by 5.}$$

$$t \log 2 = 5 \log(20) \qquad \textit{Divide both sides by log 2.}$$

$$t = \boxed{\frac{5 \log(20)}{\log 2}} \qquad \textit{Using a calculator, we get}$$

$$t \approx \boxed{21.6}$$

This answer is close to the answer we estimated on the graph in Example 3 of Section 10.1.

EXAMPLE 6

Solve for y: $\quad 2^y = 3^{y+4}$

Solution

$$2^y = 3^{y+4} \qquad \textit{Since we cannot express both 2 and 3 as powers of the same base, we take the log of both sides.}$$

$$\log 2^y = \log 3^{y+4} \qquad \textit{Use the power rule on both sides.}$$

$$y \log 2 = (y + 4)\log 3 \qquad \textit{Solve for y. To isolate y, first multiply out the right side.}$$

$$y \log 2 = y \log 3 + 4 \log 3 \qquad \textit{Collect all y terms on one side and all non-y terms on the other side.}$$

$$y \log 2 - y \log 3 = 4 \log 3 \qquad \textit{Factor y from the left side.}$$

$$y(\log 2 - \log 3) = 4 \log 3 \qquad \textit{Divide both sides by the constant multiplier of y, log 2 - log 3.}$$

$$y = \boxed{\frac{4 \log 3}{\log 2 - \log 3}} \qquad \textit{Which we evaluate as}$$

$$y \approx \boxed{-10.84} \text{ to two decimal places.}$$

EXAMPLE 7

Chemists have defined the pH (which stands for *hydrogen potential*) of a solution to be

$$pH = -\log[H_3O^+]$$

where $[H_3O^+]$ stands for the concentration of the hydronium ion in the solution (in moles per liter). The pH is a measure of the acidity or alkalinity of a solution. Water, which is neutral, has a pH of 7. Solutions with a pH below 7 are acidic, whereas those with a pH above 7 are alkaline.

(a) What is the pH of a glass of orange juice if its hydronium ion concentration is 5.43×10^{-4} mol per liter?

(b) What is the hydronium ion concentration of a solution whose pH is 4.7?

Solution

(a) According to the definition of pH, we have

$$pH = -\log[H_3O^+]$$ *Since $[H_3O^+] = 5.43 \times 10^{-4}$, we have*

$$pH = -\log(5.43 \times 10^{-4})$$ *Using a calculator, we get*

$$pH = 3.2652$$

Since pH values are customarily rounded to the nearest tenth, $pH = \boxed{3.3}$.

(b) To find the hydronium ion concentration of a solution whose pH is 4.7, we again use the definition of pH. This time, we are solving for $[H_3O^+]$.

$$pH = -\log[H_3O^+]$$ *Since pH = 4.7, we have*

$$4.7 = -\log[H_3O^+]$$ *Multiply each side by -1 to get*

$$-4.7 = \log[H_3O^+]$$ *Rewrite in exponential form.*

$$10^{-4.7} = [H_3O^+]$$ *Evaluate $10^{-4.7}$.*

$$1.995 \times 10^{-5} = [H_3O^+]$$

The hydronium ion concentration is $\boxed{1.995 \times 10^{-5} \text{ mol per liter}}$.

As you can see, if we are solving for a variable that appears in the exponent of an equation, we use logs and then log properties to remove the variable from the exponent. In the previous examples, to remove the variable from the exponent, we needed to "take the log of both sides" of the equation. In some cases we can arrive at the same result by simply rewriting the exponential expression in logarithmic form, as demonstrated in Example 8.

EXAMPLE 8

Solve for *r*: $32 = e^{3r}$

Solution

Our first goal in this example is to remove the variable from the exponent. Recalling from Section 10.2 that $x = b^y$ is equivalent to $y = \log_b x$, we see that we can quickly remove the variable from the exponent by rewriting the expression in logarithmic form:

$$32 = e^{3r}$$ *Rewrite this equation in logarithmic form, ($32 = e^{3r}$ is equivalent to $\log_e 32 = 3r$ or simply $\ln 32 = 3r$.)*

$$\ln 32 = 3r$$ *Now solve for r.*

$$\boxed{\frac{\ln 32}{3} = 1.1552453 = r}$$

It would be instructive to approach this same problem by taking the ln of both sides to show how we can arrive at the same result using the log properties:

$$32 = e^{3r} \quad \text{\textit{Take the ln of both sides to get}}$$

$$\ln 32 = \ln e^{3r} \quad \text{\textit{But} } \ln e^{3r} = 3r \text{ \textit{(do you see why?). Hence}}$$

$$\ln 32 = 3r \quad \text{\textit{And we have}}$$

$$\frac{\ln 32}{3} = 1.1552453 = r$$

In the process of solving the equation above by taking the ln of both sides, we stated that $\ln e^{3r} = 3r$. We can arrive at this result in two ways:

Recall that $\ln e = \log_e e$

Apply the power rule for logs to $\ln e^{3r}$ ***or***   *Recall from Section* 10.2 *that*

$\ln e^{3r} = 3r \ln e$ *Since* $\ln e = 1$, $\log_b b^x = x$ *Hence,*
          we have
$\quad\quad = 3r$ $\ln e^{3r} = \log_e e^{3r} = 3r$

Let's return to the ideas discussed in the bacteria example given at the beginning of this chapter.

> A bacteria culture initially has 1,000 bacteria of a type whose number doubles every 12 hours.

We developed a model for this problem and found how the number of bacteria, N, is related to the time, t, in hours. We found that if we are given the number of hours that have elapsed, we can compute the number of bacteria by simply substituting for t in the formula, and using the calculator to compute the value of N.

Let's modify this problem and see whether we can do the "opposite" (actually the inverse); that is, can we find t if we are given N? Let's restate the problem posed at the beginning of this section.

EXAMPLE 9

Lab technicians are trying to develop a fast lab test for doctors to use in their office to determine whether patients have a certain bacterial infection. They found that when this strain of infectious bacteria is placed in a certain medium, the population doubles every hour. They estimate that an average throat culture of an infected person usually contains about 1,000 bacteria, and that in the medium, the culture must grow to at least 2 million for it to be clearly seen with the naked eye. If a doctor takes a throat culture from a patient and puts it in the medium, how long will the doctor and patient have to wait before the bacteria can be clearly seen?

Solution

We stated at the beginning of this section that if we start with 1,000 bacteria and the bacteria count doubles every hour, then after the first hour there are 1,000(2) bacteria; after the second hour there are $1,000(2)(2) = 1,000(2^2)$ bacteria; after the third hour there are $1,000(2^2)(2) = 1,000(2^3)$ bacteria, and so on. Hence, the model is

$$N(t) = 1,000(2^t)$$

where $N(t)$ is the number of bacteria present after t hours.

To determine how long it will take before we can see the bacteria with the naked eye, or how long it will take before there are 2,000,000 bacteria, we need to solve the equation $2,000,000 = 1,000(2^t)$ for t.

$$2,000,000 = 1,000(2^t) \quad \text{\textit{This is an exponential equation: First isolate}}$$
$$\text{\textit{2}}^t \text{ \textit{by dividing both sides by 1,000.}}$$

$$\frac{2,000,000}{1,000} = \frac{1,000(2^t)}{1,000}$$

$$2,000 = 2^t \quad \text{\textit{Take the log of both sides of the equation.}}$$

$$\log 2,000 = \log 2^t \quad \text{\textit{Use the power rule on the right side to get}}$$

$$\log 2{,}000 = t \log 2 \qquad \textit{Divide both sides by log 2.}$$

$$\frac{\log 2{,}000}{\log 2} = t$$

$$t = \frac{\log 2{,}000}{\log 2} \approx 10.97$$

Hence, it would take about $\boxed{11 \text{ hours}}$ before the bacteria can be seen with the naked eye.

EXERCISES 10.5

In Exercises 1–48, solve the equations.

1. $\log_3 5 + \log_3 x = 2$

2. $\log_4 x + \log_4 5 = 1$

3. $\log_2 x = 2 + \log_2 3$

4. $\log_5 x = 2 - \log_5 3$

5. $2 \log_5 x = \log_5 36$

6. $3 \log_4 x = \log_4 125$

7. $\log_3 x + \log_3(x - 8) = 2$

8. $\log_6 x + \log_6(x - 5) = 1$

9. $\log_2 a + \log_2(a + 2) = 3$

10. $\log_3 a + \log_3(a - 2) = 1$

11. $\log_2 y - \log_2(y - 2) = 3$

12. $\log_2 x - \log_2(x + 3) = 2$

13. $\log_3 x - \log_3(x + 3) = 5$

14. $\log_3(2x) - \log_3(x - 2) = 4$

15. $\log_b 5 + \log_b x = \log_b 10$

16. $\log_b 6 + \log_b y = \log_b 18$

17. $\log_p x - \log_p 2 = \log_p 7$

18. $\log_p 2 - \log_p x = \log_p 8$

19. $\log_5 x + \log_5(x + 1) = \log_5 2$

20. $\log_3 y + \log_3(y - 2) = \log_3 3$

21. $\log_3 x - \log_3(x - 2) = \log_3 4$

22. $\log_3 a + \log_3(a + 2) = \log_3 15$

23. $\log_4 x - \log_4(x - 4) = \log_4(x - 6)$

24. $\log_9(2x + 7) - \log_9(x - 1) = \log_9(x - 7)$

25. $2 \log_2 x = \log_2(2x - 1)$

26. $2 \log_4 y = \log_4(y + 2)$

27. $\dfrac{1}{2} \log_3 x = \log_3(x - 6)$

28. $\dfrac{1}{2} \log_4 x = \log_4(2x + 1)$

29. $2 \log_b x = \log_b(6x - 5)$

30. $2 \log_b x = \log_b(2x + 1)$

31. $2^x = 5$

32. $3^x = 7$

33. $2^{x+1} = 6$

34. $5^{x+1} = 6$

35. $4^{2x+3} = 5$

36. $5^{2x+1} = 9$

37. $7^{y+1} = 3^y$

38. $6^{y+2} = 5^y$

39. $6^{2x+1} = 5^{x+2}$

40. $5^{2x-1} = 3^{x-3}$

41. $8^{3x-2} = 9^{x+2}$

42. $10^{3x+2} = 5^{x+3}$

43. $3^x = 5 \cdot 2x$

44. $5^x = 7 \cdot 3^x$

45. $2^y 5^y = 3$

46. $3^a 5^a = 10$

47. $4^a 3^{a+1} = 2$

48. $6^a 5^{a-1} = 8$

49. What is the pH of a solution if its hydronium ion concentration is 3.98×10^{-6} mol per liter? (Refer to Example 7.)

50. What is the hydronium ion concentration of a solution whose pH is 7.4? (Refer to Example 7.)

51. What is the hydronium ion concentration of water, which has a pH of 7? (See Example 7.)

52. Compute the pH of a solution for which $[H_3O^+] = 4.82 \times 10^{-3}$. (See Example 7.)

53. The unit of measurement frequently used to measure sound levels is the *decibel*. Decibels are measured on a logarithmic scale. The number of decibels, N, of a sound with intensity I (usually measured in watts per square centimeter) is defined to be

$$N = 10 \log I + 160$$

What is the intensity of sound of 200 dB?

54. What is the decibel rating of a jet plane that is emitting a sound whose intensity, I, is 10^{-1} W/cm^2?

55. The population of a certain species of mouse doubles every month. Suppose the population of mice starts at 10.

(a) Develop a mathematical model describing the number of mice at any given month.

(b) Use this model to determine the size of the mouse population in 2 years.

(c) How long (to the nearest month) does it take for the population to grow to 10,000?

56. Lab technicians are trying to develop a fast lab test for a certain strain of infectious bacteria. When this strain of bacteria is placed in a certain medium, the population triples every hour. Suppose the population of bacteria starts at 1,000.

(a) Develop a mathematical model describing the number of bacteria present in this medium at any given hour.

(b) How long (to the nearest tenth of an hour) does it take for the population to grow to 2 million?

 Mini-Review

57. Write an equation of the circle with center $(2, -3)$ and radius 6.

58. Use the graph of the function $y = f(x) = x^2 - 4$ to determine whether it has an inverse. Explain.

59. *Solve for x:* $2x^2 - 5x \leq 3$

60. *Solve for x and y:* $\begin{cases} 3x - 5y = 10 \\ 4x + 2y = 8 \end{cases}$

 # Applications: Exponential and Logarithmic Functions as Mathematical Models

In this section we will see how the ideas discussed in this chapter can be used to deal with a variety of applications.

We begin with an example from finance and will see how the results can be generalized to other areas such as biology and physics. Let's examine a situation we have discussed previously and attempt to generalize the results.

Suppose that $1,000 is invested in an account at an interest rate of 5% per year. The $1,000 is called the *principal* and is often designated as P. At the end of 1 year the amount in the account would be

(1) $$\text{Principal} + \text{Interest} \quad = \text{New principal}$$

$$1,000 \quad + 0.05(1,000) = 1,000(1 + 0.05) = (1,000)(1.05)$$

(Rather than computing the amount, we factored 1,000 from the left-hand expression and left it in factored form.)

If the interest earned each year remains in the account so that in subsequent years the interest itself earns interest, the account is said to be earning *compound interest*.

Let's start with P dollars in an account paying an interest rate of r (expressed as a decimal) compounded annually (meaning that the interest is computed at the end of each year), and develop a formula for the amount, A, present in an account after t years.

Our computation will be simplified if we recognize that, as in equation (1) above, *the amount in the account at the end of any year is $(1 + r)$ times the principal* (where again r is the interest rate expressed as a decimal). We therefore have the following:

At the end of 1 year: $A = P(1 + r)$ *This is now the new principal, as in equation (1).*

At the end of 2 years: $A = \underbrace{P(1 + r)} \cdot [1 + r] = P(1 + r)^2$
Previous principal

At the end of 3 years: $A = P(1 + r)^2[1 + r] = P(1 + r)^3$

$$\vdots$$

(2) At the end of t years: $A = P(1 + r)^t$

EXAMPLE 1

Suppose $1,000 is placed into a savings account paying 5% interest per year.

(a) How much money will be in the account after 12 years?

(b) How long will it take the initial principal of $1,000 to double?

Solution

(a) Using the formula in equation (2) above, we substitute $P = 1,000$, $r = 0.05$, and $t = 12$ to get

$$A = 1,000(1 + 0.05)^{12} = 1,000(1.05)^{12} \quad \text{\textit{Using a calculator, we get}}$$

$$= \boxed{\$1,795.86} \text{ in the account at the end of 12 years.}$$

(b) Again we use the formula in equation (2). Since we are looking for the number of years it takes the money to double, we substitute $A = 2,000$, $P = 1,000$, $r = 0.05$, and solve for t.

$2,000 = 1,000(1.05)^t$ *First we isolate the expression containing t by dividing both sides of the equation by 1,000 to get*

$$\frac{2,000}{1,000} = \frac{1,000(1.05)^t}{1,000}$$

$2 = 1.05^t$ *Next, take ln (or log) of both sides.*

$\ln 2 = \ln 1.05^t$ *Apply the power rule.*

$\ln 2 = t \ln 1.05$ *Keep in mind that ln 2 and ln 1.05 are numbers; to isolate t, divide both sides by ln 1.05.*

$$t = \frac{\ln 2}{\ln 1.05} \approx 14.2 \text{ years}$$

Since interest is paid at the end of each year, it will take $\boxed{15 \text{ years}}$ for the money to double.

Most banks do not compute the interest on an account once yearly; rather, they compound the interest a number of times per year. For example, if a bank compounds the interest semiannually, it computes and credits the interest twice a year (every 6 months) using one-half of the annual interest rate each time. Similarly, banks may compound monthly by computing and adding the interest to the account 12 times per year using one-twelfth the annual interest rate each time.

Essentially, compound interest calculations are made using shorter interest periods with smaller interest rates. If we are compounding $1,000 quarterly at an annual interest rate of 5%, we are computing a new principal four times per year by using an interest rate that is one-quarter of the yearly rate, or $\frac{0.05}{4} = 0.0125$. (This situation will yield the same amount of money as if $1,000 were invested at 1.25% per year for 4 years compounded annually.)

Let's examine what happens at the end of a year to $1,000 invested at an annual interest rate of 5% if the interest is compounded quarterly, monthly, and daily:

$$\text{Compounding quarterly:} \quad A = 1{,}000\left(1 + \frac{0.05}{4}\right)^4 = \$1{,}050.95$$

$$\text{Compounding monthly:} \quad A = 1{,}000\left(1 + \frac{0.05}{12}\right)^{12} = \$1{,}051.16$$

$$\text{Compounding daily:} \quad A = 1{,}000\left(1 + \frac{0.05}{365}\right)^{365} = \$1{,}051.27$$

Note that increasing the number of interest periods from 12 per year to 365 per year increased the annual interest by only 11 cents.

Also note that if we were compounding the annual interest monthly for a period of 8 years, there would be 8(12) = 96 interest periods and so

$$A = 1{,}000\left(1 + \frac{0.05}{12}\right)^{8(12)} = \$1{,}490.59$$

As a result of the above discussion, we state the following formula.

Compound Interest Formula	If P dollars is invested at an annual interest rate r, compounded n times per year, then A, the amount of money present after t years, is given by $$A = P\left(1 + \frac{r}{n}\right)^{nt}$$

EXAMPLE 2

In Example 1 we saw that it takes 15 (actually 14.2) years for $1,000 invested at 5% per year to double. If the $1,000 is invested at 5% compounded daily, how long will it take to double?

Solution

Using the compound interest formula, we substitute $A = 2{,}000$, $P = 1{,}000$, $r = 0.05$, $n = 365$, and we solve for t:

$$2{,}000 = 1{,}000\left(1 + \frac{0.05}{365}\right)^{365t}$$

$2{,}000 = 1{,}000(1.00013699)^{365t}$ *Divide both sides by* **1,000** *to get*

$2 = 1.00013699^{365t}$ *Take* **ln** *of both sides (or* **log** *of both sides).*

$\ln 2 = \ln 1.00013699^{365t}$ *Apply the power rule.*

$\ln 2 = 365t(\ln 1.00013699)$ *Divide both sides by* **365 ln 1.00013699.**

$$t = \frac{\ln 2}{365 \ln 1.00013699} \approx 13.86$$

Thus, with daily compounding, the doubling time has been reduced from 14.2 years to ⎡ 13.86, or a little less than 14 years ⎤.

A natural question one might ask is what happens if we start with P dollars and compound an interest rate of r every minute? Every second? Will the amount A keep growing larger? Is there any limit to how much the amount A will grow as we compound more frequently? As we saw above, increasing the number of interest periods does increase the amount A, but not by as much as we might have expected. It turns out there is a limit to how much A can grow, regardless of how many times we compound per year.

Let's suppose that $1 is invested at an annual rate of 100% per year compounded n times per year. By the formula for compound interest, the amount A at the end of one year is given by

$$A = \left(1 + \frac{1}{n}\right)^n$$ *We are using the compound interest formula with $r = 1$ (a rate of 100%).*

If we compound it

annually, then $n = 1$ $A = \left(1 + \frac{1}{1}\right)^1 = 2$

semiannually, then $n = 2$ $A = \left(1 + \frac{1}{2}\right)^2 = 2.25$

quarterly, then $n = 4$ $A = \left(1 + \frac{1}{4}\right)^4 \approx 2.44$

monthly, then $n = 12$ $A = \left(1 + \frac{1}{12}\right)^{12} \approx 2.61$

daily, then $n = 365$ $A = \left(1 + \frac{1}{365}\right)^{365} \approx 2.7146$

each minute, then $n = 525{,}600$ $A = \left(1 + \frac{1}{525{,}600}\right)^{525{,}600} \approx 2.718279$

each second, then $n = 31{,}536{,}000$ $A = \left(1 + \frac{1}{31{,}536{,}000}\right)^{31{,}536{,}000} \approx 2.718282$

As soon as n gets larger than 365, the amount remains steady at $2.72, assuming that the bank will round off to the nearest cent. If we allow the number of interest periods to get larger and larger, we are approaching what is called *continuous compounding*.

Note that as n increases, the value of $\left(1 + \frac{1}{n}\right)^n$ seems to be approaching e. (In fact, this is one way of defining e.)

The following formula is derived in calculus.

Compound Interest Formula (Continuous Compounding)	If P dollars is invested at an annual interest rate r compounded *continuously*, then A, the amount of money present after t years, is given by $$A = Pe^{rt}$$

EXAMPLE 3

At what annual rate, compounded continuously, must $5,000 be invested to grow to $15,000 in 10 years?

Solution   We use the formula $A = Pe^{rt}$ with $A = 15{,}000$, $P = 5{,}000$, and $t = 10$, and solve for r.

$15{,}000 = 5{,}000e^{10r}$ *Divide both sides by 5,000.*

$\dfrac{15{,}000}{5{,}000} = \dfrac{5{,}000e^{10r}}{5{,}000}$

$3 = e^{10r}$ *Rewrite this equation in logarithmic form. (Remember that $e^{10r} = 3$ is equivalent to $\log_e 3 = 10r$.)*

$$\ln 3 = 10r \qquad \textit{Then solve for r.}$$

$$r = \frac{\ln 3}{10} \approx 0.1099 \qquad \textit{Writing r as a percent, we have}$$

The annual rate must be $\boxed{10.99\%}$.

Growth and Decay

There are many relationships in the real world that can be fairly accurately described by exponential functions. In such a case, we often say that the equation is a *model* of the situation.

We will now describe two models that involve the number e, the base for natural logarithms mentioned in the previous section. First, if we start with a population A_0 that is growing continuously at a rate of $r\%$ per year, then after t years the population will have grown to A, where

$$A = A_0 e^{rt}$$

where r is written as a decimal. This is called an ***exponential growth model***.

EXAMPLE 4

The population of the earth in 1992 was estimated to be about 5.5 billion. Assume that the world's population is growing at a continuous rate of 1.9% per year.

(a) What will the population be in the year 2008?

(b) How long will it take for the earth's population to grow to 10 billion?

Solution

By "a continuous rate of 1.9% per year," we mean "a rate of 1.9% per year compounded continuously." We use the exponential growth model $A = A_0 e^{rt}$ with $A_0 = 5.5$ and $r = 0.019$ (1.9% written as a decimal). Thus, A represents the population (in billions) t years after 1992, and A_0 represents the population (in billions) of the earth in 1992 (this is the initial population). Thus, we have

$$A = 5.5 e^{0.019t}$$

(a) Since we want to estimate the earth's population in the year 2008, 16 years will have elapsed since the initial year of 1992. Therefore, we substitute $t = 16$ into the growth equation:

$$A = 5.5 e^{0.019t} \qquad \textit{Substitute t = 16.}$$
$$= 5.5 e^{0.019(16)}$$
$$A = 7.45$$

rounded to two places. Thus the earth's population in the year 2008 will be approximately $\boxed{\text{7.45 billion people}}$.

(b) To determine in what year the earth's population will be 10 billion, we again use the equation $A = 5.5 e^{0.019t}$, but this time we want to know what value of t will make $A = 10$.

$$A = 5.5 e^{0.019t} \qquad \textit{We let A = 10 and solve for t.}$$

$$10 = 5.5 e^{0.019t} \qquad \textit{Divide both sides by 5.5.}$$

$$\frac{10}{5.5} = e^{0.019t} \qquad \textit{Take ln of both sides.}$$

$$\ln\!\left(\frac{10}{5.5}\right) = \ln e^{0.019t} \qquad \textit{But ln } e^{0.019t} = 0.019t \textit{ (do you see why?), so we get}$$

$$\ln\left(\frac{10}{5.5}\right) = 0.019t$$

$$\frac{\ln(10/5.5)}{0.019} = t \qquad \textit{Using a calculator, we get}$$

$$t \approx 31.47$$

It would take about $\boxed{31\frac{1}{2} \text{ years}}$ or until the middle of the year 2023 ($31\frac{1}{2}$ years after 1992) for the earth's population to reach 10 billion.

If we start with a certain amount of radioactive material, A_0, that is decaying continuously at the rate of $r\%$ per year, then after t years the amount of radioactive material still present is A, where

$$A = A_0 e^{-rt}$$

where r is written as a decimal. This is called an ***exponential decay model***.

Note that in the exponential decay model the exponent is negative, which means that A gets smaller as t increases.

EXAMPLE 5

Suppose we have 100 g of a radioactive substance that decays continuously at the rate of 8% per hour.

(a) How long will it take until only 50 g of radioactive substance remain?

(b) Find the time it takes for an initial amount A_0 of this radioactive substance to decay to half this amount. This time is called the *half-life* of the substance.

Solution   The exponential decay equation for this substance is

$$A = 100e^{-0.08t} \qquad \textit{Note the negative exponent for the decay model.}$$

(a) To determine how long it will take the 100 g to decay to 50 g, we set $A = 50$, and solve for t.

$$50 = 100e^{-0.08t} \qquad \textit{Divide both sides by 100.}$$

$$0.5 = e^{-0.08t} \qquad \textit{Rewrite in logarithmic form to get}$$

$$\ln 0.5 = -0.08t$$

$$t = \frac{\ln 0.5}{-0.08} \approx 8.66$$

Thus, it takes approximately $\boxed{8.66 \text{ hours}}$ for the 100 g to decay to 50 g.

(b) To determine the half-life of this substance, we want to know how long it takes the initial amount to decay to half its size. In other words, we want to find t when $A = \frac{1}{2}A_0$.

$$\frac{1}{2}A_0 = A_0 e^{-0.08t} \qquad \textit{Divide both sides by } A_0.$$

$$\frac{1}{2} = e^{-0.08t}$$

$$0.5 = e^{-0.08t} \qquad \textit{Which is rewritten in logarithmic form as}$$

$$\ln 0.5 = -0.08t \qquad \textit{Divide both sides by } -0.08 \textit{ to get}$$

$$t = \frac{\ln 0.5}{-0.08} \approx 8.66$$

Thus, the half-life of this substance is approximately $\boxed{8.66 \text{ hours}}$.

Note that as part **(a)** suggests, this answer is independent of the initial amount. This fact is crucial for scientists' ability to use the radioactive decay of carbon-14 to determine the age of fossils and archeological artifacts. (See Exercise 29.)

EXAMPLE 6

Suppose a colony of bacteria grows from a population of approximately 400 to 2,600 in 6 hours.

(a) Find the exponential growth model for these bacteria.

(b) Using this model, determine how many bacteria are present after 20 hours.

Solution

(a) We start with the exponential growth model, $A = A_0 e^{rt}$, and note that we are given $A = 2,600$, $A_0 = 400$, and $t = 6$, and we need to solve for r.

$$2,600 = 400e^{6r} \qquad \textit{Divide both sides by 400.}$$

$$6.5 = e^{6r} \qquad \textit{Rewrite in logarithmic form.}$$

$$\ln 6.5 = 6r$$

$$r = \frac{\ln 6.5}{6} \qquad \frac{\ln 6.5}{6} \approx 0.3120, \textit{ but for accuracy we leave } r = \frac{\ln 6.5}{6}.$$

Thus, the exponential growth model for these bacteria is

$$\boxed{A = A_0 e^{(t \ln 6.5)/6}}$$

(b) To find the size of the population after 20 hours, we use the model we developed in part **(a)** for these bacteria:

$$A = A_0 e^{(t \ln 6.5)/6}$$

and note that the initial amount, A_0, is still 400, $t = 20$, and now we want to compute A:

$$A = 400e^{(20 \ln 6.5)/6} \qquad \textit{Which we compute on a calculator as}$$

$$A = 205,008$$

Thus, there are approximately $\boxed{205,000}$ bacteria present after 20 hours.

EXERCISES 10.6

1. If $8,000 is deposited in a bank account paying an interest rate of 6% per year compounded semiannually, how much money will be in the account in 5 years?

2. If $8,000 is deposited in a bank account paying an interest rate of 6% per year compounded quarterly, how much money will be in the account in 5 years?

3. If $8,000 is deposited in a bank account paying an interest rate of 6% per year compounded monthly, how much money will be in the account in 5 years?

4. If $8,000 is deposited in a bank account paying an interest rate of 6% per year compounded daily, how much money will be in the account in 5 years?

5. If $9,000 is deposited in a bank account paying an interest rate of 6.2% per year compounded monthly, how long would it take for the money to double?

6. If \$9,000 is deposited in a bank account paying an interest rate of 6.2% per year compounded daily, how long would it take for the money to double?

7. For any amount of money deposited in a bank account paying an interest rate of 6.2% per year compounded monthly, how long would it take for the money to double?

8. For any amount of money deposited in a bank account paying an interest rate of 6.2% per year compounded daily, how long would it take for the money to double?

9. How long would it take for a sum of money to double in a bank account paying an interest rate of 10% per year compounded monthly?

10. How long would it take for a sum of money to double in a bank account paying an interest rate of 10% per year compounded daily?

11. How long would it take for \$5,000 to grow to \$20,000 in a bank account paying an interest rate of 7.3% per year compounded continuously?

12. How long would it take money to quadruple in a bank account paying an interest rate of 7.3% per year compounded continuously?

13. In 1990, the population of Rabbittville was growing at a rate of 8% per year. If there were 2,000 inhabitants of Rabbittville in 1990, use the exponential growth model to determine the size of the population in 2005.

14. In 1990, the population of a city was growing at a rate of 2% per year. If there were 2,000 inhabitants of the city in 1990, use the exponential growth model to determine the size of the population in 2005.

15. In a certain environment, a colony of bacteria grows according to the model

$$A = 10,000e^{0.0542t}$$

where A is the number of bacteria present at time t (in hours) after a culture is taken. How many bacteria are present 5 hours after a culture is taken?

16. In a certain environment, a colony of bacteria grows according to the model

$$A = 10,000e^{0.0542t}$$

where A is the number of bacteria present at time t (in hours) after a culture is taken. In another environment, the growth model for the same bacteria is

$$A = 10,000e^{0.122t}$$

For each model determine how many bacteria are present after 24 hours.

17. In a certain environment, a colony of bacteria grows according to the model

$$A = 10,000e^{0.0542t}$$

where A is the number of bacteria present at time t (in hours) after a culture is taken. In another environment, the growth model for the same bacteria is

$$A = 10,000e^{0.122t}$$

For each environment determine how long it takes for 100,000 bacteria to be present.

18. In a certain environment, a colony of bacteria grows according to the model

$$A = 10,000e^{0.0542t}$$

where A is the number of bacteria present at time t (in hours) after a culture is taken. In another environment, the growth model for the same bacteria is

$$A = 10,000e^{0.122t}$$

When the bacteria count reaches 200,000 the colony can be seen with the naked eye. For each environment determine how long it would take for it to be seen with the naked eye.

19. In a certain environment, a colony of bacteria grows at a continuous rate of 8% per hour. How long would it take 1,000 bacteria to grow to 10,000 at this rate?

20. In a certain environment, a colony of bacteria grows at a continuous rate of 4% per hour. How long would it take 1,000 bacteria to grow to 10,000 at this rate?

21. In a certain environment, a colony of bacteria grows from 1,000 to 3,000 in 5 hours. Develop a model to describe this growth, and determine how long it takes for the bacteria colony to grow to 50,000.

22. In a certain environment, a colony of bacteria grows from 1,000 to 4,000 in 2 hours. Develop a model to describe this growth, and determine how long it takes for the bacteria colony to grow to 50,000.

23. The formula for the radioactive decay of radium is $A = A_0 e^{-0.0004t}$. Use this formula to determine the half-life of radium. [*Hint:* Find the value for t for which $A = \frac{1}{2}A_0$.]

24. Find the half-life for carbon-14, whose decay formula is $A = A0e^{-0.000121t}$.

25. Radon has a half-life of approximately 4 days and satisfies the exponential decay model. Find the value of r in the formula for radon. (Keep in mind that the value for r is usually given as an *annual rate* of decay. However, the formula applies to any unit of time.)

26. Certain man-made radioactive materials have extremely short half-lives. Repeat Exercise 25 for a substance whose half-life is 3 minutes.

27. The Richter scale is used in measuring the intensity of an earthquake. On the Richter scale, the magnitude of an earthquake, R, is defined to be

$$R = \log \frac{I}{I_0}$$

where I is the intensity of the earthquake, and I_0 is a standard intensity. For the sake of simplicity we will assume that $I_0 = 1$. Compare the intensities I_1 and I_2 of two earthquakes whose magnitudes on the Richter scale are $R_1 = 3.6$ and $R_2 = 7.2$.

28. What is the intensity, I, of an earthquake whose magnitude on the Richter scale is 6.85? (See Exercise 27.)

29. All living plant and animal tissue contains carbon-12 and carbon-14. Carbon-12 is not radioactive, but carbon-14 is radioactive with a half-life of approximately 5,730 years. While an organism is living, the ratio of carbon-14 to carbon-12 remains constant. After the organism dies, however, the carbon-14 begins to decay. Thus, the smaller the ratio of carbon-14 to carbon-12, the older the remains of living tissue. We start $t = 0$ at the time the organism dies since carbon-14 begins to decay at that time.

    (a) Verify that the decay model for carbon-14 is $A = A_0 e^{-0.000121t}$.

    (b) Suppose a fossil now contains 35% of the carbon-14 it originally contained. How old is the fossil?

 Mini-Review

30. Sketch the graph of $y = 6x - x^2$. Label the intercepts.

31. Sketch the graph of $y = 6 - x$. Label the intercepts.

32. Given $f(x) = 2x^2 - 3x + 1$ and $g(x) = 5x + 2$, find $f[g(x)]$ and $g[f(x)]$.

33. *Solve:* $(3x - 4)^2 = 5$

CHAPTER 10 SUMMARY

After having completed this chapter you should be able to:

1. Sketch the graph of an exponential function (Section 10.1).

   For example: Figure 10.12 shows the graph of $y = f(x) = 4^x$.

   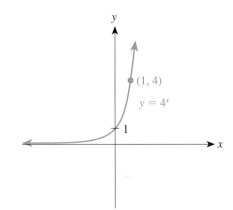

   Figure 10.12
   Graph of $y = 4^x$

2. Sketch the graph of a logarithmic function (Section 10.2).

   For example: Figure 10.13 shows the graph of $f(x) = \log_4 x$.

   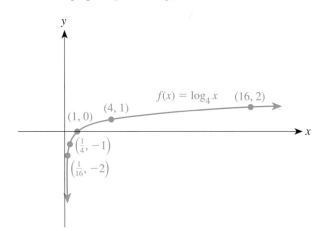

   Figure 10.13
   Graph of $f(x) = \log_4 x$

3. Translate logarithmic statements into exponential form and vice versa (Section 10.2).

   For example:

   (a) $\log_3 \dfrac{1}{9} = -2$ is equivalent to $3^{-2} = \dfrac{1}{9}$.

   (b) $2^5 = 32$ is equivalent to $\log_2 32 = 5$.

4. Evaluate certain logarithms (Section 10.2).

For example: Find $\log_8 \dfrac{1}{16}$.

Solution: Let

$$\log_8 \frac{1}{16} = t \qquad \text{\textit{Write in exponential form.}}$$

$$8^t = \frac{1}{16} \qquad \text{\textit{Express both sides in terms of the same base.}}$$

$$(2^3)^t = \frac{1}{2^4}$$

$$2^{3t} = 2^{-4}$$

$$3t = -4$$

$$t = -\frac{4}{3}$$

Therefore, $\log_8 \dfrac{1}{16} = -\dfrac{4}{3}$.

5. Use the properties of logarithms to rewrite logarithmic expressions (Section 10.3).

For example:

$$\log_b \frac{\sqrt{xy}}{z^3} = \log_b \frac{(xy)^{1/2}}{z^3} \qquad \text{\textit{Use the quotient rule.}}$$

$$= \log_b (xy)^{1/2} - \log_b z^3 \qquad \text{\textit{Now use the power rule.}}$$

$$= \frac{1}{2} \log_b (xy) - 3 \log_b z \qquad \text{\textit{Now use the product rule.}}$$

$$= \frac{1}{2}(\log_b x + \log_b y) - 3 \log_b z$$

$$= \frac{1}{2} \log_b x + \frac{1}{2} \log_b y - 3 \log_b z$$

6. Use a calculator to compute logarithms (Section 10.4).

For example:

(a) $\log 0.063 = -1.2007$

(b) $\ln 800 = 6.6846$

(c) $\log_6 934 = \dfrac{\log 934}{\log 6}$ *We are using the change of base formula:* $\log_b a = \dfrac{\log a}{\log b}$.

$$\approx 3.817 \quad \text{(to three decimal places)}$$

7. Solve exponential and logarithmic equations (Sections 10.1 and 10.5).

For example: Solve the following equations:

(a) $81^x = \dfrac{1}{9}$

$$(9^2)^x = 9^{-1}$$

$$9^{2x} = 9^{-1}$$

$$2x = -1$$

$$x = -\frac{1}{2}$$

(b) $\log_2(x + 5) + \log_2(x - 1) = 4$ *Use the product rule.*

$\log_2(x + 5)(x - 1) = 4$ *Write in exponential form.*

$(x + 5)(x - 1) = 2^4$

$x^2 + 4x - 5 = 16$

$x^2 + 4x - 21 = 0$

$(x + 7)(x - 3) = 0$

$x = -7$ or $x = 3$

Check: $x = -7$: $\log_2(x + 5) + \log_2(x - 1) = 4$

$\log_2(-7 + 5) + \log_2(-7 - 1) \overset{?}{=} 4$

$\log_2(-2) + \log_2(-8) \overset{?}{=} 4$

Since the logarithm of a negative number is undefined, $x = -7$ is not a solution.

Check: $x = 3$: $\log_2(x + 5) + \log_2(x - 1) = 4$

$\log_2(3 + 5) + \log_2(3 - 1) \overset{?}{=} 4$

$\log_2(8) + \log_2(2) \overset{?}{=} 4$

$3 + 1 \overset{\checkmark}{=} 4$

Thus, the solution is $x = 3$.

(c) $7^{y+5} = 5^{2y}$ *Take the log (base 10) of both sides.*

$\log 7^{y+5} = \log 5^{2y}$ *Use the power rule.*

$(y + 5)\log 7 = (2y)\log 5$ *Solve for y.*

$y \log 7 + 5 \log 7 = 2y \log 5$

$y \log 7 - 2y \log 5 = -5 \log 7$

$y(\log 7 - 2 \log 5) = -5 \log 7$

$y = \dfrac{-5 \log 7}{\log 7 - 2 \log 5}$ *Which, when evaluated, is*

$y = 7.64$ to two decimal places

8. Apply logarithms to solve problems involving exponential and logarithmic expressions (Sections 10.1, 10.5, and 10.6).

CHAPTER 10 REVIEW EXERCISES

In Exercises 1–6, sketch the graph of the given function.

1. $y = 6^x$

2. $f(x) = \log_6 x$

3. $f(x) = \log_{10} x$

4. $y = \left(\dfrac{2}{3}\right)^x$

5. $y = 2^{-x}$

6. $f(x) = \left(\dfrac{1}{4}\right)^{-x}$

In Exercises 7–18, write exponential statements in logarithmic form and logarithmic statements in exponential form.

7. $\log_3 81 = 4$

8. $6^3 = 216$

9. $4^{-3} = \dfrac{1}{64}$

10. $\log_2 \dfrac{1}{4} = -2$

11. $\log_8 4 = \dfrac{2}{3}$

12. $10^{-3} = 0.001$

13. $25^{1/2} = 5$

14. $\left(\dfrac{1}{3}\right)^{-2} = 9$

15. $\log_7 \sqrt{7} = \dfrac{1}{2}$

16. $\log_{11} \sqrt[3]{11} = \dfrac{1}{3}$

17. $\log_6 1 = 0$

18. $12^1 = 12$

In Exercises 19–32, evaluate the given logarithm.

19. $\log_{10} 1{,}000$

20. $\log_{10} 0.0001$

21. $\log_3 \dfrac{1}{9}$

22. $\log_2 64$

23. $\log_2 \dfrac{1}{4}$

24. $\log_7 \dfrac{1}{7}$

25. $\log_{1/3} 9$

26. $\log_{1/10} 0.01$

27. $\log_9 \dfrac{1}{3}$

28. $\log_4 \dfrac{1}{8}$

29. $\log_b \sqrt{b}$

30. $\log_p \sqrt[3]{p}$

31. $\log_{16} 32$

32. $\log_{32} 16$

In Exercises 33–42, write the given logarithm as a sum of simpler logarithms where possible.

33. $\log_b(x^3 y^7)$

34. $\log_b \dfrac{x^3}{y^7}$

35. $\log_b \dfrac{u^2 v^5}{w^3}$

36. $\log_b \dfrac{u^2}{v^5 w^3}$

37. $\log_b \sqrt[3]{xy}$

38. $\log_b \sqrt{\dfrac{x}{y}}$

39. $\log_b(x^3 + y^4)$

40. $\log_b \sqrt{\dfrac{x^3 y^5}{z^7}}$

41. $\log_b \sqrt{\dfrac{x^6 y^2}{z^2}}$

42. $\dfrac{\log_b x^4}{\log_b y^9}$

In Exercises 43–44, evaluate the logarithm given $\log_b 7 = 1.32$ and $\log_b 2 = 1.1$.

43. $\log_b 28$

44. $\log_b\!\left(\dfrac{2}{49}\right)$

In Exercises 45–56, solve the given exponential or logarithmic equation.

45. $9^x = \dfrac{1}{81}$

46. $\left(\dfrac{1}{2}\right)^x = 4$

47. $16^x = 32$

48. $32^x = 16$

49. $5^{x+1} = 3$

50. $2^x = 7^{x+4}$

51. $\log(x + 10) - \log(x + 1) = 1$

52. $\log_2 x + \log_2(x + 1) = 1$

53. $\log_2(t + 1) + \log_2(t - 1) = 3$

54. $\log_3(2x + 1) + \log_3(x - 1) = 3$

55. $\log_b 3x + \log_b(x + 2) = \log_b 9$

56. $\log_p x - \log_p(x + 1) = \log_p 2$

In Exercises 57–64, use a calculator to compute the given expression.

57. $\log 783$

58. $\log 0.584$

59. $\log 0.00499$

60. $\ln 2.5$

61. $\ln 0.0063$

62. $e^{-4.2}$

63. $e^{7.8}$

64. $\ln 194$

In Exercises 65–66, find x, accurate to three decimal places.

65. $\log x = -3$

66. $\log x = 3.6571$

In Exercises 67–72, use the change of base formula to compute the given logarithm to four decimal places.

67. $\log_5 73$

68. $\log_9 0.074$

69. $\log_{12} 764$

70. $\log_7 0.0063$

71. $\log_{0.2} 190$

72. $\log_{0.5} 834$

Use logarithms to solve each of the following exercises.

73. If \$6,000 is invested in an account at 8.2% compounded semiannually, how much money will there be in the account after 8 years?

74. How long will it take a sum of money invested at 7.2% and compounded quarterly to triple?

75. If a radioactive substance decays according to the formula $A = A_0 e^{-0.045t}$, how long will it take for 100 g of radioactive material to decay to 25 g of radioactive material?

76. If the population of a small town satisfies the exponential growth model with $r = 0.015$, how long will it take the town's population to increase from 8,500 to 15,000?

77. If the pH of a solution is defined to be pH $= -\log[H_3O^+]$, find the pH of a solution for which $[H_3O^+] = 6.21 \times 10^{-9}$.

78. If the pH of a solution is defined to be pH $= -\log[H_3O^+]$, where $[H_3O^+]$ is the hydronium ion concentration, find the hydronium ion concentration of a solution whose pH is 8.4.

CHAPTER 10 PRACTICE TEST ⊙

1. Sketch the graph of $y = \left(\dfrac{1}{3}\right)^x$.

2. Sketch the graph of $y = \log_7 x$.

3. Write in exponential form:

(a) $\log_2 16 = 4$

(b) $\log_9 \dfrac{1}{3} = -\dfrac{1}{2}$

4. Write in logarithmic form:

(a) $8^{2/3} = 4$

(b) $10^{-2} = 0.01$

5. Evaluate each of the following logarithms:

   (a) $\log_3 \dfrac{1}{3}$ **(b)** $\log_{81} 9$

   (c) $\log_8 32$

6. Write as a sum of simpler logarithms:

   (a) $\log_b(x^3 y^5 z)$ **(b)** $\log_4 \dfrac{64\sqrt{x}}{yz}$

7. Use a calculator to compute each of the following:

   (a) $\log 27{,}900$ **(b)** $\ln 0.004$

   (c) $e^{-0.02}$

8. Find x accurate to four decimal places. $\log x = 0.135$

9. Use the change of base formula to compute the following to three decimal places.

   (a) $\log_5 67$ **(b)** $\log_8 0.0034$

10. Solve each of the following equations:

   (a) $3^x = \dfrac{1}{81}$ **(b)** $\log_2(x + 4) + \log_2(x - 2) = 4$

   (c) $4^{5x} = 32^{3x-4}$ **(d)** $\log(5x) - \log(x - 5) = 1$

   (e) $9^{x+3} = 5$

11. If \$5,000 is invested in an account paying 8.4% interest compounded quarterly, how much money will there be in the account after 7 years?

12. Assuming the exponential decay model $A = A_0 e^{-0.04t}$, how long will it take for 100 g of a radioactive material to decay to 25 g of radioactive material?

11

More Systems of Equations and Systems of Inequalities

© JG Photography/Alamy

In Section 4.3, we discussed solving systems of two linear equations in two variables by manipulating equations. In this chapter we will generalize and apply these methods to solving other systems of equations, as well as discuss other methods of solving linear systems.

11.1 3 × 3 Linear Systems

Let's consider the following situation.

A hospital administrator must allocate $140,400 in funds for supplies, 3,480 hours of doctors' services, and 5,600 hours of nurses' services among an emergency room, an outpatient clinic, and a surgery ward. The table indicates how much of each of these resources is needed for a single day's operation of each location. Determine how many days each of these locations can operate if the administrator allocates all of the resources.

	Emergency Room	Outpatient Clinic	Surgery Ward
Supplies	$3,000	$1,800	$2,000
Doctors' hours	75	50	45
Nurses' hours	100	80	90

Since we are being asked to find how many days each of the locations can operate, we begin by letting e = the number of days the emergency room can operate, c = the number of days the outpatient clinic can operate, and s = the number of days the surgery ward can operate. We use the table to construct a system of equations in e, c, and s, where each equation represents the conditions on supplies and nurses' and doctors' hours:

$$\begin{cases} 3,000e + 1,800c + 2,000s = 140,400 \\ 75e + 50c + 45s = 3,480 \\ 100e + 80c + 90s = 5,600 \end{cases}$$

This equation represents the amount of money allocated for supplies.

This equation represents the number of hours of doctors' services.

This equation represents the number of hours of nurses' services.

In order to solve this problem, we need to be able to solve this system of three equations in three unknowns. The remainder of this section is devoted to describing procedures for solving such a system. We will return to this problem at the end of this section.

A **3 × 3** (read "3 by 3") *system* is a system of three equations in three variables. Whereas the graph of a first-degree equation in two variables is a straight line in a two-dimensional coordinate system, a first-degree equation in three variables is a *plane* in a three-dimensional coordinate system. Instead of solutions being ordered pairs, they are *ordered triplets* (x, y, z) for a 3 × 3 system.

For example, the triplet $(1, -3, -2)$, which means $x = 1$, $y = -3$, $z = -2$, satisfies the system

$$\begin{cases} x + y + z = -4 \\ 4x - y + z = 5 \\ 3x + y - z = 2 \end{cases}$$

For a 2 × 2 system of linear equations, we can represent the unique solution geometrically as the point of intersection of the two lines described by the equations in the system. For a 3 × 3 system, the solution is represented geometrically by the intersection of the three planes described by each equation in the system. As is the case with a 2 × 2 system, a 3 × 3 system also has three possibilities. Geometric examples are illustrated in Figure 11.1(**a–c**).

1. There is one unique solution to the system (see Figure 11.1(**a**)).

2. There are no solutions to the system (see Figure 11.1(**b**)).

3. There are infinitely many solutions to the system (see Figure 11.1(**c**)).

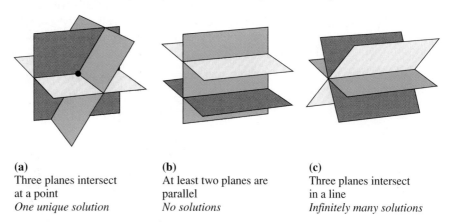

(a)
Three planes intersect
at a point
One unique solution

(b)
At least two planes are
parallel
No solutions

(c)
Three planes intersect
in a line
Infinitely many solutions

Figure 11.1

Let's recall how we solve 2 × 2 systems and see how we can generalize the procedures to solve 3 × 3 systems. In solving 2 × 2 systems of linear equations, the first algebraic method we discussed was the elimination method. The goal of this method, as implied by its name, is to eliminate variables (and equations). This was accomplished by combining equations. For example, to solve the system

$$\begin{cases} 2x - 3y = 19 \\ 3x + y = 12 \end{cases}$$

we can transform the second equation by using the multiplication property of equality to change the coefficients of y so that they are opposites. Then we add the two equations and eliminate the variable with opposite coefficients.

$$
\begin{array}{llll}
2x - 3y = 19 & \textit{As is} \rightarrow & 2x - 3y = 19 & \textit{Add.} \\
3x + y = 12 & \textit{Multiply by 3} \rightarrow & 9x + 3y = 36 \\
& & \overline{11x = 55} \\
& & x = 5
\end{array}
$$

Next we find y by substituting $x = 5$ into the first equation:

$$
\begin{array}{ll}
2x - 3y = 19 & \textit{Substitute } x = 5. \\
2(5) - 3y = 19 & \\
y = -3 &
\end{array}
$$

Hence, the solution is $(5, -3)$. The student should check this solution in both equations.

The elimination method generalizes quite naturally to systems of equations with more than two variables. Let's illustrate how we can use the elimination method to solve a 3 × 3 system.

EXAMPLE 1

Solve the following system of equations:

$$\begin{cases} x + y + 2z = 11 & \textbf{(1)} \\ 2x - y - z = 0 & \textbf{(2)} \\ 3x + y + 4z = 25 & \textbf{(3)} \end{cases}$$

Note: We number the equations for ease of reference.

Solution

The elimination method for a 3×3 system requires us to take the 3×3 system, eliminate one of the variables to produce a 2×2 system, and then use the elimination process again to solve this 2×2 system.

We must first decide which variable we want to eliminate, and for a 3×3 system, making the "right" choice can significantly simplify the procedure. If we take a moment to look at the system, we can see that y can be eliminated most easily. We proceed as follows:

$$\begin{array}{ll} x + y + 2z = 11 & \textit{Add equations (1) and (2).} \\ \underline{2x - y - z = 0} & \\ 3x \qquad + z = 11 & \text{Let's call this equation \textbf{(4)}.} \end{array}$$

If we can now get another equation involving x and z only, we will have a 2×2 system.

$$\begin{array}{ll} 2x - y - z = 0 & \textit{Add equations (2) and (3).} \\ \underline{3x + y + 4z = 25} & \\ 5x \qquad + 3z = 25 & \text{Let's call this equation \textbf{(5)}.} \end{array}$$

Equations (4) and (5) together are a 2×2 system that we can solve:

$$\begin{cases} 3x + z = 11 & \textit{Multiply by } -3 \rightarrow \\ 5x + 3z = 25 & \textit{As is} \rightarrow \end{cases} \qquad \begin{array}{l} -9x - 3z = -33 \\ \underline{5x + 3z = \quad 25} \\ -4x \qquad = -8 \\ \qquad x = 2 \end{array}$$

Add the two equations.

Now we can substitute $x = 2$ into either equation (4) or (5) to solve for z. We will substitute $x = 2$ into equation (4):

$$3x + z = 11$$
$$3(2) + z = 11$$
$$6 + z = 11$$
$$z = 5$$

Having the values for x and z, we can go back to any of the original equations and solve for y. We will substitute $x = 2$, $z = 5$ into equation (1):

$$x + y + 2z = 11$$
$$2 + y + 2(5) = 11$$
$$y + 12 = 11$$
$$y = -1$$

Thus, our solution is $\boxed{x = 2, y = -1, z = 5, \quad \text{or} \quad (2, -1, 5)}$.

Check: We check in all three original equations.

$$\begin{array}{ccc} x + y + 2z = 11 & 2x - y - z = 0 & 3x + y + 4z = 25 \\ 2 + (-1) + 2(5) \overset{?}{=} 11 & 2(2) - (-1) - 5 \overset{?}{=} 0 & 3(2) + (-1) + 4(5) \overset{?}{=} 25 \\ 2 - 1 + 10 \overset{\checkmark}{=} 11 & 4 + 1 - 5 \overset{\checkmark}{=} 0 & 6 - 1 + 20 \overset{\checkmark}{=} 25 \end{array}$$

The following outline summarizes the elimination method for a 3 × 3 system.

Elimination Method for a 3 × 3 System

1. Look over the system and choose the most convenient variable to eliminate (it may be that there is no preference).

2. Use any two of the equations to eliminate one of the variables. This gives an equation in at most two variables; call it equation (4).

3. Using a *different* pair of equations, eliminate the *same* variable as in step 2. This gives another equation in the same two variables; call it equation (5).

4. Use the elimination process to solve the system of equations (4) and (5).

5. Substitute the two values obtained in step 4 into one of the original equations to solve for the remaining variable.

6. Check the solution in all three original equations.

EXAMPLE 2

Solve the following system of equations:
$$\begin{cases} 3x - 2y - 3z = 3 & \textbf{(1)} \\ x - 2y = 2 & \textbf{(2)} \\ 11x + 6y + 2z = 1 & \textbf{(3)} \end{cases}$$

Solution   Noticing that equation (2) already involves only x and y, we use equations (1) and (3) to eliminate z as well.

2 times equation (1) → $6x - 4y - 6z = 6$

3 times equation (3) → $\underline{33x + 18y + 6z = 3}$ *Add these two equations.*

$\qquad\qquad\qquad\quad 39x + 14y \quad\;\; = 9$ **(4)**

We now have the resulting 2 × 2 system consisting of equations (2) and (4):

$$\begin{cases} x - 2y = 2 & \text{\textit{Multiply by 7}} \rightarrow & 7x - 14y = 14 \\ 39x + 14y = 9 & \text{\textit{As is}} \rightarrow & \underline{39x + 14y = \;\; 9} \end{cases}$$

$$46x \qquad\quad = 23$$
$$x = \frac{1}{2}$$

Now we substitute $x = \dfrac{1}{2}$ into equation (2):

$$x - 2y = 2$$
$$\frac{1}{2} - 2y = 2$$
$$-2y = \frac{3}{2}$$
$$y = -\frac{3}{4}$$

Finally, we substitute $x = \dfrac{1}{2}$ and $y = -\dfrac{3}{4}$ into equation (1):

$$3x - 2y - 3z = 3$$
$$3\left(\frac{1}{2}\right) - 2\left(-\frac{3}{4}\right) - 3z = 3$$
$$\frac{3}{2} + \frac{3}{2} - 3z = 3$$
$$-3z = 0$$
$$z = 0$$

Thus, the solution is $\boxed{\; x = \dfrac{1}{2}, y = -\dfrac{3}{4}, z = 0 \;}$. This check is left to the student.

EXAMPLE 3

Solve the following system of equations:

$$\begin{cases} 4x - 5y + 8z = 12 & \textbf{(1)} \\ -4x + 7y + z = 9 & \textbf{(2)} \\ x - \dfrac{7}{4}y - \dfrac{1}{4}z = 1 & \textbf{(3)} \end{cases}$$

Solution

We choose to eliminate x. Adding equations (1) and (2), we get

$$2y + 9z = 21 \qquad \textbf{(4)}$$

We already used equations (1) and (2). Now we must use a different pair: equations (2) and (3).

Next we eliminate x again:

Equation (2) $\rightarrow$	$-4x + 7y + z = 9$	
4 times equation (3) $\rightarrow$	$4x - 7y - z = 4$	*Add the two equations.*
	$0 = 13$	*This is always false.*

Therefore, the system of equations has $\boxed{\text{no solutions}}$.

EXAMPLE 4

Solve the following system of equations:

$$\begin{cases} x + 2y + 3z = 5 & \textbf{(1)} \\ \dfrac{x}{6} + \dfrac{y}{3} + \dfrac{z}{2} = \dfrac{5}{6} & \textbf{(2)} \\ 4x + 8y + 12z = 20 & \textbf{(3)} \end{cases}$$

Solution

Whichever variable we choose to eliminate, any pair of equations will always yield an identity. (Try it.) In fact, if we look carefully at the system, we can see that if we multiply equation (2) by 6 we get equation (1), and if we divide equation (3) by 4 we get equation (1). Thus, all three equations are equivalent. All three equations represent the same plane.

The solution set to this system is the set of all points that satisfy any one of the equations. Thus, the solution set is

$$\boxed{\{(x, y, z) \mid x + 2y + 3z = 5\}}$$

We now return to the problem posed at the beginning of this section.

EXAMPLE 5

A hospital administrator must allocate \$140,400 in funds for supplies, 3,480 hours of doctors' services, and 5,600 hours of nurses' services among an emergency room, an outpatient clinic, and a surgery ward. The table indicates how much of each of these resources is needed for a single day's operation of each location. Determine how many days each of these locations can operate if the administrator allocates all of the resources.

	Emergency Room	**Outpatient Clinic**	**Surgery Ward**
Supplies	\$3,000	\$1,800	\$2,000
Doctors' hours	75	50	45
Nurses' hours	100	80	90

Solution   Let e = the number of days the emergency room can operate.

Let c = the number of days the outpatient clinic can operate.

Let s = the number of days the surgery ward can operate.

Each equation in our system represents a particular constraint or condition that must be met by the hospital administrator.

$$
\begin{cases}
3{,}000e + 1{,}800c + 2{,}000s = 140{,}400 & \textbf{(1)} \\
\phantom{3{,}00}75e + 50c + 45s = 3{,}480 & \textbf{(2)} \\
\phantom{3{,}0}100e + 80c + 90s = 5{,}600 & \textbf{(3)}
\end{cases}
$$

This equation represents the amount of money allocated for supplies.

This equation represents the number of hours of doctors' services.

This equation represents the number of hours of nurses' services.

We choose to eliminate e:

Equation (1) $\rightarrow$

$$3{,}000e + 1.800c + 2{,}000s = 140{,}400$$

−30 times equation (3) $\rightarrow$

$$\underline{-3{,}000e - 2{,}400c - 2{,}700s = -168{,}000}$$

$$-600c - 700s = -27{,}600 \qquad \textit{Divide by 100.}$$

$$-6c - 7s = -276 \qquad \textbf{(4)}$$

Let's eliminate e again

Equation (1) $\rightarrow$

$$3{,}000e + 1{,}800c + 2{,}000s = 140{,}400$$

−40 times equation (2) $\rightarrow$

$$\underline{-3{,}000e - 2{,}000c - 1{,}800s = -139{,}200}$$

$$-200c + 200s = 1{,}200 \qquad \textit{Divide by 200.}$$

$$-c + s = 6 \qquad \textbf{(5)}$$

Thus, our 2 × 2 system is

$$
\begin{array}{lll}
-6c - 7s = -276 & \textbf{(4)} & \textit{As is} \rightarrow \\
-c + s = 6 & \textbf{(5)} & \textit{Multiply by 7} \rightarrow
\end{array}
\qquad
\begin{array}{l}
-6c - 7s = -276 \\
\underline{-7c + 7s = 42} \\
-13c = -234 \\
c = 18
\end{array}
$$

Substitute $c = 18$ into equation (5):

$$-c + s = 6$$
$$-18 + s = 6$$
$$s = 24$$

Substitute $c = 18$ and $s = 24$ into equation (3):

$$100e + 80c + 90s = 5{,}600$$
$$100e + 80(18) + 90(24) = 5{,}600$$
$$100e + 1{,}440 + 2{,}160 = 5{,}600$$
$$100e = 2{,}000$$
$$e = 20$$

The check is left to the student.

Thus, the emergency room can operate for 20 days, the outpatient clinic can operate for 18 days, and the surgery ward can operate for 24 days.

EXERCISES 11.1

In Exercises 1–26, solve the system of equations.

1. $\begin{cases} x + y + z = 9 \\ 2x - y + z = 9 \\ x - y + z = 3 \end{cases}$

2. $\begin{cases} x + y + z = 6 \\ 3x + y - z = 6 \\ 2x + y - z = 4 \end{cases}$

3. $\begin{cases} -x + 2y + z = 0 \\ x - y + 2z = 1 \\ x + 3y + z = 5 \end{cases}$

4. $\begin{cases} 2x + y - z = 8 \\ 3x - y - z = 11 \\ 2x - y + 2z = 2 \end{cases}$

5. $\begin{cases} x + y - z = 1 \\ 2x + 2y + 2z = 0 \\ x - y + z = 3 \end{cases}$

6. $\begin{cases} 2x - 3y + z = 1 \\ -2x + y - 2z = 6 \\ 2x - y - z = 3 \end{cases}$

7. $\begin{cases} x + 2y + 3z = 1 \\ 3x + 6y + 9z = 3 \\ 4x + 8y + 12z = 4 \end{cases}$

8. $\begin{cases} 3x + 2y + z = 2 \\ 6x + 4y + 2z = 4 \\ 15x + 10y + 5z = 10 \end{cases}$

9. $\begin{cases} 3x - 2y + 5z = 2 \\ 4x - 7y - z = 19 \\ 5x - 6y + 4z = 13 \end{cases}$

10. $\begin{cases} 5x + 4y - 3z = 3 \\ 6x + 3y - 2z = 7 \\ x - 2y + 4z = 8 \end{cases}$

11. $\begin{cases} 2a + b - 3c = -6 \\ 4a - 4b + 2c = 10 \\ 6a - 7b + c = 12 \end{cases}$

12. $\begin{cases} 2a + 3b - c = 2 \\ 5a - 6b + 4c = 3 \\ 6a - 9b + 5c = 4 \end{cases}$

13. $\begin{cases} x + 3y + 2z = 3 \\ x + 3z = 4 \\ x - 4y - z = 0 \end{cases}$

14. $\begin{cases} 2x - 3y + 4z = 1 \\ 5y - 2z = 5 \\ 3x + 2y - 5z = 8 \end{cases}$

15. $\begin{cases} \dfrac{1}{2}s + \dfrac{1}{3}t + u = 3 \\ \dfrac{1}{3}s - \dfrac{1}{2}t - 2u = 1 \\ \dfrac{2}{3}s - \dfrac{1}{6}t + \dfrac{1}{2}u = 6 \end{cases}$

16. $\begin{cases} \dfrac{s}{4} + \dfrac{t}{6} - \dfrac{u}{3} = 1 \\ \dfrac{s}{2} + \dfrac{t}{3} + u = -3 \\ \dfrac{s}{8} + \dfrac{t}{4} - u = 5 \end{cases}$

17. $\begin{cases} p + q + r = 6 \\ 2q + r - p = 6 \\ r - p + q = 4 \end{cases}$

18. $\begin{cases} 2r - s + t = 6 \\ s - 2r + t = 0 \\ 3t - r + s = 20 \end{cases}$

19. $\begin{cases} x + y = 0 \\ y + z = 0 \\ x + z = 2 \end{cases}$

20. $\begin{cases} 2x - y = 0 \\ y - z = 0 \\ 3x + 2z = 0 \end{cases}$

21. $\begin{cases} a + b = 2b + c \\ a - 2b = c + 3 \\ 2a - b = 3c - 9 \end{cases}$

22. $\begin{cases} m + 3n = p + 8 \\ 2m - 4n = 2p + 6 \\ 3m - n = -p + 2 \end{cases}$

23. $\begin{cases} 12a + 5b + 3c = 24{,}000 \\ 10a + 6b + 4c = 13{,}300 \\ 8a + 7b + 5c = 8{,}700 \end{cases}$

24. $\begin{cases} 9a + 12b + 15c = 26{,}100 \\ 6a + 8b + 10c = 17{,}400 \\ 3a + 4b + 5c = 8{,}700 \end{cases}$

25. $\begin{cases} 0.06x + 0.07y + 0.08z = 440 \\ 0.05x + 0.06y + 0.08z = 410 \\ 0.04x + 0.05y + 0.06z = 320 \end{cases}$

26. $\begin{cases} 0.08x + 0.10y + 0.12z = 17{,}000 \\ 0.06x + 0.05y + 0.08z = 12{,}300 \\ 0.04x + 0.10y + 0.06z = 11{,}000 \end{cases}$

In Exercises 27–38, solve the problems algebraically. Clearly label what each variable represents.

27. A collection of 48 coins consists of dimes, quarters, and half-dollars, and has a total value of $10.55. If there are 2 fewer dimes than quarters and half-dollars combined, how many of each type of coin are there?

28. A bank teller gave a customer change for a $500 bill in $5, $10, and $20 bills. The number of $5 and $20 bills combined was 5 less than twice the number of $10 bills. If there were 40 bills in all, how many of each type of bill were there?

29. Martha splits up a total of $12,000 into three investments. She has a bank account paying 8.7%, a bond paying 9.3%, and a stock paying 12.66%. Her annual interest from the three investments is $1,266. If the interest from the stock is equal to the interest from the bank account and the bond combined, how much is invested at each rate?

30. George wants to divide $17,000 into three investments so that the interests from the three investments are equal. If he is going to invest part in a certificate of deposit paying 8.5%, part in a corporate bond paying 13.6%, and the rest in a high-risk real estate deal paying 17%, how much should he invest at each rate?

31. A theater group plans to sell 750 tickets for a play. They are charging $12 for orchestra seats, $8 for mezzanine seats, and $6 for balcony seats, and they plan to collect $7,290. If there are 100 more orchestra tickets than mezzanine and balcony tickets combined, how many tickets of each type are there?

32. An opera house is planning to put on a performance and wants to determine its ticket prices. The theater has an orchestra section that holds 550 people, a mezzanine that holds 140 people, and a balcony that holds 275 people. If they want a balcony seat to cost 20% less than a mezzanine seat, an orchestra seat to cost $5 more than a mezzanine seat, and a full house to bring in $16,400, how much should they charge for each type of seat?

33. A computer manufacturer produces three models of personal computers—model A, model B, and model C. The company knows how much production, assembly, and testing time is needed for each model. This information is found in the accompanying table. If the company has allocated 721 hours for production, 974 hours for assembly, and 168 hours for testing, how many of each model can the company produce if it wants to use up all the time allocated for each phase of the process?

Model	Hours Required for Production	Hours Required for Assembly	Hours Required for Testing
Model A	2.1	3.2	0.5
Model B	2.8	3.6	0.6
Model C	3.2	4.0	0.8

34. A nutritionist wants to create a food supplement out of three substances: A, B, and C. She wants the food supplement to have the following characteristics: 5 g of the supplement should supply 1.54 g of iron and cost 48¢. The iron content and cost of substances A, B, and C are entered in the accompanying table. How many grams of each substance should be used to make such a food supplement?

Substance	Iron Content per Gram	Cost per Gram
Substance A	0.3 g	10¢
Substance B	0.28 g	8¢
Substance C	0.4 g	12¢

35. A nutritionist wishes to mix three foods—food A, food B, and food C—whose contents (as percentages of fat, carbohydrates, and protein) are described in the accompanying table. How many grams of each should be used to produce a mixture with 9 g of fat, 28.5 g of carbohydrates, and 97 g of protein?

	Food A	Food B	Food C
Fat	5%	6%	4%
Carbohydrates	20%	15%	10%
Protein	40%	60%	70%

36. A farmer wants to plant 300 acres with corn, potatoes, and melons. She will plant 20 more acres of corn than potatoes. She knows that she will earn a profit of $300 per acre for corn, $150 per acre for potatoes, and $200 per acre for melons. If she earns a total of $66,250, how many acres were allotted to each crop?

37. A land manager needs to determine how many acres of farmland to plant based on the following limitations of resources. There is $94,000 available for fertilizer, 7,200 hours available for labor, and 555,000 gallons of water available for irrigation. Using the accompanying table, which indicates how much of each of these resources is needed to plant one acre of each of three crops, A, B, and C, determine how many acres of each crop should be planted to use all of the resources available.

	Crop A	Crop B	Crop C
Fertilizer	$90	$110	$75
Hours of labor	6	10	5
Gallons of water	400	500	600

38. A manufacturer produces three types of thread. Each type uses a different amount of three raw materials: cotton, rayon, and polyester. The accompanying table contains the relevant information on the amount of each raw material used to produce one spool of each type of thread. If the manufacturer has 27 kilograms (kg) of cotton, 10.8 kg of rayon, and 8.2 kg of polyester in stock and wants to use it all, how many spools of each type should be produced?

Type of Thread	Amount of Cotton	Amount of Rayon	Amount of Polyester
Type A	15 g	3 g	2 g
Type B	10 g	6 g	4 g
Type C	8 g	6 g	6 g

Mini-Review

39. Find the domain of the function $f(x) = \dfrac{3x}{2x^2 - x - 3}$.

40. If z varies directly as x and inversely as y, what happens to z if x is tripled and y is halved?

41. Sketch the graph of $y = 8 - 3x - x^2$.

42. *Simplify:* $\sqrt{18x^5y} - x^2\sqrt{50xy}$

11.2 Solving Linear Systems Using Augmented Matrices

In the process of using the elimination method for solving systems of equations, we concentrated mainly on the coefficients of the variables and the constants. We were concerned only with the actual variables when we wanted to make sure we were "matching up" the correct variables to be eliminated. Hence, as long as we have some way of keeping track of the variables, we should be able to solve systems focusing primarily on the coefficients and constants of the system. In this and the next section we discuss methods of solving systems using the constants and the coefficients of the variables of the system.

A rectangular array of numbers is called a *matrix.* The numbers in a matrix are called the *elements* or *entries* of the matrix. The horizontal arrays of entries are called the *rows,* and the vertical arrays of entries are called the *columns.* The following are two examples of matrices:

$$\begin{bmatrix} 1 & -2 & 3 & 4 \\ 0 & 5 & 0 & -3 \end{bmatrix}$$

$$\begin{array}{l} \text{Row 1} \rightarrow \\ \text{Row 2} \rightarrow \\ \text{Row 3} \rightarrow \end{array} \begin{bmatrix} 2 & 3 & -1 \\ 0 & -5 & 2 \\ 1 & 3 & 5 \end{bmatrix}$$

We usually specify the size of a matrix by first giving the number of rows and then the number of columns. The first example is a 2×4 (read "2 by 4") matrix; the second is a 3×3 matrix. A *square matrix* is a matrix that contains an equal number of rows and columns; the 3×3 matrix above is a square matrix.

We will now describe how we can use matrices to solve systems of equations. We begin by taking a system and rewriting it as an *augmented matrix* as follows:

$$\begin{cases} 2x - 3y = 4 \\ 5x + y = -1 \end{cases}$$

is written as the augmented matrix

$$\left[\begin{array}{cc|c} 2 & -3 & 4 \\ 5 & 1 & -1 \end{array} \right]$$

The matrix $\begin{bmatrix} 2 & -3 \\ 5 & 1 \end{bmatrix}$, made up of the coefficients of the variables of the system, is called the *coefficient matrix.* The *augmented matrix* of a system of equations consists of the coefficient matrix of the system with the constants of the system adjoined to the right. Consider another example:

$$\begin{cases} 3x + 5y - z = 7 \\ 2x + 3z = -2 \\ x - 2y - 2z = 4 \end{cases} \quad \text{is written as} \quad \left[\begin{array}{ccc|c} 3 & 5 & -1 & 7 \\ 2 & 0 & 3 & -2 \\ 1 & -2 & -2 & 4 \end{array} \right]$$

The relative positions of the column entries in the coefficient matrix indicate the variable: The first column consists of the coefficients of the x variable; the second column, the coefficients of the y variable; and the third column, the coefficients of the z variable. The constants are on the augmented side. Thus, it is important that the variables be lined up columnwise before we represent a system by its augmented matrix.

Notice that a 0 is entered in the second row of the last augmented matrix for the missing y variable in the second equation. Also observe that 1 is entered in row 3 as the coefficient of x in the third equation of the system.

As we do in solving single equations, we define *equivalent systems of equations* as systems with the same solution(s). For example, the three systems on the left side of the following display are equivalent systems. When we solve each system we find that each system has the unique solution $(2, -1, 3)$. Look carefully at each system's augmented matrix on the right side.

| **System** | | **Augmented matrix** |

1. $\begin{cases} 3x - y + z = 10 \\ x + y + z = 4 \\ x + 2y + 3z = 9 \end{cases}$ $\qquad \begin{bmatrix} 3 & -1 & 1 & | & 10 \\ 1 & 1 & 1 & | & 4 \\ 1 & 2 & 3 & | & 9 \end{bmatrix}$

2. $\begin{cases} 3x - y + z = 10 \\ -x - y - z = -4 \\ y + 2z = 5 \end{cases}$ $\qquad \begin{bmatrix} 3 & -1 & 1 & | & 10 \\ -1 & -1 & -1 & | & -4 \\ 0 & 1 & 2 & | & 5 \end{bmatrix}$

3. $\begin{cases} 3x - y + z = 10 \\ -4y - 2z = -2 \\ 6z = 18 \end{cases}$ $\qquad \begin{bmatrix} 3 & -1 & 1 & | & 10 \\ 0 & -4 & -2 & | & -2 \\ 0 & 0 & 6 & | & 18 \end{bmatrix}$

The main diagonal

Notice that the last system, system 3, is easiest to solve. We simply start at the bottom equation and solve for z. Then we substitute the value we found for z in the second equation and solve for y. Finally, we substitute the values we found for y and z into the first equation to find the value for x. This process of substitution is called ***back substitution.***

System 3 and its augmented matrix are said to be in *triangular* (or *echelon*) *form.** The elements $3, -4, 6$ in this augmented matrix are called the ***main diagonal.*** Note that the main diagonal of an augmented matrix is the diagonal of the coefficient matrix. A matrix is in ***triangular*** (or ***echelon***) ***form*** if it has all zero entries below the main diagonal.

This will be our goal: Change an augmented matrix into triangular form, convert the matrix to its associated system, and then solve the simpler system by back substitution. We begin with the following definition:

DEFINITION

Two matrices are ***row-equivalent*** if their associated systems of equations are equivalent.

Thus, all of the matrices shown on the right in the previous display are row-equivalent since their corresponding systems are all equivalent.

In the previous section, we used the properties of equality to transform a given system of equations into a "simpler" *equivalent* system of equations. Now that we have defined row-equivalence for matrices, the procedures we illustrated in the previous section translate into the following list of transformations that we can apply to a matrix to produce a row-equivalent matrix. We call these transformations the *elementary row operations.*

Elementary Row Operations

1. Multiply each entry in a given row by any nonzero constant.
2. Interchange any two rows.
3. Add a multiple of one row to another row.

Let's illustrate how these elementary row operations can be used to transform a matrix into a second row-equivalent matrix.

---

*Sometimes the term *triangular form* is restricted to the system of equations rather than its augmented matrix; however, we will use the term *triangular form* to describe both the system and its augmented matrix.

For example,

1.
$$\begin{bmatrix} 2 & 3 & -2 \\ 3 & 2 & 5 \end{bmatrix}$$
$-3R_1 \to R_1$

This notation means multiply each entry in row 1 by -3 to get the new row 1.

$$\begin{bmatrix} -6 & -9 & 6 \\ 3 & 2 & 5 \end{bmatrix}$$

2.
$$\begin{bmatrix} 2 & 5 & 0 \\ 3 & 1 & 2 \\ 1 & 0 & 4 \end{bmatrix}$$
$R_1 \leftrightarrow R_2$

This notation means interchange row 1 and row 2. (Row 3 remains unchanged.)

$$\begin{bmatrix} 3 & 1 & 2 \\ 2 & 5 & 0 \\ 1 & 0 & 4 \end{bmatrix}$$

3.
$$\begin{bmatrix} 2 & -3 & 3 \\ 3 & 0 & 2 \\ 1 & 2 & -1 \end{bmatrix}$$
$2R_1 + R_3 \to R_3$

This notation means multiply each entry in row 1 by 2 and add the resulting entries to each corresponding entry in row 3 to get a new row 3. (Rows 1 and 2 remain unchanged.)

$$\begin{bmatrix} 2 & -3 & 3 \\ 3 & 0 & 2 \\ 5 & -4 & 5 \end{bmatrix}$$

Since this last illustration is a bit more complex than the first two, we will demonstrate how we found the new row 3.

Multiply each entry in row 1 by 2:

$$2 \quad -3 \quad 3 \quad \longrightarrow \quad 4 \quad -6 \quad 6 \quad \text{\textit{This is } } 2R_1.$$

Then add each entry in this multiple of row 1 to each entry in row 3: $+$ $\underline{\quad 1 \quad\quad 2 \quad -1}$ *This is R_3.*

This gives the new row 3: $5 \quad -4 \quad\quad 5$ *This is the new R_3.*

Notice that the elementary row operations are equivalent to the transformations performed on *systems of equations:* Row operation 1 is equivalent to multiplying both sides of any equation by a nonzero constant; row operation 2 is equivalent to interchanging any two equations in a system; and row operation 3 is equivalent to adding a multiple of one equation to another.

Since elementary row operations produce associated systems that are equivalent to each other, we can conclude:

> A matrix can be transformed into a row-equivalent matrix by performing any elementary row operation on the matrix.

As we mentioned previously, our intention is to solve a system of equations by using the elementary row operations on its augmented matrix to produce row-equivalent matrices that are closer to triangular form. Once we get the system into triangular form, we can solve the system by back substitution.

The method we will use in solving systems of equations by matrices, called **Gaussian elimination,** is described in the next box.

Method for Solving Linear Systems Using Augmented Matrices (Gaussian Elimination)

1. Set up the augmented matrix of the system.
2. Use the elementary row operations to transform the augmented matrix into a row-equivalent matrix in triangular form.
3. For the augmented matrix in triangular form, write the corresponding system of equations.
4. Solve this system by back substitution.
5. Check your solution in the original equations.

EXAMPLE 1

Solve the following system by Gaussian elimination:

$$\begin{cases} x + 2y = 8 \\ 4x - 3y = 21 \end{cases}$$

Solution

The first step is to set up the augmented matrix of the system:

$$\left[\begin{array}{cc|c} 1 & 2 & 8 \\ 4 & -3 & 21 \end{array}\right]$$ *Now we use the elementary row operations to transform this matrix into a row-equivalent matrix in triangular form.*

For a 2×2 matrix to be in triangular form, all we need is one 0 in the bottom left corner of the matrix (below the main diagonal: $1, -3$):

$$\left[\begin{array}{cc|c} 1 & 2 & 8 \\ 4 & -3 & 21 \end{array}\right] \qquad -4R_1 + R_2 \to R_2 \qquad \left[\begin{array}{cc|c} 1 & 2 & 8 \\ 0 & -11 & -11 \end{array}\right]$$

Multiply the entries in row 1 by -4 and add the result to the entries in row 2 to get the new row 2. *This matrix is now in triangular form.*

Now that the matrix is in triangular form, we can convert the new (row-equivalent) matrix into its associated system:

$$\begin{cases} x + 2y = 8 \\ -11y = -11 \end{cases}$$ *Now we can solve this system by back substitution.*

First solve for y using the last equation:

$$-11y = -11$$ *Divide both sides of the equation by -11.*
$$y = 1$$

Then substitute 1 for y in the first equation and solve for x:

$$x + 2y = 8 \qquad \text{Substitute } y = 1.$$
$$x + 2(1) = 8 \qquad \text{Solve for } x.$$
$$x = 6$$

The solution is $\boxed{x = 6, y = 1}$. *You should check that this is the solution to the original system of equations.*

EXAMPLE 2

Solve the following system by Gaussian elimination:

$$\begin{cases} x + 2y + 2z = 10 \\ 2x + 3y - z = 6 \\ 3x + y + 5z = 8 \end{cases}$$

Solution

First, set up the augmented matrix of the system:

$$\left[\begin{array}{ccc|c} 1 & 2 & 2 & 10 \\ 2 & 3 & -1 & 6 \\ 3 & 1 & 5 & 8 \end{array}\right]$$ *Now use the elementary row operations to transform this matrix into a row-equivalent matrix in triangular form.*

Begin by getting 0 below 1 in the first column:

$$\left[\begin{array}{ccc|c} 1 & 2 & 2 & 10 \\ 2 & 3 & -1 & 6 \\ 3 & 1 & 5 & 8 \end{array}\right] \qquad -2R_1 + R_2 \to R_2 \qquad \left[\begin{array}{ccc|c} 1 & 2 & 2 & 10 \\ 0 & -1 & -5 & -14 \\ 3 & 1 & 5 & 8 \end{array}\right]$$

Multiply row 1 by -2 and add it to row 2 to get the new row 2.

Next, we need to get another 0 in the bottom left corner:

$$\begin{bmatrix} 1 & 2 & 2 & | & 10 \\ 0 & -1 & -5 & | & -14 \\ 3 & 1 & 5 & | & 8 \end{bmatrix} \qquad -3R_1 + R_3 \rightarrow R_3 \qquad \begin{bmatrix} 1 & 2 & 2 & | & 10 \\ 0 & -1 & -5 & | & -14 \\ 0 & -5 & -1 & | & -22 \end{bmatrix}$$

Multiply row 1 by −3 and add it to row 3 to get the new row 3.

The last step is to get 0 under −1 in the second column. Note that if we try to add a multiple of row 1 to row 3, we will lose the 0 in row 3, column 1. Because there is a 0 in row 2, column 1, adding a multiple of row 2 will not affect the 0 in row 3, column 1.

$$\begin{bmatrix} 1 & 2 & 2 & | & 10 \\ 0 & -1 & -5 & | & -14 \\ 0 & -5 & -1 & | & -22 \end{bmatrix} \qquad -5R_2 + R_3 \rightarrow R_3 \qquad \begin{bmatrix} 1 & 2 & 2 & | & 10 \\ 0 & -1 & -5 & | & -14 \\ 0 & 0 & 24 & | & 48 \end{bmatrix}$$

Multiply row 2 by −5 and add it to row 3 to get the new row 3.   *This matrix is in triangular form.*

Now that the matrix is in triangular form, we can convert this last (row-equivalent) matrix into its associated system:

$$\begin{cases} x + 2y + 2z = 10 \\ \quad\;\; -y - 5z = -14 \\ \qquad\qquad 24z = 48 \end{cases}$$ *Now we can solve this system by back substitution.*

First solve for z using the last equation:

$$24z = 48 \qquad \textit{Divide both sides of the equation by 24.}$$
$$z = 2$$

Then substitute 2 for z in the second equation and find y:

$$-y - 5z = -14 \qquad \textit{Substitute } z = 2.$$
$$-y - 5(2) = -14 \qquad \textit{Solve for y.}$$
$$-y - 10 = -14$$
$$y = 4$$

Then substitute 2 for z, and 4 for y in the first equation, and solve for x:

$$x + 2y + 2z = 10 \qquad \textit{Substitute } z = 2 \textit{ and } y = 4.$$
$$x + 2(4) + 2(2) = 10 \qquad \textit{Now solve for x.}$$
$$x = -2$$

Hence, the solution is $\boxed{(-2, 4, 2)}$.   *Check this solution in the original system of equations.*

EXAMPLE 3

Solve the following system by Gaussian elimination:

$$\begin{cases} 6x - 4y = 7 \\ 3x - 2y = 4 \end{cases}$$

Solution   First, identify its augmented matrix:

$$\begin{bmatrix} 6 & -4 & | & 7 \\ 3 & -2 & | & 4 \end{bmatrix}$$

Again, for a 2×2 matrix to be in triangular form, all we need is one 0 in the bottom left corner of the matrix. To get 0 in the bottom left corner, we would have to multiply row 1 by $-\frac{1}{2}$ and add it to row 2. In this example we can avoid computations with fractions by first applying elementary row operation 2—interchanging rows 1 and 2:

$$\begin{bmatrix} 6 & -4 & | & 7 \\ 3 & -2 & | & 4 \end{bmatrix} \qquad R_1 \leftrightarrow R_2 \qquad \begin{bmatrix} 3 & -2 & | & 4 \\ 6 & -4 & | & 7 \end{bmatrix}$$

Now we can get 0 in the bottom left corner:

$$\begin{bmatrix} 3 & -2 & | & 4 \\ 6 & -4 & | & 7 \end{bmatrix} \qquad -2R_1 + R_2 \rightarrow R_2 \qquad \begin{bmatrix} 3 & -2 & | & 4 \\ 0 & 0 & | & -1 \end{bmatrix}$$

Multiply the entries in row 1 by −2 and add this to the entries in row 2 to get the new row 2.

This matrix is now in triangular form. We write the associated system of the new matrix:

$$\begin{cases} 3x - 2y = 4 \\ \qquad\quad 0 = -1 \end{cases}$$

Since the second equation is a contradiction, the system is inconsistent. We conclude that there is $\boxed{\text{no solution}}$.

EXAMPLE 4

Solve the following system by Gaussian elimination:

$$\begin{cases} 2x + 6y + 4z = 8 \\ 4x + 12y + 10z = 20 \\ 3x + 9y + 6z = 12 \end{cases}$$

Solution

First, identify its augmented matrix:

$$\begin{bmatrix} 2 & 6 & 4 & | & 8 \\ 4 & 12 & 10 & | & 20 \\ 3 & 9 & 6 & | & 12 \end{bmatrix}$$

Now, we want to get 0 below 2 in the first column:

$$\begin{bmatrix} 2 & 6 & 4 & | & 8 \\ 4 & 12 & 10 & | & 20 \\ 3 & 9 & 6 & | & 12 \end{bmatrix} \qquad -2R_1 + R_2 \rightarrow R_2 \qquad \begin{bmatrix} 2 & 6 & 4 & | & 8 \\ 0 & 0 & 2 & | & 4 \\ 3 & 9 & 6 & | & 12 \end{bmatrix}$$

Multiply the entries in row 1 by −2 and add the result to the entries in row 2 to get a new row 2.

Then we want to get 0 in the bottom left corner. We could multiply the entries in row 1 by $-\frac{3}{2}$ and add the result to the entries in row 3 to get 0 in this spot. However, we can avoid fractions by first noting that the LCM (least common multiple) of 2 and 3 is 6, and multiply row 1 by 3 and row 3 by 2 to get 6's in the first column. We will do both operations in this step:

$$\begin{bmatrix} 2 & 6 & 4 & | & 8 \\ 0 & 0 & 2 & | & 4 \\ 3 & 9 & 6 & | & 12 \end{bmatrix} \qquad \begin{matrix} 3R_1 \rightarrow R_1 \\ \\ 2R_3 \rightarrow R_3 \end{matrix} \qquad \begin{bmatrix} 6 & 18 & 12 & | & 24 \\ 0 & 0 & 2 & | & 4 \\ 6 & 18 & 12 & | & 24 \end{bmatrix}$$

Now we can easily get 0 in the bottom left corner:

$$\begin{bmatrix} 6 & 18 & 12 & | & 24 \\ 0 & 0 & 2 & | & 4 \\ 6 & 18 & 12 & | & 24 \end{bmatrix} \qquad -R_1 + R_3 \rightarrow R_3 \qquad \begin{bmatrix} 6 & 18 & 12 & | & 24 \\ 0 & 0 & 2 & | & 4 \\ 0 & 0 & 0 & | & 0 \end{bmatrix}$$

We convert the matrix into its associated system:

$$\begin{cases} 6x + 18y + 12z = 24 \\ 2z = 24 \end{cases}$$ *Note that the third equation, which would be 0 = 0, need not appear in the system.*

We solve the system by first solving for z in the last equation to get

$$2z = 4$$
$$z = 2$$

Then we substitute 2 for z in the first equation:

$6x + 18y + 12z = 24$ *Substitute z = 2 to get*

$6x + 18y + 12(2) = 24$ *Which is equivalent to*

$6x + 18y = 0$ *or equivalently:*

$x + 3y = 0$

There is no unique solution for this system. The first and third equations of the original system are equivalent and we end up with two planes intersecting in a line similar to Figure 11.1(c) in Section 11.1. The best we can do is list the set of all points that satisfy this system in the following way:

$$\boxed{\{(x, y, 2) \mid x + 3y = 0\}}$$ *Notice z = 2.*

Gauss–Jordan Elimination

In using an augmented matrix to solve a linear system, we still have to back substitute to arrive at the system's solution. Also, keep in mind that triangular matrices for a system are not unique; that is, a system can have different (but equivalent) triangular matrices.

If we continued to apply the elementary row operations to a matrix in triangular form in an attempt to get all 1's on the main diagonal and 0's in the rest of the coefficient matrix as pictured in the 3 × 3 matrix below (where a, b, and c are constants), then we can read off the solution directly from the matrix. In the matrix below, we can see that $x = a$, $y = b$, and $z = c$:

$$\begin{bmatrix} 1 & 0 & 0 & | & a \\ 0 & 1 & 0 & | & b \\ 0 & 0 & 1 & | & c \end{bmatrix}$$

The form of this 3 × 3 matrix is called *reduced echelon form.*

A matrix is in *reduced echelon form* when

1. Row consisting entirely of 0 entries are on the bottom of the matrix.
2. The first nonzero entry in any row is 1, called a *pivot element.*
3. All nonzero rows are arranged so that the pivot element occurs farther to the right than the pivot element in the preceding row.
4. In each column with a pivot element, the remaining entries are all 0.

We can put a matrix into reduced echelon form using the elementary row operations. The method of solving systems by putting a system's augmented matrix into reduced echelon form is called *Gauss–Jordan elimination.* The next example illustrates an efficient approach for putting a matrix into reduced echelon form.

EXAMPLE 5

Solve the following system by Gauss–Jordan elimination:

$$\begin{cases} 2x - y + z = 3 \\ x + y + z = 4 \\ x + 2y + 3z = 4 \end{cases}$$

Solution

We identify the system's augmented matrix:

$$\begin{bmatrix} 2 & -1 & 1 & 3 \\ 1 & 1 & 1 & 4 \\ 1 & 2 & 3 & 4 \end{bmatrix}$$

1. First, we want to get a 1 in the upper left corner. Rather than divide row 1 by 2 and end up with fractions in the first row, we can interchange row 1 with row 2:

$$\begin{bmatrix} 2 & -1 & 1 & 3 \\ 1 & 1 & 1 & 4 \\ 1 & 2 & 3 & 4 \end{bmatrix} \quad R_1 \leftrightarrow R_2 \quad \begin{bmatrix} 1 & 1 & 1 & 4 \\ 2 & -1 & 1 & 3 \\ 1 & 2 & 3 & 4 \end{bmatrix}$$

2. Now we want to get 0's below the 1 in the first column. (This is called *sweeping out the column*.) We will take two steps at once; observe the transformations.

$$\begin{bmatrix} 1 & 1 & 1 & 4 \\ 2 & -1 & 1 & 3 \\ 1 & 2 & 3 & 4 \end{bmatrix} \quad \begin{matrix} -2R_1 + R_2 \to R_2 \\ -R_1 + R_3 \to R_3 \end{matrix} \quad \begin{bmatrix} 1 & 1 & 1 & 4 \\ 0 & -3 & -1 & -5 \\ 0 & 1 & 2 & 0 \end{bmatrix}$$

3. Next we want to get 1 where -3 is (in the middle of the main diagonal). Again, rather than divide row 2 by -3 and get fractions, we can interchange row 2 and row 3:

$$\begin{bmatrix} 1 & 1 & 1 & 4 \\ 0 & -3 & -1 & -5 \\ 0 & 1 & 2 & 0 \end{bmatrix} \quad R_2 \leftrightarrow R_3 \quad \begin{bmatrix} 1 & 1 & 1 & 4 \\ 0 & 1 & 2 & 0 \\ 0 & -3 & -1 & -5 \end{bmatrix}$$

4. Now we want to get 0's in the first and third entries of the second column (sweep out the rest of the second column). Again, we will take two steps:

$$\begin{bmatrix} 1 & 1 & 1 & 4 \\ 0 & 1 & 2 & 0 \\ 0 & -3 & -1 & -5 \end{bmatrix} \quad \begin{matrix} -R_2 + R_1 \to R_1 \\ 3R_2 + R_3 \to R_3 \end{matrix} \quad \begin{bmatrix} 1 & 0 & -1 & 4 \\ 0 & 1 & 2 & 0 \\ 0 & 0 & 5 & -5 \end{bmatrix}$$

5. Next we want to get 1 in the lower right corner of the coefficient matrix:

$$\begin{bmatrix} 1 & 0 & -1 & 4 \\ 0 & 1 & 2 & 0 \\ 0 & 0 & 5 & -5 \end{bmatrix} \quad \tfrac{1}{5}R_3 \to R_3 \quad \begin{bmatrix} 1 & 0 & -1 & 4 \\ 0 & 1 & 2 & 0 \\ 0 & 0 & 1 & -1 \end{bmatrix}$$

6. Finally, we want to sweep out the rest of the third column:

$$\begin{bmatrix} 1 & 0 & -1 & 4 \\ 0 & 1 & 2 & 0 \\ 0 & 0 & 1 & -1 \end{bmatrix} \quad \begin{matrix} R_3 + R_1 \to R_1 \\ -2R_3 + R_2 \to R_2 \end{matrix} \quad \begin{bmatrix} 1 & 0 & 0 & 3 \\ 0 & 1 & 0 & 2 \\ 0 & 0 & 1 & -1 \end{bmatrix}$$

The matrix is in reduced echelon form and we can directly read off the matrix by columns to get the solution to the original system:

$$\boxed{x = 3, y = 2, \text{ and } z = -1}$$

A few comments should be made about trying to put a matrix into triangular form versus reduced echelon form. Unlike triangular form, one advantage to reduced echelon form is that it is unique for a system. The key advantage to Gauss–Jordan elimination is that it lends itself well to computer programming. However, Gauss–Jordan elimination requires more elementary row operations than does Gaussian elimination and you may feel that it would be quicker to stop and begin back substitution when you have a matrix in triangular form. This is especially true when you cannot avoid extensive computations with fractions.

EXERCISES 11.2

In Exercises 1–4, set up the augmented matrices of the systems of equations.

1. $\begin{cases} 3x - 2y = 5 \\ x - y = 8 \end{cases}$

2. $\begin{cases} 2x + y = -6 \\ 7x + 2y = 0 \end{cases}$

3. $\begin{cases} x - 2y + 3z = 4 \\ y - z = -3 \\ 2x + 3y = 8 \end{cases}$

4. $\begin{cases} x - 2y + z = -1 \\ 3y - z = 2 \\ x - 2z = 0 \end{cases}$

In Exercises 5–26, solve the systems of equations by Gaussian elimination.

5. $\begin{cases} x - 2y = 7 \\ 2x - 3y = 12 \end{cases}$

6. $\begin{cases} x + 2y = 3 \\ 5x - 3y = -11 \end{cases}$

7. $\begin{cases} x - 3y = 6 \\ 3x + 5y = -10 \end{cases}$

8. $\begin{cases} 5x - 3y = 4 \\ 10x - 6y = 2 \end{cases}$

9. $\begin{cases} 2x - 5y = -8 \\ 7x + 3y = -28 \end{cases}$

10. $\begin{cases} 5x - 2y = 18 \\ 3x + 4y = -10 \end{cases}$

11. $\begin{cases} 6x + 2y = 9 \\ 4x - y = -1 \end{cases}$

12. $\begin{cases} 2x - 3y = 7 \\ 3x + 6y = 14 \end{cases}$

13. $\begin{cases} 6x + 2y = 5 \\ 3x - 4y = 0 \end{cases}$

14. $\begin{cases} 4x + 6y = 0 \\ 8x - 2y = 7 \end{cases}$

15. $\begin{cases} x + 3y + z = 8 \\ x + 2y + z = 7 \\ x - 2y + 2z = 6 \end{cases}$

16. $\begin{cases} x + y + z = 4 \\ 3x + 2y - z = 13 \\ 2x - y + 2z = -1 \end{cases}$

17. $\begin{cases} x + y - z = 2 \\ 3x + y - z = -2 \\ 4x - 2y + z = -13 \end{cases}$

18. $\begin{cases} x + 3y - z = 0 \\ 2x + y - z = 4 \\ 3x - y + 2z = 5 \end{cases}$

19. $\begin{cases} x - 2y + 3z = 7 \\ 2x + 3y - z = 0 \\ x + y + z = 1 \end{cases}$

20. $\begin{cases} x + 2y - 2z = 4 \\ 3x + 3y - z = 7 \\ 5x - 2y + 2z = 8 \end{cases}$

21. $\begin{cases} 4x - y + 2z = 6 \\ 2x + 3y - z = 4 \\ 2x - 2y + z = 0 \end{cases}$

22. $\begin{cases} 2x - 3y + 2z = 11 \\ x - 6y - z = 1 \\ 3x - 3y + z = 10 \end{cases}$

23. $\begin{cases} 2x + 3y + 2z = 4 \\ 4x + 6y + 4z = 8 \\ 2x + 3y + 5z = 13 \end{cases}$

24. $\begin{cases} x + y - z = 6 \\ 2x + 2y - 2z = 12 \\ 3x + 3y - 3z = 18 \end{cases}$

25. $\begin{cases} w - 2x + y - z = 2 \\ w + 2x + 2y + z = 0 \\ 2w - 2x + y - z = 3 \\ 2w - 2y + z = 5 \end{cases}$

26. $\begin{cases} w - x + y = -2 \\ x - z = -3 \\ -2x + 2y - z = -7 \\ w + 2y + z = -1 \end{cases}$

In Exercises 27–30, put the augmented matrix in reduced echelon form.

27. $\begin{bmatrix} 2 & 3 & | & 5 \\ 1 & 4 & | & 0 \end{bmatrix}$

28. $\begin{bmatrix} 3 & 2 & | & 0 \\ 2 & 1 & | & 1 \end{bmatrix}$

29. $\begin{bmatrix} 1 & 1 & 2 & | & 1 \\ 2 & 4 & 2 & | & 6 \\ 3 & 1 & 2 & | & 5 \end{bmatrix}$

30. $\begin{bmatrix} 2 & 1 & 3 & | & 13 \\ 1 & 2 & 1 & | & 2 \\ 2 & 3 & 2 & | & 6 \end{bmatrix}$

In Exercises 31–38, solve by the Gauss–Jordan method.

31. $\begin{cases} x - 3y = 5 \\ 3x + 5y = 1 \end{cases}$

32. $\begin{cases} x - 2y = 0 \\ 6x - 4y = 8 \end{cases}$

33. $\begin{cases} x + 2y + z = 8 \\ x + 4y - z = 12 \\ x - 2y + z = -4 \end{cases}$

34. $\begin{cases} x + y - 4z = -3 \\ 2x - y + z = 9 \\ 2x + 2y - 2z = 0 \end{cases}$

35. $\begin{cases} x - 2y = -4 \\ 2y + 2z = 4 \\ x + z = 1 \end{cases}$

36. $\begin{cases} x + z = 2 \\ 2x - 2y = -6 \\ 2y - z = 4 \end{cases}$

37. $\begin{cases} w + x + y + z = 7 \\ 2w + x - y + 2z = 6 \\ 3w + 2x + y - z = 3 \\ w - x - y - 2z = -10 \end{cases}$

38. $\begin{cases} w - 2x + y - z = 3 \\ w + y + z = -1 \\ 2w + x - y = 5 \\ 3w - x + z = 4 \end{cases}$

Mini-Review

39. Given $f(x) = \dfrac{3}{x + 1}$ and $g(x) = \dfrac{x - 1}{x}$, find $f[g(x)]$ and $g[f(x)]$.

40. *Simplify:* $\dfrac{(x^{-2}y^{-2/3})^{1/2}}{9(x^{-1/4}y^{-1/6})^2}$

41. Write an equation of the line passing through the point $(-2, 3)$ that is perpendicular to the line whose equation is $5x + 3y = 7$.

11.3   The Algebra of Matrices

In the previous section we introduced matrices to develop methods for solving linear systems of equations. In this section we expand our discussion of matrices and matrix operations.

Recall that a matrix is a rectangular array of numbers. The numbers in the matrix are called the **elements** or **entries** of the matrix. When we talk about the size of a matrix, we specify the number of rows and columns in the matrix. The following matrix A is called a 3×4 (read "3 by 4") matrix, which means that it has 3 rows and 4 columns:

$$A = \begin{bmatrix} a_{11} & a_{12} & a_{13} & a_{14} \\ a_{21} & a_{22} & a_{23} & a_{24} \\ a_{31} & a_{32} & a_{33} & a_{34} \end{bmatrix}$$

Look carefully at the subscripts in the matrix A. The subscript indicates the position of the entry in the matrix. The first number indicates the row; the second number indicates the column. For example, a_{23} is the entry in the second row and third column. In general, a_{ij} is the entry in row i and column j.

As we have done previously after defining a new type of algebraic object, we define the various arithmetic operations on matrices. We must first ensure that this new object, called a matrix, is well defined—that is, that we can recognize when two matrices are equal.

DEFINITION Two matrices are *equal* if and only if the corresponding entries of the two matrices are equal.

Thus, for example, the following two matrices are equal:

$$\begin{bmatrix} 4 & 7 \\ 1 & -5 \end{bmatrix} \quad \text{and} \quad \begin{bmatrix} \sqrt{16} & 4+3 \\ 2\left(\frac{1}{2}\right) & -4-1 \end{bmatrix}$$

We begin by defining addition and subtraction of matrices.

Addition and Subtraction of Matrices

If A and B are both $m \times n$ matrices, then

The matrix $A + B$ is obtained by adding the corresponding entries of the two matrices.

The matrix $A - B$ is obtained by subtracting each entry in B from the corresponding entry in A.

EXAMPLE 1

Given matrices $A = \begin{bmatrix} 4 & -3 & 2 & 5 \\ -1 & 6 & 0 & 1 \end{bmatrix}$ and $B = \begin{bmatrix} -8 & 2 & 4 & 7 \\ 0 & 5 & 1 & 8 \end{bmatrix}$, find $A + B$ and $A - B$.

Solution We must first ensure that it is possible to add and subtract these two matrices. (If the matrices are not the same size, then addition and subtraction are not defined.) We note that both A and B are 2×4 matrices and so we can add and subtract them according to the rule given above.

$$A + B = \begin{bmatrix} 4 & -3 & 2 & 5 \\ -1 & 6 & 0 & 1 \end{bmatrix} + \begin{bmatrix} -8 & 2 & 4 & 7 \\ 0 & 5 & 1 & 8 \end{bmatrix}$$

$$= \begin{bmatrix} 4+(-8) & -3+2 & 2+4 & 5+7 \\ -1+0 & 6+5 & 0+1 & 1+8 \end{bmatrix} = \begin{bmatrix} -4 & -1 & 6 & 12 \\ -1 & 11 & 1 & 9 \end{bmatrix}$$

$$A - B = \begin{bmatrix} 4 & -3 & 2 & 5 \\ -1 & 6 & 0 & 1 \end{bmatrix} - \begin{bmatrix} -8 & 2 & 4 & 7 \\ 0 & 5 & 1 & 8 \end{bmatrix}$$

$$= \begin{bmatrix} 4-(-8) & -3-2 & 2-4 & 5-7 \\ -1-0 & 6-5 & 0-1 & 1-8 \end{bmatrix} = \begin{bmatrix} 12 & -5 & -2 & -2 \\ -1 & 1 & -1 & -7 \end{bmatrix}$$

Using the definition of matrix addition, we can show that matrix addition is commutative and associative. We can also show that there is an additive identity matrix and an additive inverse matrix. See Exercise 43.

Matrices are very useful for organizing and combining data, as illustrated in the following example.

EXAMPLE 2

A small computer consulting firm keeps track of its costs and revenues for both hardware and software and records these figures for the four quarters of 2005. For hardware, the company recorded quarterly costs of 0.7, 0.9, 0.5, and 1.2 million dollars, respectively; and quarterly revenues of 2.8, 3.2, 2.1, and 3.7 million dollars, respectively. For software, the company recorded quarterly costs of 0.9, 1.2, 0.6, and 1.4 million dollars, respectively; and quarterly revenues of 3.1, 2.9, 2.4, and 3.9 million dollars, respectively. Organize the given data in matrix form, and express the company's total cost and revenue by quarter.

Solution   For both hardware and software, we define a 4×2 matrix in which the 4 rows represent the 4 quarters of the year and the 2 columns represent the cost and revenue. If we let H and S represent hardware and software, respectively, we have

$$H = \begin{matrix} \text{Cost} & \text{Rev.} \\ \begin{bmatrix} 0.7 & 2.8 \\ 0.9 & 3.2 \\ 0.5 & 2.1 \\ 1.2 & 3.7 \end{bmatrix} \end{matrix} \qquad S = \begin{matrix} \text{Cost} & \text{Rev.} \\ \begin{bmatrix} 0.9 & 3.2 \\ 1.2 & 2.9 \\ 0.6 & 2.4 \\ 1.4 & 3.9 \end{bmatrix} \end{matrix}$$

The total quarterly cost and revenue are obtained by adding the two matrices:

$$H + S = \begin{matrix} \text{Cost} & \text{Rev.} \\ \begin{bmatrix} 0.7 + 0.9 & 2.8 + 3.1 \\ 0.9 + 1.2 & 3.2 + 2.9 \\ 0.5 + 0.6 & 2.1 + 2.4 \\ 1.2 + 1.4 & 3.7 + 3.9 \end{bmatrix} \end{matrix} = \begin{matrix} \text{Cost} & \text{Rev.} \\ \begin{bmatrix} 1.6 & 5.9 \\ 2.1 & 6.1 \\ 1.1 & 4.5 \\ 2.6 & 7.6 \end{bmatrix} \end{matrix}$$

Alternatively, we can display the entries of this matrix in a table:

Quarter	Total Cost	Total Revenue
1	1.6	5.9
2	2.1	6.1
3	1.1	4.5
4	2.6	7.6

We next define *scalar multiplication,* which means multiplying a matrix by a constant.

Scalar Multiplication

If A is a matrix and k is a constant, then kA is the matrix obtained by multiplying each entry of A by k.

For example, if $A = \begin{bmatrix} 2 & -3 & 0 \\ -1 & 4 & 2 \end{bmatrix}$, then

$$5A = 5 \begin{bmatrix} 2 & -3 & 0 \\ -1 & 4 & 2 \end{bmatrix} = \begin{bmatrix} 5 \cdot 2 & 5 \cdot (-3) & 5 \cdot 0 \\ 5 \cdot (-1) & 5 \cdot 4 & 5 \cdot 2 \end{bmatrix} = \begin{bmatrix} 10 & -15 & 0 \\ -5 & 20 & 10 \end{bmatrix}$$

EXAMPLE 3

If $A = \begin{bmatrix} -3 & 4 \\ 1 & -2 \end{bmatrix}$ and $B = \begin{bmatrix} 2 & 5 \\ -1 & 1 \end{bmatrix}$, find $2A - 3B$.

Solution   We begin by noting that scalar multiplication does not affect the size of a matrix, so that $2A$ and $3B$ are both 2×2 matrices and can be subtracted.

$$2A - 3B = 2 \begin{bmatrix} -3 & 4 \\ 1 & -2 \end{bmatrix} - 3 \begin{bmatrix} 2 & 5 \\ -1 & 1 \end{bmatrix}$$

$$= \begin{bmatrix} 2(-3) - 3(2) & 2(4) - 3(5) \\ 2(1) - 3(-1) & 2(-2) - 3(1) \end{bmatrix} = \begin{bmatrix} -12 & -7 \\ 5 & -7 \end{bmatrix}$$

Matrix Multiplication

The product of two matrices is not defined as the product of the corresponding entries of the two matrices. Instead, we begin by defining an operation called the *inner product,* defined for matrices for a single row and a single column. A matrix with one row is often called a **row vector,** and a matrix with one column is often called a **column vector.**

The Inner Product

The **inner product** of a $1 \times p$ row vector and a $p \times 1$ column vector is

$$[a_{11} \, a_{12} \, a_{13} \cdots a_{1p}] \cdot \begin{bmatrix} b_{11} \\ b_{21} \\ b_{31} \\ \vdots \\ b_{p1} \end{bmatrix} = a_{11}b_{11} + a_{12}b_{21} + a_{13}b_{31} + \cdots + a_{1p}b_{p1}$$

The inner product of a row vector and a column vector is the sum of the products obtained by multiplying each entry in the row vector by its corresponding entry in the column vector. For example,

$$[4 \, -3 \, -2 \quad 0] \cdot \begin{bmatrix} 6 \\ 3 \\ -4 \\ 5 \end{bmatrix} = 4(6) + (-3)(3) + (-2)(-4) + 0(5) = 23$$

Two important points must be emphasized. First, the inner product of a row vector with a column vector is a **number;** second, the inner product is defined only for a $1 \times p$ row vector multiplied *left* of a $p \times 1$ column vector. Also keep in mind that the "·" we use in an inner product is not the usual dot used to indicate numerical multiplication.

We can now define matrix multiplication.

Matrix Multiplication

If the number of columns in matrix A equals the number of rows in matrix B, then the product AB is the matrix whose entry c_{ij} in row i and column j is obtained as the inner product of row i in matrix A and column j in matrix B:

$$c_{ij} = (\text{row } i \text{ of } A) \cdot (\text{column } j \text{ of } B)$$

Alternatively, we can say that the product of an $m \times n$ matrix A with an $n \times p$ matrix B is the $m \times p$ matrix C, where the entry in the ith row and jth column of C is the inner product of the ith row vector of A with the jth column vector of B.

Keep in mind that this definition of matrix multiplication requires that the number of columns of A, the matrix on the left, equal the number of rows of B, the matrix on the right.

EXAMPLE 4

Find the product AB for $A = \begin{bmatrix} 3 & -1 & 4 \\ 1 & 5 & 7 \end{bmatrix}$ and $B = \begin{bmatrix} 4 & 2 \\ -3 & 0 \\ 1 & 6 \end{bmatrix}$

Solution

We begin by recognizing that multiplying A and B makes sense because the number of columns of A is equal to the number of rows of B. We follow the multiplication rule given in the box, and illustrate how each entry in the product is computed.

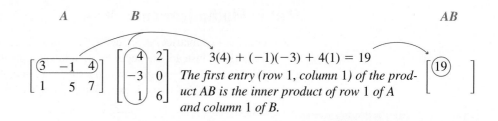

$$3(4) + (-1)(-3) + 4(1) = 19$$
The first entry (row 1, column 1) of the product AB is the inner product of row 1 of A and column 1 of B.

$$3(2) + (-1)(0) + 4(6) = 30$$
The entry for row 1, column 2 of the product AB is the inner product of row 1 of A and column 2 of B.

The entry for row 2, column 1 of the product AB is the inner product of row 2 of A and column 1 of B:
$$1(4) + 5(-3) + 7(1) = -4$$

Finally, the entry in row 2, column 2 of the product AB is the inner product of row 2 of A and column 2 of B:
$$1(2) + 5(0) + 7(6) = 44$$

Matrix multiplication:
$$(A) \quad (B) \quad = \quad (C)$$
$$(m \times n)(n \times p) = (m \times p)$$
same

The product $AB = \begin{bmatrix} 3 & -1 & 4 \\ 1 & 5 & 7 \end{bmatrix} \begin{bmatrix} 4 & 2 \\ -3 & 0 \\ 1 & 6 \end{bmatrix} = \begin{bmatrix} 19 & 30 \\ -4 & 44 \end{bmatrix}$. Notice that we end up

with a 2×2 matrix; the resulting matrix AB should have the same number of rows as A and the same number of columns as B.

On the other hand, we find a different result computing the matrix BA for the same A and B, as demonstrated in the next example.

EXAMPLE 5

If $A = \begin{bmatrix} 3 & -1 & 4 \\ 1 & 5 & 7 \end{bmatrix}$ and $B = \begin{bmatrix} 4 & 2 \\ -3 & 0 \\ 1 & 6 \end{bmatrix}$, find BA.

Solution

$$4(3) + 2(1) = 14$$
The entry for row 1, column 1 of the product BA is the inner product of row 1 of B and column 1 of A.

The entry for row 2, column 3 of the product BA is the inner product of row 2 of B and column 3 of A:
$$-3(4) + 0(7) = -12$$

The remaining entries are as follows:

$$BA = \begin{bmatrix} 4 & 2 \\ -3 & 0 \\ 1 & 6 \end{bmatrix} \begin{bmatrix} 3 & -1 & 4 \\ 1 & 5 & 7 \end{bmatrix} = \begin{bmatrix} 4(3) + 2(1) & 4(-1) + 2(5) & 4(4) + 2(7) \\ -3(3) + 0(1) & -3(-1) + 0(5) & -3(4) + 0(7) \\ 1(3) + 6(1) & 1(-1) + 6(5) & 1(4) + 6(7) \end{bmatrix}$$

$$= \begin{bmatrix} 14 & 6 & 30 \\ -9 & 3 & -12 \\ 9 & 29 & 46 \end{bmatrix}$$

Notice that BA is a 3×3 matrix (the number of rows of B by the number of columns of A).

As the previous two examples illustrate, if A is a 2×3 matrix and B is a 3×2 matrix, both products AB and BA are defined. But the product AB is a 2×2 matrix, whereas the product BA is a 3×3 matrix. This illustrates that, in general, matrix multiplication is not commutative. In fact, it may be that we can multiply two matrices in one order but not in the reverse order. For instance, we can multiply a 4×2 by a 2×3 matrix to produce a 4×3 matrix. However, we cannot multiply a 2×3 matrix by a 4×2 matrix because that number of columns of the first matrix does not match the number of rows in the second.

Although matrix multiplication is not commutative, matrix multiplication does have a number of useful properties, some of which are listed in the following box.

Properties of Matrix Multiplication	Suppose that A, B, and C are matrices for which the following matrix addition and multiplication operations are defined, and k is constant. Then 1. $A(BC) = (AB)C$ 2. $A(B + C) = AB + AC$ 3. $(B + C)A = BA + CA$ 4. $(kA)B = k(AB)$

Matrices can be used where we need to perform operations on tables of numbers, as illustrated in the next example.

EXAMPLE 6   Jane adopts the following monthly investment plan. She will invest \$500 each month, of which she will place \$360 in a growth fund and \$140 in an income fund. The growth fund allocates 50% of her investment in high-tech stocks, 20% in automotive stocks, and 30% in medical stocks. The income fund allocates 65% of her investment in high-tech stocks, 15% in automotive stocks, and 20% in medical stocks. During a particular month, high-tech stocks have a rate of return of 3%, automotive stocks have a rate of return of 1.8%, and medical stocks have a rate of return of 4.2%. Organize these data into a matrix format and use matrix multiplication to determine the earnings from one month's investments.

Solution   We first define a 2×3 matrix A, which represents the amounts invested monthly. The rows correspond to the two funds, and the columns correspond to the three types of stocks she has chosen:

$$A = \begin{bmatrix} 50\% \text{ of } 360 & 20\% \text{ of } 360 & 30\% \text{ of } 360 \\ 65\% \text{ of } 140 & 15\% \text{ of } 140 & 20\% \text{ of } 140 \end{bmatrix} = \begin{bmatrix} 180 & 72 & 108 \\ 91 & 21 & 28 \end{bmatrix} \quad \begin{matrix} \textit{Growth} \\ \textit{Income} \end{matrix}$$

Next we define a 3×1 matrix R, which represents the rates returned on each investment:

$$R = \begin{bmatrix} 0.03 \\ 0.018 \\ 0.042 \end{bmatrix}$$

The interest earned on all the investments can be found by computing the matrix product AR. We leave this to the student and give the result, which we computed by calculator:

$$AR = \begin{bmatrix} 180 & 72 & 108 \\ 91 & 21 & 28 \end{bmatrix} \begin{bmatrix} 0.03 \\ 0.018 \\ 0.042 \end{bmatrix} = \begin{bmatrix} 11.23 \\ 4.28 \end{bmatrix}$$

which means that from one month's investment of $500 she would earn $11.23 from the growth fund and $4.28 from the income fund.

In the next section we will discuss how the matrix operations offer another approach to solving a linear system of equations.

EXERCISES 11.3

In Exercises 1–8, perform the operations if possible.

1. $\begin{bmatrix} 1 & -3 \\ 5 & 2 \end{bmatrix} + \begin{bmatrix} 2 & 5 \\ 3 & 0 \end{bmatrix}$

2. $\begin{bmatrix} 2 & 9 \\ 15 & -23 \end{bmatrix} + \begin{bmatrix} -14 & 6 \\ 0 & -20 \end{bmatrix}$

3. $\begin{bmatrix} 1 & 3 \\ -5 & 0 \\ 2 & -7 \end{bmatrix} + \begin{bmatrix} -1 & -3 \\ 5 & 0 \\ -2 & 7 \end{bmatrix}$

4. $\begin{bmatrix} 2 & -3 & 8 \\ 5 & -1 & 0 \\ -2 & 7 & 18 \end{bmatrix} + \begin{bmatrix} -1 & -3 \\ 5 & 0 \\ -2 & 7 \end{bmatrix}$

5. $\begin{bmatrix} 0 & 3 & 6 \\ 80 & 0 & -19 \\ 2 & 12 & 4 \end{bmatrix} + \begin{bmatrix} 2 & 3 \\ 5 & 0 \end{bmatrix}$

6. $\begin{bmatrix} 2 & -3 & 8 \\ 5 & -1 & 0 \\ -2 & 7 & 18 \end{bmatrix} + \begin{bmatrix} 0 & 0 & 0 \\ 0 & 0 & 0 \\ 0 & 0 & 0 \end{bmatrix}$

7. $3 \begin{bmatrix} 0 & 3 \\ 17 & 0 \end{bmatrix}$

8. $-4 \begin{bmatrix} 2 & -3 & 8 \\ 5 & -1 & 0 \\ -2 & 7 & 18 \end{bmatrix}$

In Exercises 9–16, compute the matrix, if possible, given matrices A, B, C, and D.

$$A = \begin{bmatrix} 2 & -3 \\ -1 & 0 \end{bmatrix} \quad B = \begin{bmatrix} 5 & 1 \\ -2 & 3 \end{bmatrix} \quad C = \begin{bmatrix} 1 & -1 & 3 \\ -2 & 3 & 0 \\ 5 & 8 & 2 \end{bmatrix} \quad D = \begin{bmatrix} 3 & 1 & 0 \\ -2 & 4 & 6 \\ 3 & 8 & -10 \end{bmatrix}$$

9. $-A$

10. $2C$

11. $\dfrac{2}{3}B$

12. $-3D$

13. $A - B$

14. $B - A$

15. $2A - B$

16. $C - 3D$

In Exercises 17–32, compute the matrix products, if possible.

17. $[3 \quad -3 \quad 1] \begin{bmatrix} 5 \\ 1 \\ -2 \end{bmatrix}$

18. $[5 \quad 0 \quad -1] \begin{bmatrix} 3 \\ -9 \\ 0 \end{bmatrix}$

19. $[0 \quad -1 \quad 1 \quad 2] \begin{bmatrix} 1 \\ -1 \\ -2 \\ 2 \end{bmatrix}$

20. $\begin{bmatrix} 1 \\ -1 \\ -2 \\ 2 \end{bmatrix} \begin{bmatrix} 3 & 2 & 5 & 1 \\ 0 & -1 & 1 & 2 \end{bmatrix}$

21. $\begin{bmatrix} 0 & -1 \\ 3 & 2 \end{bmatrix} \begin{bmatrix} -1 & 3 \\ 0 & -2 \end{bmatrix}$

22. $\begin{bmatrix} 1 & 0 \\ 0 & 1 \end{bmatrix} \begin{bmatrix} 3 & 5 \\ -4 & 6 \end{bmatrix}$

23. $\begin{bmatrix} 3 & -1 & 4 \\ 3 & -2 & 5 \end{bmatrix} \begin{bmatrix} -1 & 3 & 5 \\ 0 & -2 & 1 \end{bmatrix}$

24. $\begin{bmatrix} 1 & 0 & -3 \\ 2 & 1 & -4 \end{bmatrix} \begin{bmatrix} 3 & 5 \\ -4 & 6 \\ 3 & -1 \end{bmatrix}$

25. $\begin{bmatrix} 0 & 1 & -5 \\ -3 & 1 & 6 \end{bmatrix} \begin{bmatrix} -1 & 3 \\ 0 & -2 \\ 6 & 0 \end{bmatrix}$

26. $\begin{bmatrix} -1 & 3 \\ 0 & -2 \\ 6 & 0 \end{bmatrix} \begin{bmatrix} 0 & 1 & -5 \\ -3 & 1 & 6 \end{bmatrix}$

27. $\begin{bmatrix} 1 & 0 & -2 \\ 3 & 1 & -1 \\ 2 & 0 & 1 \end{bmatrix} \begin{bmatrix} 1 & -3 & 0 \\ 0 & 2 & -2 \\ -6 & 0 & 5 \end{bmatrix}$

28. $\begin{bmatrix} 1 & -3 & 0 \\ 0 & 2 & -2 \\ -6 & 0 & 5 \end{bmatrix} \begin{bmatrix} 1 & 0 & -2 \\ 3 & 1 & -1 \\ 2 & 0 & 1 \end{bmatrix}$

29. $\begin{bmatrix} 2 & 0 & -2 & 1 \\ -2 & 3 & -1 & 1 \end{bmatrix} \begin{bmatrix} 1 & 3 \\ 1 & 0 \\ -1 & 5 \\ 1 & -3 \end{bmatrix}$

30. $\begin{bmatrix} 1 & 3 \\ 1 & 0 \\ -1 & 5 \\ 1 & -3 \end{bmatrix} \begin{bmatrix} 2 & 0 & -2 & 1 \\ -2 & 3 & -1 & 1 \end{bmatrix}$

31. $\begin{bmatrix} 1 & 0 & 0 \\ 0 & 1 & 0 \\ 0 & 0 & 1 \end{bmatrix} \begin{bmatrix} 3 & 1 & 0 \\ -2 & 4 & 6 \\ 3 & 8 & -10 \end{bmatrix}$

32. $\begin{bmatrix} 3 & 1 & 0 \\ -2 & 4 & 6 \\ 3 & 8 & -10 \end{bmatrix} \begin{bmatrix} 1 & 0 & 0 \\ 0 & 1 & 0 \\ 0 & 0 & 1 \end{bmatrix}$

In Exercises 33–36, given matrices A, B, C, and D, demonstrate whether the given statement is true or false.

$$A = \begin{bmatrix} 2 & 1 & 0 \\ 3 & 1 & -2 \end{bmatrix} \quad B = \begin{bmatrix} 3 & 2 \\ -5 & 0 \end{bmatrix} \quad C = \begin{bmatrix} 2 & -2 \\ 4 & 1 \end{bmatrix} \quad D = \begin{bmatrix} 5 & -1 \\ 0 & -2 \end{bmatrix}$$

33. $CB = BC$

34. $A(B + C) = AB + AC$

35. $(B + C)A = BA + CA$

36. $D(BC) = (DB)C$

37. Show that

$$\left(\frac{1}{2} \begin{bmatrix} 3 & 5 \\ 2 & 4 \end{bmatrix} \right) \begin{bmatrix} 0 & 2 \\ 1 & -1 \end{bmatrix} = \frac{1}{2} \left(\begin{bmatrix} 3 & 5 \\ 2 & 4 \end{bmatrix} \begin{bmatrix} 0 & 2 \\ 1 & -1 \end{bmatrix} \right)$$

38. If A is a 3×4 matrix and B is a 4×3 matrix, what size is the matrix AB? What size is the matrix BA?

39. If A is a 5×4 matrix and B is a 3×5 matrix, what size is the matrix AB? What size is the matrix BA?

40. When does it make sense to discuss A^2 if A is a matrix?

41. Over a 6-month period, the political science and the sociology departments use office supplies as illustrated by the accompanying table.

Supplies	Sociology Department	Political Science Department
Cases of legal pads	7	4
Cases of duplicating paper	6	10
Gross of pencils	2	9

Tom's Office Supply Company charges $230 for a case of legal pads, $65 for a case of duplicating paper, and $18 for a gross of pencils. Kuma's Office Supply Company charges $250 for a case of legal pads, $56 for a case of duplicating paper, and $16 for a gross of pencils. If supplies are ordered for each department, which office supply company is the less expensive for each 6-month order?

42. Three scientific supply companies each manufacture their own brand of pneumatic articulators: brand A, brand B, and brand C. Each brand contains a certain amount of plastic, rubber, and glass tubing, as illustrated by the accompanying table.

Type of Tubing Required	Brand A	Brand B	Brand C
Plastic tubing	12 in.	18 in.	14 in.
Rubber tubing	14 in.	6 in.	12 in.
Glass tubing	18 in.	12 in.	5 in.

Only two companies in the country manufacture tubing suitable for the pneumatic articulators: George's Tubing Company and Raju's Tubing Company. George's Tubing Company charges 1¢ per inch for plastic tubing, 1.5¢ per inch for rubber tubing, and 1.8¢ per inch for glass tubing. Raju's Tubing Company charges 1.2¢ per inch for plastic tubing, 1.4¢ per inch for rubbing tubing, and 1.6¢ per inch for glass tubing. If tubing supplies are ordered for all three brands, which tubing company offers the less expensive tubing for each brand?

? Questions for Thought

43. Given matrices $A = \begin{bmatrix} a & b \\ c & d \end{bmatrix}$, $B = \begin{bmatrix} e & f \\ g & h \end{bmatrix}$, and $C = \begin{bmatrix} r & s \\ t & u \end{bmatrix}$, show that

(a) $A + B = B + A$

(b) $A + (B + C) = (A + B) + C$

(c) $A(BC) = (AB)C$

(d) The 2×2 matrix with all zero entries is the additive identity for addition of 2×2 matrices.

(e) Matrix $-A$ is the additive inverse of matrix A.

(f) The matrix $I_2 = \begin{bmatrix} 1 & 0 \\ 0 & 1 \end{bmatrix}$ is the multiplicative identity for 2×2 matrices.

Mini-Review

44. Sketch the graph of each of the following:

(a) $y = x^2$ **(b)** $y = 2^x$

45. *Solve for x:* $(2x - 1)(x + 3) = (3x - 2)(x + 4)$

46. Express in $a + bi$ form: $\dfrac{5 + 2i}{3 + i}$

47. Write an equation of the circle with a diameter having endpoints $(-3, 4)$ and $(5, -6)$.

 Solving Linear Systems Using Matrix Inverses

Our goal in this section is to develop a method for solving a linear system of equations as though it were a simple linear equation such as $5x = 3$. Let's begin by examining a special matrix product.

EXAMPLE 1	Compute the matrix products AB and BA for

Throughout this section, we will be restricting our attention to square matrices.

$$A = \begin{bmatrix} 2 & -3 & 5 \\ 1 & 0 & 2 \\ -3 & 4 & 1 \end{bmatrix} \quad \text{and} \quad B = \begin{bmatrix} 1 & 0 & 0 \\ 0 & 1 & 0 \\ 0 & 0 & 1 \end{bmatrix}$$

Solution   We note that both A and B are 3×3 matrices and so we can multiply them in either order. We will compute the product AB:

$$AB = \begin{bmatrix} 2 & -3 & 5 \\ 1 & 0 & 2 \\ -3 & 4 & 1 \end{bmatrix} \begin{bmatrix} 1 & 0 & 0 \\ 0 & 1 & 0 \\ 0 & 0 & 1 \end{bmatrix}$$

$$= \begin{bmatrix} 2(1) + (-3)(0) + 5(0) & 2(0) + (-3)(1) + 5(0) & 2(0) + (-3)(0) + 5(1) \\ 1(1) + 0(0) + 2(0) & 1(0) + 0(1) + 2(0) & 1(0) + 0(0) + 2(1) \\ -3(1) + 4(0) + 1(0) & -3(0) + 4(1) + 1(0) & -3(0) + 4(0) + 1(1) \end{bmatrix}$$

$$= \begin{bmatrix} 2 & -3 & 5 \\ 1 & 0 & 2 \\ -3 & 4 & 1 \end{bmatrix}$$

Note that $AB = A$.

Similarly,

$$BA = \begin{bmatrix} 1 & 0 & 0 \\ 0 & 1 & 0 \\ 0 & 0 & 1 \end{bmatrix} \begin{bmatrix} 2 & -3 & 5 \\ 1 & 0 & 2 \\ -3 & 4 & 1 \end{bmatrix}$$

$$= \begin{bmatrix} 1(2) + 0(1) + 0(-3) & 1(-3) + 0(0) + 0(4) & 1(5) + 0(2) + 0(1) \\ 0(2) + 1(1) + 0(-3) & 0(-3) + 1(0) + 0(4) & 0(5) + 1(2) + 0(1) \\ 0(2) + 0(1) + 1(-3) & 0(-3) + 0(0) + 1(4) & 0(5) + 0(2) + 1(1) \end{bmatrix}$$

$$= \begin{bmatrix} 2 & -3 & 5 \\ 1 & 0 & 2 \\ -3 & 4 & 1 \end{bmatrix}$$

Note that BA = A.

Note that both products *AB* and *BA* yield the original matrix *A*.

As Example 1 illustrates, multiplying matrix *A* by matrix *B* leaves matrix *A* unchanged. In fact, it is not difficult to prove that any 3 × 3 matrix when multiplied by matrix *B* (on either side) will remain unchanged. Recall that the number 1 is called the multiplicative identity because $a \cdot 1 = 1 \cdot a = a$ for all real numbers *a*. Similarly, the matrix *B* with entries of 1 down the main diagonal and entries of 0 elsewhere acts as the multiplicative identity for 3 × 3 matrices.

In general, the square $n \times n$ matrix with 1's in the main diagonal and 0's elsewhere acts as the multiplicative identity for all $n \times n$ matrices. We call this matrix the $n \times n$ *identity matrix* and designate it I_n.

Next, let's compute another special matrix product.

EXAMPLE 2

Compute the matrix products *AB* and *BA* for $A = \begin{bmatrix} 1 & 2 \\ 2 & 0 \end{bmatrix}$ and $B = \begin{bmatrix} 1 & \frac{1}{2} \\ \frac{1}{2} & -\frac{1}{4} \end{bmatrix}$

Solution

$$AB = \begin{bmatrix} 1 & 2 \\ 2 & 0 \end{bmatrix} \begin{bmatrix} 0 & \frac{1}{2} \\ \frac{1}{2} & -\frac{1}{4} \end{bmatrix} = \begin{bmatrix} 1(0) + 2(\frac{1}{2}) & 1(\frac{1}{2}) + 2(-\frac{1}{4}) \\ 2(0) + 0(\frac{1}{2}) & 2(\frac{1}{2}) + 0(-\frac{1}{4}) \end{bmatrix} = \begin{bmatrix} 1 & 0 \\ 0 & 1 \end{bmatrix}$$

Note that this is I_2, the 2 × 2 identity matrix.

It is left to the student to verify that *BA* gives us the same result.

It is also true that a is the inverse of b, and that matrix A is the inverse of matrix B.

Throughout this section, whenever we refer to the inverse of matrix A, we mean its multiplicative *inverse,* A^{-1}.

When we multiply two real numbers *a* and *b* such that $ab = 1$, then we have $b = \frac{1}{a}$, and *b* is called the (multiplicative) inverse of *a*. In the previous example, we have two matrices that multiply to the identity matrix. In such a case, we say that matrix *B* is the inverse of matrix *A*, which we denote as A^{-1}, read as "*A* inverse."

It would be nice if we would find that A^{-1} is simply some sort of reciprocal of the elements of *A*; unfortunately, because of the complex way matrix multiplication is defined, finding the inverse of a matrix takes a bit more work. A discussion of the derivation of matrix inverses is beyond the scope of this textbook. We will, however, demonstrate how to find inverses of square matrices in the next few examples.

EXAMPLE 3

Find the inverse of the matrix $A = \begin{bmatrix} 1 & 1 & 2 \\ 1 & 2 & 4 \\ 3 & 4 & 10 \end{bmatrix}$.

Solution

1. We are looking for inverse, A^{-1}, of this 3 × 3 matrix. We first form a new matrix in the following way:

   We put the 3 × 3 matrix *A* together with the identity matrix of order 3, I_3, where the elements of I_3 are right of the elements of *A* to form a **grafted** 3 × 6 matrix [*A* | *I*]:

$$\begin{bmatrix} 1 & 1 & 2 & | & 1 & 0 & 0 \\ 1 & 2 & 4 & | & 0 & 1 & 0 \\ 3 & 4 & 10 & | & 0 & 0 & 1 \end{bmatrix}$$

This is known as *grafting* I_3 *onto a 3 × 3 matrix.*

2. Our goal is to try to transform the left square matrix (the A portion of the grafted matrix) into the identity matrix I_3. We use the elementary row operations discussed in Section 11.2 as we did when we transformed a matrix into reduced echelon form. *All transformations performed on the A portion are applied to the entire 3×6 grafted matrix.*

   We already have 1 in the upper left corner of the main diagonal of A, so our first step is to sweep out (get zeros in the remaining entries of) the first column.

$$\left[\begin{array}{ccc|ccc} 1 & 1 & 2 & 1 & 0 & 0 \\ 1 & 2 & 4 & 0 & 1 & 0 \\ 3 & 4 & 10 & 0 & 0 & 1 \end{array}\right] \quad \begin{array}{c} -R_1 + R_2 \rightarrow R_2 \\ -3R_1 + R_3 \rightarrow R_3 \end{array} \quad \left[\begin{array}{ccc|ccc} 1 & 1 & 2 & 1 & 0 & 0 \\ 0 & 1 & 2 & -1 & 1 & 0 \\ 0 & 1 & 4 & -3 & 0 & 1 \end{array}\right]$$

We already have 1 in the second row, second column; now we sweep out the second column.

$$\left[\begin{array}{ccc|ccc} 1 & 1 & 2 & 1 & 0 & 0 \\ 0 & 1 & 2 & -1 & 1 & 0 \\ 0 & 1 & 4 & -3 & 0 & 1 \end{array}\right] \quad \begin{array}{c} -R_2 + R_1 \rightarrow R_1 \\ -R_2 + R_3 \rightarrow R_3 \end{array} \quad \left[\begin{array}{ccc|ccc} 1 & 0 & 0 & 2 & -1 & 0 \\ 0 & 1 & 2 & -1 & 1 & 0 \\ 0 & 0 & 2 & -2 & -1 & 1 \end{array}\right]$$

Next, get 1 in row 3, column 3.

$$\left[\begin{array}{ccc|ccc} 1 & 0 & 0 & 2 & -1 & 0 \\ 0 & 1 & 2 & -1 & 1 & 0 \\ 0 & 0 & 2 & -2 & -1 & 1 \end{array}\right] \quad \tfrac{1}{2}R_3 \rightarrow R_3 \quad \left[\begin{array}{ccc|ccc} 1 & 0 & 0 & 2 & -1 & 0 \\ 0 & 1 & 2 & -1 & 1 & 0 \\ 0 & 0 & 1 & -1 & -\frac{1}{2} & \frac{1}{2} \end{array}\right]$$

Finally, sweep out the third column.

$$\left[\begin{array}{ccc|ccc} 1 & 0 & 0 & 2 & -1 & 0 \\ 0 & 1 & 2 & -1 & 1 & 0 \\ 0 & 0 & 1 & -1 & -\frac{1}{2} & \frac{1}{2} \end{array}\right] \quad -2R_3 + R_2 \rightarrow R_2 \quad \left[\begin{array}{ccc|ccc} 1 & 0 & 0 & 2 & -1 & 0 \\ 0 & 1 & 0 & 1 & 2 & -1 \\ 0 & 0 & 1 & -1 & -\frac{1}{2} & \frac{1}{2} \end{array}\right]$$

3. The right 3×3 matrix in the grafted matrix is A^{-1}.

$$A^{-1} = \left[\begin{array}{ccc} 2 & -1 & 0 \\ 1 & 2 & -1 \\ -1 & -\frac{1}{2} & \frac{1}{2} \end{array}\right]$$

The student should verify that $AA^{-1} = I_3$.

Procedure for Finding A^{-1}

In general, if A is an $n \times n$ matrix, to find A^{-1} we

1. Graft I_n onto A to get $[A \mid I_n]$.

2. Use the elementary row operations on the entire matrix to transform the A portion of the grafted matrix into I_n.

3. When the left portion of the grafted matrix is I_n, the right $n \times n$ matrix is A^{-1}, that is; if $[A \mid I_n]$ is row-equivalent to $[I_n \mid B]$, then B is the inverse of A.

By now we are very comfortable with solving an equation of the form $ax = b$. We usually describe the process of solving this equation as "divide both sides of the equation by a." However, we can also view the process as "multiply both sides of the equation by $\frac{1}{a}$," as illustrated by the following:

$$ax = b \qquad \textit{Multiply both sides by } \frac{1}{a}.$$

$$\frac{1}{a}(ax) = \frac{1}{a}(b)$$

$$\left(\frac{1}{a} \cdot a\right)x = \frac{1}{a}(b)$$

$$x = \frac{1}{a}(b)$$

Let's examine a typical system of linear equations from a matrix viewpoint and see how the various ideas we have discussed thus far in this section can be applied.

Consider a typical 2×2 system of linear equations:

$$\begin{cases} 10x - 5y = 3 \\ 8x - 3y = 2 \end{cases} \quad (*)$$

If we let

$$A = \begin{bmatrix} 10 & -5 \\ 8 & -3 \end{bmatrix} \quad X = \begin{bmatrix} x \\ y \end{bmatrix} \quad \text{and} \quad K = \begin{bmatrix} 3 \\ 2 \end{bmatrix}$$

then the matrix equation

$$A \quad \cdot X = K \qquad \textit{becomes}$$

$$\begin{bmatrix} 10 & -5 \\ 8 & -3 \end{bmatrix}\begin{bmatrix} x \\ y \end{bmatrix} = \begin{bmatrix} 3 \\ 2 \end{bmatrix} \qquad \textit{Do the matrix multiplication.}$$

$$\begin{bmatrix} 10x - 5y \\ 8x - 3y \end{bmatrix} = \begin{bmatrix} 3 \\ 2 \end{bmatrix} \qquad \textit{The equality of these two matrices means that } 10x - 5y = 3 \textit{ and } 8x - 3y = 2, \textit{ which is the same as the system in } (*).$$

Thus, solving the system in $(*)$ is equivalent to finding a matrix X that satisfies the matrix equation $AX = K$. We call A the ***coefficient matrix,*** X the ***variable matrix,*** and K the ***constant matrix.*** Using matrix properties, we can solve the matrix equation $AX = K$ as follows:

Note how similar this process is to solving ax = b.

$$AX = K \qquad \textit{Multiply both sides of the equation on the left by } A^{-1}.$$

$$A^{-1}(AX) = A^{-1}K \qquad \textit{Use the associative property of matrix multiplication.}$$

$$(A^{-1}A)X = A^{-1}K \qquad \textit{Since } A^{-1}A \textit{ gives the identity matrix, we get where } I_2 \textit{ is the } 2 \times 2 \textit{ identity matrix}$$

$$I_2X = A^{-1}K \qquad \textit{Since } I_2 \textit{ is the identity matrix, } I_2X = X, \textit{ and so we get}$$

$$X = A^{-1}K$$

In other words, to solve a linear system of equations, we take the constant matrix *left-multiplied* by the inverse of the coefficient matrix to get the solution matrix. In this way we approach solving a system of equations as though we were solving a simple linear equation. However, it is important to keep in mind that since matrix multiplication is not commutative, we must multiply A^{-1} to the *left* of K.

EXAMPLE 4

Using matrix inverses, find the solution to the following system of equations:

$$\begin{cases} 10x - 5y = 3 \\ 8x - 3y = 2 \end{cases}$$

Solution   We identify matrices A, X, and K:

$$A = \begin{bmatrix} 10 & -5 \\ 8 & -3 \end{bmatrix} \quad X = \begin{bmatrix} x \\ y \end{bmatrix} \quad K = \begin{bmatrix} 3 \\ 2 \end{bmatrix}$$

Hence, $AX = K$, and our solution is found by $X = A^{-1}K$, which means that we must find the inverse of the coefficient matrix and multiply it by the constant matrix to find the solution. (We said that we can view it as a simple linear equation, but we did not say the computations weren't substantial.) First we find A^{-1}:

1. Graft I_2 onto A:

$$A = \begin{bmatrix} 10 & -5 & | & 1 & 0 \\ 8 & -3 & | & 0 & 1 \end{bmatrix}$$

2. Transform the left 2×2 portion into I_2:

$$\begin{bmatrix} 10 & -5 & | & 1 & 0 \\ 8 & -3 & | & 0 & 1 \end{bmatrix} \quad \frac{1}{10}R_1 \to R_1 \quad \begin{bmatrix} 1 & -\frac{1}{2} & | & \frac{1}{10} & 0 \\ 8 & -3 & | & 0 & 1 \end{bmatrix}$$

$$\begin{bmatrix} 1 & -\frac{1}{2} & | & \frac{1}{10} & 0 \\ 8 & -3 & | & 0 & 1 \end{bmatrix} \quad -8R_1 + R_2 \to R_2 \quad \begin{bmatrix} 1 & -\frac{1}{2} & | & \frac{1}{10} & 0 \\ 0 & 1 & | & -\frac{4}{5} & 1 \end{bmatrix}$$

$$\begin{bmatrix} 1 & -\frac{1}{2} & | & \frac{1}{10} & 0 \\ 0 & 1 & | & -\frac{4}{5} & 1 \end{bmatrix} \quad \frac{1}{2}R_2 + R_1 \to R_1 \quad \begin{bmatrix} 1 & 0 & | & -\frac{3}{10} & \frac{1}{2} \\ 0 & 1 & | & -\frac{4}{5} & 1 \end{bmatrix}$$

3. Since the left portion is now I_2, the right portion is A^{-1}. Hence,

$$A^{-1} = \begin{bmatrix} -\frac{3}{10} & \frac{1}{2} \\ -\frac{4}{5} & 1 \end{bmatrix}$$

Some calculators have the capability of computing the inverse of a matrix. If you have access to this type of calculator or have computer software that performs this function, then you may find this method to be the easiest of the methods discussed in this chapter.

Now that we have A^{-1}, we can solve for X:

$$X = A^{-1}K$$

$$= \begin{bmatrix} -\frac{3}{10} & \frac{1}{2} \\ -\frac{4}{5} & 1 \end{bmatrix} \begin{bmatrix} 3 \\ 2 \end{bmatrix} \quad \textit{Perform matrix multiplication to get}$$

$$= \begin{bmatrix} -(\frac{3}{10})(3) + (\frac{1}{2})(2) \\ (-\frac{4}{5})(3) + (1)(2) \end{bmatrix} = \begin{bmatrix} \frac{1}{10} \\ -\frac{2}{5} \end{bmatrix}$$

Hence $X = \begin{bmatrix} x \\ y \end{bmatrix} = \begin{bmatrix} \frac{1}{10} \\ -\frac{2}{5} \end{bmatrix}$, which means

$$\boxed{x = \frac{1}{10}, \quad y = -\frac{2}{5}}$$

If we were to use Gaussian elimination with back substitution on the last example, we would get to the solution much more quickly than by using matrix inverses. So a natural question to ask is, Why bother with this method? We answer this question by considering the following example.

EXAMPLE 5

Danny invests $10,000, part at 6% per year in a conservative investment and part at 12% per year in a riskier investment. How much should he invest at each rate if he wants a return on his total investment of **(a)** 8%? **(b)** 9%? **(c)** 10.5%?

Solution   If we let x = the amount invested at 6% and y = the amount invested at 12%, the first equation for all three parts of this problem is

$$x + y = 10,000 \qquad \textit{The total amount invested}$$

The second equation involves the interest received on each investment. In general, if Danny invests x dollars at 6% and y dollars at 12%, then his total interest for the year is $0.06x + 0.12y$. The three parts of this example differ on what this amount should be.

For part **(a)**, interest is expressed in the equation

$$0.06x + 0.12y = 0.08(10,000) = 800$$

For part **(b),** interest is expressed in the equation

$$0.06x + 0.12y = 0.09(10,000) = 900$$

For part **(c),** interest is expressed in the equation

$$0.06x + 0.12y = 0.105(10,000) = 1,050$$

Hence, we are being asked to solve three systems of equations:

(a) $\begin{cases} x + \quad y = 10,000 \\ 0.06x + 0.12y = 800 \end{cases}$

(b) $\begin{cases} x + \quad y = 10,000 \\ 0.06x + 0.12y = 900 \end{cases}$

(c) $\begin{cases} x + \quad y = 10,000 \\ 0.06x + 0.12y = 1,050 \end{cases}$

Notice that all three parts of this example have the same matrix of coefficients. We could solve each system separately using Gaussian elimination; however, once we have the inverse of the coefficient matrix, we can solve all three quickly.

The coefficient matrix of all three systems above is the same:

$$A = \begin{bmatrix} 1 & 1 \\ 0.06 & 0.12 \end{bmatrix}$$

The student should verify that the inverse of this matrix is $A^{-1} = \begin{bmatrix} 2 & -\frac{50}{3} \\ -1 & \frac{50}{3} \end{bmatrix}$.

Given that we have the inverse of the coefficient matrix, we can find the solution to all three systems easily:

(a) $\begin{cases} x + \quad y = 10,000 \\ 0.06x + 0.12y = 800 \end{cases}$

The constant matrix for this problem is $\begin{bmatrix} 10,000 \\ 800 \end{bmatrix}$. We have

$$X = A^{-1}K$$

$$\begin{bmatrix} x \\ y \end{bmatrix} = \begin{bmatrix} 2 & -\frac{50}{3} \\ -1 & \frac{50}{3} \end{bmatrix} \begin{bmatrix} 10,000 \\ 800 \end{bmatrix} = \begin{bmatrix} 6,666.67 \\ 3,333.33 \end{bmatrix}$$

Hence, $x = 6,666.67$ and $y = 3,333.33$; therefore, Danny should invest $\boxed{\$6,666.67 \text{ at } 6\%}$ and $\boxed{\$3,333.33 \text{ at } 12\%}$ in order to gain a return of 8% on his total investment.

(b) $\begin{cases} x + \quad y = 10,000 \\ 0.06x + 0.12y = 900 \end{cases}$

The constant matrix for this problem is $\begin{bmatrix} 10,000 \\ 900 \end{bmatrix}$. We have

$$X = A^{-1}K$$

$$\begin{bmatrix} x \\ y \end{bmatrix} = \begin{bmatrix} 2 & -\frac{50}{3} \\ -1 & \frac{50}{3} \end{bmatrix} \begin{bmatrix} 10,000 \\ 900 \end{bmatrix} = \begin{bmatrix} 5,000 \\ 5,000 \end{bmatrix}$$

Hence, $x = 5,000$ and $y = 5,000$; therefore, Danny should invest $\boxed{\$5,000 \text{ at } 6\%}$ and $\boxed{\$5,000 \text{ at } 12\%}$ in order to gain a return of 9% on his total investment.

(c) $\begin{cases} x + y = 10,000 \\ 0.06x + 0.12y = 1,050 \end{cases}$

The constant matrix for this problem is $\begin{bmatrix} 10,000 \\ 1,050 \end{bmatrix}$. We have

$$X = A^{-1}K$$

$$\begin{bmatrix} x \\ y \end{bmatrix} = \begin{bmatrix} 2 & -\frac{50}{3} \\ -1 & \frac{50}{3} \end{bmatrix} \begin{bmatrix} 10,000 \\ 1,050 \end{bmatrix} = \begin{bmatrix} 2,500 \\ 7,500 \end{bmatrix}$$

Hence, $x = 2,500$ and $y = 7,500$; therefore, Danny should invest $\boxed{\$2,500 \text{ at } 6\%}$ and $\boxed{\$7,500 \text{ at } 12\%}$ in order to gain a return of 10.5% on his total investment.

The method we have used in this section requires that the coefficient matrix A have an inverse, which is not always the case. Systems for which there is no solution will correspond to those for which the coefficient matrix has no inverse. See Exercise 33.

In the next section we will discuss additional methods for solving a system of equations.

EXERCISES 11.4

In Exercises 1–6, find the inverse of the matrix if it exists.

1. $\begin{bmatrix} 1 & 2 \\ 2 & 0 \end{bmatrix}$

2. $\begin{bmatrix} 3 & -2 \\ 1 & 1 \end{bmatrix}$

3. $\begin{bmatrix} 2 & -1 \\ 0 & 1 \end{bmatrix}$

4. $\begin{bmatrix} 3 & 0 \\ -5 & 1 \end{bmatrix}$

5. $\begin{bmatrix} 1 & 1 & 2 \\ 2 & 0 & 1 \\ 3 & -1 & 1 \end{bmatrix}$

6. $\begin{bmatrix} 1 & -1 & 0 \\ -2 & 2 & 1 \\ 4 & -1 & 0 \end{bmatrix}$

In Exercises 7–24, solve the given system using matrix inverses.

7. $\begin{cases} x + 2y = 3 \\ x - y = 6 \end{cases}$

8. $\begin{cases} x + y = 5 \\ 3x - y = 8 \end{cases}$

9. $\begin{cases} x - 2y = 4 \\ 2x + y = 13 \end{cases}$

10. $\begin{cases} 3x + 2y = 3 \\ 3x + y = 0 \end{cases}$

11. $\begin{cases} 5x - 2y = 11 \\ x - 5y = -7 \end{cases}$

12. $\begin{cases} x = 3y - 1 \\ 2x - 5y = -3 \end{cases}$

13. $\begin{cases} y = 3x - 15 \\ 8x - 2y = 10 \end{cases}$

14. $\begin{cases} 2x + 7y = -1 \\ 3x + 21y = -3 \end{cases}$

15. $\begin{cases} \dfrac{3}{2}x + \dfrac{1}{3}y = 1 \\ \phantom{\dfrac{3}{2}}x + \dfrac{1}{12}y = \dfrac{1}{4} \end{cases}$

16. $\begin{cases} 5x + 2y = -8 \\ 3x + y = -4 \end{cases}$

17. $\begin{cases} x + y + z = 6 \\ 3x + y - z = 6 \\ 2x + y - z = 4 \end{cases}$

18. $\begin{cases} x + y - z = 1 \\ 2x + 2y + 2z = 0 \\ x - y + z = 3 \end{cases}$

19. $\begin{cases} x - 3y + z = 0 \\ 3x - y + z = 2 \\ x + 2y - z = 2 \end{cases}$

20. $\begin{cases} x + y + 2z = 7 \\ x - 2y - z = 1 \\ x + y + z = 4 \end{cases}$

21. $\begin{cases} x + y + z = 6 \\ 3x + y - z = 4 \\ 2x + 2y + z = 4 \end{cases}$

22. $\begin{cases} x + y - 2z = 3 \\ 2x + 2y - 2z = 6 \\ x - y + z = 5 \end{cases}$

23. $\begin{cases} x + 0.2y + z = 0.02 \\ 0.3x - y + 0.2z = 0 \\ x + y + z = 0.1 \end{cases}$

24. $\begin{cases} x + y = 2x + z \\ x - 2y = z - 1 \\ 4x - 2y = 6z - 5 \end{cases}$

25. Jenna has $20,000 to split between a low-risk certificate of deposit yielding 6% per year and a high-risk oil stock yielding 18% per year. How much should she invest at each rate if she wants a return on her total investment of **(a)** 8%? **(b)** 10%? **(c)** 12%?

26. Sam has $10,000 to split between a low-risk certificate of deposit yielding 5% per year and a high-risk Internet stock with an average yield of 25% per year. How much should she invest in each if she wants a return on her total investment of **(a)** 8%? **(b)** 12%? **(c)** 18%?

27. Josh has $12,000 to split among three investments: a low-risk certificate of deposit yielding 4% per year, a medium-risk bond that pays 6% per year, and a high-risk utility stock with an average yield of 20% per year. He would like to invest twice as much money in the low-risk CD as in the high-risk stock. How much should he invest in each if he wants a return on his total investment of **(a)** 8%? **(b)** 9%? **(c)** 12%?

28. Rita has $20,000 to split among three investments: a low-risk certificate of deposit yielding 5% per year, a medium-risk bond that pays 10% per year, and a high-risk utility stock with an average yield of 25% per year. She would like to invest as much in the high-risk certificate as in the total of the remaining two investments. How much should she invest in each if she wants a return on her total investment of **(a)** 8%? **(b)** 15%? **(c)** 16%?

29. A garment factory manufactures shirts, blouses, and skirts. Each type of garment requires attention from each of three departments, as indicated in the accompanying table. The design, cutting, and shipping departments have available a maximum of 400, 980, and 160 labor-hours per week, respectively. How many of each type of garment should be manufactured each week for the factory to work at full capacity?

	Shirts	**Blouses**	**Skirts**
Design	0.1	0.2	0.2
Cutting	0.3	0.4	0.5
Shipping	0.08	0.06	0.05

30. A dietitian is trying to create a food supplement that supplies exactly 57.5 mg of calcium, 37.5 g of fiber, and 50 units of vitamin E. The supplement will be a

combination of three foods—food A, food B, and food C. The amounts of these nutrients contained in 1 oz of each of the foods are given in the following table.

	mg of Calcium	g of Fiber	Units of Vitamin E
Food A	2.5	5	0
Food B	5	2.5	5
Food C	5	5	2.5

How many ounces of each of these foods must be combined to create a supplement that satisfies the dietitian's nutritional requirement?

31. A nutritionist wants to create a food supplement out of three substances: *X, Y,* and *Z.* The iron content and cost of the substances *X, Y,* and *Z* are entered in the accompanying table.

Substance	Iron Content per gram	Cost per gram
X	0.5	10¢
Y	0.2	6¢
Z	0.4	8¢

How many grams of each substance should be used to make a food supplement if she wants 10 g of the food supplement to supply 3.5 g of iron and cost **(a)** 77¢? **(b)** 78¢? **(c)** 79¢?

32. Jean wants to supplement her diet with 300 mg of vitamin C, 200 mg of vitamin E, and 400 mg of niacin. The accompanying table indicates the amount of each of these nutrients contained in 1 oz of each of three foods. Determine how many ounces of each food substance should be mixed to produce the required supplement.

	Vitamin C	Vitamin E	Niacin
Food A	15	20	40
Food B	30	15	30
Food C	5	5	0

? **Question for Thought**

33. To solve a system of equations using the matrix inverse process discussed in this section, it is necessary that the coefficient matrix of the system have an inverse. In the case of an inconsistent system, which has no solutions, the coefficient matrix will have no inverse. A matrix that has no inverse is called a *singular matrix.*

 Solve the following system by an algebraic method to verify that it has no solutions. Try to find the inverse of the coefficient matrix.

$$\begin{cases} 3x - 2y = 5 \\ 6x - 4y = 9 \end{cases}$$

Mini-Review

34. *Solve for x:* $\sqrt{3x + 1} + 3 = x$

35. If *y* varies directly as the square of *x* and inversely as the square root of *z*, and $y = 10$ when $x = 3$ and $z = 4$, find *y* when $x = 2$ and $z = 9$.

36. *Solve for x:* $\dfrac{x}{x + 1} + 1 = \dfrac{x}{x + 2}$

37. *Solve for x:* $\dfrac{x}{x + 1} = \dfrac{x}{x + 2}$

11.5   Determinants and Cramer's Rule

In the first two sections we solved many systems of equations. It is quite likely that you noticed the repetitive nature of the procedure. We kept repeating the same basic process—it was just the numbers that changed each time.

This type of situation leads us quite naturally to ask whether we can apply the procedure to the general case and produce a "formula" for the solutions of linear systems. You may recall that this is exactly the same type of question we asked in Chapter 8 regarding the method of completing the square, where the answer turned out to be the quadratic formula. The answer to the question here, regarding systems of linear equations, is called ***Cramer's rule.***

EXAMPLE 1

Solve the following system of equations for *x* and *y:*

$$\begin{cases} a_1x + b_1y = k_1 \\ a_2x + b_2y = k_2 \end{cases}$$

Solution

This is the "general" 2×2 linear system. We solve the system by the elimination method. We first eliminate *y*.

$a_1x + b_1y = k_1$	*Multiply by b_2* →	$a_1b_2x + b_1b_2y = k_1b_2$
$a_2x + b_2y = k_2$	*Multiply by $-b_1$* →	$\underline{-a_2b_1x - b_1b_2y = -k_2b_1}$ *Add.*

$$a_1b_2x - a_2b_1x = k_1b_2 - k_2b_1 \qquad \textit{Solve for x.}$$
$$(a_1b_2 - a_2b_1)x = k_1b_2 - k_2b_1$$
$$x = \frac{k_1b_2 - k_2b_1}{a_1b_2 - a_2b_1}$$

This solution is valid provided $a_1b_2 - a_2b_1 \neq 0$.

Similarly, we can solve for *y* by eliminating *x:*

$a_1x + b_1y = k_1$	*Multiply by $-a_2$* →	$-a_1a_2x - a_2b_1y = -a_2k_1$
$a_2x + b_2y = k_2$	*Multiply by a_1* →	$\underline{a_1a_2x + a_1b_2y = a_1k_2}$ *Add.*

$$a_1b_2y - a_2b_1y = a_1k_2 - a_2k_1 \qquad \textit{Solve for y.}$$
$$(a_1b_2 - a_2b_1)y = a_1k_2 - a_2k_1$$
$$y = \frac{a_1k_2 - a_2k_1}{a_1b_2 - a_2b_1}$$

Again, this solution for *y* is valid provided $a_1b_2 - a_2b_1 \neq 0$.

Thus, the solution to this general 2×2 system is

$$\boxed{x = \frac{k_1b_2 - k_2b_1}{a_1b_2 - a_2b_1}, \quad y = \frac{a_1k_2 - a_2k_1}{a_1b_2 - a_2b_1}}$$

While the solution obtained in Example 1 is "general," it can be very confusing to remember. To make this general solution easier to remember, we interrupt our discussion to introduce the idea of a *determinant*. We will return to the general solution in a moment.

DEFINITION

The symbol $\begin{vmatrix} a & c \\ b & d \end{vmatrix}$ is called a **2×2 determinant.** Its value is defined to be

$$\begin{vmatrix} a & c \\ b & d \end{vmatrix} = ad - bc,$$ which we may indicate as follows:

$$\begin{vmatrix} a & c \\ b & d \end{vmatrix} = ad - bc$$

EXAMPLE 2

Evaluate each of the following:

(a) $\begin{vmatrix} 7 & 3 \\ 5 & 4 \end{vmatrix}$ **(b)** $\begin{vmatrix} 8 & -2 \\ 3 & -4 \end{vmatrix}$

Solution

Using this definition is sometimes called *expanding the determinant.*

(a) $\begin{vmatrix} 7 & 3 \\ 5 & 4 \end{vmatrix} = 7(4) - 5(3) = 28 - 15 = \boxed{13}$

(b) $\begin{vmatrix} 8 & -2 \\ 3 & -4 \end{vmatrix} = 8(-4) - (3)(-2) = -32 + 6 = \boxed{-26}$

EXAMPLE 3

Solve for x: $\begin{vmatrix} x^2 & 3 \\ x & 1 \end{vmatrix} = 10$

Solution

$\begin{vmatrix} x^2 & 3 \\ x & 1 \end{vmatrix} = 10$ *Expand the determinant.*

$x^2 - 3x = 10$ *This is a quadratic equation. Solve by factoring.*

$x^2 - 3x - 10 = 0$

$(x - 5)(x + 2) = 0$

$x - 5 = 0 \quad \text{or} \quad x + 2 = 0$

$\boxed{x = 5 \quad \text{or} \quad x = -2}$

Check:

$\begin{vmatrix} 5^2 & 3 \\ 5 & 1 \end{vmatrix} = 25 \cdot 1 - 5 \cdot 3 \overset{\checkmark}{=} 10 \qquad \begin{vmatrix} (-2)^2 & 3 \\ -2 & 1 \end{vmatrix} = 4 \cdot 1 - (-2) \cdot (3) \overset{\checkmark}{=} 10$

We now return to the result of Example 1. We found the solution to the general system

$$\begin{cases} a_1x + b_1y = k_1 \\ a_2x + b_2y = k_2 \end{cases}$$

to be $x = \dfrac{k_1b_2 - k_2b_1}{a_1b_2 - a_2b_1}$ and $y = \dfrac{a_1k_2 - a_2k_1}{a_1b_2 - a_2b_1}$

which we can write using determinant notation as follows:

$$x = \frac{\begin{vmatrix} k_1 & b_1 \\ k_2 & b_2 \end{vmatrix}}{\begin{vmatrix} a_1 & b_1 \\ a_2 & b_2 \end{vmatrix}} \quad \text{and} \quad y = \frac{\begin{vmatrix} a_1 & k_1 \\ a_2 & k_2 \end{vmatrix}}{\begin{vmatrix} a_1 & b_1 \\ a_2 & b_2 \end{vmatrix}}$$

It is useful to note that the denominator in each case is the determinant of the coefficients of x and y in our general system. This determinant is usually denoted by D:

$$D = \begin{vmatrix} a_1 & b_1 \\ a_2 & b_2 \end{vmatrix}$$

The numerator of the solution for each variable is obtained by taking D and replacing the column of coefficients of that variable by the column of constant terms. That is, the numerators are denoted as

$$D_x = \begin{vmatrix} k_1 & b_1 \\ k_2 & b_2 \end{vmatrix} \quad \text{and} \quad D_y = \begin{vmatrix} a_1 & k_1 \\ a_2 & k_2 \end{vmatrix}$$

We can summarize our discussion thus far in the following theorem.

THEOREM   **Cramer's Rule (part 1)**

The solution to the system

$$\begin{cases} a_1x + b_1y = k_1 \\ a_2x + b_2y = k_2 \end{cases}$$

is given by

$$x = \frac{D_x}{D} \quad \text{and} \quad y = \frac{D_y}{D}$$

where

$$D = \begin{vmatrix} a_1 & b_1 \\ a_2 & b_2 \end{vmatrix} \neq 0, \qquad D_x = \begin{vmatrix} k_1 & b_1 \\ k_2 & b_2 \end{vmatrix}, \qquad D_y = \begin{vmatrix} a_1 & k_1 \\ a_2 & k_2 \end{vmatrix}$$

If $D = 0$, there is no unique solution: The system is either inconsistent and has no solutions (if $D_x \neq 0$ or $D_y \neq 0$), or dependent and has infinitely many solutions (if $D_x = 0$ and $D_y = 0$).

The next example illustrates the use of Cramer's rule.

EXAMPLE 4   Solve the following system of equations:

$$\begin{cases} 5x - 3y = 7 \\ 7x - 8y = 4 \end{cases}$$

Solution   We first evaluate D, D_x, and D_y:

$$D = \begin{vmatrix} 5 & -3 \\ 7 & -8 \end{vmatrix} = 5(-8) - 7(-3) = -40 + 21 = -19$$

$$D_x = \begin{vmatrix} 7 & -3 \\ 4 & -8 \end{vmatrix} = 7(-8) - 4(-3) = -56 + 12 = -44$$

$$D_y = \begin{vmatrix} 5 & 7 \\ 7 & 4 \end{vmatrix} = 5(4) - 7(7) = 20 - 49 = -29$$

According to Cramer's rule, our solution is

$$x = \frac{D_x}{D} = \frac{-44}{-19} = \frac{44}{19}, \qquad y = \frac{D_y}{D} = \frac{-29}{-19} = \frac{29}{19}$$

$$\boxed{x = \frac{44}{19}, \quad y = \frac{29}{19}}$$

The check is left to the student.

In a similar manner, Cramer's rule can be extended to 3×3 linear systems. To do this we must define what we mean by a 3×3 determinant.

DEFINITION   A **3 × 3 determinant** is defined as follows:

$$\begin{vmatrix} a_1 & b_1 & c_1 \\ a_2 & b_2 & c_2 \\ a_3 & b_3 & c_3 \end{vmatrix} = a_1 \begin{vmatrix} b_2 & c_2 \\ b_3 & c_3 \end{vmatrix} - a_2 \begin{vmatrix} b_1 & c_1 \\ b_3 & c_3 \end{vmatrix} + a_3 \begin{vmatrix} b_1 & c_1 \\ b_2 & c_2 \end{vmatrix}$$

The 2×2 determinants in this definition are called the ***minors*** of their coefficients.

A minor is obtained by choosing an element of the determinant and crossing off the row and column that contain it. Thus, the minor of a_1 is obtained as follows:

$$\begin{vmatrix} a_1 & b_1 & c_1 \\ a_2 & b_2 & c_2 \\ a_3 & b_3 & c_3 \end{vmatrix} \quad \text{which gives} \quad \begin{vmatrix} b_2 & c_2 \\ b_3 & c_3 \end{vmatrix}$$

The other minors are found in a similar manner.

The procedure specified in the definition is called *expanding* the 3×3 determinant *down its first column.*

EXAMPLE 5   Evaluate the following determinant.

$$\begin{vmatrix} 3 & 2 & 2 \\ 4 & 1 & -1 \\ -2 & -3 & 5 \end{vmatrix}$$

Solution   Using the definition, we get

$$\begin{vmatrix} 3 & 2 & 2 \\ 4 & 1 & -1 \\ -2 & -3 & 5 \end{vmatrix} = 3 \begin{vmatrix} 3 & 2 & 2 \\ 4 & 1 & -1 \\ -2 & -3 & 5 \end{vmatrix} - 4 \begin{vmatrix} 3 & 2 & 2 \\ 4 & 1 & -1 \\ -2 & -3 & 5 \end{vmatrix} + (-2) \begin{vmatrix} 3 & 2 & 2 \\ 4 & 1 & -1 \\ -2 & -3 & 5 \end{vmatrix}$$

$$= 3 \begin{vmatrix} 1 & -1 \\ -3 & 5 \end{vmatrix} - 4 \begin{vmatrix} 2 & 2 \\ -3 & 5 \end{vmatrix} + (-2) \begin{vmatrix} 2 & 2 \\ 1 & -1 \end{vmatrix}$$

$$= 3(5 - 3) - 4(10 + 6) - 2(-2 - 2)$$

$$= 6 - 64 + 8 = \boxed{-50}$$

In fact, a 3×3 determinant may be expanded across any row or down any column provided we *prefix* each entry with its proper sign, which is determined by its position. The **sign array,** as it is called, is

$$\begin{vmatrix} + & - & + \\ - & + & - \\ + & - & + \end{vmatrix}$$

which is obtained by putting a $+$ sign in the upper left position and then alternating signs along each row and column.

EXAMPLE 6

Evaluate: $\begin{vmatrix} 3 & 2 & 2 \\ 4 & 1 & -1 \\ -2 & 3 & 5 \end{vmatrix}$ by expanding across the second row.

Solution

$$\begin{vmatrix} 3 & 2 & 2 \\ 4 & 1 & -1 \\ -2 & 3 & 5 \end{vmatrix} = -4 \begin{vmatrix} 2 & 2 \\ 3 & 5 \end{vmatrix} + 1 \begin{vmatrix} 3 & 2 \\ -2 & 5 \end{vmatrix} - (-1) \begin{vmatrix} 3 & 2 \\ -2 & 3 \end{vmatrix}$$

Note the signs from the second row of the sign array.

$$= -4(10 - 6) + 1(15 + 4) - (-1)(9 + 4)$$

$$= -4(4) + 19 + 13 = \boxed{16}$$

EXAMPLE 7

Evaluate: $\begin{vmatrix} 1 & 0 & 3 \\ 2 & 0 & 5 \\ -1 & 4 & 6 \end{vmatrix}$

Solution

We can expand the determinant using any row or column, so we might as well choose a row or column that contains zeros (if there is one) to make the computation easier. We will expand down the second column.

$$\begin{vmatrix} 1 & 0 & 3 \\ 2 & 0 & 5 \\ -1 & 4 & 6 \end{vmatrix} = -0 \begin{vmatrix} 2 & 5 \\ -1 & 6 \end{vmatrix} + 0 \begin{vmatrix} 1 & 3 \\ -1 & 6 \end{vmatrix} - 4 \begin{vmatrix} 1 & 3 \\ 2 & 5 \end{vmatrix}$$

Signs from the second column of the sign array.

$$= 0 + 0 - 4(-1) = \boxed{4}$$

We can now state Cramer's rule for 3×3 linear systems.

THEOREM

Cramer's Rule (part 2)

The solution to the system

$$\begin{cases} a_1 x + b_1 y + c_1 z = k_1 \\ a_2 x + b_2 y + c_2 z = k_2 \\ a_3 x + b_3 y + c_3 z = k_3 \end{cases}$$

(Continued)

is given by

$$x = \frac{D_x}{D}, \quad y = \frac{D_y}{D}, \quad z = \frac{D_z}{D}$$

where

$$D = \begin{vmatrix} a_1 & b_1 & c_1 \\ a_2 & b_2 & c_2 \\ a_3 & b_3 & c_3 \end{vmatrix} \neq 0 \qquad D_x = \begin{vmatrix} k_1 & b_1 & c_1 \\ k_2 & b_2 & c_2 \\ k_3 & b_3 & c_3 \end{vmatrix}$$

$$D_y = \begin{vmatrix} a_1 & k_1 & c_1 \\ a_2 & k_2 & c_2 \\ a_3 & k_3 & c_3 \end{vmatrix} \qquad D_z = \begin{vmatrix} a_1 & b_1 & k_1 \\ a_2 & b_2 & k_2 \\ a_3 & b_3 & k_3 \end{vmatrix}$$

EXAMPLE 8

Solve the following system of equations by using Cramer's rule.

$$\begin{cases} x + y + z = 1 \\ 3x + 2y - 6z = 1 \\ 9x - 4y + 12z = 3 \end{cases}$$

Solution   We begin by computing D, since if $D = 0$ there is no unique solution. We compute D by expanding across the first row:

$$D = \begin{vmatrix} 1 & 1 & 1 \\ 3 & 2 & -6 \\ 9 & -4 & 12 \end{vmatrix} = 1 \begin{vmatrix} 2 & -6 \\ -4 & 12 \end{vmatrix} - 1 \begin{vmatrix} 3 & -6 \\ 9 & 12 \end{vmatrix} + 1 \begin{vmatrix} 3 & 2 \\ 9 & -4 \end{vmatrix}$$

$$= (24 - 24) - (36 + 54) + (-12 - 18)$$
$$= 0 - 90 - 30$$
$$= -120$$

We compute D_x by expanding across the first row:

$$D_x = \begin{vmatrix} 1 & 1 & 1 \\ 1 & 2 & -6 \\ 3 & -4 & 12 \end{vmatrix} = 1 \begin{vmatrix} 2 & -6 \\ -4 & 12 \end{vmatrix} - 1 \begin{vmatrix} 1 & -6 \\ 3 & 12 \end{vmatrix} + 1 \begin{vmatrix} 1 & 2 \\ 3 & -4 \end{vmatrix}$$

$$= (24 - 24) - (12 + 18) + (-4 - 6)$$
$$= 0 - 30 - 10$$
$$= -40$$

Just for variety, we compute D_y by expanding down the third column:

$$D_y = \begin{vmatrix} 1 & 1 & 1 \\ 3 & 1 & -6 \\ 9 & 3 & 12 \end{vmatrix} = 1 \begin{vmatrix} 3 & 1 \\ 9 & 3 \end{vmatrix} - (-6) \begin{vmatrix} 1 & 1 \\ 9 & 3 \end{vmatrix} + 12 \begin{vmatrix} 1 & 1 \\ 3 & 1 \end{vmatrix}$$

$$= (9 - 9) + 6(3 - 9) + 12(1 - 3)$$
$$= 0 - 36 - 24$$
$$= -60$$

We expand D_z across the second row:

$$D_z = \begin{vmatrix} 1 & 1 & 1 \\ 3 & 2 & 1 \\ 9 & -4 & 3 \end{vmatrix} = -3\begin{vmatrix} 1 & 1 \\ -4 & 3 \end{vmatrix} + 2\begin{vmatrix} 1 & 1 \\ 9 & 3 \end{vmatrix} - 1\begin{vmatrix} 1 & 1 \\ 9 & -4 \end{vmatrix}$$

Note the signs from the second row of the sign array.

$$= -3(3 + 4) + 2(3 - 9) - (-4 - 9)$$
$$= -21 - 12 + 13$$
$$= -20$$

Therefore, the solution to the system is

$$x = \frac{D_x}{D} = \frac{-40}{-120} = \frac{1}{3}, \quad y = \frac{D_y}{D} = \frac{-60}{-120} = \frac{1}{2}, \quad z = \frac{D_z}{D} = \frac{-20}{-120} = \frac{1}{6}$$

$$\boxed{x = \frac{1}{3}, \quad y = \frac{1}{2}, \quad z = \frac{1}{6}}$$

The check is left to the student.

Keep in mind that in order to use Cramer's rule, the system of equations must be in standard form.

EXERCISES 11.5

In Exercises 1–18, compute the value of the determinant.

1. $\begin{vmatrix} 1 & 2 \\ 3 & 4 \end{vmatrix}$
    2. $\begin{vmatrix} 3 & 1 \\ 4 & 2 \end{vmatrix}$
    3. $\begin{vmatrix} 5 & 1 \\ -2 & 4 \end{vmatrix}$
    4. $\begin{vmatrix} 4 & -3 \\ 2 & 6 \end{vmatrix}$

5. $\begin{vmatrix} -3 & -1 \\ 2 & -2 \end{vmatrix}$
    6. $\begin{vmatrix} -5 & -4 \\ -3 & 2 \end{vmatrix}$
    7. $\begin{vmatrix} 1 & 0 \\ 5 & 4 \end{vmatrix}$
    8. $\begin{vmatrix} -1 & 7 \\ 8 & 0 \end{vmatrix}$

9. $\begin{vmatrix} 4 & 6 \\ 6 & 9 \end{vmatrix}$
    10. $\begin{vmatrix} 2 & -3 \\ -6 & -9 \end{vmatrix}$
    11. $\begin{vmatrix} 1 & 2 & -2 \\ 3 & -3 & 1 \\ -4 & 2 & -1 \end{vmatrix}$
    12. $\begin{vmatrix} 5 & 2 & -1 \\ -3 & 4 & 1 \\ -2 & 2 & -2 \end{vmatrix}$

13. $\begin{vmatrix} 1 & 2 & 3 \\ 2 & 4 & 6 \\ 1 & 1 & 1 \end{vmatrix}$
    14. $\begin{vmatrix} 3 & -1 & 2 \\ 4 & 5 & 6 \\ -3 & 1 & -2 \end{vmatrix}$
    15. $\begin{vmatrix} -3 & 0 & 4 \\ 5 & 2 & -3 \\ 7 & 0 & 6 \end{vmatrix}$
    16. $\begin{vmatrix} 1 & -2 & 6 \\ 0 & 8 & 0 \\ 5 & -3 & 4 \end{vmatrix}$

17. $\begin{vmatrix} 1 & 0 & 0 \\ 0 & 1 & 0 \\ 0 & 0 & 1 \end{vmatrix}$
    18. $\begin{vmatrix} 2 & 0 & 0 \\ 0 & 2 & 0 \\ 0 & 0 & 2 \end{vmatrix}$

In Exercises 19–24, expand the determinant and solve the resulting equation for x.

19. $\begin{vmatrix} 2x & 4 \\ 3 & 5 \end{vmatrix} = 18$
    20. $\begin{vmatrix} x + 3 & -2 \\ 5 & 4 \end{vmatrix} = 30$
    21. $\begin{vmatrix} x^2 & 5 \\ x & 1 \end{vmatrix} = 14$
    22. $\begin{vmatrix} 2 & -x \\ -1 & x^2 \end{vmatrix} = 15$

23. $\begin{vmatrix} x & 3 \\ 2 & x + 2 \end{vmatrix} = 9$

24. $\begin{vmatrix} x - 5 & 3 \\ 2 & 3x \end{vmatrix} = 2x$

In Exercises 25–54, *solve the system of equations by using Cramer's rule.*

25. $\begin{cases} 9x + 2y = 1 \\ 5x + \ y = 0 \end{cases}$

26. $\begin{cases} 3x - 5y = 2 \\ 2x - 3y = 2 \end{cases}$

27. $\begin{cases} 2x - 4y = 5 \\ -3x + 6y = 7 \end{cases}$

28. $\begin{cases} -4x + 6y = 11 \\ 6x - 9y = 1 \end{cases}$

29. $\begin{cases} 3x - 5y - \ 2 = 0 \\ 4x - 3y - 10 = 0 \end{cases}$

30. $\begin{cases} 5x - 4y + \ 3 = 0 \\ 6x + 7y - 20 = 0 \end{cases}$

31. $\begin{cases} 2x - 7y = -4 \\ 3y - 4x = 8 \end{cases}$

32. $\begin{cases} 6y - 11x = -6 \\ 5x - \ 3y = \ \ 3 \end{cases}$

33. $\begin{cases} 6x = 5y - 7 \\ 4y = 3x - 5 \end{cases}$

34. $\begin{cases} 9a = 5b + 2 \\ 3b = 5a - 2 \end{cases}$

35. $\begin{cases} 2s - 9t = 4 \\ 3s + 5t = 6 \end{cases}$

36. $\begin{cases} 4m + 7n = 3 \\ 5m - 6n = 1 \end{cases}$

37. $\begin{cases} 6u + 7v = 3 \\ 5u + 8v = 9 \end{cases}$

38. $\begin{cases} 9w - 5z = 10 \\ 8w - 2z = 5 \end{cases}$

39. $\begin{cases} \dfrac{1}{2}x - 12y = 6 \\[2mm] \dfrac{1}{3}x - \ 8y = 6 \end{cases}$

40. $\begin{cases} \dfrac{2}{3}x + \dfrac{3}{4}y = 1 \\[2mm] 6x + \ 9y = 7 \end{cases}$

41. $\begin{cases} \ \ x + y + z = 3 \\ \ \ x - y + z = 2 \\ -x + y + z = 4 \end{cases}$

42. $\begin{cases} x + y + 2z = 4 \\ x - y + 2z = 2 \\ x - y - \ z = 2 \end{cases}$

43. $\begin{cases} 3x + 4y + 2z = 1 \\ 2x + 3y + \ z = 1 \\ 6x + \ y + 5z = 1 \end{cases}$

44. $\begin{cases} 4x + \ y + \ z = 0 \\ 2x - 3y + 4z = -7 \\ 3x + 4y - 2z = 6 \end{cases}$

45. $\begin{cases} \ r + 2s + 3t = 10 \\ 6r + 5s + 4t = 20 \\ 7r + 8s + 9t = 30 \end{cases}$

46. $\begin{cases} 2r - 3s + \ t = 5 \\ 3r - 2s + 4t = 5 \\ 5r - 3s + 2t = 5 \end{cases}$

47. $\begin{cases} 3x - \ y = 4 \\ 2y + \ z = 6 \\ 3x + 4z = 14 \end{cases}$

48. $\begin{cases} 5x - 3y = -1 \\ 3x + 4z = 11 \\ 5y + 2z = 16 \end{cases}$

49. $\begin{cases} \ \ 6x - 2y + 4z = 5 \\ \ 12x - 4y + 8z = 2 \\ -3x + \ y - 2z = 2 \end{cases}$

50. $\begin{cases} \ \ r - 2s + 3z = 3 \\ -3r + 6s - 9z = 1 \\ \ \ 2r - 4s + 6z = 9 \end{cases}$

51. $\begin{cases} x = y + z + 2 \\ y = x - z + 3 \\ z = x + y + 4 \end{cases}$

52. $\begin{cases} r = 2s - t + 1 \\ s = 3r + t + 4 \\ t = 5r - s - 1 \end{cases}$

53. $\begin{cases} 2a + b = 8 \\ -3a + c = -13 \\ 2b + 5c = 0 \end{cases}$

54. $\begin{cases} 3u + 2v = 0 \\ 5v + 6w = 5 \\ 4u - 3w = 0 \end{cases}$

In Exercises 55–56, compute the value of the determinant.

55. $\begin{vmatrix} 68 & 85 \\ 920 & 743 \end{vmatrix}$

56. $\begin{vmatrix} 0.34 & 0.6 & -0.09 \\ 2 & -0.81 & 0.4 \\ 9 & 12 & 0.5 \end{vmatrix}$

In Exercises 57–60, solve the system of equations using Cramer's rule. Express your answer rounded to three decimal places.

57. $\begin{cases} 5x - 3y = -0.181 \\ 0.2x + 4y = 0.2482 \end{cases}$

58. $\begin{cases} 64x - 2y = 85 \\ 0.02x + 0.04y = 2 \end{cases}$

59. $\begin{cases} 0.02x + 0.04y - 0.06z = 12 \\ 0.5x + 0.6y - 0.3z = 16 \\ 0.1x - 0.02y + z = 15 \end{cases}$

60. $\begin{cases} 1{,}000x - 3y + z = 200 \\ 0.02x - y + 0.01z = 20 \\ 10x - 0.5y + 0.2z = 60 \end{cases}$

? Question for Thought

61. Expand the determinant

$$\begin{vmatrix} a_1 & b_1 & c_1 \\ a_2 & b_2 & c_2 \\ a_3 & b_3 & c_3 \end{vmatrix}$$

across the first row and also down the first column. Show by regrouping terms and factoring that the results are the same.

◆ Mini-Review

62. *Express in exponential form:* $\log_4 64 = 3$

63. *Express in logarithmic form:* $3^{-4} = \dfrac{1}{81}$

64. *Evaluate:* $\log_8 16$

65. Which is greater, $\log_4 60$ or $\log_3 40$? Explain.

11.6 Systems of Linear Inequalities

Let's examine the following problem:

This problem is solved in Example 4 in this section.

The Cal-Q-Late Electronics Co. makes two types of calculators—a printing model and a programmable scientific model. The printing model requires $30 in materials and 6 minutes to assemble and package. The scientific model requires $10 in materials and 12 minutes to assemble and package. The company decides to spend a maximum of $5,000 on materials and to allot a maximum of 35 hours for packing and assembly. How many of each type of calculator can the company produce under these restrictions?

We'll write a system of inequalities to describe this situation: Letting x = the number of printing calculators the company can produce and y = the number of scientific calculators the company can produce, we can translate the problem into the following system of inequalities:

$$\begin{cases} 30x + 10y \le 5{,}000 \\ 6x + 12y \le 2{,}100 \\ x \ge 0 \\ y \ge 0 \end{cases}$$

Since the total cost for materials is $\le$ \$5,000

Since the total number of minutes for packaging and assembly is $\le$ 2,100 minutes (35 hours)

Because the number of calculators produced cannot be negative

Recall that in Section 4.4 we discussed solving linear inequalities in two variables by graphing the solution set. We found that the graph of a linear inequality is a half-plane. Since we have four inequalities in the problem above, when we graph them we will have four half-planes. In this section we will see how we can put this information together in a meaningful manner. We will begin with a simpler case—solving a 2×2 system of inequalities—and return to the problem above later on in this section.

EXAMPLE 1

Solve the following system of inequalities:

$$\begin{cases} x + y \le 4 \\ 2x + y \ge 2 \end{cases}$$

Solution

As we outlined in Section 4.4, we graph the *equations* $x + y = 4$ and $2x + y = 2$, choose a test point, and shade the appropriate region.

The equation $x + y = 4$ is graphed in Figure 11.2**(a)**. Choose $(0, 0)$ as the test point. Since $(0, 0)$ satisfies the inequality $x + y \le 4$, we shade the region containing $(0, 0)$.

The equation $2x + y = 2$ is graphed in Figure 11.2**(b)**. Choose $(0, 0)$ as the test point. Since $(0, 0)$ does not satisfy the inequality $2x + y \ge 2$, we shade the region *not* containing $(0, 0)$.

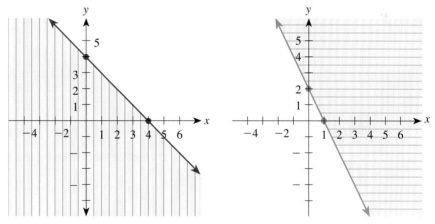

Figure 11.2   **(a)** Graph of $x + y \le 4$ **(b)** Graph of $2x + y \ge 2$

Combining the two graphs, we obtain the solution set to the system of inequalities—the *intersection* of the two shaded regions (see Figure 11.3 on page 696).

We may choose a check point in this region, such as $(3, 0)$, to see whether it satisfies the system.

$$\begin{array}{ccc} x + y \le 4 & \leftarrow \textit{Substitute } (3, 0) \rightarrow & 2x + y \ge 2 \\ 3 + 0 \overset{?}{\le} 4 & & 2(3) + 0 \overset{?}{\ge} 2 \\ 3 \overset{\checkmark}{\le} 4 & & 6 + 0 \overset{\checkmark}{\ge} 2 \end{array}$$

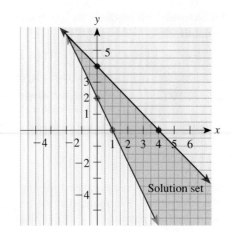

Figure 11.3
Graph of the system
$$\begin{cases} x + y \le 4 \\ 2x + y \ge 2 \end{cases}$$

EXAMPLE 2

Solve the following system of inequalities:

$$\begin{cases} y - x > 4 \\ x - y > 3 \end{cases}$$

Solution   We proceed as in the last example, except that since the inequalities are strict, we draw dashed rather than solid lines (see Figure 11.4**(a)** and **(b)**).

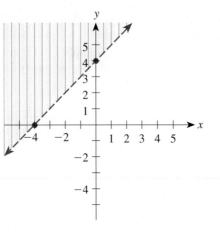

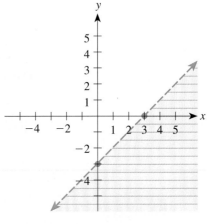

Figure 11.4   **(a)** Graph of $y - x > 4$ **(b)** Graph of $x - y > 3$

Putting the graphs together, we obtain Figure 11.5.

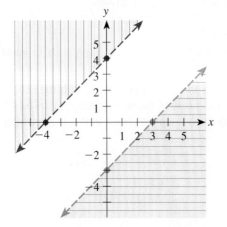

Figure 11.5
Graph of the system
$$\begin{cases} y - x > 4 \\ x - y > 3 \end{cases}$$

Looking at the equations

$$y - x = 4 \;\rightarrow\; y = x + 4$$
$$x - y = 3 \;\rightarrow\; y = x - 3$$

we can see that the lines are parallel (they both have slope equal to 1). The two lines never cross and the regions do not intersect.

There are ⎡ no solutions ⎤ to this system.

We may use the same procedure to solve systems involving more than two inequalities.

EXAMPLE 3   Solve the following system of inequalities:

$$\begin{cases} 2x + 3y \le 12 \\ \qquad x \ge 0 \\ \qquad y \le 2 \end{cases}$$

Solution   We graph each inequality, as shown in Figure 11.6.

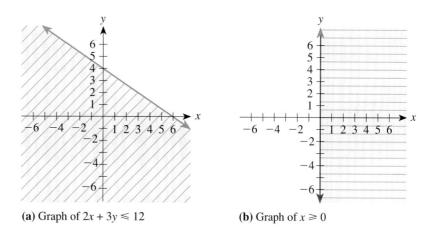

(a) Graph of $2x + 3y \le 12$ **(b)** Graph of $x \ge 0$

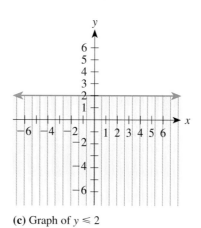

Figure 11.6   **(c)** Graph of $y \le 2$

The solution set, which is the intersection of these three regions, is indicated in Figure 11.7.

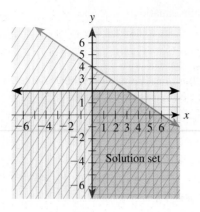

Figure 11.7
Graph of the system
$$\begin{cases} 2x + 3y \le 12 \\ \quad\;\; x \ge 0 \\ \quad\;\; y \le 2 \end{cases}$$

We may choose $(1, 1)$ as a check point and verify that it satisfies the system.

Now we can complete the problem we began at the beginning of this section.

EXAMPLE 4

The Cal-Q-Late Electronics Co. makes two types of calculators—a printing model and a programmable scientific model. The printing model requires \$30 in materials and 6 minutes to assemble and package. The scientific model requires \$10 in materials and 12 minutes to assemble and package. The company decides to spend a maximum of \$5,000 on materials and to allot a maximum of 35 hours for packing and assembly. How many of each type of calculator can the company produce under these restrictions? Write a system of inequalities to describe this situation and sketch the solution set of the system.

Solution

Let x = Number of printing calculators the company can produce.

Let y = Number of scientific calculators the company can produce.

We can translate the information given in the problem into the following system of inequalities:

$$\begin{cases} 30x + 10y \le 5,000 \\ 6x + 12y \le 2,100 \\ \qquad\quad x \ge 0 \\ \qquad\quad y \ge 0 \end{cases}$$

Since the total cost for materials is $\le$ \$5,000

Since the total number of minutes for packaging and assembly is $\le$ 2,100 minutes (35 hours)

Because the number of calculators produced cannot be negative

We can now sketch the solution set of the first two inequalities (see Figures 11.8(**a**) and (**b**)).

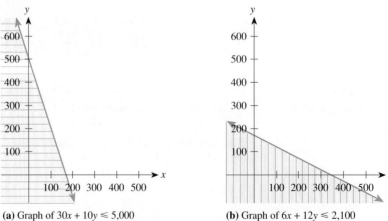

Figure 11.8　　(**a**) Graph of $30x + 10y \le 5,000$　　(**b**) Graph of $6x + 12y \le 2,100$

We note that the inequalities $x \geq 0$ and $y \geq 0$ restrict us to the first quadrant and its boundary. Thus, the solution set of this system of inequalities is as shown in Figure 11.9.

This graph indicates that any point (x, y) in the darkest shaded region satisfies the system. However, since we cannot produce fractional parts of a calculator, only the ordered pairs in the darkest region with whole-number coordinates are actual choices for the company.

Figure 11.9
Graph of the system
$$\begin{cases} 30x + 10y \leq 5{,}000 \\ 6x + 12y \leq 2{,}100 \\ x \geq 0 \\ y \geq 0 \end{cases}$$

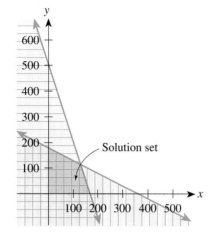

EXERCISES 11.6

In Exercises 1–24, *sketch the solution set of the system of inequalities.*

1. $\begin{cases} x + y \leq 6 \\ x + 2y \geq 3 \end{cases}$

2. $\begin{cases} x + y \geq 4 \\ 3x + y < 6 \end{cases}$

3. $\begin{cases} 2y + x \geq 6 \\ 3y + x \geq 9 \end{cases}$

4. $\begin{cases} x - 2y \leq 8 \\ x - 3y \leq 12 \end{cases}$

5. $\begin{cases} x - y \geq 2 \\ y - x > -1 \end{cases}$

6. $\begin{cases} x - 2y \leq 2 \\ 4y - 2x \leq 0 \end{cases}$

7. $\begin{cases} x + y \leq 5 \\ 2x + y \leq 8 \end{cases}$

8. $\begin{cases} y - x \leq 4 \\ y + x \leq 4 \end{cases}$

9. $\begin{cases} 3x + 2y \leq 12 \\ y \leq x \end{cases}$

10. $\begin{cases} 3y - x > 6 \\ y \leq 5 \end{cases}$

11. $\begin{cases} 3x + 2y \geq 12 \\ y \geq x \end{cases}$

12. $\begin{cases} 3y - x < 6 \\ y \geq 5 \end{cases}$

13. $\begin{cases} 5x - 3y \leq 15 \\ x < 3 \end{cases}$

14. $\begin{cases} 2x + 7y > 14 \\ x \geq 7 \end{cases}$

15. $\begin{cases} 5x - 3y \geq 15 \\ x > 3 \end{cases}$

16. $\begin{cases} 2x + 7y < 14 \\ x \leq 7 \end{cases}$

17. $\begin{cases} x - 2y < 10 \\ x \geq 2 \\ y \leq 2 \end{cases}$

18. $\begin{cases} 3y - x > 6 \\ y \leq 5 \\ x \geq -3 \end{cases}$

19. $\begin{cases} 4x + 3y \leq 12 \\ x \geq 0 \\ y \geq 0 \end{cases}$

20. $\begin{cases} 2x - 5y \leq 10 \\ x \geq 0 \\ y \geq 0 \end{cases}$

21. $\begin{cases} x + 3y \geq 6 \\ 3x + 2y \leq 18 \\ y \leq 3 \\ x \geq 0 \end{cases}$

22. $\begin{cases} 2x + y \leq 8 \\ 3x - 2y \leq 12 \\ x \geq 0 \\ y \geq -4 \end{cases}$

23. $\begin{cases} x + 2y \geq 4 \\ 2x - 4y \leq -8 \\ x \geq 0 \\ y \geq 0 \end{cases}$

24. $\begin{cases} x > y \\ x + y < 8 \\ 3x - 4y < 24 \end{cases}$

In Exercises 25–30, write a system of inequalities to describe the given situation and sketch the graph of its solution set.

25. The Bedding Store needs to order a shipment of single and double mattresses. A single mattress costs $80 and takes up 16 cubic feet of storage space; a double mattress costs $120 and takes up 36 cubic feet of storage space. If the store manager wants to order no more than $8,000 worth of mattresses and has at most 2,000 cubic feet of storage space, how many of each mattress can she order? Write a system of inequalities to describe this situation and sketch the solution set of the system.

26. Megan has no more than $50,000 available for investment. She can invest in two types of certificates of deposit: a 6-month certificate that yields a 5% yearly return on investment, and a 1-year certificate that yields a 6.5% yearly return. How much should she invest in each if she wants to earn at least $2,800 interest in one year?

27. A jogger wants to establish a more balanced diet. She reads the nutritional information on cereal boxes and finds out that 1 oz of brand X cereal contains 10 g of carbohydrates and 0.33 g of sodium, whereas 1 oz of brand Y cereal contains 13 g of carbohydrates and 0.37 g of sodium. If she wants to create a mixture of brands X and Y that will contain at least 21 g of carbohydrates and no more than 1 g of sodium, how much of each cereal can she use?

28. A discount appliance store wants to order a shipment of refrigerators and televisions. A refrigerator costs $400 and uses 30 cubic feet of storage space, whereas a TV costs $275 and uses 12 cubic feet of storage space. If the store wants to order at least $10,000 worth of merchandise but has at most 10,000 cubic feet of storage space in its warehouse, how many of each can it order?

29. A shoe manufacturer uses 1 square meter of leather and $\frac{1}{3}$ square meter of crepe rubber to make each pair of men's shoes, and $\frac{3}{4}$ square meter of leather and $\frac{1}{4}$ square meter of crepe rubber to make each pair of women's shoes. He has 500 square meters of leather and 150 square meters of crepe rubber on hand. If he cannot get any more leather or crepe rubber, how many pairs of each type of shoe can he produce?

30. A carpenter makes two kinds of bookcases—one type is custom-made to specifications and the other type is made from prefabricated parts. On the average, it takes 5 hours to make a custom-made bookcase and 3 hours to make one from prefabricated parts. He can make a profit of $175 on each custom-made bookcase and $100 on each prefabricated bookcase. If he can spend no more than 20 hours per week on bookcases and wants to earn at least $500 per week from making bookcases, how many of each type can he produce?

 ## Mini-Review

31. *Express as a sum or difference of simpler logarithms:* $\log_b \dfrac{\sqrt{x^3y^4}}{5b}$

32. *Solve for x:* $\log_2 x + \log_2(x - 7) = 3$

33. If $6,000 is invested at an annual rate of 7.2% compounded weekly, how much money will be in the account after 100 weeks?

34. *Solve for x:* $2^{5x} = 8^{x^2}$

 11.7 # Nonlinear Systems of Equations

In this section we will discuss systems of equations involving second-degree equations. To help us visualize the solutions, in the course of our discussion we will refer back to our work on graphing straight lines, parabolas, and circles.

As was the case with a linear system, when we solve a nonlinear system we often interpret the solution(s) as the point(s) of intersection of the graphs. Since we cannot graph complex-number solutions to systems of equations, we are going to restrict our attention to *real-number* solutions.

The next two examples illustrate that the two methods we used to solve linear systems—the elimination and substitution methods—can be applied to nonlinear systems as well.

EXAMPLE 1

Solve the following system of equations:

$$\begin{cases} x^2 + y^2 = 5 \\ 2x^2 + y = 0 \end{cases}$$

Solution

Although it is not necessary, it is often helpful to sketch the graphs of the equations in the system, so we know what to expect. Using the methods developed in Chapter 8, we get the graphs in Figure 11.10. The graph of $x^2 + y^2 = 5$ is a circle with center $(0, 0)$ and radius $\sqrt{5}$. The graph of $2x^2 + y = 0$ or $y = -2x^2$ is a parabola with vertex $(0, 0)$ that opens downward.

In addition to a parabola and a circle, there are two other possibilities for the graph of a second-degree equation in two variables. They are called an ellipse and a hyperbola. See Appendix B.

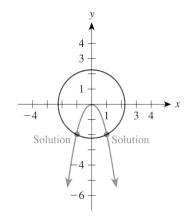

Figure 11.10

Graphs of the circle $x^2 + y^2 = 5$ and the parabola $2x^2 + y = 0$

As these graphs clearly show, we expect to find *two* solutions to this system of equations (one solution for each point of intersection).

Since both equations contain an "x^2" term, we can use the elimination method to eliminate x^2 and get an equation involving the variable y only.

$$\begin{array}{lll} x^2 + y^2 = 5 & \textit{Multiply by } -2 \rightarrow & -2x^2 - 2y^2 = -10 \\ 2x^2 + y = 0 & \textit{As is} \rightarrow & \underline{2x^2 + y = 0} \qquad \textit{Add.} \\ & & -2y^2 + y = -10 \end{array}$$

We now have a quadratic equation, which we put in standard form and solve:

$$\begin{array}{ll} -2y^2 + y + 10 = 0 & \textit{Multiply both sides of the equation by } -1. \\ 2y^2 - y - 10 = 0 & \textit{We can factor or use the quadratic formula.} \\ (2y - 5)(y + 2) = 0 & \\ \quad 2y - 5 = 0 \quad \text{or} \quad y + 2 = 0 & \\ \qquad\quad y = \dfrac{5}{2} \quad \text{or} \qquad y = -2 & \end{array}$$

Now we substitute each y value into the second equation and solve for x:

Substitute $y = \dfrac{5}{2}$	*Substitute* $y = -2$
$2x^2 + y = 0$	$2x^2 + y = 0$
$2x^2 + \dfrac{5}{2} = 0$	$2x^2 - 2 = 0$
$2x^2 = -\dfrac{5}{2}$	$2x^2 = 2$
$x^2 = -\dfrac{5}{4}$	$x^2 = 1$
$x = \pm\sqrt{-\dfrac{5}{4}}$	$x = \pm 1$
No real solutions.	There are *two* x values for this *one* y value.

Thus, we have two real solutions to this system: $\boxed{(1, -2) \text{ and } (-1, -2)}$.

This result agrees quite well with what we saw in Figure 11.10. We check each solution in *both* equations.

Check: $(1, -2)$:

$$\begin{array}{ll} x^2 + y^2 = 5 & 2x^2 + y = 0 \\ (1)^2 + (-2)^2 \stackrel{?}{=} 5 & 2(1)^2 + (-2) \stackrel{?}{=} 0 \\ 1 + 4 \stackrel{\checkmark}{=} 5 & 2 - 2 \stackrel{\checkmark}{=} 0 \end{array}$$

Check: $(-1, -2)$:

$$\begin{array}{ll} x^2 + y^2 = 5 & 2x^2 + y = 0 \\ (-1)^2 + (-2)^2 \stackrel{?}{=} 5 & 2(-1)^2 + (-2) \stackrel{?}{=} 0 \\ 1 + 4 \stackrel{\checkmark}{=} 5 & 2 - 2 \stackrel{\checkmark}{=} 0 \end{array}$$

If we reflect a moment on second-degree systems of equations and analyze the possibilities, we can see that such a system can have as many as four solutions or as few as none. That is, the graphs can have as many as four points of intersection or as few as none. For instance, if we consider the case of a circle and a parabola, we saw in Example 1 the situation in which they intersect in two points. Figure 11.11 illustrates five possibilities.

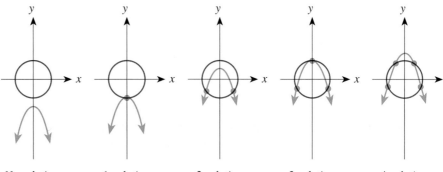

Figure 11.11
Possibilities for solutions of second-degree systems of equations

No solutions   *1 solution*   *2 solutions*   *3 solutions*   *4 solutions*

| **EXAMPLE 2** | Solve the following system of equations: |

$$\begin{cases} x^2 + y^2 = 17 \\ 2x + y = 7 \end{cases}$$

Solution   Figure 11.12 illustrates the graphs associated with this system of equations.

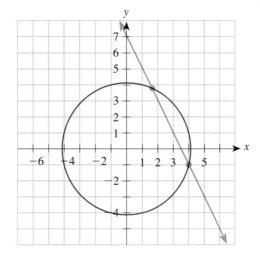

Figure 11.12
The graphs of the circle $x^2 + y^2 = 17$ and the straight line $2x + y = 7$

Unlike Example 1, we cannot use the elimination method to solve this system because we do not have "like" terms to eliminate. In such a case, we can use the substitution method. We begin by solving the second equation for y.

From the second equation we get $y = 7 - 2x$, which we can substitute into the first equation.

$$x^2 + y^2 = 17 \qquad \textit{Substitute } y = 7 - 2x.$$
$$x^2 + (7 - 2x)^2 = 17$$
$$x^2 + 49 - 28x + 4x^2 = 17 \qquad \textit{We put this quadratic equation into standard form.}$$
$$5x^2 - 28x + 32 = 0 \qquad \textit{We can solve this equation by factoring.}$$
$$(5x - 8)(x - 4) = 0$$
$$x = \frac{8}{5} \qquad \text{or} \quad x = 4 \qquad \textit{Now we substitute these x values into the equation } y = 7 - 2x \textit{ to get the y values.}$$
$$y = 7 - 2\left(\frac{8}{5}\right) \quad \text{or} \quad y = 7 - 2(4)$$
$$y = \frac{19}{5} \qquad \text{or} \quad y = -1$$

Thus, there are two solutions, $\left(\frac{8}{5}, \frac{19}{5}\right)$ and $(4, -1)$, which agree very well with the graph in Figure 11.12. The algebraic check of these solutions is left to the student.

EXAMPLE 3

Solve the following system of equations:

$$\begin{cases} x^2 + y^2 = 0 & \textbf{(1)} \\ \qquad y = 3 - x^2 & \textbf{(2)} \end{cases}$$

Solution

We offer two methods of solution.

Using the Substitution Method:

We use equation (2) to substitute for y in equation (1):

$$x^2 + y^2 = 9 \qquad \textit{Replace } y \textit{ with } 3 - x^2.$$
$$x^2 + (3 - x^2)^2 = 9$$
$$x^2 + 9 - 6x^2 + x^4 = 9$$
$$x^4 - 5x^2 = 0$$
$$x^2(x^2 - 5) = 0$$
$$x^2 = 0 \quad \text{or} \quad x^2 - 5 = 0$$
$$x = 0 \quad \text{or} \qquad x = \pm\sqrt{5}$$

We can now substitute these x values into equation (2) to get their associated y values.

Substitute $x = 0$	**Substitute $x = \sqrt{5}$**	**Substitute $x = -\sqrt{5}$**
$y = 3 - x^2$	$y = 3 - \left(\sqrt{5}\right)^2$	$y = 3 - \left(-\sqrt{5}\right)^2$
$y = 3 - 0^2$	$y = 3 - 5$	$y = 3 - 5$
$y = 3$	$y = -2$	$y = -2$
$\boxed{(0, 3)}$	$\boxed{(\sqrt{5}, -2)}$	$\boxed{(-\sqrt{5}, -2)}$

There is an important point to be made here. Suppose we had decided to substitute our x values into equation (1) instead of equation (2). We would have obtained another "solution." That is, if we substitute $x = 0$ into equation (1), we get

$$x^2 + y^2 = 9$$
$$0^2 + y^2 = 9$$
$$y^2 = 9$$
$$y = \pm 3$$

which gives another "solution," $(0, -3)$. ***However,*** $(0, -3)$ does *not* satisfy equation (2):

$$y = 3 - x^2 \qquad \textit{Substitute } (0, -3).$$
$$-3 \overset{?}{=} 3 - (0)^2$$
$$-3 \neq 3$$

Remember that a solution to a system of equations must satisfy *every* equation in the system. Thus, we should always check our solutions in *all* the equations in the system.

Using the Elimination Method:

We can rewrite equation (2) and eliminate x^2:

(1) $x^2 + y^2 = 9$	$\rightarrow x^2 + y^2 = 9$	*As is* $\rightarrow$	$x^2 + y^2 = 9$	
(2) $\quad y = 3 - x^2$	$\rightarrow x^2 + y = 3$	*Multiply by* -1	$-x^2 - y = -3$	*Add.*

$$y^2 - y = 6$$
$$y^2 - y - 6 = 0$$
$$(y - 3)(y + 2) = 0$$
$$y - 3 = 0 \quad \text{or} \quad y + 2 = 0$$
$$y = 3 \quad \text{or} \qquad y = -2$$

Substitute these *y* values into equation (1) and solve for *x*:

Substitute $y = 3$	**Substitute $y = -2$**
$x^2 + y^2 = 9$	$x^2 + y^2 = 9$
$x^2 + 3^2 = 9$	$x^2 - (-2)^2 = 9$
$x^2 + 9 = 9$	$x^2 + 4 = 9$
$x^2 = 0$	$x^2 = 5$
$x = 0$	$x = \pm\sqrt{5}$
$\boxed{(0, 3)}$	$\boxed{(\sqrt{5},\, -2) \text{ and } (-\sqrt{5},\, -2)}$

In terms of choosing a method of solution, we suggest the following. If the elimination method allows us to eliminate one of the variables completely, then it is usually the easier method; otherwise, use the substitution method.

EXAMPLE 4

Solution

Find the dimensions of a rectangle whose perimeter is 33 cm and whose area is 65 sq cm.

We represent our rectangle as shown in Figure 11.13.

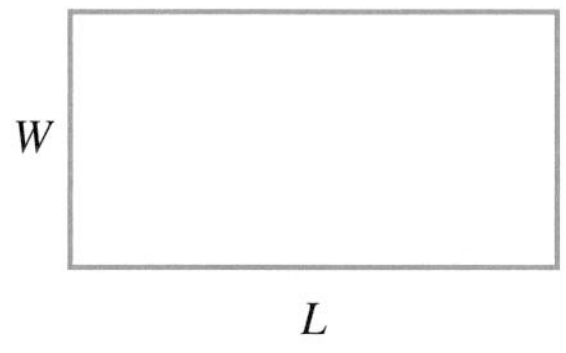

W

L

Figure 11.13

Let W = Width.
Let L = Length.

The information given in the problem allows us to write the following system of equations:

$$\begin{cases} 2W + 2L = 33 & \textbf{(1)} \quad \textit{Perimeter of the rectangle is 33 cm.} \\ WL = 65 & \textbf{(2)} \quad \textit{Area of the rectangle is 65 sq cm.} \end{cases}$$

Since we do not have "like" terms to eliminate, we will use the substitution method.

We can choose to solve either equation for one of the variables. We choose to solve the second equation for *L*.

$$WL = 65 \rightarrow L = \frac{65}{W} \qquad \textit{Substitute into equation (1).}$$

$$2W + 2\left(\frac{65}{W}\right) = 33 \qquad \textit{Now solve for W.}$$

$$2W + \frac{130}{W} = 33 \qquad \textit{Clear fractions; multiply each side by W.}$$

$$2W^2 + 130 = 33W$$

$$2W^2 - 33W + 130 = 0$$

$$(2W - 13)(W - 10) = 0$$

$$2W - 13 = 0 \qquad \text{or} \quad W - 10 = 0$$

$$W = \frac{13}{2} \quad \text{or} \quad W = 10$$

If $W = \frac{13}{2}$, then $L = \frac{65}{W} = \frac{65}{\frac{13}{2}} = \frac{65}{1} \cdot \frac{2}{13} = 10$.

Thus, the rectangle is $\frac{13}{2}$ by 10 or $\boxed{6.5 \text{ cm by } 10 \text{ cm}}$.

Note that if we substitute $W = 10$, then $L = \frac{65}{W} = \frac{65}{10} = 6.5$, giving us the same dimensions.

Check:

The perimeter is $2(6.5) + 2(10) = 13 + 20 \overset{\checkmark}{=} 33$ cm.

The area is $(6.5)(10) \overset{\checkmark}{=} 65$ sq cm.

EXERCISES 11.7

In Exercises 1–36, solve the system of equations.

1. $\begin{cases} x^2 + y^2 = 10 \\ x^2 - y^2 = 8 \end{cases}$

2. $\begin{cases} 2x^2 + y^2 = 6 \\ -x^2 + y^2 = 3 \end{cases}$

3. $\begin{cases} x^2 + y^2 = 25 \\ x - y = 5 \end{cases}$

4. $\begin{cases} x^2 - y^2 = 9 \\ x + y = 3 \end{cases}$

5. $\begin{cases} 16x^2 - 4y^2 = 64 \\ x^2 + y^2 = 9 \end{cases}$

6. $\begin{cases} 4x^2 - 16y^2 = 64 \\ x^2 - 5y^2 = 8 \end{cases}$

7. $\begin{cases} x^2 + y^2 = 9 \\ 2x - y = 3 \end{cases}$

8. $\begin{cases} x^2 + y^2 = 9 \\ 2x^2 - y^2 = -6 \end{cases}$

9. $\begin{cases} x^2 + y^2 = 13 \\ 3x^2 - y^2 = 3 \end{cases}$

10. $\begin{cases} x^2 + y^2 = 13 \\ 3x - y = 3 \end{cases}$

11. $\begin{cases} x^2 - y = 0 \\ x^2 - 2x + y = 6 \end{cases}$

12. $\begin{cases} y - 2x^2 = 0 \\ x^2 + 6x - y = 6 \end{cases}$

13. $\begin{cases} y = 1 - x^2 \\ x + y = 2 \end{cases}$

14. $\begin{cases} y = x^2 - 4 \\ x - y = 5 \end{cases}$

15. $\begin{cases} y = x^2 - 6x \\ y = x - 12 \end{cases}$

16. $\begin{cases} y = x^2 + 8x - 10 \\ y = 3x + 4 \end{cases}$

17. $\begin{cases} x^2 + y^2 = 16 \\ x = y^2 - 16 \end{cases}$

18. $\begin{cases} 4x^2 + y^2 = 16 \\ y = x - 4 \end{cases}$

19. $\begin{cases} x^2 + y^2 - 25 = 0 \\ x + y - 7 = 0 \end{cases}$

20. $\begin{cases} x^2 = y \\ 2x - y = 1 \end{cases}$

21. $\begin{cases} x^2 - y^2 = 4 \\ 2x^2 + y^2 = 16 \end{cases}$

22. $\begin{cases} x^2 + y^2 = 25 \\ 4x^2 + 3y^2 = 36 \end{cases}$

23. $\begin{cases} 9x^2 + y^2 = 9 \\ 3x + y = 3 \end{cases}$

24. $\begin{cases} x^2 + 4y^2 = 16 \\ x + 4y = 4 \end{cases}$

25. $\begin{cases} x^2 + y = 9 \\ 3x + 2y = 16 \end{cases}$

26. $\begin{cases} x^2 - y = 10 \\ 2x - 3y = -10 \end{cases}$

27. $\begin{cases} x = y^2 - 3 \\ x + y = 9 \end{cases}$

28. $\begin{cases} x - y^2 = 3 \\ x - y = 3 \end{cases}$

29. $\begin{cases} x^2 + 4x + y^2 - 6y = 7 \\ 2x + y = 5 \end{cases}$

30. $\begin{cases} x^2 - 2x + y^2 + 4y = 15 \\ x - 2y = -5 \end{cases}$

31. $\begin{cases} x^2 + y^2 = 10 \\ xy = 4 \end{cases}$

32. $\begin{cases} 2x^2 + 6 = y^2 \\ xy = 6 \end{cases}$

33. $\begin{cases} x^2 - y^2 = 4 \\ xy = 2\sqrt{3} \end{cases}$

34. $\begin{cases} x^2 - 2y^2 = 0 \\ xy = 3\sqrt{2} \end{cases}$

35. $\begin{cases} x^2 + y^2 = 25 \\ x^2 - xy + y^2 = 13 \end{cases}$

36. $\begin{cases} x^2 - y^2 = 21 \\ x^2 + xy - y^2 = 31 \end{cases}$

[*Hint for Exercises 35–36: First eliminate x^2 or y^2, then solve for x and y and use the substitution method.*]

In Exercises 37–46, *solve the system of equations and sketch the graphs.*

37. $\begin{cases} x^2 + y^2 = 29 \\ x - y = 3 \end{cases}$

38. $\begin{cases} y = x^2 - 4x + 3 \\ y = 3x - 3 \end{cases}$

39. $\begin{cases} x^2 + y^2 = 29 \\ y = x^2 - 9 \end{cases}$

40. $\begin{cases} y = x^2 + 4 \\ y = 16 - x^2 \end{cases}$

41. $\begin{cases} x - y = 0 \\ 3y + x^2 = 4 \end{cases}$

42. $\begin{cases} x^2 + y^2 = 18 \\ x - y = 6 \end{cases}$

43. $\begin{cases} x^2 + y^2 = 5 \\ y = 8x^2 - 6 \end{cases}$

44. $\begin{cases} y = x \\ y = x^2 \end{cases}$

45. $\begin{cases} x^2 + y^2 = 1 \\ x^2 + 2x + y^2 = 0 \end{cases}$

46. $\begin{cases} x^2 + y^2 = 9 \\ x^2 + 2x + y^2 = 9 \end{cases}$

Solve each of the following problems algebraically.

47. Find the dimensions of a rectangle if the area is 210 sq in. and the perimeter is 59 in.

48. Find the dimensions of a rectangle if the perimeter is 57 cm and the area is 189 sq cm.

49. Maria makes a round trip totaling 120 miles by car. Her return trip took 15 minutes less than her trip going because she drove 12 mph faster returning than she did going. Find her rate and time in each direction.

50. An electronic mail sorter sorts two batches of 500 letters. It sorts the second batch, set at high speed, in 10 minutes less than it took to sort the first batch at medium speed. If high speed is 150 letters per hour faster than medium speed, find the rates at medium and high speed and the time needed to sort each batch.

Question for Thought

51. Discuss what is ***wrong*** (if anything) with the following "solution."

$$\begin{cases} 4x^2 + y^2 = 25 \\ \quad x + y = 4 \end{cases}$$ *Square both sides of the second equation.*

$$\begin{array}{r} 4x^2 + y^2 = 25 \\ \underline{x^2 + y^2 = 16} \qquad \textit{Subtract.} \\ 3x^2 \qquad\quad = 9 \\ x^2 \qquad\quad = 3 \\ x = \pm\sqrt{3} \end{array}$$

◆ **Mini-Review**

52. *Solve for x:* $4^{3x+2} = 5$. *Round your answer to four decimal places.*

53. A bacteria culture is growing according to the growth model $A = 2{,}000e^{0.0316t}$, where A is the number of bacteria t hours after the culture is taken.

   (a) How many bacteria are there after 3 hours?

   (b) How long will it take until there are 5,000 bacteria?

54. *Solve for x:* $\log_5 x - \log_5(x - 1) = 1$.

55. Sketch the graph of $y = \left(\dfrac{2}{3}\right)^x$.

CHAPTER 11 SUMMARY

After having completed this chapter you should be able to:

1. Solve a 3×3 system of linear equations by using the elimination method (Section 11.1).

   For example: Solve for x, y, and z:

$$\begin{cases} x - y + 2z = 4 & \textbf{(1)} \\ 3x + 2y - z = 3 & \textbf{(2)} \\ 5x - 3y + 3z = 8 & \textbf{(3)} \end{cases}$$

   Solution: We choose to eliminate z:

$$\begin{array}{lr} \textit{Equation (1)} & x - y + 2z = 4 \\ \textit{2 times equation (2)} & \underline{6x + 4y - 2z = 6} \qquad \textit{Add.} \\ & 7x + 3y \qquad\;\; = 10 \qquad \text{(4)} \end{array}$$

$$\begin{array}{lr} \textit{3 times equation (2)} & 9x + 6y - 3z = 9 \\ \textit{Equation (3)} & \underline{5x - 3y + 3z = 8} \qquad \textit{Add.} \\ & 14x + 3y \qquad\;\; = 17 \qquad \text{(5)} \end{array}$$

Using equations (4) and (5), we have the following 2×2 system:

$$\begin{cases} 7x + 3y = 10 & \textit{Multiply by } -1 \rightarrow & -7x - 3y = -10 \\ 14x + 3y = 17 & \textit{As is} \rightarrow & \underline{14x + 3y = \quad 17} \\ & & 7x \qquad\;\; = \quad 7 \\ & & x = \quad 1 \end{cases}$$

Substitute $x = 1$ into equation (4): Substitute $x = 1$, $y = 1$ into equation (1):

$$7x + 3y = 10 \qquad\qquad x - y + 2z = 4$$
$$7(1) + 3y = 10 \qquad\qquad 1 - 1 + 2z = 4$$
$$y = 1 \qquad\qquad\qquad z = 2$$

Thus, our solution is $x = 1$, $y = 1$, $z = 2$.

Check in all three original equations.

2. Solve a linear system by Gaussian elimination (Section 11.2).

For example: Solve the following system using Gaussian elimination:

$$\begin{cases} 2x + y - z = 4 \\ x - 2y + z = 7 \\ 3x - y + 2z = 11 \end{cases}$$

Solution: Set up the augmented matrix, and transform the augmented matrix into a (row-equivalent) matrix in triangular form:

$$\begin{bmatrix} 2 & 1 & -1 & | & 4 \\ 1 & -2 & 1 & | & 7 \\ 3 & -1 & 2 & | & 11 \end{bmatrix} \quad R_1 \leftrightarrow R_2 \quad \begin{bmatrix} 1 & -2 & 1 & | & 7 \\ 2 & 1 & -1 & | & 4 \\ 3 & -1 & 2 & | & 11 \end{bmatrix}$$

$$\begin{bmatrix} 1 & -2 & 1 & | & 7 \\ 2 & 1 & -1 & | & 4 \\ 3 & -1 & 2 & | & 11 \end{bmatrix} \quad \begin{matrix} -2R_1 + R_2 \rightarrow R_2 \\ -3R_1 + R_3 \rightarrow R_3 \end{matrix} \quad \begin{bmatrix} 1 & -2 & 1 & | & 7 \\ 0 & 5 & -3 & | & -10 \\ 0 & 5 & -1 & | & -10 \end{bmatrix}$$

$$\begin{bmatrix} 1 & -2 & 1 & | & 7 \\ 0 & 5 & -3 & | & -10 \\ 0 & 5 & -1 & | & -10 \end{bmatrix} \quad -R_2 + R_3 \rightarrow R_3 \quad \begin{bmatrix} 1 & -2 & 1 & | & 7 \\ 0 & 5 & -3 & | & -10 \\ 0 & 0 & 2 & | & 0 \end{bmatrix} \quad \textit{This matrix is now in triangular form.}$$

Set up the associated system of the matrix in triangular form:

$$\begin{cases} x - 2y + z = 7 \\ 5y - 3z = -10 \\ 2z = 0 \end{cases}$$

Solve by back substitution:

First solve for z in the last equation: $2z = 0$ | *Substitute z = 0 in the second equation and solve for y:* $5y - 3z = -10$ | *Substitute z = 0 and y = −2 in the first equation and solve for x:* $x - 2y + z = 7$
$z = 0$ | $5y - 3(0) = -10$ | $x - 2(-2) + 0 = 7$
| $y = -2$ | $x = 3$

The solution is $(3, -2, 0)$.

3. Solve a linear system by Gauss–Jordan elimination (Section 11.2).

For example: Solve the following system by Gauss–Jordan elimination:

$$\begin{cases} 2x + 4y = 24 \\ 3x + y = 11 \end{cases}$$

Set up the augmented matrix and transform the augmented matrix into reduced echelon form:

$$\begin{bmatrix} 2 & 4 & | & 24 \\ 3 & 1 & | & 11 \end{bmatrix} \quad \tfrac{1}{2}R_1 \rightarrow R_1 \quad \begin{bmatrix} 1 & 2 & | & 12 \\ 3 & 1 & | & 11 \end{bmatrix}$$

$$\begin{bmatrix} 1 & 2 & | & 12 \\ 3 & 1 & | & 11 \end{bmatrix} \qquad -3\,R_1 + R_2 \rightarrow R_2 \qquad \begin{bmatrix} 1 & 2 & | & 12 \\ 0 & -5 & | & -25 \end{bmatrix}$$

$$\begin{bmatrix} 1 & 2 & | & 12 \\ 0 & -5 & | & -25 \end{bmatrix} \qquad -\tfrac{1}{5}R_2 \rightarrow R_2 \qquad \begin{bmatrix} 1 & 2 & | & 12 \\ 0 & 1 & | & 5 \end{bmatrix}$$

$$\begin{bmatrix} 1 & 2 & | & 12 \\ 0 & 1 & | & 5 \end{bmatrix} \qquad -2R_2 + R_1 \rightarrow R_1 \qquad \begin{bmatrix} 1 & 0 & | & 2 \\ 0 & 1 & | & 5 \end{bmatrix}$$

This matrix is in reduced echelon form.

We can see from the matrix in reduced echelon form that the solution is $x = 2$, $y = 5$.

4. Perform matrix operations (Section 11.3).

For example: Given $A = \begin{bmatrix} -2 & 3 \\ 4 & 0 \end{bmatrix}$, $B = \begin{bmatrix} 1 & 5 \\ 2 & -1 \end{bmatrix}$, and $C = \begin{bmatrix} 1 & 5 & -3 \\ 2 & -1 & 0 \end{bmatrix}$, find

(a) $A - 2B$ (b) BC

(a) We first note that since matrices A and B are of the same size we can compute $A - 2B$:

$$A - 2B = \begin{bmatrix} -2 & 3 \\ 4 & 0 \end{bmatrix} - 2\begin{bmatrix} 1 & 5 \\ 2 & -1 \end{bmatrix} = \begin{bmatrix} -2 - 2(1) & 3 - 2(5) \\ 4 - 2(2) & 0 - 2(-1) \end{bmatrix} = \begin{bmatrix} -4 & -7 \\ 0 & 2 \end{bmatrix}$$

(b) Again we recognize that we can multiply matrices B and C because the number of columns in matrix B is equal to the number of rows in matrix C. Thus,

$$BC = \begin{bmatrix} 1 & 5 \\ 2 & -1 \end{bmatrix}\begin{bmatrix} 1 & 5 & -3 \\ 2 & -1 & 0 \end{bmatrix}$$

$$= \begin{bmatrix} 1(1) + 5(2) & 1(5) + 5(-1) & 1(-3) + 5(0) \\ 2(1) + (-1)(2) & 2(5) + (-1)(-1) & 2(-3) + (-1)(0) \end{bmatrix} = \begin{bmatrix} 11 & 0 & -3 \\ 0 & 11 & -6 \end{bmatrix}$$

5. Use matrix inverses to solve a system of equations (Section 11.4).

For example:

Using a matrix inverse, find the solution to the following system of equations:

$$\begin{cases} x - 2y = 4 \\ x + 5y = -3 \end{cases}$$

Solution:

We identify matrices A, X, and K:

$$A = \begin{bmatrix} 1 & -2 \\ 1 & 5 \end{bmatrix} \qquad X = \begin{bmatrix} x \\ y \end{bmatrix} \qquad K = \begin{bmatrix} 4 \\ -3 \end{bmatrix}$$

Then $AX = K$. We find A^{-1} to be

$$A^{-1} = \begin{bmatrix} \tfrac{5}{7} & \tfrac{2}{7} \\ -\tfrac{1}{7} & \tfrac{1}{7} \end{bmatrix} = \tfrac{1}{7}\begin{bmatrix} 5 & 2 \\ -1 & 1 \end{bmatrix}$$

Now we can solve for X:

$$X = A^{-1}K$$

$$= \tfrac{1}{7}\begin{bmatrix} 5 & 2 \\ -1 & 1 \end{bmatrix}\begin{bmatrix} 4 \\ -3 \end{bmatrix} = \begin{bmatrix} 2 \\ -1 \end{bmatrix}$$

Hence $X = \begin{bmatrix} x \\ y \end{bmatrix} = \begin{bmatrix} 2 \\ -1 \end{bmatrix}$, which means $x = 2$ and $y = -1$.

6. Evaluate 2×2 and 3×3 determinants (Section 11.5).

For example: Evaluate each of the following:

(a) $\begin{vmatrix} 3 & -2 \\ 4 & 5 \end{vmatrix} = 3(5) - 4(-2) = 15 + 8 = 23$

(b) $\begin{vmatrix} 2 & -1 & 3 \\ 1 & 4 & 2 \\ 3 & 1 & 2 \end{vmatrix} = 2\begin{vmatrix} 4 & 2 \\ 1 & 2 \end{vmatrix} - 1\begin{vmatrix} -1 & 3 \\ 1 & 2 \end{vmatrix} + 3\begin{vmatrix} -1 & 3 \\ 4 & 2 \end{vmatrix}$ *Expanding down the first column*

$$= 2(8 - 2) - (-2 - 3) + 3(-2 - 12)$$

$$= 12 + 5 - 42 = -25$$

7. Use Cramer's rule to solve 2×2 and 3×3 linear systems (Section 11.5).

For example: Solve the following system using Cramer's rule:

$$\begin{cases} 2x - 7y = 4 \\ 5x + 3y = 2 \end{cases}$$

Solution:

$$D = \begin{vmatrix} 2 & -7 \\ 5 & 3 \end{vmatrix} = 6 + 35 = 41$$

$$D_x = \begin{vmatrix} 4 & -7 \\ 2 & 3 \end{vmatrix} = 12 + 14 = 26$$

$$D_y = \begin{vmatrix} 2 & 4 \\ 5 & 2 \end{vmatrix} = 4 - 20 = -16$$

According to Cramer's rule,

$$x = \frac{D_x}{D} = \frac{26}{41} \quad \text{and} \quad y = \frac{D_y}{D} = -\frac{16}{41}$$

Thus, the solution is $\left(\dfrac{26}{41}, -\dfrac{16}{41} \right)$.

8. Solve a system of linear inequalities by the graphical method (Section 11.6).

For example: Solve the following system of inequalities:

$$\begin{cases} x + 2y \le 6 \\ \quad y < x - 6 \end{cases}$$

Solution: We graph the line $x + 2y = 6$ (with a solid line) and $y = x - 6$ (with a dashed line), as shown in Figure 11.14.

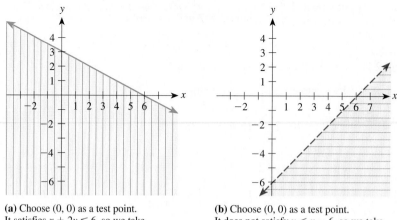

(a) Choose $(0, 0)$ as a test point.
It satisfies $x + 2y \leq 6$, so we take
the region containing $(0, 0)$.

(b) Choose $(0, 0)$ as a test point.
It does not satisfy $y < x - 6$, so we take
the region not containing $(0, 0)$.

Figure 11.14

The solution set is the intersection of the two regions, as indicated in Figure 11.15.

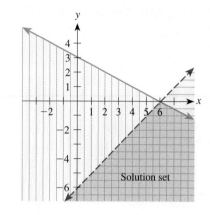

Figure 11.15
Graph of the system
$$\begin{cases} x + 2y \leq 6 \\ \quad y < x - 6 \end{cases}$$

We may choose $(6, -3)$ as a check point in the shaded region and verify that it satisfies the system.

9. Solve 2×2 nonlinear systems of equations and sketch their graphs (Section 11.7).

For example: Solve for x and y:

$$\begin{cases} x^2 + y^2 = 13 & \textbf{(1)} \\ x \ + y \ = \ 1 & \textbf{(2)} \end{cases}$$

Solution: Figure 11.16 illustrates the graphs of the equations in the system.

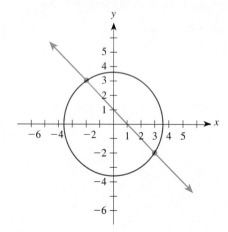

Figure 11.16
Graphs of $x^2 + y^2 = 13$ and
$x + y = 1$

Since we do not have "like" terms to eliminate, we use the substitution method. We solve equation **(2)** for *y:*

$$y = 1 - x \qquad \text{Call this equation (3).}$$

Substitute $y = 1 - x$ into equation (1):

$$x^2 + y^2 = 13$$
$$x^2 + (1 - x)^2 = 13$$
$$x^2 + 1 - 2x + x^2 = 13$$
$$2x^2 - 2x - 12 = 0$$
$$2(x^2 - x - 6) = 0$$
$$2(x - 3)(x + 2) = 0$$
$$x - 3 = 0 \quad \text{or} \quad x + 2 = 0$$
$$x = 3 \quad \text{or} \quad x = -2$$

Substitute *x* values into equation (3):

Substitute $x = 3$ **Substitute $x = -2$**

$$y = 1 - x \qquad\qquad y = 1 - x$$
$$y = 1 - 3 \qquad\qquad y = 1 - (-2)$$
$$y = -2 \qquad\qquad\quad y = 3$$

The solutions are $(3, -2)$ and $(-2, 3)$.

CHAPTER 11 REVIEW EXERCISES

In Exercises 1–6, solve the system of equations.

1. $\begin{cases} x + 2y - z = 2 \\ 2x + 3y + 4z = 9 \\ 3x + y - 2z = 2 \end{cases}$

2. $\begin{cases} 3x - 2y + 5z = 1 \\ 4x + 3y - 6z = 14 \\ 2x + 4y + 4z = 0 \end{cases}$

3. $\begin{cases} 2x - 3z = 7 \\ 3x + 4y = -11 \\ 3y - 2z = 0 \end{cases}$

4. $\begin{cases} a + 2b = 3c - 3 \\ 2a - 3b = c - 1 \\ 3b - 4c = a - 4 \end{cases}$

5. $\begin{cases} 3x + y - z = 2 \\ 3x + y - z = 5 \\ x - y + z = 1 \end{cases}$

6. $\begin{cases} 2x + 10y + z = 1 \\ x - 100y - 4z = -1 \\ x - 10y + z = -2 \end{cases}$

In Exercises 7–10, put the matrices into reduced echelon form.

7. $\left[\begin{array}{cc|c} 1 & 2 & 6 \\ 2 & 6 & 8 \end{array}\right]$

8. $\left[\begin{array}{cc|c} 3 & 4 & 2 \\ 1 & 2 & 6 \end{array}\right]$

9. $\left[\begin{array}{ccc|c} 2 & 2 & 6 & 4 \\ 2 & 5 & 9 & -2 \\ 1 & 2 & 3 & -1 \end{array}\right]$

10. $\left[\begin{array}{ccc|c} 1 & 3 & -1 & -3 \\ 2 & 7 & -3 & -9 \\ 3 & 11 & -1 & -3 \end{array}\right]$

In Exercises 11–14, solve by Gaussian elimination.

11. $\begin{cases} x - 2y = 1 \\ 2x + 3y = 9 \end{cases}$

12. $\begin{cases} 3x - y = 3 \\ x + 2y = 1 \end{cases}$

13. $\begin{cases} x + 2y - z = 5 \\ x - y + z = 0 \\ x + y + 2z = 1 \end{cases}$

14. $\begin{cases} x - 2y - z = -4 \\ 2x + y - z = 1 \\ 3x - y + 2z = -7 \end{cases}$

In Exercises 15–16, solve by Gauss–Jordan elimination.

15. $\begin{cases} x - 3y = -1 \\ 2x + y = 5 \end{cases}$

16. $\begin{cases} x + y + 9z = -6 \\ -x - 6z = 7 \\ 2x + y + 12z = -10 \end{cases}$

In Exercises 17–26, use the following matrices to perform the indicated matrix operations if possible.

$$A = \begin{bmatrix} 1 \\ 0 \\ -2 \end{bmatrix} \quad B = [2 \quad -3 \quad 1] \quad C = \begin{bmatrix} 2 & -1 \\ 0 & 5 \end{bmatrix} \quad D = \begin{bmatrix} -3 & 0 \\ 2 & 1 \end{bmatrix} \quad E = \begin{bmatrix} 1 & 5 & -2 \\ -1 & 0 & 6 \end{bmatrix}$$

17. $5C$

18. $-4E$

19. $C - D$

20. $2C + 3D$

21. ED

22. BA

23. CE

24. EC

25. $(C + D)E$

26. $E(C + D)$

In Exercises 27–30, use the method of matrix inverses to solve the system of equations.

27. $\begin{cases} 2x + 3y = 14 \\ 3x + \dfrac{1}{2}y = 1 \end{cases}$

28. $\begin{cases} 7x + 4y = 0 \\ 9x + 5y = -1 \end{cases}$

29. $\begin{cases} 2x - 3y + 4z = 18 \\ 5x + 4y - 3z = -4 \\ 3x - 2y + z = 12 \end{cases}$

30. $\begin{cases} x + y + z = 1 \\ 2x + y - 2z = 1 \\ 3x - 2y + 3z = 12 \end{cases}$

In Exercises 31–34, evaluate the given determinant.

31. $\begin{vmatrix} 2 & 3 \\ 4 & -1 \end{vmatrix}$

32. $\begin{vmatrix} 8 & 5 \\ 2 & 4 \end{vmatrix}$

33. $\begin{vmatrix} 1 & 3 & -2 \\ 2 & 4 & 3 \\ 5 & -1 & -3 \end{vmatrix}$

34. $\begin{vmatrix} 2 & -4 & 1 \\ 0 & 6 & 0 \\ 5 & 3 & 7 \end{vmatrix}$

In Exercises 35–40, solve the system using Cramer's rule.

35. $\begin{cases} 3x + 7y = 4 \\ 5x + 4y = 2 \end{cases}$

36. $\begin{cases} 5x - 2y = 3 \\ 6x - 7y = 1 \end{cases}$

37. $\begin{cases} 4a + 2b = 7 \\ 6a + 3b = 9 \end{cases}$

39. $\begin{cases} 9m + 6n = 12 \\ 3m + 2n = 4 \end{cases}$

39. $\begin{cases} 3s + 4t + u = 3 \\ 5s - 3t + 6u = 2 \\ 4s - 5t - 5u = 1 \end{cases}$

40. $\begin{cases} -4u + 3v + 6w = 4 \\ 3u - 5v - 7w = -1 \\ 5u - 4v + 2w = 3 \end{cases}$

In Exercises 41–46, solve the nonlinear system of equations.

41. $\begin{cases} x^2 + y^2 = 8 \\ x + y = 4 \end{cases}$

42. $\begin{cases} x^2 - 3y^2 = 4 \\ 2x^2 + 5y^2 = 12 \end{cases}$

43. $\begin{cases} 2x^2 - y^2 = 14 \\ 2x - 3y = -8 \end{cases}$

44. $\begin{cases} x^2 + y^2 = 9 \\ x^2 + y = 3 \end{cases}$

45. $\begin{cases} x^2 + y^2 = 4 \\ 2x^2 + 3y^2 = 18 \end{cases}$

46. $\begin{cases} 2x^2 + 3y^2 = 4 \\ x^2 - 4y^2 = 9 \end{cases}$

In Exercises 47–50, graph the system of inequalities.

47. $\begin{cases} x + y \le 2 \\ 2x - 3y \le 6 \end{cases}$

48. $\begin{cases} y - 2x \le 8 \\ y > 8 - x \end{cases}$

49. $\begin{cases} 2x + 3y \le 12 \\ y < x \\ x \ge 0 \\ y \ge 0 \end{cases}$

50. $\begin{cases} y - x > 3 \\ x - y > 2 \end{cases}$

51. A total of \$20,000 is split into three investments. Part is invested in a bank account paying 7.7% interest per year, part in a corporate bond paying 8.6% interest per year, and the rest in a municipal bond paying 9.8% interest per year. The sum of the amounts invested in the two bonds is \$3,000 more than the amount invested in the bank account. If the annual interest from all three investments is $1,719.10, how much is invested at each rate?

52. Jane, June, and Jean go shopping at a flea market. Jane buys one sweater, two blouses, and one pair of jeans. June buys two sweaters, one blouse, and one pair of jeans. Jean buys two sweaters, three blouses, and two pairs of jeans. If Jane spends $28.50, June spends $30, and Jean spends $52, what are the prices of an individual sweater, blouse, and pair of jeans?

53. Find the dimensions of a rectangle with perimeter 41 ft and area 100 sq ft.

54. A farmer has 50 acres available to plant crops. It costs $200 an acre for labor to plant crop X, and $100 an acre to plant crop Y. She does not want to spend more than $8,000 for planting crops. How many acres of each crop should she plant? Write a system of inequalities to describe this situation and sketch the solution set of the system.

CHAPTER 11 PRACTICE TEST

1. Solve the following system of simultaneous equations:

$$\begin{cases} 2x - 3y + 4z = 2 \\ 3x + 2y - z = 10 \\ 2x - 4y + 3z = 3 \end{cases}$$

2. Solve the following systems using augmented matrices:

(a) $\begin{cases} 5x + 2y = 1 \\ x + 3y = 8 \end{cases}$

(b) $\begin{cases} x - y = 2 \\ x + 2z = 7 \\ -2x + 3y + 4z = 5 \end{cases}$

3. Perform the indicated matrix operations using

$$A = \begin{bmatrix} 1 & 5 & -2 \\ -1 & 0 & 6 \\ 2 & 2 & 0 \end{bmatrix} \quad \text{and} \quad B = \begin{bmatrix} 2 & -3 & 1 \\ 0 & 1 & -1 \\ 4 & 0 & 1 \end{bmatrix}.$$

(a) $3A - 4B$ (b) AB

4. Solve the following system of equations using matrix inverses.

$$\begin{cases} 2x - 5y = 11 \\ 3x + 2y = 7 \end{cases}$$

5. Evaluate the following determinants:

(a) $\begin{vmatrix} 2 & 3 \\ -1 & 4 \end{vmatrix}$

(b) $\begin{vmatrix} 5 & 0 & 2 \\ 2 & 3 & 1 \\ 1 & 1 & 2 \end{vmatrix}$

6. Solve the following using Cramer's rule:

(a) $\begin{cases} 2x - 6y = -1 \\ 4x = 8y + 5 \end{cases}$

(b) $\begin{cases} 5x + y = 9 \\ 3x - z = 3 \\ y + 3z = 8 \end{cases}$

7. A bank teller gave a customer change for a $500 bill in $5, $10, and $20 bills. There were twice as many $10 bills as $20 bills and altogether there were 40 bills. How many of each did the customer receive for this $500 bill?

8. Solve the following nonlinear system of equations:

$$\begin{cases} x^2 + 2y^2 = 6 \\ x - y = 3 \end{cases}$$

9. Sketch the solution set of the following system of inequalities: $\begin{cases} y < 4 \\ x \le 2 \\ x + y > 3 \end{cases}$

CHAPTERS 10–11 CUMULATIVE REVIEW

In Exercises 1–6, solve the given system of equations.

1. $\begin{cases} x + y + z = 6 \\ 2x + y - 2z = 6 \\ 3x - y + 3z = 10 \end{cases}$

2. $\begin{cases} 2x - 3y + 4z = 22 \\ 3x - y + 2z = 10 \\ 4x + 2y - z = -8 \end{cases}$

3. $\begin{cases} 3x - 4y + 5z = 1 \\ 2x - y + 3z = 2 \\ x + 2y + z = 3 \end{cases}$

4. $\begin{cases} \dfrac{x}{2} + \dfrac{y}{3} + \dfrac{z}{4} = 13 \\ x - \dfrac{y}{2} + \dfrac{z}{3} = 10 \\ \dfrac{x}{3} + \dfrac{y}{4} - \dfrac{z}{2} = 1 \end{cases}$

5. $\begin{cases} x - 2y + 3z = 4 \\ \dfrac{3}{2}x - 3y + \dfrac{9}{2}z = 6 \\ -3x + 6y - 9z = -12 \end{cases}$

6. $\begin{cases} 3x - y = 0 \\ 4x + z = 1 \\ 6y - 3z = 2 \end{cases}$

In Exercises 7–14, evaluate the given determinant.

7. $\begin{vmatrix} 3 & 2 \\ 4 & 5 \end{vmatrix}$

8. $\begin{vmatrix} 1 & -2 \\ 3 & -1 \end{vmatrix}$

9. $\begin{vmatrix} 1 & 2 \\ 2 & 4 \end{vmatrix}$

10. $\begin{vmatrix} 3 & 5 \\ 3 & -2 \end{vmatrix}$

11. $\begin{vmatrix} 4 & -1 & 2 \\ 2 & 1 & 0 \\ -1 & 2 & -3 \end{vmatrix}$

12. $\begin{vmatrix} 3 & 2 & -2 \\ 6 & 4 & -4 \\ 5 & 1 & -1 \end{vmatrix}$

13. $\begin{vmatrix} 5 & 4 & 3 \\ 2 & 0 & 0 \\ 3 & 1 & 1 \end{vmatrix}$

14. $\begin{vmatrix} 2 & 2 & 1 \\ 2 & 2 & 1 \\ 3 & 4 & 1 \end{vmatrix}$

In Exercises 15–16, solve using augmented matrices:

15. $\begin{cases} x + 2y = 0 \\ 2x - 3y = 7 \end{cases}$

16. $\begin{cases} x + y - z = -2 \\ x + 2y + 2z = 9 \\ 2x + z = 1 \end{cases}$

In Exercises 17–20, solve the given system of equations using Cramer's rule.

17. $\begin{cases} 3x + 7y = 2 \\ 10x + 5y = 11 \end{cases}$

18. $\begin{cases} 5x - 6y = 4 \\ 11x + 8y = 5 \end{cases}$

19. $\begin{cases} 2x + 3y + 4z = 3 \\ 5x + 2y - 3z = 2 \\ 3x - 7y + 5z = -7 \end{cases}$

20. $\begin{cases} 4x - 6y - z = 8 \\ 2x + 3y + 3z = 9 \\ 6x + 9y + 5z = 10 \end{cases}$

In Exercises 21–28, solve the given system of equations.

21. $\begin{cases} x + y = 1 \\ x^2 + y^2 = 5 \end{cases}$

22. $\begin{cases} x - 2y = 2 \\ x^2 + 4y^2 = 20 \end{cases}$

23. $\begin{cases} y - x^2 = 4 \\ x^2 + y = 1 \end{cases}$

24. $\begin{cases} y = x^2 - 2x + 1 \\ y = x - x^2 \end{cases}$

25. $\begin{cases} x^2 + y^2 = 10 \\ 3x^2 - 4y^2 = 23 \end{cases}$

26. $\begin{cases} x^2 + 4y^2 = 40 \\ 4x^2 + y^2 = 25 \end{cases}$

27. $\begin{cases} x - y^2 = 3 \\ 3x - 2y = 9 \end{cases}$

28. $\begin{cases} x^2 - y^2 = 5 \\ y^2 - x = 7 \end{cases}$

In Exercises 29–34, sketch the solution set to each system of inequalities.

29. $\begin{cases} x + y \le 6 \\ 2x + y \ge 4 \end{cases}$

30. $\begin{cases} 3x + y > 6 \\ 2x - y < 4 \end{cases}$

31. $\begin{cases} 2x + 3y < 12 \\ x < y \end{cases}$

32. $\begin{cases} 4x - y \le 6 \\ x \le -2 \end{cases}$

33. $\begin{cases} x + y \le 4 \\ x - y \le 4 \\ x \ge 0 \end{cases}$

34. $\begin{cases} 3x + y \le 9 \\ x + 3y \le 9 \\ x \ge 0 \\ y \ge 0 \end{cases}$

In Exercises 35–40, use the following matrices to perform the indicated matrix operations, if possible.

$$A = \begin{bmatrix} 3 \\ 1 \\ -4 \end{bmatrix} \qquad B = \begin{bmatrix} 5 & 0 & 2 \end{bmatrix} \qquad C = \begin{bmatrix} -1 & 2 \\ 5 & 0 \end{bmatrix} \qquad D = \begin{bmatrix} -4 & 1 \\ 0 & 3 \end{bmatrix}$$

35. $-3C$

36. $D - C$

37. $3C + 5D$

38. BA

39. CD

40. DC

In Exercises 41–42, use the method of matrix inverses to solve the system of equations.

41. $\begin{cases} 3x + 2y = 10 \\ \dfrac{1}{2}x + 3y = 1 \end{cases}$

42. $\begin{cases} 2x - y + 2z = 2 \\ y - 2z = 2 \\ 2x + 2y + z = 4 \end{cases}$

In Exercises 43–46, sketch the graph of the given function.

43. $f(x) = 3^{x+1}$

44. $f(x) = \left(\dfrac{1}{2}\right)^x - 1$

45. $f(x) = \log_5 x$

46. $f(x) = \log_2 x + 1$

In Exercises 47–52, translate logarithmic statements into exponential form and vice versa.

47. $\log_2 64 = 6$

48. $3^{-4} = \dfrac{1}{81}$

49. $\sqrt[3]{125} = 5$

50. $\log_{10} 0.01 = -2$

51. $\log_{27} 81 = \dfrac{4}{3}$

52. $4^{-1/2} = \dfrac{1}{2}$

In Exercises 53–60, find the given logarithm (assume $b > 0$ and $b \ne 1$).

53. $\log_3 81$

54. $\log_8 2$

55. $\log_4 \dfrac{1}{16}$

56. $\log_{16} 32$

57. $\log_9 \dfrac{1}{3}$

58. $\log_7 7^{2/3}$

59. $\log_b 1$

60. $\log_b b$

In Exercises 61–64, write each logarithm as a sum of simpler logarithms, if possible.

61. $\log_b \sqrt[3]{5xy}$

62. $\log_b \dfrac{x^3 y}{z^5}$

63. $\log_3 \dfrac{x^2\sqrt{y}}{9wz}$

64. $\dfrac{\log_b u^8}{\log_b v^4}$

In Exercises 65–68, use a calculator to find the following.

65. $\log 73{,}600$

66. $\ln 968$

67. Find x such that $\log x = 0.6085$.

68. $e^{0.4}$

In Exercises 69–70, use the change of base formula to compute the following.

69. $\log_9 384$

70. $\log_5 0.006$

In Exercises 71–80, solve the given equation.

71. $\dfrac{1}{2}\log_8 x = \log_8 5$

72. $2\log_b x = \log_b(6x - 9)$

73. $5^x = \dfrac{1}{25}$

74. $16^x = \dfrac{1}{2}$

75. $\log_6 x + \log_6 4 = 3$

76. $\log_4 x + \log_4(x - 6) = 2$

77. $\dfrac{4^{x^2}}{2x} = 64$

78. $9^{\sqrt{x}} = 3^{3x-8}$

79. $9^x = 7^{x+3}$

80. $\log_b(x + 4) - \log_b(x - 4) = \log_b 5$

81. How long will it take $3,000 invested at an annual rate of 8% and compounded quarterly to grow to $5,000?

82. Based on the exponential growth model, how long will it take an initial population of 30,000 growing at 3% per year to double?

83. Based on the exponential decay model with $r = 0.002$, how long will it take 20 g of radioactive material to decay to 2 g of radioactive material?

84. If the pH of a solution is defined to be $pH = -\log[H_3O^+]$, find the pH of a solution for which $[H_3O^+] = 3.6 \times 10^{-4}$.

85. If the pH of a solution is defined to be $pH = -\log[H_3O^+]$, where $[H_3O^+]$ is the hydronium ion concentration, find the hydronium ion concentration of a solution whose pH is 8.2.

CHAPTERS 10–11 CUMULATIVE PRACTICE TEST

1. *Solve the following system of equations:* $\begin{cases} x + y + z = 6 \\ 3x + 2y - z = 11 \\ 2x - 4y - z = 12 \end{cases}$

2. *Solve the following system using augmented matrices:* $\begin{cases} x + 2y \quad\;\; = 4 \\ x + y + z = 0 \\ x - y + 2z = -6 \end{cases}$

3. *Solve the following systems of equations by using Cramer's rule:*

   (a) $\begin{cases} 6x + 5y = 13 \\ 7x + 8y = 26 \end{cases}$

   (b) $\begin{cases} 4x - 5y + 2z = 17 \\ 3x + 7y - 5z = 2 \\ 5x - 6y + 3z = 21 \end{cases}$

4. *Solve the system of equations:*

   $$\begin{cases} x^2 - y^2 = 7 \\ x - y = 1 \end{cases}$$

5. *Sketch the solution set of the following system of inequalities:*

   $$\begin{cases} x - y \le 5 \\ x - 2y \le 0 \\ y \ge 0 \end{cases}$$

6. Given the following matrices A and B,

   $$A = \begin{bmatrix} 3 & -4 \\ 0 & 2 \end{bmatrix} \qquad B = \begin{bmatrix} -2 & 0 \\ 1 & 5 \end{bmatrix}$$

   compute

   (a) $4A - 3B$

   (b) AB

7. Use matrix inverses to solve the following system of equations.

   $$\begin{cases} 2x + 4y = -4 \\ 3x + 2y = -8 \end{cases}$$

8. *Sketch the graphs of the following equations:*

   (a) $y = \left(\dfrac{1}{4}\right)^x$

   (b) $y = |x + 5|$

9. *Write in exponential form:* $\log_8 \dfrac{1}{4} = -\dfrac{2}{3}$

10. *Find each of the following logarithms:*

    (a) $\log_3 \dfrac{1}{27}$

    (b) $\log_4 8$

11. *Express as a sum of simpler logarithms, if possible:*

    (a) $\log_b(x \sqrt[3]{y})$

    (b) $\log_b \dfrac{x^3}{\sqrt{xy}}$

12. *Use a calculator to find the following.*

    (a) $\log 0.00637$ (b) Find x such that $\log x = 3.9263$.   (c) $\ln 94$

13. Use the change of base formula to compute $\log_8 985$.

14. *Solve the following equations:*

    (a) $2^{x^2} = 8^{3x}$

    (b) $\log_6 x + \log_6(x + 5) = 2$

15. How long must \$5,000 be invested so that if the interest is compounded quarterly at 10%, the \$5,000 will have grown to \$10,000?

APPENDIX A

Sets

A *set* is simply a *well-defined* collection of objects. The phrase *well-defined* means that there are clearly determined criteria for membership in the set. The criteria can be a list of objects in the set, called the ***elements*** or members of the set, or it can be a description of the objects in the set.

For example, it is not sufficient to say "the set of all tall people in our class." "Tall" is not well defined; we may not agree on whether a $5'11''$ person belongs in the set or not. To make the set well defined, we could say "the set of people in our class over 6 feet tall."

One way to represent a set is to list the elements of the set, separated by commas, and to enclose the list in *set brackets,* which look like { }.

We often designate sets by using capital letters such as A, B, and C. For example,

$$A = \{1, 6, 17, 45\} \qquad B = \{k, l, m, n, o\}$$

We say that the set A is the set consisting of the numerals 1, 6, 17, and 45.

The symbol we use to indicate that an object is a member of a particular set is "$\in$". Thus, $x \in S$ is a symbolic way of writing that x is a member or an element of the set S.

We use the symbol "$\notin$" to indicate that an object is *not* an element of a set. (In general, when we put a "/" through a math symbol it means *not;* thus, "$\neq$" means "not equal to.") For example, using the sets A and B listed above, we have:

$$17 \in A \qquad \text{17 is an element of } A.$$
$$p \notin B \qquad \text{p is not an element of } B.$$

Sometimes, in order to exhibit a set that contains many elements or a set that contains an infinite number of elements, we use a variation on the listing method. We list a few elements followed by a comma and three dots. For example, two sets of numbers to which we frequently refer are N and W. The set of numbers we use for counting is called the set of ***natural numbers,*** and is usually denoted by the letter N:

$$N = \{1, 2, 3, 4, 5, \ldots\}$$

When we include the number 0 with this set it is called the set of ***whole numbers*** and is denoted by the letter W:

$$W = \{0, 1, 2, 3, 4, 5, \ldots\}$$

Of course, this method of listing a set can be used only when the first few elements clearly show the pattern for *all* the elements in the set.

Often when we describe a set we use the word *between,* which can be ambiguous. When we say "the numbers between 7 and 15," do we mean to include or exclude 7 and 15? Therefore, let's agree that unless we specifically indicate otherwise, when we say *between* and *in between,* we do *not* include the first and last numbers.

EXAMPLE 1

List the elements of the following sets:

(a) The set *A* of even numbers between 6 and 54
(b) The set *B* of odd numbers greater than 19

Solution

(a) An even number is (an integer) exactly divisible by 2.

$$A = \{8, 10, 12, 14, \ldots, 52\}$$

Note that 6 and 54 are not included. Also note that if the set is large but not infinite, we use the three-dot notation, but list the last element of the set preceded by a comma.

(b) Since no upper limit to this set is given, our answer is

$$B = \{21, 23, 25, \ldots\}$$

Note that 19 is not included.

Another way of writing a set is by using set-builder notation. **Set-builder notation** consists of the set braces, a variable that acts as a placeholder, a vertical bar (|) that is read "such that," and a sentence that describes what the variable can be. This last part is called the **condition** on the variable. For example,

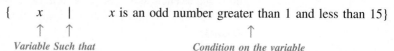

$$\{ \quad x \quad | \qquad x \text{ is an odd number greater than 1 and less than 15}\}$$

Variable   Such that   Condition on the variable

This is read "the set of all *x* such that *x* is an odd number greater than 1 and less than 15," which is the set $\{3, 5, 7, 9, 11, 13\}$.

It is possible to place a condition on the variable that cannot be satisfied, as for example

$$F = \{x \,|\, x \text{ is a whole number less than } 0\}$$

Since it is impossible for a whole number to be less than 0, this set *F* has no members. A set with no members is called the **empty set** or the **null set** and is symbolized by "$\varnothing$". Thus, we have $F = \varnothing$.

In dealing with numbers, we use the equals symbol, "=," to indicate that two expressions represent the same number. For example, $6 + 2 = 8$ since both $6 + 2$ and 8 represent the same number. For sets, the same symbol of equality is used to indicate that two sets are identical (they contain the same elements). Hence,

$$\{1, 3, 8\} = \{3, 8, 1\}$$

because both sets contain the same elements. Note that the order of the elements is unimportant.

DEFINITION

The set *B* is a **subset** of *A*, written $B \subset A$, if all elements of *B* are also contained in *A*.

Hence, if $B = \{3, 7, 9\}$ and $A = \{2, 3, 4, 5, 7, 9\}$ then *B* is a subset of *A*, since all elements of *B* (3, 7, and 9) are contained in *A*.

Note that $N \subset W$ since all natural numbers are contained in the set of whole numbers.

We can also create new sets from old sets by certain *operations* on sets.

DEFINITION

The **union** of two sets A and B, written $A \cup B$, is the set made up of elements in A or in B, or in both A and B.

For example, if $S = \{a, b, c\}$ and $T = \{$red, white, blue$\}$, then $S \cup T$ is the new set formed by combining elements of S and T:

$$S \cup T = \{a, b, c, \text{red, white, blue}\}$$

We often see $A \cup B$ defined symbolically as follows:

$$A \cup B = \{x \mid x \in A \text{ or } x \in B\}$$

(In mathematics, the word *or* includes the possibility that x is an element of *both* A and B.)

DEFINITION

The **intersection** of two sets A and B, written $A \cap B$, is the set made up of all elements common to both A and B.

For example, if $S = \{1, 3, 8, 9, 15\}$ and $T = \{3, 7, 9, 11\}$, then $S \cap T$ is the new set made up of elements that are common to both sets:

$$S \cap T = \{3, 9\}$$

We often see $A \cap B$ defined symbolically as follows:

$$A \cap B = \{x \mid x \in A \text{ and } x \in B\}$$

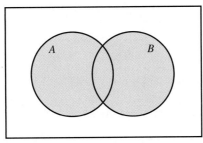
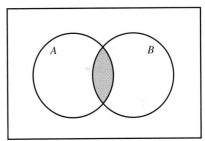

Figure A.1
Union and intersection

(a) $A \cup B$ is the set of all elements in A or B, or in both A and B.

(b) $A \cap B$ is the set of all elements common to both A and B.

The shaded areas in Figure A.1 represent the union and intersection of sets A and B, respectively.

EXAMPLE 2

List the elements of the following sets:

(a) $A = \{y \mid y \in N, y$ is a multiple of 6, and y is less than 30$\}$
(b) $B = \{s \mid s \in N, s$ is a multiple of 4, and s is less than 28$\}$
(c) $A \cap B$
(d) $A \cup B$

Solution

(a) $A = \{y \mid y \in N, y$ is a multiple of 6, and y is less than 30$\}$. Hence,

$$A = \{6, 12, 18, 24\}$$

(b) $B = \{s \mid s \in N, s$ is a multiple of 4, and s is less than 28$\}$. Hence,

$$B = \{4, 8, 12, 16, 20, 24\}$$

(c) *A* ∩ *B*: Since 12 and 24 belong to both sets *A* and *B*, we have

$$A \cap B = \{12, 24\}$$

(d) *A* ∪ *B*: Looking at sets *A* and *B*, we have

$$A \cup B = \{4, 6, 8, 12, 16, 18, 20, 24\}$$

Note that in *A* ∪ *B*, even though the elements 12 and 24 appear in both sets, we write them only once.

EXAMPLE 3

If $A = \{x \mid x \leq 4\}$, $B = \{x \mid x > -1\}$, $C = \{x \mid -2 \leq x < 3\}$, and $D = \{x \mid x < -3\}$, find:

(a) *A* ∩ *B*
(b) *A* ∪ *B*
(c) *A* ∩ *C*
(d) *B* ∩ *D*
(e) *B* ∪ *D*

Solution   If we graph the sets on the real number line, it is easier to "see" the solutions:

(a) *A* ∩ *B* is the intersection of sets *A* and *B*:

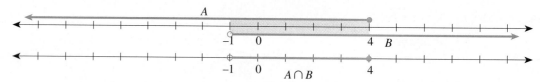

Hence, $A \cap B = \{x \mid -1 < x \leq 4\}$

(b) *A* ∪ *B* is the union of sets *A* and *B*:

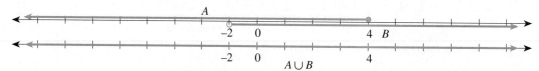

Hence, $A \cup B = \mathbb{R}$

(c) *A* ∩ *C* is the intersection of sets *A* and *C*:

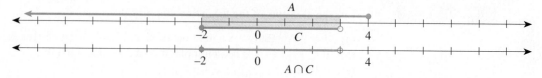

Hence, $A \cap C = \{x \mid -2 \leq x < 3\} = C$

(d) *B* ∩ *D* is the intersection of sets *B* and *D*:

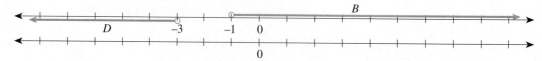

Since there are no real numbers common to both *B* and *D*, $B \cap D = \emptyset$.

(e) $B \cup D$ is the union of sets B and D:

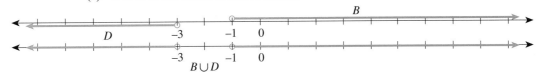

Hence, $B \cup D = \{x \mid x < -3 \text{ or } x > -1\}$

APPENDIX A EXERCISES

In Exercises 1–12, indicate whether the statement is true or false.

1. $3 \in N$ **2.** $-3 \in N$ **3.** $0 \notin W$ **4.** $0 \in N$

5. $\{a, b\} \subset \{a, b, c\}$ **6.** $e \in \{a, b, c, d\}$ **7.** $\{1, 2, 3\} = \{3, 2, 1\}$ **8.** $12 \in \{1, 2, 3\}$

9. $a \in \{x \mid x \text{ is a letter of the alphabet}\}$ **10.** $\{a, b, d\} \subset \{a, b, c, d\}$

11. $\{a \mid a \text{ is a multiple of } 6\} \subset \{a \mid a \text{ is a multiple of } 3\}$ **12.** $\{x \mid x \text{ is a multiple of } 3\} \subset \{x \mid x \text{ is a multiple of } 6\}$

In Exercises 13–30, list the elements in the specified set.

13. $\{n \mid n \in N, n \text{ is less than } 12\}$ **14.** $\{x \mid x \in W, x \text{ is less than } 12\}$

15. $\{m \mid m \in N, m \text{ is greater than } 6 \text{ and less than } 13\}$ **16.** $\{m \mid m \in N, m \text{ is greater than } 2 \text{ and less than or equal to } 8\}$

17. $\{a \mid a \in W, a \text{ is greater than or equal to } 3 \text{ and less than } 18\}$ **18.** $\{t \mid t \in W, t \text{ is greater than } 16 \text{ and less than or equal to } 28\}$

19. $\{n \mid n \in W, n \text{ is greater than } 5 \text{ and less than } 4\}$ **20.** $\{n \mid n \in W, n \text{ is greater than } 5 \text{ and less than } 5\}$

21. $\{n \mid n \text{ is a prime number between } 20 \text{ and } 40, n \in N\}$ **22.** $\{n \mid n \text{ is a composite number between } 20 \text{ and } 40, n \in N\}$

23. $\{m \mid m \text{ is both a prime and a composite number}, m \in N\}$ **24.** $\{m \mid m \in N \text{ is neither a prime nor a composite number}\}$

25. $\{t \mid t \in W \text{ and } t \text{ is a multiple of } 12\}$ **26.** $\{t \mid t \in W \text{ and } t \text{ is a multiple of } 2 \text{ but not of } 10\}$

27. $\{n \mid n \in W \text{ and } n \text{ is a multiple of } 5\}$ **28.** $\{n \mid n \in W \text{ and } n \text{ is a multiple of } 8 \text{ but not of } 4\}$

29. $\{n \mid n \in W \text{ and } n \text{ is a factor of } 48\}$ **30.** $\{t \mid t \in W \text{ and } t \text{ is a factor of } 58\}$

In Exercises 31–38, let

$A = \{1, 2, 4, 5, 8\}$
$B = \{x \mid x \in W, x \text{ is a multiple of } 4, \text{ and } x \text{ is less than } 36\}$
$C = \{p \mid p \text{ is a prime number}\}$
$D = \{x \mid x \text{ is between } 5 \text{ and } 10, x \in N\}$
$E = \{x \mid x \in N \text{ and } x \text{ is less than or equal to } 12\}$

List the elements in the indicated set.

31. $A \cap B$ **32.** $A \cap C$ **33.** $A \cup B$ **34.** $B \cap C$

35. $D \cap C$ **36.** $D \cap E$ **37.** $A \cap D$ **38.** $A \cap E$

In Exercises 39–44, let

$$A = \{x \mid -2 \le x < 5\}, \quad B = \{x \mid x \ge 8\}, \quad C = \{x \mid x < 0\}, \quad \text{and} \quad D = \{x \mid x \le 10\},$$

Find

39. $A \cap C$ **40.** $A \cup C$ **41.** $A \cap B$ **42.** $A \cup B$

43. $B \cap D$ **44.** $B \cup D$

Conic Sections

In Chapter 3 we graphed first-degree equations in two variables, that is, equations of the form $Ax + By = C$, on the rectangular coordinate system. Graphs of equations of this form are straight lines and hence we called them *linear equations*.

In this appendix we will be concerned with graphs of second-degree equations—in particular, equations of the form

$$Ax^2 + By^2 + Cx + Dy + E = 0$$

where A and B are not both zero.

If the equation can be graphed, it can be shown that (with a few exceptions) the graph will be one of four figures: the circle, the parabola, the ellipse, or the hyperbola. These figures are called **conic sections** since they describe the intersection of a plane and a double-napped cone, as illustrated in Figure B.1.

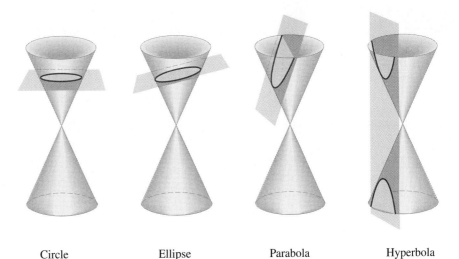

Figure B.1   Circle   Ellipse   Parabola   Hyperbola

In Section 8.8 we derived the distance formula using the Pythagorean theorem (see Figure B.2).

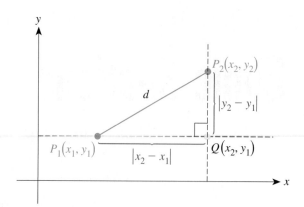

Figure B.2

The distance, d, between points $P_1(x_1, y_1)$ and $P_2(x_2, y_2)$ is

$$d = \sqrt{(x_2 - x_1)^2 + (y_2 - y_1)^2}$$

This formula is useful in deriving the general equations for conic sections. For example, in Section 8.8 we applied the distance formula to the definition of a circle and derived the standard form of a circle (see Figure B.3).

The Circle

The equation

$$(x - h)^2 + (y - k)^2 = r^2$$

is the standard form of the equation of a circle with a center of (h, k) and radius r.

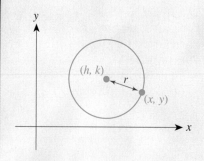

Figure B.3

Recall that with completing the square, we can take any equation of the form

$$Ax^2 + By^2 + Cx + Dy + E = 0 \text{ (where } A \text{ and } B \text{ are not both 0)}$$

and put it in the standard form of a circle if A and B, the coefficients of the squared terms, are equal.

EXAMPLE 1   Find the center and radius of the circle $2x^2 + 2y^2 + 8x - 4y - 12 = 0$.

Solution $2x^2 + 2y^2 + 8x - 4y - 12 = 0$ *Divide each side by 2.*

$x^2 + y^2 + 4x - 2y - 6 = 0$ *Separate terms containing x from terms containing y.*

$$(x^2 + 4x \quad) + (y^2 - 2y \quad) = 6$$

Complete the square for each quadratic expression:

$$\left[\frac{1}{2}(4)\right]^2 = 4; \quad \left[\frac{1}{2}(-2)\right]^2 = 1$$

Add $4 + 1$ to each side.

$$(x^2 + 4x + 4) + (y^2 - 2y + 1) = 6 + 4 + 1$$

Rewrite each quadratic expression in factored form.

$$(x + 2)^2 + (y - 1)^2 = 11$$

We have a circle with center $(-2, 1)$ and radius $\sqrt{11}$.

The Parabola

A *parabola* is defined to be the set of points whose distance from a fixed point is equal to its distance from a fixed line. That is, for any point (x, y) on the parabola, the distance from the point (x, y) to a fixed point (a, b) is equal to the distance from the point (x, y) to a fixed line, $y = c$ (see Figure B.4).

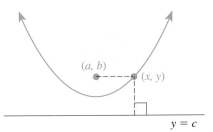

Figure B.4

As with the circle, we can derive the general equation of the parabola using the distance formula in terms of a, b, c, x, and y, and arrive at an equation of the form $y = Ax^2 + Bx + C$, where A, B, and C are constants in terms of a, b, and c (see Exercise 73 at the end of this appendix). In Chapter 8 we found that the graph of a quadratic function, a function that can be put in the form $y = A(x - h)^2 + k$, is a parabola. We examined the graph of this function in great detail in Section 8.6. We refer to this parabola as a "vertical" parabola because its axis of symmetry is a vertical line, and therefore the parabola opens either upward or downward. We can generalize the results from that section to the "horizontal" parabola, which has a horizontal line as its axis of symmetry and therefore opens to the left or to the right.

The Graph of $x = Ay^2 + By + C$

If the second-degree term involves y rather than x, then the roles of x and y are switched. That is,

---

An equation of the form $x = Ay^2 + By + C$ (where $A \neq 0$) is a parabola.

The equation of the axis of symmetry is $y = -\dfrac{B}{2A}$.

The y-coordinate of the vertex is $-\dfrac{B}{2A}$.

The parabola opens to the *right* if $A > 0$, and to the *left* if $A < 0$.

Figure B.5 shows the graph of the equation $x = y^2 - 10y + 9$. We leave it to the student to check the vertex, axis of symmetry, and intercepts for this graph.

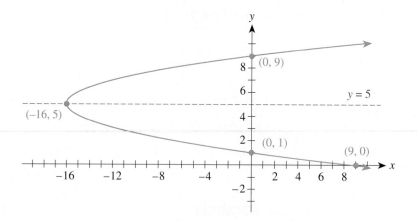

Figure B.5
Graph of $x = y^2 - 10y + 9$

Observe that if an equation is in the form $Ax^2 + By^2 + Cx + Dy + E = 0$, then if either A or B but not both is 0, the equation will usually describe a parabola. (We will discuss the exceptions at the end of this appendix.)

The Ellipse: Centered at the Origin

An *ellipse* is the set of all points whose sum of the distances from two fixed points is a constant (see Figure B.6).

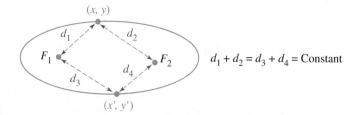

Figure B.6

Instead of deriving the equation of the ellipse from the distance formula (see Exercise 75 at the end of this appendix), we will start with the standard form of the equation of an ellipse centered at the origin.

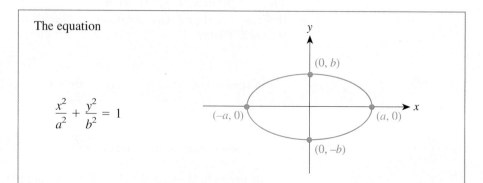

The equation

$$\frac{x^2}{a^2} + \frac{y^2}{b^2} = 1$$

is the standard form of the equation of an *ellipse* centered at the origin, with intercepts $(-a, 0)$, $(a, 0)$, $(0, b)$, and $(0, -b)$, where a and b are *positive* constants.

If we set $x = 0$ in $\dfrac{x^2}{a^2} + \dfrac{y^2}{b^2} = 1$ and solve for y, we obtain the y-intercepts:

$$\dfrac{(0)^2}{a^2} + \dfrac{y^2}{b^2} = 1 \qquad \textit{Set } x = 0.$$

$$\dfrac{y^2}{b^2} = 1 \qquad \textit{Multiply both sides by } b^2.$$

$$y^2 = b^2$$

$$y = \pm b \qquad \textit{The y-intercepts}$$

If we set $y = 0$, then we can find the x-intercepts, $\pm a$, as we would expect.

For the ellipse centered at the origin, the **axes of the ellipse** are the line segments joining the opposite pairs of intercepts. The longer axis is called the **major axis;** the shorter axis is called the **minor axis.**

For the ellipse centered at the origin, the axes lie on the x- and y-coordinate axes. The endpoints of the major (longer) axis are called the **vertices of the ellipse.**

EXAMPLE 2

Graph the following. Label the intercepts and vertices.

(a) $\dfrac{x^2}{9} + \dfrac{y^2}{16} = 1$ **(b)** $\dfrac{x^2}{25} + \dfrac{y^2}{20} = 1$

Solution

(a) This is the standard form of the equation of an ellipse. Therefore, we have an ellipse centered at the origin with the x- and y-coordinate axes as the axes of the ellipse.

Since the intercepts of the equation in standard form are $(a, 0)$, $(-a, 0)$, $(0, b)$, and $(0, -b)$, all we need to do now is to identify a and b by simply observing the denominator of the squared terms:

$$a^2 = 9 \text{ (the denominator of } x^2) \quad \text{and} \quad b^2 = 16 \text{ (the denominator of } y^2)$$

Thus, $a = 3$ and $b = 4$. (Remember that a and b are *positive* constants.) Therefore, the intercepts of the ellipse in this example are $(3, 0)$, $(-3, 0)$, $(0, 4)$, and $(0, -4)$.

The ellipse is graphed in Figure B.7. Since the longer (major) axis lies on the y-axis, and the vertices are the endpoints of the major axis, the vertices are $(0, 4)$ and $(0, -4)$.

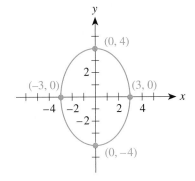

Figure B.7

Graph of $\dfrac{x^2}{9} + \dfrac{y^2}{16} = 1$

(b) This is the standard form of the equation of an ellipse. Thus, we have an ellipse centered at the origin. The axes of the ellipse lie on the x- and y-coordinate axes.

Since the denominator of x^2 is 25, we have $a^2 = 25$. Therefore, $a = 5$. The denominator of y^2 is 20; therefore, $b^2 = 20$ and $b = \sqrt{20} = 2\sqrt{5}$ (≈ 4.47).

Thus, the intercepts are $(5, 0)$, $(-5, 0)$, $(0, 2\sqrt{5})$, and $(0, -2\sqrt{5})$. The major axis lies on the x-axis, and therefore the vertices are $(-5, 0)$ and $(5, 0)$. See Figure B.8.

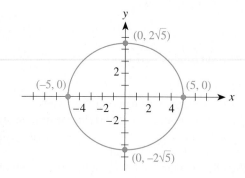

Figure B.8

Graph of $\dfrac{x^2}{25} + \dfrac{y^2}{20} = 1$

By algebraically manipulating the equation, we can identify the intercepts and vertices, and sketch a graph of the equation of an ellipse not in standard form, as demonstrated in the next example.

EXAMPLE 3

Graph the following. Label the vertices and identify the axes.

(a) $4x^2 + 25y^2 = 100$ **(b)** $100x^2 + y^2 = 25$

Solution

(a) This equation is not in standard form. To put the equation in standard form, we observe that the right-hand side of the equation in standard form is 1. Hence, we must divide both sides of $4x^2 + 25y^2 = 100$ by 100 in order to get 1 on the right-hand side.

$$4x^2 + 25y^2 = 100$$ *To get 1 on the right-hand side, divide both sides of the equation by 100.*

$$\frac{4x^2 + 25y^2}{100} = \frac{100}{100}$$ *Separate fractions.*

$$\frac{4x^2}{100} + \frac{25y^2}{100} = 1$$ *Reduce each fraction.*

$$\frac{x^2}{25} + \frac{y^2}{4} = 1$$

Now the equation is in standard form.

a^2 (the denominator of x^2) is 25; thus, $a = 5$.

b^2 (the denominator of y^2) is 4; thus, $b = 2$.

The intercepts of the ellipse are $(-5, 0)$, $(5, 0)$, $(0, -2)$, and $(0, 2)$. The major (longer) axis lies on the x-axis and therefore the vertices are $(-5, 0)$ and $(5, 0)$. See Figure B.9.

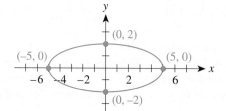

Figure B.9

Graph of

$4x^2 + 25y^2 = 100$

(b) This equation is not in standard form.

$$100x^2 + y^2 = 25$$ *Divide both sides of the equation by 25 and the right-hand side will be 1.*

$$\frac{100x^2 + y^2}{25} = \frac{25}{25}$$ *Separate fractions.*

$$\frac{100x^2}{25} + \frac{y^2}{25} = 1$$ *Reduce.*

$$4x^2 + \frac{y^2}{25} = 1$$

Now we can easily identify b^2 (the denominator of y^2) as 25, but what is a^2, the denominator of x^2? It is surely not 4, since 4 is not in the denominator of x^2. We have to rewrite $4x^2$ as $\dfrac{x^2}{a^2}$. For this we use our knowledge of quotients of fractions and rewrite

$$4x^2 \quad \text{as} \quad \frac{x^2}{\frac{1}{4}} \qquad \text{Since} \qquad \frac{x^2}{\frac{1}{4}} = x^2\left(\frac{4}{1}\right) = 4x^2$$

Hence, $4x^2 + \dfrac{y^2}{25} = 1$ becomes $\dfrac{x^2}{\frac{1}{4}} + \dfrac{y^2}{25} = 1$.

We have

$$a^2 = \frac{1}{4} \rightarrow a = \sqrt{\frac{1}{4}} = \frac{1}{2} \quad \text{and} \quad b^2 = 25 \rightarrow b = 5$$

The equation is that of an ellipse centered at $(0, 0)$ with intercepts $\left(\frac{1}{2}, 0\right)$, $\left(-\frac{1}{2}, 0\right)$, $(0, 5)$, and $(0, -5)$, as illustrated in Figure B.10. The vertices are $(0, 5)$ and $(0, -5)$.

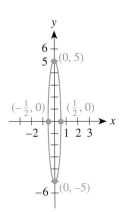

Figure B.10
Graph of
$100x^2 + y^2 = 25$

EXAMPLE 4

The arch of a bridge over a highway is semielliptical. The base of the bridge covers the entire 60-ft width of the highway, and the highest part of the bridge is 20 ft directly above the center of the highway. What is the height of the bridge 10 ft from the center of the road? (See Figure B.11.)

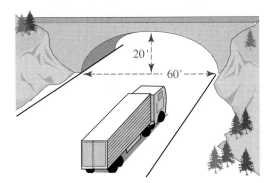

Figure B.11

Solution   We can draw an ellipse, centering it at the origin of a set of coordinate axes, with the base of the semiellipse (the horizontal axis of the ellipse) lying on the x-axis as shown in Figure B.12.

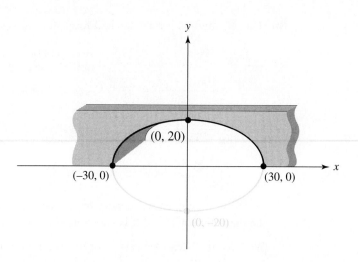

Figure B.12

Note that since the total length of the base is 60 ft, if we center the ellipse at the origin, the horizontal vertices are $(-30, 0)$ and $(30, 0)$. Since the bridge is 20 ft high above the center, the vertical intercepts of the ellipse are $(0, 20)$ and $(0, -20)$. Hence we have an ellipse centered at the origin with $a = 30$ and $b = 20$. Therefore, the equation of the ellipse is

$$\frac{x^2}{30^2} + \frac{y^2}{20^2} = 1 \quad \text{or} \quad \frac{x^2}{900} + \frac{y^2}{400} = 1$$

We are looking for the height of the bridge 10 ft from the center line. Looking at the graph, this translates into finding y (the height) when $x = 10$. (See Figure B.13.)

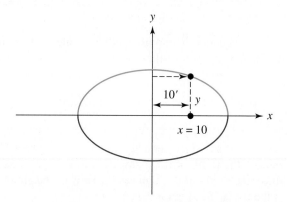

Figure B.13

Hence, we substitute $x = 10$ in the equation of the ellipse and solve for y:

$$\frac{x^2}{900} + \frac{y^2}{400} = 1 \qquad \qquad \textit{Substitute } x = 10.$$

$$\frac{10^2}{900} + \frac{y^2}{400} = 1$$

$$\frac{100}{900} + \frac{y^2}{400} = 1 \qquad \qquad \textit{Solve for } y\text{: First isolate } y^2.$$

$$\frac{y^2}{400} = 1 - \frac{100}{900} = 1 - \frac{1}{9} \qquad \qquad \textit{Hence,}$$

$$\frac{y^2}{400} = \frac{8}{9}$$

$$y^2 = 400\left(\frac{8}{9}\right) = \frac{3,200}{9} \qquad \qquad \textit{Take square roots.}$$

$$y = \pm\sqrt{\frac{3,200}{9}} = \pm\frac{40\sqrt{2}}{3}$$

Since we are interested only in the height of the semiellipse, that is, the top of the semiellipse, we ignore the negative y value and get $\boxed{\dfrac{40\sqrt{2}}{3} \approx 18.86 \text{ ft}}$.

The Hyperbola: Centered at the Origin

A **hyperbola** is the set of all points such that the absolute value of the *difference* of their distances from two fixed points is a constant (see Figure B.14).

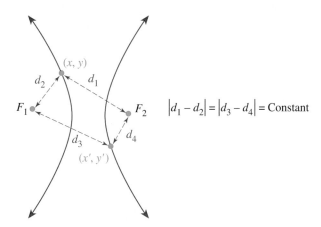

Figure B.14
A hyperbola

Again, instead of deriving the equation of a hyperbola from the distance formula (see Exercise 77 at the end of this appendix), we will start with the standard form of the equation of a hyperbola.

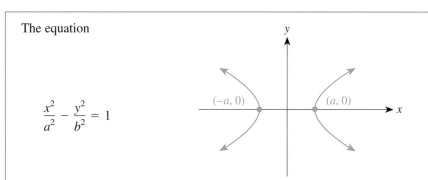

The equation

$$\frac{x^2}{a^2} - \frac{y^2}{b^2} = 1$$

is the standard form of the equation of the *hyperbola* centered at the origin, with vertices $(-a, 0)$ and $(a, 0)$, where a and b are positive constants.

The axes of a hyperbola centered at the origin are the x- and y-axes.

Note the differences and similarities between the standard forms of the equations of the hyperbola and the ellipse. For example, in graphing the hyperbola $\dfrac{x^2}{9} - \dfrac{y^2}{16} = 1$, we note that

a^2 (the denominator of x^2) $= 9 \;\rightarrow\; a = 3$

b^2 (the denominator of y^2) $= 16 \rightarrow b = 4$

Hence, by the standard form given in the box, the vertices are $(-3, 0)$ and $(3, 0)$.

Notice that if we set $y = 0$ in $\dfrac{x^2}{9} - \dfrac{y^2}{16} = 1$, we obtain $\dfrac{x^2}{9} = 1$, and hence

$$x^2 = 9 \rightarrow x = \pm 3$$

Thus, the x-intercepts and vertices are identical for hyperbolas of the form $\dfrac{x^2}{a^2} - \dfrac{y^2}{b^2} = 1$: $(-a, 0)$ and $(a, 0)$.

On the other hand, if we set $x = 0$ in $\dfrac{x^2}{9} - \dfrac{y^2}{16} = 1$, then we have $-\dfrac{y^2}{16} = 1$. This yields $y^2 = -16$, which has no real solutions (since y^2 must be positive).

Hence, x cannot be 0, which means that

the graph of the hyperbola $\dfrac{x^2}{a^2} - \dfrac{y^2}{b^2} = 1$ does not intersect the y-axis.

In graphing a hyperbola, we are guided by two intersecting lines that the hyperbola approaches but never touches. These lines are called **asymptotes.** For example, in graphing $\dfrac{x^2}{9} - \dfrac{y^2}{16} = 1$, we are guided by the asymptotes

$$y = \frac{4}{3}x \quad \text{and} \quad y = -\frac{4}{3}x$$

The hyperbola gets very close to, but never touches, the lines $y = \frac{4}{3}x$ and $y = -\frac{4}{3}x$ (see Figure B.15).

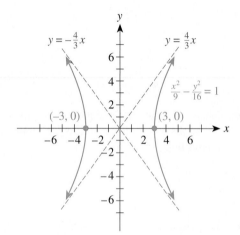

Figure B.15

Graph of $\dfrac{x^2}{9} - \dfrac{y^2}{16} = 1$

The asymptotes guide us in determining the shape of the hyperbola. We can manipulate the equation of the hyperbola (see Exercise 78 at the end of this appendix) to determine the asymptotes, as indicated in the box.

The asymptotes of the hyperbola $\dfrac{x^2}{a^2} - \dfrac{y^2}{b^2} = 1$ are

$$y = \pm \frac{b}{a}x$$

EXAMPLE 5

Sketch a graph of the hyperbola $\dfrac{x^2}{4} - \dfrac{y^2}{9} = 1$.

Solution

Since the hyperbola is in standard form, we can identify a and b to find the vertices and the asymptotes.

$$a^2 = 4 \quad \text{and} \quad b^2 = 9; \quad \text{therefore,} \quad a = 2 \quad \text{and} \quad b = 3$$

Since $a = 2$, the vertices are $(2, 0)$ and $(-2, 0)$.

We can sketch the asymptotes using a and b by the following procedure:

First *plot* the vertices—the points $(2, 0)$ and $(-2, 0)$—and *locate* the points $(0, 3)$ and $(0, -3)$.

Then draw a rectangle such that the points just found are the midpoints of the sides of the rectangle.

The lines containing the diagonals of the rectangle are the asymptotes of the hyperbola (see Figure B.16).

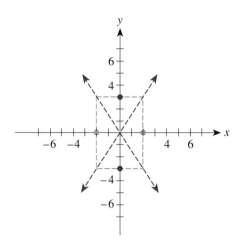

Figure B.16

Now we sketch the hyperbola, using the asymptotes as a guide (see Figure B.17).

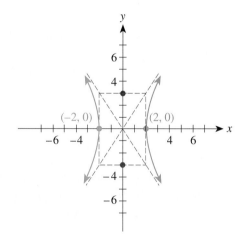

Figure B.17

Graph of $\dfrac{x^2}{4} - \dfrac{y^2}{9} = 1$

Why do the lines containing the diagonals of the rectangles have the same equation as the asymptotes? First, notice in Figure B.17 that the slopes of the diagonals are $\pm\frac{3}{2}$. Then note that the diagonals pass through the origin, and hence their y-intercepts are 0. Therefore, by the slope–intercept form (Section 4.2), the equations of the lines containing the diagonals are

$$y = \pm\frac{3}{2}x$$

which are the asymptotes.

In general, to graph a hyperbola, we suggest the outline given in the box.

It is important for us to realize that the rectangle and the asymptotes are only guides to help us draw the hyperbola; they are not actually part of the graph of the hyperbola.

To graph the hyperbola $\dfrac{x^2}{a^2} - \dfrac{y^2}{b^2} = 1$:

1. Plot the points $(a, 0)$ and $(-a, 0)$, and locate the points $(0, b)$ and $(0, -b)$.
2. Draw a rectangle such that the points found in step 1 are the midpoints of the sides of the rectangle.
3. The lines containing the diagonals of the rectangle are the asymptotes.
4. The vertices are $(a, 0)$ and $(-a, 0)$.
5. Draw the hyperbola using the asymptotes and vertices as a guide.

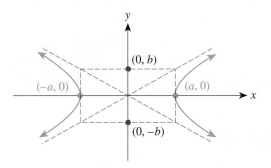

EXAMPLE 6

Sketch a graph of the hyperbola $x^2 - 36y^2 = 36$.

Solution

The equation $x^2 - 36y^2 = 36$ is not in standard form. We divide both sides of the equation by 36 to get

$$\frac{x^2 - 36y^2}{36} = \frac{36}{36}$$

$$\frac{x^2}{36} - \frac{y^2}{1} = 1$$

Therefore, $a^2 = 36 \rightarrow a = 6$ and $b^2 = 1 \rightarrow b = 1$.

We plot the points $(6, 0)$ and $(-6, 0)$ and locate points $(0, 1)$ and $(0, -1)$ as shown in Figure B.18. We then draw the rectangle and the diagonals. Finally, we graph the hyperbola centered at the origin with vertices $(-6, 0)$ and $(6, 0)$ and asymptotes $y = \pm\frac{1}{6}x$.

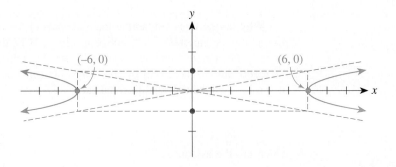

Figure B.18
Graph of $x^2 - 36y^2 = 36$

We have just discussed equations of the form $\dfrac{x^2}{a^2} - \dfrac{y^2}{b^2} = 1$. Note that the term $\dfrac{y^2}{b^2}$ is being subtracted from the term $\dfrac{x^2}{a^2}$. What if the situation were reversed, that is, what if $\dfrac{x^2}{a^2}$ were being subtracted from $\dfrac{y^2}{b^2}$? Then the form would be

$$\frac{y^2}{b^2} - \frac{x^2}{a^2} = 1$$

Note that if $x = 0$, then $\dfrac{y^2}{b^2} = 1$ and $y = \pm b$. Hence, we would have y-intercepts $(0, b)$ and $(0, -b)$. However, if $y = 0$, then we arrive at the impossible situation $-\dfrac{x^2}{a^2} = 1$. Therefore, the graph does not cross the x-axis.

The equation

$$\frac{y^2}{b^2} - \frac{x^2}{a^2} = 1$$

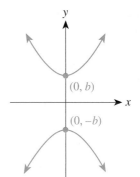

is the standard form of the equation of the hyperbola centered at the origin, with vertices $(0, -b)$ and $(0, b)$, where a and b are positive constants.

Fortunately, the asymptotes still remain the same:

The asymptotes of the hyperbola $\dfrac{y^2}{b^2} - \dfrac{x^2}{a^2} = 1$ are $y = \pm\dfrac{b}{a}x$.

Thus, we simply follow the same guidelines in graphing $\dfrac{y^2}{b^2} - \dfrac{x^2}{a^2} = 1$ as we did for graphing $\dfrac{x^2}{a^2} - \dfrac{y^2}{b^2} = 1$, except that the vertices, $(0, b)$ and $(0, -b)$, now lie on the y-axis and the hyperbola opens up and down rather than side to side. *Note that a^2 is still the denominator of x^2 and b^2 is still the denominator of y^2.*

EXAMPLE 7

Graph $9y^2 - 4x^2 = 144$.

Solution

We can put the equation in standard form by dividing both sides of the equation by 144 to get

$$\frac{y^2}{16} - \frac{x^2}{36} = 1$$

Since a^2 (the denominator of x^2) is 36 and b^2 (the denominator of y^2) is 16, we have

$$a = 6 \quad \text{and} \quad b = 4$$

We plot the points $(0, 4)$ and $(0, -4)$, locate the points $(6, 0)$ and $(-6, 0)$, and draw our rectangle with these points as midpoints of its sides (see Figure B.19).

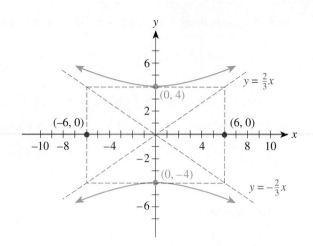

Figure B.19
Graph of $9y^2 - 4x^2 = 144$

The vertices for this hyperbola are on the y-axis. Note that the asymptotes are still $y = \pm\dfrac{b}{a}x$, which are

$$y = \pm\frac{2}{3}x$$

Also, the hyperbola now opens up and down, as shown in the figure.

We can recognize the equation of an ellipse or hyperbola by the signs of the co-efficients of the squared terms: The coefficients of the squared terms of a hyperbola have opposite signs, whereas the coefficients of the squared terms of an ellipse have the same sign.

Once we recognize a hyperbola, we can also determine what type of hyperbola we have by locating its x- and y-intercepts. A hyperbola centered at the origin will have only one set of intercepts. By knowing which axis it crosses we know what type of hyperbola we have.

By now it may have occurred to you that (with the help of completing the square) we may be able to convert any equation of the form

$$Ax^2 + By^2 + Cx + Dy + E = 0 \quad \text{(where } A \text{ and } B \text{ are not both zero)}$$

into one of the conic section forms discussed in this appendix.

However, that is not necessarily possible, since the coefficients (A, B, and C) and h, k, and r must be real. For example, consider the equation $x^2 + y^2 = -4$. Since x and y must be real numbers, their squares must be nonnegative and can never sum to the negative number, -4. Since we require all constants and variables to be real, we can never put the equation in the form $(x - h)^2 + (y - k)^2 = r^2$, where h, k, and r are real numbers. Therefore, the equation $x^2 + y^2 = -4$ has no graph.

If the equation can be graphed, the next question to ask is, will we get a conic section? Here we run into another problem. The equation $x^2 + y^2 = 0$ is in the form of the equation of a circle with center $(0, 0)$ and radius $= 0$. Therefore its graph is not a circle, it is the point $(0, 0)$. As another example, the equation $\dfrac{x^2}{16} + \dfrac{y^2}{25} = 0$ *cannot* be put into one of the standard forms. It *can* be graphed, but it yields only the point $(0, 0)$. We call such forms ***degenerate forms.*** The degenerate form of a circle and ellipse is a point. (See Figure B.20.)

Figure B.20
The degenerate form of a circle

The degenerate form of the hyperbola is two intersecting lines. (Try solving the equation $\dfrac{x^2}{9} - \dfrac{y^2}{4} = 0$ for y and describe its graph.) See Figure B.21.

The equation $x^2 - 4 = 0$ looks like it may be a parabola. However, this equation is equivalent to the equations $x = -2$ and $x = +2$, which, when graphed on a rectangular coordinate system, yield two vertical lines. This is a degenerate form of a parabola. The graph of $x^2 - 6x + 9 = 0$ is a single vertical line, which is also a degenerate form of a parabola.

Thus, we *can* say the following:

If the equation $Ax^2 + By^2 + Cx + Dy + E = 0$ (where A and B are not both 0) can be graphed, the graph will yield a conic section or one of its degenerate forms.

Actually, we can say more. If the figure can be graphed, we can identify what the figure may be by simply looking at the coefficients of the squared terms.

Figure B.21
The degenerate form
of a hyperbola

Assume that the equation $Ax^2 + By^2 + Cx + Dy + E = 0$, where A and B are not both 0, can be graphed.

1. If either $A = 0$ or $B = 0$, but not both, the equation will yield a parabola or one of its degenerate forms—a line or two parallel lines.

2. If the signs of A and B are the same, the equation will yield a circle if $A = B$, an ellipse if $A \neq B$, or their degenerate form—a point.

3. If the signs of A and B are different, the equation will yield a hyperbola or its degenerate form—two intersecting lines.

EXAMPLE 8

Identify the type of conic section by its coefficients, assuming the equation can be graphed.

(a) $3x^2 + 4y^2 - 2x - 3y = 0$ (b) $5x^2 - 3x + 2y = 4$

(c) $7y^2 - 2x^2 + 3x = 7$ (d) $-6x^2 - 6y^2 - x + y = 8$

(e) $3x + 2y = 5$

Solution

(a) Since the coefficients of the squared terms have the same sign but are not identical, the equation will yield an ellipse or a point if it can be graphed.

(b) Since there is no y^2 term ($B = 0$), the equation will yield a parabola or one of its degenerate forms if it can be graphed.

(c) Since the signs of the coefficients of the squared terms are different (7 and -2), the equation will yield a hyperbola or its degenerate form if it can be graphed.

(d) Since the coefficients of the squared terms are identical, the equation will yield a circle or a point if it can be graphed.

(e) Since it is a first degree equation in two variables, we know that the graph is a straight line.

APPENDIX B EXERCISES ⊙

In Exercises 1–4, *identify the center and radius of each circle.*

1. $x^2 + y^2 - 4x + 12y - 18 = 0$ **2.** $3x^2 + 3y^2 + 12x - 6y = 21$

3. $-x^2 - y^2 - 8x + 12y + 15 = 0$ **4.** $2x^2 + 2y^2 + 12x - 6y = 22$

In Exercises 5–12, sketch a graph of each parabola.

5. $y = 3x^2 + 2x$

6. $y - 25 = x^2 - 10x$

7. $y - x^2 = 6x - 8$

8. $y = -x^2 + 3x + 28$

9. $x = 2y^2 + 4y + 1$

10. $x - 3y^2 = 12y + 3$

11. $x = 2y^2 + y + 4$

12. $x - 2 = 3y^2 - y$

In Exercises 13–18, identify the intercepts and vertices of the ellipse described by the given equation.

13. $\dfrac{x^2}{16} + \dfrac{y^2}{9} = 1$

14. $\dfrac{x^2}{4} + \dfrac{y^2}{9} = 1$

15. $\dfrac{y^2}{36} + \dfrac{x^2}{25} = 1$

16. $\dfrac{y^2}{49} + \dfrac{x^2}{81} = 1$

17. $\dfrac{x^2}{24} + \dfrac{y^2}{20} = 1$

18. $\dfrac{x^2}{20} + \dfrac{y^2}{18} = 1$

In Exercises 19–26, sketch a graph of the ellipse. Label the vertices.

19. $x^2 + \dfrac{y^2}{16} = 1$

20. $\dfrac{x^2}{9} + y^2 = 1$

21. $4x^2 + 25y^2 = 100$

22. $4y^2 + 25x^2 = 100$

23. $8x^2 + 7y^2 = 56$

24. $7x^2 + 8y^2 = 56$

25. $25x^2 + 16y^2 = 1$

26. $36x^2 + 16y^2 = 1$

27. The arch of a bridge over a highway is semielliptical. The base of the bridge covers the entire 50-ft width of the highway, and the highest part of the bridge is 15 ft directly above the center of the highway. Can a truck 11 ft wide and 14 ft high pass through the bridge staying right of the center line?

28. The arch of a bridge over a road is semielliptical. The base of the bridge covers the entire 70-ft width of the road, and the highest part of the bridge is 20 ft directly above the center of the road. How high is the bridge 20 ft from the edge of the road?

In Exercises 29–32, identify the vertices and write the equations of the asymptotes for the hyperbolas described by the equation.

29. $\dfrac{x^2}{9} - \dfrac{y^2}{16} = 1$

30. $\dfrac{x^2}{25} - \dfrac{y^2}{16} = 1$

31. $\dfrac{x^2}{36} - \dfrac{y^2}{25} = 1$

32. $\dfrac{y^2}{9} - \dfrac{x^2}{16} = 1$

In Exercises 33–40, sketch a graph of the hyperbola. Label the vertices and asymptotes.

33. $\dfrac{y^2}{4} - \dfrac{x^2}{12} = 1$

34. $\dfrac{y^2}{8} - \dfrac{x^2}{12} = 1$

35. $2y^2 - x^2 = 4$

36. $y^2 - 2x^2 = 4$

37. $12y^2 - 5x^2 = 60$

38. $8y^2 - 9x^2 = 72$

39. $225x^2 - y^2 = 25$

40. $16x^2 - 4y^2 = 1$

In Exercises 41–58, assume the following equations can be graphed. Identify what figure the equation will yield without putting the equation in standard form.

41. $x^2 + y^2 = 12$

42. $9x^2 + 25y^2 = 220$

43. $6x^2 + 7y^2 = 42$

44. $4x^2 - 5y^2 = 1$

45. $3x^2 + 2x + 3y = 2$

46. $-7y^2 + 6x - 3y = -9$

47. $\dfrac{x}{3} + \dfrac{y}{2} = 1$

48. $5x^2 + 5y^2 = 12$

49. $x^2 - y^2 = 9$

50. $5y - 3x = 2$

51. $5y^2 - 9x^2 - 30y - 36x = 36$

52. $y^2 - 3x^2 - 4y + 24x = 45$

53. $x^2 + y^2 - 2x + 6y = -2$

54. $\dfrac{(x - 1)^2}{3} + \dfrac{(y - 1)^2}{3} = 1$

55. $3x^2 + 3y^2 + 18x - 12y = -24$

56. $5x^2 - 3x + 4y = -7$

57. $6x^2 - 7y^2 = 12$

58. $-x^2 - y^2 - 12x - 8y = 37$

In Exercises 59–72, identify and sketch a graph of the equations, if possible. Label the important aspects of the figure.

59. $x^2 + y^2 = 16$

60. $16x^2 + 9y^2 = 144$

61. $x^2 + 2x - y = 9$

62. $16x^2 + 5y^2 = -80$

63. $x^2 - 100y^2 = 25$

64. $2y^2 + 16y - x = -24$

65. $16x^2 - 9y^2 = 0$

66. $100x^2 - 16y^2 = 1{,}600$

67. $3x^2 + 3y^2 = 24$

68. $2x^2 + 2y^2 = 0$

69. $2x^2 + 4x - y = -6$

70. $y^2 - 8y - x = -7$

71. $2x^2 + y^2 = 8$

72. $16x^2 + 5y^2 = 80$

73. A parabola is defined to be the set of all points whose distance from a fixed point is equal to its distance from a fixed line. Suppose we let the fixed point be the point $F(0, a)$ and the fixed line be the line $y = -a$ (see the accompanying figure). Pick any point on the parabola and call it $P(x, y)$. Drop a perpendicular line down from the point (x, y) to the line $y = -a$. Then the perpendicular line intersects the line $y = -a$ at point D, which has coordinates $(x, -a)$ (why?).

Call d_1 the distance from P to F, and d_2 the distance from P to D. By definition of the parabola, d_1 must equal d_2. Using the distance formula, write an equation in terms of x, y, and a. What is the difference between this equation and the equation of the parabola $y = Ax^2 + Bx + C$?

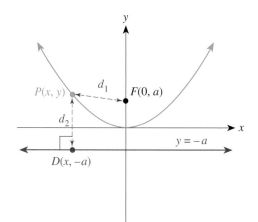

74. What type of figure is created by graphing $\dfrac{x^2}{a^2} + \dfrac{y^2}{a^2} = 1$?

75. An ellipse is defined to be the set of all points whose sum of the distances from two fixed points is a constant. The accompanying figure shows two fixed points, $F_1(-s, 0)$ and $F_2(s, 0)$. Pick any point $P(x, y)$ on the ellipse. We will call d_1 the distance from P to F_1, and d_2 the distance from P to F_2.

   By definition of the ellipse, the sum of the distances, $d_1 + d_2$, must be constant. Let's call this constant k. Then we have $d_1 + d_2 = k$. Using the distance formula, write this as an equation in x, y, s, and k.

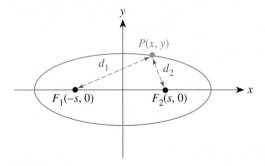

76. Look at the standard form of the equation of a circle with its center at the origin, and compare it with the standard form of a circle with center (h, k). Then generalize the standard form of an ellipse centered at the origin to the case where the center is at (h, k).

77. A hyperbola is defined to be the set of all points such that the absolute value of the difference of their distances from two fixed points is a constant. The accompanying figure shows two fixed points, $F_1(-s, 0)$ and $F_2(s, 0)$, on the x-axis. Pick any point $P(x, y)$ on the hyperbola. We will call d_1 the distance from P to F_1, and d_2 the distance from P to F_2.

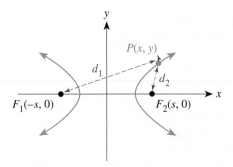

   By definition of the hyperbola, the difference of the distances $|d_1 - d_2|$ must be constant. Let's call it k. Then we have $|d_1 - d_2| = k$. Using the distance formula, write this as an equation in x, y, s, and k.

78. **(a)** Show that if we start with the equation $\dfrac{x^2}{4} - \dfrac{y^2}{9} = 1$ and solve for y, we obtain

$$y = \pm\frac{3}{2}\sqrt{x^2 - 4}$$

   (b) Compute y from the following two equations for these values of x: $x = 4$, 10, 20, 100, 200, 1000, 2000:

$$y = \pm\frac{3}{2}\sqrt{x^2 - 4} \quad \text{and} \quad y = \pm\frac{3}{2}x$$

   Do you see why the graph of the hyperbola "gets closer" to $y = \pm\dfrac{3}{2}x$?

79. The graph of the equation $xy = k$, where k is a positive constant, is given here for $k = 1$. This figure is known as a *rectangular hyperbola.* What are the asymptotes?

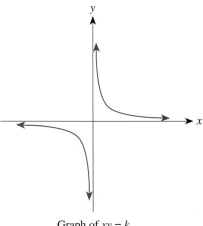

Graph of $xy = k$

80. Look at the standard form of the equation of a circle with its center at the origin, and compare it to the standard form of a circle with center (h, k). Then generalize the standard form of a hyperbola centered at the origin to the case where the center is at (h, k).

Sequences and Series: The Binomial Theorem

At the very heart of mathematics is the attention directed to recognizing and identifying patterns. In the first four sections of this appendix, we discuss sequences and series, which are two particularly important types of patterns in mathematics. They play a fundamental role in calculus and ultimately even in the way our scientific calculators compute values of certain functions. We conclude this appendix with a discussion of the binomial theorem (Section C.5), which provides us with a pattern for finding $(x + y)^n$, perfect nth powers of the binomial $x + y$, without multiplying out the binomials.

C.1 Sequences

When we use the word *sequence* in everyday conversation, we are usually referring to a list of things occurring in a particular order. For example, if a student wishes to go on a cross-country trip during summer vacation, she might consider the following sequence of events, which we have labeled with subscripts to indicate that they follow a particular order:

E_1: Get a part-time job.

E_2: Earn \$3,000.

E_3: Buy a used car.

E_4: Take a cross-country trip.

Note that the student is concerned not only with the things on the list, but also with the order in which they occur—which is first, which is second, and so on.

In mathematics, a sequence is viewed in much the same way—an *ordered* set of things. For example, we could have the sequence of numbers 2, 4, 6, 8—where 2 is the first, 4 is the second, 6 is the third, and 8 is the fourth number. The following are examples of sequences:

$$\{1, 4, 9, 16, 25\}$$

$$\{2, 4, 6, 8, 10\}$$

$$\{2, 4, 6, 8, 10, \ldots\}$$

$$\{3, 9, 27, 81, \ldots\}$$

The first two sequences are examples of *finite sequences* because they terminate. The third and fourth sequences are examples of *infinite sequences* because, as the three-dot notation indicates, they continue on forever in the same established pattern. For the most part, we will focus our attention on infinite sequences.

While describing a sequence as an ordered list of things may serve us intuitively, in mathematics we need to define our terminology more precisely. Since we are concerned with order, we can associate the first term in a sequence with the number 1, the second with the number 2, the third with the number 3, and so on. This approach lends itself quite naturally to using function notation to define a sequence.

DEFINITION

A *sequence* is a function whose domain is a set of consecutive positive integers, usually beginning with 1.

The sequence $\{3, 9, 27, 81, \ldots\}$ can be viewed as a function, f, with domain $\{1, 2, 3, 4, \ldots\}$ and range $\{3, 9, 27, 81, \ldots\}$, where

$$f(1) = 3 \qquad \text{\textit{The first number in the sequence is 3.}}$$

$$f(2) = 9 \qquad \text{\textit{The second number in the sequence is 9.}}$$

$$f(3) = 27 \qquad \text{\textit{The third number in the sequence is 27.}}$$

$$f(4) = 81 \qquad \text{\textit{The fourth number in the sequence is 81.}}$$

$$\vdots \qquad \text{\textit{And so on}}$$

Instead of writing $f(1)$, $f(2)$, $f(3)$, and $f(4)$, it is a more common practice with sequences to use subscript notation instead and write f_1, f_2, f_3, and f_4, respectively. Thus, $f_1 = 3, f_2 = 9, f_3 = 27$, and $f_4 = 81$, etc. Thus, the entire sequence can be denoted $\{f_i\}$.

In a sequence $\{a_i\}$, i is called the *index.* Unless otherwise indicated, we shall start our indices at 1. Thus, if we write $\{a_i\} = \{1, 4, 9, 16, 25\}$, this notation means $a_1 = 1$, $a_2 = 4, a_3 = 9, a_4 = 16, a_5 = 25$. The functional values a_1, a_2, a_3, a_4, and a_5 are called the *terms* of the sequence.

Often, rather than writing out the terms of the sequence, we may simply define the sequence by writing the general term of the sequence, $\{a_i\}$, in terms of i. For example, a general term for the sequence $\{a_i\} = \{2, 4, 6, 8, 10, \ldots\}$ can be written as $a_i = 2i$. We find each term of the sequence by substituting successive positive integers for i, beginning with $i = 1$. Thus, for $a_i = 2i$,

$$a_1 = 2(1) = 2$$

$$a_2 = 2(2) = 4$$

$$a_3 = 2(3) = 6$$

$$a_4 = 2(4) = 8$$

$$\vdots$$

$$\text{\textit{And so on}}$$

EXAMPLE 1

Write the first four terms and the eighth term of the sequence whose nth term is given by the following:

(a) $a_n = 2^n$ **(b)** $x_n = 2n - 5$ **(c)** $b_n = \dfrac{n+1}{n}$ **(d)** $y_n = \dfrac{(-1)^n}{n+1}$

Solution

(a) Since $a_n = 2^n$, the first term of the sequence, a_1, is found by substituting 1 for n. We get

$a_1 = 2^1 = 2$ *The second term, a_2, is found by substituting 2 for n in a_n.*

Hence,

$$a_2 = 2^2 = 4 \qquad \textit{The third term is}$$
$$a_3 = 2^3 = 8 \qquad \textit{And the fourth term is}$$
$$a_4 = 2^4 = 16$$

So the first four terms of the sequence are $\boxed{2, 4, 8, \text{ and } 16}$.

The eighth term is $a_8 = 2^8 = \boxed{256}$.

(b) Since $x_n = 2n - 5$ *We substitute 1 for n in x_n to get the first term.*

$$x_1 = 2(1) - 5 = -3 \qquad \textit{Similarly, for } x_2, x_3, \textit{ and } x_4, \textit{ we substitute 2, 3, and 4,}$$
$$\textit{respectively, for n in } x_n \textit{ to get}$$
$$x_2 = 2(2) - 5 = -1$$
$$x_3 = 2(3) - 5 = 1$$
$$x_4 = 2(4) - 5 = 3$$

Hence, the first four terms of the sequence are $\boxed{-3, -1, 1, 3}$.

The eighth term is $x_8 = 2(8) - 5 = \boxed{11}$.

(c) Since $b_n = \dfrac{n+1}{n}$,

$$b_1 = \frac{1+1}{1} = 2 \qquad b_2 = \frac{2+1}{2} = \frac{3}{2}$$

$$b_3 = \frac{3+1}{3} = \frac{4}{3} \qquad b_4 = \frac{4+1}{4} = \frac{5}{4}$$

Hence, the first four terms are $\boxed{2, \dfrac{3}{2}, \dfrac{4}{3}, \dfrac{5}{4}}$.

The eighth term is $b_8 = \dfrac{8+1}{8} = \boxed{\dfrac{9}{8}}$.

(d) Since $y_n = \dfrac{(-1)^n}{n+1}$,

$$y_1 = \frac{(-1)^1}{1+1} = \frac{-1}{2} \qquad y_2 = \frac{(-1)^2}{2+1} = \frac{1}{3}$$

$$y_3 = \frac{(-1)^3}{3+1} = \frac{-1}{4} \qquad y_4 = \frac{(-1)^4}{4+1} = \frac{1}{5}$$

Thus, the first four terms of the sequence are $\boxed{-\dfrac{1}{2}, \dfrac{1}{3}, -\dfrac{1}{4}, \dfrac{1}{5}}$.

The eighth term is $y_8 = \dfrac{(-1)^8}{8+1} = \boxed{\dfrac{1}{9}}$.

Given a short sequence of numbers, we may be able to identify a pattern and guess at the general term, but here we must be careful: Two different sequences may start with the same terms. For example, the sequence $\{3, 5, 7, \ldots\}$ could represent a list of odd numbers starting with the number 3, or it could represent a list of prime numbers starting with the number 3.

EXAMPLE 2

Find a possible general term for the following sequences:

(a) $\{5, 10, 15, 20, 25, \ldots\}$ **(b)** $\{3, 9, 27, 81, 243, \ldots\}$

(c) $\left\{1, \dfrac{1}{2}, \dfrac{1}{3}, \dfrac{1}{4}, \ldots\right\}$

Solution

We find the general term by examining the given terms and looking for a pattern.

(a) We can see that

$$a_1 = 5 = 5 \cdot 1, \quad a_2 = 10 = 5 \cdot 2, \quad a_3 = 15 = 5 \cdot 3, \quad \text{etc.,}$$

so it appears that a general term can be written as $\boxed{a_n = 5n}$.

(b) We can see that

$$a_1 = 3 = 3^1, \quad a_2 = 9 = 3^2, \quad a_3 = 27 = 3^3, \quad \text{etc.,}$$

so it appears that a general term can be written as $\boxed{a_n = 3^n}$.

(c) We can see that

$$a_1 = 1 = \frac{1}{1}, \quad a_2 = \frac{1}{2}, \quad a_3 = \frac{1}{3}, \quad \text{etc.,}$$

so it appears that a general term can be written as $\boxed{a_n = \dfrac{1}{n}}$.

Sequences often come up in real-life applications.

EXAMPLE 3

Virginia deposits $1,000 in a bank that gives 6% interest compounded annually. Write a sequence showing the amount of money Virginia has in the bank at the end of each year for 4 years.

Solution

In this problem, interest is calculated at a rate of 6% per year. Virginia starts out with $1,000; hence, at the end of the first year, Virginia earns interest of

$$\text{Interest} = 6\% \text{ of } \$1,000 = (0.06)(\$1,000) = \$60.00$$

So at the end of the first year, she has

$$\text{Her original principal} + \text{Interest} = \$1,000.00 + \$60.00 = \$1,060.00$$

in the bank. Now her principal for the second year is $1,060.00, so interest for the second year is calculated using this new principal:

$$\text{Interest} = 6\% \text{ of } \$1,060 = (0.06)(\$1,060) = \$63.60$$

So at the end of the second year, she has

$$\left(\begin{array}{c}\text{Her principal} \\ \text{for the second year}\end{array}\right) + \left(\begin{array}{c}\text{Her interest} \\ \text{for the second year}\end{array}\right)$$

$$\$1,060.00 \quad + \quad \$63.60 \quad = \$1,123.60$$

in the bank at the end of the second year. Continuing, we get

$$\begin{pmatrix} \text{Her principal} \\ \text{for the third year} \end{pmatrix} + \begin{pmatrix} \text{Her interest} \\ \text{for the third year} \end{pmatrix}$$

$$\$1{,}123.60 \quad + \quad 0.06(\$1{,}123.60) \quad = \$1{,}191.02$$

in the bank at the end of the third year. Finally,

$$\begin{pmatrix} \text{Her principal} \\ \text{for the fourth year} \end{pmatrix} + \begin{pmatrix} \text{Her interest} \\ \text{for the fourth year} \end{pmatrix}$$

$$\$1{,}191.02 \quad + \quad 0.06(\$1{,}191.02) \quad = \$1{,}262.48$$

in the bank at the end of the fourth year.

Therefore, the sequence of amounts in the bank at the end of each year for 4 years would be

$$\boxed{\$1{,}060.00, \ \$1{,}123.60, \ \$1{,}191.02, \ \text{and} \ \$1{,}262.48}$$

In Section C.4 we will derive a general term for this sequence.

A *recursive sequence* is one in which later terms are defined by earlier terms. For example, the sequence

$$a_1 = 3 \quad \text{and} \quad a_n = 2a_{n-1} + 7$$

is a recursively defined sequence. The recursive relationship $a_n = 2a_{n-1} + 7$ means that each term after the first equals 7 more than twice the previous term. Hence,

$$a_n = 2a_{n-1} + 7 \qquad \textit{Gives}$$
$$a_2 = 2a_1 + 7 \qquad \textit{For n = 2}$$
$$a_3 = 2a_2 + 7 \qquad \textit{For n = 3}$$
$$a_4 = 2a_3 + 7 \qquad \textit{For n = 4}$$

Notice that we have to find a_1 before finding a_2, a_2 before finding a_3, a_3 before finding a_4, and so on; we are usually given at least one numerical term to start. In the recursively defined sequence above, we are given $a_1 = 3$, so we have

$$a_2 = 2a_1 + 7 = 2(3) + 7 = 13 \qquad \textit{Since } a_2 = 13, \textit{ we get}$$
$$a_3 = 2a_2 + 7 = 2(13) + 7 = 33 \qquad \textit{Since } a_3 = 33, \textit{ we get}$$
$$a_4 = 2a_3 + 7 = 2(33) + 7 = 73$$

Thus, the first four terms of this sequence are 3, 13, 33, and 73.

EXAMPLE 4

Find the fourth term of the recursive sequence:

$$a_1 = 2 \quad \text{and} \quad a_n = 5a_{n-1} - 6$$

Solution   We have $a_1 = 2$ and $a_n = 5a_{n-1} - 6$; thus,

$$a_2 = 5a_1 - 6 = 5(2) - 6 = 4 \qquad \textit{Since } a_2 = 4, \textit{ we get}$$
$$a_3 = 5a_2 - 6 = 5(4) - 6 = 14 \qquad \textit{Since } a_3 = 14, \textit{ we get}$$
$$a_4 = 5a_3 - 6 = 5(14) - 6 = 64$$

Hence, the fourth term is $\boxed{a_4 = 64}$.

APPENDIX C.1 EXERCISES ⊙

In Exercises 1–16, write the first four terms and the tenth term of the given sequence.

1. $a_n = 3n - 5$

2. $a_n = 2n + 4$

3. $b_n = 4n + 3$

4. $b_n = 6n + 4$

5. $c_i = 4^i$

6. $c_i = 3^i$

7. $x_j = \dfrac{j - 1}{j + 1}$

8. $x_j = \dfrac{j + 2}{j + 1}$

9. $x_n = \dfrac{(-1)^n}{n + 2}$

10. $y_n = \dfrac{(-1)^{n+1}}{3n - 4}$

11. $a_n = \dfrac{(-1)^n n}{n + 1}$

12. $a_n = \dfrac{(-1)^n n^2}{2n - 3}$

13. $a_n = 2^n + |n - 5|$

14. $a_n = 2^n + |n + 1|$

15. $b_n = 5 + (-0.1)^n$

16. $b_n = 3 - (0.01)^n$

In Exercises 17–26, find a possible general term.

17. $\{4, 8, 12, 16, \ldots\}$

18. $\{6, 12, 18, 24, \ldots\}$

19. $\{1, 3, 5, 7, 9, \ldots\}$

20. $\{4, 6, 8, 10, 12, \ldots\}$

21. $\left\{\dfrac{1}{2}, \dfrac{2}{3}, \dfrac{3}{4}, \dfrac{4}{5}, \ldots\right\}$

22. $\left\{\dfrac{1}{4}, \dfrac{2}{5}, \dfrac{3}{6}, \dfrac{4}{7}, \ldots\right\}$

23. $\left\{\dfrac{1}{2}, -\dfrac{1}{3}, \dfrac{1}{4}, -\dfrac{1}{5}, \ldots\right\}$

24. $\left\{\dfrac{1}{2}, -\dfrac{2}{3}, \dfrac{3}{4}, -\dfrac{4}{5}, \ldots\right\}$

25. $\{-0.5, 0.05, -0.005, 0.0005, \ldots\}$

26. $\{-0.06, 0.0006, -0.000006, 0.00000006, \ldots\}$

27. Beth gets a raise of \$850 per year for the first 3 years she works for her firm. If her starting salary is \$26,000, write a sequence to show her income at the end of each year for the first 3 years.

28. Susan gets a raise of \$600 every 6 months for the first 2 years she works for her firm. If her starting salary is \$24,500, write a sequence to show her income at the end of each 6-month period for the first 2 years.

29. The population of Portertown increases at a rate of 6% per year. If the population is currently 20,000 people, write a sequence to show the population at the end of each year for 3 years.

30. The population of Harborville increases at a rate of 5% per year. If the population is currently 100,000 people, write a sequence to show the population at the end of each year for 4 years.

31. José deposits \$5,000 in a bank that yields 6% interest compounded annually. Write a sequence to show how much José has in the bank at the end of each year for 4 years.

32. Karl deposits \$8,000 in a bank that yields 8% interest compounded annually. Write a sequence to show how much Karl has in the bank at the end of each year for 3 years.

33. Jennifer gets a 6% raise each year. If her starting salary is \$30,000 per year, write a sequence to show how much she is paid at the beginning of each year for the first 4 years.

34. Louise gets a 7% raise each year. If her starting salary is $30,000 per year, write a sequence to show how much she is paid at the beginning of each year for the first 3 years.

35. A ball bounces to $\frac{1}{2}$ the distance it falls on a rebound. If it is dropped from a height of 60 ft, write a sequence to show how high it bounces for the first five bounces.

36. A ball bounces to $\frac{2}{3}$ the distance it falls on a rebound. If it is dropped from a height of 60 ft, write a sequence to show how high it bounces for the first five bounces.

37. Harold borrows $5,000 and promises to pay back $200 per month plus, as interest, 1% of the remaining balance each month. Write a sequence showing how much Harold must pay each month for the first 6 months.

38. Paige borrows $10,000 and promises to pay back $500 per month plus, as interest, 1% of the remaining balance each month. Write a sequence showing how much Paige must pay each month for the first 6 months.

In Exercises 39−44, find the first five terms of the given recursive sequence.

39. $a_1 = 4$; $a_n = 5a_{n-1}$

40. $a_1 = 3$; $a_n = a_{n-1} - 4$

41. $a_1 = 6$; $a_n = 5a_{n-1} - 2$

42. $a_1 = 7$; $a_n = 3a_{n-1} + 5$

43. $a_1 = 4, a_2 = 5$; $a_n = a_{n-1} - a_{n-2}$

44. $a_1 = 3, a_2 = 8$; $a_n = 2a_{n-1} + a_{n-2}$

 Questions for Thought

45. In Example 3 we wrote a sequence showing the amount of money Virginia had in the bank at the end of each year for 4 years. Express the amount of money Virginia has in the bank at the end of each year as a recursive relationship. [*Hint:* How is the amount in the bank at the end of the year related to the amount in the bank at the end of the previous year?]

C.2   Series and Sigma Notation

Given a sequence, $\{a_i\}$, we can associate with it a sum, obtained by adding the first n terms of the sequence. For example, for the sequence

$$\{2, 4, 6, 8, 10, 12, \ldots , 2i, \ldots\}$$

we can associate the sum S_n, where S_n is *the sum of the first n terms of the sequence*, $\{a_i\}$. Thus,

$S_1 = a_1$ $= 2$ S_1 *is the first term of the sequence.*

$S_2 = a_1 + a_2$ $= 2 + 4$ S_2 *is the sum of the first* **two** *terms of the sequence.*

$S_3 = a_1 + a_2 + a_3$ $= 2 + 4 + 6$ S_3 *is the sum of the first* **three** *terms of the sequence.*

$S_4 = a_1 + a_2 + a_3 + a_4$ $= 2 + 4 + 6 + 8$ S_4 *is the sum of the first* **four** *terms of the sequence.*

$$\vdots$$

$S_n = a_1 + a_2 + a_3 + a_4 + \cdots + a_n = 2 + 4 + 6 + 8 + \cdots + 2n$ S_n *is the sum of the first n terms of the sequence.*

We have the following definition:

DEFINITION   The series S_n associated with the sequence $\{a_i\}$ is defined by

$$S_n = a_1 + a_2 + a_3 + \cdots + a_n$$

In words, this says that S_n is the sum of the first n terms of the sequence $\{a_i\}$. Note that we use uppercase letters to designate series, and lowercase letters to designate sequences.

To find the series associated with the sequence $\{a_n\}$, where $a_n = 3n - 5$, we first write out some of the terms of the sequence: $-2, 1, 4, 7, 10, \ldots$ (Check for yourself that these are the first few terms of $\{a_n\}$.) Then we find the indicated sums of the given sequence. Hence,

$$
\begin{aligned}
S_1 &= a_1 & &= -2 \\
S_2 &= a_1 + a_2 & &= -2 + 1 & &= -1 \\
S_3 &= a_1 + a_2 + a_3 & &= -2 + 1 + 4 & &= 3 \\
S_4 &= a_1 + a_2 + a_3 + a_4 & &= -2 + 1 + 4 + 7 = 10
\end{aligned}
$$

EXAMPLE 1   Find S_5 for the sequences whose nth term is given by

(a) $a_n = 4n - 1$ **(b)** $b_n = 2^n$ **(c)** $x_n = \dfrac{(-1)^n}{n + 1}$

Solution   **(a)** S_5 is the sum of the first five terms of the sequence $\{a_n\}$ where $a_n = 4n - 1$. We start by finding the first five terms:

$$
\begin{aligned}
a_1 &= 4(1) - 1 = 3 \\
a_2 &= 4(2) - 1 = 7 \\
a_3 &= 4(3) - 1 = 11 \\
a_4 &= 4(4) - 1 = 15 \\
a_5 &= 4(5) - 1 = 19
\end{aligned}
$$

Thus, $S_5 = a_1 + a_2 + a_3 + a_4 + a_5 = 3 + 7 + 11 + 15 + 19 = \boxed{55}$.

(b) S_5 is the sum of the first five terms of the sequence $\{b_n\}$ where $b_n = 2^n$. Again, we find the first five terms:

$$
\begin{aligned}
b_1 &= 2^1 = 2 \\
b_2 &= 2^2 = 4 \\
b_3 &= 2^3 = 8 \\
b_4 &= 2^4 = 16 \\
b_5 &= 2^5 = 32
\end{aligned}
$$

Hence, $S_5 = b_1 + b_2 + b_3 + b_4 + b_5 = 2 + 4 + 8 + 16 + 32 = \boxed{62}$.

(c) We begin by finding the first five terms:

$$x_1 = \frac{(-1)^1}{1 + 1} = \frac{-1}{2} \qquad x_2 = \frac{(-1)^2}{2 + 1} = \frac{1}{3} \qquad x_3 = \frac{(-1)^3}{3 + 1} = \frac{-1}{4}$$

$$x_4 = \frac{(-1)^4}{4 + 1} = \frac{1}{5} \qquad x_5 = \frac{(-1)^5}{5 + 1} = \frac{-1}{6}$$

Therefore, we have $S_5 = \dfrac{-1}{2} + \dfrac{1}{3} + \dfrac{-1}{4} + \dfrac{1}{5} + \dfrac{-1}{6} = \boxed{-\dfrac{23}{60}}$.

Sigma Notation

When we know the general term of a sequence, we can express its series in a more compact form by using the Greek letter (uppercase) sigma, Σ, often called the summation symbol, along with the general term. Sigma notation is used to express a *sum*.

DEFINITION

$$\sum_{i=k}^{N} a_i = a_k + a_{k+1} + a_{k+2} + a_{k+3} + \cdots + a_N \qquad \text{for } k \le N$$

The subscript i is called the *index of summation* or simply the *index*.

The symbol Σ indicates that we are going to *add* the terms. The number k, beneath the symbol, is the number substituted for the index, i, in order to find the *first* term of the sum. Then we increase the index by 1 and evaluate each subsequent term until the index reaches the number above the sigma, N, which is the last number used for i in the sum. In other words, we are adding the terms of the sequence $\{a_n\}$ starting with a_k up to and including a_N.

EXAMPLE 2

Evaluate each of the following:

(a) $\displaystyle\sum_{i=1}^{5} 3i$ **(b)** $\displaystyle\sum_{i=1}^{6} \dfrac{1}{i+1}$ **(c)** $\displaystyle\sum_{i=2}^{5} 10i^2$ **(d)** $\displaystyle\sum_{j=3}^{7} (5j-1)$

Solution

(a) $\displaystyle\sum_{i=1}^{5} 3i$ *Each term in the sum is found by substituting consecutive integers for i in the general term 3i starting with i = 1 and ending with i = 5. Remember: The Σ is telling us to add the terms.*

$\qquad = 3(1) + 3(2) + 3(3) + 3(4) + 3(5)$

$\qquad = \boxed{45}$

(b) $\displaystyle\sum_{i=1}^{6} \dfrac{1}{i+1}$ *Find each term in the sum by substituting consecutive integers for i in the general term, $\dfrac{1}{i+1}$, starting with i = 1 and ending with i = 6.*

$\qquad = \dfrac{1}{1+1} + \dfrac{1}{2+1} + \dfrac{1}{3+1} + \dfrac{1}{4+1} + \dfrac{1}{5+1} + \dfrac{1}{6+1}$

$\qquad = \dfrac{1}{2} + \dfrac{1}{3} + \dfrac{1}{4} + \dfrac{1}{5} + \dfrac{1}{6} + \dfrac{1}{7}$

$\qquad = \dfrac{669}{420} = \boxed{\dfrac{223}{140}}$ *Try computing this sum with your calculator.*

(c) $\displaystyle\sum_{i=2}^{5} 10i^2$ *This time the first term is found by substituting i = 2 in $10i^2$. The next terms are found by substituting i = 3, 4, and 5. Remember that the Σ tells us to add these terms.*

$\qquad = 10(2)^2 + 10(3)^2 + 10(4)^2 + 10(5)^2$

$\qquad = 40 + 90 + 160 + 250$

$\qquad = \boxed{540}$

(d) $\displaystyle\sum_{j=3}^{7} (5j - 1)$ *This time we are using j as the index of summation. The first term is found by evaluating the expression $5j - 1$ for $j = 3$. The next terms are found by evaluating the expression $5j - 1$ for $j = 4, 5, 6,$ and 7, respectively.*

$$= [5(3) - 1] + [5(4) - 1] + [5(5) - 1] + [5(6) - 1] + [5(7) - 1]$$
$$= 14 + 19 + 24 + 29 + 34$$
$$= \boxed{120}$$

Note that the index is merely a placeholder and so can be replaced with any other letter. Hence,

$$\sum_{i=k}^{N} a_i = \sum_{j=k}^{N} a_j = \sum_{h=k}^{N} a_h$$

Since we denote the sum of the first N terms of the sequence $\{a_i\}$ by S_N, we may write

$$S_N = a_1 + a_2 + a_3 + \cdots + a_N = \sum_{i=1}^{N} a_i$$

EXAMPLE 3

Given a sequence for which $a_j = j^3$, evaluate S_4.

Solution

$$S_4 = \sum_{j=1}^{4} j^3$$
$$= 1^3 + 2^3 + 3^3 + 4^3 = 1 + 8 + 27 + 64 = \boxed{100}$$

APPENDIX C.2 EXERCISES

In Exercises 1–18, use the given sequence (or general term) to find the indicated sum, S_n.

1. $\{2, 5, 8, 11, 14, 17, 20, \ldots\}$; S_5

2. $\{3, 7, 11, 15, 19, 23, 27, \ldots\}$; S_4

3. $\{-2, 0, 2, 4, 6, 8, \ldots\}$; S_6

4. $\{-5, -2, 1, 4, 7, 10, \ldots\}$; S_4

5. $\{7, 4, 1, -2, -5, -8, \ldots\}$; S_3

6. $\{6, 2, -2, -6, -10, -14, \ldots\}$; S_5

7. $a_n = 3n + 1$; S_2

8. $a_n = 2n + 1$; S_3

9. $b_n = 5n - 1$; S_3

10. $b_n = 4n + 3$; S_2

11. $x_j = 2^j$; S_5

12. $x_j = 3^j$; S_6

13. $y_i = 2^i + 1$; S_6

14. $y_i = 3^i - 1$; S_5

15. $a_n = \dfrac{n}{n + 1}$; S_4

16. $a_n = \dfrac{n + 2}{n}$; S_4

17. $b_n = \dfrac{(-1)^n}{n + 1}$; S_5

18. $b_n = \dfrac{(-1)^n n}{n + 3}$; S_3

In Exercises 19−32, rewrite each sum without using sigma notation, then compute each sum.

19. $\displaystyle\sum_{i=1}^{5} i$

20. $\displaystyle\sum_{i=1}^{8} 2i$

21. $\displaystyle\sum_{i=1}^{7} 5i$

22. $\displaystyle\sum_{i=1}^{4} 3i$

23. $\displaystyle\sum_{i=1}^{6} i^2$

24. $\displaystyle\sum_{j=1}^{4} j^3$

25. $\displaystyle\sum_{m=2}^{5} 2m^3$

26. $\displaystyle\sum_{k=3}^{6} 2k^2$

27. $\displaystyle\sum_{n=4}^{7} (2n+3)$

28. $\displaystyle\sum_{j=3}^{8} (3j-1)$

29. $\displaystyle\sum_{j=2}^{8} (4j+1)$

30. $\displaystyle\sum_{m=3}^{7} (5m+1)$

31. $\displaystyle\sum_{i=2}^{6} (2i^2-1)$

32. $\displaystyle\sum_{i=3}^{7} (3i^2-2)$

In Exercises 33−36, use sigma notation to represent the sum of the first n terms of the given sequence.

33. $\{2, 4, 6, \ldots, 2k, \ldots\}$ for $n = 5$

34. $\{3, 6, 9, \ldots, 3k, \ldots\}$ for $n = 8$

35. $\{1, 4, 9, \ldots, k^2, \ldots\}$ for $n = 7$

36. $\{7, 9, 11, \ldots, (2k+5), \ldots\}$ for $n = 8$

? Questions for Thought

37. Show that

(a) $\displaystyle\sum_{i=1}^{n} ca_i = c\left(\sum_{i=1}^{n} a_i\right)$

(b) $\displaystyle\sum_{i=1}^{n} (a_i + b_i) = \sum_{i=1}^{n} a_i + \sum_{i=1}^{n} b_i$

(*Hint:*   Write out each sum and use the real number properties.)

38. Show that

$$\sum_{i=1}^{n} a_i = \sum_{i=2}^{n+1} a_{i-1}$$

C.3   Arithmetic Sequences and Series

Let's examine the following sequences of numbers:

$$\{a_i\} = \{3, 5, 7, 9, 11, \ldots\}$$
$$\{b_i\} = \{1, 4, 7, 10, 13, \ldots\}$$
$$\{c_i\} = \{8, 5, 2, -1, -4, \ldots\}$$

Can you predict the next three terms in each sequence?

In previous sections, we defined and examined sequences in general and found it easy to predict terms in a sequence once a general term for the sequence had been uncovered or defined. For example, the sequence $\{3, 5, 7, 9, 11, \ldots\}$ might have a general term of $a_n = 2n + 1$. But trying to uncover a general term can be a difficult task.

An easier way to describe the sequences above is to examine the relationship between successive terms in each given sequence. In the first sequence, $\{a_i\}$, note that the difference between any two successive terms is 2, and so it would be reasonable to predict that each successive term following 11 is found by adding 2 to the previous term. Hence, 13 follows 11, 15 follows 13, and so on. Symbolically, we can represent this relationship by

$$a_i = a_{i-1} + 2$$

Note that this means that any term (say the "*i*th" term) is 2 more than the previous term [the "$(i - 1)$th" term]. If $i = 8$, we would get

$$a_8 = a_7 + 2$$

Equivalently, we could write $a_i = a_{i-1} + 2$ as a difference:

$$a_i - a_{i-1} = 2$$

This says that the *difference* between any two successive terms is 2.

In the second sequence, 1, 4, 7, 10, 13, . . . , note that the *common* difference between successive terms is 3; thus, it would be reasonable to conclude that adding 3 will produce the next term in the sequence. Algebraically we write

$$b_i = b_{i-1} + 3$$ *This says that any term is found by adding 3 to the preceding term.*

Thus, the terms following 13 would be 16, 19, 22, and so on.

Notice that in $\{c_i\}$, the third sequence given above, the difference between two successive terms is -3 because, for example, $5 - 8 = -3$ and $-4 - (-1) = -3$. Hence, we could describe the relationship between successive terms in the sequence as

$$c_i = c_{i-1} - 3$$

So in $\{c_i\}$, -4 would be followed by -7, -10, -13, and so on.

Sequences in which each successive term can be found by adding the same constant to the previous term are called ***arithmetic sequences.***

DEFINITION   An ***arithmetic sequence*** is one in which the difference between successive terms (called the *common difference*) is constant. This common difference is usually denoted by d.

An arithmetic sequence is sometimes called an *arithmetic progression*.

Hence, referring back to the sequences given at the beginning of this section, the common difference in the first sequence, $\{a_i\}$, is 2; the common difference in the second sequence, $\{b_i\}$, is 3; and the common difference in the third sequence, $\{c_i\}$, is -3.

If $\{a_1, a_2, a_3, a_4, \ldots, a_{k-1}, a_k, \ldots\}$ is an arithmetic sequence, then $a_k - a_{k-1} = d$ (the common difference) for all k and so

$$a_2 - a_1 = a_3 - a_2 = a_4 - a_3 = a_5 - a_4 \quad \text{and so on}$$

because they are all equal to d.

Similarly, if the common difference of a sequence is d, and we start with a_1, then the next term in the sequence, a_2, is

$$a_2 = a_1 + d$$ *The common difference is added to the first term.*

The next term, a_3, is

$$a_3 = a_2 + d = (a_1 + d) + d = a_1 + 2d$$
$$a_4 = a_3 + d = (a_1 + 2d) + d = a_1 + 3d$$

Note the pattern.

$$a_5 = a_4 + d = a_1 + 4d$$ *To get the fifth term, we add the common difference to a_1 four times.*

$$a_6 = a_5 + d = a_1 + 5d$$ *To get the sixth term, we add the common difference to a_1 five times.*

$$\vdots$$

In general, we have

$$a_n = a_{n-1} + d = a_1 + (n-1)d$$

We have thus derived the following formula:

The *n*th Term of an Arithmetic Sequence	If $\{a_n\}$ is an arithmetic sequence with common difference d and first term a_1, then the *n*th term is given by $$a_n = a_1 + (n-1)d \qquad \textbf{(C.1)}$$

For example, in an arithmetic sequence, $a_8 = a_1 + 7d$. In words, this says that to obtain the eighth term in an arithmetic sequence, we add the common difference seven times to the first term.

EXAMPLE 1

Given an arithmetic sequence with the first term 4 and the common difference 7, find the first five terms and the 19th term.

Solution

The first term of the arithmetic sequence is 4; hence,

$a_1 = 4$ *In an arithmetic sequence with common difference d, each term is d more than the previous term.*

$a_2 = 4 + 7 = 11$ *Since the common difference, d, is 7, a_2 is 7 more than $a_1 = 4$.*

$a_3 = 11 + 7 = 18$ *a_3 is 7 more than $a_2 = 11$.*

$a_4 = 18 + 7 = 25$

$a_5 = 25 + 7 = 32$

Thus, the first five terms are $\boxed{4,\ 11,\ 18,\ 25,\ \text{and } 32}$.

To find the 19th term, we can use formula (C.1):

$a_n = a_1 + (n-1)d$ *Substituting n = 19, $a_1 = 4$, and d = 7, we get*

$a_{19} = 4 + (19 - 1)7 = 4 + 18(7) = \boxed{130}$

EXAMPLE 2

Given an arithmetic sequence whose first two terms are 8 and 14, find the next three terms and the 20th term.

Solution

Since the sequence starts with 8, 14, . . . and is an *arithmetic* sequence, the difference between the first two terms is also the common difference between any two successive terms. Hence, $d = 14 - 8 = 6$. Since $a_1 = 8$ and $a_2 = 14$ we have

$a_3 = 14 + 6 = 20$ *Because $a_3 = a_2 + d$*

Thus, we have

$a_4 = 20 + 6 = 26$ *Because $a_4 = a_3 + d$*

$a_5 = 26 + 6 = 32$

Therefore, the three terms following 8, 14, are $\boxed{20,\ 26,\ \text{and } 32}$.

The 20th term is found by using formula (C.1) as follows:

$a_n = a_1 + (n-1)d$ *Where n = 20, $a_1 = 8$, and d = 6*

$a_{20} = 8 + (20 - 1)(6) = \boxed{122}$

Given an arithmetic sequence $\{a_i\}$, let's examine the difference between the third term, a_3, and the eighth term, a_8. First, we point out that since the difference between successive terms in an arithmetic sequence is a constant, which we are calling d, the difference between *any* two terms in an arithmetic sequence is a multiple of d. Because we have to add d each time we want the next term, when we start with a_3, we have to add d to get a_4, another d to get a_5, another d to get a_6, and so on, until we get to a_8. Hence to get from a_3 to a_8 we have to add five d's, or

$$a_8 = a_3 + 5d$$

We can derive a formula for the relationship between *any* two terms in an arithmetic sequence in general as follows. We begin with formula (C.1), which gives the nth term of an arithmetic sequence with first term a_1 and common difference d:

$$a_n = a_1 + (n - 1)d$$

Now suppose we have two terms in an arithmetic sequence, a_j and a_k. Let's apply this formula to a_j and a_k by substituting $n = k$ and $n = j$; we get

$$a_k = a_1 + (k - 1)d$$
$$a_j = a_1 + (j - 1)d \qquad \textit{Subtracting the second equation from the first, we get}$$
$$a_k - a_j = (k - 1)d - (j - 1)d$$
$$= kd - d - jd + d$$
$$= kd - jd \qquad \textit{Factor out d.}$$
$$= (k - j)d$$

which gives us the following formula.

The difference between any two terms, a_j and a_k, in an arithmetic sequence with common difference d is

$$a_k - a_j = (k - j)d \qquad\qquad \textbf{(C.2)}$$

EXAMPLE 3

Given an arithmetic sequence with $a_4 = 13$ and $a_7 = 31$, find a_1 and a_{20}.

Solution

We are given that the fourth term, a_4, is 13, and the seventh term, a_7, is 31 in an arithmetic sequence. Using formula (C.2) with $k = 7$ and $j = 4$, the difference between these two terms is

$$a_7 - a_4 = (7 - 4)d \qquad \textit{Substitute for } a_7 \textit{ and } a_4.$$
$$31 - 13 = (7 - 4)d$$
$$18 = 3d$$
$$6 = d$$

Once we know the value of d, we find a_1 by substituting $d = 6$ in formula (C.1) for the nth term of an arithmetic sequence:

$$a_n = a_1 + (n - 1)d \qquad \textit{Because we know the value of } a_4, \textit{ we use } n = 4 \textit{ to obtain}$$
$$a_4 = a_1 + (4 - 1)d \qquad \textit{Substituting } a_4 = 13 \textit{ and } d = 6, \textit{ we get}$$
$$13 = a_1 + 3(6)$$
$$13 = a_1 + 18$$
$$\boxed{-5 = a_1}$$

Knowing $a_1 = -5$, which was the first term we needed to find, and $d = 6$, we have the information we need to find a_{20}. We use formula (C.1) for the nth term of an arithmetic sequence:

$$a_n = a_1 + (n - 1)d \qquad \textit{Substituting } a_1 = -5, \, d = 6, \textit{ and } n = 20, \textit{ we get}$$
$$a_{20} = -5 + (20 - 1)6 = -5 + 19(6) = -5 + 114$$

$$\boxed{a_{20} = 109}$$

Arithmetic Series

An arithmetic sequence has the interesting property that not only do we have a formula for a_k, but we can also derive a formula for S_k, the sum of the first k terms of an arithmetic sequence.

The series obtained from an arithmetic sequence is called an ***arithmetic series.***

Let's first consider a special arithmetic sequence and its series. If $a_1 = 1$ and $d = 1$, then $a_2 = 2, a_3 = 3, a_4 = 4, \ldots, a_k = k$. In other words, this sequence is the sequence of positive integers. The series for this sequence will be denoted by

$$A_k = 1 + 2 + 3 + \cdots + (k - 2) + (k - 1) + k$$

which is the sum of the first k positive integers.

Let us write the sequence A_k and below it, rewrite the same sequence in reverse:

$$A_k = 1 + \quad 2 \quad + \quad 3 \quad + \cdots + (k - 2) + (k - 1) + k$$
$$A_k = k + (k - 1) + (k - 2) + \cdots + \quad 3 \quad + \quad 2 \quad + 1$$

On the right side note that there are k columns, and each column sums to $k + 1$.

Adding the two equations, we get

$$2A_k = \underbrace{(k + 1) + (k + 1) + (k + 1) + \cdots + (k + 1) + (k + 1) + (k + 1)}_{k \text{ times}}$$

Therefore,

$$2A_k = k(k + 1)$$

$$A_k = \frac{k(k + 1)}{2}$$

Thus, we have derived the following formula:

The Sum of the First k Positive Integers	The sum of the first k positive integers is given by $$A_k = 1 + 2 + 3 + \cdots + k = \frac{k(k + 1)}{2} \qquad \text{(C.3)}$$

For example, using this formula for $k = 10$, we find that the sum of the first 10 positive integers is

$$A_{10} = 1 + 2 + 3 + \cdots + 10 = \frac{10(10 + 1)}{2} = \frac{10(11)}{2} = 55$$

We can now derive a formula for the general arithmetic series, S_n. The sum $S_n = a_1 + a_2 + a_3 + a_4 + \cdots + a_{n-2} + a_{n-1} + a_n$ is called the arithmetic series associated with the sequence $\{a_n\}$. By using the formula $a_n = a_1 + (n - 1)d$, we can rewrite S_n in terms of a_1, d, and n as follows:

$$S_n = a_1 + \quad a_2 \quad + \quad a_3 \quad + \quad a_4 \quad + \cdots + \quad a_{n-1} \quad + a_n \qquad \textit{Becomes}$$

$$S_n = a_1 + (a_1 + d) + (a_1 + 2d) + (a_1 + 3d) + \cdots + [a_1 + (n - 2)d] + [a_1 + (n - 1)d]$$

We collect all the a_1 terms (there are n of them):

$$S_n = na_1 + [d + 2d + 3d + \ldots + (n - 2)d + (n - 1)d] \qquad \textit{Now factor out d from within the brackets.}$$

$$= na_1 + d[1 + 2 + 3 + \ldots + (n - 2) + (n - 1)]$$

Inside the brackets we have the sum of the first $n - 1$ positive integers. We use formula (C.3) to find the sum of the first k positive integers, substituting $n - 1$ for k:

$$S_n = na_1 + d\left[\frac{(n - 1)n}{2}\right] \qquad \textit{Combine into a single fraction.}$$

$$= \frac{2na_1}{2} + \frac{n(n - 1)d}{2} = \frac{2na_1 + n(n - 1)d}{2} \qquad \textit{Factor out n in the numerator.}$$

$$= \frac{n[2a_1 + (n - 1)d]}{2}$$

Finally, we rewrite this formula by separating out the factor of $\frac{n}{2}$.

The Sum of the First n Terms of an Arithmetic Sequence	The sum of the first n terms of an arithmetic sequence is given by $$S_n = \sum_{i=1}^{n} a_i = \frac{n}{2}[2a_1 + (n - 1)d] \qquad \textbf{(C.4)}$$

Note that since $a_n = a_1 + (n - 1)d$, we can rewrite $2a_1$ as $a_1 + a_1$, and S_n becomes

$$S_n = \frac{n}{2}[2a_1 + (n - 1)d] = \frac{n}{2}[a_1 + \underbrace{a_1 + (n - 1)d}_{a_n}] = \frac{n}{2}(a_1 + a_n)$$

Hence, we have derived an alternate (and simpler) version of formula (C.4).

The Sum of the First n Terms of an Arithmetic Sequence (Alternative Form)	An alternative form for the sum of the first n terms of an arithmetic sequence is given by $$S_n = \frac{n}{2}(a_1 + a_n) \qquad \textbf{(C.5)}$$

Note that formula (C.4) is most useful when we know n, a_1 (the first term), and d (the common difference), whereas formula (C.5) is most useful when we know n, a_1, and a_n (the last term).

EXAMPLE 4

Given the arithmetic sequence $\{5, 9, 13, 17, \ldots\}$, find the following:

(a) S_8 **(b)** The sum of the first 20 terms

Solution

(a) We could calculate S_8 by writing out the first eight terms in the sequence and adding them together, or we could use one of the formulas for S_n, where $n = 8$. Using formula (C.4), we can find S_n by knowing a_1, d, and n. The sequence

$\{5, 9, 13, 17, \ldots\}$ has first term $a_1 = 5$, and the common difference $d = 4$ (why?), and for S_8, $n = 8$. We can substitute these values into formula (C.4) to get

$$S_n = \frac{n}{2}[2a_1 + (n-1)d] \qquad \textit{We substitute } n = 8, a_1 = 5, \textit{ and } d = 4.$$

$$S_8 = \frac{8}{2}[2 \cdot 5 + (8-1)4]$$

$$= 4[10 + 7(4)] = 4(38)$$

$$= \boxed{152} \qquad \textit{You may want to check this answer by writing out the first eight terms in the sequence and finding their sum.}$$

(b) To find the sum of the first 20 terms of the sequence, we are looking for the series S_{20}. Thus, we use formula (C.4) again, this time with $n = 20$:

$$S_{20} = \frac{20}{2}[2 \cdot 5 + (20-1)4]$$

$$= 10[10 + 19(4)]$$

$$= \boxed{860}$$

EXAMPLE 5

Find the sum of the first 30 multiples of 5 beginning with 5.

Solution

We want to compute the sum of the terms $5, 10, 15, \ldots, 150$, which is an arithmetic sequence with $n = 30$, $d = 5$, and $a_1 = 5$. Since we can easily identify the first and last terms, we use the alternate formula for an arithmetic series [formula (C.5)]:

$$S_n = \frac{n}{2}(a_1 + a_n)$$

$$S_{30} = \frac{30}{2}(5 + 150) = 15(155)$$

$$= \boxed{2{,}325}$$

EXAMPLE 6

Jake gets a job offer from each of two firms. Firm A offers him a job with a starting annual salary of $48,000 and a guaranteed raise of $3,000 per year. On the other hand, firm B offers him a job at a starting annual salary of $54,000 but will guarantee a raise of only $2,000 per year.

(a) What would his annual salary be at each firm in the 10th year?
(b) Over the first 10 years, how much would he earn at each firm?

Solution

(a) For firm A, the annual salaries form the arithmetic sequence 48,000, 51,000, 54,000, $\ldots$, with first term $a_1 = \$48{,}000$ and common difference $d = 3{,}000$. Letting a_n be the salary in year n, we use formula (C.1), $a_n = a_1 + (n-1)d$. Since we want the salary for the 10th year, we have $n = 10$ and so

$$a_{10} = a_1 + 9d = 48{,}000 + 9(3{,}000) = 75{,}000$$

For firm B, the annual salaries form the arithmetic sequence 56,000, 58,000, 60,000, $\ldots$, with first term $b_1 = 54{,}000$ and common difference $d = 2{,}000$. Letting b_n be the salary in year n, we have $b_n = b_1 + (n-1)d$. Hence,

$$b_{10} = b_1 + 9d = 54{,}000 + 9(2{,}000) = 72{,}000$$

In the 10th year Jake's annual salary would be

$$\boxed{\$75{,}000 \text{ at firm A and } \$72{,}000 \text{ at firm B}}$$

(b) To find out how much he would earn at each firm over the first 10 years, we add the first 10 annual salaries: For firm A, the total would be $a_1 + a_2 + \ldots + a_{10}$; for firm B, the total would be $b_1 + b_2 + \ldots + b_{10}$. In other words, we are looking for the sum of the first ten terms for each arithmetic sequence. We use formula (C.5) since we already have the first and last terms of each sequence:

$$a_1 + a_2 + a_3 + \cdots + a_{10} = \frac{10}{2}(a_1 + a_{10})$$ *Since $a_1 = 48,000$*
and $a_{10} = 75,000$

$$= 5(48,000 + 75,000) = 615,000$$

$$b_1 + b_2 + b_3 + \cdots + b_{10} = \frac{10}{2}(b_1 + b_{10})$$ *Since $b_1 = 54,000$*
and $b_{10} = 72,000$

$$= 5(54,000 + 72,000) = 630,000$$

Over the first 10 years Jake would have earned

> $615,000 working in firm A and $630,000 working in firm B as on left

Notice that although Jake would have a higher salary during (and after) the 10th year at firm A, the total amount of money he makes over the 10-year period is higher at firm B.

APPENDIX C.3 EXERCISES ⊙

*In Exercises 1−18, the given sequences are **arithmetic**. Find the next two terms and the 15th term.*

1. a_i, where the first term is 3 and the common difference is 2

2. a_j, where the first term is 8 and the common difference is 5

3. x_n, where the first term is 2 and the common difference is -4

4. y_m, where the first term is -5 and the common difference is -4

5. z_n, where the first term is -2 and the common difference is $\frac{1}{2}$

6. z_n, where the first term is 3 and the common difference is $\frac{1}{10}$

7. $\{1, 4, 7, 10, \ldots\}$

8. $\{5, 7, 9, 11, \ldots\}$

9. $\{-2, 0, 2, 4, \ldots\}$

10. $\{8, 13, 18, \ldots\}$

11. $\{2, -3, -8, \ldots\}$

12. $\{5, 2, -1, -4, \ldots\}$

13. $\{8, 16, 24, \ldots\}$

14. $\{10, 20, 30, \ldots\}$

15. $\left\{\frac{1}{2}, \frac{5}{6}, \frac{7}{6}, \ldots\right\}$

16. $\left\{\frac{1}{4}, \frac{7}{12}, \frac{11}{12}, \ldots\right\}$

17. $\left\{-\frac{1}{3}, -\frac{5}{6}, \ldots\right\}$

18. $\left\{-\frac{1}{2}, 1, \ldots\right\}$

19. Given the arithmetic sequence $\{a_i\}$ with $a_5 = 25$ and $a_{15} = 85$, find a_1 and d.

20. Given the arithmetic sequence $\{a_i\}$ with $a_5 = 19$ and $a_{15} = 59$, find a_1 and d.

21. Given the arithmetic sequence $\{a_i\}$ with $a_7 = -21$ and $a_{12} = -41$, find a_5.

22. Given the arithmetic sequence $\{a_i\}$ with $a_3 = -8$ and $a_{17} = -50$, find a_{12}.

23. Harry gets a raise of $800 every 6 months for the first 5 years of his job. If his starting salary is $36,500, what is his yearly salary at the end of 3 years?

24. Joan gets a raise of $400 every 3 months for the first 3 years of her job. If her starting salary is $36,500, what is her yearly salary at the end of 2 years?

25. Cindy bench-presses 65 lb and wants to increase the amount of weight she bench-presses by 5 lb every 2 weeks. If she increases weights consistently, how much should she be bench-pressing during the 20th week? (Start at 65 lb for the first 2 weeks.)

26. Janice saves $1,000 a year and wants to consistently increase her savings by an additional $200 each year. How much should she be saving during the eighth year? (Start at $1,000 for the first year.)

27. Suppose that an object dropped from the top of a building falls 16 ft the first second, 48 ft the second second, 80 ft the third second, and so on. The distances it falls each second form an arithmetic sequence. How many feet will the object fall during the sixth second?

28. How many feet will the object in Exercise 27 fall *during* the 10th second?

In Exercises 29−40, find the indicated S_n for the given arithmetic sequences.

29. $\{a_i\}$, where the first term is 8 and the common difference is 2; S_6

30. $\{a_i\}$, where the first term is -2 and the common difference is 3; S_4

31. $\{x_n\}$, where the first term is 5 and the common difference is -1; S_5

32. $\{y_n\}$, where the first term is 0 and the common difference is -3; S_7

33. $\{b_j\}$, where the first term is 4 and the common difference is $\frac{1}{2}$; S_5

34. $\{b_k\}$, where the first term is -1 and the common difference is $\frac{1}{3}$; S_4

35. $\{2, -3, -8, \ldots\}$; S_5

36. $\{8, 13, 18, \ldots\}$; S_3

37. $\{-2, 0, 2, 4, \ldots\}$; S_6

38. $\{5, 2, -1, -4, \ldots\}$; S_6

39. $\left\{-\frac{1}{3}, -\frac{5}{6}, \ldots\right\}$; S_4

40. $\left\{-\frac{1}{2}, -1, \ldots\right\}$; S_4

In Exercises 41−48, find the indicated sums.

41. $\sum_{i=1}^{6} 3i$

42. $\sum_{k=1}^{8} 4k$

43. $\sum_{j=1}^{10} 10j$

44. $\sum_{m=1}^{7} 6m$

45. $\sum_{i=1}^{5} (4i + 3)$

46. $\sum_{n=1}^{6} (9n - 4)$

47. $\sum_{i=30}^{80} i$

48. $\sum_{n=50}^{100} n$

49. Find the sum of the first 50 positive even numbers.

50. Find the sum of the first 50 positive odd numbers.

51. Find the sum of the first 10 positive multiples of 8.

52. Find the sum of the first 30 positive multiples of 7.

53. Given the arithmetic sequence $\{a_i\}$ with $a_4 = 3$ and $a_{14} = 23$, find S_{15}.

54. Given the arithmetic sequence $\{a_i\}$ with $a_7 = 17$ and $a_{14} = 31$, find S_{12}.

55. Given the arithmetic sequence $\{a_i\}$ with $a_8 = -38$ and $a_{12} = -58$, find S_{15}.

56. Given the arithmetic sequence $\{a_i\}$ with $a_5 = -28$ and $a_{14} = -91$, find S_{20}.

57. Suppose that an object dropped from the top of a building falls 16 ft the first second, 48 ft the second second, 80 ft the third second, and so on. The distances it falls each second form an arithmetic sequence. How many total feet will the object have fallen after (at the end of) 6 seconds?

58. If the object in Exercise 57 is dropped from the top of the same building and hits the ground at the end of 8 seconds, how high is the building?

59. Karen gets a job offer from each of two firms. Firm A offers her a job with a starting annual salary of $40,000 and a guaranteed raise of $2,000 per year. On the other hand, firm B offers her a job at a starting annual salary of $42,000 but will guarantee a raise of only $1,500 per year.

(a) What would her annual salary be at each firm in the eighth year?

(b) Over the first 8 years, how much would she earn at each firm?

60. Darren gets a job offer from each of two firms. Firm A offers him a job with a starting annual salary of $48,000 and a guaranteed raise of $1,500 per year. On the other hand, firm B offers him a job at a starting annual salary of $44,000 but will guarantee a raise of $2,500 per year.

(a) What would his annual salary be at each firm in the seventh year?

(b) Over the first 7 years, how much would he earn at each firm?

? Questions for Thought

61. Refer to Exercise 60. Darren considers the advantages and disadvantages of working for either firm A or B, and decides that they are fairly evenly matched except for salary. He decides to work for one of the firms based on salary alone. As demonstrated in Example 6 (and Exercise 60), the financial advantages of one salary offer over the other are not clear-cut. Write a few paragraphs discussing under what conditions one salary would be better than the other. Support your arguments mathematically.

C.4 Geometric Sequences and Series

In the previous section we examined arithmetic sequences and series. In this section we discuss another type of special sequence and series.

To begin, let's examine the following group of sequences:

$$\{a_i\} = \{3, 6, 12, 24, 48, \ldots\}$$
$$\{b_i\} = \{5, 15, 45, 135, 405, \ldots\}$$

Neither of these sequences is arithmetic because, in each sequence, the differences between each successive pair of terms vary. (For sequence $\{a_i\}$, $6 - 3 = 3$, $12 - 6 = 6$, $24 - 12 = 12$, and so on.) However, on closer inspection of the terms in each sequence, we find that there is a pattern: The *quotient* of two successive terms (each term divided by its preceding term), is constant. In the case of $\{a_i\}$, the quotient is equal to 2 since

$$\frac{6}{3} = \frac{12}{6} = \frac{24}{12} = \frac{48}{24} = 2$$

This type of sequence—one in which the ratio formed by any two successive terms is constant—is called a **geometric sequence.** In a geometric sequence, each successive term is found by *multiplying* the previous term by a constant. Each successive term in $\{a_i\}$ is found by multiplying the previous term by 2. Hence, 3 is followed by 6, 6 is followed by 12, 12 is followed by 24, and if this is a geometric sequence, then 48 would be followed by 96, 96 would be followed by 192, and so on. Each term is twice the preceding term.

Symbolically, we can represent the relationship in sequence $\{a_i\}$ by the following:

$$a_i = 2a_{i-1}$$

Note that this means that any term (the "ith" term) is 2 times the previous term (the "$(i-1)$th" term). If $i = 12$, we would get

$$a_{12} = 2a_{11}$$

Equivalently, we could write the above relationship as a ratio. That is,

$$a_i = 2a_{i-1} \quad \text{is equivalent to} \quad \frac{a_i}{a_{i-1}} = 2$$

This says that the ratio between any two successive terms is 2. This ratio is called the *common ratio*.

Sequences in which each successive term is a constant *multiple* of the previous term are called *geometric sequences*.

DEFINITION

A *geometric sequence* is one in which the ratio between successive terms (called the common ratio) is constant. This common ratio is usually denoted by r.

A geometric sequence is sometimes called a *geometric progression*.

Note the similarities and differences between an arithmetic sequence and a geometric sequence: In an arithmetic sequence, the *differences* between successive terms are constant, whereas in a geometric sequence the *quotients* of successive terms are constant.

In the geometric sequence $\{a_1, a_2, a_3, a_4, \ldots, a_{k-1}, a_k, \ldots\}$, where r is the common ratio, we have

$$r = \frac{a_2}{a_1} = \frac{a_3}{a_2} = \frac{a_4}{a_3} = \cdots = \frac{a_k}{a_{k-1}} = \cdots$$

For the geometric sequence $\{b_i\} = \{5, 15, 45, 135, 405, \ldots\}$ the common ratio is 3, since

$$\frac{15}{5} = \frac{45}{15} = \frac{135}{45} = \frac{405}{135} = 3$$

Hence, it follows that, if the sequence is geometric, the next two terms in the sequence would be $405 \cdot 3 = 1,215$ and $1,215 \cdot 3 = 3,645$.

In general, in a geometric sequence, if the common ratio is r and the first term is a_1, then the second term is

$a_2 = a_1 r$	*To get the next term, we multiply by the common ratio, r.*
$a_3 = a_2 r = (a_1 r)r = a_1 r^2$	*Because $a_2 = a_1 r$; similarly*
$a_4 = a_3 r = (a_1 r^2)r = a_1 r^3$	*Continuing the pattern, we have*
$a_5 = a_4 r = a_1 r^4$	
$a_6 = a_5 r = a_1 r^5$	
$\vdots$	

In general, we have

$$a_n = a_{n-1} r = a_1 r^{n-1}$$

This gives us the following general formula:

The nth Term of a Geometric Sequence

The nth term of a geometric sequence with common ratio r is

$$a_n = a_1 r^{n-1} \qquad (n > 1) \tag{C.6}$$

EXAMPLE 1

Given the geometric sequence $\{2, 10, 50, \ldots\}$, find the next two terms and the eighth term.

Solution

Since we are given that the sequence is geometric, we first find the common ratio, r, which is $10/2 = 5$. Therefore, the next two terms following 50 in the sequence are found by multiplying successively by the common ratio 5; that is,

$$50 \cdot 5 = \boxed{250} \quad \text{and} \quad 250 \cdot 5 = \boxed{1{,}250}$$

The eighth term is found by using formula (C.6) for the nth term of a geometric sequence:

$$a_n = a_1 r^{n-1} \qquad \text{\textit{Here we substitute } } a_1 = 2, n = 8, \text{\textit{ and } } r = 5.$$
$$a_8 = a_1 r^7$$
$$= 2 \cdot 5^7$$
$$= 2(78{,}125)$$
$$= \boxed{156{,}250}$$

EXAMPLE 2

Find the next term in the geometric sequence $\left\{ 1, \dfrac{1}{2}, \dfrac{1}{4}, \dfrac{1}{8}, \dfrac{1}{16}, \ldots \right\}$.

Solution

The geometric sequence $\left\{ 1, \dfrac{1}{2}, \dfrac{1}{4}, \dfrac{1}{8}, \dfrac{1}{16}, \ldots \right\}$ has a common ratio of $\dfrac{1}{2}$ since

$$\frac{\frac{1}{2}}{1} = \frac{\frac{1}{4}}{\frac{1}{2}} = \frac{\frac{1}{8}}{\frac{1}{4}} = \frac{\frac{1}{16}}{\frac{1}{8}} = \frac{1}{2}$$

Actually, if we know that a sequence is geometric, we can use any two successive terms to find that $r = \dfrac{1}{2}$.

Given the term $\dfrac{1}{16}$, we can use the fact that $a_n = a_{n-1} r$ to find the term following $\dfrac{1}{16}$. We simply multiply it by the common ratio $r = \dfrac{1}{2}$ to get $\dfrac{1}{16} \cdot \dfrac{1}{2} = \dfrac{1}{32}$.

EXAMPLE 3

Find the sixth term of the geometric sequence whose first term is 2 and whose fourth term is 54.

Solution

First we need to find the common ratio, r. Given $a_1 = 2$ and $a_4 = 54$, we use formula (C.6) for the nth term of a geometric sequence as follows:

$$a_n = a_1 r^{n-1} \qquad \text{\textit{We use the fact that we are given } } a_1 \text{ \textit{and} } a_4 \text{ \textit{and let} } n = 4.$$
$$a_4 = a_1 r^3 \qquad \text{\textit{Substitute } } a_1 = 2 \text{ \textit{and} } a_4 = 54.$$
$$54 = 2r^3 \qquad \text{\textit{Now we solve for } } r.$$
$$27 = r^3 \qquad \text{\textit{Take cube roots.}}$$
$$3 = r \qquad \text{\textit{Now we use formula (C.6) again to find } } a_6.$$
$$a_6 = a_1 r^5$$
$$a_6 = 2(3)^5 = 2(243)$$
$$a_6 = \boxed{486}$$

EXAMPLE 4

Virginia receives 6% interest annually on the principal in her bank account. If she deposits $1,000, how much will she have after the end of 4 years? At the end of 8 years?

Solution

This situation is the same as the one encountered in Example 3 of Section C.1. This time we will use what we know about geometric sequences and develop a shortcut for handling the problem. When we solved this problem before, we noted that at the end of each year the amount Virginia had in the bank is

$$\left(\begin{matrix} \text{Her principal for the} \\ \text{beginning of the year} \end{matrix}\right) + \left(\begin{matrix} \text{Interest on the principal at} \\ \text{the beginning of the year} \end{matrix}\right)$$

If we call P her original principal, then at the end of the *first* year, Virginia had in the bank:

$$P + 0.06P \qquad \textit{Which simplifies to}$$
$$= 1.06P \qquad \textit{In the bank at the end of the first year}$$

Note that this says that the new principal (or the amount at the end of the first year) is equal to 106% of the original principal. In general, we find this to be true for each year: The new principal (or the amount at the end of each year) is 106% of the principal of the previous year.

At the end of the *second* year she has

106% of the amount in the bank at the end of the first year

$$= 1.06(1.06P) \qquad \textit{Which simplifies to}$$
$$= P(1.06)^2$$

At the end of the *third* year she has

106% of the amount in the bank at the end of the second year

$$= 1.06[P(1.06)^2] \qquad \textit{Which simplifies to}$$
$$= P(1.06)^3$$

In general, the amount she has in the bank at the end of the ith year is

$$P(1.06)^i$$

If we call A_i the amount Virginia has in the bank at the end of year i, then

$$A_i = P(1.06)^i$$

This pattern fits the definition of a geometric sequence: Each successive term is a constant (the constant is 1.06) times the previous term.

Since Virginia starts out with $1,000, call this P. Hence, $A_i = 1,000(1.06)^i$ is the principal at the end of the ith year.

At the end of the first year, $i = 1$ and

$$A_1 = 1,000(1.06)^1 = \$1,060.00$$

is the amount she has in the bank at the end of the first year (as we saw earlier).

At the end of the fourth year, $i = 4$ and

$$A_4 = 1,000(1.06)^4 = \$1,262.48$$

Thus, Virginia has $\boxed{\$1,262.48 \text{ in the bank at the end of the fourth year}}$.

At the end of the eighth year, $i = 8$ and

$$A_8 = 1,000(1.06)^8 = \$1,593.85$$

Virginia has $\boxed{\$1,593.85 \text{ in the bank at the end of the eighth year}}$.

EXAMPLE 5	A radioactive substance loses $\frac{1}{2}$ its mass per year. How much is left of 100 g of the substance after 10 years?

Solution

If we list how much is left after each year and note that each term is $\frac{1}{2}$ the previous term, we obtain a geometric sequence as follows.

Let A_i be the amount left at the end of year i. Then

$$A_1 = 100\left(\frac{1}{2}\right) \qquad \textit{Amount left at the end of year 1}$$

$$A_2 = \frac{1}{2} \text{ of } A_1 = \frac{1}{2}\left[100\left(\frac{1}{2}\right)\right] = 100\left(\frac{1}{2}\right)^2 \qquad \textit{Amount left at the end of year 2}$$

$$A_3 = \frac{1}{2} \text{ of } A_2 = \frac{1}{2}\left[100\left(\frac{1}{2}\right)^2\right] = 100\left(\frac{1}{2}\right)^3 \qquad \textit{Amount left at the end of year 3}$$

In general,

$$A_i = 100\left(\frac{1}{2}\right)^i \qquad \textit{Amount left at the end of year i}$$

Hence, at the end of 10 years there is

$$A_{10} = 100\left(\frac{1}{2}\right)^{10}$$

$$= 100\left(\frac{1}{1,024}\right) = \frac{100}{1,024}$$

$$= \boxed{0.09766 \text{ g}}$$

Geometric Series

If $\{a_i\}$ is a geometric sequence then its associated series, S_n, is

$$S_n = a_1 + a_2 + a_3 + \cdots + a_{n-1} + a_n$$

and S_n is called the *geometric series* associated with the geometric sequence $\{a_i\}$. Throughout this discussion n is, as usual, a positive integer.

As with arithmetic sequences, we can find a formula to describe a geometric series associated with it. Before we begin to derive the formula, however, we must first make note of the following product:

$$(1 - r)(1 + r + r^2 + r^3 + \cdots + r^{n-2} + r^{n-1}) = 1 - r^n \qquad \textbf{(C.7)}$$

We leave the proof as an exercise (see Exercise 60), but will demonstrate this product for $n = 6$:

$$(1 - r)(1 + r + r^2 + r^3 + r^4 + r^5) \qquad \textit{Use the distributive property to get}$$

$$= 1(1 + r + r^2 + r^3 + r^4 + r^5) - r(1 + r + r^2 + r^3 + r^4 + r^5)$$

$$= 1 + r + r^2 + r^3 + r^4 + r^5 - r - r^2 - r^3 - r^4 - r^5 - r^6$$

$$\qquad \textit{Note how all but the first and last terms drop out.}$$

$$= 1 - r^6$$

Hence,

$$(1 - r)(1 + r + r^2 + r^3 + r^4 + r^5) = 1 - r^6 \qquad \textit{As required}$$

If we divide both sides of equation (C.7) by $1 - r$ (assuming $r \neq 1$), we get

$$1 + r + r^2 + r^3 + \cdots + r^{n-2} + r^{n-1} = \frac{1 - r^n}{1 - r} \qquad \textbf{(C.8)}$$

We will use this fact to find the formula for the sum of n terms of a geometric sequence as follows. Suppose $\{a_i\}$ is a geometric sequence with its associated geometric series S_n. Writing out the terms of S_n, we get

$$S_n = a_1 + a_2 + a_3 + a_4 + \cdots + a_{n-1} + a_n$$

Since $\{a_i\}$ is a geometric sequence, we can use formula (C.6), which says that $a_n = a_1 r^{n-1}$, to rewrite each term using a_1 and r as follows:

$$S_n = a_1 + a_1 r + a_1 r^2 + a_1 r^3 + \cdots + a_1 r^{n-2} + a_1 r^{n-1} \qquad \text{\textit{Factoring out } } a_1, \text{ \textit{we get}}$$

$$= a_1(1 + r + r^2 + r^3 + \cdots + r^{n-2} + r^{n-1}) \qquad \textit{Using the result in formula (C.8) on the expression inside the parentheses, we get}$$

$$= a_1\left(\frac{1 - r^n}{1 - r}\right)$$

Thus, we have derived the following formula:

The Sum of the First n Terms of a Geometric Sequence	The formula for the sum of the first n terms of a geometric sequence is $$S_n = \frac{a_1(1 - r^n)}{1 - r} \qquad \text{for } r \neq 1 \qquad \textbf{(C.9)}$$

Equivalently, we can write the formula for S_n using sigma notation:

> If $\{a_i\}$ is a geometric sequence with $r \neq 1$, then
> $$\sum_{i=1}^{n} a_i = \frac{a_1(1 - r^n)}{1 - r}$$

EXAMPLE 6

Given the geometric sequence $\{a_i\} = \{5, 15, 45, \ldots\}$, find the following.

(a) S_6 **(b)** $\displaystyle\sum_{i=1}^{10} a_i$

Solution

(a) To find S_6 we could simply find the next three terms of the sequence and add the first six terms together. (Try this!) Instead, we use formula (C.9), noting that the first term is $a_1 = 5$, and $r = 3$:

$$S_n = \frac{a_1(1 - r^n)}{1 - r}$$

$$S_6 = \frac{5(1 - 3^6)}{1 - 3}$$

$$= \frac{5(1 - 729)}{-2} = \frac{5(-728)}{-2}$$

$$= \boxed{1,820}$$

Does this answer agree with what you obtained by finding the first six terms in the sequence and adding them together?

(b) Here, we are looking for the sum of the first 10 terms in the sequence. Again, $a_1 = 5$ and $r = 3$, but this time $n = 10$.

$$\sum_{i=1}^{10} a_i = S_{10} = \frac{5(1 - 3^{10})}{1 - 3}$$

$$= \frac{5(1 - 59{,}049)}{-2} = \frac{5(-59{,}048)}{-2}$$

$$= \boxed{147{,}620}$$

In order to appreciate the usefulness of this formula, try to compute all 10 terms and find their sum.

Infinite Geometric Series

So far we have examined the sums of finite geometric sequences—that is, sequences that terminate. Can we find the sum or a formula for the sum of an infinite sequence? Let's examine two examples of infinite geometric sequences:

$$\{x_i\} = \{3, 6, 12, 24, \ldots\}$$

$$\{y_i\} = \left\{\frac{1}{2}, \frac{1}{4}, \frac{1}{8}, \ldots\right\}$$

For $\{x_i\}$, note that $x_1 = 3$ and $r = 2$. If we calculate S_5, S_{10}, and S_{20} using formula (C.9), we get:

$$S_5 = \frac{3(1 - 2^5)}{1 - 2} = 93$$

$$S_{10} = \frac{3(1 - 2^{10})}{1 - 2} = 3,069$$

$$S_{20} = \frac{3(1 - 2^{20})}{1 - 2} = 3,145,725$$

The sums, S_i, are increasing quickly because the terms, x_i, are increasing. (Here $r = 2$ and the expression 2^n continues to increase quickly as n increases.) Hence, it would make no sense to find the sum of the *infinite* sequence $\{x_i\}$.

Looking at the sequence $\{y_i\}$, we have $y_1 = \frac{1}{2}$ and $r = \frac{1}{2}$; we find

$$S_n = \frac{\frac{1}{2}\left[1 - \left(\frac{1}{2}\right)^n\right]}{1 - \frac{1}{2}} = \frac{\frac{1}{2}\left[1 - \left(\frac{1}{2}\right)^n\right]}{\frac{1}{2}} = 1 - \left(\frac{1}{2}\right)^n$$

If we calculate y_n and S_n for a few values of n, a pattern does emerge. (See Table C.1.)

Table C.1

n	1	2	3	4	5	10	20
y_n	0.5	0.25	0.125	0.0625	0.03125	0.0009766	0.0000009537
S_n	0.5	0.75	0.875	0.9375	0.96875	0.9990234	0.9999990463

Observe the following:

1. As n increases, y_n approaches, or gets close to, 0.
2. As n increases, S_n approaches, or gets close to, 1.

This seems to suggest that as n gets larger and larger, S_n gets closer and closer to 1. In such a situation we let S be the sum of the entire infinite sequence (called an *infinite series*). In the example we have just been discussing, we have $S = 1$.

Indeed, we find that there are some cases in which we can find the sum of an infinite sequence and some cases in which we cannot: This is dependent on the size of r.

If we start with a geometric sequence, $\{a_i\}$, we know that the associated series S_n can be found by formula (C.9) (provided $r \neq 1$):

$$S_n = \frac{a_1(1 - r^n)}{1 - r} \qquad \textit{Which we can rewrite as}$$

$$= \frac{a_1 - a_1 r^n}{1 - r}$$

Let's suppose $-1 < r < 1$ (usually written as $|r| < 1$), and take a closer look at what happens to the right side of the last equation as n gets very large.

If $|r| < 1$, then r^n will get closer and closer to 0 as n gets larger and larger. (Table C.1 illustrates this for $r = \frac{1}{2}$.) If n is really large, then r^n is so small that it becomes "negligible." That is, it is so small that we can ignore the expression $a_1 r^n$ being subtracted from the term a_1 in the numerator. Thus, we have the result given in the next box.

The Sum of an Infinite Geometric Sequence	If $|r| < 1$, the sum S of all terms of an infinite geometric sequence $\{a_i\}$ can be calculated by $$S = \frac{a_1}{1-r} \qquad \textbf{(C.10)}$$ If $|r| > 1$ then r^n increases (we say *increases without bound*) and thus we cannot find the sum S.

We can write S using sigma notation by writing the symbol ∞ above the Σ.

> For $|r| < 1$ and $\{a_i\}$ a geometric sequence,
> $$S = \sum_{i=1}^{\infty} a_i = \frac{a_i}{1-r}$$
> The ∞ symbol means that we keep increasing the index i without ever reaching a final value.

EXAMPLE 7

Find the sum of the given infinite geometric sequences, if possible.

(a) $\left\{5, \dfrac{5}{3}, \dfrac{5}{9}, \dfrac{5}{27}, \ldots\right\}$ **(b)** $\{5, 10, 20, 40, 80, \ldots\}$

Solution

(a) $\left\{5, \dfrac{5}{3}, \dfrac{5}{9}, \dfrac{5}{27}, \ldots\right\}$ is an infinite geometric sequence with first term 5 and common ratio

$$r = \frac{\dfrac{5}{3}}{5} = \frac{1}{3}$$

Since $|r| < 1$, the sum S is given by

$$S = \frac{5}{1 - \dfrac{1}{3}} = \frac{5}{\dfrac{2}{3}} = \frac{15}{2}$$

Hence, $S = 5 + \dfrac{5}{3} + \dfrac{5}{9} + \dfrac{5}{27} + \cdots = \boxed{\dfrac{15}{2}}$.

(b) $\{5, 10, 20, 40, 80, \ldots\}$ is an infinite sequence with first term 5 and common ratio $r = \frac{10}{5} = 2$. Since $|r| > 1$, $\boxed{\text{the sum does not exist}}$.

EXAMPLE 8

A ball returns $\frac{1}{2}$ the distance it falls on a rebound. (See Figure C.1.) If you could let the ball bounce forever, what is the total distance it would travel if you dropped it from a height of 40 ft?

Solution

First we observe that the total distance that the ball travels is a result of two motions: falling and rising. We will examine each motion separately. We start with the falling motion (the solid blue lines in Figure C.1).

First, the ball is dropped (or falls) 40 ft. Then the ball returns half the distance, 20 ft, and then *falls* 20 ft to the ground. Continuing in this manner, we find that the distances the ball *falls* on each rebound form an infinite geometric sequence:

$$a_1 = 40$$

$$a_2 = \frac{1}{2} \cdot 40$$

$$a_3 = \frac{1}{2}\left(\frac{1}{2} \cdot 40\right) = 40\left(\frac{1}{2}\right)^2$$

$$a_4 = \frac{1}{2}\left[\frac{1}{2}\left(\frac{1}{2} \cdot 40\right)\right] = 40\left(\frac{1}{2}\right)^3$$

$$\vdots$$

We have an infinite geometric sequence with $r = \frac{1}{2}$ and $a_1 = 40$. The sum of the distances the ball falls is found by using formula (C.10) for the sum of an infinite series:

$$S = \frac{a_1}{1 - r} \qquad \text{We substitute } r = \tfrac{1}{2} \text{ and } a_1 = 40.$$

$$= \frac{40}{1 - \frac{1}{2}}$$

$$= \frac{40}{\frac{1}{2}} = 80$$

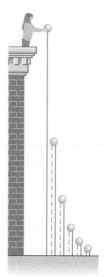

Figure C.1

Hence, the ball *falls* a total distance of 80 ft.

Next, we consider the rising motion of the ball (the dashed blue lines in Figure C.1). We note that the first time the ball rises is after it hits the ground: It has already fallen 40 ft. Its first ascent begins with 20 ft (half the 40-ft distance it fell). The next time, the ball rises 10 ft (half the 20-ft distance it fell). We could continue in this way and come up with a geometric sequence similar to the falling motion. (For the rising motion, the first term would be 20 ft.) An easier way is to realize that after the first 40 ft, the ball rises the same distance it falls. Since the ball falls a total of 80 ft, then the ball rises $80 - 40 = 40$ ft. Hence, the total distance the ball travels is

$$\text{Falling distance} + \text{Rising distance} = \text{Total distance traveled}$$
$$80 \qquad + \qquad 40 \qquad = \qquad 120$$

The total distance traveled by the ball is $\boxed{120 \text{ ft}}$.

EXAMPLE 9

Express the repeating decimal $0.26\overline{26}$ as a fraction.

Solution

In the Questions for Thought in Section 1.1, we demonstrated one approach for solving this problem. Now we will demonstrate a different approach using what we now know about infinite sequences.

We can rewrite $0.26\overline{26}$ as an infinite sum in the following way:

$$0.26\overline{26} = 0.26 + 0.0026 + 0.000026 + \cdots$$

Note that the terms of this sum, $0.26, 0.0026, 0.000026, \cdots$, form an infinite geometric sequence; the first term is 0.26 and $r = 0.01$ (each term is 0.01 times the preceding term). Thus,

$$0.26\overline{26} = \sum_{i=1}^{\infty} a_i \qquad \text{where } a_1 = 0.26 \quad \text{and} \quad r = 0.01$$

Since $|r| < 1$, we can find the sum using formula (C.10) to get

$$\sum_{i=1}^{\infty} a_i = \frac{a_1}{1-r} = \frac{0.26}{1-0.01} = \frac{0.26}{0.99} = \frac{26}{99}$$

Thus, $0.26\overline{26} = \boxed{\dfrac{26}{99}}$.

APPENDIX C.4 EXERCISES

For the geometric sequences in Exercises 1–14, find the required term.

1. $\{2, 6, 18, 54, \ldots\}$; a_6

2. $\{2, 8, 32, 128, \ldots\}$; a_5

3. $\{1, 6, 36, \ldots\}$; a_5

4. $\{1, 8, 64, \ldots\}$; a_6

5. $\left\{3, 1, \dfrac{1}{3}, \dfrac{1}{9}, \ldots\right\}$; a_8

6. $\left\{2, 1, \dfrac{1}{2}, \dfrac{1}{4}, \ldots\right\}$; a_9

7. $\left\{5, \dfrac{5}{6}, \dfrac{5}{36}, \ldots\right\}$; a_6

8. $\left\{7, \dfrac{7}{10}, \dfrac{7}{100}, \ldots\right\}$; a_6

9. $\left\{\dfrac{1}{2}, \dfrac{1}{6}, \dfrac{1}{18}, \ldots\right\}$; a_5

10. $\left\{3, \dfrac{3}{2}, \dfrac{3}{4}, \ldots\right\}$; a_6

11. Sequence $\{a_i\}$, with $a_1 = 3$ and $r = 2$; find a_4.

12. Sequence $\{a_i\}$, with $a_1 = 2$ and $r = 5$; find a_6.

13. Sequence $\{b_i\}$, with $b_1 = \dfrac{1}{2}$ and $r = \dfrac{1}{5}$; find b_5.

14. Sequence $\{c_i\}$, with $c_1 = 6$, and $r = \dfrac{1}{3}$; find c_6.

15. Find the fifth term of the geometric sequence whose first term is 1 and whose third term is 16.

16. Find the eighth term of the geometric sequence whose first term is 3 and whose fourth term is 24.

17. Find the seventh term of the geometric sequence whose first term is 1 and whose fifth term is $\dfrac{1}{16}$.

18. Find the sixth term of the geometric sequence whose first term is 2 and whose fourth term is $\dfrac{2}{27}$.

19. A car loses $\dfrac{1}{6}$ of its value each year. If the car is worth \$16,000 today, how much will it be worth 4 years from now?

20. A boat is now worth \$14,000, and loses 10% of its value each year. What would it be worth 5 years from now?

21. A machine depreciates in value by $\dfrac{1}{5}$ each year. If the machine is now worth \$24,000, how much will it be worth 3 years from now?

22. An antique coin is now worth \$8,000. If it appreciates in value at the rate of 10% per year, how much would it be worth 5 years from now?

23. The population of Davisville is increasing at a rate of 5% per year. If the population is currently 100,000, what will the population be 4 years from now?

24. The population of Sivadville is increasing at a rate of 10% per year. If the population is currently 100,000, what will the population be 4 years from now?

25. Roberta invests \$10,000 in a bank that pays 8% interest compounded annually. How much will Roberta have in the bank at the end of 5 years?

26. How much would Roberta of Exercise 25 have in the bank at the end of 5 years if the bank paid 6% interest compounded annually?

27. Beverly invests $5,000 in a bond that pays 10% interest compounded annually. How much is her bond worth at the end of 5 years?

28. How much would the bond of Exercise 27 be worth at the end of 10 years?

29. A ball bounces up one-half the distance that it falls on a rebound. How far does the ball bounce up on the fifth rebound if it is dropped from a height of 20 ft? (Do not count the first drop as a rebound.)

30. A ball bounces up one-third the distance it falls on a rebound. How far does the ball bounce up on the fourth rebound if it is dropped from a height of 27 ft? (Do not count the first drop as a rebound.)

In Exercises 31−36, find S_6 for the given geometric sequences.

31. $\{3, 9, 27, \ldots\}$ **32.** $\{4, 12, 36, \ldots\}$

33. $\{5, 10, 20, \ldots\}$ **34.** $\{1, 4, 16, \ldots\}$

35. $\left\{\dfrac{1}{3}, \dfrac{2}{3}, \dfrac{4}{3}, \ldots\right\}$ **36.** $\left\{3, \dfrac{3}{10}, \dfrac{3}{100}, \ldots\right\}$

In Exercises 37−40, find the following sums for the geometric sequence

$$\{a_i\} = \left\{1, \frac{1}{10}, \frac{1}{100}, \ldots\right\}.$$

37. $\displaystyle\sum_{i=1}^{5} a_i$ **38.** $\displaystyle\sum_{j=1}^{8} a_j$ **39.** $\displaystyle\sum_{k=4}^{8} a_k$ **40.** $\displaystyle\sum_{n=1}^{\infty} a_n$

41. If a ball rebounds $\frac{1}{3}$ the distance it falls, find the total distance the ball has rebounded at the end of the fifth rebound if it is dropped from a height of 30 ft. (Do not count the first drop as a rebound.)

42. Jack promises to pay Pete a debt in the following way: He would get 1 cent on the first day, 2 cents on the second day, 4 cents on the third day, and so on, where the payment on a particular day is double that of the previous day. At this rate, how much will Jack have paid Pete at the end of 30 days?

43. Manoj gets a job offer from each of two firms. Firm A offers him a job with a starting annual salary of $30,000 and a guaranteed raise of 6% per year. On the other hand, firm B offers him a job at a starting salary of $34,000 but will guarantee a raise of only 4% per year.

   (a) What would his annual salary be at each firm in the eighth year?

   (b) Over the first 8 years, how much would he earn at each firm?

44. Kara gets a job offer from two firms. Firm A offers her a job with a starting annual salary of $28,000 and a guaranteed raise of 4.5% per year. On the other hand, firm B offers her a job at a starting salary of $26,000 but will guarantee a raise of 6% per year.

   (a) What would her annual salary be at each firm in the tenth year?

   (b) Over the first 10 years, how much would she earn at each firm?

In Exercises 45−52, find S, the sum of the infinite sequence $\{a_i\}$ (if it exists), for the given geometric sequences.

45. $\left\{2, 1, \dfrac{1}{2}, \dfrac{1}{4}, \ldots\right\}$ **46.** $\left\{3, 1, \dfrac{1}{3}, \dfrac{1}{9}, \ldots\right\}$

47. $\left\{\dfrac{1}{5}, \dfrac{1}{10}, \dfrac{1}{20}, \cdots\right\}$

48. $\left\{1, \dfrac{2}{3}, \dfrac{4}{9}, \cdots\right\}$

49. $\left\{5, \dfrac{5}{3}, \dfrac{5}{9}, \cdots\right\}$

50. $\left\{\dfrac{1}{2}, 1, 2, 4, \cdots\right\}$

51. $\left\{\dfrac{1}{5}, \dfrac{2}{5}, \dfrac{4}{5}, \cdots\right\}$

52. $\left\{7, \dfrac{7}{10}, \dfrac{7}{100}, \cdots\right\}$

53. Express $0.35\overline{3535}$ as a fraction.

54. Express $0.243\overline{243}$ as a fraction.

55. Express $0.0245\overline{0245}$ as a fraction.

56. Express $0.7891\overline{7891}$ as a fraction.

57. A ball returns $\frac{2}{3}$ the distance it falls on a rebound. If you could let the ball bounce forever, what is the total distance it would travel if you dropped it from a height of 60 ft?

58. If the ball of Exercise 57 is dropped from a height of 20 ft, what is the total distance it would travel?

59. A pendulum sweeps out a 10-ft arc on its first pass and each time it passes it travels $\frac{2}{3}$ of the distance of the previous arc. How far would it travel before coming to rest?

Questions for Thought

60. Show that $(1 - r)(1 + r + r^2 + r^3 + \cdots + r^{n-2} + r^{n-1}) = 1 - r^n$.

61. Show that the terms of the sum $\displaystyle\sum_{i=1}^{N} ab^i$ form a geometric sequence. What is the common ratio?

62. Given the result of Exercise 61, find the following:

(a) $\displaystyle\sum_{i=1}^{10} 3 \cdot 2^i$ **(b)** $\displaystyle\sum_{i=1}^{10} 5 \cdot 3^i$

C.5

The Binomial Theorem

Early in this textbook we discussed finding the product $(x + y)^n$ for small n. For $(x + y)^2$ we pointed out that if we knew the pattern (which we called the *perfect square of the sum*), we could write out the product in simplified form quickly, without using the distributive property.

To find products $(x + y)^3$, we needed to find $(x + y)^2$ first and then multiply the result by $(x + y)$:

$$
\begin{aligned}
(x + y)^3 &= (x + y)(x + y)^2 \\
&= (x + y)(x^2 + 2xy + y^2) \\
&= x^3 + 2x^2y + xy^2 + x^2y + 2xy^2 + y^3 \\
&= x^3 + 3x^2y + 3xy^2 + y^3
\end{aligned}
$$

To find $(x + y)^4$ we needed to find $(x + y)^3$ first, and then multiply the result by $(x + y)$:

$$
\begin{aligned}
(x + y)^4 &= (x + y)(x + y)^3 \\
&= (x + y)(x^3 + 3x^2y + 3xy^2 + y^3) \qquad \textit{which yields} \\
&= x^4 + 4x^3y + 6x^2y^2 + 4xy^3 + y^4
\end{aligned}
$$

In this section we will demonstrate how to find $(x + y)^n$, perfect nth powers of the binomial $x + y$, for any n without multiplying out the binomials. As with the perfect square, we will examine the patterns and then generalize these patterns.

First we examine the product $(x + y)^n$ multiplied out. When $(x + y)^n$ is multiplied out and simplified (terms are combined where possible) we call it the *expanded form* of $(x + y)^n$. We will expand $(x + y)^n$ for $n = 0$ through $n = 5$. We will assume that neither x nor y is 0.

$$
\begin{array}{lll}
n = 0 & (x + y)^0 = & 1 \\
n = 1 & (x + y)^1 = & x + y \\
n = 2 & (x + y)^2 = & x^2 + 2xy + y^2 \\
n = 3 & (x + y)^3 = & x^3 + 3x^2y + 3xy^2 + y^3 \\
n = 4 & (x + y)^4 = & x^4 + 4x^3y + 6x^2y^2 + 4xy^3 + y^4 \\
n = 5 & (x + y)^5 = & x^5 + 5x^4y + 10x^3y^2 + 10x^2y^3 + 5xy^4 + y^5
\end{array}
$$

Let's first ignore the coefficients of the terms and look at the powers of the variables for the terms in each expansion. Note the pattern of the exponents for each expansion:

$$
\begin{array}{ll}
n = 1 & x, \quad y \\
n = 2 & x^2, \quad xy, \quad y^2 \\
n = 3 & x^3, \quad x^2y, \quad xy^2, \quad y^3 \\
n = 4 & x^4, \quad x^3y, \quad x^2y^2, \quad xy^3, \quad y^4 \\
n = 5 & x^5, \quad x^4y, \quad x^3y^2, \quad x^2y^3, \quad xy^4, \quad y^5
\end{array}
$$

First notice that for each n, the expansion starts with x^n and ends with y^n.

Now observe the terms on the line where $n = 5$. The leftmost x term has an exponent of 5 (which is n), and as we move to the right, the exponent of x decreases (for successive terms) by exactly 1. If we write the last term, y^5, as x^0y^5 (remember that $x^0 = 1$), then the pattern is complete, with each term having a power of x.

On the other hand, the pattern is reversed for the powers of y in each term. The first term, x^5, can be written as x^5y^0 and hence each subsequent term to the immediate right contains a power of y increased by 1, until the last term contains a fifth (or nth) power of y.

Observe the same pattern for $n = 1$ through $n = 4$: Moving to the right, each term contains consecutively decreasing powers of x and consecutively increasing powers of y.

In general, this pattern of descending powers of x and ascending powers of y for the terms in the expansion of $(x + y)^n$ holds true for all n.

Notice as well that the sum of the exponents of each term in the expansion of $(x + y)^5$ is 5. Again, this is true in general for n: In the expansion of $(x + y)^n$, the sum of the exponents of each term is n. Hence, in the expansion of $(x + y)^n$, the powers of x and y in the terms will have the following pattern:

$$
x^n, \quad x^{n-1}y, \quad x^{n-2}y^2, \quad x^{n-3}y^3, \quad \ldots, \quad x^3y^{n-3}, \quad x^2y^{n-2}, \quad xy^{n-1}, \quad y^n
$$

Note that the exponents in each term add up to n.

We can now write the powers of the variables of the binomial expansion of $(x + y)^8$ with this pattern in mind:

1. The first term is x^8.
2. The exponent of x is decreased by 1 in each subsequent term to the right until we reach x^0.
3. Starting with y^0 in the first term, the exponent of y is increased by 1 in each subsequent term to the right until we reach y^8. Consequently, we get

$$
x^8, \quad x^7y, \quad x^6y^2, \quad x^5y^3, \quad x^4y^4, \quad x^3y^5, \quad x^2y^6, \quad xy^7, \quad y^8
$$

4. Note that the exponents in each term sum to 8.

We now know what the powers of the variables used in each term will be when we expand $(x + y)^n$. Let's look at the coefficients.

First, we examine the coefficients of the terms in the binomial expansions of $(x + y)^n$ for $n = 0$ through $n = 5$. We can write them out in the form of a triangle:

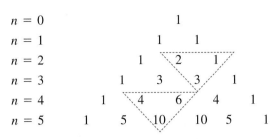

This triangle of coefficients is known as **Pascal's triangle,** *named after the 17th-century philosopher and mathematician, Blaise Pascal.*

One pattern we can identify (indicated by the dashed triangles within Pascal's triangle) is that, other than the end coefficients of 1, each coefficient is the sum of the two coefficients immediately above it. Thus, we can figure out the next two lines of coefficients for the expansion of $(x + y)^6$ and $(x + y)^7$ by adding neighboring coefficients to produce the coefficients in the line below:

$$
\begin{array}{ccccccccc}
n = 5 & & 1 & 5 & 10 & 10 & 5 & 1 & \\
n = 6 & 1 & 6 & 15 & 20 & 15 & 6 & 1 & \\
n = 7 & 1 & 7 & 21 & 35 & 35 & 21 & 7 & 1 \\
\end{array}
$$

This is one way of finding the coefficients of the terms for an expanded binomial. However, this approach is not very practical for finding the coefficients when n is large. There is another pattern in the coefficients we can discern, but before we discuss this pattern, we introduce a new notation to make the pattern more convenient to express (and remember).

We define $n!$, called *n factorial,* as follows:

> If n is an integer and $n > 0$, then
> $$n! = n \cdot (n - 1) \cdot (n - 2) \cdot (n - 3) \cdots \cdots 3 \cdot 2 \cdot 1$$

Thus, *n factorial* is the product of n consecutively decreasing integers from n to 1. For example,

$$3! = 3 \cdot 2 \cdot 1 \qquad \textit{Which, when evaluated, is}$$
$$= 6$$
$$5! = 5 \cdot 4 \cdot 3 \cdot 2 \cdot 1 \qquad \textit{Which, when evaluated, is}$$
$$= 120$$

We also define

$$\boxed{0! = 1}$$

EXAMPLE 1

Evaluate the following:

(a) $6!$ **(b)** $(5 - 1)!$ **(c)** $5! - 1$ **(d)** $\dfrac{8!}{6!}$ **(e)** $\dfrac{6!}{2!4!}$

Solution

(a) $6!$ *6! is the product of the six consecutively decreasing positive integers, from 6 to 1.*

$$= 6 \cdot 5 \cdot 4 \cdot 3 \cdot 2 \cdot 1$$
$$= \boxed{720}$$

(b) $(5 - 1)!$ *Perform the operation in parentheses first.*

$= 4!$ *4! is the product of the four consecutively decreasing positive integers, from 4 to 1.*

$= 4 \cdot 3 \cdot 2 \cdot 1$

$= \boxed{24}$

(c) $5! - 1$ *Multiply (perform factorial operation) first.*

$= (5 \cdot 4 \cdot 3 \cdot 2 \cdot 1) - 1$ *Multiply.*

$= 120 - 1$ *Then subtract.*

$= \boxed{119}$

Note the differences between parts **(b)** and **(c)**.

(d) $\dfrac{8!}{6!}$ *Do not try to reduce yet; this is not 8 divided by 6. By definition of factorial, we have*

$= \dfrac{8 \cdot 7 \cdot 6 \cdot 5 \cdot 4 \cdot 3 \cdot 2 \cdot 1}{6 \cdot 5 \cdot 4 \cdot 3 \cdot 2 \cdot 1}$ *Now reduce.*

$= \dfrac{8 \cdot 7 \cdot \cancel{6} \cdot \cancel{5} \cdot \cancel{4} \cdot \cancel{3} \cdot \cancel{2} \cdot \cancel{1}}{\cancel{6} \cdot \cancel{5} \cdot \cancel{4} \cdot \cancel{3} \cdot \cancel{2} \cdot \cancel{1}}$

$= 8 \cdot 7 = \boxed{56}$

(e) $\dfrac{6!}{2!4!}$ *By definition of factorial notation, we have*

$= \dfrac{6 \cdot 5 \cdot 4 \cdot 3 \cdot 2 \cdot 1}{2 \cdot 1 \cdot 4 \cdot 3 \cdot 2 \cdot 1}$ *Then reduce.*

$= \dfrac{\overset{3}{\cancel{6}} \cdot 5 \cdot \cancel{4} \cdot \cancel{3} \cdot \cancel{2} \cdot \cancel{1}}{2 \cdot 1 \cdot \cancel{4} \cdot \cancel{3} \cdot \cancel{2} \cdot \cancel{1}}$

$= \boxed{15}$

With factorial notation in hand, we can now reexamine the coefficients of the binomial expansion of $(x + y)^n$, where $n = 5$ and $n = 4$. First, we expand $(x + y)^5$:

$$(x + y)^5 = x^5 + 5x^4y + 10x^3y^2 + 10x^2y^3 + 5xy^4 + y^5$$

We will demonstrate that the coefficients of the terms are related to the exponents of the terms.

The third term in the expansion of $(x + y)^5$ is $10x^3y^2$. If we write the expression

$$\frac{5!}{3!2!} \quad \text{we get} \quad \frac{5 \cdot \overset{2}{\cancel{4}} \cdot 3 \cdot 2 \cdot \cancel{1}}{3 \cdot 2 \cdot \cancel{1} \cdot 2 \cdot 1} = 10$$

The fifth term is $5xy^4$. If we write

$$\frac{5!}{1!4!} \quad \text{we get} \quad \frac{5 \cdot \cancel{4} \cdot \cancel{3} \cdot \cancel{2} \cdot \cancel{1}}{1 \cdot \cancel{4} \cdot \cancel{3} \cdot \cancel{2} \cdot \cancel{1}} = 5$$

Notice that the denominator is the product of the factorials of the exponents of the variables of the term $5xy^4$.

Let's look at $(x + y)^4 = x^4 + 4x^3y + 6x^2y^2 + 4xy^3 + y^4$. The third term of this expansion is $6x^2y^2$. If we write the expression

$$\frac{4!}{2!2!} \quad \text{we get} \quad \frac{\overset{2}{\cancel{4}} \cdot 3 \cdot 2 \cdot \cancel{1}}{2 \cdot 1 \cdot 2 \cdot \cancel{1}} = 6$$

The numerator is the factorial of n and the denominator is the product of the factorials of the exponents of the variables x and y.

In general, we have the result given in the box.

> In the expansion of $(x + y)^n$, the term containing the expression $x^{n-k}y^k$ has coefficient $\dfrac{n!}{(n-k)!k!}$, $1 \le k \le n$.

Note that $n - k$ and k, the exponents of x and y, add up to n. For example, in the expansion of $(x + y)^6$, the term x^2y^4 has coefficient 15, since

$$\frac{n!}{(\text{Exponent of } x)!(\text{Exponent of } y)!} = \frac{n!}{(n-k!)k!}$$

$$= \frac{6!}{2!4!}$$

$$= \frac{\overset{3}{\cancel{6}} \cdot 5 \cdot \cancel{4} \cdot \cancel{3} \cdot \cancel{2} \cdot \cancel{1}}{2 \cdot 1 \cdot \cancel{4} \cdot \cancel{3} \cdot \cancel{2} \cdot \cancel{1}} = 15$$

We can now give the formula for the expansion of $(x + y)^n$:

The Binomial Theorem

For $n > 0$,

$$(x + y)^n = x^n + \frac{n!}{(n-1)!1!}x^{n-1}y + \frac{n!}{(n-2)!2!}x^{n-2}y^2$$

$$+ \frac{n!}{(n-3)!3!}x^{n-3}y^3 + \cdots + \frac{n!}{1!(n-1)!}xy^{n-1} + y^n$$

EXAMPLE 2

Expand the following:

(a) $(x + y)^8$ **(b)** $\left(\dfrac{m}{2} - n\right)^5$

Solution   **(a)** $(x + y)^8$ *Using the binomial theorem with n = 8, we get*

$$= x^8 + \frac{8!}{7!1!}x^7y + \frac{8!}{6!2!}x^6y^2 + \frac{8!}{5!3!}x^5y^3 + \frac{8!}{4!4!}x^4y^4 + \frac{8!}{3!5!}x^3y^5$$

$$+ \frac{8!}{2!6!}x^2y^6 + \frac{8!}{1!7!}xy^7 + y^8$$

Notice that the exponents of the variables appear in the denominators of the coefficients.

$$= \boxed{x^8 + 8x^7y + 28x^6y^2 + 56x^5y^3 + 70x^4y^4 + 56x^3y^5 + 28x^2y^6 + 8xy^7 + y^8}$$

(b) $\left(\dfrac{m}{2} - n\right)^5 = \left[\dfrac{m}{2} + (-n)\right]^5 = \left(\dfrac{m}{2}\right)^5 + \dfrac{5!}{4!1!}\left(\dfrac{m}{2}\right)^4(-n) + \dfrac{5!}{3!2!}\left(\dfrac{m}{2}\right)^3(-n)^2$

$$+ \frac{5!}{2!3!}\left(\frac{m}{2}\right)^2(-n)^3 + \frac{5!}{1!4!}\left(\frac{m}{2}\right)(-n)^4 + (-n)^5$$

$$= \frac{m^5}{32} - 5\left(\frac{m^4}{16}\right)n + 10\left(\frac{m^3}{8}\right)n^2 - 10\left(\frac{m^2}{4}\right)n^3 + 5\left(\frac{m}{2}\right)n^4 - n^5$$

$$= \boxed{\frac{1}{32}m^5 - \frac{5}{16}m^4n + \frac{5}{4}m^3n^2 - \frac{5}{2}m^2n^3 + \frac{5}{2}mn^4 - n^5}$$

EXAMPLE 3

Find the following terms:

(a) The term containing $r^7 s^3$ in the expansion of $(r + s)^{10}$

(b) The term containing $a^4 b^9$ in the expansion of $(a^2 - 2b^3)^5$

Solution

As we pointed out earlier, a general term in the expansion of $(x + y)^n$ can be written as

$$\frac{n!}{(n - k!)k!} x^{n-k} y^k$$

(a) For $(r + s)^{10}$, we can quickly identify $n = 10$. What remains is to find k, the exponent of y in the general term above. Letting $x = r$ and $y = s$, we compare the parts of the term we are given, $r^7 s^3$, to the corresponding parts of the general term, $x^{n-k} y^k = r^{n-k} s^k$, and observe that k is the exponent of s; thus, $k = 3$. Using the general term above with $n = 10$ and $k = 3$, the term must be

$$\frac{10!}{(10 - 3)!3!} r^{(10-3)} s^3 = \frac{10!}{7!3!} r^7 s^3$$

$$= \frac{10 \cdot \overset{3}{9} \cdot \overset{4}{8} \cdot 7 \cdot 6 \cdot 5 \cdot 4 \cdot 3 \cdot 2 \cdot 1}{7 \cdot 6 \cdot 5 \cdot 4 \cdot 3 \cdot 2 \cdot 1 \cdot 3 \cdot 2 \cdot 1} r^7 s^3$$

$$= \boxed{120 r^7 s^3}$$

(b) For $(a^2 - 2b^3)^5$, we again can quickly identify $n = 5$. However, finding k is a bit trickier. Letting $x = a^2$ and $y = -2b^3$, we compare the parts of the term we are given, $a^4 b^9$, with the corresponding parts of the general term $x^{n-k} y^k = (a^2)^{n-k}(-2b^3)^k$.

If we simplified $(a^2)^{n-k}(-2b^3)^k$, the exponent of b would be $3k$. Since the exponent of b of the given term $a^4 b^9$ is 9, we have $9 = 3k$, and therefore $k = 3$. Since $n = 5$ and $k = 3$, the term containing $a^4 b^9$ in the expansion of $(a^2 - 2b^3)^5$ must be

$$\frac{5!}{(5 - 3!)3!} (a^2)^{(5-3)}(-2b^3)^3 = \frac{5!}{2!3!}(a^2)^2(-2b^3)^3$$

$$= \frac{5 \cdot \overset{2}{4} \cdot 3 \cdot 2 \cdot 1}{2 \cdot 1 \cdot 3 \cdot 2 \cdot 1} a^4(-8b^9)$$

$$= (10)a^4(-8b^9)$$

$$= \boxed{-80 a^4 b^9}$$

Although we can write the terms of the binomial expansion in any order we please, when we refer to the "5th" term in an expansion, assume we are using the order given in the binomial theorem on page 781, and counting from left to right.

Look at the first five terms of the expansion of $(x + y)^9$:

$$x^9 + \frac{9!}{8!1!}x^8 y + \frac{9!}{7!2!}x^7 y^2 + \frac{9!}{6!3!}x^6 y^3 + \frac{9!}{5!4!}x^5 y^4 + \cdots$$

We now compare these terms with $\dfrac{n!}{(n - k!)k!} x^{n-k} y^k$, a general term in the expansion of $(x + y)^n$.

In the 3rd term, $\dfrac{9!}{7!2!}x^7 y^2$, k is 2.

In the 4th term, $\dfrac{9!}{6!3!}x^6 y^3$, k is 3.

In the 5th term, k is 4.

In general, in the rth term, $k = r - 1$.

Thus, if we want to find the rth term of the expansion of $(x + y)^n$, we substitute $r - 1$ for k in the expression

$$\frac{n!}{(n - k!)k!}x^{n-k}y^k$$

EXAMPLE 4

Find the following terms:

(a) The third term in the expansion of $(x + y)^7$

(b) The fourth term in the expansion of $(2a - 5)^6$

Solution

(a) For the third term, $r = 3$ and thus $k = r - 1 = 3 - 1 = 2$. We substitute $n = 7$ and $k = 2$ in the general term

$$\frac{n!}{(n - k!)k!}x^{n-k}y^k$$

to get

$$\frac{7!}{(7 - 2!)2!}x^{7-2}y^2 = \frac{7!}{5!2!}x^5y^2$$

$$= \frac{7 \cdot \overset{3}{\cancel{6}} \cdot 5 \cdot 4 \cdot 3 \cdot 2 \cdot 1}{5 \cdot 4 \cdot 3 \cdot 2 \cdot 1 \cdot 2 \cdot 1}x^5y^2$$

$$= \boxed{21x^5y^2}$$

(b) Since $r = 4$ (the fourth term), $k = r - 1 = 3$. We substitute 6 for n, 3 for k, $2a$ for x, and -5 for y in

$$\frac{n!}{(n - k!)k!}x^{n-k}y^k$$

to get

$$\frac{6!}{(6 - 3!)3!}(2a)^{6-3}(-5)^3 = \frac{6!}{3!3!}(2a)^3(-5)^3$$

$$= \frac{6 \cdot 5 \cdot 4 \cdot 3 \cdot 2 \cdot \cancel{1}}{3 \cdot 2 \cdot 1 \cdot 3 \cdot 2 \cdot \cancel{1}}(8a^3)(-125)$$

$$= 20(8a^3)(-125)$$

$$= \boxed{-20{,}000a^3}$$

APPENDIX C.5 EXERCISES

In Exercises 1–20, evaluate the expression.

1. $6!$ **2.** $2!$ **3.** $4!$ **4.** $7!$

5. $0!$ **6.** $1!$ **7.** $(7 - 2)!$ **8.** $(8 - 2)!$

9. $7! - 2$ **10.** $8! - 2$ **11.** $(2 \cdot 3)!$ **12.** $(3 \cdot 4)!$

13. $2 \cdot 3!$ **14.** $3 \cdot 4!$ **15.** $\dfrac{5!}{1!4!}$ **16.** $\dfrac{7!}{4!3!}$

17. $\dfrac{8!}{5!3!}$ **18.** $\dfrac{9!}{6!3!}$ **19.** $\dfrac{12!}{3!4!5!}$ **20.** $\dfrac{12!}{4!6!2!}$

In Exercises 21−36, expand the binomial.

21. $(a + b)^6$ **22.** $(x + y)^7$ **23.** $(a - b)^6$ **24.** $(x - y)^7$

25. $(2a + 1)^5$ **26.** $(3a + 1)^4$ **27.** $(2a - 1)^5$ **28.** $(3a - 1)^4$

29. $(1 - 2a)^4$ **30.** $(1 - 3a)^4$ **31.** $(a^2 + 2b)^5$ **32.** $(x + 2y^2)^5$

33. $(2x^2 - 3y^2)^4$ **34.** $(3x^2 + 2y^2)^4$ **35.** $\left(\dfrac{x}{3} + 2\right)^4$ **36.** $\left(\dfrac{a}{2} - 4\right)^4$

In Exercises 37−42, give the first four terms of the binomial expansion.

37. $(a^2 - b^4)^6$ **38.** $(x^4 - y^2)^6$ **39.** $(2a + 3b)^8$ **40.** $(3a - 2b)^7$

41. $(3a^2 - 2)^6$ **42.** $(2a^2 - 3)^8$

In Exercises 43−52, find the given terms.

43. The third term in the expansion of $(x + y)^8$

44. The fifth term in the expansion of $(x + y)^9$

45. The fourth term in the expansion of $(a - 2b)^6$

46. The fourth term in the expansion of $(2x - b)^6$

47. The third term in the expansion of $(3a^2 - 2)^7$

48. The fourth term in the expansion of $(2r^2 - 3s)^8$

49. The sixth term in the expansion of $(3a^2 - 2)^7$

50. The sixth term in the expansion of $(2r^2 - 3s)^8$

51. The ninth term in the expansion of $(3a^2 - 2)^8$

52. The tenth term in the expansion of $(2a^2 - 1)^9$

? Questions for Thought

53. In words, state the relationship among the exponents of the variables in the terms of the expansion of $(x + y)^n$, the coefficients of the terms of the expansion of $(x + y)^n$, and n.

54. Compute $(1.02)^7$ to four places using the binomial theorem. [*Hint:* Rewrite 1.02 as $1 + 0.02$.] Why do you not need to compute all the terms in the expansion?

55. The expression $\dfrac{n!}{r!(n - r)!}$ is sometimes written in the form $\dbinom{n}{r}$.

Hence, $\dbinom{7}{2} = \dfrac{7!}{2!5!} = \dfrac{7 \cdot \overset{3}{\cancel{6}} \cdot \cancel{5} \cdot \cancel{4} \cdot \cancel{3} \cdot \cancel{2} \cdot \cancel{1}}{2 \cdot 1 \cdot \cancel{5} \cdot \cancel{4} \cdot \cancel{3} \cdot \cancel{2} \cdot \cancel{1}} = 21.$

Compute the following:

(a) $\dbinom{6}{2}$ **(b)** $\dbinom{5}{0}$ **(c)** $\dbinom{9}{3}$ **(d)** $\dbinom{9}{6}$

APPENDIX C SUMMARY

After having completed this appendix you should be able to:

1. Find a specific term in a sequence (Section C.1).

For example:

Given a sequence with general term $a_n = \dfrac{2n - 1}{n^2}$, then $a_7 = \dfrac{2(7) - 1}{7^2} = \dfrac{14 - 1}{49} = \dfrac{13}{49}$.

2. Use sigma (Σ) notation (Section C.2).

For example:

$$\sum_{k = 1}^{5} (2k^2 - 1) = [2(1)^2 - 1] + [2(2)^2 - 1] + [2(3)^2 - 1] + [2(4)^2 - 1] + [2(5)^2 - 1]$$
$$= 1 + 7 + 17 + 31 + 49$$
$$= 105$$

3. Recognize arithmetic and geometric sequences (Sections C.3−C.4).

For example:

(a) The numbers {3, 8, 13, 18, 23, . . .} form an *arithmetic* sequence, since successive terms have a common difference of $d = 8 - 3 = 13 - 8 = 18 - 13 = 23 - 18 = 5$.

(b) The numbers (3, 12, 48, 192, 768, . . . } form a *geometric* sequence, since successive terms have a common ratio

$$r = \frac{12}{3} = \frac{48}{12} = \frac{192}{48} = \frac{768}{192} = 4$$

4. Find a specific term of a given arithmetic or geometric sequence (Sections C.2−C.4).

For example:

(a) To find the tenth term of the arithmetic sequence whose first term is 4 and whose common difference is 7, we use formula (C.1) for the *n*th term of an arithmetic sequence:

$a_n = a_1 + (n - 1)d$ ⠀⠀⠀*We substitute $a_1 = 4$, $n = 10$, and $d = 7$.*

$a_{10} = 4 + (10 - 1)7$

⠀⠀$= 67$

(b) To find the eighth term of the geometric sequence whose first term is 3 and whose fourth term is −24, we could use formula (C.6) for the *n*th term of a geometric sequence:

$a_n = a_1 r^{n - 1}$ ⠀⠀⠀⠀⠀⠀⠀⠀⠀*We substitute for a_1 and a_4 to find r.*

$a_4 = a_1 r^3$

$-24 = 3r^3$

$-8 = r^3$ ⠀⠀⠀⠀⠀⠀⠀⠀⠀⠀*Take cube roots.*

$-2 = r$ ⠀⠀⠀⠀⠀⠀⠀⠀⠀⠀*Now we can use the formula again to find a_8.*

$a_8 = a_1 r^7 = 3(-2)^7 = 3(-128)$ ⠀⠀*We substitute for a_1 and a_4 to find r.*

⠀⠀$= -384$

5. Find the sum of an arithmetic or geometric sequence (Sections C.3−C.4).

For example:

(a) To find the sum of the first nine terms of an *arithmetic* sequence whose first term is 1 and whose common difference is 2, we use formula (C.4):

$$S_n = \frac{n}{2}[2a_1 + (n - 1)d] \qquad \textit{We substitute } a_1 = 1, \ n = 9, \text{ and } d = 2 \text{ to get}$$

$$S_9 = \frac{9}{2}[2(1) + (9 - 1)2]$$

$$= \frac{9}{2}[2 + 16]$$

$$= 81$$

(b) To find the sum of the first nine terms of a *geometric* sequence whose first term is 1 and whose common ratio is 2, we use formula (C.9):

$$S_n = \frac{a_1(1 - r^n)}{1 - r} \qquad \textit{We substitute } a_1 = 1, \ n = 9, \text{ and } r = 2 \text{ to get}$$

$$S_9 = \frac{1(1 - 2^9)}{1 - 2}$$

$$= \frac{1 - 512}{1 - 2}$$

$$= \frac{-511}{-1}$$

$$= 511$$

(c) To find the sum of an infinite geometric sequence $\{a_i\} = \left\{2, \frac{2}{3}, \frac{2}{9}, \frac{2}{27}, \ldots\right\}$

(if it exists), we first identify a_1 and r. We are given $a_1 = 2$ and we find that

$$r = \frac{\dfrac{2}{3}}{2} = \frac{1}{3}$$

Since $|r| < 1$, we know that the sum S exists and

$$S = \frac{a_1}{1 - r} = \frac{2}{1 - \dfrac{1}{3}} = 3$$

6. Solve applications involving sequences and series (Sections C.1–C.4).

For example:

(a) Bob gets a \$1,100 raise every 6 months for the first 5 years of his job. If his starting salary is \$22,000, how much is his salary at the beginning of his fifth year of work?

Solution:

Bob's salary every 6 months forms the sequence \$22,000, \$23,100, \$24,200, . . . until the end of the fifth year. This is an arithmetic sequence with the first term $a_1 = 22,000$ and the common difference $d = 1,100$. Since we are looking for the salary at the beginning of the fifth year, we are looking for a_9. (Why?) We use formula (C.1) to find the nth term of an arithmetic sequence with $n = 9$:

$$a_n = a_1 + (n - 1)d \qquad \textit{Substitute } n = 9, \ d = 1,100, \text{ and } a_1 = 22,000.$$

$$a_9 = 22,000 + (9 - 1)1,100$$

$$= 30,800$$

Bob's salary at the beginning of the fifth year is \$30,800.

(b) Andrea invests $10,000 in a bank that pays 9% interest compounded annually. How much will Andrea have in the bank at the end of 5 years?

Solution:

Andrea starts out with $10,000. At 9% interest, her principal at the end of the ith year is

$$A_i = 10,000(1.09)^i \qquad \textit{We substitute } i = 5.$$
$$A_5 = 10,000(1.09)^5$$
$$= 15,386.24$$

Andrea's principal at the end of the fifth year is $15,386.24.

7. Evaluate expressions with factorial notation (Section C.5).

For example: $\dfrac{8!3!}{4!} = \dfrac{8 \cdot 7 \cdot 6 \cdot 5 \cdot 4 \cdot 3 \cdot 2 \cdot 1 \cdot 3 \cdot 2 \cdot 1}{4 \cdot 3 \cdot 2 \cdot 1} = 10,080$

8. Expand a binomial using the binomial theorem (Section C.5).

For example: Expand $(3x - y)^5$.

$(3x - y)^5 = [3x + (-y)]^5$ *   Using the binomial theorem with $n = 5$, we get*

$$= (3x)^5 + \frac{5!}{4!1!}(3x)^4(-y) + \frac{5!}{3!2!}(3x)^3(-y)^2 + \frac{5!}{2!3!}(3x)^2(-y)^3$$
$$+ \frac{5!}{1!4!}(3x)(-y)^4 + (-y)^5$$
$$= 243x^5 - 405x^4y + 270x^3y^2 - 90x^2y^3 + 15xy^4 - y^5$$

9. Identify the rth term in the expansion of a binomial (Section C.5).

For example: To find the fourth term in the expansion of $(x + y)^8$, we have $r = 4$ and therefore $k = r - 1 = 4 - 1 = 3$. We substitute $k = 3$ and $n = 8$ in the general term

$$\frac{n!}{(n - k)!k!}x^{n-k}y^k$$

to get

$$\frac{8!}{(8 - 3)!3!}x^{8-3}y^3 = \frac{8!}{5!3!}x^5y^3 = 56x^5y^3$$

APPENDIX C REVIEW EXERCISES

In Exercises 1−8, write the first four terms and the 12th term of the sequence whose nth term is given.

1. $a_n = 2n - 5$

2. $b_n = 3n + 1$

3. $x_n = 2n^2$

4. $y_n = 3n$

5. $a_n = \dfrac{(-1)^n}{n + 1}$

6. $b_n = \dfrac{(-1)^{n+1}}{n + 2}$

7. $x_n = 3 + (-1)^n$

8. $y_n = 2^n + |n - 3|$

In Exercises 9−12, find the possible general term for the given sequences.

9. $\{3, 8, 13, 18, \ldots\}$

10. $\{2, 6, 18, 54, \ldots\}$

11. $\left\{\dfrac{1}{2}, -\dfrac{1}{4}, \dfrac{1}{6}, -\dfrac{1}{8}, \ldots\right\}$

12. $\left\{\dfrac{1}{2}, \dfrac{1}{4}, \dfrac{1}{8}, \dfrac{1}{16}, \ldots\right\}$

13. Jack gets a raise of $700 every 6 months for the first 5 years he works at his job. If his starting salary is $23,000 per year, write a sequence to show how his annual income changes during the first 3 years at his job.

14. The population of Soraville increases at a rate of 11% per year. If the current population is 20,000, write a sequence to show how the population changes over the next 4 years.

15. Samantha invests $500 in a bond that increases in value by 12% each year. How much will her bond be worth in 5 years?

16. Jenna invests $500 in a bond that increases in value by 8% each year. How much will her bond be worth in 5 years?

Given the sequences or general terms defined in Exercises 17–22, find the indicated sum, S_n.

17. $\{3, 6, 9, 12, 15, \ldots\}$; S_5

18. $\{5, 9, 13, 17, 21, \ldots\}$; S_5

19. $\{2, 4, 8, 16, 32, 64, \ldots\}$; S_6

20. $\{3, 9, 81, 243, \ldots\}$; S_4

21. $a_i = 3i - 1$; S_6

22. $b_i = 4i + 3$; S_8

In Exercises 23–28, rewrite each sum without using sigma notation, then compute each sum.

23. $\displaystyle\sum_{i=1}^{6} i$

24. $\displaystyle\sum_{i=3}^{8} i$

25. $\displaystyle\sum_{i=1}^{5} 2i$

26. $\displaystyle\sum_{i=1}^{3} 4i$

27. $\displaystyle\sum_{n=1}^{6} 3n^2$

28. $\displaystyle\sum_{k=2}^{6} (2k + 3)$

*In Exercises 29–34, find the next two terms and the tenth term of the given **arithmetic** sequence.*

29. $\{a_n\}$, where the first term is 2 and the common difference is 3

30. $\{b_i\}$, where the first term is -5 and the common difference is 2

31. $\{3, 8, 13, 18, \ldots\}$

32. $\{-4, -2, 0, 2, \ldots\}$

33. $\{1, -3, -7, -11, \ldots\}$

34. $\left\{1, \dfrac{1}{2}, 0, -\dfrac{1}{2}, \ldots\right\}$

35. Given the arithmetic sequence $\{x_i\}$ with $x_4 = 10$ and $x_8 = 22$, find x_1 and d.

36. Given the arithmetic sequence y_i with $y_3 = -9$ and $y_5 = -13$, find y_1 and d.

37. An object dropped from the top of a building falls 16 ft the first second, 48 ft the second second, and so on, forming an arithmetic sequence. How many feet will the object fall during the eighth second?

38. How many feet will the object described in Exercise 37 fall during the tenth second?

*In Exercises 39–42, use a formula for S_n to find the indicated S_n for the given **arithmetic** sequence.*

39. $\{a_i\}$, where the first term is 5 and the common difference is 3; find S_{10}

40. $\{b_i\}$, where the first term is 9 and the common difference is -2; find S_9

41. $\{3, 10, 17, \ldots\}$; S_8

42. $\{-5, -1, 3, \ldots\}$; S_8

43. Find $\displaystyle\sum_{i=1}^{30} 2i$.

44. Find $\displaystyle\sum_{i=1}^{20} 3i$.

45. Find the sum of the first 30 positive odd integers.

46. Find the sum of the first 20 positive even integers.

47. Given an arithmetic sequence with $a_4 = 5$ and $a_8 = 8$, find S_{10}.

48. Given an arithmetic sequence with $a_3 = 2$ and $a_6 = 5$, find S_{12}.

*In Exercises 49−54, find the required a_n for the given **geometric** sequence.*

49. $\{2, 6, 18, \ldots\}$; a_5

50. $\{1, 4, 16, \ldots\}$; a_4

51. $\left\{1, \dfrac{1}{3}, \dfrac{1}{9}, \ldots\right\}$; a_6

52. $\left\{2, \dfrac{2}{3}, \dfrac{2}{9}, \ldots\right\}$; a_6

53. Sequence $\{a_i\}$ with $a_1 = 5$ and $r = 2$; find a_5.

54. Sequence $\{b_i\}$ with $b_1 = 2$ and $r = \frac{1}{2}$; find a_6.

55. Harry promises to pay Kathy a debt in the following way: He would pay her $5 the first month, $10 the second month, $20 the third month, and so on, where the payment in any month is double that of the previous month. How much should Kathy receive in the 12th month?

56. How much will Harry have paid Kathy of Exercise 55 altogether at the end of the 12th month?

*In Exercises 57−60, find the sum, S, of the following **geometric** sequences (if they exist).*

57. $\left\{3, 1, \dfrac{1}{3}, \ldots\right\}$

58. $\left\{1, \dfrac{1}{5}, \dfrac{1}{25}, \ldots\right\}$

59. $\left\{\dfrac{1}{5}, \dfrac{2}{5}, \dfrac{4}{5}, \ldots\right\}$

60. $\left\{\dfrac{1}{4}, \dfrac{3}{4}, \dfrac{9}{4}, \ldots\right\}$

61. Express $0.646\overline{464}$ as a fraction.

62. A ball bounces back $\frac{3}{5}$ the distance it falls on a rebound. If you could let the ball bounce forever, what is the total distance it would travel if you dropped it from a height of 50 ft?

In Exercises 63−66, evaluate the expression.

63. $8!$

64. $(9 - 6)!$

65. $9! - 6!$

66. $\dfrac{9!5!}{6!}$

In Exercises 67−70, expand the binomial.

67. $(x - y)^5$

68. $(3a + b)^4$

69. $(2a - 3b^2)^4$

70. $(5a - 1)^3$

In Exercises 71−72, give the first four terms of the binomial expansion.

71. $(3a - 2b)^5$

72. $(2a^2 - b^3)^6$

In Exercises 73−74, find the given terms.

73. The fourth term in the expansion of $(2a^2 - 3)^6$

74. The fifth term in the expansion of $(3x - 2y^2)^7$

APPENDIX C PRACTICE TEST ⊙

1. Identify the first three terms and the ninth term of the following sequences:

   (a) $\{x_i\}$, where $x_i = 3i + 1$

   (b) $\{y_i\}$, where $y_i = 2i^3 - 1$

2. Given the following sequences, find the indicated S_n:

   (a) $\{3, 7, 11, 15, 19, \ldots\};$ S_5

   (b) $\{a_n\}$, where $a_n = 4n + 1;$ S_4

3. Rewrite the following as a sum without using sigma notation, then compute the sum.

$$\sum_{i=3}^{6} (2i^2 + 1)$$

4. Find the 12th term of the arithmetic sequence with first term 3 and common difference 7.

5. Find the fifth term of a geometric sequence with first term 4 and common ratio 2.

6. Find S_{12} for the arithmetic sequence $\{2, 7, 12, 17, \ldots\}$.

7. Find S_4 for the geometric sequence $2, 10, \ldots$.

8. Find the sum of the first 20 positive multiples of 4.

9. A pendulum sweeps out a 12-ft arc on its first pass, and each time it passes, it covers $\frac{3}{5}$ the distance of the previous arc. How far would it travel before coming to rest?

10. Expand the binomial: $(2a^2 + 3)^4$.

11. Find the fourth term in the expansion of $(3a^2 - 1)^6$.

Answers to Selected Exercises

This answer section contains the answers to all odd-numbered exercises as well as the answers to all exercises in the Mini-Reviews, Chapter Reviews, Chapter Tests, Cumulative Reviews, and Cumulative Tests.

Exercises 1.1

1. $\{3, 4, 5, 6, 7, 8, 9, 10, 11\}$
3. $\varnothing$ 5. $\{41, 43, 47\}$
7. $\{0, 7, 14, 21, \ldots\}$
9. $\{1, 2, 3, 6, 9, 18, 27, 54\}$
11. $\{0, 3, 6\}$
13. $\{0, 1, 2, 3, 4, 5, 6, 9, 12, 15, 18, 21, 24, 27, 30, 33\}$
15. $\{0, 1, 2, 3, 4, 5, 6, 7, 8, 9, 10, 11, 12, 13\}$
17. $2 \cdot 3 \cdot 11$
19. $2 \cdot 2 \cdot 2 \cdot 2 \cdot 2 \cdot 2 \cdot 2$
21. Prime number 23. $7 \cdot 13$
25. True 27. True 29. True
31. False 33. False 35. True
37. $>, \geq, \neq$ 39. $<, \leq, \neq$
41.
43.
45.
47.
49.
51.
53. No solution
55.
57. $C \cup D = \{x \mid -4 \leq x < 9, x \in Z\}$

59. $A \cap D = \{x \mid 1 \leq x < 9, x \in Z\}$

61. $A \cup B = \{x \mid x \in R\}$

63. $C \cap D = \{x \mid 1 \leq x \leq 6\}$

65. $A \cup C = \{x \mid x \geq -4\}$

67. Associative property of addition
69. Commutative property of addition
71. Distributive property 73. False

75. Distributive property 77. False
79. Associative property of multiplication
81. Multiplicative inverse property
83. Additive inverse property
85. Commutative property of addition
87. Distributive property
89. Closure property of multiplication
91. False

Exercises 1.2

1. 5 3. -11 5. 24 7. 6.481
9. -12 11. -23 13. -60
15. -27 17. -11 19. 3
21. -16 23. 0.87 25. 4 27. $\frac{16}{5}$
29. 8 31. 4 33. 1 35. -12
37. -4 39. -3 41. 45 43. 17
45. 10 47. -16 49. 2 51. 2
53. -8 55. 36 57. 29 59. 50
61. -81 63. -40 65. -360
67. -46 69. $\frac{50}{7}$ 71. -13
73. 111 75. -6 77. 10 79. 24
81. -4 83. $1,230.42 85. 30,335 ft
87. (a) $\frac{-9}{49}$ (b) -0.18
89. 0 91. 30
93. 6 95. 1 97. Undefined
99. 1.55 101. (a) 0.070 (b) 0.110
103. (a) 285 million (b) 3 million
105. (a) $390.9 billion, $407.5 billion, $424.2 billion, $440.8 billion
(b) 4.3%, 1998–1999; 4.1%, 1999–2000; 3.9%, 2000–2001

Exercises 1.3

1. $8x$ 3. $12x^2$ 5. $-4x$ 7. $-12x^2$
9. $-6m$ 11. $60m^3$ 13. $-9t^2$
15. $-24t^6$ 17. $2x + 3y + 5z$
19. $30xyz$ 21. $x^3 + x^2 + 2x$
23. $2x^6$ 25. $-17x^2y$ 27. $30x^4y^2$
29. $x^2 + 2x - 6$ 31. $11x^2y - 7xy^2$
33. $9m + 12n$ 35. $2a - 16b$
37. $6c - 16d$ 39. $x^2 - 2xy + y^2$
41. 0 43. $40a^3b^4c^3$ 45. $72x^5$
47. $18x^5$ 49. $-32x^{11}$ 51. 0
53. $-16x^3$ 55. $-b + 10$

57. $17t + 12$ 59. $13a - 64$
61. $4x^2 - 8x$ 63. $10x - 7y$
65. $-8y^3 + 2xy^2 + 2xy + 3x$
67. $12s^2$ 69. -131.31 71. -73.38
73. (a) $P = 16.13t - 31,488.2$
(b) 868,600
75. $96 - 3x$ 77. $x^2 + 3x$

Exercises 1.4

1. Let $x =$ number; $x + 8$
3. Let $x =$ number; $2x - 3$
5. Let $x =$ number; $3x + 4 = x - 7$
7. Let x and y be the two numbers; $x + y = xy + 1$
9. Let the smaller number be x; the larger number $= 2x + 5$
11. Let $x =$ smallest number; the other two numbers are $3x$ and $3x + 12$
13. x and $x + 1$ 15. $x, x + 2, x + 4$
17. $x^3 + (x + 1)^3$
19. Let $x =$ one number; the other number is $40 - x$
21. Let $x =$ first number; second number $= 2x$; third number $= 100 - 3x$
23. Let $x =$ width; area $= (x)(3x) = 3x^2$; perimeter $= 2x + 2(3x) = 8x$
25. Let $x =$ second side; perimeter $= 2x + x + x + 4 = 4x + 4$
27. (a) 31 coins
(b) Value of coins $= $4
29. (a) Number of coins $= n + d + q$
(b) Value of coins (in cents) $= 5n + 10d + 25q$ cents
31. (a) $2w$ meters (b) $4w$ dollars
(c) $6w$ (d) $30w$ dollars
(e) $34w$ dollars
33. Let $x =$ nickels, then $20 - x =$ dimes; value of coins (in cents) $= 5x + 10(20 - x) = 200 - 5x$ cents
35. (a) $1,590 (b) $1,685.40
(c) $A = 1,500(1.06)^n$
(d) $A = 1,500(1.06)^{15} = $3,594.84

37. (a) 2 hr: $12, Standard; $16.20, Saver. 1 hr: $6.00, Standard; $12.60, Saver

(b) Standard: $P = 0.10t$; Saver: $P = 0.06t + 9$. Standard Plan is cheaper for $t < 225$ min; Saver Plan is cheaper for $t > 225$ min.

39. (a) $984.38

(b) $0.75x + 0.05(0.75x) = 0.7875x$

41. (a) $2,100

(b) $0.5x + 0.5(0.5x) = 0.75x$

Exercises 1.5

1. 3 satisfies; 0 and 5 do not

3. 2 satisfies; -3 and 0 do not

5. -1 and 5 satisfy; 2 does not

7. 1 satisfies; -1 and 3 do not

9. $x = 6$ **11.** $y = -\frac{3}{4}$ **13.** $m = 0$

15. $t = 4$ **17.** $y = -9$ **19.** $s = -\frac{9}{2}$

21. $x = -\frac{11}{9} \approx -1.22$ **23.** $x = \frac{7}{8}$

25. $x = 4$ **27.** $x = 0$ **29.** $t = 2$

31. $x = 5$ **33.** $x = \frac{3}{2}$

35. $x = 10,500$ **37.** $x = 12$

39. $x = -12$ **41.** $x = \frac{8}{57}$ **43.** $x = \frac{8}{105}$

45. $x = -\frac{5}{33}$ **47.** $x = \frac{22}{3}$ **49.** $x = \frac{1}{3}$

51. (a) $117,731

(b) In 1988

53. $x = \dfrac{4 - 7y}{5}$ **55.** $y = \dfrac{2x - 11}{9}$

57. $x = -2y - 4$

59. $x > -\frac{3}{5}$

61. $x < \frac{17}{3}$

63. $x > 4$

65. $x \geq \frac{1}{3}$

67. $x \leq 2$

69. $x \geq \frac{26}{23}$

Chapter 1 Review Exercises

1. $A = \{1, 2, 3, 4\}$

2. $B = \{6, 7, 8, 9, \ldots\}$

3. $C \cap D = \{b\}$

4. $C \cup D = \{a, b, c, e, f, g, r, s\}$

5. $A \cup B = \{0, 1, 2, 3, 4, 6, 7, \ldots\}$ or $A \cup B = \{x \mid x \in N, x \neq 5\}$

6. $\varnothing$ **7.** $A = \{1, 2, 3, 4, 6, 12\}$

8. $B = \{0, 12, 24, 36, 48, \ldots\}$

9. $A \cap B = \{12\}$

10. $A \cup B = \{0, 1, 2, 3, 4, 6, 12, 24, 36, 48, 60, \ldots, 12n, \ldots\}$, $n = 6, 7, 8, \ldots$

11. $B \cap C = \{x \mid x \in W \text{ and } x \text{ is a multiple of } 12\} = \{0, 12, 24, 36, \ldots\}$

12. $A \cap C = \{6, 12\}$

13. $A \cap B = \{r \mid 3 \leq r \leq 4, r \in Z\} = \{3, 4\}$

14. $A \cup B = \{-1, 0, 1, 2, 3, 4, 5, 6, 7, 8, 9, 10, 11, 12\}$

15.

16.

17.

18.

19.

20.

21. False   **22.** True   **23.** True

24. True   **25.** False   **26.** True

27. True   **28.** False

29. Commutative property of addition

30. Commutative property of multiplication

31. Distributive property

32. Distributive property

33. Multiplicative inverse property

34. Additive identity property

35. False   **36.** False   **37.** -6

38. -11 **39.** -8 **40.** -9 **41.** -30

42. -210 **43.** 64   **44.** -64

45. -11 **46.** 3   **47.** 63   **48.** -15

49. -34 **50.** 16   **51.** -61

52. -136 **53.** -17

54. Undefined   **55.** 1   **56.** 11   **57.** 0

58. 4   **59.** Not defined   **60.** 0

61. 3.86   **62.** 0.09

63. $-6x^3y - 3x^2y^2$ **64.** $-36a^3b^4$

65. $36x^4y^4$ **66.** $54r^8s^6$ **67.** $3y - 4x$

68. $-8a - 2b$ **69.** $-5r^2s + rs^2$

70. $-3x^2y^3 - 2xy^2$ **71.** -1

72. $3a^2 - 3ab + 3ac$

73. $10r^2s + 15rs^2$ **74.** $-x + 12$

75. $7y - 18x - 9$ **76.** $4a - 9$

77. $-2r - 3s + 36$ **78.** $-6x + 30$

79. $-21y + 31$

80. (a) $V = 1481t - 2,844,269$

(b) $107,364

81. Let the two numbers be x and y; $xy - 5 = x + y + 3$

82. Let $x =$ number; $2x + 8 = x^2 - 3$

83. Let $x =$ first odd integer; $x + (x + 2) = (x + 4) - 5$

84. Let $x =$ first even integer; $(x + 2)(x + 4) = 10x + 8$

85. Let width $= x$; area $= (4x - 5)x = 4x^2 - 5x$; perimeter $= 2x + 2(4x - 5) = 10x - 10$

86. Let $w =$ width; area $= (3w + 5)(w) = 3w^2 + 5w$; perimeter $= 2(3w + 5) + 2w = 8w + 10$

87. Let x and y be the numbers; $x^2 + y^2 = xy + 8$

88. Let x and y be the two numbers; $(x + y)^2 = xy + 8$

89. Let $x =$ number of dimes, then $40 - x =$ number of nickels; value (in cents) $= 10x + 5(40 - x) = 5x + 200$ cents

90. $18x + 10(30 - x) = 8x + 300$ dollars

91. $x = 0$ **92.** $x = 4$

93. $x = 11$ **94.** $y = \frac{4}{5}$

95. $x = -1$ **96.** $x = -2$

97. $a = \frac{7}{3}$ **98.** $b = -\frac{1}{2}$

99. No solution   **100.** All reals

101. $x = 0$ **102.** $a = 0$

103. $x = -\frac{14}{3}$ **104.** $x = \frac{5}{4}$

105. $x = -2$ **106.** $y = \frac{5}{6}$

107. $x = -12$

108. (a) 691,150 physicians

(b) 1989

109. $y = \dfrac{22 - 7x}{8}$

110. $a = \dfrac{5b + 2}{2}$

111. $y = \dfrac{7x - 2}{8}$

112. $a = \dfrac{10b + 2}{5}$

113. $x < -21$

114. $x \geq 1$

115. $x < \frac{27}{5}$

116. $x \geq -\frac{33}{17}$

117. $x \geq 0.713$

118. $x > 1.246$

Chapter 1 Practice Test

1. (a) $A \cap B = \{2, 3, 5, 7\}$
   (b) $A \cup B = \{2, 3, 5, 7, 11, 13, 17, 19, 23\}$
2. (a) False (b) True (c) False
3. (a)

   (b)
4. (a) False
   (b) Commutative property of addition
5. (a) 8 (b) 85 (c) 1 (d) 6
6. (a) 1 (b) $\frac{5}{17}$
7. (a) $10x^6 y^5$ (b) $-rs^2 - 5r^2 s - 7rs$
   (c) 0 (d) $9r - 6s - 3$
8. Let $x =$ width; perimeter $= 2x + 2(3x - 8) = 8x - 16$
9. Number of nickels $= 34 - x$; value of coins (in cents) $= 10x + 5(34 - x) = 5x + 170$
10. (a) $x = -5$ (b) $x = -\frac{1}{3}$
    (c) $x = -\frac{1}{8}$ (d) $x \geq -1$
    (e) $x < -\frac{1}{32}$
11. $t = \dfrac{3s + 4}{7}$

Exercises 2.1

1. Let $G =$ the course grade, $f =$ the final exam grade. We can use the equation $G = 0.75(82) + 0.25f$
   (a) 74
   (b) She would need a score of at least 114 on the final exam. Assuming a final exam based on 100 points, it would be impossible for her to get a 90 course grade.
3. Let $G =$ the course grade, $f =$ the final exam grade. We can use the equation $G = 0.75(78) + 0.25f$. She would need a score of at least 86 to get a course grade of 80.
5. Let $C =$ weekly commission, $G =$ gross sales. We can use the equation $C = 0.09G$. Weekly sales of \$6,666.67 would be needed.
7. Let $u =$ U.S. dollars and $x =$ eurodollar; $1.24x = u$; $1.24(300) = \$372$
9. Let $x =$ British pounds and $u =$ U.S. dollars; $1.7984x = u$; $(1.7984)400 = \$719.36$

11. Let $P =$ weekly pay, $G =$ gross sales. The strictly commission equation is $P = 0.09G$. The base salary plus commission equation is $P = 120 + 0.06G$. For gross sales greater than \$4,000, the strictly commission plan is better. For gross sales less than \$4,000, the salary plus commission plan is better.
13. Let $P =$ perimeter, $w =$ width, and $3w + 1 =$ length. We have $P = 2w + 2(3w + 1)$. The rectangle is 9.75 in. by 30.25 in.
15. Let $I =$ interest earned, $x =$ the amount invested. We have $I = 0.083x$. She should invest \$12,048.19.
17. Let $I =$ interest earned, $x =$ the amount invested at 8%. Then $12{,}500 - x =$ the amount invested at 5%. The interest earned is $I = 0.08x + 0.05(12{,}500 - x) = 0.03x + 625$. If $x = 0$ (no money invested in the long-term certificate), then the total interest is \$625 (the minimum amount). If all the money is put into the long-term certificate ($x = \$12{,}500$), the total amount of interest would be \$1,000 (the maximum amount of interest).
19. Let $n =$ total number of copies printed, $t =$ the number of minutes the faster printer works, $t - 5 =$ the number of minutes the slower printer works. The total number of copies produced by both printers working together is $n = 9t + 4(t - 5)$; it takes approximately 12.5 minutes to print 142 pages.
21. Let $x =$ number of \$850 computers sold, $58 - x =$ the number of \$600 computers sold. The total C collected on the sale of all the computers is $C = 850x + 600(58 - x)$; twenty-two \$850 computers were sold.
23. Let $x =$ the value of the stock yesterday; the value of the stock today $= \frac{47}{61}x$; $\frac{47}{61}x = 2{,}500$; $x = \$3{,}244.69$
25. Let $h =$ number of hours Lewis works, $h - \frac{1}{2} =$ number of hours Arthur works. The total number n of forms they can process is $n = 200h + 300(h - \frac{1}{2})$. It takes 6.3 hours to process 3,000 forms from the time Lewis starts.
27. Let $x =$ number of AM radios, $24 - x =$ number of AM/FM radios. The total cost C is $C = 35x + 50(24 - x) + 70$; there are 18 AM radios and 6 AM/FM radios.

Exercises 2.2

1. Contradiction 3. Identity
5. Identity 7. Contradiction
9. $x = -0.8, 0$ do not satisfy; $x = 5$ satisfies
11. $x = -5, -\frac{1}{2}$ do not satisfy; $x = 5$ satisfies
13. $x = 4$ 15. All reals
17. $t = 6$ 19. No solution
21. No solution 23. $t = \frac{5}{7}$
25. No solution 27. All reals
29. $x = 0$ 31. $x = -\frac{2}{11}$
33. $t = -\frac{1}{2}$ 35. $x = -40$
37. $y = \frac{32}{5}$ 39. $y = \frac{11}{2}$ 41. $a = \frac{19}{15}$
43. Let $x =$ first number; $x + (2x + 5) = 23$; 6 and 17
45. Let $x =$ first number; $x + (x + 1) + (x + 2) + (x + 3) = (x + 2) + 1$; $-1, 0, 1, 2$
47. Let $u =$ number of U.S. dollars; $u = 1.24(500)$; \$620
49. Let $x =$ stock value on Jan. 4; $1.25(0.75x) = 2{,}500$; \$2,666.67
51. Let $x =$ width; $2x + 2(2x) = 42$; 7 m by 14 m
53. Let $x =$ second side; $33 = (x - 5) + x + 2(x - 5)$; sides are 7, 12, 14 cm
55. Let $x =$ original width; $2(x + 2) + 2[2(3x + 1)] = 5(3x + 1) - 3$; original width $= 6$; original length $= 19$
57. Let $x =$ weekly gross sales; $0.08x = 600$; $x = \$7{,}500$
59. Let $x =$ final exam grade; $0.45x + 0.55(72) = 80$; $x = 89.8$
61. Let $x =$ number of \$5 bills; $1[25 - (2x + 1)] + 5x + 10(x + 1) = 164$; four \$1 bills, ten \$5 bills, eleven \$10 bills
63. Let $x =$ number of 20-lb packages; $20x + 25(50 - x) = 1{,}075$; number of 20-lb packages $= 35$; number of 25-lb packages $= 15$
65. Let $s =$ number of shares at \$18.825. Then $18.825s + 20.375(20 - s) = 3889$; 120 shares at \$18.825 and 80 shares at 20.375.
67. Let $x =$ number of pairs of shoes sold; $468 = 300 + 3(70 - x) + 2x$; 42 pairs of shoes
69. Let $x =$ quantity of \$4/lb coffee; $4x + 30(6) = 5.2(x + 30)$; 20 lb of \$4/lb coffee
71. Let $x =$ number of orchestra seats; $96x + 76(56 - x) = 5{,}016$; 38 orchestra seats
73. Let $x =$ number of hours the plumber worked; $42x + 24(x + 2) = 444$; the plumber worked 6 hours

75. Let t = hours until they meet;
$345 = 55t + 60t$; $t = 3$;
at 6:00 P.M.

77. $595 = 35t + 50t$, where
t = time of travel; 7 hours

79. $17t = 7(t + 3)$, where t = time it
takes to overtake; $t = 2.1$ hours
(2 hours and 6 minutes)

81. Let t = length of time to go to
convention; $48t = 54(17 - t)$;
$t = 9$ hours; distance to
convention = $48 \cdot 9 = 432$ km

83. Let t = time running;
$18t + 50(6 - t) = 124$;
$t = 5.5$ hours; 99 km

85. Let older model work for t minutes;
$35t + 50(110 - t) = 5,125$;
$t = 25$ minutes; older machine
makes $35 \cdot 25 = 875$ copies

87. Let x = number of hours;
$60x + 30x = 6,750$; 75 hours

89. Let x = number of hours by
experienced worker;
$60(x) + 30(x - 3) = 6,750$;
76 hours to complete the job

91. Let x = hours the trainee is working; $80(x - 3) + 48x = 656$;
$x = 7$ hours; finishing time =
5:00 P.M.

Exercises 2.3

1. Identity   **3.** Contradiction
5. Identity   **7.** Identity
9. $u = -3.4, 2$ do not satisfy
11. $y = -2.5, -1$ do not satisfy
13. $z = -2$ satisfies; $z = 4.6$ does not
satisfy
15. Makes sense
17. Makes sense $(-8 \leq w < -6)$
19. Makes sense
21. Does not make sense: no number
can be both less than -3 and greater
than 2.
23. $(-2, 6)$ **25.** $(-\infty, 7)$
27. $(-12, -8]$
29. $y \geq \frac{1}{4}$ or $[\frac{1}{4}, \infty)$
31. $x \geq -\frac{27}{14}$ or $[-\frac{27}{14}, \infty)$
33. $x \geq -9$ or $[-9, \infty)$
35. $y > -9$ or $(-9, \infty)$
37. $(-\infty, -\frac{8}{3})$;

39. $(-\infty, 4)$;

41. $[-\frac{10}{3}, \infty)$;

43. $(-\infty, -5)$;

45. $(-9, \infty)$;

47. $[3, \infty)$;

49. Identity $(-\infty, \infty)$;

51. $[-2, 2)$;

53. $(\frac{7}{2}, 6]$;

55. $(1, 5)$;

57. $-\frac{4}{3} \leq t \leq \frac{7}{3}$ **59.** $-5 < x < 0$
61. $3 < x \leq 6$ **63.** $4 < x \leq 7$
65. $-\frac{8}{5} \leq z \leq \frac{2}{5}$
67. $2.625 \leq x < 13.\overline{3}$
69. Let x = number; $3x - 4 < 17$;
$x < 7$
71. Let x = number; $6x + 12 > 3x$;
$x > -4$
73. Let s = length of a side; $4s \leq 72$;
maximum length = 18 feet
75. Let l = length; $16 + 2l \geq 80$;
length ≥ 32 cm
77. Let w = width;
$50 \leq 2w + 36 \leq 70$;
the range of values for
the width w is $7 \leq w \leq 17$ inches
79. $106 \leq$ perimeter ≤ 138 in.
81. Let x = shortest side;
$30 \leq x + 2x + (x + 2) \leq 50$;
$7 \leq x \leq 12$ cm
83. (a) $18.5 \leq \frac{703w}{68^2} \leq 24.9$;
122 lb $\leq w \leq 164$ lb
(b) answer varies.
85. Let x = price of reserved ticket;
$300x + 150(x - 2) \geq 3,750$;
minimum price is $9
87. Let x = number of dimes;
$10x + 5(40 - x) \leq 285$;
maximum number is 17
89. Let x = number of elephants;
$25x + 20(24 - x) \geq 575$; at least 19
91. Let x = number of hours of tutoring
time; $20x + 10(30 - x) \geq 650$;
$x \geq 35$; he cannot make $650
working 30 hours
93. Let x = number of superdogs;
$45x + 25(100 - x) \geq 3,860$;
at least 68
95. Let x = number of shares of stock B;
$2(1,000 - x) + 3x \geq 2,400$; at
least 400 shares

Exercises 2.4

1. $x = 4$ or $x = -4$
3. $-4 < x < 4$
5. $x < -4$ or $x > 4$
7. $-4 \leq x \leq 4$
9. $x \leq -4$ or $x \geq 4$
11. No solution
13. All reals
15. No solution
17. $t = 5$ or $t = 1$
19. $t = 5$ or $t = -5$
21. $n = 4$ or $n = 6$
23. $2 < a < 8$
25. $a \leq -1$ or $a \geq 3$
27. No solution
29. $-2 < a < 2$
31. $x = -\frac{2}{3}$ or $x = 2$
33. No solution
35. $x = -5$ or $x = 0$
37. $1 \leq a \leq 5$
39. $a < 2$ or $a > 3$
41. $-8 < x < 0$
43. $x = 6$ or $x = -4$;

45. $[1, 5]$;

47. $(-\infty, -4) \cup (-3, \infty)$

49. $(\frac{1}{2}, 2)$

51. $(\frac{1}{2}, 1)$

53. $t = 4$ or $t = -\frac{2}{9}$
55. $r = 6$ or $r = -1$ **57.** $a = \frac{7}{2}$
59. $x = -1$ or $x = 1$ **61.** $x = 0$
63. At 40°F the temperature ranges
between 39.25°F and 40.75°F.
At 50°F the temperature ranges
between 49.25°F and 50.75°F.
At 80°F the temperature ranges
between 79.25°F and 80.75°F.
65. Between 58% and 70%

Chapter 2 Review Exercises

1. Let s = number of single beds,
$24 - s$ = number of larger beds;
$415 = 15s + 20(24 - s)$;
13 single beds, 11 larger beds
2. Let p = pay, g = gross sales;
$500 = 120 + 0.06g$; $6,333.33
in gross sales

3. Let w = width, length = $4w - 2$; $2w + 2(4w - 2) = 100$; 10.4 in. by 39.6 in.

4. Let g = number of hours grading, $25 - g$ = number of hours tutoring; $10g + 25(25 - g) \geq 500$; she can grade papers no more than about $8\frac{1}{3}$ hours per week

5. $x = \frac{13}{3}$ **6.** $x = \frac{19}{4}$ **7.** $x = \frac{25}{3}$

8. No solution **9.** $x = 0$

10. $a = 0$ **11.** $a = \frac{62}{11}$ **12.** All reals

13. $x = \frac{65}{18}$ **14.** $x = -\frac{15}{4}$

15. $x = \frac{23}{9}$ **16.** $x = \frac{7}{2}$

17. $x = 3$ or $x = -3$

18. No solution **19.** No solution

20. $y = 7$ or $y = -7$

21. $x = 4$ or $x = -4$

22. $x = \frac{5}{3}$ or $x = -\frac{5}{3}$

23. $a = 3$ or $a = -5$

24. $a = 8$ or $a = -2$

25. $z = -\frac{5}{4}$ **26.** $x = \frac{4}{7}$ or $x = -\frac{8}{7}$

27. $y = 5$ or $y = 0$

28. $x = \frac{11}{4}$ or $x = \frac{7}{4}$ **29.** $x = 3$

30. $a = -2$ **31.** $t = 2$

32. $a = 2$ or $a = \frac{8}{3}$

33. $x \leq 3$

34. $x \leq -3$

35. $x \leq \frac{4}{3}$

36. $x < 2$

37. $z > -\frac{5}{3}$

38. $y > \frac{5}{4}$

39. $s < 4$

40. No solution **41.** All reals

42. No solution

43. $-1 \leq x \leq 3$

44. $1 \leq x \leq 6$

45. $-\frac{1}{2} \leq x \leq \frac{11}{2}$

46. $-9 \leq x \leq -1$

47. $3 \leq x \leq 6$

48. $2 \leq x \leq \frac{21}{5}$

49. $(-\infty, -3)$;
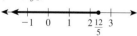

50. No solution **51.** No solution

52. $(\frac{2}{3}, \infty)$;

53. $(-\infty, \frac{12}{5}]$;
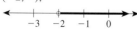

54. $(-2, \infty)$;
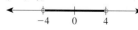

55. $(-4, 4)$;
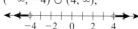

56. $(-\infty, -4) \cup (4, \infty)$;
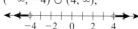

57. $(-\infty, -5] \cup [5, \infty)$;

58. $[-5, 5]$;

59. No solution

60. $t = 0$;

61. $(-1, 3)$;
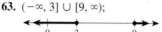

62. $(-\infty, -1) \cup (3, \infty)$;

63. $(-\infty, 3] \cup [9, \infty)$;
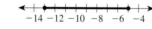

64. $[3, 9]$;

65. $[-13, -5]$;
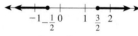

66. $(-\infty, 5] \cup [13, \infty)$;
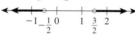

67. $(-\infty, -\frac{1}{2}] \cup [\frac{3}{2}, \infty)$;

68. $(-\infty, -\frac{1}{2}] \cup (\frac{3}{2}, \infty)$;
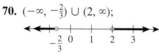

69. $(-\frac{2}{3}, 2)$;
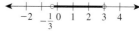

70. $(-\infty, -\frac{2}{3}] \cup (2, \infty)$;

71. $-1 \leq x \leq 4$

72. $x \leq -1$ or $x \geq 4$

73. $x < -\frac{10}{3}$ or $x > 0$

74. $-\frac{10}{3} < x < 0$

75. No solution

76. $-\frac{9}{2} \leq x \leq \frac{15}{2}$

77. Let x = number; $3x = 4x - 4$; $x = 4$

78. Let x = number; $5(x + 6) = x - 2$; $x = -8$

79. Let u = U.S. dollars; yen = $114.05u$; $114.05(800) = 91,240$ yen

80. Let x = value of stock on April 5; $0.85x + 0.2(0.85x) = 4,500$; $4,411.76

81. Let G = gross sales; $650 = 0.09G$; $G = \$7,222.22$

82. Let G = gross sales; $600 = 100 + 0.05G$; $G = \$10,000$

83. Let f = final exam grade; $(0.60)(74) + 0.40f = 80$; $f = 89$

84. Let f = final exam grade; $(0.55)(78) + 0.45f = 80$; $f = 82.4$

85. Let x = number of packages weighing 8 lb each; $8x + 5(30 - x) = 186$; twelve 8-lb packages and eighteen 5-lb packages

86. Let x = number of packages weighing 8 lb; $8x + 5(45 - x) = 276$; seventeen 8-lb packages and twenty-eight 5-lb packages

87. Let x = number of single beds; $10x + 15(23 - x) = 295$; 10 single beds and 13 larger beds

88. Let x = number of hours the plumber worked; $65x + 40(7 - x) + 27 = 369.50$; plumber worked 2.5 hours; assistant worked 4.5 hours

89. Let x = number of hours of tutoring; $20(30 - x) + 25x \geq 660$; at least 12 hours

90. Let x = number of hours tutoring; $25x + 20(40 - x) \leq 925$; no more than 25 hours tutoring

91. Let t = time for the first machine in minutes. $200t + 300(30 - t) = 8,500$; $t = 5$; 5 minutes

92. Let w = number of \$875 web sites. $875w + 2050(22 - w) = 23,950$; $w = 18$; 18 web sites at \$875 and 4 web sites at \$2,050

93. Let x = number of hits in category 3; $x + 2x + (x + 2x) = 18,576$; $x = 3,096$; 3,096 hits in category 3, 6,192 hits in category 2, and 9,288 hits in category 1

Chapter 2 Practice Test

1. Let x = amount in savings account; $I = 0.035x + 0.06(5,000 - x)$ \$3,200 in savings

2. $x = \frac{11}{2}$ **3.** No solution

4. $a = -9$ **5.** $x \geq 4$

6. $x < \frac{1}{5}$

7. $\frac{1}{2} < x < 4$

8. $x = 7$ or $x = 0$;

9. $(-\frac{1}{3}, 3)$;

10. $(-\infty, 1] \cup [\frac{9}{5}, \infty)$;

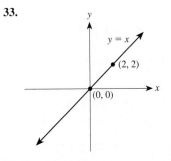

11. Let u = U.S. dollars; $\frac{d}{1.20}$ = 500; $600

12. Let x = number of boxes weighing 35 kg; $35x + 45(93 - x) = 3,465$; seventy-two 35-kg boxes and twenty-one 45-kg boxes

13. $8x = 3(x + 2)$, where x = time of jogger; $x = \frac{6}{5} = 1\frac{1}{5}$ hours to catch up

14. Let x = number of dimes; $10x + 5(32 - x) \geq 265$; at least 21 dimes

Exercises 3.1

1. No **3.** No **5.** Yes
7. Yes **9.** Yes
11. $(-1, 9), (0, 8), (1, 7), (10, -2),$ $(8, 0), (4, 4)$
13. $(-2, \frac{15}{2}), (0, 5), (4, 0), (8, -5),$ $(4, 0), (\frac{4}{5}, 4)$
15. $(-3, 8), (0, 4), (3, 0), (6, -4),$ $(3, 0), (0, 4)$
17. x-intercept: 6; y-intercept: 6
19. x-intercept: 6; y-intercept: -6
21. x-intercept: -6; y-intercept: 6
23. x-intercept: 6; y-intercept: 3
25. x-intercept: 3; y-intercept: 4
27. x-intercept: 5; y-intercept: -3
29. x-intercept: $-\frac{7}{3}$; y-intercept: $\frac{7}{2}$

31.

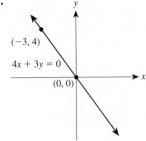

33.

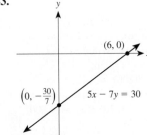

35.

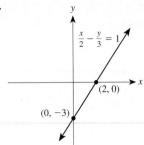

37.

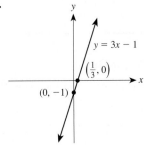

39.

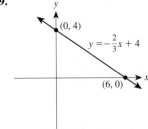

41.

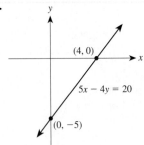

43.

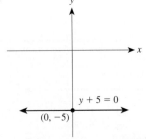

45.

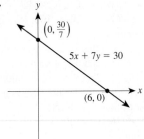

47.

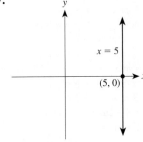

49.

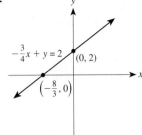

51.

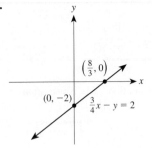

53.

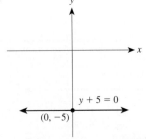

55.

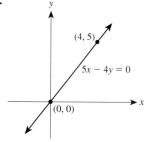
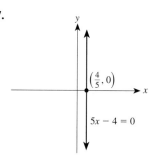

57.

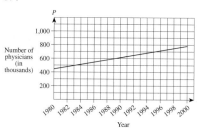

59.

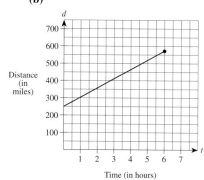

61. (a) $S = 0.05v + 220$
   (b)

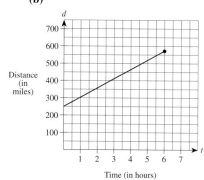

63. (a) $d = 260 + 52t$
   (b)

65. (a) $C = 29 + 0.14n$
   (b)

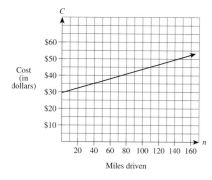

Miles driven

Exercises 3.2

1. (a) $y = -3$ when $x = -4$;
      $y = 1$ when $x = 4$
   (b) $x = 6$ when $y = 2$
3. (a) $y = -4$ when $x = -3$;
      $y = -4, 0,$ and 3 when $x = 2$
   (b) $x = 1$ when $y = 2$;
      $x = 0$ and -2 when $y = -6$
5. $y = -1$ **7.** $x = 3, x = -1$
9. (a) $(3, -3)$ **(b)** $(-4, 4)$
   (c) $(3, -3)$ and $(-2, -3)$
   (d) $(5, 4)$ and $(-4, 4)$
11. (a) $y = -3$ and 3
    (b) $x = -2$ and 2
13. (a) y-intercept $= 4$
    (b) x-intercept $= 8$
15. (a) y-intercept $= 2$
    (b) x-intercept $= -4$
    (c) Smallest x is -4; no largest x
    (d) Smallest y is 0; no largest y
17. (a) Smallest x is -3; no largest x
    (b) Largest y is 3; no smallest y
19. (a)

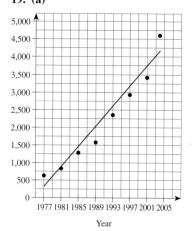

Year

21.

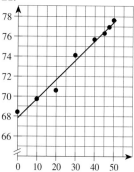

23. (a) $C = 0.10g$
   (b)

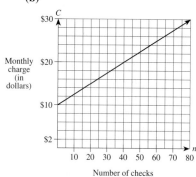

Gross sales (in dollars)

   (c) $400
   (d) $8,500
25. (a) $C = 10 + 0.25n$
   (b)

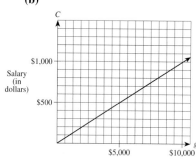

Number of checks

   (c) $22 to $23
   (d) 20 checks

Exercises 3.3

1. $\{(3, 9), (3, 7), (8, 2), (7, 2)\}$
3. $\{(3, a), (3, b), (8, b), (-1, b), (-1, c)\}$
5. Domain: $\{3, 4, 5\}$; range: $\{2, 3\}$
7. Domain: $\{-2, 3, 4\}$;
   range: $\{-2, -1, 3\}$
9. $\{x \mid x \neq 0\}$ **11.** All reals
13. $\{x \mid x \neq -\frac{3}{2}\}$ **15.** All reals
17. $\{x \mid x \geq 4\}$ **19.** $\{x \mid x \leq \frac{5}{4}\}$
21. $\{x \mid x \geq 0\}$ **23.** $\{x \mid x > 3\}$
25. Function   **27.** Not a function
29. Function   **31.** Not a function
33. Function   **35.** Function

37. Not a function   **39.** Function
41. Not a function   **43.** Not a function
45. Function   **47.** Not a Function
49. Function   **51.** Not a function   **53.** 0
55. $(7, -6)$ **57.** Three; $x = -6, -2, 4$
59. One   **61.** $\{x \mid -7 \le x \le 7\}$
63. 3   **65.** $(7, -3)$
67. Three: $x = -3, 2, 6$
69. One: $y = 2$ **71.** $\{x \mid -5 \le x \le 7\}$
73. Domain $= \{x \mid -4 \le x \le 5\}$;
    range $= \{y \mid -4 \le y \le 6\}$
75. Domain $= \{x \mid 0 \le x \le 4\}$;
    range $= \{y \mid -3 \le y \le 3\}$
77. Domain $= \{x \mid -5 \le x \le -2$ or
    $1 \le x \le 5\}$;
    range $= \{y \mid -3 \le y \le 1$ or
    $3 \le y \le 4\}$
79. $A = s^2$ **81.** $A = 2a^2 - 15; a \ge 3$
83. $C = 116 + 0.22m$
85. $C = 22d + 60$
87. $20,200; \$30,800; \$65,000;$
    $\$130,000$
89. (a) $C = 29.95 + 0.33m$
    (b) $54.70; \$89.35; \$128.95$
91. (a) Yes
    (b) Domain: The rent varies
      between $400 and $600;
      Range: The profit varies
      between $35,000 and $50,000.
    (c) $500; \$50,000$

Exercises 3.4

1. -3 **3.** 11   **5.** 15
7. $\sqrt{8} = 2\sqrt{2}$ **9.** $\sqrt{2}$
11. $\sqrt{a + 5}$ **13.** 33   **15.** 6
17. $2x + 1$ **19.** $4x^2 + 2$
21. $6x + 1$ **23.** 4   **25.** 9
27. $x^2 + 2x - 7$ **29.** $x^2 + 2x - 24$
31. $9x^2 + 6x - 3$ **33.** $3x^2 + 6x - 9$
35. $-\dfrac{3}{2}$ **37.** 0   **39.** $\dfrac{x + 5}{x + 10}$
41. (a) 1   (b) -1 (c) -2
    (d) $-\dfrac{1}{2}$ (e) 2
43. (a) 3   (b) 5   (c) 0
    (d) $x = -5, -3, 5$ (e) Two
45. $C = 87 + 0.14m$
47. $A = 3L^2 - 12L$
49. $C = 30x + 24(600 - x)$
51. $N = 20m + 22(m + 35)$
53. $A = 65h + 45(h - 2)$
55. $I = 30t + 9.35(15 - t)$

Exercises 3.5

1. (a) 7,500   (b) 4 A.M.   (c) 30,000
    (d) Between 4 P.M. and 4 A.M.
    (e) Between 10 A.M. and 4 P.M.
    (f) Between 10 A.M. and 4 P.M.

3. (a) 5.5%   (b) 1997–2001
    (c) The unemployment rate
      decreased from 1994 until 2000.
      From 2000 until 2003, the
      unemployment rate increased.
      From 2003 through 2004, the
      unemployment rate decreased
      again.
5. (a) Consumed: 80 quadrillion Btu;
      produced: 65 quadrillion Btu;
      imported: 15 quadrillion Btu
    (b) Production: There was a slight
      decrease in production between
      1982 and 1983, followed by
      an increase beween 1983 and
      1984. Production then slightly
      increased over the years
      1984−1998. Consumption:
      There was a decrease in
      consumption from 1980 to
      1983, then a relatively steady
      increase from 1983 to 1998.
      Imports: There was a decrease
      from 1980 to 1982 and then
      a relatively steady increase
      from 1982 to 1998.
7. (a) The graph shows that the
      temperature does *not* drop
      steadily as altitude increases.
    (b) Between 0 and 10 km, and
      between 50 and 80 km
    (c) Starting at 45 km, the
      temperature increases until it
      reaches approximately 10°C at
      an altitude of 50 km, then the
      temperature decreases.
9. (a) Approximately 30%
    (b) About 2 days
    (c) As time passes material is forgot-
      ten. Material is forgotten rapidly
      during the first 3 hours of the
      given time period (more than half
      the material is forgotten during
      this time), then the rate of forget-
      ting slows down. Only about
      18−20% of the material is
      remembered by the 6th day.
11. (a) 14 trials   (b) 3 trials
    (c) Performance with spaced
      practice is superior to
      performance with massed
      practice. Learning with spaced
      practice occurs more rapidly
      than with massed practice.
13. Starting at home, Kyle starts his trip
    by traveling 90 miles away during
    the first 2 hours of the day. His aver-
    age rate of speed during the first
    hour was 50 mph, and during the sec-
    ond hour was 40 mph. For the next
    2 hours Kyle did not travel. Then
    Kyle traveled closer to his home

between hours 4 and 5. He was
traveling an average of 30 mph
during this time. Kyle stopped again
for about 1 hour, and then drove
home in $1\frac{1}{2}$ hours at an average rate
of 40 mph.
15. The graph shows that children
    assigned to each group had approxi-
    mately the same number of aggres-
    sive behaviors before treatment.
    By the end of the treatment period,
    however, treatments A and C seemed
    to be the most effective in reducing
    these aggressive behaviors, whereas
    treatment C reduced their behaviors
    most quickly.
      After the treatment period,
    however, the children in treatment
    C reverted back to their aggressive
    behavior. Hence treatment C ends
    up being the least effective at the end
    of the time period given following
    treatment.
      The children in treatment B
    slowly reduced their aggressive
    behaviors even after the treatment
    ended. Although it seemed the least
    effective at the end of the treatment,
    it actually is shown to be the most
    effective of the three at the end of
    the time period given following
    treatment.

Chapter 3 Review Exercises

1.

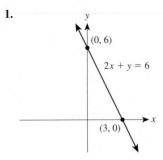

2.

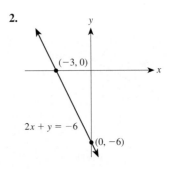

3.

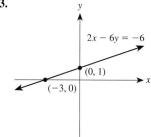

$2x - 6y = -6$
$(0, 1)$
$(-3, 0)$

4.

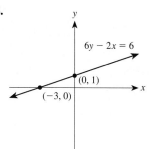

$6y - 2x = 6$
$(0, 1)$
$(-3, 0)$

5.

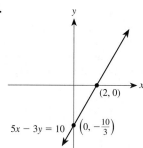

$(2, 0)$
$5x - 3y = 10$ $\left(0, -\frac{10}{3}\right)$

6.

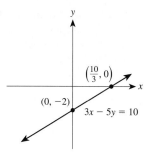

$\left(\frac{10}{3}, 0\right)$
$(0, -2)$
$3x - 5y = 10$

7.

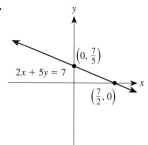

$\left(0, \frac{7}{5}\right)$
$2x + 5y = 7$
$\left(\frac{7}{2}, 0\right)$

8.

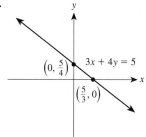

$\left(0, \frac{5}{4}\right)$
$3x + 4y = 5$
$\left(\frac{5}{3}, 0\right)$

9.

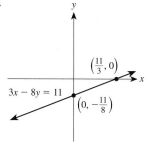

$\left(\frac{11}{3}, 0\right)$
$3x - 8y = 11$
$\left(0, -\frac{11}{8}\right)$

10.

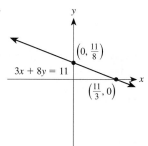

$\left(0, \frac{11}{8}\right)$
$3x + 8y = 11$
$\left(\frac{11}{3}, 0\right)$

11.

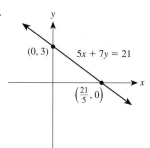

$(0, 3)$
$5x + 7y = 21$
$\left(\frac{21}{5}, 0\right)$

12.

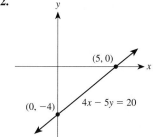

$(5, 0)$
$(0, -4)$ $4x - 5y = 20$

13.

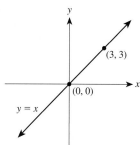

$(3, 3)$
$(0, 0)$
$y = x$

14.

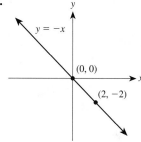

$y = -x$
$(0, 0)$
$(2, -2)$

15.

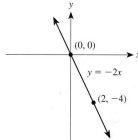

$(0, 0)$
$y = -2x$
$(2, -4)$

16.

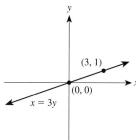

$(3, 1)$
$(0, 0)$
$x = 3y$

17.

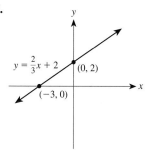

$y = \frac{2}{3}x + 2$ $(0, 2)$
$(-3, 0)$

18.

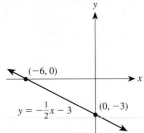

19.

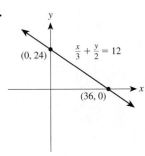

20.

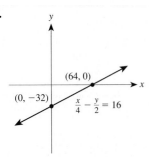

21.

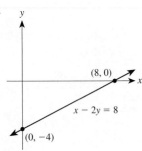

22.

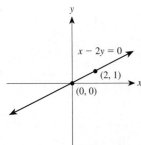

23.

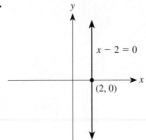

24.

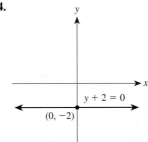

25.

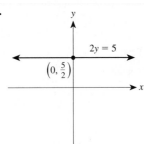

26.

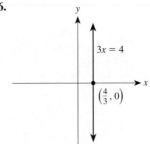

27. (a) $y = 3$ when $x = -2$;
  $y = -1$ when $x = 2$
  (b) $x = 1$ and $x = -2$ when $y = 3$
28. (a) $y = 4$ and $y = -2$ when
  $x = -3$
  (b) $x = 3$ when $y = 2$;
  $x = 2$ when $y = -5$
29. $y = 6$
30. (a) $(5, 4)$ **(b)** $(5, 4)$ and $(-1, 4)$
31. x-intercepts $= 1$ and 3
32. y-intercept $= 2$
33. $y = -4$ and 4

34. $x = -2$ and 2
35. (a) $I = 150 + 0.05g$
  (b)

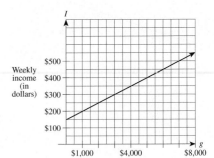

  (c) $400
  (d) $12,000
36. (a) $C = 30 + 0.20m$
  (b)

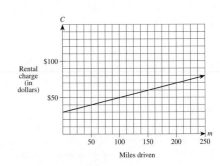

  (c) $70
  (d) 175 miles
37. All reals   **38.** All reals
39. $\{x \mid x \leq 4\}$ **40.** $\{x \mid x \geq -3\}$
41. $\{x \mid x \neq -2\}$ **42.** $\{x \mid x \neq -\frac{1}{2}\}$
43. Function   **44.** Not a function
45. Not a function   **46.** Function
47. Function   **48.** Function
49. Function   **50.** Not a function
51. Function   **52.** Function
53. Not a function   **54.** Function
55. Domain $= \{-2, 0, 3, 5\}$;
  range $= \{3, 4, 5, 7\}$; function
56. Domain $= \{-3, 0, 2\}$;
  range $= \{-8, 0, 7, 10\}$; not a
  function
57. Domain $= \{1, 4, 6, 8\}$;
  range $= \{9\}$; function
58. Domain $= \{6\}$;
  range $= \{-5, -3, 0, 2\}$; not a
  function
59. Domain $= \{x \mid -5 \leq x \leq 6\}$;
  range $= \{y \mid -2 \leq y \leq 5\}$;
  function
60. Domain $=$ all real numbers;
  range $= \{y \mid y \geq -3\}$; function
61. Domain $= \{x \mid -3 \leq x \leq 3\}$;
  range $= \{y \mid -4 \leq y \leq 4\}$;
  not a function

62. Domain = all real numbers; range = all real numbers; function
63. 2, 5, 8, 11   **64.** 9, 5, 1, −3
65. 7, 2, 1, 4   **66.** −8, −3, 2, 25
67. 1, 0, $h(4)$ is not a real number [4 not in the domain of $h(x)$]
68. $\sqrt{2}$, $g(2)$ is not a real number [2 not in the domain of $g(x)$], $\sqrt{8} = 2\sqrt{2}$
69. 0, $\frac{1}{3}$, undefined [−3 not in the domain of $h(x)$]
70. 0, undefined [0 not in the domain of $h(x)$], $\frac{5}{4}$
71. $2a^2 + 4a - 1, 2z^2 + 4z - 1$
72. $2x^2 - 4x + 2, 2z^2 - 4z + 2$
73. $5x + 12$ **74.** $5x + 17$
75. $5x + 4$ **76.** $5x + 5$
77. $5x + 14$ **78.** $5x + 19$
79. $4 - x$ **80.** $3 - x$
81. $6 - 2x$ **82.** $6 - 3x$
83. $12 - 2x$ **84.** $18 - 3x$
85. (a) 0 (b) 2 (c) −2 (d) −3
   (e) $f(5)$ (f) $x = -6, -2, 5$
86. (a) 5 (b) 1 (c) 3 (d) −1
   (e) 2 (f) $x = 3, 5$
87. Rasheed starts his trip by traveling 90 miles away from his home during the first 2 hours of the day. His average rate of speed for the first hour was 40 mph; during the second hour his average rate was 50 mph. During the next hour Rasheed did not travel. Then Rasheed drove closer to his home between hours 3 and 4. He was traveling an average of 60 mph during this time. Rasheed drove another 20 miles away from his home, traveling at 20 mph between the 4th and 5th hour. Finally, during the last hour, Rasheed drove home at a rate of 50 mph.
88. If you look carefully at the distance it travels (the vertical change) for each half-second interval (horizontal change), you see that the distance it falls is getting longer for each progressive half-second. If you compute the speed, you can see that the longer the ball stays in the air, the faster it travels; that is, its speed is increasing with time.
89. $C = 30 + 42n$
90. $M = 5,000 + 2,500s$
91. (a) 14 million (b) 1999–2002 (c) The number employed steadily increased from 1994 until the beginning of 2001. Then the number decreased from 2001 through 2003. There seemed to be a leveling off from 2003 through 2004.

Chapter 3 Practice Test

1. (a)
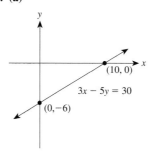

(b)
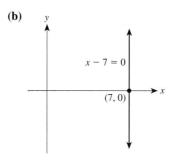

2. $y = 3$ when $x = 5$
3. (a) Domain = $\{2, 3, 4\}$;
   range = $\{-3, 5, 6\}$
   (b) Domain = $\{x \mid -4 \le x \le 5\}$;
   range = $\{y \mid -3 \le y \le 6\}$
4. (a) $\{x \mid x \ge 4\}$ (b) $\{x \mid x \ne \frac{4}{3}\}$
5. (a) Not a function (b) Function
   (c) Not a function (d) Function
6. (a) $\sqrt{5}$ (b) 23 (c) $5x - 13$
   (d) $3x^4 - 4$ (e) $3x^2 - 75$
7. (a) 2 (b) −2 (c) 0 (d) 2
   (e) $f(-4)$ (f) $x = 0, 5$
8. (a) 30 minutes (b) 8 mg
   (c) Between 10 and 20 minutes; 3 mg

Chapters 1–3 Cumulative Review

The number in parentheses after the answer indicates the section in which the material is discussed.
1. $\{11, 13, 17, 19, 23, 29, 31, 37\}$ (§1.1)
2. $\varnothing$ (§1.1)   **3.** $\varnothing$ (§1.1)
4. $\{10, 11, 13, 15, 17, 19, 20, 23, 25, 29, 30, 31, 35, 37\}$ (§1.1)
5. True (§1.2)   **6.** False (§1.2)
7. (§1.2):

8. (§1.2):
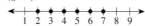
9. Closure property for addition (§1.3)

10. Multiplicative inverse property (§1.3)
11. −6 (§1.4)   **12.** −17 (§1.4)
13. 36 (§1.4)   **14.** 43 (§1.4)
15. (a) 0 (§1.4) (b) Undefined (§1.4)
16. $-3x^3 + 6x^2 - 9x$ (§1.5)
17. $-12x^4y^3$ (§1.5)
18. (a) $M = 3.058t - 5,898.49$
   (b) 202.22 million (§1.4)
19. Let x = the number; $3 + 8x$ (§1.6)
20. Let x = the first number;
   $x(x + 1) - (x + 2)$ (§1.6)
21. Let w = the width; $A = w(4w - 3)$;
   $P = 10w - 6$ (§1.6)
22. Let n = number of nickels;
   $5n + 10(35 - n)$ (in cents) (§1.6)
23. $x = -11$ (§2.2)
24. No solution (§2.2)
25. $x = 2$ (§2.2)   **26.** $x = -\frac{1}{3}$ (§2.2)
27. $x = \frac{4}{19}$ (§2.2)   **28.** $x = -\frac{4}{3}$ (§2.2)
29. $x = 4, -5$ (§2.4)
30. $x = \frac{1}{2}, \frac{3}{4}$ (§2.4)
31. $x < -\frac{3}{2}$ (§2.3):
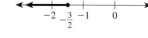
32. $x < -\frac{1}{4}$ (§2.3):

33. All reals (§2.3):

34. $-3 < x < 4$ (§2.3):

35. $1 < x < 9$ (§2.4):

36. $x < -4$ or $x > 3$ (§2.4):

37. $x < -\frac{1}{2}$ or $x > 3$ (§2.4):

38. $0 < x < \frac{16}{3}$ (§2.4):

39. Let x = the final exam grade, and G = the course grade; $0.60(63) + 0.40x = G$; he would need 80.5 on the final. (§2.1)
40. Let g = her gross sales and S = her weekly salary; $S = 250 + 0.02g$; she would need \$12,500 in gross sales to make \$500 a week. (§2.1)

41. Let x = number; $4x = 2x - 3$; $-\frac{3}{2}$ (§2.2)

42. Let x = number of dimes; $10x + 25(25 - x) = 325$; 20 dimes, 5 quarters (§2.2)

43. Let p = number of packages with 100 plates; $1 \cdot p + 2(28 - p) = 33$; 23 packages of 100 plates, 5 packages of 300 plates; 3,800 plates (§2.2)

44. Let x = number of hours assistant painted; $7x + 14(58 - x) = 567$; 35 hours (§2.2)

45. Let t = time for VF-44 printer to do the job; $(30)(60)(80) = x(60)(120)$; 20 minutes (§2.2)

46. Let t = time new machine ran during the job; $90t + 60(120 - t) = 9{,}750$; $t = 85$ minutes; $(90)(85) = 7{,}650$ copies (§2.2)

47. Let w = width; perimeter = $8w + 4$; 44 ft to 100 ft (§2.3)

48. Let t = number of hours working as a mechanic; $15t + 12(30 - t) \geq 399$; at least 13 hours (§2.3)

49. $x = -5$ when $y = 3$ (§3.2)

50. $y = 3$ and $y = -3$ when $x = -5$ (§3.2)

51. Domain: $\{2, 4, 7\}$; range: $\{-1, 3, 5\}$; function (§3.3)

52. Domain: $\{-2, 1, 5\}$; range: $\{3, 6\}$; function (§3.3)

53. Domain: $\{3, 4\}$; range: $\{2, 7, 9\}$; not a function (§3.3)

54. Domain: $\{-2, -1, 1, 2\}$; range: $\{1, 4\}$; function (§3.3)

55. All reals (§3.3)

56. $\{x \,|\, x \geq 3\}$ (§3.3)

57. $\{x \,|\, x \neq \frac{4}{5}\}$ (§3.3)

58. $\{x \,|\, x \geq -4, x \neq 3\}$ (§3.3)

59. Function (§3.3)

60. Not a function (§3.3)

61. Function (§3.3)

62. Not a function (§3.3)

63. $-\frac{10}{3}$ (§3.4)   **64.** $\frac{16}{9}$ (§3.4)

65. $3x^2 + 1$ (§3.4)   **66.** $\frac{1}{x^4}$ (§3.4)

67. $a + 3 + \dfrac{1}{a + 3}$ (§3.4)

68. $a + \dfrac{1}{a} + 3$ (§3.4)

69. $3a + \dfrac{1}{3a}$ (§3.4)

70. $9a + \dfrac{1}{a}$ (§3.4)

71. 3 (§3.4)   **72.** 31 (§3.4)

73. $7\frac{1}{7}$ (§3.4)   **74.** $8\frac{1}{2}$ (§3.4)

75. (§3.1):

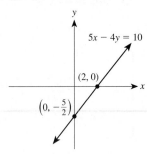

76. (§3.1):

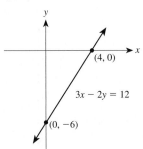

77. (§3.1):

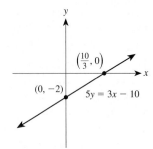

78. (§3.1):

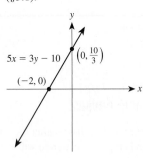

79. (§3.1):

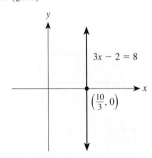

80. (§3.1):

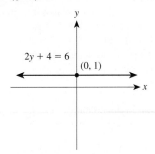

81. **(a)** Approx. 40,000
   (b) Approx. 4.7 hr
   (c) Between 3.5 and 5.5 hr
   (d) After approx. 7.5 hr (§3.5)

82. $I = 275 + 0.025s$ (§3.3)

83. $C = 350 + 75h$ (§3.3)

84. Domain = $\{x \,|\, -3 \leq x \leq 5\}$; range = $\{y \,|\, -1 \leq y \leq 4\}$ (§3.3)

85. Domain = all real numbers; range = $\{y \,|\, y \geq -3\}$ (§3.3)

86. **(a)** -3 **(b)** -3 **(c)** 5   **(d)** -4
   (e) $f(1)$ **(f)** $-6, -1, 5$ (§3.4)

87. Let h = the number of hours Jonas grades homework papers, and I = the total amount of money he makes; $I = 15h + 25(20 - h) = 500 - 10h$; he makes \$450 grading 5 hours; to make \$380, he should be grading homework for 12 hours. (§3.1):

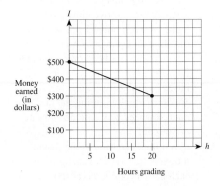

Chapters 1–3
Cumulative Practice Test

1. **(a)** $\{4, 8, 12, 16, 20\} = A$
   (b) $\{2, 4, 6, 8, 10, 12, 14, 16, 18, 20, 22\} = B$

2. Distributive property

3. **(a)** -4 **(b)** -18 **4.** 6

5. **(a)** $x = -3$ **(b)** No solution
   (c) $x = 11, -5$

6. (a) $x > \frac{9}{4}$:

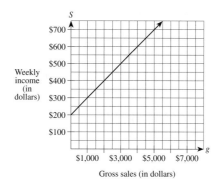

(b) $-1 < x < \frac{7}{3}$:

(c) $x \le 1$ or $x \ge \frac{3}{2}$:

7. Let p = number of packages weighing 30 lb; $30p + 10(170 - p) = 3{,}140$; seventy-two 30-lb packages, ninety-eight 10-lb packages

8. Let t = time the faster car travels; $55t = 40(t + 1)$; $2\frac{2}{3}$ hours

9. Evan starts his trip by traveling 50 miles away from his home during the first hour of the day; his average rate of speed for the first hour was 50 mph. During the next 2 hours Evan did not travel. Then Evan drove closer to his home between hours 3 and 4. He was traveling an average of 40 mph during this time, and ended up only 10 miles from his home. Evan drove another 60 miles away from his home traveling at 60 mph between the 4th and 5th hour, and stopped traveling for an hour. Finally, during the last hour and a half, Evan drove home at a rate of 47 mph.

10.

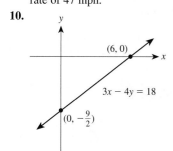

11. The x-intercepts are -4 and 4.

12. (a) $\{x \mid x \ne 9\}$ **(b)** $\left\{x \mid x \ge \frac{3}{2}\right\}$

13. (a) $\frac{7}{4}$ **(b)** $4x^2 - 1$

(c) $\frac{x}{x + 3}$ **(d)** $\frac{x - 2}{x + 1} + 2$

(e) $20x - 1$ **(f)** $20x - 5$

(g) Undefined **(h)** -9

14. (a) 2 **(b)** -2 **(c)** 0 **(d)** 3

(e) $f(-3)$ **(f)** 0 and 5

(g) Domain $= \{x \mid -6 \le x \le 7\}$; range $= \{y \mid -2 \le y \le 3\}$

15. Let g = his gross sales, and S = his weekly salary; $S = 200 + 0.10g$; he would need about $\$4{,}000$ in gross sales to make $\$600$ a week.

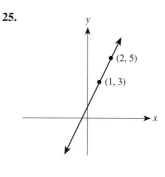

Gross sales (in dollars)

Exercises 4.1

1.

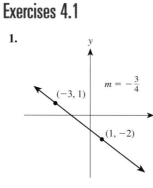

3.

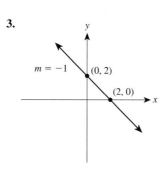

5.

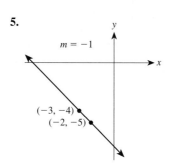

7. Slope $= 0$ **9.** Undefined

11. 1 **13.** $\frac{b^2 - a^2}{b - a} = b + a$

15. -0.531 (to 3 places)

17. -20.412 (to 3 places)

19. 1.5 **21.** 3.5 **23.** $\frac{3}{5}$

25.

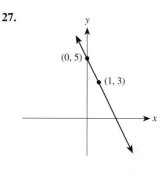

27.

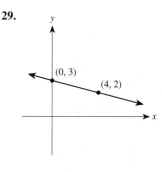

29.

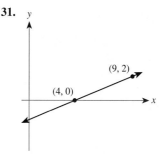

31.

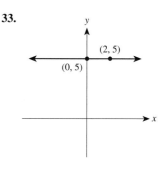

33.

35.

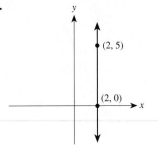

37. $m = 2$ **39.** $m = -0.4$
41. $m = 1.5$ **43.** Parallel
45. Neither **47.** Perpendicular
49. $h = 13$
51. The slopes of the parallel sides are $\frac{1}{2}$ and $-\frac{5}{2}$.
53. The slope of one side is $-\frac{1}{2}$ and of another side is 2. Hence, the two sides are perpendicular to each other.
55. 160 m $= 0.16$ km
57. Approximately 10,417 ft
59. Sprinter runs 10.8 m/sec.
61. The slope represents the average change in °F per hour. In this case, the temperature *falls* an average of 3.5°F per hour.
63. The slope gives the change in profit per change in wholesale price. Choosing points (100, 40) and (135, 100) on the line, we obtain a slope of $\frac{12}{7}$. This means the profit increases by $12,000 for every $7 increase in wholesale price (or a $1,714 increase in profit for every dollar increase in wholesale price).

Mini-Review

70. $54x^{11}$ **71.** $-6x^2y - 16xy^2$
72. 3 **73.** $t = 21$

Exercises 4.2

1. $y = 5x - 8$ **3.** $y = -3x - 13$
5. $y = \frac{2}{3}x - 3$ **7.** $y = -\frac{1}{2}x + 2$
9. $y = \frac{3}{4}x + 5$ **11.** $y = -4$
13. $x = -4$ **15.** $y = \frac{4}{3}x + \frac{1}{3}$
17. $y = 4x + 7$ **19.** $y = -x + 5$
21. $y = 4x + 6$ **23.** $y = -2x - 6$
25. $y = 3$ **27.** $x = -2$
29. $y = \frac{2}{3}x + 2$ **31.** $y = 3x - 4$
33. $y = \frac{3}{2}x + \frac{13}{2}$ **35.** $y = -x$
37. $y = \frac{3}{2}x - \frac{1}{2}$ **39.** $y = -\frac{3}{4}x - \frac{9}{4}$
41. $y = -\frac{5}{8}x - 4$ **43.** $y = -2x + 4$
45. $y = 3$ **47.** $x = -1$ **49.** $y = \frac{3}{2}x$
51. $y = \frac{7}{2}x - 7$
53. (a) $P = 0.13t - 257$
    (b) 3.65 billion

55. Parallel **57.** Neither
59. Perpendicular **61.** Parallel
63. $C = 0.12n + 29$; $m = 0.12$
65. $B = 0.825n + 23$; $m = 0.825$
67. $P = 30x - 340$; if $x = 200$, $P = \$5,660$
69. Test B score: 85
71. $V = -1,000E + 105,000$; $E = 90$ when $V = 15,000$

Mini-Review

75. -6 **76.** $x \geq -1$ **77.** $\{x \mid x \neq \frac{5}{4}\}$
78. $y = \frac{8x - 12}{3}$

Exercises 4.3

1. $x = 5, y = 2$ **3.** $x = 4, y = 3$
5. $x = 1, y = 3$ **7.** $x = 6, y = 2$
9. $x = 5, y = -4$ **11.** Dependent
13. Inconsistent **15.** $x = 5, y = 0$
17. $x = -1, y = 1$
19. $a = \frac{13}{27}, b = \frac{17}{27}$ **21.** $s = \frac{5}{2}, t = \frac{5}{2}$
23. $m = \frac{8}{7}, n = -\frac{16}{7}$
25. $p = 3, q = 1$ **27.** $u = \frac{27}{5}, v = \frac{4}{5}$
29. $w = 12, z = 6$ **31.** $x = 3, y = 2$
33. $x = 2, y = 6$ **35.** $x = 3, y = 7$
37. $x = \frac{1}{2}, y = 2$
39. Let $x =$ amount invested at 5% and $y =$ amount at 8%; $x + y = 14,000, 0.05x + 0.08y = 835$; $9,500 at 5% and $4,500 at 8%
41. Let $x =$ amount at 8% and $y =$ amount at 10%; $x + y = 10,000, 0.08x = 0.10y$; $5,555.56 at 8% and $4,444.44 at 10%
43. Let $x =$ amount invested at 6% and $y =$ amount invested at 4%; $.06x + .04y = 540, .04x + .06y = 510$; $x = \$6,000$; total $= \$10,500$
45. Let $L =$ length and $W =$ width; $L = W + 2, 2L + 2W = 36$; 8 cm by 10 cm
47. Let $T =$ cost of 35-mm roll and $M =$ cost of movie roll; $5T + 3M = 35.60, 3T + 5M = 43.60$; cost of movie roll $= \$6.95$ each; cost of 35-mm roll $= \$2.95$ each
49. Let $c =$ cost of cream-filled donut and $j =$ cost of jelly donut; $7c + 5j = 6.33, 4c + 8j = 6.08$; single cream donut $= 56¢$; single jelly donut $= 48¢$
51. Let $E =$ # of expensive models and $L =$ # of less expensive models; $6E + 5L = 730, 3E + 2L = 340$; 80 of the more expensive type and 50 of the less expensive type

53. Let $y =$ cost of the plan, $x =$ total annual medical expenses; $y = 1,680 + 0.10x, y = 1,200 + 0.25x$; plans are equal at $3,200 in annual medical expenses.
55. Let $C =$ cost of rental, $n =$ number of miles driven; $C = 29 + 0.12n, C = 22 + 0.15n$; costs are equal if 233.3 miles are driven.
57. The break-even point is 14,400 units
59. Let $C =$ cost per month, $n =$ number of calls per month; $C = 18 + 0.11n, C = 24 + 0.09(n - 50)$; plans are equal at 75 calls per month.
61. Let $f =$ # of five-dollar bills and $t =$ # of ten-dollar bills; $f + t = 43, 5f + 10t = 340$; eighteen $5 bills and twenty-five $10 bills
63. Let $f =$ flat rate and $r =$ mileage rate; $f + 85r = 44.30, f + 125r = 51.50$; $29 flat rate and 18¢ per mile
65. Let $x =$ speed of plane and $y =$ wind speed; $6(x + y) = 2,310, 6(x - y) = 1,530$; speed of plane $= 320$ mph, speed of wind $= 65$ mph
67. $\frac{y - 3}{x - 2} = -1; \frac{y + 2}{x - 1} = 2$; $x = 3, y = 2$
69. 1994

Mini-Review

73. $x = 8$ **74.** $x = 24$
75. $f(-4) = 57; f(a) = 3a^2 - 2a + 1$

Exercises 4.4

1.

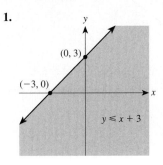

3.

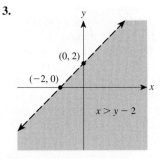

5.

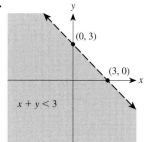

7.

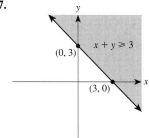

9.

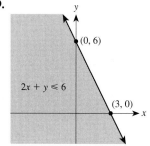

11.

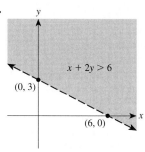

13.

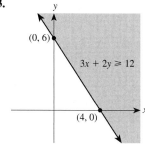

15.

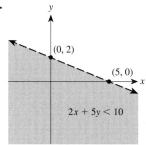

17.

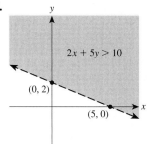

19.

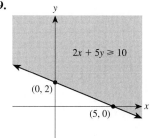

21.

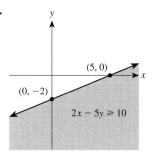

23.

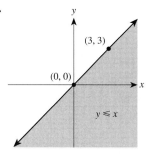

25.

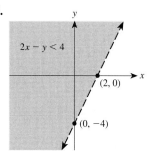

27.

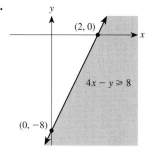

29.

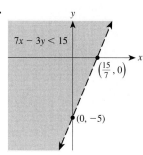

31.

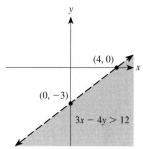

33.

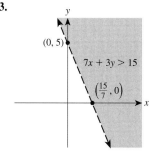

35.

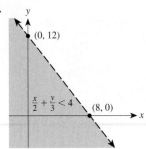

37.

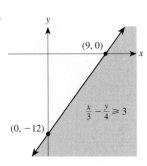

39.

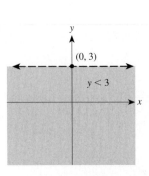

41.

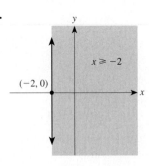

43.

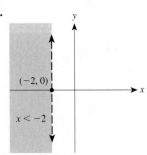

45.

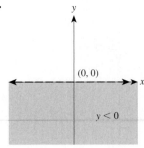

47.

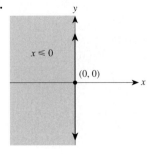

49.

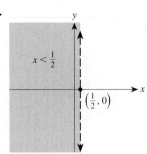

51.

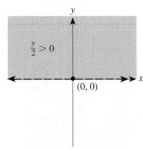

53. $2x + 2y \leq 80$;

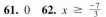

55. $20s + 28p \leq 200$;

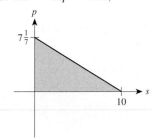

57. $12r + 8a \leq 2{,}500$;

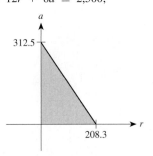

59. Let $r =$ # of regular clerks,
   $s =$ # of special service clerks;
   $15r + 10s \geq 90$;

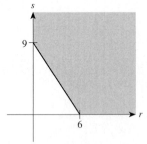

Mini-Review

61. 0   **62.** $x \geq \frac{-7}{3}$
63. Distributive property
64. $x = 0$ or $x = 5$
65. Let $t =$ the time Bobby was
   running; $6t = 5(t + 1)$; it takes
   Bobby 5 hours
66. Let $t =$ the number of minutes it
   takes for both to complete the job;
   $32t + 40t = 3{,}960$; it takes
   55 minutes

Chapter 4 Review Exercises

1. $-\frac{1}{2}$ **2.** $-\frac{3}{5}$ **3.** 3 **4.** $\frac{2}{3}$ **5.** $\frac{3}{4}$ **6.** $\frac{1}{3}$
7. $\frac{1}{2}$ **8.** $-\frac{4}{3}$ **9.** 0 **10.** -2 **11.** 0
12. Undefined **13.** 3 **14.** $-\frac{1}{5}$
15. $-\frac{3}{5}$ **16.** $\frac{3}{2}$ **17.** Undefined
18. 0 **19.** $a = 14$ **20.** $a = 0$
21. $a = \frac{21}{10}$ **22.** $a = \frac{25}{7}$

23. $y = -\frac{7}{3}x - \frac{5}{3}$ **24.** $y = 2x - 2$

25. $y = \frac{5}{3}x$ **26.** $y = -3x$

27. $y = \frac{2}{5}x + \frac{21}{5}$ **28.** $y = -4x + 4$

29. $y = 5x - 13$ **30.** $y = -\frac{3}{4}x + \frac{41}{4}$

31. $y = 5x + 3$ **32.** $y = -3x - 4$

33. $y = 3$ **34.** $y = -4$ **35.** $y = \frac{3}{2}x$

36. $y = \frac{1}{5}x + \frac{3}{5}$ **37.** $y = -\frac{2}{5}x + 6$

38. $y = \frac{6}{7}x + \frac{5}{7}$ **39.** $y = \frac{5}{3}x$

40. $y = -\frac{3}{5}x$ **41.** $y = \frac{5}{3}x - 5$

42. $y = -\frac{8}{5}x + 8$ **43.** $y = \frac{3}{2}x + \frac{3}{5}$

44. $y = -\frac{2}{5}x - 2$

45. $P = 160x - 28{,}000$;
$P = \$36{,}000$ when $x = 400$
gadgets

46. Jake's score on test B: 86;
Charles' score on test A: 37

47. (a) $P = 16.6t - 32{,}440$
(b) 826,000 physicians

48. (a) $D = 2.3t - 4{,}432$
(b) 177,200 dentists

49. $x = 5,\ y = 1$ **50.** $x = 3,\ y = -2$

51. $x = 8,\ y = 0$ **52.** $x = \frac{1}{2},\ y = \frac{1}{3}$

53. Dependent **54.** Inconsistent

55. $x = -\frac{1}{5},\ y = \frac{7}{5}$

56. $s = \frac{64}{23},\ t = -\frac{2}{23}$

57. Let $\$x$ be deposited at 4.75% and
$\$y$ be deposited at 6.65%;
$0.0475x + 0.0665y = 512.05$,
$x + y = 8{,}500$; \$5,700 at 6.65%
and \$2,800 at 4.75%

58. Let price per lb of bread $= \$x$
and price per lb of cookies $= \$y$;
$3x + 5y = 22.02,\ 2x + 3y = 13.43$; $x = \$1.09$/lb, $y = \$3.75$/lb

59.

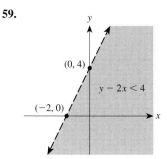

60.

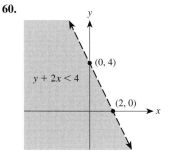

61.

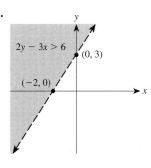

62.

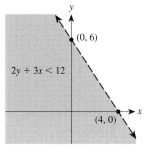

63.

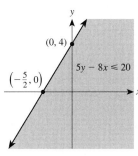

64.

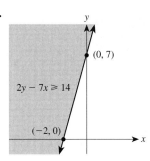

65.

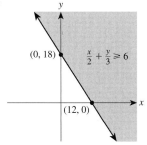

66.

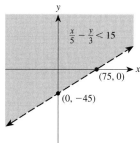

67.

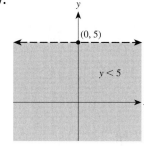

68.

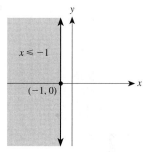

69. $2x + 2y > 100$;

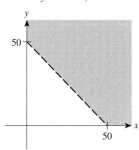

70. $9t + 6b \le 72$;

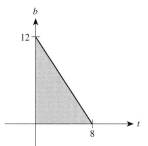

Chapter 4 Practice Test

1. (a)

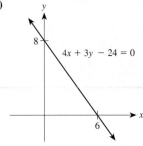

$4x + 3y - 24 = 0$

(b)

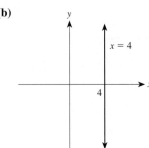

$x = 4$

(c)

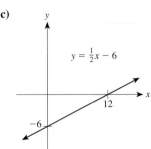

$y = \frac{1}{2}x - 6$

2. (a) 1 **(b)** $\frac{3}{2}$ **3.** $a = \frac{1}{2}$
4. (a) $y = 8x - 19$
 (b) $y = -4x + 4$
 (c) $y = x + 3$
 (d) $y = -\frac{x}{3} - \frac{7}{3}$
 (e) $y = 3x - 9$
 (f) $y = -1$
5. 165 on test B **6.** $y = \frac{1}{2}x + \frac{3}{2}$
7. (a) $x = 8, y = -5$
 (b) $a = -3, b = 6$
8. Let x = amount invested at $8\frac{1}{2}\%$
 and y = amount invested at 9%;
 $x + y = 3,500, 0.085x + 0.09y$
 $= 309$; $1,200 at $8\frac{1}{2}\%$, $2,300 at 9%
9.

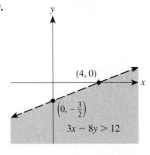

$(4, 0)$
$\left(0, -\frac{3}{2}\right)$
$3x - 8y > 12$

10. $450c + 525s \leq 9,200$

Exercises 5.1

1. 2 terms; degree 2; 1 variable;
 coefficients are 5 and 4
3. 1 term; degree 0; no variables;
 coefficient is 59
5. Not a polynomial
7. 1 term; degree 11; 3 variables;
 coefficient is -4
9. Not a polynomial
11. 4 terms; degree 3; 1 variable;
 coefficients are 4, -1, -2, and 1
13. 1970: 31.4; 1975: 35.9; 1980: 41.2;
 1985: 47.0; 1990: 53.5; 1995: 60.7;
 2000: 68.5; 2002: 71.8
15. (a) $R(d) = 4,300d - 10d^3$
 (b) $R(8) = 29,280; R(10) =$
 $33,000; R(12) = 34,320;$
 $R(14) = 32,760$
17. (a) $R(n) = 2n + 0.45n^2 - 0.001n^3$
 (b) $R(100) = 3,700; R(200) =$
 $10,400; R(300) = 14,100$ at
 prices $37, $52, and $47,
 respectively
19. $S(w) = 14w^2 - 48w;$
 $S(20) = 4,640$ sq in.;
 $S(26) = 8,216$ sq in.;
 $S(42) = 22,680$ sq in.
21. (a) $(0, 0), (5, 8.7), (10, 21.8),$
 $(15, 24.5), (20, 1.6), (25, 0),$
 $(30, 0)$
 (b) Approximately 13 minutes
 after the medication is taken

Mini-Review

23. 4 **24.** -6
25. First side is 34 in.;
 second side is 17 in.
26. 4.5 hr

Exercises 5.2

1. $5x^2 - 9x + 9$
3. $-x^2 - 3xy - 4y^2 + 3x - 2$
5. $6a^2 - 5ab + 3b^2$
7. $ab + 5b^2$ **9.** $-5b^2 - ab$
11. $-5y^2 + 3xy - 3y + 4$
13. $-11x^2 - 3xy + 2y^2$
15. $-5a^2 + ab - 4b^2$
17. $3x^4y - 2x^3y^2 + 2x^2y^3$
19. $3x^4y - 2x^3y^2 + 2x^2y^3$
21. $x^2 + 7x + 12$
23. $6x^2 + x - 1$
25. $6x^2 - x - 1$
27. $3a^2 + 2ab - b^2$
29. $4r^2 - 4rs + s^2$ **31.** $4r^2 - s^2$
33. $9x^2 - 4y^2$
35. $9x^2 - 12xy + 4y^2$

37. $ax - bx - ay + by$
39. $4ar + 6br + 6as + 9bs$
41. $2y^4 - y^2 - 3$
43. $5x^3 + 12x^2 - 5x + 12$
45. $27a^3 + 18a^2 - 12a - 5$
47. $a^3 + b^3$
49. $x^2 + y^2 + z^2 - 2xy - 2xz + 2yz$
51. $x^3 + 8$
53. $a^2 + b^2 + 2ab + ac + ad$
 $+ bc + bd$
55. $P = 2w + 2(3w + 5)$
 $= 8w + 10;$
 $A = w(3w + 5) = 3w^2 + 5w$
57. $A = \pi(8 + x)^2 - \pi(8)^2$
 $= \pi x^2 + 16\pi x$ sq. ft
59. $A = (20 + 2x)(60 + 2x)$
 $- (20)(60)$
 $= 4x^2 + 160x$ sq. ft
61. $-3x^2 + 7x - 5$
63. $x^3 + 3x^2 - 8x + 5$
65. $6x^3 - 19x^2 + 19x - 6$
67. $3x^5 - 3x^4 - 4x^3 + 3x^2 + x$
69. $3x^2 - 5x + 6$
71. $3x^2 + 7x + 4$
73. $x^3 - x - 1$

Mini-Review

77. $\frac{19}{7}$
78. 36 30-lb packages;
 24 25-lb packages
79. $-1 \leq x \leq 4$ **80.** No solution

Exercises 5.3

1. $x^2 + 9x + 20$ **3.** $x^2 - 4x - 21$
5. $x^2 - 19x + 88$ **7.** $x^2 - x - 30$
9. $6x - 30$ **11.** -3
13. $2x^2 - 7x - 4$ **15.** $25a^2 - 16$
17. $9z^2 + 30z + 25$
19. $9z^2 - 30z + 25$ **21.** $9z^2 - 25$
23. $15r^3 + 25r^2s + 9rs + 15s^2$
25. $9s^2 - 12sy + 4y^2$
27. $9y^2 + 60yz + 100z^2$
29. $9y^2 - 100z^2$
31. $9a^2 + 18ab + 8b^2$
33. $64a^2 - 16a + 1$
35. $25r^2 - 20rs + 4s^2$
37. $56x^2 - 113x + 56$
39. $25t^2 - 15s^2t + 15st - 9s^3$
41. $9y^6 - 24xy^3 + 16x^2$
43. $9y^6 - 16x^2$
45. $4r^2s^2t^2 - 28rstxyz + 49x^2y^2z^2$
47. $-9x^2 - 5x - 5$ **49.** $-4ab$
51. $-30a^3 - 5a^2 + 10a$
53. $-10a^2 + 8a + 2$ **55.** $3rs^2$
57. $-y^2 + 19y - 7$
59. $125a^3 - 225a^2b + 135ab^2 - 27b^3$

61. $-2x^3 - x^2 + x$
63. $a^2 + 2ab + b^2 - 1$
65. $a^2 - 4ab + 4b^2 + 10az$
$\qquad\qquad - 20bz + 25z^2$
67. $a^2 - 4ab + 4b^2 - 25z^2$
69. $a^2 + 2ab + b^2 - 4x^2 - 4x - 1$
71. $a^{2n} - 9$ **73.** $3x^2 + 10x + 9$
75. $3x^2 - 2x + 10$
77. $12x^2 - 4x + 1$
79. $6x^2 - 4x + 2$
81. $7x^2 - 6x + 11$
83. $V = 4x^3 - 4x^2 + x$
85. $2x^2 - 30$ **87.** $36x + 138$

Mini-Review

93. $x < -1$ or $x > \frac{7}{3}$
94. 3 hours 15 minutes
95. 2.8 in. by 18 in.
96. (a)

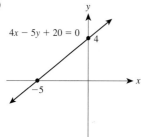

(b)

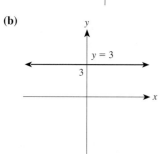

Exercises 5.4

1. $2x(2x + 1)$ **3.** $x(x + 1)$
5. $3xy^2(1 - 2xy)$
7. $3x^2(2x^2 + 3x - 7)$
9. $5x^2y^3z(7x^2y - 3xy^2 + 2z)$
11. $6r^2s^3(4rs - 3rs^2 - 1)$
13. $7ab(5ab^2 - 3a^2b + 1)$
15. $(3x + 5)(x + 2)$
17. $(3x - 5)(x + 2)$
19. $(2x + 5y)(x + 3y)$
21. $(x + 4y)(3x - 5y)$
23. $2(r + 3)(a - 2)$
25. $2(a - 3)^2(x + 1)$
27. $4a(b - 4)(4ab - 16a - 1)$
29. $(2x + 3)(x - 4)$
31. $(3x + 5y)(x - 4y)$
33. $(7x + 3y)(a - b)$
35. $(7x - 3y)(a + b)$

37. $(7x - 3y)(a - b)$
39. Not factorable
41. $(2r - s)(r + s)$
43. Not factorable
45. $(5a - 2b)(a - b)$
47. $(3a - 1)(a - 2)$
49. $(3a + 1)(a - 2)$
51. $(3a - 1)(a + 2)$
53. $(a + 2)(a^2 + 4)$

Mini-Review

58. $a = \frac{1}{2}$
59. $1 \le a \le 5$ **60.** $a \le 1$ or $a \ge 5$
61.

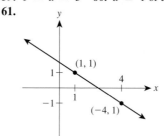

Exercises 5.5

1. $(x + 9)(x - 5)$
3. $(y - 5)(y + 2)$
5. $(x - 5y)(x + 3y)$
7. $(r + 9)(r - 9)$
9. $(x - 3y)(x + 2y)$
11. Not factorable
13. $(r - 3s)(r - 4s)$
15. $(r - 4s)(r + 3s)$
17. $(3x + 7y)(3x - 7y)$
19. $(5x + 4)(3x + 1)$
21. $(5x + 3y)(3x - 2y)$
23. $(5x - 3y)(3x + 2y)$
25. $(3x - 5y)(x + 5y)$
27. $(5a - 2b)(2a + 5b)$
29. $(5 + y)(5 - 2y)$
31. Not factorable
33. $2x(x + 4)(x - 2)$
35. $x(x + 7)(x - 4)$
37. $(x - 2)(3x + 1)$
39. $3ab(2 - 3x)(3 + 2x)$
41. $6y^2(5y - 3)(3y - 2)$
43. $(2r^2 + 1)(3r^2 - 2)$
45. $3(x + 3)(x - 3)$
47. $2xy(10y^4 + xy^2 - 4x^2)$
49. $3xy(6x - y^2)(6x + 5y^2)$
51. $ab(3a^3 - 2)(4a^3 + 3)$
53. $(5a^4 - 2)(3a^4 + 5)$
55. $(4x + 3y)(4x - 3y)$
57. $(x + 2y)^2$ **59.** $(x - 2y)^2$
61. Not factorable
63. $(3xy - 1)(2xy - 1)$
65. $(9rs - 4)(9rs + 4)$

67. $(3xy + 2z)(3xy - 2z)$
69. $a^2(5 - 2b)(5 + 2b)$
71. $(1 - 7a)^2$
73. $(2a - b)(4a^2 + 2ab + b^2)$
75. $(x + 5y)(x^2 - 5xy + 25y^2)$
77. Not factorable
79. $2(2y^2 - 10y + 5)$
81. $3ac(2a + 3c)^2$
83. $(2x^2 + 9y^2)(2x^2 - 9y^2)$
85. $(5a^3 + b)^2$ **87.** $3y^2(4y^4 + 9)$
89. $(a + b + 2)(a + b - 2)$
91. $(a - b + 4)(a - b - 4)$
93. $(x + 3 + r)(x + 3 - r)$
95. $(a + 1)(a + 2)(a - 2)$
97. $(x + 1)(x - 1)(x^2 + x + 1)$
99. $n(n - 2)(n - 1)$

Mini-Review

110. $m = 11$
111. $y = -6x + 31$
112.

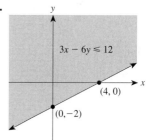

113. Let s = # of pairs of sandals and
F = the weekly temperature in °F;
$s = 2F - 128$; at 90°, 52 pairs
will be sold.

Exercises 5.6

1. $x = -2, 3$ **3.** $y = \frac{1}{2}, 4$
5. $x = \pm 5$ **7.** $x = 0, 4$
9. $x = 6, -2$ **11.** $y = 0, 7$
13. $x = \pm 4$ **15.** $c = \pm \frac{4}{3}$
17. $a = \pm \frac{3}{2}$ **19.** $x = 3, -2$
21. $y = 1, \frac{1}{2}$ **23.** $x = \frac{7}{2}, -4$
25. $x = 0, -3, 2$ **27.** $t = 0, -1$
29. $m = 0, -\frac{1}{2}, \frac{1}{2}$ **31.** $x = \frac{10}{3}$
33. $t = 5$ **35.** $t = -5, 2$
37. $x = -10, -2$ **39.** $x = -\frac{2}{3}, \frac{2}{3}$
41. $x = 3, -2$ **43.** $x = -10$
45. (a) $x = -3, 5$; **(b)** $x = -4, 6$
47. Let x = number;
$x^2 - 5 = 5x + 1$; $x = 6$
49. Let x = number; $x^2 + 64 = 68$;
$x = 2$
51. Let x = number; $(x + 6)^2 = 169$;
$x = -19$ or $x = 7$
53. Let x = number; $x + \frac{1}{x} = \frac{13}{6}$;
$x = \frac{2}{3}$ or $x = \frac{3}{2}$

55. Let x = number;
$x + \frac{2}{x} = 3$; $x = 2$ or $x = 1$
57. (a) Profit = $0 **(b)** Price = $6
59. (a) s = 48 ft **(b)** t = 2 sec
 (c) 64 ft
61. Let x = width; $(x + 2)(x) = 80$;
 length = 10 ft, width = 8 ft
63. Let x = width of walkway;
 $(20 + 2x)(55 + 2x) - (20)(55)$
 $= 400$; width of walkway is $2\frac{1}{2}$ ft

Mini-Review

67. Let h = # of hours it takes 52-mph
 car to catch up; $52h = 40(h + 1)$;
 3 hours 20 minutes
68. $[-\frac{2}{3}, \frac{7}{3})$;

69. $m = \frac{3}{2}$ **70.** $y = \frac{5}{4}x + 5$

Exercises 5.7

Note: R stands for remainder.
 1. $x + 4$ **3.** $a + 2$
 5. $7z - 2, R\ -14$
 7. $a^2 + 2a + 3, R\ -5$
 9. $3x^2 + x - 1, R\ 8$
11. $2z - 3, R\ -6$
13. $x^3 + 3x^2 + 2x + 1$
15. $3y^3 - 2y^2 + 5, R\ -5$
17. $2a^3 - 5, R\ 16$
19. $y^2 - y + 1, R\ -2$
21. $4a^2 + 2a + 1, R\ 2$
23. $2y^3 - 3y^2 - 2y + 2, R\ 3y + 2$
25. $3z^3 - 6z^2 + 11z - 25,$
 $R\ 73z - 35$
27. $7x^4 - 3, R\ x - 5$
29. $x = -2, \frac{3}{2}, 2$ **31.** $x = -\frac{1}{2}, 1, 8$

Mini-Review

33. $f(4) = 70$;
 $f(x^2) = 5x^4 - 3x^2 + 2$

Chapter 5 Review Exercises

 1. $2x^3 - 7x^2 + 3x - 5$
 2. $6x^2 - 8x + 15$
 3. $6x^3y - 9x^2y + 3x^2$
 4. $15r^3s^3 - 10r^2s^4 - 15rs^2$
 5. $-ab^2$ **6.** $-4x^2y - 11xy^2 + 18xy$
 7. $-2x + 16$ **8.** $21x - 28$
 9. $x^2 - x - 6$ **10.** $x^2 - 13x + 40$
11. $6y^2 - 7y + 2$ **12.** $25y^2 - 9a^2$

13. $9x^2 - 24xy + 16y^2$
14. $5a^2 - 7ab - 6b^2$
15. $16x^4 - 40x^2y + 25y^2$
16. $16a^2 - 9b^4$ **17.** $49x^4 - 25y^6$
18. $6x^3 - 19x^2 + 12x - 5$
19. $10y^3 - 21y^2 + 23y - 21$
20. $21x^3 - 5x^2y - 12xy^2 + 4y^3$
21. $4x^2 + 12xy + 9y^2 - 2x - 3y - 20$
22. $a^2 + 2ab + b^2 - x^2 + 2xy - y^2$
23. $x^2 - 2xy + y^2 + 10x - 10y + 25$
24. $x^2 - 6x + 9 - y^2$
25. $a^2 + 2ab + b^2 - 8a - 8b + 16$
26. $x^2 + 2xy + y^2 + 10x + 10y + 25$
27. $3xy(2x - 4y + 3)$
28. $5ab^2(3a - 2b^2 + 1)$
29. $(3x - 2)(a + b)$
30. $(y - 1)(5y + 3)$
31. $(a - b)(2a - 2b + 3)$
32. $(a - b)(4a - 4b + 7)$
33. $(x - 1)(5a + 3b)$
34. $(a + b)(2x + 3)$
35. $(5y + 3)(y - 1)$
36. $(2a - b)(7a + 3b)$
37. $2a(a - 5)$ **38.** $t(t + 3)(t - 2)$
39. Not factorable **40.** Not factorable
41. $(x - 7)(x + 5)$
42. $(a + 9)(a - 4)$
43. $(a + 7b)(a - 2b)$
44. $(y^2 + 3x^2)(y^2 - 2x^2)$
45. $(7a + 2b)(5a + b)$
46. $(x - 3y)^2$ **47.** $(a - 3b^2)^2$
48. $2x(x^2 + 3x - 27)$
49. $3a(a - 5)(a - 2)$
50. $5ab(a + 3b)(a - 2b)$
51. $2x(x + 5y)(x - 5y)$
52. $3y(y + 4)^2$
53. $(3x - 2)(2x + 3)$
54. $(5y - 4)(5y + 3)$
55. $(2x + 5y)(4x^2 - 10xy + 25y^2)$
56. $(x - 3)(x^2 + 3x + 9)$
57. $(6a + b)(a - 3b)$
58. $(6x + 5y^2)(2x + y^2)$
59. $(7a^2 + 2b^2)(3a^2 + 5b^2)$
60. $2a(3a - 2b)(9a^2 + 6ab + 4b^2)$
61. $(5x^2 - 4y^2)^2$
62. $3x(2x + 3y)(3x - 2y)$
63. $5xy(2x - 3y)^2$
64. $(2a^2 + 3b^2)(2a^2 - 3b^2)$
65. $2a^2b(3a + 2b)(a - 2b)$
66. $7x^2y(2x + 3y)(2x - 3y)$
67. $2x(x^2 - 2)(3x^2 + 1)$
68. $5xy(3x^2 - y)(2x^2 - 5y)$
69. $(a - b + 2)(a - b - 2)$
70. $(3x + 5y + 1)(3x - 5y - 5)$
71. $(3y + 5 + 3x)(3y + 5 - 3x)$
72. $(5x^2 + y + 4)(5x^2 - y - 4)$
73. $3x^2 + 2x + 11, R\ 17$
74. $x^3 - x^2 + x - 2, R\ 1$
75. $4a^2 + 6a + 9$ **76.** $y^4 + y, R\ 1$
77. $x = 0, -5, \frac{4}{3}$ **78.** $x = -2, 10$
79. $x = 0, \pm5$ **80.** $x = 0, -12, 2$
81. $x = -\frac{5}{3}, -2$ **82.** $x = -3, x = -10$

83. $6x^3 - 17x^2 + 14x - 3$
84. $2x^2 - 2x + 2$ **85.** $2x^2 - x$
86. $3x - 6$ **87.** $x = -3, -\frac{1}{2}, 3$
88. $x = -5, -2, 1$
89. $\frac{1}{6}n(2n + 1)(n + 1)$
90. $\frac{1}{4}n^2(n + 1)^2$
91. (a) $R(c) = 20,000c - 1,000c^3$
 (b) $R(1) = \$19,000$;
  $R(2) = \$32,000$;
  $R(4) = \$16,000$
92. (a) $6,000 **(b)** $4 or $7

Chapter 5 Practice Test

 1. $27a^3 - 8b^3$ **2.** $6x - 18$
 3. $x^2 - 2xy + y^2 - 4$
 4. $9x^2 + 6xy^2 + y^4$
 5. $2xy(5x^2y - 3xy^2 + 1)$
 6. $(3a - b)(2x + 5)$
 7. $(x + 15y)(x - 4y)$
 8. $(5x + 2)(2x - 3)$
 9. Not factorable
10. $3ab(a + b)(a - b)$
11. $rs(r - 5s)^2$
12. $(2a - 1)(4a^2 + 2a + 1)$
13. $(2x + y + 8)(2x + y - 8)$
14. $(a + 4)(a + 1)(a - 1)$
15. $x^2 + 3x + 16, R\ 40$
16. (a) $x = 0, -6, 2$ **(b)** $x = 1, 5$
17. (a) 48 ft **(b)** 3 seconds
18. (a) $f(5) = 12$;
 (b) $f(x + 5) = x^2 + 7x + 12$

Exercises 6.1

 1. $x \neq 2$
 3. x and y can be any number
 5. $x \neq 0$ **7.** $3x \neq y$
 9. $\{x \mid x \neq 8\}$
11. $\left\{x \mid x \neq \frac{8}{5}\right\}$
13. $\{x \mid x \neq 5, -5\}$
15. $\{x \mid x \neq -2, 3\}$
17. (a) Let w = width. $L(w) = \frac{20}{w}$
 (b) $L(0.5) = 40; L(1) = 20$;
  $L(2) = 10; L(4) = 5$;
  $L(5) = 4; L(10) = 2$;
  $L(20) = 1$.
 The larger the width, the
 smaller the length.
19. (a) $A(5) = \$1,600.40$;
  $A(10) = \$800.40$;
  $A(50) = \$160.40$;
  $A(100) = \$80.40$;
  $A(200) = \$40.40$;
  $A(300) = \$27.07$;
 The larger the width, the smaller
 the length.

(b)

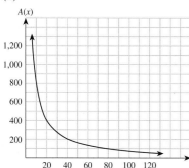

21. (a) $n(1) = 5; n(10) = 33;$
$n(50) = 71; n(100) = 83$

(b)

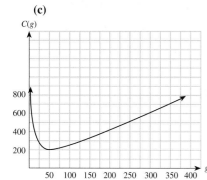

23. (a) $g > 0$
(b) $C(1) = \$5,002;$
$C(5) = \$1,010;$
$C(10) = \$520;$
$C(20) = \$290;$
$C(40) = \$205;$
$C(100) = \$250;$
$C(200) = \$425;$
$C(300) = \$616.67$

(c)

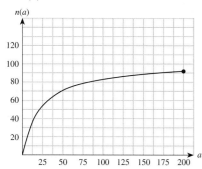

Exercises 6.2

1. $\dfrac{x}{5}$ **3.** $\dfrac{8x}{9ya^2}$ **5.** $\dfrac{x - 5}{x - 3}$

7. $\dfrac{(2x - 3)(x - 5)}{(2x + 3)(x + 3)}$ **9.** -1

---

11. $\dfrac{x - 1}{2(x + 3)}$ **13.** $\dfrac{x + 3}{x - 3}$

15. $\dfrac{(w - 7z)(w - z)}{(w + 7z)(w + z)}$ **17.** $x + 4$

19. $-(x + 4)$ **21.** $\dfrac{a - 6}{5a + 1}$

23. $\dfrac{2(x - 3)(x + 2)}{3(x - 2)}$ **25.** $\dfrac{2x^2}{3(x^2 - 2)}$

27. $-\dfrac{3a + 2}{1 + 2a}$ **29.** -1

31. $\dfrac{a^2 - ab + b^2}{(a + b)^2}$ **33.** $-\dfrac{a + b}{2x}$

35. $a + b - x - y$ **37.** $6x^3$
39. $25a^2b$ **41.** $3x(x + 5)$
43. $(a - b)(x + y)$
45. $-(y - 2)(y + 4)$
47. $(x - y)(x + y)$

Mini-Review

51. $x = 40$ **52.** $y = -\frac{5}{4}x - \frac{13}{4}$
53. $2x^6 - x^4 + 3$

Exercises 6.3

1. $\dfrac{6y}{a^2b^6}$ **3.** $\dfrac{b^3}{4}$ **5.** $\dfrac{128x^3ab^4}{3}$

7. $\dfrac{12r^2s^5}{a^2}$ **9.** $\dfrac{16r^2s}{27a^2b^2}$

11. $\dfrac{2(x + 1)^2(x - 2)}{3(x + 3)^2}$ **13.** $\dfrac{3(x - 2)}{2}$

15. $\dfrac{5a(a - b)}{3}$ **17.** $\dfrac{5a(a - b)}{3(a + b)^2}$

19. -1 **21.** $\dfrac{2a + 3}{a + 1}$ **23.** $\dfrac{1}{x - y}$

25. $3x$ **27.** $\dfrac{2(1 - x)}{(x + 1)^2}$ **29.** 1

31. $\dfrac{1}{5m(m + 5)}$ **33.** $\dfrac{2t + 1}{2t + 1}$

35. $\dfrac{1}{3a + 1}$ **37.** 1 **39.** $\dfrac{-1}{xy(x + y)^2}$

41. $2(x + 2)^2$ **43.** $\dfrac{1}{2(x - 3)^2}$

45. $\dfrac{2(x - 3)^2}{3x(x + 2)}$

Mini-Review

47. $x = \frac{1}{2}, y = -2$
48. $x^2 + 8x - 15$
49. Degree is 3
50. $x = -7, -\frac{1}{5}$

Exercises 6.4

1. 2 **3.** 10 **5.** $\dfrac{1}{a + 1}$

7. $\dfrac{x - y}{x + y}$ **9.** $\dfrac{12(x + 1)}{x + 3}$

11. $\dfrac{27x - 8x^2y}{18y^2}$ **13.** $\dfrac{21y^3 - 16x^4}{18x^2y^2}$

15. $\dfrac{28}{27y}$ **17.** $\dfrac{252a^2c^2 + 168b - 3bc^2}{7b^2c^3}$

19. $\dfrac{x^2 + 3x + 6}{x(x + 2)}$ **21.** $\dfrac{a^2 - ab + b^2}{a(a - b)}$

23. $\dfrac{3x - 29}{(x + 7)(x - 3)}$

25. $\dfrac{3(x^2 - 6x - 1)}{(x - 7)(x + 2)}$

27. $\dfrac{r^2 + 6rs - s^2}{(r - s)(r + s)}$ **29.** -1

31. 1 **33.** $\dfrac{a^2 + ab - b}{a^2 - b^2}$

35. $\dfrac{a(a + b)}{b}$

37. $\dfrac{y^3 + 10y^2 + 19y + 3}{(y + 2)(y + 3)}$

39. $\dfrac{5x + 3}{x + 2}$ **41.** $\dfrac{5x + 1}{2}$

43. $\dfrac{2x^2 - 4x - 3}{x(x - 3)^2}$

45. $\dfrac{2a^2 + 4a + 9}{(a + 4)(a - 3)}$ **47.** $\dfrac{-1}{y - 3}$

49. $\dfrac{(3x + 4)(x - 2)}{(x - 5)(x + 2)}$

51. $\dfrac{2a^2 - 15a - 2}{(a - 2)(a + 3)(a - 5)}$

53. $\dfrac{75a^3 + 51a^2 + 2a - 3}{(3a - 1)(5a + 2)}$

55. $2 + x$ **57.** $\dfrac{1}{r - s}$

59. $\dfrac{3s^2 - 3st + 2s + t}{s^3 - t^3}$

61. (a) $\dfrac{x + 4}{4}$ **(b)** $\dfrac{x}{4} + 1$

63. (a) $3x - 2y$ **(b)** $3x - 2y$

65. (a) $\dfrac{6m - 4m^2n - 9}{15n}$

(b) $\dfrac{2m}{5n} - \dfrac{4m^2}{15} - \dfrac{3}{5n}$

67. $\dfrac{x^2 - 2x + 4}{(x + 2)(x - 2)}$ **69.** $\dfrac{x}{x + 2}$

71. $\dfrac{x^2 + 4x}{x^2 - 4}$ **73.** $\dfrac{x + 2}{2}$

75. $\dfrac{-3h}{x(x + h)}$

77. $\dfrac{-3h}{(x + h - 3)(x - 3)}$

Mini-Review

80. $x = 48$ **81.** $m = -\dfrac{4}{9}$ **82.** $-12x$

83.

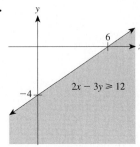

$2x - 3y \geq 12$

Exercises 6.5

1. $\dfrac{x}{5y}$ **3.** $\dfrac{9}{(x - y)^2}$ **5.** $\dfrac{a}{3a + 1}$

7. $\dfrac{x^3 + 2}{x(2x + 1)}$ **9.** $x + 2$

11. $\dfrac{1}{y + 8}$ **13.** $z(z - 2)$

15. $\dfrac{2x - 3y}{2x + 3y}$ **17.** $\dfrac{2x - 5y}{y(3x - y)}$

19. $\dfrac{x - 5}{4x}$ **21.** $\dfrac{3t - 5}{2t}$

23. $\dfrac{1}{x(x - 3)}$ **25.** $\dfrac{-h}{h^2 - 1}$

27. $\dfrac{6x - 2}{11x - 5}$ **29.** $-\dfrac{4}{x(x + 3)}$

31. $-\dfrac{4}{7}$ **33.** $-\dfrac{x}{2}$

35. $-\dfrac{6}{x(x - 3)}$ **37.** $\dfrac{2}{x(x - 2)}$

39. $\dfrac{2x^2 - x + 3}{(x - 3)(x + 2)}$ **41.** $\dfrac{x - 1}{x + 2}$

43. $\dfrac{x^2 + x + 6}{3x^2 - 9x}$

Mini-Review

45. $x = 5$ **46.** $x = -5, 2$

47. $y = \dfrac{3}{2}x + 11$

48. $4x - 20$

Exercises 6.6

1. $x = 12$ **3.** $t < -30$ **5.** $a = 4$
7. $y = 7$ **9.** $y \leq 17$ **11.** $x = -17$
13. $x \leq -17$ **15.** $x = -3$
17. $x > -\dfrac{177}{11}$ **19.** $x = 4$

21. $\dfrac{75 - 2x}{10x}$ **23.** $t = \dfrac{3}{4}$ **25.** $a = 5$

27. No solution **29.** $y = \dfrac{38}{13}$

31. No solution **33.** $\dfrac{3y^2 + y + 6}{6y(y + 3)}$

35. $x = 2$ **37.** $x = 1$ **39.** $x = \dfrac{22}{7}$

41. $\dfrac{13(x + 1)}{(x + 2)(x - 2)}$ **43.** $x = 3$

45. No solution

47. $\dfrac{3n^2 - 8n + 2}{(3n + 2)(3n - 2)}$

49. $\left\{n \mid n \neq \dfrac{4}{3}, -\dfrac{4}{3}\right\}$ **51.** $x = -\dfrac{5}{8}$

53. $x = 2, 3$ **55.** $x = 0, 3$
57. $x = -1, 2$ **59.** $x = -1, \dfrac{5}{2}$

61. No solution **63.** $x = -\dfrac{7}{2}$
65. $x = -\dfrac{22}{3}$ **67.** $x = 0, 3$

69. $x = \dfrac{5}{18}$ **71.** $n = 3143$

73. $\dfrac{x^2 + 6x - 9}{(x - 2)^2}$

75. $\dfrac{-3x^2 - 3x - 32}{(x - 2)^2}$

77. $\dfrac{3x + 1}{(x - 2)(x + 5)}$

Mini-Review

79. $6x^5y^3$ **80.** $6x^5y + 3x^2y^3$

81. $f(x) = x^2$ **82.** $\left\{x \mid x > \dfrac{5}{3}\right\}$

Exercises 6.7

1. $x = \dfrac{4 - 7y}{5}$ **3.** $y = \dfrac{2x - 11}{9}$

5. $w = 5z - 4$ **7.** $r > \dfrac{15}{2}t$

9. $n = \dfrac{3m + 4p + 8}{9}$

11. $a = \dfrac{-5b}{9}$ **13.** $x = \dfrac{21}{19}y$

15. $x = \dfrac{d - b}{a - c}$

17. $x = \dfrac{by - 2y + 6}{3 - a}$

19. $x = \dfrac{a}{y + 7} - 3 = \dfrac{a - 3y - 21}{y + 7}$

21. $u = \dfrac{-y - 1}{y - 1}$

23. $t = \dfrac{5x - 3}{3x - 2}$ **25.** $b = \dfrac{2A}{h}$

27. $b_1 = \dfrac{2A}{h} - b_2$ **29.** $r = \dfrac{A - P}{Pt}$

31. $F = \dfrac{9}{5}C + 32$ **33.** $P_2 = \dfrac{P_1V_2}{V_1}$

35. $g = \dfrac{2}{t^2}(S - s_0 - v_0t)$

37. $x < \mu + 1.96s$

39. $f_1 = \dfrac{ff_2}{f_2 - f}$

41. $h = \dfrac{S - 2\pi r^2}{2\pi r}$

Mini-Review

43. $x = 1, 5$
44. $x^3 - 6x^2 + 12x - 8$
45. 4 heavy-duty, 14 regular
46. Domain = $\{x \mid -4 \leq x \leq 3\}$;
range = $\{y \mid -3 \leq y \leq 2\}$;
$f(-3) = 2$

Exercises 6.8

1. Let x = number; $\dfrac{3}{4}x = \dfrac{2}{5}x - 7$;
$x = -20$

3. Let x = # of men; $\dfrac{7}{9} = \dfrac{x}{810}$;
630 men

5. Let x = the larger number;
$\dfrac{5}{12} = \dfrac{x - 21}{x}$; two numbers are
36 and 15

7. Let x = # inches in 52 cm;
$\dfrac{1}{2.54} = \dfrac{x}{52}$; 20.47 in.

9. Let x = # of euros;
$\dfrac{1}{1.24} = \dfrac{x}{312}$; 251.61 euros

11. Let x = # of dribbles;
$\dfrac{x}{28} = \dfrac{25}{7} \cdot \dfrac{4}{5}$; 80 dribbles

13. Let x = second side;
$P = 22 = \dfrac{1}{2}x + x + x + 2$;
lengths are 4, 8, and 10 cm

15. Let x = width; $50 = 2(2.5x + x)$;
$x = \dfrac{50}{7}$ cm (width), length = $\dfrac{125}{7}$ cm

17. Let x = distance from home;
$\dfrac{3x}{4} = 6$; 8 miles

19. Let x = distance from home to
school; $x - \dfrac{1}{5}x - \dfrac{1}{4}\left(\dfrac{4}{5}x\right) = 2$;
$3\dfrac{1}{3}$ blocks

21. $\dfrac{4}{5} = \dfrac{1}{2} + \dfrac{1}{5} + \dfrac{1}{R_3}$; R_3 = 10 ohms

23. Let $x be invested in bond; $0.10x + 0.06(x + 3,000) = 580$; $2,500 in bond; $5,500 in CD

25. Let $x be invested at $5\frac{1}{2}\%$; $0.055x + 0.07(25,000 - x) = 1,465$; $19,000 at $5\frac{1}{2}\%$, $6,000 at 7%

27. Let $x be invested at $8\frac{1}{2}\%$; $0.085x + 0.11(18,000 - x) = 0.10(18,000)$; $7,200 at $8\frac{1}{2}\%$, $10,800 at 11%

29. Let $x =$ number of hours working together; $\frac{x}{3} + \frac{x}{5} = 1$; $1\frac{7}{8}$ hours

31. Let $x =$ number of hours working together; $\frac{x}{30} + \frac{x}{20} = 1$; 12 hours

33. Let $x =$ time working together; $\frac{x}{2\frac{2}{3}} + \frac{x}{5} = 1$; $1\frac{17}{23}$ days

35. Let $x =$ # hours for the Super-Quickie Service to finish cleaning; $\frac{10}{30} + \frac{x}{20} = 1$; $x = \frac{40}{3}$; Total time $= 23\frac{1}{3}$ hours

37. Let $x =$ number of minutes to overflow; $\frac{x}{10} - \frac{x}{15} = 1$; 30 minutes

39. Let $x =$ # hours they work together; $\frac{3}{10} + \frac{x}{10} + \frac{x}{8} = 1$; $x = 3.11$, at 6:07 P.M.

41. Let $x =$ the number of hours it takes Megan to do the job herself; $\frac{1}{6} + \frac{2}{6} + \frac{2}{x} = 1$; 4 hours

43. Let $x =$ Bill's rate; $\frac{10}{x} = \frac{15}{x + 10}$; 20 kph

45. Let $x =$ number of miles at slow speed; $\frac{600 - x}{50} + \frac{x}{20} = 14$; $66\frac{2}{3}$ miles

47. Let $x =$ number of ounces of 20% alcohol; $0.20x + 0.50(5) = 0.30(x + 5)$; 10 ounces of 20% alcohol

49. Let $x =$ number of ml of 30% solution; $0.30x + 0.75(80 - x) = 0.50(80)$; $44\frac{4}{9} \approx 44.44$ ml of 30% solution, $35\frac{5}{9} \approx 35.56$ ml of 75% solution

51. Let $x =$ number of tons of 40% iron; $0.40x + 0.60(80 - x) = 0.55(80)$; $x = 20$ tons of 40% iron; 60 tons of 60% iron

53. Let $x =$ number of liters of pure alcohol; $0.60(2) + 1(x) = 0.80(x + 2)$; 2 liters of pure alcohol

55. Let $x =$ number of gallons drained off $=$ water added; $0.30(3 - x) + 0(x) = 0.20(3)$; 1 gallon

57. Let $x =$ number of advance tickets sold; $25x + 30.50(3,600 - x) = 97,700$; 2,200 advance tickets

59. Let $x =$ number of nickels; $0.05x + 0.10(x + 5) + 0.25(2x) = 20.00$; 30 nickels, 35 dimes, 60 quarters

61. Let $x =$ number of orchestra tickets; $25x + 16(2x) + 20.5(900 - 3x) = 17,325$; 250 orchestra seats, 500 general admission, 150 balcony

63. Let $x =$ score on final exam; $0.20(85) + 0.20(65) + 0.20(72) + 0.40(x) \geq 80$; he needs at least an 89

65. Let $x =$ amount invested in high-risk bond; $0.082x + 0.039(20,000 - x) \geq 1,000$; $x \approx 5,116.28$; at least $5,116.28 invested in high-risk bond

67. Let $x =$ score on the final exam; $0.20(85) + 0.20(92) + 0.20(86) + 0.40x \geq 90$; $x \geq 93.5$; at least 93.5

69. (a) 6 ohms

(b) $\frac{6A}{6 - A}$

71. (a) 61 ($x = 60.12$)

(b) $\frac{6,000}{T - 0.2}$

Mini-Review

74. $-5x + 23$ **75.** $3xy(x - y)^2$
76. $(x - a - 3)(x - a + 3)$
77. $-9 \leq x \leq 1$

Chapter 6 Review Exercises

1. $\{x \mid x \neq -5\}$ **2.** $\left\{x \mid x \neq -\frac{4}{3}\right\}$
3. $\left\{x \mid x \neq -\frac{1}{3}, 4\right\}$ **4.** $\left\{x \mid x \neq -7, \frac{1}{2}\right\}$
5. $\frac{x}{4y^2}$ **6.** $\frac{3b^3}{4a^3c^2}$ **7.** $\frac{x + 4}{x + 5}$
8. $\frac{3x + 1}{2x - 3}$ **9.** $x^2 - 2x + 3$
10. $\frac{6a^2b^4 - 8b^3 + 9}{2ab}$
11. $\frac{5x - 7}{3x - 2}$ **12.** $2a - b$ **13.** $\frac{xy^3z^2}{2}$
14. $\frac{2xz^2}{9y^5}$ **15.** $\frac{2}{x^2y}$ **16.** 0

17. $\frac{3(x + 1)}{x - 1}$ **18.** 2   **19.** 2   **20.** 3

21. $\frac{20b^3 - 21a}{12a^2b^4 - 2}$ **22.** $\frac{4y + 35}{15xy^2}$

23. $\frac{-6x^2 + 19x + 5}{10x^2}$

24. $\frac{14x^2 + 3x + 2}{6x^2}$

25. $\frac{2(x + 5)}{x - 5}$ **26.** $\frac{a - 2}{a - 4}$

27. $2x$ **28.** $\frac{a - 3}{2}$

29. $\frac{1}{a - b}$ **30.** $\frac{r + s}{3}$

31. $\frac{3x^2 - 22x - 15}{(2x + 3)(x - 4)}$

32. $\frac{13a^2 + 2a}{(a - 1)(2a + 1)}$

33. $\frac{16x^2 - 44x - 8}{(2x - 7)(2x - 3)}$

34. $\frac{5x^2 - 22x - 11}{(x - 2)(x + 3)}$

35. $\frac{10a^3 + 10a^2 + a - 3}{2a^2(a + 1)(a - 3)}$

36. $\frac{-2x - 17}{(x + 4)(x - 2)(x + 1)}$

37. $\frac{8x + 4}{(x - 2)^2(x + 2)}$

38. $\frac{-2a + 24}{(a + 3)(a - 3)^2}$

39. $\frac{x(x + 3)}{7(x - 3)(x + 1)}$

40. $\frac{5x + 5y - 3}{x^2 - y^2}$

41. $\frac{8x^2 + 2x - 3}{x^2 - 4}$

42. $\frac{-x^2 + 26x + 1}{x^2 - 25}$ **43.** $\frac{1}{y}$ **44.** b^2

45. $\frac{3(x^2 + 5x + 5)}{(x + 3)(x - 2)(x + 1)}$

46. $\frac{x^2 - 9x + 2}{(2x + 3)(x + 5)(x - 3)}$

47. $\frac{5x + 3}{(x - 2)(x + 1)}$

48. $\frac{2r^2 + 2s^2 + 3r - 6s}{(r - 2s)(r + s)(r - s)}$

49. $\frac{3x}{2x - 3}$ **50.** $\frac{3x}{2x + 3}$

51. $\frac{2x + 3y}{2}$ **52.** $\frac{2}{2x + 3y}$

53. $\dfrac{(x^2 + 6)(x + 1)}{2x^2}$ **54.** xy

55. $\dfrac{8axy}{3b}$ **56.** $\dfrac{x}{y}$ **57.** $\dfrac{1}{5}$ **58.** $\dfrac{1}{2 - x}$

59. $\dfrac{(b - 1)(2b + 5)}{b^3 + b + 2}$

60. $\dfrac{z^2 + 2z - 2}{3z}$

61. $\dfrac{x^2 - 2x - 10}{(x - 2)(x - 4)}$

62. $\dfrac{-h}{(x + h - 4)(x - 4)}$

63. $-\dfrac{3}{10}$ **64.** $-\dfrac{18}{7}$ **65.** $x = 2$
66. $x = 5$ **67.** $x > -5$ **68.** $x \geq 5$
69. $x = 4$ **70.** $x = \dfrac{11}{10}$ **71.** $x < -22$
72. $x \geq -\dfrac{28}{3}$ **73.** $x = \dfrac{1}{2}$ **74.** $x = \dfrac{3}{2}$
75. $x = 7$ **76.** $x = 4$ **77.** $x > \dfrac{8}{3}$
78. $x \leq \dfrac{6}{11}$ **79.** No solution
80. No solution **81.** $x = 4$
82. No solution

83. $x = \dfrac{10y}{3}$ **84.** $y = \dfrac{3x}{10}$

85. $y = \dfrac{4}{x}$ **86.** $b = \dfrac{a}{a - 1}$

87. $y = \dfrac{2x + 1}{x}$ **88.** $a = \dfrac{b}{c - x}$

89. $x = \dfrac{dy - b}{a - cy}$ **90.** $s = \dfrac{3 + 2r}{3r - 2}$

91. $b = \dfrac{adc}{ac - cd - ad}$

92. $c = \dfrac{ad}{b + de}$

93. Let x = # of inches in 1 cm;
$\dfrac{2.54 \text{ cm}}{1 \text{ in.}} = \dfrac{1}{x}$, $x = \dfrac{1}{254} \approx 0.39$ in.
94. Let x = # of Democrats;
$\dfrac{4}{3} = \dfrac{x}{1{,}890 - x}$; 1,080 Democrats
95. Let total distance = x;
$\dfrac{1}{2}x + \dfrac{1}{3}\left(\dfrac{1}{2}x\right) + \dfrac{3}{2} = x$; 4.5 miles
96. Let r = Carlos' rate; $\dfrac{3}{r - 5} = \dfrac{5}{r}$;
$12\dfrac{1}{2}$ mph
97. Let x = number of days required
by both; $\dfrac{x}{2\frac{1}{3}} + \dfrac{x}{2\frac{1}{3}} = 1$; $1\dfrac{6}{29}$ days
98. Let x = amount of time to clean
kitchen together; $\dfrac{x}{\frac{1}{3}} + \dfrac{x}{\frac{2}{3}} = 1$;
$\dfrac{2}{9}$ hour
99. Let x = # hours it takes John to
complete the job alone; $\dfrac{2}{6} + \dfrac{3}{x}$
$= 1$; $4\dfrac{1}{2}$ hours
100. Let x = # hours it takes for the
other to complete the job;
$\dfrac{3}{8} + \dfrac{3}{x} = 1$; $4\dfrac{4}{5}$ hours

101. Let x = amount of 35% alcohol;
$0.70(5) + 0.35x = 0.60(x + 5)$;
2 liters
102. Let x = quantity of pure water;
$0(x) + 0.65(3) = 0.30(x + 3)$;
$x = 3.5$ liters
103. Let x = number of children's
tickets; $3.50x + 6.25(980 - x) =$
$4{,}970.00$; 420 children's tickets,
560 adult tickets
104. Let x = number of dimes;
$0.05(4x) + 0.10x$
$+ 0.01(82 - 5x) = 3.32$;
40 nickels, 10 dimes, and 32 pennies
105. (a) $c(0.25) = 3.68$;
$c(0.5) = 6.76$; $c(1) = 10.13$;
$c(2) = 10.13$; $c(4) = 6.76$;
$c(10) = 2.98$

(b)

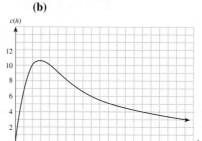

106. (a) 205 items

(b) $\dfrac{2{,}500}{A - 0.3}$ items

Chapter 6 Practice Test

1. $\left\{x \mid x \neq \dfrac{5}{2}\right\}$

2. (a) $\dfrac{3y^3}{8x}$ **(b)** $\dfrac{x + 3}{x - 3}$

(c) $\dfrac{3(2x + 1)}{5x}$

3. (a) $\dfrac{3}{4x^3y^2ab^4}$ **(b)** $\dfrac{4x^3 + 15y}{24x^2y^2}$

(c) $2(r + s)$ **(d)** 2

(e) $\dfrac{x^2 - x + 8}{(x + 2)(x - 2)(x - 3)}$

(f) $\dfrac{x + 3}{x}$

4. $\dfrac{x(1 - 2x)}{(x + 1)(x + 5)}$ **5.** $-\dfrac{4}{17}$

6. (a) $x > \dfrac{-15}{4}$ **(b)** No solution
(c) $x = 2$

7. $x = \dfrac{-y - 2}{2y - 1}$

8. Let x = amount of 30% alcohol
solution; $0.30(x) + 0.45(8) =$
$0.42(x + 8)$; 2 liters

9. Let x = time required working
together; $\dfrac{x}{3\frac{1}{2}} + \dfrac{x}{2} = 1$;
$x = 1\dfrac{3}{11}$ hours
10. 180 items

Chapters 4–6 Cumulative Review

1. (§4.1)

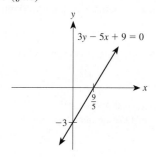

2. (§4.1)

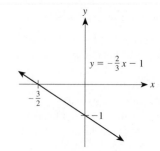

3. (§4.1)

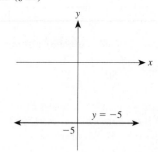

4. (§4.1)

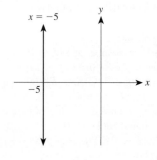

5. (§4.1)

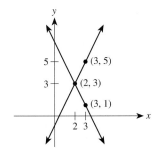

6. (§4.1)

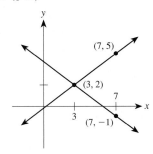

7. $m = 8$ (§4.1)   **8.** $m = -1$ (§4.1)

9. $m = 5$ (§4.1)   **10.** $m = \frac{3}{2}$ (§4.1)

11. $m = \frac{5}{4}$ (§4.1)   **12.** $m = -\frac{2}{5}$ (§4.1)

13. $a = 1$ (§4.1)   **14.** $a = \frac{1}{2}$ (§4.1)

15. $y = 3x + 13$ (§4.2)

16. $y = -2x + 9$ (§4.2)

17. $y = -6x + 19$ (§4.2)

18. $y = 5$ (§4.2)

19. $y = 4x + 2$ (§4.2)

20. $y = -x + 3$ (§4.2)

21. $y = -\frac{3}{5}x - \frac{9}{5}$ (§4.2)

22. $y = -\frac{1}{2}x - 2$ (§4.2)

23. $y = \frac{3}{2}x + 1$ (§4.2)

24. $P = \frac{125}{6}T - \frac{4,025}{3}$; $533.33 (§4.2)

25. $x = 2, y = -1$ (§4.3)

26. $x = \frac{1}{2}, y = 0$ (§4.3)

27. $u = 4, v = -1$ (§4.3)

28. $s = \frac{1}{4}, t = -\frac{1}{2}$ (§4.3)

29. Inconsistent (§4.3)

30. Dependent (§4.3)

31. $x = 8, y = 15$ (§4.3)

32. $x = 3, y = -2$ (§4.3)

33. (§4.4)

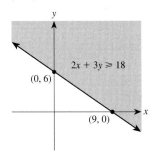

34. (§4.4)

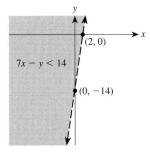

35. (§4.4)

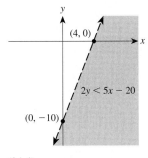

36. (§4.4)

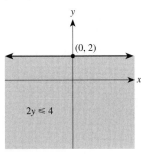

37. 5 (§5.1)   **38.** 0 (§5.1)

39. $x^2 + 5x - 9$ (§5.2)

40. $-6x^2 + 15xy - 12x$ (§5.2)

41. $x^2 - 4xy + 3y^2$ (§5.2)

42. $3x^2 - 5xy + 2y^2$ (§5.2)

43. $x^2 + 2xy + y^2 - 2x - 2y$ (§5.2)

44. $8a^3 + b^3$ (§5.2)

45. $9x^2 - 30xy + 25y^2$ (§5.3)

46. $4m^2 - 9n^2$ (§5.3)

47. $9x^2 - 25y^2$ (§5.3)

48. $4m^2 + 12mn + 9n^2$ (§5.3)

49. $4x^2 + 4xy + y^2 - 9$ (§5.3)

50. $x^2 - 4xy + 4y^2 + 6x - 12y + 9$ (§5.3)

51. $(x - 8)(x + 3)$ (§5.5)

52. $(y - 2x)(y + 2x)$ (§5.5)

53. $(y - 7x)(y - 5x)$ (§5.5)

54. $(5a + b)(2a - b)$ (§5.5)

55. $(2y + 3z)(2y + 5z)$ (§5.5)

56. $(3x - 5z)(3x + 5z)$ (§5.5)

57. $(5a + 2b)^2$ (§5.5)

58. Not factorable (§5.5)

59. $9(2x - 1)(2x + 1)$ (§5.5)

60. $(5x - 3z)^2$ (§5.5)

61. Not factorable (§5.5)

62. $y(6y + 1)(2y - 3)$ (§5.5)

63. $y(3y - 1)(y + 2)$ (§5.5)

64. $(5a^2 - 2b^2)(5a^2 + 4b^2)$ (§5.5)

65. $(7a^2 - 3z)(7a^2 + z)$ (§5.5)

66. $ab(6a^2 + b)(3a^2 - 2b)$ (§5.5)

67. $(x - y - 4)(x - y + 4)$ (§5.5)

68. $(a - 2b - 5)(a - 2b + 5)$ (§5.5)

69. $(4a^2 + b^2)(2a - b)(2a + b)$ (§5.5)

70. $(x + y + 2)(x + y + 1)$ (§5.5)

71. $(2a + 5b)(4a^2 - 10ab + 25b^2)$ (§5.5)

72. $(x - 1)(x + 5)(x - 5)$ (§5.5)

73. $x = -3, 4$ (§5.6)

74. $t = 1, \frac{5}{2}$ (§5.6)

75. $a = -4, 6$ (§5.6)

76. $c = -\frac{1}{8}$ (§5.6)

77. $r = 2, -10$ (§5.6)

78. $x = 0, \frac{5}{3}$ (§5.6)

79. $x = 1, 4$ **80.** $x = 0, 5$

81. $x - 2, R\ 11$ (§5.7)

82. $x + 3, R\ -4$ (§5.7)

83. $2x^2 - 3x + 6, R\ -17$ (§5.7)

84. $x^2 - 2x + 3, R\ -4x - 1$ (§5.7)

85. $\left\{x \mid x \neq -\frac{3}{2}\right\}$ (§6.1)

86. $\left\{x \mid x \neq -\frac{1}{3}, 2\right\}$ (§6.1)

87. $\dfrac{9b}{8a^2}$ (§6.2)   **88.** $\dfrac{x + 3y}{x - 3y}$ (§6.2)

89. $\dfrac{2a + b}{a(a + b)}$ (§6.2)

90. $\dfrac{3x - y}{2}$ (§6.2)

91. $\dfrac{5yb^2}{7a^2x^2}$ (§§6.3–6.4)

92. $(2x + y)(x - y)$ (§§6.3–6.4)

93. $\dfrac{9x^2 + 4y^2}{6xy}$ (§§6.3–6.4)

94. -3 (§§6.3–6.4)

95. $\dfrac{x - 4}{2}$ (§§6.3–6.4)

96. $3a - b$ (§§6.3–6.4)

97. $\dfrac{-x + 11}{(x - 5)(x - 2)}$ (§§6.3–6.4)

98. $\dfrac{-7x - 2y}{(x - 3y)(x + y)(2x + y)}$ (§§6.3–6.4)

99. $\dfrac{x}{x + 3y}$ (§§6.3–6.4)

100. $\dfrac{x - 2y}{x - 5y}$ (§§6.3–6.4)

101. $\dfrac{2x^2 - 4}{x - 2x^2}$ (§6.5)

102. $\dfrac{(x - 2y)(x + 2y)}{(2x - y)(x + 4y)}$ (§6.5)

103. 2 (§6.5)

104. $\dfrac{-3h}{(x + h + 1)(x + 1)}$ (§6.5)

105. $x = \frac{6}{5}$ (§6.6)   **106.** $x = -2$ (§6.6)

107. $x > 6$ (§6.6)

108. No solution (§6.6)

109. $x = 1$ (§6.6)   **110.** $x \geq -\frac{1}{5}$ (§6.6)

111. $a = \dfrac{b}{3}$ (§6.7)

112. $y = \dfrac{5x}{3x - 5}$ (§6.7)

113. $y = \dfrac{x}{x + 1}$ (§6.7)

114. $x = \dfrac{2y + 3}{y - 2}$ (§6.7)

115. Let $x = $ total # of cars;

$\dfrac{5}{6} = \dfrac{x - 1,200}{1,200}$; 2,200 cars (§6.8)

116. Let $x = $ amount of 20% solution;
$0.20x + 0.35(8) = 0.30(8 + x)$;
4 liters (§6.8)

117. Let $t = $ time they would take to

work together; $\dfrac{t}{4} + \dfrac{t}{4\frac{1}{2}} = 1$;

$2\frac{2}{17}$ hours (§6.8)

118. Let $x = $ # hours it takes Jason to
complete the job alone;

$\dfrac{2}{5} + \dfrac{1}{5} + \dfrac{1}{x} = 1$; $2\frac{1}{2}$ hours

119. Let $x = $ number of general
admission tickets sold;
$3.50x + 4.25(505 - x) = $
$1,861.25$; 380 general admission
tickets, 125 reserved seat tickets
(§6.8)

120. (a) $C(5) = 355;$
$C(10) = 332.5;$
$C(50) = 325.3;$
$C(100) = 325.08;$
$C(300) = 325.0083;$
$C(500) = 325.003;$
$C(1,000) = 325.00075$ (§6.1)

(b)

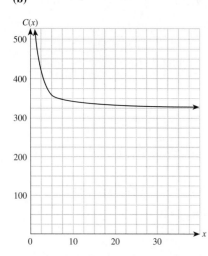

Chapters 4–6
Cumulative Practice Test

1. (a) $-3x^2 - xy + 3y^2$
(b) $15a^2 - ab - 6b^2$
(c) $4x^4 - y^2$
(d) $9y^2 - 12yz + 4z^2$
(e) $x^2 + 2xy + y^2 - 6x - 6y + 9$

2. (a) $(a - 7)(a - 2)$
(b) $(3r + 2s)(2r - 3s)$
(c) $(5a^2 + 3y)(2a^2 - 3y)$
(d) $(2a^2 - 3b^2)^2$
(e) $(a + 2b - 5)(a + 2b + 5)$
(f) $(5x - 1)(25x^2 + 5x + 1)$

3. $x = -9, 3$

4. (a) $x = -2, \frac{1}{2}$
(b) $x = -5, 4$

5. $2x^2 - 4x + 14, R\, {-23}$

6. $\left\{x \,|\, x \neq -1, \frac{2}{3}\right\}$

7. (a) $5x + 3y$
(b) -2

(c) $\dfrac{(3x - 14)(x - 1)}{(x - 5)(x - 3)(x - 7)}$

8. $\dfrac{-2x^2 - x + 1}{3x^2 + 4x}$

9. $\dfrac{h}{(2x + 2h + 1)(2x + 1)}$

10. (a) $x = \frac{9}{7}$
(b) $x < \frac{45}{2}$
(c) No solution

11. $a = \dfrac{y}{1 - y}$

12. Let $t = $ time it takes to process

80 forms together; $\dfrac{t}{6} + \dfrac{t}{5\frac{1}{2}} = 1$;

$2\frac{20}{23}$ hours

13.

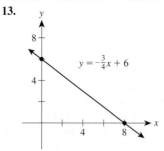

$y = -\frac{3}{4}x + 6$

14. $m = -1$
15. (a) $y = \frac{2}{3}x - \frac{13}{3}$ **(b)** $y = -\frac{3}{2}x$
16. $B = 28$ **17.** $x = 2, y = -3$
18.

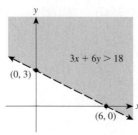

$3x + 6y > 18$

$(0, 3)$

$(6, 0)$

Exercises 7.1

1. x^{11} **3.** $-6a^7b^{10}$ **5.** a^{10} **7.** $\frac{4}{9}$

9. $x^2 + 2xy^3 + y^6$ **11.** x^2y^6

13. $2^6 \cdot 3^4$ **15.** $x^{26}y^{19}$

17. $-8a^{10}b^8$ **19.** $16r^{10}s^{11}t^7$

21. x^3 **23.** $\dfrac{x^2}{y^2}$ **25.** $\dfrac{1}{4}$ **27.** $\dfrac{ab^8}{c^4}$

29. $-\dfrac{1}{y}$ **31.** $\dfrac{1}{18}$ **33.** $\dfrac{1}{y^9}$ **35.** y^{15}

37. $\dfrac{4r^8s^9}{3}$ **39.** $\dfrac{x^5y^4}{2}$

41. $-1,728a^3b^9$ **43.** $216a^3b^6x^3y^6$

45. $r^{26}s^{28}$ **47.** 0.000003

49. 7.984000 **51.** $\dfrac{1}{x}$ **53.** $\dfrac{x^2}{y^2}$

55. 3^6 **57.** $\dfrac{1}{a^4b^6}$ **59.** $\dfrac{r^6}{s^8}$ **61.** $-\dfrac{1}{4}$

63. $\dfrac{1}{9}$ **65.** $\dfrac{2^6}{3^6}$ **67.** $\dfrac{s^{18}}{3^{12}}$ **69.** x^6

71. y^2x^2 **73.** 2 **75.** $-\dfrac{125}{27}$ **77.** $\dfrac{1}{x^4}$

79. $\dfrac{a^5}{b^4}$ **81.** x^6 **83.** 81 **85.** $\dfrac{4y^6}{9x^6}$

87. $\dfrac{1}{x^6y^9}$ **89.** $\dfrac{9b^3}{a^4}$ **91.** $\dfrac{y^3}{x^2}$

93. $\dfrac{y^3}{x^2y^3 + 1}$ **95.** $\dfrac{x^4y^4}{(x^2 + y^2)^2}$ **97.** 1

99. $\dfrac{x + y}{x^2}$ **101.** $\dfrac{s(s + r^2)}{r(s^2 + r)}$

103. $\dfrac{a(2b^2 + a)}{b^2(1 + a^2b)}$

Mini-Review

111. $\left(-3, \frac{7}{2}\right]$
112. $28w^2 - 70w$ dollars
113. $y = \frac{2}{3}x - 4$
114. $f(3) + f(2) = -1; f(5) = 24$

Exercises 7.2

1. $1,000$ **3.** $1,000,000,000$
5. 10 **7.** 0.000001 **9.** $1,000,000$
11. 16.2 **13.** $760,000,000$
15. 0.000000851 **17.** $6,000$
19. 8.24×10^2 **21.** 5.0×10^0
23. 9.3×10^{-3} **25.** 8.27546×10^8
27. 7.2×10^{-7} **29.** 7.932×10^1
31. $0.16\overline{6}$ or 0.17 **33.** 0.000001
35. Red: 5×10^6; White: 7.5×10^3
37. $\dfrac{2,000}{3}$ **39.** 5 hr 28 min 51 sec
41. 6.048×10^{13} tons in one day,
2.20752×10^{16} tons in one year
43. 10^4 Å in 1 micron

45. 0.66×10^{-8} cm $= 6.6 \times 10^{-9}$ cm

47. $\dfrac{1 \text{ mile}}{1.6 \text{ km}} = \dfrac{5.86 \times 10^{12} \text{ miles}}{x \text{ km}};$
9.376×10^{12} km

49. 60×10^{-4} cm $= 6.0 \times 10^{-3}$ cm

51. 10^{23} atoms   **53.** 240 cm

55. Approximately 63,011 AU

Mini-Review

59. $2x^2 - 18x$ **60.** $\dfrac{2x + 2}{x}$

61. -17 **62.** Approx. 11:42 A.M.

Exercises 7.3

1. 2   **3.** -2 **5.** -10 **7.** 4
9. 3   **11.** Not a real number
13. 1   **15.** -7 **17.** -6 **19.** 2
21. 5   **23.** -8 **25.** $\frac{1}{2}$ **27.** $-\frac{1}{2}$
29. $-\frac{1}{9}$ **31.** $\frac{1}{16}$ **33.** 16
35. Not a real number   **37.** -16
39. 9   **41.** $\frac{4}{3}$ **43.** $\frac{2}{3}$ **45.** 16   **47.** $x^{7/6}$
49. $a^{3/8}$ **51.** 1   **53.** $\dfrac{r^{1/3}}{s}$ **55.** $\dfrac{r}{s^{1/3}}$
57. $\dfrac{1}{x^{1/6}}$ **59.** $\dfrac{b^{2/15}}{a^{3/4}}$ **61.** x^5 **63.** $-\frac{1}{2}$
65. $\dfrac{1}{x^6 y^{47/12}}$ **67.** $x + x^{3/2}$
69. $\dfrac{x^2 - 4x + 4}{x}$
71. $(xy)^{1/3}$ **73.** $(x^2 + y^2)^{1/2}$
75. $(5a^2 b^3)^{1/5}$ **77.** $2(3xyz^4)^{1/3}$
79. $5(x - y)^{2/3}$ **81.** $(x^n - y^n)^{1/n}$
83. $(x^{5n+1} y^{2n-1})^{1/n}$
85. $\sqrt[3]{x}$ **87.** $m\sqrt[3]{n}$
89. $\sqrt[3]{(-a)^2}$ **91.** $-\sqrt[4]{a^2}$
93. $\sqrt[3]{a^2 b}$ **95.** $\sqrt{x^2 + y^2}$
97. $\sqrt{x^n - y^n}$
99. 0.1667   **101.** 2.8856

Mini-Review

110. $x = 0, \frac{3}{2}$
111. (a) -3 (b) -6 (c) 3
  (d) 0 (e) $-1, 2, 5$
112. $\dfrac{-x + 24}{6}$ **113.** $x = -24$

Exercises 7.4

1. $2\sqrt{14}$ **3.** $4\sqrt{3}$ **5.** $2\sqrt[5]{2}$
7. 12   **9.** $8x^4$ **11.** $3x^3$

13. $8x^{30}\sqrt{2}$ **15.** $2x^{15}\sqrt[4]{8}$
17. $2x^{12}\sqrt[5]{4}$ **19.** xy^2
21. $4ab^2\sqrt{2}$ **23.** $a^7 b^{15}$ **25.** $x^2 y^2$
27. $\dfrac{\sqrt{2}}{2}$ **29.** $\dfrac{\sqrt{5x}}{5}$ **31.** $\dfrac{3\sqrt{5}}{2}$
33. $\dfrac{\sqrt{3}}{15}$ **35.** $8x^2 y^4\sqrt{x}$
37. $3x^2 y^2\sqrt[3]{3x^2 y}$ **39.** $6y\sqrt[4]{2y^2}$
41. $x + y^2$ **43.** $\sqrt[4]{x^4 - y^4}$
45. $30st\sqrt{2st}$ **47.** $12a^3 b^2$
49. $\dfrac{xy^2}{2}$ **51.** $\dfrac{2x^2\sqrt[4]{2x}}{y^3}$ **53.** $3\sqrt{3}$
55. $\dfrac{y^4}{x^3}$ **57.** $\dfrac{\sqrt{15x}}{5x}$ **59.** $\dfrac{\sqrt{3xy}}{xy^2}$
61. $\dfrac{\sqrt[3]{12}}{2}$ **63.** $\dfrac{\sqrt[3]{18}}{2}$ **65.** $\dfrac{\sqrt{6}}{2}$
67. $\dfrac{3y}{2x}\sqrt[3]{12x^2}$ **69.** $\dfrac{x^2}{3a^5}$
71. $-\dfrac{6s^3\sqrt{sr}}{r}$ **73.** $\sqrt{a}$
75. $x\sqrt[12]{x}$ **77.** $\sqrt[6]{a}$ **79.** $x^5 y^3$

Mini-Review

82. $(2x - 3)(x + 5)$
83. $x^2 - 4x + 14, R\ -53$

Exercises 7.5

1. $4\sqrt{3}$ **3.** $-3\sqrt{5}$
5. $4\sqrt{3} + 2\sqrt{6}$ **7.** $\sqrt{3} - \sqrt{5}$
9. 0   **11.** $(7a - 3a^3)\sqrt{b}$
13. $(3x + 4)\sqrt[3]{x^2}$ **15.** $1 - 2\sqrt[3]{a}$
17. $-\sqrt{3}$ **19.** $5\sqrt{6} - 3\sqrt{3}$
21. $6\sqrt{3} - 36$ **23.** $5\sqrt{6} - 20\sqrt{3}$
25. $6\sqrt[3]{3} - 11\sqrt[3]{6}$
27. $2ab\sqrt{a} - 3ab\sqrt{b} - ab\sqrt{ab}$
29. 0   **31.** $4x^4 y^4\sqrt{5x}$
33. $9x\sqrt[3]{9x^2} - 3x\sqrt[3]{x^2}$
35. $(8 - 4x)\sqrt[4]{x}$ **37.** $\dfrac{\sqrt{5} + 10}{5}$
39. 0   **41.** $\dfrac{3\sqrt{2}}{2}$ **43.** $\dfrac{7\sqrt{10}}{10}$
45. $\dfrac{5\sqrt{7} - 21\sqrt{5}}{35}$ **47.** $-\dfrac{11\sqrt[3]{4}}{2}$
49. $\dfrac{y\sqrt{x} + x\sqrt{y}}{xy}$ **51.** $\dfrac{\sqrt[3]{3} - 3\sqrt[3]{9}}{3}$

53. $\dfrac{3\sqrt{2} + 21\sqrt{7}}{7}$ **55.** $\dfrac{14\sqrt{10}}{5}$
57. $4\sqrt[3]{25}$ **59.** $\dfrac{(2x - 1)\sqrt{x - 1}}{2(x - 1)}$

Mini-Review

61. (a) $\{x \mid -4 \le x \le 6\}$
  (b) $\{y \mid -1 \le y \le 5\}$
  (c) $\{x \mid -4 \le x \le -2$ and
    $3 \le x \le 6\}$
  (d) $\{x \mid -2 \le x \le 3\}$
62. $3x(x + 5)(x - 5)$
63.

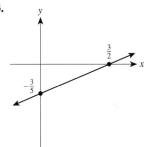

Exercises 7.6

1. $5\sqrt{5} - 15$ **3.** $-2\sqrt{3} - 6\sqrt{5}$
5. $a + \sqrt{ab}$ **7.** $\sqrt{10} + 2$
9. $6\sqrt{15} - 60$ **11.** $-5\sqrt{6} + 14$
13. $\sqrt{15} + \sqrt{5} - 2\sqrt{3} - 2$
15. 2   **17.** $8 - 2\sqrt{15}$ **19.** 3
21. 5   **23.** $4a - 4\sqrt{ab} + b$
25. $9x - 4y$ **27.** $x - 6\sqrt{x} + 9$
29. $x - 3$ **31.** $-2\sqrt{x}$ **33.** -1
35. $2\sqrt{2} - 3\sqrt{3}$ **37.** $\dfrac{5\sqrt{2} - \sqrt{7}}{4}$
39. $3\sqrt{2} + \sqrt{5}$ **41.** $\dfrac{2 + \sqrt{7}}{2}$
43. $-\dfrac{\sqrt{2} + 3}{7}$ **45.** $\dfrac{5\sqrt{5} - 5}{2}$
47. $\dfrac{2\sqrt{3} + 2\sqrt{a}}{3 - a}$ **49.** $\dfrac{x + \sqrt{xy}}{x - y}$
51. $\dfrac{\sqrt{10} + 2}{3}$ **53.** $\dfrac{2\sqrt{10} + 2}{9}$
55. $\dfrac{\sqrt{10} - 1}{2}$ **57.** $5 + 2\sqrt{6}$
59. $\dfrac{18 + 5\sqrt{10}}{2}$ **61.** $\sqrt{x} + \sqrt{y}$
63. $(x + 1)(\sqrt{x} + \sqrt{2})$ **65.** 12
67. $9\sqrt{7} - 5\sqrt{3}$ **69.** $x = -1$

71. $x = \dfrac{-3 + \sqrt{3}}{2}$

73. $x = \dfrac{-3 + \sqrt{15}}{2}$

Mini-Review

75. 12; 12; answers are identical
76. $3x^2 - 75$ **77.** $x = -\frac{5}{2}, \frac{3}{2}$
78. -0.2 **79.** $2(3x - 4)(2x - 5)$

Exercises 7.7

1. $a = 49$ **3.** $a = 31$
5. $x = \dfrac{26}{3}$ **7.** $y = 116$
9. $x = \dfrac{10}{3}$ **11.** $x = 16$
13. $a = 8.90$ **15.** No solution
17. $x = 8$ **19.** $x = -1$
21. $x = 64$ **23.** $s = -243$
25. $y = -32$ **27.** $x = 125$
29. $x = -8$ **31.** $x = 118$
33. $x = \frac{1}{256}$ **35.** $x = -11$
37. $x = 60.84$ **39.** $x = 92.28$
41. $x = 9.49$ **43.** $x = 8.60$
45. $x = 11$ **47.** $x = 512$
49. $a = \dfrac{25}{3}$ **51.** $x = 22$
53. $y = 53.76$ **55.** $x = 2.88$
57. $x = -2$ **59.** 151.25 ft
61. $L = 1{,}588.7$ cm **63.** 2.9 in.

Mini-Review

66. $x = -2, 8$ **67.** $x = 3$
68. $\dfrac{11x - 73}{20}$

Exercises 7.8

1. $-i$ **3.** $-i$ **5.** $-i$ **7.** 1
9. $-i$ **11.** -1 **13.** $i\sqrt{5}$
15. $3 - i\sqrt{2}$ **17.** $-2\sqrt{7} + 4i$
19. $\frac{3}{5} - i\dfrac{\sqrt{2}}{5}$ **21.** $2 - i\dfrac{\sqrt{3}}{3}$
23. $1 - i$ **25.** 4 **27.** $15 - 35i$
29. $-4 + 2i$ **31.** -15 **33.** -9
35. $-4 + 6i$ **37.** $10 + 6i$
39. 0 **41.** $25 - 5i$ **43.** 10
45. 53 **47.** $2i$ **49.** -5
51. -1 **53.** $2 - \frac{5}{3}i$
55. $-1 - 3i$ **57.** $-1 - \frac{5}{2}i$

59. $-\frac{4}{29} + \frac{10}{29}i$ **61.** $\frac{10}{29} + \frac{4}{29}i$
63. $\frac{3}{5} + \frac{4}{5}i$ **65.** $-\frac{21}{29} - \frac{20}{29}i$
67. $\frac{1}{29} - \frac{41}{29}i$ **69.** $\frac{5}{34} - \frac{31}{34}i$
71. $x = -2 + i$
73. $x = -1 + i\dfrac{\sqrt{6}}{2}$
75. $x^2 + 4 = 0$
Substitute $x = 2i$.
$(2i)^2 + 4 \overset{?}{=} 0$
$4i^2 + 4 \overset{?}{=} 0$
$4(-1) + 4 \overset{\checkmark}{=} 0$
77. $\hspace{2cm} x^2 - 4x + 5 = 0$
Substitute $x = 2 - i$.
$(2 - i)^2 - 4(2 - i) + 5 \overset{?}{=} 0$
$4 - 4i + i^2 - 8 + 4i + 5 \overset{?}{=} 0$
$4 - 4i - 1 - 8 + 4i + 5 \overset{\checkmark}{=} 0$

Mini-Review

82. $\dfrac{-4x}{x + 4}$ **83.** $\dfrac{1}{(x - 6)^2}$
84. $x = \dfrac{7}{a - 6}$ **85.** $\{x \mid x \neq \pm 5\}$

Chapter 7 Review Exercises

1. x^{12} **2.** $x^{11}y^9$ **3.** $-6x^4y^5$
4. $-30x^3y^6$ **5.** a^{12} **6.** b^{20}
7. $a^{14}b^{21}$ **8.** $r^{16}s^{24}$ **9.** $a^{10}b^9$
10. x^7y^{22} **11.** $a^7b^8c^7$ **12.** $-162x^8y^5$
13. $\dfrac{1}{a}$ **14.** x^5 **15.** 1 **16.** y^2
17. $\dfrac{1}{x^{16}y^{14}}$ **18.** $\dfrac{1}{a^2b^6}$ **19.** a^4
20. $\dfrac{a^4b^4}{c^4}$ **21.** $9a^4x^4$ **22.** $-3x^6y^2$
23. $\dfrac{-72y^8}{x^2}$ **24.** $\dfrac{-72b^5}{125a^5}$ **25.** $\dfrac{1}{a^2}$
26. $\dfrac{1}{x^9}$ **27.** $\dfrac{x^8}{y^{20}}$ **28.** $\dfrac{1}{x^4}$
29. $-\dfrac{1}{162}$ **30.** 1 **31.** 1 **32.** 1
33. $\dfrac{1}{x^4}$ **34.** $\dfrac{y}{x^4}$ **35.** x^8 **36.** a^8b^{12}
37. $\dfrac{r^{11}}{s^5}$ **38.** $-\dfrac{s^{11}}{3r^5}$ **39.** $\frac{25}{4}$ **40.** $\frac{256}{81}$
41. $\dfrac{27y^8}{4xz^2}$ **42.** $\dfrac{8x}{9yz^8}$ **43.** $\dfrac{x^2 - y^2}{xy}$
44. $\dfrac{y^4 + 4xy^2 + 4x^2}{x^2y^4}$ **45.** $\dfrac{y^3 + x}{y^5}$
46. $\dfrac{b(b + a^2)}{a}$ **47.** 28,300 **48.** 6.29

49. 0.0000796 **50.** 0.0000008264
51. 9.259×10^1 **52.** 5.78×10^{-3}
53. 6.25897×10^5 **54.** 7.3×10^{-6}
55. 0.00000001 **56.** 1,000
57. $x^{5/6}$ **58.** $\dfrac{1}{y^{1/6}}$ **59.** x **60.** $y^{1/12}$
61. $x^{13/30}$ **62.** $\dfrac{s^{1/3}}{r^{5/6}}$ **63.** a^5b^3
64. $\dfrac{y^3}{x}$ **65.** $\dfrac{x^{5/6}}{y^{7/6}}$ **66.** $\dfrac{1}{a^{1/6}}$ **67.** 2^{10}
68. $\dfrac{1}{a^{1/6}} - \dfrac{b^{1/2}}{a^{2/3}} = \dfrac{a^{1/2} - b^{1/2}}{a^{2/3}}$
69. $a^{2/3} - b^{2/3}$ **70.** $\dfrac{1}{a} + \dfrac{4b^{1/2}}{a^{1/2}} + 4b$
71. $\sqrt{x}$ **72.** $\sqrt[3]{x}$ **73.** $x\sqrt{y}$
74. $\sqrt{xy}$ **75.** $\sqrt[3]{m^2}$ **76.** $\sqrt{m^3}$
77. $\sqrt[3]{(5x)^3}$ **78.** $5\sqrt[4]{x^3}$ **79.** $a^{1/3}$
80. $t^{1/5}$ **81.** $-n^{4/5}$ **82.** $t^{5/3}$ **83.** $t^{-7/5}$
84. $t^{-5/7}$ **85.** 36 **86.** $3\sqrt[3]{2}$
87. x^{30} **88.** x^{15} **89.** $2xy^2\sqrt[3]{6xy^2}$
90. $2x^4y^6\sqrt{7xy}$ **91.** $15x\sqrt{y}$
92. $24ab$ **93.** $2x^4y^2\sqrt{y}$ **94.** $6x^7y^2$
95. $\frac{2}{3}$ **96.** $\frac{2}{3}$ **97.** $\dfrac{\sqrt{y}}{x}$ **98.** $\dfrac{\sqrt{xy}}{x}$
99. $\dfrac{4\sqrt{ab}}{a^2b}$ **100.** $\dfrac{9y\sqrt{2x}}{2x}$ **101.** $\dfrac{\sqrt{5a}}{a}$
102. $\dfrac{\sqrt[3]{5a^2}}{a}$ **103.** $\dfrac{2\sqrt[3]{4a^2}}{a}$
104. $\dfrac{2\sqrt[3]{4a}}{a}$ **105.** $\sqrt[3]{x^2}$ **106.** $\sqrt[8]{x^3}$
107. $\sqrt[6]{2^5}$ **108.** $\sqrt[6]{2}$ **109.** 0
110. $33\sqrt{6}$ **111.** 0
112. $2s^2t^2\sqrt{s} - 3s^2t^2\sqrt{t}$
113. $\dfrac{3\sqrt{6} + 2\sqrt{15}}{6}$
114. $\dfrac{3\sqrt{35} + 7\sqrt{6}}{21}$
115. 0 **116.** $-\sqrt{10}$
117. $27 - 7\sqrt{21}$
118. $6x + 4\sqrt{x} - 3\sqrt{5x} - 2\sqrt{5}$
119. $x - 10\sqrt{x} + 25$
120. $10 - 2\sqrt{21}$
121. $-14\sqrt{a} - 42$
122. $12 - 6\sqrt{b}$ **123.** $6\sqrt{5} - 6\sqrt{3}$
124. $\sqrt{m} + n$ **125.** $4\sqrt{2x} + 4\sqrt{5y}$
126. $-3\sqrt{2}$ **127.** $x = 27$
128. $x = \frac{4}{3}$ **129.** $x = 27$
130. $x = \frac{14}{3}$ **131.** $x = 85$

132. $x = 5$ **133.** $x = 16$
134. $x = 80$ **135.** $x = 17.5$
136. $x = 35.7$ **137.** 2.1 psi
138. 1.7 psi **139.** $-i$ **140.** i
141. 1 **142.** -1 **143.** $9 - i$
144. $1 - 5i$ **145.** $8 - 25i$
146. 41 **147.** $-6 - 8i$
148. $-5 - 10i$ **149.** $5 - 12i$
150. $-5 - 12i$ **151.** -37
152. $34 - 6i$ **153.** $\frac{9}{10} - \frac{13}{10}i$
154. $-i$

Chapter 7 Practice Test

1. $a^4 b^{11}$ **2.** $-200 x^8 y^3$
3. $-\dfrac{4b^2 y^2}{3ax^2}$ **4.** $-\dfrac{2x^4}{9y^5}$ **5.** $\dfrac{9r^4 s^{10}}{25}$

6. $\frac{1}{3}$ **7.** $x^{7/3}$ **8.** $\dfrac{1 + x^2}{xy}$

9. $x - x^{4/3}$ **10.** (a) -5 (b) $-\frac{1}{8}$
11. $3\sqrt{a^2}$ **12.** 0.0000032
13. In one hour it travels
$186{,}000 \times 60 \times 60 =$
6.696×10^8 miles.

14. $3x^2 y^3$ **15.** $2x\sqrt[3]{y^2}$ **16.** $6x^4 y$

17. $\dfrac{\sqrt{35}}{7}$ **18.** $\dfrac{2\sqrt[3]{3}}{3}$ **19.** $\dfrac{3xy^2 \sqrt{2}}{2}$

20. $\sqrt[4]{5^3}$ **21.** $5\sqrt{2}$
22. $x - 6\sqrt{x} + 9$ **23.** $3 + \sqrt{6}$
24. $x = 19$ **25.** $x = 121$
26. $x = 17.30$ **27.** $-i$ **28.** $-5 - 5i$
29. $-\frac{4}{5} - \frac{7}{5}i$

Exercises 8.1

1. (a) $P(30) = 14{,}862$;
$P(50) = 21{,}270$;
$P(70) = 24{,}878$;
$P(100) = 25{,}040$;
$P(120) = 21{,}648$ dollars. As
the number of items, x,
increases, profit rises for a while,
and when the number of items
exceeds \$100, the profit begins
to decrease.
(b) The profit seems to reach a
maximum of \$25,700, occurring
between 85 and 86 items.
3. (a) $s(5) = 3{,}225$; $s(10) = 5{,}650$;
$s(20) = 8{,}100$; $s(30) = 7{,}350$;
$s(40) = 3{,}400$; $s(45) = 225$

(b) The maximum height is about
8,200 ft, occurring at
$20 - 25$ seconds.
(c) The rocket hits the ground
somewhere between 45 and
46 seconds.
5. (a) $A(x) = 100x - x^2$
(b) $A(10) = 900$; $A(40) = 2{,}400$;
$A(60) = 2{,}400$; $A(70) = 2{,}100$;
$A(90) = 900$ sq ft.
As the length, x, increases, the
area increases until the length x
is a value between 40 and 60
feet. The area then decreases as
the length exceeds that value.
(c) The maximum area occurs at
50 ft, so the dimensions are $50'$
by $50'$, and the maximum area is
2,500 sq ft for these dimensions.
7. Let $x =$ the first number, then the
other number is $225 - x$. Their
product is $A = x(225 - x)$. A will
be a maximum when x is between
112 and 113, and hence the other
number will be between 112 and 113
(exact values are 112.5 for each
number).

9. (a) $A(x) = 2x\left(\dfrac{1{,}000 - 4x}{3}\right)$
$= \dfrac{2{,}000x - 8x^2}{3}$

(b) $A(50) = 26{,}666.67$;
$A(100) = 40{,}000$;
$A(120) = 41{,}600$;
$A(150) = 40{,}000$;
$A(200) = 26{,}666.67$ sq ft.
As the length, x, increases, the
area increases until the length x
is a value around 120 ft. The
area then decreases as the length
exceeds that value.
(c) The maximum area occurs at
about 125 ft, and the area for
$x = 125$ is about 41,667 sq ft
where $y = 166.7$ ft.
11. (a) 24 items (b) \$5 per item
(c)

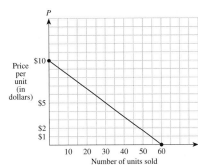

Price
per
unit
(in
dollars)

P

\$10
\$5
\$2
\$1

10 20 30 40 50 60 x
Number of units sold

(d) $R(x) = x\left(10 - \frac{x}{6}\right)$
(e) $R(5) = 45.83$; $R(15) = 112.5$;
$R(25) = 145.83$; $R(35) = 145.83$;

$R(40) = 133.33$ dollars.
As the number of units sold, x,
increases, the revenue increases
until the number of units sold is
a value between 25 and 45. Then
the revenue decreases as the
number of units sold exceeds
that value.
(f) The maximum revenue occurs at
30 units sold, and the maximum
revenue is \$150.

Mini-Review

13. -247 **14.** $-\frac{2}{3} < x < \frac{10}{3}$
15. $x \le -\frac{1}{2}$ or $x \ge \frac{11}{2}$
16. Let $x =$ # of more expensive models, $y =$ # of less expensive models;
$x + 0.75y = 150$, $0.5x + 0.25y = 60$; 60 more expensive models, 120
less expensive models

Exercises 8.2

1. $x = -2, 3$ **3.** $x = \pm 5$
5. $x = 0, 4$ **7.** $x = 6, -2$
9. $y = 0, 7$ **11.** $x = \pm 4$
13. $c = \pm\frac{4}{3}$ **15.** $a = \pm\frac{3}{2}$
17. $x = 3, -2$ **19.** $y = 1, \frac{1}{2}$
21. $x = \pm\sqrt{6}$ **23.** $x = \pm\dfrac{\sqrt{133}}{7}$
25. $x = 5, -2$ **27.** $a = -\frac{1}{3}, \frac{1}{2}$
29. $y = \frac{3}{2}, -\frac{2}{3}$ **31.** $x = \pm\dfrac{i\sqrt{15}}{3}$
33. $s = -\frac{5}{6}, 2$ **35.** $x = 3, -7$
37. $x = 5, 1$ **39.** $x = -9$
41. $x = \pm i\sqrt{2}$ **43.** $x = 8 \pm 2\sqrt{2}$
45. $y = 5 \pm 4i$ **47.** $x = 2, 3$
49. $a = \pm\sqrt{13}$ **51.** $x = -\frac{4}{3}$
53. $x = 1$ **55.** $x = 5$
57. $x = \frac{14}{3}, -1$ **59.** $a = \pm\dfrac{\sqrt{b}}{2}$
61. $y = \pm\dfrac{\sqrt{63 - 35x^2}}{7}$
63. $r = \dfrac{\sqrt{6V\pi}}{2\pi}$ **65.** $a = \pm 2b$
67. $x = 3y, -2y$
69. Let $x =$ number;
$x^2 - 5 = 5x + 1$; $x = 6$
71. Let $x =$ number; $x^2 + 64 = 68$;
$x = 2$
73. Let $x =$ number; $(x + 6)^2 = 169$;
$x = -19$ or $x = 7$
75. Let $x =$ number; $x + \dfrac{1}{x} = \dfrac{13}{6}$;
$x = \frac{2}{3}$ or $x = \frac{3}{2}$

77. (a) Profit = $0 **(b)** Price = $6
79. (a) $h(1) = 24$ ft

   (b) $t = \dfrac{\sqrt{10}}{2}$ sec ≈ 1.58 sec

   (c) 40 ft
81. Let x = width; $(x + 2)(x) = 80$; length = 10 ft, width = 8 ft
83. Let side of square = x inches; $x^2 = 60$; $2\sqrt{15} \approx 7.75$ inches (each side)
85. Let x = larger number; $(x)(x - 2) = 120$; $x = 12$ or $x = -10$; two sets of numbers: 12 and 10; -12 and -10
87. Let x = one number; $x(3x + 2) = 85$; two sets of numbers: 5 and 17, $-\frac{17}{3}$ and -15
89. Length of other leg = $\sqrt{33} \approx 5.74''$
91. Dimensions are 8, $2\frac{1}{3}$, and $8\frac{1}{3}$
93. Length of diagonal = $\sqrt{193} \approx 13.9''$
95. It reaches $2\sqrt{209} \approx 28.9$ ft high
97. Area = 32 sq in.
99. Area = $5\sqrt{119} \approx 54.54$ sq in.
101. Let t = # of hours: $(50t)^2 + (60t)^2 = 160^2$; $t = \dfrac{16\sqrt{61}}{61} \approx 2.05$ hours
103. Let t = # of hours: $(90t)^2 + (110t)^2 = 500^2$; $t = \dfrac{25\sqrt{202}}{202} \approx 2.52$ hours; 6:31 P.M.
105. $n = -3, 4$
107. Let x = width of picture; $(x + 2)(2x + 2) = 60$; dimensions of picture are 4 in. by 8 in.
109. Let r = speed of boat in still water; $\dfrac{20}{r - 5} + \dfrac{10}{r + 5} = \dfrac{4}{3}$; 25 kph

Exercises 8.3

1. $x = 3 \pm \sqrt{10}$ **3.** $c = 1 \pm \sqrt{6}$
5. $y = \dfrac{-5 \pm \sqrt{33}}{2}$
7. $x = \dfrac{-3 \pm \sqrt{5}}{2}$
9. $a = \dfrac{1 \pm \sqrt{17}}{2}$ **11.** $a = -1 \pm i$
13. $x = -2 \pm \sqrt{6}$
15. $a = \dfrac{1 \pm \sqrt{33}}{2}$

17. $x = \dfrac{5 \pm i\sqrt{19}}{2}$
19. $x = \dfrac{-5 \pm \sqrt{195}}{5}$
21. $t = -\frac{3}{2}, 1$ **23.** $x = 2 \pm \sqrt{15}$
25. $x = \frac{1}{4}, 3$ **27.** $a = \dfrac{1 \pm i\sqrt{3}}{2}$
29. $y = \dfrac{-1 \pm 3i}{2}$ **31.** $n = \dfrac{1 \pm i\sqrt{7}}{2}$
33. $t = \dfrac{-1 \pm i\sqrt{7}}{2}$
35. $x = \dfrac{-2 \pm \sqrt{19}}{5}$

Mini-Review

41. $x = 30, y = 15$ **42.** $27x^2y^4$
43. 6 **44.** $y = \frac{7}{2}x - 14$

Exercises 8.4

1. $x = 5, -1$ **3.** $a = \dfrac{3 \pm \sqrt{41}}{4}$
5. $y = \dfrac{2 \pm \sqrt{2}}{2}$
7. $y = \dfrac{-7 \pm \sqrt{77}}{2}$
9. $a = \dfrac{-1 \pm i\sqrt{23}}{6}$
11. $s = -1 \pm 3i$
13. Roots are not real.
15. Roots are real and distinct.
17. Roots are real and distinct.
19. Roots are real and distinct.

21. $a = 4, -1$ **23.** $y = \pm\dfrac{\sqrt{6}}{4}$
25. $x = \frac{4}{5}, \frac{3}{2}$ **27.** $x = 1$
29. $x = \frac{5}{2}, -\frac{1}{5}$ **31.** $a = 0, -1$
33. $x = 0.3, 5$ **35.** $x = \dfrac{3 \pm i\sqrt{11}}{2}$
37. $s = \pm\sqrt{6}$
39. $x = \dfrac{-1 \pm i\sqrt{39}}{2}$
41. $x = \frac{19}{4}$ **43.** $x = -2.7, 1.8$
45. $a = \dfrac{2 \pm i\sqrt{2}}{3}$
47. $x = \dfrac{5 \pm \sqrt{57}}{2}$
49. $a = 1, \frac{1}{2}$ **51.** $z = \frac{7}{6}$ **53.** $y = \frac{6}{5}$
55. $x = -0.05, 2{,}000.05$

57. $a = \dfrac{1 \pm \sqrt{97}}{4}$
59. $a = \dfrac{-1 \pm i\sqrt{15}}{2}$
61. Let x = number; $x(x - 3) = 40$; two sets of numbers: 8 and 5, -5 and -8
63. Hypotenuse = $\sqrt{73} \approx 8.54''$
65. Length of diagonal = $5\sqrt{2} \approx 7.07''$
67. Length of side = $4\sqrt{2} \approx 5.66''$
69. $\dfrac{c - 1}{c} = -\dfrac{c}{c - 2}$; $c = \dfrac{3 \pm i\sqrt{7}}{4}$; since c is not real, no solution.
71. 60 items
73. Let x = width of path; $\pi(10 + x)^2 - \pi(10)^2 = 44\pi$; width of path is 2 ft.
75. Let x = price per roll the first week, $x - 2$ = price per roll the second week; $(x - 2)\left(2{,}250 - \dfrac{10{,}000}{x}\right) = 10{,}000$; $10 per roll and $8 per roll

Mini-Review

80. 1 **81.** $\frac{1}{9}$ **82.** $\dfrac{6}{x(x - 3)}$
83. Let x = amount invested at 8%, y = amount invested at 10%; $0.08x + 0.10y = 730$, $0.10x + 0.08y = 710$; $8,000 altogether

Exercises 8.5

1. $x = \frac{9}{4}$ **3.** $x = 4$ **5.** $a = 13$
7. $a = 8$ **9.** $x = 8$ **11.** $a = 1$
13. $y = 1$ **15.** $a = 2$ **17.** $s = 9$
19. $a = -2$

21. (a) $\sqrt{x^2 + 4} + 9 - x$

   (b) $T = \dfrac{\sqrt{x^2 + 4}}{2} + \dfrac{9 - x}{6}$

   $= \dfrac{3\sqrt{x^2 + 4} + 9 - x}{6}$

23. (a) $\dfrac{x}{4}$ and $\dfrac{6 - x}{4}$

   (b) $A = \left(\dfrac{x}{4}\right)^2 + \left(\dfrac{6 - x}{4}\right)^2$

   $= \dfrac{x^2 - 6x + 18}{8}$

25. $L = \sqrt{x^2 + 20^2}$

$\qquad + \sqrt{30^2 + (50 - x)^2}$

$\quad = \sqrt{x^2 + 400}$

$\qquad + \sqrt{x^2 - 100x + 3400}$

27. $x = (b - a)^2$

29. $L = \dfrac{g^2 T^2}{\pi}$

31. $x = \dfrac{b^2 + 11b + 36}{5}$

33. $n = \left(\dfrac{ts}{\overline{X} - a}\right)^2$

35. $x = 0, 5, -3$

37. $a = 0, \frac{2}{3}, -\frac{1}{2}$ **39.** $y = \pm 4, \pm 1$

41. $a = \pm 2, \pm \sqrt{2}$

43. $b = \pm 4, \pm \sqrt{7}$

45. $x = \pm 2\sqrt{2}, \pm 1$

47. $x = \pm 1$ **49.** $x = \pm 8$ **51.** $x = 4$

53. $y = 125, y = -1$

55. No solution **57.** $x = \frac{1}{2}, \frac{1}{3}$

59. $x = -2, 3$ **61.** $x = \pm \frac{1}{3}, \pm \frac{1}{2}$

63. $a = 81$ **65.** $x = 625$

67. $a = -7$ **69.** $x = \frac{2}{3}, -\frac{1}{2}$

71. $x = -2, 1, \dfrac{-1 \pm i\sqrt{7}}{2}$

73. $a = 10, -1, 5, -2$ **75.** 0.77

Mini-Review

78. $x = 3$ **79.** $m = -\frac{2}{5}$

80. $a = \dfrac{3}{r + 3}$ **81.** $1{,}422{,}525$

Exercises 8.6

1.

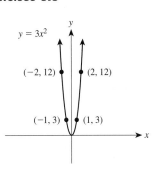

3.

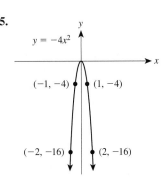

5.

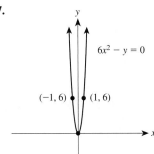

7.

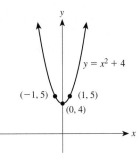

9.

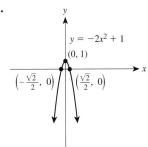

11.

13.

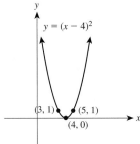

15.

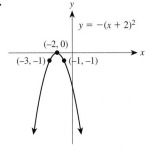

17.

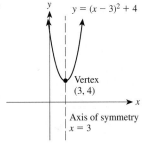

19.

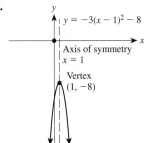

21.

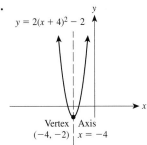

23. Vertex: $(3, 1)$; axis: $x = 3$
25. Vertex: $(2, 3)$; axis: $x = 2$
27. Vertex: $(5, 5)$; axis: $x = 5$

29.

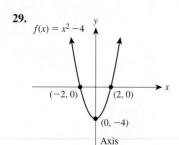

$f(x) = x^2 - 4$

$(-2, 0)$ $(2, 0)$

$(0, -4)$

Axis $x = 0$

31.

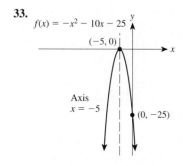

$(-1, 10)$ $(1, 10)$

$(0, 8)$

$f(x) = 2x^2 + 8$

Axis $x = 0$

33.

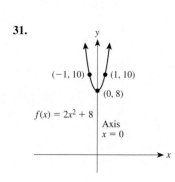

$f(x) = -x^2 - 10x - 25$

$(-5, 0)$

Axis $x = -5$

$(0, -25)$

35.

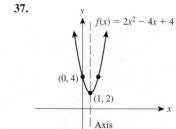

$f(x) = x^2 - 2x - 35$

$(-5, 0)$ $(7, 0)$

Axis $x = 1$

$(0, -35)$ $(1, -36)$

37.

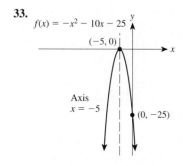

$f(x) = 2x^2 - 4x + 4$

$(0, 4)$

$(1, 2)$

Axis $x = 1$

39.

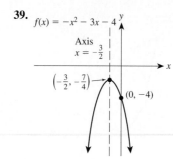

$f(x) = -x^2 - 3x - 4$

Axis $x = -\frac{3}{2}$

$\left(-\frac{3}{2}, -\frac{7}{4}\right)$

$(0, -4)$

41.

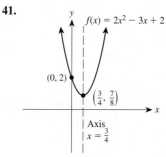

$f(x) = 2x^2 - 3x + 2$

$(0, 2)$

$\left(\frac{3}{4}, \frac{7}{8}\right)$

Axis $x = \frac{3}{4}$

43.

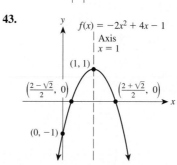

$f(x) = -2x^2 + 4x - 1$

Axis $x = 1$

$(1, 1)$

$\left(\frac{2 - \sqrt{2}}{2}, 0\right)$ $\left(\frac{2 + \sqrt{2}}{2}, 0\right)$

$(0, -1)$

45. 3,500 bagels, $P = \$1,225$

47. 1984

49. Number of cases $= 56$;
maximum $P = \$2,601$

51. $t = \frac{25}{2}$ seconds;
maximum height $= 2,500$ ft

53. Let $x =$ the length of the side
adjacent to the house, and $A =$ area;
$A = -2x^2 + 100x; x = 25$; 25 ft
by 50 ft, where 25 ft is the length of
the adjacent side

55. $R(x) = -\frac{x^2}{6} + 10x$;
maximum revenue $= \$150$,
when $x = 30$

57. (a) $R(x) = (200,000 - 7,250x)(3 +$
$0.25x) = -1,812.5x^2 +$
$28,250x + 600,000$

(b) The maximum profit occurs at
$x = 7.79$, or 8 price increases.
The maximum revenue price
at $x = 8$ is \$5, and the revenue
for this price is \$710,000.

Mini-Review

65. $x = \frac{1}{2}, 5$

66. $\dfrac{1}{(x + 3)^2}$ **67.** 12 oz

68.

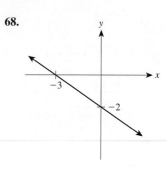

-3

-2

Exercises 8.7

1. $-4 < x < 2$;

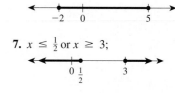

-4 0 2

3. $x < -4$ or $x > 2$;

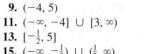

-4 0 2

5. $-2 \le x \le 5$;

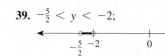

-2 0 5

7. $x \le \frac{1}{2}$ or $x \ge 3$;

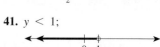

0 $\frac{1}{2}$ 3

9. $(-4, 5)$

11. $(-\infty, -4] \cup [3, \infty)$

13. $[-\frac{1}{2}, 5]$

15. $(-\infty, -\frac{1}{3}) \cup (\frac{1}{2}, \infty)$

17. $[-5, \frac{2}{3}]$

19. $(-\infty, \infty)$

21. No solution

23. $(-\infty, \frac{3}{2}) \cup (5, \infty)$

25. $(-\infty, -\frac{1}{3}] \cup [2, \infty)$ **27.** $x = -1$

29. $(-1, 2)$

31. $(-\infty, -3] \cup (5, \infty)$

33. $(-4, 6)$

35. $(4, \infty)$

37. $y < 1$ or $y > 4$;

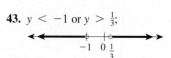

0 1 4

39. $-\frac{5}{2} < y < -2$;

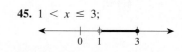

$-\frac{5}{2}$ -2 0

41. $y < 1$;

0 1

43. $y < -1$ or $y > \frac{1}{3}$;

-1 0 $\frac{1}{3}$

45. $1 < x \le 3$;

0 1 3

47. $x < -3$;

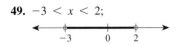

49. $-3 < x < 2$;

51. $x \le 0.7$ or $x \ge 4.3$ (endpoints are estimates)

Mini-Review

56. $\dfrac{8 - 3x}{12 - 2x}$

57. **(a)** $P = 10x + 100$
   (b) \$1,000

58. $\dfrac{x^2}{y}$ **59.** $\dfrac{x^2}{1 + x^2 y}$

Exercises 8.8

1. 5 **3.** $3\sqrt{2}$ **5.** $6\sqrt{13}$ **7.** 1

9. $\dfrac{\sqrt{145}}{6}$ **11.** 0.5 **13.** (0, 6)

15. (0, 1) **17.** (0, 0) **19.** $\left(\frac{11}{30}, \frac{11}{8}\right)$

21. Yes **23.** No

25. Yes; both diagonals have midpoint (7, 5).

27. No **29.** $x^2 + y^2 = 1$

31. $(x - 1)^2 + y^2 = 49$

33. $(x - 2)^2 + (y - 5)^2 = 36$

35. $(x - 6)^2 + (y + 2)^2 = 25$

37. $(x + 3)^2 + (y + 2)^2 = 1$

39. Center (0, 0); radius = 4

41. Center (0, 0); radius = $2\sqrt{6}$

43. Center (3, 0); radius = 4

45. Center (2, 1); radius = 1

47. Center (−1, 3); radius = 5

49.

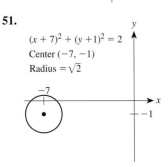

$(x + 2)^2 + (y + 3)^2 = 32$
Center (−2, −3)
Radius = $4\sqrt{2}$

51.

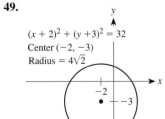

$(x + 7)^2 + (y + 1)^2 = 2$
Center (−7, −1)
Radius = $\sqrt{2}$

53.

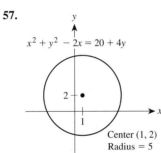

$x^2 + y^2 - 2x = 15$

Center (1, 0)
Radius = 4

55.

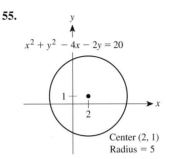

$x^2 + y^2 - 4x - 2y = 20$

Center (2, 1)
Radius = 5

57.

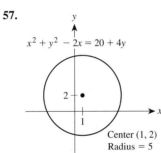

$x^2 + y^2 - 2x = 20 + 4y$

Center (1, 2)
Radius = 5

59. Center (2, −5); radius = 10

61. Center (−3, 1); radius = $2\sqrt{2}$

63. Center (1, −1); radius = $\sqrt{13}$

65. Center $\left(\frac{1}{2}, -1\right)$; radius = 4

67.

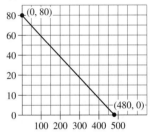

Circle
Center (0, 0)
Radius = 4

$x^2 + y^2 = 16$

69.

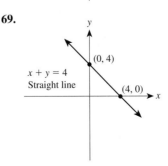

$x + y = 4$
Straight line

(0, 4)

(4, 0)

71. $(x - 1)^2 + \left(y - \dfrac{3}{2}\right)^2 = \dfrac{205}{4}$

73. $(x - 3)^2 + (y + 5)^2 = 122$

75. $C = 4\pi\sqrt{17}$

77. $(x - 3)^2 + (y + 2)^2 = 4$

79. $(x - 3)^2 + (y + 3)^2 = 9$

Mini-Review

83. 10^7 **84.** $\dfrac{y^2 + x}{x^2 y}$

85. $2\sqrt{x} - \sqrt[3]{25x^2}$ **86.** 16

Chapter 8 Review Exercises

1. **(a)** $d(30) = 369.9$,
   $d(40) = 417.5$,
   $d(50) = 427.4$,
   $d(60) = 399.5$
   (b) Approx. 35 mph and 60 mph

2. **(a)** 240 units **(b)** \$46.67
   (c)

(0, 80)

(480, 0)

   (d) $R(x) = x\left(80 - \dfrac{x}{6}\right)$
   (e) $R(150) = 8,250$; $R(200) = 9,353.33$; $R(250) = 9,583.33$;
   $R(300) = 9,000$;
   $R(350) = 7,583.33$;
   (f) The maximum revenue is \$9,600, which occurs at $x = 240$.

3. $y = -\frac{1}{2}, 1$ **4.** $a = -\frac{2}{3}, 5$

5. $x = -\frac{4}{3}, 7$ **6.** $a = \frac{1}{2}, -\frac{1}{5}$

7. $a = \pm 9$ **8.** $y = \pm\sqrt{65}$

9. $z = \pm i\sqrt{5}$ **10.** $r = \pm\dfrac{i\sqrt{10}}{2}$

11. $x = 3$ **12.** $a = 2, -1$

13. $a = 5, -2$ **14.** $y = 3, -\frac{1}{2}$

15. $x = \pm\sqrt{5}$ **16.** $a = 6, 4$

17. $x = -6, 3$ **18.** $x = -1, -\frac{11}{3}$

19. $x = -1 \pm \sqrt{5}$ **20.** $x = 1 \pm \sqrt{5}$

21. $y = \dfrac{-2 \pm \sqrt{10}}{2}$

22. $y = \dfrac{-2 \pm i\sqrt{2}}{2}$

23. $a = \dfrac{-3 \pm 2\sqrt{6}}{3}$

24. $a = \dfrac{3 \pm i\sqrt{6}}{3}$

25. $a = \dfrac{4 \pm \sqrt{43}}{2}$

26. $a = \pm \dfrac{\sqrt{22}}{2}$

27. $a = -\frac{1}{3}, \frac{5}{2}$ **28.** $x = 4, -\frac{2}{5}$

29. $a \approx -0.7, 2.5$ **30.** $y = 1, -\frac{7}{2}$

31. $a = \pm 1$ **32.** $a = 5, -3$

33. $x = 2 \pm \sqrt{6}$ **34.** $y = 1 \pm \sqrt{39}$

35. $t = -4$ **36.** $u = \frac{5}{3}$

37. $x = \pm \dfrac{\sqrt{6}}{2}$ **38.** $x = \pm \dfrac{\sqrt{57}}{3}$

39. $x = 0$ **40.** $x = \dfrac{3.8 \pm i\sqrt{145.4}}{14.8}$

41. $x = 2 \pm \sqrt{2}$

42. $x \approx -1.66, 2.16$

43. $z = 1 \pm \sqrt{10}$ **44.** $z = 3, -1$

45. $x = \dfrac{-16 \pm \sqrt{6}}{5}$

46. $x = \dfrac{-7 \pm \sqrt{201}}{4}$

47. $x = \dfrac{4 \pm \sqrt{22}}{2}$ **48.** $x = 9, -1$

49. $r = \dfrac{\sqrt{\pi A h}}{\pi h}$ **50.** $t = \dfrac{\sqrt{2lg}}{g}$

51. $x = \dfrac{-3y}{2}, x = y$

52. $y = 5x, y = -x$

53. $a = 3$ **54.** $a = 4, 1$

55. $a = 5$ **56.** $y = 0, 3$

57. $x \approx 44.8$ **58.** $x = 16$

59. $x = 7, 4$ **60.** $x = 5$

61. $y = \dfrac{x^2 - z}{3}$ **62.** $y = \dfrac{x^2 + z}{5}$

63. $y = \dfrac{(x - z)^2}{3}$ **64.** $y = \dfrac{(x + z)^2}{5}$

65. $x = 0, 5, -3$ **66.** $x = 0, 7, -5$

67. $x = 0, 3, -\frac{1}{2}$ **68.** $x = 0, -1, \frac{2}{3}$

69. $a = \pm 4, \pm 1$ **70.** $y = \pm 3$

71. $y = \pm 2, \pm i$ **72.** $a = \pm 3, \pm 2i$

73. $z = \pm\sqrt{5}, \pm 1$ **74.** $z = \pm 3, \pm\sqrt{2}$

75. $a = 81$ **76.** $a = 27, -8$

77. $x = 27, -\frac{1}{8}$ **78.** $x = 81$

79. $x = 625$ **80.** $x = 16, 81$

81. $x = -1, \frac{3}{2}$ **82.** $x = -\frac{5}{3}, 1$

83.

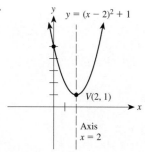

84.

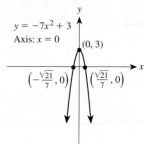

85.

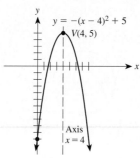

86.

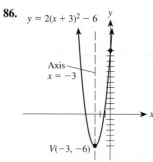

87. $y = 2(x - 3)^2 - 14$;
vertex: $(3, -14)$; axis: $x = 3$

88. $y = 2\left(x - \frac{3}{2}\right)^2 - \frac{7}{2}$;
vertex: $\left(\frac{3}{2}, -\frac{7}{2}\right)$; axis: $x = \frac{3}{2}$

89. $y = -(x - 2)^2 - 8$;
vertex: $(2, -8)$; axis: $x = 2$

90. $y = -(x + 3)^2 + 9$;
vertex: $(-3, 9)$; axis: $x = -3$

91.

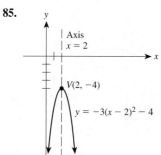

92.

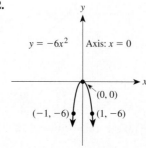

93.

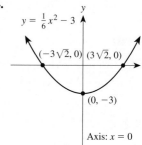

94.

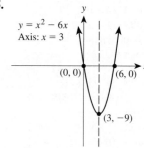

95.

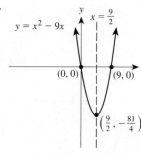

96.

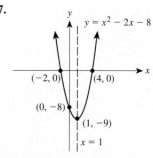

97.

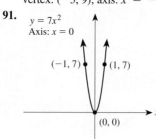

98.

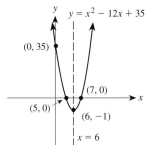

99.

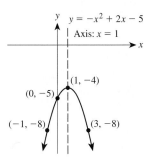

100.

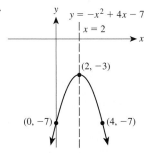

101. $x < -1$ or $x > 2$
102. $a \le -5$ or $a \ge 1$
103. $-\frac{1}{3} \le x \le 2$ **104.** $-\frac{1}{2} < y < 6$
105. $(-\infty, 1) \cup (4, \infty)$;

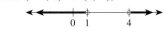

106. $[4, 9]$;

107. $(-9, 9)$;

108. $(-\infty, -9) \cup (9, \infty)$;

109. $(-\infty, -\frac{2}{5}] \cup [4, \infty)$;

110. $[-3, 5]$;

111. $(-2, 3)$;

112. $(-\infty, -4) \cup (2, \infty)$;

113. $(-\infty, -2) \cup [3, \infty)$;

114. $[-4, 2)$;

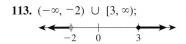

115. $x < 3$
116. $x < 2$ or $x \ge \frac{11}{4}$
117. $-4 < x \le -\frac{11}{4}$
118. $x < -5$ or $x > -\frac{8}{3}$
119. $x < -3.6$ or $x > 0.6$
120. $-3.6 < x < 0.6$
121. $2\sqrt{10}$; $(1, 3)$ **122.** $2\sqrt{5}$; $(3, 10)$
123. 4; $(0, 5)$ **124.** 16; $(3, 0)$
125. $2\sqrt{2}$; $(5, -5)$ **126.** $2\sqrt{58}$; $(0, 0)$
127. $2\sqrt{29}$; $(0, 0)$ **128.** $2\sqrt{29}$; $(0, 0)$
129. Yes   **130.** No

131.

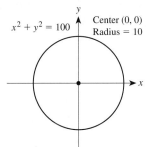

132.

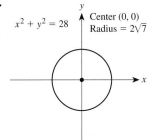

133.

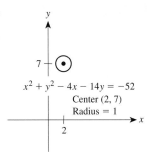

134.

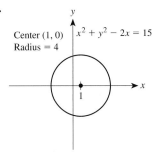

135.

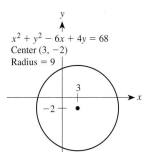

136.

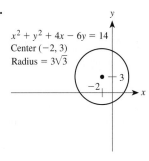

137. $(x - 2)^2 + (y - 6)^2 = 20$
138. $(x - 2)^2 + (y + 4)^2 = 101$
139. Ticket price $= \$6$
140. **(a)** Maximum height $= \frac{65}{8}$ ft
   (b) 5 ft
141. Let $x =$ number; $x^2 + 4 = 36$;
   $x = \pm 4\sqrt{2}$
142. Let $x =$ number; $(x + 4)^2 = 36$;
   $x = 2$ or -10
143. Let $x =$ number; $x + \frac{1}{x} = \frac{53}{14}$;
   $x = \frac{2}{7}$ or $x = \frac{7}{2}$
144. Let $x =$ number; $x - \frac{1}{x} = \frac{40}{21}$;
   $x = \frac{7}{3}$ or $-\frac{3}{7}$
145. Let width $= x$; $50 = x(2x)$;
   $x = 5$ ft (width); length $= 10$ ft
146. Let $w =$ width; $w(3w + 2) =$
   85; width $= 5''$, length $= 17''$
147. Let width of frame $= x$;
   $(5 + 2x)(8 + 2x) - 40 = 114$;
   $x = 3''$
148. Let $x =$ width of walkway;
   $(2x + 20)(2x + 30) - 600 =$
   216; 2 ft wide
149. Hypotenuse $= 5\sqrt{10} \approx 15.81''$
150. Length of leg $= 15\sqrt{3} \approx 25.98'$
151. Length of diagonal $= \sqrt{41} \approx 6.40''$
152. Length of side $= 10\sqrt{2} \approx 14.14''$
153. Let rate of wind $= x$ mph;
   $$\frac{300}{200 - x} + \frac{300}{200 + x} = \frac{25}{8};$$
   $x = 40$ mph
154. **(a)** $R(x) = -75n^2 + 1{,}850n$
   $+ 84{,}000$
   (b) $n = 12$ at $\$18$. Maximum
   revenue is $\$95{,}400$ for $n = 12$.
155. **(a)** $R(x) =$
   $(36{,}000 + 3{,}000x)(4 - 0.25x)$
   $= 144{,}000 + 3{,}000x - 750x^2$

(b) At fare $= \$3.50$, $x = 2$ and revenue is $147,000.

13. Center: $(4, -6)$; radius: $\sqrt{50} = 5\sqrt{2}$

Chapter 8 Practice Test

1. (a) $z = 3, -\frac{1}{2}$ **(b)** $x = \pm\sqrt{5}$

2. (a) $y = -2 \pm \sqrt{5}$

   (b) $a = \dfrac{-1 \pm \sqrt{17}}{4}$

   (c) $x = 5, -4$

3. Roots are distinct and imaginary.

4. $r = \dfrac{\sqrt{2V\pi h}}{2\pi h}$ **5.** $x = 12$

6. $x = 625$

7. (a) $x \le -4$ or $x \ge 9$;

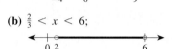

   (b) $\frac{2}{3} < x < 6$;

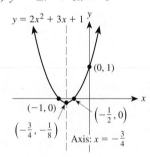

8. Length of diagonal $= 9\sqrt{2} \approx 12.73'$

9. Let speed of boat in still water $= x$ mph;
$$\frac{15}{x-3} + \frac{12}{x+3} = \frac{9}{4};$$
$x = 13$ mph

10. (a) $y = 2x^2 + 3x + 1$

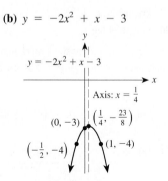

   (b) $y = -2x^2 + x - 3$

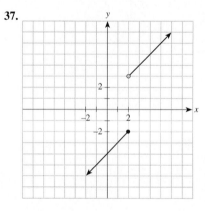

11. 500 widgets; $1,000

12. Distance $= 2\sqrt{10}$; midpoint: $(-1, 6)$

Exercises 9.1

1. $f(-2) = -16; f(0) = -2;$ $f(2) = 12$

3. -2 and 0 are not in the domain of $f(x); f(2) = 2$

5. $f(-2) = 2; f(0) = \sqrt{2}; f(2) = 0$

7. $f(-2) = -1;$ 0 is not in the domain of $f(x); f(2) = 1$

9. $f(-2) = 0; f(0) = -1;$ 2 is not in the domain of $f(x)$

11. All reals   **13.** All reals

15. All reals   **17.** $\left\{x \mid x \ge \frac{1}{2}\right\}$

19. $\left\{x \mid x \le \frac{2}{3}\right\}$ **21.** $\{x \mid x \ne 0\}$

23. $\{x \mid x \ne 4\}$ **25.** $\{x \mid x \ne 1, -2\}$
27. $\{x \mid x \ne 4, -3\}$ **29.** $\{x \mid x > 0\}$

31. $\left\{x \mid x > \frac{1}{2}\right\}$

33. (a) $2x + 4$ **(b)** $2x + 7$
   (c) $4x - 3$ **(d)** $4x - 6$
   (e) 10 **(f)** 2 **(g)** $2x + 2h - 3$
   (h) $2x + 2h - 6$ **(i)** 2

35. (a) $\dfrac{5 + 5x}{2x - 2}$ **(b)** $\dfrac{5}{x + 2}$

   (c) $\dfrac{5}{3x - 1}$ **(d)** $\dfrac{15}{x - 1}$

   (e) $\dfrac{-15}{(x + 2)(x - 1)}$

   (f) $\dfrac{-5}{(x + 2)(x - 1)}$

   (g) $\dfrac{5}{x + h - 1}$

   (h) $\dfrac{5x + 5h - 10}{(x - 1)(h - 1)}$

   (i) $\dfrac{-5}{(x + h - 1)(x - 1)}$

37.

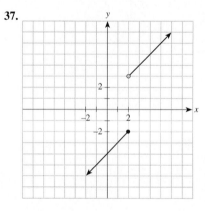

$$f(x) = \begin{cases} x - 4 & \text{if } x \le 2 \\ x + 1 & \text{if } x > 2 \end{cases}$$

$f(0) = -4; f(2) = -2;$
$f(3) = 4$

39.

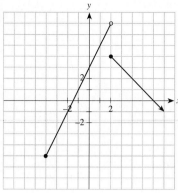

$$h(x) = \begin{cases} 2x + 3 & \text{if } -4 \le x < 2 \\ 6 - x & \text{if } x \ge 2 \end{cases}$$

$h(-1) = 1; h(2) = 4; h(5) = 1$

41.

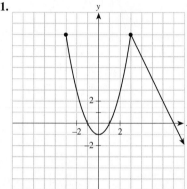

$$g(x) = \begin{cases} x^2 - 1 & \text{if } -3 \le x \le 3 \\ 14 - 2x & \text{if } x > 3 \end{cases}$$

$g(-3) = 8; g(3) = 8; g(6) = 2$

43.

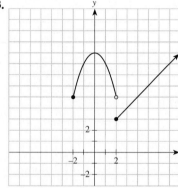

$$f(x) = \begin{cases} 9 - x^2 & \text{if } -2 \le x < 2 \\ x + 1 & \text{if } x \ge 2 \end{cases}$$

$f(-3)$ is not defined; $f(1) = 8;$
$f(5) = 6$

45. (a)
$$C(t) = \begin{cases} 0.61t & \text{if } 0 \le t \le 200 \\ 0.59(t - 200) + 122 & \text{if } t > 200 \end{cases}$$

   (b) $C(150) = \$91.50;$
   $C(200) = \$122;$
   $C(300) = \$181$

47. (a)
$$C(k) = \begin{cases} 2.50 + 0.125k & \text{if } 0 \le k \le 560 \\ 0.118(k - 560) + 72.50 & \text{if } k > 560 \end{cases}$$

(b) $C(300) = \$40;$
   $C(600) = \$77.22;$
   $C(1,000) = \$124.42$

Mini-Review

50. $x = -\dfrac{1}{2}$ **51.** $\dfrac{2x^2 - 4x - 4}{(x + 1)(x - 1)}$

52. (a) 0 **(b)** -3
   (c) $x = -7, 2,$ and 6
   (d) $f(6)$ is greater because the
      graph is falling between
      $x = 6$ and $x = 7$.

53. Let $r =$ rate going;

$\dfrac{60}{r} = \dfrac{60}{r + 12} + \dfrac{1}{4};$

48 mph going, $t = 1\frac{1}{4}$ hr;
60 mph returning, $t = 1$ hr.

Exercises 9.2

1. -4 **3.** 4 **5.** 0
7. Undefined **9.** $x^2 - x - 4$

11. $\dfrac{x - 1}{x^2 - 2x - 3}$ **13.** $5\frac{1}{3}$

15. Undefined **17.** $\frac{1}{35}$

19. $\dfrac{1}{(2x - 1)(2x + 1)}$ **21.** 0

23. $\sqrt{6}$ **25.** $x - 3 \ (x \ge 1)$

27. $\sqrt{x^2 - 3}$ **29.** $\dfrac{2\sqrt{3}}{3}$ **31.** $-\dfrac{5}{2}$

33. $\sqrt{\dfrac{x + 1}{x}} = \dfrac{\sqrt{x^2 + x}}{x}$

35. 1 is not in the domain of $g[f(x)]$

37. 15 **39.** 33 **41.** $\dfrac{141}{25}$

43. 2 is not in the domain of $g[h(x)]$

45. $\frac{41}{4}$ **47.** $9x^2 - 15x + 9$

49. $\dfrac{5x - 7}{4 - 2x}$ **51.** $9x - 4$

53. (a) $36\pi \approx 113.10$ sq in.
   (b) $9\pi t^2$ sq in.
55. (a) $V = 4,500\pi \approx 14,137$ cu in.
   (b) $V = 36\pi t^3 \approx 113.1t^3$ cu in.
57. $C = 8P$
59. Let $s =$ length; $1,000 = s^2;$
   $s = \sqrt{1,000};$ cost $= x(4s) =$
   $x(4\sqrt{1,000}) = 40x\sqrt{10}$ dollars

Mini-Review

62. Let $x =$ one side;

other side $= \dfrac{59 - 2x}{2};$

$210 = x\left(\dfrac{59 - 2x}{2}\right);$

12×17.5 in.

63. $2x^{5/8}y^2$ **64.** $\dfrac{2x + 3}{x - 3}$

65. $4x^4y^2\sqrt{2y}$

Exercises 9.3

1. Linear function
3. Quadratic function
5. Polynomial function
7. Quadratic function
9. Square root function
11. Absolute value function
13. Linear function
15. Quadratic function
17. $9, 5, 1, 3, 7$ **19.** $7, 3, -1, 3, 7$
21. $9, 5, 1, 5, 9$ **23.** $7, 3, 1, 5, 9$

25.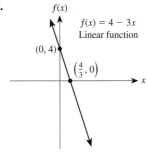

27.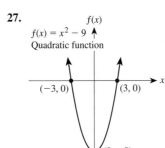

29.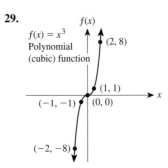

31.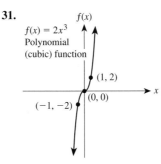

33.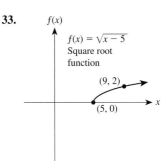

35.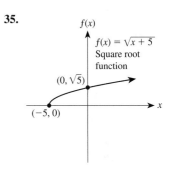

37.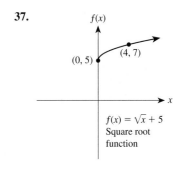

39.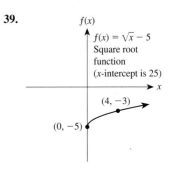

41.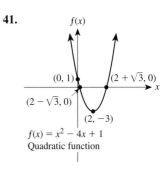

43.

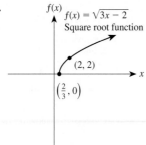

$f(x) = \sqrt{3x - 2}$
Square root function
$(2, 2)$
$\left(\frac{2}{3}, 0\right)$

45.

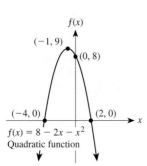

$(-1, 9)$
$(0, 8)$
$(-4, 0)$ $(2, 0)$
$f(x) = 8 - 2x - x^2$
Quadratic function

47.

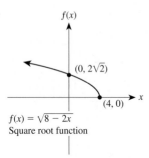

$(0, 2\sqrt{2})$
$(4, 0)$
$f(x) = \sqrt{8 - 2x}$
Square root function

49.

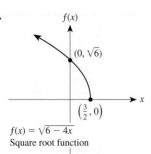

$(0, \sqrt{6})$
$\left(\frac{3}{2}, 0\right)$
$f(x) = \sqrt{6 - 4x}$
Square root function

51.

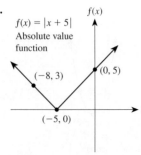

$f(x) = |x + 5|$
Absolute value function
$(-8, 3)$
$(0, 5)$
$(-5, 0)$

53.

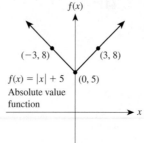

$f(x)$
$(-3, 8)$ $(3, 8)$
$f(x) = |x| + 5$ $(0, 5)$
Absolute value function

55.

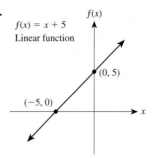

$f(x) = x + 5$
Linear function
$(0, 5)$
$(-5, 0)$

57.

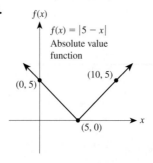

$f(x) = |5 - x|$
Absolute value function
$(0, 5)$ $(10, 5)$
$(5, 0)$

Mini-Review

65. Let r = rate of the medium speed sorter; $\dfrac{500}{r} = \dfrac{500}{r + 150} + \dfrac{1}{6}$;
$r = 600$ letters/minute, $t = 50$ minutes; high speed rate = 750 letters/minute, $t = 40$ minutes

66. -6

67. (a) $3x^2 - 14x + 21$
(b) $3x^2 - 2x - 8$

Exercises 9.4

1. $\{(3, 1), (5, 2)\}$
3. $\{(2, 3), (-1, -3), (3, -1)\}$
5. Function has no inverse.
7. Function   **9.** One-to-one function
11. Neither   **13.** Function
15. One-to-one function
17. One-to-one function
19. Function   **21.** Neither
23. One-to-one function
25. Domain: $\{2, 3\}$;
inverse: $\{(-2, 3), (-3, 2)\}$;
domain of inverse: $\{-2, -3\}$

27. Domain: $\{-3, 2, 6\}$;
inverse: $\{(-3, 6), (-4, 2), (6, -3)\}$;
domain of inverse: $\{-4, -3, 6\}$
29. Domain: $\{2, 3, 4\}$; no inverse
31. Domain: all reals,
$f^{-1}(x) = \dfrac{x - 4}{3}$;
domain of $f^{-1}(x)$: all reals
33. Domain: all reals;
$g^{-1}(x) = \dfrac{x + 3}{2}$;
domain of $g^{-1}(x)$: all reals
35. Domain: all reals;
$h^{-1}(x) = \dfrac{4 - x}{5}$;
domain of $h^{-1}(x)$: all reals
37. Domain of $f(x)$: all reals; no inverse
39. Domain: all reals;
$f^{-1}(x) = \sqrt[3]{x - 4}$;
domain of $f^{-1}(x)$: all reals
41. Domain: $\{x \mid x \neq 0\}$; $g^{-1}(x) = \dfrac{1}{x}$;
domain of $g^{-1}(x)$: $\{x \mid x \neq 0\}$
43. Domain: $\{x \mid x \neq -3\}$;
$g^{-1}(x) = \dfrac{2 - 3x}{x}$;
domain of $g^{-1}(x)$: $\{x \mid x \neq 0\}$
45. Domain: $\{x \mid x \neq 0\}$;
$g^{-1}(x) = \dfrac{1}{1 - x}$;
domain of $g^{-1}(x)$: $\{x \mid x \neq 1\}$
47. Domain: $\{x \mid x \neq 1\}$;
$h^{-1}(x) = \dfrac{x + 2}{x - 1}$
domain of $h^{-1}(x)$: $\{x \mid x \neq 1\}$

49.

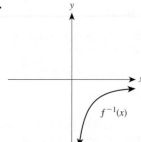

y
x
$f^{-1}(x)$

51. Graph is its own inverse (same graph).

Mini-Review

55. $x - 10\sqrt{x} + 25$

56. $x = \dfrac{3 \pm \sqrt{57}}{4}$

57. 7%

Exercises 9.5

1. $y = 6$ **3.** $y = \frac{40}{3}$ **5.** $x = \frac{15}{22}$

7. $a = 36$ **9.** $r = \frac{243}{4}$ **11.** $y = 10$

13. $x = 28$ **15.** $x = \frac{125}{6}$

17. $V = \frac{1,000}{3}\,\text{m}^3$ **19.** $V = \frac{256\pi}{3}\,\text{cm}^3$

21. $a = \frac{3}{256}$ **23.** $a = \frac{32}{3}$

25. $5\sqrt{2} \approx 7.07$ seconds

27. $E = 6.25$ foot-candles

29. $z = 15$ **31.** $z = \frac{50}{3}$ **33.** $d = \frac{5}{4}$

35. $z = 40$ **37.** $z = \frac{160}{9}$

39. $z = 160$ **41.** $z = \frac{32}{3}$

43. $6\pi\,\text{m}^3$ **45.** 3.75 ohms

47. If r is doubled, the circumference is doubled; if r is tripled, the circumference is tripled; if r is halved, the circumference is halved.

49. y is divided by 4.

51. z is multiplied by 20.

53. s is multiplied by 4.

Mini-Review

55. $(5x)^{1/2} + 5x^{1/3}$

56.

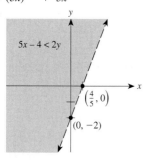

$5x - 4 < 2y$

$\left(\frac{4}{5}, 0\right)$

$(0, -2)$

57. Let $x =$ amount invested at 6.5%;
$0.065x + 0.07(x + 5,000) = 1,430;$
$8,000 at 6.5%; $13,000 at 7%

58. $\dfrac{10x - 3x^2}{25 + 4x^2}$

Chapter 9 Review Exercises

1. $f(-3) = -11; f(1) = 9;$
$f(3) = 19$

2. $f(-3) = 1; f(1) = 5; f(3) = 19$

3. $f(-3)$ is undefined; $f(1) = 0;$
$f(3) = \sqrt{2}$

4. $f(-3) = \sqrt{11};$ 1 and 3 are not in the domain of f

5. $f(-1) = \frac{3}{2}; f(1)$ is undefined;
3 is not in the domain of f

6. $f(-3) = -3 - 2\sqrt{6};$
$f(1) = 1 - 2\sqrt{2}; f(3) = 3$

7. All reals **8.** All reals

9. All reals **10.** $\{x \mid x \le 2\}$

11. $\left\{x \mid x \le \frac{3}{5}\right\}$ **12.** $\{x \mid x \ne 2\}$

13. $\left\{x \mid x > \frac{1}{2}\right\}$ **14.** $\{x \mid x > 5\}$

15. $\left\{x \mid x > \frac{3}{2}\right\}$ **16.** $5x + 12$

17. $5x + 17$ **18.** $5x + 4$

19. $5x + 5$ **20.** $5x + 14$

21. $5x + 19$ **22.** $\sqrt{-2x - 1}$

23. $\sqrt{-2x - 3}$ **24.** $\sqrt{3 - 4x}$

25. $\sqrt{3 - 6x}$ **26.** $2\sqrt{3 - 2x}$

27. $3\sqrt{3 - 2x}$ **28.** $5h$

29. $\sqrt{3 - 2x - 2h} - \sqrt{3 - 2x}$

30. $5\sqrt{3 - 2x} + 2$

31. $\sqrt{-10x - 1}$

32. 5 **33.** 1 **34.** 7 **35.** $-\frac{6}{5}$

36. $-\frac{5}{6}$ **37.** Undefined **38.** 0

39. 7 **40.** 2 **41.** -5 **42.** -2

43. $-\frac{17}{4}$ **44.** $\frac{17}{42}$ **45.** 1

46. Undefined

47. $x^2 - x - 4$ **48.** $x^2 + x - 5$

49. $\dfrac{3x + 1}{2x}$ **50.** $\dfrac{x + 2}{2x + 2}$

51. (a) $r = 4\sqrt{\dfrac{2}{\pi}} \approx 3.2$ in.

(b) $r = \sqrt{\dfrac{8t}{\pi}} = 2\sqrt{\dfrac{2t}{\pi}} = \dfrac{2\sqrt{2\pi t}}{\pi}$

52. (a) $A = 400$ sq in.
(b) $A = 16t^2$ sq in.

53. $20\pi x\sqrt{2}$ dollars

54. (a) 12.5 sq ft (b) $A = \frac{1}{2}d^2$

55.

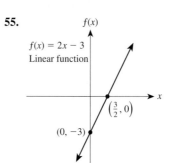

$f(x) = 2x - 3$
Linear function

$\left(\frac{3}{2}, 0\right)$

$(0, -3)$

56.

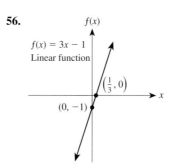

$f(x) = 3x - 1$
Linear function

$\left(\frac{1}{3}, 0\right)$

$(0, -1)$

57.

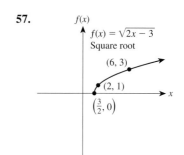

$f(x) = \sqrt{2x - 3}$
Square root

$(6, 3)$

$(2, 1)$

$\left(\frac{3}{2}, 0\right)$

58.

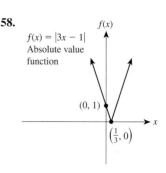

$f(x) = |3x - 1|$
Absolute value function

$(0, 1)$

$\left(\frac{1}{3}, 0\right)$

59.

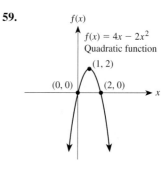

$f(x) = 4x - 2x^2$
Quadratic function

$(1, 2)$

$(0, 0)$ $(2, 0)$

60.

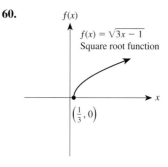

$f(x) = \sqrt{3x - 1}$
Square root function

$\left(\frac{1}{3}, 0\right)$

61.

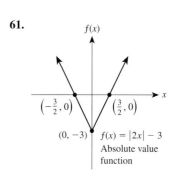

$\left(-\frac{3}{2}, 0\right)$ $\left(\frac{3}{2}, 0\right)$

$(0, -3)$ $f(x) = |2x| - 3$
Absolute value function

62.

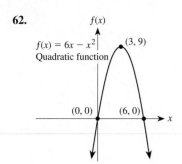

$f(x) = 6x - x^2$
Quadratic function
$(3, 9)$
$(0, 0)$ $(6, 0)$

63. Function   **64.** Neither
65. One-to-one   **66.** One-to-one
67. One-to-one   **68.** Function
69. Neither   **70.** One-to-one
71. $\{(3, 2), (4, 3)\}$
72. $\{(-8, 5), (7, 3), (6, 2)\}$
73. $y = \dfrac{x - 8}{3}$ **74.** $y = \dfrac{x + 1}{5}$
75. $y = \sqrt[3]{x}$ **76.** $y = \sqrt[5]{x}$
77. $y = \dfrac{3 - x}{x}$ **78.** $y = -\dfrac{3x}{x - 2}$

79.

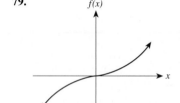

80.

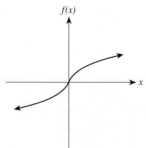

81. $x = \frac{16}{3}$ **82.** $x = 12$
83. $s = \pm 2\sqrt{2}$ **84.** $s = \sqrt[3]{2}$
85. $z = \frac{96}{7}$ **86.** $z = 36$
87. $V = 288\pi$ in.^{3} **88.** 158.73 lb
89. 60 m^3 **90.** \$11,400

Chapter 9 Practice Test

1. (a) All reals   **(b)** $\left\{x \mid x \neq \frac{3}{2}\right\}$
   (c) $\left\{x \mid x \leq \frac{8}{3}\right\}$
2. (a) Undefined   **(b)** $3x + 1$
   (c) $\dfrac{-9}{(x + 4)(x + 1)}$
3. (a) 6   **(b)** 1   **(c)** $\sqrt{31}$
   (d) $x - 5$ **(e)** $\sqrt{x^2 - 5}$

4. (a) $P = 16\sqrt{2}$ ft
   (b) $P = 4\left(\dfrac{d}{\sqrt{2}}\right) = 2d\sqrt{2}$

5.

$f(x)$
$f(x) = |2x - 1|$
$(0, 1)$ $(1, 1)$
$\left(\frac{1}{2}, 0\right)$

6. No inverse   **7.** No
8. (a) $f^{-1}(x) = \dfrac{x + 4}{2}$
   (b) $f^{-1}(x) = \dfrac{3x}{1 - x}$
9. $y = \frac{8}{3}$ **10.** $z = 405$

Chapters 7–9
Cumulative Review

1. $54x^5y^7$ (§7.1)   **2.** $\dfrac{y^4}{2x^2}$ (§7.1)
3. $\dfrac{1}{4x^2y^2}$ (§7.1)   **4.** $-6a^2b^3$ (§7.1)
5. $\dfrac{1}{x^{12}y}$ (§7.1)   **6.** $\dfrac{x^6}{4y^2}$ (§7.1)
7. $\dfrac{1}{rs^2}$ (§7.1)   **8.** $-\frac{27}{8}$ (§7.1)
9. $-\dfrac{8}{rs^7}$ (§7.1)   **10.** $\dfrac{5y^3}{x^3}$ (§7.1)
11. $\dfrac{y^2 - x^2}{xy}$ (§7.1)
12. $\dfrac{b^2 + a}{b^2 + ab^2}$ (§7.1)
13. 5.642932×10^4 (§7.2)
14. 7.52×10^{-5} (§7.2)
15. $(1.86 \times 10^5)(60)(60)(24) =$
   1.60704×10^{10} miles (§7.2)
16. 4.8×10^7 (§7.2)
17. -10 (§7.3)   **18.** $\frac{16}{9}$ (§7.3)
19. $a^{1/6}$ (§7.3)   **20.** $x^{1/15}$ (§7.3)
21. $\frac{1}{9}$ (§7.3)   **22.** $\dfrac{1}{x^{1/2}}$ (§7.3)
23. $\sqrt[4]{x^3}$ (§7.3)   **24.** $2\sqrt{x}$ (§7.3)
25. $4a^2b^4$ (§7.4)
26. $2ab^5\sqrt{2a}$ (§7.4)
27. $24x^3y^2\sqrt{x}$ (§7.4)
28. $3xy\sqrt[3]{3xy^2}$ (§7.4)   **29.** 4 (§7.4)
30. $\dfrac{\sqrt{3x}}{xy}$ (§7.4)   **31.** $\sqrt{x}$ (§7.4)

32. $\dfrac{\sqrt[3]{50x^2}}{5x}$ (§7.4)   **33.** $7\sqrt{5}$ (§7.5)
34. 0 (§7.5)   **35.** $2a^2\sqrt[3]{a}$ (§7.5)
36. $-\dfrac{\sqrt{6}}{6}$ (§7.5)   **37.** $2 + \sqrt{2}$ (§7.6)
38. $8 - 2\sqrt{15}$ (§7.6)   **39.** 2 (§7.6)
40. $-8\sqrt{x} - 12$ (§7.6)
41. $4\sqrt{3} - 5\sqrt{2}$ (§7.6)
42. $\sqrt{15} - 3$ (§7.6)
43. $x = \frac{47}{3}$ (§7.7)   **44.** $x = \frac{25}{3}$ (§7.7)
45. $x = -7$ (§7.7)   **46.** $x = -1$ (§7.7)
47. $-i$ (§7.8)   **48.** -1 (§7.8)
49. $17 - 7i$ (§7.8)   **50.** $\frac{7}{5} + \frac{1}{5}i$ (§7.8)
51. $a = 5, -3$ (§8.2)
52. $a = 6, 1$ (§8.2)
53. $x = \pm\dfrac{\sqrt{38}}{2}$ (§8.2)
54. $x = 5 \pm 2\sqrt{7}$ (§8.2)
55. $y = -3 \pm \sqrt{10}$ (§8.3)
56. $x = \dfrac{-3 \pm \sqrt{15}}{3}$ (§8.3)
57. $a = \dfrac{1 \pm \sqrt{7}}{3}$ (§8.4)
58. $x = 2 \pm i\sqrt{3}$ (§8.4)
59. $x = -1, -3$ (§8.4)
60. $y = -4, 3$ (§8.4)
61. $x = 3, -\frac{9}{2}$ (§8.4)
62. $b = 5$ (§8.4)
63. $x = 2 \pm \sqrt{7}$ (§8.4)
64. $x = \dfrac{25 \pm \sqrt{865}}{10}$ (§8.4)

65. (§8.6)

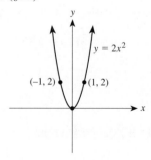

$y = 2x^2$
$(-1, 2)$ $(1, 2)$

66. (§8.6)

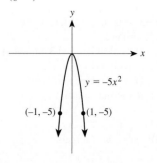

$y = -5x^2$
$(-1, -5)$ $(1, -5)$

67. (§3.1)

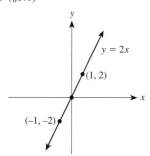

68. (§3.1)

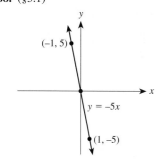

69. (§8.6)

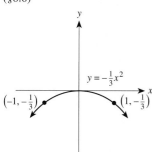

70. (§8.6)

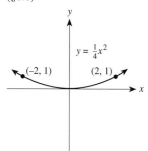

71. (§8.6)

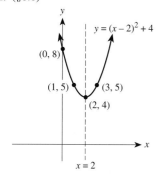

72. (§8.6)

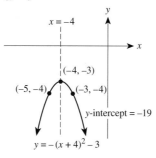

73. (§8.6)

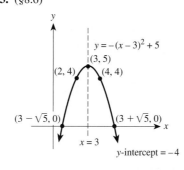

74. (§3.1)

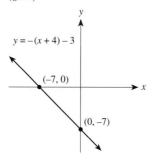

75. (§8.6)

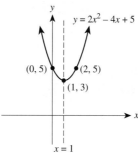

76. (§8.6)

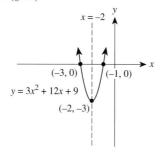

77. $\{x \mid -\frac{1}{2} \le x \le 7\}$ (§8.7)

78. $\{x \mid -2 < x < 8\}$ (§8.7)

79. $\{x \mid x < -\frac{3}{2} \text{ or } x > 2\}$ (§8.7)

80. $\{x \mid x < -\frac{1}{2} \text{ or } x > 6\}$ (§8.7)

81. $\{x \mid x \le -7 \text{ or } x > -3\}$ (§8.7)

82. $\{x \mid -1 < x < 2 \text{ or } x \ge 3\}$ (§8.7)

83. \$1.25; \$312.50 (§8.6)
84. $(x + 3)^2 + (y - 5)^2 = 16$ (§8.8)
85. $x^2 + (y - 2)^2 = 26$ (§8.8)
86. Distance $= \sqrt{82}$ (§8.8);

midpoint $= \left(\frac{5}{2}, \frac{1}{2}\right)$ (§8.8)

87. $(x - 1)^2 + (y - 3)^2 = 13$ (§8.8)

88. Center: $(0, -2)$; radius: $\sqrt{10}$ (§8.8)
89. $2x^2 - x - 6$ (§9.2)
90. $6x^3 - 24x^2 + 24x$ (§9.2)
91. $18x^2 - 84x + 96$ (§9.2)
92. $6x^2 - 12x - 6$ (§9.2)

93. $\dfrac{2}{x + 1}$ $(x \ne 0)$ (§9.2)

94. $\dfrac{2x^2 - x - 1}{x(x + 1)}$ (§9.2)

95. $\dfrac{2x}{3}$ (§9.2) **96.** $3x^2 - 6x$ (§9.2)

97. $\dfrac{2}{x + 1}$ $(x \ne 0)$ (§9.2)

98. $\dfrac{x + 1}{2x}$ (§9.2)

99.

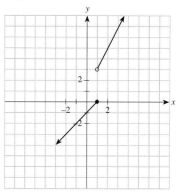

$$f(x) = \begin{cases} x - 1 & \text{if } x \le 1 \\ 2x + 1 & \text{if } x > 1 \end{cases}$$

$f(-2) = -3; f(1) = 0;$
$f(3) = 7$ (§9.1)

100.

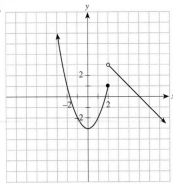

$$g(x) = \begin{cases} x^2 - 3 & \text{if } x \le 2 \\ 5 - x & \text{if } x > 2 \end{cases}$$

$g(0) = -3; g(2) = 1;$
$g(3) = 2$ (§9.1)

101. $y = \sqrt[3]{x + 1}$ (§9.4)

102. $y = \dfrac{3x + 5}{2}$ (§9.4)

103. $y = \dfrac{1}{x - 1}$ (§9.4)

104. No inverse function (§9.4)

105. (§9.3)

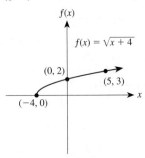

106. (§9.3)

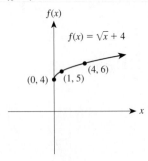

107. (§9.3)

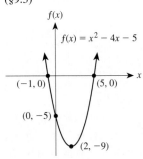

108. (§9.3)

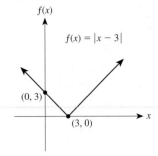

109. (§8.8)

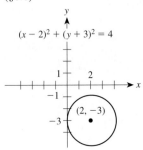

110. (§8.8)

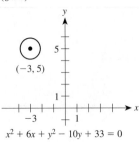

$x^2 + 6x + y^2 - 10y + 33 = 0$

111. $y = \frac{80}{3}$ (§9.5)   **112.** $y = \frac{12}{5}$ (§9.5)
113. $x = 20$ (§9.5)   **114.** $x = \frac{24}{5}$ (§9.5)
115. The area gets multiplied by 6. (§9.5)
116. The illumination quadruples. (§9.5)

Chapters 7–9
Cumulative Practice Test

1. (a) $-72x^7y^{11}$ **(b)** $-\dfrac{18x^3}{y^3}$

     (c) $\dfrac{9x^6}{y^4}$ **(d)** $\dfrac{b}{b + a}$

2. 3.4×10^{-5} **3.** $3{,}000{,}000$

4. $-\frac{1}{8}$ **5. (a)** $x^{2/3}$ **(b)** $\dfrac{1}{x^{7/6}y^{3/4}}$

6. (a) $2xy^2\sqrt{6y}$ **(b)** $2a^3b^3\sqrt{2ab}$

     (c) $\dfrac{\sqrt[3]{5a^2}}{a}$

7. (a) 0 **(b)** $\dfrac{2\sqrt{xy} - y\sqrt{2x}}{2y}$

     (c) $4\sqrt{3} - 7$ **(d)** $\dfrac{5\sqrt{10} + 10}{3}$

8. $x = \frac{1}{2}$ **9.** $\frac{7}{10} - \frac{1}{10}i$

10. (a) $x = -\frac{3}{2}, 1$ **(b)** $x = \pm 2\sqrt{3}$

     (c) $x = \dfrac{-2 \pm \sqrt{10}}{2}$

     (d) $x = \dfrac{7 \pm \sqrt{217}}{6}$

11. (a)

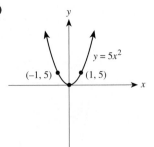

(b)

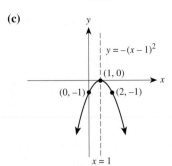

(c)

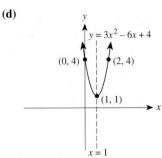

(d)

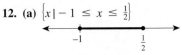

12. (a) $\{x \mid -1 \le x \le \frac{1}{2}\}$

(b) $\{x \mid -\frac{1}{3} < x < 4\}$

(c) $\{x \mid x < -5 \text{ or } x \geq 3\}$

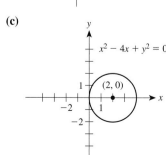

(d) $\{x \mid x < 2 \text{ or } x \geq 3\}$

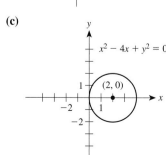

13. $(x + 2)^2 + y^2 = 50$

14. Distance $= \sqrt{29}$;
midpoint $= \left(\frac{1}{2}, 7\right)$

15.

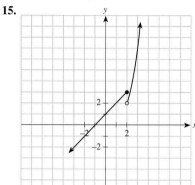

$$f(x) = \begin{cases} x + 1 & \text{if } x \leq 2 \\ x^2 - 2 & \text{if } x > 2 \end{cases}$$

$f(0) = 1; f(2) = 3; f(3) = 7$

16. (a) $9x^3 - 15x^2 + 13x - 3$
(b) $27x^2 - 30x + 10$

17. (a)

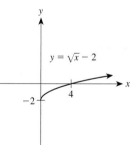

(b)

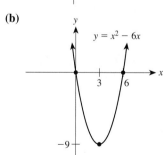

(c)

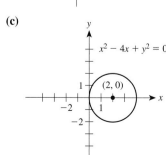

18. $y = \frac{200}{9}$

Exercises 10.1

1.

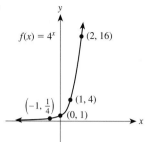

3.

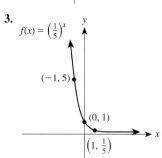

5.

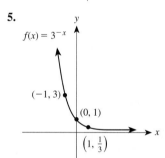

7.

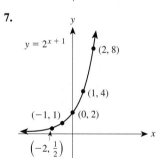

9.

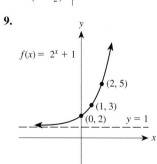

11. (a) $P(0) = 4.28$;
$P(2) = 17.57$;
$P(4) = 72.15$;
$P(6) = 296.26$;
$P(8) = 1216.39$;
$P(10) = 4994.38$

(b)

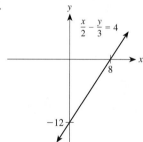

13. $x = 1$ **15.** $x = 2$ **17.** $x = -1$
19. $x = 2$ **21.** $x = 2, x = -1$
23. $x = \frac{9}{4}$ **25.** $x = 1, x = -\frac{1}{4}$
27. $5,000, 10,000, 80,000, 2,500(2^{t/10})$
29. $40,000, 80,000, 20,000(2^{(Y-2004)/14})$
31. $2,500, 1,250, 10,000\left(\frac{1}{2}\right)^t$
33. \$7,716.05 **35.** \$12,288
37. 121,551 **39.** \$14,693.28
41. \$8,052.55
43. (a) \$18,193.97 **(b)** year 18

Mini-Review

45. $\frac{72}{17}$ **46.** $\frac{15x + 6}{2x(x + 2)}$

47.

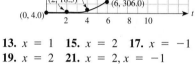

48. 52

Exercises 10.2

1. $7^2 = 49$ **3.** $\log_3 81 = 4$
5. $10^4 = 10,000$
7. $\log_{10} 1,000 = 3$
9. $9^2 = 81$ **11.** $81^{1/2} = 9$
13. $\log_6\left(\frac{1}{36}\right) = -2$ **15.** $3^{-1} = \frac{1}{3}$
17. $\log_{25} 5 = \frac{1}{2}$ **19.** $8^1 = 8$
21. $\log_8 1 = 0$ **23.** $16^{3/4} = 8$
25. $\log_{27}\left(\frac{1}{9}\right) = -\frac{2}{3}$ **27.** $8^{-1/3} = \frac{1}{2}$
29. $\left(\frac{1}{2}\right)^{-2} = 4$ **31.** $\log_3 1 = 0$
33. $7^0 = 1$ **35.** $6^{1/2} = \sqrt{6}$
37. $\log_6 \sqrt{6} = \frac{1}{2}$ **39.** 3 **41.** 2

43. -1 **45.** -3 **47.** $-\frac{1}{2}$ **49.** $\frac{2}{3}$
51. Not defined   **53.** $-\frac{3}{2}$ **55.** $\frac{1}{2}$
57. $\frac{2}{3}$ **59.** 1   **61.** 0   **63.** 7
65. $x = 125$ **67.** $y = 3$ **69.** $b = 4$
71. $x = \frac{1}{36}$ **73.** $x = 8$ **75.** $y = \frac{5}{3}$
77. $b = 2$ **79.** $x = 1$ **81.** 4.3728
83. -3.0339
85.

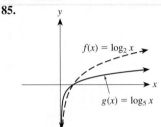

Mini-Review

91. $\sqrt{7} - 2$ **92.** $x^3 - y^3$
93. $x^2 - 3x + 9$
94. Real and distinct

Exercises 10.3

1. $\log_5 x + \log_5 y + \log_5 z$
3. $\log_7 2 - \log_7 3$ **5.** $3 \log_3 x$
7. $\left(\frac{2}{3}\right) \log_b a$ **9.** 8   **11.** $-\frac{1}{4}$
13. $2 \log_b x + 3 \log_b y$
15. $4 \log_b m - 2 \log_b n$
17. $\dfrac{\log_b x}{2} + \dfrac{\log_b y}{2}$
19. $\dfrac{2 \log_2 x}{5} + \dfrac{\log_2 y}{5} - \dfrac{3 \log_2 z}{5}$
21. $\log_b(xy + z^2)$
23. $2 \log_b x - \log_b y - \log_b z$
25. $\dfrac{1}{2} + \log_6 m + \dfrac{\log_6 n}{2}$
$\qquad - \dfrac{5 \log_6 p}{2} - \dfrac{\log_6 q}{2}$
27. $\log_b(xy)$ **29.** $\log_b\left(\dfrac{m^2}{n^3}\right)$
31. $\log_b 80$ **33.** $\log_b\left(\dfrac{x^{1/3} y^{1/4}}{z^{1/5}}\right)$
35. $\log_b \dfrac{\sqrt{xy}}{z^2}$ **37.** $\log_b\left(\dfrac{x^2}{yz^3}\right)$
39. $\log_p\left(\dfrac{x^{2/3} y^{4/3}}{z^{3/7}}\right)$ **41.** 3.3
43. -0.9 **45.** -1.42 **47.** 6
49. 6.6   **51.** 2.25   **53.** $\dfrac{2A}{3}$

55. $3A + 2B - C$
57. $\dfrac{5A}{2} + \dfrac{B}{2} - \dfrac{3C}{2}$

Mini-Review

62. Axis of symmetry: $x = \frac{3}{2}$;
vertex: (1.5, 4.75)
63. $6x - 7\sqrt{x} - 20$
64. $f^{-1}(x) = \dfrac{x + 3}{5}$;

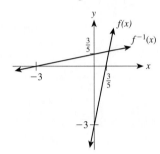

65. (a) $N = \frac{5}{3}s + 55$
(b) 105 beats per minute

Exercises 10.4

1. 2.7664   **3.** -2.4306 **5.** 5.4472
7. -4.2573 **9.** 0.9031
11. -0.0608 **13.** 2.5625
15. 3.4556   **17.** 1.5041
19. Undefined   **21.** 8.7297
23. 54.5982   **25.** $\dfrac{\log x}{\log 5}$ **27.** $\dfrac{\log 8}{\log 7}$
29. $2 \log_{25} 8$ **31.** 2.7748
33. 4.0249   **35.** 6.1082   **37.** 2.0305
41.

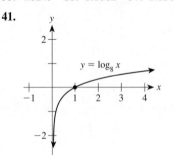

43.

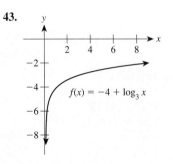

45.

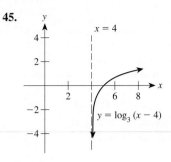

Mini-Review

46. $\sqrt{109}$ **47.** $\left(\frac{7}{2}, -1\right)$
48.

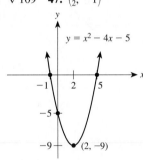

49. $x = 0, 3$

Exercises 10.5

1. $x = \frac{9}{5}$ **3.** $x = 12$ **5.** $x = 6$
7. $x = 9$ **9.** $a = 2$ **11.** $y = \frac{16}{7}$
13. No solution   **15.** $x = 2$
17. $x = 14$ **19.** $x = 1$ **21.** $x = \frac{8}{3}$
23. $x = 8$ **25.** $x = 1$ **27.** $x = 9$
29. $x = 5, x = 1$
31. $x = \dfrac{\log 5}{\log 2} \approx 2.322$
33. $x = \dfrac{\log 6 - \log 2}{\log 2} = \dfrac{\log 3}{\log 2} \approx 1.585$
35. $x = \dfrac{\log 5 - 3 \log 4}{2 \log 4} \approx -0.9195$
37. $y = \dfrac{\log 7}{\log 3 - \log 7} \approx -2.297$
39. $x = \dfrac{2 \log 5 - \log 6}{2 \log 6 - \log 5} \approx 0.723$
41. $x = \dfrac{2 \log 8 + 2 \log 9}{3 \log 8 - \log 9} \approx 2.117$
43. $x = \dfrac{\log 5}{\log 3 - \log 2} \approx 3.969$
45. $y = \dfrac{\log 3}{\log 2 + \log 5} = \dfrac{\log 3}{\log 10}$
$\qquad = \log 3 \approx 0.4771$
47. $a = \dfrac{\log 2 - \log 3}{\log 4 + \log 3} \approx -0.1632$

49. 5.4   **51.** 1×10^{-7}
53. $I = 10{,}000$ watts/cm^2
55. Let $P =$ population after
  m months.
  (a) $P = 10(2^m)$
  (b) $P = 10(2^{24}) = 167{,}772{,}160$
  (c) 10 months

Mini-Review

57. $(x - 2)^2 + (y + 3)^2 = 36$
58. No inverse; the graph fails the
  horizontal line test
59. $-\frac{1}{2} \le x \le 3$
60. $x = 2.31, y = -0.62$

Exercises 10.6

1. \$10,751.33   **3.** \$10,790.80
5. 11.21 years   **7.** 11.21 years
9. 6.96 years   **11.** 18.99 years
13. 6,640   **15.** 13,113
17. 42.5 hours for the first model,
  18.9 hours for the second
19. 28.8 hours
21. $A = A_0 e^{0.21972t}$, 17.8 hours
23. $t = 1{,}732.87$ years
25. Rate $= 17.33\%$ per day
27. $I_1 = 3{,}981.07, I_2 = 15{,}848{,}932$
29. (b) Approx. 8,676 years

Mini-Review

30.

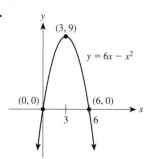

31.

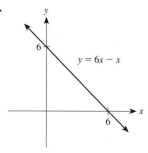

32. $f[g(x)] = 50x^2 + 25x + 3$;
  $g[f(x)] = 10x^2 - 15x + 7$

33. $x = \dfrac{4 \pm \sqrt{5}}{3}$

Chapter 10 Review Exercises

1.

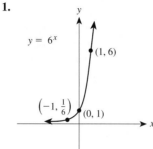

2.

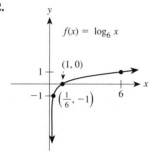

3.

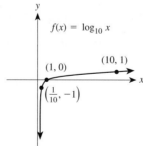

4.

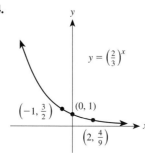

5.

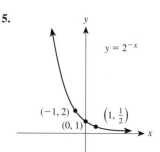

6.

7. $3^4 = 81$ **8.** $\log_6 216 = 3$
9. $\log_4\left(\frac{1}{64}\right) = -3$ **10.** $2^{-2} = \frac{1}{4}$
11. $8^{2/3} = 4$ **12.** $\log_{10}(0.001) = -3$
13. $\log_{25} 5 = \frac{1}{2}$ **14.** $\log_{1/3} 9 = -2$
15. $7^{1/2} = \sqrt{7}$ **16.** $11^{1/3} = \sqrt[3]{11}$
17. $6^0 = 1$ **18.** $\log_{12} 12 = 1$
19. 3   **20.** -4 **21.** -2 **22.** 6
23. -2 **24.** -1 **25.** -2 **26.** 2
27. $-\frac{1}{2}$ **28.** $-\frac{3}{2}$ **29.** $\frac{1}{2}$ **30.** $\frac{1}{3}$
31. $\frac{5}{4}$ **32.** $\frac{4}{5}$
33. $3 \log_b x + 7 \log_b y$
34. $3 \log_b x - 7 \log_b y$
35. $2 \log_b u + 5 \log_b v - 3 \log_b w$
36. $2 \log_b u - 5 \log_b v - 3 \log_b w$
37. $\frac{1}{3} \log_b x + \frac{1}{3} \log_b y$
38. $\frac{1}{2} \log_b x - \frac{1}{2} \log_b y$
39. $\log_b(x^3 + y^4)$
40. $\frac{3}{2} \log_b x + \frac{5}{2} \log_b y - \frac{7}{2} \log_b z$
41. $3 \log_b x + \log_b y - \log_b z$
42. $\dfrac{4 \log_b x}{9 \log_b y}$ **43.** 3.52   **44.** -1.54
45. $x = -2$ **46.** $x = -2$
47. $x = \frac{5}{4}$ **48.** $x = \frac{4}{5}$
49. $\dfrac{\log 3 - \log 5}{\log 5} \approx -0.3174$
50. $\dfrac{4 \log 7}{\log 2 - \log 7} \approx -6.213$
51. $x = 0$ **52.** $x = 1$ **53.** $t = 3$
54. $x = 4$ **55.** $x = 1$
56. No solution   **57.** 2.8938
58. -0.2336 **59.** -2.3019
60. 0.9163   **61.** -5.0672
62. 0.0150   **63.** 2,440.6020
64. 5.2679   **65.** 0.001
66. 4,540.462   **67.** 2.6658
68. -1.1850 **69.** 2.6716
70. -2.6040 **71.** -3.2602

72. -9.7039 **73.** $\$11,412.03$
74. 15.4 years   **75.** $t = 30.81$ years
76. 37.87 years   **77.** 8.2
78. 3.981×10^{-9}

Chapter 10 Practice Test

1.

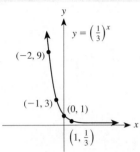

2.

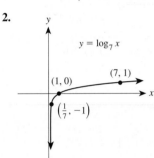

3. (a) $2^4 = 16$ (b) $9^{-1/2} = \frac{1}{3}$
4. (a) $\log_8 4 = \frac{2}{3}$
  (b) $\log_{10}(0.01) = -2$
5. (a) -1 (b) $\frac{1}{2}$ (c) $\frac{5}{3}$
6. (a) $3 \log_b x + 5 \log_b y + \log_b z$
  (b) $3 + \frac{1}{2} \log_4 x - \log_4 y - \log_4 z$
7. (a) 4.4456   (b) -5.5215
  (c) 0.9802
8. 1.3646
9. (a) 2.613   (b) -2.733
10. (a) $x = -4$ (b) $x = 4$
  (c) $x = 4$ (d) $x = 10$
  (e) $x = \dfrac{\log 5 - 3 \log 9}{\log 9} \approx -2.268$
11. $\$8,947.27$ **12.** ≈ 34.7 years

Exercises 11.1

1. $x = 6, y = 3, z = 0$
3. $x = 2, y = 1, z = 0$
5. $x = 2, y = -\frac{3}{2}, z = -\frac{1}{2}$
7. $\{(x, y, z) \mid x + 2y + 3z = 1\}$
9. $x = 1, y = -2, z = -1$
11. $a = \frac{1}{2}, b = -1, c = 2$
13. $x = 1, y = 0, z = 1$
15. $s = 6, t = -6, u = 2$
17. $p = 1, q = 2, r = 3$
19. $x = 1, y = -1, z = 1$
21. $a = 3, b = -3, c = 6$

23. No solutions
25. $x = 1,000, y = 2,000, z = 3,000$
27. Let d = # of dimes, q = # of
quarters, and h = # of half-dollars;
$h + q + d = 48, d = q + h - 2$,
$0.10d + 0.25q + 0.50h = 10.55$;
23 dimes, 17 quarters, 8 half-dollars
29. Let a = amount in the bank
account, b = amount in bonds, and
c = amount in stocks; $a + b + c$
$= 12,000, 0.087a + 0.093b +$
$0.1266c = 1,266, 0.1266c =$
$0.087a + 0.093b$; $\$3,000$ at 8.7%,
$\$4,000$ at 9.3%, $\$5,000$ at 12.66%
31. Let r = # of orchestra tickets, m = #
of mezzanine tickets, and b = # of
balcony tickets; $r + m + b =$
$750, 12r + 8m + 6b = 7,290$,
$m + b = r - 100$; 425 orchestra
tickets, 120 mezzanine tickets, and
205 balcony tickets
33. Let A = # of model A's, B = # of
model B's, and C = # of model
C's; $2.1A + 2.8B + 3.2C = 721$,
$3.2A + 3.6B + 4C = 974$,
$0.5A + 0.6B + 0.8C = 168$;
Model A: 110, model B: 95,
model C: 70
35. Let A = # of grams of food A,
B = # of grams of food B,
C = # of grams of food C;
$0.05A + 0.06B + 0.04C = 9$,
$0.2A + 0.15B + 0.1C = 28.5$,
$0.4A + 0.6B + 0.7C = 97$;
80 g of food A, 50 g of food B,
50 g of food C
37. Let A = # of acres of crop A,
B = # of acres of crop B,
C = # of acres of crop C;
$90A + 110B + 75C = 94,000$,
$6A + 10B + 5C = 7,200$,
$400A + 500B + 600C =$
$555,000$; 200 acres for crop A,
350 acres for crop B, 500 acres
for crop C

Mini-Review

39. $\{x \mid x \neq \frac{3}{2}, -1\}$
40. z is multiplied by 6.
41.

42. $-2x^2 \sqrt{2xy}$

Exercises 11.2

1. $\begin{bmatrix} 3 & -2 & | & 5 \\ 1 & -1 & | & 8 \end{bmatrix}$

3. $\begin{bmatrix} 1 & -2 & 3 & | & 4 \\ 0 & 1 & -1 & | & -3 \\ 2 & 3 & 0 & | & 8 \end{bmatrix}$

5. $(3, -2)$ **7.** $(0, -2)$ **9.** $(-4, 0)$
11. $(\frac{1}{2}, 3)$ **13.** $(\frac{2}{3}, \frac{1}{2})$ **15.** $(2, 1, 3)$
17. $(-2, 1, -3)$ **19.** $(3, -2, 0)$
21. $(\frac{1}{2}, 2, 3)$
23. $\{(x, y, 3) \mid 2x + 3y = -2\}$
25. $(1, -2, 0, 3)$

27. $\begin{bmatrix} 1 & 0 & | & 4 \\ 0 & 1 & | & -1 \end{bmatrix}$

29. $\begin{bmatrix} 1 & 0 & 0 & | & 2 \\ 0 & 1 & 0 & | & 1 \\ 0 & 0 & 1 & | & -1 \end{bmatrix}$

31. $(2, -1)$ **33.** $(1, 3, 1)$
35. $(2, 3, -1)$ **37.** $(0, 2, 2, 3)$

Mini-Review

39. $f[g(x)] = \dfrac{3x}{2x - 1} (x \neq 0)$;
  $g[f(x)] = \dfrac{2 - x}{3} (x \neq 1)$
40. $\dfrac{1}{9x^{1/2}}$ **41.** $y = \frac{3}{5}x + \frac{21}{5}$

Exercises 11.3

1. $\begin{bmatrix} 3 & 2 \\ 8 & 2 \end{bmatrix}$ **3.** $\begin{bmatrix} 0 & 0 \\ 0 & 0 \\ 0 & 0 \end{bmatrix}$

5. Not defined   **7.** $\begin{bmatrix} 0 & 9 \\ 51 & 0 \end{bmatrix}$

9. $\begin{bmatrix} -2 & 3 \\ 1 & 0 \end{bmatrix}$ **11.** $\begin{bmatrix} \frac{10}{3} & \frac{2}{3} \\ -\frac{4}{3} & 2 \end{bmatrix}$

13. $\begin{bmatrix} -3 & -4 \\ 1 & -3 \end{bmatrix}$ **15.** $\begin{bmatrix} -1 & -7 \\ 0 & -3 \end{bmatrix}$

17. $[10]$ **19.** $[3]$ **21.** $\begin{bmatrix} 0 & 2 \\ -3 & 5 \end{bmatrix}$

23. Not defined   **25.** $\begin{bmatrix} -30 & -2 \\ 39 & -11 \end{bmatrix}$

27. $\begin{bmatrix} 13 & -3 & -10 \\ 9 & -7 & -7 \\ -4 & -6 & 5 \end{bmatrix}$ **29.** $\begin{bmatrix} 5 & -7 \\ 3 & -14 \end{bmatrix}$

31. $\begin{bmatrix} 3 & 1 & 0 \\ -2 & 4 & 6 \\ 3 & 8 & -10 \end{bmatrix}$

33. False **35.** True
39. AB is not defined;
BA is a 3×4 matrix.
41. Tom's Office Supply, $3,768;
Kuma's Office Supply, $3,822;
Tom's is less expensive

Mini-Review

44. (a)

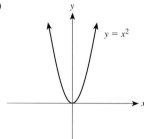

(b)

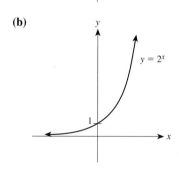

45. $x = \dfrac{-5 \pm 3\sqrt{5}}{2}$

46. $\dfrac{17}{10} + \dfrac{1}{10}i$

47. $(x - 1)^2 + (y + 1)^2 = 41$

Exercises 11.4

1. $\begin{bmatrix} 0 & \frac{1}{2} \\ \frac{1}{2} & -\frac{1}{4} \end{bmatrix}$ **3.** $\begin{bmatrix} \frac{1}{2} & \frac{1}{2} \\ 0 & 1 \end{bmatrix}$

5. $\begin{bmatrix} -\frac{1}{2} & \frac{3}{2} & -\frac{1}{2} \\ -\frac{1}{2} & \frac{5}{2} & -\frac{3}{2} \\ 1 & -2 & 1 \end{bmatrix}$

7. $(5, -1)$ **9.** $(6, 1)$ **11.** $(3, 2)$
13. $(-10, -45)$ **15.** $(0, 3)$
17. $(2, 2, 2)$ **19.** $(1, 0, -1)$
21. $(7, -9, 8)$ **23.** $(1, 0.1, -1)$

25. (a) $16,666.67 invested in the certificate and $3,333.33 invested in the stock
(b) $13,333.33 invested in the certificate and $6,666.67 invested in the stock
(c) $10,000 invested in the certificate and $10,000 invested in the stock
27. (a) CD = $4,800; bond = $4,800; utility stocks = $2,400
(b) CD = $7,200; bond = $1,200; utility stocks = $3,600
(c) impossible
29. 966 shirts, 683 blouses, 833 skirts
31. (a) X: 2 g; Y: 3.5 g; Z: 4.5 g
(b) X: 3 g; Y: 4 g; Z: 3 g
(c) X: 4 g; Y: 4.5 g; Z: 1.5 g

Mini-Review

34. $x = 8$ **35.** $y = \frac{80}{27}$
36. $x = -2 \pm \sqrt{2}$ **37.** $x = 0$

Exercises 11.5

1. -2 **3.** 22 **5.** 8 **7.** 4 **9.** 0
11. 11 **13.** 0 **15.** -92 **17.** 1
19. $x = 3$ **21.** $x = 7, x = -2$
23. $x = -5, x = 3$
25. $x = -1, y = 5$
27. No unique solution
29. $x = 4, y = 2$
31. $x = -2, y = 0$
33. $x = -\frac{53}{9}, y = -\frac{17}{3}$
35. $s = 2, t = 0$ **37.** $u = -3, v = 3$
39. No unique solution
41. $x = -\frac{1}{2}, y = \frac{1}{2}, z = 3$
43. $x = 1, y = 0, z = -1$
45. No unique solution
47. $x = 2, y = 2, z = 2$
49. No unique solution
51. No unique solution
53. $a = \frac{49}{11}, b = -\frac{10}{11}, c = \frac{4}{11}$
55. $-27,676$
57. $x = 0.001, y = 0.062$
59. $x = -1,341.25, y = 1,231.25,$
$z = 173.75$

Mini-Review

62. $4^3 = 64$ **63.** $\log_3 \frac{1}{81} = -4$
64. $\frac{4}{3}$ **65.** $\log_3 40$ is larger

Exercises 11.6s

1.

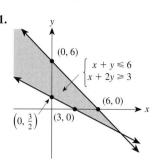

3.

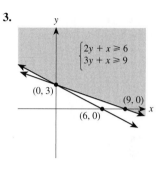

5. No solution

7.

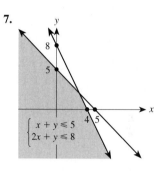

9.

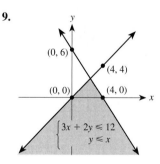

11.

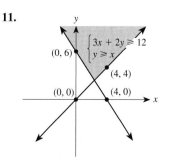

13.

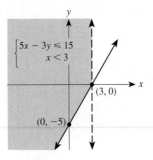

15.

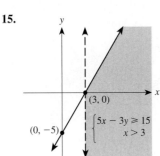

17.

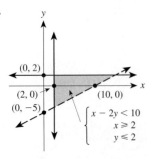

19.

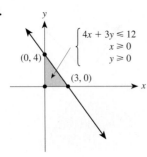

21.

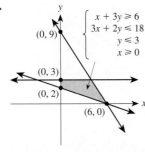

23.

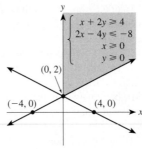

25. Let $x = $ # of single mattresses,
$y = $ # of double mattresses:
$80x + 120y \leq 8{,}000$,
$16x + 36y \leq 2{,}000$, $x \geq 0$, $y \geq 0$

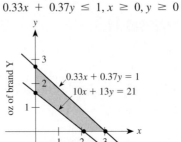

27. Let $x = $ number of oz of brand X,
$y = $ number of oz of brand Y;
$10x + 13y \geq 21$,
$0.33x + 0.37y \leq 1$, $x \geq 0$, $y \geq 0$

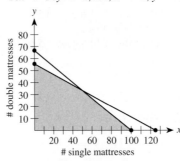

29. Let $x = $ men's shoes,
$y = $ ladies' shoes;
$1x + \frac{3}{4}y \leq 500$, $\frac{1}{3}x + \frac{1}{4}y \leq 150$,
$x \geq 0$, $y \geq 0$

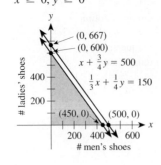

Mini-Review

31. $\frac{3}{2}\log_b x + 2\log_b y - \log_b 5 - 1$
32. $x = 8$ **33.** $6{,}890.37
34. $x = 0, \frac{5}{3}$

Exercises 11.7

1. $(3, 1)$, $(3, -1)$, $(-3, 1)$, $(-3, -1)$
3. $(5, 0)$, $(0, -5)$
5. $\left(\sqrt{5}, 2\right)$, $\left(\sqrt{5}, -2\right)$, $\left(-\sqrt{5}, 2\right)$,
$\left(-\sqrt{5}, -2\right)$
7. $(0, -3)$, $\left(\frac{12}{5}, \frac{9}{5}\right)$
9. $(2, 3)$, $(2, -3)$, $(-2, 3)$, $(-2, -3)$
11. $\left(\dfrac{1 + \sqrt{13}}{2}, \dfrac{7 + \sqrt{13}}{2}\right)$,
$\left(\dfrac{1 - \sqrt{13}}{2}, \dfrac{7 - \sqrt{13}}{2}\right)$
13. No real solutions
15. $(3, -9)$, $(4, -8)$
17. $(0, 4)$, $(0, -4)$, $\left(-1, \sqrt{15}\right)$,
$\left(-1, -\sqrt{15}\right)$
19. $(3, 4)$, $(4, 3)$
21. $\left(\dfrac{2\sqrt{15}}{3}, \dfrac{2\sqrt{6}}{3}\right)$,
$\left(\dfrac{2\sqrt{15}}{3}, \dfrac{-2\sqrt{6}}{3}\right)$
$\left(\dfrac{-2\sqrt{15}}{3}, \dfrac{2\sqrt{6}}{3}\right)$,
$\left(\dfrac{-2\sqrt{15}}{3}, \dfrac{-2\sqrt{6}}{3}\right)$
23. $(1, 0)$, $(0, 3)$
25. $\left(-\frac{1}{2}, \frac{35}{4}\right)$, $(2, 5)$
27. $(13, -4)$, $(6, 3)$
29. $(2, 1)$, $\left(-\frac{6}{5}, \frac{37}{5}\right)$
31. $\left(2\sqrt{2}, \sqrt{2}\right)$, $\left(-2\sqrt{2}, -\sqrt{2}\right)$,
$\left(\sqrt{2}, 2\sqrt{2}\right)$, $\left(-\sqrt{2}, -2\sqrt{2}\right)$
33. $\left(\sqrt{6}, \sqrt{2}\right)$, $\left(-\sqrt{6}, -\sqrt{2}\right)$
35. $(4, 3)$, $(-4, -3)$, $(3, 4)$, $(-3, -4)$
37. $(5, 2)$, $(-2, -5)$

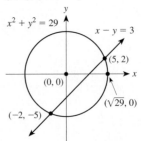

39. $(2, -5)$, $(-2, 5)$, $\left(\sqrt{13}, 4\right)$,
$\left(-\sqrt{13}, 4\right)$

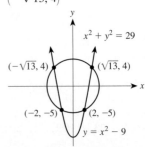

41. $(-4, -4), (1, 1)$

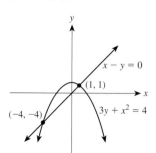

43. $\left(\dfrac{\sqrt{31}}{8}, -\dfrac{17}{8}\right), \left(-\dfrac{\sqrt{31}}{8}, -\dfrac{17}{8}\right),$
$(1, 2), (-1, 2)$

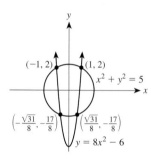

45. $\left(-\dfrac{1}{2}, \dfrac{\sqrt{3}}{2}\right), \left(-\dfrac{1}{2}, -\dfrac{\sqrt{3}}{2}\right)$

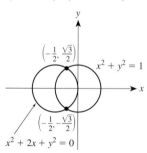

47. Let L = length and W = width;
$2L + 2W = 59, LW = 210;$
17.5 in. by 12 in.

49. Let r = rate going and t = time
going; $60 = rt,$
$60 = (r + 12)(t - \frac{1}{4});$
rate going = 48 mph and
time = 1.25 hr,
rate returning = 60 mph and
time = 1 hr

Mini-Review

52. $x = -0.2797$
53. **(a)** Approx. 2,200
   (b) Approx. 29 hours
54. $x = \dfrac{5}{4}$

55.

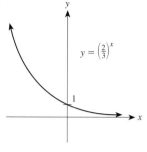

Chapter 11 Review Exercises

1. $x = 1, y = 1, z = 1$
2. $x = 2, y = 0, z = -1$
3. $x = -1, y = -2, z = -3$
4. $a = 0, b = 0, c = 1$
5. No solution   **6.** $(1, 0.1, -2)$

7. $\begin{bmatrix} 1 & 0 & | & 10 \\ 0 & 1 & | & -2 \end{bmatrix}$

8. $\begin{bmatrix} 1 & 0 & | & -10 \\ 0 & 1 & | & 8 \end{bmatrix}$

9. $\begin{bmatrix} 1 & 0 & 0 & | & 2 \\ 0 & 1 & 0 & | & -3 \\ 0 & 0 & 1 & | & 1 \end{bmatrix}$

10. $\begin{bmatrix} 1 & 0 & 0 & | & 0 \\ 0 & 1 & 0 & | & 0 \\ 0 & 0 & 1 & | & 3 \end{bmatrix}$

11. $(3, 1)$ **12.** $(1, 0)$ **13.** $(2, 1, -1)$
14. $(-1, 2, -1)$ **15.** $(2, 1)$
16. $(-1, 4, -1)$

17. $\begin{bmatrix} 10 & -5 \\ 0 & 25 \end{bmatrix}$ **18.** $\begin{bmatrix} -4 & -20 & 8 \\ 4 & 0 & -24 \end{bmatrix}$

19. $\begin{bmatrix} 5 & -1 \\ -2 & 4 \end{bmatrix}$ **20.** $\begin{bmatrix} -5 & -2 \\ 6 & 13 \end{bmatrix}$

21. Cannot be multiplied **22.** $[0]$

23. $\begin{bmatrix} 3 & 10 & -10 \\ -5 & 0 & 30 \end{bmatrix}$

24. Cannot be multiplied

25. $\begin{bmatrix} 0 & -5 & -4 \\ -4 & 10 & 32 \end{bmatrix}$

26. Cannot be multiplied
27. $x = -\frac{1}{2}, y = 5$
28. $x = -4, y = 7$
29. $x = 2, y = -2, z = 2$
30. $x = 2.1, y = -1.8, z = 0.7$
31. -14 **32.** 22 **33.** 98 **34.** 54
35. $x = -\frac{2}{23}, y = \frac{14}{23}$
36. $x = \frac{19}{23}, y = \frac{13}{23}$
37. Inconsistent **38.** Dependent
39. $s = \frac{32}{53}, t = \frac{16}{53}, u = -\frac{1}{53}$
40. $u = \frac{-95}{107}, v = \frac{-156}{107}, w = \frac{86}{107}$

41. $x = 2, y = 2$
42. $\left(\dfrac{2\sqrt{154}}{11}, \dfrac{2\sqrt{11}}{11}\right),$
$\left(\dfrac{2\sqrt{154}}{11}, -\dfrac{2\sqrt{11}}{11}\right),$
$\left(-\dfrac{2\sqrt{154}}{11}, \dfrac{2\sqrt{11}}{11}\right),$
$\left(-\dfrac{2\sqrt{154}}{11}, -\dfrac{2\sqrt{11}}{11}\right)$

43. $(5, 6), \left(\dfrac{-19}{7}, \dfrac{6}{7}\right)$
44. $(0, 3), (\sqrt{5}, -2), (-\sqrt{5}, -2)$
45. No real solutions
46. No real solutions

47.

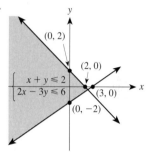

48.

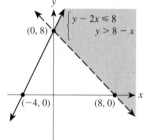

49.

50. No solution
51. Let $\$x$ be invested at 7.7%, $\$y$ be
   invested at 8.6%, and $\$z$ be invested
   at 9.8%; $y + z = 3{,}000 + x,$
   $x + y + z = 20{,}000,$
   $0.077x + 0.086y + 0.098z$
   $= 1{,}719.10;$ \$8,500 at 7.7%,
   \$5,200 at 8.6%, \$6,300 at 9.8%
52. Let s = price of a sweater,
   b = price of a blouse, and
   j = price of jeans; $s + 2b + j$
   $= 28.50, 2s + b + j = 30,$

$2s + 3b + 2j = 52$; $6.50 for each sweater, $5 for each blouse, and $12 for each pair of jeans

53. Let $w =$ the width and $l =$ the length: $2w + 2l = 41$, $lw = 100$; dimensions are 8 ft by 12.5 ft

54. Let $x =$ # of acres of crop X and $y =$ # of acres of crop Y: $x + y \le 50$, $200x + 100y \le 8{,}000$, $x \ge 0$, $y \ge 0$

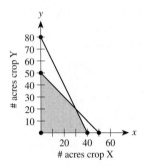

Chapter 11 Practice Test

1. $x = 3, y = 0, z = -1$

2. (a) $(-1, 3)$ **(b)** $(3, 1, 2)$

3. (a) $\begin{bmatrix} -5 & 27 & -10 \\ -3 & -4 & 22 \\ -10 & 6 & -4 \end{bmatrix}$

(b) $\begin{bmatrix} -6 & 2 & -6 \\ 22 & 3 & 5 \\ 4 & -4 & 0 \end{bmatrix}$

4. $x = 3, y = -1$

5. (a) 11   **(b)** 23

6. (a) $x = \frac{19}{4}, y = \frac{7}{4}$
(b) $x = 2, y = -1, z = 3$

7. Let $f =$ # of $5 bills, $t =$ # of $10 bills, $w =$ # of $20 bills; $2w = t$, $f + t + w = 40$, $5f + 10t + 20w = 500$; four $5 bills, twenty-four $10 bills, twelve $20 bills

8. $(2, -1)$

9.

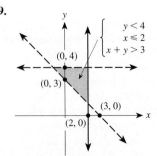

Chapters 10–11
Cumulative Review

1. $x = 3, y = 2, z = 1$ (§11.1)
2. $x = 0, y = -2, z = 4$ (§11.1)
3. $x = 7, y = 0, z = -4$ (§11.1)
4. $x = 12, y = 12, z = 12$ (§11.1)
5. $\{(x, y, z) \mid x - 2y + 3z = 4\}$ (§11.1)
6. $x = \frac{1}{6}, y = \frac{1}{2}, z = \frac{1}{3}$ (§11.1)
7. 7 (§11.5)   **8.** 5 (§11.5)
9. 0 (§11.5)   **10.** -21 (§11.5)
11. -8 (§11.5)   **12.** 0 (§11.5)
13. -2 (§11.5)   **14.** 0 (§11.5)
15. $(2, -1)$ (§11.2)
16. $(-1, 2, 3)$ (§11.2)
17. $x = \frac{67}{55}, y = -\frac{13}{55}$ (§11.5)
18. $x = \frac{31}{53}, y = -\frac{19}{106}$ (§11.5)
19. $x = 0, y = 1, z = 0$ (§11.5)
20. $x = \frac{19}{32}, y = -\frac{79}{48}, z = \frac{17}{4}$ (§11.5)
21. $(2, -1), (-1, 2)$ (§11.7)
22. $(-2, -2), (4, 1)$ (§11.7)
23. No solution (§11.7)
24. $(1, 0), \left(\frac{1}{2}, \frac{1}{4}\right)$ (§11.7)
25. $(3, 1), (3, -1), (-3, 1), (-3, -1)$ (§11.7)
26. $(2, 3), (2, -3), (-2, 3), (-2, -3)$ (§11.7)
27. $(3, 0), \left(\frac{31}{9}, \frac{2}{3}\right)$ (§11.7)
28. $\left(4, \sqrt{11}\right), \left(4, -\sqrt{11}\right), (-3, 2), (-3, -2)$ (§11.7)
29. (§11.6)

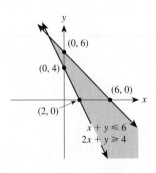

30. (§11.6)

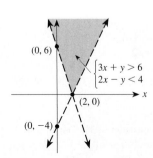

31. (§11.6)

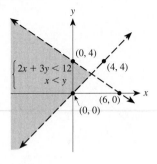

32. (§11.6)

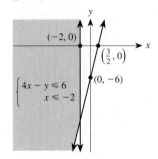

33. (§11.6)

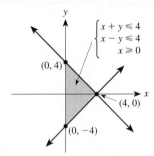

34. (§11.6)

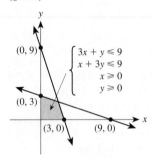

35. $\begin{bmatrix} 3 & -6 \\ -15 & 0 \end{bmatrix}$ (§11.3)

36. $\begin{bmatrix} -3 & -1 \\ -5 & 3 \end{bmatrix}$ (§11.3)

37. $\begin{bmatrix} -23 & 11 \\ 15 & 15 \end{bmatrix}$ (§11.3)

38. $[7]$ (§11.3)

39. $\begin{bmatrix} 4 & 5 \\ -20 & 5 \end{bmatrix}$ (§11.3)

40. $\begin{bmatrix} 9 & -8 \\ 15 & 0 \end{bmatrix}$ (§11.3)

41. $x = 3.5, y = -0.25$ (§11.4)

42. $x = 2, y = 0.4, z = -0.8$ (§11.4)

43. (§10.1)

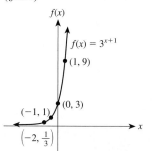

44. (§10.1)

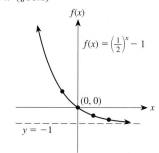

45. (§10.2)

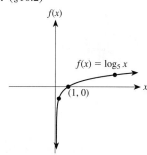

46. (§10.2)

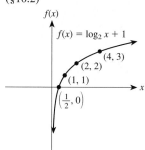

47. $2^6 = 64$ (§10.2)

48. $\log_3\left(\frac{1}{81}\right) = -4$ (§10.2)

49. $\log_{125} 5 = \frac{1}{3}$ (§10.2)

50. $10^{-2} = 0.01$ (§10.2)

51. $27^{4/3} = 81$ (§10.2)

52. $\log_4\left(\frac{1}{2}\right) = -\frac{1}{2}$ (§10.2)

53. 4 (§10.2)   **54.** $\frac{1}{3}$ (§10.2)

55. -2 (§10.2)   **56.** $\frac{5}{4}$ (§10.2)

57. $-\frac{1}{2}$ (§10.2)   **58.** $\frac{2}{3}$ (§10.2)

59. 0 (§10.2)   **60.** 1 (§10.2)

61. $\frac{1}{3}\log_b 5 + \frac{1}{3}\log_b x + \frac{1}{3}\log_b y$ (§10.3)

62. $3 \log_b x + \log_b y - 5 \log_b z$ (§10.3)

63. $2 \log_3 x + \frac{1}{2} \log_3 y - 2 - \log_3 w$
$\qquad\qquad\qquad - \log_3 z$ (§10.3)

64. $\dfrac{2 \log_b u}{\log_b v}$ (§10.3)

65. 4.8669 (§10.4)

66. 6.8752 (§10.4)

67. 4.0598 (§10.4)

68. 1.4918 (§10.4)

69. 2.7083 (§10.4)

70. -3.1787 (§10.4)

71. $x = 25$ (§10.5)

72. $x = 3$ (§10.5)

73. $x = -2$ (§10.5)

74. $x = -\frac{1}{4}$ (§10.5)

75. $x = 54$ (§10.5)

76. $x = 8$ (§10.5)

77. $x = 2, -\frac{3}{2}$ (§10.5)

78. $x = 4$ (§10.5)

79. $x = \dfrac{3 \log 7}{\log 9 - \log 7}$
$\qquad \approx 23.23$ (§10.5)

80. $x = 6$ (§10.5)

81. ≈ 6.449 yr (§10.6)

82. ≈ 23.1 yr (§10.6)

83. $\approx 1,151$ yr (§10.6)

84. 3.4 (§10.6)

85. 6.31×10^{-9} (§10.6)

Chapters 10–11
Cumulative Practice Test

1. $x = 5, y = -1, z = 2$

2. $(0, 2, -2)$

3. (a) $x = -2, y = 5$
(b) $x = 3, y = -1, z = 0$

4. $x = 4, y = 3$

5.

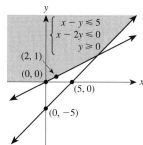

6. (a) $\begin{bmatrix} 18 & -16 \\ -3 & -7 \end{bmatrix}$

(b) $\begin{bmatrix} -10 & -20 \\ 2 & 10 \end{bmatrix}$

7. $x = -3, y = \frac{1}{2}$

8. (a)

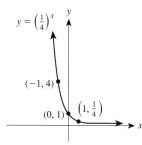

(b)

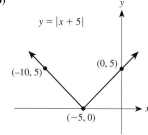

9. $8^{-2/3} = \frac{1}{4}$ **10.** (a) -3 (b) $\frac{3}{2}$

11. (a) $\log_b x + \frac{1}{3}\log_b y$

(b) $\frac{5}{2}\log_b x - \frac{1}{2}\log_b y$

12. (a) -2.1959 (b) 8,439.18
(c) 4.5433

13. 3.3147

14. (a) $x = 0, 9$ (b) $x = 4$

15. ≈ 7.018 yr

Appendix A Exercises

1. True   **3.** False   **5.** True

7. True   **9.** True   **11.** True

13. $\{1, 2, 3, 4, 5, 6, 7, 8, 9, 10, 11\}$

15. $\{7, 8, 9, 10, 11, 12\}$

17. $\{3, 4, 5, 6, 7, 8, 9, 10, 11, 12, 13, 14,$
$15, 16, 17\}$

19. $\varnothing$ **21.** $\{23, 29, 31, 37\}$ **23.** $\varnothing$

25. $\{0, 12, 24, 36, 48, 60, 72, \ldots\}$

27. $\{0, 5, 10, 15, 20, 25, 30, 35, 40, \ldots\}$

29. $\{1, 2, 3, 4, 6, 8, 12, 16, 24, 48\}$

31. $\{4, 8\}$

33. $\{1, 2, 4, 5, 8, 12, 16, 20, 24, 28, 32\}$

35. $\{7\}$ **37.** $\{8\}$

39. $\{x \mid -2 \le x < 0\}$ **41.** $\varnothing$

43. $\{x \mid 8 \le x \le 10\}$

Appendix B Exercises

1. $C = (2, -6); r = \sqrt{58}$

3. $C = (-4, 6); r = \sqrt{67}$

5.
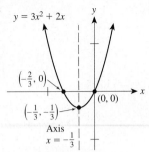

7.
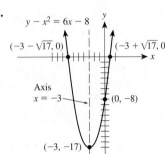

9.
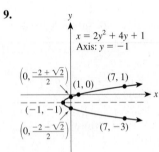

11.
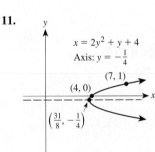

13. $(0, 3), (0, -3), (-4, 0), (4, 0)$
15. $(0, 6), (0, -6), (-5, 0), (5, 0)$
17. $(0, 2\sqrt{5}), (0, -2\sqrt{5}), (-2\sqrt{6}, 0),$
$(2\sqrt{6}, 0)$

19.
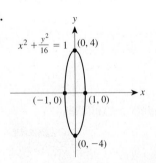

21.
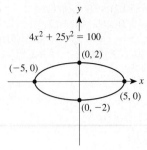

23.
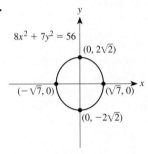

25.
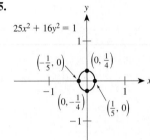

27. No; the bridge is 13.47 ft high
11 ft right of the center line.
29. Vertices: $(-3, 0), (3, 0)$;
asymptotes $y = \pm\frac{4}{3}x$
31. Vertices: $(-6, 0), (6, 0)$;
asymptotes: $y = \pm\frac{5}{6}x$

33.
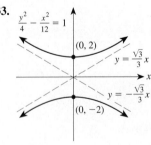

35.
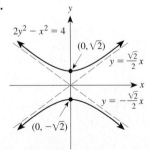

37.
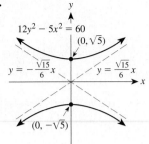

39.
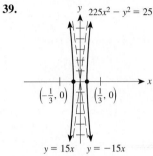

41. Circle **43.** Ellipse
45. Parabola **47.** Straight line
49. Hyperbola **51.** Hyperbola
53. Circle **55.** Circle
57. Hyperbola

59.
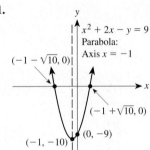

61.
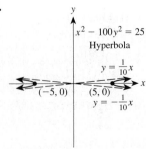

63.

65.

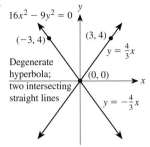

$16x^2 - 9y^2 = 0$

$(-3, 4)$ $(3, 4)$

$y = \frac{4}{3}x$

$(0, 0)$

Degenerate hyperbola; two intersecting straight lines

$y = -\frac{4}{3}x$

67.

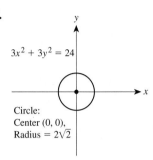

$3x^2 + 3y^2 = 24$

Circle: Center $(0, 0)$, Radius $= 2\sqrt{2}$

69.

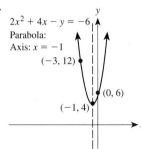

$2x^2 + 4x - y = -6$

Parabola: Axis: $x = -1$

$(-3, 12)$

$(0, 6)$

$(-1, 4)$

71.

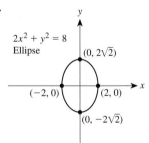

$2x^2 + y^2 = 8$

Ellipse

$(0, 2\sqrt{2})$

$(-2, 0)$ $(2, 0)$

$(0, -2\sqrt{2})$

Exercises C.1

1. $-2, 1, 4, 7; a_{10} = 25$
3. $7, 11, 15, 19; b_{10} = 43$
5. $4, 16, 64, 256; c_{10} = 1{,}048{,}576$
7. $0, \frac{1}{3}, \frac{1}{2}, \frac{3}{5}; x_{10} = \frac{9}{11}$
9. $-\frac{1}{3}, \frac{1}{4}, -\frac{1}{5}, \frac{1}{6}; x_{10} = \frac{1}{12}$
11. $-\frac{1}{2}, \frac{2}{3}, -\frac{3}{4}, \frac{4}{5}; a_{10} = \frac{10}{11}$
13. $6, 7, 10, 17; a_{10} = 1{,}029$
15. $4.9, 5.01, 4.999, 5.0001;$
$b_{10} = 5.0000000001$
17. $a_n = 4n$ **19.** $b_n = 2n - 1$
21. $a_n = \dfrac{n}{n + 1}$ **23.** $b_n = \dfrac{(-1)^{n+1}}{n + 1}$

25. $a_n = 5(-0.1)^n$
27. $\{\$26{,}000, \$26{,}850, \$27{,}700\}$
29. $\{21{,}200, 22{,}472, 23{,}820\}$
31. $\{\$5{,}300, \$5{,}618, \$5{,}955.08,$
$\$6{,}312.38\}$
33. $\{\$30{,}000, \$31{,}800, \$33{,}708,$
$\$35{,}730.48\}$
35. $\{30', 15', 7.5', 3.75', 1.875'\}$
37. $a_n = 200 + 0.01[5{,}000 - 200(n - 1)]; \{\$250, \$248, \$246,$
$\$244, \$242, \$240\}$
39. $\{4, 20, 100, 500, 2{,}500\}$
41. $\{6, 28, 138, 688, 3{,}438\}$
43. $\{4, 5, 1, -4, -5\}$

Exercises C.2

1. 40 **3.** 18 **5.** 12 **7.** 11
9. 27 **11.** 62 **13.** 132
15. $\frac{163}{60}$ **17.** $-\frac{23}{60}$
19. $1 + 2 + 3 + 4 + 5 = 15$
21. $5 + 10 + 15 + 20 + 25 + 30$
$+ 35 = 140$
23. $1 + 4 + 9 + 16 + 25 + 36 = 91$
25. $16 + 54 + 128 + 250 = 448$
27. $11 + 13 + 15 + 17 = 56$
29. $9 + 13 + 17 + 21 + 25 + 29$
$+ 33 = 147$
31. $7 + 17 + 31 + 49 + 71 = 175$
33. $\sum_{k=1}^{5} 2k$ **35.** $\sum_{k=1}^{7} k^2$

Exercises C.3

1. $5, 7; a_{15} = 31$
3. $-2, -6; x_{15} = -54$
5. $-\frac{3}{2}, -1; z_{15} = 5$
7. $13, 16; a_{15} = 43$
9. $6, 8; a_{15} = 26$
11. $-13, -18; a_{15} = -68$
13. $32, 40; a_{15} = 120$
15. $\frac{3}{2}, \frac{11}{6}; a_{15} = \frac{31}{6}$
17. $-\frac{4}{3}, -\frac{11}{6}; a_{15} = -\frac{22}{3}$
19. $a_1 = 1, d = 6$
21. $a_1 = 3, d = -4$ **23.** $\$40{,}500$
25. 110 lb **27.** 176 ft **29.** 78
31. 15 **33.** 25 **35.** -40
37. 18 **39.** $-\frac{13}{3}$ **41.** 63
43. 550 **45.** 75 **47.** 2,805
49. 2,550 **51.** 440 **53.** 165
55. -570 **57.** 576 ft
59. (a) During the eighth year, $\$54{,}000$ at firm A and $\$52{,}500$ at firm B
(b) Over 8 years, $\$376{,}000$ at firm A and $\$378{,}000$ at firm B

Exercises C.4

1. 486 **3.** 1,296 **5.** $\frac{1}{729}$
7. $\frac{5}{7{,}776}$ **9.** $\frac{1}{162}$ **11.** 24
13. $\frac{1}{1{,}250}$ **15.** 256 **17.** $\frac{1}{64}$
19. $\$7{,}716.05$ **21.** $\$12{,}288$
23. 121,551 **25.** $\$14{,}693.28$
27. $\$8{,}052.55$ **29.** $\frac{5}{8}$ ft
31. 1,092 **33.** 315 **35.** 21
37. $\frac{11{,}111}{10{,}000}$ **39.** $\frac{11{,}111}{10{,}000{,}000}$ **41.** $14\frac{76}{81}$ ft
43. (a) During the eighth year, $\$45{,}108.91$ at firm A and $\$44{,}741.68$ at firm B
(b) Over 8 years, $\$296{,}924.04$ at firm A and $\$313{,}283.69$ at firm B
45. 4 **47.** $\frac{2}{5}$ **49.** $\frac{15}{2}$
51. Does not exist **53.** $\frac{35}{99}$
55. $\frac{245}{9{,}999}$ **57.** 300 ft **59.** 30 ft

Exercises C.5

1. 720 **3.** 24 **5.** 1 **7.** 120
9. 5,038 **11.** 720 **13.** 12
15. 5 **17.** 56 **19.** 27,720
21. $a^6 + 6a^5b + 15a^4b^2 + 20a^3b^3$
$+ 15a^2b^4 + 6ab^5 + b^6$
23. $a^6 - 6a^5b + 15a^4b^2 - 20a^3b^3$
$+ 15a^2b^4 - 6ab^5 + b^6$
25. $32a^5 + 80a^4 + 80a^3 + 40a^2$
$+ 10a + 1$
27. $32a^5 - 80a^4 + 80a^3 - 40a^2$
$+ 10a - 1$
29. $1 - 8a + 24a^2 - 32a^3 + 16a^4$
31. $a^{10} + 10a^8b + 40a^6b^2 + 80a^4b^3$
$+ 80a^2b^4 + 32b^5$
33. $16x^8 - 96x^6y^2 + 216x^4y^4$
$- 216x^2y^6 + 81y^8$
35. $\frac{1}{81}x^4 + \frac{8}{27}x^3 + \frac{8}{3}x^2 + \frac{32}{3}x + 16$
37. $a^{12} - 6a^{10}b^4 + 15a^8b^8 - 20a^6b^{12}$
39. $256a^8 + 3{,}072a^7b + 16{,}128a^6b^2$
$+ 48{,}384a^5b^3$
41. $729a^{12} - 2{,}916a^{10} + 4{,}860a^8$
$- 4{,}320a^6$
43. $28x^6y^2$ **45.** $-160a^3b^3$
47. $20{,}412a^{10}$ **49.** $-6{,}048a^4$ **51.** 256

Appendix C Review Exercises

1. $-3, -1, 1, 3; a_{12} = 19$
2. $4, 7, 10, 13; b_{12} = 37$
3. $2, 8, 18, 32; x_{12} = 288$

4. 3, 6, 9, 12; $y_{12} = 36$

5. $-\frac{1}{2}, \frac{1}{3}, -\frac{1}{4}, \frac{1}{5}$; $a_{12} = \frac{1}{13}$

6. $\frac{1}{3}, -\frac{1}{4}, \frac{1}{5}, -\frac{1}{6}$; $b_{12} = -\frac{1}{14}$

7. 2, 4, 2, 4; $x_{12} = 4$

8. 4, 5, 8, 17, $y_{12} = 4{,}105$

9. $a_i = 5i - 2$ **10.** $a_i = 2 \cdot 3^{i-1}$

11. $a_n = \dfrac{(-1)^{n+1}}{2n}$ **12.** $a_n = \dfrac{1}{2^n}$

13. $23{,}000, \$23{,}700, \$24{,}400,$
$\$25{,}100, \$25{,}800, \$26{,}500$

14. 20,000, 22,200, 24,642, 27,353

15. \$881.17 **16.** \$734.66

17. 45 **18.** 65 **19.** 126

20. 336 **21.** 57 **22.** 168

23. $1 + 2 + 3 + 4 + 5 + 6 = 21$

24. $3 + 4 + 5 + 6 + 7 + 8 = 33$

25. $2 + 4 + 6 + 8 + 10 = 30$

26. $4 + 8 + 12 = 24$

27. $3 + 12 + 27 + 48 + 75$
$+ 108 = 273$

28. $7 + 9 + 11 + 13 + 15 = 55$

29. 5, 8; $a_{10} = 29$

30. $-3, -1; b_{10} = 13$

31. 23, 28; $a_{10} = 48$

32. 4, 6; $a_{10} = 14$

33. $-15, -19; a_{10} = -35$

34. $-1, -\frac{3}{2}; a_{10} = -\frac{7}{2}$

35. $x_1 = 1, d = 3$

36. $y_1 = -5, d = -2$ **37.** 240 ft

38. 304 ft **39.** 185 **40.** 9

41. 220 **42.** 72 **43.** 930

44. 630 **45.** 900 **46.** 420

47. $\frac{245}{4}$ **48.** 66 **49.** 162

50. 64 **51.** $\frac{1}{243}$ **52.** $\frac{2}{243}$

53. 80 **54.** $\frac{1}{16}$ **55.** \$10,240

56. \$20,475 **57.** $\frac{9}{2}$ **58.** $\frac{5}{4}$

59. Does not exist

60. Does not exist **61.** $\frac{64}{99}$

62. 200 ft **63.** 40,320 **64.** 6

65. 362,160 **66.** 60,480

67. $x^5 - 5x^4y + 10x^3y^2 - 10x^2y^3$
$+ 5xy^4 - y^5$

68. $81a^4 + 108a^3b + 54a^2b^2$
$+ 12ab^3 + b^4$

69. $16a^4 - 96a^3b^2 + 216a^2b^4$
$- 216ab^6 + 81b^8$

70. $125a^3 - 75a^2 + 15a - 1$

71. $243a^5 - 810a^4b + 1{,}080a^3b^2$
$- 720a^2b^3$

72. $64a^{12} - 192a^{10}b^3 + 240a^8b^6$
$- 160a^6b^9$

73. $-4{,}320a^6$ **74.** $15{,}120x^3y^8$

Appendix C Practice Test

1. **(a)** 4, 7, 10; $x_9 = 28$
 (b) 1, 15, 53; $y_9 = 1{,}457$

2. **(a)** 55 **(b)** 44

3. $19 + 33 + 51 + 73 = 176$

4. 80 **5.** 64 **6.** 354 **7.** 312

8. 840 **9.** 30 ft

10. $16a^8 + 96a^6 + 216a^4 + 216a^2 + 81$

11. $-540a^6$

Subject Index

Applications Index